IMPORTANT:

HERE IS YOUR REGISTRATION CODE TO ACCESS YOUR PREMIUM McGRAW-HILL ONLINE RESOURCES.

For key premium online resources you need THIS CODE to gain access. Once the code is entered, you will be able to use the Web resources for the length of your course.

If your course is using **WebCT** or **Blackboard**, you'll be able to use this code to access the McGraw-Hill content within your instructor's online course.

Access is provided if you have purchased a new book. If the registration code is missing from this book, the registration screen on our Website, and within your WebCT or Blackboard course, will tell you how to obtain your new code.

Registering for McGraw-Hill Online Resources

To gain access to your McGraw-Hill web resources simply follow the steps below:

1. USE YOUR WEB BROWSER TO GO TO: www.mhhe.com/raven7
2. CLICK ON **FIRST TIME USER**.
3. ENTER THE REGISTRATION CODE* PRINTED ON THE TEAR-OFF BOOKMARK ON THE RIGHT.
4. AFTER YOU HAVE ENTERED YOUR REGISTRATION CODE, CLICK **REGISTER**.
5. FOLLOW THE INSTRUCTIONS TO SET-UP YOUR PERSONAL UserID AND PASSWORD.
6. WRITE YOUR UserID AND PASSWORD DOWN FOR FUTURE REFERENCE. KEEP IT IN A SAFE PLACE.

TO GAIN ACCESS to the McGraw-Hill content in your instructor's **WebCT** or **Blackboard** course simply log in to the course with the UserID and Password provided by your instructor. Enter the registration code exactly as it appears in the box to the right when prompted by the system. You will only need to use the code the first time you click on McGraw-Hill content.

Thank you, and welcome to your McGraw-Hill online Resources!

* YOUR REGISTRATION CODE CAN BE USED ONLY ONCE TO ESTABLISH ACCESS. IT IS NOT TRANSFERABLE.

0-07-291843-8 T/A RAVEN/JOHNSON : BIOLOGY, 7/E

ONLINE RESOURCES

REGISTRATION CODE

KEPJ-SJLV-DB21-T1WX-1X3S

How's Your Math?

Do you have the math skills you need to succeed?

Why risk not succeeding because you struggle with your math skills?

Get access to a web-based, personal math tutor:

- Available 24/7, unlimited use
- Driven by artificial intelligence
- Self-paced
- An entire month's subscription **for much less** than the cost of one hour with a human tutor

ALEKS is an inexpensive, private, infinitely patient math tutor that's accessible any time, anywhere you log on.

ALEKS® McGraw Hill

Log On for a
FREE 48-hour Trial

www.highedstudent.aleks.com

ALEKS is a registered trademark of ALEKS Corporation.

BIOLOGY

Seventh Edition

Peter H. Raven
Director, Missouri Botanical Gardens;
Engelmann Professor of Botany
Washington University

George B. Johnson
Professor Emeritus of Biology
Washington University

Jonathan B. Losos
Professor of Biology
Washington University

Susan R. Singer
Professor of Biology
Carleton College

Illustration Authors
William C. Ober, M.D.
and
Claire W. Garrison, R.N.

McGraw Hill **Higher Education**

Boston   Burr Ridge, IL   Dubuque, IA   Madison, WI   New York   San Francisco   St. Louis
Bangkok   Bogotá   Caracas   Kuala Lumpur   Lisbon   London   Madrid   Mexico City
Milan   Montreal   New Delhi   Santiago   Seoul   Singapore   Sydney   Taipei   Toronto

The McGraw·Hill Companies

McGraw Hill Higher Education

BIOLOGY, SEVENTH EDITION

Published by McGraw-Hill, a business unit of The McGraw-Hill Companies, Inc., 1221 Avenue of the Americas, New York, NY 10020. Copyright © 2005, 2002, 1999, 1996 by The McGraw-Hill Companies, Inc. All rights reserved. No part of this publication may be reproduced or distributed in any form or by any means, or stored in a database or retrieval system, without the prior written consent of The McGraw-Hill Companies, Inc., including, but not limited to, in any network or other electronic storage or transmission, or broadcast for distance learning.

Some ancillaries, including electronic and print components, may not be available to customers outside the United States.

♻ This book is printed on recycled, acid-free paper containing 10% postconsumer waste.

International 1 2 3 4 5 6 7 8 9 0 VNH/VNH 0 9 8 7 6 5 4 3
Domestic   1 2 3 4 5 6 7 8 9 0 VNH/VNH 0 9 8 7 6 5 4 3

ISBN 0–07–243731–6
ISBN 0–07–111182–4 (ISE)

Publisher: *Martin J. Lange*
Senior sponsoring editor: *Patrick E. Reidy*
Developmental editor: *Anne L. Winch*
Marketing manager: *Tami Petsche*
Lead project manager: *Peggy J. Selle*
Production supervisor: *Kara Kudronowicz*
Senior media project manager: *Jodi K. Banowetz*
Senior media technology producer: *John J. Theobald*
Senior coordinator of freelance design: *Michelle D. Whitaker*
Cover/interior designer: *Christopher Reese*
Senior photo research coordinator: *Lori Hancock*
Photo research: *Meyers Photo-Art*
Supplement producer: *Brenda A. Ernzen*
Compositor: *Carlisle Communications, Ltd.*
Typeface: *10/12 Janson Text Roman*
Printer: *Von Hoffmann Corporation*

Cover images: DNA: © *Doug Struthers/Getty Images*; Pollen: *S. Lowry, Univ. Ulster/Getty Images*; Beetle: © *Davies + Starr/Getty Images*; Leopard: © *Ryan McVay/Getty Images*; Man's Profile: © *Suza Scalora/Getty Images*; Leaf: *Christopher Reese*

The credits section for this book begins on page C-1 and is considered an extension of the copyright page.

Library of Congress Cataloging-in-Publication Data

Biology / Peter H. Raven . . . [et al.]. 7th ed.
    p.   cm.
  Rev. ed. of: Biology / Peter H. Raven and George B. Johnson. 6th ed., © 2002.
  ISBN 0–07–243731–6 (hard : alk. paper)
  1. Biology. I. Raven, Peter H. II. Raven, Peter H. Biology.

QH308.2.R38   2005
570—dc22   2003016998
                                                      CIP

INTERNATIONAL EDITION ISBN 0–07–111182–4
Copyright © 2005. Exclusive rights by The McGraw-Hill Companies, Inc., for manufacture and export. This book cannot be re-exported from the country to which it is sold by McGraw-Hill. The International Edition is not available in North America.

www.mhhe.com

Brief Contents

Part I The Origin of Living Things 1
1. The Science of Biology 1
2. The Nature of Molecules 19
3. The Chemical Building Blocks of Life 35
4. The Origin and Early History of Life 61

Part II Biology of the Cell 79
5. Cell Structure 79
6. Membranes 105
7. Cell-Cell Interactions 125
8. Energy and Metabolism 143
9. How Cells Harvest Energy 159
10. Photosynthesis 185
11. How Cells Divide 207

Part III Genetic and Molecular Biology 227
12. Sexual Reproduction and Meiosis 227
13. Patterns of Inheritance 241
14. DNA: The Genetic Material 279
15. Genes and How They Work 301
16. Gene Technology 319
17. Genomes 343
18. Control of Gene Expression 361
19. Cellular Mechanisms of Development 381
20. Cancer Biology and Cell Technology 405

Part IV Evolution 433
21. Genes Within Populations 433
22. The Evidence for Evolution 453
23. The Origin of Species 471
24. Evolution of Genomes and Developmental Mechanisms 491

Part V Diversity of Life on Earth 509
25. Systematics and the Phylogenetic Revolution 509
26. Viruses 531
27. Prokaryotes 545
28. Protists 561
29. Overview of Plant Diversity 579
30. Fungi 599
31. Overview of Animal Diversity 617
32. Noncoelomate Invertebrates 633
33. Coelomate Invertebrates 651
34. Vertebrates 683

Part VI Plant Form and Function 729
35. Plant Form 729
36. Vegetative Plant Development 755
37. Transport in Plants 767
38. Plant Nutrition 781
39. Plant Defense Responses 795
40. Sensory Systems in Plants 807
41. Plant Reproduction 831

Part VII Animal Form and Function 855
42. The Animal Body and How It Moves 855
43. Fueling Body Activities: Digestion 887
44. Circulation and Respiration 907
45. The Nervous System 939
46. Sensory Systems 969
47. The Endocrine System 991
48. The Immune System 1013
49. Maintaining the Internal Environment 1039
50. Sex and Reproduction 1061
51. Vertebrate Development 1081

Part VIII Ecology and Behavior 1105
52. Behavioral Biology 1105
53. Population Ecology 1137
54. Community Ecology 1161
55. Dynamics of Ecosystems 1183
56. The Biosphere 1203
57. Conservation Biology 1227

Contents

Part I The Origin of Living Things

1 The Science of Biology 1
1.1 Biology is the science of life.
1.2 Scientists form generalizations from observations.
1.3 Darwin's theory of evolution illustrates how science works.
1.4 Four themes unify biology as a science.

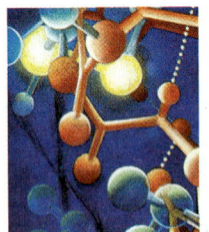

2 The Nature of Molecules 19
2.1 Atoms are natures building material.
2.2 The atoms of living things are among the smallest.
2.3 Chemical bonds hold molecules together.
2.4 Water is the cradle of life.

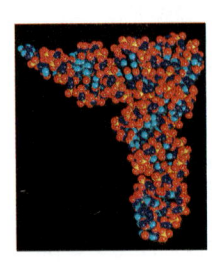

3 The Chemical Building Blocks of Life 35
3.1 Molecules are the building blocks of life.
3.2 Proteins perform the chemistry of the cell.
3.3 Nucleic acids store and transfer genetic information.
3.4 Lipids make membranes and store energy.
3.5 Carbohydrates store energy and provide building materials.

4 The Origin and Early History of Life 61
4.1 All living things share key characteristics.
4.2 There is considerable disagreement about the origin of life.
4.3 There are many hypotheses about the origin of cells.
4.4 Cells became progressively more complex as they evolved.
4.5 Scientists are beginning to take the possibility of extraterrestrial life seriously.

Part II Biology of the Cell

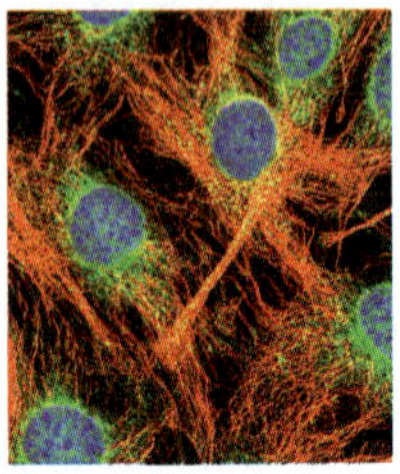

5 Cell Structure 79
5.1 All organisms are composed of cells.
5.2 Eukaryotic cells are more structurally complex than prokaryotic cells.
5.3 Take a tour of a eukaryotic.
5.4 Not all eukaryotic cells are the same.

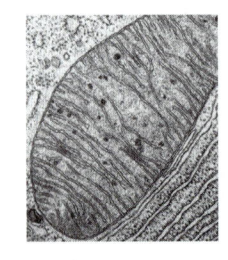

6 Membranes 105
6.1 Biological membranes are fluid layers of lipid.
6.2 Proteins embedded within cell membranes determine their character.
6.3 Passive transport across membranes moves down the concentration gradient.
6.4 Bulk transport utilizes endocytosis.
6.5 Active transport across membranes requires energy.

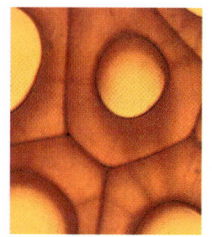

7 Cell-Cell Interactions 125

7.1 Cells signal one another with chemicals.
7.2 Proteins in the cell and on its surface receive signals from other cells.
7.3 Follow the journey of information into the cell.
7.4 Cell surface proteins mediate cell-cell interactions.

8 Energy and Metabolism 143

8.1 The laws of thermodynamics describe how energy changes.
8.2 Enzymes are biological catalysts.
8.3 ATP is the energy currency of life.
8.4 Metabolism is the chemical life of a cell.

9 How Cells Harvest Energy 159

9.1 Cells harvest the energy in chemical bonds.
9.2 Cellular respiration oxidizes food molecules.
9.3 Catabolism of proteins and fats can yield considerable energy.
9.4 Cells can metabolize food without oxygen.
9.5 The stages of cellular respiration evolved over time.

10 Photosynthesis 185

10.1 What is photosynthesis?
10.2 Learning about photosynthesis: An experimental journey.
10.3 Pigments capture energy from sunlight.
10.4 Cells use the energy and reducing power captured by the light-dependent reactions to make organic molecules.

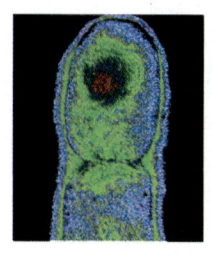

11 How Cells Divide 207

11.1 Prokaryotes divide far more simply than do eukaryotes.
11.2 The chromosomes of eukaryotes are highly ordered structures.
11.3 Mitosis is a key phase of the cell cycle.
11.4 The cell cycle is carefully controlled.

Part III Genetic and Molecular Biology

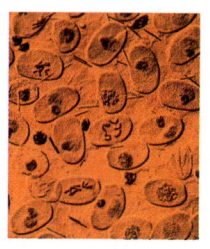

12 Sexual Reproduction and Meiosis 227

12.1 Meiosis produces haploid cells from diploid cells.
12.2 Meiosis has unique features.
12.3 The sequence of events during meiosis involves two nuclear divisions.
12.4 The evolutionary origin of sex is a puzzle.

13 Patterns of Inheritance 241

13.1 Mendel solved the mystery of heredity.
13.2 Human genetics follows Mendelian principles.
13.3 Genes are on chromosomes.

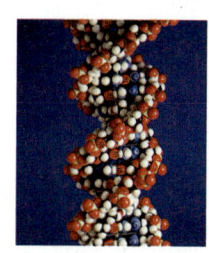

14 DNA: The Genetic Material 279

14.1 What is the genetic material?
14.2 What is the structure of DNA?
14.3 How does DNA replicate?
14.4 What is a gene?

15 Genes and How They Work 301

15.1 The Central Dogma traces the flow of gene-encoded information.
15.2 Genes encode information in three-nucleotide code words.
15.3 Genes are first transcribed, then translated.
15.4 Eukaryotic gene transcripts are spliced.

16 Gene Technology 319

16.1 Molecular biologists can manipulate DNA to clone genes.
16.2 Genetic engineering involves easily understood procedures.
16.3 Biotechnology is producing a scientific revolution.

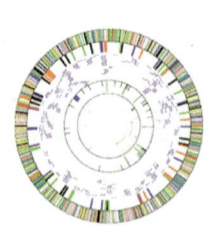

17 Genomes 343

17.1 Genomes can be mapped both genetically and physically.
17.2 Genome sequencing produces the ultimate physical map.
17.3 Being more complex does not necessarily require more genes.
17.4 Genomics is opening a new window on life.

18 Control of Gene Expression 361

18.1 Gene expression is controlled by regulating transcription.
18.2 Regulatory proteins read DNA without unwinding it.
18.3 Prokaryotes regulate genes by controlling transcription initiation.
18.4 Transcriptional control in eukaryotes operates at a distance.

19 Cellular Mechanisms of Development 381

19.1 Development is a regulated process.
19.2 Multicellular organisms employ the same basic mechanisms of development.
19.3 Aging can be considered a developmental process.

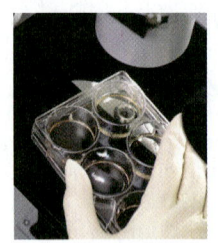

20 Cancer Biology and Cell Technology 405

20.1 Recombination alters gene location.
20.2 Mutations are changes in the genetic message.
20.3 Most cancer results from mutation of growth-regulating genes.
20.4 Reproductive cloning of animals, once thought impossible, isn't.
20.5 Therapeutic cloning is a promising but controversial possibility.

Part IV Evolution

21 Genes Within Populations 433

21.1 Genes vary in natural populations.
21.2 Why do allele frequencies change in populations?
21.3 Selection can act on traits affected by many genes.

22 The Evidence for Evolution 453

22.1 Evidence indicates that natural selection can produce evolutionary change.
22.2 Fossil evidence indicated that evolution has occurred.
22.3 Evidence for evolution can be found in other fields of biology.
22.4 The theory of evolution has proven controversial.

vi Contents

23 The Origin of Species 471

23.1 Species are the basic units of evolution.
23.2 Species maintain their genetic distinctiveness through barriers to reproduction.
23.3 We have learned a great deal about how species form.
23.4 Clusters of species reflect rapid evolution.

24 Evolution of Genomes and Developmental Mechanisms 491

24.1 Evolutionary history is written in genomes.
24.2 Developmental mechanisms are evolving.

Part V Diversity of Life on Earth

25 Systematics and the Phylogenetic Revolution 509

25.1 Biologists name organisms in a systematic way.
25.2 Scientists construct phylogenies to understand the evolutionary relationships among species.
25.3 Phylogenetics is the basis of all comparative biology.
25.4 All living organisms are grouped into one of a few major categories.
25.5 Molecular data are revolutionizing taxonomy.

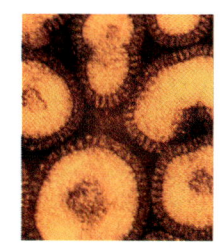

26 Viruses 531

26.1 Viruses are stands of nucleic acid encased within a protein coat.
26.2 Bacterial viruses exhibit two sorts of reproductive cycles.
26.3 HIV is a complex animal virus.
26.4 Nonliving infectious agents are responsible for many human diseases.

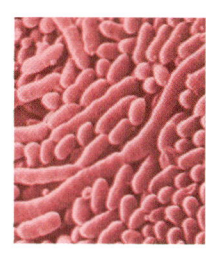

27 Prokaryotes 545

27.1 Prokaryotes are the smallest and most numerous organisms.
27.2 Prokaryotes exhibit considerable diversity in both structure and metabolism.
27.3 Prokaryotes are more complex than commonly supposed.
27.4 Prokaryotes are responsible for many diseases, but they also make important contributions to ecosystems.

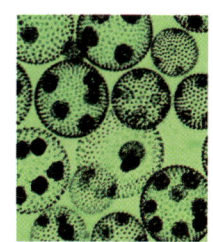

28 Protists 561

28.1 Eukaryotes probably arose by endosymbiosis.
28.2 The kingdom Protista is by far the most diverse of any kingdom.
28.3 Protists can be categorized into six groups.

29 Overview of Plant Diversity 579

29.1 Plants have multicellular haploid and diploid stages in their life cycles.
29.2 Nonvascular plants are relatively unspecialized but successful in many terrestrial environments.
29.3 Seedless vascular plants have well-developed conducting tissues in their sporophytes.
29.4 Seeds protect and aid in the dispersal of plant embryos.

30 Fungi 599

30.1 The fungi share several key characteristics.
30.2 There are four major groups of fungi.
30.3 Fungi participate in many symbioses.

31 Overview of Animal Diversity 617

31.1 Animals are multicellular heterotrophs without cell walls.
31.2 Animals are a very diverse kingdom.
31.3 The animal body plan has undergone many changes.
31.4 The way we classify animals is being reevaluated.

32 Noncoelomate Invertebrates 633

32.1 The classification of invertebrates is currently being reevaluated.
32.2 The simplest animals are not bilaterally symmetrical.
32.3 Acoelomates are solid worms that lack a body cavity.
32.4 Pseudocoelomates have a simple body cavity.

33 Coelomate Invertebrates 651

33.1 Mollusks were among the first coelomates.
33.2 Annelids were the first segmented animals.
33.3 Lophophorates appear to be a transitional group.
33.4 Arthropods are the most diverse of all animal groups.
33.5 Echinoderms are radially symmetrical as adults.

34 Vertebrates 683

34.1 Attaching muscles to an internal framework greatly improves movement.
34.2 Nonvertebrate chordates have a notochord but no backbone.
34.3 The evolution of vertebrates involved invasions of sea, land, and air.
34.4 Evolution among the primates has focused on brain size and locomotion.

Part VI Plant Form and Function

35 Plant Form 729

35.1 Meristems elaborate the plant body plan after germination.
35.2 Plants have three basic tissues, each composed of several cell types.
35.3 Root cells differentiate as they become distanced from the dividing root apical meristem.
35.4 Stems are the backbone of the shoot, transporting nutrients and supporting the aerial plant organs.
35.5 Leaves are adapted to support basic plant functions.

36 Vegetative Plant Development 755

36.1 Plant embryo development establishes a basic body plan.
36.2 The seed protects the dormant embryo from water loss.
36.3 Fruit formation enhances the dispersal of seeds.
36.4 Germination initiates post-seed development.

37 Transport in Plants 767

37.1 Water and minerals move upward through the xylem.
37.2 Dissolved sugars and hormones are transported in the phloem.

38 Plant Nutrition 781

38.1 Plants require a variety of nutrients in addition to the direct products of photosynthesis.
38.2 Global change could alter the balance among photosynthesis, respiration, and use of nutrients acquired through the soil.
38.3 Some plants have novel strategies for obtaining nutrients.
38.4 Plants can remove harmful chemicals from the soil.

39 Plant Defense Responses 795

39.1 Morphological and physiological features protect plants from invasion.
39.2 Some plant defenses act by poisoning the invader.
39.3 Some plants have coevolved with bodyguards.
39.4 Systemic responses also protect plants from invaders.

40 Sensory Systems in Plants 807

40.1 Plants respond to light.
40.2 Plants respond to gravity.
40.3 Plants respond to touch.
40.4 Water and temperature elicit plant responses.
40.5 The hormones that guide growth are keyed to the environment.

41 Plant Reproduction 831

41.1 The environment influences reproduction.
41.2 Flowers are highly evolved for reproducing.
41.3 Many plants can clone themselves by asexual reproduction.
41.4 How long do plants and plant organs live?

Part VII   Animal Form and Function

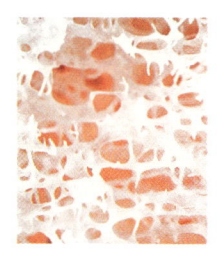

42 The Animal Body and How It Moves 855

42.1 The bodies of vertebrates are organized into functional systems.
42.2 Epithelial tissue forms membranes and glands.
42.3 Connective tissues contain abundant extracellular material.
42.4 Muscle tissue provides for movement, and nerve tissue provides for control.
42.5 Coordinated efforts of organ systems are necessary for locomotion.
42.6 Muscle contraction powers animal locomotion.

43 Fueling Body Activities: Digestion 887

43.1 Animals employ a digestive system to prepare food for assimilation by cells.
43.2 Food is ingested, swallowed, and transported to the stomach.
43.3 The small and large intestines have very different functions.
43.4 Neural, hormonal, and accessory organ regulation controls digestion.
43.5 All animals require food energy and essential nutrients.

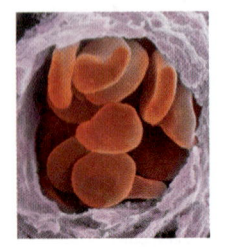

44 Circulation and Respiration 907

44.1 Circulatory systems are the transportation highways of the animal body.
44.2 The circulatory and respiratory systems evolved together in vertebrates.
44.3 The cardiac cycle drives the cardiovascular system.
44.4 Respiration has evolved to maximize the rate of gas diffusion.
44.5 Mammalian breathing is a dynamic process.

45 The Nervous System 939

45.1 The nervous system consists of neurons and supporting cells.
45.2 Nerve impulses are produced on the axon membrane.
45.3 Neurons form junctions called synapses with other cells.
45.4 The central nervous system consists of the brain and spinal cord.
45.5 The peripheral nervous system consists of sensory and motor neurons.

46 Sensory Systems 969

46.1 Animals employ a wide variety of sensory receptors.
46.2 Mechanical and chemical receptors sense the body's condition.
46.3 Auditory receptors detect pressure waves in the air.
46.4 Optic receptors detect light over a broad range of wavelengths.
46.5 Some vertebrates use heat, electricity, or magnetism for orientation.

47 The Endocrine System 991

47.1 Regulation is often accomplished by chemical messengers.
47.2 Lipophilic and polar hormones regulate their target cells by different means.
47.3 The hypothalamus controls the secretions of the pituitary gland.
47.4 Endocrine glands secrete hormones that regulate many body functions.

48 The Immune System 1013

48.1 Many body's most effective defenses are nonspecific.
48.2 Specific immune defenses require the recognition of antigens.
48.3 T cells organize attacks against invading microbes.
48.4 B cells label specific cells for destruction.
48.5 The immune system can be defeated.

49 Maintaining the Internal Environment 1039

49.1 The regulatory systems of the body maintain homeostasis.
49.2 The extracellular fluid concentration is constant in most vertebrates.
49.3 The functions of the vertebrate kidney are performed by nephrons.

50 Sex and Reproduction 1061

50.1 Animals employ both sexual and asexual reproductive strategies.
50.2 The evolution of reproduction among the vertebrates has led to internalization of fertilization and development.
50.3 Male and female reproductive systems are specialized for different functions.

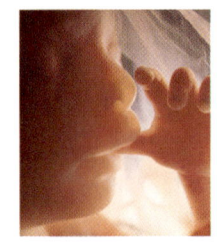

51 Vertebrate Development 1081

51.1 Fertilization is the initial event in development.
51.2 Cell cleavage and the formation of a blastula set the stage for later development.
51.3 Gastrulation forms the three germ layers of the embryo.
51.4 Body architecture is determined during the next stages of embryonic development.
51.5 Human development is divided into trimesters.

Part VIII Ecology and Behavior

52 Behavioral Biology 1105
52.1 Many behavioral patterns are innate.
52.2 Learning influences behavior.
52.3 Communication is a key element of many animal behaviors.
52.4 Evolutionary forces shape behavior.

53 Population Ecology 1137
53.1 Organisms must cope with a varied environment.
53.2 Populations are individuals of the same species that live in the same place.
53.3 Population dynamics depend critically upon age distribution.
53.4 Life histories often reflect trade-offs between reproduction and survival.
53.5 Population growth is limited by the environment.
53.6 The human population has grown explosively in the last three centuries.

54 Community Ecology 1161
54.1 Biological communities are composed of species that occur together.
54.2 Interactions among competing species shape ecological niches.
54.3 Predation has ecological and evolutionary effects.
54.4 Species within a community interact in many ways.
54.5 Ecological succession may increase the species richness of communities.

55 Dynamics of Ecosystems 1183
55.1 Chemicals cycle within ecosystems.
55.2 Energy flows through ecosystems.
55.3 Interactions occur among different trophic levels.
55.4 Biodiversity promotes ecosystem stability.

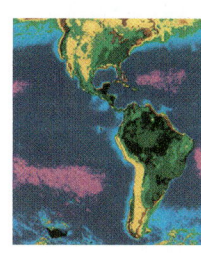

56 The Biosphere 1203
56.1 Climate shapes the character of ecosystems.
56.2 Biomes are widespread terrestrial ecosystems.
56.3 Aquatic ecosystems cover much of the earth.
56.4 Human activity is placing the biosphere under increasing stress.

57 Conservation Biology 1227
57.1 The new science of conservation biology is focused on conserving biodiversity.
57.2 The extinction crisis is a result of many factors.
57.3 Successful recovery efforts need to be multidimensional.

Glossary G-1
Credits C-1
Index I-1

About the Authors

Dr. Peter Raven is director of the Missouri Botanical Garden and Engelmann Professor of Botany at Washington University. A distinguished scientist, Dr. Raven is a member of the National Academy of Sciences, the National Research Council, and is a MacArthur and a Guggenheim fellow. He has received numerous honors and awards for his botanical research and work in tropical conservation, including the National Medal of Science. In addition to coauthoring this text, Raven has authored twenty other books and several hundred scientific articles.

George Johnson is professor emeritus of biology at Washington University in St. Louis, where he has taught genetics and general biology to undergraduates for 30 years. Also professor of genetics at Washington University School of Medicine, he is a student of population genetics and evolution. He has authored more than fifty scientific publications, and several high school and college texts, including *The Living World*, a very successful non-majors college biology text. He has pioneered the development of interactive CD-ROM and web-based investigations for biology teaching.

Jonathan Losos is a professor in the Department of Biology at Washington University and is also chair of the undergraduate Environmental Studies program. An evolutionary biologist, Losos's research has focused on studying patterns of adaptive radiation and evolutionary diversification in lizards. The recipient of several awards including the prestigious Theodosius Dobzhanksy and David Starr Jordan Prizes for outstanding young evolutionary biologists, Losos has published more than eighty scientific articles. He is currently the editor of the *American Naturalist*, a leading journal integrating the fields of evolutionary biology, behavior and ecology.

Susan Singer is professor of biology at Carleton College in Northfield, Minnesota, where she has taught introductory biology, plant biology, plant development, and developmental genetics for 18 years. Her research interests are focused on the development and evolution of flowering plants. Singer has authored numerous scientific publications on plant development, contributed chapters to developmental biology texts, and been actively involved with the education efforts of several professional societies. She serves on the NRC Committee on Undergraduate Science Education.

Preface

We first began work on this text in 1982, over twenty years ago. We set out to write a text that explained biology the way we taught it in the classroom—as the product of evolution. Most texts in 1982 relegated evolution to a few chapters in the diversity section. But evolution pervades biology, and is just as evident in the bacterial character of the mitochondria within your cells, in the biochemical similarities of photosynthesis and glycolysis, in the evolution of genes that control development—everywhere you look in biology, you see Darwin staring back at you. This evolutionary approach has proven popular among our nation's biology faculty, and most texts to greater or lesser degrees now adopt it.

Our text has changed a lot over twenty years, reflecting great changes in biology. The book has become more molecular, as biology has. In particular, a lot more is said about cell biology and development, areas where biology has made enormous strides. But our text remains fundamentally an evolutionary explanation of biology. In this edition, for example, while there is a new chapter on genomes, there is also a new "evo-devo" chapter that examines how genomes and developmental control mechanisms have evolved. This is just one example of our efforts to integrate biological questions and approaches at multiple levels of organization throughout the text, as we strive to guide students toward a connected understanding of biology.

This new seventh edition marks perhaps the greatest change in our text: the addition of two new biologists to the author team, Jonathan Losos, also at Washington University, and Susan Singer of Carleton College. Both made major contributions to the previous edition—Jonathan to the chapter on evolution and ecology, and Susan to the chapters on plant biology—and we are delighted to welcome them as full-fledged authors. In this edition their responsibilities have broadened to include the revolution that is ongoing in our understanding of systematics and evolution at the DNA level, matters that affect many chapters of this book.

Text development today involves an even greater number of people, as instructors from across the country are continually invited to share their knowledge and experience with us through reviews and focus groups. All of the feedback we have received has shaped this edition, resulting in new chapters, reorganization of the table of contents, and expanded coverage in key areas. This edition also incorporated the expertise of three consultants: Randy Di Domenico, University of Colorado—Boulder; Kenneth Mason, Purdue University; and Randall Phillis, University of Massachusetts—Amherst, who provided detailed suggestions for improving the clarity, flow and accuracy of large portions of the text.

How We Have Responded to You

Perhaps more than any other text on the market, this text has continued to evolve as a result of feedback from instructors teaching majors' biology. Overwhelmingly, they have told us that up-to-date content, a clear writing style, quality illustrations and dynamic presentation materials are the most important factors they consider when evaluating textbooks. We have let those values guide our revision of the text, as McGraw-Hill Education worked with those same instructors to create supplements that will help them in the classroom.

Up-to-date Treatment

The core of any majors' biology course is the exploration of cells and genetics, always covered in the first half of any majors' text. This book has been particularly aggressive in keeping its treatment of cell biology and genetics comprehensive and up-to-date. It was the first to present a chapter on cell communication, for example (other books soon followed). We are continuing in that tradition by incorporating such cutting edge topics as the structure of ATP synthetase, small RNAs and RNA editing. This edition also includes a new chapter 17 that explores what we can learn about genomes, covering topics ranging from human health issues and concerns about privacy, agricultural applications, and the potential of genomics in minimizing bioterrorism.

We did not contain this revision to a few select chapters in the first half of the text, however, and it's possible to point to many areas where treatment of recent breakthroughs has been integrated. By concluding our evolution section with a comparative approach to genomes and evolution of development, we were able to connect new breakthroughs in these areas and provide a springboard into the diversity section. Major changes in our understanding of phylogenetic relationships among land plants, protists, and fungi along with other major groups, are reflected in the extensive revision of the diversity chapters. Rapid advances in our knowledge of plant defense responses led to a new chapter on this topic.

Writing Style

Students of biology are responsible for an ever-growing volume of information, and that amount of detail is reflected in today's textbooks, which are increasingly becoming encyclopedic references as opposed to teaching texts.

But students are more likely to succeed with a text that they enjoy reading, that gives them a sense of the wonderment that inspired their own instructors to study biology. For this reason, we have endeavored to strike a balance; an inviting and accessible writing style with the level of authority and rigor expected of a majors' level text.

To further aid the student, every page or two-page spread in this book functions as a semi-independent learning module, organized under its own heading at the top of the left-hand page, with its own summary at the bottom of the right-hand page. This modular presentation makes the conceptual organization of the chapter clear, greatly enhancing student learning.

Illustrations

This book is set apart form others in that its artists, William Ober, M.D. and Claire Garrison, R.N., are part of the author team. Their respective backgrounds as a practicing physician and pediatric registered nurse, and their experience creating art for highly successful anatomy and physiology, zoology and marine biology textbooks, bring an invaluable contribution to the text. The close collaboration between text and illustration authors results in dynamic, accurate figures that aid student understanding and instructor presentation.

- **Combination Figures** These pieces combine a photo or micrograph with a line drawing, to make the connection between conceptual figures and what the student may encounter in lab (Figure 5.10, page 88).
- **Biochemistry Pathway Icons** Found in the discussion of metabolism, these icons help students follow complex processes by highlighting the step currently under discussion (Figure 9.15, page 174).
- **Phylogeny Guideposts** This icon is presented as each group is introduced, to remind students of relationships among diverse organisms (TA 32.1, page 636).
- **Process Boxes** These figures include step-by-step descriptions to walk the student through a compact summary of important concepts (Figure 6.18, page 121)

We have also been fortunate in that this collaboration has allowed us to carefully integrate explanatory text into the figures. The end results are uncluttered, easy to follow illustrations that guide a student through a concept. They also benefit the instructor, as figures without distracting captions can be used for presentation while still allowing instructors to tell their own story.

What Sets this Book Apart

Those who have not used or reviewed previous editions will want to know how this book differs from others.

Evolutionary Focus

The treatment of evolution in this book differs from others in a simple but very important way: Evolution is the organizing principle guiding the teaching of each chapter. Instead of leaping from chemistry directly to cell structure as in other books, this book uses the chemistry of the first chapters to examine the origin of life and the evolution of cells; the cell chapters that follow can then be seen in a broader evolutionary context. Similarly, the treatment of animal anatomy and physiology in other books is largely limited to structure and function—this is the organ and this is how it functions. This book examines each animal body system in terms of how it has evolved. Every section of this book, whether it is genetics or plant biology, presents biology from an evolutionary perspective.

Chemistry in a Biological Context

In talking to students over 30 years of teaching freshman biology, a consistent student complaint has been that the introductory biology course begins with a heavy jolt of chemistry. In other books, only after as many as 100 pages of chemistry do students encounter any biology. This is very off-putting for many students, and gets the course off to a rocky start. This book, by contrast, integrates the chemistry of the first section with biological themes. The treatment of macromolecules in Chapter 3 starts with proteins, which can be easily understood without detailed knowledge of carbohydrates. This arrangement has the distinct advantage of starting the student off with material of obvious relevance to biology.

A Modern Approach

Some of the most obvious differences between this book and others can be seen in the second half of the book, that part devoted to coverage of evolution, diversity, plant biology, anatomy and physiology, and ecology.

Evolution. Our approach to evolutionary biology is unique in two respects. First, we strongly emphasize the role of experiments in studying evolution. Although much of evolutionary biology concerns the study of what happened in the past, that does not mean that experimental approaches are impossible. We emphasize the role that experiments play in studying evolutionary phenomena. More generally, like any detective story, we point out how various approaches must be integrated to fully understand evolutionary diversity.

Second, our book devotes an entire chapter to the evidentiary basis for evolutionary biology. Unlike other aspects of biology, or science in general, the factual basis of evolutionary biology is disputed by some segments of society. Thus, we feel that it is important to clearly present the diversity and depth of evolution that leads almost all biologists to conclude that evolution has occurred. We feel that it is essential for all college biology students—regardless of their own opinions—to understand the scientific basis for this view.

Diversity. Our text has been organized so the diversity section is framed by a discussion of the revolution in taxonomy and phylogenetics (Chapter 25). Complete with vivid examples of dramatic changes in our understanding of relationships among organisms, this chapter can be used alone as an abbreviated approach to diversity or as a foundation for a more comprehensive evolutionary investigation of the diversity of life in the chapters that follow.

The book also differs from other majors' biology textbooks in that its coverage of diversity is more extensive. Consider for example the invertebrates. Other books devote as few as 30 pages to the invertebrates, presenting only the briefest of sketches of what used to be the core of traditional biology courses. This book devotes more than twice as many pages to the invertebrates, followed by a more comprehensive chapter on the vertebrates than is found in other books. Why is this more extensive treatment of diversity important? Even in courses that don't cover diversity in detail, it is important that students be able to uncover for themselves the relations among animal groups.

Plant Biology. The plant biology chapters have undergone extensive revision, and are now organized to lead the student through the plant life cycle. In addition, we have carefully integrated both developmental and genetic perspectives, a fusion not found in other texts. For example, Chapter 36, Vegetative Plant Development, explores root formation in the context of the *monopterous* mutant of *Arabidopsis* that fails to make a root. The shift from its developmental role to its functional role as an auxin receptor begins to move students toward a physiological understanding of plant function.

Anatomy and Physiology. Most books devote nearly the same amount of space to anatomy and physiology, about 250 pages. The differences lie primarily in approach, this book having a more evolutionary focus than others, and in its emphasis on fundamentals.

Ecology and Behavior. We take an integrative view to understanding how the environment functions and how organisms interact with it. This section is broken into different chapters, such as behavioral ecology, population ecology, and community ecology, but the topics are carefully integrated. Moreover, we apply this information extensively in Chapters 56 and 57 (The Biosphere and Conservation Biology) to address the environmental issues facing our planet. We believe it is of the utmost importance that students understand the scientific bases to current problems so they can evaluate efforts to solve them.

Changes to the Seventh Edition

The seventh edition of *Biology* is the result of extensive analysis of the text and evaluation of input from biology instructors who conscientiously reviewed chapters during various stages of this revision. We have utilized the constructive comments provided by these professionals in our continuing efforts to enhance the strengths of the text. Listed first are general changes that have been made to the entire text, which is then followed by specific changes for each part.

End-of-chapter Pedagogy

The end-of-chapter student review has been greatly expanded, offering students a full-page chapter review and three assessment tools: Self Test, multiple choice questions; Test Your Visual Understanding, questions based on a figure from the chapter; and Apply Your Knowledge, critical thinking questions. The assessment doesn't end there, however. These tools are carried over to the web where the student can take an interactive version of the test and receive instructional feedback.

Inquiry Questions

In this edition we have developed Inquiry Questions, which follow the legend in figures presenting graphed data. These questions require the student to think about the information contained in the graph in even greater depth, increasing their understanding of, and facility with, the material.

Answers to both end of chapter questions and Inquiry Questions are found on the Online Learning Center at www.mhhe.com/raven7.

Volumes

We recognize that instructors don't always use the entire text, so we now offer *Biology* in the following volumes:

Volume 1 Chapters 1–20 Chemistry, Cell and Genetics
Volume 2 Chapters 35–51 Plant Biology and Animal Biology
Volume 3 Chapters 21–34, 52–57 Evolution, Diversity and Ecology

Content Changes by Part
Part I The Origin of Living Things

Part I was revised with the intention of creating a more solid foundation of key concepts in biology, which students can then build on in later chapters. Discussions are now clearer and better supported with illustrations.

New Topics and Revised Treatments

Chapter 1 Properties of life, hierarchical organization *Revised*; Additional topics in evolution *New*

Chapter 2 The Nature of Molecules *Entire chapter revised*

Chapter 3 Figures on chaperons and protein denaturing *New*; Protein folding, lipids *Revised*

Chapter 4 Figures on endosymbiosis, Domains/Kingdoms, phylogenetic tree of life *New* Bacteria and archaebacteria, microfossils *Revised*

Part II Biology of the Cell

Randall Phillis assisted in the revision of Part II by directing the authors to concepts that needed additional detail, and by providing suggestions for improving the accuracy and parsimony of the narrative. Concepts that were covered too briefly in previous editions are now supported with more extensive discussion and new illustrations.

New Topics and Revised Treatments

Chapter 6 Membrane microdomains *New* Osmosis; coupled transport *Revised*

Chapter 7 Signal amplification, expression of cellular identification *Revised*

Chapter 8 Redox reactions, ATP functioning *Revised*

Chapter 9 ATP synthetase *New*; Electron transport; reducing power; chemiosmosis *Revised*

Chapter 10 The Calvin cycle *Revised*

Chapter 11 Cell cycle control *New*; Chromosome structure *Revised*

Part III Genetic and Molecular Biology

With the help of Kenneth Mason, Part III was carefully updated to incorporate the most current research. Chapter 13, Patterns of Inheritance, was rewritten for better organization and clearer presentation. Two new chapters provide expanded discussion in fields where our knowledge has grown exponentially.

Chapter 17, "Genomes" integrates plant and animal genomics, functional genomics and proteomics in a chapter that is inquiry driven rather than a discussion of techniques.

Chapter 20, "Cancer Biology and Cell Technology" explores two areas where recent advances in cell and molecular biology have the potential of revolutionizing medicine. The first is cancer, where research into the molecular events leading to cancer is beginning to suggest effective therapies. The second is cell technology, including cloning, embryonic stem cells, and the exciting and controversial proposal of therapeutic cloning.

New Topics and Revised Treatments

Chapter 12 Meiotic prophase *Revised*

Chapter 13 Patterns of inheritance *Entire chapter revised*

Chapter 14 Eukaryotic DNA replication *Revised*

Chapter 15 Eukaryotic transcription *New*

Chapter 16 The tools of genetic engineering *New*

Chapter 18 Small RNAs, iRNA, RNA editing *New* Transcriptional control in prokaryotes *Revised*

Chapter 19 Vertebrate embryonic axis formation, evolution of homeotic genes *New* Cell movement; cell induction; embryonic determination; pattern formation *Revised*

Part IV Evolution

The Evolution section has been revised to bring more experimental data and analysis into the discussions. Because presentation of the experimental data used to derive conclusions and concepts is key to understanding how the concepts arose from the research, you will see that graphs and charts have become more plentiful in these chapters. The evolution of many groups is reassessed in light of new molecular data.

Chapter 24, "Evolution of Genomes and Developmental Mechanisms" is a new comparative genomics chapter that addresses our emerging understanding of the evolution of development, and helps to provide a conceptual framework for the diversity chapters that follow. The chapter was developed in conjunction with Chapter 17, Genomes, to first provide students with an understanding of what we can learn about genomes, and then having learned about evolution, delve into a deeper discussion of how development has evolved to yield novel phenotypes.

New Topics and Revised Treatments

Chapter 21 Measuring fitness, components of fitness, role of selection in maintaining variation *New* Hardy-Weinberg *Revised*

Chapter 22 Darwin's finches, industrial melanism *New* Evidence from developmental biology for evolution *Revised*

Chapter 23 Plant speciation by chromosomal change, the future of evolution *New*

Part V Diversity of Life on Earth

The fungi chapter has been moved to the phylogenetically appropriate place in the diversity section following plant diversity. Where appropriate, chapters in the diversity unit have been updated to reflect phylogenetic changes. The thoroughly revised and rewritten **Chapter 25, "Systematics and the Phylogenetic Revolution"** addresses the current tension between taxonomy and systematics. The chapter can be used alone, to teach the basic concepts of diversity, or can be used as a starting place for a more in-depth study of this area of biology.

New Topics and Revised Treatments

Chapter 25 Phylogenetics and classification *Entire chapter revised*

Chapter 26 Virus genomes *New;* Viral diseases and HIV *Revised*

Chapter 27 Term "eubacteria" replaced with "bacteria," figures of cell structure and clades *New*

Chapter 28 Phylogenetic approach *New;* Protist disease in South, Central and North America, relationships between algae and land plants *Revised*

Chapter 29 Fossil evidence of ancient angiosperm Archaefructacea, evolution of triploid endosperm *New*
Monophyletic relationships between ferns and horsetails *Revised*

Chapter 30 Fungi *Entire chapter revised*

Chapter 31 Protostomes and deuterostomes *New and expanded*
Classification *Revised*

Chapter 32 Protostome phylogeny, rotifers and cycliophora *Revised*

Chapter 33 Mollusks, annelids, arthropods, and echinoderms combined into one chapter *New*

Chapter 34 Characteristics and phylogeny illustration, primate evolution *New*

Part VI Plant Form and Function

The plant biology chapters have been revised so that the traditional discussion of evolutionary influences on plant form and function are brought into a developmental context. Evolution is still presented as the underlying explanation for the character of vascular tissues, seeds, flowers, and fruits, however the developmental processes that produce these organs are now given more emphasis. Two previously combined topics, transport and nutrition, have been split into separate chapters allowing for more in-depth discussion of both topics.

Chapter 39, "Plant Defense Responses" is a new chapter that provides a thorough discussion of secondary compounds and their roles in both plant defense and human applications. Wound responses and R gene mediated responses are explored in depth with an emphasis on signaling pathways.

New Topics and Revised Treatments

Chapter 35 Updated photographs, discussion of genetic regulation of trichomes *Revised*

Chapter 36 Discussion of signal transduction in germination, comparison of roles of *Hox* genes in plant and animal development *Revised*

Chapter 37 Water relations problems, mRNA transport in phloem *New*

Chapter 38 Newly expanded chapter on plant nutrition. Effects of global change on photosynthesis and balance of plant nutrients, phytoremediation *New*
Nutritional symbioses *Revised and expanded*

Chapter 40 Signal transduction mediated by light including phot1 *New*
Light responses *Revised*

Chapter 41 Plant reproduction *Entire chapter revised and reorganized*

Part VII Animal Form and Function

With the assistance of Randy DiDomenico, many discussions were rewritten for better organization and clarity. Previous chapters on circulation and respiration were combined into one chapter, as were chapters on body organization and locomotion.

New Topics and Revised Treatments

Chapter 42 Combined organization of the animal body and locomotion into one chapter, coordination of organ systems *New*

Chapter 43 Neural, hormonal and accessory organ regulation *New*
Small intestine discussion reorganized to group all functions together *Revised*

Chapter 44 Maximizing rate of gas diffusion *New*
Integration of circulation and respiration chapter *Revised*

Chapter 45 Graded potentials *New;* Membrane and action potentials, synapses and drug addiction *Revised*

Chapter 46 Sensory transduction *Revised*

Chapter 47 The Endocrine System *Entire chapter revised*

Chapter 48 Immunoglobulins, illustrated table *New*
AIDS *Revised*

Chapter 49 Discussion of ammonia, urea and uric acid reorganized, nephron *Revised*

Chapter 50 Sex determination, reptiles and birds *Revised*
Human intercourse *Omitted*

Chapter 51 Combined discussion of chick and mammalian extraembryonic membranes, neurulation *Revised*

Part VIII Ecology and Behavior

The ecology and behavior chapters were moved to follow diversity and physiology, where these chapters are more often taught. There is now an even greater emphasis on experimental data and analysis.

New Topics and Revised Treatments

Chapter 52 Animal cognition *New;* Integration of animal behavior and behavioral ecology chapters *Revised*

Chapter 53 Introduction to ecology, integration of autoecology and population ecology *New* Population regulation and limitation, human population growth *Revised*

Chapter 54 Introduction, definition of community *New* Parasitism, succession, disturbance *Revised*

Chapter 55 Geochemical cycles, energy flow, species richness *Revised*

Chapter 56 Differences between aquatic and terrestrial ecosystems *New* Integration of biosphere and future of the biosphere chapters, global climate change, El Niño *Revised*

Chapter 57 Chapter organization, biodiversity hotspots, amphibian extinctions, invasive species *New* Economic benefits of biodiversity *Revised*

Overview of Changes to BIOLOGY, Seventh Edition

All Cell & Genetics Chapters Extensively Revised
In addition to discussing important advances, many sections have been reworked for improved clarity.

Chapter 17, "Genomes"
This new chapter describes how researchers sequence entire genomes, and how the information is being used.

Chapter 20, "Cancer Biology and Cell Technology"
This new chapter updates progress in understanding cancer, and introduces many new advances in cloning and stem cell technology.

Evolution & Diversity Sections Extensively Revised
New RNA and genomic information is leading to a reassessment of traditional evolutionary phylogenies.

Chapter 24, "Evolution of Genomes and Developmental Mechanisms"
This new chapter explores the wealth of new information on genome sequences, and introduces the new and exciting field of "evo/devo", the evolution of development.

Treatment of Plant Biology Expanded
A total of seven plant chapters provide extensive plant biology coverage with a molecular development point of view. Chapters have been organized to lead the student through the life cycle of a plant.

Chapter 39, "Plant Defense Responses"
This new chapter captures the excitement of this area of biology, which has seen rapid advances and recent breakthroughs in understanding plant defense responses.

Ecology Chapters Updated and Expanded
Up-to-date examples have been integrated into all chapters; note the use of case histories in Chapter 57.

Physiology Chapters Reworked
Discussions of processes like nervous conduction reworked for increased clarity, and related subjects like circulation and respiration treated together.

End-of-Chapter Assessment
Two pages are now devoted to student review and assessment. A full-page chapter summary is followed by multiple choice, illustration-based and application questions.

Illustrations
Many new illustrations clarify difficult concepts; others illustrate tables to aid understanding. Wherever data are presented in graphs, the figure is accompanied by an Inquiry Question to test the student's understanding.

Teaching and Learning Supplements

McGraw-Hill offers various tools and technology products to support *Biology*. Students can order supplemental study materials by contacting their local bookstore or by calling 800-262-4729. Instructors can obtain teaching aids by calling the Customer Service Department at 800-338-3987, visiting our website at www.mhhe.com/biology, or contacting their local McGraw-Hill sales representative.

For the Instructor:

Digital Content Manager CD-ROM

This multimedia collection of visual resources allows instructors to utilize artwork from the text in multiple formats to create customized classroom presentations, visually based tests and quizzes, dynamic course website content, or attractive printed support material. The digital assets on this cross-platform CD-ROM include:

Art Library Color-enhanced, digital files of all illustrations in the book, plus the same art saved in unlabeled and gray scale versions, can be readily incorporated into lecture presentations, exams, or custom-made classroom materials. Upsized labels make the images appropriate for use in large lecture halls.

TextEdit Art Library Every line art piece is placed into a PowerPoint presentation that allows the user to revise, move, or delete labels as desired for creation of customized presentations or for testing purposes.

Active Art Library Active Art consists of art files that have been converted to a format that allows the artwork to be edited inside of PowerPoint. Each piece can be broken down to its core elements, grouped or ungrouped, and edited to create customized illustrations.

Animations Library Full color presentations involving key process figures in the book have been brought to life via animation. These animations offer flexibility for instructors and were designed to be used in lecture. Instructors can pause, rewind, fast forward, and turn audio off/on to create dynamic lecture presentations.

PowerPoint Lecture Outlines These ready-made presentations combine art and lecture notes for each of the 57 chapters of the book. The presentations can be used as they are, or can be customized to reflect your preferred lecture topics and organization.

PowerPoint Outlines The art, photos, and tables for each chapter are inserted into blank PowerPoint presentations to which you can add your own notes.

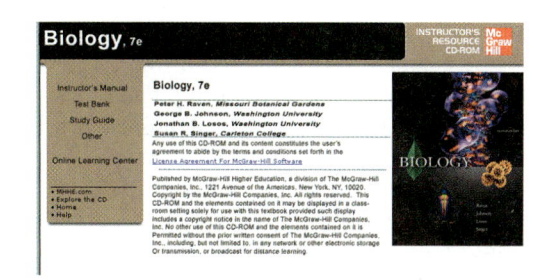

Photo Library Like the Art Library, digital files of all photographs from the book are available.

Table Library Every table that appears in the book is provided in electronic form.

Video Library Contains digitized video clips that can be inserted into a PowerPoint lecture.

Additional Photo Library Over 700 photos, not found in *Biology*, are available for use in creating lecture presentations.

Instructor's Testing and Resource CD-ROM

The cross-platform CD-ROM contains the Instructor's Manual and Test Item File, both available in both Word and PDF formats. The manual contains chapter synopses, objectives, key terms, outlines, instructional strategies and sources for additional visual resources. The Test Bank offers questions that can be used for homework assignments or the preparation of exams. The computerized test bank utilizes Brownstone Diploma testing software, which allows the user to quickly create customized exams. This user-friendly program allows instructors to search questions by topic, format, or difficulty level; edit existing questions or add new ones; and scramble questions and answer keys for multiple versions of the same test.

Transparencies

A set of 1300 transparency overheads includes every piece of line art and table in the text. The images are printed with better visibility and contrast than ever before, and labels are large and bold for clear projection.

Online Learning Center
www.mhhe.com/raven7

Instructor resources at this site include access to online laboratories, course-specific current articles, real-time news feeds, course updates and research links.

Course Delivery Systems

With help from our partners, WebCT, Blackboard, TopClass, eCollege, and other course management systems, instructors can take complete control over their course content. These course cartridges also provide online testing and powerful student tracking features. The *Biology* Online Learning Center is available within all of these platforms.

For the Student:

Online Learning Center
www.mhhe.com/raven7

The site includes quizzes for each chapter, interactive activities, and answers to questions from the text. Turn to the inside cover of the text to learn more about the exciting features provided for students through the enhanced *Biology* Online Learning Center.

Student Study Guide

This student resource contains activities and questions to help reinforce chapter concepts. The guide provides students with tips and strategies for mastering the chapter content, concept outlines, concept maps, key terms and sample quizzes.

Acknowledgements

Our goal for *Biology* has always been to present the science in an interesting and engaging way while maintaining a comprehensive and authoritative text. This is a lofty goal considering the mountain of information and research we must go through just to update the text from one edition to the next. This seventh edition would not have been possible without the contributions of many. We are indebted to our colleagues across the country and around the globe that provided numerous suggestions on how to improve on the sixth edition. We wish particularly to thank Kenneth Mason of Purdue University, Randy DiDomenico of the University of Colorado, Boulder, and Randall Phyllis of the University of Massachusetts, Amherst for very detailed advice on how to improve large sections of the text.

As any author knows, a textbook is made not only by an author team aided by their colleagues, but also by a publishing team, a group of people that guide the raw book written by the authors through a yearlong process of reviewing, editing, fine-tuning and production. This edition was particularly fortunate in its book team, led by Patrick Reidy, sponsoring editor; Anne Winch, developmental editor; Tami Petsche, marketing manager; Peggy Selle, project manager; Michelle Whitaker, designer; Megan Jackman and Elizabeth Sievers, off-site editors; Kennie Harris, copy editor, and many more people behind the scenes.

The illustrations are critically important to a biology text, and ours continue to be superbly conceived and rendered by Bill Ober and Claire Garrison.

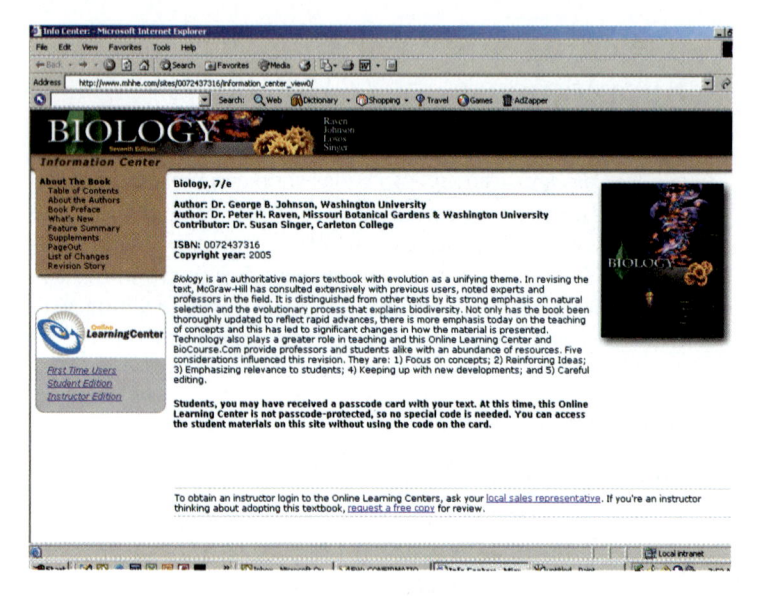

As always, we have had the support of spouses and children, who have seen less of us than they might have liked because of the pressures of getting this revision completed. They have adapted to the many hours this book draws us away from them and, even more than us, look forward to its completion.

As with every edition, acknowledgements would not be complete without thanking the generations of students who have used the many editions of this text. They have taught us as least as much as we have taught them, and, thanks to e-mail, are an increasing part of our lives.

Finally, we need to thank our reviewers. Every text owes a great deal to those instructors across the country who review it. Serving as sensitive antennae for errors and omissions, and as sounding boards for new approaches, reviewers are among the most valuable tools at an author's disposal. Many improvements in this edition are the direct result of their suggestions. Every one of them has our heartfelt thanks.

Reviewers of the Seventh Edition

Heather Addy   *University of Calgary*
Lawrence A. Alice   *Western Kentucky University*
Terry C. Allison   *The University of Texas–Pan American*
Loran C. Anderson   *Florida State University*
Mohammad Ashraf   *City Colleges of Chicago*
Ellen Baker   *Santa Monica College*
R. Neal Band   *Michigan State University*
Dale L. Barnard   *Utah State University*
Diane C. Bassham   *Iowa State University*
Wayne M. Becker   *University of Wisconsin–Madison*
Robert L. Beckmann   *North Carolina State University*
Gerald Bergtrom   *University of Wisconsin–Milwaukee*
Cheryl Briggs   *University of California–Berkeley*
Trey Broadhurst   *Montgomery College*
Arthur L. Buikema, Jr.   *Virginia Tech*
Ann B. Burgess   *University of Wisconsin–Madison*
Carol A. Burkart   *Mountain Empire Community College*
D. Brent Burt   *Stephen F. Austin State University*
David Byres   *Florida Community College at Jacksonville*
Les Chappell   *University of Aberdeen, UK*

Jung Choi *Georgia Institute of Technology*
Don Cipollini *Wright State University*
Richard J. Cogdell *University of Glasgow*
Jerry Cook *Sam Houston State University*
David T. Corey *Midlands Technical College*
George Cornwall *University of Colorado–Boulder/Metropolitan State College of Denver*
Francie Smith Cuffney *Meredith College*
Paul V. Cupp, Jr. *Eastern Kentucky University*
James A. Danoff-Burg *Columbia University*
Sandra L. Davis *University of Louisiana at Monroe*
Mark D. Decker *University of Minnesota*
Mary B. Dettman *Seminole Community College*
John Dickerman *Northern Illinois University*
Cathy Donald-Whitney *Collin County Community College*
Thomas W. Dreschel *Brevard Community College, Kennedy Space Center*
Carolyn S. Dunn *University of North Carolina at Wilmington*
Roland R. Dute *Auburn University*
Frederick B. Essig *University of South Florida*
Bruce E. Felgenhauer *University of Louisiana at Lafayette*
James Franzen *University of Pittsburgh*
Andrea Gargas *University of Wisconsin–Madison*
John V. Gartner, Jr. *St. Petersburg College*
John R. Geiser *Western Michigan University*
Florence K. Gleason *University of Minnesota*
John S. Graham *Bowling Green State University*
John S. Greenwood *University of Guelph*
Peggy J. Guthrie *University of Central Oklahoma*
Adrian Hailey *University of Bristol*
Dana Brown Haine *Central Piedmont Community College*
Robert O. Hall *University of Wyoming*
Robert W. Hamilton *Loyola University Chicago*
David S. Hibbett *Clark University*
Leland N. Holland, Jr. *Pasco-Hernando Community College*
Eva A. Horne *Kansas State University*
Jeffrey Jack *University of Louisville*
Lee F. Johnson *The Ohio State University*
Gregory A. Jones *Santa Fe Community College*
Walter S. Judd *University of Florida*
Richard R. Jurin *University of Northern Colorado*
Thomas C. Kane *University of Cincinnati*
Ronald Keiper *Valencia Community College*
John J. Kelly *Loyola University Chicago*
Cheryl A. Kerfeld *University of California, Los Angeles*
David J. Kittlesen *University of Virginia*
William Kroll *Loyola University Chicago*
Harry D. Kurtz, Jr. *Clemson University*
Roberta Lammers-Campbell *Loyola University Chicago*
Peter Lavrentyev *The University of Akron*
Michael Lawson *Missouri Southern State College*
Roger M. Lloyd *Florida Community College at Jacksonville*
David Magrane *Morehead State University*
Richard Malkin *University of California–Berkeley*
Terry C. Maxwell *Angelo State University*
Michael McLeod *Belmont Abbey College*
Frank J. Messina *Utah State University*
Sandra Millward *University of Cincinnati*
Jacalyn S. Newman *University of Pittsburgh*
Janice Moore *Colorado State University*
Deborah A. Neher *University of Toledo*

Erik T. Nilsen *Virginia Tech*
T. Mark Olsen *University of Notre Dame*
John C. Osterman *University of Nebraska–Lincoln*
Daniel M. Pavuk *Bowling Green State University*
Andrew J. Pease *Villa Julie College*
Rhoda E. Perozzi *Virginia Commonwealth University*
Carolyn Peters *Spoon River College*
Susan Phillips *Brevard Community College*
Eric R. Pianka *University of Texas at Austin*
Aleksandar Popadic *Wayne State University*
Angela R. Porta *Kean University*
Calvin A. Porter *Xavier University of Louisiana*
Elena Pravosudova *Sierra College*
Linda R. Richardson *Blinn College*
Laurel Roberts *University of Pittsburgh*
Charles L. Rutherford *Virginia Tech University*
Erik P. Scully *Towson University*
Wendy E. Sera *Seton Hall University*
Alison M. Shakarian *Salve Regina University*
Neil F. Shay *University of Notre Dame*
Shree R. Singh *Alabama State University*
David A. Smith *Lock Haven University of Pennsylvania*
Willie Smith *Brevard Community College*
Nancy G. Solomon *Miami University*
Alan J. Spindler *Brevard Community College*
Ann Springer *Hillsborough Community College*
Amy C. Sprinkle *Jefferson Community College Southwest*
Bruce Stallsmith *University of Alabama in Huntsville*
John D. Story *North West Arkansas Community College*
Robert Sullivan *Marist College*
Marshall D. Sundberg *Emporia State University*
Pamela S. Thomas *University of Central Florida*
Patrick A. Thorpe *Grand Valley State University*
Rani Vajravelu *University of Central Florida*
Carol M. F. Wake *South Dakota State University*
Jane Waterman *University of Central Florida*
Cindy Martinez Wedig *University of Texas–Pan American*
Olivia Masih White *University of North Texas*
Lance R. Williams *The Ohio State University*
Michael Zimmerman *University of Wisconsin–Oshkosh*

General Biology Symposium

Each year McGraw-Hill holds a General Biology Symposium, which is attended by instructors from across the country. These events are an opportunity for editors from McGraw-Hill to gather information about the needs and challenges of instructors teaching the major's biology course, however it also offers professors a forum for exchanging ideas and experiences with colleagues they might not have otherwise met. The feedback we have received has been invaluable, and has contributed to the success of *Biology* and its supplements.

2003

Marc Ammerlaan *University of Michigan–Ann Arbor*
Scott Chandler *University of California, Los Angeles*
Bill Collins *SUNY at Stony Brook*
Elizabeth Connor *University of Massachusetts–Amherst*
Steve Connor *University of South Florida*

Robert Fulginiti   *Xavier University*
Florence Gleason   *University of Minnesota*
Carla Haas   *Penn State University*
David Julian   *University of Florida*
Steve Kelso   *University of Illinois at Chicago*
Bob Locy   *Auburn University*
Kenneth Mason   *Purdue University*
Nancy Solomon   *Miami University–Oxford*
Bill Stein   *SUNY at Binghamton*
Sally Swain   *Middle Tennessee State University*
Linda Waters   *University of Central Florida*

2002

Richard J. Cyr   *Penn State University*
Randy DiDomenico   *University of Colorado–Boulder*
Doug Gaffin   *University of Oklahoma*
Marielle Hoefnagels   *University of Oklahoma*
Jan Jenner   *Science Writer*
Cheryl A. Kerfeld   *University of California, Los Angeles*
Kenneth Mason   *Purdue University*
Michael Meighan   *University of California–Berkeley*
Jane Phillips   *University of Minnesota*
Randall W. Phillis   *University of Massachusetts–Amherst*
Joelle Presson   *University of Maryland*
Leslie Winemiller   *Texas A & M University*
Denise Woodward   *Penn State University*

2001

Mark Ammerlaan   *University of Michigan–Ann Arbor*
Doug Gaffin   *University of Oklahoma*
Jon C. Glase   *Cornell University*
Richard Hallik   *University of Arizona*
Marielle Hoefnagels   *University of Oklahoma*
Fernan Jaramillo   *Carleton College*
Randall Johnson   *University of California, San Diego*
Kenneth Mason   *Purdue University*
Sally Frost-Mason   *Purdue University*
Jorge Moreno   *University of Colorado–Boulder*
Tom Owens   *Cornell University*
Deanna Raineri   *University of Illinois at Urbana-Champaign*
Jon Ruehle   *University of Central Arkansas*
Steven A. Wasserman   *University of California, San Diego*

Reviewers of the Sixth Edition

Michael Adams   *Pasco-Hernando Community College*
Sylvester Allred   *Northern Arizona University*
Lon Alterman   *Clarke College*
Elena Amesbury   *University of Florida*
William Anyonge   *University of California–Los Angeles*
Amir Assadirad   *Delta College*
Gary I. Baird   *Brigham Young University*
Ellen Baker   *Santa Monica College*
Stephen W. Banks   *Louisiana State University–Shreveport*
Ruth Beattie   *University of Kentucky*
Samuel N. Beshers   *University of Illinois*
Christine Konicki Bieszczad   *Saint Joseph College*
John Birdsell   *University of Arizona*
Brenda C. Blackwelder   *Central Piedmont Community College*

Sandra Bobrick   *Community College of Allegheny County Allegheny Campus*
Randall Breitwisch   *University of Dayton*
Mark Browning   *Purdue University*
Roger Buckanan   *Arkansas State University*
Theodore Burk   *Creighton University*
John S. Campbell   *Northwest College*
John R. Capeheart   *University of Houston–Downtown*
Michael S. Capp   *Carlow College*
Jeff Carmichael   *University of North Dakota*
George P. Chamuris   *Bloomsburg University*
Susan Cockayne   *Brigham Young University*
William Cohen   *University of Kentucky*
W. Wade Cooper   *Shelton State Community College*
Lisa M. Coussens   *University of California–San Francisco, Cancer Research Institute*
Wilson Crone   *Hudson Valley Community College*
Paul V. Cupp Jr.   *Eastern Kentucky University*
Richard Cyr   *The Pennsylvania State University*
Grayson Davis   *Trinity University*
Mark A. DeCrosta   *University of Tampa*
David L. Denlinger   *Ohio State University*
C. Lynn Dorn   *Valencia Community College*
Charles D. Drewes   *Iowa State University*
Sondra Dubowsky   *Allen County Community College*
Peter I. Ekechukwu   *Horry-Georgetown Technical College*
Dennis Emery   *Iowa State University*
Frederick B. Essig   *University of South Florida*
Bruce Evans   *Huntington College*
Deborah Fahey   *Wheaton College*
Linda E. Fisher   *University of Michigan–Dearborn*
Rob Fitch   *Wenatchee Valley College*
Robert Fogel   *University of Michigan*
James Franzen   *University of Pittsburgh–Pittsburgh Campus*
William Friedman   *University of Colorado*
Lawrence Fritz   *Northern Arizona University*
Bernard Frye   *University of Texas at Arlington*
Robert J. Full   *University of California–Berkeley*
Warren Gallin   *University of Alberta*
Darrell Galloway   *The Ohio State University*
Ted Gish   *St. Mary's College*
Donald Glassman   *Des Moines Area Community College*
Jim Glenn   *Red Deer College*
Jim R. Goetze   *Laredo Community College*
Jack M. Goldberg   *University of California–Davis*
Elizabeth Godrick   *Boston University*
Dalton Gossett   *Louisiana State University–Shreveport*
John Griffis   *Joliet Junior College*
Kathryn Gronlund   *New Mexico State University–Carlsbad*
Elizabeth L. Gross   *The Ohio State University*
Patricia A. Grove   *College of Mount St. Vincent*
Randolph Hampton   *University of California–San Diego*
Sehoya E. Harris   *The Pennsylvania State University*
Carla Ann Hass   *The Pennsylvania State University*
Chris Haynes   *Shelton State Community College*
Albert A. Herrera   *University of Southern California*
Pamela Higgins   *Allentown College of St. Francis DeSales*
Richard Hill   *Michigan State University*
Phyllis Hirsch   *East Los Angeles College*
Victoria Hittinger   *Rhode Island College*
Nan Ho   *Las Positas College*

Leland N. Holland, Jr. *Pasco-Hernando Community College–West Campus*
Elisabeth A. Hooper *Truman State University*
Terry L. Hufford *The George Washington University*
Allen Hunt *Elizabethtown Community College*
Sobrasua E. M. Ibin *Morris Brown College*
Louis Irwin *University of Texas at El Paso*
Laurie E. Iten *Purdue University*
Jeffrey Jack *College of Arts & Sciences*
James B. Jensen *Brigham Young University*
Judy Jernstedt *University of California - Davis*
George P. Johnson *Arkansas Tech University*
Kenneth V. Kardong *Washington State University*
Cheryl Kerfeld *University of California–Los Angeles*
Joanne M. Kilpatrick *Auburn University at Montgomery*
Peter King *Francis Marion University*
Edward C. Kisailus *Canisius College*
Robert M. Kitchin *University of Wyoming*
Will Kleinelp *Middlesex County College*
Kenton Ko *Queen's University*
Ross E. Koning *Eastern Connecticut State University*
Karen L. Koster *University of South Dakota*
V.A. Langman *Louisiana State University–Shreveport*
Simon Lawrance *Otterbein College*
Jeffrey N. Lee *Essex County College*
Laura G. Leff *Kent State University*
Mary E. Lehman *Longwood College*
Niles Lehman *University at Albany SUNY*
Michael Lema *Midlands Technical College*
Charles Kingsley Levy *Boston University*
Leslie Lichtenstein *Massasoit Community College*
Harvey Liftin *Broward Community College*
Richard Londraville *University of Akron*
Sonja L. Maki *Clemson University*
Bradford D. Martin *La Sierra University*
Barbara Maynard *Colorado State University*
Deanna McCullough *University of Houston Downtown*
L. R. McEdward *University of Florida*
Michael Ray Meighan *University of California–Berkeley*
John Merrill *Michigan State University*
Harry A. Meyer *McNeese State University*
Dennis J. Minchella *Purdue University*
Jonathan D. Monroe *James Madison University*
David L. Moore *Utica College of Syracuse University*
Tony E. Morris *Fairmont State College*
Roger N. Morrissette *Framingham State College*
Richard Mortensen *Albion College*
William H. Nelson *Morgan State University*
Peter H. Niewiarowski *University of Akron*
Colleen J. Nolan *St. Mary's University*
John C. Osterman *University of Nebraska–Lincoln*
Thomas G. Owens *Cornell University*
Bruce Parker *Utah Valley State University*
Dustin Penn *University of Utah*
Stacia Pieffer-Schneider *Marquette University*
Carl S. Pike *Franklin and Marshall College*
Nancy A. Perigo *Willamette University*
Greg Phillips *Blinn College–Brenham Campus*
Jon Pigage *University of Colorado at Colorado Springs*
Barbara Pleasants *Iowa State University*
John Pleasants *Iowa State University*

Peggy Pollack *Northern Arizona University*
Mitch Price *The Pennsylvania State University*
Margene Ranieri *Bob Jones University*
Arthur Raske *Northland Baptist Bible College*
Keith Redetzke *University of Texas at El Paso*
Peter J. Rizzo *Texas A&M University*
Ellison Robinson *Midlands Technical College*
Lyndell P. Robinson *Lincoln Land Community College*
Angel M. Rodriguez *Broward Community College*
June R. P. Ross *Western Washington University*
Patricia Rugaber *Coastal Georgia Community College*
Connie Rye *Bevill State Community College*
Nancy K. Sanders *Truman State University*
Robert B. Sanders *University of Kansas–Main Campus*
Lisa M. Sardinia *Pacific University*
Brian W. Schwartz *Columbus State University*
Bruce S. Serlin *DePauw University*
Mark A. Sheridan *North Dakota State University*
Janet Anne Sherman *Penn College of Technology*
Louis Sherman *Purdue University*
Jim Shinkle *Trinity University*
Richard Shippee *Vincennes University*
Brian Shmaefsky *Kingwood College*
Michele Shuster *University of Pittsburgh*
Robert C. Sizemore *Alcorn State University*
Mark Smith *Victor Valley College*
Nancy Solomon *Miami University*
Norm Stacey *University of Alberta*
Ruth Stutts-Moseley *Bishop State Community College*
Kathy Sympson *Florida Keys Community College*
Stan Szarek *Arizona State University*
Robert H. Tamarin *University of Massachusetts Lowell*
Michael Tenneson *Evangel University*
Sharon Thoma *Edgewood College*
Joanne Kivela Tillotson *Purchase College State University of New York*
Maurice Thomas *Palm Beach Atlantic College*
Thomas Tomasi *Southwest Missouri State University*
Leslie Towill *Arizona State University*
Akif Uzman *University of Houston–Downtown*
Thomas J. Volk *University of Wisconsin–La Crosse*
Keith D. Waddington *University of Miami*
D. Alexander Wait *Southwest Missouri State University*
Timothy S. Wakefield *Auburn University*
Charles Walcott *Cornell University*
Eileen Walsh *Westchester Community College*
Frederick Wasserman *Boston University*
Steven A. Wasserman *University of California–San Diego*
Robert F. Weaver *University of Kansas*
Andrew N. Webber *Arizona State University*
Harold J. Webster *Penn State DuBois*
Mark Wheelis *University of California–Davis*
Lynn D. Wike *University of South Carolina at Aiken*
William Williams *Saint Mary's College of Maryland*
Mary L. Wilson *Gordon College*
Kevin Winterling *Emory & Henry College*
E. William Wischusen *Louisiana State University and Agricultural and Mechanical College*
Kenneth Wunch *Tulane University*
Mark L. Wygoda *McNeese State University*
Roger Young *Drury College*

Instructive Art Program

The core of every biology textbook is its art program, and the text and illustration authors of *Biology* have worked together to create a dynamic program of full-color illustrations and photographs that support and further clarify the text explanations. Brilliantly rendered and meticulously reviewed for accuracy and consistency, the carefully conceived illustrations and accompanying photos provide concrete, visual reinforcement of the topics discussed throughout the text.

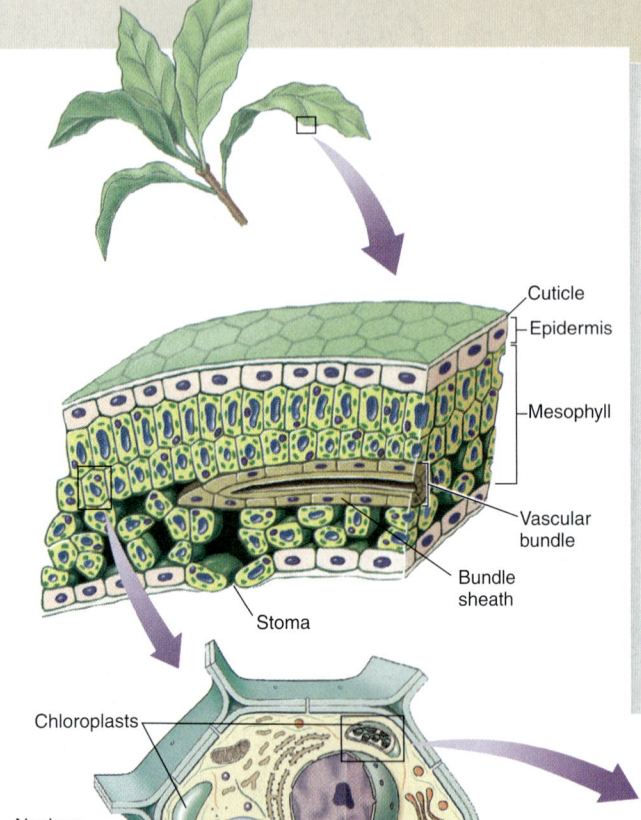

Multi-Level Perspective

Illustrations depicting complex structures or processes combine macroscopic and microscopic views to help you see the relationship between increasingly detailed images.

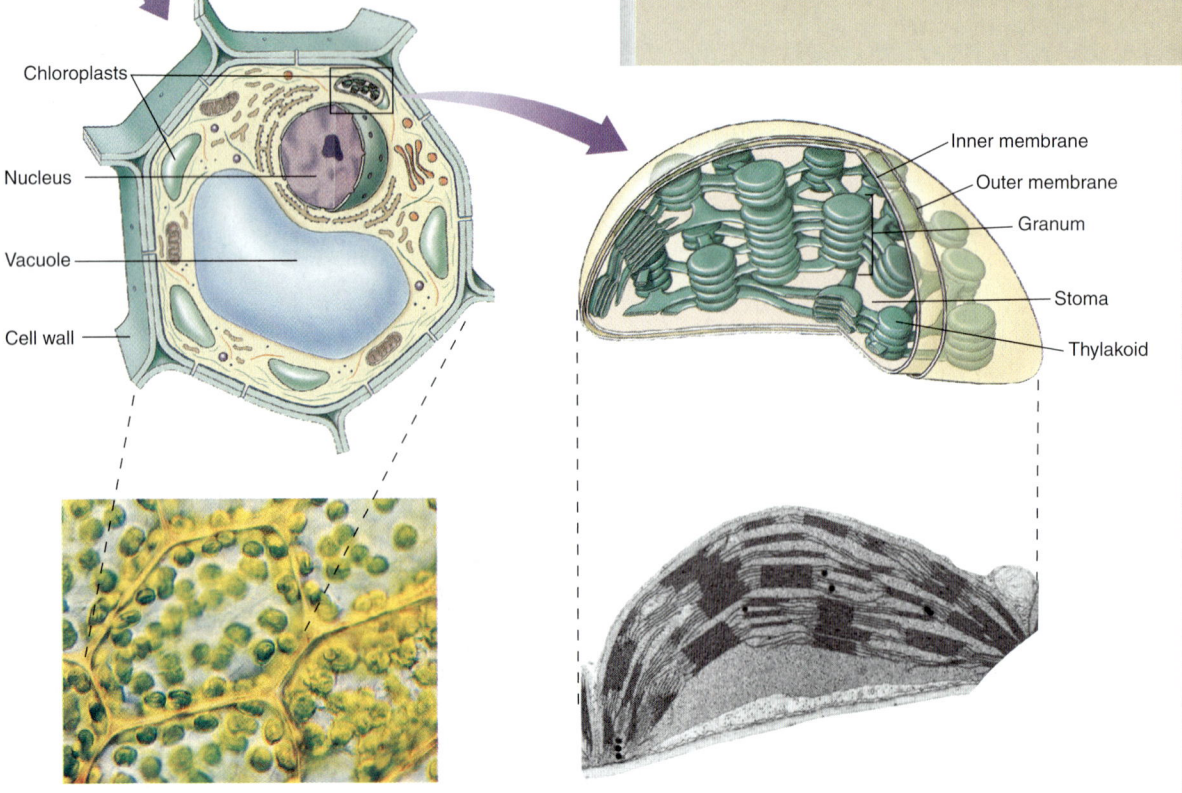

Light micrographs, as well as scanning and transmission electron micrographs, are used in conjunction with illustrations to present a true picture of what you would encounter in lab. A micron bar is added whenever the magnification is known.

Combination Figures
Line drawings are often combined with photographs to facilitate visualization of structures.

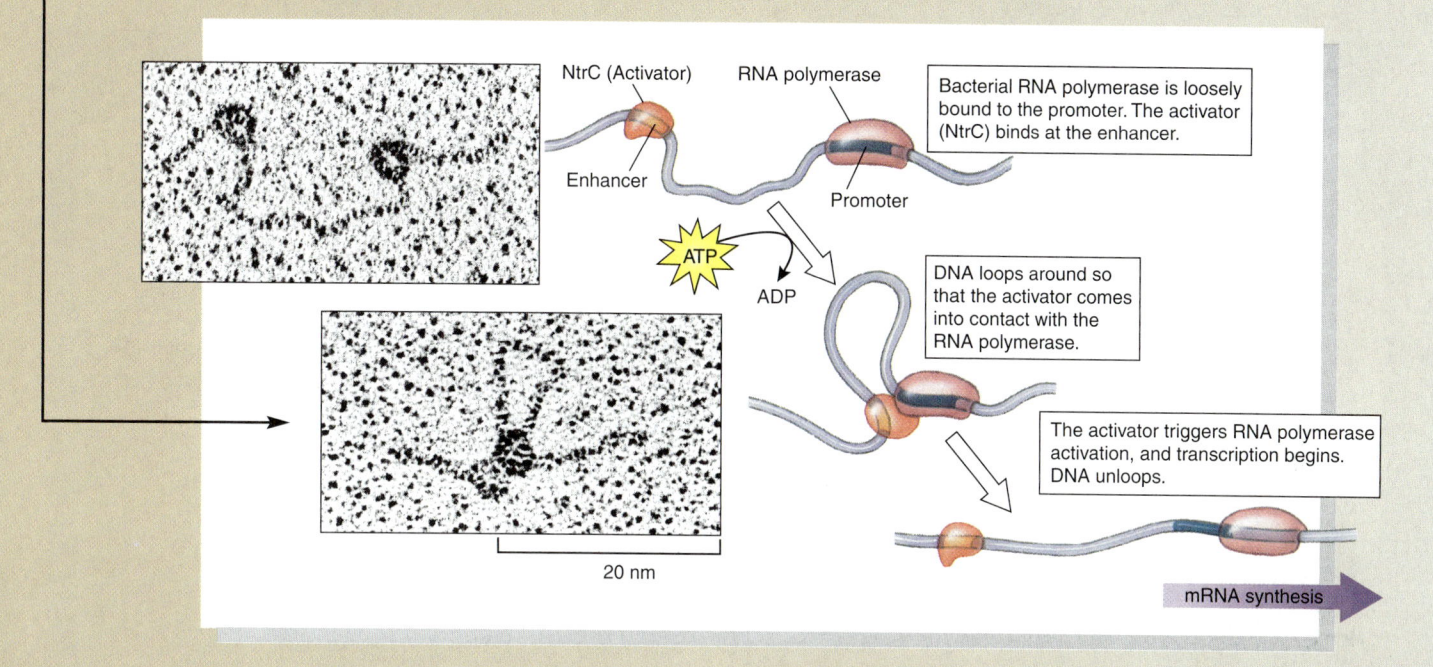

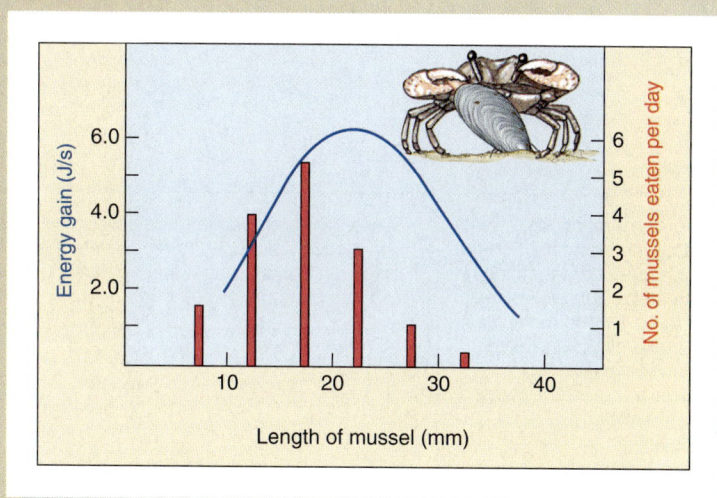

FIGURE 52.26
Optimal diet. The shore crab selects a diet of energetically profitable prey. The curve describes the net energy gain (equal to energy gained minus energy expended) derived from feeding on different sizes of mussels. The bar graph shows the numbers of mussels of each size in the diet. Shore crabs tend to feed on those mussels that provide the most energy.
What factors might be responsible for the slight difference in peak prey length relative to the length optimal for maximum energy gain?

Inquiry Questions
These questions follow figure legends in illustrations that present graphed data. The questions require that you think more carefully about the information presented, and apply that information in new ways. Answers to the Inquiry Questions are found on the *Biology* Online Learning Center at *www.mhhe.com/raven7*.

Instructive Art Program

Explanatory text boxes describe the action depicted in each step. The discrete, carefully placed boxes guide you through the process, without cluttering the image.

Instructors benefit from this style as well, as the images can be used for presentation without the distraction of extraneous captions.

THE CALVIN CYCLE

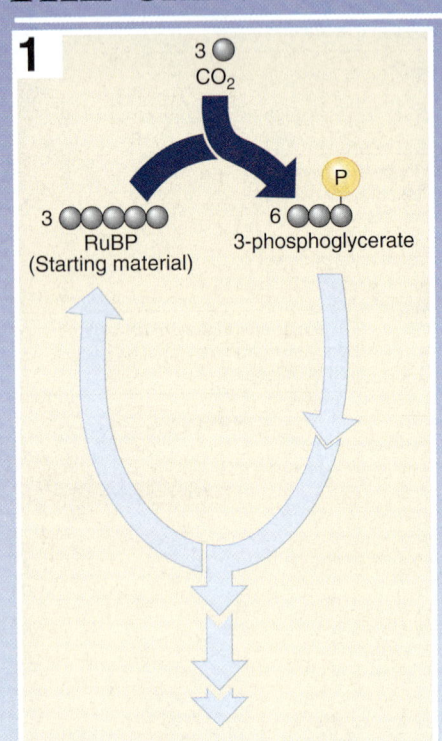

1

3 CO_2 is added to 3 RuBP (Starting material), producing 6 3-phosphoglycerate.

The Calvin cycle begins when a carbon atom from a CO_2 molecule is added to a five-carbon molecule (the starting material). The resulting six-carbon molecule is unstable and immediately splits into three-carbon molecules.

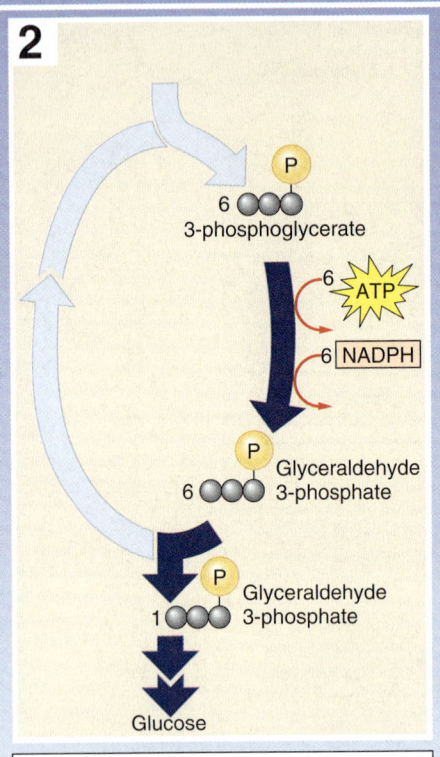

2

6 3-phosphoglycerate, with 6 ATP and 6 NADPH, becomes 6 Glyceraldehyde 3-phosphate; 1 Glyceraldehyde 3-phosphate → Glucose.

Then, through a series of reactions, energy from ATP and hydrogens from NADPH (the products of the light-dependent reactions) are added to the three-carbon molecules. The now-reduced three-carbon molecules either combine to make glucose or are used to make other molecules.

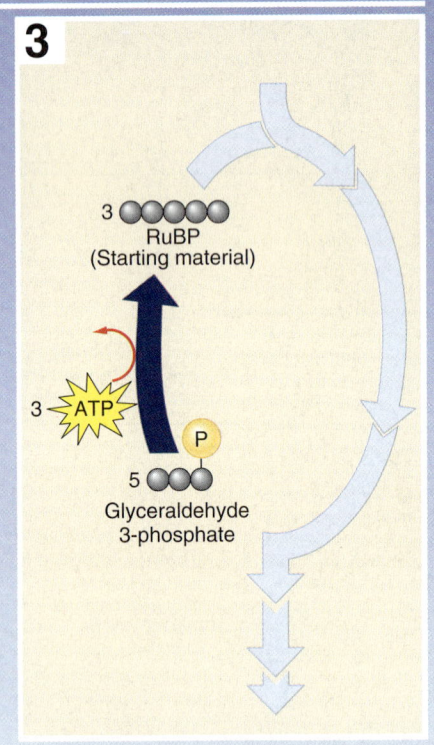

3

5 Glyceraldehyde 3-phosphate + 3 ATP → 3 RuBP (Starting material).

Most of the reduced three-carbon molecules are used to regenerate the five-carbon starting material, thus completing the cycle.

FIGURE 10.17
How the Calvin cycle works.

Process Boxes
Process Boxes break down complex processes into a series of small steps, allowing you to track the key occurrences and learn them as you go.

Phylogeny Guideposts

Phylogeny Guideposts are used in the diversity chapters to help you track relationships among diverse organisms. As each group is introduced, the appropriate branch on the phylogenetic tree is highlighted.

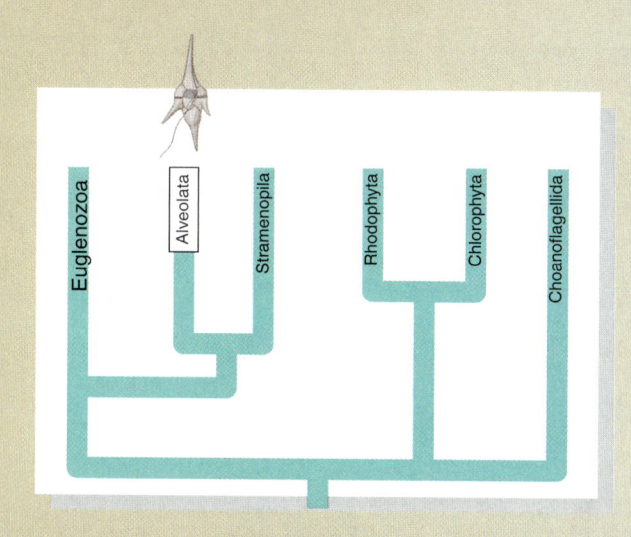

Biochemistry Pathway Icons

These icons are paired with more detailed illustrations to assist you in keeping the big picture in mind when learning complex metabolic processes. The icon highlights which step the main illustration represents, and where that step occurs in the complete process.

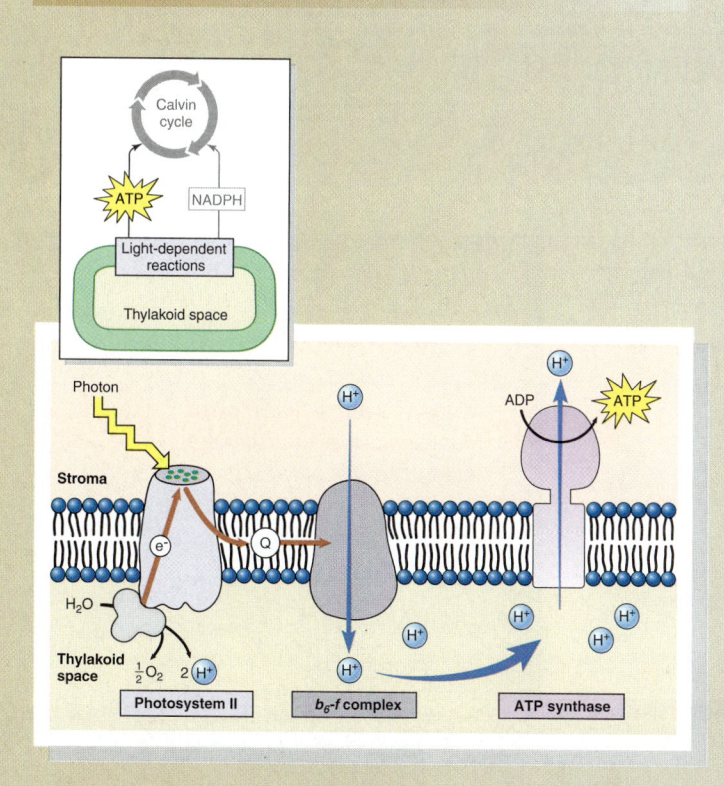

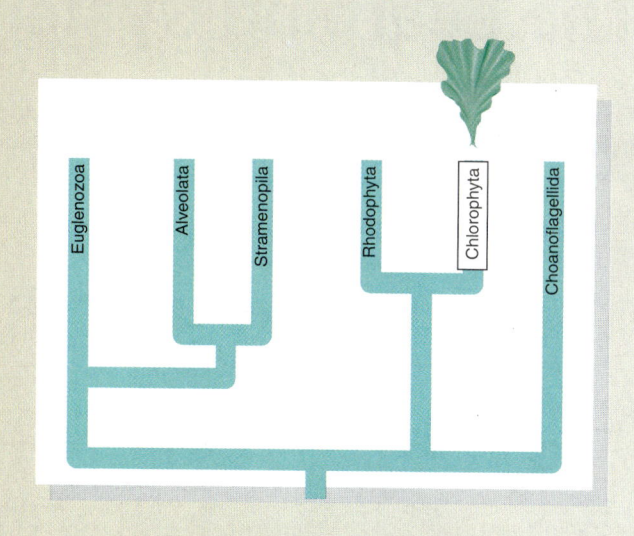

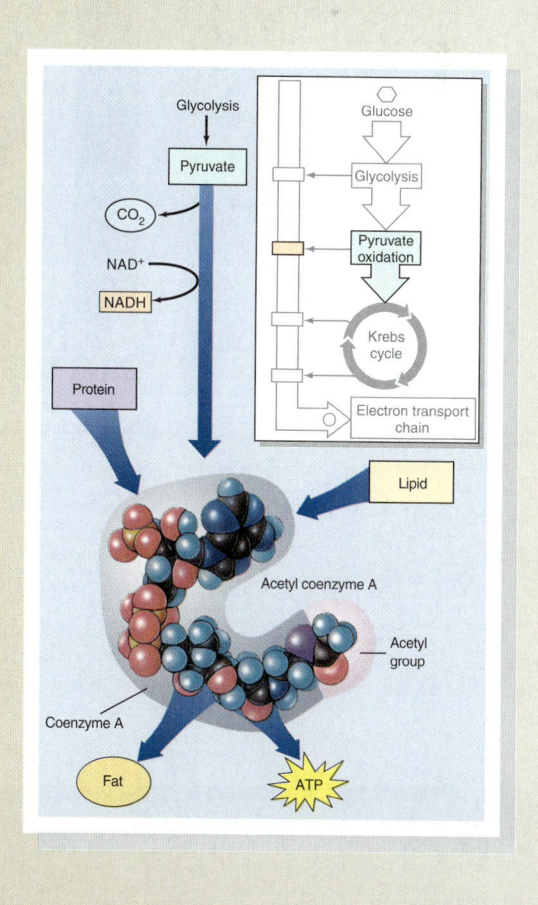

The Learning System

This text is designed to help you learn in a systematic fashion. Simple facts are the building blocks for developing explanations of more complex concepts. The text discussion is presented within a supporting framework of learning aids that help organize studying, reinforce learning, and promote problem-solving skills.

Numbered Headings
The numbered headings employed in the modules form the backbone of the Concept Outline. This consistency makes it easier to identify the key concepts for each chapter, and to then manage the supporting details for each concept.

11
How Cells Divide

Concept Outline

11.1 Prokaryotes divide far more simply than do eukaryotes.

Cell Division in Prokaryotes. Prokaryotic cells divide by splitting in two.

11.2 The chromosomes of eukaryotes are highly ordered structures.

Discovery of Chromosomes. All eukaryotic cells contain chromosomes, but different organisms possess differing numbers of chromosomes.

The Structure of Eukaryotic Chromosomes. Proteins play an important role in packaging DNA in chromosomes.

11.3 Mitosis is a key phase of the cell cycle.

The Cell Cycle. The cell cycle consists of three growth phases, a nuclear division phase, and a cytoplasmic division stage.

Interphase: Preparing for Mitosis. In interphase, the cell grows, replicates its DNA, and prepares for cell division.

Mitosis. In prophase, the chromosomes condense, and microtubules attach sister chromosomes to opposite poles of the cell. In metaphase, the chromosomes align along the center of the cell. In anaphase, the chromosomes separate; in telophase, the spindle dissipates and the nuclear envelope re-forms.

Cytokinesis. In cytokinesis, the cytoplasm separates into two roughly equal halves.

11.4 The cell cycle is carefully controlled.

General Strategies of Cell Cycle Control. At three points in the cell cycle, feedback from the cell determines whether the cycle will continue.

Molecular Mechanisms of Cell Cycle Control. Special proteins regulate the checkpoints of the cell cycle.

Cancer and the Control of Cell Proliferation. Cancer results from damage to genes encoding proteins that regulate the cell division cycle.

FIGURE 11.1
Cell division in prokaryotes. It's hard to imagine fecal coliform bacteria as being beautiful, but here is *Escherichia coli*, inhabitant of the large intestine and the biotechnology lab, spectacularly caught in the act of fission.

All species of organisms—bacteria, alligators, the weeds in a lawn—grow and reproduce. From the smallest creature to the largest, all species produce offspring like themselves and pass on the hereditary information that makes them what they are. In this chapter, we examine how cells divide and reproduce (figure 11.1). The mechanism of cell reproduction and its biological consequences have changed significantly during the evolution of life on earth. The process is complex in eukaryotes, involving both the replication of chromosomes and their separation into daughter cells. Much of what we are learning about the causes of cancer relates to how cells control this process, and in particular their propensity to divide, a mechanism that in broad outline remains the same in all eukaryotes.

207

11.1 Prokaryotes divide far more simply than do eukaryotes.

Cell Division in Prokaryotes

The end result of cell division in both prokaryotic and eukaryotic cells is two daughter cells, each with the same genetic information as the original cell. The differences between these two basic cell types lead to large differences in how this process occurs. Despite these differences, the essentials of the process are the same: duplication and segregation of genetic information into daughter cells, and division of cellular contents. We will begin by looking at the simpler process, which occurs in prokaryotes: division by **binary fission**.

Most prokaryotes have a genome made up of a single, circular DNA molecule. Despite its apparent simplicity, the DNA molecule of the bacterium *Escherichia coli* is actually on the order of 500 times longer than the cell itself! Thus, this "simple" structure is actually exquisitely packaged to fit into the cell. Although not found in a nucleus, the DNA is in a compacted form called a *nucleoid* that is distinct from the cytoplasm around it.

For many years, it was believed that the *E. coli* DNA molecule was passively segregated by attachment to the membrane and growth of the membrane as the cell elongates. More recently, a more complex picture is emerging that involves both active partitioning of the DNA and formation of a septum that divides the elongated cell in half. Although the details differ, species as different as *E. coli* and *Bacillus subtilis* both exhibit active partitioning of the newly replicated DNA molecules during the division process. This requires both specific sites on the chromosomes and a number of proteins actively involved in the process.

Binary fission begins with the replication of the prokaryotic DNA at a specific site—the origin of replication (see chapter 15)—and proceeds bidirectionally around the circular DNA to a specific site of termination (figure 11.2). Growth of the cell results in elongation, and the newly replicated DNA molecules are actively partitioned to one-fourth and three-quarters of the cell length. This process requires sequences near the origin of replication and results in these sequences being attached to the membrane. The cell itself is partitioned by the growth of new membrane and cell material called a septum (see figure 11.2). This process of septation is complex and under control of the cell as well.

The site of septation is usually the midpoint of the cell and begins with the formation of a ring composed of the molecule FtsZ (figure 11.3). This then results in the accumulation of a number of other proteins, including ones embedded in the membrane. The exact mechanism of septation is not known, but this structure grows inward radially until the cells pinch off into new cells.

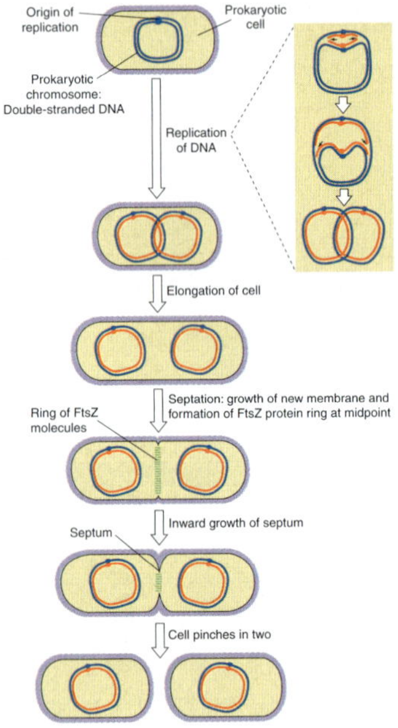

FIGURE 11.2
Binary fission. Prior to cell division, the prokaryotic DNA molecule replicates. The replication of the double-stranded, circular DNA molecule (*blue*) that constitutes the genome of a prokaryote begins at a specific site, called the origin of replication. The replication enzymes move out in both directions from that site and make copies (*red*) of each strand in the DNA duplex. The enzymes continue until they meet at another specific site, the terminus of replication. After the DNA is replicated, the cell elongates, and the DNA is partitioned in the cell. Septation then begins, in which new cell membrane material begins to grow and form a septum at approximately the midpoint of the cell. A protein molecule called FtsZ facilitates this process. When the septum is complete, the cell pinches in two, and two daughter cells are formed, each containing a prokaryotic DNA molecule.

208   Part II   Biology of the Cell

Concept Outline
Each chapter begins with an outline that gives you an overview of the content contained within that chapter. Reviewing the concept outline before reading the chapter will help focus your attention on the major concepts you should take away from the chapter.

Modular Format

Each page or two-page spread in *Biology* is organized as an independent module, with its own numbered heading at the top of the left-hand page, and a highlighted summary at the bottom of the right-hand page. This system organizes the information in the chapter within a clear conceptual framework, which in turn helps you learn and retain the material.

Section Summaries

Each module ends with a summary intended to reinforce the key concepts from that section. Reviewing the summary after reading the section will indicate whether you learned the main ideas presented in the module.

Vocabulary Boxes

These boxes are found throughout the text, in chapters that require you to learn many new terms. This saves you time when studying, by placing the definitions you need in one location. It is also a handy study tool, as it reinforces the key terms for that chapter.

FIGURE 11.3
The FtsZ protein. In these dividing *E. coli* bacteria, the FtsZ protein is fluorescent, and its location during binary fission can be seen. The protein assembles into a ring at approximately the midpoint of the cell, where it facilitates septation and cell division. Bacteria in which the *ftsZ* gene is mutated cannot divide.

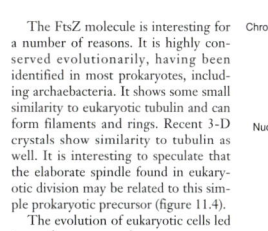

The FtsZ molecule is interesting for a number of reasons. It is highly conserved evolutionarily, having been identified in most prokaryotes, including archaebacteria. It shows some small similarity to eukaryotic tubulin and can form filaments and rings. Recent 3-D crystals show similarity to tubulin as well. It is interesting to speculate that the elaborate spindle found in eukaryotic division may be related to this simple prokaryotic precursor (figure 11.4).

The evolution of eukaryotic cells led to much more complex genomes composed of multiple linear chromosomes housed in a membrane-bounded nucleus. These chromosomes contain even more DNA, and thus pose packaging problems that are solved by DNA being complexed with protein and packaged into functionally distinct chromosomes. This creates more challenges both for the replication of the genome and for its accurate segregation during cell division. The process that evolved to accomplish this segregation of chromosomes is called mitosis.

> Prokaryotes divide by binary fission. Fission begins in the middle of the cell. An active partitioning process ensures that one genome will end up in each daughter cell.

FIGURE 11.4
A comparison of protein assemblies during cell division among different organisms. The prokaryotic protein FtsZ has a structure that is similar to that of the eukaryotic protein tubulin. Tubulin is the protein component of microtubules, which are fibers that play an important role in eukaryotic cell division.

Prokaryotes
No nucleus; single circular chromosome. After DNA is replicated, it is partitioned in the cell. After cell elongation, FtsZ protein assembles into a ring and facilitates septation and cell division.

Some protists
Nucleus present and nuclear envelope remains intact during cell division. Chromosomes linear. Fibers called microtubules, composed of the protein tubulin, pass through tunnels in the nuclear membrane and set up an axis for separation of replicated chromosomes, and cell division.

Other protists
A spindle of microtubules forms between two pairs of centrioles at opposite ends of the cell. The spindle passes through one tunnel in the intact nuclear envelope. Kinetochore microtubules form between kinetochores on the chromosomes and the spindle poles and pull the chromosomes to each pole.

Yeasts
Nuclear envelope remains intact; spindle microtubules form inside the nucleus between spindle pole bodies. A single kinetochore microtubule attaches to each chromosome and pulls each to a pole.

Animals
Spindle microtubules begin to form between centrioles outside of nucleus. As these centrioles move to the poles, the nuclear envelope breaks down, and kinetochore microtubules attach kinetochores of chromosomes to spindle poles. Polar microtubules extend toward the center of the cell and overlap.

Chapter 11 How Cells Divide 209

A Vocabulary of Cell Division

binary fission Reproduction of a cell by division into two equal or nearly equal parts. Prokaryotes divide by binary fission.

centromere A constricted region of a chromosome about 220 nucleotides in length, composed of highly repeated DNA sequences. During mitosis, the centromere joins the two sister chromatids and is the site to which the kinetochores are attached.

chromatid One of the two copies of a replicated chromosome, joined by a single centromere to the other strand.

chromatin The complex of DNA and proteins of which eukaryotic chromosomes are composed.

chromosome The structure within cells that contains the genes. In eukaryotes, it consists of a single linear DNA molecule associated with proteins. The DNA replicates during S phase, and the replicas separate during M phase.

cytokinesis Division of the cytoplasm of a cell after nuclear division.

euchromatin The portion of a chromosome that is extended except during cell division, and from which RNA is transcribed.

heterochromatin The portion of a chromosome that remains permanently condensed and, therefore, is not transcribed into RNA. Most centromere regions are heterochromatic.

homologues Homologous chromosomes; in diploid cells, one of a pair of chromosomes that carry equivalent genes.

kinetochore A disk of protein bound to the centromere and attached to microtubules during mitosis, linking each chromatid to the spindle apparatus.

microtubule A hollow cylinder, about 25 nanometers in diameter, composed of subunits of the protein tubulin. Microtubules lengthen by the addition of tubulin subunits to their end(s) and shorten by the removal of subunits.

mitosis Nuclear division in which replicated chromosomes separate to form two genetically identical daughter nuclei. When accompanied by cytokinesis, it produces two identical daughter cells.

nucleosome The basic packaging unit of eukaryotic chromosomes, in which the DNA molecule is wound around a cluster of histone proteins. Chromatin is composed of long strings of nucleosomes that resemble beads on a string.

214 Part II Biology of the Cell

The Learning System

Concept Review
An expanded version of the Concept Outline, the Concept Review details each numbered section head followed by its supporting ideas. Each supporting idea is page referenced to allow you to focus your time on areas where you need additional study.

Concept Review
For interactive testing, visit the Online Learning Center with PowerWeb at www.mhhe.com/Raven7

11.1 Prokaryotes divide far more simply than do eukaryotes.
Cell Division in Prokaryotes
- Most prokaryotes have a genome made up of a single, circular DNA molecule, and replicate via binary fission. (p. 208)
- Binary fission begins with DNA replication, which starts at the origin site and proceeds bidirectionally around the circular DNA to a specific site of termination. (p. 208)
- The evolution of eukaryotic cells led to much more complex genomes and, thus, new and different ways to replicate and segregate the genome during cell division. (p. 209)

11.2 The chromosomes of eukaryotes are highly ordered structures.
Discovery of Chromosomes
- Chromosomes were first discovered in 1882 by Walther Fleming. (p. 210)
- The number of chromosomes varies from one species to another. Humans have 23 nearly identical pairs for a total of 46 chromosomes. (p. 210)

The Structure of Eukaryotic Chromosomes
- The DNA is a very long, double-stranded fiber extending unbroken through the entire length of the chromosome. A typical human chromosome contains about 140 million nucleotides. (p. 211)
- Every 200 nucleotides, the DNA duplex is coiled around a core of eight histone proteins, forming a nucleosome. (p. 211)
- The particular array of chromosomes an individual possesses is its karyotype. (p. 212)
- The number of different chromosomes a species contains is known as its haploid (*n*) number, and is considered one complete set of chromosomes. (p. 212)
- Humans are diploid, with homologues coming from both

Interphase: Preparing for Mitosis
- The cell grows throughout interphase. The G_1 and G_2 phases are periods of protein synthesis and organelle production, while the S phase is when DNA replication occurs. (p. 214)

Mitosis
- Chromatin condensation continues into prophase. The spindle apparatus is assembled, and sister chromatids are linked to opposite poles of the cell by microtubules. The nuclear envelope breaks down. (p. 215)
- During metaphase, chromosomes align in the center of the cell along the metaphase plate. (p. 215)
- Anaphase begins when centromeres divide, freeing the two sister chromatids from each other. Sister chromatids are pulled to opposite poles as the attached microtubules shorten. (pp. 216–217)
- In telophase, the spindle apparatus disassembles, and the nuclear membrane begins to re-form. (p. 217)

Cytokinesis
- Cytokinesis is the phase of the cell cycle when the cell actually divides. Cytokinesis generally involves the cleavage of the cell into roughly equal halves, forming two daughter cells. (p. 218)

11.4 The cell cycle is carefully controlled.
General Strategies of Cell Cycle Control
- A cell uses three main checkpoints to both assess the internal state of the cell and integrate external signals. The G_1/S checkpoint is the primary point at which the cell decides to divide; the G_2/M checkpoint represents a commitment to mitosis; and the spindle checkpoint ensures that all chromosomes are attached to the spindle in preparation for anaphase. (p. 219)

Molecular Mechanisms of Cell Cycle Control
- Two groups of proteins, cyclins and Cdk's, interact and regulate the cell cycle. (p. 220)
- Cells also receive protein signals (growth factors) that affect cell division. (p. 222)

Cancer and the Control of Cell Proliferation
- Cancer is failure of cell division control. (p. 223)
- It is believed that a malfunction in the *p53* gene may allow cells to go through repeated cell division without being stopped at the appropriate checkpoints. (p. 223)
- Proto-oncogenes are normal cellular genes that become oncogenes when mutated. Proto-oncogenes can encode growth factors, protein relay switches, and kinase enzymes. (p. 224)
- Tumor-suppressor genes can also lead to cancer when they are mutated. (p. 224)

Test Your Understanding
For interactive testing, visit the Online Learning Center with PowerWeb at www.mhhe.com/Raven7

Self Test
1. Bacterial cells divide by
   a. mitosis.
   b. replication.
   c. cytokinesis.
   d. binary fission.
2. Most eukaryotic organisms have _____ chromosomes in their cells.
   a. 1–5
   b. 10–50
   c. 100–500
   d. over 1000
3. Replicate copies of each chromosome are called _____ and are joined at the _____.
   a. homologues/centromere
   b. sister chromatids/kinetochore
   c. sister chromatids/centromere
   d. homologues/kinetochore
4. During which phase of the cell cycle is DNA synthesized?
   a. G_1
   b. G_2
   c. S
   d. M
5. Chromosomes are visible under a light microscope
   a. during mitosis.
   b. during interphase.
   c. when they are attached to their sister chromatids.
   d. All of these are correct.
6. During mitosis, the sister chromatids are separated and pulled to opposite poles during which stage?
   a. interphase
   b. metaphase
   c. anaphase
   d. telophase
7. Cytokinesis is
   a. the same process in plant and animal cells.
   b. the separation of cytoplasm and the formation of two cells.
   c. the final stage of mitosis.
   d. the movement of kinetochores.
8. The eukaryotic cell cycle is controlled at several points; which of these statements is *not* true?
   a. Cell growth is assessed at the G_1/S checkpoint.
   b. DNA replication is assessed at the G_2/M checkpoint.
   c. Environmental conditions are assessed at the G_0 checkpoint.
   d. The chromosomes are assessed at the spindle checkpoint.
9. What proteins are used to control cell growth specifically in *multicellular* eukaryotic organisms?
   a. Cdk
   b. MPF
   c. cyclins
   d. growth factors
10. What causes cancer in cells?
    a. damage to genes
    b. chemical damage to cell membranes
    c. UV damage to transport proteins
    d. All of these cause cancer in cells.

Test Your Visual Understanding

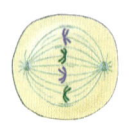

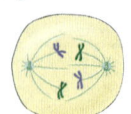

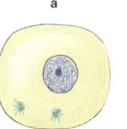

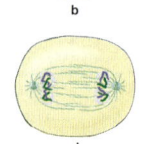

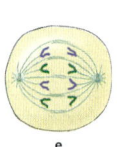

a   b   c   d   e

1. Match the mitotic and cell cycle phases with the appropriate figure.
   anaphase
   interphase
   metaphase
   prophase
   telophase

Apply Your Knowledge
1. An ancient plant called horsetail contains 216 chromosomes. How many homologous pairs of chromosomes does it contain? How many chromosomes are present in its cells during metaphase?
2. Colchicine is a poison that binds to tubulin and prevents its assembly into microtubules; cytochalasins are compounds that bind to the ends of actin filaments and prevent their elongation. What effects would these two substances have on cell division in animal cells?
3. If you could construct an artificial chromosome, what elements would you introduce into it, at a minimum, so that it could function normally in mitosis?

226   Part II   Biology of the Cell

Testing Yourself
Each chapter concludes with a set of questions designed to test your knowledge of the content, including multiple choice questions, illustration-based questions, and application questions. Answers to these questions are found on the *Biology* Online Learning Center at: www.mhhe.com/raven7. At the site you can take an interactive version of the end-of-chapter quiz that provides you with hints and instructional feedback.

BIOLOGY

21
Genes Within Populations

Concept Outline

21.1 Genes vary in natural populations.

Genetic Variation Is the Raw Material of Evolution. Selection acts on the genetic variation present in populations, favoring variants that increase the likelihood of survival and reproduction.

Genetic Variation in Nature. Natural populations contain considerable amounts of variation, present at the DNA level and expressed in proteins.

21.2 Why do allele frequencies change in populations?

The Hardy–Weinberg Principle. The proportion of homozygotes and heterozygotes in a population should remain unchanged in the absence of agents of evolutionary change.

Five Agents of Evolutionary Change. The frequency of alleles in a population can be changed by evolutionary forces such as mutation, gene flow, nonrandom mating, genetic drift, and selection.

Measuring Fitness. An organism's reproductive success is affected by how long it survives, how often it mates, and how many offspring it produces per mating.

Interactions Among Evolutionary Forces. Allele frequencies sometimes reflect a balance between opposing evolutionary processes.

Natural Selection Can Maintain Variation in Populations. Frequency-dependent selection, oscillating selection, and heterozygote advantage can all maintain genetic variation in populations.

21.3 Selection can act on traits affected by many genes.

Forms of Selection. Selection on traits affected by many genes can favor both extremes of the trait, or intermediate values, or only one extreme.

Selection on Color in Guppies. Experiments in natural settings can test evolutionary hypotheses and reveal the workings of natural selection.

Limits to What Selection Can Accomplish. Selection cannot act on traits lacking genetic variation and can be affected by gene interactions.

FIGURE 21.1
Genetic variation. Natural populations contain abundant genetic variation for many traits, such as flower color.

No other human being is exactly like you (unless you have an identical twin). Often the particular characteristics of an individual have an important bearing on its survival, on its chances to reproduce, and on the success of its offspring. Evolution is driven by such consequences. Genetic variation that influences these characteristics provides the raw material for natural selection, and natural populations contain a wealth of such variation. In plants (figure 21.1), insects, and vertebrates, many genes exhibit some level of variation. In this chapter, we explore genetic variation in natural populations and consider the evolutionary forces that cause allele frequencies in natural populations to change. These deceptively simple matters lie at the core of evolutionary biology, which is the topic of this chapter and chapters 22–24.

Part IV Evolution

21.1 Genes vary in natural populations.

Genetic Variation Is the Raw Material of Evolution

Evolution: Descent with Modification

The word *evolution* is widely used in the natural and social sciences. It refers to how an entity—be it a social system, a gas, or a planet—changes through time. Although development of the modern concept of evolution in biology can be traced to Darwin's *On the Origin of Species*, the first five editions of this book never actually used the term! Rather, Darwin used the phrase "descent with modification." Although many more complicated definitions have been proposed, Darwin's phrase probably best captures the essence of biological evolution: Through time, species accumulate differences; as a result, when new species are formed, the descendant species differ from their ancestors.

Natural Selection: An Important Mechanism of Evolutionary Change

Darwin was not the first to propose a theory of evolution. Rather, he followed a long line of earlier philosophers and naturalists who deduced that the many kinds of organisms around us were produced by a process of evolution. Unlike his predecessors, however, Darwin proposed **natural selection** as the mechanism of evolution. Natural selection produces evolutionary change when some individuals in a population possess certain inherited characteristics and produce more surviving offspring than individuals lacking these characteristics. As a result, the population gradually comes to include more and more individuals with the advantageous characteristics. In this way, the population evolves and becomes better adapted to its local circumstances.

Natural selection was by no means the only evolutionary mechanism proposed. A rival theory, championed by the prominent biologist Jean-Baptiste Lamarck, was that evolution occurred by the **inheritance of acquired characteristics**. According to Lamarck, individuals passed on to their offspring body and behavior changes acquired during their lives. Thus, Lamarck proposed that ancestral giraffes with short necks tended to stretch their necks to feed on tree leaves, and this extension of the neck was passed on to subsequent generations, leading to the long-necked giraffe (figure 21.2*a*). In Darwin's theory, by contrast, the variation is not created by experience, but is the result of preexisting genetic differences among individuals (figure 21.2*b*).

Although the efficacy of natural selection is now widely accepted, it is not the only process that can lead to changes in the genetic makeup of populations. Allele frequencies can also change when mutations occur repeatedly from one allele to another and when migrants bring alleles into a population. In addition, when populations are small, the frequencies of alleles can change randomly as the result of chance events. The relationship between natural selection and other processes varies. Sometimes, natural selection overwhelms the effects of other processes, but as we shall see later in this chapter, this is not always the case.

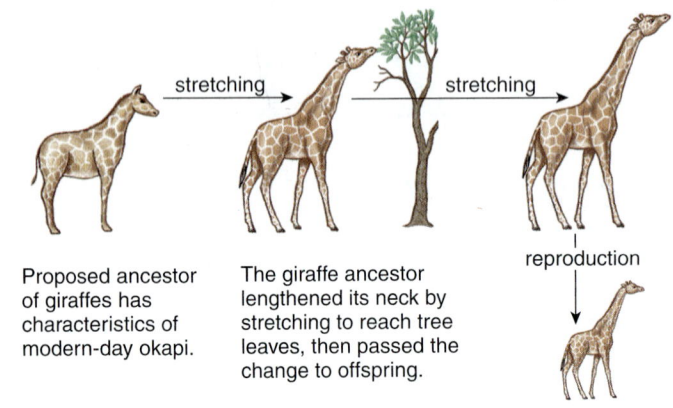

(a) Lamarck's theory: variation is acquired.

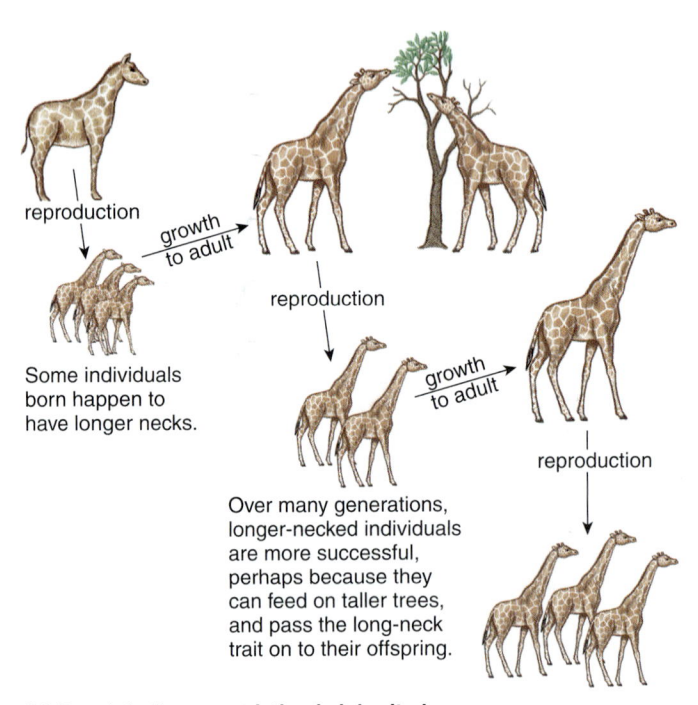

(b) Darwin's theory: variation is inherited.

FIGURE 21.2
Two ideas of how giraffes might have evolved long necks.

> **Darwin proposed that natural selection on variants within populations leads to evolutionary change.**

434   Part IV   Evolution

Genetic Variation in Nature

Evolution within a species can result from any process that causes a change in the genetic composition of a population, and thus we cannot talk about evolution without also considering **population genetics,** the study of the properties of genes in populations. It is best to start by looking at the genetic variation present among individuals within a species. This is the raw material available for the selective process.

Measuring Levels of Genetic Variation

As we saw in chapter 13, a natural population can contain a great deal of genetic variation. This is true not only of humans, but of all organisms. How much variation usually occurs? Biologists have looked at many different genes in an effort to answer this question:

1. **Genes that influence blood groups.** Chemical analysis has revealed the existence of more than 30 blood group genes in humans, in addition to the ABO locus. At least one-third of these genes are routinely found in several alternative allelic forms in human populations. In addition to these, more than 45 variable genes encode other proteins in human blood cells and plasma, which are not considered blood groups. Thus, many genetically variable genes are present in this one system alone.
2. **Genes that influence enzymes.** Alternative alleles of genes specifying particular enzymes are easy to distinguish by measuring how fast the alternative proteins migrate in an electrical field (a process called **electrophoresis**). A great deal of variation exists at enzyme-specifying loci. About 5% of the enzyme loci of a typical human are heterozygous: If you picked an individual at random, and in turn selected one of the enzyme-encoding genes of that individual at random, the chances are 1 in 20 (5%) that the gene you selected would be heterozygous in that individual.

Considering the entire genome, it is fair to say that almost all humans are different from one another. This is also true of other organisms, except for those that reproduce asexually. In nature, genetic variation is the rule.

Enzyme Polymorphism

Many loci in a given population have more than one allele at frequencies significantly greater than would occur due to mutation alone. Researchers refer to a locus with more variation than can be explained by mutation as **polymorphic** (Greek *poly,* "many," + *morphe,* "forms") (figure 21.3). The extent of such variation within natural populations was not even suspected a few decades ago, until modern techniques such as protein electrophoresis made it possible to examine enzymes and other proteins directly. We now know that most populations of insects and plants are polymorphic (that

FIGURE 21.3
Polymorphic variation. These Australian snails, all of the species *Bankivia fasciata,* exhibit considerable variation in pattern and color. Individual differences are inherited and passed on to offspring.

is, have more than one allele occurring at a frequency greater than 5%) at more than half of their enzyme-encoding loci, although vertebrates are somewhat less polymorphic. **Heterozygosity** (the probability that a randomly selected gene will be heterozygous for a randomly selected individual) is about 15% in *Drosophila* and other invertebrates, between 5% and 8% in vertebrates, and around 8% in outcrossing plants. These high levels of genetic variability provide ample supplies of raw material for evolution.

DNA Sequence Polymorphism

The advent of gene technology has made it possible to assess genetic variation even more directly by sequencing the DNA itself. In a pioneering study in 1989, Martin Kreitman sequenced ADH genes isolated from 11 individuals of the fruit fly *Drosophila melanogaster.* He found 43 variable sites, only one of which had been detected by protein electrophoresis! Since then, numerous other studies of variation at the DNA level have confirmed these findings: Abundant variation exists in both the coding regions of genes and in their nontranslated introns—considerably more variation than we can detect by examining enzymes with electrophoresis.

Natural populations contain considerable amounts of genetic variation—more than can be accounted for by mutation alone.

21.2 Why do allele frequencies change in populations?

Genetic variation within natural populations was a puzzle to Darwin and his contemporaries. The way in which meiosis produces genetic segregation among the progeny of a hybrid had not yet been discovered. Selection, scientists then thought, should always favor an optimal form, and so tend to eliminate variation. Moreover, the theory of **blending inheritance**—in which offspring were expected to be phenotypically intermediate relative to their parents—was widely accepted. If blending inheritance were correct, then the effect of any new genetic variant would quickly be diluted to the point of disappearance in subsequent generations.

The Hardy–Weinberg Principle

Following the rediscovery of Mendel's research, two people in 1908 independently solved the puzzle of why genetic variation persists—G. H. Hardy, an English mathematician, and W. Weinberg, a German physician. They pointed out that the original proportions of the genotypes in a population will remain constant from generation to generation, as long as the following assumptions are met:

1. The population size is very large.
2. Random mating is occurring.
3. No mutation takes place.
4. No genes are input from other sources (no immigration takes place).
5. No selection occurs.

Because their proportions do not change, the genotypes are said to be in **Hardy–Weinberg equilibrium.**

In algebraic terms, the Hardy–Weinberg principle is written as an equation. Consider a population of 100 cats in which 84 are black and 16 are white. The frequencies of the two phenotypes would be 0.84 (or 84%) black and 0.16 (or 16%) white. Based on these phenotypic frequencies, can we deduce the underlying frequency of genotypes? If we assume that the white cats are homozygous recessive for an allele we designate b, and the black cats are therefore either homozygous dominant BB or heterozygous Bb, we can calculate the **allele frequencies** of the two alleles in the population from the proportion of black and white individuals, assuming the population is in Hardy–Weinberg equilibrium. Let the letter p designate the frequency of the B allele and the letter q the frequency of the alternative allele. Because there are only two alleles, p plus q must always equal 1.

The Hardy–Weinberg equation can now be expressed in the form of what is known as a binomial expansion:

$$(p + q)^2 = \underbrace{p^2}_{\substack{\text{(Individuals} \\ \text{homozygous} \\ \text{for allele } B\text{;} \\ \text{color black)}}} + \underbrace{2pq}_{\substack{\text{(Individuals} \\ \text{heterozygous} \\ \text{with alleles } B + b\text{;} \\ \text{color black)}}} + \underbrace{q^2}_{\substack{\text{(Individuals} \\ \text{homozygous} \\ \text{for allele } b\text{;} \\ \text{color white)}}}$$

If $q^2 = 0.16$ (the frequency of white cats), then $q = 0.4$. Therefore, p, the frequency of allele B, would be 0.6 (1.0 − 0.4 = 0.6). We can now easily calculate the **genotype frequencies**: There are $p^2 = (0.6)^2 = 0.36$, or 36 homozygous dominant BB individuals in a population of 100 cats. The heterozygous individuals have the Bb genotype, and would have a frequency of $2pq$, or $(2 \times 0.6 \times 0.4) = 0.48$, or 48 heterozygous Bb individuals.

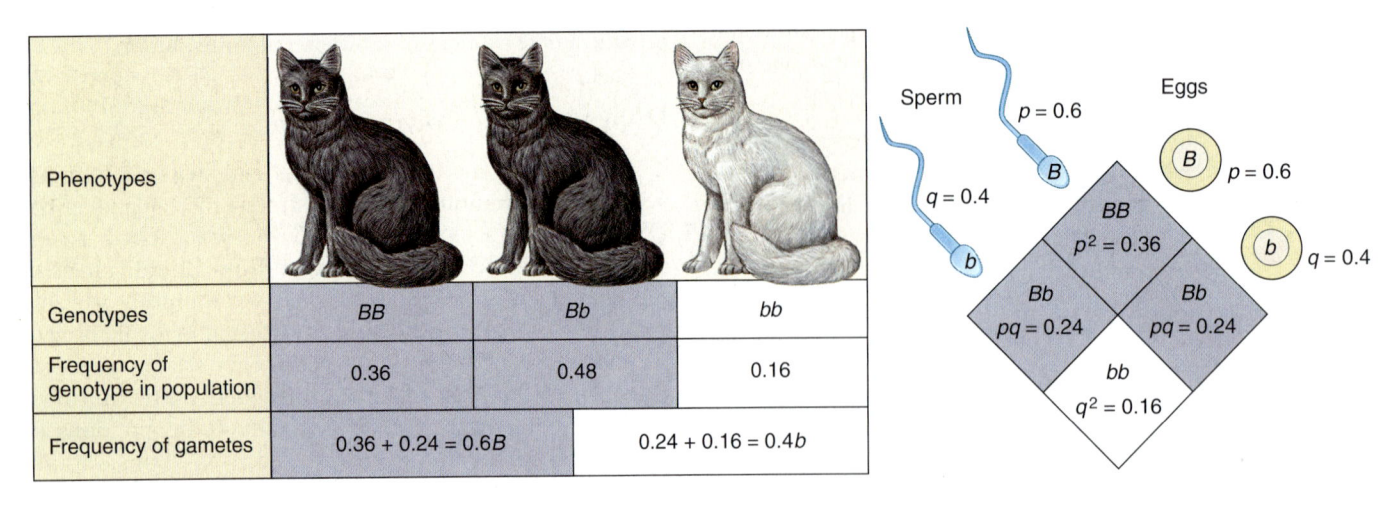

Phenotypes	(black)	(black)	(white)
Genotypes	BB	Bb	bb
Frequency of genotype in population	0.36	0.48	0.16
Frequency of gametes	0.36 + 0.24 = 0.6B		0.24 + 0.16 = 0.4b

Sperm: $p = 0.6$ (B), $q = 0.4$ (b)
Eggs: $p = 0.6$ (B), $q = 0.4$ (b)

- BB: $p^2 = 0.36$
- Bb: $pq = 0.24$
- Bb: $pq = 0.24$
- bb: $q^2 = 0.16$

FIGURE 21.4
The Hardy–Weinberg equilibrium. In the absence of factors that alter them, the frequencies of gametes, genotypes, and phenotypes remain constant generation after generation.
If all white cats died, what proportion of the kittens in the next generation would be white?

Using the Hardy–Weinberg Equation

The Hardy–Weinberg equation is a simple extension of the Punnett square described in chapter 13, with two alleles assigned frequencies, p and q. Figure 21.4 allows you to trace genetic reassortment during sexual reproduction and see how it affects the frequencies of the B and b alleles during the next generation. In constructing this diagram, we have assumed that the union of sperm and egg in these cats is random, so that all combinations of b and B alleles occur. For this reason, the alleles are mixed randomly and represented in the next generation in proportion to their original representation. Each individual egg or sperm in each generation has a 0.6 chance of receiving a B allele ($p = 0.6$) and a 0.4 chance of receiving a b allele ($q = 0.4$).

In the next generation, therefore, the chance of combining two B alleles is p^2, or 0.36 (that is, 0.6×0.6), and approximately 36% of the individuals in the population will continue to have the BB genotype. The frequency of bb individuals is q^2 (0.4×0.4) and so will continue to be about 16%, and the frequency of Bb individuals will be $2pq$ ($2 \times 0.6 \times 0.4$), or on average, 48%. Phenotypically, if the population size remains at 100 cats, we will still see approximately 84 black individuals (with either BB or Bb genotypes) and 16 white individuals (with the bb genotype) in the population. Allele, genotype, and phenotype frequencies have remained unchanged from one generation to the next, despite the reshuffling of genes that occurs during meiosis and sexual reproduction. Dominance and recessiveness of alleles thus only affect how an allele is expressed in an individual and not how allele frequencies will change through time.

Of course, populations need not be in Hardy–Weinberg equilibrium. Suppose, for example, that the frequencies of the BB and bb genotypes were each 0.45, and that of Bb was 0.10. How could such an excess of homozygotes and absence of heterozygotes be explained? Two possibilities are that heterozygotes do not survive for long or that individuals choose to mate with genetically similar individuals. (Because $BB \times BB$ and $bb \times bb$ always produce homozygous offspring, but only half of $Bb \times Bb$ produce heterozygous offspring, such mating patterns would lead to an excess of homozygotes.)

Thus, it is important to remember that Hardy–Weinberg equilibrium is a null hypothesis: When all assumptions are met, populations will be in equilibrium. Situations in which populations are not at equilibrium are the most interesting, because they indicate that one or more evolutionary processes are at work.

Why Do Allele Frequencies Change?

According to the Hardy–Weinberg principle, both the allele and genotype frequencies in a large, random-mating population will remain constant from generation to generation if no mutation, no gene flow, and no selection occur. The stipulations tacked onto the end of this statement are important. In fact, they are the key to the importance of the Hardy–Weinberg principle, because individual allele frequencies often change in natural populations, with some alleles becoming more common and others decreasing in frequency. The Hardy–Weinberg principle establishes a convenient baseline against which to measure such changes. By looking at how various factors alter the proportions of homozygotes and heterozygotes, we can identify the forces affecting particular situations we observe.

Many factors can alter allele frequencies. Only five, however, alter the proportions of homozygotes and heterozygotes enough to produce significant deviations from the proportions predicted by the Hardy–Weinberg principle: mutation, gene flow (including both immigration into and emigration out of a given population), nonrandom mating, genetic drift (random change in allele frequencies, which is more likely in small populations), and selection (table 21.1). Of these, only selection produces adaptive evolutionary change because only in selection does the result depend on the nature of the environment. The other factors operate relatively independently of the environment, so the changes they produce are not shaped by environmental demands.

Table 21.1 Agents of Evolutionary Change

Factor	Description
Mutation	The ultimate source of variation. Individual mutations occur so rarely that mutation alone usually does not change allele frequency much.
Gene flow	A very potent agent of change. Populations exchange members or gametes.
Nonrandom mating	Inbreeding is the most common form. It does not alter allele frequency but changes the proportion of heterozygotes.
Genetic drift	Statistical accidents. The random fluctuation in allele frequencies increases as population size decreases.
Selection	The only agent that produces *adaptive* evolutionary changes.

> The Hardy–Weinberg principle states that in a large population mating at random and in the absence of other forces that would change the proportions of the different alleles at a given locus, the process of sexual reproduction (meiosis and fertilization) alone will not change these proportions.

Five Agents of Evolutionary Change

1. Mutation

Mutation from one allele to another can obviously change the proportions of particular alleles in a population. Mutation rates are generally so low that they have little effect on the Hardy–Weinberg proportions of common alleles. A typical gene mutates about once per 100,000 cell divisions. Because this rate is so low, other evolutionary processes are usually more important in determining how allele frequencies change. Nonetheless, mutation is the ultimate source of genetic variation and thus makes evolution possible (figure 21.5a). It is important to remember, however, that the likelihood of a particular mutation occurring is not affected by natural selection; that is, mutations do not occur more frequently in situations in which they would be favored by natural selection.

2. Gene Flow

Gene flow is the movement of alleles from one population to another. It can be a powerful agent of change because members of two different populations may exchange genetic material. Sometimes gene flow is obvious, as when an animal moves from one place to another. If the characteristics of the newly arrived animal differ from those of the animals already there, and if the newcomer is adapted well enough to the new area to survive and mate successfully, the genetic composition of the receiving population may be altered. Other important kinds of gene flow are not as obvious. These subtler movements include the drifting of gametes or immature stages of plants or marine animals from one place to another (figure 21.5b). Pollen, the male gamete of flowering plants, is often carried great distances by insects and other animals that visit flowers. Seeds may also blow in the wind or be carried by animals to new populations far from their place of origin. In addition, gene flow may also result from the mating of individuals belonging to adjacent populations.

Consider two populations initially different in allele frequencies: In population 1, $p = 0.2$ and $q = 0.8$; in population 2, $p = 0.8$ and $q = 0.2$. Gene flow will tend to bring the rarer allele into each population. Thus, allele frequencies will change from generation to generation, and the populations will not be in Hardy–Weinberg equilibrium. Only when allele frequencies reach 0.5 for both alleles in both populations will equilibrium be attained. This example also indicates that gene flow tends to homogenize allele frequencies among populations.

3. Nonrandom Mating

Individuals with certain genotypes sometimes mate with one another more commonly than would be expected on a random basis, a phenomenon known as nonrandom mating (fig-

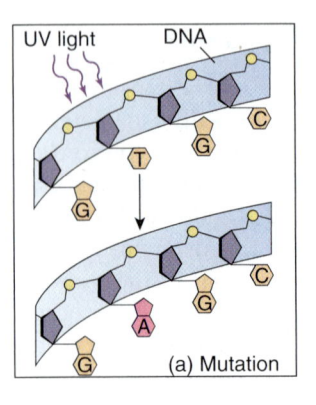

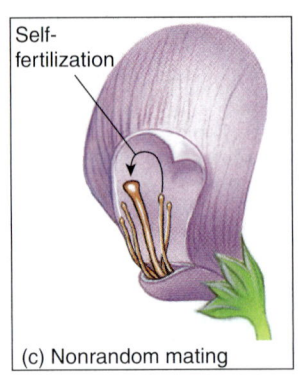

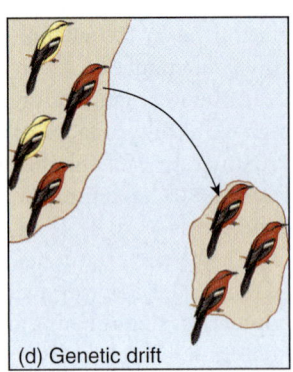

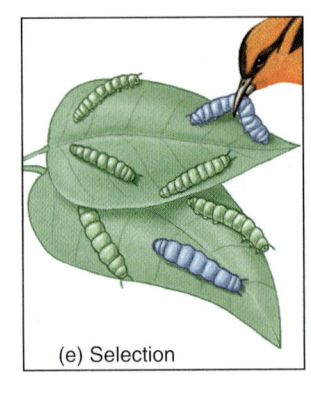

FIGURE 21.5
Five agents of evolutionary change. (*a*) Mutation, (*b*) gene flow, (*c*) nonrandom mating, (*d*) genetic drift by founder effect, and (*e*) selection.

ure 21.5c). **Assortative mating,** in which phenotypically similar individuals mate, is a type of nonrandom mating that causes the frequencies of particular genotypes to differ greatly from those predicted by the Hardy–Weinberg principle. Assortative mating does not change the frequency of the alleles, but rather increases the proportion of homozygous individuals because phenotypically similar individuals are likely to be genetically similar and thus produce offspring with two copies of the same allele. This is why populations of self-fertilizing plants consist primarily of homozygous individuals. By contrast, **disassortative mating,** in which phenotypically different individuals mate, produces an excess of heterozygotes.

4. Genetic Drift

In small populations, frequencies of particular alleles may change drastically by chance alone. Such changes in allele frequencies occur randomly, as if the frequencies were drifting, and are thus known as **genetic drift** (figure 21.5d). For this reason, a population must be large to be in Hardy–Weinberg equilibrium. If the gametes of only a few individuals form the next generation, the alleles they carry may by chance not be representative of the parent population from which they were drawn, as illustrated in figure 21.6, where a small number of individuals are removed from a bottle containing many. By chance, most of the individuals removed are blue, so the new population has a much higher population of blue individuals than the parent generation had.

A set of small populations that are isolated from one another may come to differ strongly as a result of genetic drift even if the forces of natural selection do not differ between the populations. Indeed, because of genetic drift, harmful alleles may increase in frequency in small populations, despite selective disadvantage, and favorable alleles may be lost even though they are selectively advantageous. It is interesting to realize that humans have lived in small groups for much of the course of their evolution; consequently, genetic drift may have been a particularly important factor in the evolution of our species.

Even large populations may feel the effect of genetic drift. Large populations may have been much smaller in the past, and genetic drift may have greatly altered allele frequencies at that time. Imagine a population containing only two alleles of a gene, B and b, in equal frequency (that is, $p = q = 0.5$). In a large Hardy–Weinberg population, the genotype frequencies are expected to be 0.25 BB, 0.50 Bb, and 0.25 bb. If only a small sample produces the next generation, large deviations in these genotype frequencies can occur by chance. Imagine, for example, that four individuals form the next generation, and that by chance they are two Bb heterozygotes and two BB homozygotes—that is, the allele frequencies in the next generation are $p = 0.75$ and $q = 0.25$! If you were to replicate this experiment 100 times, each time randomly drawing four individuals from the parental population, one of the two alleles would be missing entirely from about 8 of the 100 populations. This leads to an important conclusion: Genetic drift results in the loss of alleles in isolated populations. Although genetic drift occurs in any population, it is particularly likely in situations with founder effects and bottlenecks.

Founder Effects. Sometimes one or a few individuals disperse and become the founders of a new, isolated population at some distance from their place of origin. These pioneers are not likely to have all the alleles present in the source population. Thus, some alleles may be lost from the new population, and others may change drastically in frequency. In some cases, previously rare alleles in the source population may be a significant fraction of the new population's genetic endowment. This phenomenon is called the **founder effect**. Founder effects are not rare in nature. Many self-pollinating plants start new populations from a single seed.

Founder effects have been particularly important in the evolution of organisms on distant oceanic islands, such as the Hawaiian Islands and the Galápagos Islands visited by Darwin. Most of the organisms in such areas probably derive from one or a few initial "founders." In a similar way, isolated human populations founded by relatively few individuals are often dominated by genetic features characteristic of their particular founders. Amish populations in the United States, for example, have unusually high frequencies of a number of conditions, such as polydactylism (the presence of a sixth finger).

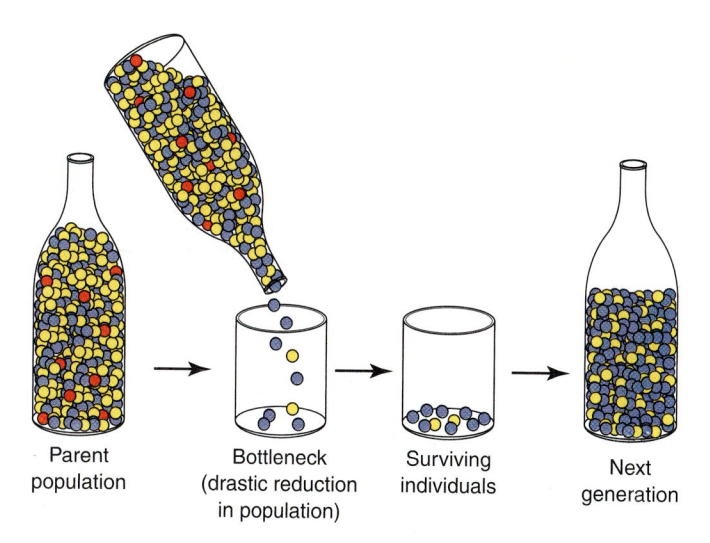

FIGURE 21.6
Genetic drift: A bottleneck effect. The parent population contains roughly equal numbers of blue and yellow individuals and a small number of red individuals. By chance, the few remaining individuals that contribute to the next generation are mostly blue. The bottleneck occurs because so few individuals form the next generation, as might happen after an epidemic or a catastrophic storm.

The Bottleneck Effect. Even if organisms do not move from place to place, occasionally their populations may be drastically reduced in size. This may result from flooding, drought, epidemic disease, and other natural forces, or from progressive changes in the environment. The few surviving individuals may constitute a random genetic sample of the original population (unless some individuals survive specifically because of their genetic makeup). The resultant alterations and loss of genetic variability have been termed the bottleneck effect.

Some living species appear to be severely depleted genetically and have probably suffered from a bottleneck effect in the past. For example, the northern elephant seal, which breeds on the western coast of North America and nearby islands, was nearly hunted to extinction in the nineteenth century and was reduced to a single population containing perhaps no more than 20 individuals on the island of Guadalupe off the coast of Baja, California. As a result of this bottleneck, even though the seal populations have rebounded and now number in the tens of thousands, this species has lost almost all of its genetic variation.

Chapter 21 Genes Within Populations 439

5. Selection

As Darwin pointed out, some individuals leave behind more progeny than others, and the rate at which they do so is affected by phenotype and behavior. We describe the results of this process as **selection** (figure 21.5*e*). In **artificial selection**, a breeder selects for the desired characteristics. In **natural selection**, environmental conditions determine which individuals in a population produce the most offspring. For natural selection to occur and result in evolutionary change, three conditions must be met:

a. **Variation must exist among individuals in a population.** Natural selection works by favoring individuals with some traits over individuals with alternative traits. If no variation exists, natural selection cannot operate.

b. **Variation among individuals results in differences in the number of offspring surviving in the next generation.** This is the essence of natural selection. Because of their phenotype or behavior, some individuals are more successful than others in producing offspring. Although many traits are phenotypically variable, varying individuals do not always differ in survival and reproductive success.

c. **Variation must be genetically inherited.** For natural selection to result in evolutionary change, the selected differences must have a genetic basis. However, not all variation has a genetic basis—even genetically identical individuals may be phenotypically quite distinctive if they grow up in different environments. Such environmental effects are common in nature. In many turtles, for example, individuals that hatch from eggs laid in moist soil are heavier, with longer and wider shells, than individuals from nests in drier areas. When phenotypically different individuals do not differ genetically, then differences in the number of their offspring will not alter the genetic composition of the population in the next generation, and thus, no evolutionary change will have occurred.

It is important to remember that natural selection and evolution are not the same—the two concepts often are incorrectly equated. Natural selection is a process, whereas evolution is the historical record of change through time. Evolution is an outcome, not a process. Natural selection (the process) can lead to evolution (the outcome), but natural selection is only one of several processes that can produce evolutionary change. Moreover, natural selection can occur without producing evolutionary change; only if variation is genetically based will natural selection lead to evolution.

Selection to Avoid Predators. Many of the most dramatic documented instances of adaptation involve genetic changes that decrease the probability of capture by a predator. The caterpillar larvae of the common sulphur butterfly *Colias eurytheme* usually exhibit a dull, kelly green color, providing excellent camouflage against the alfalfa plants on which they feed. An alternative bright blue color morph is kept at very low frequency because this color renders the larvae highly visible on the food plant, making it easier for bird predators to see them (see figure 21.5*e*). In a similar fashion, the way the shell markings in the land snail *Cepaea nemoralis* match its background habitat reflects the same pattern of avoiding predation by camouflage.

One of the most dramatic examples of background matching involves ancient lava flows in the middle of deserts in the American Southwest. In these areas, the black rock formations produced when the lava cooled contrast starkly with the surrounding bright glare of the desert sand. Populations of many species of animals occurring on these rocks—including lizards, rodents, and a variety of insects—are dark in color, whereas sand-dwelling populations in surrounding areas are much lighter (figure 21.7). Predation is the likely cause selecting for these differences in color. Laboratory studies have confirmed that predatory birds are adept at picking out individuals occurring on backgrounds to which they are not adapted.

FIGURE 21.7
Pocket mice from the Tularosa Basin of New Mexico whose color matches their background. Black lava formations are surrounded by desert, and selection favors coat color in pocket mice that matches their surroundings.

Selection to Match Climatic Conditions. Many studies of selection have focused on genes encoding enzymes, because in such cases the investigator can directly assess the consequences to the organism of changes in the frequency of alternative enzyme alleles. Often investigators find that enzyme allele frequencies vary latitudinally, with one allele more common in northern populations but progressively less common at more southern locations. A superb example is seen in studies of a fish, the mummichog *(Fundulus heteroclitus)*, which ranges along the eastern coast of North America. In this fish, allele frequencies vary geographically for the gene that produces the enzyme lactate dehydrogenase, which catalyzes the conversion of pyruvate to lactate (figure 21.8). Biochemical studies show that the enzymes formed by these alleles function differently at different temperatures, thus explaining their geographic distributions. For example, the form of the enzyme that is more frequent in the north is a better catalyst at low temperatures than the enzyme from the south. Moreover, studies indicate that at low temperatures, individuals with the northern allele swim faster, and presumably survive better, than individuals with the alternative allele.

Selection for Pesticide Resistance. A particularly clear example of selection in natural populations is provided by studies of pesticide resistance in insects. The widespread use of insecticides has led to the rapid evolution of resistance in more than 500 pest species. For example, in the housefly, the resistance allele at the *pen* gene decreases the uptake of insecticide, whereas alleles at the *kdr* and *dld-r* genes decrease the number of target sites, thus decreasing the binding ability of the insecticide (figure 21.9). Other alleles enhance the ability of the insects' enzymes to identify and detoxify insecticide molecules.

Single genes are also responsible for resistance in other organisms. The pigweed, *Amaranthus hybridus*, is one of about 28 agricultural weeds that have evolved resistance to the herbicide triazine. Triazine inhibits photosynthesis by binding to a protein in the chloroplast membrane, causing the plant to die. Single amino acid substitutions in the gene encoding the protein diminish the ability of triazine to decrease the plant's photosynthetic capabilities. Similarly, Norway rats are normally susceptible to the pesticide warfarin, which diminishes the clotting ability of the rat's blood and leads to fatal hemorrhaging. However, a resistance allele at a single gene alters a metabolic pathway and renders warfarin ineffective.

Five factors can bring about a deviation from the proportions of homozygotes and heterozygotes predicted by the Hardy–Weinberg principle. Only selection regularly produces adaptive evolutionary change, but the genetic constitution of populations, and thus the course of evolution, can also be affected by mutation, gene flow, nonrandom mating, and genetic drift.

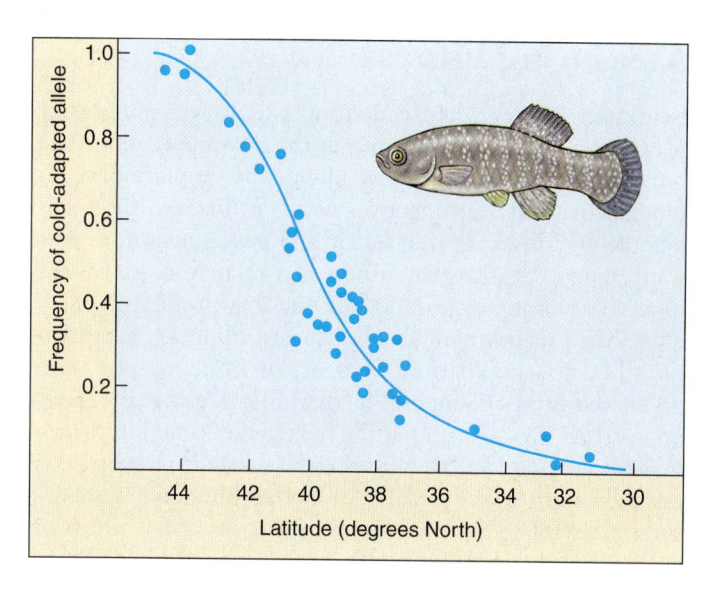

FIGURE 21.8
Selection to match climatic conditions. The frequency of the cold-adapted allele for lactate dehydrogenase in a type of fish (the mummichog, *Fundulus heteroclitus*) decreases at lower latitudes, which are warmer.
Why does the allele frequency change from north to south?

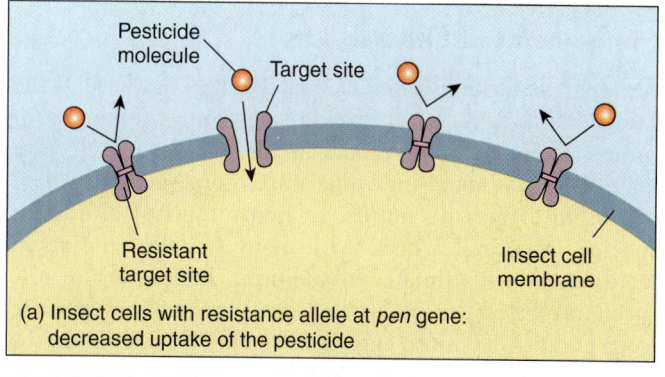

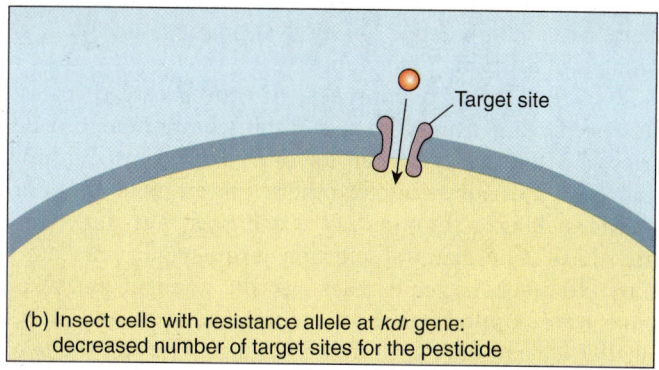

FIGURE 21.9
Selection for pesticide resistance. Resistance alleles at genes such as *pen* and *kdr* allow insects to be more resistant to pesticides. Insects that possess these resistance alleles have become more common through selection.

Chapter 21 Genes Within Populations 441

Measuring Fitness

Selection occurs when individuals with one phenotype leave more surviving offspring in the next generation than individuals with an alternative phenotype. Evolutionary biologists quantify reproductive success as **fitness,** the number of surviving offspring left in the next generation. It is important to realize that fitness is a relative concept; the most fit phenotype is simply the one that produces, on average, the greatest number of offspring. Suppose, for example, that in a population of toads, two phenotypes exist: green and brown. Suppose, further, that green toads leave, on average, 4.0 offspring in the next generation, but brown toads leave only 2.5. By custom, the most fit phenotype is assigned a fitness value of 1.0, and other phenotypes are expressed as relative proportions. In this case, the fitness of the brown phenotype would be 2.5/4.0 = 0.625. A difference in fitness of 0.375 is quite large; natural selection in this case strongly favors the green phenotype. If differences in color have a genetic basis, then we would expect evolutionary change to occur; the frequency of green toads should be substantially greater in the next generation. Further, if the fitness of two phenotypes remained unchanged, we would expect alleles for the brown phenotype to disappear from the population.

Components of Fitness

Although selection is often characterized as "survival of the fittest," differences in survival are only one component of fitness. Even if no differences in survival occur, selection may operate if some individuals are more successful than others in attracting mates. In many territorial animal species, large males mate with many females, and small males rarely get to mate. In addition, the number of offspring produced per mating is also important. Large female frogs and fish lay more eggs than smaller females and thus may leave more offspring in the next generation. Fitness is therefore a combination of survival, mating success, and number of offspring per mating.

Selection favors phenotypes with the greatest fitness, but predicting fitness from a single component can be tricky because traits favored for one component of fitness may be at a disadvantage for others. For example, in water striders, larger females lay more eggs per day (figure 21.10). Thus, natural selection at this stage favors large size. However, larger females also die younger and thus have fewer opportunities to reproduce; consequently, smaller females have a survival advantage. Overall, the two opposing directions of selection cancel each other out, and intermediate-sized females, on average, leave the most offspring in the next generation.

> An organism's reproductive success is affected by how long it survives, how often it mates, and how many offspring it produces per mating.

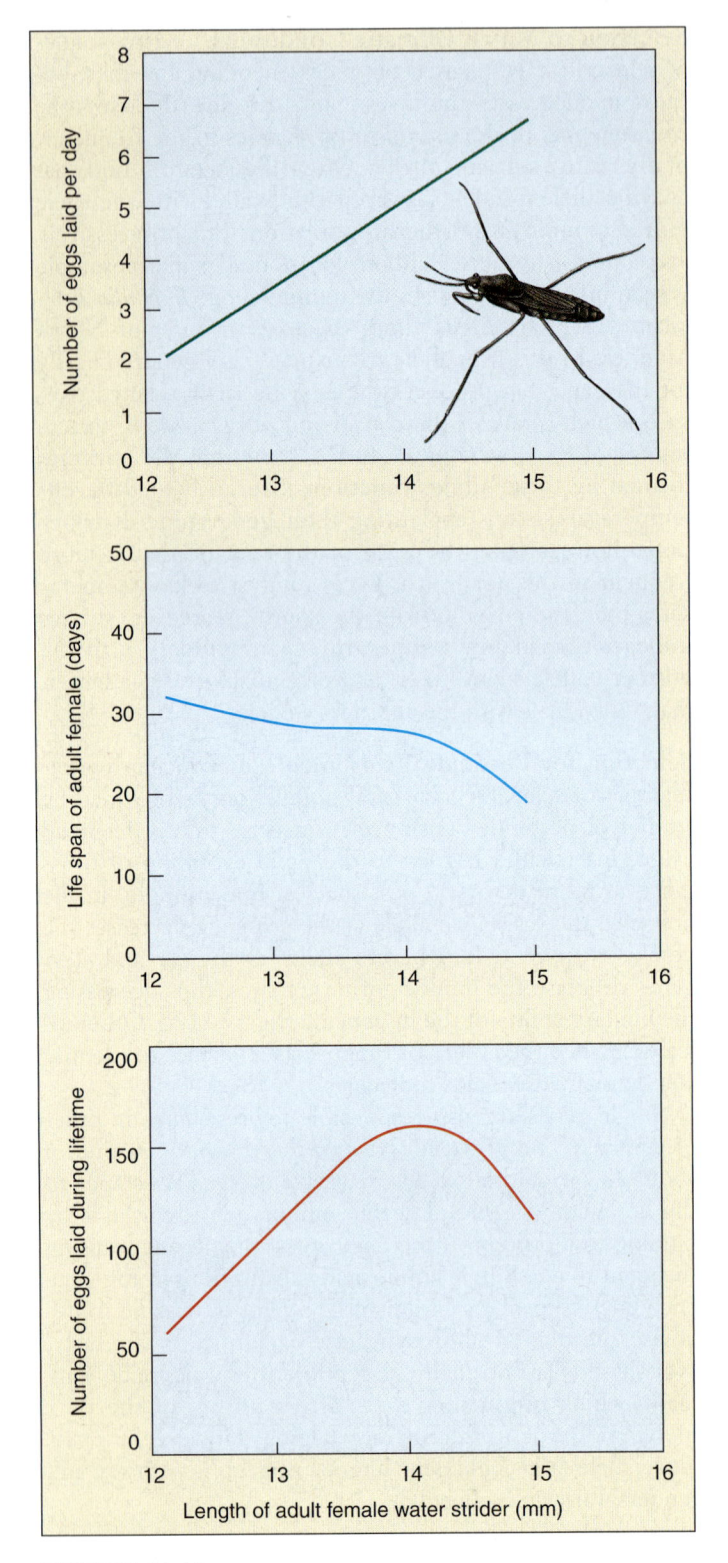

FIGURE 21.10
Body size and egg-laying in water striders. Larger female water striders lay more eggs per day, but also survive for a shorter period of time. As a result, intermediate-sized females produce the most offspring over the course of their entire lives and thus have the highest fitness.
What evolutionary change in body size might you expect? If the number of eggs laid per day was not affected by body size, would your prediction change?

Interactions Among Evolutionary Forces

Levels of variation retained in a population may be determined by the relative strength of different evolutionary processes. In theory, for example, if allele *B* mutates to allele *b* at a high enough rate, allele *b* could be maintained in the population even if natural selection strongly favored allele *B*. In nature, however, mutation rates are rarely high enough to counter the effects of natural selection.

The effect of natural selection also may be countered by genetic drift. Both processes may act to remove variation from a population. However, whereas selection is a deterministic process that operates to increase the representation of alleles that enhance survival and reproductive success, drift is a random process. Thus, in some cases, drift may lead to a decrease in the frequency of an allele that is favored by selection. In some extreme cases, drift may even lead to the loss of a favored allele from a population. Remember, however, that the magnitude of drift is negatively related to population size; consequently, natural selection is expected to overwhelm drift except when populations are very small.

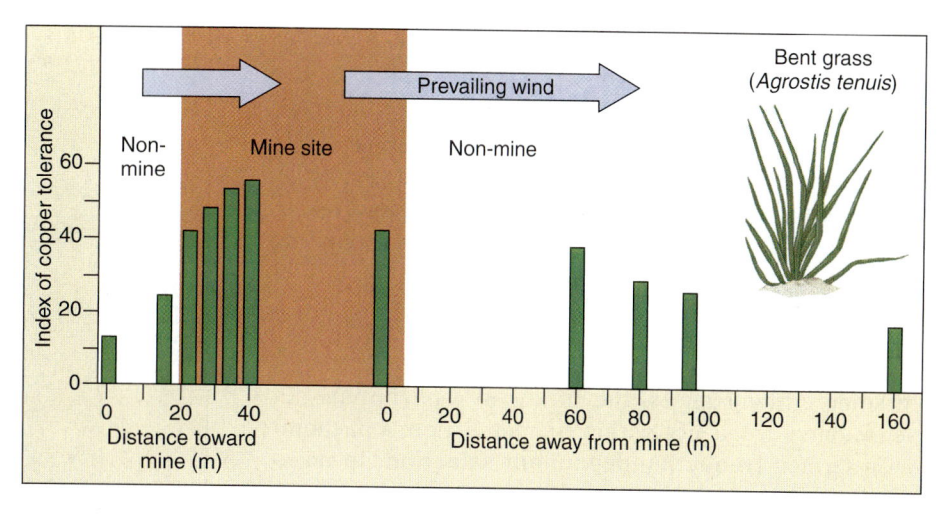

Gene Flow Versus Natural Selection

Gene flow can be either a constructive or a constraining force. On one hand, gene flow can increase the adaptedness of a species by spreading a beneficial mutation that arises in one population to other populations within a species. On the other hand, gene flow can impede adaptation within a population by continually importing inferior alleles from other populations. Consider two populations of a species that live in different environments. In this situation, natural selection might favor different alleles—*B* and *b*—in the different populations. In the absence of gene flow and other evolutionary processes, the frequency of *B* would be expected to reach 100% in one population and 0% in the other. However, if gene flow were occurring between the two populations, then the less favored allele would continually be reintroduced into each population. As a result, the frequency of the two alleles in each population would reflect a balance between the rate at which gene flow brings the inferior allele into a population and the rate at which natural selection removes it.

A classic example of gene flow opposing natural selection occurs on abandoned mine sites in Great Britain. Although mining activities ceased hundreds of years ago, the concentration of metal ions in the soil is still much greater than in surrounding areas. Heavy metal concentrations are generally toxic to plants, but alleles at certain genes confer resistance. The ability to tolerate heavy metals comes at a price, however; individuals with the resistance allele exhibit lower growth rates on nonpolluted soil. Consequently, we would expect the resistance allele to occur with a frequency of 100% on mine sites and 0% elsewhere. Heavy metal tolerance has been studied particularly intensively in the slender bent grass *Agrostis tenuis*, in which researchers have found that the resistance allele occurs at intermediate levels in many areas (figure 21.11). The explanation relates to the reproductive system of this grass in which pollen, the male gamete (that is, the floral equivalent of sperm), is dispersed by the wind. As a result, pollen—and the alleles it carries—can be blown for great distances, leading to levels of gene flow between mine sites and unpolluted areas high enough to counteract the effects of natural selection.

In general, the extent to which gene flow can hinder the effects of natural selection should depend on the relative strengths of the two processes. In species in which gene flow is generally strong, such as birds and wind-pollinated plants, the frequency of the less favored allele may be relatively high, whereas in more sedentary species that exhibit low levels of gene flow, such as salamanders, the favored allele should occur at a frequency near 100%.

FIGURE 21.11
Degree of copper tolerance in grass plants on and near ancient mine sites.
Individuals with tolerant alleles have decreased growth rates on unpolluted soil. Thus, we would expect copper tolerance to be 100% on mine sites and 0% on non-mine sites. However, prevailing winds blow pollen containing nontolerant alleles onto the mine site and tolerant alleles beyond the site's borders.
Would you expect the frequency of copper tolerance to be affected by distance from the mine site? How would your answer change depending on whether you were upwind or downwind from the mine site?

> Allele frequencies sometimes reflect a balance between opposing processes, such as gene flow and natural selection. In such cases, observed frequencies will depend on the relative strength of the processes.

Natural Selection Can Maintain Variation in Populations

In the previous pages, natural selection has been discussed as a process that removes variation from a population by favoring one allele over others at a gene locus. However, in some circumstances, selection can do exactly the opposite and actually maintain population variation.

Frequency-Dependent Selection

In some circumstances, the fitness of a phenotype depends on its frequency within the population, a phenomenon termed **frequency-dependent selection**. In negative frequency-dependent selection, rare phenotypes are favored by selection. Assuming a genetic basis for phenotypic variation, such selection will have the effect of making rare alleles more common, thus maintaining variation.

Negative frequency-dependent selection can occur for many reasons. For example, it is well-known that animals or people searching for something form a "search image." That is, they become particularly adept at picking out certain objects. Consequently, predators may form a search image for common prey phenotypes. Rare forms may thus be preyed upon less frequently. An example is fish predation on an insect, the water boatman, which occurs in three different colors. Experiments indicate that each of the color types is preyed upon disproportionately when it is the most common one (figure 21.12*a*). Another cause of negative frequency dependence is resource competition. If genotypes differ in their resource requirements, as occurs in many plants, then the rarer genotype will have fewer competitors. If the different resource types are equally abundant, the rarer genotype will be at an advantage relative to the more common genotype.

Positive frequency-dependent selection has the opposite effect; by favoring common forms, it tends to eliminate variation from a population. For example, predators don't always select common individuals. In some cases, "oddballs" stand out from the rest and attract attention (figure 21.12*b*).

Oscillating Selection

In some cases, selection favors one phenotype at one time and another phenotype at another time, a phenomenon called **oscillating selection**. If selection repeatedly oscillates in this fashion, the effect will be to maintain genetic variation in the population. One example, discussed in chapter 22, concerns the medium ground finch in the Galápagos Islands. In times of drought, the supply of small, soft seeds is depleted, but there are still enough large seeds around. Consequently, birds with big bills are favored. However, when wet conditions return, the ensuing abundance of small seeds favors birds with smaller bills.

Oscillating selection and frequency-dependent selection are similar because in both cases, the form of selection

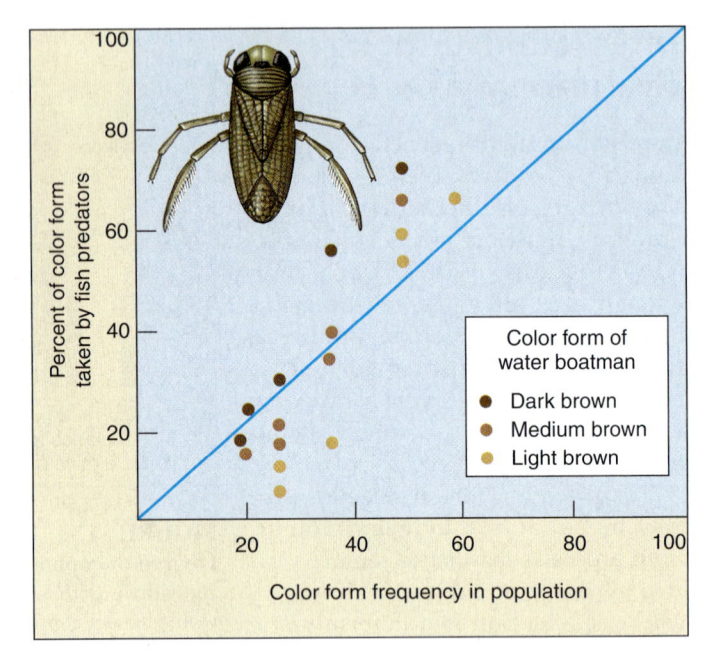

FIGURE 21.12
Frequency-dependent selection. (*a*) Predators often form search images for the most common prey. In one experiment, fish were placed in an enclosure with water boatmen (aquatic insects) of three different colors. When each color was common, the fish disproportionately captured boatmen of that color. By contrast, when each color form was uncommon, it was rarely taken (negative frequency-dependent selection). (*b*) However, in some cases, rare individuals stand out from the rest and draw the attention of predators; thus, in these cases, common phenotypes have the advantage (positive frequency-dependent selection).

changes through time. However, it is important to recognize that they are not the same: In oscillating selection, at any one time, the fitness of a phenotype does not depend on its frequency; rather, environmental changes lead to the oscillation in selection. In contrast, in frequency-dependent selection, it is the change in frequencies themselves that leads to the changes in fitness of the different phenotypes.

Heterozygote Advantage

If heterozygotes are favored over homozygotes, then natural selection actually tends to maintain variation in the population. Such **heterozygote advantage** will favor individuals with copies of both alleles, and thus will work to maintain both alleles in the population. Some evolutionary biologists believe that heterozygote advantage is pervasive and can explain the high levels of polymorphism observed in natural populations. Others, however, believe that it is relatively rare.

The best-documented example of heterozygote advantage is sickle cell anemia, a hereditary disease affecting hemoglobin in humans. Individuals with sickle cell anemia exhibit symptoms of severe anemia and abnormal red blood cells that are irregular in shape, with a great number of long, sickle-shaped cells (figure 21.13*a*). Chapter 13 discusses why the sickle cell mutation (*S*) causes red blood cells to sickle.

The average incidence of the *S* allele in central African populations is about 0.12, far higher than that found among African Americans. From the Hardy–Weinberg principle, you can calculate that 1 in 5 central African individuals is heterozygous at the *S* allele, and 1 in 100 is homozygous and develops the fatal form of the disorder. People who are homozygous for the sickle cell allele almost never reproduce because they usually die before they reach reproductive age. Why is the *S* allele not eliminated from the central African population by selection, rather than being maintained at such high levels? People who are heterozygous for the sickle cell allele (and thus do not suffer from sickle cell anemia) are much less susceptible to malaria—one of the leading causes of illness and death in central Africa, especially among young children. The reason is that when the parasite that causes malaria, *Plasmodium falciparum*, enters a red blood cell, it causes extremely low oxygen tension in the cell, which leads to cell sickling even in heterozygotes (but not in individuals that do not have the sickle cell allele). Such cells are quickly filtered out of the bloodstream by the spleen, thus eliminating the parasite. (The spleen's filtering effect is what leads to anemia in homozygotes as large numbers of red blood cells are removed; in the case of malaria, only cells with the *Plasmodium* parasite sickle).

Consequently, even though most homozygous recessive individuals die before they have children, the sickle cell allele is maintained at high levels in these populations (it is selected for) because it is associated with resistance to malaria in heterozygotes and also, for reasons not yet fully understood, with increased fertility in female heterozygotes.

For people living in areas where malaria is common, having the sickle cell allele in the heterozygous condition has adaptive value (figure 21.13*b*). Among African Americans, however, many of whose ancestors have lived for some 15 generations in a country where malaria has been relatively rare and is now essentially absent, the environment does not place a premium on resistance to malaria. Consequently, no adaptive value counterbalances the ill effects of the disease; in this nonmalarial environment, selection is acting to eliminate the *S* allele. Only 1 in 375 African Americans develops sickle cell anemia, far less than in central Africa.

Selection can maintain variation within populations in a number of ways.

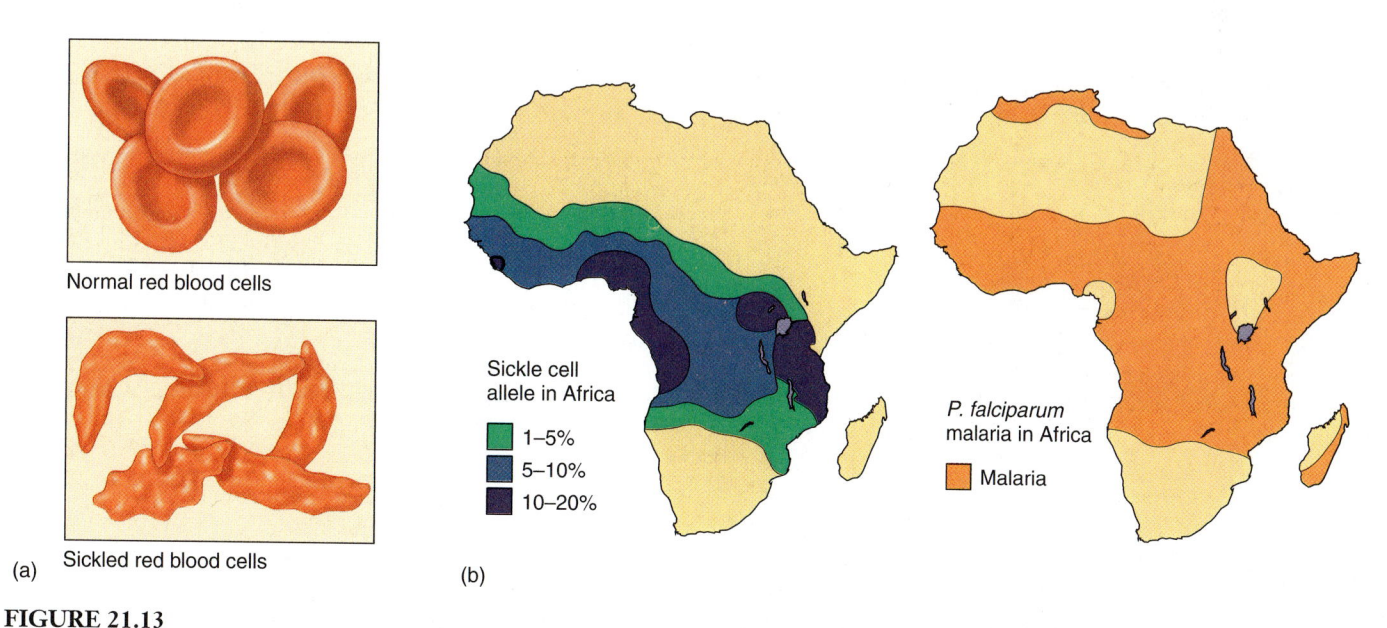

FIGURE 21.13
Frequency of sickle cell allele and distribution of *Plasmodium falciparum* malaria. (*a*) The red blood cells of people homozygous for the sickle cell allele collapse into sickled shapes when the oxygen level in the blood is low. (*b*) The distribution of the sickle cell allele in Africa coincides closely with that of *P. falciparum* malaria.

21.3 Selection can act on traits affected by many genes.

Forms of Selection

In nature, many traits—perhaps most—are affected by more than one gene. The interactions between genes are typically complex, as you saw in chapter 13. For example, alleles of many different genes play a role in determining human height (see figure 13.16). In such cases, selection operates on all the genes, influencing most strongly those that make the greatest contribution to the phenotype. How selection changes the population depends on which genotypes are favored.

Disruptive Selection

In some situations, selection acts to eliminate intermediate types, a phenomenon called **disruptive selection** (figure 21.14a). A clear example is the different beak sizes of the African fire-bellied seedcracker finch *Pyronestes ostrinus* (figure 21.15). Populations of these birds contain individuals with large and small beaks, but very few individuals with intermediate-sized beaks. As their name implies, these birds feed on seeds, and the available seeds fall into two size categories: large and small. Only large-beaked birds can open the tough shells of large seeds, whereas birds with the smaller beaks are more adept at handling small seeds. Birds with intermediate-sized beaks are at a disadvantage with both seed types: unable to open large seeds and too clumsy to efficiently process small seeds. Consequently, selection acts to eliminate the intermediate phenotypes, in effect partitioning (or "disrupting") the population into two phenotypically distinct groups.

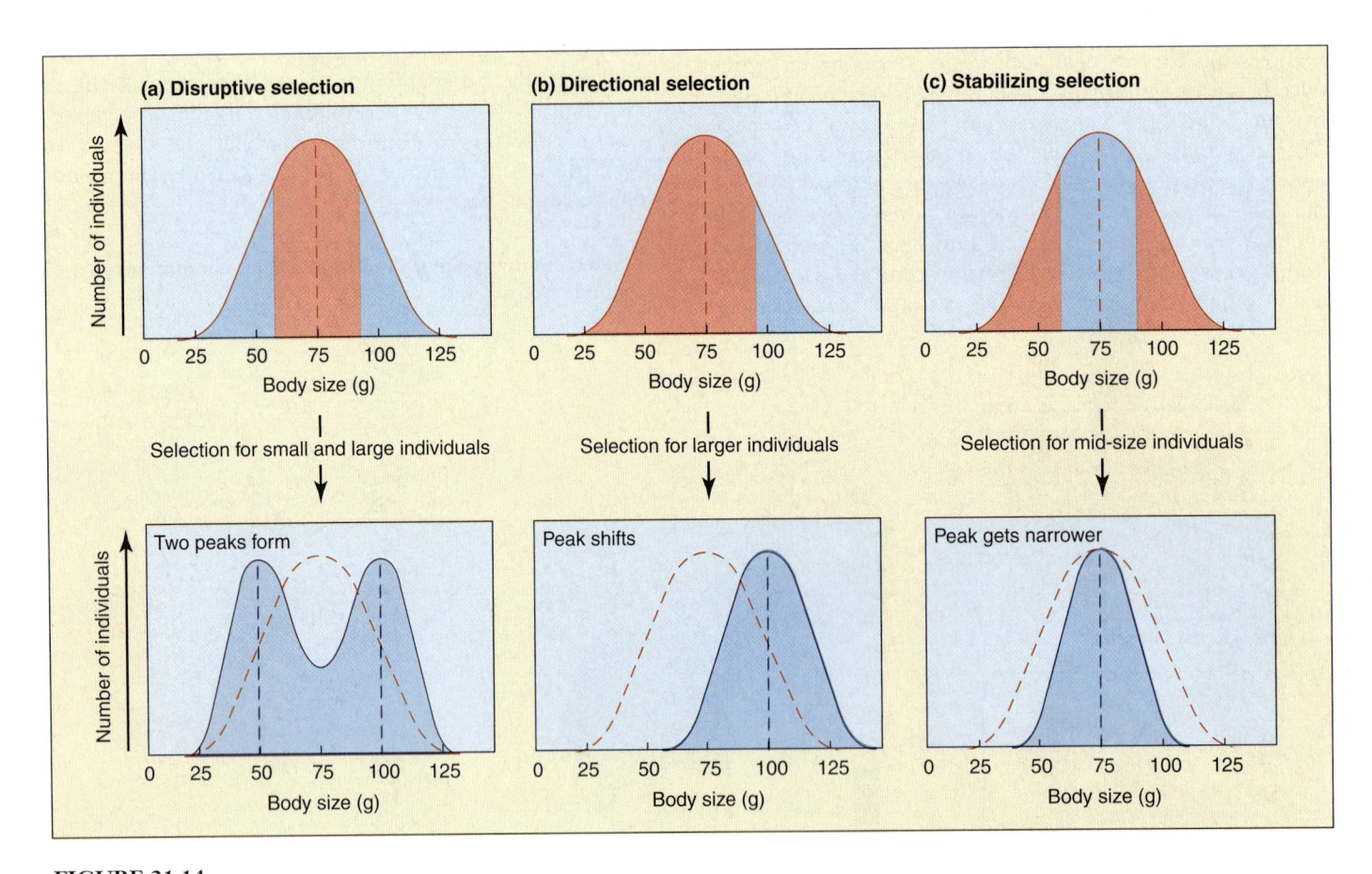

FIGURE 21.14
Three kinds of selection. The top panels show the populations before selection has occurred, with the forms that will be selected against shaded red and the forms that will be favored shaded blue. The bottom panels indicate what the populations will look like after selection has occurred, assuming that variation in the traits is genetically based. (*a*) In disruptive selection, individuals in the middle of the range of phenotypes of a certain trait are selected against (*red*), and the extreme forms of the trait are favored (*blue*). (*b*) In directional selection, individuals concentrated toward one extreme of the array of phenotypes are favored. (*c*) In stabilizing selection, individuals with midrange phenotypes are favored, with selection acting against both ends of the range of phenotypes.

FIGURE 21.15
Disruptive selection for large and small beaks. Differences in beak size in the fire-bellied seedcracker finch of west Africa are the result of disruptive selection.

Directional Selection

When selection acts to eliminate one extreme from an array of phenotypes, the genes promoting this extreme become less frequent in the population. This form of selection is called **directional selection** (figure 21.14b). Thus, in the *Drosophila* population illustrated in figure 21.16, the elimination of flies that move toward light causes the population to contain fewer individuals with alleles promoting such behavior. If you were to pick an individual at random from the new fly population, there is a smaller chance that it would spontaneously move toward light than if you had selected a fly from the old population. Artificial selection has changed the population in the direction of lower light attraction.

Stabilizing Selection

When selection acts to eliminate both extremes from an array of phenotypes, the result is to increase the frequency of the already common intermediate type. This form of selection is called **stabilizing selection** (figure 21.14c). In effect, selection is operating to prevent change away from this middle range of values. Selection does not change the most common phenotype of the population, but rather makes it even more common by eliminating extremes. Many examples are known. In humans, infants with intermediate weight at birth have the highest survival rate (figure 21.17). In ducks and chickens, eggs of intermediate weight have the highest hatching success.

> Selection on traits affected by many genes can favor both extremes of the trait, or intermediate values, or only one extreme.

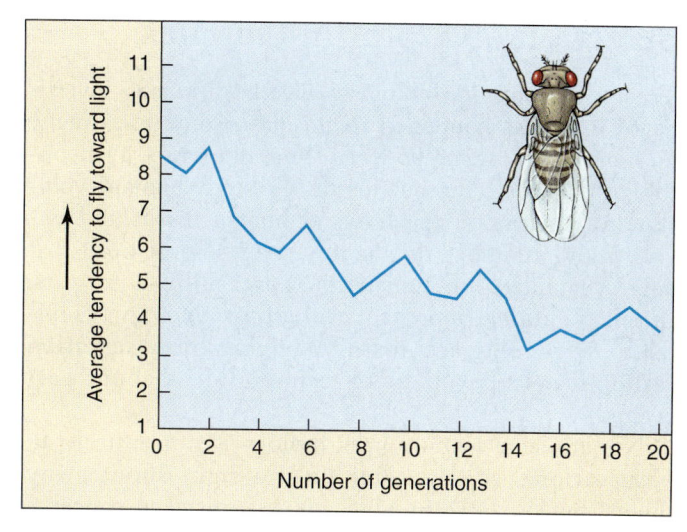

FIGURE 21.16
Directional selection for negative phototropism in *Drosophila*. Flies that moved toward light were discarded, and only flies that moved away from light were used as parents for the next generation. This procedure was repeated for 20 generations, producing substantial evolutionary change.
What would happen if after 20 generations, experimenters started keeping flies that moved *toward* the light and discarded the others?

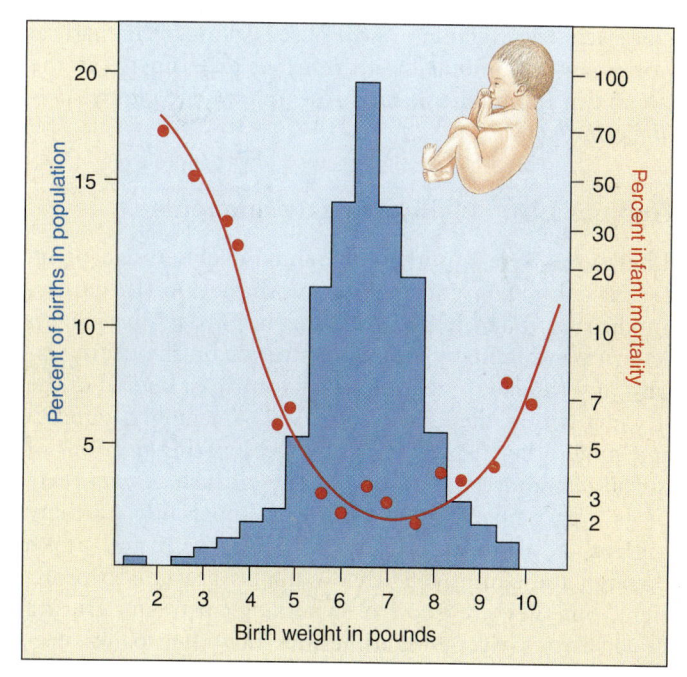

FIGURE 21.17
Stabilizing selection for birth weight in human beings. The death rate among babies (*red curve; right y-axis*) is lowest at an intermediate birth weight; both smaller and larger babies have a greater tendency to die than those around the most frequent weight (*blue area; left y-axis*) of between 7 and 8 pounds.
As improved medical technology leads to decreased infant mortality rates, how would you expect the distribution of birthrates in the population to change?

Chapter 21 Genes Within Populations 447

Selection on Color in Guppies

To study evolution, biologists have traditionally investigated what has happened in the past, sometimes many millions of years ago. To learn about dinosaurs, a paleontologist looks at dinosaur fossils. To study human evolution, an anthropologist looks at human fossils and, increasingly, examines the "family tree" of mutations that have accumulated in human DNA over millions of years. In this traditional approach, evolutionary biology is similar to astronomy and history, relying on observation rather than experiment to examine ideas about past events.

Nonetheless, evolutionary biology is not entirely an observational science. Darwin was right about many things, but one area in which he was mistaken concerns the pace at which evolution occurs. Darwin thought that evolution occurred at a very slow, almost imperceptible pace. However, in recent years many case studies have demonstrated that in some circumstances evolutionary change can occur rapidly. Consequently, it is possible to establish experimental studies to test evolutionary hypotheses. Although laboratory studies on fruit flies and other organisms have been common for more than 50 years, it has only been recently that scientists have started conducting experimental studies of evolution in nature. One excellent example of how observations of the natural world can be combined with rigorous experiments in the lab and in the field concerns research on the guppy, *Poecilia reticulata*.

Guppies Live in Different Environments

The guppy is a popular aquarium fish because of its bright coloration and prolific reproduction. In nature, guppies are found in small streams in northeastern South America and in many mountain streams on the nearby island of Trinidad. One interesting feature of several of the streams is that they have waterfalls. Amazingly, guppies and some other fish are capable of colonizing portions of the stream above the waterfall. The killifish, *Rivulus hartii*, is a particularly good colonizer; apparently on rainy nights, it will wriggle out of the stream and move through the damp leaf litter. Guppies are not so proficient, but they are good at swimming upstream. During flood seasons, rivers sometimes overflow their banks, creating secondary channels that move through the forest. On these occasions, guppies may be able to move upstream and invade the pools above waterfalls. By contrast, some species are not capable of such dispersal and thus are only found in streams below the first waterfall. One species whose distribution is restricted by waterfalls is the pike cichlid, *Crenicichla alta*, a voracious predator that feeds on other fish, including guppies.

Because of these barriers to dispersal, guppies can be found in two very different environments. In pools just

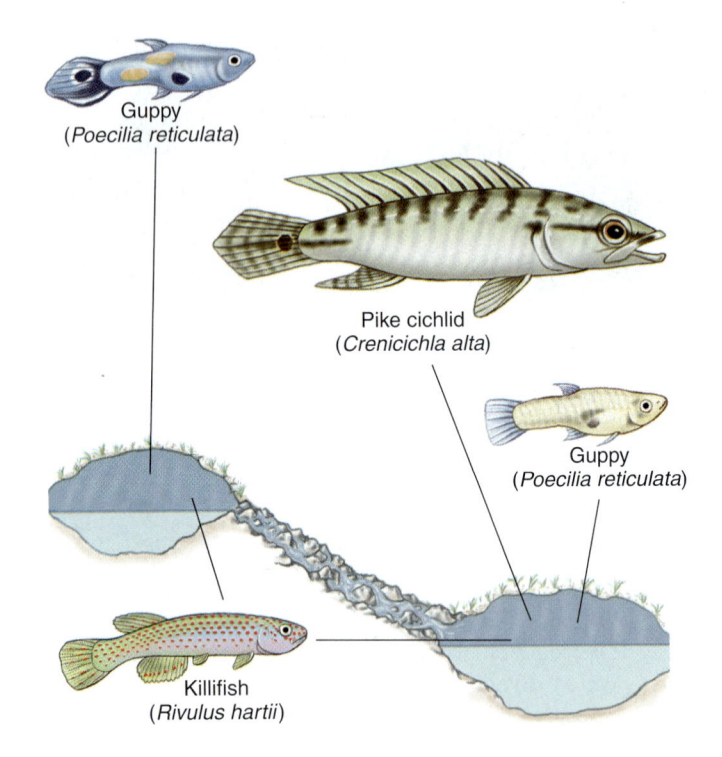

FIGURE 21.18
The evolution of protective coloration in guppies. In pools below waterfalls where predation is high, male guppies (*Poecilia reticulata*) are drab colored. In the absence of the highly predatory pike cichlid (*Crenicichla alta*), male guppies in pools above waterfalls are much more colorful and attractive to females. The killifish (*Rivulus hartii*) is also a predator, but it only rarely eats guppies. The evolution of these differences in guppies can be experimentally tested.

below the waterfalls, predation by the pike cichlid is a substantial risk, and rates of survival are relatively low. By contrast, in similar pools just above the waterfall, the only predator present is the killifish, which rarely preys on guppies. Guppy populations above and below waterfalls exhibit many differences. In the high-predation pools, guppies exhibit drab coloration. Moreover, they tend to reproduce at a younger age and attain relatively smaller adult sizes. By contrast, male fish above the waterfall display gaudy colors that they use to court females (figure 21.18). Adults mature later and grow to larger sizes.

These differences suggest the function of natural selection. In the low-predation environment, males display gaudy colors and spots that they use to court females. Moreover, larger males are most successful at holding territories and mating with females, and larger females lay more eggs. Thus, in the absence of predators, larger and more colorful fish may have produced more offspring, leading to the evolution of those traits. In pools below the waterfall, however, natural selection would favor different traits. Colorful males are likely to attract the attention of

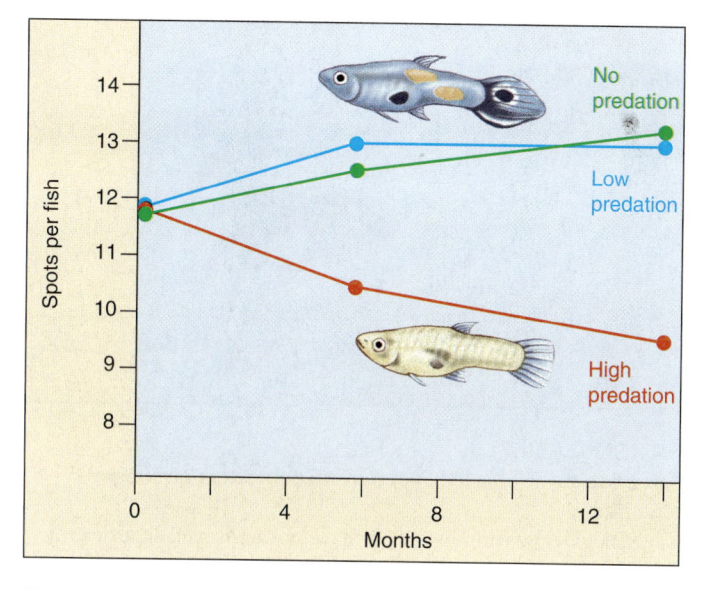

FIGURE 21.19
Evolutionary change in spot number. Guppies raised in low-predation or no-predation environments in laboratory greenhouses had a greater number of spots, whereas selection in more dangerous environments, such as the pools with the highly predatory pike cichlid, led to less conspicuous fish. The same results are seen in field experiments conducted in pools above and below waterfalls (*photo*).
How do these results depend on the manner by which the guppy predators locate their prey?

the pike cichlid, and high predation rates mean that most fish live short lives; thus, individuals that are more drab and shunt energy into early reproduction, rather than growth to a larger size, are likely to be favored by natural selection.

The Experiments

Although the differences between guppies living above and below the waterfalls suggest that they represent evolutionary responses to differences in the strength of predation, alternative explanations are possible. Perhaps, for example, only very large fish are capable of crawling past the waterfall to colonize pools. If this were the case, then a founder effect would occur in which the new population was established solely by individuals with genes for large size.

Laboratory Experiment. The only way to rule out such alternative possibilities is to conduct a controlled experiment. John Endler, now at the University of California, Santa Barbara, conducted the first experiments in large pools in laboratory greenhouses. At the start of the experiment, a group of 2000 guppies was divided equally among 10 large pools. Six months later, pike cichlids were added to four of the pools and killifish to another four, with the remaining two pools left to serve as "no-predation" controls. Fourteen months later (which corresponds to 10 guppy generations), the scientists compared the populations. The guppies in the killifish and control pools were indistinguishable—brightly colored and large. In contrast, the guppies in the pike cichlid pools were smaller and drab in coloration. These results established that predation can lead to rapid evolutionary change, but do these laboratory experiments reflect what occurs in nature?

Field Experiment. To find out, Endler and colleagues—including David Reznick, now at the University of California, Riverside—located two streams that had guppies in pools below a waterfall, but not above it. As in other Trinidadian streams, the pike cichlid was present in the lower pools, but only the killifish was found above the waterfalls. The scientists then transplanted guppies to the upper pools and returned at several-year intervals to monitor the populations. Despite originating from populations in which predation levels were high, the transplanted populations rapidly evolved the traits characteristic of low-predation guppies: They matured late, attained greater size, and had brighter colors. Control populations in the lower pools, by contrast, continued to be drab and to mature early and at a smaller size (figure 21.19). Laboratory studies confirmed that the variations between the populations were the result of genetic differences. These results demonstrate that substantial evolutionary change can occur in less than 12 years. More generally, these studies indicate how scientists can formulate hypotheses about how evolution occurs and then test these hypotheses in natural conditions. The results give strong support to the theory of evolution by natural selection.

> Evolutionary biology is a historical science. Nonetheless, in some circumstances, experiments can be conducted in nature to test hypotheses about how evolution occurs. Such studies often reveal that natural selection can lead to rapid evolutionary change.

Chapter 21 Genes Within Populations 449

Limits to What Selection Can Accomplish

Genes Have Multiple Effects

Although selection is the most powerful of the principal agents of genetic change, there are limits to what it can accomplish. These limits arise for a variety of reasons. For example, alleles often affect multiple aspects of the phenotype (the phenomenon of pleiotropy; see chapter 13). These multiple effects tend to set limits on how much a phenotype can be altered. For example, selecting for large clutch size in chickens eventually leads to eggs with thinner shells that break more easily. For this reason, we could never produce chickens that lay eggs twice as large as the best layers do now, gigantic cattle that yield twice as much meat as our leading strains, or corn with an ear at the base of every leaf, instead of just at the base of a few leaves.

Evolution Requires Genetic Variation

Over 80% of the gene pool of the thoroughbred horses racing today goes back to 31 known ancestors from the late eighteenth century. Despite intense directional selection on thoroughbreds, their performance times have not improved for more than 50 years (figure 21.20). Years of intense selection presumably have removed variation from the population at a rate greater than it could be replenished by mutation, such that now no genetic variation remains and evolutionary change is not possible.

In some cases, phenotypic variation for a trait may never have had a genetic basis. The compound eyes of insects are made up of hundreds of visual units, termed ommatidia (see chapter 33). In some individuals, the left eye contains more ommatidia than the right eye. In other individuals, the right eye contains more than the left. However, despite intense selection in the laboratory, scientists have never been able to produce a line of fruit flies that consistently have more ommatidia in the left eye than in the right. The reason is that separate genes do not exist for the left and right eyes. Rather, the same genes affect both eyes, and differences in the number of ommatidia result from differences that occur as the eyes are formed in the development process (figure 21.21). Thus, despite the existence of phenotypic variation, no genetic variation is available for selection to favor.

Gene Interactions Affect Fitness of Alleles

As discussed in chapter 13, **epistasis** is the phenomenon in which an allele for one gene may have different effects, depending on the alleles present at other genes. Because of epistasis, the selective advantage of an allele at one gene may vary among genotypes. If a population is polymorphic for a second gene, then selection on the first gene may be constrained because different alleles are favored in different individuals of the same population.

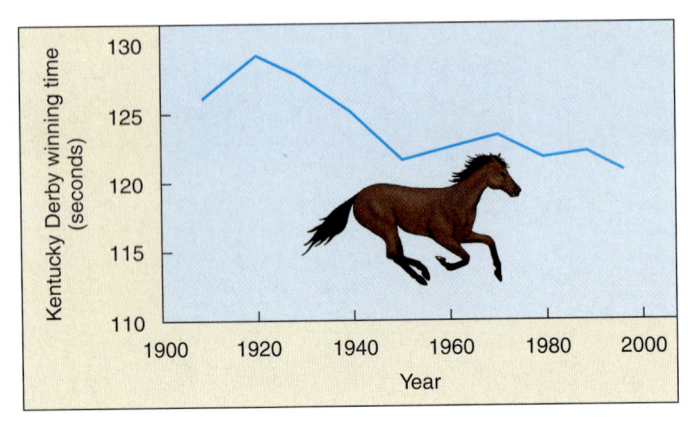

FIGURE 21.20
Selection for increased speed in racehorses is no longer effective.
Kentucky Derby winning speeds have not improved significantly since 1950.
What might explain the lack of change in winning speeds?

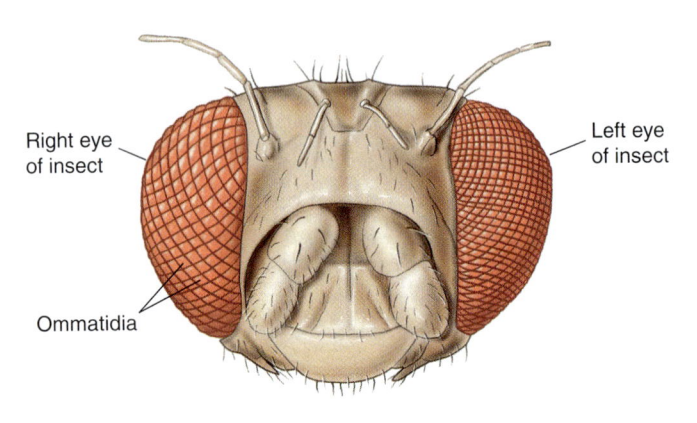

FIGURE 21.21
Phenotypic variation in insect ommatidia. In some individuals, the number of ommatidia in the left eye is greater than the number in the right eye.

Dan Hartl, now of Harvard University, and colleagues grew bacteria with different alleles of the enzyme 6-PGD on gluconate, the enzyme's substrate. Under normal conditions, Hartl found that bacteria with the different alleles all grew at the same rate; natural selection thus did not favor one allele over another. However, a second gene controls an alternative biochemical pathway for the metabolism of gluconate. An uncommon allele at this gene disables this second pathway so that only 6-PGD mediates the utilization of gluconate. When Hartl studied bacteria with this uncommon allele, he found that bacteria with several 6-PGD alleles were markedly superior to others. Thus, epistatic interactions exist between the two genes, and the outcome of natural selection on one gene depends on which alleles are present at the second gene.

The ability of selection to produce evolutionary change is hindered by a variety of factors.

Concept Review

21.1 Genes vary in natural populations.

Genetic Variation Is the Raw Material of Evolution
- Evolution refers to change in a species over time as genetic differences accumulate. (p. 434)
- Darwin proposed that natural selection was the mechanism of evolution, while Jean-Baptiste Lamarck believed individuals acquired characteristics during their lifetime, and then passed those onto their offspring. (p. 434)

Genetic Variation in Nature
- Evolution can result from any process causing change in a population's genetic variation. (p. 435)
- DNA analysis has found abundant evidence of polymorphic genes (loci with more than one allele). (p. 435)

21.2 Why do allele frequencies change in populations?

The Hardy–Weinberg Principle
- Original genotypic proportions in a population will remain constant as long as the population is large, mates at random, experiences no mutation or immigration, and is not affected by selection. (p. 436)

Five Agents of Evolutionary Change
- Mutation, gene flow, nonrandom mating, genetic drift (including founder effects and bottleneck effects), and selection (either natural or artificial) are the major factors driving evolutionary change. (pp. 438–441)

Measuring Fitness
- Fitness is measured as the number of living offspring in the next generation. Maximum fitness is defined as 1.0. (p. 442)

Interactions Among Evolutionary Forces
- Population variation may be determined by counterbalancing forces such as gene flow and natural selection. The extent of interaction depends on the relative strength of each force. (p. 443)

Natural Selection Can Maintain Variation in Populations
- Negative frequency-dependent selection will favor rare phenotypes, while positive frequency-dependent selection will favor common phenotypes. (p. 444)
- Oscillating selection will favor different phenotypes at different times. (p. 444)
- Variation can be maintained if heterozygotes are favored over homozygotes, as in the case of sickle cell anemia and malaria. (p. 445)

21.3 Selection can act on traits affected by many genes.

Forms of Selection
- Disruptive selection acts to eliminate intermediate phenotypes. (p. 446)
- Directional selection acts to eliminate one phenotypic extreme. (p. 447)
- Stabilizing selection acts to eliminate both phenotypic extremes. (p. 447)

Selection on Color in Guppies
- Guppies live in different natural environments. Populations under predation pressure tend to exhibit drab colors and to be smaller than guppies not under predation pressure. (p. 448)
- Laboratory and field experiments have shown that predation can lead to rapid evolutionary change and that the differences between populations are genetically based. (p. 449)

Limits to What Selection Can Accomplish
- Multiple effects of the same gene can limit the level to which a phenotype can be altered. (p. 450)
- Selection on phenotypic variation cannot lead to evolutionary change without genetic variation. (p. 450)
- Due to epistasis, the effect of an allele at one gene may depend on what alleles are present at other genes. (p. 450)

Test Your Understanding

For interactive testing, visit the Online Learning Center with PowerWeb at www.mhhe.com/Raven7

Self Test

1. Which of the following is *not* an assumption of the Hardy–Weinberg equilibrium?
   a. Mating occurs preferentially.
   b. The size of the population is large.
   c. There is no migration.
   d. There are no mutations.
2. In a population of red (dominant) and white flowers, the frequency of red flowers is 91%. What is the frequency of the red allele?
   a. 9%
   b. 30%
   c. 91%
   d. 70%
3. Which of the following best describes gene flow?
   a. random mating
   b. migration
   c. genetic drift
   d. selection
4. Which of the following conditions is *not* needed for natural selection to occur in a population?
   a. Individuals must be able to move between populations.
   b. Variation must be genetically inherited.
   c. Certain variations allow an individual to produce more offspring that survive in the next generation.
   d. There must be variations in the phenotypes of individuals.
5. Which of the following is the ultimate source of genetic variation in a population?
   a. gene flow
   b. assortative mating
   c. mutation
   d. selection
6. Natural selection can be countered by which of the following?
   a. genetic drift
   b. gene flow
   c. mutation
   d. All of these can counter natural selection in some way.
7. The maintenance of the sickle cell allele in human populations in central Africa is an example of
   a. gene flow.
   b. heterozygote advantage.
   c. genetic drift.
   d. nonrandom mating.
8. What would happen in the U.S. if malaria once again became a widespread disease?
   a. Over time, the sickle cell allele would become more prevalent in the population.
   b. Many individuals in the population would die of malaria.
   c. Individuals who were heterozygous for the sickle cell allele would be less susceptible to malaria.
   d. All of these events would occur.
9. _____ operates to eliminate intermediate phenotypes.
   a. Directional selection
   b. Disruptive selection
   c. Stabilizing selection
   d. Random chance
10. When transplanted to streams above waterfalls, guppy populations evolved more spots because
    a. predators are not present.
    b. they consumed different sources of food.
    c. spots make guppies harder to see against their background.
    d. all of the above

Test Your Visual Understanding

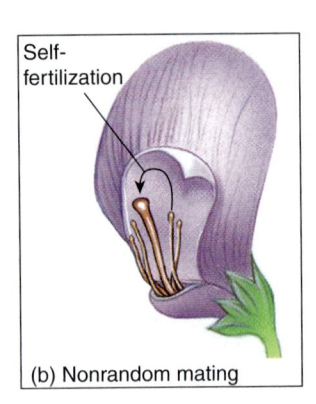

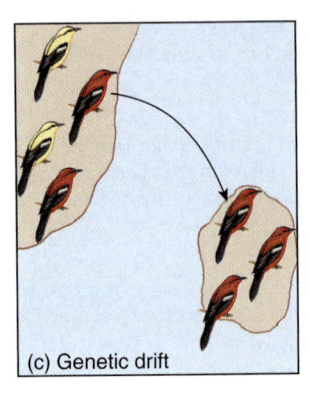

(a) Gene flow

(b) Nonrandom mating

(c) Genetic drift

(d) Selection

1. Match the following descriptions with the correct panels in the figure.
   a. Phenotypically similar individuals mate.
   b. Individuals migrate.
   c. Space exploration expands with the settlement of a population of 100 individuals on Mars.
   d. A nuclear power plant dumps boiling water into a reservoir, killing all bacteria except those that contain a heat shock gene.

Applying Your Knowledge

1. Consider a human population that is similar to the ideal Hardy–Weinberg population in that it is very large and generally random-mating. Although mutations occur, they alone do not lead to great changes in allele frequencies. However, migration occurs at relatively high levels—perhaps 1% per year. The following data describe relative numbers of individuals bearing the two alleles of the MN blood group:

	MM	MN	NN	Total
Individuals	1787	3037	1305	6129

Do these data suggest that some factor is disrupting the Hardy–Weinberg proportions of the three genotypes? What are the allele frequencies of M and N?

22
The Evidence for Evolution

Concept Outline

22.1 Evidence indicates that natural selection can produce evolutionary change.

The Beaks of Darwin's Finches. Natural selection favors stouter bills in dry years, when large tough-to-crush seeds are the only food available to finches.

Peppered Moths and Industrial Melanism. Natural selection favors dark-colored moths in areas of heavy pollution, while light-colored moths survive better in unpolluted areas.

Artificial Selection. Artificial selection practiced in laboratory studies, agriculture, and domestication demonstrate that selection can produce substantial evolutionary change.

22.2 Fossil evidence indicates that evolution has occurred.

The Fossil Record. When fossils are arranged in the order of their age, a continual series of change is seen, with new changes added at each stage.

The Evolution of Horses. The record of horse evolution is particularly well-documented and instructive.

22.3 Evidence for evolution can be found in other fields of biology.

The Anatomical Record. When the anatomical features of living animals are examined, evidence of shared ancestry is often apparent.

The Molecular Record. When gene or protein sequences from organisms are arranged, species thought to be closely related based on fossil evidence are seen to be more similar than species thought to be distantly related.

Convergent Evolution and the Biogeographical Record. Natural selection favors similar forms under similar circumstances, and the geographic distribution of species reflects evolutionary divergence.

22.4 The theory of evolution has proven controversial.

Darwin's Critics. Critics have raised seven objections to Darwin's theory of evolution by natural selection.

FIGURE 22.1
Fossil remains of trilobites. Trilobites were a diverse class of arthropods that lived 245–540 million years ago. The rich fossil record of these primarily bottom-dwelling marine organisms clearly documents the origins of new species and reveals how species evolve through time.

As we discussed in chapter 1, when Darwin proposed his revolutionary theory of evolution by natural selection, little actual evidence existed to bolster his case. Instead, Darwin relied on observations of the natural world, logic, and results obtained by breeders working with domestic animals. Since his day, however, the evidence for Darwin's theory has become overwhelming. The case is built upon two pillars: first, evidence that natural selection can produce evolutionary change, and second, evidence from the fossil record that evolution has occurred (figure 22.1). In addition, information from many different areas of biology—fields as different as anatomy, molecular biology, and biogeography—is only interpretable scientifically as the outcome of evolution.

22.1 Evidence indicates that natural selection can produce evolutionary change.

As we saw in chapter 21, a variety of processes produce evolutionary change. Nonetheless, in agreement with Darwin, most evolutionary biologists believe that natural selection is the primary process responsible for major evolutionary changes. Although we cannot travel back through time, a variety of modern-day evidence confirms the power of natural selection as an agent of evolutionary change. These data come from both the field and the laboratory and from both natural and human-altered situations.

The Beaks of Darwin's Finches

Darwin's finches are a classic example of evolution by natural selection. When he visited the Galápagos Islands off the coast of Ecuador in 1835, Darwin collected 31 specimens of finches from three islands. Darwin, not an expert on birds, had trouble identifying the specimens, believing by examining their bills that his collection contained wrens, "gross-beaks," and blackbirds.

The Importance of the Beak

Upon Darwin's return to England, ornithologist John Gould recognized that Darwin's collection was in fact a closely related group of distinct species, all similar to one another except for their bills. In all, 14 species are now recognized. The ground finches with the larger bills in figure 22.2 feed on seeds that they crush in their beaks, whereas the ones with narrower bills eat insects. Other species include fruit and bud eaters, insect eaters, and species that feed on cactus fruits and the insects they attract; some populations of the sharp-beaked ground finch even include "vampires" that creep up on seabirds and use their sharp beaks to drink their blood. Perhaps most remarkable are the tool users, woodpecker finches that pick up a twig, cactus spine, or leaf stalk, trim it into shape with their bills, and then poke it into dead branches to pry out grubs.

The correspondence between the beaks of the 14 finch species and their food source immediately suggested to Darwin that natural selection had shaped them. In *The*

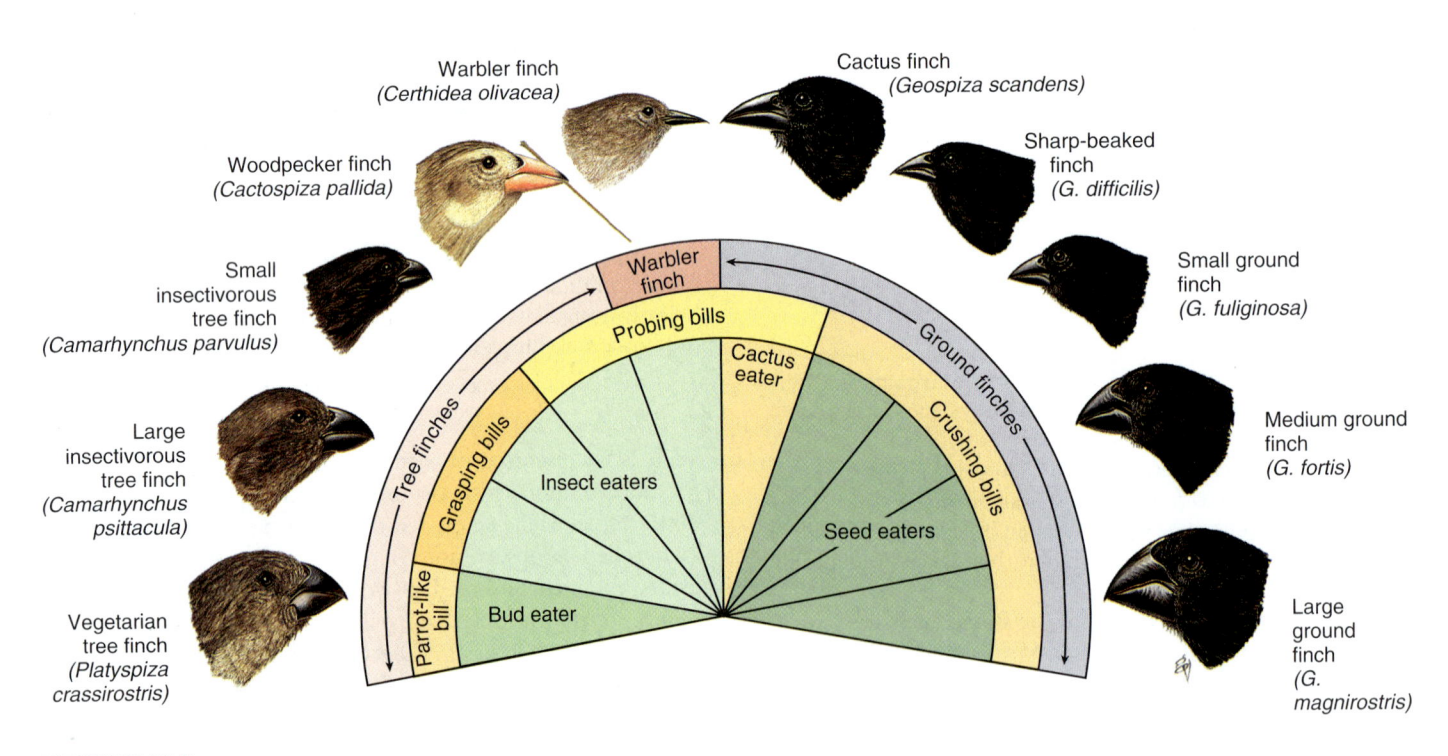

FIGURE 22.2
Darwin's finches. Ten species of Darwin's finches occur on Santa Cruz, one of the Galápagos Islands. These species show differences in bills and feeding habits. This condition presumably arose when the finches evolved new species in habitats lacking small birds. The bills of several species resemble those of distinct families of birds on the mainland. For example, the woodpecker finch uses cactus spines to probe in crevices of bark and rotten wood for food. All of these species are derived from a single common ancestor.

Voyage of the Beagle, Darwin wrote, "Seeing this gradation and diversity of structure in one small, intimately related group of birds, one might really fancy that from an original paucity of birds in this archipelago, one species has been taken and modified for different ends."

A Closer Look

Darwin's observations suggest that differences among species in beak size and shape have evolved as the species adapted to use different food resources, but can this hypothesis be tested? As we discussed in chapter 21, the theory of evolution by natural selection requires that three conditions be met: (1) Variation must exist in the population; (2) this variation must lead to differences among individuals in survival and reproductive success; and (3) variation among individuals must be genetically transmitted to the next generation.

The key to successfully testing Darwin's proposal proved to be patience. Starting in 1973, Peter and Rosemary Grant of Princeton University and generations of their students have studied the medium ground finch, *Geospiza fortis*, on a tiny island in the center of the Galápagos called Daphne Major. These finches feed preferentially on small, tender seeds, produced in abundance by plants in wet years. The birds resort to larger, drier seeds, which are harder to crush, only when small seeds become depleted during long periods of dry weather, when plants produce few seeds.

The Grants quantified beak shape among the medium ground finches of Daphne Major by carefully measuring beak depth (width of beak, from top to bottom, at its base) on individual birds. Measuring many birds every year, they were able to assemble for the first time a detailed portrait of evolution in action. The Grants found that not only did a great deal of variation in beak depth exist among members of the population, but the average beak depth changed from one year to the next in a predictable fashion. During droughts, plants produced few seeds, and all available small seeds were quickly eaten, leaving large seeds as the major remaining source of food. As a result, birds with large beaks survived better, because they were better able to break open these large seeds. Consequently, the average beak depth of birds in the population increased the next year, only to decrease again when wet seasons returned (figure 22.3*a*).

Could these changes in beak dimension reflect the action of natural selection? An alternative possibility might be that the changes in beak depth do not reflect changes in gene frequencies, but rather are simply a response to diet; for example, perhaps crushing large seeds causes a growing bird to develop a larger beak. To rule out this possibility, the Grants measured the relation of parent beak size to offspring beak size, examining many broods over several years. The depth of the beak was passed down faithfully from one generation to the next, regardless of environ-

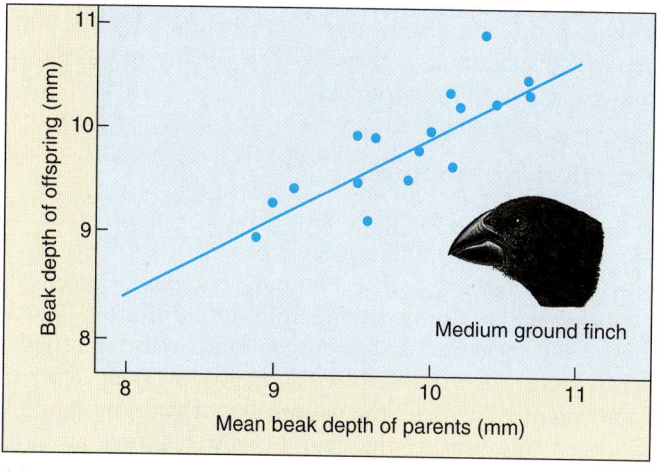

FIGURE 22.3
Evidence that natural selection alters beak size in *Geospiza fortis*. (*a*) In dry years, when only large, tough seeds are available, the mean beak size increases. In wet years, when many small seeds are available, smaller beaks become more common. (*b*) Beak depth is inherited from parents to offspring.
Suppose a bird with a large bill mates with a bird with a small bill. Would the bills of the pair's offspring tend to be larger or smaller than the bills of offspring from a pair of birds with medium-sized bills?

mental conditions (figure 22.3*b*), suggesting that the differences among individuals in beak size reflect genetic differences, and thus, that the year-to-year changes in average beak depth represent evolutionary change resulting from natural selection.

> Among Darwin's finches, natural selection adjusts the shape of the beak in response to the nature of the available food supply. Such adjustments can be seen to occur even today.

Peppered Moths and Industrial Melanism

When the environment changes, natural selection often may favor new traits in a species. One classic example concerns the peppered moth, *Biston betularia*. Adults come in a range of shades, from light gray with black speckling (hence the name "peppered" moth) to jet black (melanic). Extensive genetic analysis has shown that the moth's body color is a genetic trait, reflecting different alleles of a single gene. Black individuals have a dominant allele, one that was present but very rare in populations before 1850. From that time on, dark individuals increased in frequency in moth populations near industrialized centers until they made up almost 100% of these populations. Biologists soon noticed that in industrialized regions where the dark moths were common, the tree trunks were darkened almost black by the soot of pollution, which also killed many of the light-colored lichens on tree trunks.

Selection for Melanism

Why did dark moths gain a survival advantage around 1850? An amateur moth collector named J. W. Tutt proposed in 1896 what became the most commonly accepted hypothesis explaining the decline of the light-colored moths. He suggested that peppered forms were more visible to predators on sooty trees that have lost their lichens. Consequently, birds ate the peppered moths resting on the trunks of trees during the day. The black forms, in contrast, were at an advantage because they were camouflaged (figure 22.4). Although Tutt initially had no evidence, British ecologist Bernard Kettlewell tested the hypothesis in the 1950s by releasing equal numbers of dark and light individuals into two sets of woods: one near heavily polluted Birmingham, and the other in unpolluted Dorset. Kettlewell then set up traps in the woods to see how many of both kinds of moths survived. To evaluate his results, he had marked the released moths with a dot of paint on the underside of their wings, where birds could not see it.

In the polluted area near Birmingham, Kettlewell trapped only 19% of the light moths, but 40% of the dark ones. This indicated that dark moths had a far better chance of surviving in these polluted woods, where the tree trunks were dark. In the relatively unpolluted Dorset woods, Kettlewell recovered 12.5% of the light moths but only 6% of the dark ones. This indicated that where the tree trunks were still light-colored, light moths had a much better chance of survival. Kettlewell later solidified his argument by placing moths on trees and filming birds looking for food. Sometimes the birds actually passed right over a moth that was the same color as its background. Kettlewell's finding—that birds more frequently detect moths whose color does not match their background—has subsequently been confirmed in eight separate field studies, with a variety of experimental designs and corrections for deficiencies in Kettlewell's initial experimental design. These results, combined with the recapture studies, provide strong evidence for the action of natural selection and implicate birds as the agent of selection in the case of the peppered moth.

FIGURE 22.4
Tutt's hypothesis explaining industrial melanism. These photographs show dead specimens of the peppered moth (*Biston betularia*) placed on trees. Tutt proposed that the dark melanic variant of the moth is more visible to predators on unpolluted trees (*top*), while the light "peppered" moth is more visible to predators on bark blackened by industrial pollution (*bottom*).

Industrial Melanism

The term **industrial melanism** refers to the process by which darker individuals come to predominate over lighter individuals. In industrialized areas throughout Eurasia and North America, dozens of other species of moths have changed in the same way as the peppered moth.

Could it be that melanism confers an inherent physiological advantage over lighter-colored forms? While consistent with the worldwide success of industrial melanism, this possibility can be eliminated because of the subsequent reversal of melanism. That is, natural selection is now acting *against* the melanic allele.

Selection Against Melanism

In the second half of the twentieth century, with the widespread implementation of pollution controls, the trend toward melanism began reversing for many species of moths throughout the northern continents.

In England, the air pollution promoting industrial melanism began to reverse following enactment of the Clean Air Act in 1956. Beginning in 1959, the *Biston* population at Caldy Common outside Liverpool has been sampled each year. The frequency of the melanic (dark) form has dropped from a high of 93% in 1959 to a low of 15% in 1995 (figure 22.5). The drop correlates well with a significant drop in air pollution, particularly with a lowering of the levels of sulfur dioxide and suspended particulates, both of which act to darken trees. The drop is consistent with a 15% selective disadvantage acting against moths with the dominant melanic allele.

Interestingly, the same reversal of melanism occurred in the United States. Of 576 peppered moths collected at a field station near Detroit from 1959 to 1961, 515 were melanic, a frequency of 89%. The American Clean Air Act, passed in 1963, led to significant reductions in air pollution. Resampled in 1994, the Detroit field station peppered moth population had only 15% melanic moths (figure 22.5)! The moths in Liverpool and Detroit, both part of the same natural experiment, exhibit strong evidence for natural selection.

Reconsidering the Agent of Natural Selection

While the evidence for natural selection in the case of the peppered moth is strong, Tutt's hypothesis about the agent of selection is currently being reevaluated. Researchers have noted that the recent selection against melanism does not appear to correlate with changes in tree lichens. At Caldy Common, the light form of the peppered moth began to increase in frequency long before lichens began to reappear on the trees. At the Detroit field station, the lichens never changed significantly as the dark moths first became dominant and then declined over the last 30 years. In fact, investigators have not been able to find peppered moths on Detroit trees at all, whether covered with lichens or not. Wherever the moths rest during the day, it does not appear to be on tree bark. Some evidence suggests they rest on leaves on the treetops, but no one is sure. Could poisoning by pollution rather than predation by birds be the agent of natural selection on the moths? Perhaps—but to date, only predation by birds is backed by experimental evidence.

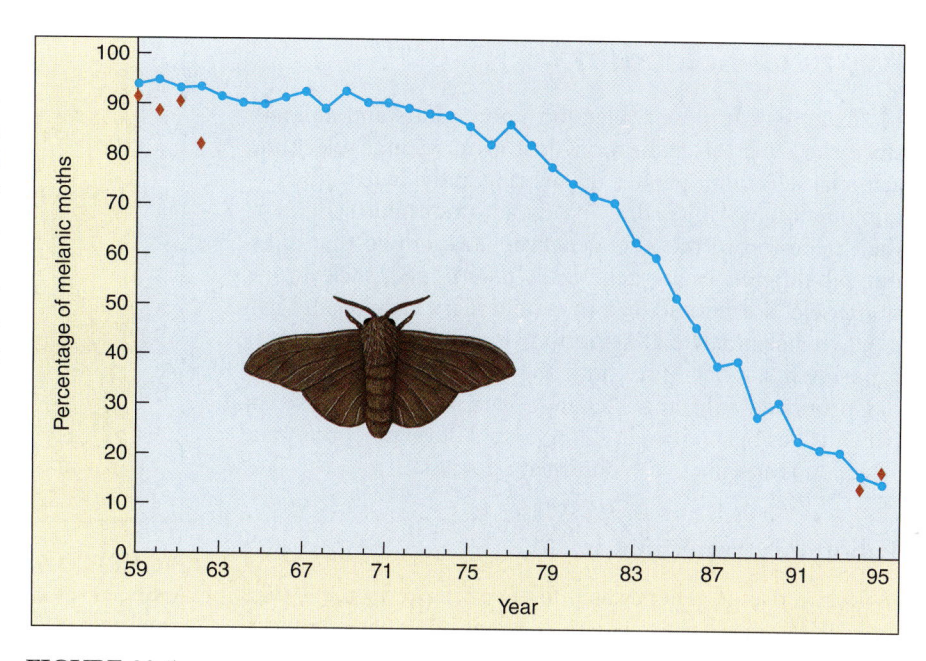

FIGURE 22.5

Selection against melanism. The circles indicate the frequency of melanic *Biston betularia* moths at Caldy Common in England, sampled continuously from 1959 to 1995. Diamonds indicate frequencies of melanic *B. betularia* in Michigan from 1959 to 1962 and from 1994 to 1995.

What can you conclude from the fact that the frequency of melanic moths decreased to the same degree in the two locations?

Researchers supporting bird predation as the agent of natural selection point out that a bird's ability to detect moths may depend less on the presence or absence of lichens, and more on other ways in which the environment is darkened by industrial pollution. Pollution tends to cover all objects in the environment with a fine layer of particulate dust, which tends to decrease how much light surfaces reflect. In addition, pollution has a particularly severe effect on birch trees, which are light in color. Both effects would tend to make the environment darker, and thus would favor darker moths by protecting them from predation by birds.

Despite this uncertainty over the agent of selection, the overall pattern is clear. Kettlewell's experiments establish indisputably that selection favors dark moths in polluted habitats and light moths in pristine areas. The increase and subsequent decrease in the frequency of melanic moths, correlated with levels of pollution independently on two continents, demonstrates clearly that this selection drives evolutionary change. The current reconsideration of the agent of natural selection illustrates well the way in which scientific progress is achieved: Hypotheses, such as Tutt's, are put forth and then tested. If rejected, new hypotheses are formulated, and the process begins anew.

Natural selection has favored the dark form of the peppered moth in areas subject to severe air pollution, perhaps because on darkened trees they are less easily seen by moth-eating birds. Selection has in turn favored the light form as pollution has abated.

Artificial Selection

Humans have imposed selection upon plants and animals since the dawn of civilization. Just as in natural selection, artificial selection operates by favoring individuals with certain phenotypic traits, allowing them to reproduce and pass their genes on to the next generation. Assuming that phenotypic differences are genetically determined, such directional selection should lead to evolutionary change, and indeed, it has. Artificial selection, imposed in laboratory experiments, agriculture, and the domestication process, has produced substantial change in almost every case in which it has been applied. This success is strong proof that selection is an effective evolutionary process.

Laboratory Experiments

With the rise of genetics as a field of science in the 1920s and 1930s, researchers began conducting experiments to test the hypothesis that selection can produce evolutionary change. A favorite subject was the laboratory fruit fly, *Drosophila melanogaster*. Geneticists have imposed selection on just about every conceivable aspect of the fruit fly—including body size, eye color, growth rate, life span, and exploratory behavior—with a consistent result: Selection for a trait leads to strong and predictable evolutionary response.

In one classic experiment, scientists selected for fruit flies with many bristles (stiff, hairlike structures) on their abdomen. At the start of the experiment, the average number of bristles was 9.5. Each generation, scientists picked out the 20% of the population with the greatest number of bristles and allowed them to reproduce, thus establishing the next generation. After 86 generations of such selection, the average number of bristles had quadrupled, to nearly 40! In another experiment, fruit flies were selected for either the most or the fewest numbers of bristles. Within 35 generations, the populations did not overlap at all in range of variation (figure 22.6).

Similar experiments have been conducted on a wide variety of other laboratory organisms. For example, by selecting for rats that were resistant to tooth decay, scientists were able to increase in less than 20 generations the average time for onset of decay from barely over 100 days to greater than 500 days.

Agriculture

Similar methods have been practiced in agriculture for many centuries. Familiar livestock, such as cattle and pigs, and crops, such as corn and strawberries, are greatly different from their wild ancestors (figure 22.7). These differences have resulted from generations of selection for desirable traits, such as milk production and corn ear size.

An experiment with corn demonstrates the ability of artificial selection to rapidly produce major change in crop plants. In 1896, agricultural scientists began selecting on

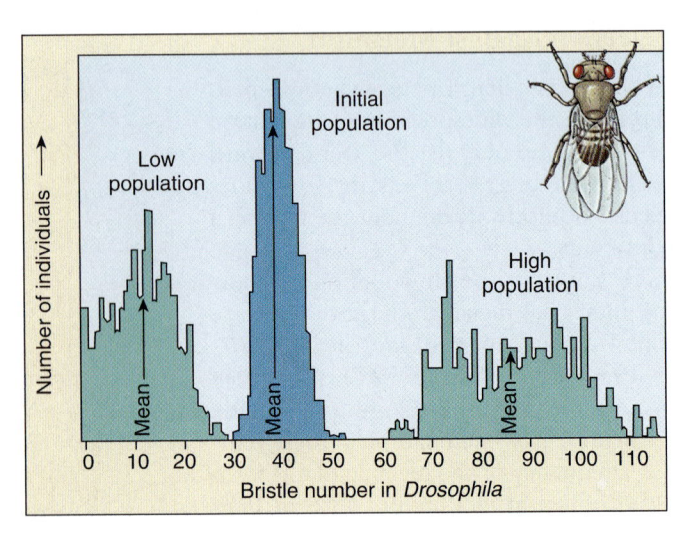

FIGURE 22.6
Artificial selection in the laboratory. In this experiment, one population of *Drosophila* was selected for low numbers of bristles and the other for high numbers. Note that not only did the means of the populations change greatly in 35 generations, but also all individuals in both experimental populations lie outside the range of the initial population.
What would happen if, within a population, both small and large individuals were allowed to breed, but middle-sized ones were not?

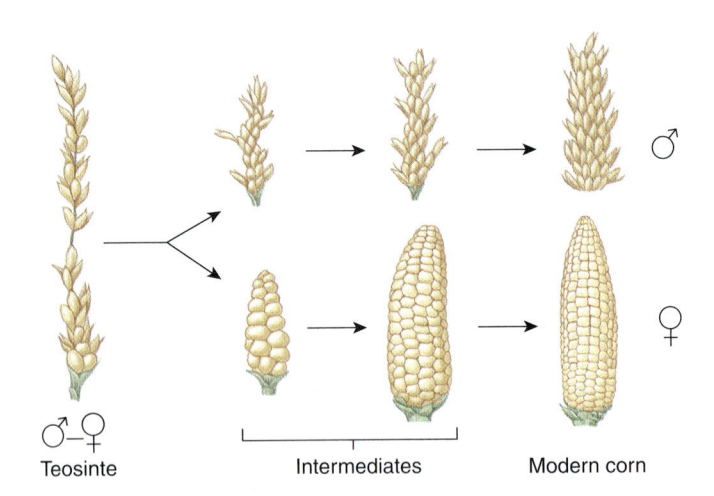

FIGURE 22.7
Corn looks very different from its ancestor. The tassels and seeds of a wild grass, such as teosinte, evolved into the male tassels and female ears of modern corn, which are on different parts of the plant, in contrast to the ancestral condition in teosinte.

the oil content of corn kernels, which initially was 4.5%. As in the fruit fly experiments, the top 20% of all individuals were allowed to reproduce. By 1986, at which time 90 generations had passed, average oil content had increased approximately 450% in the corn kernels.

Domestication

Artificial selection has also been responsible for the great variety of breeds of cats, dogs (figure 22.8), pigeons, and other domestic animals. In some cases, breeds have been developed for particular purposes. Greyhound dogs, for example, were bred by selecting for maximal running abilities, resulting in an animal with long legs, a long tail (for use as a rudder), an arched back (to increase stride length), and great muscle mass. By contrast, the odd proportions of the ungainly dachshund resulted from selection for dogs that could enter narrow holes in pursuit of badgers. In other cases, breeds have been developed primarily for their appearance, such as the many colorful and ornamented breeds of pigeons or cats.

Domestication also has led to unintentional selection for some traits. In recent years, as part of an attempt to domesticate the silver fox, Russian scientists have chosen the most docile animals in each generation and allowed them to reproduce. Within 40 years, most foxes were exceptionally docile, not only allowing themselves to be petted, but also whimpering to get attention and sniffing and licking their caretakers (figure 22.9). In many respects, they had become no different than domestic dogs! However, it was not only their behavior that changed. These foxes also began to exhibit different color patterns, such as the floppy ears, curled tails, and shorter legs and tails seen in some dog breeds. Presumably, the genes responsible for docile behavior either affect these traits as well or are closely linked to the gene for these other traits (the phenomena of pleiotropy and linkage discussed in chapter 13). As selection has favored docile animals, it has also led to the evolution of these other traits.

Can Selection Produce Major Evolutionary Changes?

Given that we can observe the results of selection operating over relatively short periods of time, most scientists believe that natural selection is the process responsible for the evolutionary changes documented in the fossil record. Some critics of evolution accept that selection can lead to changes within a species, but contend that such changes are relatively minor in scope and not equivalent to the substantial changes documented in the fossil record. In other words, it is one thing to change the number of bristles on a fruit fly or the size of an ear of corn, and quite another to produce an entirely new species.

This argument does not fully appreciate the extent of change produced by artificial selection. Consider, for example, the existing breeds of dogs, all of which have been produced since wolves were first domesticated, perhaps 10,000 years ago. If the various dog breeds did not exist and a paleontologist found fossils of animals similar to dachshunds, greyhounds, mastiffs, and chihuahuas, there is no question that they would be considered different species. Indeed, the differences in size and shape exhibited by these breeds are greater than those between members of different genera in the family Canidae—such as coyotes, jackals, foxes, and wolves—which have been evolving separately for 5–10 million years. Consequently, the claim that artificial selection produces only minor changes is clearly incorrect. If selection operating over a period of only 10,000 years can produce such substantial differences, it should be powerful enough, over the course of many millions of years, to produce the diversity of life we see around us today.

FIGURE 22.8
Breeds of dogs. The differences between these dogs are greater than the differences displayed in any wild species of canids.

FIGURE 22.9
Domesticated foxes. After 40 years of selectively breeding the tamest individuals, artificial selection has produced silver foxes that are not only as friendly as domestic dogs, but also exhibit many physical traits seen in dog breeds.

> **Artificial selection often leads to rapid and substantial results over short periods of time, thus demonstrating the power of selection to produce major evolutionary change.**

22.2 Fossil evidence indicates that evolution has occurred.

The Fossil Record

The most direct evidence that evolution has occurred is found in the fossil record. Today we have a far more complete understanding of this record than was available in Darwin's time. Fossils are the preserved remains of once-living organisms. They include specimens preserved in amber, Siberian permafrost, and dry caves, as well as the more common fossils preserved as rocks. Rock fossils are created when three events occur. First, the organism must become buried in sediment; then, the calcium in bone or other hard tissue must mineralize; and finally, the surrounding sediment must eventually harden to form rock. The process of fossilization probably occurs rarely. Usually, animal or plant remains decay or are scavenged before the process can begin. In addition, many fossils occur in rocks that are inaccessible to scientists. When they do become available, they are often destroyed by erosion and other natural processes before they can be collected. As a result, only a very small fraction of the species that have ever existed (estimated by some to be as many as 500 million) are known from fossils. Nonetheless, the fossils that have been discovered are sufficient to provide detailed information on the course of evolution through time.

Dating Fossils

By dating the rocks in which fossils occur, we can get an accurate idea of how old the fossils are. In Darwin's day, rocks were dated by their position with respect to one another (*relative dating*); rocks in deeper strata are generally older. Knowing the relative positions of sedimentary rocks and the rates of erosion of different kinds of sedimentary rocks in different environments, geologists of the nineteenth century derived a fairly accurate idea of the relative ages of rocks.

Today, geologists take advantage of radioactive decay to age rocks (*absolute dating*). Many types of rock, such as the igneous rocks formed when lava cools, contain radioactive elements such as uranium-238. These isotopes transform at a precisely known rate into nonradioactive forms. For example, for U^{238} the *half-life* (that is, the amount of time for one-half of the original amount to be transformed) is 4.5 billion years. Once a rock is formed, no additional radioactive isotopes are added. Thus, by measuring the ratio of the radioactive isotope to its derivative, "daughter" isotope (figure 22.10), geologists can determine the age of the rock. If a fossil is found between two layers of rock, each of which can be dated, then the age at which the fossil formed can be determined.

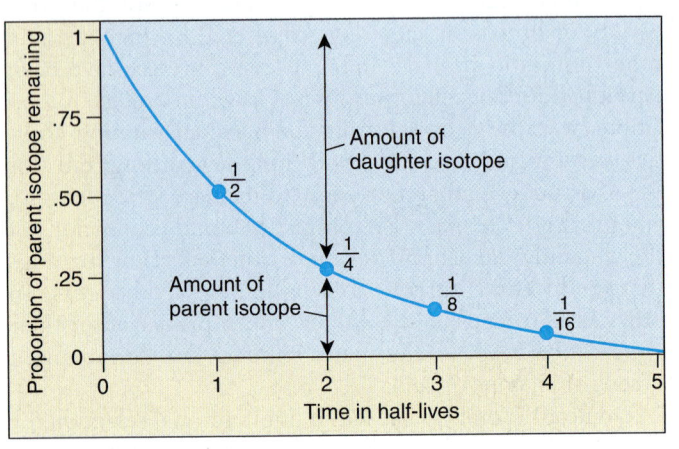

FIGURE 22.10
Radioactive decay. Radioactive elements decay at a known rate, called their half-life. After one half-life, one-half of the original amount of parent isotope has transformed into a nonradioactive daughter isotope. After each successive half-life, one-half of the remaining amount of parent isotope is transformed.

A History of Evolutionary Change

When fossils are arrayed according to their age (figure 22.11), from oldest to youngest, they often provide evi-

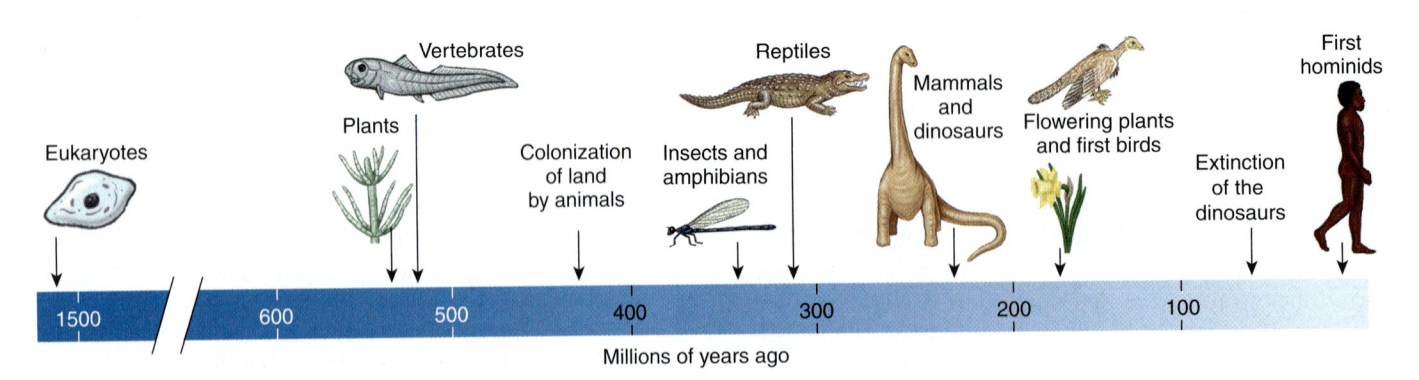

FIGURE 22.11
Timeline of the history of life as revealed by the fossil record.

dence of successive evolutionary change. At the largest scale, the fossil record documents the course of life through time, from the origin of eukaryotic organisms, through the evolution of fishes, the rise of land-living organisms, the reign of the dinosaurs, and on to the origin of humans.

Gaps in the Fossil Record

Given the low likelihood of fossil preservation and recovery, it is not surprising that there are gaps in the fossil record. While many gaps interrupted the fossil record in Darwin's era, even then, scientists knew of the *Archaeopteryx* fossil, which is transitional between dinosaurs and birds. Since then, paleontologists (the scientists who study fossils) have continued to fill in the gaps in the fossil record. Today, the fossil record is far more complete, particularly among the vertebrates; fossils have been found linking all the major groups. Recent years have seen spectacular discoveries, closing some of the major remaining gaps in our understanding of vertebrate evolution. For example, recently a four-legged aquatic mammal was discovered that provides important insights concerning the evolution of whales and dolphins from land-living, hoofed ancestors (figure 22.12). Similarly, a fossil snake with legs has shed light on the evolution of snakes, which are descended from lizards that gradually became more and more elongated with the simultaneous reduction and eventual disappearance of the limbs.

On a finer scale, evolutionary change within some types of animals is known in exceptional detail. For example, about 200 million years ago, oysters underwent a change from small, curved shells to larger, flatter ones, with progressively flatter fossils seen in the fossil record over a period of 12 million years (figure 22.13). A host of other examples illustrate a record of successive change. The demonstration of this successive change is one of the strongest lines of evidence that evolution has occurred.

> The fossil record provides a clear record of the major evolutionary transitions that have occurred through time.

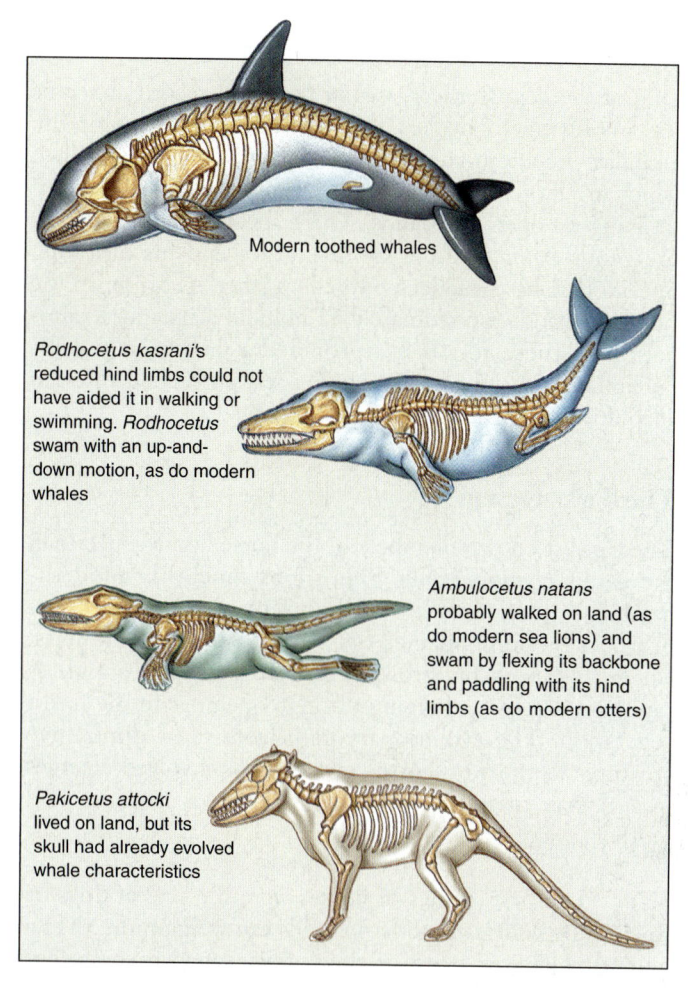

FIGURE 22.12
Whale "missing links." The recent discoveries of *Ambulocetus*, *Rodhocetus*, and *Pakicetus* have filled in the gaps between whales and their hoofed mammal ancestors. The features of *Pakicetus* illustrate that intermediate forms are not intermediate in all characteristics; rather, some traits evolve before others. In the case of the evolution of whales, changes occurred in the skull prior to evolutionary modification of the limbs. All three fossil forms occurred in the Eocene period, 45–55 million years ago.

FIGURE 22.13
Evolution of shell shape in oysters. Over 12 million years of the Early Jurassic period, the shells of this group of coiled oysters became larger, thinner, and flatter. These animals rested on the ocean floor, and the larger, flatter shells may have evolved because they were more stable in disruptive water movements.

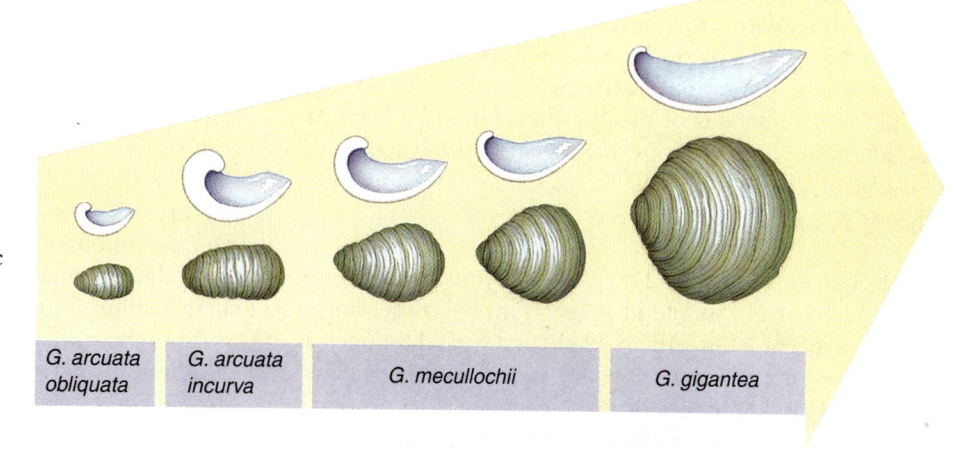

G. arcuata obliquata | G. arcuata incurva | G. mecullochii | G. gigantea

Chapter 22 The Evidence for Evolution 461

The Evolution of Horses

One of the best-studied cases in the fossil record concerns the evolution of horses. Modern-day members of the Equidae include horses, zebras, donkeys, and asses, all of which are large, long-legged, fast-running animals adapted to living on open grasslands. These species, all classified in the genus *Equus*, are the last living descendants of a long lineage that has produced 34 genera since its origin in the Eocene period, approximately 55 million years ago. Examination of these fossils has provided a particularly well-documented case of how evolution has proceeded through adaptation to changing environments.

The First Horse

The earliest known members of the horse family, species in the genus *Hyracotherium*, didn't look much like modern-day horses at all. Small, with short legs and broad feet (figure 22.14), these species occurred in wooded habitats, where they probably browsed on leaves and herbs and escaped predators by dodging through openings in the forest vegetation. The evolutionary path from these diminutive creatures to the workhorses of today has involved changes in a variety of traits, including size, toe reduction, and tooth size and shape.

Size. The first species of horses were the size of dogs or smaller. By contrast, modern equids can weigh more than a half ton. Examination of the fossil record reveals that horses changed little in size for their first 30 million years, but since then, a number of different lineages have exhibited rapid and substantial increases. However, trends toward decreased size were also exhibited among some branches of the equid evolutionary tree (figure 22.15).

Toe Reduction. The feet of modern horses have a single toe enclosed in a tough, bony hoof. By contrast, *Hyracotherium* had four toes on its front feet and three on its hind feet. Rather than hooves, these toes were encased in fleshy pads like those of dogs and cats. Examination of the fossils clearly shows the transition through time: increase in length of the central toe, development of the bony hoof, and reduction and loss of the other toes (figure 22.16). As with body size, these trends occurred concurrently on several different branches of the horse evolutionary tree. At the same time, horses were evolving changes in the length and skeletal structure of their limbs, leading to animals capable of running long distances at high speeds.

Tooth Size and Shape. The teeth of *Hyracotherium* were small and relatively simple in shape. Through time, horse teeth have increased greatly in length and have developed a complex pattern of ridges on their molars and premolars (figure 22.16). The effect of these changes is to produce teeth better capable of chewing tough and gritty vegetation, such as grass, which tends to wear teeth down. Accompanying these changes have been alterations in the shape of the skull that strengthened the skull to withstand the stresses imposed by continual chewing. As with body size, evolutionary change has not been constant through time. Rather, much of the change in tooth shape has occurred within the past 20 million years.

FIGURE 22.14
Hyracotherium sandrae, one of the earliest horses. This species was the size of a housecat.

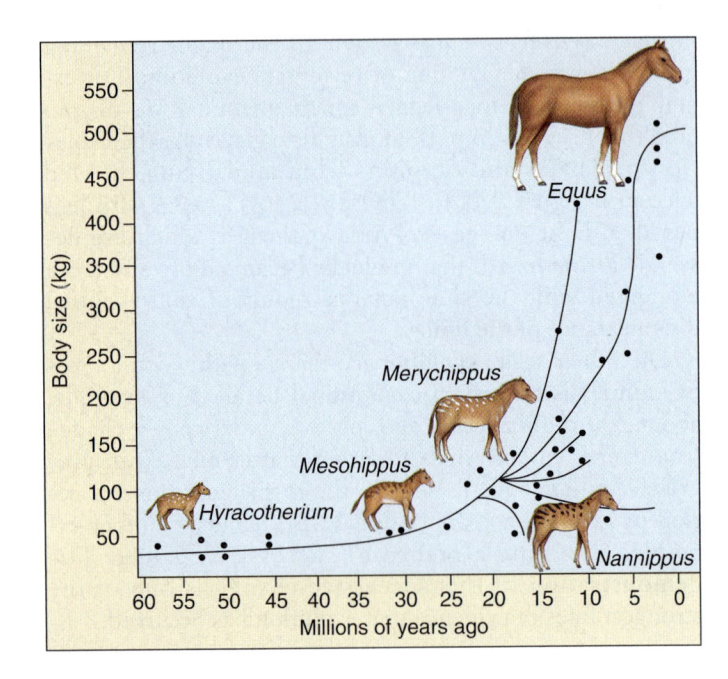

FIGURE 22.15
Evolutionary change in body size of horses. Lines indicate evolutionary relationships and reveal that although most of the change involved increases in size, some decreases also occurred. **Why might the evolutionary line leading to *Nannippus* have experienced an evolutionary decrease in body size?**

All of these changes may be understood as adaptations to changing global climates. In particular, during the late Miocene and early Oligocene epochs (approximately 20 to 25 million years ago), grasslands became widespread in

North America, where much of horse evolution occurred. As horses adapted to these habitats, high-speed locomotion probably became more important to escape predators. By contrast, the greater flexibility provided by multiple toes and shorter limbs, which was advantageous for ducking through complex forest vegetation, was no longer beneficial. At the same time, horses were eating grasses and other vegetation that contained more grit and other hard substances, thus favoring teeth and skulls better suited for withstanding such materials.

Evolutionary Trends

For many years, horse evolution was held up as an example of constant evolutionary change through time. Some even saw in the record of horse evolution evidence for a progressive, guiding force, consistently pushing evolution in a single direction. We now know that such views are misguided; evolutionary change over millions of years is rarely so simple.

Rather, the fossils demonstrate that, although overall trends have been evident in a variety of characteristics, evolutionary change has been far from constant and uniform through time. Instead, rates of evolution have varied widely, with long periods of little observable change and some periods of great change. Moreover, when changes happen, they often occur simultaneously in different lineages of the horse evolutionary tree. Finally, even when a trend exists, exceptions, such as the evolutionary decrease in body size exhibited by some lineages, are not uncommon. These patterns, evident in our knowledge of horse evolution, are usually discovered for any group of plants and animals for which we have an extensive fossil record, as we shall see when we discuss human evolution in chapter 34.

Horse Diversity

One reason that horse evolution was originally conceived of as linear through time may be that modern horse diversity is relatively limited. Thus, it is easy to mentally picture a straight line from *Hyracotherium* to modern-day *Equus*. However, today's limited horse diversity—only one surviving genus—is unusual. Indeed, at the peak of horse diversity in the Miocene epoch, 13 genera of horses could be found in North America alone. These species differed in body size and in a wide variety of other characteristics. Presumably, they lived in different habitats and exhibited different dietary preferences. Had this diversity existed to modern times, early evolutionary biologists presumably would have had a different outlook on horse evolution.

The extensive fossil record for horses provides a detailed view of the evolutionary diversification of this group, from small forest dwellers to large and fast modern grassland species.

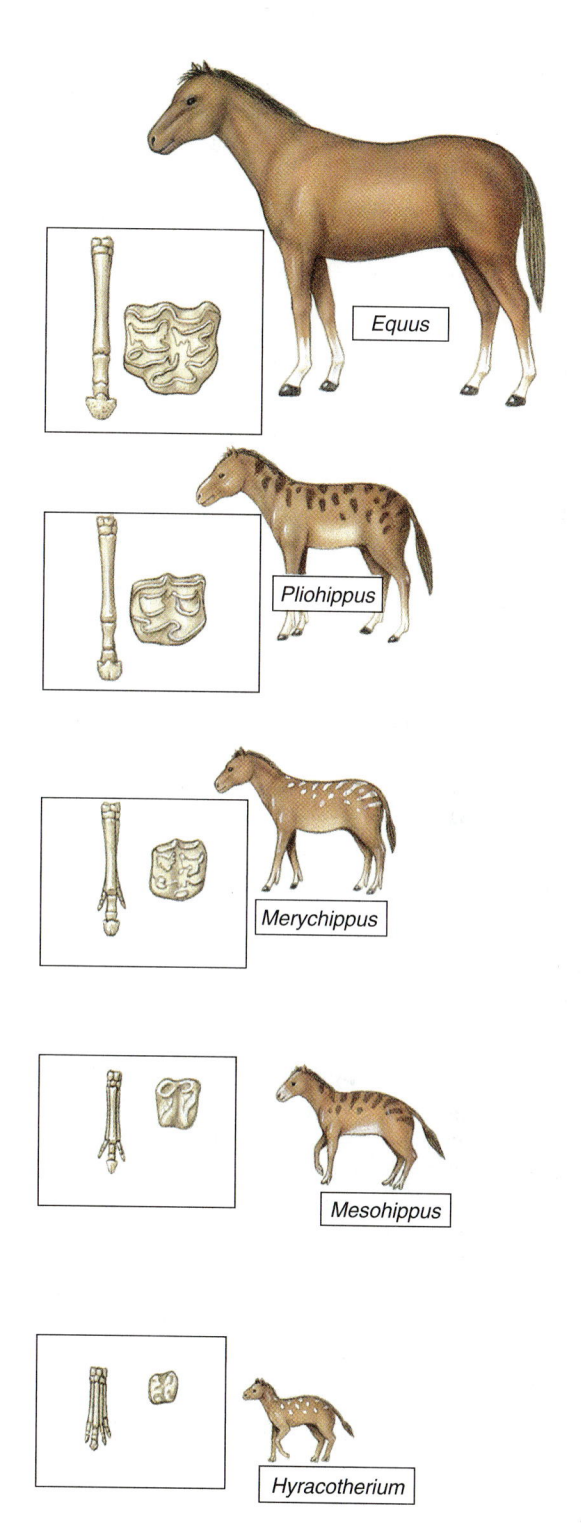

FIGURE 22.16
Evolutionary trends in horses through time. Five representative species illustrate that body size increased over time in horses. The inset boxes show the decrease in number of toes, producing the single hoof of modern horses, and the increased size and complexity of the molar teeth (viewed from the top). As figure 22.15 demonstrates, not all evolutionary changes conformed to these trends.

22.3 Evidence for evolution can be found in other fields of biology.

The Anatomical Record

Much of the power of the theory of evolution is its ability to provide a sensible framework for understanding the diversity of life. Many observations from throughout biology simply cannot be understood in any meaningful way except as a result of evolution.

Homology

As vertebrates evolved, the same bones were sometimes put to different uses. Yet the bones are still seen, their presence betraying their evolutionary past. For example, the forelimbs of vertebrates are all **homologous structures**—structures with different appearances and functions that all derived from the same body part in a common ancestor. You can see in figure 22.17 how the bones of the forelimb have been modified in different ways for different mammals. Why should these very different structures be composed of the same bones? If evolution had not occurred, this would indeed be a riddle. But when we consider that all of these animals are descended from a common ancestor, it is easy to understand that natural selection has modified the same initial starting blocks to serve very different purposes.

Development

Some of the strongest anatomical evidence supporting evolution comes from comparisons of how organisms develop. Embryos of different types of vertebrates, for example, often are similar early on, but become more different as they develop. Early in their development, human and fish embryos both possess pharyngeal pouches, which in humans develop into various glands and ducts and in fish turn into gill slits. At a later stage, every human embryo has a long bony tail, the vestige of which we carry to adulthood as the coccyx at the end of our spine. Human fetuses even possess a fine fur (called *lanugo*) during the fifth month of development. These relict developmental forms suggest strongly that our development has evolved, with new instructions modifying ancestral developmental patterns.

Imperfect Structures

Because natural selection can only work on the variation present in a population, it should not be surprising that some organisms do not appear perfectly adapted to their environments. For example, most animals with long necks have many neck vertebrae for enhanced flexibility: Geese have up to 25, and plesiosaurs, the long-necked reptiles that patrolled the seas during the age of dinosaurs, had as many as 76. By contrast, almost all mammals have only 7 neck vertebrae, and so does the giraffe! In the absence of variation in vertebrae number, selection led to evolutionary increase in vertebrae size to produce the long neck of the giraffe.

An excellent example of an imperfect design is the eye of vertebrate animals, in which the photoreceptors face backward, toward the wall of the eye (figure 22.18*a*). As a result, the nerve fibers extend not backward, toward the brain, but forward into the eye chamber, where they slightly obstruct light. Moreover, these fibers bundle together to form the optic nerve, which exits through a hole at the back of the eye, creating a blind spot. By contrast, the eye of mollusks—such as squid and octopuses—are more optimally designed: The photoreceptors face for-

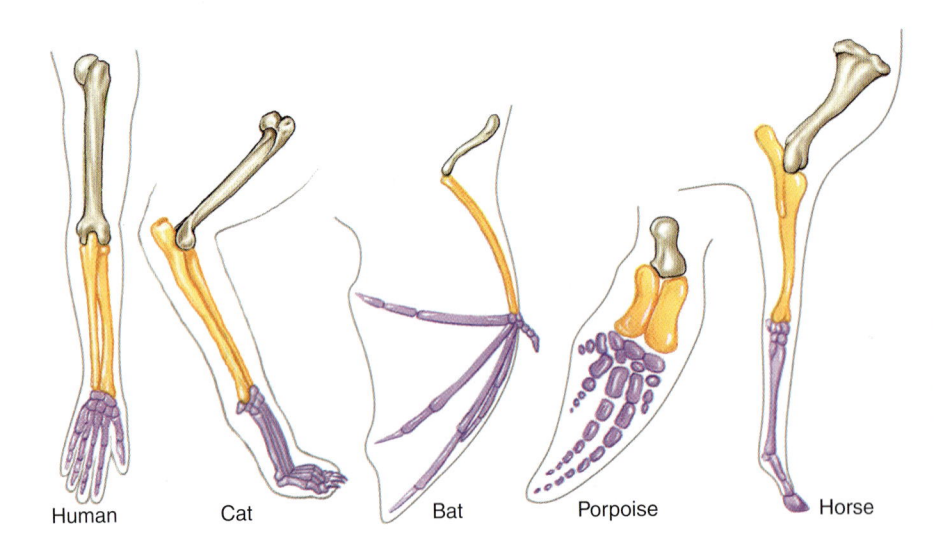

FIGURE 22.17
Homology of the bones of the forelimb of mammals. Although these structures show considerable differences in form and function, the same basic bones are present in the forelimbs of humans, cats, bats, porpoises, and horses.

ward, and the nerve fibers exit at the back, neither obstructing light nor creating a blind spot (figure 22.18*b*). Such examples illustrate that natural selection is like a tinkerer, working with whatever material is available to craft a workable solution, rather than like an engineer, who can design and build the best possible structure for a given task. Workable, but imperfect structures such as the vertebrate eye are an expected outcome of evolution by natural selection.

Vestigial Structures

Many organisms possess **vestigial structures** that have no apparent function but resemble structures their presumed ancestors had. Humans, for example, possess a complete set of muscles for wiggling their ears, just like many other mammals do. Boa constrictors have hip bones and rudimentary hind legs. Manatees (a type of aquatic mammal often referred to as "sea cows") have fingernails on their fins (which evolved from legs). Blind cave fish, which never see the light of day, have small, nonfunctional eyes. Figure 22.19 illustrates the skeleton of a baleen whale, which contains pelvic bones, as other mammal skeletons do, even though such bones serve no known function in the whale. The human vermiform appendix is apparently vestigial; it represents the degenerate terminal part of the cecum, the blind pouch or sac in which the large intestine begins. In other mammals, such as mice, the cecum is the largest part of the large intestine and functions in storage—usually of bulk cellulose in herbivores. Although some functions have been suggested, it is difficult to assign any current function to the vermiform appendix. In many respects, it is a dangerous organ: Quite often, it becomes infected, leading to an inflammation called appendicitis. Without surgical removal, the appendix may burst, allowing the contents of the gut to come in contact with the lining of the body cavity, a potentially fatal event.

It is difficult to understand vestigial structures such as these as anything other than evolutionary relicts, holdovers from the past. However, the existence of vestigial structures argues strongly for the common ancestry of the members of the groups that share them, regardless of how different those groups have subsequently become.

> Comparisons of the anatomy of different living animals often reveal evidence of shared ancestry. In some instances, the same organ has evolved to carry out different functions; in others, an organ loses its function altogether.

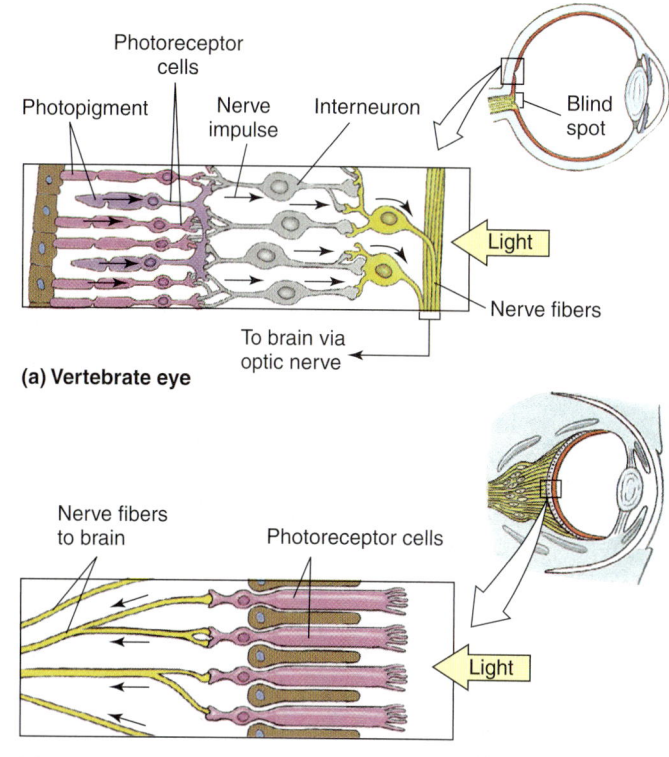

FIGURE 22.18
The eyes of vertebrates and mollusks. (*a*) Photoreceptors of vertebrates point backward, whereas (*b*) those of mollusks face forward. As a result, vertebrate nerve fibers pass in front of the photoreceptor; where they bundle together and exit the eye, a blind spot is created. Mollusks' eyes have neither of these problems.

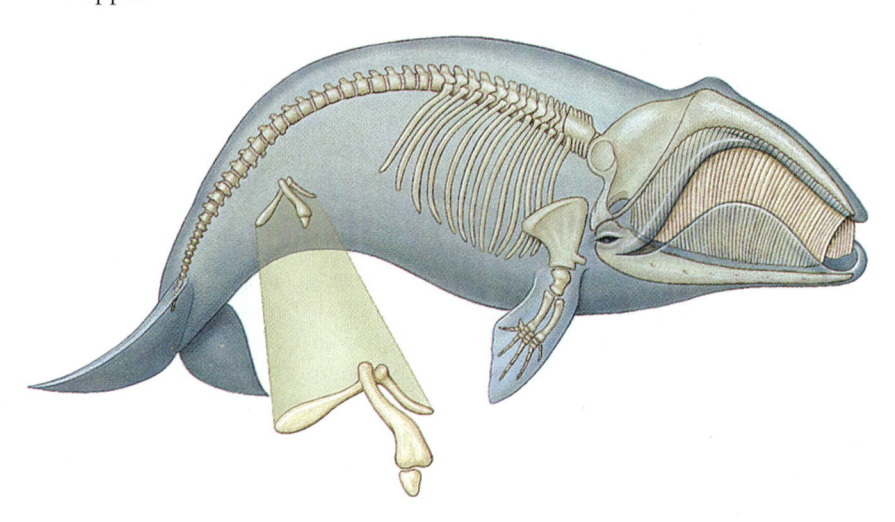

FIGURE 22.19
Vestigial structures. The skeleton of a whale reveals the presence of pelvic bones. These bones resemble those of other mammals, but are only weakly developed in the whale and have no apparent function.

Chapter 22 The Evidence for Evolution 465

The Molecular Record

Traces of our evolutionary past are also evident at the molecular level. The fact that organisms have evolved successively from relatively simple ancestors implies that a record of evolutionary change is present in the cells of each of us—in our DNA. When an ancestral species gives rise to two or more descendants, those descendants initially exhibit fairly high overall similarity in their DNA. However, as the descendants evolve independently, they accumulate more and more differences in their DNA. Consequently, organisms that are more distantly related would be expected to accumulate a greater number of evolutionary differences, whereas two species that are more closely related should share a greater portion of their DNA.

To examine this hypothesis, we need an estimate of evolutionary relationships that has been developed from data other than DNA. (It would be a circular argument to use DNA to estimate relationships and then conclude that closely related species are more similar in their DNA than are distantly related species.) Such a hypothesis of evolutionary relationships is provided by the fossil record, which indicates when particular types of organisms evolved. In addition, by comparing the anatomical structures of fossils and of modern species, we can infer how closely species are related to each other.

When degree of genetic similarity is compared with our ideas of evolutionary relationships based on fossils, a close match is evident. For example, when the human hemoglobin polypeptide is compared to the corresponding molecule in other species, closely related species are found to be more similar. Chimpanzees, gorillas, orangutans, and macaques, vertebrates thought to be more closely related to humans, have fewer differences from humans in the 146-amino-acid hemoglobin β chain than do more distantly related mammals, such as dogs. Nonmammalian vertebrates differ even more, and nonvertebrate hemoglobins are the most different of all (figure 22.20). Similar patterns are also evident when the DNA itself is compared. For example, chimps and humans, which are thought to have descended from a common ancestor that lived approximately 6 million years ago, exhibit few differences in their DNA.

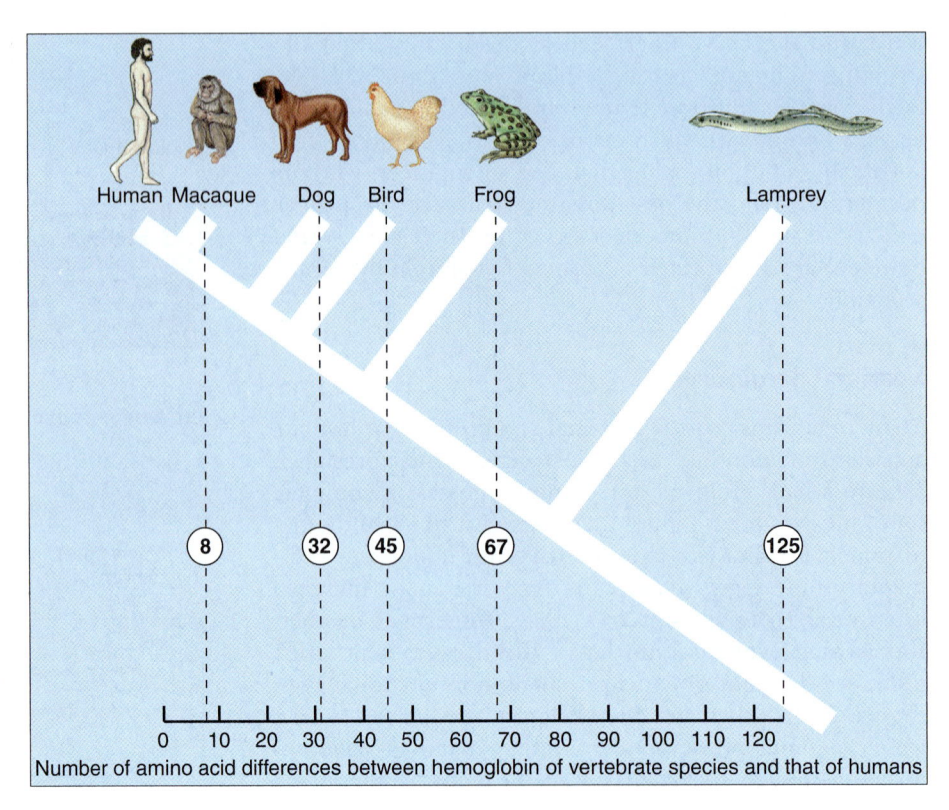

FIGURE 22.20
Molecules reflect evolutionary divergence. The greater the evolutionary distance from humans (as revealed by the white evolutionary tree which is based on the fossil record), the greater is the number of amino acid differences in the vertebrate hemoglobin polypeptide.

Why should closely related species be similar in DNA? Because DNA is the genetic code that produces the structure of living organisms, one might expect species that are similar in overall appearance and structure, such as humans and chimpanzees, to be more similar in DNA than are more dissimilar species, such as humans and frogs. This expectation would hold true even if evolution had not occurred. However, as we saw in chapter 17, large portions of the genome are composed of noncoding stretches of DNA (sometimes called "junk DNA") that have no known function and appear to serve no purpose. If evolution had not occurred, there would be no reason to expect similar-appearing species to also be similar in their junk DNA. However, comparisons of such stretches of DNA provide the same results as for other parts of the genome: More closely related species are more similar, an observation that only makes sense if evolution has occurred.

> Comparison of the DNA of different species provides strong evidence for evolution. Species deduced from the fossil record to be closely related are more similar in their DNA than are species thought to be more distantly related.

Convergent Evolution and the Biogeographical Record

Biogeography, the study of the geographic distribution of species, reveals that different geographical areas sometimes exhibit groups of plants and animals of strikingly similar appearance, even though the organisms may be only distantly related. It is difficult to explain so many similarities as the result of coincidence. Instead, natural selection appears to have favored parallel evolutionary adaptations in similar environments. Because selection in these instances has tended to favor changes that made the two groups more alike, their phenotypes have converged. This form of evolutionary change is referred to as **convergent evolution.**

The Marsupial-Placental Convergence

In the best-known case of convergent evolution, two major groups of mammals, marsupials and placentals, have evolved in a very similar way, even though the two lineages have been living independently on separate continents. Australia separated from the other continents more than 70 million years ago, after marsupials had evolved but before placental mammals had appeared. As a result, the only placental mammals in Australia are bats and a few colonizing rodents, and Australia is dominated by marsupials, members of a group in which the young are born in a very immature condition and held in a pouch until they are ready to emerge into the outside world. Thus, even though placental mammals are the dominant mammalian group throughout most of the world, marsupials retained supremacy in Australia.

What are the Australian marsupials like? To an astonishing degree, they resemble the placental mammals living today on the other continents (figure 22.21). The similarity between some individual members of these two sets of mammals argues strongly that they are the result of convergent evolution, similar forms having evolved in different, isolated areas because of similar selective pressures in similar environments.

Island Evolution

The geographical distribution of species provides evidence for evolution in other ways. Darwin was the first to present evidence that animals and plants living on oceanic islands resemble most closely the forms on the nearest continent—a relationship that only makes sense as reflecting common ancestry. For example, Galápagos tortoises and finches are

FIGURE 22.21
Convergent evolution. Marsupials in Australia resemble placental mammals occupying similar ecological niches elsewhere in the rest of the world. They evolved in isolation after Australia separated from other continents.

more similar to South American tortoises and finches than to those of more distant continents or islands, even though the environment in the Galápagos is quite different from nearby parts of South America. This relationship strongly suggests that the island forms evolved from individuals that came from the adjacent mainland at some time in the past. In the absence of evolution, there seems to be no logical explanation for why individual kinds of island plants and animals would be clearly related to others on the nearest mainland, rather than to species occupying similar habitats on more distant landmasses.

> Convergence is the evolution of similar forms in different lineages when exposed to the same selective pressures. The biogeographical distribution of species often reflects the outcome of evolutionary diversification.

22.4 The theory of evolution has proven controversial.

Darwin's Critics

In the century since he proposed it, Darwin's theory of evolution by natural selection has become nearly universally accepted by biologists, but has proven controversial among some of the general public. Darwin's critics raise seven principal objections to teaching evolution:

1. **Evolution is not solidly demonstrated.** *"Evolution is just a theory,"* Darwin's critics point out, as if "theory" meant lack of knowledge, some kind of guess. Scientists, however, use the word theory in a very different sense than the general public does. Theories are the solid ground of science—that about which we are most certain. Few of us doubt the theory of gravity because it is "just a theory."

2. **There are no fossil intermediates.** *"No one ever saw a fin on the way to becoming a leg,"* critics claim, pointing to the many gaps in the fossil record in Darwin's day. Since then, however, many fossil intermediates in vertebrate evolution have indeed been found. A clear line of fossils now traces the transition between hoofed mammals and whales, between reptiles and mammals, between dinosaurs and birds, and between apes and humans. The fossil evidence of evolution between major forms is compelling.

3. **The intelligent design argument.** *"The organs of living creatures are too complex for a random process to have produced—the existence of a clock is evidence of the existence of a clockmaker."* The intermediates in the evolution of the mammalian ear can be seen in fossils, and many intermediate "eyes" are known in various invertebrates. These intermediate forms arose because they have value—being able to detect light a little is better than not being able to detect it at all. Complex structures such as eyes evolved as a progression of slight improvements.

4. **Evolution violates the Second Law of Thermodynamics.** *"A jumble of soda cans doesn't by itself jump neatly into a stack—things become more disorganized due to random events, not more organized."* Biologists point out that this argument ignores what the second law really says: Disorder increases in a closed system, which the earth most certainly is not. Energy continually enters the biosphere from the sun, fueling life and all the processes that organize it.

5. **Proteins are too improbable.** *"Hemoglobin has 141 amino acids. The probability that the first one would be leucine is 1/20, and that all 141 would be the ones they are by chance is $(1/20)^{141}$, an impossibly rare event."* This argument illustrates a lack of understanding of probability and statistics—you cannot use probability to argue backwards. The probability that a student in a classroom has a particular birthday is 1/365; arguing this way, the probability that everyone in a class of 50 would have the birthdays they do is $(1/365)^{50}$, and yet there the class sits.

6. **Natural selection does not imply evolution.** *"No scientist has come up with an experiment in which fish evolve into frogs and leap away from predators."* Is microevolution (evolution within a species) the mechanism that has produced macroevolution (evolution among species)? Most biologists who have studied the problem think so. The differences between breeds produced by artificial selection—such as chihuahuas, mastiffs, and greyhounds—are more distinctive than the differences between some wild species, and laboratory selection experiments sometimes create forms that cannot interbreed and thus would in nature be considered different species. Thus, production of radically different forms has indeed been observed, repeatedly. To object that evolution still does not explain really major differences, such as those between fish and amphibians, simply takes us back to point 2. These changes take millions of years, and are seen clearly in the fossil record.

7. **The irreducible complexity argument.** *The intricate molecular machinery of the cell cannot be explained by evolution from simpler stages. For example, because each part of a complex cellular process such as blood clotting is essential to the overall process, how can natural selection fashion any one part?* What's wrong with this argument is that each part of a complex molecular machine evolves as part of the system. Natural selection can act on a complex system because at every stage of its evolution, the system functions. Parts that improve function are added, and because of later changes, become essential. The mammalian blood clotting system, for example, has evolved from much simpler systems. The core clotting system evolved at the dawn of the vertebrates more than 500 million years ago, and is found today in primitive fish such as lampreys. One hundred million years later, as vertebrates evolved, proteins were added to the clotting system, making it sensitive to substances released from damaged tissues. Fifty million years later, a third component was added, triggering clotting by contact with the jagged surfaces produced by injury. At each stage, as the clotting system evolved to become more complex, its overall performance came to depend on the added elements. Thus, blood clotting has become "irreducibly complex" as the result of Darwinian evolution.

> **Darwin's theory of evolution has proven controversial among some of the general public, although the commonly raised objections are without scientific merit.**

Concept Review

For interactive testing, visit the Online Learning Center with PowerWeb at www.mhhe.com/Raven7

22.1 Evidence indicates that natural selection can produce evolutionary change.

The Beaks of Darwin's Finches
- Darwin collected 31 finch specimens from three Galápagos Islands. (p. 454)
- Darwin found correspondence between 14 finch species and their food sources (p. 454)
- Observations suggest that beak differences evolved as adaptations to different food sources. (p. 455)

Peppered Moths and Industrial Melanism
- Adults range in color from light gray to black (dominant). (p. 456)
- Black moths were rare before 1850, but common after that. (p. 456)
- The Industrial Revolution caused trees to become sooty, making light-colored moths more visible to predators. Thus, darker moths became more prevalent than lighter moths. (p. 456)
- The second half of the twentieth century saw widespread pollution control, and subsequently a reversed trend away from melanism. (p. 457)
- No matter the agent or the selective force, selection favored dark moths in polluted habitats and light moths in pristine areas. (p. 457)

Artificial Selection
- Artificial selection has produced substantial change in almost every case to which it has been applied. (p. 458)
- Selection for a trait in a lab animal such as the fruit fly leads to strong and predictable evolutionary response. (p. 458)
- Differences in agricultural crops and the domestication of animals have resulted from generations of selection for desirable traits. (pp. 458–459)
- By extrapolation, selection, over the course of many millions of years, likely has the power to produce the current diversity of life. (p. 459)

22.2 Fossil evidence indicates that evolution has occurred.

The Fossil Record
- Rock fossils are created when an organism becomes buried in sediments, and hard tissue mineralizes. (p. 460)
- Fossils can be dated in a relative manner by looking at their relative position in rock strata, and dated in an absolute manner using radioactive isotopes and the ratio of derivative isotopes. (p. 460)
- At its largest scale, the fossil record documents the course of life through time. (p. 460)
- Demonstration of successive change is one of the strongest lines of evolutionary evidence. (p. 461)

The Evolution of Horses
- The evolution of horses is one of the best-studied cases in the fossil record. (p. 462)
- The earliest horses were small, with short legs and broad feet. (p. 462)
- Most changes to horses have been explained as adaptations to changing global climates. (p. 463)
- Modern horse diversity is relatively limited. (p. 463)

22.3 Evidence for evolution can be found in other fields of biology.

The Anatomical Record
- Homologous structures have different appearances and functions but are derived from the same body part in a common ancestor. (p. 464)
- Relict developmental forms in a wide variety of vertebrate embryos suggest that evolution has modified ancestral developmental patterns. (p. 464)
- Vestigial structures have no apparent function, but resemble structures of presumed ancestors. (p. 465)

The Molecular Record
- Recent descendants from a common ancestor initially exhibit relatively high DNA similarity. (p. 466)
- The more distantly related two organisms are, the larger the number of DNA differences, in both coding and noncoding regions, that evolve. (p. 466)

Convergent Evolution and the Biogeographical Record
- Different areas can exhibit groups of organisms that resemble one another. (p. 467)
- Natural selection appears to have favored parallel evolutionary adaptations within similar environments.
- Island forms are often closely related to species on the nearest mainland. (p. 467)

22.4 The theory of evolution has proven controversial.

Darwin's Critics
- Darwin's critics raise several objections to teaching evolution: evolution is not solidly demonstrated; there is a lack of fossil intermediates; the intelligent design argument; evolution violates the Second Law of Thermodynamics; proteins are too improbable; natural selection does not imply evolution; and the irreducible complexity argument. None of these objections has scientific merit. (p. 468)

Test Your Understanding

Self Test

1. Which of the following best describes the correlation between beak size and the amount of rain on Daphne Major?
   a. Birds with small beaks are favored in dry years.
   b. All birds are favored equally in wet years.
   c. Birds with large beaks are favored during wet years.
   d. Birds with large beaks are favored during dry years.
2. In peppered moths, the black coloration is selected when soot covers tree bark; this phenomenon is called
   a. artificial selection.
   b. convergent evolution.
   c. industrial melanism.
   d. none of these.
3. Evolutionary change through artificial selection has been demonstrated in all but which of the following?
   a. Galápagos finches
   b. *Drosophila*
   c. corn
   d. dog breeding
4. Darwin's examinations of fossils relied on _____ dating to determine the evolution of species.
   a. absolute
   b. carbon
   c. relative
   d. radioactive isotope
5. The missing links between whales and their hoofed ancestors include
   a. *Pakicetus*.
   b. *Archaeopteryx*.
   c. *Equus*.
   d. all of these.
6. Evolution has occurred in the horse as seen by
   a. a reduction in body size.
   b. an increase in complexity of the ridges on teeth.
   c. an increase in the number of toes.
   d. all of these.
7. Over time, the same bones in different vertebrates were put to different uses. This falls under the category of
   a. missing links.
   b. vestigial structures.
   c. analogous structures.
   d. homologous structures.
8. After examining the evidence related to the evolution of hemoglobin, you might conclude that
   a. bird hemoglobin evolved prior to lamprey hemoglobin.
   b. frogs are more closely related to lampreys than to birds.
   c. evolutionary changes occur at the molecular level.
   d. only DNA can be examined for establishing evolutionary differences.
9. An example of convergent evolution is
   a. Australian marsupials and placental mammals.
   b. the flippers in fish, penguins, and dolphins.
   c. the wings in birds, bats, and insects.
   d. all of these.
10. The shape of the beaks of Darwin's finches, industrial melanism, and the changes in horse teeth are all examples of
    a. artificial selection.
    b. natural selection.
    c. convergent evolution.
    d. homologous structures.

Test Your Visual Understanding

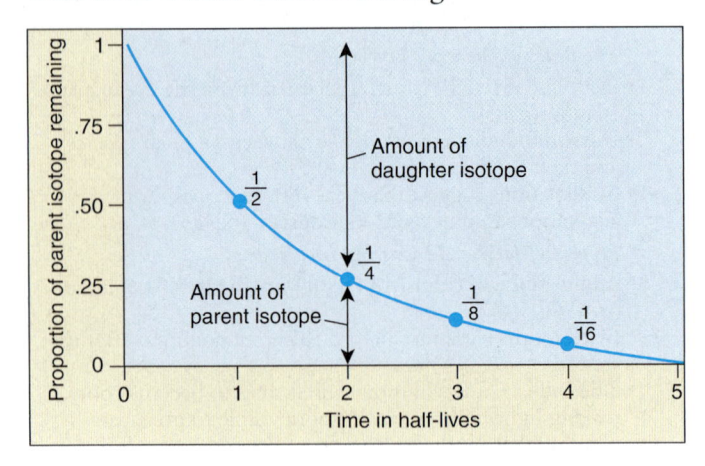

1. The graph illustrates how a radioactive isotope decays over time. Some isotopes decay more quickly than others do (they have shorter half-lives), but all isotope decay follows this same scale—that is, half of the isotope atoms decay with each half-life. For each of the isotopes in the following list, calculate how long it will take each parent sample to decay to 12.5% of the original amount. Also, graph three of the isotopes, plotting the proportion of parent isotope remaining to the number of half-lives, and compare these three graphs with the figure. Are they similar, or are they different?

Isotope	Half-life
a. beryllium-11	13.81 seconds
b. oxygen-15	2 minutes
c. sodium-24	15 hours
d. phosphorus-32	14.3 days
e. carbon-14	5,730 years
f. plutonium-239	24,110 years

Apply Your Knowledge

1. In a laboratory experiment, researchers selected for an increase and a decrease in protein content of corn seeds. The initial population contained an average of 9.5% protein by weight. As with the artificial selection experiments described in this chapter, corn seeds with the top 20% protein content were crossed and corn seeds with the lowest 20% protein contents were crossed. After 50 generations, the high-protein offspring averaged 19.2% protein, and the low-protein offspring averaged 5.4% protein.
   a. What percentage of change was recorded for the high-protein and low-protein populations?
   b. Which trait, the high- or the low-protein level, was modified more because of selection? Can you explain why one trait was modified more?
2. Why is it incorrect to think of evolution as progressive (i.e., proceeding from lowest or simplest to highest or most complex)?

23
The Origin of Species

Concept Outline

23.1 Species are the basic units of evolution.

The Nature of Species. Species are groups of actually or potentially interbreeding natural populations that are reproductively isolated from other such groups and that maintain connectedness over geographic distances.

23.2 Species maintain their genetic distinctiveness through barriers to reproduction.

Prezygotic Isolating Mechanisms. Some breeding barriers prevent the formation of zygotes.
Postzygotic Isolating Mechanisms. Other breeding barriers prevent the proper development or reproduction of the zygote after it forms.
Problems with the Biological Species Concept. Hybridization and other difficulties have prompted some scientists to propose alternative species concepts.

23.3 We have learned a great deal about how species form.

Reproductive Isolation May Evolve as a By-Product of Evolutionary Change. Speciation can occur in the absence of natural selection, but reproductive isolation generally occurs more quickly when populations are adapting to different environments.
The Geography of Speciation. Speciation occurs most readily when populations are geographically isolated.

23.4 Clusters of species reflect rapid evolution.

Hawaiian *Drosophila*. More than one-quarter of the world's fruit fly species are found on the Hawaiian Islands.
Darwin's Finches. Thirteen species of finches, all descendants of one ancestral finch, occupy diverse niches.
Lake Victoria Cichlid Fishes. Isolation has led to extensive species formation among these small fishes.
New Zealand Alpine Buttercups. Repeated glaciations have fostered waves of species formation.
The Pace of Evolution. Evolutionary change can be slow and gradual or rapid and discontinuous.
Speciation and Extinction Through Time. The number of species has increased through time.
The Future of Evolution. Evolution continues in human-altered environments.

FIGURE 23.1
A Galápagos marine iguana basks in the sun on its isolated island. How does geographic isolation contribute to the formation of new species?

Although Darwin titled his book *On the Origin of Species*, he never actually discussed what he referred to as that "mystery of mysteries"—how one species gives rise to another. Rather, his argument concerned evolution by natural selection; that is, how one species evolves through time to adapt to its changing environment. Although of fundamental importance to evolutionary biology, the process of adaptation does not explain how one species becomes another (figure 23.1). Much less can it explain how one species can give rise to many descendant species, a process we call **speciation**. As we shall see, adaptation may be involved in the speciation process, but it need not be.

471

23.1 Species are the basic units of evolution.

The Nature of Species

Before we can discuss how one species gives rise to another, we need to understand exactly what a species is. Even though the definition of a species is of fundamental importance to evolutionary biology, this issue has still not been completely settled and is currently the subject of considerable research and debate. However, any concept of a species must account for two phenomena: the distinctiveness of species that occur together at a single locality, and the connection that exists among populations of the same species that are geographically separated.

The Distinctiveness of Sympatric Species

Put out a birdfeeder on your balcony or back porch and you will attract a wide variety of different types of birds (especially if you put out a variety of different kinds of foods). In the midwestern United States, for example, you might routinely see cardinals, blue jays, downy woodpeckers, house finches—even hummingbirds in the summer. Although it might take a few days of careful observation, you would soon be able to readily distinguish the many different species. The reason is that species that occur together (termed **sympatric**) are distinctive entities that are phenotypically different, utilize different parts of the habitat, and behave separately. This observation is generally true not only for birds, but also for most other types of organisms.

Occasionally, two species occur together that appear to be nearly identical. In most cases, however, our inability to distinguish the two reflects our own reliance on vision as our primary sense. When the mating calls or chemicals exuded by such species are examined, they usually reveal great differences. In other words, even though we have trouble separating them, the animals themselves have no such difficulties!

Geographic Variation Within Species

Within the units classified as species, populations that occur in different areas may be more or less distinct from one another. Such groups of distinctive individuals may be classified taxonomically as **subspecies,** or **varieties.** (The vague term "race" has a similar connotation, but is no longer commonly used.) In areas where these populations approach one another, individuals often exhibit combinations of features characteristic of both populations (fig-

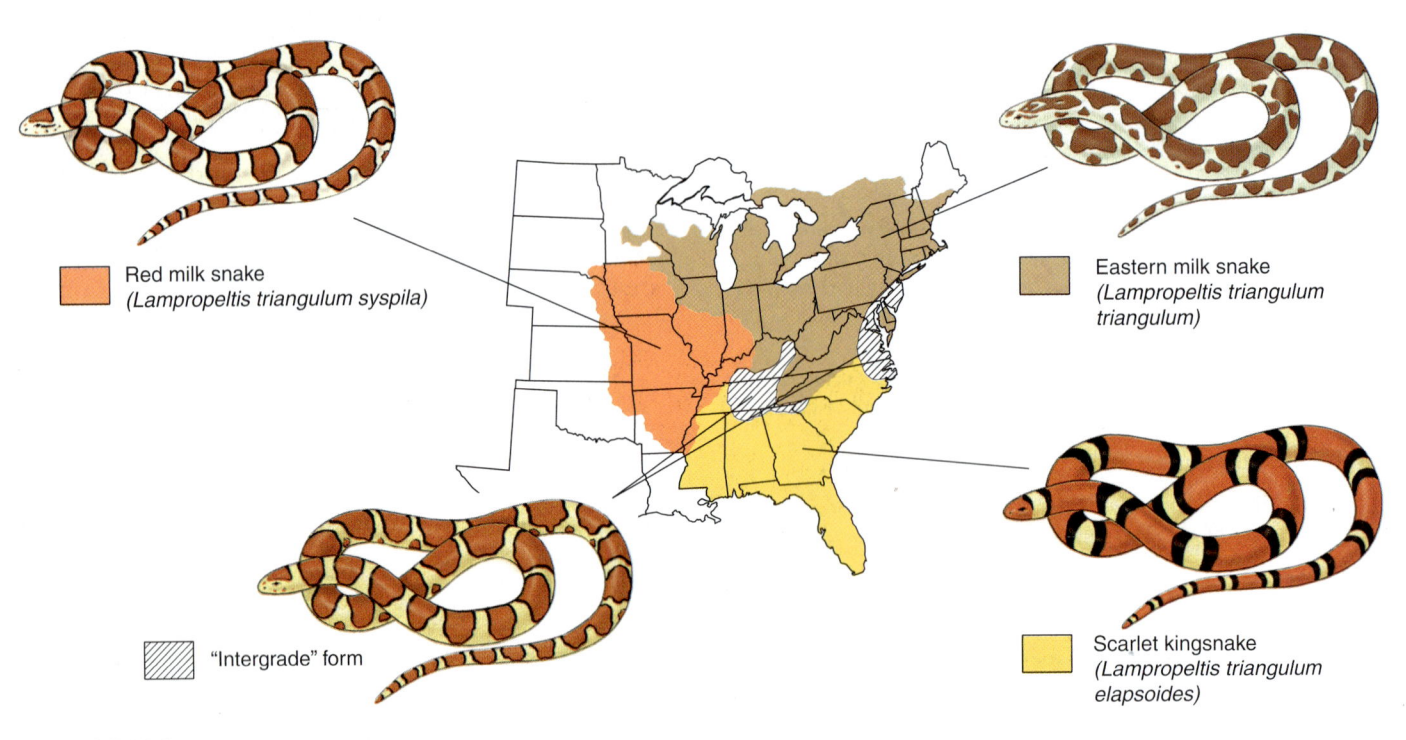

FIGURE 23.2
Geographic variation in the milk snake, *Lampropeltis triangulum*. Although subspecies appear phenotypically quite distinctive from each other, they are connected by populations that are phenotypically intermediate.

ure 23.2). In other words, even though geographically distant populations may appear distinct, they are usually connected by intervening populations that are intermediate in their characteristics.

The Biological Species Concept

What can account both for the distinctiveness of sympatric species and for the connectedness of geographic populations of the same species? One obvious possibility is that each species exchanges genetic material only with other members of its species. If sympatric species commonly exchanged genes, we might expect such species to rapidly lose their distinctions as the **gene pools** (that is, all of the alleles present in a species) of the different species became homogenized. Thus, the inability of sympatric species to exchange genes may allow them to remain distinct. Conversely, the ability of geographically distant populations of the same species to share genes through the process of gene flow may keep these populations integrated as members of the same species. Based on these ideas, the evolutionary biologist Ernst Mayr coined the **biological species concept,** which defines species as ". . . groups of actually or potentially interbreeding natural populations which are reproductively isolated from other such groups."

In other words, the biological species concept says that a species is composed of populations whose members mate with each other and produce fertile offspring—or would do so if they came into contact. Conversely, populations whose members do not mate with each other or who cannot produce fertile offspring are said to be **reproductively isolated** and, thus, members of different species.

What causes reproductive isolation? If organisms cannot interbreed or cannot produce fertile offspring, they clearly belong to different species. However, some populations that are considered separate species can interbreed and produce fertile offspring, but they ordinarily do not do so under natural conditions. They are still considered reproductively isolated in that genes from one species generally will not be able to enter the gene pool of the other species. Table 23.1 summarizes the steps at which barriers to successful reproduction may occur. Such barriers are termed **reproductive isolating mechanisms** because they prevent genetic exchange between species. We will discuss examples of these, beginning with those that prevent the formation of zygotes, which are called prezygotic isolating mechanisms. Postzygotic isolating mechanisms prevent the proper functioning of zygotes after they form.

> Species are groups of organisms that are distinct from other, co-occurring species and that are interconnected geographically. The ability to exchange genes appears to be a hallmark of such species.

Table 23.1 Reproductive Isolating Mechanisms

Mechanism	Description
PREZYGOTIC ISOLATING MECHANISMS	
Geographic isolation	 Species occur in different areas, which are often separated by a physical barrier such as a river or mountain range.
Ecological isolation	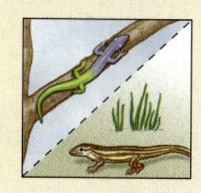 Species occur in the same area, but they occupy different habitats and rarely encounter each other.
Temporal isolation	 Species reproduce in different seasons or at different times of the day.
Behavioral isolation	 Species differ in their mating rituals.
Mechanical isolation	 Structural differences between species prevent mating.
Prevention of gamete fusion	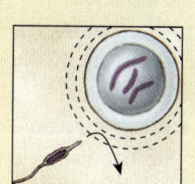 Gametes of one species function poorly with the gametes of another species or within the reproductive tract of another species.
POSTZYGOTIC ISOLATING MECHANISMS	
Hybrid inviability or infertility	 Hybrid embryos do not develop properly, hybrid adults do not survive in nature, or hybrid adults are sterile or have reduced fertility.

23.2 Species maintain their genetic distinctiveness through barriers to reproduction.

How do species keep their separate identities? Reproductive isolating mechanisms fall into two categories: **prezygotic isolating mechanisms,** which prevent the formation of zygotes, and **postzygotic isolating mechanisms,** which prevent the proper functioning of zygotes after they form. In the following discussion, we examine various isolating mechanisms in these two categories and offer examples that illustrate how the isolating mechanisms operate to help species retain their identities.

Prezygotic Isolating Mechanisms

Ecological Isolation

Even if two species occur in the same area, they may utilize different portions of the environment and thus not hybridize because they do not encounter each other. For example, in India, the ranges of lions and tigers overlapped until about 150 years ago. Even so, there were no records of natural hybrids. Lions stayed mainly in the open grassland and hunted in groups called prides; tigers tended to be solitary creatures of the forest (figure 23.3). Because of their ecological and behavioral differences, lions and tigers rarely came into direct contact with each other, even though their ranges overlapped over thousands of square kilometers.

In another example, the ranges of two toads, *Bufo woodhousei* and *B. americanus*, overlap in some areas. Although these two species can produce viable hybrids, they usually do not interbreed because they utilize different portions of the habitat for breeding. Whereas *B. woodhousei* prefers to breed in streams, *B. americanus* breeds in rainwater puddles. Similarly, the ranges of two species of dragonflies overlap in Florida. However, the dragonfly *Progomphus obscurus* lives near rivers and streams, while *P. alachuenis* lives near lakes.

Similar situations occur among plants. Two species of oaks occur widely in California: the valley oak, *Quercus lobata*, and the scrub oak, *Q. dumosa*. The valley oak, a graceful deciduous tree that can be as tall as 35 meters, occurs in the fertile soils of open grassland on gentle slopes and valley floors. In contrast, the scrub oak is an evergreen shrub, usually only 1 to 3 meters tall, which often forms the kind of dense scrub known as chaparral. The scrub oak is found on steep slopes in less fertile soils. Hybrids between these different oaks do occur and are fully fertile, but they are rare. The sharply distinct habitats of their parents limit their occurrence together, and there is no intermediate habitat where the hybrids might flourish.

FIGURE 23.3
Lions and tigers are ecologically isolated. The ranges of lions and tigers used to overlap in India. However, lions and tigers do not hybridize in the wild because they utilize different portions of the habitat. Lions live in open grassland, whereas tigers are solitary animals that live in the forest. Hybrids, such as this tiglon, have been successfully produced in captivity, but hybridization does not occur in the wild.

Behavioral Isolation

Chapter 52 describes the often elaborate courtship and mating rituals of some groups of animals. Related species of organisms such as birds often differ in their courtship rituals, which tends to keep these species distinct in nature even if they inhabit the same places (figure 23.4). For example, mallard and pintail ducks are perhaps the two most common freshwater ducks in North America. In captivity, they produce completely fertile offspring, but in nature they nest side-by-side and only rarely hybridize.

Sympatric species avoid mating with members of the wrong species in a variety of ways; every mode of communication imaginable appears to be used by some species. Differences in visual signals, as we have just discussed, are common. However, other types of animals rely more on other sensory modes for communication. Many species, such as frogs, birds and a variety of insects, use vocalizations to attract mates. Predictably, sympatric species of these animals produce different calls (figure 23.5).

Other species rely on the detection of chemical signals, called **pheromones.** The use of pheromones in moths has been particularly well studied. When they are ready to mate, female moths emit a pheromone that males can detect at great distances. Sympatric species differ in the pheromone they produce: Either they use different chemical compounds, or if using the same compounds, they differ in the proportions used. Laboratory studies indicate that males are remarkably adept at distinguishing the pheromones of their own species from those of other species or even from synthetic compounds similar, but not identical, to that of their own species.

Some species even use electroreception. African electric fish have specialized organs in their tails that produce electrical discharges and electroreceptors on their skins that detect electrical charges. These discharges are used to communicate in social interactions; field experiments indicate that males can distinguish between signals produced by their own and other species, probably on the basis of differences in the timing of the electrical pulses.

Other Prezygotic Isolating Mechanisms

Temporal Isolation. *Lactuca graminifolia* and *L. canadensis*, two species of wild lettuce, grow together along roadsides throughout the southeastern United States. Hybrids between these two species are easily made experimentally and are completely fertile. But such hybrids are rare in nature because *L. graminifolia* flowers in early spring and *L. canadensis* flowers in summer. When their blooming periods overlap, as happens occasionally, the two species do form hybrids, which may become locally abundant.

Many species of closely related amphibians have different breeding seasons that prevent hybridization between the species. For example, five species of frogs of the genus *Rana* occur together in most of the eastern United States, but hybrids are rare because the peak breeding time is different for each of them.

Mechanical Isolation. Structural differences prevent mating between some related species of animals. Aside from such obvious features as size, the structure of the male and female copulatory organs may be incompatible. In many insect and other arthropod groups, the sexual organs, particularly those of the male, are so diverse that they are used as a primary basis for distinguishing species.

Similarly, flowers of related species of plants often differ significantly in their proportions and structures. Some of these differences limit the transfer of pollen from one plant species to another. For example, bees may carry the pollen of one species on a certain place on their bodies; if this area does not come into contact with the receptive structures of the flowers of another plant species, the pollen is not transferred.

Prevention of Gamete Fusion. In animals that shed their gametes directly into water, eggs and sperm derived from different species may not attract one another. Many land animals may not hybridize successfully because the sperm of one species function so poorly within the reproductive tract of another that fertilization never takes place. In plants, the growth of pollen tubes may be impeded in

FIGURE 23.4
Differences in courtship rituals can isolate related bird species. These Galápagos blue-footed boobies select their mates only after an elaborate courtship display. This male is lifting his feet in a ritualized high-step that shows off his bright blue feet. The display behavior of the two other species of boobies that occur in the Galápagos is much different.

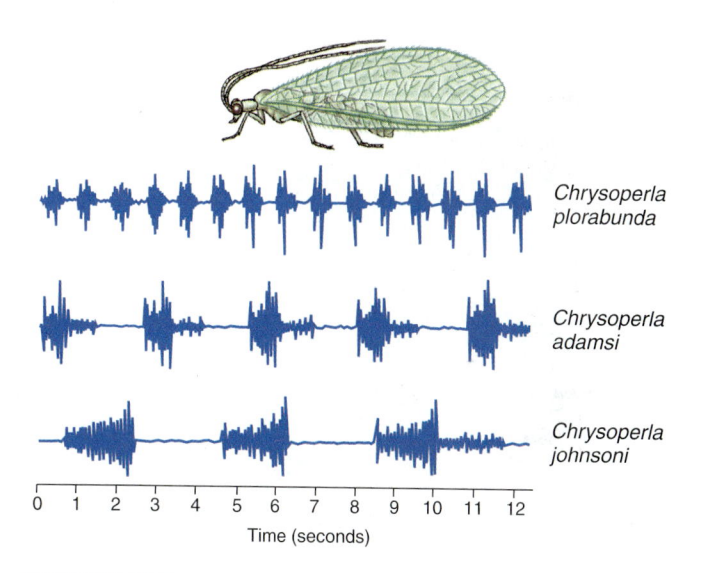

FIGURE 23.5
Differences in courtship song of sympatric species of lacewings. Lacewings are small insects that rely on auditory signals produced by vibrating their abdomens together to attract mates. As these sound recordings indicate, the sounds produced by sympatric species differ greatly. Females, which detect the calls as they are transmitted through solid surfaces such as branches, are able to distinguish calls of different species and only respond to individuals producing their own species' call.

hybrids between different species. In both plants and animals, such isolating mechanisms prevent the union of gametes, even following successful mating.

> **Prezygotic isolating mechanisms lead to reproductive isolation by preventing the formation of hybrid zygotes.**

Postzygotic Isolating Mechanisms

All of the factors we have discussed so far tend to prevent hybridization. If hybrid matings do occur and zygotes are produced, many factors may still prevent those zygotes from developing into normally functioning, fertile individuals. As we saw in chapter 19, development in any species is a complex process. In hybrids, the genetic complements of two species may be so different that they cannot function together normally in embryonic development. For example, hybridization between sheep and goats usually produces embryos that die in the earliest developmental stages.

Leopard frogs (*Rana pipiens* complex) of the eastern United States are a group of similar species, assumed for a long time to constitute a single species (figure 23.6). However, careful examination revealed that although the frogs appear similar, successful mating between them is rare because of problems that occur as the fertilized eggs develop. Many of the hybrid combinations cannot be produced even in the laboratory.

Examples of this kind, in which similar species have been recognized only as a result of hybridization experiments, are common in plants. Sometimes the hybrid embryos can be removed at an early stage and grown in an artificial medium. When these hybrids are supplied with extra nutrients or other supplements that compensate for their weakness or inviability, they may complete their development normally.

Even if hybrids survive the embryo stage, however, they may not develop normally. If the hybrids are weaker than their parents, they will almost certainly be eliminated in nature. Even if they are vigorous and strong, as in the case of the mule, a hybrid between a female horse and a male donkey, they may still be sterile and thus incapable of contributing to succeeding generations. Hybrids may be sterile because the development of sex organs is abnormal, because the chromosomes derived from the respective parents cannot pair properly, or due to a variety of other causes.

> Postzygotic isolating mechanisms are those in which hybrid zygotes fail to develop or develop abnormally, or in which hybrids cannot become established in nature.

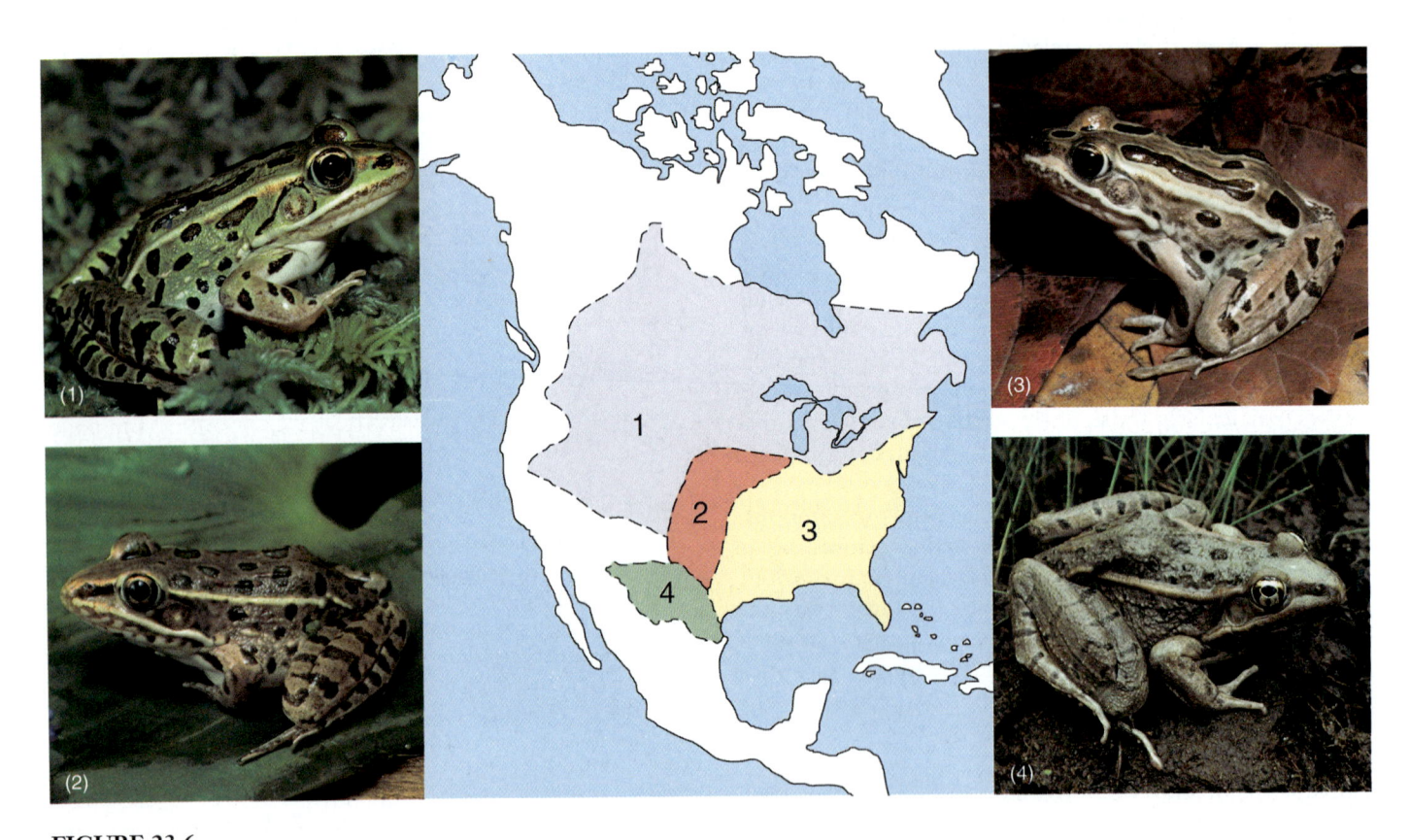

FIGURE 23.6
Postzygotic isolation in leopard frogs. Numbers indicate the following species in the geographical ranges shown: (*1*) *Rana pipiens*; (*2*) *Rana blairi*; (*3*) *Rana sphenocephala*; (*4*) *Rana berlandieri*. These four species resemble one another closely in their external features. Their status as separate species was first suspected when hybrids between some pairs of these species were found to produce defective embryos in the laboratory. Subsequent research revealed that the mating calls of the four species differ substantially, indicating that the species have both pre- and postzygotic isolating mechanisms.

Problems with the Biological Species Concept

The biological species concept has proven to be an effective way of understanding the existence of species in nature. Nonetheless, it has a number of problems that have led some scientists to propose alternative species concepts.

One criticism concerns the extent to which all species truly are reproductively isolated. By definition, under the biological species concept, species should not interbreed and produce fertile offspring. Nonetheless, in recent years biologists have detected much greater amounts of hybridization than previously realized between populations that seem to coexist as distinct biological entities. Botanists have always been aware that species can often experience substantial amounts of hybridization. For example, more than 50% of California plant species included in one study were not well defined by genetic isolation. Such coexistence without genetic isolation can be long-lasting: Fossil data show that balsam poplars and cottonwoods have been phenotypically distinct for 12 million years but have routinely produced hybrids throughout this time. Consequently, many botanists have long felt that the biological species concept only applies to animals.

What is becoming increasingly evident, however, is that hybridization is not all that uncommon in animals, either. In recent years, many cases of substantial hybridization between animal species have been documented. One recent survey indicated that almost 10% of the world's 9500 bird species are known to have hybridized in nature. Galápagos finches provide a particularly well-studied example. Three species on the island of Daphne Major—the medium ground finch, the cactus finch, and the small ground finch—are clearly distinct morphologically, and occupy different ecological niches. Studies over the past 20 years by Peter and Rosemary Grant found that, on average, 2% of the medium ground finches and 1% of the cactus ground finches mated with other species every year. Furthermore, hybrid offspring appeared to be at no disadvantage in terms of survival or subsequent reproduction. This is not a trivial amount of genetic exchange, and one might expect to see the species coalesce into one genetically variable population, but the species are maintaining their distinctiveness.

This is not to say that hybridization is rampant throughout the animal world. Most bird species do not hybridize, and even fewer probably experience significant amounts of hybridization. Still, it is common enough to cast doubt about whether reproductive isolation is the only force maintaining the integrity of species.

Natural Selection and the Ecological Species Concept

An alternative hypothesis proposes that the distinctions among species are maintained by natural selection. The idea is that each species has adapted to its own specific part of the environment. Stabilizing selection then maintains the species' adaptations; hybridization has little effect because alleles introduced into the gene pool from other species are quickly eliminated by natural selection.

We have already seen in chapter 21 that the interaction between gene flow and natural selection can have many outcomes. In some cases, strong selection can overwhelm any effects of gene flow, but in other situations, gene flow can prevent populations from eliminating less successful alleles from a population. As a general explanation, then, natural selection is not likely to have any fewer exceptions than the biological species concept, although it may prove more successful for certain types of organisms or habitats.

Other Problems with the Biological Species Concept

The biological species concept has been criticized for other reasons as well. For example, it can be difficult to apply the concept to populations that do not occur together in nature. Because individuals of these populations do not encounter each other, it is not possible to observe whether they would interbreed naturally. Although experiments can determine whether fertile hybrids can be produced, this information is not enough because many species that coexist without interbreeding in nature will readily hybridize in the artificial settings of the laboratory or zoo. Consequently, evaluating whether such populations constitute different species is ultimately a judgment call. In addition, the concept is more limited than its name would imply. Many organisms are asexual and reproduce without mating; reproductive isolation has no meaning for such organisms.

For these reasons, a variety of other ideas have been put forward to establish criteria for defining species. Many of these are specific to a particular type of organism, and none has universal applicability. In truth, there may be no single explanation for what maintains the identity of species. Given the incredible variation evident in plants, animals, and microorganisms in all aspects of their biology, it is perhaps not surprising that different processes are operating in different organisms. In addition, some scientists have turned from emphasizing the processes that maintain species distinctions to examining the evolutionary history of populations. These genealogical species concepts are currently a topic of great debate. The study of species concepts is thus an area of active research that demonstrates the dynamic nature of the field of evolutionary biology.

The surprisingly high incidence of hybridization in plants and animals is causing some researchers to seek alternative species concepts. Because of the diversity of living organisms, no single definition of what constitutes a species may be universally applicable.

23.3 We have learned a great deal about how species form.

One of the oldest questions in the field of evolution is: How does one ancestral species become divided into two descendant species? If species are defined by the existence of reproductive isolation, then the process of speciation equates with the evolution of reproductive isolating mechanisms. How do reproductive isolating mechanisms evolve?

Reproductive Isolation May Evolve as a By-Product of Evolutionary Change

Most reproductive isolating mechanisms initially arise for some reason other than to provide reproductive isolation. For example, a population that colonizes a new habitat may evolve adaptations for living in that habitat. As a result, individuals from that population might never encounter individuals from the ancestral population. Even if they do meet, the population in the new habitat may have evolved new phenotypes or behaviors so that members of the two populations no longer recognize each other as potential mates. For this reason, some biologists believe that the term "isolating mechanisms" is misguided, because it implies that the traits evolved specifically for the purpose of genetically isolating a species, which in most cases is probably incorrect.

Selection May Reinforce Isolating Mechanisms

The formation of species is a continuous process, one that we can understand because of the existence of intermediate stages at all levels of differentiation. If populations that are partly differentiated come into contact with one another, they may still be able to interbreed freely, and the differences between them may disappear over the course of time as genetic exchange homogenizes the populations. Conversely, if the populations are reproductively isolated, then no genetic exchange will occur, and the two populations will be different species.

However, there is an intermediate situation in which reproductive isolation has partially evolved, but is not complete. As a result, hybridization will occur at least occasionally. If the hybrids are partly sterile, or not as well adapted to the existing habitats as their parents, they will be at a disadvantage. As a result, selection would favor any alleles in the parental populations that prevented hybridization because individuals that avoided hybridizing would produce more successful offspring and thus pass more of their genes on to subsequent generations. The result would be the continual improvement of prezygotic isolating mechanisms until the two populations were completely reproductively isolated. This process is

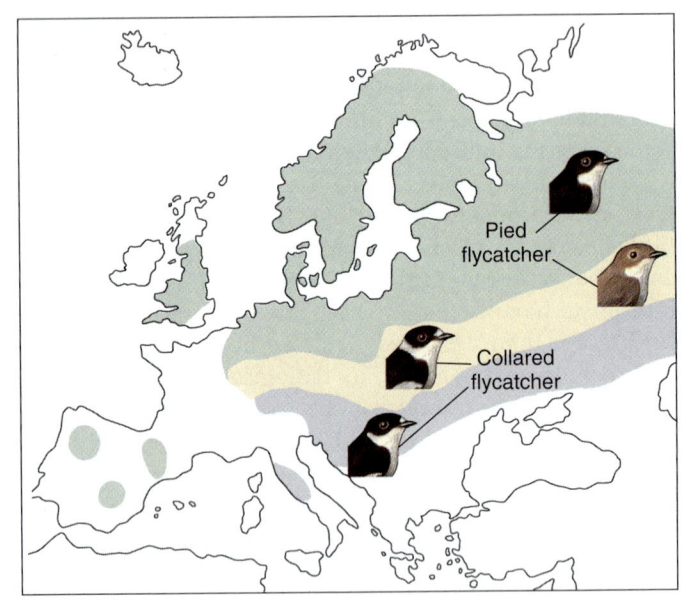

FIGURE 23.7
Reinforcement in European flycatchers. The pied flycatcher and the collared flycatcher appear very similar when they occur alone. However, in places where the two species occur sympatrically (indicated by the tan color), they have evolved differences in color, which allow individuals to choose mates from their own species and thus avoid hybridizing.

termed **reinforcement** because initially incomplete isolating mechanisms are reinforced by natural selection until they are completely effective.

An example of reinforcement is provided by pied and collared flycatchers. Throughout much of eastern and central Europe, these two species are geographically separated (allopatric) and are very similar in color (figure 23.7). However, in the Czech Republic and Slovakia, the two species occur together and occasionally hybridize, producing offspring that usually have very low fertility. At those sites, the species have evolved to look very different, and birds prefer to mate with individuals with their own species' color, in contrast to birds from allopatric populations, which prefer the allopatric color pattern. As a consequence of the color differences in sympatry, the rate of hybridization is extremely low. These results indicate that when populations of the two species came into contact, natural selection led to differences in color patterns, resulting in the evolution of pre-mating isolation.

Reinforcement is by no means inevitable, however. When incompletely isolated populations come together, gene flow immediately begins to occur between the species. Although hybrids may be inferior, they are not completely inviable or infertile (if they were, the species would already be completely reproductively isolated); hence, when these hybrids reproduce with members of either population, they will serve as a conduit of genetic exchange from one population to the other. As a result, the two populations will tend to lose their genetic distinctiveness. Thus, a race ensues: Can reproductive isolation be perfected before gene flow destroys the differences between the populations? Experts disagree on the likely outcome, but many believe that reinforcement is the much less common outcome.

The Role of Natural Selection in Speciation

What role does natural selection play in the speciation process? Certainly, the process of reinforcement is driven by natural selection favoring the perfection of reproductive isolation. But, as we have seen, reinforcement may not be common. Is natural selection necessarily involved in the initial evolution of isolating mechanisms?

Random Changes May Cause Reproductive Isolation

As we discussed in chapter 21, populations may diverge for purely random reasons. Genetic drift in small populations, founder effects, and population bottlenecks all may lead to changes in traits that cause reproductive isolation. For example, in the Hawaiian Islands, closely related species of *Drosophila* often differ greatly in their courtship behavior. Colonization of new islands by these fruit flies probably involves a founder effect, in which one or a few fruit flies—perhaps only a single pregnant female—is blown by strong winds to a new island. Changes in courtship behavior between ancestor and descendant populations may be the result of such founder events. Given long enough periods of time, any two isolated populations will diverge due to genetic drift. (Remember that even large populations experience drift, but at a lower rate than small populations do.) In some cases, this random divergence may affect traits responsible for reproductive isolation, and speciation will have occurred.

Adaptation and Speciation

Nonetheless, adaptation and speciation are probably related in many cases. As species adapt to different circumstances, they will accumulate many differences that may lead to reproductive isolation. For example, if one population of flies adapts to wet conditions and another to dry conditions, the populations will evolve a variety of differences in physiological and sensory traits; these differences may promote ecological and behavioral isolation,

FIGURE 23.8
Dewlaps of different species of Caribbean *Anolis* lizards. Males use their dewlaps in both territorial and courtship displays. Coexisting species almost always differ in their dewlaps, which are used in species recognition. Some dewlaps are easier to see in open habitats, whereas others are more visible in shaded environments.

and may cause any hybrids they produce to be poorly adapted to either habitat.

Selection might also act directly on mating behavior. Male *Anolis* lizards, for example, court females by extending a colorful flap of skin, called a "dewlap," located under their throat (figure 23.8). The ability of one lizard to see the dewlap of another lizard depends not only on the color of the dewlap, but on the environment in which the lizards occur. Thus, a light-colored dewlap is most effective in reflecting light in a dim forest, whereas dark colors are more apparent in the bright glare of open habitats. As a result, when these lizards occupy new habitats, natural selection will favor evolutionary change in dewlap color because males whose dewlaps cannot be seen will attract few mates. However, the lizards also distinguish members of their own species from those of other species by the color of the dewlap. Hence, adaptive change in mating behavior could have the incidental consequence of causing speciation.

Laboratory scientists have conducted experiments on fruit flies and other organisms in which they isolate populations in different laboratory chambers and measure how much reproductive isolation evolves. These experiments indicate that genetic drift by itself can lead to some degree of reproductive isolation, but in general, reproductive isolation evolves more rapidly when the populations are forced to adapt to different laboratory environments (such as temperature or food type).

> **Reproductive isolating mechanisms can evolve either through random changes or as an incidental by-product of adaptive evolution. Under some circumstances, however, natural selection can directly select for traits that increase the reproductive isolation of a species.**

Chapter 23 The Origin of Species 479

The Geography of Speciation

Speciation is a two-part process. First, initially identical populations must diverge, and second, reproductive isolation must evolve to maintain these differences. The difficulty with this process, as we have seen, is that the homogenizing effect of gene flow between populations will constantly be acting to erase any differences that may arise, either by genetic drift or natural selection. However, gene flow only occurs between populations that are in contact, and populations can become isolated for a variety of reasons (figure 23.9). Consequently, evolutionary biologists have long recognized that speciation is much more likely in geographically isolated, or allopatric, populations.

Allopatric Speciation

Ernst Mayr was the first biologist to demonstrate that geographically separated, or **allopatric,** populations appear much more likely to have evolved substantial differences leading to speciation. Marshalling data from a wide variety of organisms and localities, Mayr made a strong case for allopatric speciation as the primary means of speciation. For example, the Papuan kingfisher, *Tanysiptera hydrocharis,* varies little throughout its wide range in New Guinea despite the great variation in the island's topography and climate. By contrast, isolated populations on nearby islands are strikingly different from each other and from the mainland population (figure 23.10).

Many other examples indicate that speciation can occur in allopatry. Given that one would expect isolated populations to diverge over time by either drift or selection, this result is not surprising. Rather, the question becomes: Is geographic isolation *required* for speciation to occur?

Sympatric Speciation

For decades, biologists have debated whether one species could split into two at a single locality, without the two new species ever having been geographically separated. Scientists have suggested that such sympatric speciation could occur either instantaneously or over the course of multiple generations. Although most of the hypotheses suggested so far are highly controversial, one type of instantaneous sympatric speciation is known to occur commonly, as the result of polyploidy.

Instantaneous Speciation Through Polyploidy. Instantaneous sympatric speciation occurs when an individual is born that is reproductively isolated from all other members of its species. In most cases, a mutation that would cause an individual to be so different from others of its species would have many adverse pleiotropic side-effects, and the individual would not survive. One exception, however, occurs through the process of **polyploidy,** commonly seen in plants. A polyploid individual has more than two sets of chromosomes. Polyploids can arise in two ways. In **autopolyploidy,** all of the chromosomes may arise from a single species. This might happen, for example, due to an error in meiosis that causes an individual to have four sets of chromosomes. Such individuals, termed tetraploids, could fertilize themselves or mate with other tetraploids, but could not mate and produce fertile offspring with normal diploids. The reason is that the offspring from such a mating would be triploid (having three sets of chromosomes) and would be sterile due to problems with chromosome pairing during meiosis.

A more common type of polyploid speciation is **allopolyploidy,** which occurs sometimes when two species hybridize. The resulting offspring, having one copy of the chromosomes of each species, is usually infertile because the chromosomes do not pair correctly in meiosis. However, such individuals are often otherwise healthy, can reproduce asexually, and can even become fertile through a variety of events. For example, if the chromosomes of such an individual were to spontaneously

FIGURE 23.9

Populations can become geographically isolated for a variety of reasons. (*a*) Colonization of distant areas by one or a few individuals can establish populations in a distant place. (*b*) Barriers to movement can split an ancestral population into two isolated populations. (*c*) Extinction of intermediate populations can leave the remaining populations isolated from each other.

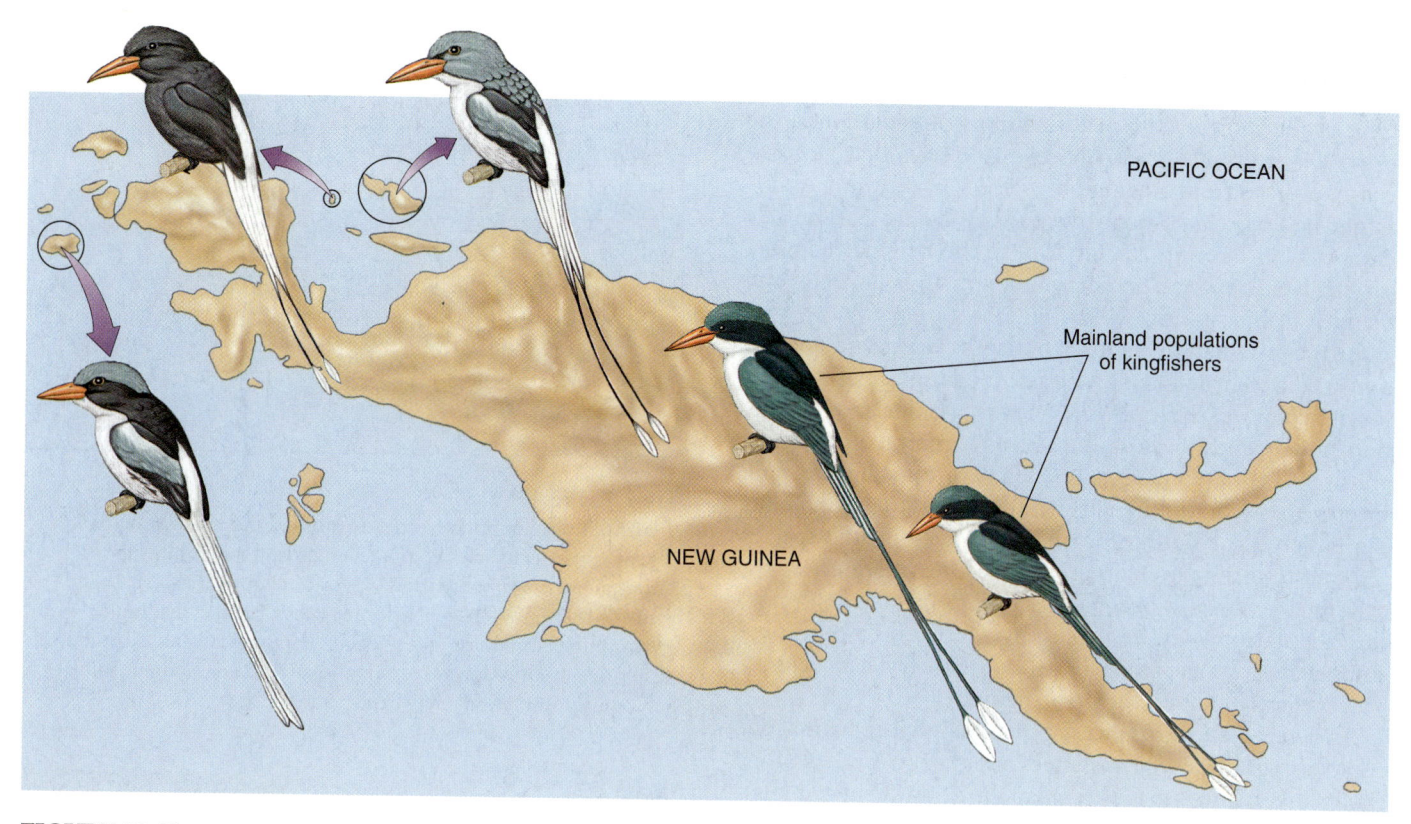

FIGURE 23.10
Phenotypic differentiation in the Papuan kingfisher in New Guinea. Isolated island populations (*left*) are quite distinctive, showing variation in tail feather structure and length, plumage coloration, and bill size, whereas kingfishers on the mainland (*right*) show little variation.

double, as just described, the resulting tetraploid would have two copies of each set of chromosomes. Consequently, pairing would no longer be a problem in meiosis: Each set of chromosomes could pair with itself. As a result, such tetraploids would be able to intermate, and a new species would have been created.

It is estimated that about half of the approximately 260,000 species of plants have a polyploid episode in their history, including many of great commercial importance, such as bread wheat, cotton, tobacco, sugarcane, bananas, and potatoes. Speciation by polyploidy is also known to occur in a variety of animals, including insects, fish, and salamanders, although much more rarely than in plants.

Sympatric Speciation by Disruptive Selection. Some investigators believe that sympatric speciation can occur over the course of multiple generations through the process of disruptive selection. As we saw in chapter 21, disruptive selection can cause a population to contain individuals exhibiting two different phenotypes. One might think that if selection were strong enough, these two phenotypes would evolve over a number of generations into different species. However, before the two phenotypes could become different species, they would have to evolve reproductive isolating mechanisms. Because the two phenotypes would initially not be reproductively isolated at all, genetic exchange between individuals of the two phenotypes would tend to prevent genetic divergence in mating preferences or other isolating mechanisms. As a result, the two phenotypes would be retained as polymorphisms within a single population. For this reason, most biologists consider sympatric speciation of this type to be a rare event.

Nonetheless, in recent years, a number of cases have appeared that are difficult to interpret in any way other than as sympatric speciation. For example, Lake Barombi Mbo in Cameroon is an extremely small and ecologically homogeneous crater lake, with no opportunity for within-lake isolation. Nonetheless, 11 species of closely related cichlid fish occur in the lake; all of the species are more closely related evolutionarily to each other than to any species outside of the crater. The most reasonable explanation is that an ancestral species colonized the crater and subsequently speciated in sympatry multiple times.

Speciation occurs much more readily in the absence of gene flow among populations. However, speciation can occur in sympatry by means of polyploidy and perhaps by disruptive selection.

Chapter 23   The Origin of Species   481

23.4 Clusters of species reflect rapid evolution.

One of the most visible manifestations of evolution is the existence of groups of closely related species that have recently evolved from a common ancestor by adapting to different habitats. Such **adaptive radiations** are particularly common on oceanic islands, where the original colonist probably encountered an environment with few species and many available resources.

Adaptive radiation requires both speciation and adaptation to different habitats. A classic model postulates that a species colonizes multiple islands in an archipelago. Speciation subsequently occurs allopatrically, and then the newly arisen species colonize other islands, producing multiple species per island (figure 23.11). Adaptation to new habitats can either occur during the allopatric phase as the species respond to different environments on the different islands, or after two species become sympatric. In the latter case, this adaptation may be driven by the need to minimize competition for available resources with other species, a process termed **character displacement**.

An alternative possibility is that adaptive radiation occurs through repeated instances of sympatric speciation, producing a suite of species adapted to different habitats. As we just discussed, such scenarios are hotly debated.

Hawaiian *Drosophila*

More than one-third of the world's species in the fly genus *Drosophila* occur on the Hawaiian Islands. New species of *Drosophila* are still being discovered in Hawaii, although the rapid destruction of the native vegetation is making the search more difficult. Aside from their sheer number, Hawaiian *Drosophila* species are unusual because of their incredible diversity of morphology and behavioral traits (figure 23.12). Evidently, when their ancestors first reached these islands, they encountered many "empty" habitats that other kinds of insects and other animals occupied elsewhere. As a result, the species have adapted to all manners of fruit fly life, and include predators, parasites, and herbivores, as well as species specialized for eating the detritus in leaf litter and the nectar of flowers. The larvae of various species live in rotting stems, fruits, bark, leaves, or roots, or feed on sap. No comparable diversity of *Drosophila* species is found anywhere else in the world.

A second, closely related genus of flies, *Scaptomyza*, also forms a species cluster in Hawaii, where it is represented by as many as 300 species. The genera *Scaptomyza* and *Drosophila* are so closely related that scientists have suggested that all of the estimated 800 species of these two genera that occur in Hawaii may have derived from a single common ancestor.

The great diversity of Hawaiian species is a result of the geological history of these islands. New islands have continually arisen from the sea in the region of the

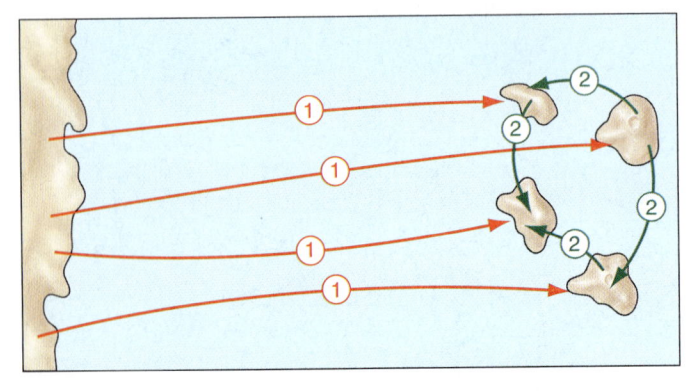

FIGURE 23.11
Classic model adaptive radiation on island archipelagoes. (*1*) An ancestral species colonizes islands in an archipelago. Subsequently, the populations speciate in allopatry. (*2*) Then some of these new species colonize other islands, leading to local communities of two or more species. Ecological specialization can either occur when species are in allopatry (*1*) or as the result of ecological interactions between species after (*2*).

FIGURE 23.12
Hawaiian *Drosophila*. The hundreds of species that have evolved on the Hawaiian Islands are extremely variable in appearance, although genetically almost identical. (*a*) *Drosophila heteroneura*. (*b*) *Drosophila digressa*.

Hawaiian Islands. As they have done so, they appear to have been invaded successively by the various *Drosophila* groups present on the older islands. New species thus have evolved as new islands have been colonized. In addition, the Hawaiian Islands are among the most volcanically active islands in the world. Periodic lava flows often have created patches of habitat within an island, termed *kipukas*, in a sea of barren rock. *Drosophila* populations isolated in these kipukas often speciate. In these ways, rampant speciation combined with ecological opportunity has led to an unparalleled diversity of insect life.

> The adaptive radiation of about 800 species of the flies *Drosophila* and *Scaptomyza* on the Hawaiian Islands, probably from a single common ancestor, is one of the most remarkable examples of intensive species formation found anywhere on earth.

Darwin's Finches

We have already mentioned the diversity of Darwin's finches on the Galápagos Islands in chapter 22. Presumably, the ancestor of Darwin's finches reached these islands before other land birds, and many of the types of habitats where birds occur on the mainland were unoccupied. As the new arrivals moved into these vacant ecological niches and adopted new lifestyles, they were subjected to diverse sets of selective pressures. Under these circumstances, and aided by the geographic isolation afforded by the many islands of the Galápagos archipelago, the ancestral finches rapidly split into a series of diverse populations, some of which evolved into separate species. These species now occupy many different kinds of habitats on the Galápagos Islands, which are comparable to the habitats several distinct groups of birds occupy on the mainland. As illustrated in figure 23.13, the 14 species comprise four groups:

1. **Ground finches.** There are six species of *Geospiza* ground finches. Most of the ground finches feed on seeds. The size of their bills is related to the size of the seeds they eat. Some of the ground finches feed primarily on cactus flowers and fruits, and have a longer, larger, and more pointed bill than the others.
2. **Tree finches.** There are five species of insect-eating tree finches. Four species have bills suitable for feeding on insects. The woodpecker finch has a chisel-like beak. This unusual bird carries around a twig or a cactus spine, which it uses to probe for insects in deep crevices.
3. **Warbler finches.** These unusual birds play the same ecological role in the Galápagos woods that warblers play on the mainland, searching continually over the leaves and branches for insects. They have slender, warbler-like beaks.
4. **Vegetarian finch.** The very heavy bill of this bud-eating bird is used to wrench buds from branches.

Recently, scientists have examined the DNA of Darwin's finches to study their evolutionary history. These studies suggest that the deepest branches in the finch evolutionary tree lead to warbler finches, which implies that warbler finches were among the first types to evolve after colonization of the islands.

> **Darwin's finches, all derived from one similar mainland species, have radiated widely on the Galápagos Islands in the absence of competition from other types of birds.**

FIGURE 23.13
An evolutionary tree of Darwin's finches. Their position at the base of the evolutionary tree of Darwin's finches suggests that warbler finches were among the first ecological types to evolve in the Galápagos.

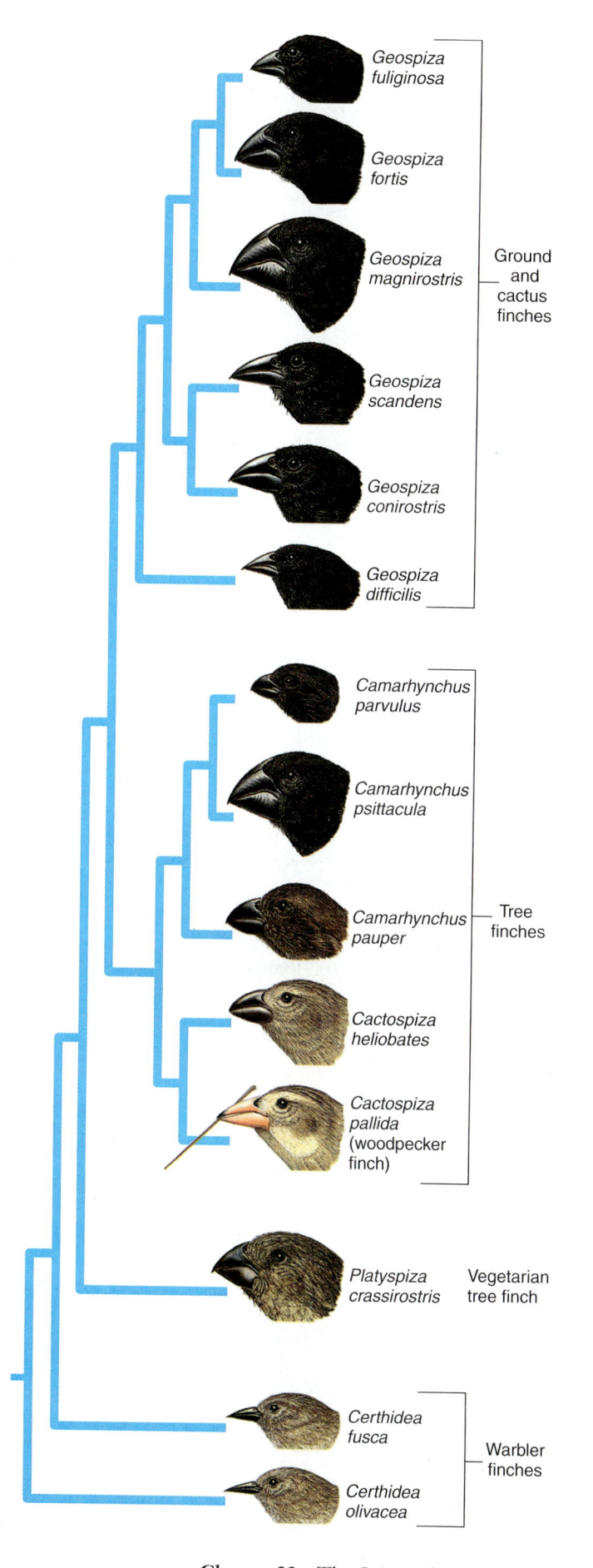

Lake Victoria Cichlid Fishes

Lake Victoria is an immense, shallow, freshwater sea about the size of Switzerland in the heart of equatorial East Africa. Until recently, the lake was home to an incredibly diverse collection of over 300 species of cichlid fishes.

Recent Radiation

The cluster of cichlid species appears to have evolved recently and quite rapidly. By sequencing the cytochrome *b* gene in many of the lake's fish, scientists have been able to estimate that the first cichlids entered Lake Victoria only 200,000 years ago, colonizing from the Nile. Dramatic changes in water level encouraged species formation. As the lake rose, it flooded new areas and opened up new habitats. Many of the species may have originated after the lake dried down 14,000 years ago, isolating local populations in small lakes until the water level rose again.

Cichlid Diversity

Cichlids are small, perchlike fishes ranging from 5 to 25 centimeters in length, and the males come in endless varieties of colors. The ecological and morphological diversity of these fish is remarkable, particularly given the short span over which they have evolved. We can gain some sense of the vast range of types by looking at how different species eat. There are mud biters, algae scrapers, leaf chewers, snail crushers, zooplankton eaters, insect eaters, prawn eaters, and fish eaters. Snail shellers pounce on slow-crawling snails and spear their soft parts with long, curved teeth before the snail can retreat into its shell. Scale scrapers rasp slices of scales off other fish. There are even cichlid species that are "pedophages," eating the young of other cichlids.

Cichlid fish have a remarkable trait that may have been instrumental in their evolutionary radiation: A second set of functioning jaws occurs in the throats of cichlid fish (figure 23.14)! The ability of these jaws to manipulate and process food has freed the oral jaws to evolve for other purposes, and the result has been the incredible diversity of ecological roles filled by these fish.

Abrupt Extinction

Recently, much of the cichlid radiation has disappeared. In the 1950s, the Nile perch, a commercial fish with a voracious appetite, was introduced on the Ugandan shore of Lake Victoria. Since then, it has spread through the lake, eating its way through the cichlids. By 1990, many of the open-water cichlid species had become extinct, as well as others living in rocky shallow regions. Over 70% of all the named Lake Victoria cichlid species had disappeared, as well as untold numbers of species that had yet to be described.

Very rapid speciation occurred among cichlid fishes isolated in Lake Victoria, but widespread extinction followed with the introduction of a predator into the lake.

FIGURE 23.14
Cichlid fishes of Lake Victoria. These fishes have evolved adaptations to use a variety of different habitats. The second set of jaws located in the throat of these fish has provided evolutionary flexibility, allowing oral jaws to be modified in many ways.

484 Part IV Evolution

New Zealand Alpine Buttercups

Adaptive radiations such as those we have described in Galápagos finches, Hawaiian *Drosophila*, and cichlid fishes seem to be favored by *periodic isolation*. For example, finches and *Drosophila* invade new islands, local species evolve, and they in turn reinvade the home island, in a cycle of expanding diversity. Similarly, cichlids become isolated by falling water levels, evolving separate species in isolated populations that later are merged when the lake's water level rises again.

A clear example of the role periodic isolation plays in species formation can be seen in the alpine buttercups (genus *Ranunculus*) that grow among the glaciers of New Zealand (figure 23.15). More species of alpine buttercup grow on the two islands of New Zealand than in all of North and South America combined. Detailed studies by the Canadian taxonomist Fulton Fisher revealed that the evolutionary mechanism responsible for inducing this diversity is recurrent isolation associated with the recession of glaciers. The 14 species of alpine *Ranunculus* occupy five distinctive habitats within glacial areas: *snowfields* (rocky crevices among outcrops in permanent snowfields at 2130 to 2740 meters elevation); *snowline fringe* (rocks at lower margin of snowfields between 1220 and 2130 m); *stony debris* (slopes of exposed loose rocks at 610 to 1830 m); *sheltered situations* (shaded by rock or shrubs at 305 to 1830 m); and *boggy habitats* (sheltered slopes and hollows, poorly drained tussocks at elevations between 760 and 1525 m).

Ranunculus speciation and diversification have been promoted by repeated cycles of glacial advance and retreat. As the glaciers retreat, populations become isolated on mountain peaks, permitting speciation (figure 23.16). In the next advance, these new species can expand throughout the mountain range, coming into contact with their close relatives. In this way, one initial species could give rise to many descendants. Moreover, on isolated mountaintops during glacial retreats, species have convergently evolved to occupy similar habitats; these distantly related but ecologically similar species have then been brought back into contact in subsequent glacial advances.

FIGURE 23.15
A New Zealand alpine buttercup. Fourteen species of alpine *Ranunculus* grow among the glaciers and mountains of New Zealand, including this *R. lyallii*, the giant buttercup.

Recurrent isolation promotes species formation.

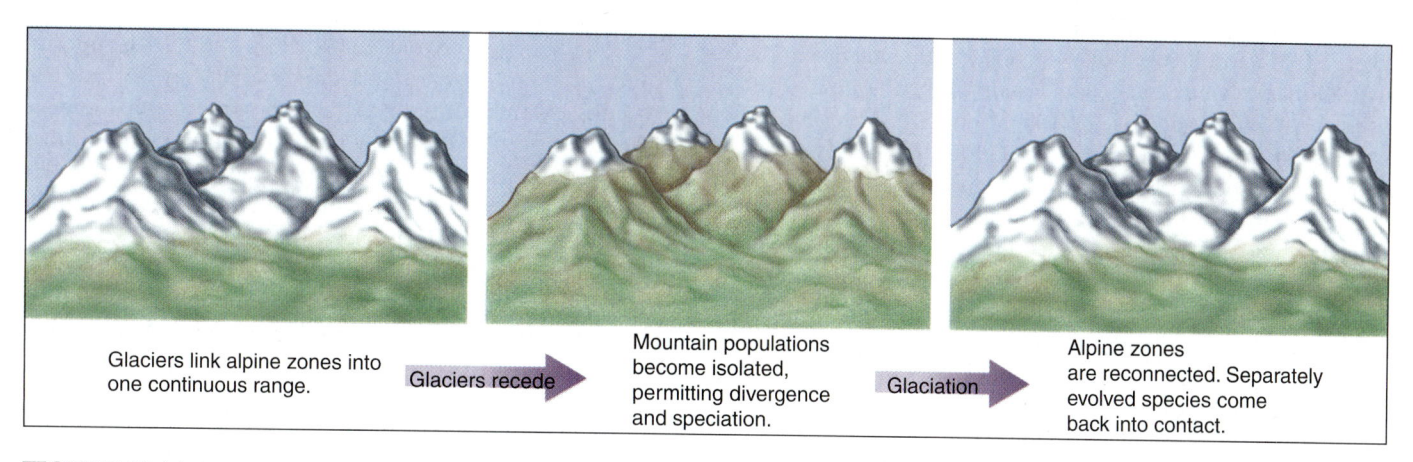

FIGURE 23.16
Periodic glaciation encouraged species formation among alpine buttercups in New Zealand. The formation of extensive glaciers during the Pleistocene epoch linked the alpine zones (*white*) of many mountains together. When the glaciers receded, these alpine zones were isolated from one another, only to become reconnected with the advent of the next glacial period. During periods of isolation, populations of alpine buttercups diverged in the isolated habitats.

The Pace of Evolution

We have discussed the manner in which speciation may occur, but we haven't yet considered the relationship between speciation and the evolutionary change that occurs within a species. For more than a century after the publication of *On the Origin of Species*, the standard view was that evolutionary change occurred extremely slowly. Such change would be nearly imperceptible from generation to generation, but would accumulate such that, over the course of thousands and millions of years, major changes could occur. This view is termed **gradualism** (figure 23.17*a*).

Evolution in Spurts?

Gradualism was challenged in 1972 by paleontologists Niles Eldredge of the American Museum of Natural History in New York and Stephen Jay Gould of Harvard University, who argued that species experience long periods of little or no evolutionary change (termed **stasis**), punctuated by bursts of evolutionary change occurring over geologically short time intervals. They called this phenomenon **punctuated equilibrium** (figure 23.17*b*), and argued that these periods of rapid change occurred only during the speciation process.

Initial criticism of the punctuated equilibrium hypothesis focused on whether rapid change could occur over short periods of time. As we have seen in the last two chapters, however, when natural selection is strong, rapid and substantial evolutionary change can occur. A more difficult question involves the long periods of lack of change: Why would species exist for thousands, even millions, of years without changing? Although a number of possible reasons have been suggested, most researchers now believe that a combination of stabilizing and oscillating selection is responsible for stasis. If the environment does not change over long periods of time, or if environmental changes oscillate back-and-forth, then selection may favor stasis, even for long periods of time. One factor that may enhance this stasis is the ability of species to shift their ranges; for example, during the ice ages, when the global climate cooled, the geographic ranges of many species shifted southward, so that the species continued to experience similar environmental conditions.

Eldredge and Gould's proposal prompted a great deal of research. Some well-documented groups, such as African mammals, clearly have evolved gradually, not in spurts. Other groups, such as marine bryozoa, seem to show the irregular pattern of evolutionary change predicted by the punctuated equilibrium model. It appears, in fact, that gradualism and punctuated equilibrium are two ends of a continuum. Although some groups appear to have evolved solely in a gradual manner and others only in a punctuated mode, many other groups show evidence of both gradual and punctuated episodes at different times in their evolutionary history. However, the idea that speciation is necessarily linked to phenotypic change has not been supported: It is now clear that speciation can occur without phenotypic change, and that phenotypic change can occur within species in the absence of speciation.

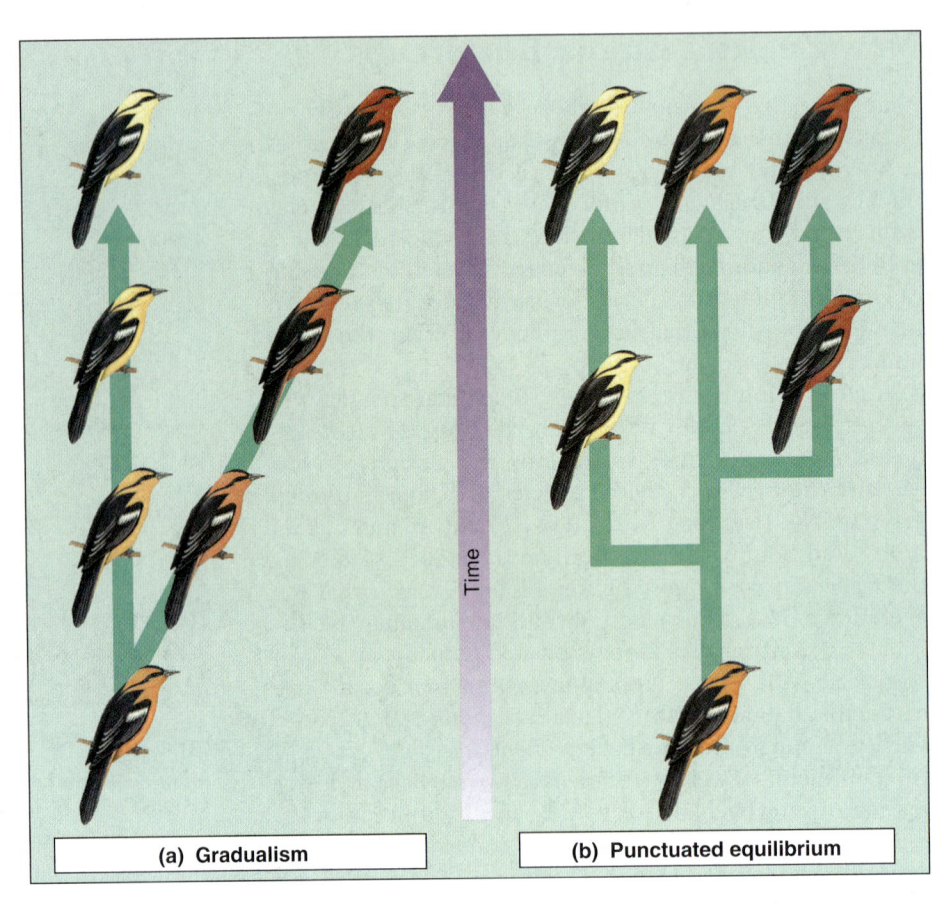

FIGURE 23.17
Two views of the pace of macroevolution. (*a*) Gradualism suggests that evolutionary change occurs slowly through time and is not linked to speciation, whereas (*b*) punctuated equilibrium surmises that phenotypic change occurs in bursts associated with speciation, separated by long periods of little or no change.

(a) Gradualism
(b) Punctuated equilibrium

> Evolutionary change can be slow and gradual. Or it may be rapid and discontinuous, separated by long periods of stasis that may result from a combination of stabilizing and oscillating selection.

Speciation and Extinction Through Time

Biological diversity has increased vastly since the Cambrian period. However, the trend has been far from consistent (figure 23.18). After a rapid rise, diversity reached a plateau for about 200 million years, but since then has risen steadily. Because changes in the number of species reflect the rate of origin of new species relative to the rate at which existing species disappear, this long-term trend reveals that speciation has, in general, outpaced extinction.

Nonetheless, speciation has not always outpaced extinction. In particular, interspersed in the long-term increase in species diversity have been a number of sharp declines, termed **mass extinctions.** Five major mass extinctions have been identified, the most severe one occurring at the end of the Permian period, approximately 250 million years ago. At that time, more than half of all families and as many as 96% of all species may have perished.

The most famous and well-studied extinction, though not as drastic, occurred at the end of the Cretaceous period (65 million years ago), at which time the dinosaurs and a variety of other organisms went extinct. Recent findings have supported the hypothesis that this extinction event was triggered when a large asteroid slammed into the earth, perhaps causing global forest fires and obscuring the sun for months by throwing particles into the air. This mass extinction did have one positive effect, though: Once the dinosaurs disappeared, mammals, which previously had been small and inconspicuous, diversified explosively, ultimately producing a wide variety of organisms, including elephants, tigers, whales, and humans.

Although species diversity rebounds after mass extinctions, the recovery is not rapid. Examination of the fossil record indicates that rates of speciation do not immediately increase after an extinction pulse, but rather take about 10 million years to reach their maximum. As a result, species diversity may require 10 million years, or even much longer, to attain its previous level.

A Sixth Extinction

The number of species in the world in recent times is greater than it has ever been. Unfortunately, that number is decreasing at an alarming rate due to human activities (see chapter 56). Some estimate that as many as one-fourth of all species will become extinct in the near future, a rate of extinction not seen on earth since the Cretaceous mass extinction. Moreover, the rebound in species diversity may be even slower than in previous mass extinction events because, instead of the ecologically depauperate, energy-rich environment that has occurred in the aftermath of previous events, a large proportion of the world's resources will be already taken up by human activities.

> **The number of species has increased through time, although not at a constant rate. Several major extinction events have substantially, though briefly, reduced the number of species. Diversity rebounds, but the recovery is not rapid, and the organisms making up that diversity are not the same as those that existed before the extinction event.**

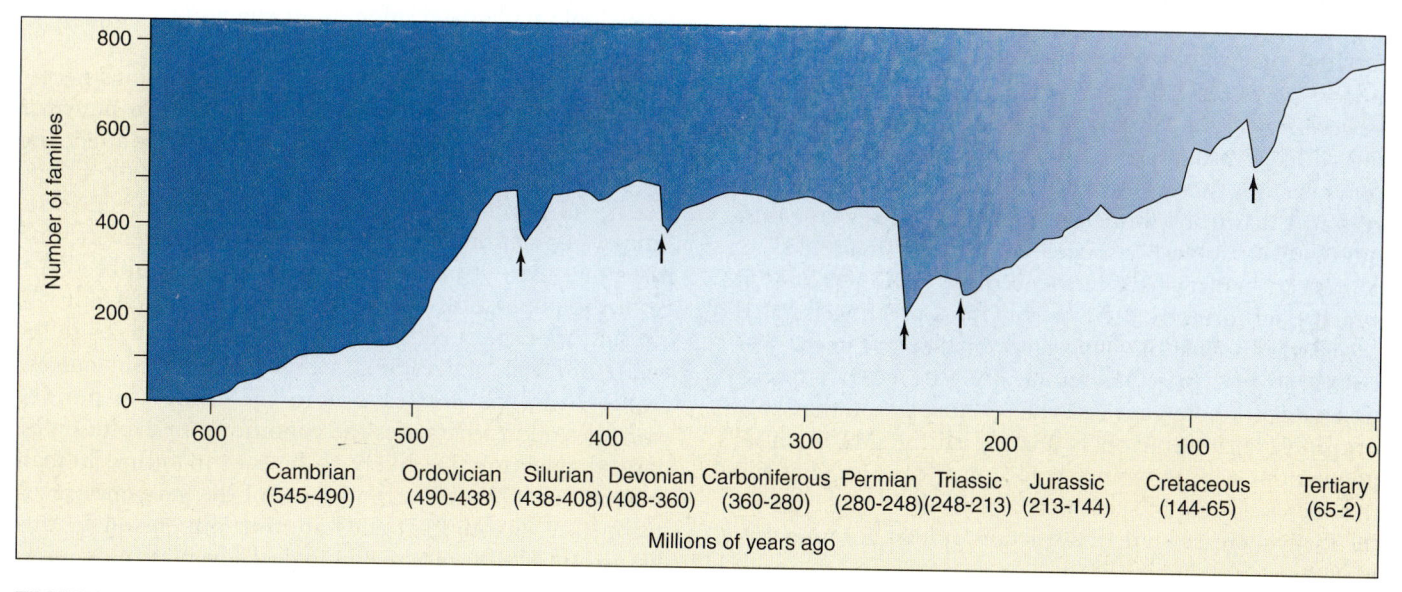

FIGURE 23.18
Diversity through time. The taxonomic diversity of families of marine animals has increased since the Cambrian period, although occasional dips have occurred. The fossil record is most complete for marine organisms because they are more readily fossilized than terrestrial species. Families are shown, rather than species, because many species are known from only one specimen, thus introducing error into estimates of the time of extinction. Arrows indicate the five major mass extinction events.

The Future of Evolution

In this chapter and chapters 21 and 22, we have discussed the results of evolution through time. What does the future hold? As mentioned on the previous page and discussed in greater detail in chapters 56 and 57, global biodiversity seems headed for a major extinction event from which recovery will be slow. Does this mean the end of evolution? We can use what we know about evolutionary processes to predict how evolution will occur in the future, both for diversity in general and for the human species in particular.

The Future Operation of Evolutionary Processes

Human impacts on the environment will affect the evolutionary process in many ways. Most obviously, by changing the environment, humans are changing the patterns of natural selection. In many cases, these changes will be so drastic that populations will be unable to adapt. But for those species that can survive, natural selection will act on genetic variation to produce evolutionary change. Global climate change, in particular, will be a major challenge, leading either to evolutionary change or extinction for many species.

Other factors will also lead to evolutionary change. Decreased population sizes will increase the likelihood of genetic drift, and geographic isolation of formerly connected populations will remove the homogenizing effect of gene flow, allowing these populations to evolve differences as adaptations to local environments. Chemicals and radiation released into the environment could even increase the mutation rate.

Consequently, for those species able to survive, evolutionary processes will continue and in some cases will even be accelerated. But what about species diversity? Extinction rates are increasing vastly, but it is also possible that speciation rates may increase, at least in some circumstances. The reason is that many formerly widespread species now exist only as geographically isolated populations (figure 23.19). Moreover, humans have introduced species to localities in which they formerly didn't occur, thus creating isolated populations. Given the importance of allopatry in the speciation process, these actions are likely to increase speciation rates for some species. This is not to say that geographic fragmentation is a good thing: Many small populations will go extinct before they can speciate, and this increased rate of speciation is unlikely to make up for the vastly increased rate of extinction, at least not for a very long time.

Human Evolutionary Future

Many science fiction writers have speculated on where evolution will take the human species, but consideration of the evolutionary process suggests that these ideas are fanciful.

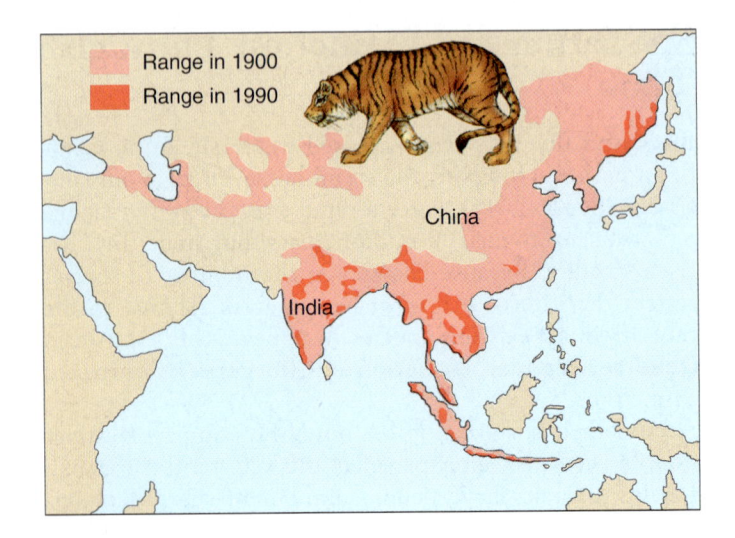

FIGURE 23.19
Tigers now exist in geographically isolated populations. Human activities, such as hunting and habitat destruction, have greatly reduced populations of the tiger, and have fragmented the species' range into many small and isolated populations.

In recent times, the movement of people around the globe has begun to erase human differences, a clear example of the homogenizing effect of gene flow. Moreover, to an ever-increasing extent, people with different ethnic origins are intermating, further diminishing differentiation among human populations. Because of the huge size of the human population, genetic drift is unlikely to be very important. Assuming that mutations don't increase too much, that leaves only natural selection as an engine of evolutionary change in humans.

The question then becomes: Do the conditions necessary for evolution by natural selection exist in humans? That is, are there phenotypic traits that both affect the number of surviving offspring and are genetically inherited from parent to offspring? Certainly, in one way, human populations will evolve because many genetic diseases that used to be fatal, and would thus remove alleles from the population, can now be treated successfully. As a result, we can expect the frequency of such alleles to increase in future generations. Other than such obvious examples, though, we leave it to the reader to ponder whether and in what cases the conditions for evolution by natural selection are likely to be met in future human populations. Of course, the advent of the genomics revolution (see chapter 17) adds another dimension to this discussion. Will future technological advances allow us to alter the human gene pool directly? And, if so, will that be a good idea?

> For those species able to avoid extinction, human-caused changes should lead to evolutionary adaptation and, in some cases, to speciation.

Concept Review

For interactive testing, visit the Online Learning Center with PowerWeb at www.mhhe.com/Raven7

23.1 Species are the basic units of evolution.

The Nature of Species
- Any species concept must be able to account for species that occur together and for populations of the same species that are geographically separated. (p. 472)
- Although sympatric species occur together, they are usually phenotypically different and utilize different parts of the shared habitat. (p. 472)
- Ernst Mayr developed the biological species concept which defines species as: "groups of actually or potentially interbreeding natural populations which are reproductively isolated from other such groups." (p. 473)
- Reproductive isolating mechanisms prevent genetic exchange between species. (p. 473)

23.2 Species maintain their genetic distinctiveness through barriers to reproduction.

Prezygotic Isolating Mechanisms
- Prezygotic isolating mechanisms prevent the formation of zygotes. (p. 474)
- Two species may not hybridize because they occupy different portions of an area and do not encounter each other. (p. 474)
- Related species often maintain distinctiveness due to behavioral differences such as courtship rituals. (p. 474)
- Sympatric species may utilize visual and sensory communication, pheromone reception, and electroreception to avoid mating with the wrong species. (pp. 474–475)
- Temporal isolation, mechanical isolation, and prevention of gamete fusion are other types of prezygotic isolating mechanisms. (p. 475)

Postzygotic Isolating Mechanisms
- Postzygotic isolating mechanisms prevent proper functioning of zygotes after they form. (p. 476)
- Hybrids that are weaker than their parents will almost certainly be eliminated in nature. (p. 476)
- Some hybrids are healthy, but infertile. (p. 476)

Problems with the Biological Species Concept
- Hybridization is more common than previously believed. (p. 477)
- Reproductive isolation may not be the only force maintaining species integrity. (p. 477)
- An alternative explanation is that species distinctions are maintained by natural selection. (p. 477)
- Due to extreme diversity, no single species definition may be adequate. (p. 477)

23.3 We have learned a great deal about how species form.

Reproductive Isolation May Evolve as a By-Product of Evolutionary Change
- Partial reproductive isolation might lead to reinforcement. Individuals in parental populations that avoid hybridization would be more successful at transmitting their genes to future generations. This would lead to continual improvement of prezygotic isolating mechanisms until the populations are completely reproductively isolated. (p. 478)
- Partial isolation might also lead to the loss of genetic distinctiveness due to gene flow. (pp. 478–479)
- Genetic drift may also cause reproductive isolation, but such isolation usually evolves more rapidly in the presence of selective pressures. (p. 479)

The Geography of Speciation
- In order for speciation to occur, similar populations must diverge, and then reproductive isolation must evolve. (p. 480)
- Allopatric populations are much more likely to diverge into separate species. (p. 480)
- Polyploidy, either by autopolyploidy or allopolyploidy, leads to immediate sympatric speciation in many plants and some animals. (pp. 480–481)
- Disruptive selection may lead to sympatric speciation, but this is controversial. (p. 481)

23.4 Clusters of species reflect rapid evolution.
- Adaptive radiation occurs when groups of closely related species evolve from a common ancestor by adapting to different habitats. (p. 482)

Hawaiian *Drosophila*
- When the ancestors of *Drosophila* reached the Hawaiian Islands, they encountered many empty habitats. Rampant speciation and ecological opportunity led to intense insect diversity. (p. 482)

Darwin's Finches
- Darwin's finches are derived from a single mainland species. The 14 species found on the islands comprise four groups: ground finches, tree finches, warbler finches, and vegetarian finches. (p. 483)

Lake Victoria Cichlid Fishes
- Many of the varied species originated after the lake dried down and isolated local populations. (p. 484)
- Ecological and morphological diversity was quite high. (p. 484)
- Widespread extinction occurred after the predatory Nile perch was introduced into the lake. (p. 484)

New Zealand Alpine Buttercups
- *Ranunculus* speciation and diversification have been fostered by repeated cycles of glacial advance and retreat. (p. 485)

The Pace of Evolution
- Darwin advocated slow evolutionary change (gradualism). (p. 486)
- Niles Eldredge and Stephen Jay Gould argued that species went through long periods of little change (stasis) and then experienced rapid bursts of evolutionary change over short periods of time during speciation events. (p. 486)

Speciation and Extinction Through Time
- Over the long term, speciation has outpaced extinction. (p. 487)
- Five major mass extinctions have interspersed the long-term increase, and have been followed by large-scale, although slow, periods of diversification. (p. 487)
- Human activities may be leading to a sixth major extinction. (p. 487)

The Future of Evolution
- Humans are changing the patterns of natural selection, and some of these changes will be so drastic that many species will not be able to adapt quickly enough to survive. (p. 488)
- Decreased population size and increased geographic isolation could also remove the homogenizing effect of gene flow and lead to moderate to high levels of speciation in species that are able to avoid extinction. (p. 488)

Test Your Understanding

For interactive testing, visit the Online Learning Center with PowerWeb at www.mhhe.com/Raven7

Self Test

1. Prezygotic isolating mechanisms include all of the following except
   a. hybrid sterility.
   b. courtship rituals.
   c. habitat separation.
   d. seasonal reproduction.
2. Which of the following is an example of mechanical isolation?
   a. Two species of birds live in the same habitat; one mates in spring and the other in summer.
   b. Two species of frogs have different mating calls.
   c. The flower structure of one species prevents the transfer of pollen from another species.
   d. One species of lizards inhabits the trees, and another species inhabits the ground cover.
3. _____ isolating mechanisms include improper development of hybrids and failure of hybrids to become established in nature.
   a. Prezygotic
   b. Postzygotic
   c. Temporal
   d. Mechanical
4. Reproductive isolation and the evolution of species could occur through which of the following?
   a. founder effect
   b. reinforcement
   c. adaptation
   d. all of these
5. Speciation occurs most frequently in populations that are
   a. sympatric.
   b. undergoing disruptive selection.
   c. allopatric.
   d. not geographically separated.
6. The large number of Hawaiian *Drosophila* species has likely resulted from
   a. adaptive radiation.
   b. a single common ancestor.
   c. geographic isolation.
   d. all of these.
7. The finch species of the Galápagos Islands are grouped according to their food sources. Which of the following is *not* a finch food source?
   a. seeds
   b. carrion
   c. insects
   d. tree buds
8. Cichlid diversity can be attributed to
   a. adaptive radiation.
   b. new habitats and geographic isolation.
   c. a second set of jaws in the throat of the fish.
   d. All of these factors contributed to cichlid diversity.
9. The hypothesis that evolution occurs in spurts, with great amounts of evolutionary change followed by periods of stasis, is
   a. punctuated equilibrium.
   b. allopatric speciation.
   c. gradualism.
   d. Hardy–Weinberg equilibrium.
10. Biological diversity through time has
    a. gradually increased.
    b. been constant.
    c. increased overall despite periodic drops.
    d. both increased and decreased with no overall change.

Test Your Visual Understanding

1. In all of the examples in the figure, one interbreeding population has been divided, which results in two or more geographically isolated populations. Over time, these isolated populations can undergo little or no evolutionary change, can undergo speciation, or can become extinct. Explain under what conditions each scenario can occur:
   a. Population undergoes little or no evolutionary change.
   b. Population undergoes speciation.
   c. Population becomes extinct.

Applying Your Knowledge

1. Adaptive radiation results when an ancestral species gives rise to many descendants which are adapted to many different parts of the environment. How would scenarios for adaptive radiation differ if speciation occurred allopatrically versus sympatrically?
2. Polyploid animals are far less common than polyploid plants. Why do you think this might be so? (*Hint:* Refer to the discussion of nondisjunction in chapter 13).

24
Evolution of Genomes and Developmental Mechanisms

Concept Outline

24.1 Evolutionary history is written in genomes.

Comparative Genomics. Comparisons among genomes are addressing the differences among genomes and what those differences mean in terms of genome evolution.

Origins of Genomic Differences. In addition to single genes mutating, whole genomes, large chunks of DNA, and individual genes have all duplicated over evolutionary time.

24.2 Developmental mechanisms are evolving.

Evolution of Development. The fields of evolution and development are intersecting to address common questions in light of growing information about the genetic and genomic bases of both evolution and development. Comparative genomics can identify genes with similar sequences, but actual experiments are needed to test the function of similar genes in different organisms.

Diversity of Eyes in the Natural World. Eyes appear to have evolved independently multiple times based on morphological traits. On the other hand, the *Pax6* gene has been found to initiate eye development in organisms as far-ranging as ribbon worms, fruit flies, and mice.

FIGURE 24.1
Frogs without polliwogs. The Puerto Rican tree frog, *Eleutheradactylus coqui*, develops on land in a huge egg. There is no tadpole stage, and an adult frog emerges from the egg. How the tree frog lost its tadpole is a fascinating evolutionary question.

How have complex traits evolved? Many clues are hidden in genomes. As more genomes are sequenced, the field of comparative genomics is yielding some surprising answers and many, many questions. Genetic and genomic tools make it possible to investigate how development has evolved, yielding novel phenotypes built from highly conserved gene families. How is it that sister species of frogs can have completely different patterns of development (figure 24.1)? One frog goes from fertilized egg to adult frog with no intermediate tadpole stage. The sister species has an extra developmental stage neatly slipped in between early development and the formation of limbs—the tadpole stage. Genomic findings are accentuating the biological paradox that many genes are highly conserved, and the tremendous diversity of life shares, at some point, a common ancestral genome.

24.1 Evolutionary history is written in genomes.

Comparative Genomics

A key challenge of modern evolutionary biology is to find a way to link the evolution of DNA sequences, which we are now able to study in great detail, with the evolution of the complex morphological characters used to construct a traditional phylogeny. Many different genes contribute to complex characters, such as feathers, and making the connection between a specific change in one of the genes and a modification in a morphological character is particularly difficult. Comparing genomes (entire DNA sequences) of different species provides a powerful new tool to explore these relationships. Genomes are more than instruction books for building and maintaining an organism; they contain vast amounts of information on the history of life. As you saw in chapter 17, the growing number of fully sequenced genomes in all kingdoms is leading to a revolution in comparative evolutionary biology (table 24.1). It is now possible to explore the genetic differences between species in a very direct way, examining one-by-one the footprints on the evolutionary path between different species.

Over 100 different prokaryotic genomes have been completed, and at least 18 different eukaryotic genomes have

Table 24.1 Milestones for Comparative Eukaryotic Genomics

Organism	Estimated Genome Size (Mb)	Estimated Number of Genes	Year Sequenced
VERTEBRATES			
Homo sapiens (human)	3,200	30,000	2001
Mus musculus (mouse)	2,500	30,000	2002
Fugu rubripes (pufferfish)	365	33,609	2002
INVERTEBRATES			
Drosophila melanogaster (fruit fly)	137	13,600	2000

Table 24.1 Milestones for Comparative Eukaryotic Genomics *Continued*

Organism	Estimated Genome Size (Mb)	Estimated Number of Genes	Year Sequenced
Anopheles gambiae (mosquito)	278	46,000–56,000	2002
FUNGI			
Schizosaccharomyces pombe (fission yeast)	13.8	4,824	2002
Saccharomyces cerevisiae (brewer's yeast)	12.7	5,805	1997
PLANTS			
Arabidopsis thaliana (wall cress)	125	25,498	2000
Oryza sativa (rice)	430	32,000–55,000	2002
PROTISTS			
Plasmodium falciparum (malaria parasite)	23	5,300	2002

been or are being sequenced. As the draft (preliminary) sequences of these and other genomes become available in the next few years, our view of the evolution of life on earth should become far clearer and our knowledge of phylogenetic relationships far more certain. The few genomic sequences that are already completed give exciting clues as to what is to come.

The Tiger Pufferfish

The draft sequence of the tiger pufferfish (*Fugu rubripes*) was completed in 2002, only the second vertebrate genome to be fully sequenced. For the first time, we were able to compare the genomes of two vertebrates. Some human and pufferfish genes have been conserved during 450 million years of evolution, while other genes are unique to each species. About 25% of human genes have no counterparts in *Fugu*. Also, extensive rearrangements have occurred during the 450 million years since mammals and teleost fish diverged, indicating a considerable scrambling of gene order. Finally, the human genome has more repetitive DNA, which counts for less than one-sixth of the *Fugu* sequence.

The pufferfish genome, 365 million base-pairs (Mb) in length, has only one-ninth of the DNA of humans, although both vertebrate species have approximately the same number of genes. Why do humans have so much extra DNA? Much of it appears to be in the form of introns that are substantially bigger than those in pufferfish. The *Fugu* genome has a handful of "giant" genes containing long introns; studying them should provide insight into the evolutionary forces that have driven the change in genome size during vertebrate evolution.

Sequences that are conserved between humans and pufferfish provide valuable clues for understanding the genetic basis of many human diseases. Amino acids critical to protein function tend to be preserved over the course of evolution, and changes at such sites within genes are more likely to cause disease. It is difficult to distinguish functionally conserved sites in the protein sequences of humans when comparing human proteins with those of other mammals because not enough time has elapsed for sufficient changes to accumulate at nonconserved sites. Because the pufferfish genome is only distantly related to humans, conserved sequences are far more easily distinguished.

The Mouse

Later in 2002, a draft sequence of the mouse (*Mus musculus*) genome was completed by an international consortium of investigators, allowing for the first time a comparison of two mammalian genomes. The human genome has about 400 million more nucleotides than that of the mouse. A comparison of the two genomes reveals that both have about 30,000 genes, and they share the bulk of them; in fact, the human genome shares 99% of its genes with mice. Humans and mice diverged about 75 million years ago, too little time for many evolutionary differences to accumulate. There are only 300 genes unique to either organism, comprising about 1% of the genome. Most of the nearly 150 genes unique to mice are linked with either the sense of smell, which is highly developed in rodents, or with reproduction. Comparing these similar genomes is already yielding information on the function of genes. It is much easier to design experiments to identify gene function in an experimental system like the mouse than in humans. Once the mouse sequence was known, the function of 1000 previously unidentified human genes was understood.

The genomes of humans and mice are so similar that one wonders why mice and humans are so different. The best explanation for why a mouse develops into a mouse and not a human is that the genes are expressed at different times and possibly in different tissues. This may be the case for the cystic fibrosis gene, which has been identified in both species. Defects in the human cystic fibrosis gene cause especially devastating effects in the lungs, but mice with cystic fibrosis do not have lung symptoms.

Comparison of the mouse and human genomes reveals that since mice and humans last shared a common ancestor about 75 million years ago (MYA), mouse DNA has mutated about twice as fast as human DNA. This is a fascinating observation in search of an explanation. The difference in generation time between mice and humans could account for some of this distinct mutation rate, because mice would have had more opportunities to mix and match genomic components during meiosis.

Perhaps the most unexpected finding in comparing the mouse and human genomes lies in the similarities between the "junk" DNA, mostly retrotransposons (refer to chapter 17), in the two species. This DNA does not code for proteins. A survey of the location of retrotransposon DNA in both species shows that it has independently ended up in comparable regions of the genome. It's beginning to look like this "junk" DNA may have more of a function than was previously assumed. The possibility that it is rich in regulatory RNA sequences, such as those described in chapter 18, is being actively investigated. In one such study, researchers collected almost all of the RNA transcripts made by mouse cells taken from every tissue. While most of the transcripts code for mouse proteins, as many as 4280 could not be matched to any known mouse protein. This suggests that a large part of the transcribed genome consists of nonprotein-encoding genes—that is, transcripts that function as RNA. Perhaps this can explain why a single retrotransposon can cause heritable differences in coat color in mice.

A draft of the rat genome has just been completed, and even more exciting news about the evolution of mammalian genomes may come forth. One of the most exciting aspects of comparing rat and mice genomes is the potential to capitalize on the extensive research on rat physiology, especially heart disease, and the long history of genetics in mice. Linking genes to disease just became easier.

Variation in the organization of genomes is as intriguing as gene sequence differences. Over long segments of chromosomes, the linear order of mouse and human genes is the same—the common ancestral sequence has been preserved in both species. This **conservation of synteny** (see chapter 17) was anticipated from earlier gene mapping studies, and provides strong evidence that evolution actively shapes the organization of the mammalian genome.

The Chimpanzee

The chimpanzee genome project is still under way. Humans and chimps diverged from a common ancestor only about 5 million years ago. This is too little time for much genetic differentiation to evolve between the two species, but enough for significant morphological and behavioral differences to have evolved. Preliminary sequence comparisons indicate that chimp DNA is 98.7% identical to human DNA. If just the gene sequences encoding proteins are considered, the similarity increases to 99.2%. How could two species differ so much in body and behavior, and yet have almost equivalent sets of genes?

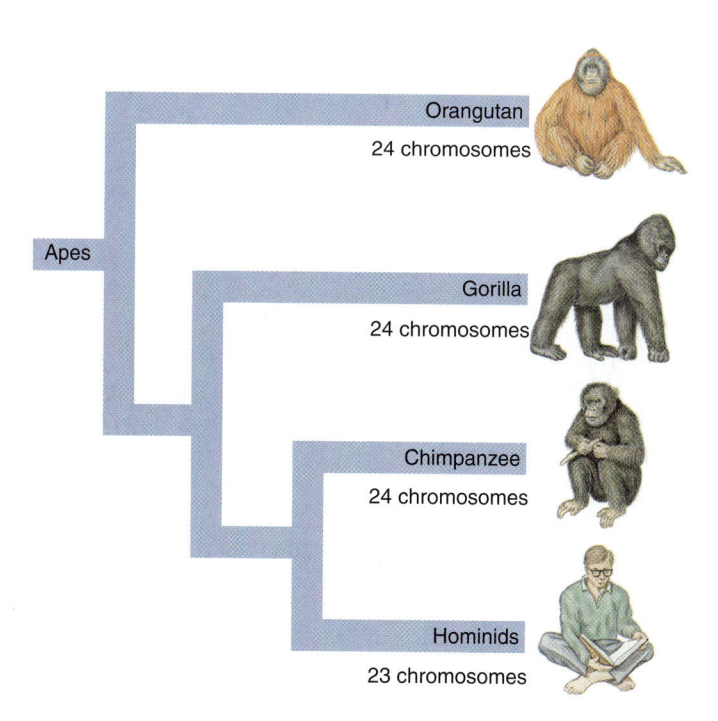

One potential answer to this question is based on the observation that chimp and human genomes show very different patterns of gene transcription activity, at least in brain cells. Investigators used microarrays (see Fig. 17.10) containing up to 18,000 human genes. Fluid extracted from living brain cells that contained expressed RNA transcripts was washed over the array of genes. Each gene lights up if a transcript of that gene is present in the fluid. The more copies of the gene, the more intense the signal. Because the chimp genome is so similar to that of humans, the microarray could detect the activity of chimp genes reasonably well. While the same genes were transcribed in chimp and human brain cells, the levels of transcription varied widely. It would seem that much of the difference between humans and chimps lies in which genes are transcribed, and when.

Humans have one fewer chromosome than chimpanzees, gorillas, and orangutans (figure 24.2). It's not that we have lost a chromosome. Rather, at some point in time, two mid-sized ape chromosomes fused to make what is now human chromosome 2, the second largest chromosome in our genome.

The fusion leading to human chromosome 2 is an example of the sort of genome reorganization that has occurred in many species. Rearrangements like this can provide evolutionary clues, but are not always definitive proof of how closely related two species are. Consider the organization of known **orthologs** (that is, genes with the same ancestral sequence) shared by humans, chickens, and mice. One study estimated that 72 chromosome arrangements had occurred since the chicken and human last shared a common ancestor. This is substantially less than the estimated 128 rearrangements between chicken and mouse or 171 between mouse and human. Does this mean that chickens and humans are more closely related than mice and humans or mice and chickens? No, what these data actually show is that chromosome rearrangements have occurred at a much lower frequency in humans and chickens than in mice. Chromosomal rearrangements in mice seem to have occurred at twice the rate seen in humans.

This finding of marked differences in the rate of chromosomal rearrangement among vertebrates raises new questions about genome evolution that are currently being explored. Identifying genomes that have undergone relatively slow chromosome change is most helpful in reconstructing the hypothetical genomes of ancestral vertebrates. If regions of chromosomes have changed little in distantly related vertebrates over the last 300 million years, it is reasonable to hypothesize that the common ancestor had genomic similarities.

FIGURE 24.2
Living great apes. All living great apes, with the exception of humans, have a haploid chromosome number of 24. Humans have not lost a chromosome; rather, two smaller chromosomes fused to make a single chromosome.

Insects: *Drosophila* and *Anopheles*

The insects are the most species-rich and morphologically diverse animal group on earth. Two insect genomes have been sequenced. *Drosophila melanogaster* has been a laboratory model for genetic studies for much of the last century, and is arguably the best-understood gene system in biology. *Anopheles gambiae*, the malaria mosquito, along with *Plasmodium falciparum*, the protistan parasite it transmits, have together caused an enormous impact on the world's health, resulting in 1.7 to 2.5 million deaths each year. The genomes of both *Anopheles* and *Plasmodium* were sequenced in 2002.

The fruit fly *Drosophila* and the mosquito *Anopheles* are separated by approximately 250 million years of evolution, and appear to have evolved more rapidly over that interval than vertebrates. The extent of similarity between these two insects is comparable to that between humans and pufferfish, which diverged 450 million years ago. The organization of genes on the chromosomes has undergone significant shuffling between the two insect species. Interestingly, *Drosophila* exhibits less noncoding DNA than *Anopheles*, although the evolutionary force driving this reduction in noncoding regions is not clear.

The Protist *Plasmodium*

The parasitic protist *Plasmodium falciparum*, which causes malaria, has a relatively small genome of 24.6 million base-pairs that proved very difficult to sequence. It has an unusually high proportion of adenine and thymine, making it hard to distinguish one portion of the genome from the next. The project took five years to complete. *P. falciparum* appears to have about 5300 genes, with those of related function clustered together, suggesting that they might share the same regulatory DNA.

P. falciparum is a particularly crafty organism that hides from our immune system inside red blood cells, regularly changing the proteins it presents on the surface of the red blood cell. This has made it particularly difficult to develop a vaccine or other treatment for malaria. Now, a link to chloroplast-like structures in *P. falciparum* has raised other possibilities for treatment. An odd subcellular component called the apicoplast, found only in *Plasmodium* and its relatives, appears to be derived from a chloroplast appropriated from algae consumed by the parasite's ancestor. Analysis of the *Plasmodium* genome reveals that about 12% of all the parasite's proteins, encoded by the nuclear genome, head for the apicoplast. These proteins act there to produce fatty acids. The apicoplast is the only place the parasite makes the fatty acids it needs to survive, suggesting that drugs targeted at this biochemical pathway might be very effective against malaria. Another possibility is to look at chloroplast-specific herbicides, which might kill *Plasmodium* by targeting the chloroplast-derived apicoplast.

Flowering Plants: Rice and *Arabidopsis*

Few plant genomes have been sequenced, the first being *Arabidopsis thaliana*, the wall cress, a tiny member of the mustard family often used as a model organism for studying plant molecular genetics and development. Its genome sequence, largely completed in 2000, revealed 25,948 genes, about as many as humans have.

Arabidopsis is mainly an experimental model, with no commercial significance. However, the second plant genome for which a draft sequence has been prepared—rice—is of enormous economic significance. *Oryza sativa* belongs to the grass family, which includes maize (corn), wheat, barley, sorghum, and sugarcane. Together, these crops provide most of the world's food and animal feed. Unlike most grasses, rice has a relatively small genome of 430 million base-pairs (the maize genome is 2500 Mb, and barley's is an enormous 4900 Mb). Two different subspecies of rice were sequenced, yielding similar results. The proportion of the rice nuclear genome devoted to repetitive DNA, for example, was 42% in one variety, and 45% in the other. Retrotransposons, the most numerous large repeats, account for more than 15% of the rice genome in each study.

The rice genome has proven to contain a surprisingly large number of genes. Both rice draft sequences place the gene number for *Oryza* higher than for any other genome yet sequenced. One study suggests 53,000 to 63,000 genes; the other, using more conservative criteria, indicates 33,000 to 50,000 genes. These numbers will become more precise as genome annotation continues.

More than 80% of the genes found in rice are also found in *Arabidopsis*. Among the other 20% must be the genes responsible for the many physiological and morphological differences between rice (a monocot) and *Arabidopsis* (a dicot), two very different kinds of flowering plants. About one-third of the genes in *Arabidopsis* and rice appear to be in some sense "plant" genes—that is, genes not found in any animal or fungal genome sequenced so far. These include the many thousands of genes involved in photosynthesis and photosynthetic anatomy. Among the other plant genes are many that are very similar to those found in animal and fungi genomes, particularly genes involved in basic intermediary metabolism, in genome replication and repair, and in protein synthesis.

Both rice and *Arabidopsis* have higher *copy numbers* for gene families (multiple slightly divergent copies of a gene) than are seen in animals or fungi, spread among different chromosomes amid clusters of other duplicated genes. This suggests that these plants have undergone numerous episodes of polyploidy and/or segmental duplication during the 150 to 200 million years since rice and *Arabidopsis* diverged from a common ancestor.

Genome sequences are being determined for a wide variety of organisms, leading to the new field of comparative genomics.

Origins of Genomic Differences

Comparison of even the limited number of genomes discussed in this chapter reveals that genomes evolve dynamically. The genome changes that provide the raw material for this evolution arise in at least six ways:

1. mutation of a single gene;
2. when regions of DNA duplicate;
3. when large chunks of chromosomes rearrange;
4. when individual chromosomes duplicate;
5. when whole genomes duplicate or combine with the genome of another species to create polyploids (cells with three or more complete copies of a genome);
6. when DNA from other species becomes integrated into genomes.

We will explore genome changes beginning with large-scale changes, including duplication of whole genomes.

Polyploidization

We will start our exploration of genome evolution with gene duplication. Duplication of entire genomes by polyploidization obviously represents a huge change in genomes. There is growing evidence that duplication of fewer genes in organisms has also been a major contributor to morphological diversity. In both cases, the duplicated genes have the opportunity to acquire new functions because the duplication provides a functioning backup gene.

Humans have nine times as much DNA as pufferfish because of longer introns, but why do tulips contain over 170 times as much DNA as the small weed *Arabidopsis thaliana*? Large-scale evolutionary change, including duplication of entire genomes, certainly provides at least part of the explanation (figure 24.3). Because meiosis requires an even number of chromosome sets, species with ploidy levels that are multiples of two can reproduce sexually. However, meiosis would be a disaster in a $3n$ organism such as the banana, since three sets of chromosomes can't be evenly divided between two cells. (Sketch out what would happen in meiosis in a $3n$ banana cell, referring back to chapter 12 if necessary.) Bananas have to rely on asexual means of propagation.

Polyploidy ($3n$ or greater) can result from either genome duplication in one species or from hybridization of two different species. Hybridization may be followed by genome duplication, providing an even number of chromosome sets and so allowing meiosis to occur. For example, bread wheat arose from two hybridization and whole-genome duplication events (figure 24.4). Such major changes in the genome have often resulted in new species.

Whole-genome duplication, however, is insufficient to explain the size of some genomes. Wheat and rice are very closely related and have similar gene content. Yet, the wheat genome is 40 times larger than the rice genome.

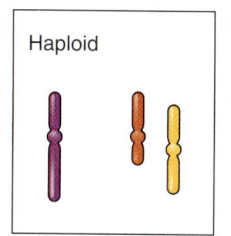

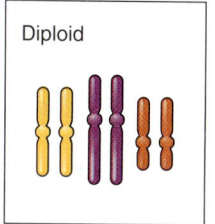

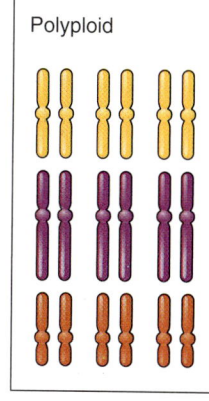

FIGURE 24.3
Chromosome numbers possible in plant genomes. *Haploid:* a set of chromosomes without their pairs; for example, the chromosome number present in a gamete. *Diploid:* a single set of chromosome pairs. *Polyploid:* multiple sets of chromosome pairs; for example, bananas have a triple set of chromosomes and are therefore polyploid.

This difference in genome size cannot be explained solely by the fact that bread wheat is a hexaploid ($6n$) and rice is a diploid ($2n$). The fact that the wheat genome contains lots of repetitive DNA has increased its DNA content, but not necessarily its gene content. Now that the rice genome is fully sequenced, attention has shifted to sequencing the other cereal grains, especially wheat. Comparisons between the rice and wheat genomes should provide clues about the genome of their common ancestor.

Segmental Duplication

One of the greatest sources of novel traits in genomes is duplication of segments of DNA. When a gene duplicates, the two most likely fates of the duplicate gene are: (1) losing function through subsequent mutation, and (2) gaining a novel function through subsequent mutation. In fact, most duplicate genes lose function, and the average half-life of duplicates is about 4 million years. This is a huge amount of time on a human scale, but very short in terms of evolutionary change. How then can researchers claim that gene duplication is a major evolutionary force for gene innovation (genes gaining new function)? One piece of evidence can be found by asking where in the genome gene duplication is most likely to occur. In humans, the highest rates of duplication have occurred in the three most gene-rich chromosomes of the genome. The seven chromosomes with the fewest genes also had the least amount of duplication. Remember, having fewer genes does not mean that there is less total DNA. Even more compelling, certain types of human genes were more likely to be duplicated: growth and development genes, immune system genes, and cell-surface receptors. About 5% of the human genome

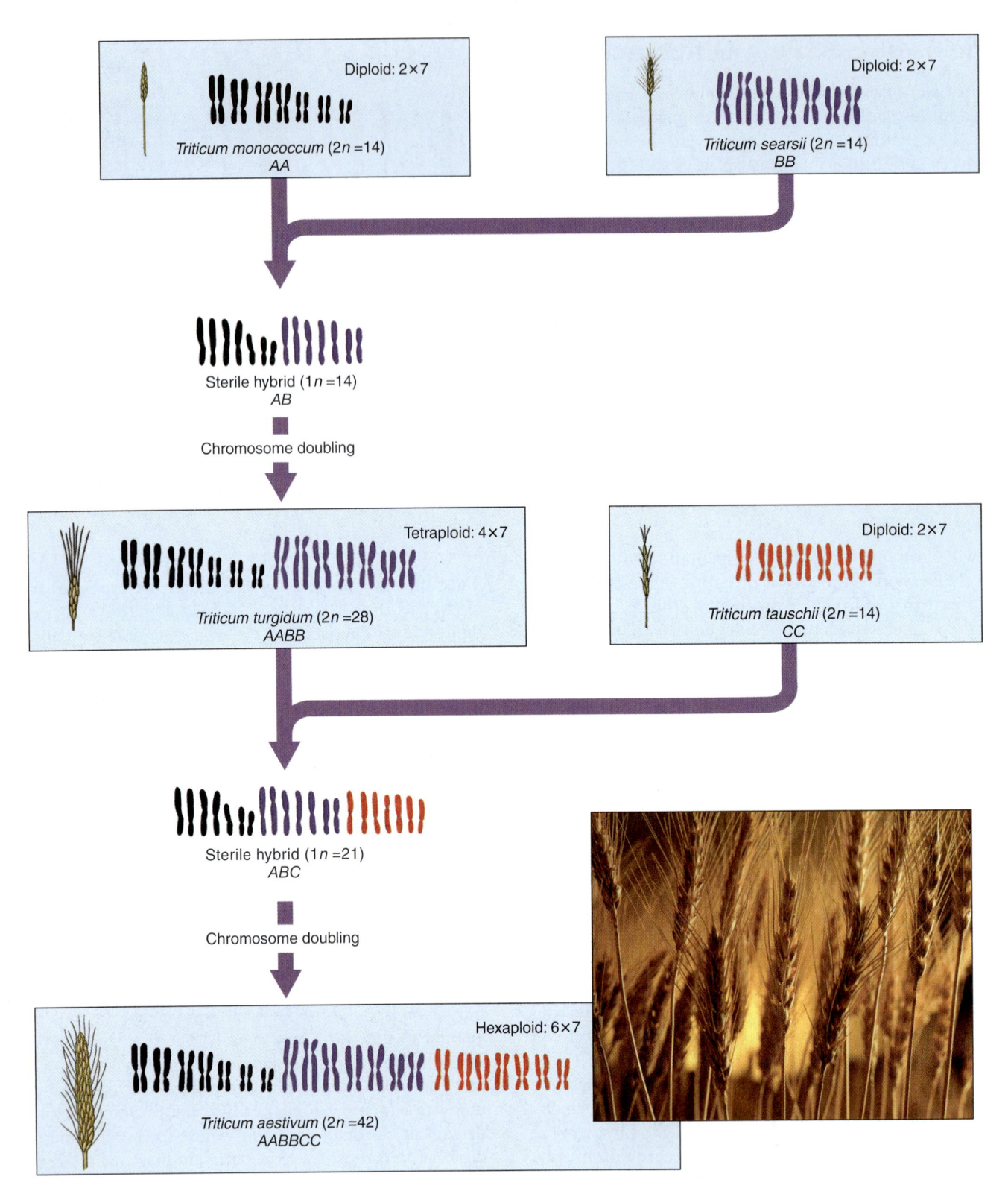

FIGURE 24.4
Evolutionary history of wheat. Domestic wheat arose in southwestern Asia in the hilly country of what is now Iraq. This region contains a rich assembly of grasses of the genus *Triticum*. Domestic wheat (*T. aestivum*) is a polyploid species of *Triticum* that arose through two so-called "allopolyploid" events. (*1*) Two different diploid species, symbolized here as *AA* and *BB*, hybridized to form an *AB* polyploid; the species were so different that *A* and *B* chromosomes could not pair in meiosis, so the *AB* polyploid was sterile. However, in some plants the chromosome number spontaneously doubled due to a failure of chromosomes to separate in meiosis, producing a fertile tetraploid species, *AABB*. This wheat is used in the production of pasta. (*2*) In a similar fashion, the tetraploid species *AABB* hybridized with another diploid species, *CC*, to produce, after another doubling event, the hexaploid *T. aestivum*, *AABBCC*. This bread wheat is commonly used throughout the world.

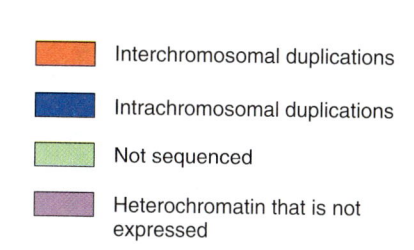

FIGURE 24.5

Segmental duplication on the human Y chromosome. Each red region has 98% sequence similarity with a sequence on a different human chromosome. Each blue region has 98% sequence similarity with a sequence elsewhere on the Y chromosome. The Y chromosome is one continuous piece of DNA shown in segments here for illustrative purposes only.

consists of segmental duplications (figure 24.5). As more species are compared, gene duplication rates appear to vary. *Drosophila* has about 31 new duplicates per genome per million years, which is equivalent to 0.0023 duplications per gene per million years. The rate is about 10 times faster for the nematode *Caenorhabditis elegans*. Two genes within an organism that arose from the duplication of one gene in an ancestor are called **paralogs**. Orthologs, as defined earlier, reflect the conservation of a single gene from a common ancestor.

Gene Inactivation

The loss of gene function is another important way genomes evolve. Consider the olfactory receptor (OR) genes that are responsible for our sense of smell. These genes code for receptors that bind odorants, initiating a cascade of signaling events that eventually lead to our perception of scents. Gene inactivation seems to have been the most frequent explanation for our reduced sense of smell relative to that of the great apes and other mammals. Primate genomes have over 1000 copies of OR genes (figure 24.6). An estimated 70% of human OR genes are inactive **pseudogenes** (sequences of DNA that are very similar to functional genes, but do not produce a functional product). Half the chimpanzee and gorilla OR genes function effectively, while half are pseudogenes. Over 95% of New World monkey OR genes and probably all mouse OR genes are working quite well. What can we conclude from this high rate of inactivation of human OR genes? Most likely, humans came to rely on other senses, reducing the selection pressure against loss of OR gene function by random mutation.

Lateral Gene Transfer

Evolutionary biologists build phylogenies on the assumption that genes are passed from generation to generation, a process called **vertical gene transfer**. Hitchhiking

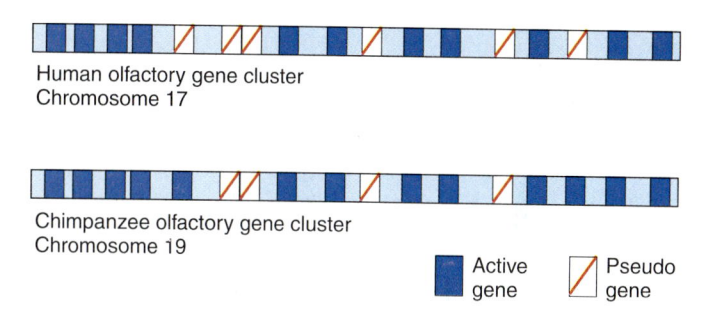

FIGURE 24.6

Gene inactivation. While almost all mouse olfactory receptor genes are functional, an evolutionary loss of olfactory receptors has occurred in primates, which rely less on their sense of smell. Comparisons of the olfactory receptor genes in humans and chimpanzees reveal that humans have more pseudogenes (inactive genes) than chimpanzees have.

genes from other species, referred to as **lateral gene transfer**, lead to phylogenetic complexity. Lateral gene transfer was most prevalent very early in the history of life, when the boundaries between individual cells and species seem to have been less firm than they are now. Earlier in the history of life, gene swapping between species was rampant.

The extensive gene swapping among early organisms is causing many researchers to reexamine the base of the tree of life. Early phylogenies based on ribosomal RNA (rRNA) sequences indicate that an early prokaryote gave rise to two major domains, the Bacteria and the Archaea. From one of these lineages, the domain Eukarya emerged, its organelles produced by engulfing specialized prokaryotes (figure 24.7).

This somewhat straightforward rRNA phylogeny is being revised as more microbial genomes are sequenced; by 2002, 87 microbial genomes had been sequenced. Gene swapping appears to have occurred frequently early in the history of life. Exchanges between organisms were

Chapter 24 Evolution of Genomes and Developmental Mechanisms

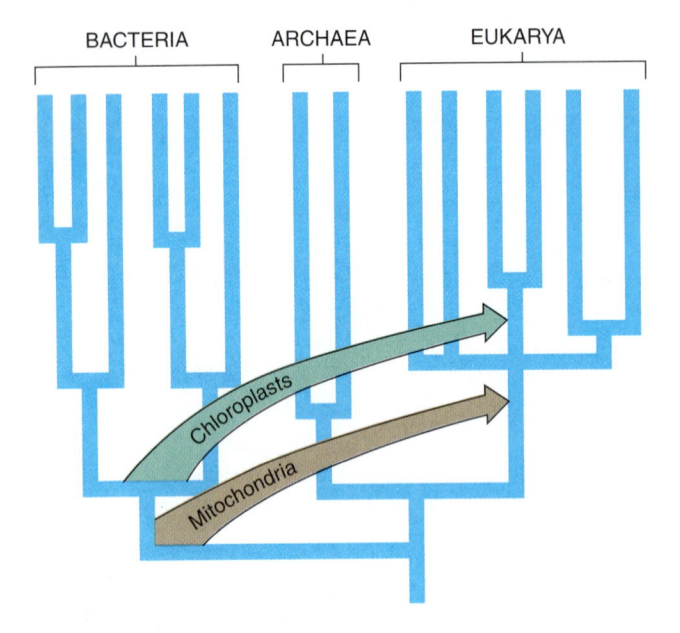

FIGURE 24.7
Phylogeny based on a universal common ancestor. The three domains share a common ancestor, and the tree of life is firmly rooted.

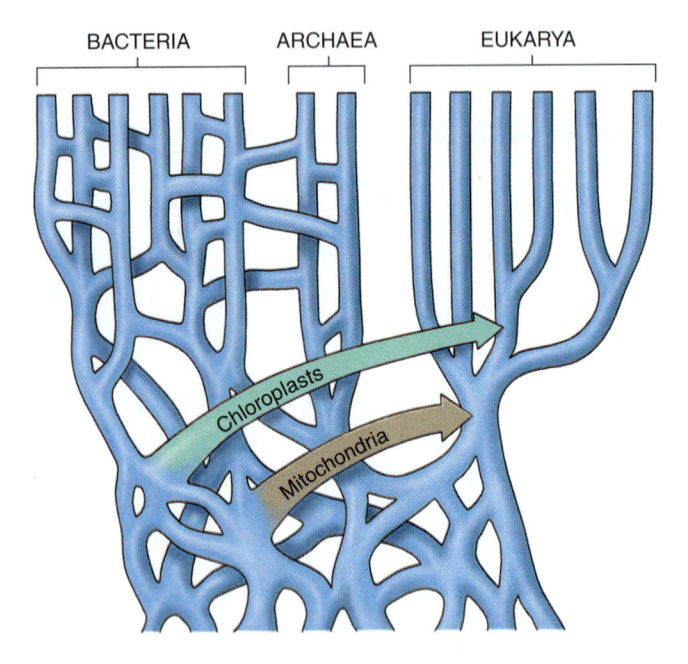

FIGURE 24.8
Lateral gene transfer. Early in the history of life, organisms may have freely exchanged genes. To a lesser extent, this transfer continues today. The tree of life may be more like a web or a net.

so substantial then that the base and the deepest branches of the tree of life are being reevaluated. Phylogenies built with rRNA sequences suggest that the domain Archaea is more closely related to the Eukarya than to the Bacteria. However, as more microbial genomes are sequenced, bacterial and archaeal genes are showing up in the same organism! The most likely conclusion is that organisms swapped genes, possibly even absorbing DNA obtained from a food source. Perhaps the base of the tree of life is better viewed as a web than a branch (figure 24.8).

Let's move a bit closer to home and look at the human genome, which is riddled with foreign DNA, often in the form of transposons (jumping genes, discussed in chapter 20). The many transposons of the human genome provide a paleontological record over several hundred million years. Comparisons among versions of a transposon that has duplicated many times allow researchers to construct a "family tree" to identify the ancestral form of the transposon. The percent sequence divergence of duplicates allows the researcher to estimate the time when that particular transposon originally invaded the human genome. In humans, most of the DNA hitchhiking seems to have occurred millions of years ago. Our genome carries many ancient transposons, making it quite different from other genomes that have been studied, such as those of *Drosophila*, *C. elegans*, and *Arabidopsis*. One explanation for the observed low level of transposons in *Drosophila* is that fruit flies somehow eliminate unnecessary DNA from their genome 75 times faster than humans do. Our genome has simply hung on to hitchhiking DNA more often. While the human genome has had minimal transposon activity in the past 50 million years, mice by contrast are continuing to acquire new transposable elements. This may explain in part the more rapid change in chromosome organization in mice than in humans that we discussed earlier in this chapter.

> **Whole-genome duplication, segmental duplication, and loss of gene function have all contributed to the evolution of genomes. Lateral gene transfer has led to an unexpected mixing of genes among organisms. This creates phylogenetic dilemmas at the base of the tree of life and leaves humans wondering whether their genome is really their own or a set of DNA shared with others.**

24.2 Developmental mechanisms are evolving.

Evolution of Development

The era of comparative genomics is making it possible to bring distinct fields of biology together to address fundamental questions about evolution and development. Closely related sea urchins have been discovered that have very distinctive developmental patterns (figure 24.9). For example, the direct-developing urchin never makes a pluteus larva—it just jumps ahead to its adult form. Sequencing the sea urchin genome, coupled with new genomics techniques such as microarrays (discussed in chapter 17), should help us unravel the changes in patterns of gene expression that most likely account for these dramatic differences in the development of complex organisms that end up with nearly identical adult bodies. These and other questions about the relationship between development and evolution are leading to exciting experiments and findings. Next, we will investigate ways that new traits have arisen by adapting a gene for a new function or by having different genes converge on a similar function.

Same Gene, New Function

If all but 300 of humans' 30,000 genes are shared with mice, why are mice and humans so different? Part of the answer is that genes with similar sequences in two different species may work in slightly or even dramatically different ways. For example, the evolution of vertebrates can partially be explained by the co-option of an existing gene for a new function. Ascidians are basal chordates that have a notochord but no vertebrae. The *Brachyury* gene of ascidians codes for a transcription factor and is expressed in the developing notochord (figure 24.10). *Brachyury* is not a novel gene that appeared in animals as vertebrates evolved. It is found in invertebrates as well, where it has a different function. Most likely, an ancestral *Brachyury* gene was co-opted for a new role in notochord development.

How does a gene gain a different function? One scenario is that it is a regulatory gene that turns on a different set of genes in different organisms. A particularly intriguing example of this is eye development. Phylogenetic analysis shows that the eyes of vertebrates and insects are analogous, not homologous, structures. Molecular genetic analysis, however, reveals that the highly conserved *Pax6* gene launches eye development in vertebrates and insects alike. In this case, the compound eye of a fruit fly and the simple eye of a mouse develop from cells that express *Pax6*. Jumping from gene to phenotype leaves out an important step—development. Development is the black box where linear sequences of DNA are translated into three-dimensional form. To fully understand evolution, we must consider development.

The use of genomics to investigate how development has evolved is a newly emerging field that combines genetics, genomics, evolution, and development to question how the diversity of life has arisen. There are some two dozen conserved gene families that regulate development in animals. Some of these gene families have ancestral genes that predate the origins of animals. These genes serve as a toolkit to build an animal.

The paradox that puzzles evolutionary developmental biologists is how this same toolkit can be used to build an insect, a bird, a bat, a whale, or a human. This paradox is illustrated with the *Brachyury* and *Pax6* genes just described. One explanation is that these genes turn on different genes or combinations of genes in different animals. For example, all tetrapods

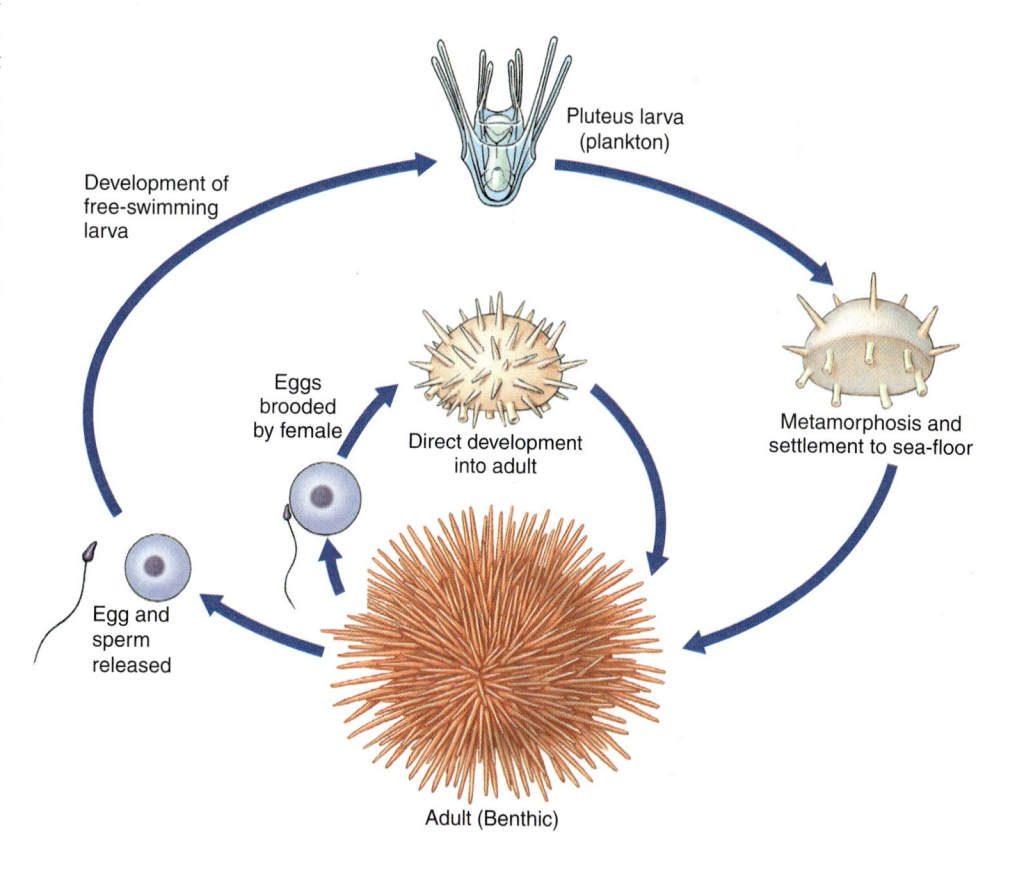

**FIGURE 24.9
Direct and indirect sea urchin development.** Phylogenetic analysis shows that indirect development was the ancestral state. Direct-developing sea urchins have lost an intermediate stage of development.

have four limbs—two hindlimbs and two forelimbs. The forelimb in a bird is actually the wing. Our forelimb is the arm. Clearly, these are two very different structures, but they have a common evolutionary origin. That is, they are homologous structures.

At the genetic level, humans and birds both express the *Tbx5* gene in developing limb buds. *Tbx5* is a member of a gene family with a specific *domain*, a conserved sequence of base-pairs. *Tbx5* encodes a protein domain called the T-box, which is a transcription factor. So, *Tbx5*-encoded protein turns on a gene or genes that are needed to make a limb. What seems to have changed as birds and humans evolved are the genes that are transcribed because of the Tbx5 protein (figure 24.11). In the ancestral tetrapod, perhaps Tbx5 protein bound to only one gene and triggered transcription. In humans and birds, a few genes are expressed in response to Tbx5 protein, but they are different genes.

The story of the evolution of limb development is far more complex than *Tbx5*, which gets limb development started. The genes that *Tbx5* regulate in turn may affect the expression of other genes. Protein

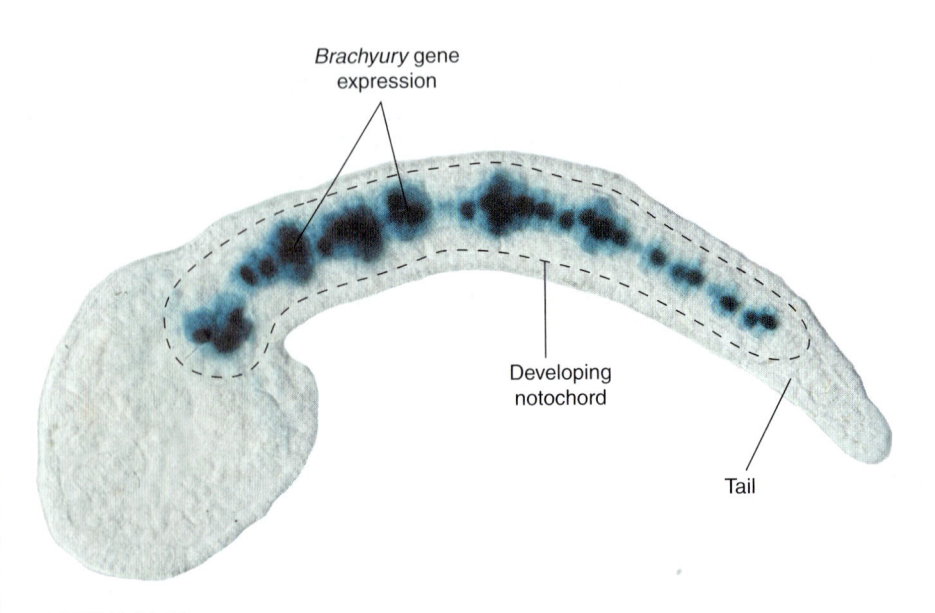

FIGURE 24.10
Co-opting a gene for a new function. *Brachyury* is a gene found in invertebrates that has been used for notochord development in this ascidian, a basal chordate. By attaching the *Brachyury* promoter to a gene with a protein product that stains blue, it is possible to see that *Brachyury* gene expression in ascidians is associated with the development of the notochord, a novel function compared to its function in organisms lacking a notochord.

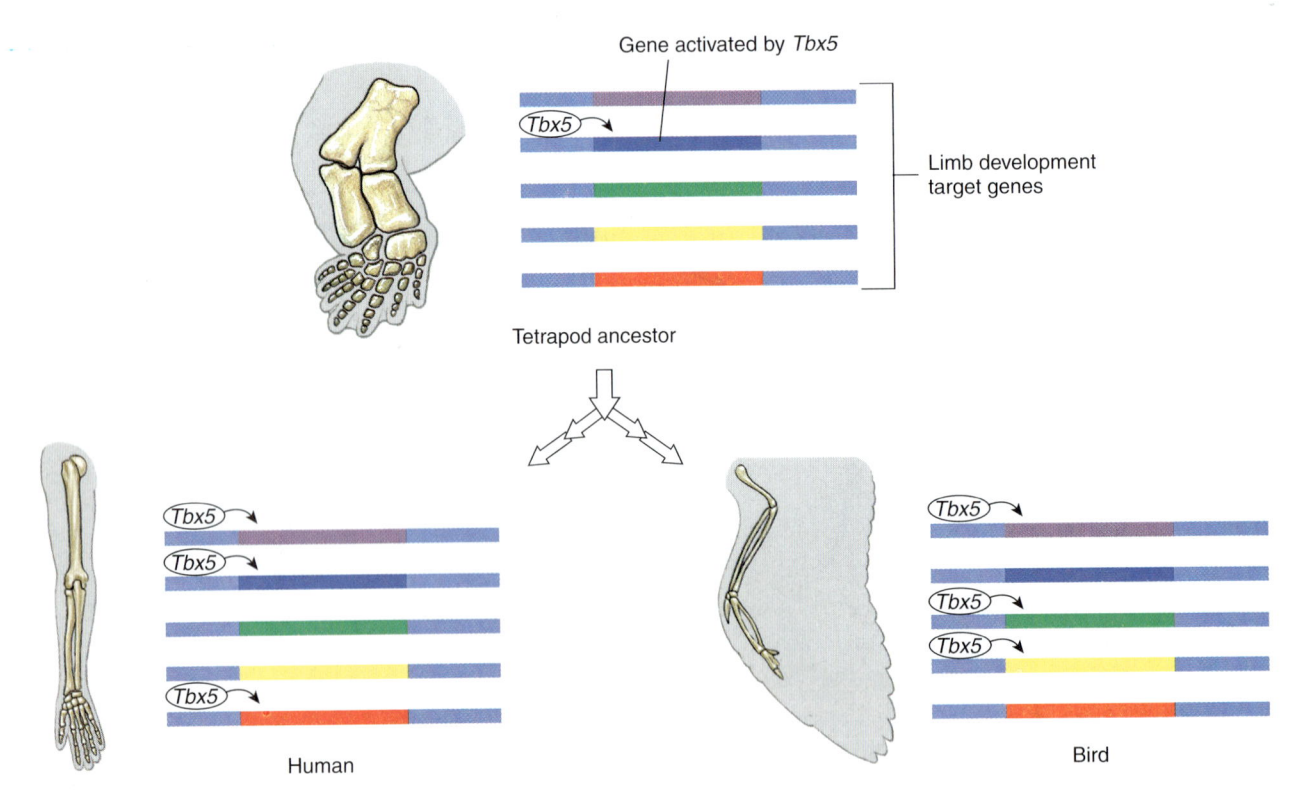

FIGURE 24.11
***Tbx5* regulates wing and arm development.** Wings and arms are very different, but the development of each depends on *Tbx5*. Why the difference? *Tbx5* turns on different genes in birds and humans.

502   Part IV   Evolution

products serve as enzymes in biochemical pathways and provide the building blocks for other protein structures. Mining genome sequences for different organisms will be essential in identifying all the genes involved. Also, development occurs in four dimensions—three-dimensional space over time. Changing the timing of gene expression, as well as the genes that are expressed, can result in dramatic changes in shape.

Different Genes, Convergent Function

Insect wings, especially those of moths and butterflies, have beautiful patterns that can protect them from predation and allow them to thermoregulate (figure 24.12). The origins of these patterns are best explained by the recruitment of existing regulatory programs for new functions. In butterflies, bristles that were linked to neurons and played a sensory role have become scales on wings that produce amazing colors. Structures that have their origins in bristles are inverted, and the daughter cell that gave rise to neuronal connections dies in the development of scales. Later in development, pigment production is triggered. Not all insects have co-opted the same sets of genes for new functions, but all the evolutionary pathways have converged around these novel, highly patterned wings.

Functional Genomics

Sequence comparisons among organisms are essential for both phylogenetic and comparative developmental studies. Careful analysis is needed to distinguish paralogs from orthologs. Rapidly evolving research on bioinformatics, which utilizes computer programming to analyze DNA and protein data, leads to hypotheses that can be tested experimentally. You have already seen how this can work with highly conserved genes such as *Pax6* and *Tbx5*. However, a single base mutation can change an active gene into an inactive or pseudogene. This means that, although function can be inferred from sequence data, experiments are necessary to demonstrate the actual function of the gene. This type of work is called functional genomics and is explained in chapter 17.

Tools for functional analysis exist in model systems, but need to be developed in other organisms on the tree of life if we are going to piece together evolutionary history. Model systems such as yeast, *Arabidopsis*, the nematode worm, the fruit fly, and the mouse have been selected because they are easy to manipulate in some ways. These models all have good genetic systems in which mutant genes can be studied by making crosses. They have fairly short life cycles (imagine trying to study genetics in redwood trees or elephants!). They can be maintained and will reproduce in the laboratory. Also, it is possible to visualize gene expression within parts of the organism using labeled markers and to create transgenic organisms that contain and may express foreign genes.

Many factors have contributed to the evolution of development, including the acquisition of new functions by the same genes. Functional genomic analysis is necessary to determine the actual function of similar genes in different species.

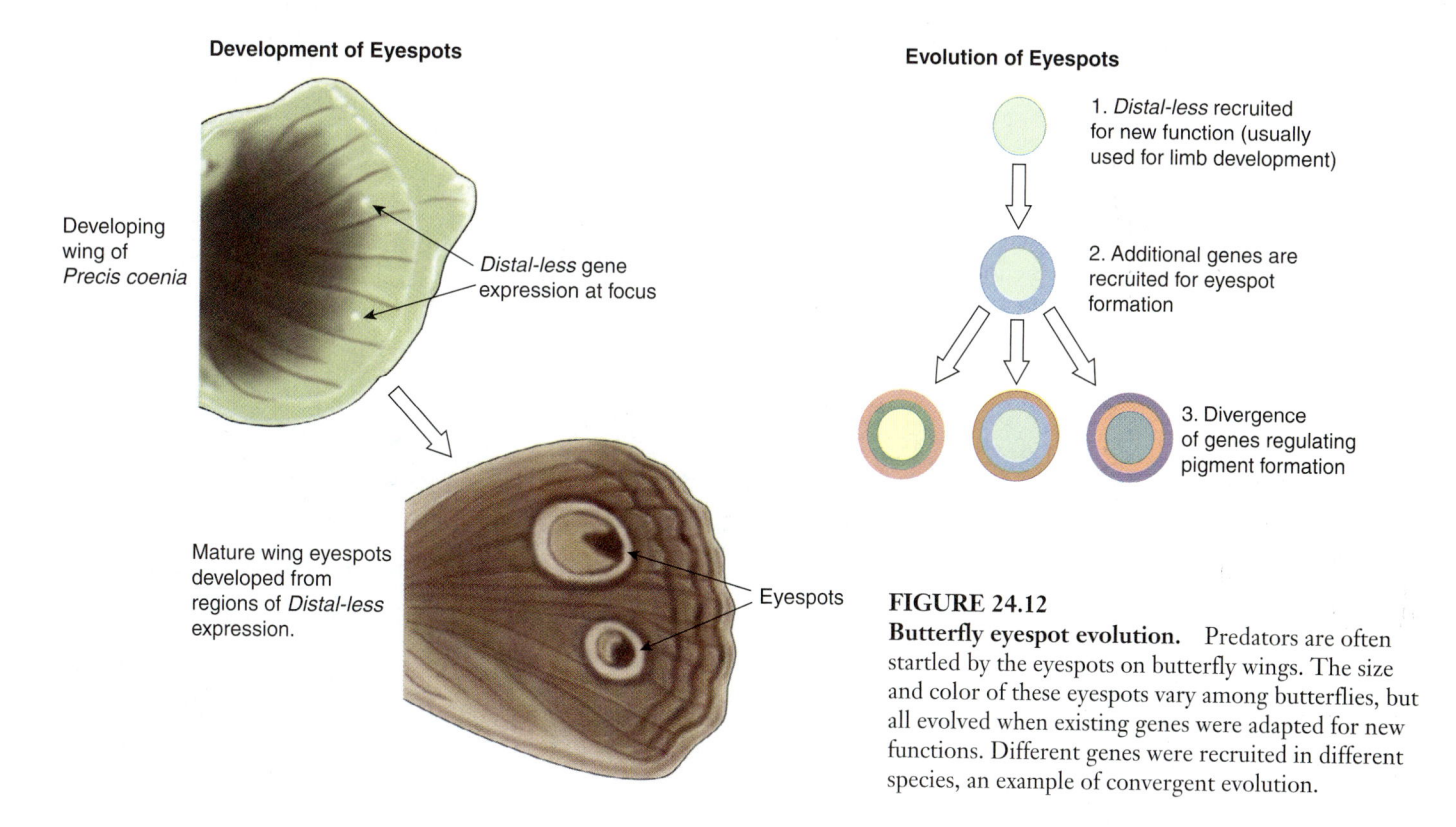

Development of Eyespots

Developing wing of *Precis coenia*

Distal-less gene expression at focus

Mature wing eyespots developed from regions of *Distal-less* expression.

Eyespots

Evolution of Eyespots

1. *Distal-less* recruited for new function (usually used for limb development)

2. Additional genes are recruited for eyespot formation

3. Divergence of genes regulating pigment formation

FIGURE 24.12
Butterfly eyespot evolution. Predators are often startled by the eyespots on butterfly wings. The size and color of these eyespots vary among butterflies, but all evolved when existing genes were adapted for new functions. Different genes were recruited in different species, an example of convergent evolution.

Diversity of Eyes in the Natural World

The eye is one of the most complex organs, and biologists have studied it for centuries. Indeed, explaining how such a complicated structure could evolve was one of the great challenges facing Darwin: If all parts of a structure such as an eye are required for proper functioning, how could natural selection build such a structure? Darwin's response was that even intermediate structures—which provide, for example, the ability to distinguish light from dark—would be advantageous compared with the ancestral state of no visual capability whatsoever, and thus would be favored by natural selection. In this way, by incremental improvements in function, natural selection could build a complicated structure.

Comparative anatomists long have noted that the structures of the eyes of different types of animals are quite different. Consider, for example, the difference in the eyes of a vertebrate, an insect, a mollusk (octopus), and a planarian (figure 24.13). The eyes of these organisms are extremely different in many ways, ranging from compound eyes, to simple eyes, to mere eyespots. Consequently, these eyes are examples of convergent evolution and are analogous, rather than homologous (that is, they are constructed from different ancestral structures). For this reason, evolutionary biologists traditionally viewed the eyes of different organisms as independently evolved, perhaps as many as 20 times! Moreover, this view holds that the most recent common ancestor of these forms was a primitive animal with no ability to detect light.

Unexpected Findings from Molecular Developmental Biology

In the early 1990s, biologists studied the development of the eye in both vertebrates and insects. In each case, a gene was discovered that codes for a transcription factor important in lens formation; the mouse gene was given the name *Pax6*, whereas the fly gene was called *eyeless* (because a mutation in that gene led to a lack of production of the transcription factor and thus the absence of eye development). When these genes were sequenced, it became apparent that they were highly similar; in essence, the homologous gene was responsible for triggering lens formation in both insects and vertebrates.

A stunning demonstration of the homology of these genes was conducted by the Swiss biologist Walter Gehring, who inserted the mouse version of the *Pax6* into the genome of a fruit fly, creating a transgenic fly. In the fly, the *Pax6* gene was turned on by regulatory factors in the leg of the fly. *Pax6* was expressed, and an eye formed on the leg of the fly (figure 24.14)!

These results were truly shocking to the evolutionary biology community. Insects and vertebrates diverged from a common ancestor more than 500 million years ago. Moreover, given the differences in structure of the vertebrate and insect eye, the standard assumption was that the eyes were independently evolved, and thus that the development of those eyes would be controlled by completely different genes. That eye development was affected by the same gene, and that the genes were so similar that the vertebrate gene seemed to function normally in the insect genome, was completely unexpected.

The *Pax6* story extends to eyeless fish found in caves (figure 24.15). Fish that live in dark caves need to rely on senses other than sight. In cavefish, *Pax6* gene expression is greatly reduced. Eyes start to develop, but then degenerate.

How ancient is the *Pax6* gene? Has it always been responsible for initiating eye development, or did it have an

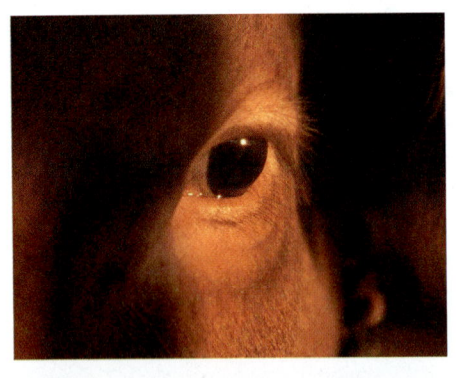

FIGURE 24.13
A diversity of eyes. Morphological and anatomical comparisons of eyes are consistent with the hypothesis of independent, convergent evolution of eyes in diverse species such as flies and humans.

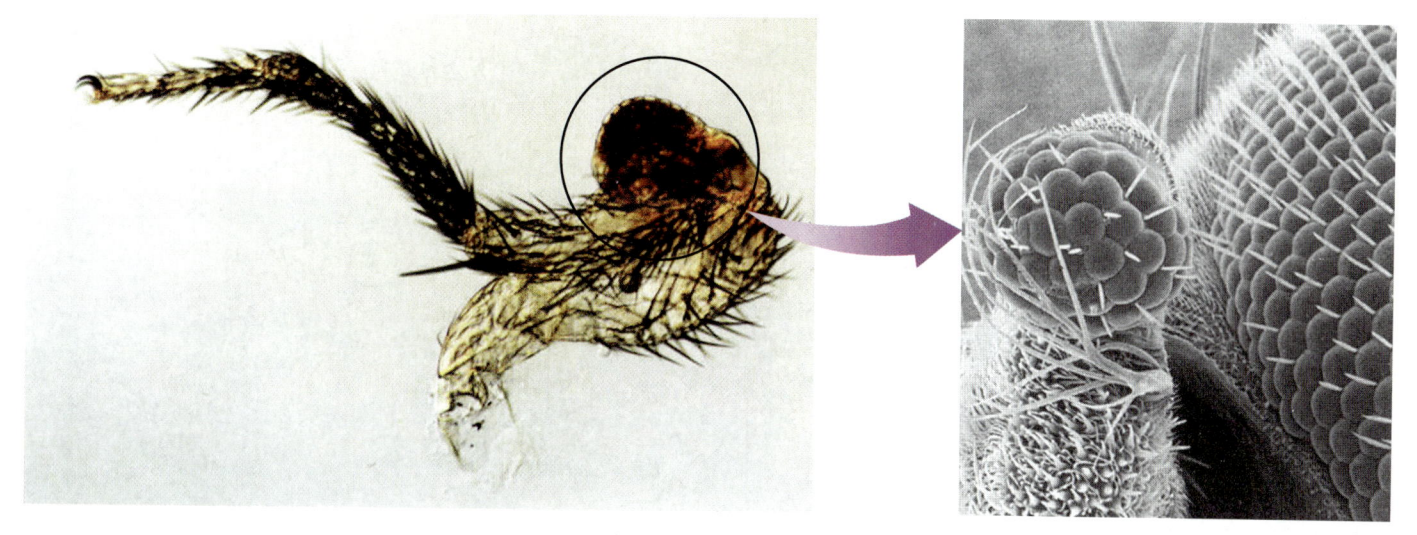

FIGURE 24.14
Mouse *Pax6* makes an eye on the leg of a fly. *Pax6* and *eyeless* are functional homologs. The *Pax6* master regulator gene can initiate compound eye development in a fruit fly or simple eye development in a mouse.

(a)   (b)

FIGURE 24.15
Cavefish have lost their sight. Mexican tetras (*Astyanax mexicanus*) have (*a*) surface-dwelling members and (*b*) cave-dwelling members of the same species. The cavefish have very tiny eyes, partly because of reduced expression of *Pax6*.

ancestral function? Recent discoveries have yielded further surprises. Even the very simple ribbon worm, *Lineus sanguineus*, relies on *Pax6* for development of its eyespots. A *Pax6* homolog has been cloned and shown to be expressed at the sites where eyespots develop. This simple marine invertebrate evolved later than the flatworm planaria. Just like planaria, ribbon worms can regenerate their head region if it is removed. In an elegant experiment, the head of the ribbon worm was removed, and the regeneration of eyespots was followed while the expression of the *Pax6* homolog was observed using in situ hybridization. To observe *Pax6* gene expression, an antisense RNA sequence of the *Pax6* was made and labeled with a color marker. When the regenerating ribbon worms were exposed to the antisense *Pax6* probe, the antisense RNA paired with expressed *Pax6* RNA transcripts and could be seen as colored spots under the microscope (figure 24.16).

Similar experiments were tried with planaria species, but the conclusion was quite different from in ribbon worms. If a planaria is cut in half lengthwise, it can regenerate its missing half, including the second eyespot, but no *Pax6* gene expression is associated with regenerating the eyespots. Planaria have *Pax6*-related genes, but inactivating those genes does not stop eye regeneration (figure 24.17). These *Pax6*-related genes are, however, expressed in the central nervous system. Perhaps some clues as to the origin of *Pax6*'s role in eye development will be uncovered as comparisons between ribbon worm and planaria regeneration continue.

Evolution of the Eye Reconsidered

How can these findings be explained? One possibility is that eyes in different types of animals are truly independently

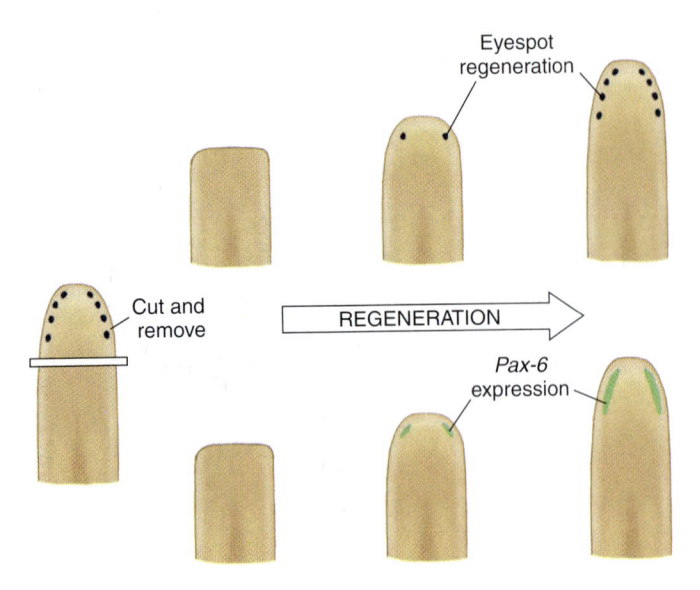

FIGURE 24.16
***Pax6* needed for ribbon worm eyespot regeneration.** *Pax6* is expressed at the same time and place as eyespots on regenerating ribbon worms. The presence of *Pax6* transcripts was visualized through in situ hybridization with a colored antisense probe.

evolved as originally believed. If this is the case, why is *Pax6* so similar and able to play a similar role in the development of the eye in so many different groups? Proponents of this viewpoint point out that *Pax6* is involved not only in development of the eye, but also in development of the entire forehead region of many organisms. Consequently, it is possible that if *Pax6* had a regulatory role in the forehead of early animals, perhaps it has been independently co-opted time and time again to serve a role in eye development. This role would be consistent with the data on planarians (figure 24.17).

Many other biologists find this interpretation unlikely. The consistent use of *Pax6* in eye development in so many organisms, the fact that it functions in the same role in each case, and the great similarity in DNA sequence and even functional replaceability suggest to many that *Pax6* only acquired its evolutionary role in eye development a single time, in the common ancestor of all extant organisms that use *Pax6* in eye development.

Given the great dissimilarity among eyes of different groups, how can this be? One hypothesis is that the common ancestor of these groups was not completely blind, as traditionally assumed. Rather, that organism may have had some sort of rudimentary visual system—maybe no more than a pigmented photoreceptor cell, maybe a slightly more elaborate organ that could distinguish light from dark. Whatever the exact phenotype, the important point is that some sort of basic visual system existed that used *Pax6* in its development. Subsequently, the descendants of this ancestor diversified independently, evolving

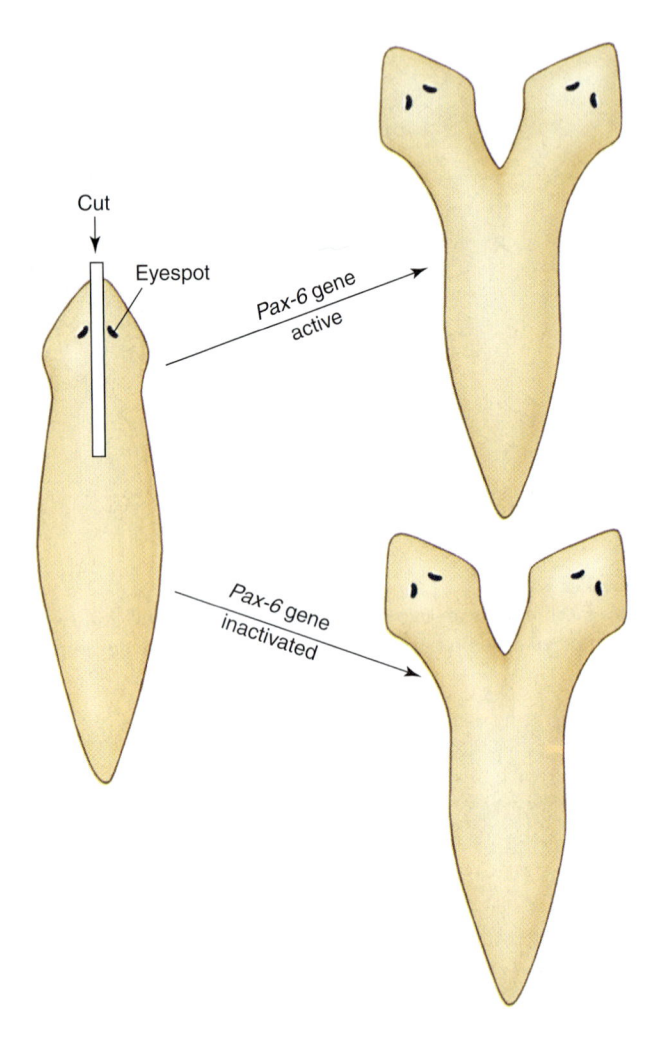

FIGURE 24.17
***Pax6* not required for planaria eyespot regeneration.** Planaria can regenerate their heads and eyespots when cut in half longitudinally. Unlike ribbon worms, *Pax6* does not appear to play a role in planaria eyespot regeneration. When both *Pax6*-related genes in planaria were prevented from producing a protein product, eyespots still formed.

the sophisticated and complex image-forming eyes exhibited by different animal groups today.

Most evolutionary and developmental biologists today support some form of this hypothesis. Nonetheless, it is important to keep in mind that there is no independent evidence that the common ancestor of most of today's animal groups, a primitive form that lived probably more than 500 million years ago, had any ability to detect light. The reason for this belief comes, not from the fossil record, but from a synthesis of phylogenetic and molecular developmental data.

Understanding the evolution of eyes illustrates the power of multidisciplinary approaches to elucidate the evolutionary history of the world's biological diversity.

Concept Review

24.1 Evolutionary history is written in genomes.

Comparative Genomics

- An important challenge of modern evolutionary biology is to link the evolution of DNA sequences with the evolution of complex morphological characters used to construct phylogenies. (p. 492)
- Over 100 prokaryotic genomes and at least 18 eukaryotic genomes have either been completed or are currently being sequenced. (pp. 492–494)
- The pufferfish genome has 365 million base-pairs—basically the same number of genes as the human genome, but only one-ninth of the DNA. (p. 494)
- Humans share 99% of their genes with mice, and thus only diverged about 75 MYA. (p. 494)
- Differences in development are explained by genes being expressed at different times and/or in different tissues. (p. 494)
- Humans and chimps diverged from a common ancestor about 5 MYA, and their sequences are 98.7% identical. (p. 495)
- If only small chromosomal differences exist between distantly related vertebrates over the past 300 million years, a reasonable hypothesis is that the common ancestor had a similar genome. (p. 495)
- The fruit fly (*Drosophila*) and the mosquito (*Anopheles*) are separated by about 250 million years of evolution, and appear to have evolved more rapidly during that time than vertebrates did. (p. 496)

Origins of Genomic Differences

- Genomic changes can be made by at least six factors: the mutation of a single gene, duplicated regions of DNA, rearrangements of large chunks of DNA, chromosome duplication, polyploidy, and interspecies gene integration. (p. 497)
- Growing evidence indicates that duplication of just a few genes has been a major contributor to morphological diversity. (p. 497)
- Whole-genome duplication is insufficient to explain the size of some genomes. (p. 497)
- One of the greatest sources of novel genomic traits is the duplication of DNA segments. The fate of the duplicate gene is most likely either the gaining of a novel function or the loss of function through subsequent mutations. (p. 497)
- About 5% of the human genome consists of segmental duplications. (pp. 497–499)
- Genomes also evolve due to the loss of gene function. (p. 499)
- Vertical gene transfer refers to genes passing from generation to generation, while lateral gene transfer refers to genes moving from one species to another. (p. 499)
- Gene swapping appears to have occurred frequently early in the history of life. (pp. 499–500)
- Most human lateral gene transfer appears to have occurred millions of years ago. (p. 500)

24.2 Developmental mechanisms are evolving.

Evolution of Development

- Genes with similar sequences in two different species may work in different ways. (p. 501)
- Regulatory genes may turn on different sets of genes in different organisms. (p. 501)
- About two dozen conserved gene families regulate animal development. (p. 501)
- Changing the timing of gene expression, and/or the genes expressed, can result in dramatic changes in form. (p. 503)
- New functions can arise for existing structures by the recruitment of existing regulatory programs. (p. 503)
- Although gene function can be inferred from sequence data, actual gene function must be exhibited through experimentation. (p. 503)

Diversity of Eyes in the Natural World

- Natural selection can build very complicated structures via incremental improvements in function. (p. 504)
- Eye development in many different animal groups is an example of convergent evolution and represents analogous structures, but a common gene appears to initiate eye development in many of these animals. (p. 504)

Test Your Understanding

For interactive testing, visit the Online Learning Center with PowerWeb at www.mhhe.com/Raven7

Self Test

1. Humans and pufferfish have a similar number of genes, yet the human genome is approximately nine times larger than the pufferfish genome. In what form is much of this extra DNA?
   a. introns
   b. exons
   c. retrotransposons
   d. RNA
2. Genome comparisons have suggested that mouse DNA has mutated about twice as fast as human DNA. What is a possible explanation for this discrepancy?
   a. Mice are much smaller than humans.
   b. Mice live in much less sanitary conditions than humans and are therefore exposed to a wider range of mutation-causing substances.
   c. Mice have a smaller genome size.
   d. Mice have a much shorter generation time.
3. How many pairs of chromosomes do chimpanzees carry?
   a. 23
   b. 46
   c. 24
   d. 48
4. Why was the genome of the protist *P. falciparum* difficult to sequence?
   a. This protist has a large genome.
   b. This organism hides inside red blood cells, making it difficult to obtain enough DNA for the sequencing project.
   c. The apicoplast structures in this protist inhibit sequencing reactions.
   d. The genome contains a high proportion of adenine and thymine.
5. All of the following are believed to contribute to genomic diversity among various species, *except*
   a. gene duplication.
   b. gene transcription.
   c. lateral gene transfer.
   d. chromosomal rearrangements.
6. What is the fate of *most* duplicated genes?
   a. gene inactivation
   b. gain of a novel function through subsequent mutation
   c. They are transferred to a new organism using lateral gene transfer.
   d. They become orthologs.
7. Which of the following best describes pseudogenes?
   a. two functional genes within an organism that arose from the duplication of one gene
   b. genes that share the same ancestral sequence, but are found in different organisms
   c. sequences of DNA that are very similar to functional genes, but do not produce a functional product
   d. sequences of DNA that are very similar to inactive genes, but do produce a functional product
8. The *Tbx5* gene is known to play a role in which process?
   a. notochord development
   b. limb formation
   c. eye formation
   d. sexual reproduction
9. Which of the following organisms is not considered a model genetic system?
   a. mice
   b. fruit flies
   c. humans
   d. yeast
10. Which of the following statements about *Pax6* is false?
    a. *Pax6* has a similar function in mice and flies.
    b. *Pax6* is involved in eyespot formation in ribbon worms.
    c. *Pax6* is required for eye formation in *Drosophila*.
    d. *Pax6* is required for eyespot formation in planaria.

Test Your Visual Understanding

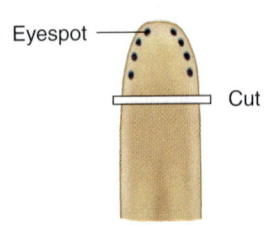

1. *Pax6* is known to play a role in the formation of eyespots during regeneration of the ribbon worm. Beginning with the removal of the head in the above diagram, outline the regeneration process and eyespot formation in a normal, wild-type ribbon worm and one that lacks *Pax6* expression. How are they different?

Apply Your Knowledge

1. How might knowledge of the *Oryza sativa* genome help combat world hunger?
2. Can a human embryo that exhibits polyploidy survive until birth? What is the difference between polyploidy and trisomy?
3. In a paper by Halder et al., 1995, the authors use the inducible GAL4-UAS expression system to artificially overexpress the *eyeless* (*Pax6*) gene in tissues where it is not normally expressed. Why does this system use the yeast transcriptional activator, GAL4, which is not normally present in *Drosophila*?

25
Systematics and the Phylogenetic Revolution

Concept Outline

25.1 Biologists name organisms in a systematic way.

The Classification of Organisms. Biologists group organisms based on shared characteristics.

25.2 Scientists construct phylogenies to understand the evolutionary relationships among species.

Systematics. Evolutionary relationships can be reconstructed.

25.3 Phylogenetics is the basis of all comparative biology.

Analogy Versus Homology. Two species may possess similar traits either because both inherited them from a common ancestor or because they have independently evolved them.

25.4 All living organisms are grouped into one of a few major categories.

The Kingdoms of Life. Organisms are grouped into three great groups called domains, and within domains into kingdoms.
Domain Archaea (Archaebacteria). The Archae are primitive prokaryotes that may live in extreme environments.
Domain Bacteria (Bacteria). Bacteria are more numerous than any other organism.
Domain Eukarya (Eukaryotes). There are four kingdoms of eukaryotes, three of them entirely or predominantly multicellular.
Viruses: A Special Case. Viruses are not organisms, and thus do not belong in any kingdom.

25.5 Molecular data are revolutionizing taxonomy.

The Impact of Molecular Cladistics. DNA sequence comparisons among organisms have led to new phylogenies.
Making Sense of the Protists. The protists are paraphyletic.
Origin of Land Plants. Molecular phylogenetics has identified the closest living relatives of land plants.
Sorting Out the Animals. Relationships among animal groups are being redefined.

FIGURE 25.1
Biological diversity. For more than 2000 years, living things have been categorized based on common characteristics. Amid the diverse life-forms in the Great Barrier Reef in Australia, different corals share a similar anatomy, pattern of development, mode of nutrition, level of organization, and biochemical composition.

All organisms share many biological characteristics. They are composed of one or more cells, carry out metabolism and transfer energy with ATP, and encode hereditary information in DNA. Yet, there is also a tremendous diversity of life, ranging from bacteria and amoebas to blue whales and sequoia trees (figure 25.1). For generations, biologists have tried to group organisms based on shared characteristics. The most meaningful groupings are based on the study of evolutionary relationships among organisms. New methods for constructing evolutionary trees and a sea of molecular sequence data are leading to new evolutionary hypotheses to explain life's diversification.

25.1 Biologists name organisms in a systematic way.

The Classification of Organisms

Organisms were first classified more than 2000 years ago by the Greek philosopher Aristotle, who categorized living things as either plants or animals. The Greeks and Romans expanded this simple system and grouped animals and plants into basic units such as cats, horses, and oaks. Eventually, these units began to be called **genera** (singular, *genus*), the Latin word for "groups." Starting in the Middle Ages, these names began to be systematically written down in Latin, the language used by scholars at that time. Thus, cats were assigned to the genus *Felis*, horses to *Equus*, and oaks to *Quercus*.

Species Names

Until the mid-1700s, whenever biologists wanted to refer to a particular kind of organism, which they called a **species**, they added a series of descriptive terms to the name of the genus; this was a *polynomial*, or "many names" system. A much simpler system of naming organisms stems from the work of the Swedish biologist Carolus Linnaeus (1707–1778). In the 1750s, Linnaeus used the polynomial names *Apis pubescens, thorace subgriseo, abdomine fusco, pedibus posticis glabris utrinque margine ciliates* to denote the European honeybee. But as a kind of shorthand, he also included a two-part name for each species. For example, the honeybee became *Apis mellifera*. These two-part names, or *binomials* (*bi*, "two") have become our standard way of designating species.

Taxonomy (Greek, "arranging rules") is the science of classifying living things, and a group of organisms at a particular level in a classification system is called a **taxon** (plural, *taxa*). By agreement among taxonomists throughout the world, no two organisms can have the same name, and all names are in Latin. The scientific name of an organism is the same anywhere in the world and avoids the confusion caused by common names (figure 25.2). Also by agreement, the first word of the binomial name is the genus to which the organism belongs. This word is always capitalized. The second word refers to the particular species and is not capitalized. The two words together are called the species name (or scientific name) and are written in italics or distinctive

FIGURE 25.2
Common names make poor labels. In North America, the common names "corn" (*a*) and "bear" (*b*) bring clear images to our minds (*photos on left*), but the images are very different for someone living in Europe or Australia (*photos on right*).

FIGURE 25.3
The hierarchical system used in classifying an organism. The organism is first recognized as a eukaryote (domain Eukarya). Within this domain, it is an animal (kingdom Animalia). Among the different phyla of animals, it is a vertebrate (phylum Chordata, subphylum Vertebrata). The organism's fur characterizes it as a mammal (class Mammalia). Within this class, it is distinguished by its gnawing teeth (order Rodentia). Next, because it has four front toes and five back toes, it is a squirrel (family Sciuridae). Within this family, it is a tree squirrel (genus *Sciurus*), with gray fur and white-tipped hairs on the tail (species *Sciurus carolinensis*, the eastern gray squirrel).

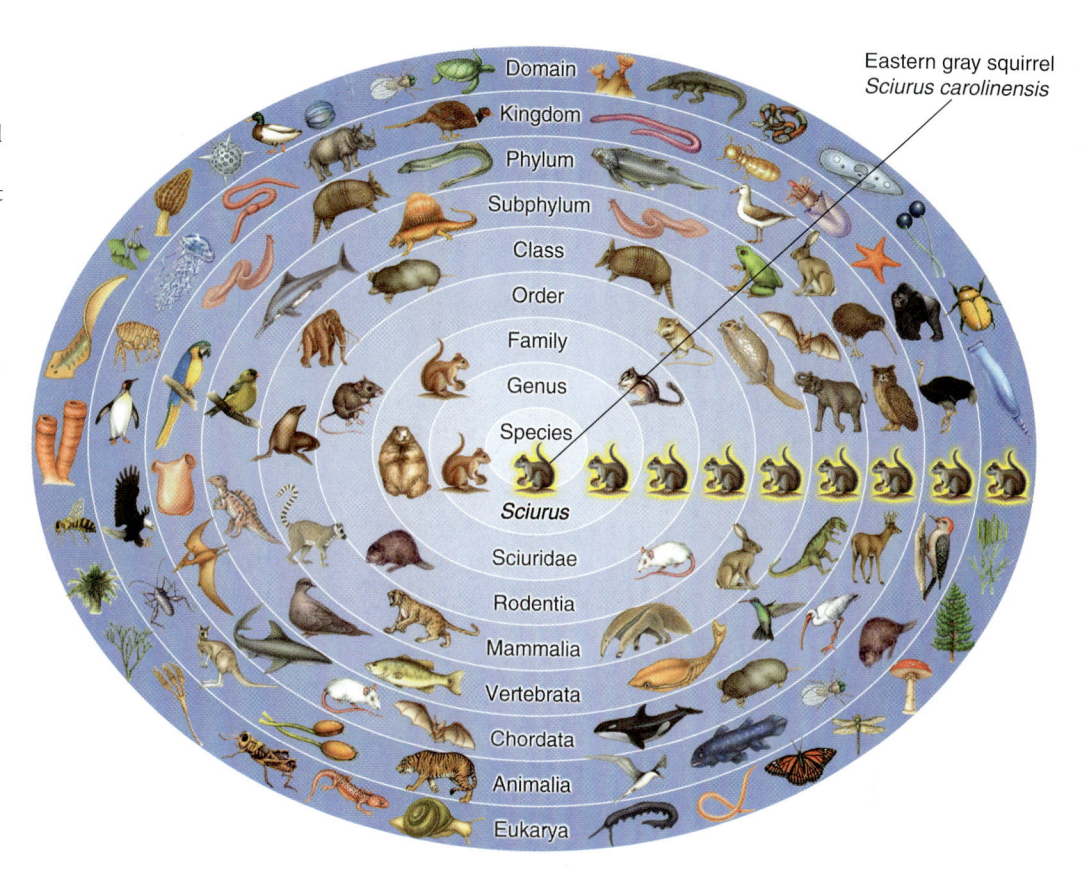

print—for example, *Homo sapiens*. Once a genus has been used in the body of a text, it is often abbreviated in later uses. For example, the dinosaur *Tyrannosaurus rex* becomes *T. rex*.

The Taxonomic Hierarchy

In the decades following Linnaeus, taxonomists began to group organisms into larger, more inclusive categories. Genera with similar properties were grouped into a cluster called a **family,** and similar families were placed into the same **order** (figure 25.3). Orders with common properties were placed into the same **class,** and classes with similar characteristics into the same **phylum** (plural, *phyla*). Finally, the phyla were assigned to one of several great groups, the **kingdoms.** Biologists currently recognize six kingdoms: two kinds of prokaryotes (Archaebacteria and Bacteria), a largely unicellular group of eukaryotes (Protista), and three multicellular groups (Fungi, Plantae, and Animalia).

In addition, an eighth level of classification, called a *domain*, is sometimes used. Biologists recognize three domains, which will be discussed in section 25.4. The names of the taxonomic units higher than the genus level are capitalized but *usually* not printed distinctively, italicized, or underlined.

The categories at the different levels may include many, a few, or only one taxon. For example, there is only one living genus of the family Hominidae, but several living genera of Fagaceae. To someone familiar with classification or having access to the appropriate reference books, each taxon implies both a set of characteristics and a group of organisms belonging to the taxon. For example, a honeybee has the species (level 1) name *Apis mellifera*. Its genus (level 2), *Apis*, is a member of the family Apidae (level 3). All members of this family are bees—some solitary, others living in hives as *A. mellifera* does. Knowledge of its order (level 4), Hymenoptera, tells you that *A. mellifera* is likely able to sting and may live in colonies. Its class (level 5), Insecta, indicates that *A. mellifera* has three major body segments, with wings and three pairs of legs attached to the middle segment. Its phylum (level 6), Arthropoda, tells us that the honeybee has a hard cuticle of chitin and jointed appendages. Its kingdom (level 7), Animalia, indicates that *A. mellifera* is a multicellular heterotroph whose cells lack cell walls.

> By convention, the first part of a binomial species name identifies the genus to which the species belongs, and the second part distinguishes that particular species from other species in the genus. Genera are grouped into families, families into orders, orders into classes, and classes into phyla. Phyla are the basic units within kingdoms; such a system is hierarchical.

Chapter 25 Systematics and the Phylogenetic Revolution   **511**

25.2 Scientists construct phylogenies to understand the evolutionary relationships among species.

One of the great challenges of modern science is to understand the history of life on earth, from the earliest single-celled organisms to the complex organisms we see around us today. Unfortunately, we cannot go back in time to retrace the course of evolution. If the fossil record were perfect, we could trace the evolutionary history of species and examine how each species arose and proliferated. However, as we discussed in chapter 22, the fossil record is far from complete. Although it answers many questions about life's diversification, it leaves many others unsettled.

Consequently, scientists must rely on other types of evidence to establish the best hypothesis of evolutionary relationships. It is important to bear in mind that the outcomes of such studies *are* hypotheses. As such, they require further testing with new data; if disproven, they lead to the formulation of new, better-established hypotheses, and the cycle begins anew.

Systematics

The reconstruction and study of evolutionary relationships is called **systematics**. By looking at the similarities and differences between species, systematics can construct an evolutionary tree, or **phylogeny**, which represents a hypothesis about which species are closely related and in what order related species evolved.

Because species evolve from other species, we would expect descendant species to be fairly similar to their ancestors. More generally, we would expect that the greater the time since two species shared a common ancestor, the more different they would be. Early systematists relied on this reasoning and constructed phylogenies based on overall similarity. If, in fact, species diverged at a constant rate, then the amount of divergence between two species would be a function of how long they had been diverging, and thus phylogenies based on degree of similarity would be accurate.

However, as chapter 21 revealed, evolution can occur very rapidly at some times and very slowly at others. Moreover, evolution is not unidirectional—sometimes species evolve in one direction and then back the other way (the phenomenon of oscillating selection). Species invading new habitats are likely to experience new selective pressures and may change greatly; those staying in the same habitats as their ancestors may change only a little. For this reason, similarity is not necessarily a good predictor of how long it has been since two species shared a common ancestor.

A second fundamental problem exists as well: Evolution is not always divergent. In chapter 22, we discussed **convergent evolution**, in which two species independently evolve to become more similar. Often, this is because species evolving the same adaptation use similar habitats. As a result, two species that are not closely related may end up being more similar to each other than they are to close relatives. Similarly, some species may reverse evolutionary course and evolve to be similar to distant ancestral species, rather than to more closely related species.

Cladistics

For these reasons, most systematists no longer construct their phylogenetic hypotheses solely on the basis of similarity. Rather, they distinguish similarity that is inherited from the common ancestor of an entire group, and is called **ancestral**, from similarity that arose within the group and thus is termed **derived**. In this method, termed **cladistics**, only shared derived characters are considered informative in determining evolutionary relationships.

To employ this method, systematists first gather data on a number of characters for all the species in the analysis. Characters can be any aspect of the phenotype, including morphology, physiology, behavior, and DNA. As chapters 17 and 24 showed, the revolution in genomics will soon provide a vast body of data that may revolutionize our ability to identify and study character variation across species.

To be useful, species must vary in *character states*. For example, consider the character "teeth" in amniote vertebrates (birds, reptiles, and mammals). This character has two states: presence in most mammals and reptiles and absence in birds and a few other groups (e.g., turtles). Once the data are assembled, the first step is to *polarize* the characters—that is, to determine whether particular character states are ancestral or derived. To polarize the character "teeth," we must determine which state was exhibited by the most recent common ancestor of this group.

Usually, we do not have a fossil that we are confident is the most recent ancestor. As a result, the method of **outgroup comparison** is used to assign character polarity. To use this method, a species or group of species that is closely related to, but not a member of, the group under study is designated as the **outgroup**. Character states exhibited by the outgroup are assumed to be ancestral, and other states are considered derived. Polarity assignments are most effective when several different outgroups are used. In the preceding example, teeth are generally present in the closest relatives of amniotes—amphibians and fish. Consequently, the presence of teeth in mammals and reptiles is considered ancestral, and their absence in birds is considered derived.

Once all characters have been polarized, systematists use this information to construct a **cladogram,** which depicts a hypothesis of evolutionary relationships. Species that share derived characters belong to a **clade.** A derived character shared by clade members is called a **synapomorphy** of that clade. Clades are thus evolutionary units and refer to all the descendants of a particular common ancestor.

Figure 25.4 illustrates that a simple cladogram is a nested set of clades, each characterized by its own synapomorphies. For example, amniotes are a clade for which the evolution of an amniotic membrane is a synapomorphy. Within that clade, mammals are a clade, with hair as a synapomorphy, and so on.

Several points need to be kept in mind when examining a cladogram. First, although the cladogram does not contain ancestral species, each node (the point where two branches of the tree come together) represents a hypothetical ancestral species. Thus, to tell how closely related two species are, we simply look to see how recently they shared a common ancestor. For example, humans and gorillas are relatively closely related because they share a recent common ancestor. By contrast, humans and sharks are not closely related because their most recent common ancestor occurs deep in the cladogram. A second consequence of this way of portraying relationships is that species that occur next to each other at the top of the cladogram are not necessarily closely related. Lizards and salamanders are right next to each other in figure 25.4, yet lizards share a more recent ancestor with—and thus are more closely related to—humans than they are to salamanders. Finally, each ancestral node has two branches, and it is arbitrary which one is put on the left and which on the right. For example, the cladogram in figure 25.4 could just as correctly have placed the branch containing the shark on the far right and the branch leading to the salamander, lizard, tiger, gorilla, and human on the left.

Ancestral states are also called **plesiomorphies,** and shared ancestral states are called **symplesiomorphies.** In contrast to synapomorphies, symplesiomorphies are not informative about phylogenetic relationships. Consider, for example, the character state "presence of a tail," which is exhibited by lampreys, sharks, salamanders, lizards, and tigers. Does this mean that tigers are more closely related—and shared a more recent common ancestor with—lizards and sharks than with apes and humans, their fellow mammals? The answer, of course, is no: Because symplesiomorphies only reflect character states inherited from a distant ancestor, they do not imply that species with that state are closely related.

In real examples, however, phylogenetic studies are rarely so simple. The reason is that in some cases, shared derived characters have evolved independently and thus are false signals of close evolutionary relationship. In addition, derived characters may sometimes be lost as

Traits: Organism	Jaws	Lungs	Amniotic membrane	Hair	No tail	Bipedal
Lamprey	0	0	0	0	0	0
Shark	1	0	0	0	0	0
Salamander	1	1	0	0	0	0
Lizard	1	1	1	0	0	0
Tiger	1	1	1	1	0	0
Gorilla	1	1	1	1	1	0
Human	1	1	1	1	1	1

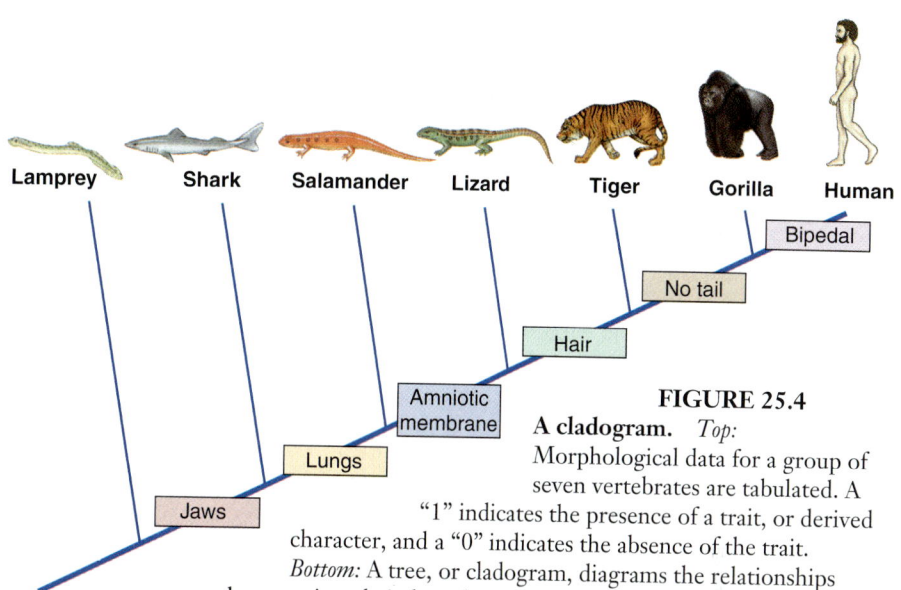

**FIGURE 25.4
A cladogram.** *Top:* Morphological data for a group of seven vertebrates are tabulated. A "1" indicates the presence of a trait, or derived character, and a "0" indicates the absence of the trait. *Bottom:* A tree, or cladogram, diagrams the relationships among the organisms based on the presence of derived characters. The derived characters between the cladogram branch points are shared by all organisms above the branch points and are not present in any below it. The outgroup (in this case, the lamprey) does not possess any of the derived characters.

species within a clade re-evolve the ancestral state. Such characteristics are referred to as **homoplasies.** For example, frogs do not have a tail. Thus, absence of a tail is a synapomorphy that unites not only gorillas and humans, but also frogs. However, frogs have neither an amniotic membrane nor hair, both of which are synapomorphies for clades that contain gorillas and humans. In cases such as this, when there are conflicts among the characters, systematists rely on the **principle of parsimony,** which favors the hypothesis that requires the fewest assumptions. As a result, the phylogeny that requires the fewest evolutionary events is considered the best hypothesis of phylogenetic relationships. Thus, in the example just stated, grouping frogs with salamanders is favored because it requires only one instance of homoplasy (the multiple origins of taillessness), whereas a phylogeny in which frogs were most closely related to humans and gorillas would require two homoplastic evolutionary events (the loss of both amniotic membranes and hair in frogs).

If characters evolve from one state to another at a slow rate compared to the frequency of speciation events, the principle of parsimony works well in reconstructing evolutionary relationships because its underlying assumption—that shared derived similarity is indicative of recent common ancestry—is usually correct. In fact, this assumption seems likely to be correct for many types of data.

However, in recent years systematists have realized that some characters evolve so rapidly that the principle of parsimony may be misleading. Of particular interest is the rate at which some parts of the DNA evolve. As discussed in chapter 17, some stretches of DNA do not appear to have any function. As a result, mutations that occur in these parts of the DNA are not removed by natural selection, and thus the rate of evolution can be quite high. Moreover, because only four character states are possible for any nucleotide base (A, C, G, or T), the probability that two species may independently evolve the same derived character state is quite high. If, in fact, such homoplasy dominates the character data set, then the assumptions of the principle of parsimony are violated, and phylogenies inferred using this method are unlikely to be accurate.

For this reason, systematists in recent years have been exploring other methods based on statistical approaches, such as maximum likelihood, to infer phylogenies. These methods start with an assumption of the rate at which characters evolve and then fit the data to these models to derive the phylogeny that best accords with these assumptions. One advantage of these methods is that different assumptions of rate of evolution can be used for different characters. Thus, if some DNA characters are more constrained to evolve slowly than are other parts of the DNA, the methods can employ different models of evolution for the different character states.

In general, cladograms such as the one in figure 25.4 only indicate the order of evolutionary branching events; they do not contain information about the timing of these events. In some cases, however, branching events can be timed, either by reference to fossils, or by making assumptions about the rate at which characters change. One widely used, but controversial, method is the **molecular clock,** which states that the rate of evolution of a molecule is constant through time, such that divergence in DNA in this model can be used to calibrate when branching events occur. Although the molecular clock appears to hold in some cases, in many others the data indicate that rates of evolution are not constant enough to be used in timing evolutionary events. Systematists are still debating how often, and under what circumstances, the molecular clock holds.

Systematics and Classification

While systematics is the reconstruction and study of evolutionary relationships, *classification* refers to how we place species and higher groups in the taxonomic hierarchy. To understand why the two are not always congruent, we need to consider how species may be grouped based on their phylogenetic relationships. A **monophyletic** group includes the most recent common ancestor of the group and all of its descendants. By definition, a clade is a monophyletic group. A **paraphyletic** group includes the most recent common ancestor of the group, but not all its descendants, and a **polyphyletic** group does not include the most recent common ancestor of all members of the group (figure 25.5).

Taxonomic hierarchies are based on shared traits and ideally should reflect evolutionary relationships. Traditional taxonomic groups, however, do not always fit well with new understandings of phylogenetic relationships. Historically, birds are placed in the class Aves, whereas other dinosaurs are in the Reptilia. But recent phylogenetic advances make clear that birds evolved from dinosaurs. The last common ancestor of all birds and a dinosaur was a meat-eating dinosaur. There is no way to arrange the evolutionary tree to support separate monophyletic groups for reptiles (including dinosaurs and crocodiles, as well as lizards, snakes, and turtles) and for birds. Yet, the terms are so familiar that suddenly referring to birds as a type of dinosaur, and thus a type of reptile, is confusing to some. Biologists are currently grappling with the extent to which our classification system should be changed to accommodate new understandings of phylogenetic relationships.

> Cladistics is the most commonly used method to reconstruct evolutionary relationships. In cladistics, shared derived characters are used to construct a cladogram, which indicates phylogenetic relationships among species.

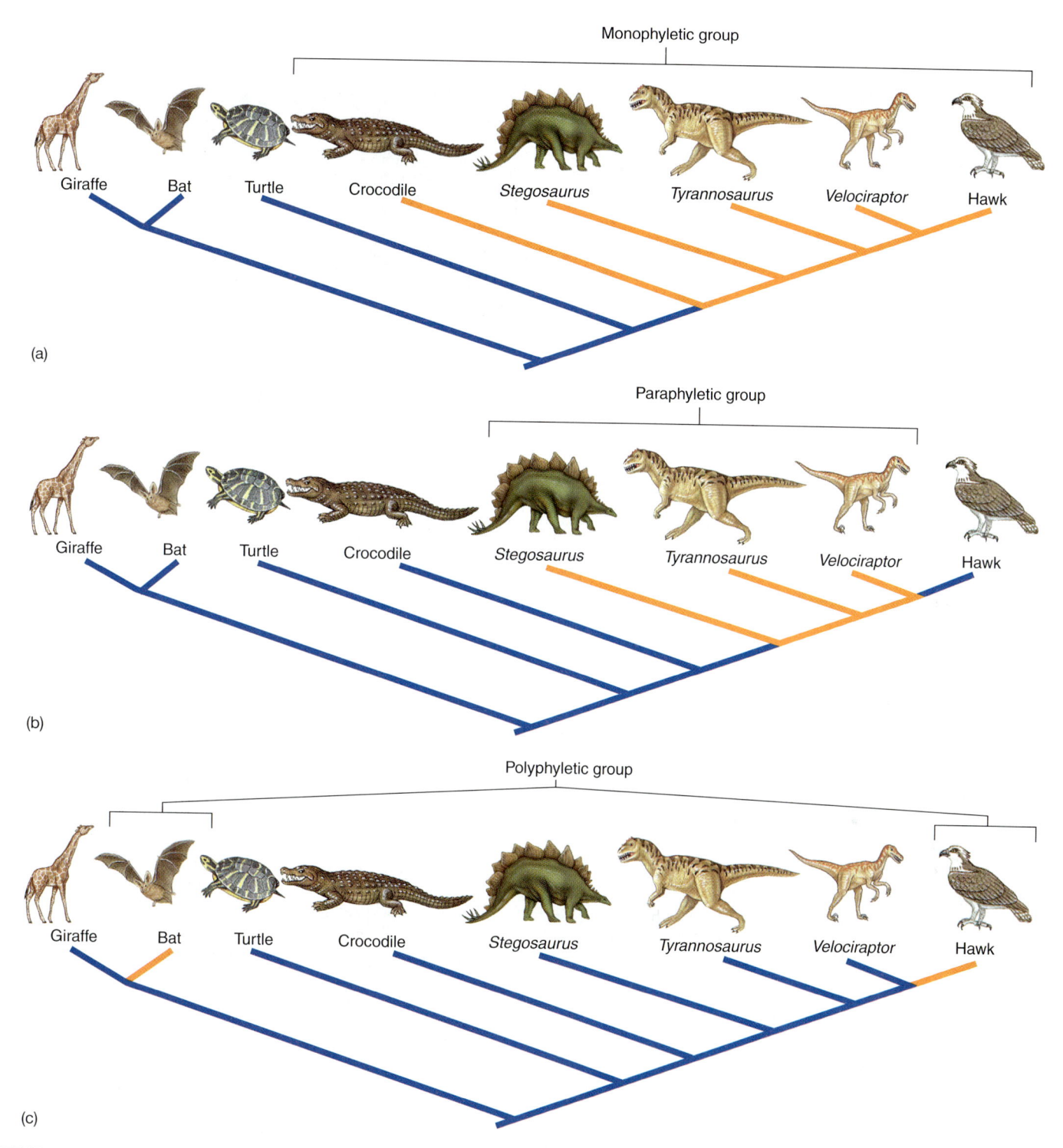

FIGURE 25.5

Monophyletic, paraphyletic, and polyphyletic groups. (*a*) A monophyletic group consists of the most recent common ancestor and all of its descendants. For example, the name "Archosaurs" is given to the monophyletic group that includes a crocodile, *Stegosaurus*, *Tyrannosaurus*, *Velociraptor*, and a hawk. (*b*) A paraphyletic group consists of the most recent common ancestor and some of its descendants. For example, some, but not all, taxonomists traditionally give the name "dinosaurs" to the paraphyletic group that includes *Stegosaurus*, *Tyrannosaurus*, and *Velociraptor*. This group is paraphyletic because one descendant of the most recent ancestor of these species, the bird, is not included in the group. Other taxonomists include birds within the Dinosauria because *Tyrannosaurus* and *Velociraptor* are more closely related to birds than to other dinosaurs. (*c*) A polyphyletic group does not contain the most recent common ancestor of the group, and taxonomists do not assign taxa to polyphyletic groups. For example, bats and birds could be classified in the same group because they have similar shapes, anatomical features, and habitats. However, their similarities reflect convergent evolution, not common ancestry.

Chapter 25   Systematics and the Phylogenetic Revolution   515

25.3 Phylogenetics is the basis of all comparative biology.

Phylogenies not only provide information about evolutionary relationships among species, but also are indispensable for understanding how evolution has occurred. By examining the distribution of traits among species in the context of the phylogenetic relationships of these species, much can be learned about how and why evolution proceeds.

Analogy Versus Homology

In chapter 22, we pointed out that homologous structures are those that are derived from the same body part in a common ancestor. Thus, the flipper of a dolphin and the leg of a horse are homologous because they are derived from the same bone in an ancestral vertebrate. By contrast, the wings of birds and dragonflies are analogous, because they are derived from different ancestral structures. Phylogenetic analysis can help determine whether structures are homologous or analogous.

For example, recent fossil discoveries have revealed that many species of dinosaurs exhibited parental care. They incubated eggs laid in nests and took care of growing baby dinosaurs, many of which could not have fended for themselves; in fact, some recent fossils show dinosaurs sitting on a nest in exactly the same posture used by birds today (figure 25.6)! Initially, these discoveries were treated as remarkable and unexpected—dinosaurs apparently had independently evolved behaviors similar to those of modern-day organisms. However, examination of the phylogenetic position of dinosaurs indicates that they are most closely related to two living groups of animals—crocodiles and birds—both of which exhibit parental care (see figure 25.5). Thus, it would appear likely that the similar parental care exhibited by crocodiles, dinosaurs, and birds did not evolve convergently; rather, the behaviors are homologous, inherited by each of these groups from their common ancestor.

In other cases, by contrast, phylogenetic analysis can indicate that similar traits have evolved independently in different clades and thus are analogous. For example, the fossil record reveals that extremely elongated canines (saber teeth) occurred in a number of different groups of extinct carnivorous mammals. Examination of saber-toothed-ness in a phylogenetic context reveals that it most likely evolved independently at least three times (figure 25.7). Although how these teeth were actually used is still debated, all saber-toothed carnivores had body proportions similar to those of cats, which suggests that these different types of carnivores all evolved a similar predatory lifestyle.

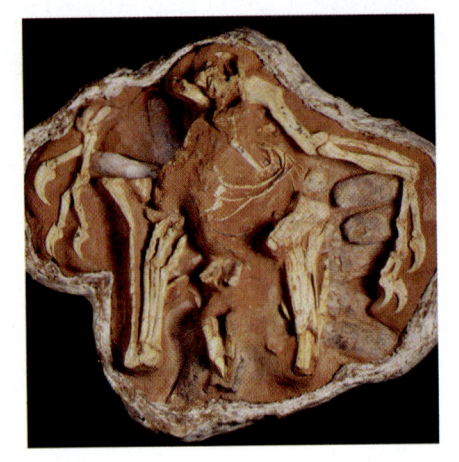

**FIGURE 25.6
Fossil dinosaur incubating its eggs.** This remarkable fossil of *Oviraptor* shows the dinosaur sitting on its nest of eggs just as chickens do today. Not only is the dinosaur squatting on the nest, but its forelimbs are outstretched, perhaps to shade the eggs.

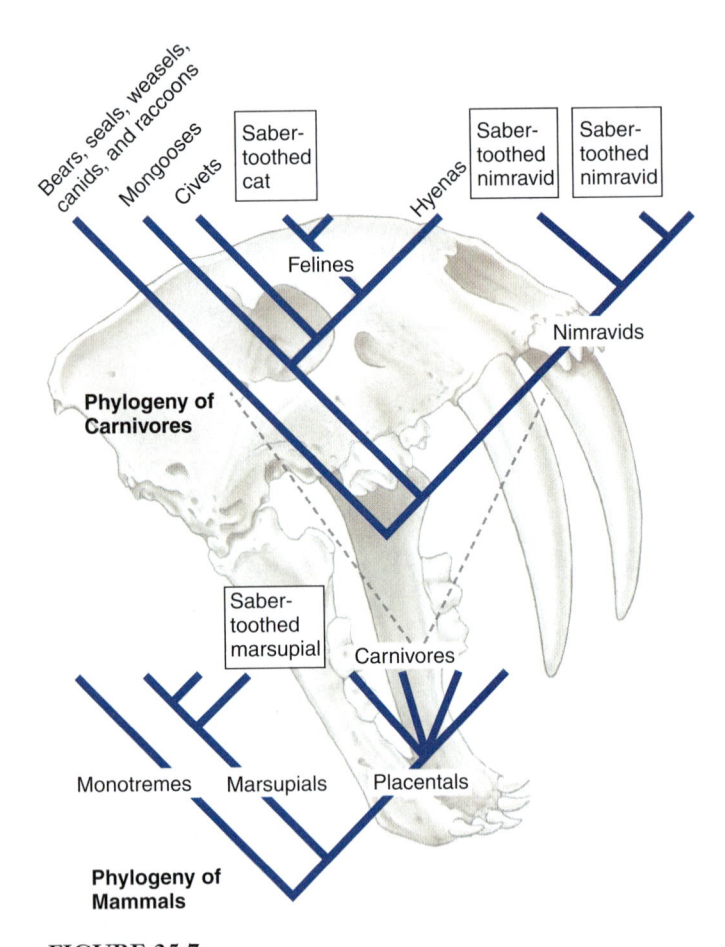

**FIGURE 25.7
Distribution of saber-toothed mammals.** Saber teeth have evolved at least three times in mammals—once within marsupials, once in felines, and at least once in a now-extinct group of catlike carnivores called nimravids. It is possible that the condition evolved twice in nimravids, but another, less likely, possibility is that saber teeth evolved only once and were subsequently lost in other taxa. (Not all of the branches within marsupials and placentals are shown).

516 Part V Diversity of Life on Earth

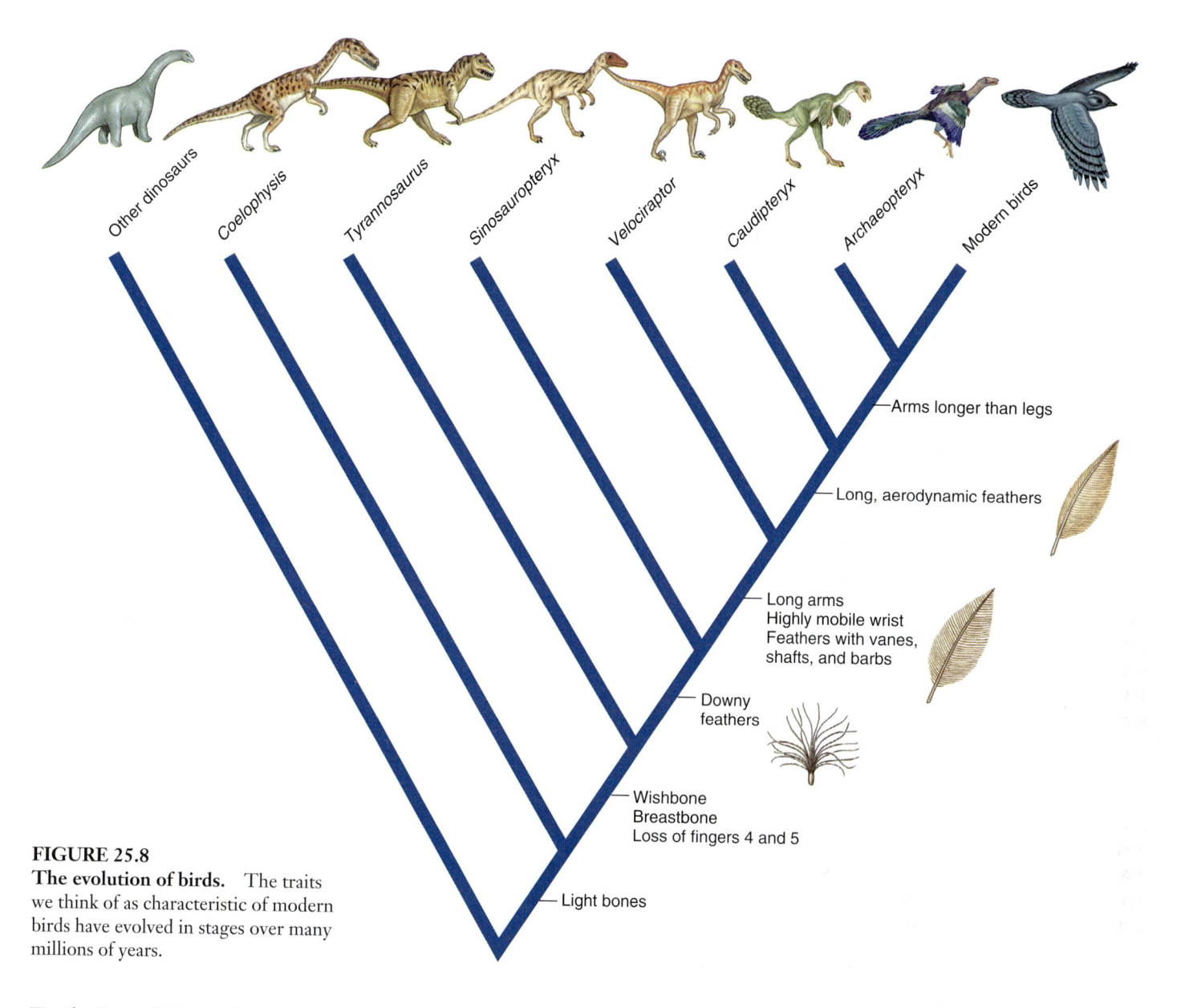

**FIGURE 25.8
The evolution of birds.** The traits we think of as characteristic of modern birds have evolved in stages over many millions of years.

Evolution of Complex Characters

Most complex characters do not evolve, fully formed, in one step. Rather, they are often built up, step-by-step, in a series of evolutionary transitions. Phylogenetic analysis can help discover these evolutionary sequences. Modern-day birds—with their wings, feathers, light bones, and breastbone—appear to be exquisitely adapted flying machines. Fossil discoveries in recent years now allow us to reconstruct the evolution of these features. When the fossils are arranged phylogenetically, it becomes clear that the features characterizing living birds did not appear simultaneously. Figure 25.8 shows how the features important to flight evolved sequentially, probably over a long period of time, in the ancestors of modern birds.

One important finding often revealed by studies of the evolution of complex characters is that the initial stages of a character evolved as an adaptation to some environmental selective pressure other than that for which the character is currently adapted. Examination of figure 25.8 reveals that the first feathery structures evolved deep in theropod phylogeny, in animals whose forearms clearly were not modified for flight. Thus, these structures must have evolved for some other purpose, perhaps to serve as insulation. Through time, these structures were modified, to the extent that modern feathers are well designed to maximize aerodynamic performance.

> Examination of the characters on a cladogram can provide great insight on how they evolved, including whether they have evolved multiple times and how complex characters were constructed evolutionarily.

Chapter 25 Systematics and the Phylogenetic Revolution 517

25.4 All living organisms are grouped into one of a few major categories.

The Kingdoms of Life

The earliest classification systems recognized only two kingdoms of living things: animals and plants. But as biologists discovered microorganisms and learned more about other organisms, they added kingdoms in recognition of certain fundamental differences discovered among organisms. Most biologists now use a six-kingdom system first proposed by Carl Woese of the University of Illinois (figure 25.9a).

In this system, four kingdoms consist of eukaryotic organisms. The two most familiar kingdoms, **Animalia** and **Plantae**, contain only organisms that are multicellular during most of their life cycle. The kingdom **Fungi** contains multicellular forms and single-celled yeasts, which are thought to have multicellular ancestors. Fundamental differences divide these three kingdoms. Plants are mainly stationary, but some have motile sperm; most fungi lack motile cells; animals are mainly motile. Animals ingest their food, plants manufacture it, and fungi digest it by means of secreted extracellular enzymes. Each of these kingdoms probably evolved from a different single-celled ancestor.

The large number of eukaryotes that do not fit in any of the three eukaryotic kingdoms are arbitrarily grouped into a single kingdom called **Protista** (see chapter 28). Most protists are unicellular or, in the case of some algae, have a unicellular phase in their life cycle. This kingdom reflects the current controversy between taxonomic and phylogenetic approaches. Growing knowledge about evolutionary relationships challenges classification systems, even at as high a level of organization as the kingdom. Protists are polyphyletic, meaning that the group does not contain the most recent ancestor of all members of the group. Another way to look at this is that there are several protist lineages with distinct evolutionary origins. These groups are not directly related to each other. Some green algae, the streptophytes (formerly the charophyceans), are more closely related to members of the kingdom Plantae than to most other protist lineages. Most systematists concur that the kingdom Plantae should be extended to include the green algae. To accomodate this new phylogentic understanding, Kingdom Plantae has been renamed the green plant kingdom, or the Virdiplantae.

The remaining two kingdoms, **Archaebacteria** and **Bacteria**, consist of prokaryotic organisms, which are vastly different from all other living things (see chapter 27). Archaebacteria are a diverse group that includes the methanogens and extreme thermophiles, and its members differ from the other prokaryotes, Bacteria. Archaebacteria are not necessarily more ancient than bacteria, as their name erroneously implies. Initially thought to inhabit only extreme environments, archaebacteria have since been found in more diverse habitats.

Domains

As biologists have learned more about the archaebacteria, it has become increasingly clear that this group is very different from all other organisms. When the full genomic DNA sequences of an archaebacterium and a bacterium were first compared in 1996, the differences proved striking. Archaebacteria are as different from bacteria as bacteria are from eukaryotes. Recognizing this, biologists are increasingly adopting a classification of living organisms that recognizes three **domains**, a taxonomic level higher than kingdom (figure 25.9b). Archaebacteria are in one domain, bacteria in a second, and eukaryotes in the third.

> Living organisms are grouped into three general categories called domains. One of the domains, the eukaryotes, is subdivided into four kingdoms: protists, fungi, plants, and animals.

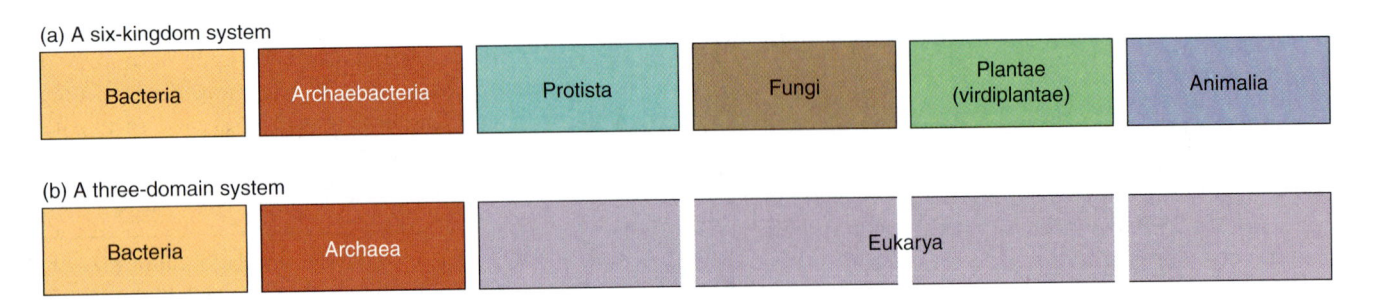

(a) A six-kingdom system

| Bacteria | Archaebacteria | Protista | Fungi | Plantae (virdiplantae) | Animalia |

(b) A three-domain system

| Bacteria | Archaea | Eukarya |

FIGURE 25.9
Different approaches to classifying living organisms. (a) Bacteria and Archaebacteria are so distinct that they have been assigned to separate kingdoms. (b) Even at the domain level, these two groups of prokaryotes are distinct.

Domain Archaea (Archaebacteria)

The archaebacteria (Greek *archaio*, "ancient") seem to have diverged very early from the bacteria and are more closely related to eukaryotes than to bacteria (figure 25.10). This conclusion comes largely from comparisons of genes that encode ribosomal RNAs. The last several years have seen an explosion of DNA sequence information from microorganisms, information that paints a more complex picture. It had been thought that by sequencing numerous microbes we could eventually come up with an accurate picture of the phylogeny of the earliest organisms on earth. The new whole-genome DNA sequence data described in chapter 24 tells us that it will not be that simple. Comparing whole-genome sequences leads evolutionary biologists to a variety of phylogenetic trees, some of which contradict each other. It appears that, during their early evolution, microorganisms have swapped genetic information (lateral gene transfer), making constructing phylogenetic trees very difficult.

As an example of the problem, we can look at *Thermotoga*, a thermophile found on Volcano Island off the coast of Italy. The sequence of one of its RNAs places it squarely within the bacteria near an ancient microbe called *Aquifex*. Recent DNA sequencing, however, fails to support any consistent relationship between the two microbes. There is disagreement as to the effect of lateral gene transfer on the ability of evolutionary biologists to provide accurate phylogenies from molecular data. For now, we will provisionally accept the tree presented in figure 25.10. Over the next few years, we can expect to see considerable change in accepted viewpoints as more and more data are brought to bear.

Although they are a diverse group, all archaebacteria share certain key characteristics (see table 25.1). Their cell walls lack peptidoglycan (an important component of the cell walls of bacteria); the lipids in the cell membranes of archaebacteria have a different structure from those in all other organisms; and archaebacteria have distinctive ribosomal RNA sequences. Some of their genes possess introns, unlike those of bacteria.

The archaebacteria are grouped into three general categories—methanogens, extremophiles, and nonextreme archaebacteria—based primarily on the environments in which they live or their specialized metabolic pathways.

Methanogens obtain their energy by using hydrogen gas (H_2) to reduce carbon dioxide (CO_2) to methane gas (CH_4). They are strict anaerobes, poisoned by even traces of oxygen. They live in swamps, marshes, and the intestines of mammals. Methanogens release about 2 billion tons of methane gas into the atmosphere each year.

Extremophiles are able to grow under conditions that seem extreme to us. There are several types of extremophiles:

Thermophiles ("heat lovers") live in very hot places, typically in temperatures ranging from 60° to 80°C. Many thermophiles are autotrophs, and their metabolisms are based on sulfur. Some thermophilic archaebacteria form the basis of food webs around deep-sea thermal vents, where they must withstand extreme temperatures and pressures. Other types, such as *Sulfolobus*, inhabit the hot sulfur springs of Yellowstone National Park at 70° to 75°C. The recently described *Pyrolobus fumarii* holds the current record for heat stability, with a 106°C temperature optimum and 113°C maximum. It is so heat-tolerant that it is not killed by a one-hour treatment in an autoclave (121°C)!

Halophiles ("salt lovers") live in very salty places, including the Great Salt Lake in Utah, Mono Lake in California, and the Dead Sea in Israel. Whereas the salinity of seawater is around 3%, these archaebacteria thrive in, and indeed require, water with a salinity of 15 to 20%.

pH-tolerant archaebacteria grow in highly acidic (pH = 0.7) and very basic (pH = 11) environments.

Pressure-tolerant archaebacteria, isolated from ocean depths, require at least 300 atmospheres of pressure to survive; they tolerate up to 800 atmospheres!

Nonextreme archaebacteria grow in the same environments bacteria do. As the genomes of archaebacteria have become better known, microbiologists have been able to identify **signature sequences** of DNA present only in archaebacteria. The newly discovered microbe *Nanoarchaeum equitens* was identified as an archaebacterium based on a signature sequence. This odd Icelandic microbe may have the smallest known genome, only 500 base-pairs.

Archaebacteria are poorly understood prokaryotes that inhabit diverse environments, some of them extreme.

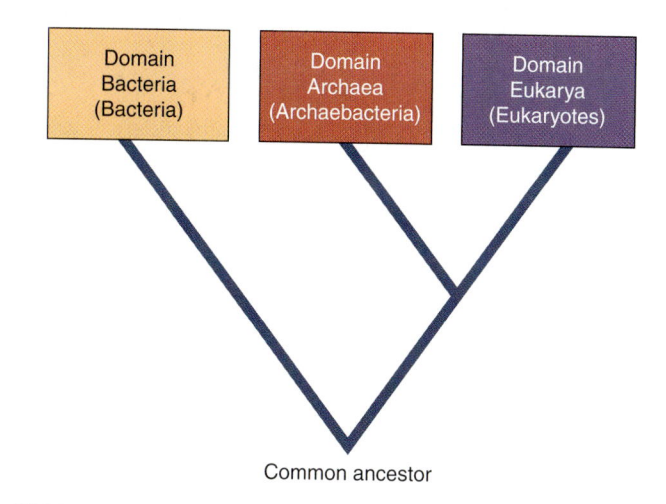

FIGURE 25.10

An evolutionary relationship among the three domains. Members of the domain Bacteria are thought to have diverged early from the evolutionary line that gave rise to the archaebacteria and eukaryotes.

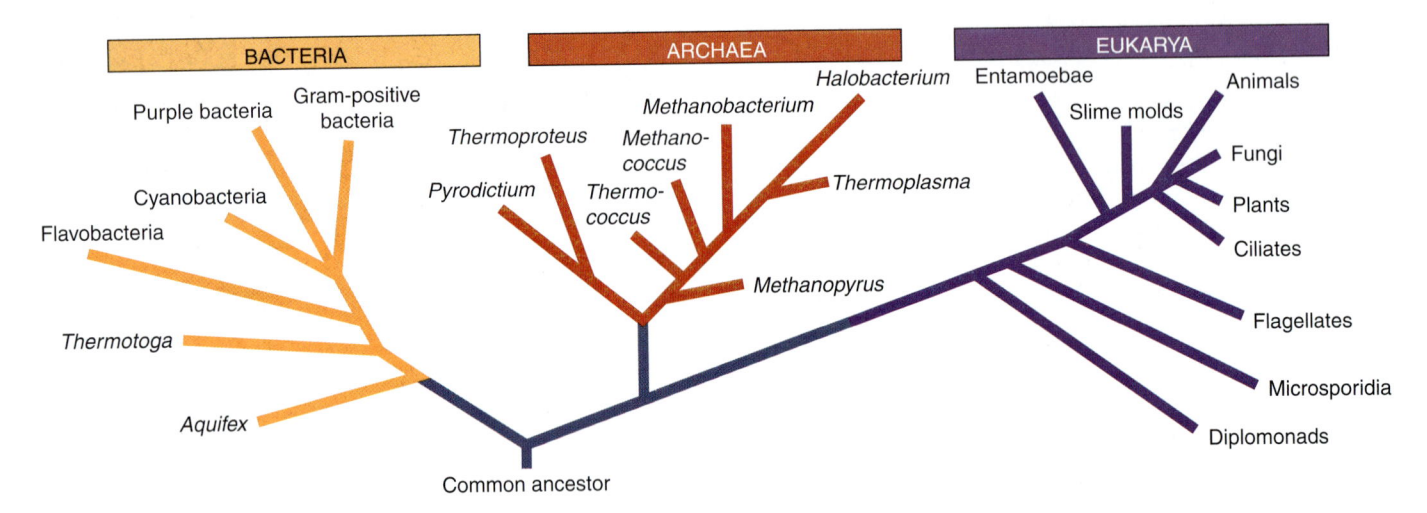

FIGURE 25.11
A tree of life. This phylogeny, prepared from rRNA analyses, shows the evolutionary relationships among the three domains. The base of the tree was determined by examining genes that are duplicated in all three domains, the duplication presumably having occurred in the common ancestor. Archaebacteria and eukaryotes diverged later than bacteria, and are more closely related to each other than either is to bacteria. Bases of trees constructed with other traits are often less clear because of lateral gene transfer (see chapter 24).

Domain Bacteria (Bacteria)

The bacteria are the most abundant organisms on earth. There are more living bacteria in your mouth than there are mammals living on earth. Although too tiny to see with the unaided eye, bacteria play critical roles throughout the biosphere. They extract from the air all the nitrogen used by organisms, and play key roles in cycling carbon and sulfur. Much of the world's photosynthesis is carried out by bacteria. However, certain groups of bacteria are also responsible for many forms of disease. Understanding their metabolism and genetics is a critical part of modern medicine.

There are many different kinds of bacteria, and the evolutionary links between them are not well understood. While taxonomists disagree about the details of bacterial classification, most recognize 12 to 15 major groups of bacteria. Comparisons of the nucleotide sequences of ribosomal RNA (rRNA) molecules are beginning to reveal how these groups are related to one another and to the other two domains. One view of our current understanding of the "tree of life" is presented in figure 25.11. The oldest divergences represent the deepest-rooted branches in the tree. The archaebacteria and eukaryotes are more closely related to each other than to bacteria and are on a separate evolutionary branch of the tree, even though archaebacteria and bacteria are both prokaryotes.

Bacteria are as different from archaebacteria as they are from eukaryotes.

Table 25.1 Features of the Domains of Life

Feature	Archaea	Bacteria	Eukarya
Amino acid that initiates protein synthesis	Methionine	Formyl-methionine	Methionine
Introns	Present in some genes	Absent	Present
Membrane-bounded organelles	Absent	Absent	Present
Membrane lipid structure	Branched	Unbranched	Unbranched
Nuclear envelope	Absent	Absent	Present
Number of different RNA polymerases	Several	One	Several
Peptidoglycan in cell wall	Absent	Present	Absent
Response to the antibiotics streptomycin and chloramphenicol	Growth not inhibited	Growth inhibited	Growth not inhibited

Domain Eukarya (Eukaryotes)

For at least 1 billion years, prokaryotes ruled the earth. No other organisms existed to eat them or compete with them, and their tiny cells formed the world's oldest fossils. Members of the third great domain of life, the eukaryotes, appear in the fossil record much later, only about 2.5 billion years ago. However, despite the metabolic similarity of eukaryotic cells to prokaryotic cells, their structure and function enabled cells to be larger, and eventually, allowed multicellular life to evolve.

Four Kingdoms of Eukaryotes

The first eukaryotes were unicellular organisms. A wide variety of unicellular eukaryotes exist today, grouped together in the kingdom Protista (along with some multicellular descendants) on the basis that they do not fit into any of the other three kingdoms of eukaryotes. Protists vary from the relatively simple, single-celled amoeba to multicellular organisms such as kelp, which can be 20 meters long.

Fungi, plants, and animals are largely multicellular kingdoms, each a distinct evolutionary line from a single-celled ancestor that would be classified in the kingdom Protista. Because of the size and ecological dominance of plants, animals, and fungi, and because they are predominantly multicellular, we recognize them as kingdoms distinct from Protista, even though the amount of diversity among the protists is much greater than that within or between the fungi, plants, and animals.

Endosymbiosis and the Origin of Eukaryotes

The hallmark of eukaryotes is complex cellular organization, highlighted by an extensive endomembrane system that subdivides the eukaryotic cell into functional compartments. Not all cellular compartments, however, are derived from the endomembrane system. With few exceptions, modern eukaryotic cells possess energy-producing organelles called mitochondria, and some eukaryotic cells possess chloroplasts, which are energy-harvesting organelles. Mitochondria and chloroplasts are both believed to have entered early eukaryotic cells by a process called **endosymbiosis**. We discussed the theory of the endosymbiotic origin of mitochondria and chloroplasts in chapters 4 and 5; their origin is also diagrammed in figure 25.12. Both organelles contain their own ribosomes, which are more similar to bacterial ribosomes than to eukaryotic cytoplasmic ribosomes. They manufacture their own inner membranes. They divide independently of the cell and contain chromosomes similar to those in bacteria. Comparison of the nucleotide sequence of this DNA with that of a variety of organisms indicates clearly that mitochondria are the descendants of purple nonsulfur bacteria that were incorporated into eukaryotic cells early in the history of the group, while chloroplasts are derived from cyanobacteria.

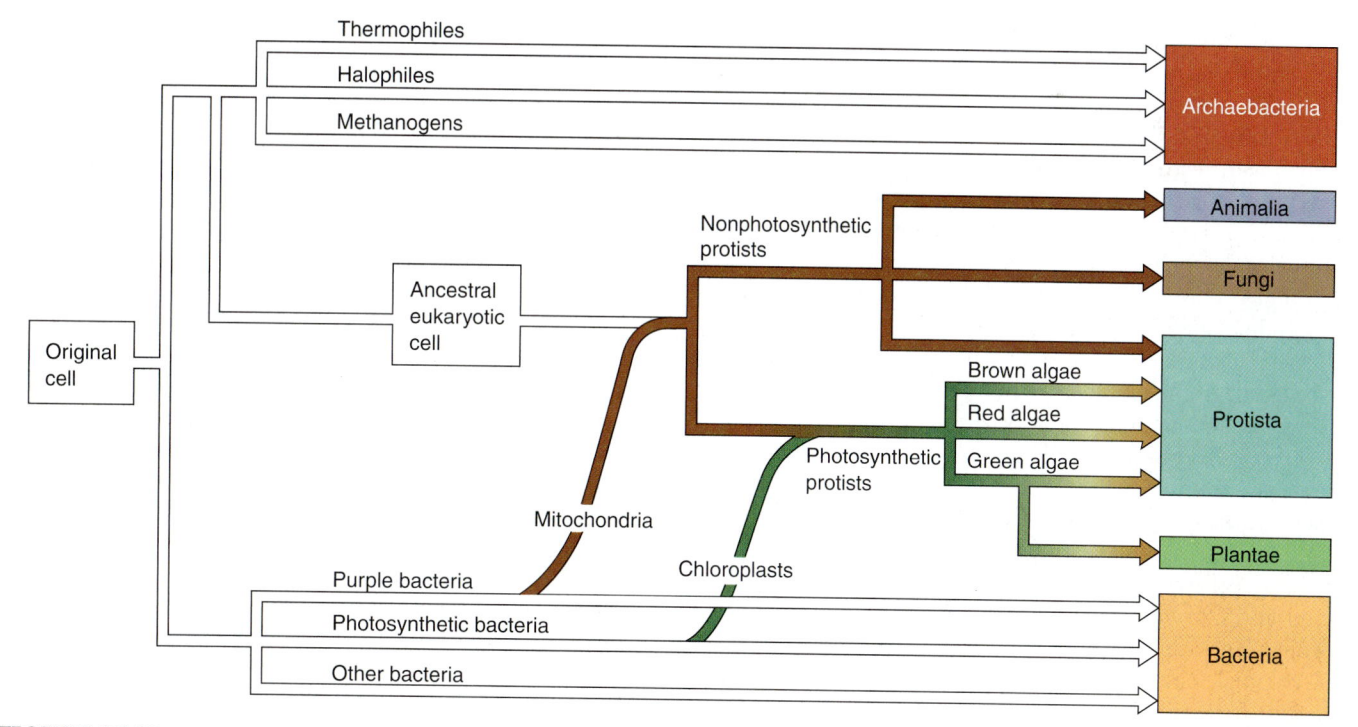

FIGURE 25.12
Diagram of the evolutionary relationships among the six kingdoms of organisms. The colored lines indicate symbiotic events.

Key Characteristics of Eukaryotes

Although eukaryotic organisms are extraordinarily diverse, they share three characteristics that distinguish them from prokaryotes: compartmentalization, multicellularity in many, but not all, eukaryotes, and sexuality.

Compartmentalization. Discrete compartments provide evolutionary opportunities for increased specialization within the cell, as we see with chloroplasts and mitochondria. The evolution of a nuclear membrane also accounts for increased complexity in eukaryotes. For example, a discrete nucleus led to new ways to control gene expression. In prokaryotes, both transcription and translation occur at the same time, and often a protein is being translated from a strand of RNA that is still being transcribed from its template DNA (see figure 15.13). In eukaryotes, RNA transcripts are processed and transported across the nuclear membrane into the cytosol, where ribosomes and other translational machinery are present to initiate the translation of the RNA message into protein. The physical separation of transcription and translation in eukaryotes adds additional levels of control to the process of gene expression.

Multicellularity. The unicellular body plan has been tremendously successful, with unicellular prokaryotes and eukaryotes constituting about half of the biomass on earth. Yet, a single cell has limits. The evolution of multicellularity allowed organisms to deal with their environments in novel ways. Distinct types of cells, tissues, and organs can be differentiated within the complex bodies of multicellular organisms. With such a functional division within its body, a multicellular organism can do many things, including protect itself, resist drought efficiently, regulate its internal conditions, move about, seek mates and prey, and carry out other activities on a scale and with a complexity that would be impossible for its unicellular ancestors. With all these advantages, it is not surprising that multicellularity has arisen independently so many times.

True multicellularity, in which the activities of individual cells are coordinated and the cells themselves are in contact, occurs only in eukaryotes and is one of their major characteristics. The cell walls of bacteria occasionally adhere to one another, and bacterial cells may also be held together within a common sheath. Some bacteria form filaments, sheets, or three-dimensional aggregates, but the individual cells remain independent of each other, reproducing and carrying on their metabolic functions without coordinating with the other cells. Such bacteria are considered colonial, but none are truly multicellular. Many protists also form similar colonial aggregates of many cells with little differentiation or integration.

Other protists—the red, brown, and green algae, for example—have independently attained multicellularity. One lineage of multicellular green algae was the ancestor of the plants (see chapters 28 and 29), and most taxonomists now place its members in the green plant kingdom, (called the Virdiplantae). Historically, the plant kingdom included only multicellular land plants, a group that arose from a single ancestor in terrestrial habitats and that has a unique set of characteristics. Aquatic plants are recent derivatives of land plants.

The multiple origins of multicellularity are also seen in the fungi and animals that arose from unicellular protist ancestors with different characteristics. As we will see in subsequent chapters, the groups that seem to have given rise to each of these kingdoms are still in existence.

Sexuality. Another major characteristic of eukaryotic organisms as a group is sexuality. Although some interchange of genetic material occurs in bacteria (see chapter 27), it is certainly not a regular, predictable mechanism in the same sense that sex is in eukaryotes. The sexual cycle characteristic of eukaryotes alternates between **syngamy**, the union of male and female gametes producing a cell with two sets of chromosomes, and **meiosis**, cell division producing daughter cells with one set of chromosomes. This cycle differs sharply from any exchange of genetic material occurring in prokaryotes. The evolution of meiosis correlates with a huge increase in the number of species on earth.

As we have seen, in diploid cells, one set of chromosomes comes from the male parent and one from the female parent. These chromosomes segregate during meiosis. Because crossing over frequently occurs during meiosis (see chapter 12), no two products of a single meiotic event are ever identical. As a result, the offspring of sexual, eukaryotic organisms vary widely, thus providing the raw material for evolution. Sexual reproduction, with its regular alternation between syngamy and meiosis, produces genetic variation. Sexual organisms can adapt to the demands of their environments because they produce a variety of progeny.

In many of the unicellular phyla of protists, sexual reproduction occurs only occasionally. The first eukaryotes were probably haploid. Diploids seem to have arisen on a number of separate occasions by the fusion of haploid cells, which then eventually divided by meiosis.

The characteristics of the six kingdoms are outlined in table 25.2; note that the archaebacteria and bacteria are grouped in the same column.

> **Eukaryotic cells are highly compartmentalized, and they acquired mitochondria and chloroplasts by endosymbiosis. The complex differentiation we associate with many life-forms depends on multicellularity and sexuality, which must have been highly advantageous to have evolved independently so often.**

Table 25.2 Characteristics of the Six Kingdoms

	Archaebacteria and Bacteria	Protista	Plantae	Fungi	Animalia
Cell Type	Prokaryotic	Eukaryotic	Eukaryotic	Eukaryotic	Eukaryotic
Nuclear Envelope	Absent	Present	Present	Present	Present
Transcription and Translation	Occur in same compartment	Occur in different compartments	Occur in different compartments	Occur in different compartments	Occur in different compartments
Histone Proteins Associated with DNA	Absent	Present	Present	Present	Present
Cytoskeleton	Absent	Present	Present	Present	Present
Mitochondria	Absent	Present (or absent)	Present	Present	Present
Chloroplasts	None (photosynthetic membranes in some types)	Present (some forms)	Present	Absent	Absent
Cell Wall	Noncellulose (polysaccharide plus amino acids)	Present in some forms, various types	Cellulose and other polysaccharides	Chitin and other noncellulose polysaccharides	Absent
Means of Genetic Recombination, If Present	Conjugation, transduction, transformation	Fertilization and meiosis	Fertilization and meiosis	Fertilization and meiosis	Fertilization and meiosis
Mode of Nutrition	Autotrophic (chemosynthetic, photosynthetic) or heterotrophic	Photosynthetic or heterotrophic, or combination of both	Photosynthetic, chlorophylls a and b	Absorption	Ingestion
Motility	Bacterial flagella, gliding or nonmotile	9 + 2 cilia and flagella; amoeboid, contractile fibrils	None in most forms; 9 + 2 cilia and flagella in gametes of some forms	Both motile and nonmotile	9 + 2 cilia and flagella, contractile fibrils
Multicellularity	Absent	Absent in most forms	Present in all forms	Present in most forms	Present in all forms
Nervous System	None	Primitive mechanisms for conducting stimuli in some forms	A few have primitive mechanisms for conducting stimuli	None	Present (except sponges), often complex

Viruses: A Special Case

Viruses possess only a portion of the properties of organisms. **Viruses** are literally "parasitic" chemicals, segments of DNA or RNA wrapped in a protein coat. They cannot reproduce on their own, and for this reason they are not considered alive by biologists. They can, however, reproduce within cells, often with disastrous results to the host organism. Earlier theories that viruses represent a kind of halfway point between life and nonlife have largely been abandoned. Instead, viruses are now viewed as detached fragments of the genomes of organisms due to the high degree of similarity found among some viral and eukaryotic genes.

Viruses thus present a special classification problem. Because they are not organisms, we cannot logically place them in any of the kingdoms. Viruses are really just complicated associations of molecules, bits of nucleic acids usually surrounded by a protein coat. But, despite their simplicity, viruses are able to invade cells and direct the genetic machinery of these cells to manufacture more of the molecules that make up the virus. Viruses can infect organisms at all taxonomic levels.

Viruses vary greatly in appearance and size. The smallest are only about 17 nanometers in diameter, and the largest are up to 1000 nanometers (1 micrometer) in their greatest dimension (figure 25.13). The largest viruses are barely visible with a light microscope, but viral morphology is best revealed using the electron microscope. Viruses are so small that they are comparable to molecules in size; a hydrogen atom is about 0.1 nanometer in diameter, and a large protein molecule is several hundred nanometers in its greatest dimension.

Biologists first began to suspect the existence of viruses near the end of the nineteenth century. European scientists attempting to isolate the infectious agent responsible for hoof-and-mouth disease in cattle concluded that it was smaller than a bacterium. Investigating the agent further, the scientists found that it could not multiply in solution—it could only reproduce itself within living host cells that it infected. The infecting agents were called viruses.

The true nature of viruses was discovered in 1933, when the biologist Wendell Stanley prepared an extract of a plant virus called tobacco mosaic virus (TMV) and attempted to purify it. To his great surprise, the purified TMV preparation precipitated (that is, separated from solution) in the form of crystals. This was surprising because precipitation is something that only chemicals do—the TMV virus was acting like a chemical off the shelf rather than like an organism. Stanley concluded that TMV is best regarded as just that—chemical matter rather than a living organism.

Within a few years, scientists disassembled the TMV virus and found that Stanley was right. TMV was not cellular but chemical. Each particle of TMV virus is in fact a mixture of two chemicals: RNA and protein. The TMV virus has the structure of a twinkie, a tube made of an RNA core surrounded by a coat of protein. Later workers were able to separate the RNA from the protein and purify and store each chemical. Then, when they reassembled the two components, the reconstructed TMV particles were fully able to infect healthy tobacco plants, so clearly the particles were the virus itself, not merely chemicals derived from it. Further experiments carried out on other viruses yielded similar results.

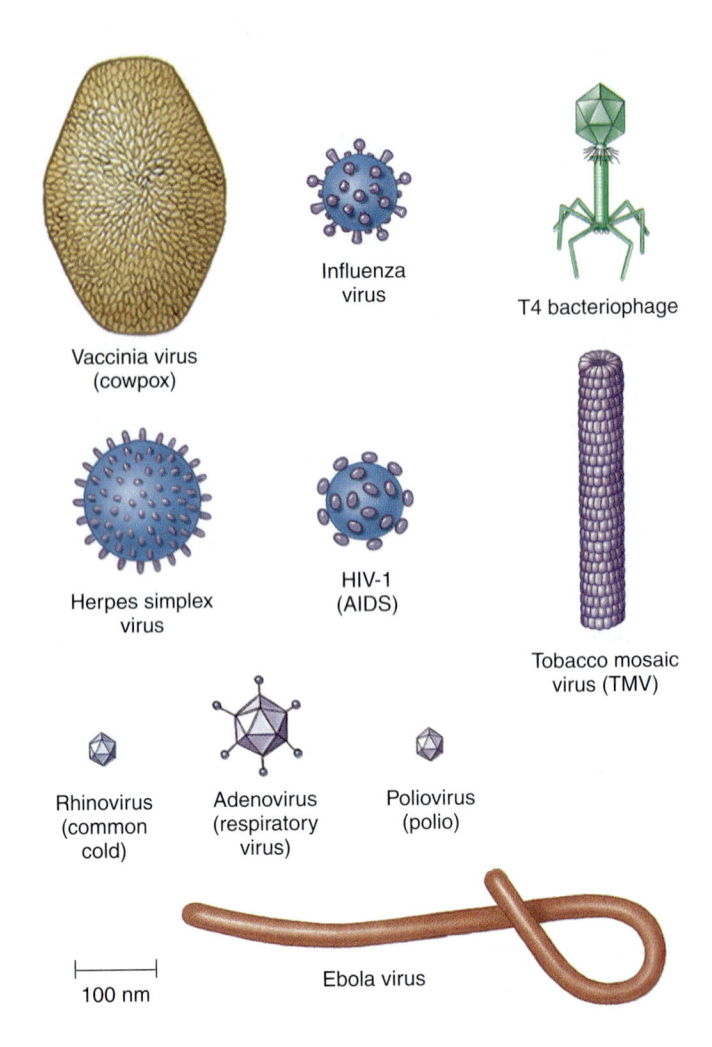

FIGURE 25.13
Viral diversity. Viruses exhibit extensive diversity in shape and size. At the scale these sample viruses are shown, a human hair would be nearly 8 meters thick.

> **Viruses are chemical assemblies that can infect cells and replicate within them. They are not organisms, and are not classified in the kingdoms of life.**

25.5 Molecular data are revolutionizing taxonomy.

The Impact of Molecular Cladistics

In reading this chapter and chapter 24, you may have sensed some tension between traditional classification systems and systems based on evolutionary relationships at the molecular as well as the cellular and organismal levels. Traditional classification systems group organisms based on morphological similarities that seem the most important, but there is no clear basis for determining which traits matter most. As a result of this taxonomic tradition, some species have widely used scientific names that do not reflect accurate evolutionary relationships. In chapter 24, you saw how molecular data are being added to character sets and used in phylogenetic analyses aimed at identifying common ancestry. As more is learned about organisms, taxonomic groupings are changing rapidly.

Systematic phylogeneticists use cladistics to build taxonomic hierarchies. Each clade has a single common ancestor. A taxonomic hierarchy arises as sytematists nest cladograms together into a phylogeny. It is important to remember that, even with exponentially increasing molecular sequence data, these phylogenies are hypotheses. Cladistic approaches emphasize the order of evolutionary events, but do not provide information about how highly diverged two groups are or are not. Lively debate about classification continues and is fueled not only by DNA sequences, but by efforts to integrate fossil data into phylogenies. In this chapter, we highlight a few of the major changes in classification. As you proceed with the subsequent chapters, you will gain an appreciation for both the diversity of living organisms and the challenges in understanding their evolutionary relationships.

> **Traditional classification systems are based on similar traits, but do not take into account evolutionary relationships. Systematic phylogenetics is based on evolutionary relationships established using cladistics. Traits can be morphological or molecular.**

Making Sense of the Protists

The weakest part of the six-kingdom classification system shown in figure 25.9 is the kingdom Protista. Eukaryotes diverged rapidly in a world that was shifting from anaerobic to aerobic conditions. We may never be able to completely sort out the relationships among different lineages during this major evolutionary transition. However, molecular systematics clearly shows that the protists are paraphyletic (figure 25.14). While we continue to use the term protist as a catchall for any eukaryote that is not a plant, fungus, or animal, this grouping is not based on evolutionary relationships.

What to do with the 200,000 protists? The six main branchings of protists, shown at the base of figure 25.14, represent a current working hypothesis, although at least 60 protists do not seem to fit in any of the six groups. Choanoflagellates are most closely related to sponges, and indeed to all animals. The green algae can be split into two monophyletic groups, one of which gave rise to land plants (the evolution of land plants is explored in more detail on the next page). Many sytematists are calling for the new kingdom called Virdiplantae, or the green plant kingdom, includes all the green algae (not the red or brown algae) and the land plants. Thus the definition of a plant has been expanded beyond those species that made it onto land. Although the kingdom Protista is in ruins, our understanding of the evolutionary relationships among these early eukaryotes is growing exponentially.

> **Molecular systematics and cladistics have led to a new understanding of the relationships among organisms formerly classified as members of kingdom Protista.**

FIGURE 25.14
The fall of kingdom Protista. Systematists have shown that protists as a group are not monophyletic. Note how some lineages are actually more closely related to plants or animals than they are to other protists.

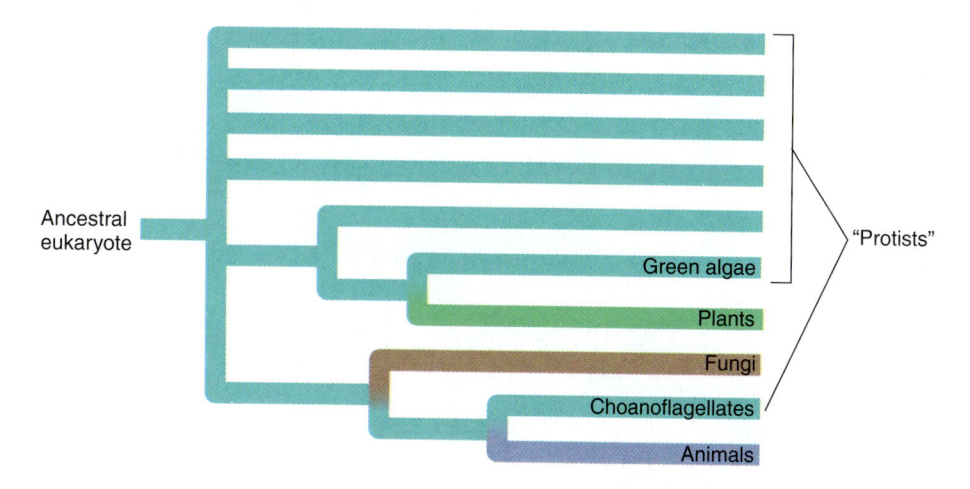

Origin of Land Plants

The origin of land plants from a green algal ancestor has long been recognized as a major evolutionary event. The phylogenetic relationships among the algae and the first land plants have been fuzzy and subject to long debate. Cell biology, biochemistry, and molecular systematics have provided surprising answers. The green algae consist of two monophyletic groups, the Chlorophyta and the Streptophyta. The land plants are actually members of the Streptophyta, rather than a separate kingdom. Within the Linnaean hierarchy, this means that new phylogenetic information demoted the land plants from a kingdom to a branch within the algal group Streptophyta. The Streptophyta, along with the sister green algal clade Chlorophyta, are now considered by most to make up the kingdom Virdiplantae, mentioned previously (figure 25.15).

In the past decade, new hypotheses about relationships within the Streptophyta have emerged, with the current phylogeny composed of seven clades as shown in figure 25.15. What was the earliest streptophyte? Conflicting answers have been obtained with different phylogenetic analyses, but growing evidence supports the hypothesis that the scaly, unicellular flagellate *Mesostigma* (order Mesostigmatales) represents the earliest streptophyte branch.

Which of the Streptophyta clades contains the closest living relative of land plants? The two contenders have been the Charales, with about 300 species, and the Coleochaetales, with about 30 species. Both lineages are freshwater algae, but the Charales are huge compared to the microscopic Coleochaetales. At the moment, the Charales appear to be the sister clade to land plants, with the Coleochaetales the next closest relatives. Charales fossils dating back 420 million years indicate that the common ancestor of land plants was a relatively complex freshwater alga.

> Molecular phylogenetics has changed the way algae and plants are classified. The challenges to existing taxonomies have been sweeping—from kingdom-level changes to a reexamination of the relationships among orders.

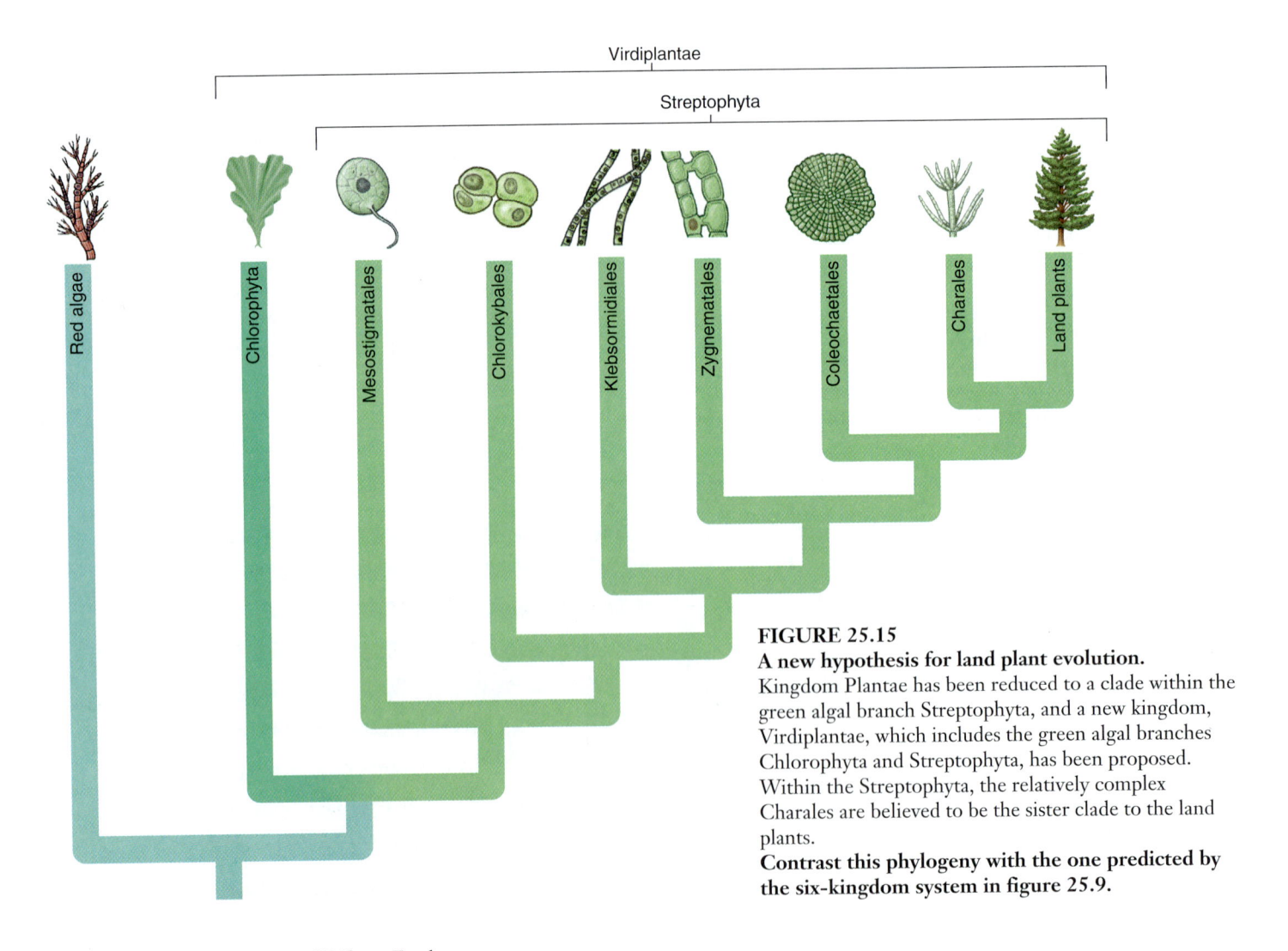

FIGURE 25.15
A new hypothesis for land plant evolution.
Kingdom Plantae has been reduced to a clade within the green algal branch Streptophyta, and a new kingdom, Virdiplantae, which includes the green algal branches Chlorophyta and Streptophyta, has been proposed. Within the Streptophyta, the relatively complex Charales are believed to be the sister clade to the land plants.
Contrast this phylogeny with the one predicted by the six-kingdom system in figure 25.9.

Sorting Out the Animals

Molecular systematics is leading to a revision of evolutionary history in all kingdoms, including the animals. Some phylogenies are changing, and others, including mammalian phylogenies, are actually being written for the first time. Next, we will explore three examples: the relationship between annelids and arthropods, relationships within the arthropods, and the discovery of phylogenetic relationships among mammals.

The Problem Posed by Segmentation

Morphological traits such as segmentation have been used to group arthropods and annelids together, but comparisons of rRNA sequences are raising questions about how closely the segmented worms (annelids) are related to arthropods, which include insects, spiders, and crustaceans (e.g., lobsters). As rRNA sequences are obtained, it is becoming increasingly clear that annelids and arthropods are more distantly related than we previously believed.

Chordates, including humans, also have segmented body plans that evolved independently of segmentation in annelids and arthropods. The vertebrae in your backbone reflect segmentation. However, developmental differences have long been used to separate chordates from the annelids and arthropods. For example, annelids and arthropods are called *protostomes* because the mouth forms before the anus, while chordates are grouped with the *deuterostomes*, in which the mouth develops second. Until molecular data became available, segmentation was thought to have evolved twice, once in the protostomes and once in the deuterostomes.

With the addition of molecular traits, annelids and arthropods fell into two distinct protostome branches (figure 25.16): lophotrochozoans and ecdysozoans. These two branches have been evolving independently since ancient times. Lophotrochozoans include flatworms, mollusks, and annelids. Two ecdysozoan phyla have been particularly successful, roundworms (nematodes) and arthropods.

In the new protostome phylogeny, annelids and arthropods do not constitute a monophyletic group. This means that segmentation arose twice, not once, in the protostomes and then once again in the chordates (figure 25.16). How did this happen? The most likely explanation is that members of the same family of genes were coopted at least three times to create segmented body plans. Segmentation is regulated by the *Hox* gene family that contains a homeodomain region (review chapter 19). The *Hox* ancestral genes predate the ecdysozoans and lophotrochozoans. The ancestor of the lophotrochozoans, ecdysozoans, and deuterostomes most likely already had seven *Hox* genes. Some of these genes appear to have evolved a role in segmentation.

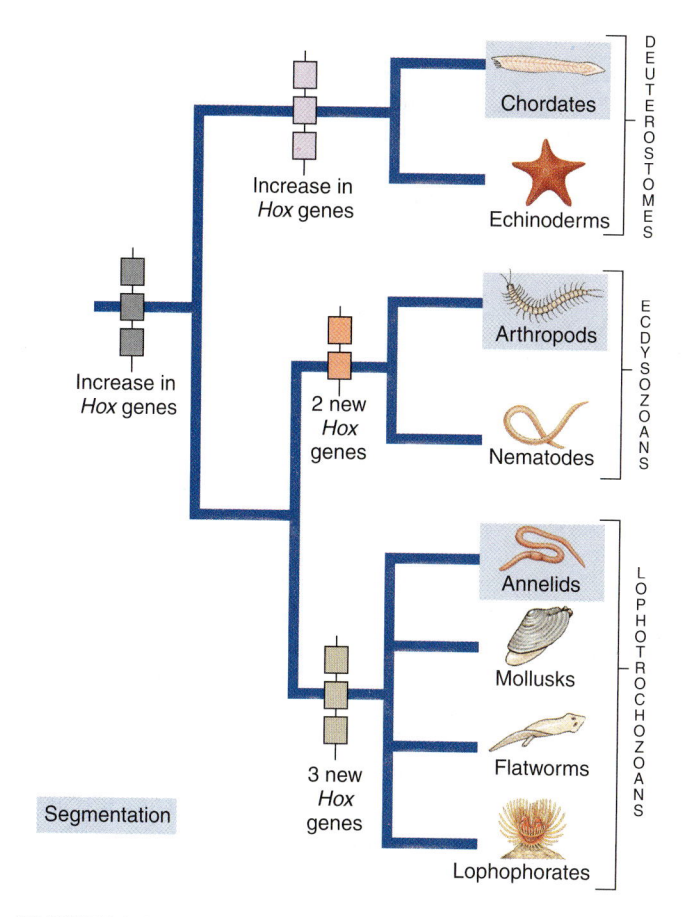

FIGURE 25.16
Multiple origins of segmentation. New phylogenies based on ribosomal RNA show that segmentation in arthropods and annelids arose independently. Segmentation in both appears to be regulated by some of the *Hox* genes.

Is an Insect a Flying Crustacean?

Arthropods are the most diverse of all the animal phyla, with more species than all other animal phyla combined, most of them insects. Insects have traditionally been set apart from the crustaceans, grouped instead with the myriapods (centipedes and millipedes) in a taxon called Tracheata. This phylogeny, still widely employed, dates back to benchmark work by Robert Snodgrass in the 1930s. He pointed out that insects, centipedes and millipedes are united by several seemingly powerful attributes, including uniramous (single-branched) legs. All crustacean appendages, by contrast, are basically biramous, or "two-branched" (figure 25.17), although some of these appendages have become single-branched by reduction in the course of their evolution.

Taxonomists have traditionally assumed a character such as two-branched appendages to be a fundamental one, conserved over the course of evolution, and thus suitable for making taxonomic distinctions. The patterning of appendages among arthropods is orchestrated by *Hox* genes. A single one of these *Hox* genes, called *Distal-less*, has been shown to initiate development of unbranched limbs in insects and branched limbs in crustaceans. The same *Distal-less* gene is found in many animal phyla, including the vertebrates. *Distal-less* appears to be necessary to initiate limb development, and it turns on genes that are more directly involved in the development of the limb itself. Evolutionary changes in the genes that *Distal-less* acts upon most likely account for differences in limb morphology.

In recent years, a mass of accumulating morphological and molecular data has led many taxonomists to suggest new arthropod phylogenies. The most revolutionary of these, championed by Richard Brusca of Columbia University, considers crustaceans to be the basic arthropod group, and insects a close sister group. Molecular phylogenies all place insects as a sister group to crustaceans, not myriapods. In conflict with 150 years of morphology-based thinking, these conclusions are certain to engender lively discussion.

Sorting Out the Mammalian Family Tree

In the preceding arthropod examples, evolutionary history has been rewritten. In mammals, however, parts of the phylogeny are just emerging, based on molecular data. Among the vertebrate classes, mammals are unique because they have mammary glands to feed their young. Mammals fall into three major groups. Monotremes, such as the platypus, are the only egg-laying mammals. Marsupials, such as kangaroos, are born quite early in development and continue development while nursing in a pouch. The majority of mammals—over 90%—are eutherians, or placental mammals. There are at least 18 extant (living) orders of eutherians, which tend to be divided into four major groups. The first major split occurred between the African clade (figure 25.18) and the other placental mammals when South America and Africa separated about 103 million years ago. Aardvarks and elephants are part of this African lineage, a clade we did not even know existed a decade ago. In South America, anteaters and armadillos soon appeared. Then two other branches arose—one including whales, horses, and carnivores, and the other, primates and rodents. Sorting out the relationships within these branches is an ongoing challenge.

Where does an aquatic mammal such as the whale fit? Whale fossils from 55 million years ago have hindlimbs. Whales were initially thought to be relatives of horses, camels, deer, antelope, pigs, and hippopotamuses based on morphological information from fossils and extant animals. DNA sequence data, however, revealed a particularly close relationship between whales and hippopotamuses. With this new phylogenetic information, the possibility arises that some adaptations to aquatic environments in both species had a common origin. Understanding evolutionary

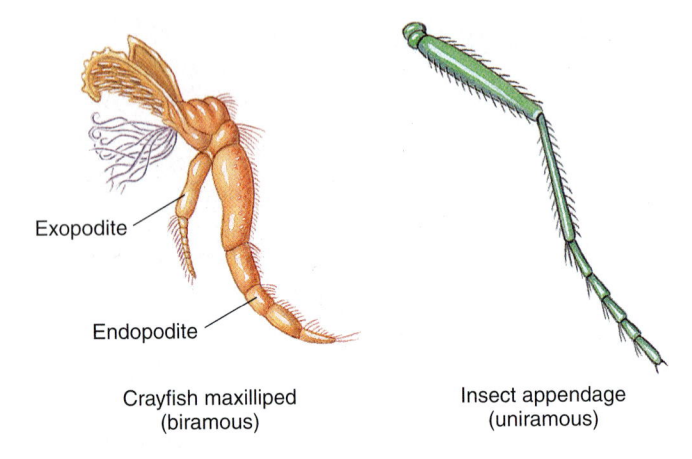

FIGURE 25.17
Branched and single appendages. A biramous leg in a crustacean (crayfish) and a uniramous leg in an insect.

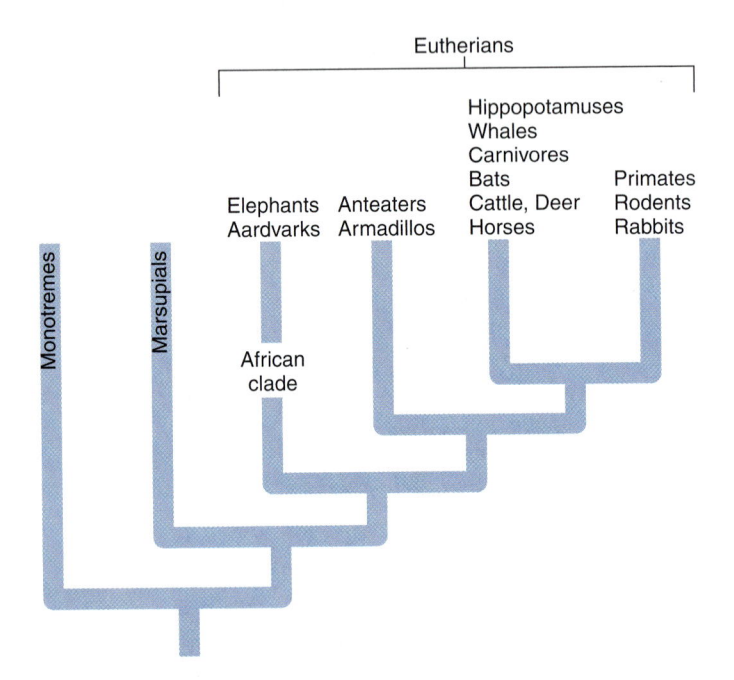

FIGURE 25.18
Major groups of mammals.

relationships among organisms does more than provide biologists with a sense of order and a logical way to name organisms. A phylogenetically based taxonomy allows researchers to ask important questions about physiology, behavior, and development using information already known about a related species.

> **Molecular systematics and cladistic approaches are providing new information about the evolutionary relationships among animals, including members of our own class, the mammals.**

Concept Review

For interactive testing, visit the Online Learning Center with PowerWeb at www.mhhe.com/Raven7

25.1 Biologists name organisms in a systematic way.
The Classification of Organisms
- Organisms were first classified by Aristotle over 2000 years ago. (p. 510)
- Similar individuals were eventually classified into units called genera (groups). (p. 510)
- Many descriptive terms were used in naming until the mid-1700s, when Linnaeus began using a binomial system. (p. 510)
- Taxonomy is the science of classifying living organisms; a group of similar organisms is put into a shared taxon. (p. 510)
- Taxonomic classifications are composed of (in ascending order): species, genus, family, order, class, phylum, kingdom, and domain. Each level may include anywhere from one to several taxa. (p. 511)

25.2 Scientists construct phylogenies to understand the evolutionary relationships among species.
Systematics
- The field of systematics constructs and studies evolutionary relationships (phylogenies). (p. 512).
- Descendant species should be relatively similar to their ancestors, and the amount of time since separation should be positively correlated with the degree of genetic divergence. But evolution is not constant in its rate, not unidirectional, and not always divergent. (p. 512)
- Cladistics distinguishes between ancestral and derived similarity, and only considers shared derived characters informative. (p. 512)
- Cladograms are constructed to show hypothesized evolutionary relationships, with species sharing derived characteristics put into the same clade. (p. 513)
- Systematists use the principle of parsimony and favor hypotheses with the fewest assumptions and the fewest evolutionary events. (p. 514)
- A monophyletic group contains the most recent common ancestor and all its descendants; a paraphyletic group does not include all the descendants; and a polyphyletic group does not include the most recent ancestor or all the members of the group. (p. 514)

25.3 Phylogenetics is the basis of all comparative biology.
Analogy Versus Homology
- Homologous structures are derived from the same body part in a common ancestor, while analogous structures are derived from different ancestral structures. (p. 516)
- Most complex characters are built up in a stepwise fashion, often with a series of evolutionary transitions. (p. 517)

25.4 All living organisms are grouped into one of a few major categories.
The Kingdoms of Life
- Most biologists now use a six-kingdom system. The kingdoms Animalia, Plantae (Virdiplantae), Fungi, and Protista contain eukaryotic organisms, while Archaebacteria and Bacteria contain prokaryotic organisms. (p. 518)
- Three different domains, a classification level higher than kingdom, are now recognized as well: Archaebacteria, Bacteria, and Eukarya. (p. 518)

Domain Archaea (Archaebacteria)
- Archaebacteria appear to have diverged from bacteria early, and are more closely related to eukaryotes than to bacteria. (p. 519)
- The archaebacteria are grouped into three general categories based on their environment: methanogens, extremophiles (e.g., thermophiles and halophiles), and nonextreme archaebacteria. (p. 519)

Domain Bacteria (Bacteria)
- Bacteria are the most abundant organisms on earth, and they carry out much of the earth's photosynthesis. (p. 520)
- Most taxonomists recognize 12–15 major groups of bacteria. (p. 520)

Domain Eukarya (Eukaryotes)
- Eukaryotes appear in the fossil record only about 2.5 billion years ago. (p. 521)
- Complex cellular organization with an extensive system of intracellular membranes is the most distinctive characteristic of eukaryotes. (p. 521)
- Compartmentalization, multicellularity (in many eukaryotes), and sexuality are all key characteristics of eukaryotes. (p. 522)

Viruses: A Special Case
- Viruses are segments of DNA or RNA wrapped in a protein coat. They can infect cells and use cellular machinery to replicate but are not considered living organisms. (p. 524)
- Viruses defy taxonomic efforts. (p. 524)

25.5 Molecular data are revolutionizing taxonomy.
The Impact of Molecular Cladistics
- Within traditional classification systems, no clear basis exists for determining the most important traits. (p. 525)
- Systematic phylogenies use cladistics to build taxonomic hierarchies. (p. 525)

Making Sense of the Protists
- Molecular systematics demonstrates that protists are paraphyletic. (p. 525)
- "Protist" is a catchall word encompassing 200,000 species. (p. 525)

Origin of Land Plants
- Although the origin of land plants from a green algal ancestor has long been recognized, new evolutionary relationships have been hypothesized. (p. 526)

Sorting Out the Animals
- New RNA investigations are raising questions about how closely annelids are related to arthropods. (p. 527)
- Arthropods are the most diverse of all the animal phyla. (p. 527)
- New DNA evidence is also bringing into question the traditional evolutionary paths of some mammals. (p. 528)
- Over 90% of mammals are placental mammals. (p. 528)

Test Your Understanding

For interactive testing, visit the Online Learning Center with PowerWeb at www.mhhe.com/Raven7

Self Test

1. There are over 300 different species of fiddler crabs. In all of these species, the adult males possess an enlarged front claw. Based on this feature, biologists group these different species together in the same genus, *Uca*. Therefore, the large claw is
   a. an outgroup.
   b. an analogous trait.
   c. a homoplasy.
   d. a synapomorphy.
2. The domestic dog, *Canis familiaris*, is related to wild dog species such as the gray wolf, *Canis lupus*. To which most exclusive category do the wolf and dog both belong?
   a. order
   b. family
   c. genus
   d. species
3. The simplest animals, the sponges (Parazoa), lack true tissues. All other animals (Eumetazoa) have distinct tissues. The presence of true tissues is a
   a. useful character for distinguishing among the eumetazoans.
   b. shared ancestral state.
   c. shared derived state.
   d. plesiomorphy.
4. Originally, organisms were classified according to a two-domain system that consisted of the bacteria and the eukaryotes. Recently, a three-domain system of classification has been proposed in which domain Bacteria was divided into domain Archaea and domain Bacteria. This is because
   a. there are so many different species of prokaryotes.
   b. of the large genetic differences between the archaebacteria and the bacteria.
   c. the archaebacteria have a nucleus like that of eukaryotes.
   d. the archaebacteria live in inhospitable environments, so they should be put in a different domain.
5. Which of the following is the most abundant group of organisms on earth?
   a. archaebacteria
   b. bacteria
   c. protists
   d. fungi
6. Biologists do not consider viruses to be alive. This is because viruses
   a. are unable to reproduce on their own.
   b. do not contain genetic material.
   c. do not contain proteins.
   d. are smaller than living organisms.
7. As more is learned about the molecular and genetic aspects of organisms, taxonomic groupings are changing. Which group of eukaryotes is the focus of the most controversy related to their taxonomic groupings?
   a. bacteria
   b. protists
   c. animals
   d. fungi
8. According to new molecular and biochemical information, should the plants still be recognized as a distinct kingdom?
   a. Yes, the green algae are now considered to be in the kingdom Plantae.
   b. Yes, the green algae and the plants are now recognized to be unrelated.
   c. No, the plants and the green algae are now classified together in the kingdom Virdiplantae.
   d. No, the plants are now classified together within the Chlorophyta.
9. The relationship between the annelids and the arthropods is now thought to be
   a. monophyletic.
   b. polyphyletic.
   c. paraphyletic
   d. unrelated.

Test Your Visual Understanding

1. Using the image from figure 25.8, fill in the following table to indicate the presence or absence of the derived character.

	Light Bones	Breastbone	Downy Feathers	Feathers with vanes, shafts, and barbs	Aerodynamic feathers	Arms longer than legs
Other Dinosaurs						
Coelophysis						
Tyrannosaurus						
Sinosauropteryx						
Velociraptor						
Caudipteryx						
Archaeopteryx						
Modern Birds						

Apply Your Knowledge

1. Most biologists consider viruses to be nonliving because they are unable to replicate outside of a host cell. How might you argue against this position? What characteristics do viruses have them make them lifelike?
2. Phylogenetic relationships are established based on assumptions about evolutionary relatedness. It is assumed that the more similar two organisms are, the more closely they are related. What problems does the process of convergent evolution cause for systematists?
3. Ethnobotanical research involves the exploration of plants that may have uses, particularly medical, for humans. Of what value is knowledge of systematics to an ethnobotanist?

26
Viruses

Concept Outline

26.1 Viruses are strands of nucleic acid encased within a protein coat.

The Nature of Viruses. Viruses occur in all organisms. Able to reproduce only within living cells, viruses are not themselves alive.

26.2 Bacterial viruses exhibit two sorts of reproductive cycles.

Bacteriophages. Some bacterial viruses, called bacteriophages, rupture the cells they infect, while others integrate themselves into the bacterial chromosome to become a stable part of the bacterial genome.

Cell Transformation and Phage Conversion. Integrated bacteriophages sometimes modify the host bacterium they infect.

26.3 HIV is a complex animal virus.

AIDS. The animal virus HIV infects certain key cells of the immune system, destroying the body's ability to defend itself from cancer and disease. The HIV infection cycle is typically a lytic cycle, in which the HIV RNA first directs the production of a corresponding DNA, and this DNA then directs the production of progeny virus particles.

The Future of HIV Treatment. Combination therapies and chemokines offer promising avenues of AIDS therapy.

26.4 Nonliving infectious agents are responsible for many human diseases.

Disease Viruses. Some of the most serious viral diseases have only recently infected human populations, as the result of transfer from other hosts.

Prions and Viroids. In some instances, proteins and "naked" RNA molecules can also transmit diseases.

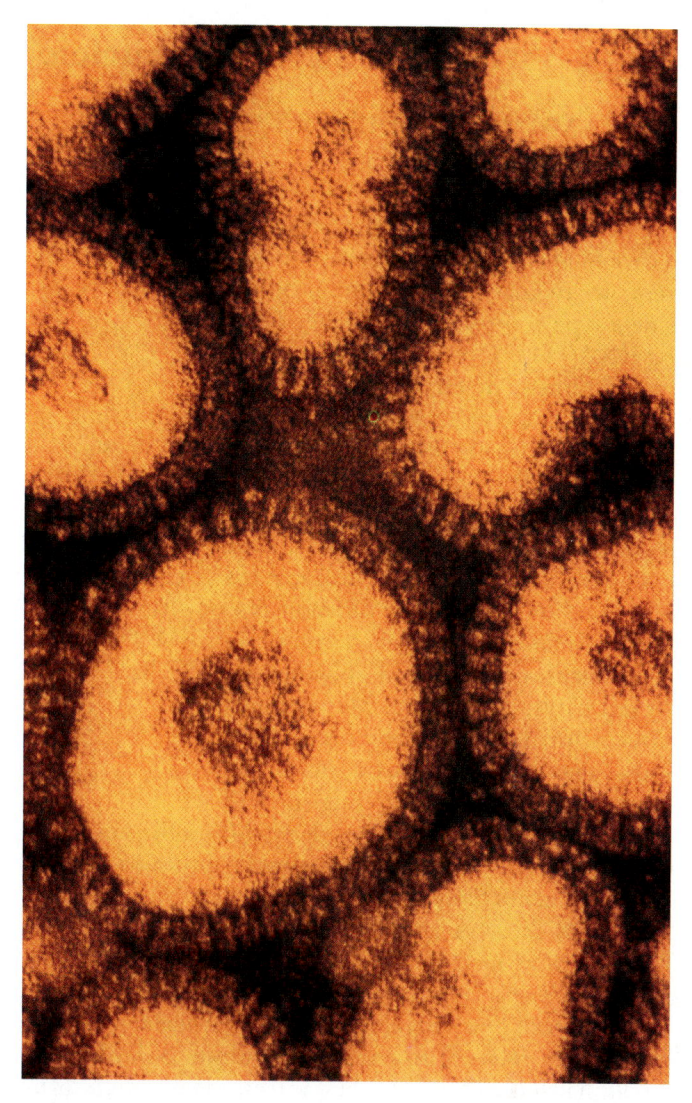

FIGURE 26.1
Influenza viruses (30,000×). A virus has been referred to as "a piece of bad news wrapped up in a protein." How can something as "simple" as a virus have such a profound effect on living organisms?

We start our exploration of the diversity of life with viruses. Viruses are genetic elements enclosed in protein and are not considered organisms, because they cannot reproduce independently. However, due to their disease-producing potential, viruses are important biological entities. The virus particles you see in figure 26.1 produce the important disease influenza. Other viruses cause such diseases as AIDS, polio, and smallpox, and some can lead to cancer. Many scientists have attempted to unravel the nature of viral genes and to discover how they work. For more than four decades, viral studies have been thoroughly intertwined with those of genetics and molecular biology. In the future, viruses are expected to be one of the principal tools used to experimentally carry genes from one organism to another. Already, viruses are being employed in the treatment of human genetic diseases.

26.1 Viruses are strands of nucleic acid encased within a protein coat.

The Nature of Viruses

Viral Structure

All viruses have the same basic structure: a core of nucleic acid surrounded by protein. Individual viruses contain only a single type of nucleic acid, either DNA or RNA. The DNA or RNA genome may be linear or circular, and single-stranded or double-stranded. Viruses are frequently classified by the nature of their genomes.

Nearly all viruses form a protein sheath, or **capsid,** around their nucleic acid core (figure 26.2). The capsid is composed of one to a few different protein molecules repeated many times. In some viruses, specialized enzymes are stored within the capsid. Many animal viruses form an **envelope** around the capsid that is rich in proteins, lipids, and glycoprotein molecules. While some of the material of the envelope is derived from the host cell's membrane, the envelope does contain proteins derived from viral genes as well.

Viruses occur in virtually every kind of organism that has been investigated for their presence. However, each type of virus can replicate in only a very limited number of cell types. The suitable cells for a particular virus are collectively referred to as its **host range.** Some viruses wreak havoc on the cells they infect; many others produce no disease or other outward sign of their infection. Still other viruses remain dormant for years until a specific signal triggers their expression. A given organism often has more than one kind of virus. This suggests that there may be many more kinds of viruses than there are kinds of organisms—perhaps millions of them. Only a few thousand viruses have been described at this point.

Viral Replication

An infecting virus can be thought of as a set of instructions, not unlike a computer program. A computer's operation is directed by the instructions in its operating program, just as a cell is directed by DNA-encoded instructions. A new program introduced into the computer can cause the computer to cease what it is doing and devote all of its energies to another activity, such as making copies of the introduced program. The new program is not itself a computer and cannot make copies of itself when it is outside the computer. Like a virus, the introduced program is simply a set of instructions.

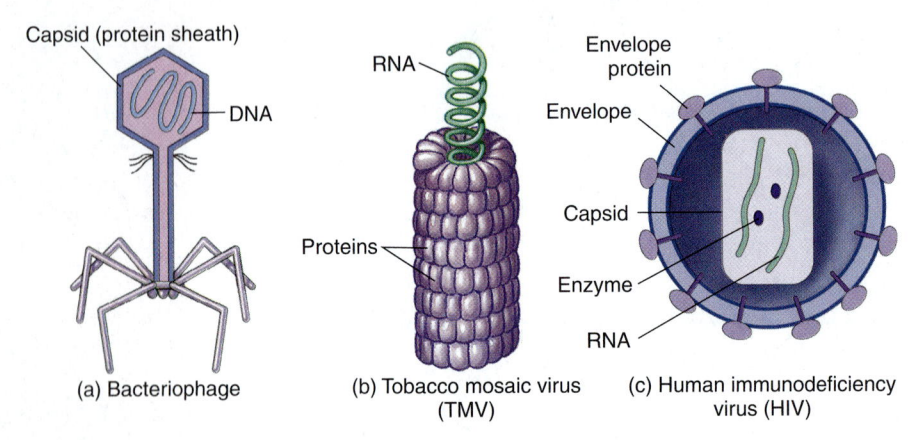

FIGURE 26.2
The structure of a bacterial, plant, and animal virus. (*a*) Bacterial viruses, called bacteriophages, often have a complex structure. (*b*) TMV infects plants and consists of 2130 identical protein molecules (*purple*) that form a cylindrical coat around the single strand of RNA (*green*). The RNA backbone determines the shape of the virus and is protected by the identical protein molecules packed tightly around it. (*c*) In the human immunodeficiency virus (HIV), the two-stranded RNA core is held within a capsid that is encased by a protein envelope.

Viruses can reproduce only when they enter cells and utilize the cellular machinery of their hosts. Viruses code their genes on a single type of nucleic acid, either DNA or RNA, but viruses lack ribosomes and the enzymes necessary for protein synthesis. Viruses are able to reproduce because their genes are translated into proteins by the cell's genetic machinery. These proteins lead to the production of more viruses.

Viral Shape

Most viruses have an overall structure that is either **helical** or **isometric.** Helical viruses, such as the tobacco mosaic virus in figure 26.2*b*, have a rodlike or threadlike appearance. Isometric viruses have a roughly spherical shape whose geometry is revealed only under the highest magnification.

The only structural pattern found so far among isometric viruses is the **icosahedron,** a structure with 20 equilateral triangular facets. Most viruses are icosahedral in basic structure. The icosahedron is the basic design of the geodesic dome and is the most efficient symmetrical arrangement that linear subunits can take to form a shell with maximum internal capacity.

Viral Genome Structure

Viral genomes vary greatly (table 26.1). Some viruses, including those that cause flu, measles, and AIDS, possess RNA genomes. Other viruses have DNA genomes, such as the viruses causing smallpox and herpes. The nucleic acid

Table 26.1 Important Human Viral Diseases

Disease	Pathogen	Genome	Vector/Epidemiology
Chicken pox	Varicella zoster	Double-stranded DNA	Spread through contact with infected individuals. No cure. Rarely fatal. Vaccine approved in U.S. in early 1995.
Hepatitis B (viral)	Hepadnavirus	Double-stranded DNA	Highly infectious through contact with infected body fluids. Approximately 1% of U.S. population infected. Vaccine available. No cure. Can be fatal.
Herpes	Herpes simplex virus	Double-stranded DNA	Fever blisters; spread primarily through contact with infected saliva. Very prevalent worldwide. No cure. Exhibits latency—the disease can be dormant for several years.
Mononucleosis	Epstein-Barr virus	Double-stranded DNA	Spread through contact with infected saliva. May last several weeks; common in young adults. No cure. Rarely fatal.
Smallpox	Variola virus	Double-stranded DNA	Historically a major killer; the last recorded case of smallpox was in 1977. A worldwide vaccination campaign wiped out the disease completely.
AIDS	HIV	(+) Single-stranded RNA (two segments)	Destroys immune defenses, resulting in death by infection or cancer. Over 42 million cases worldwide by 2002.
Polio	Enterovirus	(+) Single-stranded RNA	Acute viral infection of the CNS that can lead to paralysis and is often fatal. Prior to the development of Salk's vaccine in 1954, 60,000 people a year contracted the disease in the U.S. alone.
Yellow fever	Flavivirus	(+) Single-stranded RNA	Spread from individual to individual by mosquito bites; a notable cause of death during the construction of the Panama Canal. If untreated, this disease has a peak mortality rate of 60%.
Ebola	Filoviruses	(−) Single-stranded RNA	Acute hemorrhagic fever; virus attacks connective tissue, leading to massive hemorrhaging and death. Peak mortality is 50–90% if untreated. Outbreaks confined to local regions of central Africa.
Influenza	Influenza viruses	(−) Single-stranded RNA	Historically a major killer (22 million died in 18 months in 1918–19); wild Asian ducks, chickens, and pigs are major reservoirs. The ducks are not affected by the flu virus, which shuffles its antigen genes while multiplying within them, leading to new flu strains.
Measles	Paramyxoviruses	(−) Single-stranded RNA	Extremely contagious through contact with infected individuals. Vaccine available. Usually contracted in childhood, when it is not serious; more dangerous to adults.
SARS	Coronavirus	(−) Single-stranded RNA	Acute respiratory infection; an emerging disease, can be fatal, especially in the elderly.
Pneumonia	Influenza virus	(−) Single-stranded RNA	Acute infection of the lungs; often fatal without treatment.
Rabies	Rhabdovirus	(−) Single-stranded RNA	An acute viral encephalomyelitis transmitted by the bite of an infected animal. Fatal if untreated.

in some viruses is single-stranded, while in others, it is double-stranded. In single-stranded RNA viruses, the genome can contain the same base sequence as the mRNA used to produce the viral proteins, in which case the RNA strand can serve as the mRNA, and the virus is called a positive-strand virus. Or, the genome can contain bases complementary to the viral mRNA, in which case the virus is called a negative-strand virus.

Viruses occur in all organisms and can only reproduce within living cells.

26.2 Bacterial viruses exhibit two sorts of reproductive cycles.

Bacteriophages

Bacteriophages are viruses that infect bacteria. They are diverse both structurally and functionally, and are united solely by their occurrence in bacterial hosts. Many of these bacteriophages, called *phages* for short, are large and complex, with relatively large amounts of DNA and proteins. Some of them have been named as members of a "T" series (T1, T2, and so forth); others have been given different kinds of names. To illustrate the diversity of these viruses, T3 and T7 phages are icosahedral and have short tails. In contrast, the so-called T-even phages (T2, T4, and T6) have an icosahedral head, a capsid that consists primarily of three proteins, a connecting neck with a collar and long "whiskers," a long tail, and a complex base plate (figure 26.3).

The Lytic Cycle

During the process of bacterial infection by T4 phage, at least one of the tail fibers of the phage—they are normally held near the phage head by the "whiskers"—contacts the lipoproteins of the host bacterial cell wall. The other tail fibers set the phage perpendicular to the surface of the bacterium and bring the base plate into contact with the cell surface. The tail contracts, and the tail tube passes through an opening that appears in the base plate, piercing the bacterial cell wall. The contents of the head, mostly DNA, are then injected into the host cytoplasm.

When a virus kills the infected host cell in which it is replicating, the reproductive cycle is referred to as a **lytic cycle** (figure 26.4, *top*). The T-series bacteriophages are all **virulent viruses,** multiplying within infected cells and eventually lysing (rupturing) them. However, they vary considerably as to when they become virulent within their host cells.

The Lysogenic Cycle

Many bacteriophages do not immediately kill the cells they infect, instead integrating their nucleic acid into the genome of the infected host cell. While residing there, the virus's DNA is called a **prophage.** Among the bacteriophages that do this is the lambda (λ) phage of *Escherichia coli.* We know as much about this bacteriophage as we do about virtually any other biological particle; the complete sequence of its 48,502 bases has been determined. At least 23 proteins are associated with the development and maturation of lambda phage, and many other enzymes are involved in integrating these viruses into the host genome.

The integration of a virus into a cellular genome is called **lysogeny.** At a later time, the prophage may exit the genome and initiate virus replication. This sort of reproductive cycle, involving a period of genome integration, is called a **lysogenic cycle** (figure 26.4, *bottom*). Viruses that become stably integrated within the genome of their host cells are called **lysogenic viruses,** or **temperate viruses.**

> Bacteriophages are a diverse group of viruses that attack bacteria. Some kill their host in a lytic cycle; others integrate into the host's genome, initiating a lysogenic cycle.

(a) .05 μm

(b)

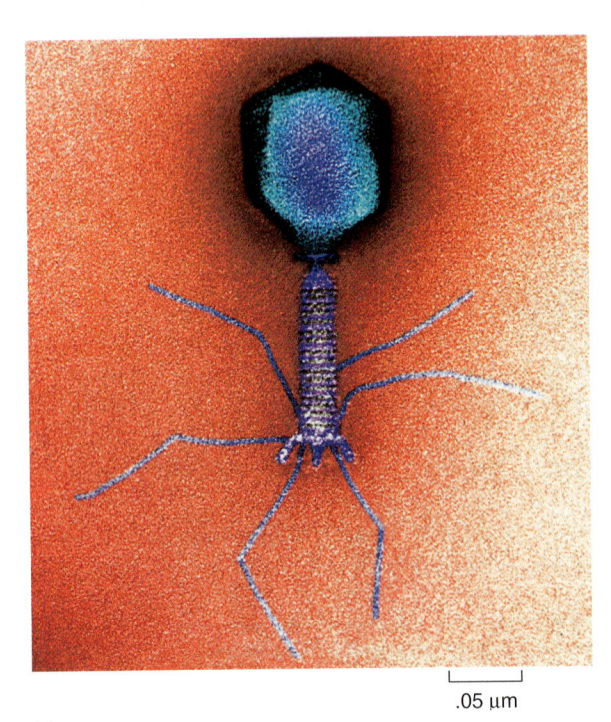

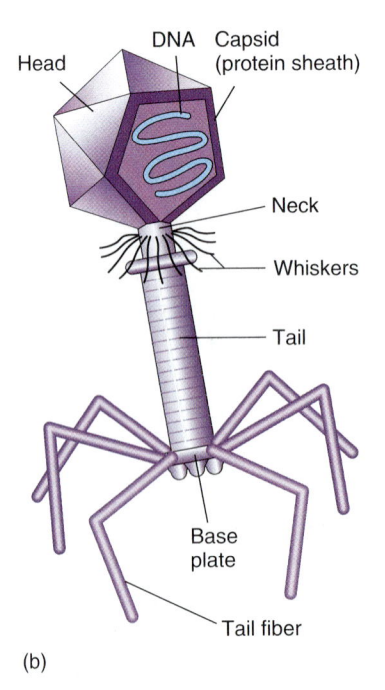

**FIGURE 26.3
A bacterial virus.** Bacteriophages exhibit a complex structure. (*a*) Electron micrograph and (*b*) diagram of the structure of a T4 bacteriophage (some facets removed to reveal the interior).

534   Part V   Diversity of Life on Earth

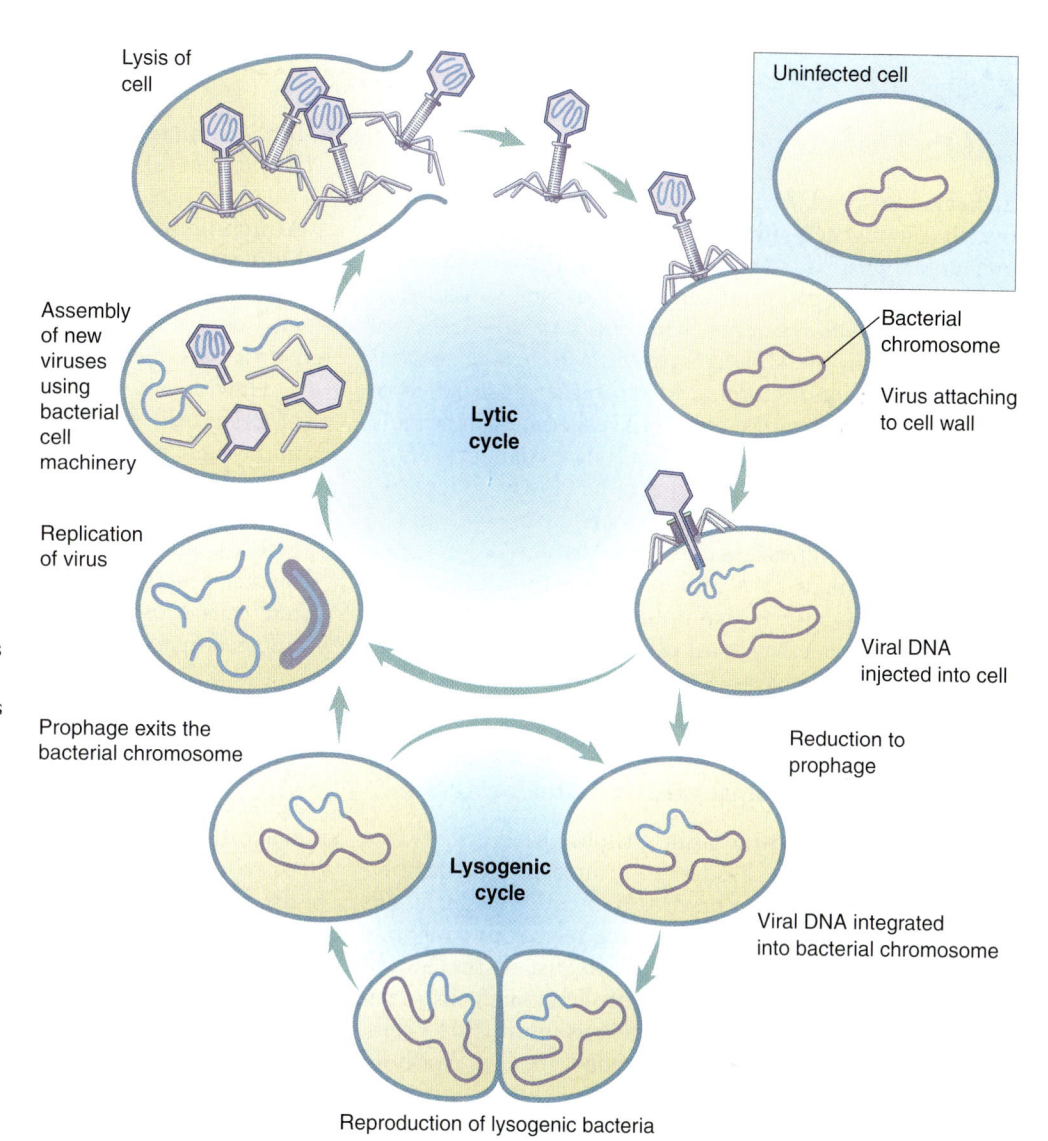

FIGURE 26.4
Lytic and lysogenic cycles of a bacteriophage. In the lytic cycle, the bacteriophage exists as viral DNA free in the bacterial host cell's cytoplasm; the viral DNA directs the production of new viral particles by the host cell until the virus kills the cell by lysis. In the lysogenic cycle, the bacteriophage DNA is integrated into the large, circular DNA molecule of the host bacterium and is reproduced along with the host DNA as the bacterium replicates. It may continue to replicate and produce lysogenic bacteria or enter the lytic cycle and kill the cell. Bacteriophages are much smaller relative to their hosts than illustrated in this diagram.

Cell Transformation and Phage Conversion

During the integrated portion of a lysogenic reproductive cycle, viral genes are often expressed. The RNA polymerase of the host cell reads the viral genes just as if they were host genes. Sometimes, expression of these genes has an important effect on the host cell, altering it in novel ways. The genetic alteration of a cell's genome by the introduction of foreign DNA is called **transformation**. When the foreign DNA is contributed by a bacterial virus, the alteration is called **phage conversion**.

Phage Conversion of the Cholera-Causing Bacterium

An important example of the sort of phage conversion directed by viral genes is provided by the bacterium responsible for an often-fatal human disease. The disease-causing bacterium *Vibrio cholerae* usually exists in a harmless form, but a second disease causing, virulent form also occurs. In this latter form, the bacterium causes the deadly disease cholera, but how the bacteria changed from harmless to deadly was not known until recently. Research now shows that a bacteriophage that infects *V. cholerae* introduces into the host bacterial cell a gene that codes for the cholera toxin. This gene becomes incorporated into the bacterial chromosome, where it is translated along with the other host genes, thereby converting the benign bacterium to a disease-causing agent. The transfer occurs through bacterial pili (see chapter 27); in further experiments, mutant bacteria that did not have pili were resistant to infection by the bacteriophage. This discovery has important implications in efforts to develop vaccines against cholera, which have been unsuccessful up to this point.

Bacteriophages convert *Vibrio cholerae* bacteria from a harmless form into disease-causing agents.

Chapter 26 Viruses 535

26.3 HIV is a complex animal virus.

AIDS

A diverse array of viruses occur among animals. A good way to gain a general idea of what they are like is to look at one animal virus in detail. Here we will examine the virus responsible for a comparatively new and fatal viral disease, acquired immunodeficiency syndrome (AIDS). AIDS was first reported in the United States in 1981. It was not long before the infectious agent, a **retrovirus** (an RNA virus that transcribes its genes into DNA) called human immunodeficiency virus (HIV), was identified by laboratories in France. Study of HIV revealed it to be closely related to a chimpanzee virus, suggesting a recent host expansion to humans from chimpanzees in central Africa.

Infected humans have little resistance to infection, and nearly all of them eventually die of diseases that noninfected individuals easily ward off. Few who contract AIDS survive more than a few years untreated. The AIDS epidemic is discussed further in chapter 48.

How HIV Compromises the Immune System

In normal individuals, an army of specialized cells (white blood cells) patrols the bloodstream, attacking and destroying any invading bacteria or viruses. In AIDS patients, this army of defenders is vanquished. One special kind of white blood cell, called a $CD4^+$ cell (discussed further in chapter 48), is required to rouse the defending cells to action. In AIDS patients, the virus homes in on $CD4^+$ cells, infecting and killing them until none are left. Without these crucial immune system cells, the body cannot mount a defense against invading bacteria or viruses. AIDS patients die of infections that a healthy person could fight off.

Clinical symptoms typically do not begin to develop until after a long latency period, generally 8 to 10 years after the initial infection with HIV. Although carriers of HIV have no clinical symptoms during this long interval, they are apparently fully infectious, which makes the spread of HIV very difficult to control. The reason HIV remains hidden for so long seems to be that its infection cycle continues throughout the 8- to 10-year latent period without doing serious harm to the infected person. Eventually, however, a random mutational event in the virus allows it to quickly overcome the immune defense, starting AIDS.

The HIV Infection Cycle

The HIV virus infects and eliminates key cells of the immune system, destroying the body's ability to defend itself from cancer and infection. The way HIV infects humans (figure 26.5) provides a good example of how animal viruses replicate. Most other viral infections follow a similar course, although the details of entry and replication differ in individual cases.

Attachment. When HIV is introduced into the human bloodstream, the virus particle circulates throughout the entire body but will only infect $CD4^+$ cells. Most other animal viruses are similarly narrow in their requirements; hepatitis goes only to the liver, and rabies to the brain.

How does a virus such as HIV recognize a specific kind of target cell? Recall from chapter 7 that every kind of cell in the human body has a specific array of cell-surface glycoprotein markers that serve to identify them to other, similar cells. Each HIV particle possesses a glycoprotein (called gp120) on its surface that precisely fits a cell-surface marker protein called CD4 on the surfaces of immune system cells called macrophages and T cells. Macrophages are infected first.

Entry into Macrophages. After docking onto the CD4 receptor of a macrophage, HIV requires a second macrophage receptor, called CCR5, to pull itself across the cell membrane. After gp120 binds to CD4, it goes through a conformational change that allows it to bind to CCR5. The current model suggests that after the conformational change, the second receptor passes the gp120-CD4 complex through the cell membrane, triggering passage of the contents of the HIV virus into the cell by endocytosis, with the cell membrane folding inward to form a deep cavity around the virus.

Replication. Once inside the macrophage, the HIV particle sheds its protective coat. This leaves virus RNA floating in the cytoplasm, along with a virus enzyme that was also within the virus shell. This enzyme, called **reverse transcriptase,** synthesizes a double strand of DNA complementary to the virus RNA, often making mistakes and so introducing new mutations. This double-stranded DNA may integrate itself into the host cell's DNA; it directs the host cell machinery to produce many copies of the virus. HIV does not rupture and kill the macrophage cells it infects. Instead, the new viruses are released from the cell by *budding*, a process much like exocytosis. HIV synthesizes large numbers of viruses in this way, challenging the immune system over a period of years.

Entry into T Cells. During this time, HIV is constantly replicating and mutating. Eventually, by chance, HIV alters the gene for gp120 in a way that causes the gp120 protein to change its second-receptor allegiance. This new form of gp120 protein prefers to bind instead to a different second receptor, CXCR4, a receptor that occurs on the surface of T lymphocyte $CD4^+$ cells. Soon the body's T lymphocytes become infected with HIV. This has deadly consequences, as new viruses exit the cell by rupturing the cell, thus

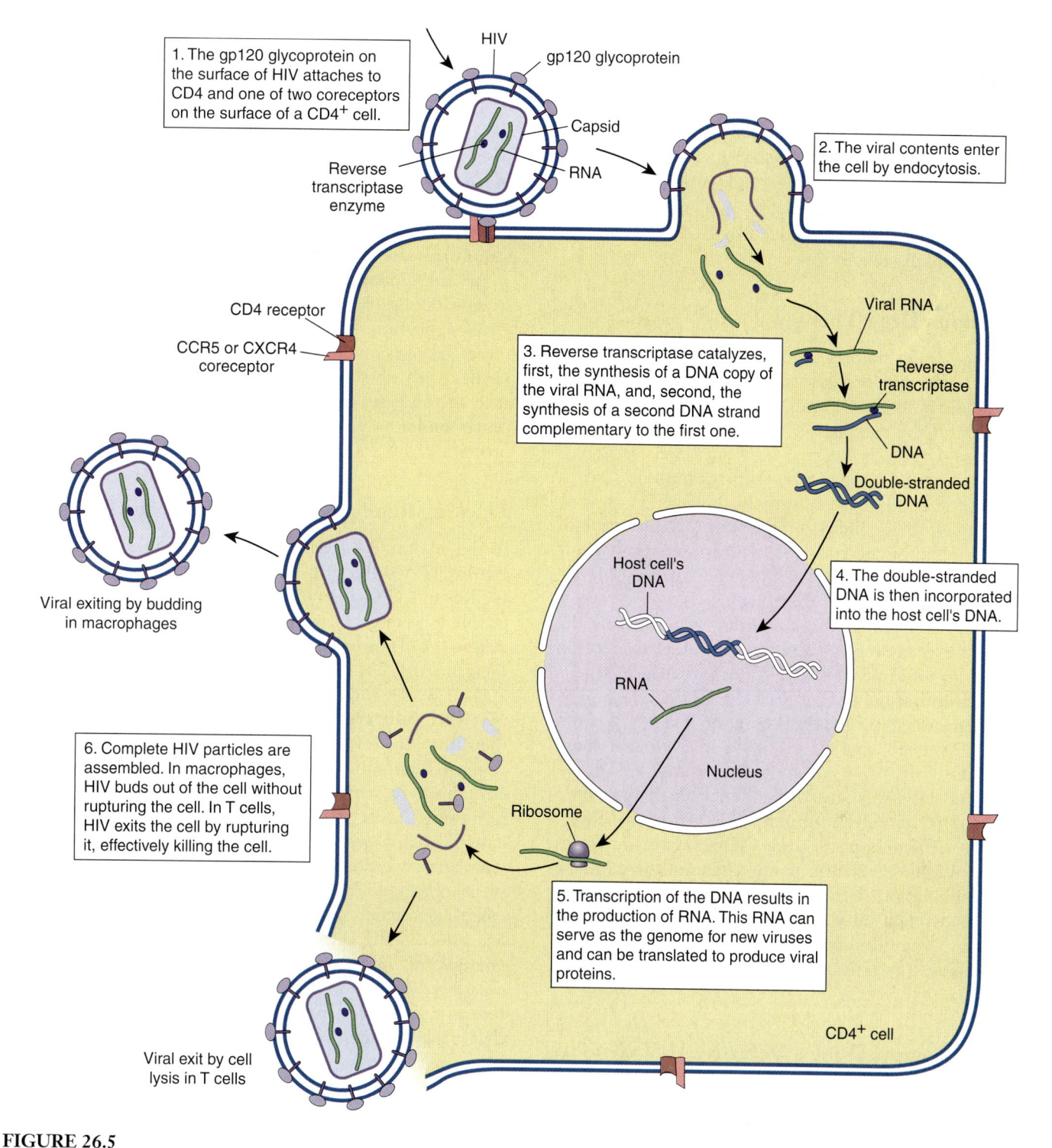

FIGURE 26.5
The HIV infection cycle. The cycle begins and ends with free HIV particles present in the bloodstream of its human host. These free viruses infect white blood cells that have CD4 receptors, cells called CD4+ cells.

killing the T cell. The shift to the CXCR4 second receptor is followed swiftly by a steep drop in the number of T cells. This destruction of the body's T cells blocks the immune response and leads directly to the onset of AIDS, with cancers and opportunistic infections free to invade the defenseless body.

> **HIV, the virus that causes AIDS, is an RNA virus that replicates inside human cells by first making a DNA copy of itself. It is only able to gain entrance to those cells possessing a particular cell-surface marker recognized by a glycoprotein on its own surface.**

Chapter 26 Viruses 537

The Future of HIV Treatment

New discoveries about how HIV works continue to fuel research on devising ways to counter HIV. For example, scientists are testing drugs and vaccines that act on HIV receptors, researching the possibility of blocking CCR5, and looking for defects in the structures of HIV receptors in individuals who are infected with HIV but have not developed AIDS. Figure 26.6 summarizes some of the recent developments and discoveries.

Combination Drug Therapy

A variety of drugs inhibit HIV in the test tube. These include AZT and similar nucleoside analogs (which inhibit viral nucleic acid replication) and protease inhibitors (which inhibit the cleavage of the large polyproteins encoded by *gag*, *pol*, and *env* genes into functional capsid, enzyme, and envelope segments). When combinations of these drugs were administered to people with HIV in controlled studies, their condition improved. A combination of a protease inhibitor and two AZT-type drugs entirely eliminated the HIV virus from many of the patients' bloodstreams. Importantly, all of these patients began to receive the drug therapy within three months of contracting the virus, before their bodies had an opportunity to develop tolerance to any one of them. Widespread use of this **combination therapy** has cut the U.S. AIDS death rate by three-fourths since its introduction in the mid-1990s, from 49,000 AIDS deaths in 1995 to 36,000 in 1996, and about 9000 in 2001.

Unfortunately, this sort of combination therapy does not appear to actually eliminate HIV from the body. While the virus disappears from the bloodstream, traces of it can still be detected in lymph tissue of the patients. When combination therapy is discontinued, virus levels in the bloodstream once again rise. Because of demanding therapy schedules and many side effects, long-term combination therapy does not seem a promising approach.

Vaccine Therapy: Using a Defective HIV Gene to Combat AIDS

Recently, five people in Australia who are HIV-positive but have not developed AIDS in 14 years were found to have all received a blood transfusion from the same HIV-positive person, who also has not developed AIDS. This led scientists to believe that the strain of virus transmitted to these people has some sort of genetic defect that prevents it from effectively disabling the human immune system. In subsequent research, a defect was found in one of the nine genes present in this strain of HIV. This gene is called *nef*, named for "negative factor," and the defective version of *nef* in the HIV strain that infected the six Australians seems to be missing some pieces. Viruses with the defective gene may have reduced reproductive capability, allowing the immune system to keep the virus in check.

This finding has exciting implications for developing a vaccine against AIDS. Before this, scientists have been unsuccessful in trying to produce a harmless strain of AIDS that can elicit an effective immune response. The Australian strain with the defective *nef* gene has the potential to be used in a vaccine that would arm the immune system against this and other strains of HIV.

Another potential application of this discovery is its use in developing drugs that inhibit HIV proteins that speed virus replication. It seems that the protein produced from the *nef* gene is one of these critical HIV proteins, because viruses with defective forms of *nef* do not reproduce, as seen in the cases of the six Australians. Research is currently under way to develop a drug that targets the *nef* protein.

Blocking Replication: Chemokines and CAF

In the laboratory, chemicals called **chemokines** appear to inhibit HIV infection by binding to and blocking the CCR5 and CXCR4 coreceptors. As you might expect, people long infected with the HIV virus who have not developed AIDS prove to have high levels of chemokines in their blood.

The search for HIV-inhibiting chemokines is intense. Not all results are promising. Researchers report that in their tests, the levels of chemokines were not different between patients in which the disease was not progressing and those in which it was rapidly progressing. More promising, levels of another factor called CAF (CD8+ cell antiviral factor) *are* different between these two groups. Researchers have not yet succeeded in isolating CAF, which seems not to block receptors that HIV uses to gain entry to cells, but instead, to prevent replication of the virus once it has infected the cells. Research continues on the use of chemokines in treatments for HIV infection, either increasing the amount of chemokines or disabling the CCR5 receptor. However, promising research on CAF suggests that it may be an even better target for treatment and prevention of AIDS.

One problem with using chemokines as drugs is that they are also involved in the inflammatory response of the immune system. Chemokines attract white blood cells to areas of infection. Chemokines work beautifully in small amounts and in local areas, but chemokines in mass numbers can cause an inflammatory response that is worse than the original infection. Injections of chemokines may hinder the immune system's ability to respond to local chemokines, or may even trigger an out-of-control inflammatory response. Thus, scientists caution that injection of chemokines could make patients *more* susceptible to infections, and they continue to research other methods of using chemokines to treat AIDS.

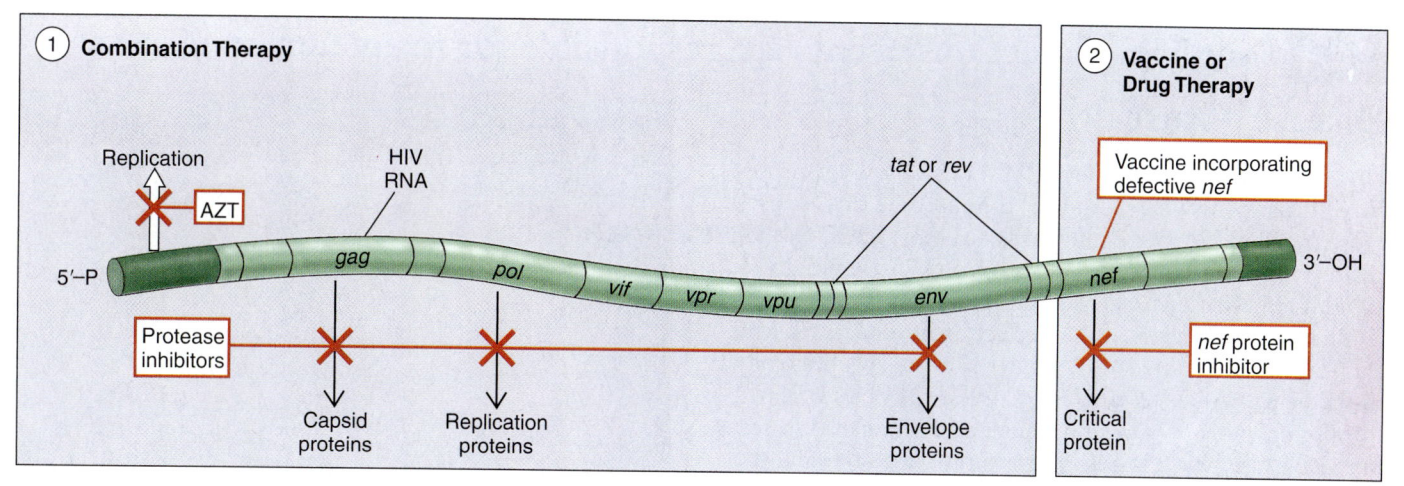

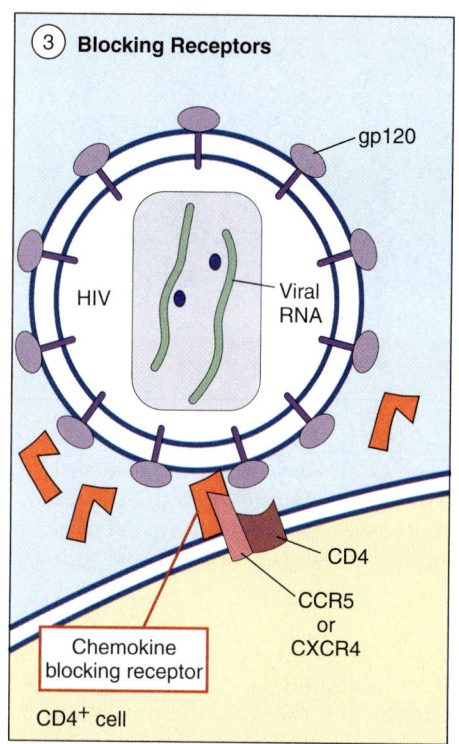

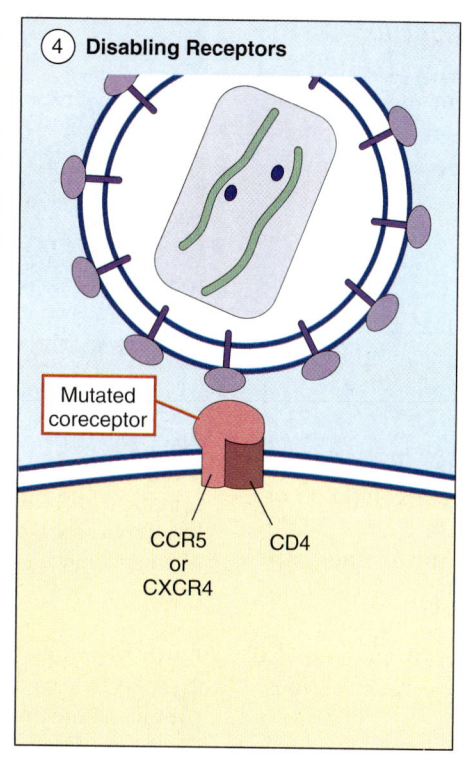

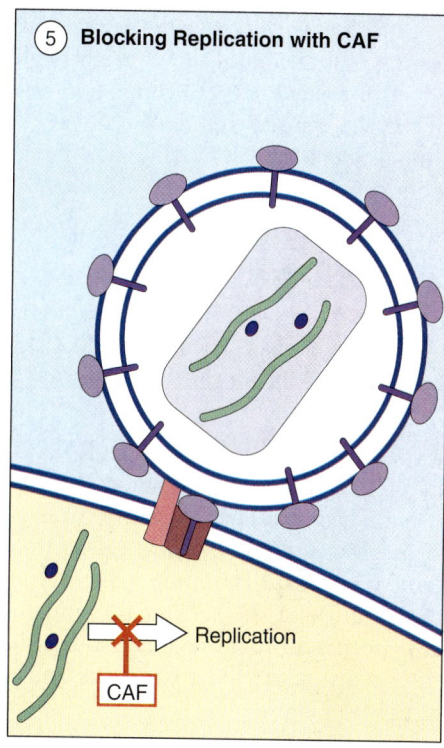

FIGURE 26.6
Potential new treatments for HIV. Research is currently under way in the following five areas: (*1*) Combination therapy involves using two drugs, AZT to block replication of the virus and protease inhibitors to block the production of critical viral proteins. (*2*) Using a defective form of the viral gene *nef*, scientists may be able to construct an HIV vaccine. Also, drug therapy that inhibits *nef*'s protein product is being tested. (*3*) Other research focuses on the use of chemokine chemicals to block receptors (CXCR4 and CCR5), thereby disabling the mechanism HIV uses to enter CD4+ cells. (*4*) Producing mutations that will disable receptors may also be possible. (*5*) Finally, CAF, an antiviral factor that acts inside the CD4+ cell, may be able to block replication of HIV.

Blocking or Disabling Receptors

A 32-base-pair deletion in the gene that codes for the CCR5 receptor appears to block HIV infection. Individuals at high risk of HIV infection who are homozygous for this mutation do not seem to develop AIDS. In one study of 1955 people, scientists found no individuals who were infected and homozygous for the mutated allele. The allele seems to be more common in Caucasian populations (10 to 11%) than in African-American populations (2%), and absent in African and Asian populations. Treatment for AIDS involving disruption of CCR5 looks promising, because research indicates that people live perfectly well without CCR5. Attempts to block or disable CCR5 are being sought in numerous laboratories.

A cure for AIDS is not yet in hand, but many new approaches look promising.

Chapter 26 Viruses 539

26.4 Nonliving infectious agents are responsible for many human diseases.

Disease Viruses

Humans have known and feared diseases caused by viruses for thousands of years. Among the diseases that viruses cause (see table 26.1) are influenza, smallpox, infectious hepatitis, yellow fever, polio, AIDS, and SARS (figure 26.7). In addition, viruses have been implicated in some cancers and leukemias. In view of their effects, it is easy to see why the late Sir Peter Medawar, Nobel laureate in physiology or medicine, wrote, "A virus is a piece of bad news wrapped in protein." Viruses not only cause many human diseases, but also cause major losses in agriculture, forestry, and the productivity of natural ecosystems.

Polio and smallpox were very serious human diseases, affecting millions, until they were brought under control. An effective vaccine largely eliminated polio in the latter half of the last century and has totally eradicated natural infections of smallpox. The smallpox virus now exists only in laboratories.

Influenza

Perhaps the most lethal virus in human history has been the influenza virus. Some 22 million Americans and Europeans died of flu within 18 months in 1918 and 1919.

Types. Flu viruses are animal RNA viruses. An individual flu virus resembles a rod studded with spikes composed of two kinds of protein. There are three general "types" of flu virus, distinguished by their capsid (inner membrane) protein, which is different for each type: Type A flu virus causes most of the serious flu epidemics in humans, and also occurs in mammals and birds. Type B and type C viruses are restricted to humans and rarely cause serious health problems.

Subtypes. Different strains of flu virus, called subtypes, differ in their protein spikes. One of these proteins, hemagglutinin (H), aids the virus in gaining access to the cell interior. The other, neuraminidase (N), helps the daughter virus break free of the host cell once virus replication has been completed. Parts of the H molecule contain "hotspots" that display an unusual tendency to change as a result of mutation of the viral RNA during imprecise replication. Point mutations cause changes in these spike proteins in 1 of 100,000 viruses during the course of each generation. These highly variable segments of the H molecule are targets against which the body's antibodies are directed. The high variability of these targets improves the reproductive capacity of the virus and hinders our ability to make perfect vaccines. Because of accumulating changes in the H and N molecules, different flu vaccines are required to protect against different subtypes. Type A flu viruses are currently classified into 13 distinct H subtypes and 9 distinct N subtypes, each of which requires a different vaccine to protect against infection. Thus, the type A virus that caused the Hong Kong flu epidemic of 1968 has type 3 H molecules and type 2 N molecules, and is called A(H3N2).

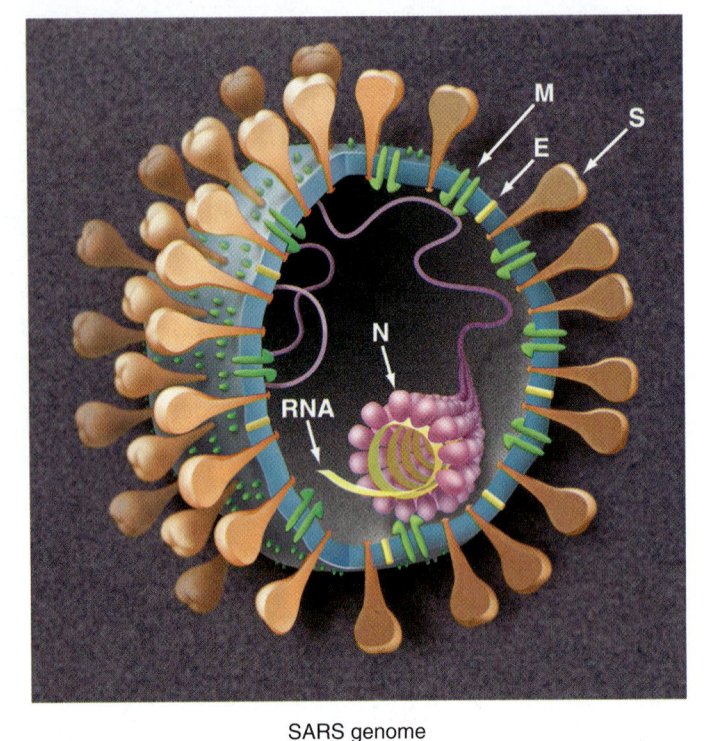

FIGURE 26.7
SARS coronavirus. The 29,751-nucleotide SARS genome is composed of RNA and contains six principle genes: R_A and R_B—replicases; S—spike proteins; E—envelope glycoproteins; M—membrane glycoprotein; N—nucleocapsid protein.

Importance of Recombination. The greatest problem in combating flu viruses arises not through mutation, but through recombination. Viral genes are readily reassorted by genetic recombination, sometimes putting together novel combinations of H and N spikes unrecognizable by human antibodies specific for the old configuration. Viral recombination of this kind seems to have been responsible for the three major flu pandemics (that is, worldwide epidemics) that occurred in the twentieth century, by producing drastic shifts in H-N combinations. The "killer flu" of 1918, A(H1N1), killed 40 million people worldwide. The Asian flu of 1957, A(H2N2), killed over 100,000 Americans. The Hong Kong flu of 1968, A(H3N2), infected 50 million people in the United States alone, of which 70,000 died.

It is no accident that new strains of flu usually originate in the Far East. The most common hosts of influenza virus are ducks, chickens, and pigs, which in Asia often live in close proximity to each other and to humans. Pigs are subject to infection by both bird and human strains of the virus, and individual animals are often simultaneously infected with multiple strains. This creates conditions favoring genetic recombination between strains, producing new combinations of H and N subtypes. The Hong Kong flu, for example, arose from recombination between A(H3N8) from ducks and A(H2N2) from humans. The new strain of influenza, in this case A(H3N2), then passed back to humans, creating an epidemic because the human population has never experienced that H-N combination before.

Emerging Viruses

Sometimes viruses that originate in one organism pass to another, thus expanding their host range. Often, this expansion is deadly to the new host. HIV, for example, arose in chimpanzees and relatively recently passed to humans. Influenza is fundamentally a bird virus. Viruses that originate in one organism and then pass to another and cause disease are called **emerging viruses.** They represent a considerable threat in an age when airplane travel potentially allows infected individuals to move about the world quickly, spreading an infection.

An emerging virus caused a sudden outbreak of a hemorrhagic-type infection in the southwestern United States in 1993. This highly fatal disease was soon attributed to a species of **hantavirus,** a single-stranded RNA virus associated with rodents. This species was eventually traced to deer mice. The deer mouse hantavirus is transmitted to humans through fecal contamination in areas of human habitation. Controlling the deer mouse population limited the disease.

Sometimes the origin of an emerging virus is unknown, making an outbreak more difficult to control. Among the most lethal of emerging viruses are a collection of filamentous viruses arising in central Africa that cause severe hemorrhagic fever. With lethality rates in excess of 50%, these so-called filoviruses are among the most lethal infectious diseases known. One, **Ebola virus** (figure 26.8), has exhibited lethality rates in excess of 90% in isolated outbreaks in central Africa. The outbreak of Ebola virus in the summer of 1995 in Zaire killed 245 people out of 316 infected—a mortality rate of 78%. The latest outbreak occurred in Gabon, West Africa, in February 1996. The natural host of Ebola is unknown.

A recently emerged species of coronavirus was responsible for a worldwide outbreak in 2003 of **severe acute respiratory syndrome (SARS),** a respiratory infection with pneumonia-like symptoms that is fatal in over 8% of cases. When the 29,751-nucleotide RNA genome of the SARS coronavirus was sequenced, it proved to be a completely new form of coronavirus, not closely related to any of the three previously described forms. Virologists suspect that the SARS coronavirus most likely came from civets (a weasel-like mammal) and other wild animals that live in China and are eaten as delicacies. If the SARS virus indeed exists in natural populations, it will be difficult to prevent future outbreaks without an effective vaccine. Genome sequences have been analyzed from SARS patients at various stages of the outbreak, and these analyses indicate that the virus's mutation rate is low, in marked contrast to another RNA virus, HIV. The stable genome of the SARS virus should make development of a SARS vaccine practical.

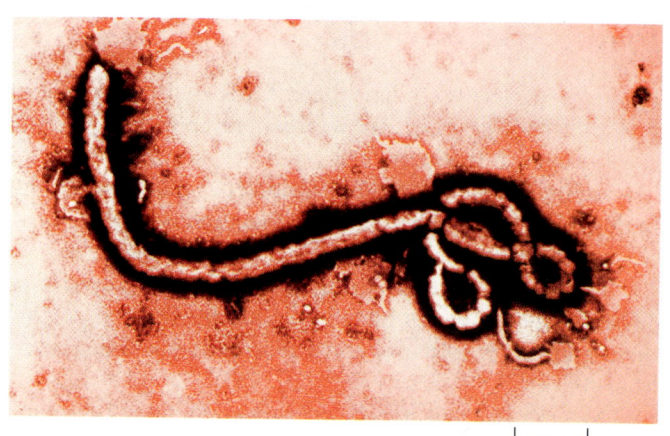

FIGURE 26.8
The Ebola virus. This virus, with a fatality rate that can exceed 90%, appears sporadically in West Africa. Health professionals are scrambling to identify the natural host of the virus, which is unknown, so they can devise strategies to combat transmission of the disease.

Viruses and Cancer

Through epidemiological studies and research, scientists have established a link between some viral infections and the subsequent development of cancer. Examples include the association between chronic hepatitis B infections and the development of liver cancer and the development of cervical carcinoma following infections with certain strains of papillomaviruses. It has been suggested that viruses contribute to about 15% of all human cancer cases worldwide. Viruses are capable of altering the growth properties of human cells they infect by triggering the expression of cancer-causing genes (see chapter 20). Virus-induced cancer is not simply a matter of infection; it involves complex interactions with cellular genes and requires a series of events in order to develop.

> Some of the most lethal diseases of humans are caused by viruses, particularly viruses that have transferred to humans from some other host, like those causing influenza (a bird virus), Ebola (animal host unknown), and SARS (small mammal virus).

Prions and Viroids

For decades, scientists have been fascinated by a peculiar group of fatal brain diseases. These diseases have an unusual property: It is years and often decades after infection before the disease is detected in infected individuals. The brains of infected individuals develop numerous small cavities as neurons die, producing a marked spongy appearance. Called **transmissible spongiform encephalopathies (TSEs),** these diseases include scrapie in sheep, mad cow disease in cattle, and kuru and Creutzfeldt-Jakob disease in humans.

TSEs can be transmitted experimentally by injecting infected brain tissue into a recipient animal's brain. TSEs can also spread via tissue transplants and, apparently, food. The disease kuru was common in the Fore people of Papua, New Guinea, because they practiced ritual cannibalism, literally eating the brains of infected individuals. Mad cow disease spread widely among the cattle herds of England in the 1990s because cows were fed bonemeal prepared from cattle carcasses to increase the protein content of their diet. Like the Fore, the British cattle were literally eating the tissue of cattle that had died of the disease.

A Heretical Suggestion

In the 1960s, British researchers T. Alper and J. Griffith noted that infectious TSE preparations remained infectious even after exposure to radiation that would destroy DNA or RNA. They suggested that the infectious agent was a protein. Perhaps, they speculated, the protein usually preferred one folding pattern, but could sometimes misfold, and then catalyze other proteins to do the same, the misfolding spreading like a chain reaction. This heretical suggestion was not accepted by the scientific community, because it violates a key tenet of molecular biology: Only DNA or RNA act as hereditary material, transmitting information from one generation to the next.

Prusiner's Prions

In the early 1970s, physician Stanley Prusiner, moved by the death of a patient from Creutzfeldt-Jakob disease, began to study TSEs. Prusiner became fascinated with Alper and Griffith's hypothesis. Try as he might, Prusiner could find no evidence of nucleic acids or viruses in the infectious TSE preparations. He concluded, as Alper and Griffith had, that the infectious agent was a *protein*, which in a 1982 paper he named a **prion,** for "proteinaceous infectious particle."

Prusiner went on to isolate a distinctive prion protein, and for two decades continued to amass evidence that prions play a key role in triggering TSEs. The scientific community resisted Prusiner's renegade conclusions, but eventually experiments done in Prusiner's and other laboratories began to convince many. For example, when Prusiner in-

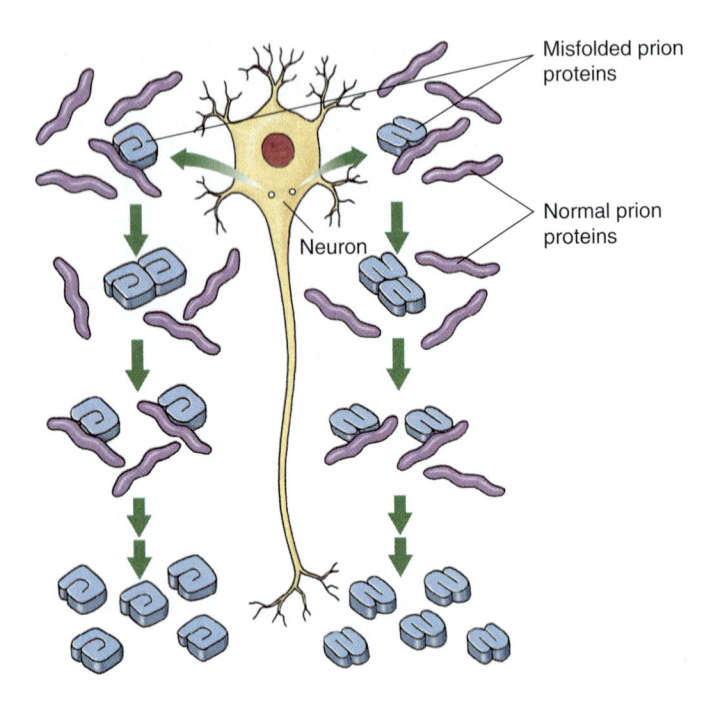

FIGURE 26.9
How prions arise. Misfolded prions seem to cause normal prion protein to misfold simply by contacting them. When prions misfolded in different ways (*blue*) contact normal prion protein (*purple*), the normal prion protein misfolds in the same way.

jected prions of a different abnormal conformation into several different hosts, these hosts developed prions with the same abnormal conformations as the parent prions (figure 26.9). In another important experiment, Charles Weissmann showed that mice genetically engineered to lack Prusiner's prion protein are immune to TSE infection. However, if brain tissue with the prion protein is grafted into the mice, the grafted tissue—but not the rest of the brain—can then be infected with TSE. In 1997, Prusiner was awarded the Nobel Prize in physiology or medicine for his work on prions.

Viroids

Viroids are tiny, naked molecules of RNA, only a few hundred nucleotides long, that are important infectious disease agents in plants. A recent viroid outbreak killed over ten million coconut palms in the Philippines. It is not clear how viroids cause disease. One clue is that viroid nucleotide sequences resemble the sequences of introns within ribosomal RNA genes. These sequences are capable of catalyzing excision from DNA—perhaps the viroids are catalyzing the destruction of chromosomal integrity.

> **Prions are infectious proteins that some scientists believe are responsible for serious brain diseases. In plants, naked RNA molecules called viroids can also transmit disease.**

Concept Review

For interactive testing, visit the Online Learning Center with PowerWeb at www.mhhe.com/Raven7

26.1 Viruses are strands of nucleic acid encased within a protein coat.

The Nature of Viruses
- All viruses have the same basic structure: a nucleic acid core surrounded by a protein. (p. 532)
- Nearly all viruses form a capsid. (p. 532)
- Each virus can only replicate in a limited number of cell types, and the suitable cells are referred to as the host range. (p. 532)
- Viruses must enter cells and utilize cellular machinery to reproduce. (p. 532)
- The overall structure of a virus is either helical or isometric. (p. 532)

26.2 Bacterial viruses exhibit two sorts of reproductive cycles.

Bacteriophages
- Bacteriophages are viruses that infect bacteria. (p. 534)
- The lytic cycle refers to the reproductive cycle when a virus kills an infected host. Virulent viruses multiply within infected cells and eventually lyse the cells. (p. 534)
- Most bacteriophages do not immediately kill their hosts; these are referred to as prophages during lysogeny, the integration of a virus into a cellular genome. (p. 534)

Cell Transformation and Phage Conversion
- Transformation refers to the alteration of a cell's genome by the introduction of foreign DNA. (p. 535)

26.3 HIV is a complex animal virus.

AIDS
- In AIDS patients, the human immunodeficiency virus infects and kills CD4+ cells, eventually causing the patients to die of an infection that would normally be fought off by a healthy immune system. (p. 536)
- HIV can only enter cells that possess a cell-surface marker recognized by a glycoprotein on the HIV particle surface. (p. 536)
- The HIV infection cycle includes the following steps: attachment, entry into cells, replication, integration into the host cell's DNA, production of new viruses, and finally exit from the cells. (p. 536)

The Future of HIV Treatment
- The future of HIV treatment includes using combination drug therapy such as AZT analogs and protease inhibitors, using a defective HIV gene, using chemokines and CAF, and blocking or disabling receptors. (pp. 538–539)

26.4 Nonliving infectious agents are responsible for many human diseases.

Disease Viruses
- Viruses are known to cause many human diseases, including influenza, smallpox, infectious hepatitis, yellow fever, polio, AIDS, and SARS. (p. 540)
- Viruses have also been implicated in some cancers and leukemias. (p. 540)
- Emerging viruses originate in one organism and then pass to another. (p. 541)

Prions and Viroids
- Brain diseases known as transmissible spongiform encephalopathies are known to be caused by a protein called a prion (proteinaceous infectious particle). (p. 542)
- Viroids, naked RNA molecules, can transmit diseases in plants. (p. 542)

Test Your Understanding

For interactive testing, visit the Online Learning Center with PowerWeb at www.mhhe.com/Raven7

Self Test

1. Viruses consist of a _____ core surrounded by a protein coat.
   a. RNA
   b. DNA
   c. chromosome
   d. nucleic acid
2. Phages infect bacterial cells by
   a. poking holes in the cell and injecting their DNA.
   b. destroying the bacterial cell wall.
   c. receptor-mediated endocytosis.
   d. exocytosis.
3. The result of the viral lytic cycle is
   a. the release of new viruses.
   b. the incorporation of the viral genome into the host genome.
   c. the conversion of the virus into a prophage.
   d. Both b and c are correct.
4. The alteration of a cell's genome by the incorporation of foreign DNA is called
   a. genetic conversion.
   b. mutation.
   c. transformation.
   d. reverse transcription.
5. Which of the following is a viral glycoprotein that plays a role in the infection of human cells by HIV?
   a. gp120
   b. CD4
   c. CCR5
   d. Both b and c are correct.
6. The HIV enzyme, _____, produces a DNA copy of the viral genome once it is inside the host cell.
   a. DNA polymerase
   b. reverse transcriptase
   c. RNA polymerase
   d. helicase
7. The drug AZT and its analogs function by
   a. inhibiting the replication of viral nucleic acid.
   b. blocking the production of envelope proteins.
   c. blocking the production of capsid proteins.
   d. blocking the binding of the virus to human cell receptors.
8. What is the function of the CD8$^+$ cell antiviral factor (CAF) in human white blood cells?
   a. disables the CD4 receptor
   b. blocks replication of the HIV virus
   c. interferes with the production of viral proteins
   d. blocks the CCR5 or CXCR4 receptors
9. New strains of flu viruses are most likely to arise in the Far East because
   a. of overpopulation in these areas.
   b. adequate sanitation is lacking.
   c. common hosts (ducks, chickens, pigs) live in close proximity to humans.
   d. of environmental mutagens that cause changes in viral RNA.
10. Which of the following infectious agents does *not* contain protein?
    a. viruses
    b. viroids
    c. prions
    d. none of these

Test Your Visual Understanding

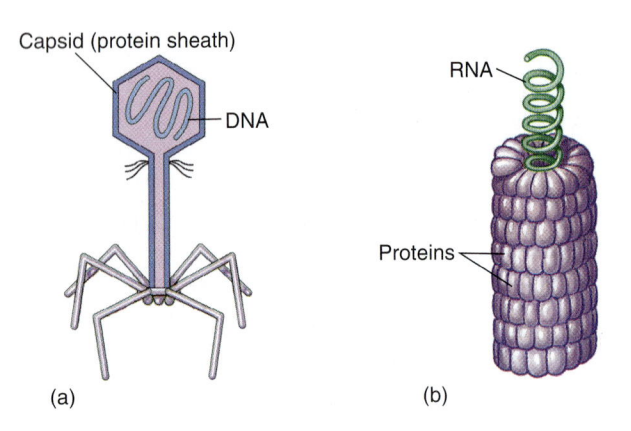

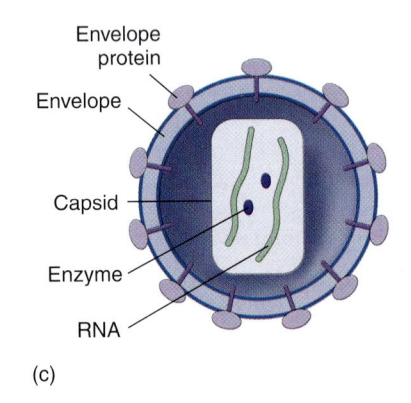

1. Describe the overall shape of each of the viruses in the figure, either as helical, isometric, or icosahedral.
2. List which of these viruses attacks:
   a. animals
   b. plants
   c. bacteria

Apply Your Knowledge

1. AIDS was first identified in 1981 in the United States, and in 2002, an estimated 495,000 cases of AIDS and/or HIV had been reported. The total U.S. population at the end of 2002 was approximately 290 million. What percent of the population was living with AIDS and/or HIV?
2. In what ways might the early, self-replicating particles that gave rise to the first organisms have resembled, or differed from, viruses?
3. If a viral disease such as influenza could kill 22 million people over 18 months in 1918–1919, why do you suppose the killing ever stopped? What prevented the flu virus from continuing its lethal assault on humanity?

27
Prokaryotes

Concept Outline

27.1 Prokaryotes are the smallest and most numerous organisms.

The Prevalence of Prokaryotes. The simplest and most abundant of organisms, prokaryotes are also thought to be the most ancient. Prokaryotes lack the high degree of internal compartmentalization characteristic of eukaryotes.

27.2 Prokaryotes exhibit considerable diversity in both structure and metabolism.

Prokaryotic Diversity. There are at least 17 major groups of prokaryotes, although many more remain to be discovered.

27.3 Prokaryotes are more complex than commonly supposed.

The Prokaryotic Cell Surface. Prokaryotes have a cell wall and may contain structures such as pili and flagella.
The Cell Interior. While prokaryotes lack extensive internal compartments, they may have complex internal membranes.
Prokaryotic Variation. Mutation and recombination generate enormous variation within bacterial populations.
Prokaryotic Metabolism. Prokaryotes obtain carbon atoms and energy from a wide array of sources, including other organisms.

27.4 Prokaryotes are responsible for many diseases, but they also make important contributions to ecosystems.

Human Bacterial Diseases. Many serious human diseases are caused by bacteria, and some of these diseases are responsible for millions of deaths each year.
Benefits of Prokaryotes. Prokaryotes have had a profound impact on the world's ecology, and play a major role in modern medicine and agriculture.

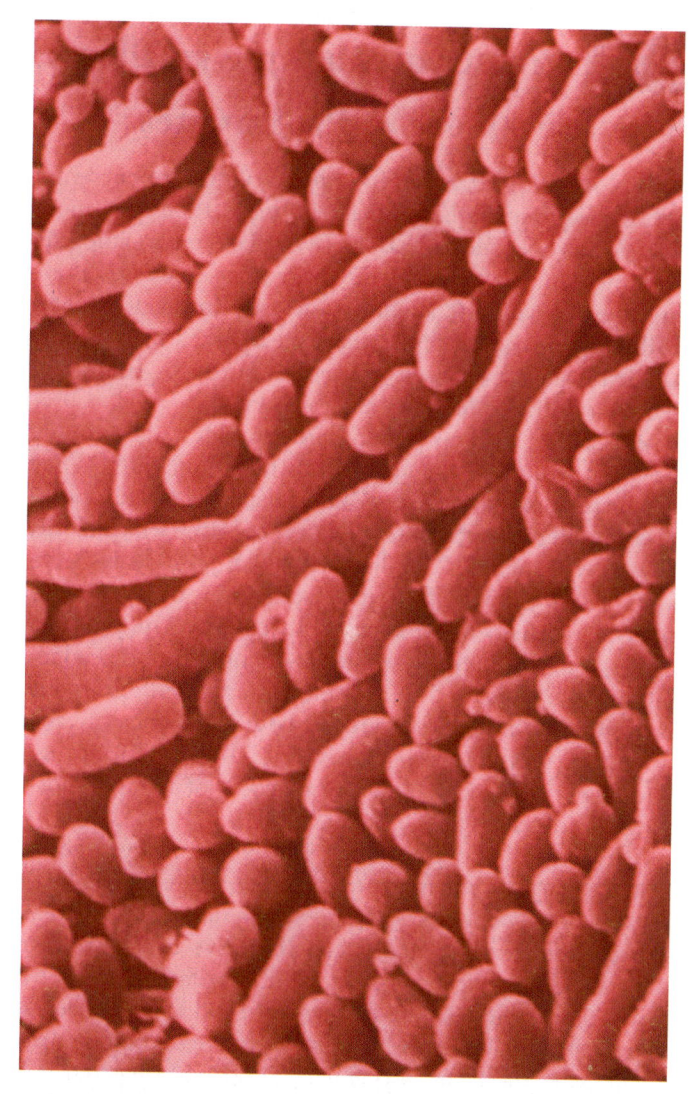

FIGURE 27.1
A colony of prokaryotes. With their enormous adaptability and metabolic versatility, prokaryotes are found in every habitat on earth, carrying out many of the vital processes of ecosystems, including photosynthesis, nitrogen fixation, and decomposition.

The simplest organisms living on earth today are prokaryotes, and biologists think they closely resemble the first organisms to evolve on earth. Too small to see with the unaided eye, prokaryotes are the most abundant of all organisms (figure 27.1) and are the only ones characterized by prokaryotic cellular organization. Life on earth could not exist without prokaryotes because they make possible many of the essential functions of ecosystems, including the capture of nitrogen from the atmosphere, decomposition of organic matter, and in many aquatic communities, photosynthesis. Indeed, prokaryotic photosynthesis is thought to have been the source for much of the oxygen in the earth's atmosphere. Prokaryotic research continues to provide extraordinary insights into genetics, ecology, and disease. Thus, an understanding of prokaryotes is essential.

27.1 Prokaryotes are the smallest and most numerous organisms.

The Prevalence of Prokaryotes

About 5000 different kinds of prokaryotes are currently recognized, but there are doubtless many thousands more awaiting proper identification. Everyplace microbiologists look, new species are being discovered, in some cases altering the way we think about prokaryotes. In the 1970s and 1980s, a new type of prokaryote was analyzed that eventually led to the division of prokaryotes into two groups, the archaebacteria (or Archaea) and the bacteria.

Archaebacteria and bacteria are the oldest, structurally simplest, and most abundant forms of life on earth. They are also the only organisms with prokaryotic cellular organization. Represented in the oldest rocks from which fossils have been obtained, 2.5 billion years old, prokaryotes were abundant for over a billion years before eukaryotes appeared in the world. Early photosynthetic bacteria (cyanobacteria) altered the earth's atmosphere by producing oxygen, which led to extreme bacterial and eukaryotic diversity. Bacteria play a vital role both in productivity and in cycling the substances essential to all other life-forms. Bacteria are the only organisms capable of fixing atmospheric nitrogen.

Prokaryotes are ubiquitous on earth, and live everywhere eukaryotes do. Many of the other, more extreme environments in which prokaryotes are found would be lethal to any other form of life. Prokaryotes live in hot springs that would cook other organisms, in hypersaline environments that would dehydrate other cells, and in atmospheres rich in such toxic gases as methane or hydrogen sulfide that would kill most other organisms. These harsh environments may be similar to the conditions present on the early earth when life first began. It is likely that prokaryotes evolved to dwell in these harsh conditions early on and have retained the ability to exploit these areas as the rest of the atmosphere has changed.

Prokaryotic Form

Prokaryotes are mostly simple in form and exhibit one of three basic structures: **bacillus** (plural, *bacilli*), straight and rod-shaped; **coccus** (plural, *cocci*), spherical-shaped; and **spirillum** (plural, *spirilla*), long and helical-shaped, also called spirochetes. Spirilla generally do not form associations with other cells; they swim singly through their environments. A complex structure within their cell wall allows them to spin their corkscrew-shaped bodies, propelling them along. Some rod-shaped and spherical bacteria form colonies, adhering end-to-end after they have divided, forming chains. Some bacterial colonies change into stalked structures, grow long, branched filaments, or form erect structures that release **spores,** single-celled bodies that grow into new bacterial individuals. Some filamentous bacteria are capable of a gliding motion, often combined with rotation around a longitudinal axis. Biologists have not yet determined the mechanism by which they move.

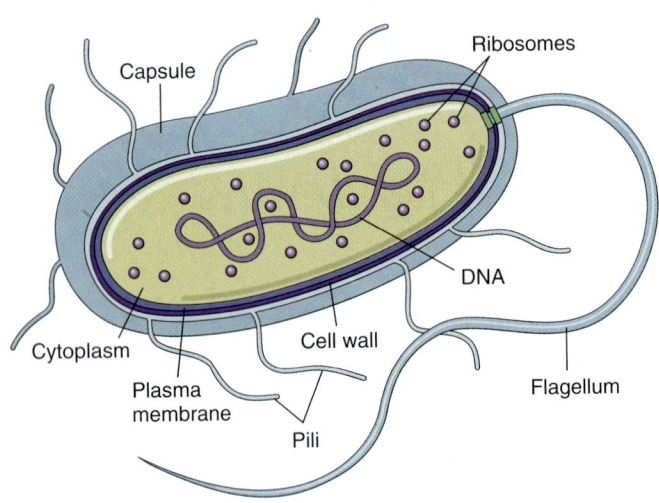

FIGURE 27.2
The structure of a prokaryotic cell. Prokaryotic cells, such as this bacterium, are small and lack interior organization. The plasma membrane is encased within a rigid cell wall, and DNA is not contained in a membrane-bounded nucleus. In addition to a flagellum, prokaryotes may also possess outgrowths called pili that aid in attachment to surfaces or other cells.

Prokaryotes Versus Eukaryotes

Prokaryotes differ from eukaryotes in numerous important features (table 27.1). These differences represent some of the most fundamental distinctions that separate any groups of organisms.

1. **Unicellularity.** All prokaryotes are fundamentally single-celled (figure 27.2). In some types, individual cells adhere to each other within a matrix and form filaments; however, the cells retain their individuality. Cyanobacteria, in particular, are likely to form such associations, but their cytoplasm is not directly interconnected, as often is the case in multicellular eukaryotes. The activities of a bacterial colony are less integrated and coordinated than those of multicellular eukaryotes. A primitive form of colonial organization occurs in gliding bacteria, which move together and form spore-bearing structures. Such coordinated multicellular forms are rare among prokaryotes.
2. **Cell size.** As new species of prokaryotes are discovered, we are finding that the size of prokaryotic cells varies tremendously, by as much as five orders of

546   Part V   Diversity of Life on Earth

magnitude. Most prokaryotic cells are only 1 micrometer or less in diameter. Most eukaryotic cells are well over 10 times that size.

3. **Chromosomes.** Eukaryotic cells have a membrane-bounded nucleus containing chromosomes made up of both nucleic acids and proteins. Prokaryotes do not have membrane-bounded nuclei, nor do they have chromosomes of the kind present in eukaryotes, in which DNA forms a structural complex with proteins. Instead, their naked circular DNA is localized in a zone of the cytoplasm called the **nucleoid.**

4. **Cell division and genetic recombination.** Cell division in eukaryotes takes place by mitosis and involves spindles made up of microtubules. Cell division in prokaryotes takes place mainly by binary fission (see chapter 11). True sexual reproduction occurs only in eukaryotes and involves syngamy and meiosis, with an alternation of diploid and haploid forms. Despite their asexual mode of reproduction, prokaryotes do have mechanisms that lead to the transfer of genetic material. These mechanisms are far less regular than those of eukaryotes and do not involve the equal participation of the individuals between which the genetic material is transferred (see chapter 20).

5. **Internal compartmentalization.** In eukaryotes, the enzymes for cellular respiration are packaged in mitochondria. In prokaryotes, the corresponding enzymes are not packaged separately but are bound to the cell membranes (see chapters 5 and 9). The cytoplasm of prokaryotes, unlike that of eukaryotes, contains no internal compartments or cytoskeleton and no organelles except ribosomes.

6. **Flagella.** Prokaryotic flagella are simple in structure, composed of a single fiber of the protein flagellin (see chapter 5). Eukaryotic flagella and cilia are complex and have a 9 + 2 structure of microtubules (see figure 5.27). Bacterial flagella also function differently, spinning like propellers, while eukaryotic flagella have a whiplike motion.

7. **Metabolic diversity.** Only one kind of photosynthesis occurs in eukaryotes, and it involves the release of oxygen. Photosynthetic bacteria have several different patterns of anaerobic and aerobic photosynthesis, involving the formation of end products such as sulfur, sulfate, and oxygen (see chapter 10). Prokaryotic cells can also be chemoautotrophic, using the energy stored in chemical bonds of inorganic molecules to synthesize carbohydrates; eukaryotes are not capable of this metabolic process.

Prokaryotes are the oldest and most abundant organisms on earth. Prokaryotes differ from eukaryotes in a wide variety of characteristics, a degree of difference as great as that separating any other groups of organisms.

Table 27.1 Prokaryotes Compared to Eukaryotes

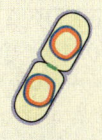

Feature	Example
Unicellularity. All prokaryotes are basically single-celled. Even though some bacteria may adhere together or form filaments, their cytoplasm is not directly interconnected, and their activities are not integrated and coordinated as is the case in multicellular eukaryotes.	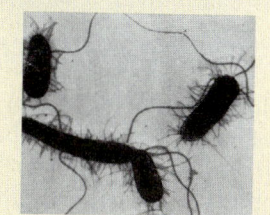 Unicellular bacteria
Cell Size. Most bacterial cells are only about 1 micrometer in diameter, while most eukaryotic cells are over 10 times that size.	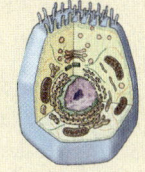 Bacterial cell / Eukaryotic cell
Chromosomes. Prokaryotic DNA exists as a single circle in the cytoplasm, while in eukaryotes, proteins are complexed with the DNA into multiple chromosomes.	 Bacterial genome / Eukaryotic chromosomes
Cell Division. Prokaryotic cells divide by binary fission (see chapter 11). The cells simply pinch in two. In eukaryotes, microtubules pull chromosomes to opposite poles during the cell division process, called mitosis.	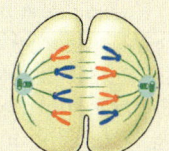 Binary fission in bacteria / Mitosis in eukaryotes
Internal Compartmentalization. Unlike eukaryotic cells, bacterial cells contain no internal compartments, no internal membrane system, and no cell nucleus.	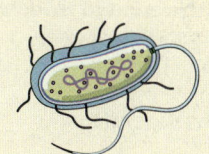 Bacterial cell
Flagella. Prokaryotic flagella are simple, composed of a single fiber of protein that spins like a propeller. Flagella in eukaryotes are complex structures that whip back and forth rather than rotating.	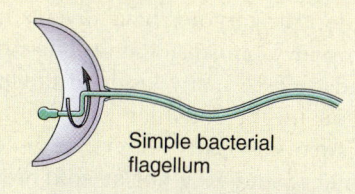 Simple bacterial flagellum
Metabolic Diversity. Prokaryotes possess many metabolic abilities that eukaryotes do not; some prokaryotes can perform several different kinds of anaerobic and aerobic photosynthesis, obtain their energy from oxidizing inorganic compounds, or fix atmospheric nitrogen.	 Chemoautotrophs

Chapter 27 Prokaryotes 547

27.2 Prokaryotes exhibit considerable diversity in both structure and metabolism.

Prokaryotic Diversity

Prokaryotes are not easily classified according to their forms, and only recently has enough been learned about their biochemical and metabolic characteristics to develop a satisfactory overall classification comparable to that used for other organisms. Early systems for classifying prokaryotes relied on differential stains such as the Gram stain. Key characteristics used in classifying prokaryotes were:

1. photosynthetic or nonphotosynthetic;
2. motile or nonmotile;
3. unicellular or colony-forming or filamentous;
4. formation of spores or division by transverse binary fission.

With the development of genetic and molecular approaches, prokaryotic classifications can at last reflect true evolutionary relatedness. Molecular approaches include: (1) the analysis of the amino acid sequences of key proteins; (2) the analysis of nucleic acid base sequences by establishing the percent of guanine (G) and cytosine (C); (3) nucleic acid hybridization, which is essentially the mixing of single-stranded DNA from two species and determining the amount of base-pairing (closely related species will have more bases pairing); (4) gene and RNA sequencing, especially looking at ribosomal RNA; and (5) whole-genome sequencing.

Based on these sorts of molecular data, several groupings of prokaryotes have been proposed. The most widely accepted is that presented in *Bergey's Manual*, second edition, 2001 (figure 27.3).

Kinds of Prokaryotes

Although they lack the structural complexity of eukaryotes, prokaryotes have diverse internal chemistries, varying metabolisms, and unique functions. Prokaryotes have adapted to many kinds of environments, including some you might consider harsh. They have successfully invaded very salty waters, very acidic or alkaline environments, and very hot or cold areas. They are found in hot springs where the temperatures exceed 78°C (172°F) and have been recovered living beneath 435 meters of ice in Antarctica!

Much of what we know about bacteria has been learned from studies in the laboratory, which has placed limits on our knowledge. Field studies suggest that bacteria able to be cultured in the lab represent but a small fraction of the kinds of prokaryotes that occur in soil, most of which cannot be cultured with existing techniques. We clearly have only scraped the surface of prokaryotic diversity.

Prokaryotes split into two lines early in the history of life. Archaebacteria and bacteria are as different in structure and metabolism from each other as either is from the eukaryotes. The domain Archaea consists of the archaebacteria ("ancient bacteria"—although they are actually not as ancient as the other prokaryotic domain). It was once thought that survivors of this group were confined to extreme environments that may resemble habitats on the early earth. However, the use of genetic screening has revealed that these "ancient" prokaryotes live in nonextreme environments as well. The other, more ancient domain, the Bacteria, includes nearly all of the named species of prokaryotes.

Comparing Archaebacteria and Bacteria

Archaebacteria and bacteria are similar in that both have a prokaryotic cellular structure, but they vary considerably at the biochemical and molecular levels. There are four key areas in which they differ:

1. **Plasma membranes.** All prokaryotes have plasma membranes with a lipid-bilayer architecture (as described in chapter 6). The plasma membranes of bacteria and archaebacteria, however, are made of very different kinds of lipids.
2. **Cell wall.** Both kinds of prokaryotes typically have cell walls covering the plasma membrane that strengthen the cell. The cell walls of bacteria are constructed of carbohydrate-protein complexes called peptidoglycan, which link together to create a strong mesh that gives the bacterial cell wall great strength. The cell walls of archaebacteria lack peptidoglycan.
3. **Gene translation machinery.** Bacteria possess ribosomal proteins and an RNA polymerase that are distinctly different from those of eukaryotes. However, the ribosomal proteins and RNA of archaebacteria are very similar to those of eukaryotes.
4. **Gene architecture.** The genes of bacteria are not interrupted by introns, while at least some of the genes of archaebacteria do possess introns.

While superficially similar, archaebacteria and bacteria differ from one another in a wide variety of characteristics.

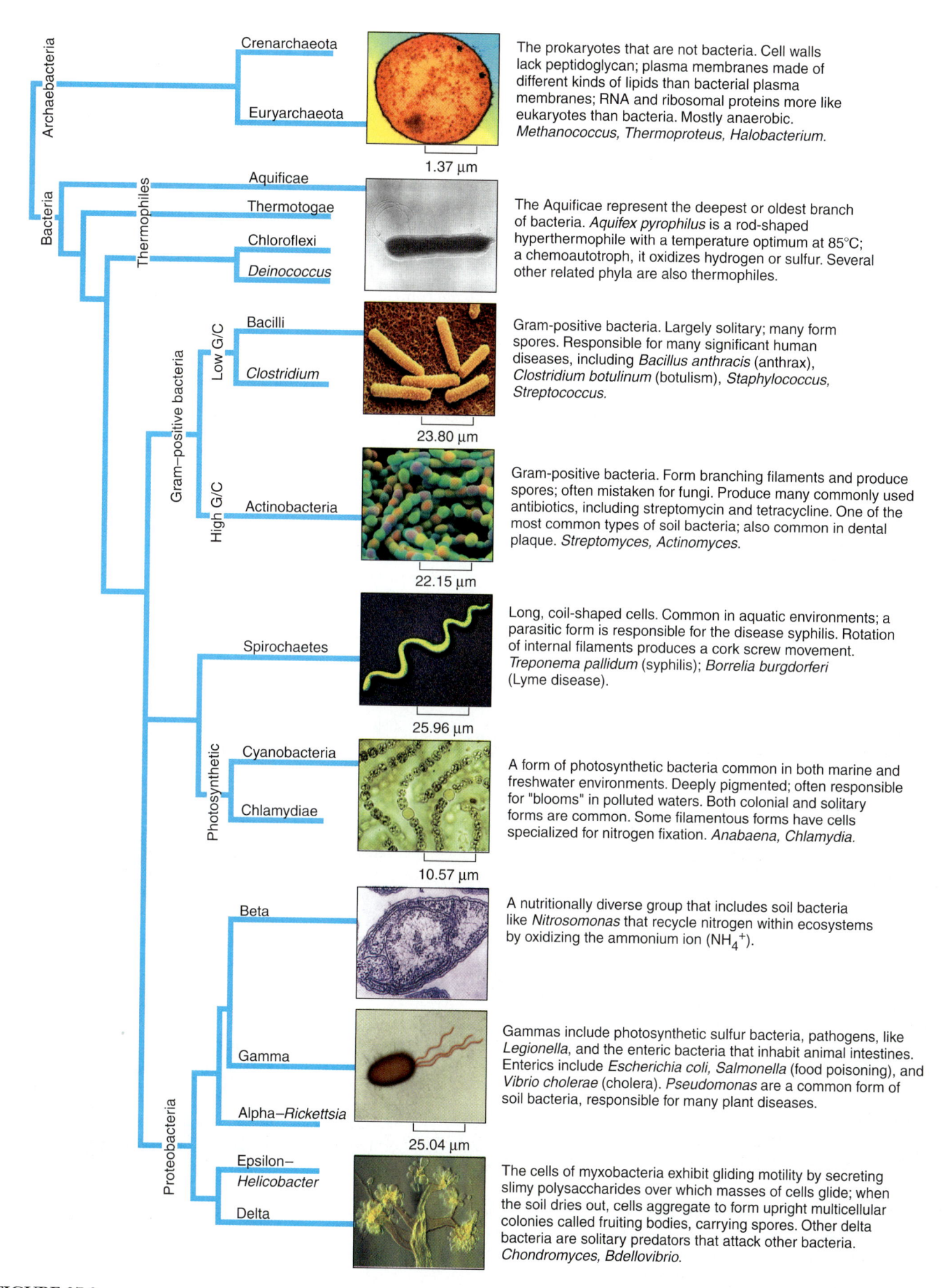

FIGURE 27.3
The major clades of prokaryotes. The classification adopted here is that of *Bergey's Manual of Systematic Bacteriology*, second edition, 2001.

Archaebacteria (Crenarchaeota, Euryarchaeota): The prokaryotes that are not bacteria. Cell walls lack peptidoglycan; plasma membranes made of different kinds of lipids than bacterial plasma membranes; RNA and ribosomal proteins more like eukaryotes than bacteria. Mostly anaerobic. *Methanococcus, Thermoproteus, Halobacterium*.

Bacteria — Thermophiles (Aquificae, Thermotogae, Chloroflexi, *Deinococcus*): The Aquificae represent the deepest or oldest branch of bacteria. *Aquifex pyrophilus* is a rod-shaped hyperthermophile with a temperature optimum at 85°C; a chemoautotroph, it oxidizes hydrogen or sulfur. Several other related phyla are also thermophiles.

Gram-positive bacteria — Low G/C (Bacilli, *Clostridium*): Gram-positive bacteria. Largely solitary; many form spores. Responsible for many significant human diseases, including *Bacillus anthracis* (anthrax), *Clostridium botulinum* (botulism), *Staphylococcus, Streptococcus*.

Gram-positive bacteria — High G/C (Actinobacteria): Gram-positive bacteria. Form branching filaments and produce spores; often mistaken for fungi. Produce many commonly used antibiotics, including streptomycin and tetracycline. One of the most common types of soil bacteria; also common in dental plaque. *Streptomyces, Actinomyces*.

Spirochaetes: Long, coil-shaped cells. Common in aquatic environments; a parasitic form is responsible for the disease syphilis. Rotation of internal filaments produces a cork screw movement. *Treponema pallidum* (syphilis); *Borrelia burgdorferi* (Lyme disease).

Photosynthetic (Cyanobacteria, Chlamydiae): A form of photosynthetic bacteria common in both marine and freshwater environments. Deeply pigmented; often responsible for "blooms" in polluted waters. Both colonial and solitary forms are common. Some filamentous forms have cells specialized for nitrogen fixation. *Anabaena, Chlamydia*.

Proteobacteria — Beta: A nutritionally diverse group that includes soil bacteria like *Nitrosomonas* that recycle nitrogen within ecosystems by oxidizing the ammonium ion (NH_4^+).

Proteobacteria — Gamma, Alpha–*Rickettsia*: Gammas include photosynthetic sulfur bacteria, pathogens, like *Legionella*, and the enteric bacteria that inhabit animal intestines. Enterics include *Escherichia coli, Salmonella* (food poisoning), and *Vibrio cholerae* (cholera). *Pseudomonas* are a common form of soil bacteria, responsible for many plant diseases.

Proteobacteria — Epsilon–*Helicobacter*, Delta: The cells of myxobacteria exhibit gliding motility by secreting slimy polysaccharides over which masses of cells glide; when the soil dries out, cells aggregate to form upright multicellular colonies called fruiting bodies, carrying spores. Other delta bacteria are solitary predators that attack other bacteria. *Chondromyces, Bdellovibrio*.

27.3 Prokaryotes are more complex than commonly supposed.

The Prokaryotic Cell Surface

The prokaryotic **cell wall** is an important structure because it maintains the shape of the cell and protects the cell from swelling and rupturing. The bacterial cell wall usually consists of **peptidoglycan,** a network of polysaccharide molecules connected by polypeptide crosslinks. In some bacteria, the peptidoglycan forms a thick, complex network around the outer surface of the cell. This network is interlaced with peptide chains. In other bacteria, a thin layer of peptidoglycan is sandwiched between two plasma membranes. The outer membrane contains large molecules of lipopolysaccharide, lipids with polysaccharide chains attached. These two major types of bacteria can be identified using a staining process called a **Gram stain.** **Gram-positive** bacteria have the thicker peptidoglycan wall and stain a purple color, while the more common **gram-negative** bacteria contain less peptidoglycan and do not retain the purple-colored dye (figure 27.4). Gram-negative bacteria stain red. The outer membrane layer makes gram-negative bacteria resistant to many antibiotics that interfere with cell wall synthesis in gram-positive bacteria. In some kinds of bacteria, an additional gelatinous layer, the capsule, surrounds the cell wall.

Many kinds of prokaryotes have slender, rigid, helical **flagella** (singular, *flagellum*) composed of the protein flagellin (figure 27.5). These flagella range from 3 to 12 micrometers in length and are very thin—only 10 to 20 nanometers thick. They are anchored in the cell wall and spin like a propeller, pulling the cell through the water.

Pili (singular, *pilus*) are other hairlike structures that occur on the cells of some prokaryotes (see figure 27.2). They are shorter than prokaryotic flagella (up to several micrometers long) and about 7.5 to 10 nanometers thick. Pili help the prokaryotic cells attach to appropriate substrates and exchange genetic information (see chapter 20).

Some prokaryotes form thick-walled **endospores** around their genome and a small portion of the surrounding cytoplasm when they are exposed to nutrient-poor conditions. These endospores are highly resistant to environmental stress, especially heat, and can germinate to form new individuals after decades or even centuries.

Prokaryotes are encased within a cell wall composed of one or more layers. They also may contain external structures such as flagella and pili.

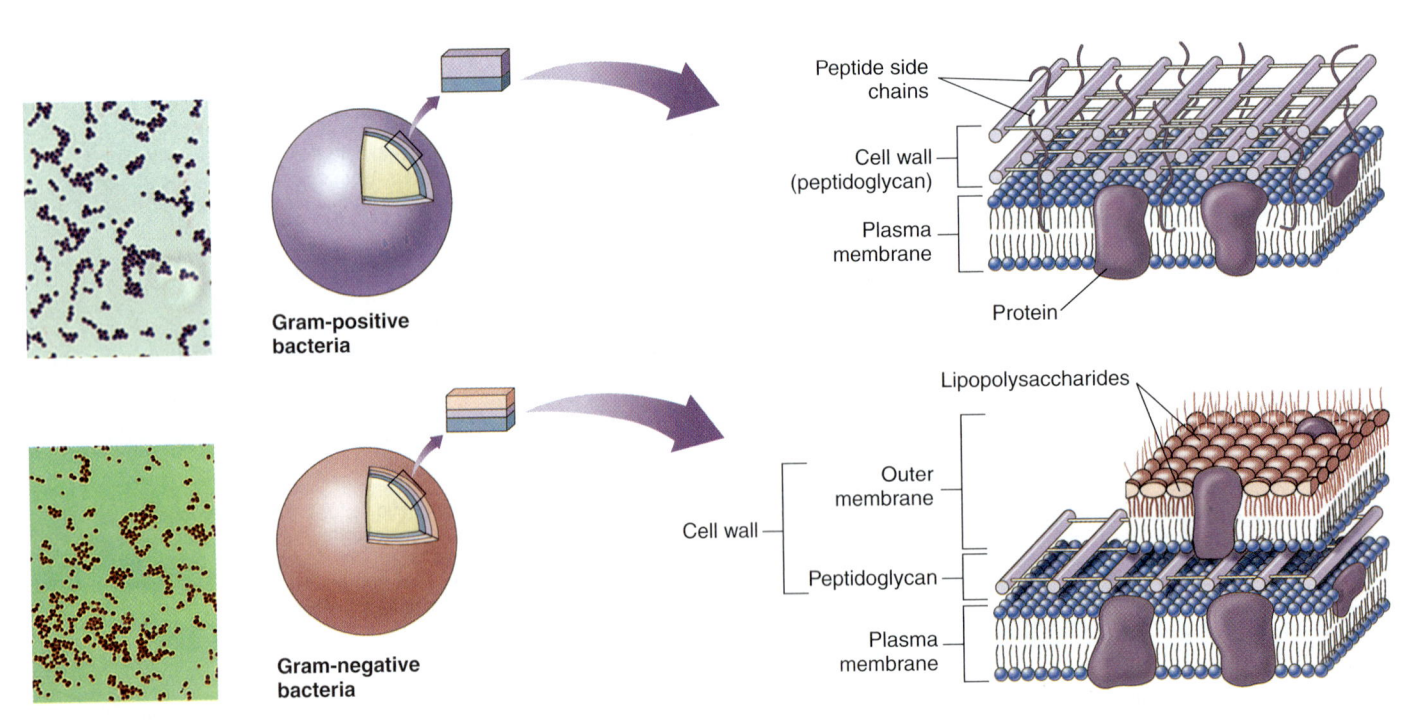

FIGURE 27.4
The Gram stain. The peptidoglycan layer encasing gram-positive bacteria traps crystal violet dye, so the bacteria appear purple in a Gram-stained smear (named after Hans Christian Gram, who developed the technique). Because gram-negative bacteria have much less peptidoglycan (located between the plasma membrane and an outer membrane), they do not retain the crystal violet dye and so exhibit the red background stain (usually a safranin dye).

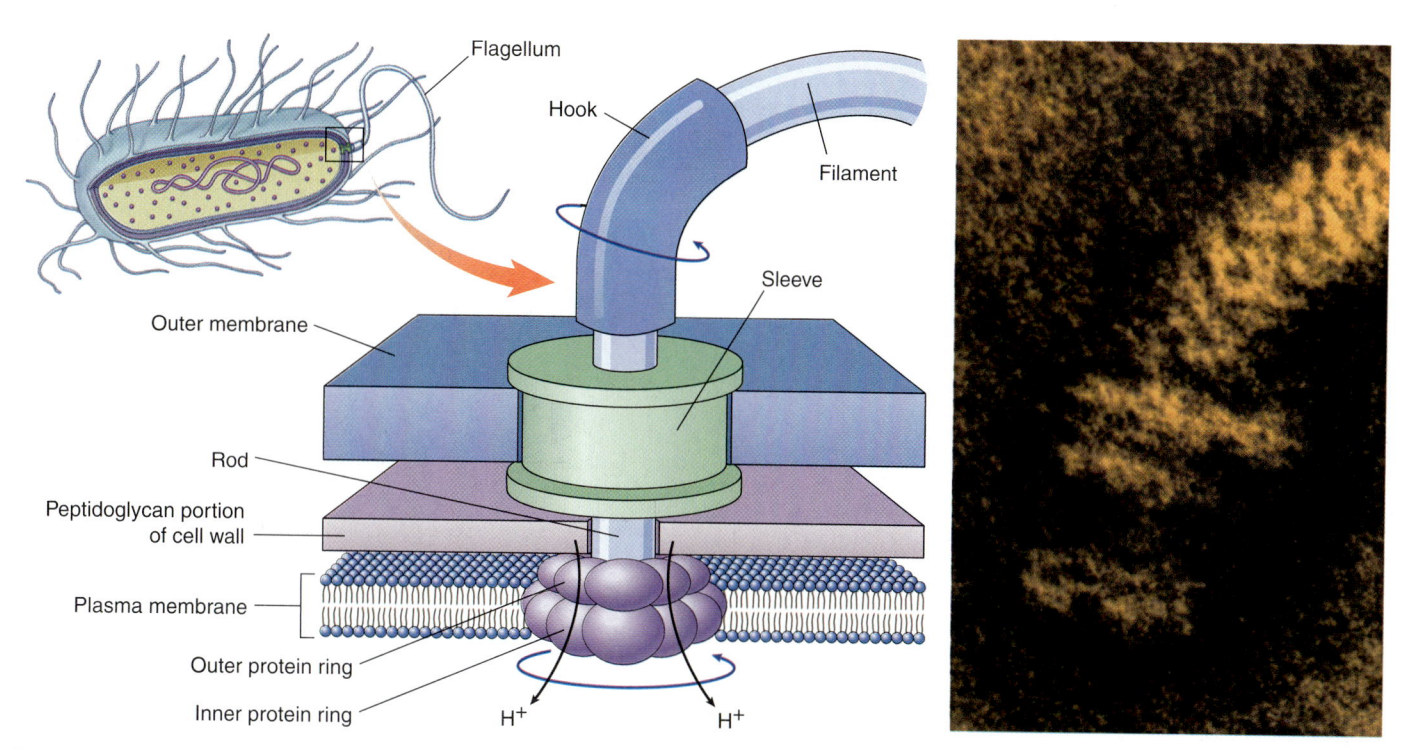

FIGURE 27.5
The flagellar motor of a gram-negative bacterium. A protein filament, composed of the protein flagellin, is attached to a protein rod that passes through a sleeve in the outer membrane and through a hole in the peptidoglycan layer to rings of protein anchored in the cell wall and plasma membrane, like rings of ballbearings. The rod rotates when the inner protein ring attached to the rod turns with respect to the outer ring fixed to the cell wall. The inner ring is an H^+ ion channel, a proton pump that uses the passage of protons into the cell to power the movement of the inner ring past the outer one.

The Cell Interior

The most fundamental characteristic of prokaryotic cells is their simple interior organization. Prokaryotic cells lack the extensive functional compartmentalization seen within eukaryotic cells, but they do have the following structures:

Internal membranes. Many prokaryotes possess invaginated regions of the plasma membrane that function in respiration or photosynthesis (figure 27.6).

Nucleoid region. Prokaryotes lack nuclei and do not possess the complex chromosomes characteristic of eukaryotes. Instead, their genes are encoded within a single double-stranded ring of DNA that is crammed into one region of the cell known as the **nucleoid region.** Many prokaryotic cells also possess small, independently replicating circles of DNA called plasmids. Plasmids contain only a few genes, which are usually not essential for the cell's survival. They are best thought of as an excised portion of the prokaryotic genome.

Ribosomes. Prokaryotic ribosomes are smaller than those of eukaryotes and differ in protein and RNA content. Antibiotics such as tetracycline and chloramphenicol can tell the difference—they bind to prokaryotic ribosomes and block protein synthesis, but do not bind to eukaryotic ribosomes.

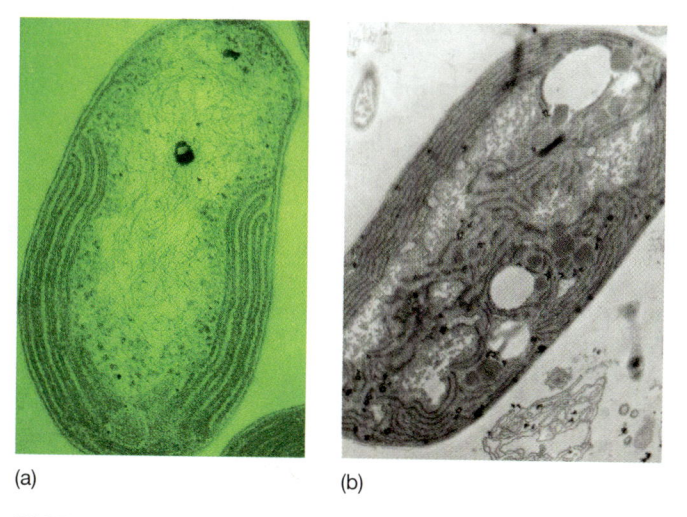

FIGURE 27.6
Prokaryotic cells often have complex internal membranes. (*a*) This aerobic bacterium exhibits extensive respiratory membranes within its cytoplasm not unlike those seen in mitochondria. (*b*) This cyanobacterium has thylakoid-like membranes that provide a site for photosynthesis.

The interior of a prokaryotic cell may possess functional membranes and a nucleoid region.

Chapter 27 Prokaryotes 551

Prokaryotic Variation

Prokaryotes reproduce rapidly, allowing genetic variations to spread quickly through a population. Two processes create variation among prokaryotes; mutation and genetic recombination.

Mutation

Mutations can arise spontaneously in bacteria as errors in DNA replication occur. Certain factors tend to increase the likelihood of errors occurring, such as radiation, ultraviolet light, and various chemicals. A typical bacterium such as *Escherichia coli* contains about 5000 genes. It is highly probable that one mutation will occur by chance in one out of every million copies of a gene. With 5000 genes in a bacterium, the laws of probability predict that 1 out of every 200 bacteria will have a mutation (figure 27.7). A spoonful of soil typically contains over a billion bacteria and therefore should contain something on the order of 5 million mutant individuals!

With adequate food and nutrients, a population of *E. coli* can double in under 20 minutes. Because bacteria multiply so rapidly, mutations can spread rapidly in a population and can change the characteristics of that population.

The ability of prokaryotes to change rapidly in response to new challenges often has adverse effects on humans. Recently, a number of strains of the bacterium *Staphylococcus aureus* associated with serious infections in hospitalized patients have appeared, some of them with alarming frequency. Unfortunately, these strains have acquired resistance to penicillin and a wide variety of other antibiotics, so that infections caused by them are very difficult to treat. *Staphylococcus* infections provide an excellent example of the way in which mutation and intensive selection can bring about rapid change in bacterial populations. Such changes have serious medical implications when, as in the case of *Staphylococcus*, strains of bacteria emerge that are resistant to a variety of antibiotics.

Recently, concern has arisen over the prevalence of antibacterial soaps in the marketplace. They are marketed as a means of protecting your family from harmful bacteria; however, it is likely that their routine use will favor bacteria that have mutations, making them immune to the antibiotics. Ultimately, extensive use of antibacterial soaps could have an adverse effect on our ability to treat common bacterial infections.

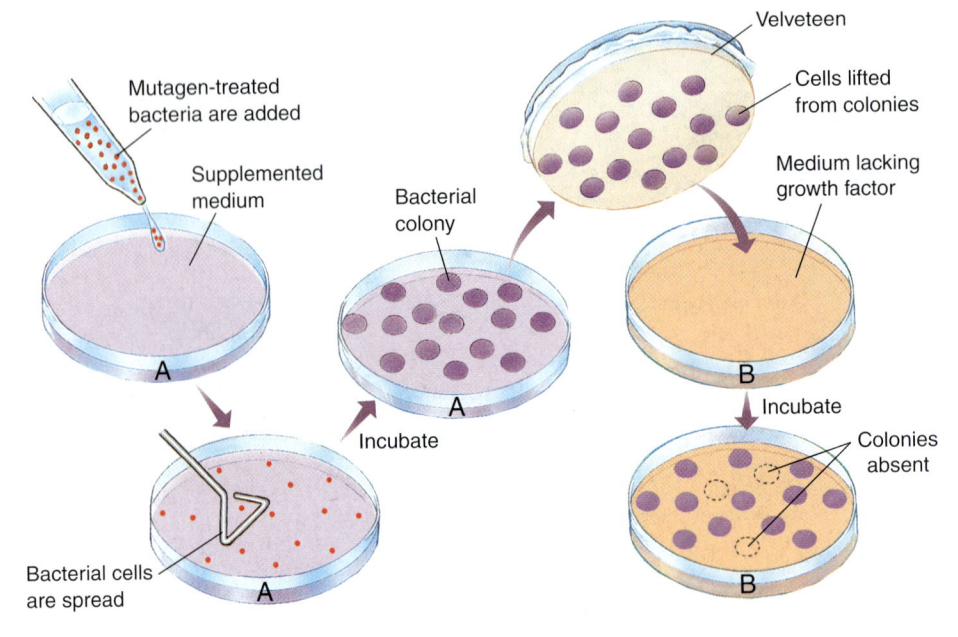

FIGURE 27.7
A mutant hunt in bacteria. Mutations in bacteria can be detected by a technique called replica plating, which allows the genetic characteristics of the colonies to be investigated without destroying them. The bacterial colonies, growing on a semisolid agar medium, are transferred from A to B using a sterile velveteen disc pressed on the plate. Plate A has a medium that includes special growth factors, while plate B has a medium that lacks some of these growth factors. Bacteria that are not mutated can produce their own growth factors and do not require them to be added to the medium. The colonies absent in B were unable to grow on the deficient medium and were thus mutant colonies; they were already present but undetected in A.

Genetic Recombination

Another source of genetic variation in populations of prokaryotes is recombination, discussed in detail in chapter 20. Recombination occurs in prokaryotes when genes are transferred from one cell to another by viruses, or through conjugation. The rapid transfer of newly produced, antibiotic-resistant genes by plasmids has been an important factor in the appearance of the resistant strains of *S. aureus* discussed earlier. An even more important example in terms of human health involves the Enterobacteriaceae, the family of bacteria to which the common intestinal bacterium *Escherichia coli* belongs. In this family are many important pathogenic bacteria, including the organisms that cause dysentery, typhoid, and other major diseases. At times, some of the genetic material from these pathogenic species is exchanged with or transferred to *E. coli* by plasmids. Because of its abundance in the human digestive tract, *E. coli* poses a special threat if it acquires harmful traits.

The short generation time of prokaryotes allows much mutation and recombination to occur, leading to genetic diversity.

Part V Diversity of Life on Earth

Prokaryotic Metabolism

Prokaryotes have evolved many mechanisms to acquire the energy and nutrients they need for growth and reproduction. Many are **autotrophs,** organisms that obtain their carbon from inorganic CO_2. Autotrophs that obtain their energy from sunlight are called *photoautotrophs*, while those that harvest energy from inorganic chemicals are called *chemoautotrophs*. Other prokaryotes are **heterotrophs,** organisms that obtain at least some of their carbon from organic molecules such as glucose. Heterotrophs that obtain their energy from sunlight are called *photoheterotrophs*, while those that harvest energy from organic molecules are called *chemoheterotrophs*.

Photoautotrophs. Many bacteria carry out photosynthesis, using the energy of sunlight to build organic molecules from carbon dioxide. The cyanobacteria use chlorophyll *a* as the key light-capturing pigment and H_2O as an electron donor, releasing oxygen gas as a by-product. Other bacteria use bacteriochlorophyll as their pigment and H_2S as an electron donor, leaving elemental sulfur as the by-product.

Chemoautotrophs. Some prokaryotes obtain their energy by oxidizing inorganic substances. Nitrifiers, for example, oxidize ammonia or nitrite to obtain energy, producing the nitrate that is taken up by plants. This process is called nitrification and is essential in terrestrial ecosystems because plants can only absorb nitrogen in the form of nitrate. Other prokaryotes oxidize sulfur, hydrogen gas, and other inorganic molecules. On the dark ocean floor at depths of 2500 meters, entire ecosystems subsist on prokaryotes that oxidize hydrogen sulfide as it escapes from thermal vents.

Photoheterotrophs. The so-called purple nonsulfur bacteria use light as their source of energy but obtain carbon from organic molecules, such as carbohydrates or alcohols that have been produced by other organisms.

Chemoheterotrophs. Most prokaryotes obtain both carbon atoms and energy from organic molecules. These include decomposers and most pathogens.

How Heterotrophs Infect Host Organisms

In the 1980s, researchers studying the disease-causing species of *Yersinia*, a group of gram-negative bacteria, found that they produced and secreted large amounts of proteins. Most proteins secreted by gram-negative bacteria have special signal sequences that allow them to pass through the bacterium's double membrane. The proteins secreted by *Yersinia* lacked a key signal sequence that two known secretion mechanisms require for transport across the double membrane of gram-negative bacteria. The proteins must therefore have been secreted by a third type of system, which researchers called the *type III system*.

As more bacterial species are studied, the genes coding for the type III system are turning up in other gram-negative animal pathogens, and even in more distantly related plant pathogens. The genes seem more closely related to one another than do the bacteria. Furthermore, the genes are similar to those that code for bacterial flagella.

The role of these proteins is still under investigation, but it seems that some of the proteins are used to transfer other virulence proteins into nearby eukaryotic cells. Given the similarity of the type III genes to the genes that code for flagella, some scientists hypothesize that the transfer proteins may form a flagellum-like structure that shoots virulence proteins into the host cells. Once in the eukaryotic cells, the virulence proteins may determine the host's response to the pathogen. In *Yersinia*, proteins secreted by the type III system are injected into macrophages; they disrupt signals that tell the macrophages to engulf bacteria. *Salmonella* and *Shigella* use their type III proteins to enter the cytoplasm of eukaryotic cells and thus are protected from the immune system of their host. The proteins secreted by *E. coli* alter the cytoskeleton of nearby intestinal eukaryotic cells, resulting in a bulge onto which the bacterial cells can tightly bind.

Currently, researchers are looking for a way to disarm the bacteria using knowledge of their internal machinery, possibly by causing the bacteria to release the virulence proteins before they are near eukaryotic cells. Others are studying the eukaryotic target proteins and the process by which they are affected.

Bacteria as Plant Pathogens

Many costly diseases of plants are associated with particular heterotrophic bacteria. Almost every kind of plant is susceptible to one or more kinds of bacterial disease, including blights, soft rots, and wilts. The symptoms of these plant diseases vary, but they are commonly manifested as spots of various sizes on the stems, leaves, flowers, or fruits. Fire blight, which destroys pear and apple trees and related plants, is a well-known example of bacterial disease. Most bacteria that cause plant diseases are members of the group of rod-shaped bacteria known as pseudomonads.

While prokaryotes obtain carbon and energy in many ways, most are chemoheterotrophs. Some heterotrophs have evolved sophisticated ways to infect their hosts.

27.4 Prokaryotes are responsible for many diseases, but they also make important contributions to ecosystems.

Human Bacterial Diseases

Bacteria cause many diseases in humans, including cholera, leprosy, tetanus, bacterial pneumonia, whooping cough, diphtheria, and Lyme disease (table 27.2). Members of the genus *Streptococcus* are associated with scarlet fever, rheumatic fever, pneumonia, and other infections. Tuberculosis (TB), another bacterial disease, is still a leading cause of death in humans. TB and some of the other bacterial diseases are mostly spread through the air in water vapor, while such diseases as typhoid fever, paratyphoid fever, and bacillary dysentery are dispersed in food or water. Typhus is spread among rodents and humans by insect vectors.

Tuberculosis

Tuberculosis has been one of the great killer diseases for thousands of years. TB afflicts the respiratory system and is easily transmitted from person to person through the air. Currently, about one-third of all people worldwide are infected with *Mycobacterium tuberculosis*, the tuberculosis bacterium (figure 27.8). Eight million new cases crop up each year, with about 3 million people dying from the disease annually; the World Health Organization predicts 4 million deaths a year by 2005. In fact, in 1997, TB was the leading cause of death from a *single infectious agent* worldwide. Since the mid-1980s, the United States has been experiencing a dramatic resurgence of tuberculosis. The causes of this resurgence include social factors such as poverty, crowding, homelessness, and incarceration (the same factors that have always promoted the spread of TB). The increasing prevalence of HIV infections is also a significant contributing factor. People with AIDS are much more likely to develop TB than people with healthy immune systems.

In addition to the increased numbers of cases—more than 15,900 nationally as of 2001—alarming outbreaks of multidrug-resistant strains of tuberculosis have occurred. These strains are resistant to the best available anti-TB medications. Thus, multidrug-resistant TB is particularly concerning because it requires much more time to treat, is more expensive to treat, and may prove fatal.

The basic principles of TB treatment and control are to make sure all patients complete a full course of medication so that all of the bacteria causing the infection are killed and drug-resistant strains do not develop. Great efforts are being made to ensure that high-risk individuals who are infected but not yet sick receive preventative therapy, which is 90% effective in reducing the likelihood of developing active TB.

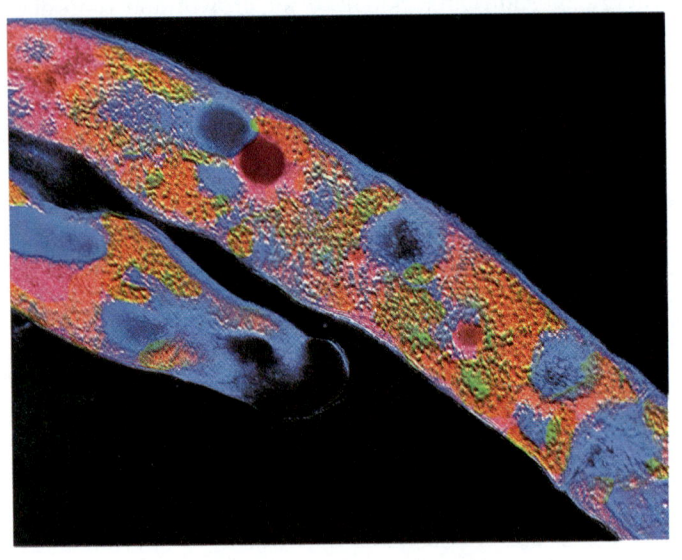

FIGURE 27.8
Mycobacterium tuberculosis. This color-enhanced image shows the rod-shaped bacterium responsible for tuberculosis in humans.

Dental Caries

One human disease we do not usually consider bacterial in origin arises in the film on our teeth. This film, or plaque, consists largely of bacterial cells surrounded by a polysaccharide matrix. Most of the bacteria in plaque are filaments of rod-shaped cells classified as various species of *Actinomyces*, which extend out perpendicular to the surface of the tooth. Many other bacterial species are also present in plaque. Tooth decay, or **dental caries,** is caused by the bacteria present in the plaque, which persists especially in places that are difficult to reach with a toothbrush. Diets that are high in sugars are especially harmful to teeth because certain bacteria (especially *Streptococcus sanguis* and *S. mutans*) ferment the sugars to lactic acid, a substance that reduces the pH of the mouth, causing the local loss of calcium from the teeth. Eating sugary snacks frequently or sucking on candy over a period of time keeps the pH level of the mouth low, resulting in the steady degeneration of the tooth enamel. As the calcium is removed from the tooth, the remaining soft matrix of the tooth becomes vulnerable to attack by bacteria, which begin to break down its proteins, and tooth decay progresses rapidly. Fluoride makes the teeth more resistant to decay because it retards the loss of calcium. It was first realized that bacteria cause tooth decay when germ-free animals were raised. Their teeth do not decay even if they are fed sugary diets.

Table 27.2 Important Human Bacterial Diseases

Disease	Pathogen	Vector/Reservoir	Epidemiology
Anthrax	*Bacillus anthracis*	Animals, including processed skins	Bacterial infection that can be transmitted through contact or ingestion. Rare except in sporadic outbreaks. May be fatal.
Botulism	*Clostridium botulinum*	Improperly prepared food	Contracted through ingestion or contact with wound. Produces acute toxic poison; can be fatal.
Chlamydia	*Chlamydia trachomatis*	Humans, STD	Urogenital infections with possible spread to eyes and respiratory tract. Occurs worldwide; increasingly common over past 20 years.
Cholera	*Vibrio cholerae*	Human feces, plankton	Causes severe diarrhea that can lead to death by dehydration; 50% peak mortality if the disease goes untreated. A major killer in times of crowding and poor sanitation; over 100,000 died in Rwanda in 1994 during a cholera outbreak.
Dental caries	*Streptococcus*	Humans	A dense collection of this bacteria on the surface of teeth leads to secretion of acids that destroy minerals in tooth enamel; sugar alone will not cause caries.
Diphtheria	*Corynebacterium diphtheriae*	Humans	Acute inflammation and lesions of mucous membranes. Spread through contact with infected individual. Vaccine available.
Gonorrhea	*Neisseria gonorrhoeae*	Humans only	STD, on the increase worldwide. Usually not fatal.
Hansen disease (leprosy)	*Mycobacterium leprae*	Humans, feral armadillos	Chronic infection of the skin; worldwide incidence about 10–12 million, especially in southeast Asia. Spread through contact with infected individuals.
Lyme disease	*Borrelia burgdorferi*	Ticks, deer, small rodents	Spread through bite of infected tick. Lesion followed by malaise, fever, fatigue, pain, stiff neck, and headache.
Peptic ulcers	*Helicobacter pylori*	Humans	Originally thought to be caused by stress or diet, most peptic ulcers now appear to be caused by this bacterium; good news for ulcer sufferers because it can be treated with antibiotics.
Plague	*Yersinia pestis*	Fleas of wild rodents: rats and squirrels	Killed ¼ of the population of Europe in the 14th century; endemic in wild rodent populations of the western U.S. today.
Pneumonia	*Streptococcus, Mycoplasma, Chlamydia*	Humans	Acute infection of the lungs; often fatal without treatment.
Tuberculosis	*Mycobacterium tuberculosis*	Humans	An acute bacterial infection of the lungs, lymph, and meninges. Its incidence is on the rise, complicated by the development of new strains of the bacterium that are resistant to antibiotics.
Typhoid fever	*Salmonella typhi*	Humans	A systemic bacterial disease of worldwide incidence. Less than 500 cases a year are reported in the U.S. The disease is spread through contaminated water or foods (such as improperly washed fruits and vegetables). Vaccines are available for travelers.
Typhus	*Rickettsia typhi*	Lice, rat fleas, humans	Historically a major killer in times of crowding and poor sanitation; transmitted from human to human through the bite of infected lice and fleas. Typhus has a peak untreated mortality rate of 70%.

Sexually Transmitted Diseases

A number of bacteria cause sexually transmitted diseases (STDs). Three that are particularly important are gonorrhea, syphilis, and chlamydia (figure 27.9).

Gonorrhea. Gonorrhea is one of the most prevalent communicable diseases in North America. Caused by the bacterium *Neisseria gonorrhoeae*, gonorrhea can be transmitted through sexual intercourse or any other sexual contacts in which body fluids are exchanged, such as oral or anal intercourse. Gonorrhea can infect the throat, urethra, cervix, or rectum and can spread to the eyes and internal organs, causing conjunctivitis (a severe infection of the eyes) and arthritic meningitis (an infection of the joints). Left untreated in women, gonorrhea can cause pelvic inflammatory disease (PID), a condition in which the fallopian tubes become scarred and blocked. PID can eventually lead to sterility. The incidence of gonorrhea has been on the decline in the United States, but it remains a serious threat worldwide.

Syphilis. Syphilis, a very destructive STD, was once prevalent but is now less common due to the advent of blood-screening procedures and antibiotics. Syphilis is caused by a spirochete bacterium, *Treponema pallidum*, that is transmitted during sexual intercourse or through direct contact with an open syphilis sore. The bacterium can also be transmitted from a mother to her fetus, often causing damage to the heart, eyes, and nervous system of the baby.

Once inside the body, the disease progresses in four distinct stages. The first, or primary stage, is characterized by the appearance of a small, painless, often unnoticed sore called a chancre. The chancre resembles a blister and occurs at the location where the bacterium entered the body about three weeks following exposure. This stage of the disease is highly infectious, and an infected person may unwittingly transmit the disease to others.

The second stage of syphilis is marked by a rash, a sore throat, and sores in the mouth. The bacteria can be transmitted at this stage through kissing or contact with an open sore. The third stage of syphilis is symptomless. This stage may last for several years, and at this point, the person is no longer infectious, but the bacteria are still present in the body, attacking the internal organs. The fourth and final stage of syphilis is the most debilitating, as the damage done by the bacteria in the third stage becomes evident. Sufferers at this stage of syphilis experience heart disease, mental deficiency, and nerve damage, which may include loss of motor functions or blindness.

Chlamydia. Sometimes called the "silent STD," chlamydia is caused by an unusual bacterium, *Chlamydia trachomatis*, that has both bacterial and viral characteristics. Like a bacterium, it is susceptible to antibiotics, and like a virus, it depends on its host to replicate its genetic material, meaning that it is an obligate internal parasite.

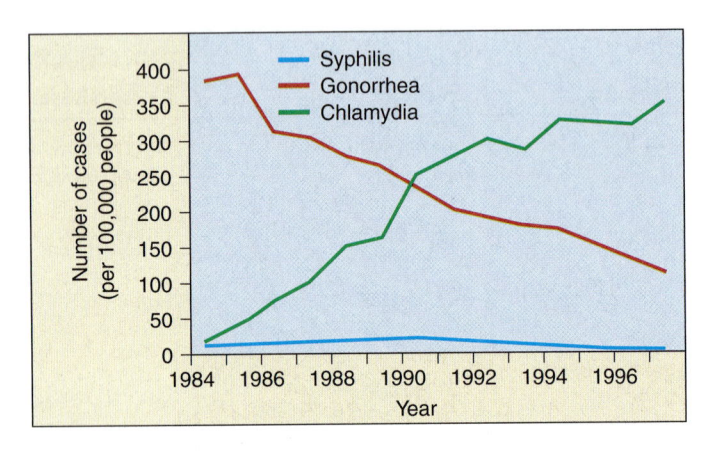

FIGURE 27.9
Trends in sexually transmitted diseases in the United States. How is it possible for the incidence of one STD (chlamydia) to rise as another (gonorrhea) falls?

The bacterium is transmitted through vaginal, anal, or oral intercourse with an infected person.

Chlamydia is called the "silent STD" because women usually experience no symptoms until after the infection has become established. In part because of this symptomless nature, the incidence of chlamydia has skyrocketed, increasing by more than sevenfold nationally since 1984. The effects of an established chlamydia infection on the female body are extremely serious. Chlamydia can cause pelvic inflammatory disease (PID), which can lead to sterility.

It has recently been established that infection of the reproductive tract by *Chlamydia* can cause heart disease. *Chlamydia* produce a peptide similar to one produced by cardiac muscle. As the body's immune system tries to fight off the infection, it recognizes this peptide. The similarity between the bacterial and cardiac peptides confuses the immune system, and T cells attack cardiac muscle fibers, inadvertently causing inflammation of the heart and other problems.

Within the last few years, two types of tests for chlamydia have been developed that look for the presence of the bacteria in the discharge from men and women. The treatment for chlamydia is antibiotics, usually tetracycline (penicillin is not effective against chlamydia). Any woman who experiences the symptoms associated with this STD should be tested for the presence of the chlamydia bacterium; otherwise, her fertility may be at risk.

This discussion of STDs may give the impression that sexual activity is fraught with danger, and in a way, it is. It is folly not to take precautions to avoid STDs. The best way to do this is to know one's sexual partners well enough to discuss the possible presence of an STD. Condom use can also prevent transmission of most of the diseases. Responsibility for protection lies with each individual.

Bacterial diseases have a major impact worldwide. Sexually transmitted diseases (STDs) are becoming increasingly widespread among Americans as sexual activity increases.

Benefits of Prokaryotes

Prokaryotes, the first organisms to evolve on earth, were largely responsible for creating the properties of the atmosphere and the soil over billions of years. Today, they affect our lives in many important ways.

Prokaryotes and the Environment

Life on earth depends critically on the cycling of chemical elements between organisms and the physical environments in which they live—that is, between the biological and physical elements of ecosystems. Prokaryotes and fungi play many key roles in this chemical cycling, a process discussed in detail in chapter 55.

Decomposition. The carbon, nitrogen, phosphorus, sulfur, and other atoms of which your body is built all have come from the physical environment, and when you die and decay, they all return to it. The prokaryotes and fungi that carry out the decomposition portion of chemical cycles, releasing a dead body's atoms to the environment, are called decomposers.

Fixation. Other prokaryotes play important roles in fixation, the other half of chemical cycles, helping to return elements from the nonliving environment to organisms. The role of photosynthetic prokaryotes in recycling carbon is obvious. The organic compounds that plants, algae, and photosynthetic prokaryotes produce from CO_2 pass up through food chains to form the bodies of all the ecosystem's heterotrophs—all the animals and fungi and nonphotosynthetic protists. Cyanobacteria are thought to have added oxygen to the earth's atmosphere as a by-product of their photosynthesis.

Less obvious, but no less critical to life on earth, is the role of prokaryotes in recycling nitrogen. The nitrogen in the earth's atmosphere is in the form of N_2 gas. A triple covalent bond links the two nitrogen atoms and is not easy to break. Among the earth's organisms, only prokaryotes are able to accomplish this feat, reducing N_2 to ammonia (NH_3), which is used to build amino acids and other nitrogen-containing biological molecules. When the organisms that contain these molecules die, other prokaryotes, called *denitrifiers*, return the nitrogen to the atmosphere, completing the cycle.

To fix atmospheric nitrogen, prokaryotes employ an enzyme complex called nitrogenase, encoded by a set of genes called *nif* ("nitrogen fixation") genes. The *nif* gene complex is found in a wide range of free-living prokaryotes. Under anaerobic conditions, the most important free-living nitrogen fixers are members of the gram-positive bacterial genus *Clostridium*. In aquatic environments, nitrogen fixation is carried out largely by cyanobacteria such as *Anabaena*. Because the nitrogen fixation process is strictly anaerobic and extremely sensitive to oxygen, cyanobacteria possess heterocysts, specialized cells impermeable to oxygen (figure 27.10). In soil, nitrogen fixation occurs in the roots of plants that harbor symbiotic colonies of nitrogen-fixing bacteria. These associations include *Rhizobium* (a genus of proteobacteria; see figure 27.3) with legumes, *Frankia* (an actinomycete) with many woody shrubs, and *Anabaena* with water ferns.

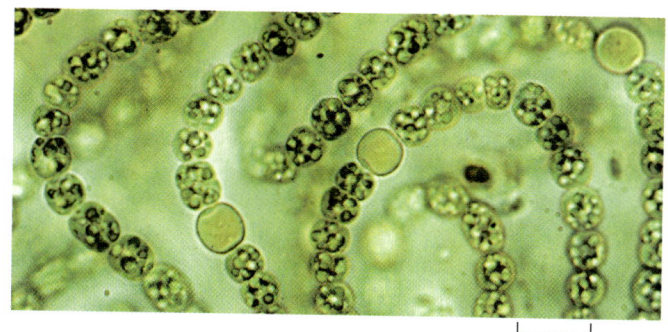

22.70 μm

FIGURE 27.10
Cyanobacteria carry out nitrogen fixation. Cells of this cyanobacterium, *Anabaena*, adhere in chains, and nitrogen is fixed in the enlarged, specialized cells, which are called heterocysts.

Symbiotic Prokaryotes

Many prokaryotes live in symbiotic association with eukaryotes. Symbiosis (Greek, "living together") refers to the ecological relationship between different species that live in direct contact with each other. The symbiotic association of nitrogen-fixing bacteria with plant roots is an example of *mutualism*, a form of symbiosis in which both parties benefit. The bacteria supply the plant with useful nitrogen, and the plant supplies the bacteria with sugars and other organic nutrients.

Many bacteria live symbiotically within the digestive tracts of animals. Cows and other grazing mammals are unable to digest cellulose (much of the bulk of the grass they eat) because they lack the required cellulase enzyme. However, living in a special section of the cow's gut called the rumen are colonies of cellulase-producing bacteria that make the cellulase enzyme, allowing the cow to digest its food (see chapter 43 for a fuller account). Similarly, humans maintain large colonies of bacteria in the large intestine that produce vitamins—particularly B_{12} and K—which the human body cannot make itself.

Many bacteria inhabit the outer surfaces of animals and plants without doing damage. These are examples of *commensalism*, in which one organism (the bacterium) receives benefits while the animal or plant is neither benefited nor harmed. As we have seen in this chapter, some pathogenic bacteria do harm their hosts. *Parasitism* is a form of symbiosis in which one member (in this case, the bacterium) is benefited, and the other (the infected animal or plant) is harmed.

Bacteria and Genetic Engineering

Applying genetic engineering methods to produce improved strains of bacteria for commercial use holds enormous promise for the future. For example, bacteria are under intense investigation as nonpolluting insect control agents. *Bacillus thuringiensis* attacks insects in nature, and improved, highly specific strains of *B. thuringiensis* have greatly increased its usefulness as a biological control agent. Genetically modified bacteria have also been extraordinarily useful in producing insulin and other therapeutic proteins.

The use of organisms to remove pollutants from water, air, and soil is called *bioremediation*. In sewage treatment plants, the solid matter from raw sewage is allowed to settle, and then gradually mixed with a growing culture of bacteria and archaebacteria. Genetically modified bacteria—in this case, pseudomonads (proteobacteria)—are being employed successfully to remove petroleum products from beaches after oil tanker spills (figure 27.11).

Bacteria are now widely used as "biofactories" in the commercial production of a variety of enzymes, vitamins, and antibiotics, an approach discussed in detail in chapter 16. Most of the insulin used to treat diabetes is produced this way. Immense cultures of bacteria, often genetically modified to enhance performance, are used to produce commercial acetone, and of course to ferment sugars into alcohol in the production of beer.

Bacteria and Bioweapons

With the success of antibiotics in treating bacterial killer diseases such as typhus and cholera, many of us have been lulled into thinking that the battle against infectious bacteria has been won. However, to defeat an infectious disease, you must control its transmission, and unfortunately the new century has seen the introduction of a more deadly way for disease to spread—by the deliberate action of people using disease as a weapon. For several decades, the United States and Russia carried out extensive bioweapons programs, and while the U.S. program was discontinued in 1969, the Russian bioweapons program continued for another two decades. In 2001, bioterrorists struck at the United States with anthrax spores developed in the American bioweapons program, adding them to letters sent through the mail. While few died, the attack points out the dangerous potential of biological weapons. Anthrax and the smallpox virus pose the greatest immediate threats, although gene-modified pathogens offer an even greater future danger. Bioweapons are discussed at length in the enhancement chapter *Infectious Disease and Bioterrorism*, which you can read at www.mhhe.com/raven7.

Prokaryotes make many important contributions to the world ecosystem, playing key roles as cyclers of carbon and nitrogen, as symbionts, as bioremediation agents, and as aids in the commercial production of medicinal drugs and other products.

FIGURE 27.11
Using bacteria to clean up oil spills. Bacteria can often be used to remove environmental pollutants, such as petroleum hydrocarbons and chlorinated compounds. In rocky areas contaminated by the *Exxon Valdez* oil spill (*left*), oil-degrading bacteria produced dramatic results (*right*).

Concept Review

27.1 Prokaryotes are the smallest and most numerous organisms.

The Prevalence of Prokaryotes

- Archaebacteria and bacteria are the oldest, structurally simplest, and most abundant forms of life on earth. (p. 546)
- About 5000 different kinds of prokaryotes are currently recognized. (p. 546)
- In the 1980s, a new discovery led to the division of prokaryotes into two groups: archaebacteria and bacteria. (p. 546)
- Prokaryotes are ubiquitous on earth and exist in some of the most hostile environments on the planet. (p. 546)
- Bacteria exhibit one of three basic structures: bacillus (straight and rod-shaped), coccus (spherical-shaped), and spirillum (long and helical-shaped). (p. 546)
- Prokaryotes differ from eukaryotes in many important features, including unicellularity, cell size, chromosomes, cell division and genetic recombination, internal compartmentalization, flagella, and metabolic diversity. (pp. 546–547)

27.2 Prokaryotes exhibit considerable diversity in both structure and metabolism.

Prokaryotic Diversity

- Key characteristics used in classifying bacteria are photosynthetic or nonphotosynthetic, motile or nonmotile, unicellular or colony-forming or filamentous, and formation of spores or division by transverse binary fission. (p. 548)
- Molecular approaches to classification include analysis of amino acid sequences of key proteins, analysis of nucleic acid base sequences, nucleic acid hybridization, ribosomal RNA sequencing, and whole-genome sequencing. (p. 548)
- Archaebacteria and bacteria differ in four key areas: plasma membrane, cell wall, gene translation machinery, and gene architecture. (p. 548)

27.3 Prokaryotes are more complex than commonly supposed.

The Prokaryotic Cell Surface

- A prokaryotic cell wall usually consists of peptidoglycan, forming either thick or thin walls; prokaryotes can be classified and identified by Gram staining as either gram-positive or gram-negative bacteria. (p. 550)
- Many prokaryotes have slender, rigid flagella for propulsion, and some prokaryotes have pili to help in attachment and genetic exchange. (p. 550)

The Cell Interior

- Prokaryotic cells lack the extensive functional compartmentalization found in eukaryotic cells, but do possess internal membranes, a nucleoid region, and ribosomes. (p. 551)

Prokaryotic Variation

- Bacteria have a very short generation time, thus mutation and genetic recombination play an important role in producing and maintaining genetic diversity. (p. 552)

Prokaryotic Metabolism

- Autotrophs can be broken down into photoautotrophs and chemoautotrophs; heterotrophs can be broken down into photoheterotrophs and chemoheterotrophs. Chemoheterotrophs appear to be the most common. (p. 553)
- Almost every kind of plant is susceptible to one or more bacterial diseases. (p. 553)

27.4 Prokaryotes are responsible for many diseases, but they also make important contributions to ecosystems.

Human Bacterial Diseases

- Tuberculosis and dental caries are just two of the many human diseases caused by bacteria. A few others are cholera, leprosy, tetanus, bacterial pneumonia, whooping cough, diptheria, and Lyme disease. (p. 554)
- A number of bacteria also cause sexually transmitted diseases such as gonorrhea, syphilis, and chlamydia. (p. 556)

Benefits of Prokaryotes

- Prokaryotes affect everyday life in many ways, including decomposition of dead organisms, chemical fixation, symbiotic relationships (such as that between nitrogen-fixing bacteria and plant roots), genetic engineering improvements in agricultural crops, commercial production of pharmaceuticals, and bioremediation. (pp. 557–558)
- Some disease-causing bacteria can even be used as forms of bioterrorism. (p. 558)

Test Your Understanding

Self Test

1. Once they evolved, _____ forever changed the atmosphere on earth.
   a. bacteria
   b. archaebacteria
   c. cyanobacteria
   d. eukaryotes
2. Which of the following statements is *not* true of prokaryotic cells?
   a. Prokaryotic cells are multicellular.
   b. Prokaryotic cells do not have a nucleus.
   c. Prokaryotic cells have circular DNA.
   d. Prokaryotic cells have flagella.
3. When bacterial cell walls are covered with an outer membrane of lipopolysaccharide, they are
   a. gram-positive.
   b. gram-negative.
   c. encapsulated.
   d. endospores.
4. The prokaryotic genome is contained in the
   a. plasmid.
   b. endospore.
   c. pilus.
   d. nucleoid region.
5. Archaebacteria and bacteria differ in all of the following ways except
   a. the structure of the cell wall.
   b. their presence in nonextreme environments.
   c. the structure of the plasma membrane.
   d. the kinds of ribosomal proteins they possess.
6. Genetic recombination has led to antibiotic resistance through the transfer of
   a. pili.
   b. endospores.
   c. plasmids.
   d. bacterial chromosomes.
7. Prokaryotic organisms that obtain their energy by oxidizing inorganic substances are called
   a. chemoautotrophs.
   b. photoautotrophs.
   c. chemoheterotrophs.
   d. photoheterotrophs.
8. What disease is experiencing new outbreaks because of antibiotic resistance?
   a. smallpox
   b. cholera
   c. tuberculosis
   d. diphtheria
9. The disease sometimes referred to as the "silent STD" because it is usually asymptomatic in women early on is
   a. chlamydia.
   b. gonorrhea.
   c. syphilis.
   d. pelvic inflammatory disease.
10. Which of the following can be attributed to bacteria?
    a. decomposition of dead organic matter
    b. increasing oxygen levels in the atmosphere
    c. production of antibiotics
    d. all of these

Test Your Visual Understanding

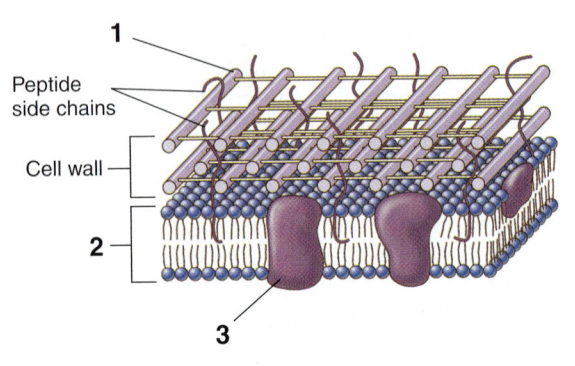

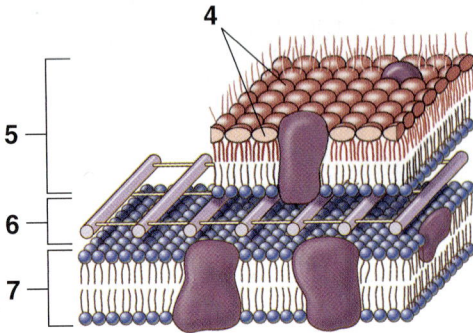

1. This figure shows two kinds of bacterial cell walls. Match the following labels with the appropriate numbered structures.
   lipopolysaccharides
   outer membrane
   peptidoglycan
   plasma membrane
   protein
2. What type of bacteria would have the cell wall structure shown in the upper figure? What type of bacteria would have the cell wall structure shown in the lower figure?

Apply Your Knowledge

1. Bacterial populations grow through binary fission, which means the bacteria divide in half such that each bacterium gives rise to two bacteria. Assume that a bacterium is placed in culture and undergoes binary fission every 30 minutes. The addition of a competing bacterium reduces the population size by 25%. How many bacteria from the original culture will be present after 24 hours?
2. Justify the assertion that life on earth could not exist without prokaryotes.
3. What are the functions of antibiotics in the prokaryotes that produce them?
4. What are endospores? Why would they be an advantage? Would a bacterium with the ability to form endospores have a greater or lesser chance of extinction? Why?

28
Protists

Concept Outline

28.1 Eukaryotes probably arose by endosymbiosis.

Endosymbiosis. Mitochondria and chloroplasts are thought to have arisen by engulfing aerobic and photosynthetic bacteria.

28.2 The kingdom Protista is by far the most diverse of any eukaryotic kingdom.

The Challenge of Classifying the Protists. Groups of protists are evolutionarily unrelated, which presents many challenges for taxonomists trying to classify them.

General Biology of the Protists. The kingdom Protista contains members exhibiting a wide range of methods of locomotion, nutrition, and reproduction.

28.3 Protists can be categorized into six groups.

Euglenozoa. Euglenoids and kinetoplastids were among the first free-living eukaryotes to contain mitochondria.

Alveolata. Characterized by a space under the plasma membrane, the dinoflagellates, apicomplexes and ciliates, have varied modes of locomotion.

Stramenopila and Rhodophyta. These brown and red algae have origins that are distinct from those of the green plants.

Chlorophyta. Chlorophyta is one of two green plant lineages. The other lineage (Streptophyta) gave rise to the land plants.

Choanoflagellida and Protists That Are Difficult to Categorize. Choanoflagellates are most like the common ancestor of all animals. Around 60 other lineages of protists have been identified, but there is not yet enough information to determine their relationships to other protist lineages.

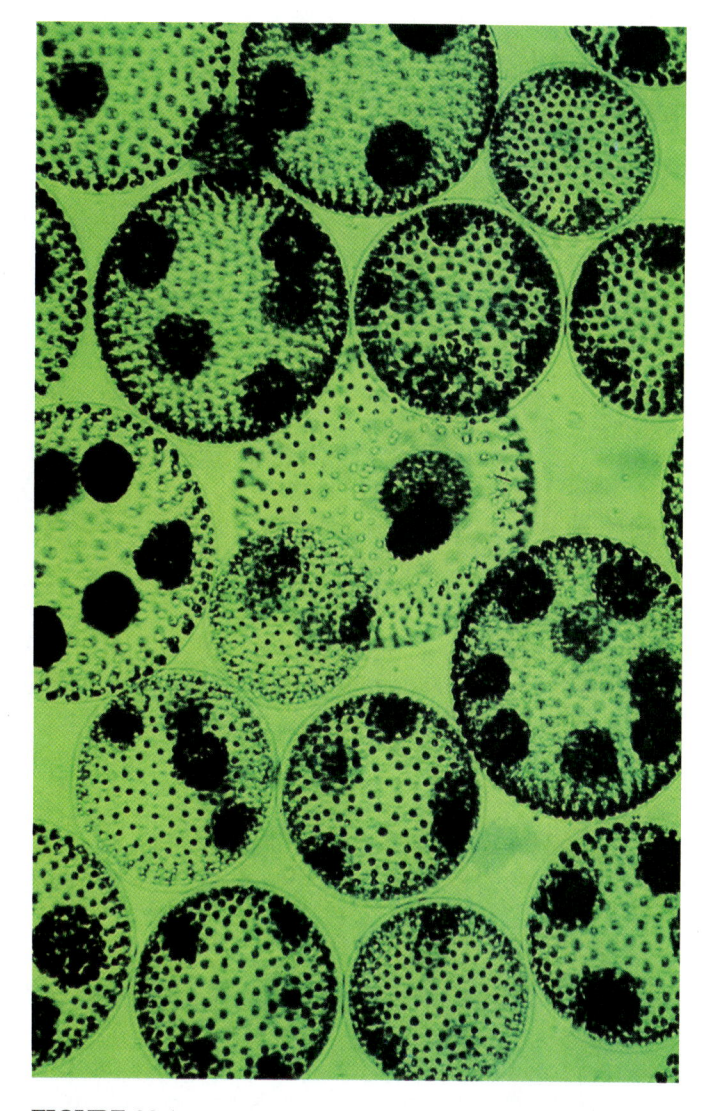

FIGURE 28.1
Volvox, **a colonial protist.** The protists are a large, diverse group of primarily single-celled organisms, a group from which the other three eukaryotic kingdoms each evolved.

For more than half of the long history of life on earth, all life was microscopic in size. The biggest organisms that existed for over 2 billion years were single-celled bacteria fewer than 6 micrometers thick. These prokaryotes lacked internal membranes, except for invaginations of surface membranes in photosynthetic bacteria. The first evidence of a different kind of organism is found in tiny fossils in rock 2.5 billion years old. These fossil cells are much larger than bacteria (up to ten times larger) and contain internal membranes and what appear to be small, membrane-bounded structures. The complexity and diversity of form among these single cells is astonishing. The step from relatively simple to quite complex cells marks one of the most important events in the evolution of life, the appearance of a new kind of organism, the eukaryote (figure 28.1). Eukaryotes that are clearly not animals, plants, or fungi have been lumped together and called protists.

28.1 Eukaryotes probably arose by endosymbiosis.

Endosymbiosis

What was the first eukaryote like? We cannot be sure, but a good model is *Pelomyxa palustris*, a single-celled, nonphotosynthetic organism that appears to represent an early stage in the evolution of eukaryotic cells (figure 28.2). The cells of *Pelomyxa* are much larger than bacterial cells and contain a complex system of internal membranes. Although they resemble some of the largest early fossil eukaryotes, these cells are unlike those of any other eukaryote because *Pelomyxa* lacks mitochondria. No clear alternative to nuclear mitosis (chromosome replication) has been described for *Pelomyxa*. Its nuclei divide by pinching apart into two daughter nuclei, around which new membranes form. Although *Pelomyxa* cells lack mitochondria, two kinds of bacteria living within them may play the same role that mitochondria do in most other eukaryotes.

Biologists know very little about the origin of *Pelomyxa*, except that in many of its fundamental characteristics it resembles the archaebacteria far more than the bacteria. Because of this general resemblance, it is widely assumed that the first eukaryotic cells were nonphotosynthetic descendants of archaebacteria.

What about the wide gap between *Pelomyxa* and all other eukaryotes? Where did mitochondria come from? Most biologists agree with the theory of **endosymbiosis**, which proposes that mitochondria originated as symbiotic, aerobic (oxygen-requiring) bacteria (figure 28.3). Symbiosis (Greek *syn*, "together with," + *bios*, "life") means living together in close association. Recall from chapter 5 that mitochondria are sausage-shaped organelles 1 to 3 micrometers long, about the same size as most bacteria. Mitochondria are bounded by two membranes. Aerobic bacteria are thought to have become mitochondria when they were engulfed by ancestral eukaryotic cells, much like *Pelomyxa*, early in the history of eukaryotes.

The most similar bacteria to mitochondria today are the nonsulfur purple bacteria, which are able to carry out oxidative metabolism (described in chapter 9). In mitochondria, the outer membrane is smooth and is thought to be derived from the endoplasmic reticulum of the host cell, which, like *Pelomyxa*, may have already contained a complex system of internal membranes. The inner membrane of a mitochondrion is folded into numerous layers, resembling the folded membranes of nonsulfur purple bacteria; embedded within this membrane are the proteins that carry out oxidative metabolism. The engulfed bacteria became the interior portion of the mitochondria we see today. Host cells were unable to carry out the Krebs cycle or other metabolic reactions necessary for living in an atmosphere that contained increasing amounts of oxygen before they had acquired these bacteria.

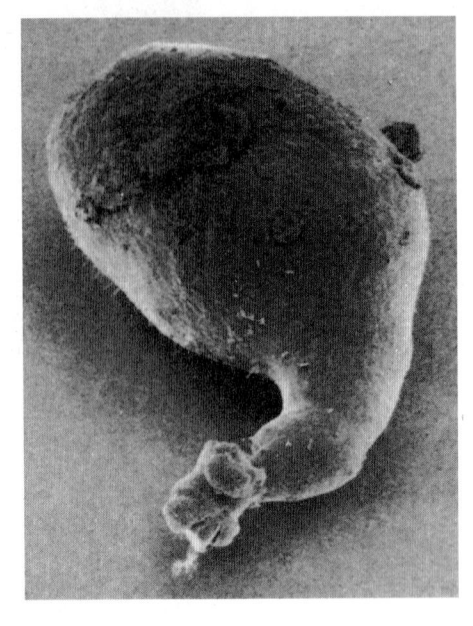

FIGURE 28.2
Pelomyxa palustris. This unique, amoeba-like protist lacks mitochondria. No alternative to nuclear mitosis in *Pelomyxa* has been described. *Pelomyxa* may represent a very early stage in the evolution of eukaryotic cells.

Pelomyxa is not the only protist to lack mitochondria. These anaerobic protists are not necessarily closely related. *Giardia*, another protist without mitochondria, can pass from human to human via contaminated water and cause diarrhea. Mitochondrial genes are found in their nuclei, leading to the conclusion that *Giardia* evolved from aerobes. Thus, *Giardia* is unlikely to represent an early protist.

During the billion and a half years in which mitochondria have existed as endosymbionts within eukaryotic cells, most of their genes have been transferred to the chromosomes of the host cells—but not all. Each mitochondrion still has its own genome, a circular, closed molecule of DNA similar to that found in bacteria, on which are located genes encoding the essential proteins of oxidative metabolism. These genes are transcribed within the mitochondrion, using mitochondrial ribosomes that are smaller than those of eukaryotic cells, very much like bacterial ribosomes in size and structure. Mitochondria divide by simple fission, just as bacteria do, and they also replicate and sort their DNA much as bacteria do. However, nuclear genes direct the process, and mitochondria cannot be grown outside of the eukaryotic cell, in cell-free culture.

The mechanisms of mitosis and cytokinesis, now so common among eukaryotes, did not evolve all at once. Traces of very different, and possibly intermediate, mechanisms survive today in some of the eukaryotes. In fungi and some groups of protists, for example, the nuclear membrane does not dissolve, and mitosis is confined to the nucleus. When mitosis is complete in these organisms, the nucleus divides into two daughter nuclei, and only then does the rest of the cell divide. In most protists, or in plants or animals, the nuclear membrane breaks down during mitosis. We do not know if mitosis without nuclear membrane dissolution represents an intermediate step on the

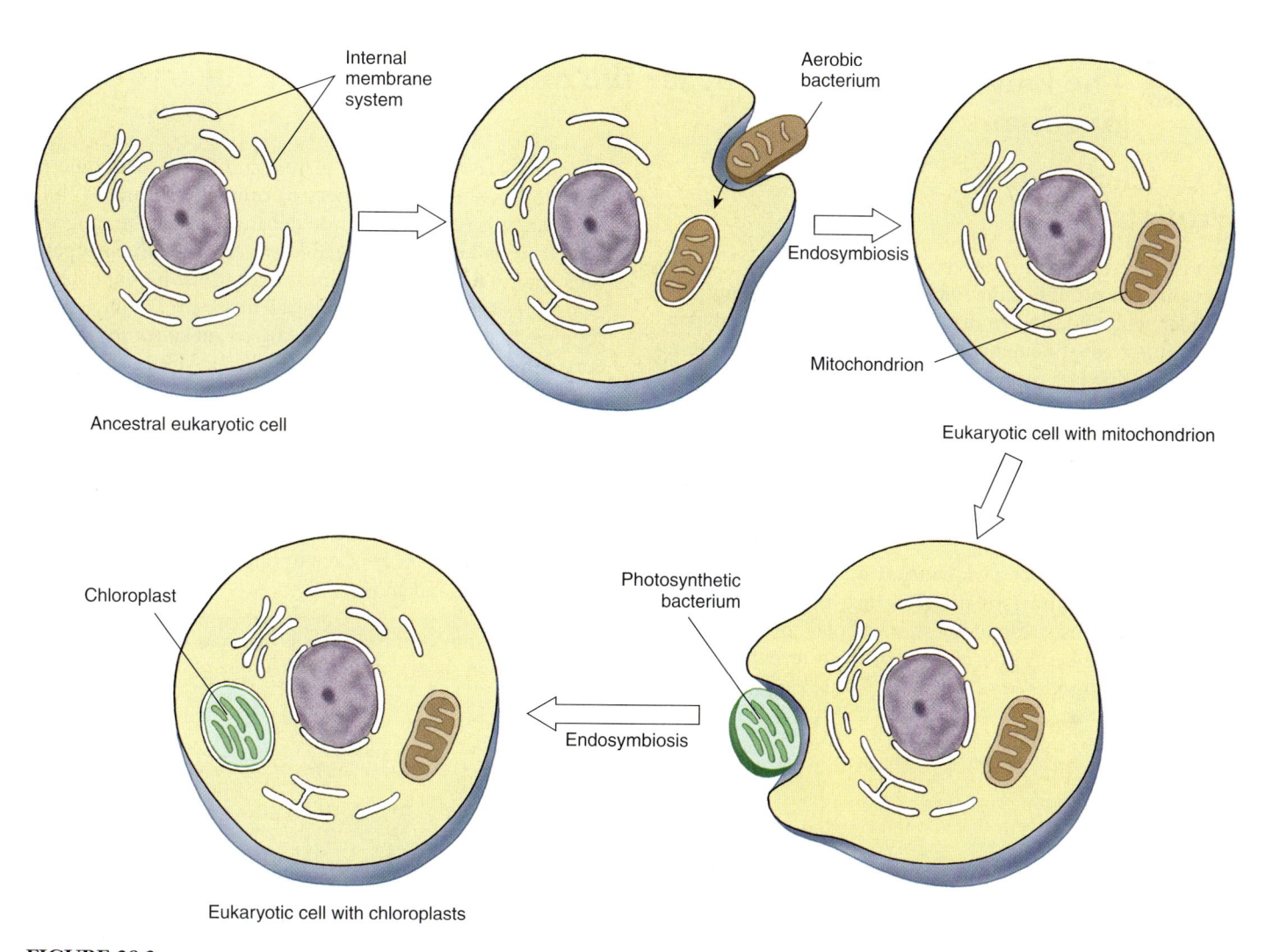

FIGURE 28.3
The theory of endosymbiosis. Scientists propose that ancestral eukaryotic cells, which already had an internal system of membranes, engulfed aerobic bacteria, which then became mitochondria in the eukaryotic cell. Chloroplasts may also have originated this way, with eukaryotic cells engulfing photosynthetic bacteria.

evolutionary journey to the form of mitosis that is characteristic of most eukaryotes today, or if it is simply a different way of solving the same problem. There are no fossils in which we can see the interiors of dividing cells well enough to be able to trace the history of mitosis.

Endosymbiosis Is Not Rare

Many eukaryotic cells contain other endosymbiotic bacteria in addition to mitochondria. Plants and algae contain chloroplasts, bacteria-like organelles that were apparently derived from symbiotic photosynthetic bacteria. Chloroplasts have a complex system of inner membranes and a circle of DNA. Centrioles, organelles associated with the assembly of microtubules, resemble in many respects spirochete bacteria, and they contain bacteria-like DNA involved in the production of their structural proteins. While all mitochondria are thought to have arisen from a single symbiotic event, it is difficult to be sure with chloroplasts. Three biochemically distinct classes of chloroplasts exist, each resembling a different bacterial ancestor. Red algae possess pigments similar to those of cyanobacteria; plants and green algae more closely resemble the photosynthetic bacterium *Prochloron*; and brown algae and other photosynthetic protists resemble a third group of bacteria. This diversity of chloroplasts has led to the widely held belief that eukaryotic cells acquired chloroplasts by endosymbiosis at least three different times. Recent comparisons of chloroplast DNA sequences, however, suggest a single origin of chloroplasts, followed by very different post-endosymbiotic histories. For example, in each of the three main lines, different genes became relocated to the nucleus, lost, or modified.

> The theory of endosymbiosis proposes that mitochondria and chloroplasts originated as symbiotic bacteria.

Chapter 28 Protists **563**

28.2 The kingdom Protista is by far the most diverse of any eukaryotic kingdom.

The Challenge of Classifying the Protists

Protists are the most diverse of the four kingdoms in the domain Eukarya. The kingdom Protista contains many unicellular, colonial, and multicellular groups. Probably the most important statement we can make about the kingdom Protista is that it is paraphyletic and not a kingdom at all; as a matter of convenience, single-celled eukaryotic organisms have typically been grouped together and called protists. This lumps 200,000 different and only distantly related forms together. The "single-kingdom" classification of the Protista is artificial and not representative of any evolutionary relationships.

New applications of a wide variety of molecular methods are providing important insights into the relationships among the protists. Many questions about how to classify the protists are being addressed with these techniques. Are protists best considered as several different kingdoms, each of equal rank with animals, plants, and fungi? Are some of the protists actually members of other kingdoms? While these questions continue to be debated, ever-increasing information is becoming available concerning which organisms among the protists are most likely to be monophyletic.

In this chapter, we group the 15 major protist phyla into six major monophyletic groups, based on our current understanding of phylogeny (figure 28.4). While these lineages may change, this approach allows us to examine groups with many shared traits. Keep in mind that about 60 of the protist lineages cannot yet be placed on the tree of life with any confidence! Protists exemplify the challenges and excitement of the revolutionary changes in taxonomy and phylogeny we explored in chapter 25. Understanding the evolution of protists is key to understanding the origins of plants, fungi, and animals.

The taxonomy of the protists is in a state of flux as new information shapes our understanding of this kingdom.

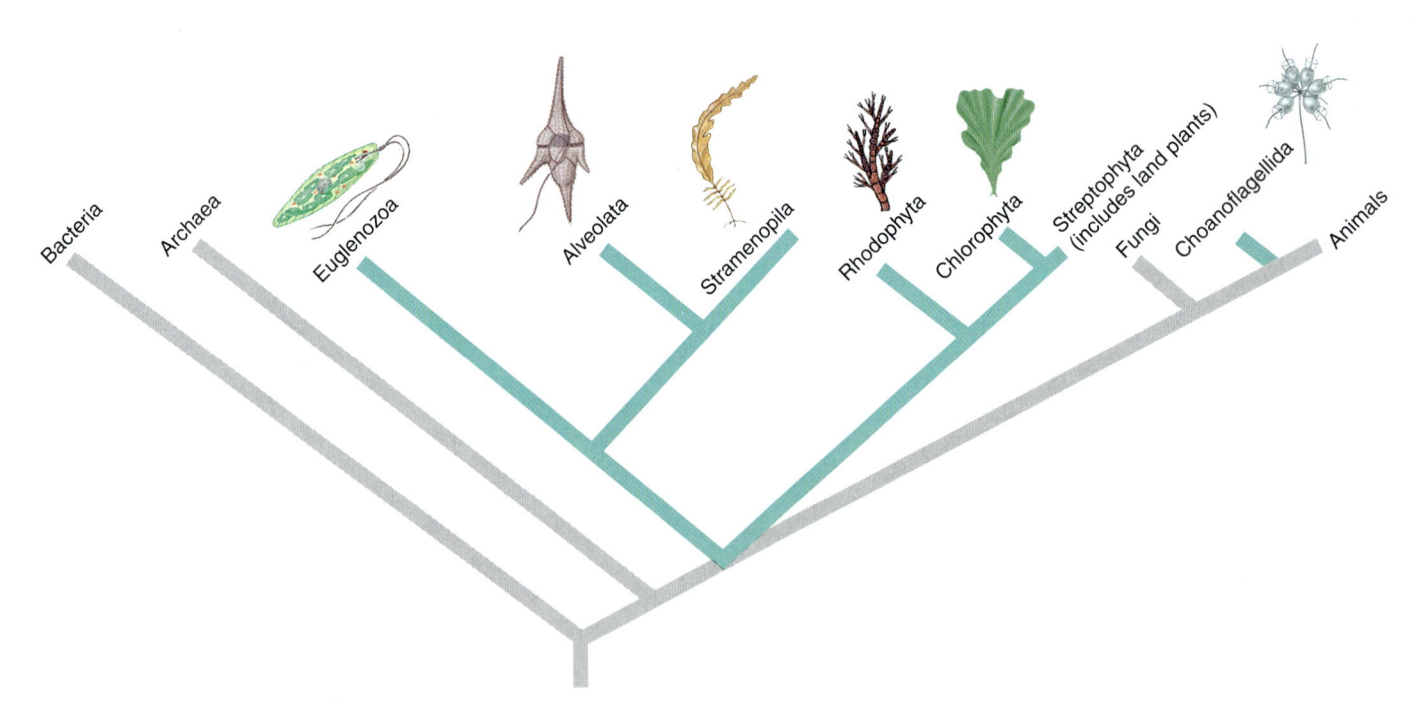

FIGURE 28.4
The challenge of protistan classification. Our understanding of the evolutionary relationships among protists is currently in flux. The most recent data support six major, monophyletic groups within the protists. Consider this a working model, not fact. The green algae (Chlorophyta) are not truly monophyletic in that another branch, Streptophyta, gave rise to the land plants. Protist lineages are shaded in blue.

General Biology of the Protists

Protists are united on the basis of a single negative characteristic: They are eukaryotes that are not fungi, plants, or animals. In all other respects, they are highly variable, with no uniting features. Many are unicellular (figure 28.5), but numerous colonial and multicellular groups exist. Most are microscopic, but some are as large as trees. They represent all symmetries, and exhibit all types of nutrition.

The Cell Surface

Protists possess a varied array of cell surfaces. Some protists, such as amoebas, are surrounded only by their plasma membrane. All other protists have a plasma membrane with extracellular material (ECM) deposited on the outside of the membrane. Some ECM forms strong cell walls. Diatoms and forams secrete glassy shells of silica.

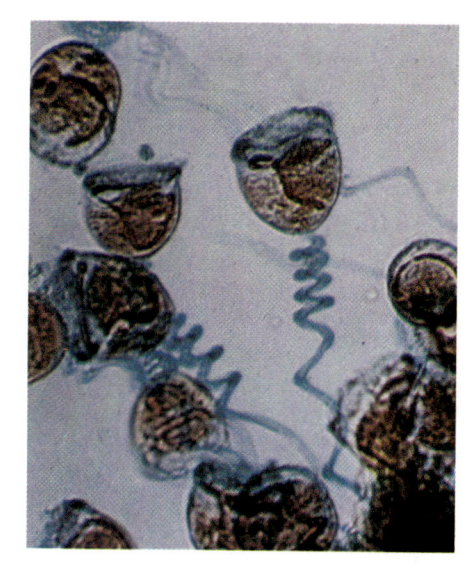

Locomotor Organelles

Movement in protists is also accomplished by diverse mechanisms. Protists move chiefly by either flagellar rotation or pseudopodial movement. Many protists wave one or more flagella to propel themselves through the water, while others use banks of short, flagella-like structures called cilia to create water currents for their feeding or propulsion. Pseudopods (Greek *pseudo-*, "false," + *podos*, "foot") are the chief means of locomotion among amoebas (see chapter 5), whose pseudopods are large, blunt extensions of the cell body called lobopodia. Other related protists extend thin, branching protrusions called filopodia. Still other protists extend long, thin pseudopods called axopodia supported by axial rods of microtubules. Axopodia can be extended or retracted. Because the tips can adhere to adjacent surfaces, the cell can move by a rolling motion, shortening the axopodia in front and extending those in the rear.

Cyst Formation

Many protists with delicate surfaces are successful in quite harsh habitats. How do they manage to survive so well? They form cysts, dormant forms of a cell with resistant outer coverings in which cell metabolism is more or less completely shut down. Not all cysts are so sturdy. Vertebrate parasitic amoebas, for example, form cysts that are quite resistant to gastric acidity, but will not tolerate desiccation or high temperature.

FIGURE 28.5
A unicellular protist. Kingdom Protista is a catch-all kingdom for many different groups of unicellular organisms, such as this *Vorticella* (a ciliate), which is heterotrophic, feeds on bacteria, and has a retractable stalk.

Nutrition

Protists employ every form of nutritional acquisition except the chemoautotrophic form, which has so far been observed only in prokaryotes. Some protists are photosynthetic and are called **phototrophs.** Others are heterotrophs that obtain energy from organic molecules synthesized by other organisms. Among the heterotrophic protists, those that ingest visible particles of food are called **phagotrophs.** Phagotrophs ingest food particles into intracellular vesicles called food vacuoles or phagosomes. Lysosomes fuse with the food vacuoles, introducing enzymes that digest the food particles within. Digested molecules are absorbed across the vacuolar membrane. Protists that ingest food in soluble form are called **osmotrophs.**

Reproduction

Protists typically reproduce asexually, although some protists have an obligate sexual reproductive phase and others undergo sexual reproduction at times of stress, including food shortages. Asexual reproduction involves mitosis, but the process is often somewhat different from the mitosis that occurs in multicellular animals. For example, the nuclear membrane often persists throughout mitosis, with the microtubular spindle forming within it. One category of asexual reproduction is fission. The most common type of fission is **binary fission,** in which a cell simply splits into nearly equal halves. When the progeny cell is considerably smaller than its parent and then grows to adult size, the fission is called **budding.** In multiple fission, or **schizogony,** common among some protists, fission is preceded by several nuclear divisions, so that fission produces several individuals almost simultaneously.

Sexual reproduction also takes place in many forms among the protists. In ciliates and some flagellates, meiosis results in gamete formation (gametic meiosis). Thus, these protists are diploid for most of their life cycle, as are metazoans. In spore-producing protists, meiosis occurs directly *after* fertilization, and all the individuals that are produced are haploid until the next zygote is formed. In algae, both haploid and diploid cells undergo mitosis, producing an alternation of generations similar to that seen in plants.

Protists exhibit a wide range of forms, locomotion, nutrition, and reproduction.

28.3 Protists can be categorized into six groups.

Six lineages of protists have been identified—Euglenozoa, Alveolata, Stramenopila, Rhodophyta, Chlorophyta, and Choanoflagellida. However, many other protists are not close relatives of any of these groups.

Euglenozoa

Euglenoids

Euglenoids diverged early and were among the earliest free-living eukaryotes to possess mitochondria. Euglenoids clearly illustrate the impossibility of distinguishing "plants" from "animals" among the protists. About one-third of the approximately 40 genera of euglenoids have chloroplasts and are fully autotrophic; the others lack chloroplasts, ingest their food, and are heterotrophic.

Some euglenoids with chloroplasts may become heterotrophic in the dark; the chloroplasts become small and nonfunctional. If they are put back in the light, they may become green within a few hours. Photosynthetic euglenoids may sometimes feed on dissolved or particulate food.

Individual euglenoids range from 10 to 500 micrometers long and are highly variable in form. Interlocking proteinaceous strips arranged in a helical pattern form a flexible structure called the pellicle, which lies within the plasma membrane of the euglenoids. Because its pellicle is flexible, a euglenoid is able to change its shape.

Reproduction in this phylum occurs by mitotic cell division. The nuclear envelope remains intact throughout the process of mitosis. No sexual reproduction is known to occur in this group.

In *Euglena* (figure 28.6), the genus for which the phylum is named, two flagella are attached at the base of a flask-shaped opening called the reservoir, which is located at the anterior end of the cell. One of the flagella is long and has a row of very fine, short, hairlike projections along one side. A second, shorter flagellum is located within the reservoir but does not emerge from it. Contractile vacuoles collect excess water from all parts of the organism and empty it into the reservoir, which apparently helps regulate the osmotic pressure within the organism. The stigma, an organ that also occurs in the green algae (phylum Chlorophyta), is light-sensitive and helps these photosynthetic organisms move toward light.

Cells of *Euglena* contain numerous small chloroplasts. These chloroplasts, like those of the green algae and plants, contain chlorophylls *a* and *b*, together with carotenoids. Although the chloroplasts of euglenoids differ somewhat in structure from those of green algae, they probably had a common origin. It seems likely that euglenoid chloroplasts ultimately evolved from a symbiotic relationship through ingestion of green algae. Recent phylogenetic evidence indicates that *Euglena* had multiple origins within the Euglenoids, and the concept of a single *Euglena* genus is now being debated.

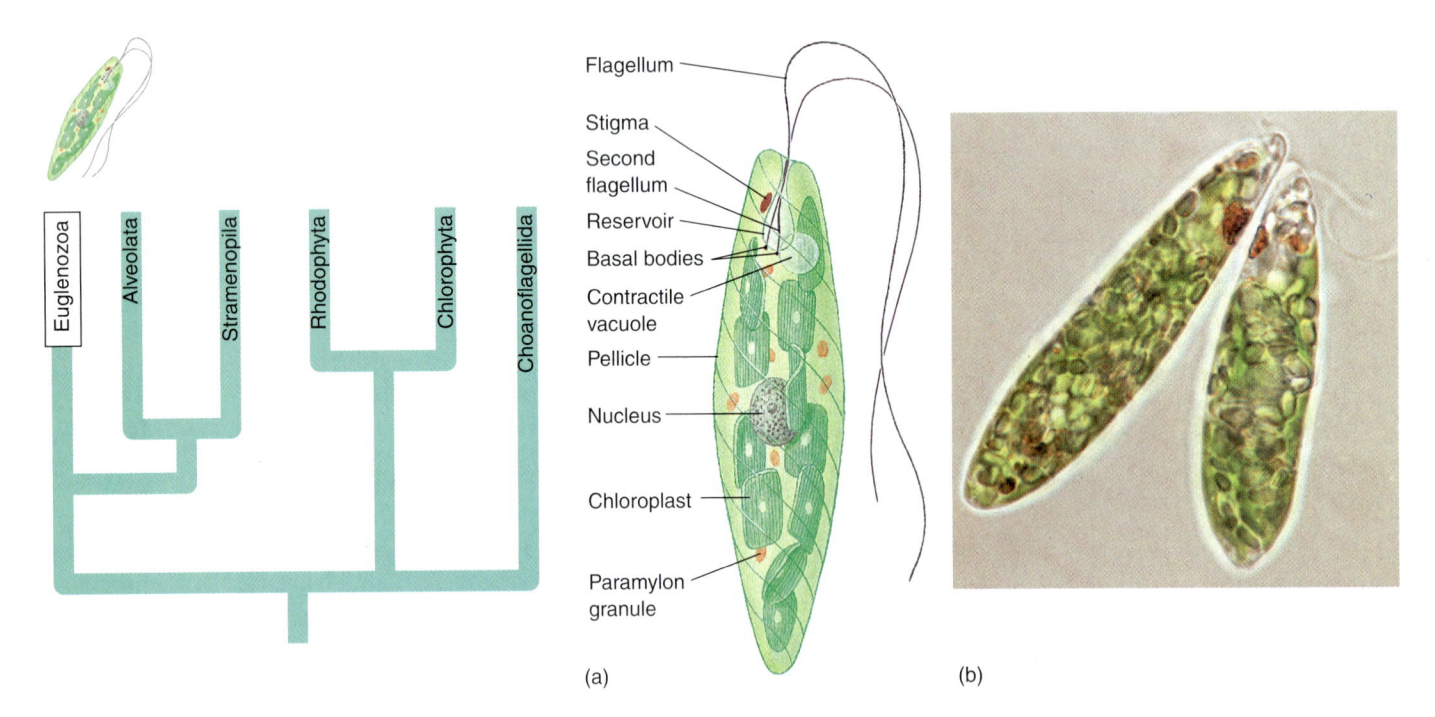

FIGURE 28.6
Euglenoids. (*a*) Diagram of *Euglena*. Paramylon granules are areas where food reserves are stored. (*b*) Micrograph of individuals of the genus *Euglena*.

Kinetoplastids

A second major group within the Euglenozoa is the kinetoplastids. The name kinetoplastid refers to a unique, single mitochondrion in each cell. The mitochondria have two types of DNA—mini-circles and maxi-circles. (Remember that prokaryotes have circular DNA, and mitochondria had prokaryotic origins.) This mitochondrial DNA is responsible for very rapid glycolysis and also for an unusual kind of editing of the DNA by guide RNAs encoded in the mini-circles.

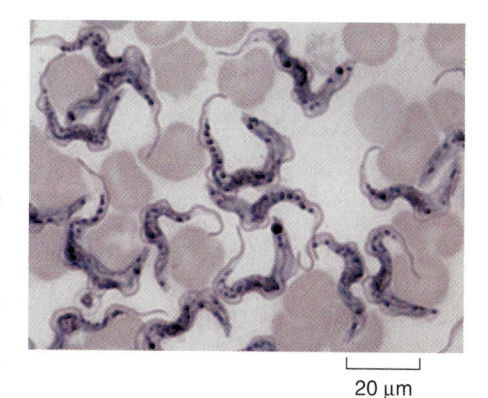

Parasitism has evolved multiple times within the kinetoplastids. Trypanosomes are kinetoplastids that cause many serious human diseases, the most familiar being trypanosomiasis, also known as African sleeping sickness, which causes extreme lethargy and fatigue (figure 28.7). Other diseases caused by trypanosomes include East Coast fever, leishmaniasis, and Chagas disease, all of great importance in tropical areas where they afflict millions of people each year. Leishmaniasis, which is transmitted by sand flies, causes skin sores and in some cases can affect internal organs, leading to death. About 1.5 million new cases are reported each year. The rise in leishmaniasis in South America correlates with the move of infected individuals from rural to urban environments, where there is a greater chance of spreading the parasite. Chagas disease is caused by *Trypanosoma cruzi*. At least 90 million people, from the southern United States to Argentina, are at risk of contracting *T. cruzi* from small wild mammals that carry the parasite and can spread it to other mammals and humans through skin contact with urine and feces. Blood transfusions have also increased the spread of the infection. Chagas disease can lead to severe cardiac and digestive problems in humans and domestic animals, but appears to be tolerated in the wild mammals.

These diseases make it impossible to raise domestic cattle for meat or milk in a large portion of Africa and are resulting in high medical expenses and loss of time from the workforce in South America. Control is especially difficult because of the unique attributes of these organisms. For example, tsetse fly–transmitted trypanosomes have evolved an elaborate genetic mechanism for repeatedly changing the antigenic nature of their protective glycoprotein coat, thus dodging the antibodies their hosts produce against them (see chapter 48). Only a single one out of some 1000 to 2000 variable antigen genes is expressed at a time. Rearrangements of these genes during the asexual cycle of the organism allow for the expression of a seemingly endless variety of different antigen genes that maintain infectivity by the trypanosomes.

When the trypanosomes are ingested by a tsetse fly, they embark on a complicated cycle of development and multiplication, first in the fly's gut and later in its salivary glands. It is their position in the salivary glands that allows them to move into their vertebrate host. Recombination has been observed between different strains of trypanosomes introduced into a single fly, suggesting that mating, syngamy, and meiosis occur, even though they have not been observed directly. Although most trypanosome reproduction is asexual, this sexual cycle, reported for the first time in 1986, affords still further possibilities for recombination in these organisms.

In the guts of the flies that spread them, trypanosomes are noninfective. When they are ready to transfer to the skin or bloodstream of their host, trypanosomes migrate to the salivary glands and acquire the thick coat of glycoprotein antigens that protect them from the host's antibodies. When they are taken up by a fly, the trypanosomes again shed their coats. The production of vaccines against such a system is complex, but tests are under way. Releasing sterilized flies to impede the reproduction of populations is another technique attempted to control the fly population. Traps made of dark cloth and scented like cows, but poisoned with insecticides, have likewise proved effective. Research is proceeding rapidly because the presence of tsetse flies with their associated trypanosomes blocks the use of some 11 million square kilometers of potential grazing land in Africa.

FIGURE 28.7
A kinetoplastid. (*a*) *Trypanosoma* among red blood cells. The nuclei (dark-staining bodies), anterior flagella, and undulating, changeable shape of the trypanosomes are visible in this photograph (500×). (*b*) The tsetse fly, shown here sucking blood from a human arm, can carry trypanosomes.

Euglenozoa include free-living and parasitic protists that move with flagella. *Euglena* have chloroplasts obtained via endosymbiosis, and trypanosomes have unusual mitochondria that use RNA editing.

Alveolata

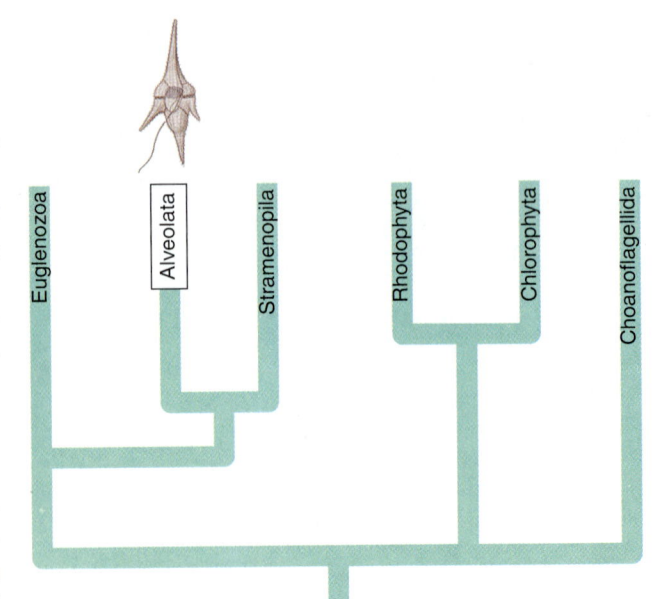

Members of the Alveolata include the dinoflagellates, apicomplexes, and ciliates, all of which have a common lineage despite their diverse modes of locomotion. One common trait is the presence of either a space or alveoli (hence the name alveolata) below their plasma membranes.

Dinoflagellates

Most dinoflagellates are photosynthetic unicells with two flagella. Dinoflagellates live in both marine and freshwater environments. Some dinoflagellates are luminous and contribute to the twinkling or flashing effects we sometimes see in the sea at night, especially in the tropics.

The flagella, protective coats, and biochemistry of dinoflagellates are distinctive, and the dinoflagellates do not appear to be directly related to any other phylum. Plates made of a cellulose-like material, often encrusted with silica, encase the dinoflagellate cells (figure 28.8). Grooves at the junctures of these plates usually house the flagella, one encircling the cell like a belt, and the other perpendicular to it. By beating in their grooves, these flagella cause the dinoflagellate to spin as it moves. Most have chlorophylls *a* and *c*, in addition to carotenoids, so that in the biochemistry of their chloroplasts, they resemble the diatoms and the brown algae, possibly acquiring such chloroplasts by forming endosymbiotic relationships with members of those groups.

The poisonous and destructive "red tides" that occur frequently in coastal areas are often associated with great population explosions, or "blooms," of dinoflagellates, whose pigments color the water. Red tides have a profound, detrimental effect on the fishing industry in the United States. Some 20 species of dinoflagellates produce powerful toxins that inhibit the diaphragm and cause respiratory failure in many vertebrates. When the toxic dinoflagellates are abundant, many fishes, birds, and marine mammals may die.

Although sexual reproduction occurs under starvation conditions, dinoflagellates reproduce primarily by asexual cell division. Asexual cell division relies on a unique form of mitosis in which the permanently condensed chromosomes divide longitudinally within a permanent nuclear envelope. After the numerous chromosomes duplicate, the nucleus divides into two daughter nuclei. Also, the dinoflagellate chromosome is unique among eukaryotes in that the DNA is not generally complexed with histone proteins. In all other eukaryotes, the chromosomal DNA is complexed with histones to form nucleosomes, structures that represent the first order of DNA packaging in the nucleus. How dinoflagellates maintain distinct chromosomes with a small amount of histones remains a mystery.

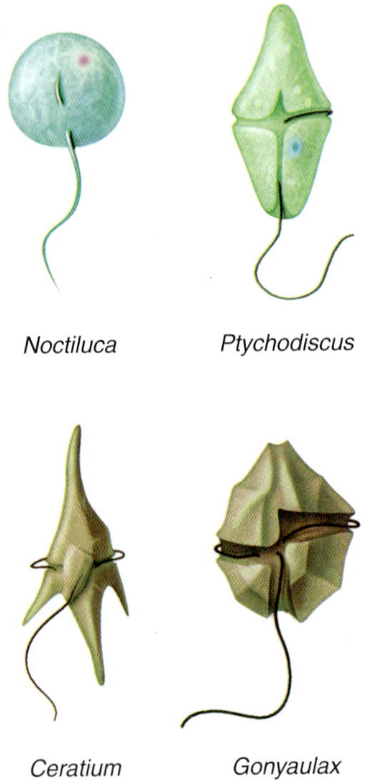

Apicomplexes

Apicomplexes are spore-forming parasites of animals. They are called apicomplexes (short for apical complex) because of a unique arrangement of fibrils, microtubules, vacuoles, and other cell organelles at one end of the cell. The best-known apicomplex is the malarial parasite *Plasmodium*. *Plasmodium* glides inside the red blood cells of its host with amoeboid-like contractility. Like other apicomplexes, *Plasmodium* has a complex life cycle involving sexual and asexual phases and alternation between different hosts, in this case mosquitoes and humans (figure 28.9). Even though *Plasmodium* has mitochondria, it is a *microaerophil*, growing best in a low-O_2, high-CO_2 environment.

Efforts to eradicate malaria have focused on (1) eliminating the mosquito vectors; (2) developing drugs to poison the parasites that have entered the human body; and (3) develop-

FIGURE 28.8
Some dinoflagellates: *Noctiluca, Ptychodiscus, Ceratium,* and *Gonyaulax*. *Noctiluca*, which lacks the heavy cellulose armor characteristic of most dinoflagellates, is one of the bioluminescent organisms that cause the waves to sparkle in warm seas. In the other three genera, the shorter, encircling flagellum is seen in its groove, with the longer one projecting away from the body of the dinoflagellate. (Not drawn to scale.)

568 Part V Diversity of Life on Earth

ing vaccines. From the 1940s to the 1960s, wide-scale applications of DDT killed mosquitoes in the United States, Italy, Greece, and certain areas of Latin America. For a time, the worldwide elimination of malaria appeared possible. But this hope was soon crushed by the development of DDT-resistant mosquitoes in many regions. Further, there are serious environmental concerns about the use of DDT. In addition to the problems with resistant strains of mosquitoes, strains of *Plasmodium* have appeared that are resistant to the drugs historically used to kill them, including quinine.

An experimental vaccine containing a surface protein of one malaria-causing parasite, *P. falciparum*, seems to induce the immune system to defend against future infections. In tests, six out of seven vaccinated people did not get malaria after being fed upon by mosquitoes that carried *P. falciparum*. Many are hopeful that this new vaccine may be able to fight malaria.

Gregarines are another group of apicomplexes that use their distinctive apical complex to attach themselves in the intestinal epithelium of arthropods, annelids, and mollusks. Most of the gregarine body, aside from the apical complex, is in the intestinal cavity, and nutrients appear to be obtained through the apicomplex attachment to the cell. One of the larger gregarines is frequently hosted by earthworms.

FIGURE 28.9
The life cycle of *Plasmodium*. *Plasmodium*, the apicomplex that causes malaria, has a complex life cycle that alternates between mosquitoes and mammals.

Ciliates

As the name indicates, most ciliates feature large numbers of cilia. These heterotrophic, unicellular protists are 10 to 3000 micrometers long. Their cilia are usually arranged either in longitudinal rows or in spirals around the cell. Cilia are anchored to microtubules beneath the plasma membrane, and they beat in a coordinated fashion. In some groups, the cilia have specialized functions, becoming fused into sheets, spikes, and rods that may then function as mouths, paddles, teeth, or feet. The ciliates have a tough but flexible outer covering called the *pellicle* that enables them to squeeze through or move around obstacles.

All known ciliates have two different types of nuclei within their cells—small micronuclei and larger macronuclei (figure 28.10). Macronuclei divide by mitosis and are essential for the physiological function of the well-known ciliate *Paramecium*. The micronucleus of some *Tetrahymena pyriformis*, a common laboratory species, was experimentally removed in the 1930s, and their descendants continue to reproduce asexually to this day! *Paramecium*, however, is not immortal. The cells divide asexually for about 700 generations and then die if sexual reproduction has not occurred. The micronucleus in ciliates is needed only for sexual reproduction.

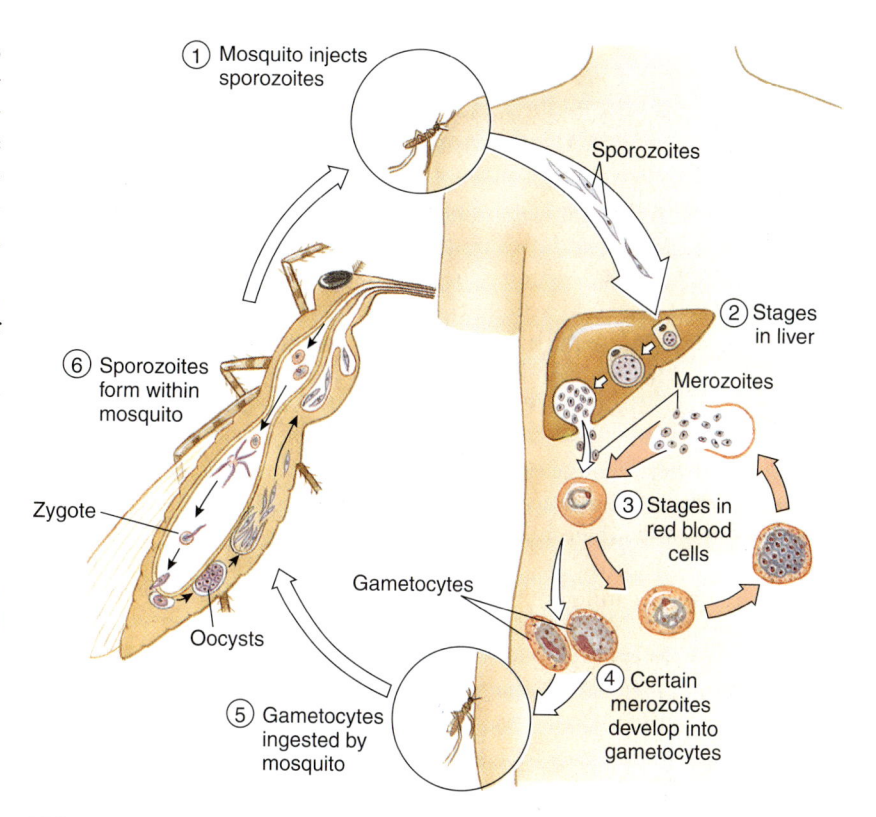

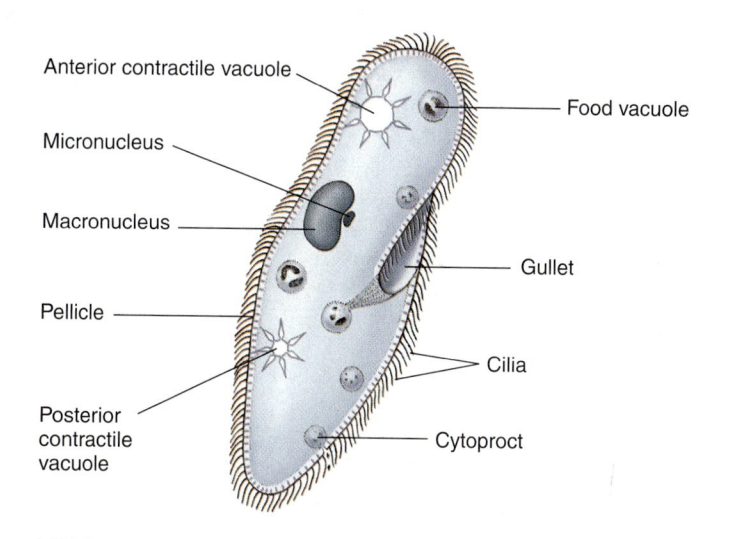

FIGURE 28.10
Paramecium. The main features of this familiar ciliate include cilia, two nuclei, and numerous specialized organelles.

Ciliates form vacuoles for ingesting food and regulating water balance. Food first enters the gullet, which in *Paramecium* is lined with cilia fused into a membrane (figure 28.10). From the gullet, the food passes into food vacuoles, where enzymes and hydrochloric acid aid in its digestion. Afterward, the vacuole empties its waste contents through a special pore

Chapter 28 Protists 569

in the pellicle called the **cytoproct,** which is essentially an exocytotic vesicle that appears periodically when solid particles are ready to be expelled. The contractile vacuoles, which regulate water balance, periodically expand and contract as they empty their contents to the outside of the organism.

Like most ciliates, *Paramecium* undergoes a sexual process called *conjugation*, in which two individual cells remain attached to each other for up to several hours (figure 28.11). Paramecia have multiple mating types. Only cells of two different genetically determined mating types can conjugate. Meiosis in the micronuclei of each individual produces several haploid micronuclei, and the two partners exchange a pair of these micronuclei through a cytoplasmic bridge between the two partners.

In each conjugating individual, the new micronucleus fuses with one of the micronuclei already present in that individual, resulting in the production of a new diploid micronucleus. After conjugation, the macronucleus in each cell disintegrates, while the new diploid micronucleus undergoes mitosis, thus giving rise to two new identical diploid micronuclei in each individual. One of these micronuclei becomes the precursor of the future micronuclei of that cell, while the other micronucleus undergoes multiple rounds of DNA replication, becoming the new macronucleus. This complete segregation of the genetic material is unique to the ciliates and makes them ideal organisms for the study of certain aspects of genetics.

Paramecium strains that kill other, sensitive strains of *Paramecium* long puzzled researchers. Initially, killer strains were believed to have genes coding for a substance toxic to sensitive strains. The true source of the toxin turned out to be an endosymbiotic bacterium in the "killer" strains. If this bacterium is engulfed by a "nonkiller" strain, the toxin is released, and the sensitive *Paramecium* dies.

Alveolata comprise what is believed to be a monophyletic group of organisms with varied forms of locomotion and reproduction.

(a)

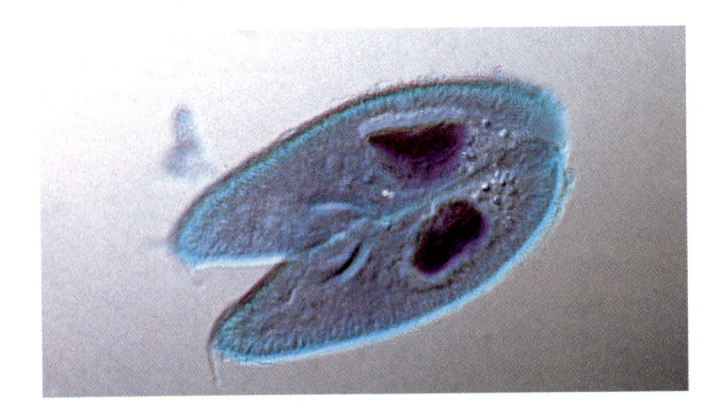

FIGURE 28.11
Life cycle of *Paramecium*. (*a,b*) In sexual reproduction, two mature cells fuse in a process called conjugation (100×).

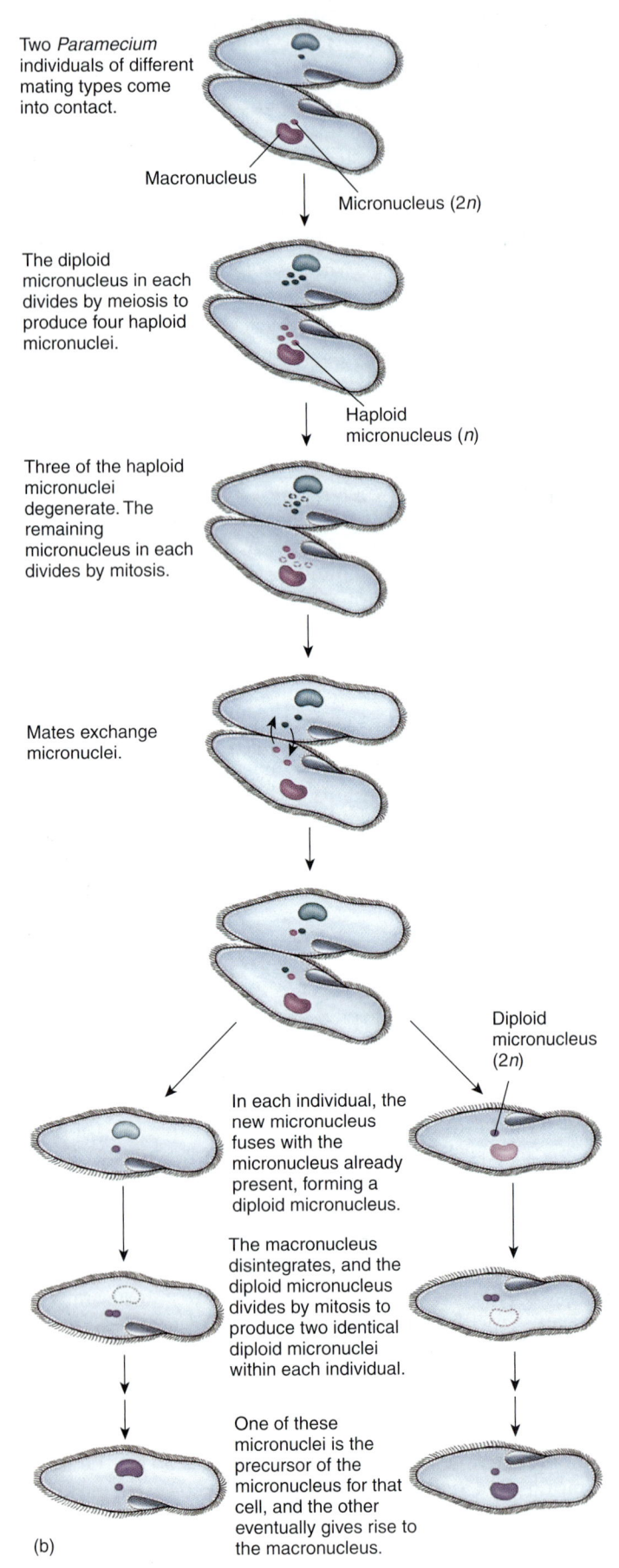

Two *Paramecium* individuals of different mating types come into contact.

Macronucleus

Micronucleus (2*n*)

The diploid micronucleus in each divides by meiosis to produce four haploid micronuclei.

Haploid micronucleus (*n*)

Three of the haploid micronuclei degenerate. The remaining micronucleus in each divides by mitosis.

Mates exchange micronuclei.

Diploid micronucleus (2*n*)

In each individual, the new micronucleus fuses with the micronucleus already present, forming a diploid micronucleus.

The macronucleus disintegrates, and the diploid micronucleus divides by mitosis to produce two identical diploid micronuclei within each individual.

One of these micronuclei is the precursor of the micronucleus for that cell, and the other eventually gives rise to the macronucleus.

(b)

Stramenopila and Rhodophyta

Stramenopila

Stramenopiles include **brown algae, diatoms,** and the **oomycetes** (water molds). Brown algae are the most conspicuous seaweeds in many northern regions (figure 28.12). The life cycle of the brown alga is marked by an alternation of generations between a sporophyte (diploid) and a gametophyte (haploid). Some sporophyte cells go through meiosis and produce spores. These spores germinate and undergo mitosis to produce the large individuals we recognize, such as the kelps. The gametophytes are often much smaller, filamentous individuals, perhaps a few centimeters in width.

Diatoms, members of the phylum Chrysophyta, are photosynthetic, unicellular organisms with unique double shells made of opaline silica, which are often strikingly and characteristically marked (figure 28.13). The shells of diatoms are like small boxes with lids, one half of the shell fitting inside the other. Their chloroplasts, containing chlorophylls *a* and *c*, as well as carotenoids, resemble those of the brown algae and dinoflagellates. Diatoms produce a unique carbohydrate called chrysolaminarin. Some diatoms move by using two long grooves, called raphes, which are lined with vibrating fibrils. The exact mechanism is still being unraveled and may involve the ejection of mucopolysaccharide streams from the raphe that propel the diatom. Pencil-shaped diatoms can slide back and forth on each other, creating an ever-changing shape.

All oomycetes are either parasites or saprobes (organisms that live by feeding on dead organic matter). They are distinguished from other protists by the structure of

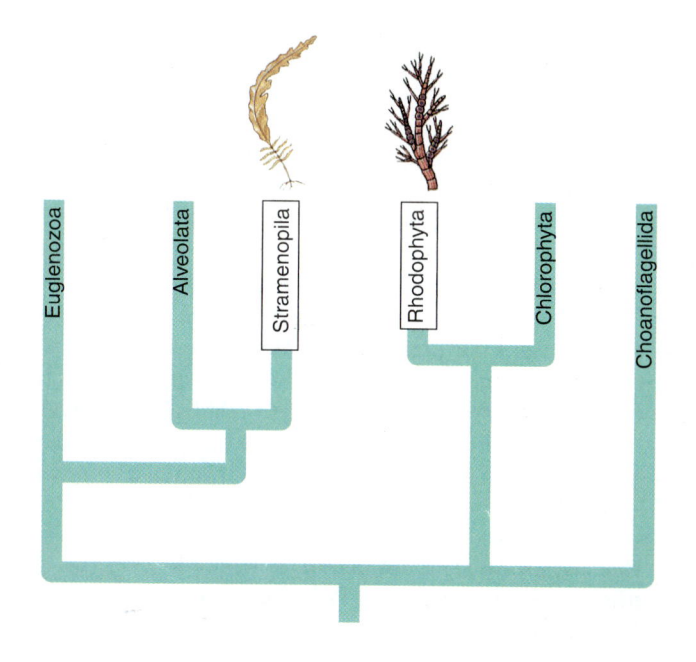

FIGURE 28.12
Brown algae. The massive "groves" of giant kelp that occur in relatively shallow water along the coasts throughout the world provide food and shelter for many different kinds of organisms.

Chapter 28 Protists 571

FIGURE 28.13
Diatoms. These different radially symmetrical diatoms have unique silica double shells.

their motile spores, or zoospores, which bear two unequal flagella, one pointed forward and the other backward. Zoospores are produced asexually in a sporangium. Sexual reproduction involves the formation of male and female reproductive organs that produce gametes. Most oomycetes are found in water, but their terrestrial relatives are plant pathogens. *Phytophthora infestans*, which causes late blight of potatoes, was responsible for the Irish Potato famine of 1845 and 1847. During the famine, about 400,000 people starved to death or died of diseases complicated by starvation. Millions of Irish people emigrated to the United States and elsewhere as a result of this disaster.

Rhodophyta

Rhodophyta, the red algae, range from microscopic organisms to those rivaling the brown algae in size. Sushi rolls are wrapped in nori, a red alga. Red algal polysaccharides are used commercially to thicken ice cream and cosmetics. This lineage lacks flagella and centrioles, and has the accessory photosynthetic pigments phycoerythrin, phycocyanin, and allophycocyanin, which are arranged within structures called phycobilisomes. They reproduce using alternation of generations.

The origin of the over 7000 species of Rhodophyta has been a source of controversy. Evidence supporting both very early eukaryotic origins and a common ancestry with green algae has been considered. Molecular comparisons of the chloroplasts in red and green algae support a single endosymbiotic origin for both. This would have been an ancient event whereby a cyanobacterium was engulfed by a host cell. Chloroplast genome comparisons tell us about the cyanobacteria symbiont, but nuclear DNA contains information about host cell origins. Comparisons of the nuclear DNA coding for the large subunit of RNA polymerase II from two red algae, a green alga, and another protist support the conclusion that the Rhodophyta emerged before the evolutionary lineage that led to plants, animals, and fungi. How can we reconcile the data from plastid and nuclear DNA? The host cells and the cyanobacterial symbionts probably did not follow congruent evolutionary pathways. The host cell that gave rise to red algae may have been distinct from the one that gave rise to plants. One possibility is that different host cells engulfed the same bacterial symbiont. What other evolutionary scenarios could explain these results? Tentatively, we will treat Rhodophyta and Chlorophyta (the green algae) as sister clades based on the substantial amount of chloroplast data.

Red and brown algae are marine or freshwater organisms. The oomycetes lack photosynthetic capabilities.

Chlorophyta

Green algae are of special interest, because of their unusual diversity and because the ancestors of the plant kingdom were clearly multicellular green algae. There are two distinct lineages of green algae—the chlorophytes discussed here, and another lineage (Streptophyta) that gave rise to the land plants (see chapter 29). The chlorophytes have an extensive fossil record dating back 900 million years. Modern chlorophytes closely resemble land plants, especially in their chloroplasts, which are biochemically similar to those of the plants. They contain chlorophylls *a* and *b*, as well as carotenoids.

Chlamydomonas probably represents a primitive state for green algae (figure 28.14). Individuals are microscopic (usually less than 25 micrometers long), green, and rounded, and they have two flagella at the anterior end. They move rapidly in water by beating their flagella in opposite directions. Each individual has an eyespot, which contains about 100,000 molecules of rhodopsin, the same pigment employed in vertebrate eyes. Light received by this eyespot is used by the alga to help direct its swimming. Most individuals of *Chlamydomonas* are haploid. *Chlamydomonas* reproduces asexually as well as sexually (figure 28.14).

Several lines of evolutionary specialization have been derived from organisms such as *Chlamydomonas*. The first is the evolution of nonmotile, unicellular green algae. *Chlamydomonas* is capable of retracting its flagella and settling down as an immobile unicellular organism if the pond in which it lives dries out. Some common algae found in soil and bark, such as *Chlorella*, are essentially like *Chlamydomonas* in this trait, but do not have the ability to form flagella. *Chlorella* is widespread in both fresh and salt water as well as in soil, and is only known to reproduce asexually.

Another major line of specialization from cells like those of *Chlamydomonas* concerns the formation of motile, colonial organisms. In these genera of green algae, the *Chlamydomonas*-like cells retain some of their individuality. The most elaborate of these organisms is *Volvox* (see figure 28.1), a hollow sphere made up of a single layer of 500 to 60,000 individual cells, each cell having two flagella. Only a small number of the cells are reproductive. Some reproductive cells may divide asexually, bulge inward, and give rise to new colonies that initially remain within the parent colony. Others produce gametes.

Nonmotile, unicellular algae and multicellular, flagellated colonies have been derived from green algae such as *Chlamydomonas*—a biflagellated, unicellular organism. Chlorophytes did not give rise to land plants.

FIGURE 28.14
Life cycle of *Chlamydomonas* (Chlorophyta). Individual cells of this microscopic, biflagellated alga, which are haploid, divide asexually, producing identical copies of themselves. At times, such haploid cells act as gametes—fusing, as shown in the lower right-hand side of the diagram, to produce a zygote. The zygote develops a thick, resistant wall, becoming a zygospore; this is the only diploid cell in the entire life cycle. Within this diploid zygospore, meiosis takes place, ultimately resulting in the release of four haploid individuals. Because of the segregation during meiosis, two of these individuals are called the (+) strain and the other two, the (−) strain. Only + and − individuals are capable of mating with each other when syngamy does take place, although both may divide asexually to reproduce themselves.

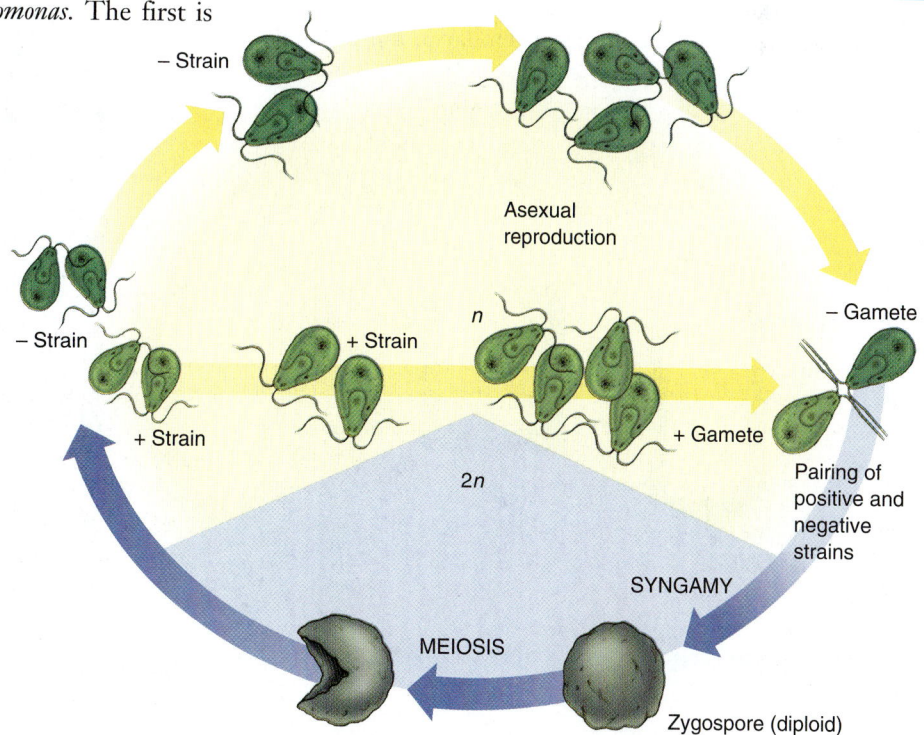

Chapter 28 Protists 573

Choanoflagellida and Protists That Are Difficult to Categorize

Choanoflagellida

Choanoflagellates are most like the common ancestor of the sponges and, indeed, all animals. Choanoflagellates have a single emergent flagellum surrounded by a funnel-shaped, contractile collar composed of closely placed filaments, a structure that is exactly matched in the sponges. Colonial forms resemble freshwater sponges. These protists feed on bacteria strained out of the water by their collar. The close relationship of choanoflagellates to animals was further demonstrated by the strong homology between a surface receptor (a tyrosine kinase receptor) found in choanoflagellates and sponges. This surface receptor initiates a signaling pathway involving phosphorylation (see chapter 7).

Amoebas

So far, we have organized the protists based on their closest relatives. Some lineages vary tremendously if you consider just a single trait. For example, the stramenopiles include autotrophic, marine algae and terrestrial plant pathogens. As seen in chapter 25, it is also possible for unrelated organisms to acquire similar traits. That is the case with amoebas, which have similar cell morphology, but are not monophyletic (figure 28.15).

Amoebas move from place to place by means of their pseudopods. Pseudopods are flowing projections of cytoplasm that extend and pull the amoeba forward or engulf food particles, a process called cytoplasmic streaming. An amoeba puts a pseudopod forward and then flows into it. Microfilaments of actin and myosin similar to those found in muscles are associated with these movements. The pseudopods can form at any point on the cell body so that it can move in any direction. The pseudopods of amoeboid cells give them truly amorphous bodies. One group, however, has more distinct structures. Members of the phylum Actinopoda, often called radiolarians, secrete glassy exoskeletons made of silica. These skeletons give the unicellular organisms a distinct shape, exhibiting either bilateral or radial symmetry. The shells of different species form many elaborate and beautiful shapes, with pseudopods extruding outward along spiky projections of the skeleton (figure 28.16). Microtubules support these cytoplasmic projections.

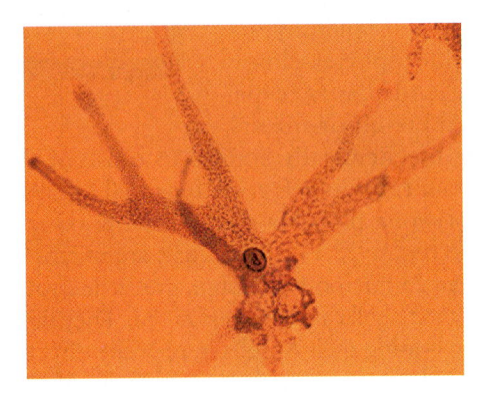

FIGURE 28.15
Amoeba proteus. The projections are pseudopods; an amoeba moves by flowing into them.

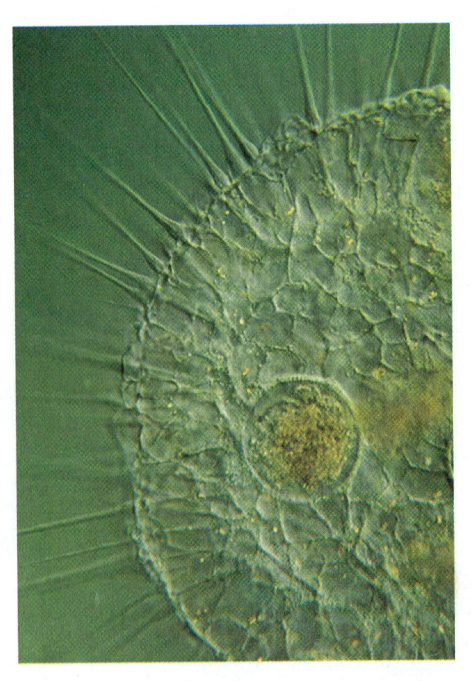

FIGURE 28.16
Actinosphaerium with needlelike pseudopods.

Foraminifera

Members of the phylum Foraminifera are heterotrophic marine protists. They range in diameter from about 20 micrometers to several centimeters. They resemble tiny snails and can form 3-meter-deep layers in marine sediments. Characteristic of the group are pore-studded shells (called *tests*) composed of organic materials usually reinforced with grains of inorganic matter. These grains may be calcium carbonate, sand, or even plates from the shells of echinoderms or spicules (minute needles of calcium carbonate) from sponge skeletons. Depending on the building materials they use, foraminifera—often informally called "forams"—may have shells of very different appearance. Some of them are brilliantly colored red, salmon, or yellow-brown.

Most foraminifera live in sand or are attached to other organisms, but two families consist of free-floating planktonic organisms. Their tests may be single-chambered, but are more often multichambered, and they sometimes have a spiral shape resembling that of a tiny snail. Thin cytoplasmic projections called *podia* emerge through openings in the tests (figure 28.17). Podia are used for swimming, gathering materials for the tests, and feeding. Forams eat a wide variety of small organisms.

The life cycles of foraminifera are extremely complex, involving alternation between haploid and diploid generations. Forams have contributed massive accumulations of their tests to the fossil record for more than 200 million years. Because of the excellent preservation of their tests and the striking differences among them, forams are very important as geological markers. The pattern of occurrence of different forams is often used as a guide in searching for oil-bearing strata. Limestones all over the world, including the famous White Cliffs of Dover in southern England, are often rich in forams (figure 28.18).

Slime Molds

Slime molds originated at least three distinct times, and the three lineages are very distantly related. We will explore two lineages—the plasmodial slime molds, which are huge, single-celled, multinucleate, oozing masses, and the cellular slime molds, in which single cells combine and differentiate, creating an early model of multicellularity.

Plasmodial slime molds stream along as a **plasmodium,** a nonwalled, multinucleate mass of cytoplasm that resembles a moving mass of slime (figure 28.19). This is called the *feeding phase*, and the plasmodia may be orange, yellow, or another color. Plasmodia show a back-and-forth streaming of cytoplasm that is very conspicuous, especially under a microscope. They are able to pass through the mesh in cloth or simply to flow around or through other obstacles. As they move, they engulf and digest bacteria, yeasts, and other small particles of organic matter. A multinucleated *Plasmodium* cell undergoes mitosis synchronously, with the

FIGURE 28.17
A representative of the foraminifera. Podia, thin cytoplasmic projections, extend through pores in the calcareous test, or shell, of this living foram (90×).

FIGURE 28.18
White Cliffs of Dover. The limestone that forms these cliffs is composed almost entirely of fossil shells of protists, including foraminifera.

FIGURE 28.19
A plasmodial protist. This multinucleate plasmodium moves about in search of the bacteria and other organic particles that it ingests.

FIGURE 28.20
Sporangia of a plasmodial slime mold. These *Arcyria* sporangia are found in the phylum Myxomycota.

nuclear envelope breaking down, but only at late anaphase or telophase. Centrioles are absent in cellular slime molds.

When either food or moisture is in short supply, the plasmodium migrates relatively rapidly to a new area. Here it stops moving and either forms a mass in which spores differentiate or divides into a large number of small mounds, each of which produces a single, mature **sporangium**, the structure in which spores are produced. These sporangia are often beautiful and extremely complex in form (figure 28.20). The spores are highly resistant to unfavorable environmental influences and may last for years if kept dry.

The cellular slime molds have become an important group for the study of cell differentiation because of their relatively simple developmental systems (figure 28.21). The individual organisms behave as separate amoebas, moving through the soil and ingesting bacteria. When food becomes scarce, the individuals aggregate to form a moving "slug." Cyclic adenosine monophosphate (cAMP) is sent out in pulses by some of the cells, and other cells move in the direction of the cAMP to form the slug. In the cellular slime mold *Dictyostelium discoideum*, this slug goes through morphogenesis to make stalk and spore cells. The spores then go on to form a new amoeba if they land in a moist habitat.

> **Choanoflagellates are the most like the common ancestor of all animals. The evolutionary origins of some protists, including amoebas and slime molds, are less well understood, and these organisms may have arisen independently more than once.**

FIGURE 28.21
Development in *Dictyostelium discoideum*, a cellular slime mold. (*a*) First, a spore germinates, forming an amoeba. The amoebas feed and reproduce until the food runs out. (*b*) The amoebas aggregate and move toward a fixed center. (*c*) Next, they form a multicellular "slug," 2 to 3 millimeters long, that migrates toward light. (*d*) The slug stops moving and begins to differentiate into (*e*) a spore-forming body called a sorocarp. (*f*) Within the heads of the sorocarps, amoebas become encysted as spores.

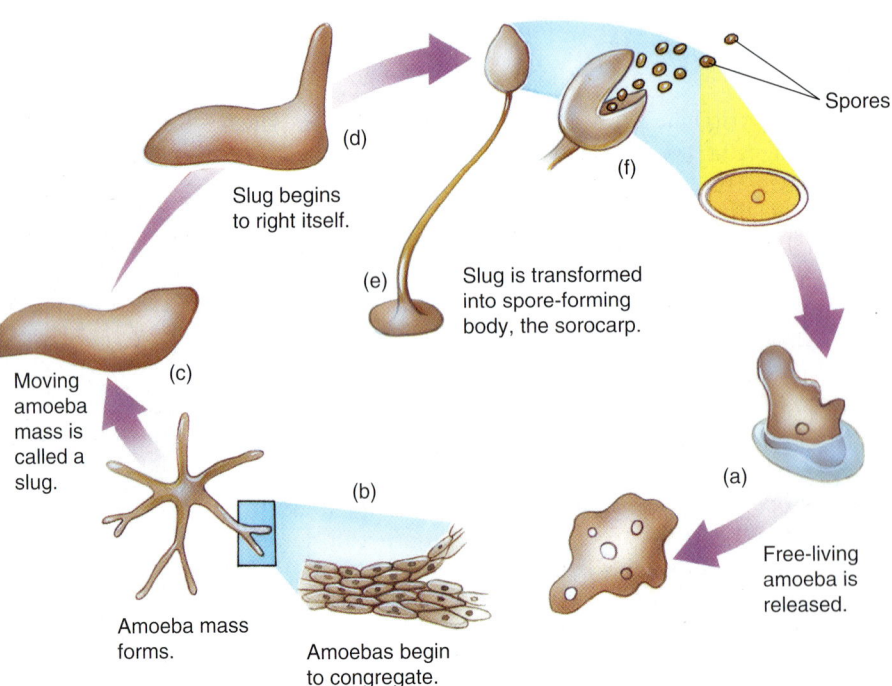

576 Part V Diversity of Life on Earth

Concept Review

For interactive testing, visit the Online Learning Center with PowerWeb at www.mhhe.com/Raven7

28.1 Eukaryotes probably arose by endosymbiosis.

Endosymbiosis

- Single-celled, nonphotosynthetic *Pelomyxa palustris* appears to represent an early stage in eukaryotic evolution. (p. 562)
- *Pelomyxa* lacks mitochondria, but contains two kinds of bacteria that may play a similar role. (p. 562)
- Most biologists believe mitochondria originated as symbiotic, aerobic bacteria; this theory is called endosymbiosis. (p. 562)
- Aerobic bacteria are believed to have evolved into mitochondria when they were engulfed by ancestral eukaryotic cells. (p. 562)
- Although most mitochondrial genes have been transferred to the host cell's chromosomes, each mitochondrion still has its own circular, closed molecule of DNA. (p. 562)
- Many eukaryotic cells contain other endosymbiotic bacteria in addition to mitochondria. (p. 563)
- Three biochemically distinct classes of chloroplasts exist, with each resembling a different bacterial ancestor. (p. 563)
- Recent investigations of chloroplast DNA suggest a single origin of chloroplasts. (p. 563)

28.2 The kingdom Protista is by far the most diverse of any eukaryotic kingdom.

The Challenge of Classifying the Protists

- Protists are the most diverse of the four kingdoms in the domain Eukarya. (p. 564)
- The classification of Protista is artificial and not representative of any evolutionary relationships. (p. 564)

General Biology of the Protists

- Protists are united on the basis that they are eukaryotes that are not fungi, plants, or animals. (p. 565)
- Protists exhibit a varied array of cell surfaces and diverse mechanisms for locomotion. (p. 565)
- Many protists form cysts with resistant coverings; they lack cell metabolism. (p. 565)
- Protists are known to be phototrophs, phagotrophs, and osmotrophs. They typically reproduce asexually through binary fission. (p. 565)

28.3 Protists can be categorized into six groups.

Euglenozoa

- Euglenoids were among the earliest free-living eukaryotes to possess mitochondria. (p. 566)
- It seems likely that chloroplasts ultimately evolved from a symbiotic relationship through ingestion of green algae. (p. 566)
- Kinetoplastids have a unique, single mitochondrion in each cell. (p. 567)
- Trypanosomes cause multiple serious diseases in humans, including trypanosomiasis (African sleeping sickness), East Coast fever, and Chagas disease. (p. 567)
- Rearrangement of genes during the asexual cycle allows for the expression of a large number of varieties of different antigen genes. (p. 567)

Alveolata

- Most dinoflagellates are photosynthetic unicells possessing two flagella. (p. 568)
- Protists cannot be grouped together based solely on the presence of flagella. (p. 568)
- At least 20 dinoflagellate species are known to produce powerful toxins that provoke respiratory failure in vertebrates. (p. 568)
- Apicomplexes are spore-forming animal parasites, such as the malarial parasite *Plasmodium*, that have a unique arrangement of organelles at one end of the cell. (pp. 568–569)
- Ciliates are extremely complex organisms with cilia usually arranged in longitudinal rows or in spirals around the cell. (pp. 569–570)
- All known ciliates have both a micro- and a macronucleus. (pp. 569–570)

Stramenopila and Rhodophyta

- Stramenopiles include brown algae, diatoms, and oomycetes. (pp. 571–572)
- Rhodophyta is made up of over 7000 species of red algae. (p. 572)
- Evidence exists supporting both early eukaryotic origins of Stramenopila and Rhodophyta and a common ancestry with green algae. (p. 572)
- Tentatively, Rhodophyta and Chlorophyta will be treated as sister clades. (p. 572)

Chlorophyta

- Chlorophytes make up one distinct lineage of green algae, while the other lineage, the streptophytes, gave rise to land plants. (p. 573)
- *Chlamydomonas* most likely represents a primitive state for green algae. (p. 573)
- Nonmotile, unicellular green algae and motile, colonial organisms are two lines of evolutionary specialization derived from early green algae. (p. 573)

Choanoflagellida and Protists That Are Difficult to Categorize

- Choanoflagellates are most similar to the sponges and to all animals. (p. 574)
- Amoebas and slime molds are very difficult to categorize because amoebas have similar cell morphology but are not monophyletic, and because slime molds originated at least three distinct times and the three lineages are only distantly related. (pp. 574–576)

Test Your Understanding

For interactive testing, visit the Online Learning Center with PowerWeb at www.mhhe.com/Raven7

Self Test

1. The mitochondria of eukaryotic cells most likely arose as a result of endosymbiosis between a eukaryotic cell and a
   a. blue-green alga.
   b. nonsulfur purple bacterium.
   c. red alga.
   d. cyanobacterium.
2. The protists are a paraphyletic group that have traditionally been grouped together because
   a. they are all genetically similar to each other.
   b. they all have very similar morphological characters.
   c. they are not fungi, animals, or plants.
   d. they all have similar nutritional modes and live in similar environments.
3. Many protists are able to resist harsh environmental conditions by
   a. forming a cyst and slowing metabolism during times of stress.
   b. utilizing a wide variety of nutritional modes.
   c. reproducing asexually through the process of budding.
   d. moving away from a harsh environment by using pseudopodia.
4. The light-sensing organ in *Euglena* is a
   a. flagellum.
   b. contractile vacuole.
   c. pellicle.
   d. stigma.
5. The parasitic kinetoplast that causes leishmaniasis must spend part of its life cycle in a nonhuman host. What organism(s) serve(s) as the vector for this life cycle?
   a. small mammals
   b. a sand fly
   c. a mosquito
   d. a tse-tse fly
6. You are examining cells from an unknown organism under the microscope. You note that the cells have a membrane-bounded nucleus and that there are small cavities in the membranes along the internal cell surface. Based upon this information alone, you conclude that these most likely are _____ cells.
   a. bacterial
   b. diatom
   c. dinoflagellate
   d. kinetoplastid
7. The parasitic protist that causes malaria, *Plasmodium*, must spend part of its life cycle in a nonhuman host. What organism(s) serve(s) as the vector for this life cycle?
   a. small mammals
   b. a sand fly
   c. a mosquito
   d. a tse-tse fly
8. *Phytopthora infestans* is the oomycete that causes
   a. Chagas disease.
   b. potato blight.
   c. red tides.
   d. African sleeping sickness.
9. A biologist discovers an alga that is marine, multicellular, lacks flagella and centrioles, and contains phycobilisomes. It probably belongs to which group?
   a. Rhodophyta
   b. brown algae
   c. Chlorophyta
   d. Foraminifera
10. The green algae gave rise to which modern group of organisms?
    a. photosynthetic bacteria
    b. plants
    c. photosynthetic euglenoids
    d. animals

Test Your Visual Understanding

1. This image shows several different types of protists that are characterized by a rigid double wall of silica. To what group do these organisms belong?

Apply Your Knowledge

1. Protists typically reproduce asexually. However, some protists undergo sexual reproduction during times of environmental stress. What advantage does sexual reproduction give a protist during these times?
2. Scientists are working to develop a vaccine against African sleeping sickness. Use what you know about the trypanosome responsible for this disease to determine why it is so difficult to develop a vaccine.

29

Overview of Plant Diversity

Concept Outline

29.1 Plants have multicellular haploid and diploid stages in their life cycles.

  The Evolutionary Origins of Plants. Plants evolved from freshwater green algae and eventually developed cuticles, stomata, conducting systems, and reproductive strategies that adapt them well for life on land.

  Plant Life Cycles. Plants have haplodiplontic life cycles. Diploid sporophytes produce haploid spores by meiosis. Spores develop into haploid gametophytes by mitosis and produce haploid gametes.

29.2 Nonvascular plants are relatively unspecialized, but successful in many terrestrial environments.

  Mosses, Liverworts, and Hornworts. The most conspicuous part of a nonvascular plant is the green photosynthetic gametophyte, which supports the smaller sporophyte nutritionally.

29.3 Seedless vascular plants have well-developed conducting tissues in their sporophytes.

  Features of Vascular Plants. In vascular plants, specialized tissue called xylem conducts water and dissolved minerals within the plant, and tissue called phloem conducts sucrose and hormones within the plant.

  Seedless Vascular Plants. Seedless vascular plants have a much more conspicuous sporophyte than nonvascular plants do, and many have well-developed conducting systems in the stems, roots, and leaves.

29.4 Seeds protect and aid in the dispersal of plant embryos.

  Seed Plants. In seed plants, the sporophyte is dominant. Male and female gametophytes develop within the sporophyte and depend on it for food. Seeds allow embryos to germinate when conditions are favorable.

  Gymnosperms. In gymnosperms, the female gametophyte (ovule) is not completely enclosed by sporophyte tissue at the time of pollination.

  Angiosperms. In angiosperms, the ovule is completely enclosed by sporophyte tissue at the time of pollination. Angiosperms, by far the most successful plant group, produce flowers.

FIGURE 29.1
An arctic tundra. Tundra is one of the harshest environments on earth, and yet many diverse plants have made it their home. These ecosystems are fragile and particularly susceptible to global change.

Plant evolution is the story of the conquest of land by green algal ancestors. For about 500 million years, algae were confined to a watery domain, limited by the need for water to reproduce, provide structural support, prevent water loss, and provide some protection from the sun's ultraviolet irradiation. Numerous evolutionary solutions to these challenges have resulted in over 300,000 species of plants dominating all terrestrial communities today, from forests to alpine tundra (figure 29.1), and from agricultural fields to deserts. Most plants are photosynthetic, converting light energy into chemical-bond energy and providing oxygen for all aerobic organisms. We rely on plants for food, clothing, wood for shelter and fuel, chemicals, and many medicines. This chapter explores the evolutionary history and strategies that have allowed plants to inhabit most terrestrial environments over millions of years.

29.1 Plants have multicellular haploid and diploid stages in their life cycles.

The Evolutionary Origins of Plants

What is a plant? We will use the term plant to refer to a group of organisms that share a freshwater algal ancestor and have evolved over a 470-million-year period. This chapter will explore land plants. The defining characteristic of land plants is the protection of their embryos, essential for survival in a terrestrial environment. It is surprising that just a single species of green algae gave rise to the entire terrestrial plant lineage, from mosses through the flowering plants (angiosperms). Exactly what this ancestral alga was is still a mystery, but close relatives, the Charales, exist in freshwater lakes today (see chapter 25 for more information on the evolution of land plants from Charales species). DNA sequence data are consistent with the claim that a single "Eve" gave rise to all plants. The shared evolutionary history with green algae has led biologists to rename kingdom Plantae as kingdom Virdiplantae to include the green algae. Fungi are not a part of this scheme. They are more closely related to metazoan animals (see chapter 31).

Land plants, though diverse, have certain characteristics in common. For example, all of them afford some protection to their embryos, and all have multicellular haploid and diploid phases. Over time, the trend has been toward more embryo protection and a smaller haploid stage in the life cycle. In addition, land plants can be compared based on the presence or absence of conducting systems, which facilitate the transport of water and nutrients. **Nonvascular plants** lack vascular tissue, while **vascular plants** have water-conducting xylem and food-conducting phloem strands of tissues in their stems, roots, and leaves. For purposes of discussion in this chapter, we subdivide the plants into the four major categories shown in figure 29.2, based on common characteristics and the order in which the types of plants are thought to have evolved not all of the groups are clades:

- The *nonvascular land plants*, which are not monophyletic, include three phyla: mosses, liverworts, and hornworts.
- Recent evidence supports two distinct monophyletic lineages of *seedless vascular plants:* (1) club mosses, and (2) ferns, whisk ferns, and horsetails. Ferns and horsetails are the closest relatives to the seed plants (the gymnosperms and angiosperms).
- *Gymnosperms* have seeds that protect their embryos. Included in this group are the conifers and cycads.
- *Angiosperms* arose about 150 million years ago with further innovations. Their distinguishing adaptations are flowers, which may attract pollinators, and fruits surrounding the seeds, protecting the embryos and aiding in seed dispersal.

FIGURE 29.2
Four major groups of land plants. In this chapter, we discuss four major groups of plants. There are actually five distinct lineages; seedless vascular plants have two different origins but are grouped here for simplicity. The ancestral green algae are discussed in chapters 25 and 28.

Adaptations to Land

Unlike their freshwater ancestors, most land plants have only limited amounts of water available. As an adaptation to living on land, most plants are protected from **desiccation**—the tendency of organisms to lose water to the air—by a waxy **cuticle** that is secreted onto their exposed surfaces. The cuticle is relatively impermeable, preventing water loss. However, this solution limits the gas exchange essential for respiration and photosynthesis. Gas diffusion into and out of a plant occurs through tiny mouth-shaped openings called **stomata** (singular, *stoma*).

Two additional adaptations allowed larger land plants to flourish. The evolution of leaves which may have occured multiple times resulted in increased photosynthetic surface area. The shift to a dominant diploid generation, accompanied by the structural support of vascular tissue, allowed plants to take advantage of the vertical dimension of the terrestrial environment, making the evolution of trees possible.

> **Plants evolved from freshwater green algae and eventually developed reproductive strategies, conducting systems, stomata, and cuticles that adapt them well for life on land.**

580 Part V Diversity of Life on Earth

Plant Life Cycles

All plants undergo mitosis after both gamete fusion and meiosis. The result is a multicellular haploid and a multicellular diploid individual, unlike the human life cycle, in which gamete fusion directly follows meiosis. Humans have a **diplontic** life cycle, meaning that only the diploid stage is multicellular, but the plant life cycle is **haplodiplontic,** having multicellular haploid and diploid stages. The basic haplodiplontic cycle is summarized in figure 29.3. Brown, red, and green algae are also haplodiplontic (see chapter 28). While humans produce gametes via meiosis, land plants actually produce gametes by mitosis in a multicellular, haploid individual. The diploid generation, or **sporophyte,** alternates with the haploid generation, or **gametophyte.** Sporophyte means "spore plant," and gametophyte means "gamete plant." These terms indicate the kinds of reproductive cells the respective generations produce.

The diploid sporophyte produces haploid spores (not gametes) by meiosis. Meiosis takes place in structures called **sporangia,** where diploid **spore mother cells (sporocytes)** undergo meiosis, each producing four haploid **spores.** Spores divide by mitosis, producing a multicellular, haploid gametophyte. Spores are the first cells of the gametophyte generation.

In turn, the haploid gametophyte is produced by mitosis and is the source of gametes. When the gametes fuse, the zygote they form is diploid and is the first cell of the next sporophyte generation. The zygote grows into a diploid sporophyte that produces sporangia in which meiosis ultimately occurs.

While all plants are haplodiplontic, the haploid generation consumes a much larger portion of the life cycle in mosses than in gymnosperms and angiosperms. In mosses, liverworts, and ferns, the gametophyte is photosynthetic and free-living; in other plants, it is usually nutritionally dependent on the sporophyte. When you look at moss, what you see is largely gametophyte tissue; the sporophytes are usually smaller, brownish or yellowish structures attached to the tissues of the gametophyte. In all vascular plants, the gametophytes are much smaller than the sporophytes. In seed plants, the gametophytes are nutritionally dependent on the sporophytes and are enclosed within their tissues. When you look at a gymnosperm or angiosperm, what you see, with rare exceptions, is a sporophyte.

While the sporophyte generation can get very large, the size of the gametophyte is limited in all plants. What we identify as a moss plant is a gametophyte, and it produces gametes at its tips. The egg is stationary, and sperm lands near the egg in a droplet of water. If the moss were the height of a sequoia, not only would it need vascular tissue for conduction and support, but the sperm would have to swim up the tree! In contrast, the fern gametophyte develops on the forest floor where gametes can meet. Fern trees are especially abundant in Australia; the haploid spores fall to the ground and develop into gametophytes.

Having completed our overview of plant life cycles, we will consider the major plant groups. As we do so, we will see a reduction of the gametophyte from group to group, a loss of multicellular **gametangia** (structures in which gametes are produced), and increasing specialization for life on land, including the remarkable structural adaptations of the flowering plants, the dominant plants today. Similar trends must have characterized the evolution of seed plants over the hundreds of millions of years since a freshwater alga first moved onto land.

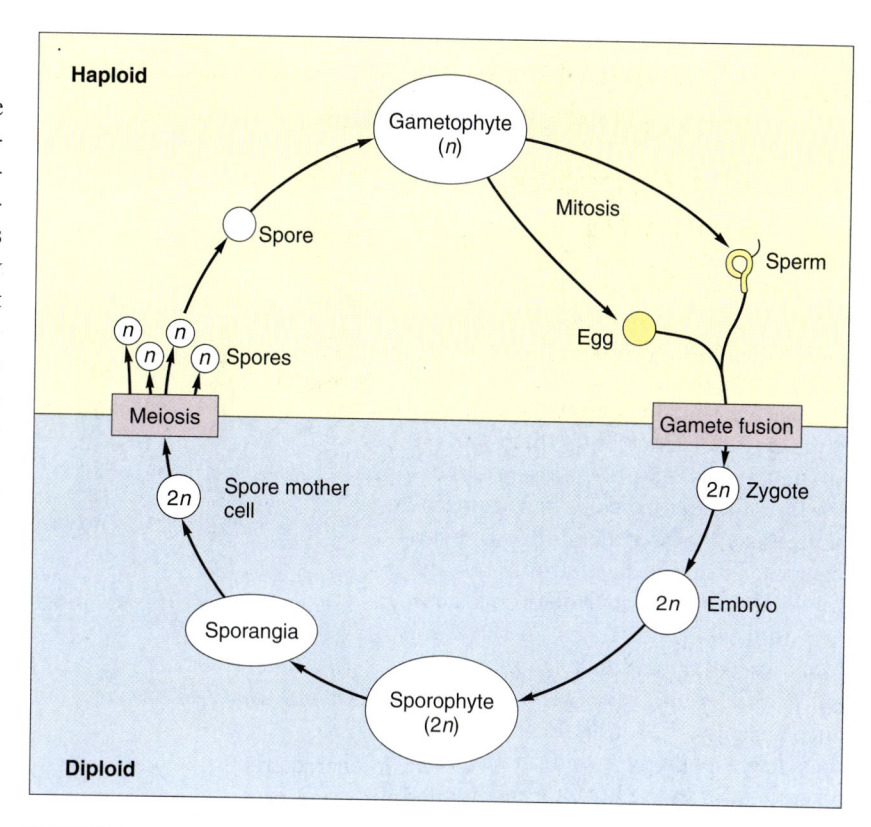

**FIGURE 29.3
A generalized plant life cycle.** Note that both haploid and diploid individuals can be multicellular. Also, spores are produced by meiosis, while gametes are produced by mitosis.

Plants have haplodiplontic life cycles. Diploid sporophytes produce haploid spores by meiosis. Spores develop into haploid gametophytes by mitosis and produce haploid gametes.

29.2 Nonvascular plants are relatively unspecialized, but successful in many terrestrial environments.

Mosses, Liverworts, and Hornworts

The approximately 24,700 species of **bryophytes** are simple but highly adapted to a diversity of terrestrial environments (even deserts!). Scientists now agree that bryophytes consist of three quite distinct phyla of relatively unspecialized plants—mosses, liverworts, and hornworts. Their gametophytes are photosynthetic. Sporophytes are attached to the gametophytes and depend on them nutritionally to varying degrees. Like ferns and certain other vascular plants, bryophytes require water (e.g., rainwater) to reproduce sexually. It is not surprising that they are especially common in moist places, both in the tropics and temperate regions.

Most bryophytes are small; few exceed 7 centimeters in height. The gametophytes are more conspicuous than the sporophytes. Some of the sporophytes are completely enclosed within gametophyte tissue; others are not and usually turn brownish or straw-colored at maturity.

Mosses (Bryophyta)

The gametophytes of mosses typically consist of small, leaflike structures (not true leaves, which contain vascular tissue) arranged spirally or alternately around a stemlike axis (figure 29.4); the axis is anchored to its substrate by means of **rhizoids.** Each rhizoid consists of several cells that absorb water, but not nearly the volume of water that is absorbed by a vascular plant root. Moss "leaves" have little in common with leaves of vascular plants, except for the superficial appearance of the green, flattened blade and slightly thickened midrib that runs lengthwise down the middle. Only one cell layer thick (except at the midrib), they lack vascular strands and stomata, and all the cells are haploid.

Water may rise up a strand of specialized cells in the center of a moss gametophyte axis. Some mosses also have specialized food-conducting cells surrounding those that conduct water.

Multicellular gametangia are formed at the tips of the leafy gametophytes (figure 29.5). Female gametangia (**archegonia**) may develop either on the same gametophyte as the male gametangia (**antheridia**) or on separate plants. A single egg is produced in the swollen lower part of an archegonium, while numerous sperm are produced in an antheridium. When sperm are released from an antheridium, they swim with the aid of flagella through a film of dew or rainwater to the archegonia. One sperm (which is haploid) unites with an egg (also haploid), forming a diploid zygote. The zygote divides by mitosis and develops into the sporophyte, a slender, basal stalk with a swollen capsule, the *sporangium*, at its tip. As the sporophyte develops, its base is embedded in gametophyte tissue, its nutritional source. The sporangium is often cylindrical or club-shaped. Spore mother cells within the sporangium undergo meiosis, each producing four haploid spores. In many mosses at maturity, the top of the sporangium pops off, and the spores are released. A spore that lands in a suitable damp location may germinate and grow into a threadlike structure, which branches to form rhizoids and "buds" that grow upright. Each bud develops into a new gametophyte plant consisting of a leafy axis.

FIGURE 29.4
A hair-cup moss, *Polytrichum* (phylum Bryophyta). The leaflike structures belong to the gametophyte. Each of the yellowish-brown stalks with a capsule, or sporangium, at its summit is a sporophyte.

In the Arctic and the Antarctic, mosses are the most abundant plants, boasting not only the largest number of individuals in these harsh regions. The greatest diversity of moss species, however, is found in the tropics. Many mosses are able to withstand prolonged periods of drought, although mosses are not common in deserts. Most are remarkably sensitive to air pollution and are rarely found in abundance in or near cities or other areas with high levels of air pollution. Some mosses, such as the peat mosses (*Sphagnum*), can absorb up to 25 times their weight in water and are valuable commercially as a soil conditioner or as a fuel when dry.

582 Part V Diversity of Life on Earth

Liverworts (Hepaticophyta)

The Old English word wyrt means "plant" or "herb." Some common liverworts have flattened gametophytes with lobes resembling those of liver—hence the name "liverwort." Although the lobed liverworts are the best-known representatives of this phylum, they constitute only about 20% of the species (figure 29.6). The other 80% are leafy and superficially resemble mosses. The gametophytes are prostrate instead of erect, and the rhizoids are one-celled.

Some liverworts have air chambers containing upright, branching rows of photosynthetic cells, each chamber having a pore at the top to facilitate gas exchange. Unlike stomata, the pores are fixed open and cannot close.

Sexual reproduction in liverworts is similar to that in mosses. Lobed liverworts may form gametangia in umbrella-like structures. Asexual reproduction occurs when lens-shaped pieces of tissue that are released from the gametophyte grow to form new gametophytes.

Hornworts (Anthocerotophyta)

The origin of hornworts is a puzzle. They are most likely among the earliest land plants, yet the earliest hornwort fossil spores date from the Cretaceous period, 65 to 145 million years ago, when angiosperms were emerging.

The small hornwort sporophytes resemble tiny green broom handles rising from filmy gametophytes usually less than 2 centimeters in diameter (figure 29.7). The sporophyte base is embedded in gametophyte tissue, from which it derives some of its nutrition. However, the sporophyte has stomata, is photosynthetic, and provides much of the energy needed for growth and reproduction. Hornwort cells usually have a single chloroplast.

> **The three major phyla of nonvascular plants are all relatively unspecialized, but well suited for diverse terrestrial environments.**

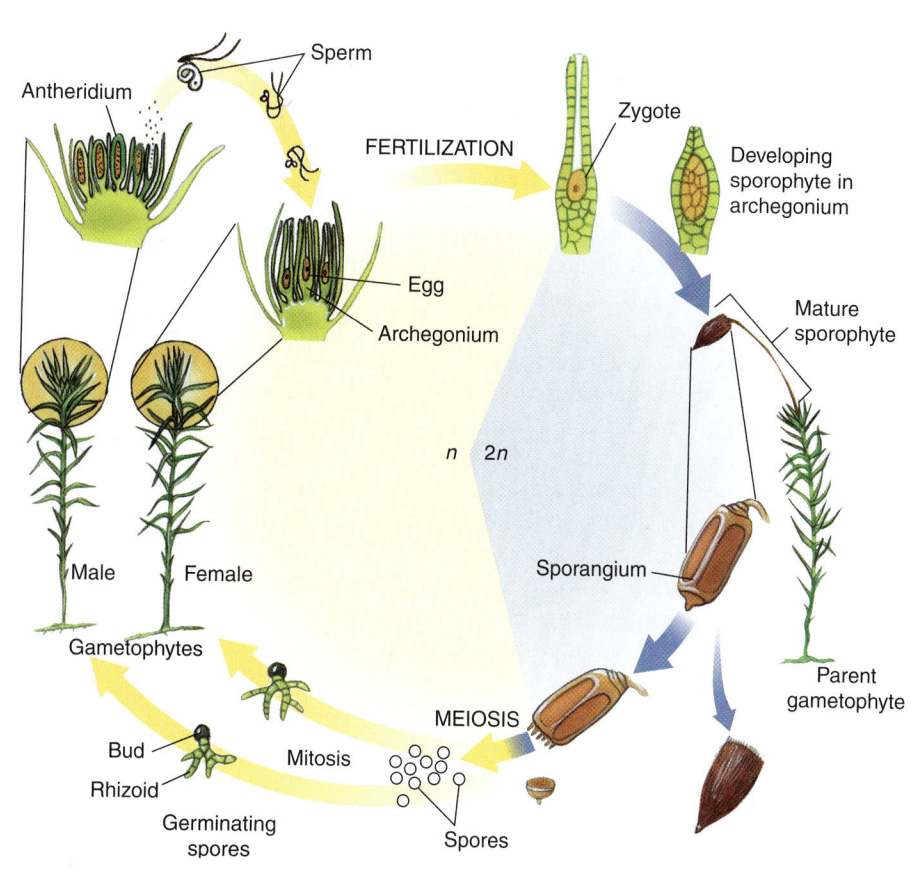

FIGURE 29.5
Life cycle of a typical moss. The majority of the life cycle of a moss is in the haploid state. The leafy gametophyte is photosynthetic, while the smaller sporophyte is not, and is nutritionally dependent on the gametophyte. Water is required to carry sperm to the egg.

FIGURE 29.6
A common liverwort, *Marchantia* **(phylum Hepaticophyta).** The sporophytes are formed by fertilization within the tissues of the umbrella-shaped structures that arise from the surface of the flat, green, creeping gametophyte.

FIGURE 29.7
Hornworts (phylum Anthocerotophyta). Hornwort sporophytes are seen in this photo. Unlike the sporophytes of other bryophytes, most hornwort sporophytes are photosynthetic.

29.3 Seedless vascular plants have well-developed conducting tissues in their sporophytes.

Features of Vascular Plants

The first vascular plants for which we have a relatively complete record belonged to the phylum Rhyniophyta. They flourished some 410 million years ago but are now extinct. We are not certain what the very earliest of these vascular plants looked like, but fossils of *Cooksonia* provide some insight into their characteristics (figure 29.8). *Cooksonia*, the first known vascular land plant, appeared in the late Silurian period about 420 million years ago. It was successful partly because it encountered little competition as it spread out over vast tracts of land. The plants were only a few centimeters tall and had no roots or leaves. They consisted of little more than a branching axis, the branches forking evenly and expanding slightly toward the tips. They were **homosporous** (producing only one type of spore). Sporangia formed at branch tips. Other ancient vascular plants that followed evolved more complex arrangements of sporangia. Leaves began to appear as protuberances from stems.

Cooksonia and the other early plants that followed it became successful colonizers of the land by developing efficient water- and food-conducting systems known as **vascular tissues** (Latin *vasculum*, "vessel" or "duct"). These tissues consist of strands of specialized cylindrical or elongated cells that form a network throughout a plant, extending from near the tips of the roots, through the stems, and into true leaves. One type of vascular tissue, *xylem*, conducts water and dissolved minerals upward from the roots; another type of tissue, *phloem*, conducts sucrose and hormonal signals throughout the plant. It is important to note that vascular tissue develops in the sporophyte, but (with few exceptions) not in the gametophyte. (See the discussion of vascular tissue structure in chapter 35.) The presence of a cuticle and stomata are also characteristic of vascular plants.

Three clades of vascular plants exist today: (1) lycophytes (club mosses), (2) pterophytes (ferns and their relatives), and (3) seed plants. Advances in molecular systematics have changed the way we view the evolutionary history of vascular plants. Whisk ferns and horsetails were long believed to be distinct phyla that were transitional between bryophytes and vascular plants. Phylogenetic evidence now shows they are the closest living relatives to ferns.

The seven living phyla of vascular plants (table 29.1) dominate terrestrial habitats everywhere, except for the highest mountains and the tundra. The haplodiplontic life cycle persists, but the gametophyte has been reduced during the evolution of some phyla. A similar reduction in multicellular gametangia has occurred.

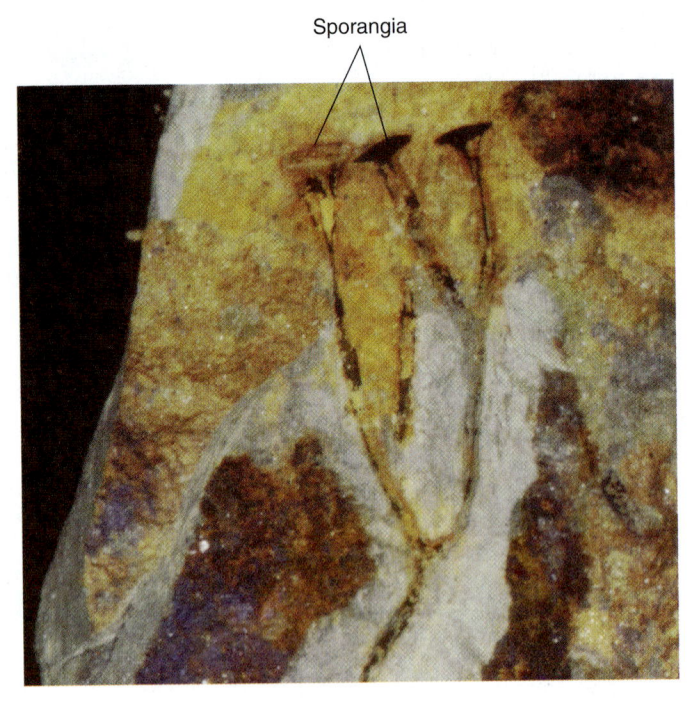

FIGURE 29.8
Cooksonia, **the first known vascular land plant.** This fossil represents a plant that lived some 410 million years ago. *Cooksonia* belongs to phylum Rhyniophyta, consisting entirely of extinct plants. Its upright, branched stems, which were no more than a few centimeters tall, terminated in sporangia, as seen here. It probably lived in moist environments such as mudflats, had a resistant cuticle, and produced spores typical of vascular plants.

Accompanying this reduction in size and complexity of the gametophytes has been the appearance of the seed. Seeds are highly resistant structures well suited to protect a plant embryo from drought and to some extent from predators. In addition, almost all seeds contain a supply of food for the young plant. Seeds occur only in **heterosporous** plants (plants that produce two types of spores). Heterospory is believed to have arisen multiple times in the plants. Fruits in the flowering plants add a layer of protection to seeds and attract animals that assist in seed dispersal, expanding the potential range of the species. Flowers, which evolved among the angiosperms, attract pollinators. Flowers allow plants to secure the benefits of wide outcrossing in promoting genetic diversity.

> Most vascular plants have well-developed conducting tissues, specialized stems, leaves, roots, cuticles, and stomata. Many have seeds, which protect embryos until conditions are suitable for further development.

Table 29.1 The Seven Phyla of Extant Vascular Plants

Phylum	Examples	Key Characteristics	Approximate Number of Living Species
SEED PLANTS			
Anthophyta	Flowering plants (angiosperms)	Heterosporous. Sperm not motile; conducted to egg by a pollen tube. Seeds enclosed within a fruit. Leaves greatly varied in size and form. Herbs, vines, shrubs, trees. About 14,000 genera.	250,000
Coniferophyta	Conifers (including pines, spruces, firs, yews, redwoods, and others)	Heterosporous seed plants. Sperm not motile; conducted to egg by a pollen tube. Leaves mostly needlelike or scalelike. Trees, shrubs. About 50 genera.	601
Cycadophyta	Cycads	Heterosporous. Sperm flagellated and motile but confined within a pollen tube that grows to the vicinity of the egg. Palmlike plants with pinnate leaves. Secondary growth slow compared to that of the conifers. 10 genera. Seeds in cones.	206
Gnetophyta	Gnetophytes	Heterosporous. Sperm not motile; conducted to egg by a pollen tube. The only gymnosperms with vessels. Trees, shrubs, vines. Three very diverse genera (*Ephedra*, *Gnetum*, *Welwitschia*).	65
Ginkgophyta	*Ginkgo*	Heterosporous. Sperm flagellated and motile but conducted to the vicinity of the egg by a pollen tube. Deciduous tree with fan-shaped leaves that have evenly forking veins. Seeds resemble a small plum with fleshy, ill-scented outer covering. One genus.	1
SEEDLESS VASCULAR PLANTS			
Pterophyta	Ferns	Primarily homosporous (a few heterosporous). Sperm motile. External water necessary for fertilization. Leaves are megaphylls that uncoil as they mature. Sporophytes and virtually all gametophytes are photosynthetic. About 365 genera.	11,000
	Horsetails	Homosporous. Sperm motile. External water necessary for fertilization. Stems ribbed, jointed, either photosynthetic or nonphotosynthetic. Leaves scalelike, in whorls; nonphotosynthetic at maturity. One genus.	15
	Whisk ferns	Homosporous. Sperm motile. External water necessary for fertilization. No differentiation between root and shoot. No leaves; one of the two genera has scalelike extensions and the other leaflike appendages.	6
Lycophyta	Club mosses	Homosporous or heterosporous. Sperm motile. External water necessary for fertilization. Leaves are microphylls. About 12–13 genera.	1,150

Seedless Vascular Plants

The earliest vascular plants lacked seeds. Members of four phyla of living vascular plants lack seeds, as do at least three other phyla known only from fossils. As we explore the adaptations of the vascular plants, we focus on both reproductive strategies and the advantages of increasingly complex transport systems.

Club Mosses (Lycophyta)

The club mosses are relicts of an ancient past when vascular plants first evolved (figure 29.9). They are the sister group to all vascular plants. Several genera of club mosses, some of them treelike, became extinct about 270 million years ago. Today, club mosses are worldwide in distribution but are most abundant in the tropics and moist temperate regions. Members of the 12–13 genera and about 1150 living species of club mosses superficially resemble true mosses, but once their internal structure and reproductive processes became known, it was clear that these vascular plants are quite unrelated to mosses. Modern club mosses are either homosporous or heterosporous. The sporophytes have leafy stems that are seldom more than 30 centimeters long.

FIGURE 29.9
A club moss. *Lycopodium clavatum*, a club moss, is representative of the vascular plant phylum most closely related to the nonvascular bryophytes.

Whisk Ferns, Horsetails, and Ferns (Pterophyta)

Whisk ferns and horsetails are close relatives of ferns. They all form antheridia and archegonia. Free water is required for the process of fertilization, during which the sperm, which have flagella, swim to and unite with the eggs. In contrast, most seed plants have nonflagellated sperm.

Whisk Ferns. Whisk ferns, which occur in the tropics and subtropics, consist merely of evenly forking green stems without roots (figure 29.10). The two or three species of the genus *Psilotum* do, however, have tiny, green, spirally arranged flaps of tissue lacking veins and stomata. Another genus, *Tmespiteris*, has more leaflike appendages. Given the simple structure of whisk ferns, it was particularly surprising to learn that they are monophyletic with ferns. The gametophytes of whisk ferns are essentially colorless and are less than 2 millimeters in diameter, but they can be up to 18 millimeters long. They form parasitic associations with fungi, which furnish their nutrients. Some develop elements of vascular tissue and have the distinction of being the only gametophytes known to do so.

FIGURE 29.10
A whisk fern. Whisk ferns have no roots or leaves.

Horsetails. The 15 living species of horsetails are all homosporous. They constitute a single genus, *Equisetum*. Fossil forms of *Equisetum* extend back 300 million years to an era when some of their relatives were treelike. Today, they are widely scattered around the world, mostly in damp places. Some that grow among the coastal redwoods of California may reach a height of 3 meters, but most are less than a meter tall (figure 29.11).

Horsetail sporophytes consist of ribbed, jointed, photosynthetic stems that arise from branching underground rhizomes with roots at their nodes. A whorl of nonphotosynthetic, scalelike leaves emerges at each node. The stems, which are hollow in the center, have silica deposits in the epidermal cells of the ribs, and the interior parts of

586   Part V   Diversity of Life on Earth

the stems have two sets of vertical, tubular canals. The larger outer canals, which alternate with the ribs, contain air, while the smaller inner canals opposite the ribs contain water. Horsetails are also called scouring rush because pioneers used them to scrub pans.

Ferns. Ferns are the most abundant group of seedless vascular plants, with about 11,000 living species. Recent research indicates that they may be the closest relatives to the seed plants. The fossil record indicates that ferns originated during the Devonian period about 350 million years ago and became abundant and varied in form during the next 50 million years. Their apparent ancestors had no broad leaves and were established on land as much as 375 million years ago. Today, ferns flourish in a wide range of habitats throughout the world; however, about 75% of the species occur in the tropics.

The conspicuous sporophytes may be less than a centimeter in diameter—as seen in small aquatic ferns such as *Azolla*—or more than 24 meters tall, with leaves up to 5 meters or longer in the tree ferns (figure 29.12). The sporophytes and the smaller gametophytes, which rarely reach 6 millimeters in diameter, are both photosynthetic. The fern life cycle differs from that of a moss primarily in the much greater development, independence, and dominance of the fern's sporophyte. The fern sporophyte is structurally more complex than the moss sporophyte; the fern sporophyte has vascular tissue and well-differentiated roots, stems, and leaves. The gametophyte, however, lacks vascular tissue.

Fern sporophytes typically have a horizontal underground stem called a *rhizome*, with roots emerging from the sides. The leaves, referred to as *fronds*, usually develop at the tip of the rhizome as tightly rolled-up coils ("fiddleheads") that unroll and expand. Many fronds are highly dissected and feathery, making the ferns that produce them prized as ornamentals. Some ferns, such as *Marsilea*, have fronds that resemble a four-leaf clover, but *Marsilea* fronds still begin as coiled fiddleheads. Other ferns produce a mixture of photosynthetic fronds and nonphotosynthetic reproductive fronds that tend to be brownish in color.

Most ferns are homosporous, producing distinctive *sporangia*, usually in clusters called *sori*, typically on the underside of the fronds. Sori are often protected during their development by a transparent, umbrella-like covering. (At first glance, one might mistake the sori for an infection on the plant.) Diploid *spore mother cells* in each sporangium undergo meiosis, producing haploid spores. At maturity, the spores are catapulted from the sporangium by a snapping action, and those that land in suitable damp locations may germinate, producing gametophytes that are often heart-shaped, are only one cell layer thick (except in the center), and have rhizoids that anchor them to their substrate. These rhizoids are not true roots because they lack vascular tissue, but they do aid in transporting water and nutrients from the soil. Flask-shaped

**FIGURE 29.11
A horsetail, *Equisetum telmateia*.** This species forms two kinds of erect stems; one is green and photosynthetic, and the other, which terminates in a spore-producing "cone," is mostly light brown.

**FIGURE 29.12
A tree fern (phylum Pterophyta) in the forests of Malaysia.** The ferns are by far the largest group of seedless vascular plants.

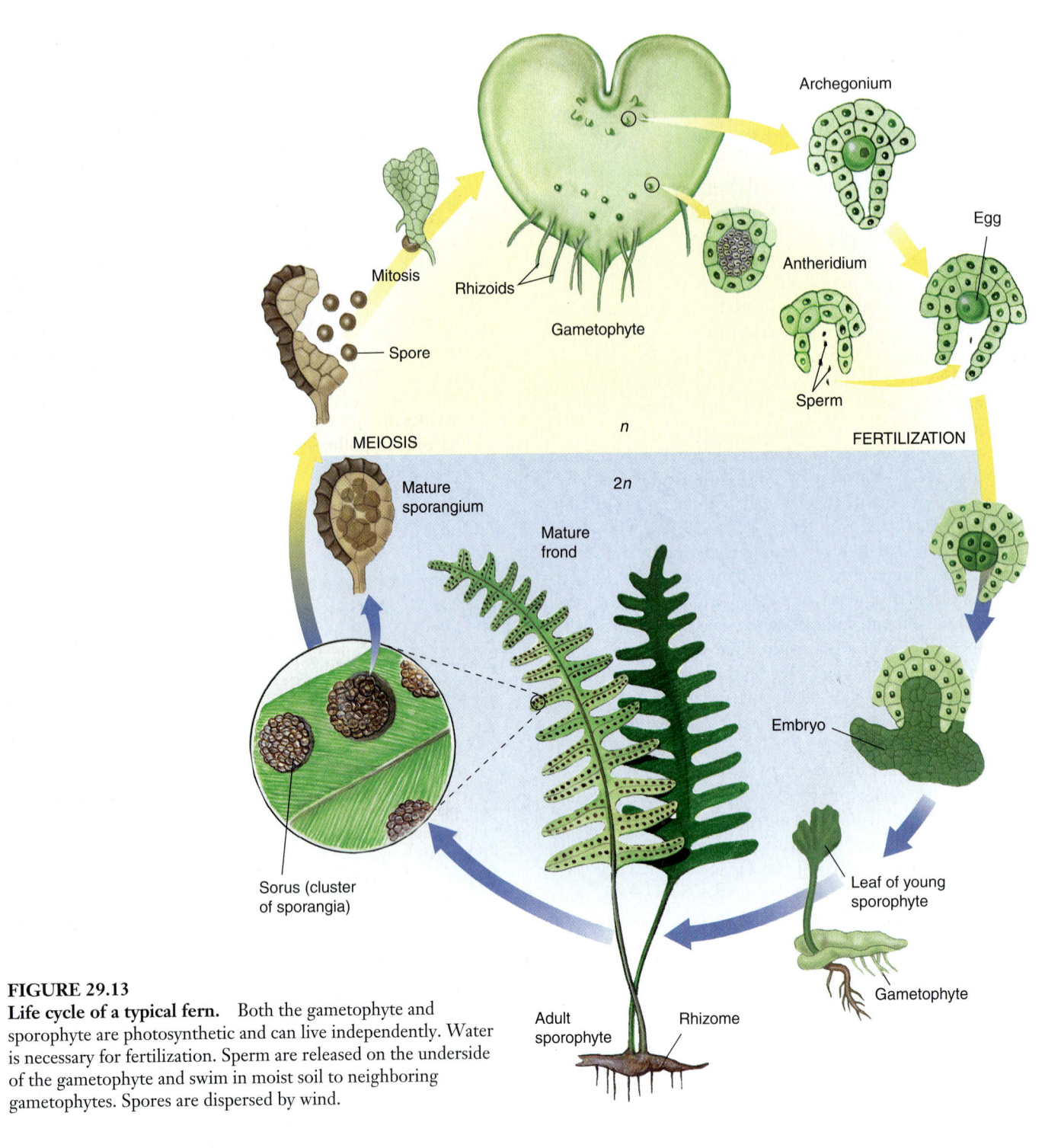

FIGURE 29.13
Life cycle of a typical fern. Both the gametophyte and sporophyte are photosynthetic and can live independently. Water is necessary for fertilization. Sperm are released on the underside of the gametophyte and swim in moist soil to neighboring gametophytes. Spores are dispersed by wind.

archegonia and globular *antheridia* are produced on either the same or a different gametophyte.

The sperm formed in the antheridia have flagella, with which they swim toward the archegonia when water is present, often in response to a chemical signal secreted by the archegonia. One sperm unites with the single egg toward the base of an archegonium, forming a *zygote*. The zygote then develops into a new sporophyte, completing the life cycle (figure 29.13). Multicellular gametangia still develop. As discussed earlier, the shift to a dominant sporophyte generation allows ferns to achieve significant height without interfering with sperm swimming to the egg. The multicellular archegonia provide some protection for the developing embryo.

> The two seedless vascular plant phyla have a large and conspicuous sporophyte, with vascular tissue. Many have well-differentiated roots, stems, and leaves. The shift to a dominant sporophyte led to the evolution of trees.

29.4 Seeds protect and aid in the dispersal of plant embryos.

Seed Plants

Seed plants first appeared about 425 million years ago. Their ancestors appear to have been spore-bearing plants known as progymnosperms. Progymnosperms shared several features with modern gymnosperms, including secondary xylem and phloem (which allows for an increase in girth later in development). Some progymnosperms had leaves. Their reproduction was very simple, and it is not certain which particular group of progymnosperms gave rise to seed plants.

From an evolutionary and ecological perspective, the seed represents an important advance. The embryo is protected by an extra layer of sporophyte tissue, creating the ovule. During development, this tissue hardens to produce the seed coat. In addition to protecting the embryo from drought, the seed can be easily dispersed. Perhaps even more significantly, the presence of seeds introduces into the life cycle a dormant phase that allows the embryo to survive until environmental conditions are favorable for further growth.

Seed plants produce two kinds of gametophytes—male and female, each of which consists of just a few cells. Pollen grains, multicellular male gametophytes, are conveyed to the egg in the female gametophyte by wind or a pollinator. In some seed plants, the sperm move toward the egg through a growing pollen tube. This eliminates the need for external water. In contrast to the seedless plants, the whole male gametophyte, rather than just the sperm, moves to the female gametophyte. A female gametophyte develops within an ovule. In flowering plants (angiosperms), the ovules are completely enclosed within diploid sporophyte tissue (ovaries that develop into the fruit). In gymnosperms (mostly cone-bearing seed plants), the ovules are not completely enclosed by sporophyte tissue at the time of pollination.

> A common ancestor that had seeds gave rise to the gymnosperms and the angiosperms. Seeds can allow for a pause in the life cycle until environmental conditions are more optimal.

A Vocabulary of Plant Terms

androecium The stamens of a flower.

anther The pollen-producing portion of a stamen. This is a sporophyte structure where microspores produced by meiosis develop into male gametophytes.

antheridium The sperm-producing structure found in the gametophytes of seedless plants and certain fungi.

archegonium The multicellular egg-producing structure in the gametophytes of seedless plants and gymnosperms.

carpel A leaflike organ in angiosperms that encloses one or more ovules; a unit of a gynoecium.

double fertilization The process in which one sperm fuses with the egg, forming a zygote, and the other sperm fuses with the two polar nuclei, forming the primary endosperm nucleus in angiosperms.

embryo sac The female gametophyte in flowering plants.

endosperm The usually triploid (although it can have a much higher ploidy level) food supply of some angiosperm seeds.

filament The stalklike structure that supports the anther of a stamen.

gametophyte The multicellular, haploid phase of a plant life cycle, in which gametes are produced by mitosis.

gynoecium The carpel(s) of a flower.

heterosporous Refers to a plant that produces two types of spores: microspores and megaspores.

homosporous Refers to a plant that produces only one type of spore.

integument The outer layer(s) of an ovule; integuments become the seed coat of a seed.

micropyle The opening in the ovule integument through which the pollen tube grows.

nucellus The tissue of an ovule in which an embryo sac develops.

ovary The basal, swollen portion of a carpel; it contains the ovules and develops into the fruit.

ovule A seed plant structure within an ovary; it contains a female gametophyte surrounded by the nucellus and one or two integuments. At maturity, an ovule becomes a seed.

pollen grain Male gametophyte in seed plants; binucleate or trinucleate seed plant structure produced from a microspore in a microsporangium.

pollination The transfer of a pollen grain from an anther to a stigma in angiosperms, or to the vicinity of the ovule in gymnosperms.

primary endosperm nucleus The triploid nucleus resulting from the fusion of a single sperm with the two polar nuclei.

seed A reproductive structure that develops from an ovule in seed plants. It consists of an embryo and a food supply surrounded by a seed coat.

seed coat The protective layer of a seed; it develops from the integument(s).

spore A haploid reproductive cell, produced when a diploid spore mother cell undergoes meiosis; it gives rise by mitosis to a gametophyte.

sporophyte The multicellular, diploid phase of a plant life cycle; it is the generation that ultimately produces spores.

stamen A unit of an androecium; it consists of a pollen-bearing anther and usually a stalklike filament.

stigma The uppermost pollen-receptive portion of a gynoecium.

Gymnosperms

There are four groups of living gymnosperms (conifers, cycads, gnetophytes, and *Ginkgo*), all of which lack the flowers and fruits of angiosperms. In all of them, the **ovule**, which becomes a seed, rests exposed on a scale (modified shoot or leaf) and is not completely enclosed by sporophyte tissues at the time of pollination. The name gymnosperm combines the Greek root *gymnos*, or "naked," with *sperma*, or "seed." In other words, gymnosperms are naked-seeded plants. However, although the ovules are naked at the time of pollination, the seeds of gymnosperms are sometimes enclosed by other sporophyte tissues by the time they are mature.

Details of reproduction vary somewhat in gymnosperms, and their forms vary greatly. For example, cycads and *Ginkgo* have motile sperm, even though the sperm are carried within a pollen tube, while conifers and gnetophytes have sperm with no flagella. The female cones range from tiny, woody structures weighing less than 25 grams and having a diameter of a few millimeters, to massive structures weighing more than 45 kilograms and growing to lengths of more than a meter.

Conifers (Coniferophyta)

The most familiar gymnosperms are conifers (phylum Coniferophyta), which include pines (figure 29.14), spruces, firs, cedars, hemlocks, yews, larches, cypresses, and others. The coastal redwood (*Sequoia sempervirens*), a conifer native to northwestern California and southwestern Oregon, is the tallest living vascular plant; it may attain nearly 100 meters (300 feet) in height. Another conifer, the bristlecone pine (*Pinus longaeva*) of the White Mountains of California, is the oldest living tree; one specimen is 4900 years of age. Conifers are found in the colder temperate and sometimes drier regions of the world. They are sources of timber, paper, resin, taxol (used to treat cancer), and other economically important products.

Pines. More than 100 species of pines exist today, all native to the northern hemisphere, although the range of one species does extend a little south of the equator. Pines and spruces are members of the vast coniferous forests that lie between the arctic tundra and the temperate deciduous forests and prairies to their south. During the past century, pines have been extensively planted in the southern hemisphere.

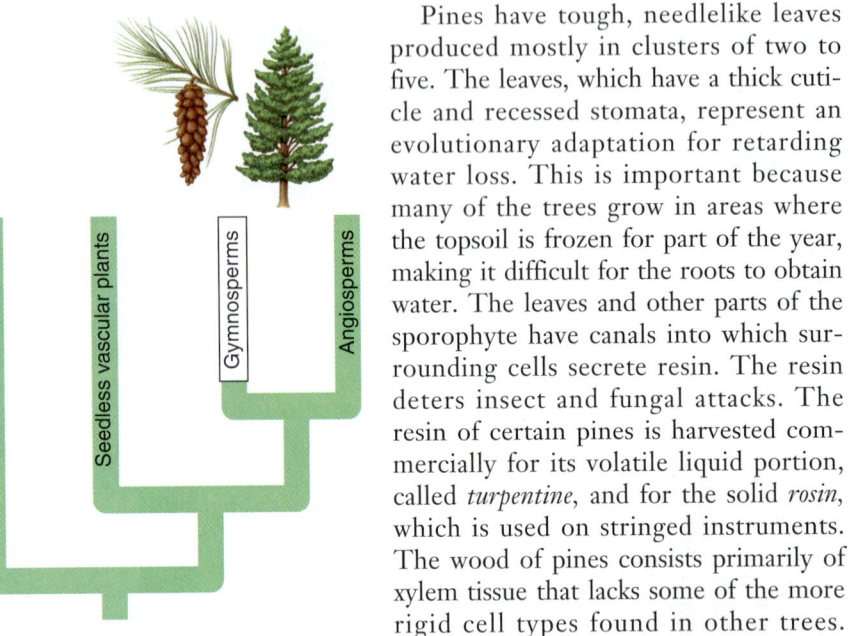

Pines have tough, needlelike leaves produced mostly in clusters of two to five. The leaves, which have a thick cuticle and recessed stomata, represent an evolutionary adaptation for retarding water loss. This is important because many of the trees grow in areas where the topsoil is frozen for part of the year, making it difficult for the roots to obtain water. The leaves and other parts of the sporophyte have canals into which surrounding cells secrete resin. The resin deters insect and fungal attacks. The resin of certain pines is harvested commercially for its volatile liquid portion, called *turpentine*, and for the solid *rosin*, which is used on stringed instruments. The wood of pines consists primarily of xylem tissue that lacks some of the more rigid cell types found in other trees. Thus, it is considered a "soft" rather than a "hard" wood. The thick bark of pines represents another adaptation for surviving fires and subzero temperatures. Some cones actually depend on fire to open them, releasing seed to reforest burnt areas.

FIGURE 29.14
Conifers. Slash pines, *Pinus palustris*, in Florida are representative of the Coniferophyta, the largest phylum of gymnosperms.

As mentioned earlier, all seed plants are heterosporous, so the spores give rise to two types of gametophytes (figure 29.15). The male gametophytes (pollen grains) of pines develop from microspores, which are produced in male cones that develop in clusters of 30 to 70, typically at the tips of the lower branches; there may be hundreds of such clusters on any single tree. The male cones generally are 1 to 4 centimeters long and consist of small, papery scales arranged in a spiral or in whorls. A pair of microsporangia form as sacs within each scale. Numerous microspore mother cells in the microsporangia undergo meiosis, each becoming four microspores. The microspores develop into four-celled pollen grains with a pair of air sacs that give them added buoyancy when released into the air. A single cluster of male pine cones may produce more than 1 million pollen grains.

Female cones typically are produced on the upper branches of the same tree that produces male cones. Female cones are larger than male cones, and their scales become woody. Two ovules develop toward the base of each scale. Each ovule contains a megasporangium called the **nucellus**. The nucellus itself is completely surrounded by a thick layer of cells called the integument that has a small opening (the **micropyle**) toward one end. One of the layers of the integument later becomes the seed coat. A single megaspore mother cell within each megasporangium undergoes meiosis, becoming a row of four megaspores. Three of the megaspores break down, but the remaining one, over the better part of a year, slowly develops into a female gametophyte. The female gametophyte at maturity may consist of thousands of cells, with two to six archegonia formed at the micropylar end. Each archegonium contains an egg so large it can be seen without a microscope.

Female cones usually take two or more seasons to mature. At first they may be reddish or purplish in color, but they soon turn green, and during the first spring, the scales spread apart. While the scales are open, pollen grains carried by the wind drift down between them, some catching in sticky fluid oozing out of the micropyle. The pollen grains within the sticky fluid are slowly drawn down through the micropyle to the top of the nucellus, and the scales close shortly thereafter. The archegonia and the rest of the female gametophyte are not mature until about a year later. While the female gametophyte is developing, a pollen tube emerges from a pollen grain at the bottom of the micropyle and slowly digests its way through the nucellus to the archegonia. While the pollen tube is growing, one of the pollen grain's four cells, the generative cell, divides by mitosis, with one of the resulting two cells dividing once more. These last two cells function as sperm. The germinated pollen grain with its two sperm is the mature male gametophyte.

About 15 months after pollination, the pollen tube reaches an archegonium, and discharges its contents into it. One sperm unites with the egg, forming a zygote. The other sperm and cells of the pollen grain degenerate. The zygote develops into an embryo within a seed. After dispersal and germination of the seed, the young sporophyte of the next generation develops into a tree.

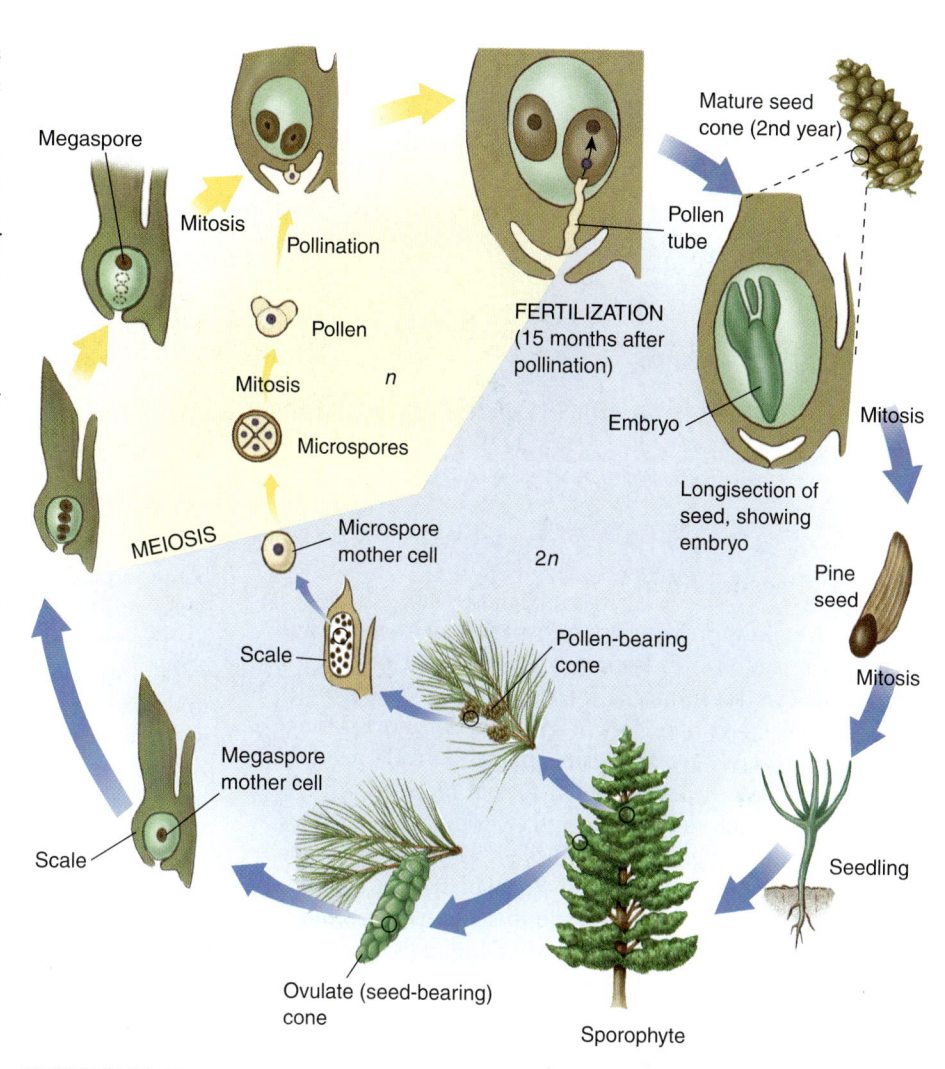

FIGURE 29.15
Life cycle of a typical pine. The male and female gametophytes are dramatically reduced in size in these plants. Wind generally disperses the male gametophyte (pollen), which produces sperm. Pollen tube growth delivers the sperm to the egg on the female cone. Additional protection for the embryo is provided by the integument, which develops into the seed coat.

Chapter 29 Overview of Plant Diversity 591

FIGURE 29.16 Three phyla of gymnosperms. (*a*) An African cycad, *Encephalartos transvenosus*. (*b*) *Welwitschia mirabilis* represents one of the three genera of gnetophytes. (*c*) Maidenhair tree, *Ginkgo biloba*, the only living representative of the phylum Ginkgophyta.

Cycads (Cycadophyta)

Cycads are slow-growing gymnosperms of tropical and subtropical regions. The sporophytes of most of the 100 known species resemble palm trees (figure 29.16*a*) with trunks that can attain heights of 15 meters or more. Unlike palm trees, which are flowering plants, cycads produce cones and have a life cycle similar to that of pines. The female cones, which develop upright among the leaf bases, are huge in some species and can weigh up to 45 kilograms. The sperm of cycads, although formed within a pollen tube, are released within the ovule to swim to an archegonium. These sperm are the largest sperm cells among all living organisms. Several species of cycads are facing extinction in the wild and soon may exist only in botanical gardens.

Gnetophytes (Gnetophyta)

There are three genera and about 65 living species of Gnetophyta. They are the only gymnosperms with vessels in their xylem, a particularly efficient conducting cell type that is a common feature in angiosperms. The members of the three genera differ greatly from one another in form. One of the most bizarre of all plants is *Welwitschia*, which occurs in the Namib and Mossamedes deserts of southwestern Africa (figure 29.16*b*). The stem is shaped like a large, shallow cup that tapers into a taproot below the surface. It has two strap-shaped, leathery leaves that grow continuously from their base, splitting as they flap in the wind. The reproductive structures of *Welwitschia* are conelike, appear toward the bases of the leaves around the rims of the stems, and are produced on separate male and female plants.

More than half of the gnetophyte species are in the genus *Ephedra*, which is common in arid regions of the western United States and Mexico. Species are found on every continent except Australia. The plants are shrubby, with stems that superficially resemble those of horsetails, being jointed and having tiny, scalelike leaves at each node. Male and female reproductive structures may be produced on the same or different plants. The drug ephedrine, widely used in the treatment of respiratory problems, was in the past extracted from Chinese species of *Ephedra*, but it has now been largely replaced with synthetic preparations. Ephedrine found in herbal remedies for weight loss has been linked to strokes and heart attacks.

The best-known species of *Gnetum* is a tropical tree, but most species are vinelike. All species have broad leaves similar to those of angiosperms. One *Gnetum* species is cultivated in Java for its tender shoots, which are cooked as a vegetable.

Ginkgo (Ginkgophyta)

The fossil record indicates that members of the Ginkgophyta were once widely distributed, particularly in the northern hemisphere; today, only one living species, the maidenhair tree (*Ginkgo biloba*), remains. The tree, which sheds its leaves in the fall, was first encountered by Europeans in cultivation in Japan and China; it apparently no longer exists in the wild. The common name comes from the resemblance of its fan-shaped leaves to the leaflets of maidenhair ferns (figure 29.16*c*). Like the sperm of cycads, those of *Ginkgo* have flagella. The ginkgo is dioecious—that is, the male and female reproductive structures are produced on separate trees. The fleshy outer coverings of the seeds of female ginkgo plants exude the foul smell of rancid butter, caused by the presence of butyric and isobutyric acids. As a result, male plants vegetatively propagated from shoots are preferred for cultivation. Because of its beauty and resistance to air pollution, *Ginkgo* is commonly planted along city streets.

Gymnosperms are mostly cone-bearing seed plants. In gymnosperms, the ovules are not completely enclosed by sporophyte tissue at pollination.

Angiosperms

The 250,000 known species of flowering plants are called angiosperms because their ovules, unlike those of gymnosperms, are enclosed within diploid tissues at the time of pollination. The name *angiosperm* derives from the Greek words *angeion*, "vessel," and *sperma*, "seed." The "vessel" in this instance refers to the carpel, which is a modified leaf that encapsulates seeds. The carpel develops into the fruit, a unique angiosperm feature. While some gymnosperms, including yew, have fleshlike tissue around their seeds, it is of a different origin and not a true fruit.

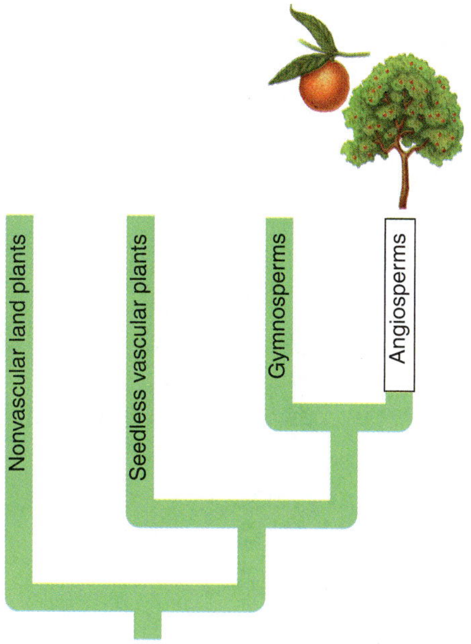

Angiosperm Origins

The origins of the angiosperms puzzled even Darwin (his "abominable mystery"). Recent fossil and molecular sequence data have provided exciting clues about basal angiosperms. In the remote Liaoning province of China, a complete angiosperm fossil that is at least 125 million years old has been found (figure 29.17). The fossil may represents a new, basal, and extinct angiosperm family, Archaefructaceae, with two species, *Archaefructus liaoningensis* and *A. sinensis*. *Archaefructus* was an herbaceous, aquatic plant. This family is proposed to be the sister clade to all other angiosperms but there is a lively debate about the validity of this claim. How do we know that *Archaefructus* is an angiosperm? The fossils have both male and female reproductive structures. They lack the sepals and petals that evolved in later angiosperms and attract pollinators. The fossils were so well preserved that fossil pollen could be examined using scanning electron microscopy. Although *Archaefructus* is ancient, it is unlikely to be the very first angiosperm. Still, the incredibly well-preserved fossils provide valuable detail on angiosperms in the Upper Jurassic/Lower Cretaceous period, when dinosaurs roamed the earth.

Consensus has also been growing on the most basal living angiosperm—*Amborella trichopoda* (figure 29.18). *Amborella*, with small, cream-colored flowers, is even more primitive than magnolias or water lilies. This small shrub, found only on the island of New Caledonia in the South Pacific, is the last remaining species of the earliest extant lineage of the angiosperms, arising about 135 million years ago. While *Amborella* is not the original angiosperm, it is sufficiently close that studying its reproductive biology may help us understand the early radiation of the angiosperms. The angiosperm phylogeny reflects an evolutionary hypothesis that is driving new research on angiosperm origins (figure 29.19).

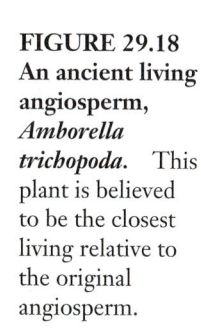

FIGURE 29.17
Fossil of basal angiosperm. *Archaefructus* fossil with multiseeded carpels (fruits) and stamens. This is the oldest known angiosperm in the fossil record.

FIGURE 29.18
An ancient living angiosperm, *Amborella trichopoda*. This plant is believed to be the closest living relative to the original angiosperm.

Monocots and Eudicots

Biologists have long grouped angiosperms into two classes, dicots and monocots. Although phylogenetic evidence indicates that much evolutionary diversity is masked by such grouping (see figure 29.19). More meaningful comparisons are made among eudicots and the lineage that gave rise to monocots and magnolias.

Eudicots (about 175,000 species) are the more primitive of the two classes, with monocots apparently having derived from early dicots. Included in the dicots are the great majority of familiar angiosperms—almost all kinds of trees and shrubs, snapdragons, mints, peas, sunflowers, and other

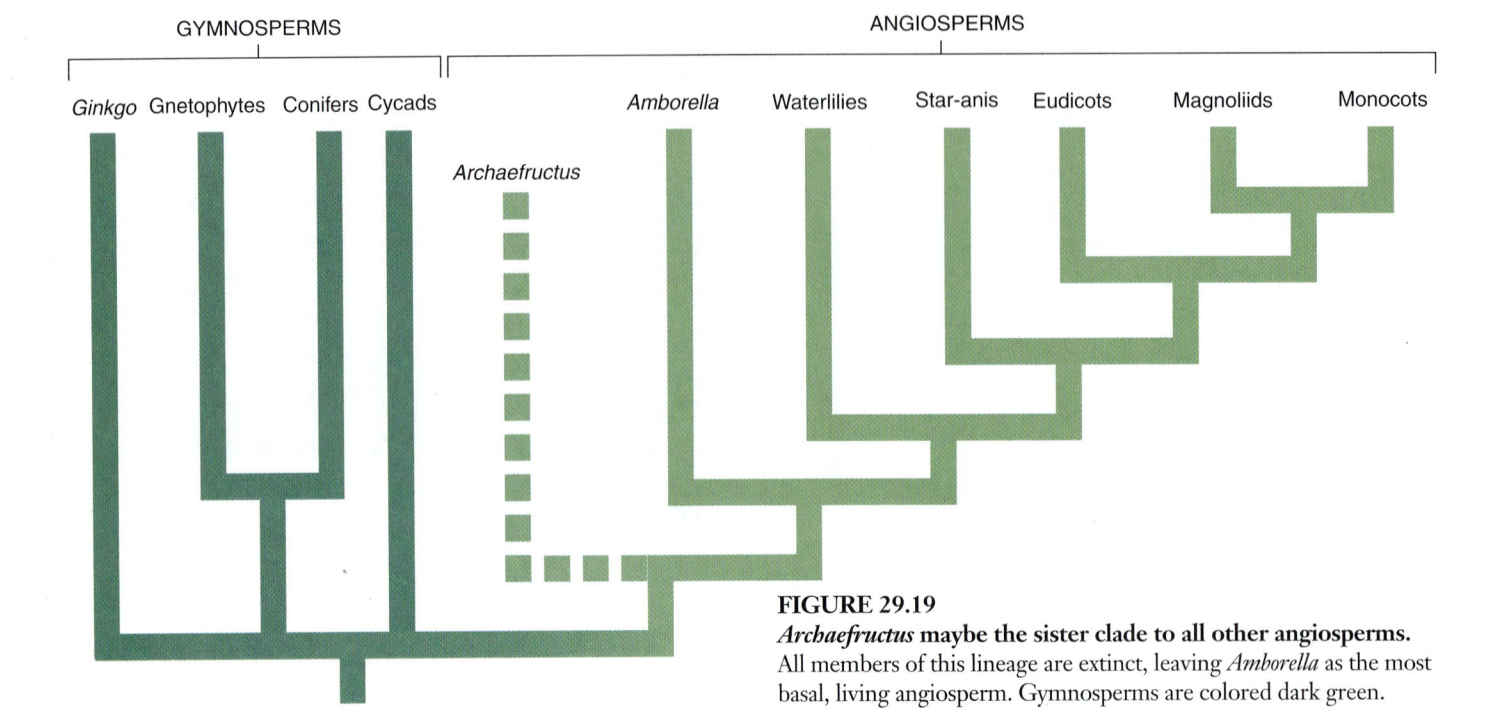

FIGURE 29.19

Archaefructus **maybe the sister clade to all other angiosperms.** All members of this lineage are extinct, leaving *Amborella* as the most basal, living angiosperm. Gymnosperms are colored dark green.

plants. Monocots (about 65,000 species) include the lilies, grasses, cattails, palms, agaves, yuccas, pondweeds, orchids, and irises.

Monocots and eudicots differ from each other in a number of features, some of which are listed in figure 29.20. Monocots and eudicots differ fundamentally in other ways as well. For example, about one-sixth of all eudicot species are **annuals** (plants that complete their entire growth cycle within a year); there are, however, very few annual monocots. Underground swollen storage organs, such as bulbs, occur much more frequently in monocots than they do in eudicots. There are many species of woody dicots (mostly trees or shrubs), but no monocots have true wood; however, a few monocots, such as *agave* and *aloe*, produce extra bundles of conducting tissues that give them a woody texture. Endosperm, which is usually present in mature monocot seeds, is largely absent in mature eudicot seeds. Other specific differences will be presented in chapter 35.

The Structure of Flowers

Flowers are considered to be modified stems bearing modified leaves. Regardless of their size and shape, they all share certain features (figure 29.21). Each flower originates as a **primordium** that develops into a bud at the end of a stalk called a **pedicel**. The pedicel expands slightly at the tip to form the **receptacle,** to which the remaining flower parts are attached. The other flower parts typically are attached in circles called *whorls*. The outermost whorl is composed of **sepals.** Most flowers have three to five sepals, which are green and somewhat leaflike; they often function in protecting the immature flower and in some species may drop off as the flower opens. The next whorl consists of **petals** that are often colored, attracting pollinators such as insects and birds. The petals, which commonly number three to five, may be separate, fused together, or missing altogether in *wind-pollinated flowers*.

The third whorl consists of **stamens,** collectively called the **androecium** (Greek *andros*, "male," + *oikos*, "house"). Each stamen consists of a pollen-bearing **anther** and a stalk called a **filament,** which may be missing in some flowers. At the center of the flower is the **gynoecium** (Greek *gynos*, "female," + *oikos*, "house"), consisting of one or more **carpels.** The first carpel is believed to have been formed from a leaflike structure with ovules along its margins. The edges of the blade then rolled inward and fused together, forming a carpel. Primitive flowers can have several to many separate carpels, but in most flowers, two to several carpels are fused together. Such fusion can be seen in an orange sliced in half; each segment represents one carpel.

MONOCOTS

1. Seed with one cotyledon ("seed leaf").
2. Leaves with parallel veins.
3. Vascular cambium rarely occur.
4. Flower parts mostly in threes or multiples of three.

EUDICOTS

1. Seed with two cotyledons ("seed leaves").
2. Leaves with a network of veins.
3. Lateral meristems (cambia) present.
4. Flower parts mostly in fours or fives or multiples of four or five.

FIGURE 29.20
Comparison of monocots and eudicots.

FIGURE 29.21
Diagram of an angiosperm flower. (*a*) The main structures of the flower are labeled. (*b*) Details of an ovule. The ovary as it matures will become a fruit; as the ovule's outer layers (integuments) mature, they will become a seed coat.

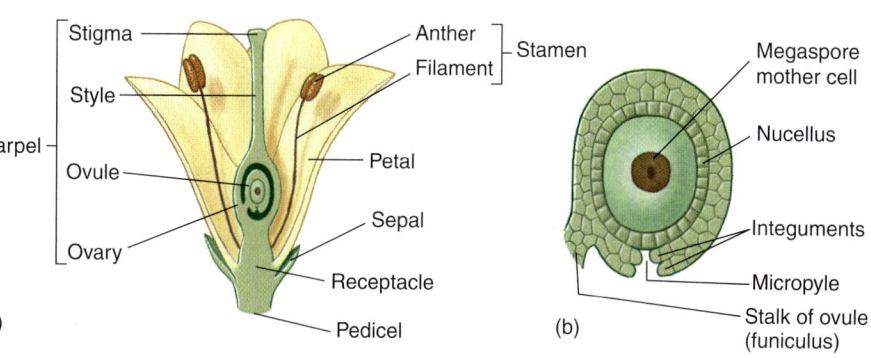

A carpel has three major regions (figure 29.21*a*). The **ovary** is the swollen base, which contains from one to hundreds of **ovules**; the ovary later develops into a **fruit**. The tip of the carpel is called a **stigma**. Most stigmas are sticky or feathery, causing pollen grains that land on them to adhere. Typically, a neck or stalk called a **style** connects the stigma and the ovary; in some flowers, the style may be very short or even missing. Many flowers have nectar-secreting glands called *nectaries*, often located toward the base of the ovary. Nectar is a fluid containing sugars, amino acids, and other molecules that attracts insects, birds, and other animals to flowers.

The Angiosperm Life Cycle

While a flower bud is developing, a single megaspore mother cell in the ovule undergoes meiosis, producing four megaspores (figure 29.22). In most flowering plants, three of the megaspores soon disappear while the nucleus of the remaining one divides mitotically, and the cell slowly expands until it becomes many times its original size. While this expansion is occurring, each of the daughter nuclei divides twice, resulting in eight haploid nuclei arranged in two groups of four. At the same time, two layers of the ovule, the **integuments**, differentiate and become the *seed coat* of a seed. The integuments, as they develop, leave a small gap or pore at one end called the *micropyle* (see figure 29.21*b*). One nucleus from each group of four migrates toward the center, where they function as **polar nuclei**. Polar nuclei may fuse together, forming a single diploid nucleus, or they may form a single cell with two haploid nuclei. Cell walls also form around the remaining nuclei. In the group closest to the micropyle, one cell functions as the **egg**; the other two nuclei are called **synergids**. At the other end, the three cells are now called **antipodals**; they have no apparent function and eventually break down and disappear. The large sac with eight nuclei in seven cells is called an **embryo sac**; it constitutes the female gametophyte. Although it is completely dependent on the sporophyte for nutrition, it is a multicellular, haploid individual.

While the female gametophyte is developing, a similar but less complex process takes place in the anthers. Most anthers have patches of tissue (usually four) that eventually become chambers lined with nutritive cells. The tissue in each patch is composed of many diploid microspore mother cells that undergo meiosis more or less simultaneously, each producing four microspores. The four microspores at first remain together as a quartet or tetrad, and the nucleus of each microspore divides once; in most species, the microspores of each quartet then separate. At the same time, a two-layered wall develops around each microspore. As the anther matures, the wall between adjacent pairs of chambers breaks down, leaving two larger sacs. At this point, the binucleate microspores have become **pollen grains**. The outer pollen grain wall layer often becomes beautifully sculptured, and it contains chemicals that may react with others in a stigma to signal whether or not development of the male gametophyte should proceed to completion. The pollen grain has areas called *apertures*, through which a pollen tube may later emerge.

Pollination is simply the mechanical transfer of pollen from its source (an anther) to a receptive area (the stigma of a flowering plant). Most pollination takes place between flowers of different plants and is brought about by insects, wind, water, gravity, bats, and other animals. In as many as one-quarter of all angiosperms, however, a pollen grain may be deposited directly on the stigma of its own flower, and self-pollination occurs. Pollination may or may not be followed by *fertilization*, depending on the genetic compatibility of the pollen grain and the flower on whose stigma it has landed. (In some species, complex, genetically controlled mechanisms prevent self-fertilization to enhance genetic diversity in the progeny.) If the stigma is receptive, the pollen grain's dense cytoplasm absorbs substances from the stigma and bulges through an aperture. The bulge develops into a *pollen tube* that responds to chemical and mechanical stimuli that guide it to the embryo sac. It follows a diffusion gradient of the chemicals and grows down through the style and into the micropyle. The pollen tube usually takes several hours to two days to reach the micropyle, but in a few instances, it may take up to a year. One of the pollen grain's two cells, the *generative cell*, lags behind. Its nucleus divides in the pollen grain or in the pollen tube, producing two sperm cells. Unlike sperm in mosses, ferns, and some gymnosperms, the sperm of flowering plants have no flagella. At this point, the pollen grain with its tube and sperm has become a mature male gametophyte.

Chapter 29 Overview of Plant Diversity

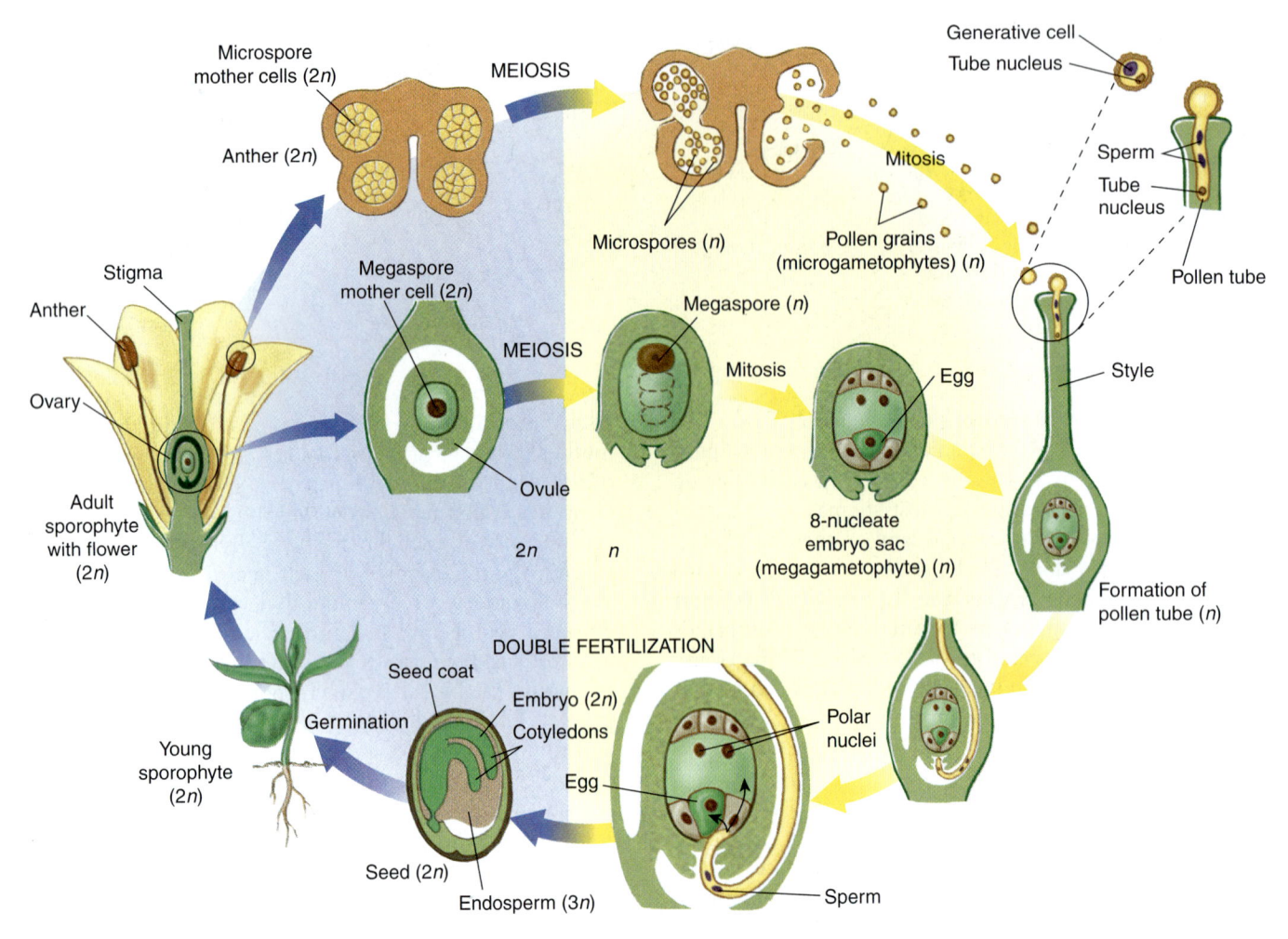

FIGURE 29.22
Life cycle of a typical angiosperm. As in pines, external water is no longer required for fertilization. In most species of angiosperms, animals carry pollen to the carpel. The outer wall of the carpel forms the fruit, which often entices animals to disperse the seed.

As the pollen tube enters the embryo sac, it destroys a synergid in the process and then discharges its contents. Both sperm are functional, and an event called **double fertilization** follows. One sperm unites with the egg and forms a zygote, which develops into an embryo sporophyte plant. The other sperm and the two polar nuclei unite, forming a triploid primary endosperm nucleus. The primary endosperm nucleus begins dividing rapidly and repeatedly, becoming triploid **endosperm tissue** that may soon consist of thousands of cells. Endosperm tissue can become an extensive part of the seed in grasses such as corn (see figure 36.7).

In most flowering plants, endosperm provides nutrients for the embryo that develops from the zygote; in many species, such as peas and beans, it disappears completely by the time the seed is mature. Following double fertilization, the integuments harden and become the seed coat of a seed. The haploid cells remaining in the embryo sac (antipodals, synergid, tube nucleus) degenerate.

Until recently, the nutritional, triploid endosperm was believed to be the ancestral state in angiosperms. A recent analysis of extant, basal angiosperms revealed that diploid endosperms were also common. The female gametophyte in these species has four, not eight nuclei. At the moment, it is unclear whether diploid or triploid endosperms are the most primitive.

Angiosperms are characterized by ovules that at pollination are enclosed within an ovary at the base of a carpel, a structure unique to the phylum; a fruit develops from the ovary. Evolutionary innovations—including flowers to attract pollinators, fruits to protect and aid in embryo dispersal, and double fertilization providing additional nutrients for the embryo—all have contributed to the widespread success of this phylum.

596 Part V Diversity of Life on Earth

Concept Review

For interactive testing, visit the Online Learning Center with PowerWeb at www.mhhe.com/Raven7

29.1 Plants have multicellular haploid and diploid stages in their life cycles.

The Evolutionary Origins of Plants

- The term plant will be used to refer to a group of organisms sharing a freshwater ancestor that have evolved over a 470-million-year period. (p. 580)
- The defining characteristic of plants is the protection of their embryos. (p. 580)
- All plants have multicellular haploid and diploid phases, with the trend over time toward increasing embryo protection and a smaller haploid stage. (p. 580)
- Most plants are protected from desiccation by a waxy cuticle, and have stomata through which gases diffuse into and out of the plant. (p. 580)

Plant Life Cycles

- Plants have haplodiplontic life cycles. (p. 581)
- The diploid sporophyte produces haploid spores by meiosis, and spores divide by mitosis, producing a multicellular, haploid gametophyte. (p. 581)

29.2 Nonvascular plants are relatively unspecialized, but successful in many terrestrial environments.

Mosses, Liverworts, and Hornworts

- Nonvascular plants are divided into three major phyla—Bryophyta, Hepaticophyta, and Anthocerophyta—with all members being relatively unspecialized, but collectively able to inhabit diverse environments. (pp. 582–583)

29.3 Seedless vascular plants have well-developed conducting tissues in their sporophytes.

Features of Vascular Plants

- The seven living phyla of vascular plants dominate almost all terrestrial habitats. (p. 584)
- In modern vascular plants, the gametophytes have been reduced in size and complexity; individuals exhibit highly specialized conductive tissues. (p. 584)

Seedless Vascular Plants

- The two seedless vascular plant phyla, Lycophyta (club mosses) and Pterophyta (whisk ferns, horsetails, and ferns), have a large and conspicuous sporophyte. Many also have well-differentiated roots, stems, and leaves. (pp. 586–588)

29.4 Seeds protect and aid in the dispersal of plant embryos.

Seed Plants

- Seed plants first appeared about 425 MYA. (p. 589)
- The seed provides an extra layer of sporophyte tissue and a dormant phase that allows a pause in the life cycle to await more favorable environmental conditions. (p. 589)
- Seed plants produce male and female gametophytes. (p. 589)

Gymnosperms

- Gymnosperms represent cone-bearing, naked-seeded plants and are composed of four living groups (conifers, cycads, gnetophytes, and ginkgo). (p. 590)
- Conifers, the most familiar gymnosperms, include pines, spruces, firs, cedars, hemlocks, yews, larches, and cypresses. (p. 590)
- During pollination, the ovules are not completely enclosed by protective sporophyte tissue. (pp. 590–592)

Angiosperms

- During pollination, ovules are enclosed within diploid tissue (an ovary). (p. 593)
- The approximately 250,000 known species of flowering plants have long been grouped into two classes—monocots and eudicots. These groups are not monophyletic. (pp. 593–594)
- Monocots and eudicots differ according to the number of cotyledons, leaf venation, presence of lateral meristems, and number of flower parts. (p. 594)
- Flowers, which may attract pollinators, are modified stems bearing modified leaves, with flower parts attached in whorls. (p. 594)
- Fruits protect embryos and seeds, and aid in seed dispersal. (pp. 593–596)
- In most flowering plants, double fertilization provides nutrients for the developing embryo. (p. 596)

Test Your Understanding

For interactive testing, visit the Online Learning Center with PowerWeb at www.mhhe.com/Raven7

Self Test

1. All plants exhibit alternation of generations. This means their life cycle
   a. includes both haploid and diploid gametes.
   b. shows only asexual reproduction.
   c. has both a multicellular haploid stage and a multicellular diploid stage.
   d. does not include meiosis.
2. The plant life cycle has both a sporophyte and a gametophyte generation. In the sporophyte stage,
   a. gametes are produced.
   b. meiosis occurs.
   c. only mitosis takes place.
   d. gametophytes form.
3. In plants,
   a. gametes are produced directly after meiosis.
   b. gametes are produced directly after mitosis.
   c. no gametes are motile.
   d. seeds are always produced.
4. Which of the following is true of the bryophytes?
   a. It is the only group that shows an alternation of generations.
   b. Bryophytes exhibit extensive vascular tissue.
   c. The sporophyte (multicellular diploid) is the conspicuous stage.
   d. The gametophyte (multicellular haploid) is the conspicuous stage.
5. A plant's vascular tissue is composed of xylem and phloem. The xylem generally transports ____, whereas the phloem transports ____.
   a. water/sugar
   b. sugar/water
   c. water/water
   d. sugar/sugar
6. In the conifers, the
   a. gametophyte is prominent, and the sporophyte is dependent upon the gametophyte.
   b. sporophyte is prominent, with the sporophyte and gametophyte living independently.
   c. sporophyte is prominent, and the gametophyte is dependent upon the sporophyte.
   d. gametophyte is prominent, and the sporophyte stage has disappeared.
7. Which of the following characters is seen in the gymnosperms, but is not seen in other seeded vascular plants?
   a. alternation of generations
   b. exposed seeds
   c. sporophyte stage
   d. pollen
8. All flowering plants (angiosperms)
   a. produce exposed seeds.
   b. are nonvascular.
   c. have flagellated sperm.
   d. have fruit.
9. In the angiosperms, the
   a. gametophyte is prominent, and the sporophyte is dependent upon the gametophyte.
   b. sporophyte is prominent, with the sporophyte and gametophyte living independently.
   c. sporophyte is prominent, and the gametophyte is dependent upon the sporophyte.
   d. gametophyte is prominent, and the sporophyte stage has disappeared.
10. Which of the following is *not* characteristic of a monocot?
    a. leaves with parallel veins
    b. flower parts usually in threes or multiples of three
    c. lateral meristems occurring rarely
    d. seed with two cotyledons

Test Your Visual Understanding

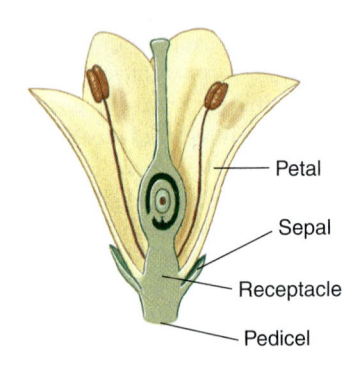

1. Label this diagram with the following terms: stigma, style, ovule, ovary, carpel, anther, filament, stamen.

Apply Your Knowledge

1. A major factor in life on land is coping with ultraviolet radiation from the sun. Early in the evolution of land plants, there was a change from having a prominent haploid gametophyte to a prominent diploid sporophyte. Explain why this change may have occurred in reference to ultraviolet radiation.
2. Plants provide many vital products for humans. Make a list of all the plant-based products that you use in a single day.
3. Many flowering plants have adapted in response to a particular animal pollinator, and they rely on that animal for proper pollination to occur. What will happen to the plant population if the animal pollinator goes extinct?

30
Fungi

Concept Outline

30.1 The fungi share several key characteristics.

Distinctive Fungal Features. All fungi are heterotrophs by absorption. Fungal cell walls include chitin.
The Body of a Fungus. Some fungi have filamentous bodies (hyphae), and others have yeast morphology.
How Fungi Reproduce. Fungi reproduce both asexually and sexually. In sexual reproduction, hyphae attracted by chemical signaling fuse.
How Fungi Obtain Nutrients. Fungi obtain nutrients by external digestion of dead or living organisms.
Metabolic Pathways. Fungal metabolic pathways provide resources for humans.
Ecology of Fungi. Fungi are among the most important decomposers in terrestrial ecosystems.

30.2 There are four major groups of fungi.

Phylogenetic Relationships. Molecular data are revising the evolutionary history of the fungi.
Chytridiomycota. Chytrids are motile fungi.
Zygomycota. In zygomycetes, the fusion of hyphae leads to the formation of a zygote inside a zygosporangium.
Basidiomycota. Basidiomycetes include mushrooms and single-celled fungi. There is only a single diploid cell in the life cycle. Meiosis immediately follows karyogamy and produces haploid spores outside a basidium.
Ascomycota. Ascomycetes range from unicells to filamentous fungi. Hyphal fusion leads to karyogamy within the ascus. The diploid cell immediately undergoes meiosis to produce spores.

30.3 Fungi participate in many symbioses.

Lichens. A lichen is a mutualistic association between a fungus and an alga or cyanobacterium.
Mycorrhizae. Mycorrhizae are mutualistic associations between fungi and the roots of plants.
Endophytes. Most likely, all plants have fungi living within them that afford protection from herbivores.
Mutualistic Animal Symbioses. Symbiotic relationships with animals range from partnerships in the guts of grazing animals to being farmed by ants.
Fungal Parasites and Pathogens. Agriculture and human health are frequently threatened by fungal parasites.

FIGURE 30.1
Spores ejected from a pore on the surface of a puffball fungus. The fungi constitute a unique kingdom of heterotrophic organisms. Along with bacteria, they are important decomposers and disease-causing organisms. Some puffballs are almost 1 meter in diameter and may contain 7 trillion spores— enough to circle the earth's equator!

The fungi, an often-overlooked group of unicellular and multicellular organisms, have a profound impact on ecology and human health. Fungi are found everywhere— from the tropics to the tundra and in both terrestrial and aquatic environments (figure 30.1). Fungi made it possible for plants to colonize land by associating with rootless stems and aiding in the uptake of nutrients and water. Mushrooms and toadstools are fungi, multicellular creatures that grow so rapidly that they seem to appear overnight on our lawns. A single *Armillaria* fungus can cover 15 hectares underground and weigh 100 tons. Some yeasts are used to make bread and beer. Other fungi cause disease in plants and animals. These fungal killers are particularly problematic because fungi are animals' closest relatives. Drugs that can kill fungi often have toxic effects on animals, including humans.

30.1 The fungi share several key characteristics.

Distinctive Fungal Features

Mycologists, scientists who study fungi, believe there may be as many as 1.5 million fungal species. Currently, taxonomists classify the fungi into four main groups: the chytrids, the zygomycetes, the basidiomycetes, and the ascomycetes (figure 30.2). Recent phylogenetic analysis of DNA sequences and protein sequences indicates that fungi are more closely related to animals than to plants. The last common ancestor was a single cell, and unique solutions to multicellularity evolved in fungi and in animals. While the fungi are incredibly diverse, they share some characteristics:

1. **Fungi are heterotrophs.** Fungi obtain their food by secreting digestive enzymes into substrates, including fallen logs and the skin of frogs. Then they absorb the organic molecules released by the enzymes. Thus, fungi actually live in their food.
2. **Fungi have several cell types.** Multicellular fungi are primarily filamentous in their growth form (that is, their bodies consist of long, slender filaments called hyphae), even though these hyphae may be packed together to form complex structures such as a mushroom. Hyphal cells have a range of morphologies. Some unicellular fungi have a flagellum.
3. **Some fungi have a dikaryon stage.** Many sexually reproducing fungi undergo a stage in which two haploid cells coexist in a single cell (dikaryon) for some period of time before fusing to make a diploid nucleus.
4. **Fungi have cell walls that include chitin.** The cell walls of fungi are built of polysaccharides (chains of sugars) and chitin, the same tough material a crab shell is made of.
5. **Fungi undergo nuclear mitosis.** Mitosis in fungi is different from that in plants and animals in one key respect: The nuclear envelope does not break down and re-form. Instead, mitosis takes place *within* the nucleus. A spindle apparatus forms there, dragging chromosomes to opposite poles of the *nucleus* (not the cell, as in most other eukaryotes). This type of mitosis is found in some protists as well (see chapter 28).

Fungi absorb their food after digesting it with secreted enzymes. Besides this mode of nutrition, the fungi are characterized by combined dikaryon stages, nuclear mitosis, and other traits.

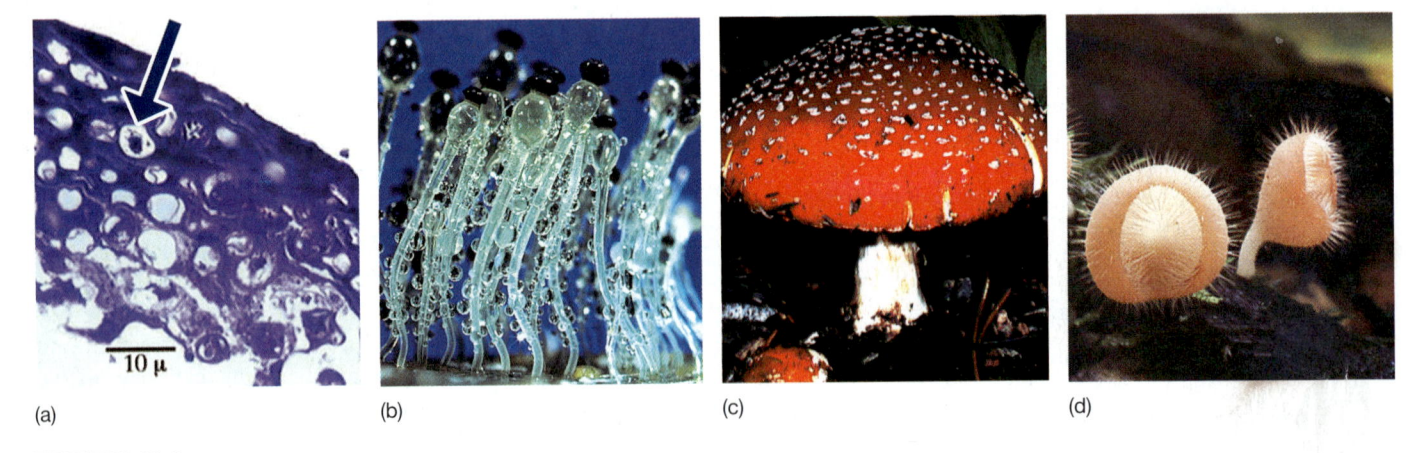

FIGURE 30.2
Representatives of the four major groups of fungi. (*a*) The parasitic chytrid *Batrachochytrium dendrobatidis* is causing a worldwide decline in amphibians. This stained skin sample from an infected frog shows sporangia of *B. dendrobatidis* (*arrow*). (*b*) *Pilobolus*, a zygomycete, grows on animal dung. Stalks about 10 millimeters long contain dark, spore-bearing sacs. (*c*) *Amanita muscaria*, the fly agaric, is a toxic basidiomycete. (*d*) The cup fungus *Cookeina tricholoma* is an ascomycete from the rain forest of Costa Rica. In the cup fungi, the spore-producing structures line the cup; in basidiomycetes that form mushrooms such as *Amanita*, they line the gills beneath the cap of the mushroom. All visible structures of fungi, including the ones shown here, arise from an extensive network of filamentous hyphae that penetrates and is interwoven with the substrate on which they grow.

The Body of a Fungus

Fungi exist either as single-celled yeasts or in multicellular form with several different cell types. We will focus on cells called **hyphae** (singular, *hypha*), which are slender filaments, barely visible to the naked eye. Some hyphae are continuous or branching tubes filled with cytoplasm and multiple nuclei. Other hyphae are typically made up of long chains of cells joined end-to-end and divided by cross-walls called **septa** (singular, *septum*). The septa rarely form a complete barrier, except when they separate the reproductive cells. Even fungi with septa can be considered one long cell. Cytoplasm characteristically flows or streams freely throughout the hyphae, passing through major pores in the septa (figure 30.3). Because of this streaming, proteins synthesized throughout the hyphae may be carried to their actively growing tips. As a result, fungal hyphae may grow very rapidly when food and water are abundant and the temperature is optimum.

A mass of connected hyphae is called a **mycelium** (plural, *mycelia*). This word and the term *mycology* are both derived from the Greek word for "fungus," *mykes*. The mycelium of a fungus (figure 30.4) constitutes a system that may, in the aggregate, be many meters long. Growth is in two dimensions, not three. This mycelium grows into the soil, wood, or other material in which the fungus is growing. Digestion begins, and all parts of such a fungus are metabolically active.

In two of the four major groups of fungi, reproductive structures formed of interwoven hyphae, such as mushrooms, puffballs, and morels, are produced at certain stages of the life cycle. These structures expand rapidly because of rapid inflation of the hyphae. For this reason, mushrooms that are ready to reproduce can appear suddenly on your lawn.

The cell walls of fungi are formed of polysaccharides and chitin, not cellulose as are those of plants and many groups of protists. Chitin is the same material that makes up the major portion of the hard shells, or exoskeletons, of arthropods, a group of animals that includes insects and crustaceans (see chapter 33). Chitin is one of the shared traits that has led scientists to believe that fungi and animals are more closely related than fungi and plants.

Mitosis in multicellular fungi differs from that in most other organisms. Because of the linked nature of the cells, the cell itself is not the relevant unit of reproduction; instead, the nucleus is. The nuclear envelope does not break down and re-form; instead, the spindle apparatus is formed *within* it. Centrioles are absent in all fungi; instead, fungi regulate the formation of microtubules during mitosis with small, relatively amorphous structures called *spindle plaques*. This unique combination of features strongly suggests that fungi originated from some unknown group of single-celled eukaryotes with these characteristics.

> **Multicellular fungi exist primarily in the form of filamentous hyphae, either with or without septa. These and other unique features support placing fungi in a separate kingdom.**

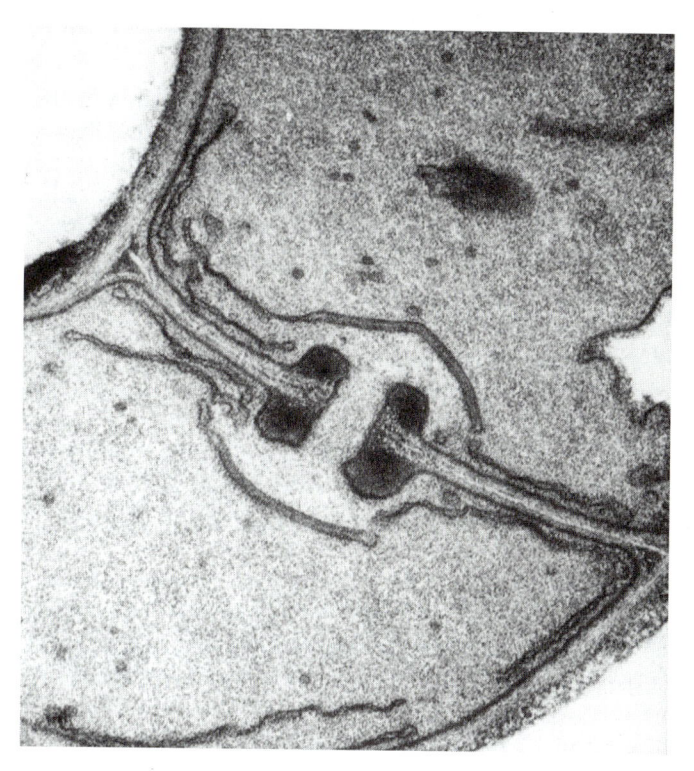

FIGURE 30.3
A septum. This transmission electron micrograph of a section through a hypha of the basidiomycete *Inonotus tomentosus* shows a pore through which the cytoplasm streams (45,000×).

FIGURE 30.4
Fungal mycelium. This mycelium, composed of hyphae, is growing through leaves on the forest floor in Maryland.

How Fungi Reproduce

Fungi are different from most animals and plants in that each cell (or hypha) can house one, two, or more nuclei. If each compartment of a hypha has only one nucleus, it is called **monokaryotic**; a hypha with two nuclei is called **dikaryotic**. In a dikaryotic cell, the two haploid nuclei exist independently. Dikaryotic hyphae have some of the genetic properties of diploids, because both genomes are transcribed.

Sometimes, many nuclei intermingle in the common cytoplasm of a fungal mycelium, which can lack distinct cells. If a dikaryotic or multinucleate hypha has nuclei that are derived from two genetically distinct individuals, the hypha is called **heterokaryotic**. Hyphae whose nuclei are genetically similar to one another are called **homokaryotic**.

These distinctions are important in understanding the life cycles of the individual groups, which can be quite complicated. Many fungi are capable of producing both sexual and asexual spores. When a fungus reproduces sexually, two haploid hyphae of compatible mating types come together and fuse. In animals, plants, and some fungi, the fusion of two haploid cells immediately results in a diploid cell ($2n$). However, in other fungi (basidiomycetes and ascomycetes), an intervening dikaryotic stage ($1n + 1n$) occurs before the parental nuclei fuse and form a diploid nucleus. In ascomycetes, this dikaryotic stage is brief, occuring in only a few cells of the sexual reproductive structure. In basidiomycetes, however, it can last for most of the life of the fungus, including both the feeding and sexual spore-producing structures.

The cytoplasm in fungal hyphae normally flows through perforated septa, or moves freely in their absence. Reproductive structures are an important exception to this general pattern. When reproductive structures form, they are cut off by complete septa that lack perforations or have perforations that soon become blocked.

Spores are the most common means of reproduction among fungi. They may form as a result of either asexual or sexual processes, and they are often dispersed by the wind. When spores land in a suitable place, they germinate, giving rise to a new fungal mycelium. Because the spores are very small, between 2 and 75 microns in diameter (figure 30.5), they can remain suspended in the air for long periods of time. Unfortunately, many of the fungi that cause diseases in plants and animals are spread rapidly by such means. The spores of other fungi are routinely dispersed by insects or other small animals. Only one group of fungi, the chytrids, retain the ancestral flagella and have motile zoospores.

It had long been believed that the worldwide presence of fungal species could be accounted for, on an evolutionary timescale, by the almost limitless, long-distance dispersal of fungal spores. However, recent biogeographic studies have examined the phylogenetic relationships among fungi in distant parts of the world and disproved this long-held assumption.

FIGURE 30.5
Fungal spores. Scanning electron micrograph of fungal spores that infect rose plants.

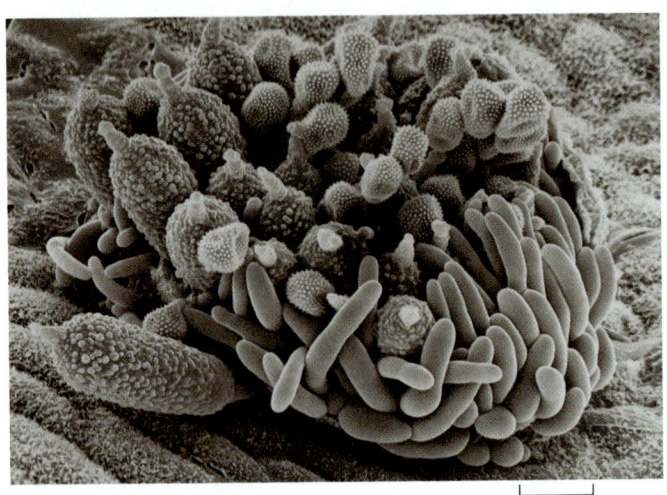

47.85 μm

> Fungi reproduce sexually after two hyphae of opposite mating types fuse. Asexual reproduction by spores is a common means of reproduction.

FIGURE 30.6
A carnivorous fungus. The oyster mushroom *Pleurotus ostreatus* not only decomposes wood but also immobilizes nematodes, which the fungus uses as a source of nitrogen.

How Fungi Obtain Nutrients

All fungi obtain their food by secreting digestive enzymes into their surroundings and then absorbing back into the fungus the organic molecules produced by this *external digestion*. The fungal body plan reflects this approach. Unicellular fungi have the greatest surface area-to-volume ratio of any fungus, maximizing the surface area for absorption. Extensive networks of hyphae provide an enormous surface area for absorption in a fungal mycelium. Many fungi are able to break down the cellulose in wood, cleaving the linkages between glucose subunits and then absorbing the glucose molecules as food. Fungi also digest lignin, which strengthens plant cell walls. Fungi's specialized metabolic pathways allow them to obtain nutrients from dead trees and from an extraordinary range of organic compounds, including tiny roundworms called nematodes.

The mycelium of the edible oyster mushroom *Pleurotus ostreatus* (figure 30.6) excretes a substance that paralyzes nematodes (see chapter 32) that feed on the fungus. When the worms become sluggish and inactive, the fungal hyphae envelop and penetrate their bodies. Then the fungus secretes digestive juices and absorbs the nematode's nutritious contents, just like it would from a plant source. The fungus usually grows within living trees or on old stumps, obtaining the bulk of its glucose through the enzymatic digestion of cellulose and lignin from plant cell walls, so that the nematodes it consumes apparently serve mainly as a source of nitrogen—a substance almost always in short supply in biological systems. Other fungi are even more active predators than *Pleurotus*, snaring, trapping, or firing projectiles into nematodes, rotifers, and other small animals on which they prey.

Fungi secrete digestive enzymes onto organic matter and then absorb the products of the digestion.

Metabolic Pathways

Humans use the metabolic pathways of fungi for commercial purposes. Anaerobic fermentation (see chapter 9) in some fungi gives wine and certain cheeses (Roquefort and Brie) their delicate flavors. Other fungi make possible the manufacture of soy sauce, miso, and other fermented foods. Vast industries depend on the biochemical manufacture of organic substances such as citric acid by fungi in culture. The ability of **yeasts** (single-celled as opposed to filamentous fungi) to ferment carbohydrates, breaking down glucose to ethanol and carbon dioxide, is fundamental in the production of bread, beer, and wine. Many different strains of yeast have been domesticated and selected for these processes. Wild yeasts—those that occur naturally in the areas where wine is made—were historically important in wine-making, but domesticated yeasts are normally used now. The most important yeast in all these processes is *Saccharomyces cerevisiae*, a yeast that has been used by humans throughout recorded history. Yeasts are now used on a large scale to produce protein for the enrichment of animal food. Some shamanic cultures have used hallucinogenic compounds produced by fungi. Many antibiotics, including the first used on a wide scale (penicillin from the asexual "mold" *Penicillium*) are derived from fungi. Fungi are also a source of steroids for medicinal use.

If water is present, some fungi can break down almost any carbon-containing compound—even jet fuel! For this reason, there is much interest in using fungi for **bioremediation,** cleaning up soils or waters that are environmentally contaminated, using organisms able to degrade the toxin. Some fungi can even remove inorganic substances from soil. For example, high levels of selenium in the soil are accumulated in local plants and can be toxic to grazing animals. At least three species of fungi have been isolated that combine selenium, accumulated at the San Luis National Wildlife Refuge in California's San Joaquin Valley, with harmless volatile chemicals—thus removing excess selenium from the soil. Bioremediation solutions will be more complex than just adding specific fungi to a waste side. For example, soils may need to have additional nutrients added to sustain restoration efforts.

Fungal metabolic pathways are used to produce food and pharmaceuticals. The fungi's amazing ability to break down almost any carbon-containing product makes members of this kingdom excellent candidates for bioremediation.

Ecology of Fungi

Fungi, together with bacteria, are the principal decomposers in the biosphere. They break down organic materials and return the substances locked in those molecules to circulation in the ecosystem. Fungi are virtually the only organisms capable of breaking down lignin, one of the major constituents of wood. By breaking down such substances, fungi release critical building blocks, such as carbon, nitrogen, and phosphorus, from the bodies of living or dead organisms and make them available to other organisms.

The interactions, or **symbioses**, between fungi and living organisms fall into a broad range of categories. In some cases, the symbiosis is obligate (essential for survival), and in other cases it is facultative (the fungus can survive without the host). Within a group of closely related fungi, several different types of symbiosis can be found. *Pathogens* and *parasites* gain resources from their host, but have a negative effect on the host that can even lead to death. The difference between pathogens and parasites is that pathogens cause disease while parasites do not. *Commensal* relationships benefit one partner and do not harm the other. Fungi that are in a *mutualistic* relationship benefit both themselves and their hosts. The varied forms of fungal symbiosis are discussed in section 30.3. The range of these interactions, along with decomposition of nonliving organic materials, contributes to the diversity and importance of fungi in ecosystems.

Fungal species cause disease in plants (figure 30.7), and they are responsible for billions of dollars in agricultural losses every year. Not only are fungi among the most harmful pests of living plants, but they also spoil food products that have been harvested and stored. In addition, fungi often secrete substances into the foods they are attacking that make these foods unpalatable, carcinogenic, or poisonous.

Human and animal diseases can also be fungal in origin. Ringworm and athlete's foot can be treated with topical antifungal ointments. *Pneumocystis jiroveci* (formerly *P. carinii*) invades the lungs, disrupting breathing, and can spread to other organs. In immune-suppressed AIDS patients, this can lead to death. Another fungus is threatening amphibian populations globally (see section 30.3).

Many mutualistic associations between fungi and autotrophic organisms are ecologically important. **Lichens** are mutualistic symbiotic associations between fungi and either green algae or cyanobacteria. They are prominent nearly everywhere in the world, especially in unusually harsh habitats such as bare rock. **Mycorrhizae**, specialized mutualistic symbiotic associations between the roots of plants and fungi, are characteristic of about 90% of all plants. In each of them, the photosynthetic organisms fix atmospheric carbon dioxide and thus make organic material available to the fungi. The metabolic activities of the fungi, in turn, enhance the overall ability of the symbiotic association to exist in a particular habitat. In the case of mycorrhizae, the fungal partner expedites the plant's absorption of essential nutrients such as phosphorus. These associations will be discussed further in section 30.3.

Fungi are key decomposers and symbionts within almost all terrestrial ecosystems and play many other important ecological roles.

FIGURE 30.7
World's largest organism? *Armillaria*, a pathogenic fungus shown here afflicting three discrete regions of coniferous forest in Montana, grows out from a central focus as a single, circular clone. The large patch at the bottom of the picture is almost 8 hectares. The largest clone measured so far has been 15 hectares—pretty impressive for a single individual!

30.2 There are four major groups of fungi.

Phylogenetic Relationships

Our understanding of fungal phylogeny is going through rapid and exciting changes, aided by increasing molecular sequence data (figure 30.8 and table 30.1). We now believe that fungi are more closely related to animals than to plants. The relationships among fungi are less clear. To illustrate the diversity within this kingdom, we will look at four groups of fungi—Chytridiomycota, Zygomycota, Basidiomycota, and Ascomycota. The Ascomycota and Basidiomycota are monophyletic, but the other two groups are not. Two other monophyletic groups are tentative in terms of phylogenetic placement. The Glomales evolved as a clade from one of the groups within the Zygomycota, and a provisional phylum, Glomeromycota, has been proposed to accommodate these organisms. Glomales are ecologically the most important fungi within the Zygomycota because they are the fungal partners in arbuscular mycorrhizal symbioses. The Microsporidia, also monophyletic, are obligate animal parasites that may belong to the kingdom Fungi. Several other groups that historically were considered fungi, such as the slime molds and water molds (see chapter 29), now are classed as protists, not fungi. As additional molecular evidence is acquired, analyzed, and integrated with other traits, a new fungal phylogeny is likely to appear.

Fungal phylogenies are changing rapidly. These close relatives of animals are provisionally placed in four groups—Chytridiomycota, Zygomycota, Basidiomycota, and Ascomycota.

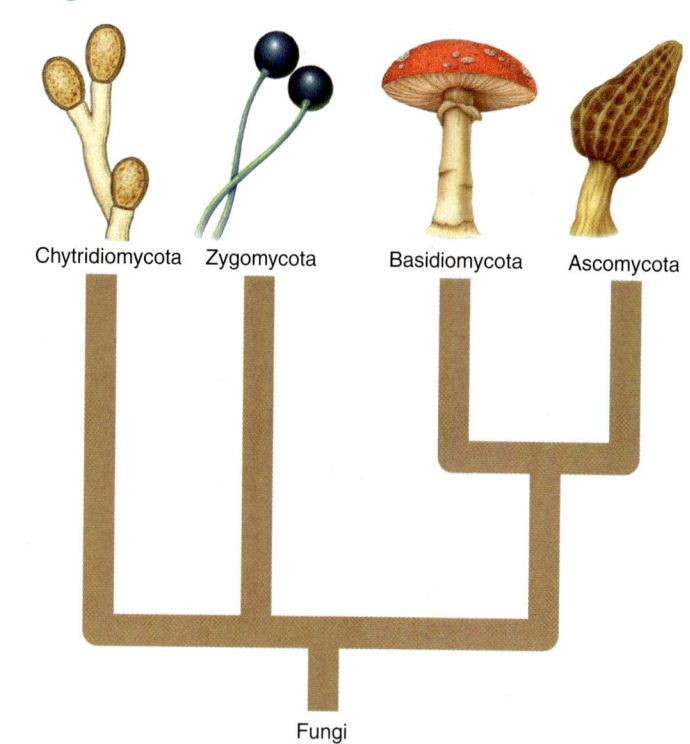

FIGURE 30.8
The four major groups of fungi. Chytridiomycota and Zygomycota are not monophyletic, but Ascomycota and Basidiomycota are. Two additional monophyletic branches have been proposed, but are still somewhat tentative and not shown here.

Table 30.1 Fungi

Group	Typical Examples	Key Characteristics	Approximate Number of Living Species
Chytridiomycota	*Allomyces*	Aquatic, flagellated fungi that produce haploid gametes in sexual reproduction or diploid zoospores in asexual reproduction	1,000
Zygomycota	*Rhizopus, Pilobolus*	Multinucleate hyphae lack septa, except for reproductive structures; fusion of hyphae leads directly to formation of a zygote in zygosporangium, in which meiosis occurs just before it germinates; asexual reproduction is most common	1,050
Basidiomycota	Mushrooms, toadstools, rusts	In sexual reproduction, basidiospores are borne on club-shaped structures called basidia; asexual reproduction occurs occasionally	22,000
Ascomycota	Truffles, morels	In sexual reproduction, ascospores are formed inside a sac called an ascus; asexual reproduction is also common	45,000

Chytridiomycota

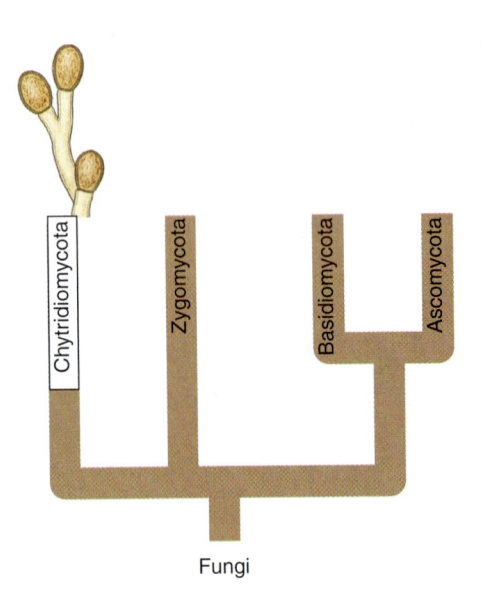

Chytridiomycota (chytrids) are aquatic, flagellated fungi that are most closely related to ancestral fungi. Motile zoospores are a distinguishing character of this fungal group (figure 30.9). Since only chytrids have flagella, this trait must have been lost among ancestors of modern groups. Also, since most chytrids are aquatic, it is likely that fungi first appeared in the water, as did plants and animals. The presence of chitin in the cell walls of chytrids is a unifying feature of all fungi.

Fossils and molecular data indicate that animals and fungi last shared a common ancestor at least 460 million years ago (MYA), but inconsistencies remain. The oldest fossil resembles extant members of the genus *Glomus* that arose within one of the Zygomycota lineages. One DNA analysis placed the animal/fungal divergence at 1500 million years ago (MYA). This latter estimate is not generally accepted, but many researchers believe that the last common ancestor existed around close to 670 million years ago (MYA) based on an analysis of multiple genes.

Regardless of the time of origin, chytrids play a major role in ecosystems. Details of their symbiotic relationships are considered in section 30.3.

Chytrids are the closest living relatives of the first fungal ancestors.

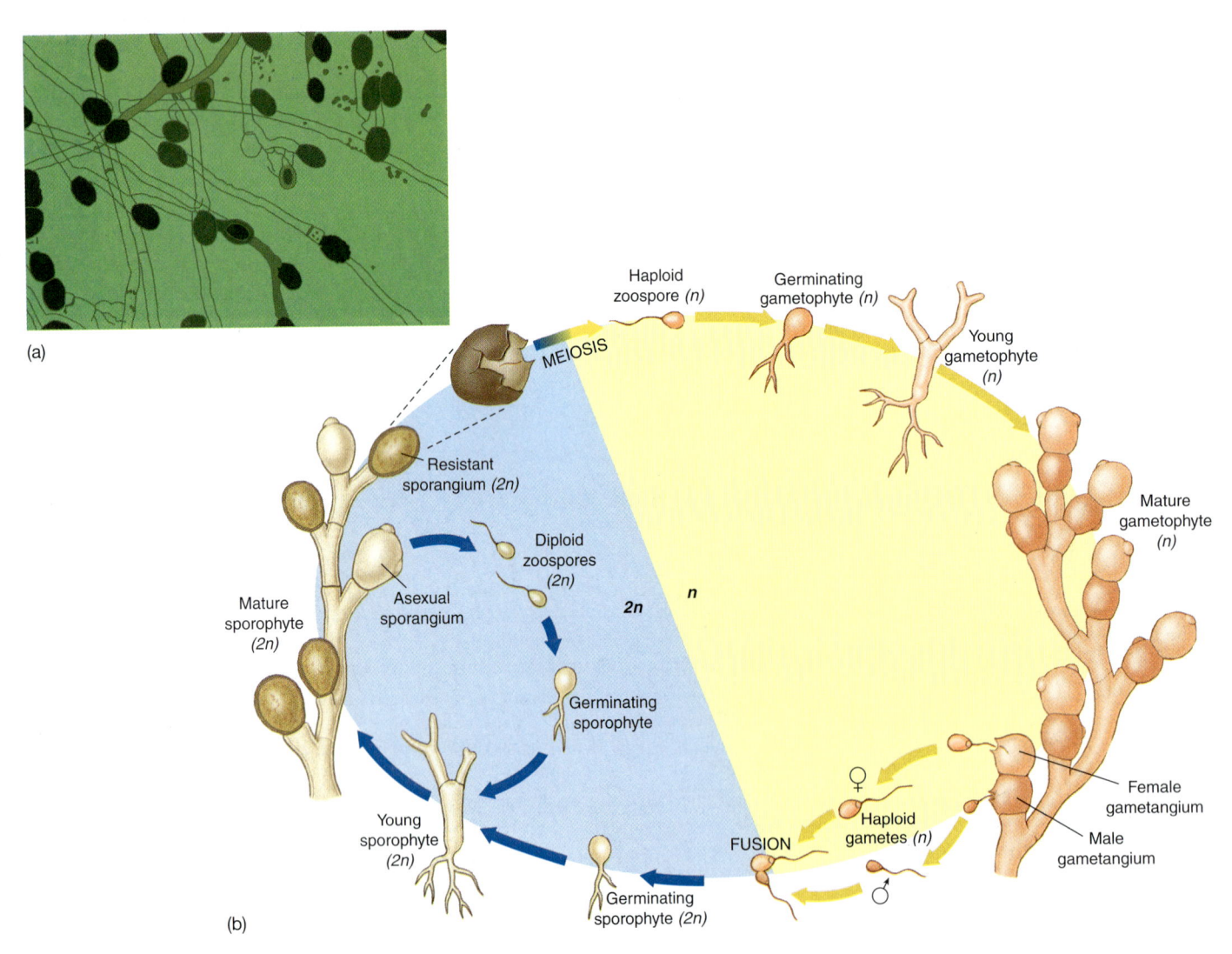

FIGURE 30.9
***Allomyces*, a chytrid that grows in the soil.** (*a*) The spherical sporangia can produce either diploid zoospores via mitosis or haploid zoospores via meiosis. (*b*) Life cycle of *Allomyces*, which has both haploid and diploid multicellular stages (alternation of generations).

Zygomycota

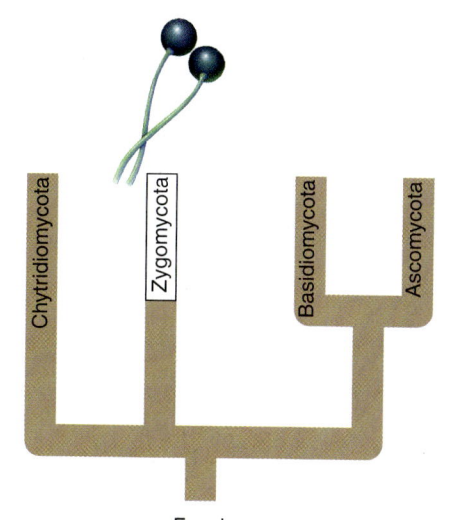

Zygomycetes include only about 1050 named species, but they are incredibly diverse. Among them are some of the more common bread molds (figure 30.10), as well as a variety of others found on decaying organic material, including strawberries and other fruits. A few human pathogens are in this group. So too are the Glomales, which perhaps made the evolution of terrestrial plants possible through arbuscular mycorrhizal symbiosis with roots, enhancing uptake of minerals and water from the soil. Today, they may help maintain plant biodiversity.

The zygomycetes lack septa in their hyphae except when they form sporangia or gametangia. The group is named after a characteristic feature of the sexual phase of the life cycle, a structure called a **zygosporangium** (figure 30.10b). Sexual reproduction occurs by the fusion of gametangia, which contain numerous nuclei. The gametangia are cut off from the hyphae by complete septa. These gametangia may be formed on hyphae of different mating types or on a single hypha. Once the haploid nuclei have fused, forming diploid zygote nuclei, the area where the fusion has taken place develops into a zygosporangium. The zygosporangium, which may contain one or more diploid nuclei, acquires a thick coat that helps the fungus survive conditions not favorable for growth. Meiosis, followed by mitosis, occurs during the germination of the zygosporangium, which releases haploid spores. Haploid hyphae grow when these haploid spores germinate. Except for the zygote nuclei, all nuclei of the zygomycetes are haploid.

Asexual reproduction occurs much more frequently than sexual reproduction in the zygomycetes. During asexual reproduction, hyphae on the surface of the fungus' food produce clumps of erect stalks, called **sporangiophores**. The tips of the sporangiophores form **sporangia**, which are separated by septa. Thin-walled haploid spores are produced within the sporangia. These spores are thus shed above the substrate, in a position where they may be picked up by the wind and dispersed to a new food source.

> Many zygomycetes form characteristic resting structures, called zygosporangia, which contain one or more diploid nuclei. The hyphae of zygomycetes are multinucleate, with septa only where gametangia or sporangia are separated.

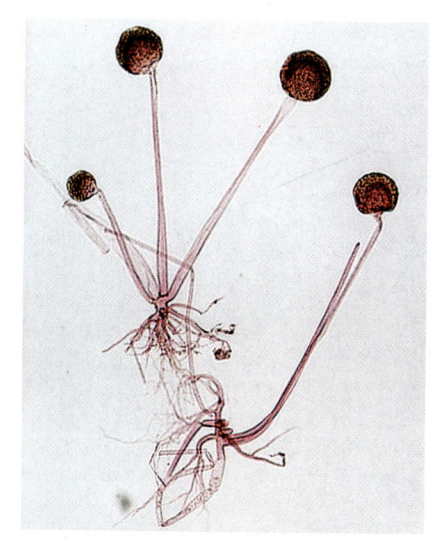

FIGURE 30.10
***Rhizopus*, a zygomycete that grows on simple sugars.** This fungus is often found on moist bread or fruit. (*a*) The dark, spherical, spore-producing sporangia are on hyphae about 1 centimeter tall. The rootlike hyphae anchor the sporangia. (*b*) Life cycle of *Rhizopus*. The Zygomycota group is named for the zygosporangia characteristic of *Rhizopus*.

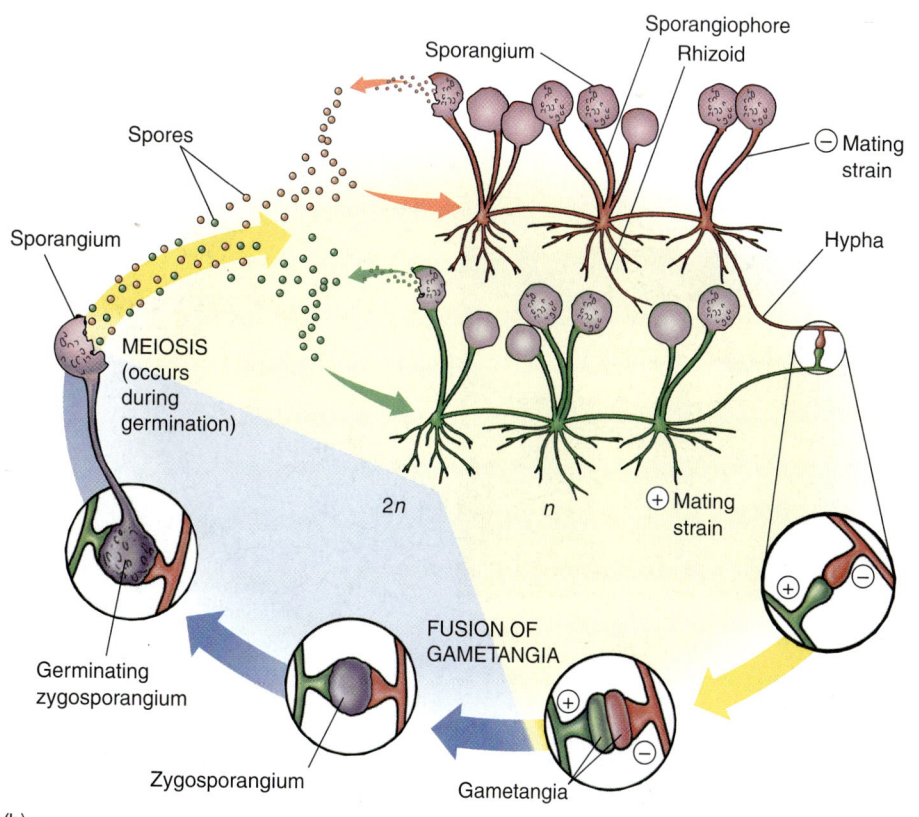

Chapter 30 Fungi 607

Basidiomycota

The basidiomycetes include some of the most familiar fungi. Among the basidiomycetes are not only the mushrooms, toadstools, puffballs, jelly fungi, and shelf fungi, but also many important plant pathogens, including rusts and smuts (figure 30.11). Rust infections resemble rusting metals, while smut infections appear black and powdery due to the spores. Many mushrooms are used as food, but others are hallucinogenic or deadly poisonous.

Basidiomycetes are named for their characteristic sexual reproductive structure, the club-shaped **basidium** (plural, *basidia*). Karyogamy occurs within the basidium, giving rise to the only diploid cell of the life cycle (figure 30.11*b*). Meiosis occurs immediately after karyogamy. In the basidiomycetes, the four haploid products of meiosis are incorporated into **basidiospores.** In most members of this phylum, the basidiospores are borne at the end of the basidia on slender projections. Thus the structure of a basidium differs from that of an ascus, although functionally the two are identical. Recall that the ascospores of the ascomycetes are borne internally in asci instead of externally as in basidiospores.

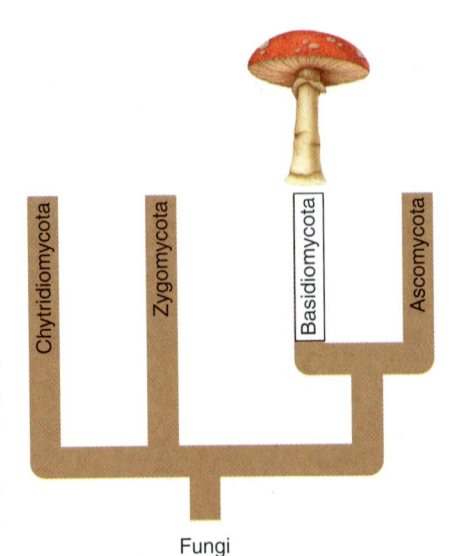

The life cycle of a basidiomycete continues with the production of monokaryotic hyphae after spore germination. These hyphae lack septa early in development. Eventually, septa form between the nuclei of the monokaryotic hyphae. A basidiomycete mycelium made up of monokaryotic hyphae is called a *primary mycelium*. Different mating types of monokaryotic hyphae may fuse, forming a dikaryotic mycelium, or *secondary mycelium*. Such a mycelium is heterokaryotic, with two nuclei representing the two different mating types, between each pair of septa. The maintenance of two genomes in the heterokaryon allows for more genetic plasticity than in a diploid cell with one nucleus. One genome may compensate for mutations in the other. The **basidiocarps,** or mushrooms, are formed entirely of secondary (dikaryotic) mycelium. Gills on the undersurface of the cap of a mushroom form vast numbers of minute spores. It has been estimated that a mushroom with a cap measuring 7.5 centimeters in diameter produces as many as 40 million spores per hour!

Most basidiomycete hyphae are dikaryotic. After karyogamy, meiosis occurs within basidia.

FIGURE 30.11
Basidiomycetes. (*a*) The death cap mushroom, *Amanita phalloides*. When eaten, these mushrooms are usually fatal. (*b*) Life cycle of a basidiomycete. The basidium is the reproductive structure.

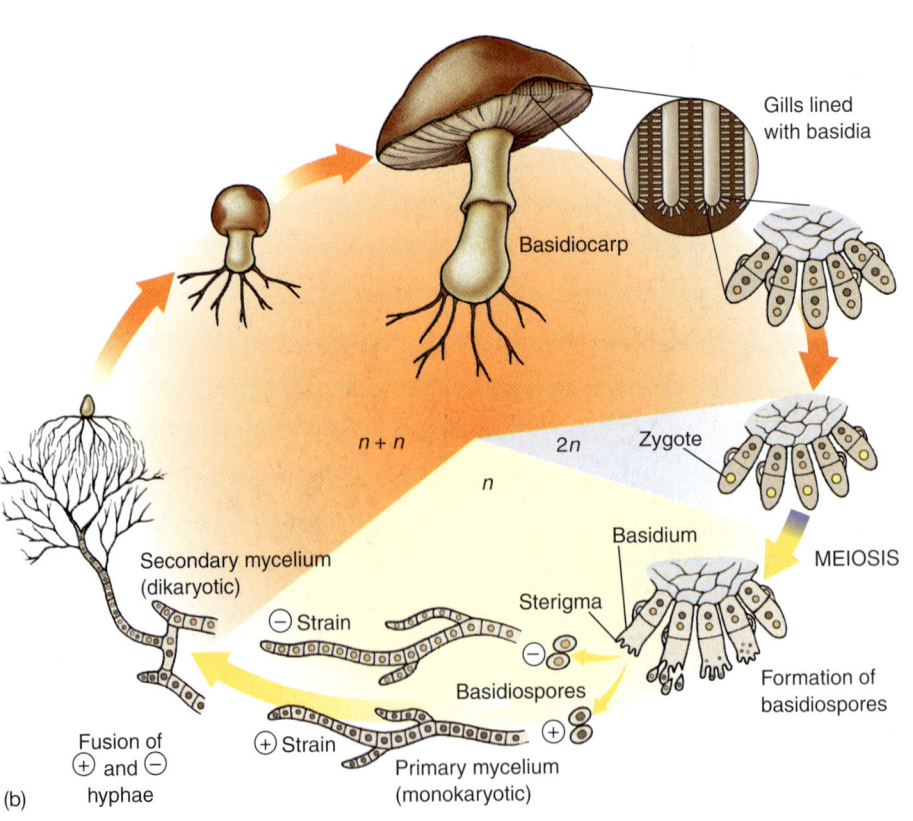

608 Part V Diversity of Life on Earth

Ascomycota

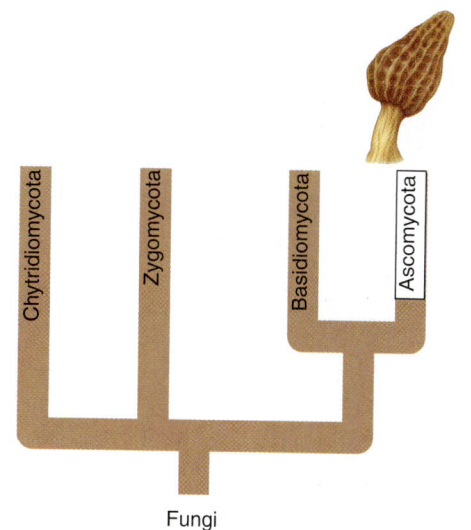

The ascomycetes (phylum Ascomycota) contain about 75% of the known fungi. Among the ascomycetes are such familiar and economically important fungi as bread yeasts, common molds, morels (figure 30.12a), and truffles. Also included in this phylum are many serious plant pathogens, including the chestnut blight, *Cryphonectria parasitica*, and Dutch elm disease, *Ophiostoma ulmi*. Penicillin-producing ascomycetes are in the genus *Penicillium*.

The ascomycetes are named for their characteristic reproductive structure, the microscopic, saclike **ascus** (plural, *asci*). Karyogamy, resulting in the only diploid nucleus of the ascomycete life cycle (figure 30.12c), occurs within the ascus. Asci are differentiated within a structure made up of densely interwoven hyphae, corresponding to the visible portions of a morel or cup fungus, called the **ascocarp**. Meiosis immediately follows karyogamy, forming four haploid daughter nuclei. These usually divide again by mitosis, producing eight haploid nuclei that become walled **ascospores**. In many ascomycetes, the ascus becomes highly turgid at maturity and ultimately bursts, often at a preformed area. When this occurs, the ascospores may be thrown as far as 30 centimeters, an amazing distance considering that most ascospores are only about 10 micrometers long. This would be equivalent to throwing a baseball (diameter 7.5 centimeters) 1.25 kilometers—about 10 times the length of a home run!

Asexual reproduction is very common in the ascomycetes. It takes place by means of **conidia** (singular, *conidium*), spores cut off by septa at the ends of modified hyphae called **conidiophores**. Conidia allow for the rapid colonization of a new food source. Many conidia are multinucleate. The hyphae of ascomycetes are divided by septa, but the septa are perforated, and the cytoplasm flows along the length of each hypha. The septa that cut off the asci and conidia are initially perforated, but later become blocked.

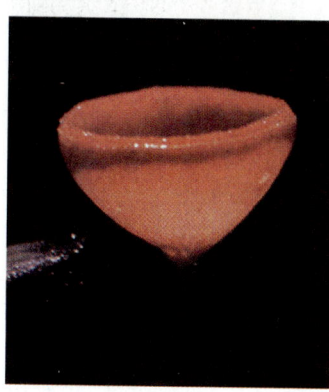

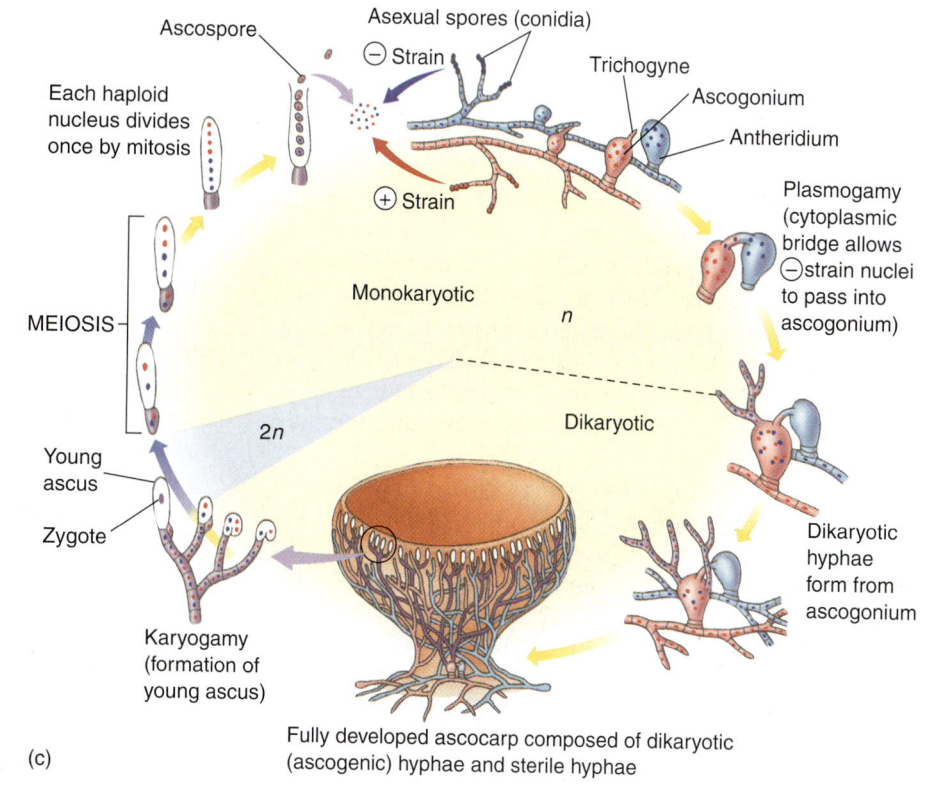

FIGURE 30.12
Ascomycetes. (*a*) This morel, *Morchella esculenta*, is a delicious edible ascomycete that appears in early spring. (*b*) A cup fungus. (*c*) Life cycle of an ascomycete. Haploid ascospores form within the ascus.

Chapter 30 Fungi 609

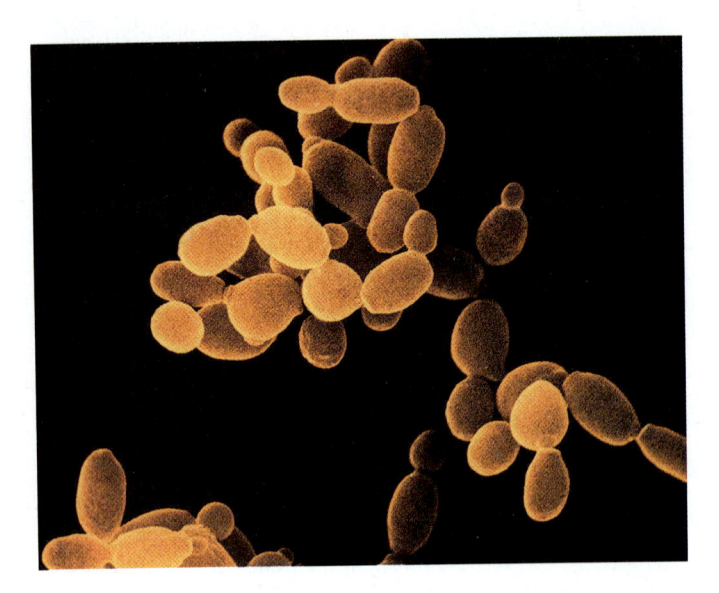

**FIGURE 30.13
Budding in *Saccharomyces*.** As shown in this scanning electron micrograph (19,000×), the cells tend to hang together in chains, a feature that calls to mind the derivation of single-celled yeasts from multicellular ancestors.

Yeast morphology is found among some ascomycetes. Most of their reproduction is asexual and takes place by cell fission or budding, when a smaller cell forms from a larger one (figure 30.13). Sometimes two yeast cells fuse, forming one cell containing two nuclei. This cell may then function as an ascus, with karyogamy followed immediately by meiosis. The resulting ascospores function directly as new yeast cells.

The ability of yeasts to ferment carbohydrates, breaking down glucose to produce ethanol and carbon dioxide, is fundamental in the production of bread, beer, and wine. Many different strains of yeast have been domesticated and selected for these processes using the sugars in rice, barley, wheat, and corn. Wild yeasts—ones that occur naturally in the areas where wine is made—were important in wine making historically, but domesticated yeasts are normally used now.

Wild yeast, often *Candida milleri*, is still important in making sourdough bread. Unlike most breads that are made with pure cultures of yeast, sourdough uses an active culture of wild yeast and bacteria that generate acid. This culture is maintained and small amounts—starter culture—are used for each batch of bread. The combination of yeast and acid-producing bacteria is needed for fermentation and gives sourdough bread its unique flavor.

Whether bread is made with wild or cultured yeast, fermentation occurs as the dough rises, releasing carbon dioxide and ethanol. The carbon dioxide is captured in the dough and causes it to rise.

The most important yeast in baking, brewing, and wine making is *Saccharomyces cerevisiae*. This yeast has been used by humans throughout recorded history. Yeast is used as a nutritional supplement because it contains high levels of B vitamins and about fifty percent of yeast is protein. Other yeasts are important pathogens and cause disease.

Over the past few decades, yeasts have become increasingly important in genetic research. They were the first eukaryotes to be manipulated extensively by the techniques of genetic engineering, and they still play the leading role as models for research in eukaryotic cells. In 1996, the genome sequence of *S. cerevisiae*, the first eukaryote to be sequenced entirely, was completed. It contains 6000 genes. With their rapid generation time and a rapidly increasing pool of genetic and biochemical information, the yeasts in general and *S. cerevisiae* in particular are becoming the eukaryotic cells of choice for many types of experiments in molecular and cellular biology. *S. cerevisiae* has become, in this respect, comparable to *Escherichia coli* among the bacteria, and is continuing to provide significant insights into the functioning of eukaryotic systems.

The fungal genome initiative is now under way to provide sequence information on other fungi. Initially fifteen fungi were selected for whole genome sequencing. These fungi were selected based on impact on human health, including plant pathogens that threaten our food supply. The ascomycete *Coccidiodes posadasii* was included because it is endemic in soil in the southwestern portion of the United States, can cause a fatal infection, and has been considered a possible bioterrorism threat. In the U.S. the annual infection rate is about 100,000 individuals, although only a small percentage of infected individuals die. A second important criterion in selecting which fungi to sequence was the potential to provide information on fungal evolution. Key evolutionary branch points were selected and additional genomes are now being considered. Extensive comparisons with the yeast genome are already yielding intriguing information. This new information will complement and expand our understanding of the diverse fungus kingdom.

Ascomycetes undergo karyogamy within a characteristic saclike structure, the ascus. Meiosis follows, resulting in the production of ascospores. Yeasts within this group generally reproduce asexually by budding.

30.3 Fungi participate in many symbioses.

Lichens

Lichens (figure 30.14) are symbiotic associations between a fungus and a photosynthetic partner. The term symbiosis was coined to describe this relationship. While many lichens are excellent examples of mutualism, the kind of symbiotic association that benefits both partners, some fungi are parasitic on their photosynthetic host. Ascomycetes are the fungal partners in all but about 20 of the approximately 15,000 species of lichens estimated to exist. Most of the visible body of a lichen consists of its fungus, but between the filaments of that fungus are cyanobacteria, green algae, or sometimes both (figure 30.15). Specialized fungal hyphae penetrate or envelop the photosynthetic cell walls within them and transfer nutrients directly to the fungal partner. Note that while fungi penetrate the cell wall, they do not penetrate the plasma membrane. Biochemical signals sent out by the fungus apparently direct its cyanobacterial or green algal component to produce metabolic substances that it does not produce when growing independently of the fungus. The fungi in lichens are unable to grow normally without their photosynthetic partners, and the fungi protect their partners from strong light and desiccation. When fungal components of lichens have been experimentally isolated from their photosynthetic partner, they survive, but grow very slowly.

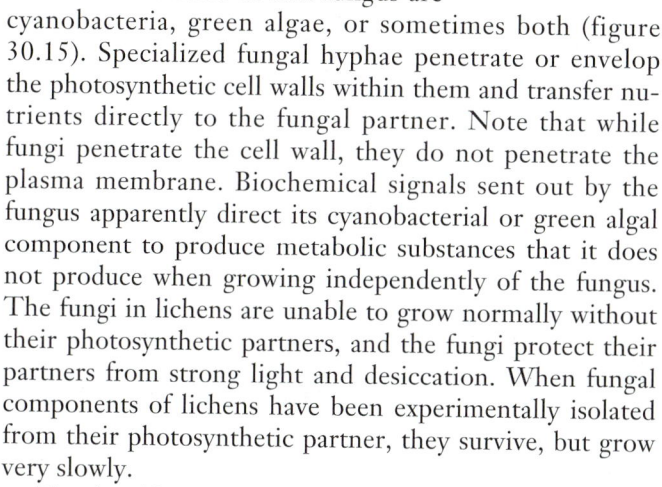

The durable construction of the fungus combined with the photosynthetic properties of its partner have enabled lichens to invade the harshest habitats—the tops of mountains, the farthest northern and southern latitudes, and dry, bare rock faces in the desert. In harsh, exposed areas, lichens are often the first colonists, breaking down the rocks and setting the stage for the invasion of other organisms.

Lichens are often strikingly colored because of the presence of pigments that probably play a role in protecting the photosynthetic partner from the destructive action of the sun's rays. These same pigments may be extracted from the lichens and used as natural dyes. The traditional method of manufacturing Scotland's famous Harris tweed used fungal dyes.

Lichens vary in sensitivity to pollutants in the atmosphere, and some species are used as bioindicators of air quality. Their sensitivity results from their ability to absorb substances dissolved in rain and dew. Lichens are generally absent in and around cities because of automobile traffic and industrial activity, but some are adapted to these conditions. As pollution decreases, lichen populations tend to increase.

FIGURE 30.14
Lichens are found in a variety of habitats. (*a*) A fruticose lichen, *Cladina evansii*, growing on the ground in Florida. (*b*) A foliose ("leafy") lichen, *Parmotrema gardneri*, growing on the bark of a tree in a mountain forest in Panama.

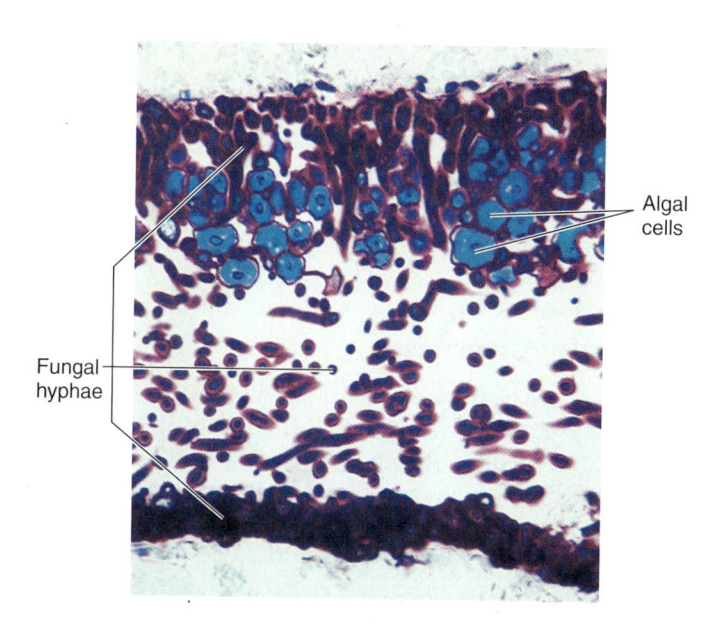

FIGURE 30.15
Stained section of a lichen. This section (250×) shows fungal hyphae (*purple*) more densely packed into a protective layer on the top and, especially, the bottom layer of the lichen. The blue cells near the upper surface of the lichen are those of a green alga. These cells supply carbohydrate to the fungus.

> Lichens are symbiotic associations between a fungus—an ascomycete in all but a very few instances—and a photosynthetic partner, which may be a green alga or a cyanobacterium or both.

Mycorrhizae

The roots of about 90% of all known species of vascular plants normally are involved in mutualistic symbiotic relationships with certain kinds of fungi. It has been estimated that these fungi probably amount to 15% of the total weight of the world's plant roots. Associations of this kind are termed **mycorrhizae** (Greek *mykes*, "fungus," + *rhiza*, "roots"). The fungi in mycorrhizal associations function as extensions of the root system. The fungal hyphae dramatically increase the amount of soil contact and total surface area for absorption. When mycorrhizae are present, they aid in the direct transfer of phosphorus, zinc, copper, and other nutrients from the soil into the roots. The plant, on the other hand, supplies organic carbon to the fungus, so the system is an example of mutualism.

There are two principal types of mycorrhizae (figure 30.16). In **arbuscular mycorrhizae,** the fungal hyphae penetrate the outer cells of the plant root, forming coils, swellings, and minute branches, and also extend out into the surrounding soil. In **ectomycorrhizae,** the hyphae surround but do not penetrate the cell walls of the roots. In both kinds of mycorrhizae, the mycelium extends far out into the soil. A single root may associate with many fungal species, dividing the root at a millimeter-by-millimeter level.

Arbuscular Mycorrhizae

Arbuscular mycorrhizae are by far the more common of the two types, involving roughly 70% of all plant species. The fungal component in them are Glomales, a monophyletic group that arose within one of the zygomycete lineages. The Glomales are associated with more than 200,000 species of plants. Unlike mushrooms, none of the Glomales produce aboveground fruiting structures, and as a result, it is difficult to arrive at an accurate count of the number of extant species. Arbuscular mycorrhizal fungi are being studied intensively because they are potentially capable of increasing crop yields with lower phosphate and energy inputs.

The earliest fossil plants often show arbuscular mycorrhizal roots. Such associations may have played an important role in allowing plants to colonize land. The soils available at such times would have been sterile and completely lacking in organic matter. Plants that form mycorrhizal associations are particularly successful in infertile soils; considering the fossil evidence, the suggestion that mycorrhizal associations found in the earliest plants helped them succeed on such soils seems reasonable. In addition, the closest living relatives of early vascular plants surviving today continue to depend strongly on mycorrhizae.

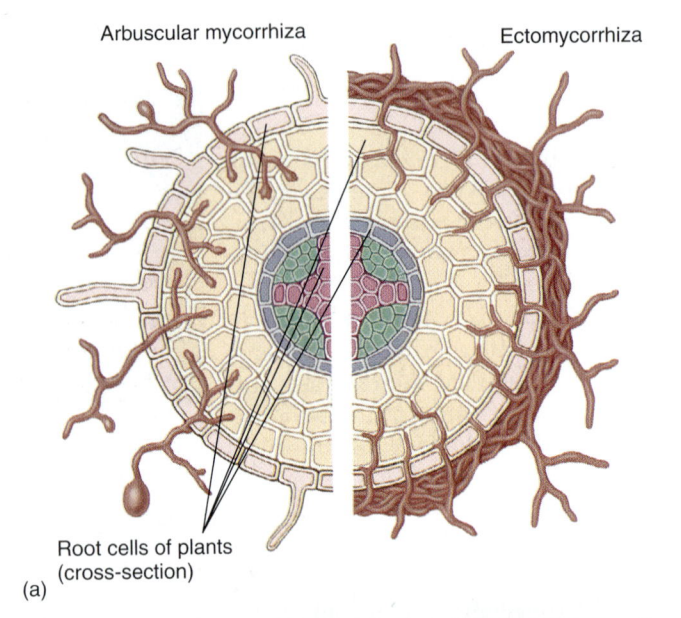

FIGURE 30.16
Arbuscular mycorrhizae and ectomycorrhizae. (*a*) In arbuscular mycorrhizae, fungal hyphae penetrate the root cells of plants. In ectomycorrhizae, fungal hyphae do not penetrate root cells, but grow around and extend between the cells.
(*b*) Ectomycorrhizae on the roots of pines. From left to right are yellow-brown mycorrhizae formed by *Pisolithus*, white mycorrhizae formed by *Rhizopogon*, and pine roots not associated with a fungus.

Some nonphotosynthetic plants cheat their fungal partner. The symbiosis is one-way because the plant has no photosynthetic resources to offer. Instead of a two-partner symbiosis, a tripartite symbiosis is established. The fungal mycelium extends between a photosynthetic plant and a nonphotosynthetic, parasitic plant. This third member of the symbiosis is called an *epiparasite*. Not only does it obtain phosphate from the fungus, but it uses the fungus to channel carbohydrates from the photosynthetic plant to itself. Epiparasitism also occurs in ectomycorrhizal symbiosis. Symbiotic relationships shift from mutualism to parasitism or from parasitism to mutualism over evolutionary time among even closely related fungi.

Ectomycorrhizae

Ectomycorrhizae (see figure 30.16b) involve far fewer kinds of plants than do arbuscular mycorrhizae—perhaps a few thousand. Most ectomycorrhizal hosts are forest trees, such as pines, oaks, birches, willows, eucalyptus, dipterocarps, and caesalpinioid legumes. The fungal components in most ectomycorrhizae are basidiomycetes, but some are ascomycetes. Most ectomycorrhizal fungi are not restricted to a single species of plant, and most ectomycorrhizal plants form associations with many ectomycorrhizal fungi. Different combinations have different effects on the physiological characteristics of the plant and its ability to survive under different environmental conditions. At least 5000 species of fungi are involved in ectomycorrhizal relationships.

Orchids are absolutely dependent on symbiosis with ectomycorrhizal fungi, especially in the early stages of development. In this relationship, the orchid has become parasitic upon the fungus.

Mycorrhizae are symbiotic associations between plants and fungi.

Endophytes

Endophytic fungi live inside plants, actually in the intercellular spaces. Found throughout the plant kingdom, many of these relationships may be examples of parasitism or commensalism. However, there is growing evidence that some of these fungi protect their hosts from herbivores by producing chemical toxins or deterrents. Most often, the fungus synthesizes alkaloids that protect the plant. As you will see in chapter 39, plants also synthesize a wide range of alkaloids, many of which serve to defend the plant.

One way to assess whether or not an endophyte is enhancing the health of its host plant is to grow plots of plants with and without the same endophyte. An experiment with Italian ryegrass, *Lolium multiflorum*, demonstrated that it is more resistant to aphid parasitism when an endophytic fungus, *Neotyphodium*, is present.

Endophytes can protect their hosts from parasites by producing chemical toxins.

Mutualistic Animal Symbioses

A range of mutualistic fungal–animal symbioses has been identified. For example, ruminant animals host fungi in their gut. Grassy diets have a high content of cellulose and lignin that cannot be digested by the grazing animal. Fungal enzymes, however, release nutrients that would otherwise be unavailable to the animal. The fungus gains a nutrient-rich environment in exchange.

To further explore fungal–animal mutualisms, we will focus on a tripartite symbiosis involving ants, plants, and fungi. Leaf-cutter ants are the dominant herbivore in the New World tropics. These ants, members of the tribe Attini, have an obligate symbiosis with specific fungi that they have domesticated and maintain in an underground garden. The ants provide fungi with leaves to eat and protection from pathogens and other predators (figure 30.17). The fungi are the ants' food source. Depending on the species of ant, the ant nest can be as small as a golf ball or as large as 50 centimeters in diameter and many feet deep. Some nests are inhabited by millions of leaf-cutter ants that maintain fungal gardens. These social insects have a caste system, and different ants have specific roles. Traveling on trails as long as 200 meters, leaf-cutter ants search for foliage for their fungi. A colony of ants can defoliate an entire tree in a day. This ant farmer–fungi symbiosis evolved multiple times and may have occurred as early as 50 million years ago.

Fungi have coevolved with animals in mutualistic relationships.

FIGURE 30.17
Ant–fungal symbiosis. Ants farming their fungal garden.

Fungal Parasites and Pathogens

Animal Parasites and Pathogens

Fungi can create devastating human diseases that are often difficult to treat because of their close phylogenetic relationship with animals. Yeast ascomycetes are important pathogens that cause diseases such as thrush; *Candida*, for example, causes common oral or vaginal infections. Mold allergies are common, and mold-infested "sick" buildings pose concerns for inhabitants. Individuals with suppressed immune systems and people undergoing steroid treatments for inflammation disorders are particularly at risk for fungal disease.

An example of a parasitic fungal–animal symbiosis is chytridiomycosis, first identified in 1998 as an emerging infectious disease of amphibians. Amphibian populations have been declining worldwide for over three decades. The decline correlates with the presence of the chytrid *Batrachochytrium dendrobatidis* (see figure 30.2a), identified after extensive studies of frog carcasses. Sick and dead frogs were more likely than healthy frogs to have flasklike structures encased in their skin (figure 30.18). The connection with *B. dendrobatidis* has been supported by DNA sequence data, by isolating and culturing the chytrid, and by infecting healthy frogs with the organism and replicating disease symptoms. Bathing frogs in antifungal drugs can halt the disease or even eliminate the chytrids. How the disease emerged simultaneously on different continents is a yet unsolved mystery. Both environmental change and carriers are being considered.

Plant Parasites and Pathogens

Pathogenic fungal–plant symbioses are numerous. Crop losses are extensive throughout the world. Fungal pathogens of plants can also harm the animals that consume the plants. *Fusarium* species growing on spoiled food produce highly toxic substances, including vomitoxin, which has been implicated in brain damage in the southwestern United States.

Aflatoxins, which are among the most carcinogenic compounds known, are produced by some *Aspergillus flavus* strains growing on corn, peanuts, and cotton seed (figure 30.19). Aflatoxins can also damage the kidneys and the nervous system. Most developed countries have legal limits on the concentration of aflatoxin permitted in different foods. More recently, aflatoxins have been considered as possible bioterrorism agents. Hot, humid summers are especially conducive to the growth of this fungus. Monoculture has enhanced its spread, while crop rotation with resistant crops can help control the spread of *Aspergillus*.

Fungi can severely harm or kill both plants and animals.

FIGURE 30.18
Frog killed by chytridiomycosis. Lesions formed by the chytrid can be seen on the abdomen of this frog.

FIGURE 30.19
***Aspergillus flavus* infects maize and can produce aflatoxins that are harmful to animals.** (*a*) Maize (corn) infected with the fungus. (*b*) A photomicrograph of *Aspergillus flavus* conidia.

Concept Review

For interactive testing, visit the Online Learning Center with PowerWeb at www.mhhe.com/Raven7

30.1 The fungi share several key characteristics.

Distinctive Fungal Features
- Recent analysis indicates that fungi are more closely related to animals than to plants. (p. 600)
- Fungi are very diverse, but share several characteristics: Fungi are heterotrophs; they have several cell types although they are primarily filamentous; some have a dikaryon stage; they have cell walls that include chitin; and they exhibit nuclear mitosis. (p. 600)

The Body of a Fungus
- Many fungi have slender filaments called hyphae that are continuous, composed of branching tubes, or made up of long chains of cells joined end-to-end and divided by septa. (p. 601)
- A mycelium is a mass of connected hyphae that grows into the material in which the fungus is feeding. (p. 601)
- Cell walls are formed of polysaccharides and chitin. (p. 601)

How Fungi Reproduce
- Fungi exhibit both sexual and asexual reproduction. (p. 602)
- Sexual reproduction occurs when the haploid hyphae of two individuals fuse. (p. 602)
- Most fungi use spores to reproduce. (p. 602)

How Fungi Obtain Nutrients
- All fungi secrete digestive enzymes into their surroundings and absorb the organic molecules produced via external digestion. (p. 603)

Metabolic Pathways
- Humans use the metabolic pathways of fungi for commercial products such as cheese, soy sauce, bread, beer, and wine. (p. 603)
- Fungi are also used in bioremediation to rid water or soils of environmental contamination. (p. 603)

Ecology of Fungi
- Fungi are the principal decomposers in the biosphere and are nearly the only organisms capable of breaking down lignin. (p. 604)
- Symbioses fall into many categories, including obligate and facultative. (p. 604)
- Fungal species are responsible for billions of dollars of agricultural losses annually, and cause many human and animal diseases, including ringworm and athlete's foot. (p. 604)
- Lichens are mutualistic associations between fungi and algae or cyanobacteria, while mycorrhizae are specialized mutualistic associations between plant roots and fungi. (p. 604)

30.2 There are four major groups of fungi.

Phylogenetic Relationships
- Although the phylogeny of fungi changes rapidly, currently they are divided into four groups (Chytridiomycota, Zygomycota, Basidiomycota and Ascomycota. (p. 605)

Chytridiomycota
- The chytrids are aquatic, flagellated fungi that are the closest living relatives to the first fungi. (p. 606)

Zygomycota
- The zygomycetes are a very diverse group that includes many of the common bread molds. (p. 607)
- Zygomycotes are named after a characteristic feature of their sexual phase, the zygosporangium. (p. 607)
- In zygomycetes, asexual reproduction occurs much more frequently than sexual reproduction. (p. 607)

Basidiomycota
- Basidiomycetes include mushrooms, toadstools, puffballs, and many plant pathogens, such as smuts and rusts. (p. 608)
- Basidiomycetes are named for their characteristic sexual reproductive structure, the basidium. (p. 608)

Ascomycota
- The ascomycetes contain about 75% of known fungi, including bread yeasts, common molds, morels and truffles, and many plant pathogens, such as chestnut blight. (p. 609)
- Ascomycetes are named for their characteristic reproductive structure, the ascus. (p. 609)
- In ascomycetes, asexual reproduction is very common and takes place by conidia produced at the ends of conidiophores. (p. 609)

30.3 Fungi participate in many symbioses.

Lichens
- Lichens are symbiotic relationships between a fungus and a photosynthetic partner that colonize even the harshest habitats on the planet. (p. 611)
- Lichens vary in their sensitivity to atmospheric pollutants, and some are used as air quality indicators. (p. 611)

Mycorrhizae
- Mycorrhizae are mutualistic relationships between plant roots and certain fungi. (p. 612)
- The fungi in mycorrhizal associations increase the amount of soil contact and the total absorption area. (p. 612)
- Two principal types of mycorrhizae are arbuscular and ectomycorrhizae. (p. 612)

Endophytes
- Endophytic fungi live in the intercellular spaces inside living plants. (p. 613)

Mutualistic Animal Symbioses
- A range of fungal–animal symbioses has been identified. These include ruminants and the fungi in their gut, and leaf-cutter ants and specific fungal species. (p. 613)

Fungal Parasites and Pathogens
- *Candida* can cause common oral or vaginal infections. (p. 614)
- A chytrid can cause chytridiomycosis in amphibians.
- Crop losses due to parasitic or pathogenic fungi are extensive throughout the world. (p. 614)
- Aflatoxins can cause kidney or nervous system damage. (p. 614)

Test Your Understanding

For interactive testing, visit the Online Learning Center with PowerWeb at www.mhhe.com/Raven7

Self Test

1. How many species of fungi are thought to exist?
   a. 1500
   b. 150,000
   c. 1.5 million
   d. 1.5 billion
2. Which of the following is *not* a characteristic of the fungi?
   a. They are all absorptive heterotrophs.
   b. They have cell walls made of chitin.
   c. Mitosis takes place within the nuclear membrane.
   d. They are all motile.
3. A mycelium is
   a. a specialized reproductive structure of a fungus.
   b. a mass of connected fungal hyphae.
   c. a mutualistic relationship between a fungus and a plant.
   d. a partition between the cells of fungal hyphae.
4. Which of the following statements best describes fungi?
   a. All are eukaryotic, multicellular autotrophs.
   b. All are eukaryotic heterotrophs that feed by absorption.
   c. All are prokaryotic, multicellular autotrophs.
   d. All are eukaryotic heterotrophs that feed by ingestion.
5. A wildlife pathologist is examining some skin tissue from a dead frog. She notes the presence of a fungus. She cultures some of the fungal cells and notices that some of the cells are flagellated. She concludes that the frog has a fungal disease caused by
   a. an ascomycete.
   b. a zygomycete.
   c. a basidiomycete.
   d. a chytrid.
6. You are walking in the woods and see a fungus that is unfamiliar to you. You remove a reproductive structure, and take it home to examine further. When you look at it under the microscope, you find a zygosporangium. Based on this information alone, this fungus is a(n)
   a. zygomycete.
   b. chytrid.
   c. basidiomycete.
   d. ascomycete.
7. A basidium is typically observed in the common
   a. bread mold.
   b. gilled mushroom.
   c. lichen.
   d. chytrid.
8. An ascomycete can be distinguished from other fungi
   a. because ascomycetes are mainly diploid.
   b. because ascomycetes lack a dikaryotic phase.
   c. by the presence of eight sexual spores in an ascus.
   d. by the presence of gills on the mycelium.
9. A lichen can be described as a mutualistic symbiosis between an ascomycete and a(n)
   a. chytrid.
   b. archaebacterium.
   c. green alga.
   d. angiosperm root.

Test Your Visual Understanding

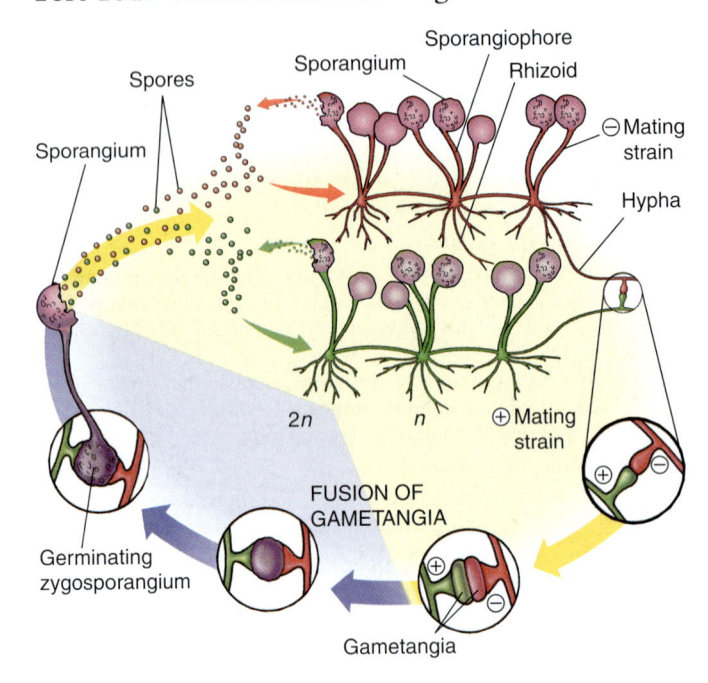

1. This is a diagram of the life cycle of a typical bread mold.
   a. Is this sexual or asexual reproduction?
   b. On this diagram, indicate where in the life cycle meiosis takes place.
   c. Label the reproductive structures of this fungus.
   d. To what group of fungi does this organism belong?

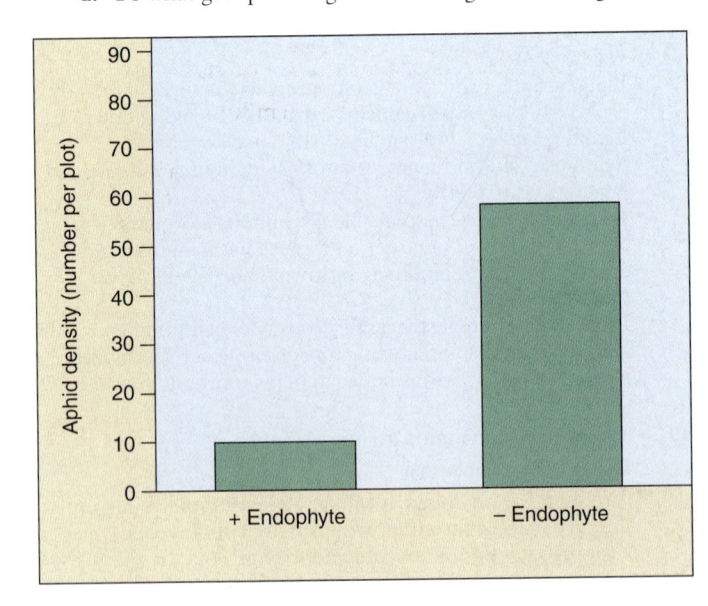

2. Two equal-sized plots of Italian ryegrass were grown, one with and one without endophytes. Each plot was exposed to aphids in the experiment shown above.
   a. What conclusion would you draw from the experimental results?
   b. What advantages and disadvantages does the ryegrass have in an endophytic relationship?

31

Overview of Animal Diversity

Concept Outline

31.1 Animals are multicellular heterotrophs without cell walls.

Some General Features of Animals. Animals lack cell walls and move more rapidly and in more complex ways than other organisms.

31.2 Animals are a very diverse kingdom.

The Traditional Classification of Animals. The 35 animal phyla have traditionally been arrayed on the animal family tree according to certain key features of the animal body plan.

31.3 The animal body plan has undergone many changes.

Five Key Transitions in Body Plan. Over the course of animal evolution, five key transitions in body organization have occurred.

31.4 The way we classify animals is being reevaluated.

A New Look at the Metazoan Family Tree. Traditional and molecular approaches lead to somewhat different views of the animal family tree.

"Evo-Devo" and the Roots of the Animal Family Tree. The 35 traditional animal phyla all evolved before the Cambrian period.

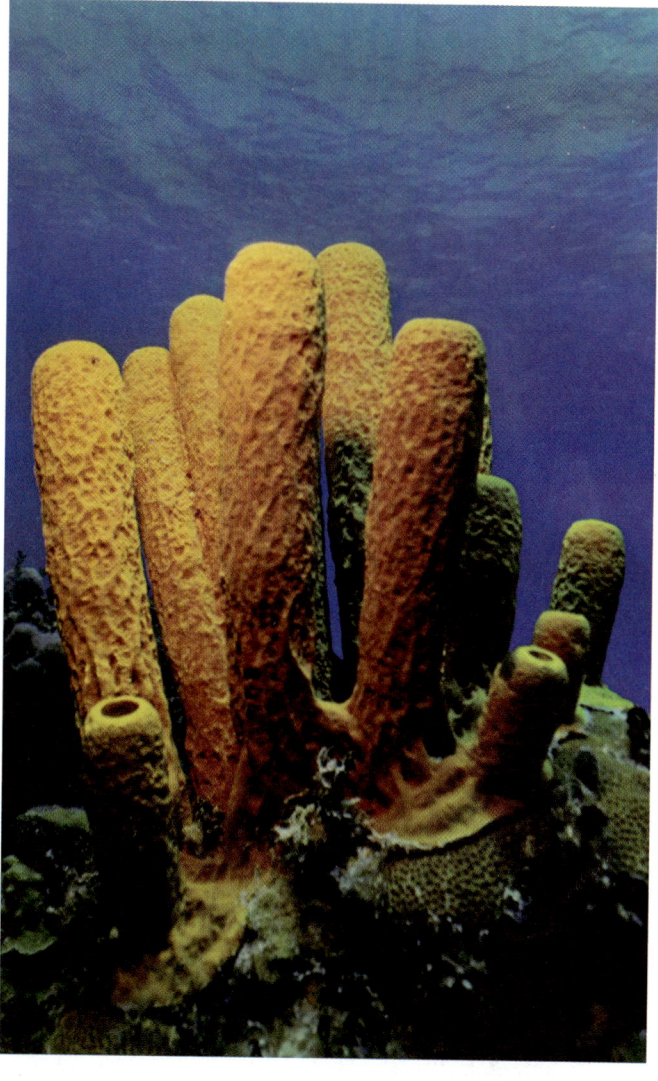

FIGURE 31.1
Biologists used to think that sponges were marine plants. Perhaps the earliest animals to evolve, sponges have motile larvae that may travel thousands of miles, but as adults, sponges are sessile and remain attached to one spot on the ocean floor.

We will now explore the great diversity of animals, the result of a long evolutionary history. Animals, constituting millions of species, are among the most abundant living things. Found in every conceivable habitat, they bewilder us with their diversity. More than a million species have been described, and several million more are thought to await discovery. Despite their great diversity, all animals have much in common. For example, locomotion is a distinctive characteristic of animals, although not all animals move about. Early naturalists thought that sponges (figure 31.1) and corals were plants because the adults seemed rooted to one place and did not move about. Biologists have long debated the evolutionary relationships among animal groups. In recent years, molecular tools are providing new insights and leading biologists to reconsider the animal tree of life.

31.1 Animals are multicellular heterotrophs without cell walls.

Some General Features of Animals

Animals are the eaters, or consumers, of the earth. Animals are a very diverse group—no one criterion fits all—but several characteristics are of major importance (table 31.1): (1) Animals are heterotrophs and must ingest plants, algae, or other animals for nourishment. (2) All animals are multicellular, and unlike plants and protists, animal cells lack cell walls. (3) Animals are able to move from place to place. (4) Animals are very diverse in form and habitat. (5) Most animals reproduce sexually. (6) Animals have a characteristic pattern of embryonic development and possess unique tissues.

> Animals are complex multicellular organisms typically characterized by high mobility and heterotrophy. Most animals also possess internal tissues and reproduce sexually.

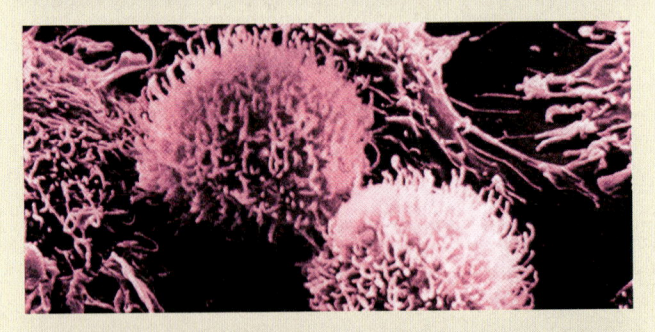

Table 31.1 General Features of Animals

Heterotrophs. Unlike autotrophic plants and algae, animals cannot construct organic molecules from inorganic chemicals. All animals are heterotrophs—that is, they obtain energy and organic molecules by ingesting other organisms. Some animals (herbivores) consume autotrophs; other animals (carnivores) consume heterotrophs; and still others (detritivores) consume decomposing organisms.

Multicellular. All animals are multicellular, often with complex bodies like that of this brittlestar *(right)*. The unicellular heterotrophic organisms called protozoa, which were at one time regarded as simple animals, are now considered members of the large and diverse kingdom Protista, discussed in chapter 28.

No Cell Walls. Animal cells are distinct among those of multicellular organisms because they lack rigid cell walls and are usually quite flexible, as are these cancer cells. The many cells of animal bodies are held together by extracellular lattices of structural proteins such as collagen. Other proteins form a collection of unique intercellular junctions between animal cells.

Active Movement. The ability of animals to move more rapidly and in more complex ways than members of other kingdoms is perhaps their most striking characteristic, one that is directly related to the flexibility of their cells and the evolution of nerve and muscle tissues. A remarkable form of movement unique to animals is flying, an ability that is well developed among vertebrates and insects such as this butterfly. The only terrestrial vertebrate group never to have evolved flight is the amphibians.

Table 31.1 General Features of Animals *Continued*.

Diverse in Form. Almost all animals (99%) are **invertebrates**, which, like this millipede, lack a backbone. Of the estimated 10 million living animal species, only 42,500 have a backbone and are referred to as **vertebrates**. Animals are very diverse in form, ranging in size from organisms too small to see with the unaided eye to enormous whales and giant squids.

Diverse in Habitat. The animal kingdom includes about 35 phyla, most of which, like these jellyfish (phylum Cnidaria), occur in the sea. Far fewer phyla occur in fresh water, and fewer still occur on land. Members of three successful marine phyla, Arthropoda (insects), Mollusca (snails), and Chordata (vertebrates), also dominate animal life on land.

Sexual Reproduction. Most animals reproduce sexually, as these tortoises are doing. Animal eggs, which are nonmotile, are much larger than the small, usually flagellated sperm. In animals, cells formed in meiosis function directly as gametes. The haploid cells do not divide by mitosis first, as they do in plants and fungi, but rather fuse directly with each other to form the zygote. Consequently, with a few exceptions, there is no counterpart among animals to the alternation of haploid (gametophyte) and diploid (sporophyte) generations characteristic of plants.

Embryonic Development. Most animals have a similar pattern of embryonic development. The zygote first undergoes a series of mitotic divisions, called *cleavage*, and like this dividing frog egg, becomes a solid ball of cells, the **morula**, and then a hollow ball of cells, the **blastula**. In most animals, the blastula folds inward at one point to form a hollow sac with an opening at one end called the **blastopore**. An embryo at this stage is called a **gastrula**. The subsequent growth and movement of the cells of the gastrula differ widely from one phylum of animals to another.

Unique Tissues. The cells of all animals except sponges are organized into structural and functional units called **tissues**, collections of cells that have joined together and are specialized to perform a specific function. Animals are unique in having two tissues associated with movement: (1) muscle tissue, which powers animal movement, and (2) nervous tissue, which conducts signals among cells. Neuromuscular junctions, where nerves connect with muscle tissue, are shown here.

Chapter 31 Overview of Animal Diversity 619

31.2 Animals are a very diverse kingdom.

The Traditional Classification of Animals

The multicellular animals, or metazoans, are traditionally divided into 35 distinct phyla. The diversity of animals can be clearly seen in table 31.2, which describes key characteristics of 20 of the more widely studied animal phyla.

How can we make sense of this great diversity? Taxonomists, attempting to sort out which phyla are related, have traditionally created phylogenies (family trees) by comparing anatomical features and aspects of embryological development. Little other data was available, and a broad consensus emerged over the last century concerning the main branches of the animal family tree.

The First Branch: Tissues

Taxonomists traditionally divide the kingdom Animalia into two main branches: **Parazoa** ("beside animals")—animals that for the most part lack a definite symmetry and possess neither tissues nor organs, mostly composed of the sponges, phylum Porifera; and **Eumetazoa** ("true animals")—animals that have a definite shape and symmetry and, in most cases, tissues organized into organs and organ systems.

The Second Branch: Symmetry

The eumetazoan branch of the animal family tree itself has two principal branches, differing in the nature of the embryonic layers that form during development and go on to differentiate into the tissues of the adult animal: Eumetazoans of the subgroup **Radiata** (having radial symmetry) have two layers, an outer *ectoderm* and an inner *endoderm*, and thus are called diploblastic. All other eumetazoans, the **Bilateria** (having bilateral symmetry), are triploblastic and produce a third layer, the *mesoderm*, between the ectoderm and endoderm (figure 31.2).

Further Branches

Further branches of the animal family tree were assigned by taxonomists by comparing traits that seemed profoundly important to the evolutionary history of phyla, key features of the body plan shared by all animals belonging to that branch. Thus, the bilateral animals were split into groups with a body cavity and those without; animals with a body cavity were split into those with a coelom (body cavity enclosed by mesoderm) and those without; animals with a coelom were split into those whose coelom derived from the digestive tube and those that did not; and so on.

Because of the either-or nature of the categories set up by traditional taxonomists, this approach has produced a family tree with a lot of paired branches (figure 31.3). The arbitrary nature of the divisions has always been obvious to biologists, but until recently most biologists felt this phylogeny faithfully represented the general nature of the evolutionary history of metazoans. Although new molecular techniques have now suggested other ways of classifying animals (see section 31.4), the traditional phylogeny is still employed by most taxonomists and will provide the framework for discussion in this book.

(a) Parazoa: no tissues
(b) Eumetazoan: tissues
(c) Radiata: radial symmetry
(d) Bilateria: bilateral symmetry

FIGURE 31.2
Broad groupings of the kingdom Animalia. (*a*, *b*) Animals are divided into Parazoa, animals that lack tissues and a definite symmetry, and Eumetazoa, animals that have symmetry and organized tissues. (*c*, *d*) Eumetazoans are further divided into Radiata, animals with radial symmetry, and Bilateria, animals with bilateral symmetry.

> Animals are traditionally classified into some 35 phyla. The evolutionary relationships among the animal phyla have been inferred until recent years by assuming relatedness of phyla that share certain fundamental morphological characters, which are assumed to have arisen only once.

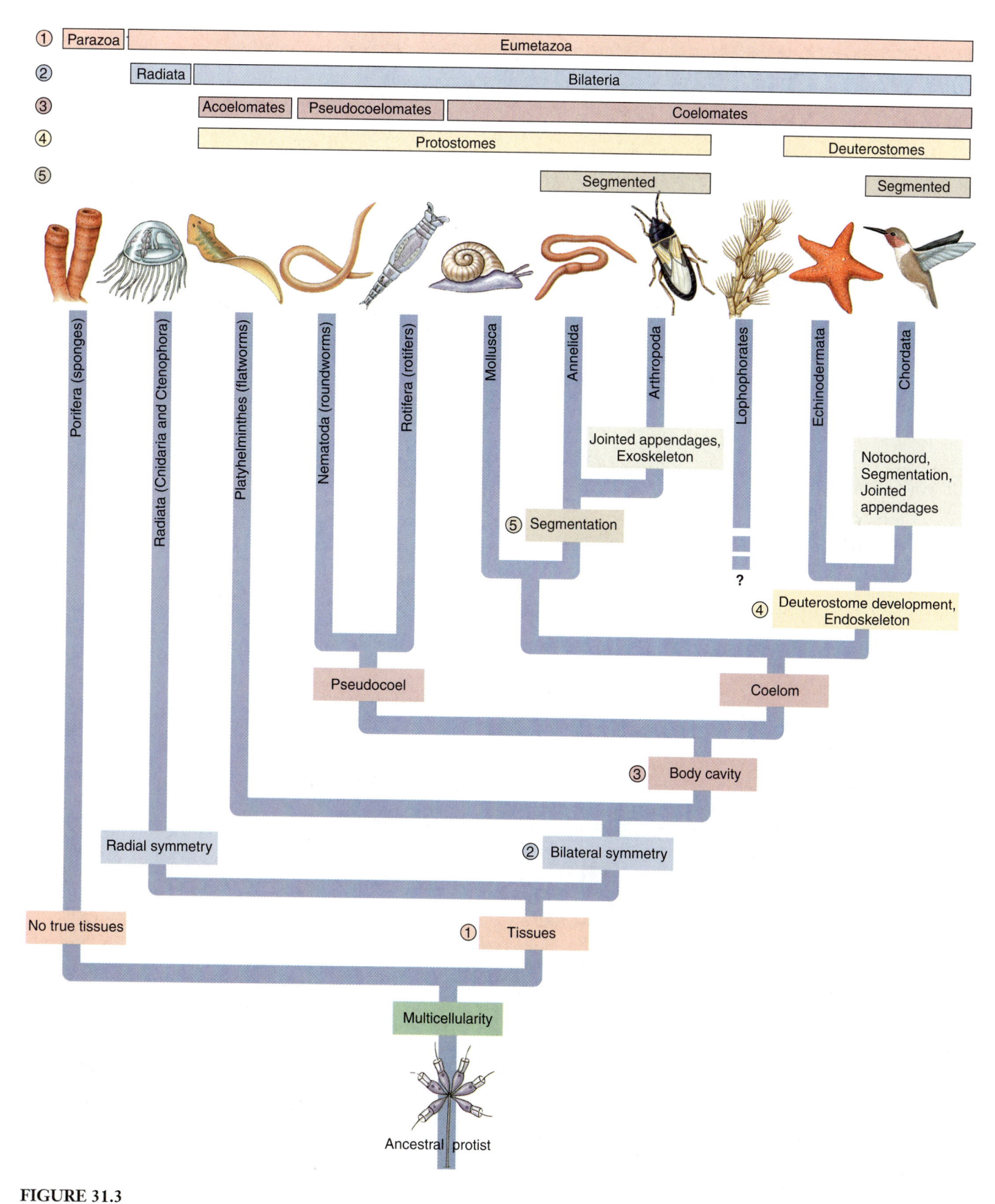

FIGURE 31.3
A possible phylogeny of the major groups of the kingdom Animalia. Transitions in the animal body plan are identified along the branches; five key advances are the evolution of tissues, bilateral symmetry, a body cavity, deuterostome development, and segmentation.

Chapter 31 Overview of Animal Diversity 621

Table 31.2 The Major Animal Phyla

Phylum	Typical Examples	Key Characteristics	Approximate Number of Named Species
Arthropoda (arthropods)	Beetles, other insects, crabs, spiders	Most successful of all animal phyla; chitinous exoskeleton covering segmented bodies with paired, jointed appendages; many insect groups have wings.	1,000,000
Mollusca (mollusks)	Snails, oysters, octopuses, nudibranchs	Soft-bodied coelomates whose bodies are divided into three parts: head-foot, visceral mass, and mantle; many have shells; almost all possess a unique rasping tongue, called a radula; 35,000 species are terrestrial.	110,000
Chordata (chordates)	Mammals, fish, reptiles, birds, amphibians	Segmented coelomates with a notochord; possess a dorsal nerve cord, pharyngeal slits, and a tail at some stage of life; in vertebrates, the notochord is replaced during development by the spinal column; 20,000 species are terrestrial.	56,000
Platyhelminthes (flatworms)	Planaria, tapeworms, liver flukes	Solid, unsegmented, bilaterally symmetrical worms; no body cavity; digestive cavity, if present, has only one opening.	20,000
Nematoda (roundworms)	*Ascaris*, pinworms, hookworms, *Filaria*	Pseudocoelomate, unsegmented, bilaterally symmetrical worms; tubular digestive tract passing from mouth to anus; tiny; without cilia; live in great numbers in soil and aquatic sediments; some are important animal parasites.	20,000
Annelida (segmented worms)	Earthworms, polychaetes, tube worms, leeches	Coelomate, serially segmented, bilaterally symmetrical worms; complete digestive tract; most have bristles called setae on each segment that anchor them during crawling.	12,000
Cnidaria (cnidarians)	Jellyfish, hydra, corals, sea anemones	Soft, gelatinous, radially symmetrical bodies whose digestive cavity has a single opening; possess tentacles armed with stinging cells called cnidocytes that shoot sharp harpoons called nematocysts; almost entirely marine.	10,000
Echinodermata (echinoderms)	Sea stars, sea urchins, sand dollars, sea cucumbers	Deuterostomes with radially symmetrical adult bodies; endoskeleton of calcium plates; five-part body plan and unique water-vascular system with tube feet; able to regenerate lost body parts; marine.	6,000
Porifera (sponges)	Barrel sponges, boring sponges, basket sponges, vase sponges	Asymmetrical bodies without distinct tissues or organs; saclike body consists of two layers breached by many pores; internal cavity lined with food-filtering cells called choanocytes; most are marine (150 species live in fresh water).	5,150

Table 31.2 The Major Animal Phyla *Continued*.

Phylum	Typical Examples	Key Characteristics	Approximate Number of Named Species
Bryozoa (moss animals)	*Bowerbankia*, *Plumatella*, sea mats, sea moss	Microscopic, aquatic deuterostomes that form branching colonies; possess circular or U-shaped row of ciliated tentacles for feeding called a lophophore that usually protrudes through pores in a hard exoskeleton; also called Ectoprocta because the anus, or proct, is external to the lophophore; marine or freshwater.	4,000
Rotifera (wheel animals)	Rotifers	Small, aquatic pseudocoelomates with a crown of cilia around the mouth resembling a wheel; almost all live in fresh water.	2,000
Five Phyla of Minor Worms	Velvet worms, acorn worms, arrow worms, giant tube worms	**Chaetognatha** (arrow worms): coelomate deuterostomes; bilaterally symmetrical; large eyes (some) and powerful jaws.	980
		Hemichordata (acorn worms): marine worms with dorsal *and* ventral nerve cords.	
		Onychophora (velvet worms): protostomes with a chitinous exoskeleton; evolutionary relicts.	
		Pogonophora (tube worms): sessile deep-sea worms with long tentacles; live within chitinous tubes attached to the ocean floor.	
		Nemertea (ribbon worms): acoelomate, bilaterally symmetrical marine worms with long, extendable proboscis.	
Brachiopoda (lamp shells)	*Lingula*	Like bryozoans, possess a lophophore, but within two clamlike shells; more than 30,000 species known as fossils.	300
Ctenophora (sea walnuts)	Comb jellies, sea walnuts	Gelatinous, almost transparent, often bioluminescent marine animals; eight bands of cilia; largest animals that use cilia for locomotion; complete digestive tract with anal pore.	100
Phoronida (phoronids)	*Phoronis*	Lophophorate tube worms; often live in dense populations; unique U-shaped gut, instead of the straight digestive tube of other tube worms.	12
Loricifera (loriciferans)	*Nanaloricus mysticus*	Tiny, bilaterally symmetrical, marine pseudocoelomates that live in spaces between grains of sand; mouthparts include a unique flexible tube. A recent addition to the traditional 35 animal phyla, loriciferans were discovered in 1983.	6

31.3 The animal body plan has undergone many changes.

Five Key Transitions in Body Plan

1. Evolution of Tissues

The simplest animals, the Parazoa, lack both defined tissues and organs. Characterized by the sponges, these animals exist as aggregates of cells with minimal intercellular coordination. All other animals, the Eumetazoa, have distinct tissues with highly specialized cells. The evolution of tissues is the first key transition in the animal body plan.

2. Evolution of Bilateral Symmetry

Sponges also lack any definite symmetry, growing asymmetrically as irregular masses. Virtually all other animals have a definite shape and symmetry that can be defined along an imaginary axis drawn through the animal's body. Animals with symmetry belong to either the Radiata, animals with radial symmetry, or the Bilateria, animals with bilateral symmetry.

Radial Symmetry. Symmetrical bodies first evolved in marine animals belonging to two phyla: Cnidaria (jellyfish, sea anemones, and corals) and Ctenophora (comb jellies). The bodies of members of these two phyla, the Radiata, exhibit **radial symmetry**, a body design in which the parts of the body are arranged around a central axis in such a way that any plane passing through the central axis divides the organism into halves that are approximate mirror images (figure 31.4a).

Bilateral Symmetry. The bodies of all other animals, the Bilateria, are marked by a fundamental **bilateral symmetry**, a body design in which the body has a right and a left half that are mirror images of each other (figure 31.4b). A bilaterally symmetrical body plan has a top and a bottom, better known respectively as the *dorsal* and *ventral* portions of the body. It also has a front, or *anterior* end, and a back, or *posterior* end. In some higher animals, such as echinoderms (starfish), the adults are radially symmetrical, but even in them the larvae are bilaterally symmetrical.

Bilateral symmetry constitutes the second major evolutionary advance in the animal body plan. This unique form of organization allows parts of the body to evolve in different ways, permitting different organs to be located in different parts of the body. Also, bilaterally symmetrical animals move from place to place more efficiently than radially symmetrical ones, which, in general, lead a sessile or passively floating existence. Due to their increased mobility, bilaterally symmetrical animals are efficient in seeking food, locating mates, and avoiding predators.

The bilaterally symmetrical eumetazoans produce three germ layers: an outer **ectoderm**, an inner **endoderm**, and a third layer, the **mesoderm**, between the ectoderm and endoderm. In general, the outer coverings of the body and the nervous system develop from the ectoderm; the digestive organs and intestines develop from the endoderm; and the skeleton and muscles develop from the mesoderm. The radially symmetrical animals have only two layers, the endoderm and the ectoderm, and sponges lack any germ layers.

Much of the nervous system in bilaterally symmetrical animals is in the form of major longitudinal nerve cords. In a very early evolutionary advance, nerve cells became grouped in the anterior end of the body. These nerve cells probably first functioned mainly to transmit impulses from the anterior sense organs to the rest of the nervous system. This trend ultimately led to the evolution of a definite head and brain area, a process called **cephalization**, as well as to the increasing dominance and specialization of these organs in the more advanced animal phyla.

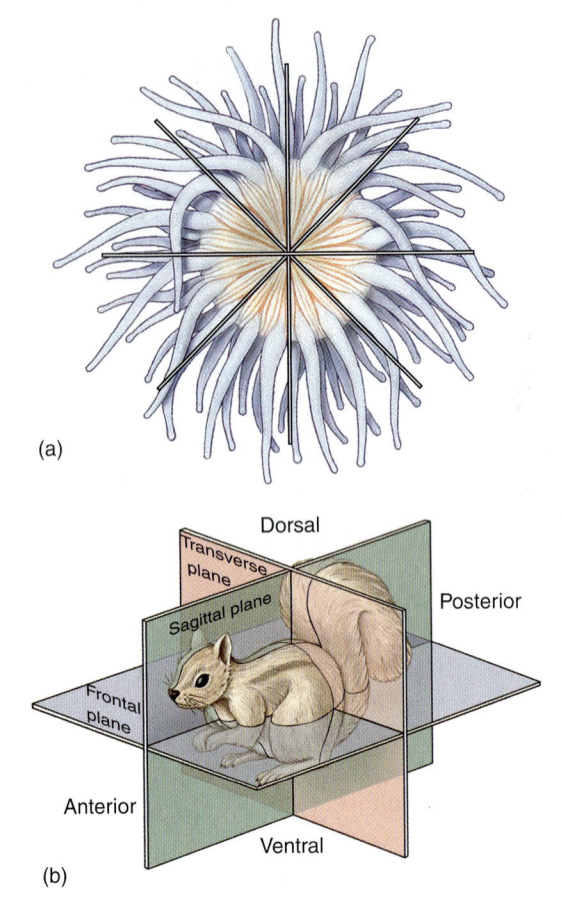

FIGURE 31.4

A comparison of radial and bilateral symmetry. (*a*) Radially symmetrical animals, such as this sea anemone, can be bisected into equal halves in any two-dimensional plane. (*b*) Bilaterally symmetrical animals, such as this squirrel, can only be bisected into equal halves in one plane (the sagittal plane).

3. Evolution of a Body Cavity

A third key transition in the animal body plan was the evolution of the body cavity. The evolution of efficient organ systems within the animal body was not possible until a body cavity evolved for supporting organs, distributing materials, and fostering complex developmental interactions.

The presence of a body cavity enables the digestive tract to be larger and longer. This longer passage allows for storage of undigested food, longer exposure to enzymes for more complete digestion, and even storage and final processing of food remnants. With such an arrangement, an animal can eat a great deal when it is safe to do so and then hide during the digestive process, thus limiting the animal's exposure to predators. The tube within the body cavity architecture is also more flexible, thus allowing the animal greater freedom to move.

An internal body cavity also provides space within which the gonads (ovaries and testes) can expand, allowing large numbers of eggs and sperm to accumulate. Such storage capacity makes possible the diverse modifications of breeding strategy that characterize the more advanced phyla of animals. Furthermore, large numbers of gametes can be stored and released when the conditions are as favorable as possible for the survival of the young animals.

Kinds of Body Cavities. Three basic kinds of body plans evolved in the Bilateria (figure 31.5). **Acoelomates** have no body cavity. **Pseudocoelomates** have a body cavity called the **pseudocoel** located between the mesoderm and endoderm. In animals having the third type of body plan, called **coelomates**, a fluid-filled body cavity develops not between the endoderm and mesoderm, but rather entirely within the mesoderm. Such a body cavity is called a **coelom**. In coelomates, the gut is suspended, along with other organ systems of the animal, within the coelom; the coelom, in turn, is surrounded by a layer of epithelial cells entirely derived from the mesoderm (figure 31.5). A major advantage of the coelomate body plan is that it allows contact between mesoderm and endoderm, so that tissues can interact during development. For example, contact between mesoderm and endoderm permits localized portions of the digestive tract to develop into complex, highly specialized regions such as the stomach. In pseudocoelomates, mesoderm and endoderm are separated by the body cavity, limiting developmental interactions between these tissues, which ultimately limits tissue specialization and development.

The development of the coelom poses a problem—circulation—which is solved in pseudocoelomates by churning the fluid within the body cavity. In coelomates, the gut is again surrounded by tissue that presents a barrier to diffusion, just as it was in the solid bodies of acoelomates. This problem is solved among coelomates by the development of a **circulatory system,** a network of vessels that carry fluids to parts of the body. The circulating fluid, or blood, carries nutrients and oxygen to the tissues and re-

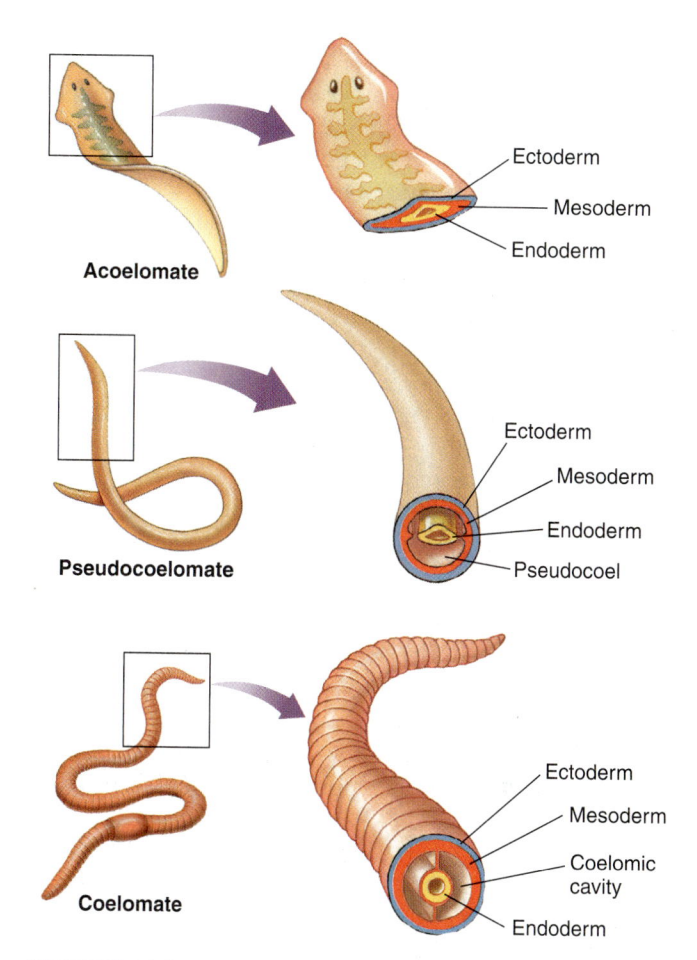

FIGURE 31.5
Three body plans for bilaterally symmetrical animals.
Acoelomates, such as flatworms, have no body cavity between the digestive tract (endoderm) and the outer body layer (ectoderm). Pseudocoelomates have a body cavity, the pseudocoel, between the endoderm and the mesoderm. Coelomates have a body cavity, the coelom, that develops entirely within the mesoderm, and so is lined on both sides by mesoderm tissue.

moves wastes and carbon dioxide. Blood is usually pushed through the circulatory system by contraction of one or more muscular hearts. In an **open circulatory system,** the blood passes from vessels into sinuses, mixes with body fluid, and then reenters the vessels later in another location. In a **closed circulatory system,** the blood is physically separated from other body fluids and can be separately controlled. Also, blood moves through a closed circulatory system faster and more efficiently than it does through an open system.

The evolutionary relationship among coelomates, pseudocoelomates, and acoelomates is not clear. Acoelomates, for example, could have given rise to coelomates, but scientists also cannot rule out the possibility that acoelomates were derived from coelomates. The different phyla of pseudocoelomates form two groups that do not appear to be closely related.

4. The Evolution of Deuterostome Development

Bilaterally symmetrical animals exhibit a pattern of embryonic development that begins with mitotic cell divisions of the egg that lead to the formation of a hollow ball of cells, called the **blastula**. The blastula indents to form a two-layer-thick ball with a **blastopore** opening to the outside and a primitive gut cavity called the **archenteron**. Bilaterians can be divided into two groups based on differences in the basic pattern of development. One group is called the **protostomes** (Greek *protos*, "first," + *stoma*, "mouth") and includes the flatworms, nematodes, mollusks, annelids, and arthropods. Two outwardly dissimilar groups, the echinoderms and the chordates, together with a few other smaller related phyla, comprise the second group, the **deuterostomes** (Greek, *deuteros*, "second," + *stoma*, "mouth"). Protostomes and deuterostomes differ in several aspects of embryo growth.

Cleavage. The progressive division of cells during embryonic growth is called *cleavage*. The cleavage pattern relative to the embryo's polar axis determines how the cells will array. In nearly all protostomes, each new cell buds off at an angle oblique to the polar axis. As a result, a new cell nestles into the space between the older ones in a closely packed array. This pattern is called **spiral cleavage** because a line drawn through a sequence of dividing cells spirals outward from the polar axis (figure 31.6 *top*). In deuterostomes, the cells divide parallel to and at right angles to the polar axis. As a result, the pairs of cells from each division are positioned directly above and below one another; this process gives rise to a loosely packed array of cells. This pattern is called **radial cleavage** because a line drawn through a sequence of dividing cells describes a radius outward from the polar axis (figure 31.6 *bottom*).

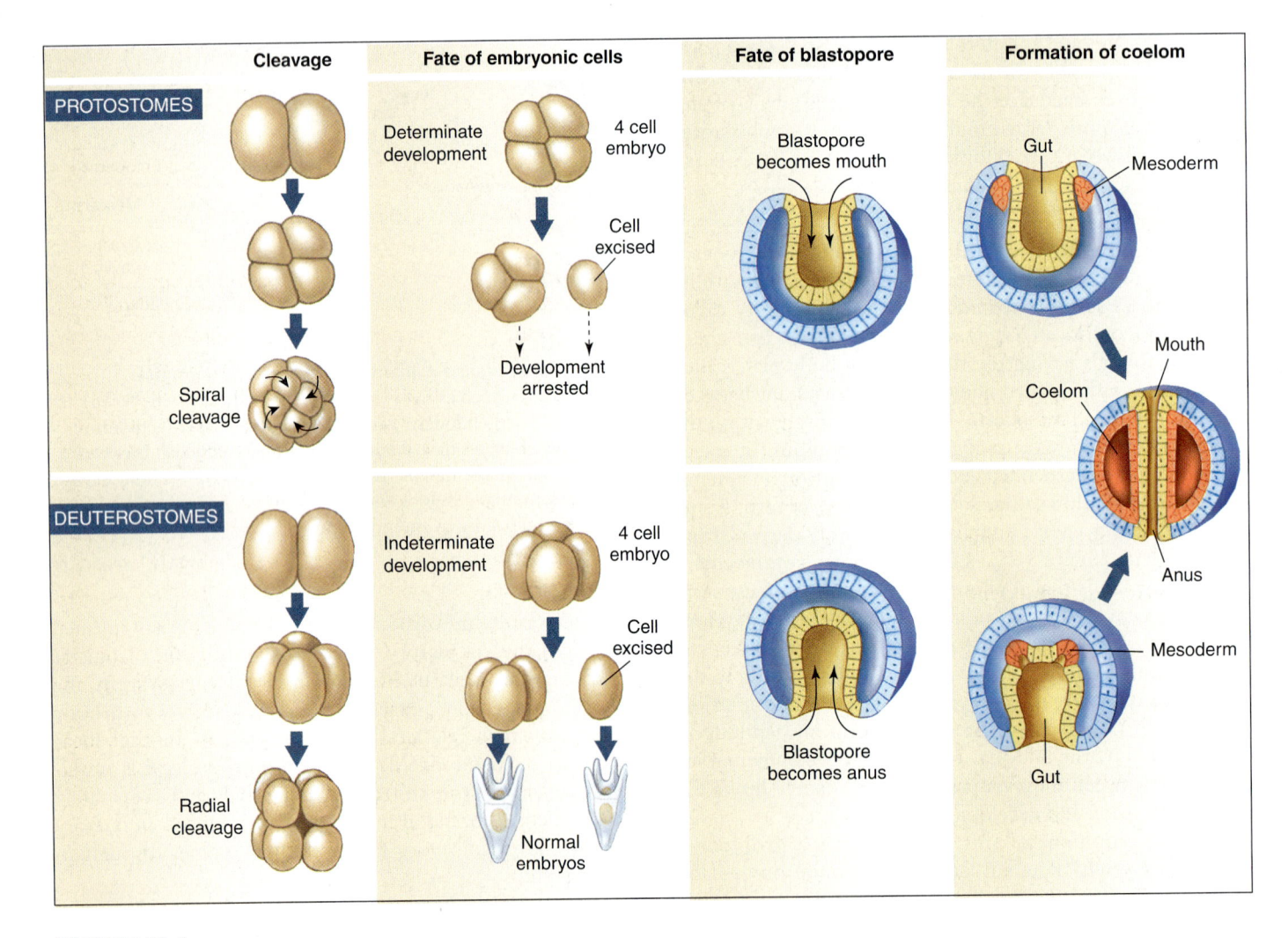

FIGURE 31.6
Embryonic development in protostomes and deuterostomes. In protostomes, embryonic cells cleave in a spiral pattern and exhibit determinate development; the blastopore becomes the animal's mouth, and the coelom originates from a mesodermal split. In deuterostomes, embryonic cells cleave radially and exhibit indeterminate development; the blastopore becomes the animal's anus, and the coelom originates from an evagination, or outpouching, of the archenteron.

Fate of Embryonic Cells. Protostomes exhibit **determinate development**. In this type of development, each embryonic cell has a predetermined fate in terms of what kind of tissue it will form in the adult. Before cleavage begins, the chemicals that act as developmental signals are localized in different parts of the egg. Consequently, the cell divisions that occur after fertilization separate different signals into different daughter cells. This process specifies the fate of even the very earliest embryonic cells. Deuterostomes, on the other hand, display **indeterminate development**. The first few cell divisions of the egg produce identical daughter cells. Any one of these cells, if separated from the others, can develop into a complete organism. This is possible because the chemicals that signal the embryonic cells to develop differently are not localized until later in the animal's development.

Fate of Blastopore. In protostomes, the mouth (stoma) of the animal develops from or near the blastopore. If such an animal has a distinct anus or anal pore, it develops later in another region of the embryo. This same pattern of development, in a general sense, is seen in all noncoelomate animals. In deuterostomes, the blastopore gives rise to the organism's anus, and the mouth develops from a second pore that arises in the blastula later in development.

Formation of the Coelom. In all coelomates, the coelom originates from mesoderm. In protostomes, this occurs simply and directly: The cells simply move away from one another as the coelomic cavity expands within the mesoderm. However, in deuterostomes, whole groups of cells usually move around to form new tissue associations. The coelom is normally produced by an evagination of the archenteron, the central tube within the gastrula, also called the primitive gut. This tube, lined with endoderm, opens to the outside via the blastopore and eventually becomes the gut cavity.

Deuterostomes evolved from protostomes more than 630 million years ago, and the consistency of deuterostome development, along with its distinctiveness from that of the protostomes, suggests that it evolved once, in a common ancestor to all of the phyla that exhibit it.

5. The Evolution of Segmentation

The fifth key transition in the animal body plan involved the subdivision of the body into segments. Just as it is efficient for workers to construct a tunnel from a series of identical prefabricated parts, so segmented animals are "assembled" from a succession of identical segments. During the animal's early development, these segments become most obvious in the mesoderm but later are reflected in the ectoderm and endoderm as well. Two advantages result from early embryonic segmentation:

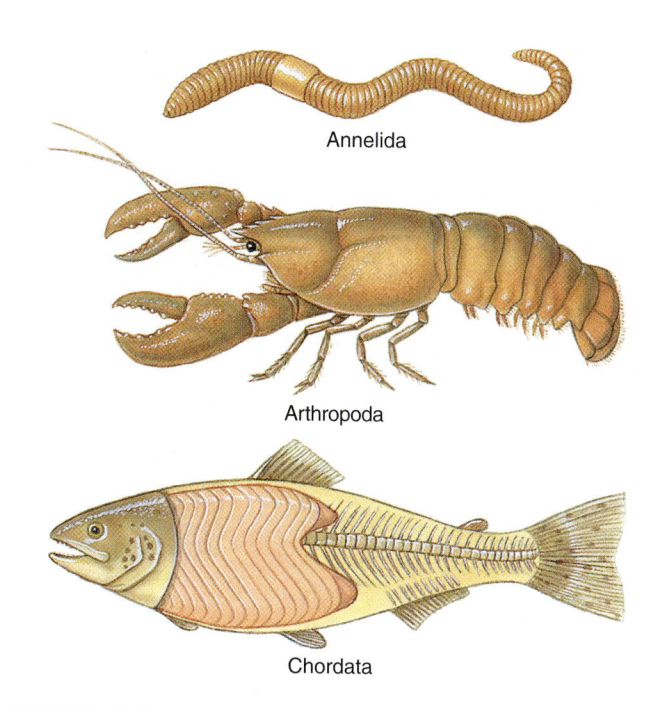

FIGURE 31.7
Segmentation. Annelids, arthropods, and chordates exhibit segmentation.

1. In annelids and other highly segmented animals, each segment may go on to develop a more or less complete set of adult organ systems. Damage to any one segment need not be fatal to the individual, since the other segments duplicate that segment's functions.
2. Locomotion is far more effective when individual segments can move independently because the animal as a whole has more flexibility of movement. Because the separations isolate each segment into an individual skeletal unit, each is able to contract or expand autonomously in response to changes in hydrostatic pressure. Therefore, a long body can move in ways that are often quite complex.

Segmentation, also referred to as *metamerism*, underlies the organization of all advanced animal body plans. In some adult arthropods, the segments are fused, but segmentation is usually apparent in their embryological development. In vertebrates, the backbone and muscular areas are segmented, although segmentation is often disguised in the adult form. True segmentation is found in only three phyla: the annelids, the arthropods, and the chordates (figure 31.7), although this trend is evident in many phyla.

> **Five key transitions in body design are responsible for most of the differences we see among the major animal phyla: the evolution of (1) tissues, (2) bilateral symmetry, (3) a body cavity, (4) deuterostome development, and (5) segmentation.**

31.4 The way we classify animals is being reevaluated.

A New Look at the Metazoan Family Tree

The traditional animal phylogeny, while accepted by a broad consensus of biologists for almost a century, is now being reevaluated. Its simple either-or organization has always presented certain problems—puzzling minor groups do not fit well into the standard scheme. For example, the myzostomids (figure 31.8), an enigmatic and anatomically bizarre group of marine animals, are parasites or symbionts of echinoderms. Myzostomid fossils are found associated with echinoderms since the Ordovician period, so the myzostomid-echinoderm relationship is a very ancient one. Their long history of obligate association has led to the loss or simplification of many myzostomid body elements, leaving them, for example, with no body cavity (they are acoelomates) and only incomplete segmentation.

This character loss has led to considerable disagreement among taxonomists. However, while taxonomists have disagreed about the details, all have generally allied myzostomids in some fashion with the annelids—sometimes within the polychaetes and sometimes as a separate phylum closely allied to the annelids.

Recently, this view has been challenged. New taxonomical comparisons using molecular data have come to very different conclusions. Researchers have examined two components of the protein synthesis machinery—the small ribosomal subunit rRNA gene and an elongation factor gene (the one called 1 alpha). The phylogeny they obtain does not place the myzostomids with the annelids. Indeed, they find that the myzostomids have no close links to the annelids at all. Instead, surprisingly, they are more closely allied with the flatworms!

This result hints strongly that the key morphological characters that biologists have traditionally used to construct animal phylogenies—segmentation, coeloms, jointed appendages, and the like—are not the conservative characters we had supposed. Among the myzostomids, these features appear to have been gained and lost again during the course of their evolution. If this unconservative evolutionary pattern should prove general, our view of the evolution of the animal body plan, and how the various animal phyla relate to one another, will soon be in need of major revision.

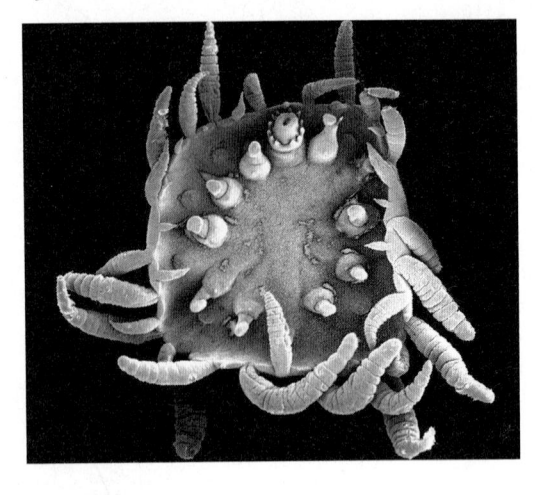

FIGURE 31.8
A taxonomic puzzle. *Myzostoma mortenensi* has no body cavity and incomplete segmentation. Such animals present a classification challenge, causing taxonomists to reconsider traditional animal phylogenies and the characters they are based on.

The last decade has seen a wealth of new molecular RNA and DNA sequence data on the various animal groups. The new field of **molecular systematics** uses unique sequences within certain genes to identify clusters of related groups. Expressed in the terminology of cladistic taxonomy, the sharing of derived sequence characters unique to a group and its ancestors defines clusters of monophyletic taxa that make up clades. The animal phylogenetic tree viewed in these terms is a hierarchy of clades nested within larger clades.

Using these sorts of molecular data, a variety of molecular phylogenies have been produced in the last decade. While differing from one another in many important respects, the new molecular phylogenies have the same deep branch structure as the traditional animal family tree. However, most agree on one revolutionary difference from the traditional phylogeny used in this text and presented in figure 31.3: The protostomes are broken into two distinct clades. Different molecular taxonomists distribute lophophorates, pseudocoelomates, and acoelomates among the two clades in different ways.

At present, molecular phylogenetic analysis of the animal kingdom is in its infancy. Phylogenies developed from different molecules sometimes suggest quite different evolutionary relationships. However, the childhood of this approach is likely to be short. Over the next few years, a mountain of additional molecular data can be anticipated. As more data are brought to bear, the confusion can be expected to lessen.

A year-2003 consensus molecular phylogeny, developed from DNA, ribosomal RNA, and protein studies, is presented in figure 31.9. In it, the traditional protostome group is broken up into Ecdysozoa and Lophotrochozoa. This new view of the metazoan Tree of Life is only a rough outline; in the future, more data should allow us to more confidently resolve relationships within groupings. Still, it is already clear that major groups are related in very different ways in molecular phylogenies than in the more traditional morphological one.

The use of molecular data to construct phylogenies is likely to significantly alter our understanding of relationships among the animal phyla.

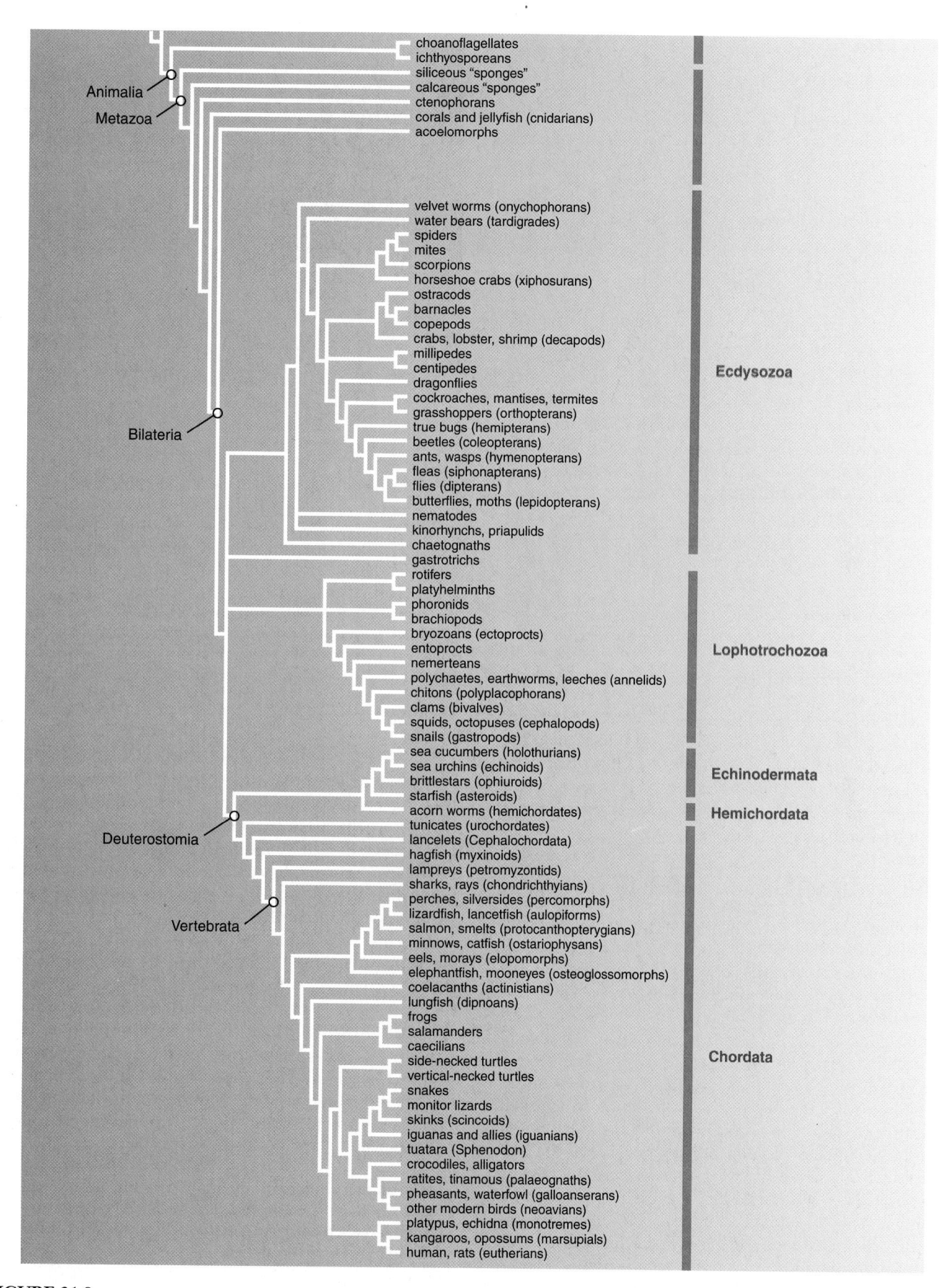

FIGURE 31.9

Proposed revision of the animal tree of life. This molecular phylogeny, based on comparisons of DNA, ribosomal RNA, and protein sequence differences, was assembled in 2003 by the journal *Science* from discussions with numerous taxonomists. It reflects an emerging consensus among molecular investigators that traditional phylogenies based on shared morphological and embryological characters do not reflect true ancestral relationships.

"Evo-Devo" and the Roots of the Animal Family Tree

Some of the most exciting contributions of molecular systematics are being made to our understanding of the base of the animal family tree.

Origin of Metazoans

Most taxonomists agree that the animal kingdom is monophyletic—that is, that parazoans and eumetazoans have a common ancestor. This ancestor was presumably a protist, but from which line of protists did animals evolve? There are three prominent hypotheses for the origin of metazoans from single-celled protists:

> The **multinucleate hypothesis** suggests that metazoans arose from a multinuclear protist similar to today's ciliates. The cells later became compartmentalized into the multicellular condition.
>
> The **colonial flagellate hypothesis,** first proposed by Haeckel in 1874, states that metazoans descended from colonial protists, hollow spherical colonies of flagellated cells. Some of the cells of sponges are strikingly like those of choanoflagellate protists.
>
> The **polyphyletic origin hypothesis** proposes that sponges evolved independently from eumetazoans.

Molecular systematics based on ribosomal RNA sequences settles this argument clearly in favor of the colonial flagellate hypothesis. The molecular evidence excludes the multinucleate ciliate hypothesis because metazoans are closer to eukaryotic algae than to ciliates. The polyphyletic origin hypothesis is excluded because metazoans represent a monophyletic assembly (the sister group of metazoans appears to be fungi).

Early Diversification of the Animal Family Tree

A second contentious issue in animal phylogeny is being addressed successfully with molecular systematics. Study of the fossil record reveals that the great diversity of animals evolved quite rapidly in geological terms around the beginning of the Cambrian period. Nearly all the major animal body plans can be seen in Cambrian rocks dating from 543 to 525 million years ago.

In rock from the earlier Ediacaran period as old as 565 million years, fossil cnidarians are found, along with what appear to be fossil mollusks and the burrows of worms. This implies that the earliest branches of the animal family tree arose before the Cambrian period.

In the half-billion years since the early Cambrian, no significant new innovations in animal body plan have occurred. Biologists have long debated what caused this **Cambrian explosion** of animal diversity (figure 31.10).

FIGURE 31.10
Diversity of animals that evolved during the Cambrian explosion. In addition to the ancestors of many present-day groups, such as insects and vertebrates, a variety of bizarre creatures evolved that left no descendants, such as: (1) *Amiskwia*, (2) *Odontogriphus*, (3) *Eldonia*, (4) *Halichondrites*, (5) *Anomalocaris canadensis*, (6) *Pikaia*, (7) *Canadia*, (8) *Marrella splendens*, (9) *Opabinia*, (10) *Ottoia*, (11) *Wiwaxia*, (12) *Yohoia*, (13) *Xianguangia*, (14) *Aysheaia*, (15) *Sidneyia*, (16) *Dinomischus*, and (17) *Hallucigenia*. The natural history of these species is open to speculation.

Many have argued that the emergence of new body plans was the consequence of the emergence of predatory lifestyles, which encouraged an arms race between defenses, such as armor, and innovations that improved mobility and hunting success. Others have attributed the rapid diversification in body plans to geological factors, such as the buildup of dissolved oxygen and minerals in the oceans. A third possibility arises from molecular studies being carried out by biologists in the new field of **"evo-devo,"** a synthesis of evolutionary biology and developmental biology. Much of the variation in animal body plan is associated with changes in the location or time of expression of *Hox* genes within developing animal embryos (see chapters 19 and 24). Perhaps the Cambrian explosion reflects the evolution of the *Hox* developmental gene complex, which provides a tool that can produce rapid changes in body plan.

> **The animal kingdom is monophyletic and arose from a colonial flagellated protist. The great diversity in the animal body plan arose quickly, possibly as the result of the evolution of *Hox* genes.**

Concept Review

For interactive testing, visit the Online Learning Center with PowerWeb at www.mhhe.com/Raven7

31.1 Animals are multicellular heterotrophs without cell walls.

Some General Features of Animals
- Animals are multicellular heterotrophs that are diverse in form and habitat, are mobile, reproduce sexually, and have characteristic embryonic development. (p. 618)

31.2 Animals are a very diverse kingdom.

The Traditional Classification of Animals
- Taxonomists have traditionally created phylogenies by comparing anatomical features and embryological development. (p. 620)
- Kingdom Animalia is traditionally divided into the Parazoa, which lack a definite symmetry and organized tissues, and the Eumetazoa, which have a definite symmetry and organized tissues. (p. 620)
- The eumetazoan branch is divided into Radiata and Bilateria. (p. 620)
- Bilateral animals further split into groups with and without a body cavity. (p. 620)

31.3 The animal body plan has undergone many changes.

Five Key Transitions in Body Plan
- The evolution of tissues involved cell specialization. (p. 624)
- The evolution of bilateral symmetry allowed organization of body parts, including cephalization, and increased motility. Bilaterally symmetrical animals produce three germ layers—ectoderm, endoderm, and mesoderm. (p. 624)
- The evolution of a body cavity allowed the subsequent development of efficient organ systems, such as an open or a closed circulatory system. Three basic types of body plans are (1) the acoelomates with no body cavity, (2) the pseudocoelomates with a cavity between the mesoderm and endoderm, and (3) the coelomates with a fluid-filled body cavity entirely within the mesoderm. (p. 625)
- The evolution of deuterostome development involved changes in cleavage patterns, determination, the fate of the blastopore, and the formation of the coelom. (pp. 626–627)
- The evolution of segmentation allowed the duplication of body organs and more effective locomotion. (p. 627)

31.4 The way we classify animals is being reevaluated.

A New Look at the Metazoan Family Tree
- Traditional characters used to construct phylogenies are not as conservative as once supposed. (p. 628)
- Molecular systematics uses unique sequences within certain genes to identify shared characteristics that define the monophyletic taxa making up clades. (p. 628)

"Evo-Devo" and the Roots of the Animal Family Tree
- The multinucleate hypothesis, the colonial flagellate hypothesis, and the polyphyletic origin hypothesis all try to account for the origin of metazoans from single-celled protists. (p. 630)
- A large diversity of animal body plans occurred around the Cambrian period, with no new innovations since. (p. 630)
- The emergence of body plans has been hypothesized to be caused by the emergence of predatory lifestyles, geological factors, and changes in the location or time of expression of *Hox* genes within developing animal embryos. (p. 630)

Test Your Understanding

Self Test

1. Which feature is *not* characteristic of the animals?
   a. multicellular heterotrophs
   b. sexual reproduction
   c. embryonic development
   d. cell walls
2. Most of the phyla of the animal kingdom are found
   a. on land.
   b. in the ocean.
   c. burrowing underground.
   d. in freshwater habitats
3. The subkingdom of animals that lack symmetry and have no true tissues or organs is the
   a. Eumetazoa.
   b. Radiata.
   c. Parazoa.
   d. Bilateria.
4. Except for the Radiata, the eumetazoans
   a. lack true tissues and organs.
   b. are monoblastic.
   c. are diploblastic.
   d. are triploblastic.
5. Cnidarians and ctenophores differ from other eumetazoans by having
   a. radial symmetry.
   b. bilateral symmetry.
   c. major organ systems.
   d. three tissue layers.
6. The advantage of bilateral symmetry is that it allowed for the evolution of
   a. appendages.
   b. cephalization.
   c. reproductive structures.
   d. parasites.
7. Members of which group are *not* deuterostomes?
   a. chordates
   b. echinoderms
   c. arthropods
   d. none of these; all are deuterostomes.
8. The evolution of an internal body cavity offered an advantage in animal body design in all areas except
   a. circulation.
   b. digestion.
   c. freedom of movement.
   d. gamete storage.
9. The segments of annelids are
   a. apparent in the embryo but not in the adult.
   b. specialized for different functions.
   c. present in the mesoderm (muscles) but not in the ectoderm.
   d. repetitive—each able to develop a complete set of adult organs.
10. Which of the following hypotheses about the origin of metazoans is supported by ribosomal RNA analysis?
    a. the multinucleate hypothesis
    b. the colonial flagellate hypothesis
    c. the polyphyletic origin hypothesis
    d. None of these; rRNA analysis doesn't support any specific hypothesis.

Test Your Visual Understanding

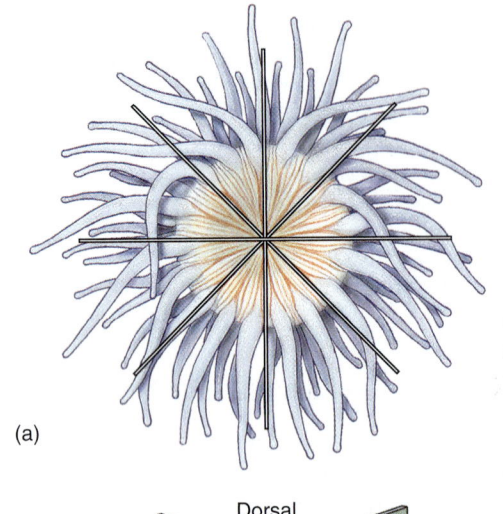

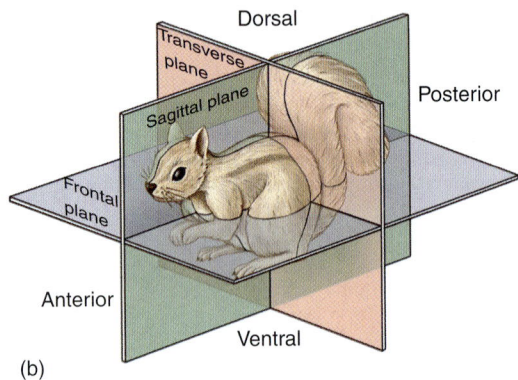

1. Which of these drawings (*a* or *b*) depicts a radially symmetrical animal? Which depicts a bilaterally symmetrical animal? Although a radially symmetrical animal can be bisected into equal halves in any two-dimensional plane, can you describe a plane of orientation of a radially symmetrical animal that produces dissimilar halves? (Refer to the planes of bisection in *b*.)

Apply Your Knowledge

1. In what ways is an earthworm more complex than a flatworm?
2. Why is it believed that echinoderms and chordates, which are so dissimilar, are members of the same evolutionary line?

32
Noncoelomate Invertebrates

Concept Outline

32.1 The classification of invertebrates is currently being reevaluated.

An Uproar Over Invertebrate Phylogeny. The bilaterally symmetrical invertebrates have traditionally been separated according to the nature of their body cavity. However, comparisons of ribosomal RNA are suggesting a very different invertebrate phylogeny.

32.2 The simplest animals are not bilaterally symmetrical.

Parazoa. Sponges are the most primitive animals, without either tissues or, for the most part, symmetry.
Radiata. Cnidarians and ctenophores have distinct tissues and radial symmetry.

32.3 Acoelomates are solid worms that lack a body cavity.

The Bilaterian Acoelomates. Flatworms are the simplest bilaterally symmetrical animals; they are solid worms that lack a body cavity, but possess true organs.

32.4 Pseudocoelomates have a simple body cavity.

The Pseudocoelomates. Nematodes and rotifers possess a simple body cavity.

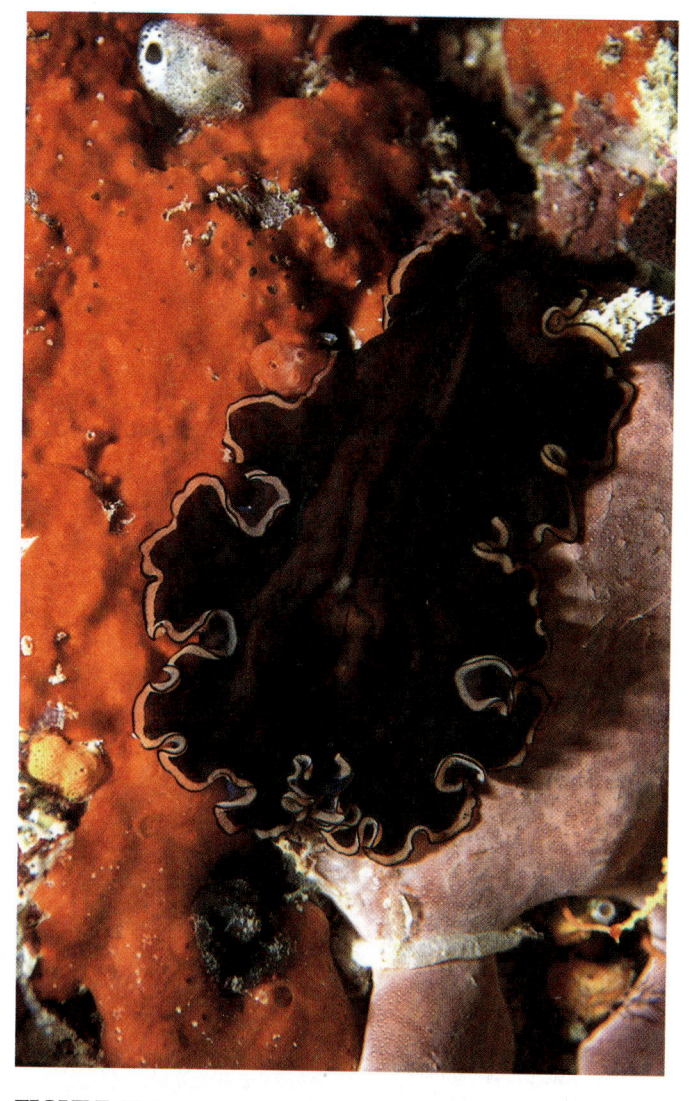

FIGURE 32.1
A noncoelomate: a marine flatworm. Some of the earliest invertebrates to evolve, marine flatworms possess internal organs but lack a mesoderm-encased body cavity called a coelom.

We will start our exploration of the great diversity of animals with the simplest members of the animal kingdom—sponges, jellyfish, and simple worms. These animals lack a body cavity called a coelom, and are thus called noncoelomates (figure 32.1). The major organization of the animal body first evolved in these animals, a basic body plan upon which all the rest of animal evolution has depended. In chapter 33, we consider the invertebrate animals that have a coelom, and in chapter 34 the vertebrates. Despite their great diversity, you will see that all animals have much in common.

633

32.1 The classification of invertebrates is currently being reevaluated.

An Uproar Over Invertebrate Phylogeny

There is little disagreement among biologists about the classification of animals. The 35 traditional animal phyla are firmly established in the biological lexicon—for example, an annelid worm would be classified in the phylum Annelida by any competent taxonomist. However, there is great disagreement about how the 35 animal phyla are related to one another. Depending on what aspects of the phyla are compared, different biologists draw quite different family trees.

Two Approaches

For many years, biologists have based their reconstructions of the animal family tree on key aspects of body architecture, lumping together those phyla that share fundamental aspects of body plan. For the better part of a century, biologists agreed on the principal aspects of this tree, basing the phylogeny largely on anatomical and embryological comparisons

However, as we learned in chapter 31, a variety of new, quite different, animal family trees have been proposed in the last decade by researchers employing molecular rather than anatomical comparisons, focusing particularly on differences in ribosomal RNA sequences.

Like the traditional tree based on body plan, these new trees based on rRNA comparisons place the sponges (phylum Porifera), the only animals without tissues, in the category Parazoa, and place all animals with tissues in the category Eumetazoa. Among the eumetazoans, both approaches place the radially symmetrical cnidarians and ctenophores in the category Radiata. All other eumetazoans are assigned to one of two groups of bilaterally symmetrical animals that differ in their embryological development—protostomes or deuterostomes.

It is in how they construct the protostome branch of the animal family tree that the new rRNA phylogenies differ radically from the traditional one.

The Traditional Protostome Phylogeny

As noted in chapter 31, the traditional animal family tree divides the bilaterally symmetrical animals into three great branches, based on the nature of the body cavity: (1) acoelomates (such as phylum Platyhelminthes), which have no body cavity; (2) pseudocoelomates (such as phylum Nematoda), which have a pseudocoel body cavity separating mesoderm from endoderm; and (3) coelomates (such as phyla Arthropoda, Annelida, and Mollusca), which have a coelom body cavity encased within mesoderm (figure 32.2). Coelomates can be either protostomes or deuterostomes.

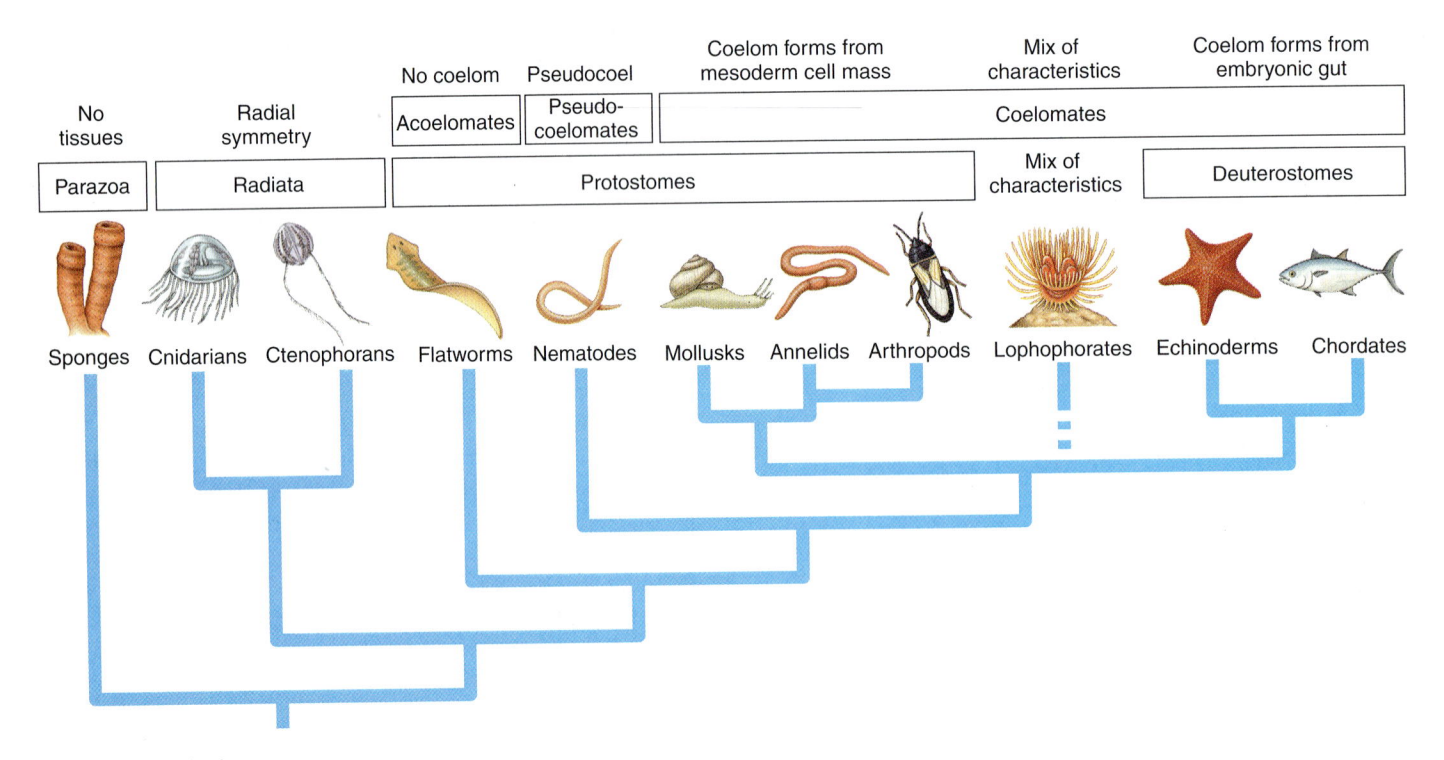

FIGURE 32.2

The traditional protostome phylogeny. Biologists have traditionally separated the bilaterally symmetrical animals into three groups that differ with respect to their body cavity: acoelomates, pseudocoelomates, and coelomates.

634 Part V Diversity of Life on Earth

The Novel rRNA Protostome Phylogenies

Phylogenies based on rRNA sequences suggest a very different lineage of the protostome phyla. Two major clades are recognized as having evolved independently since ancient times: the lophotrochozoans and the ecdysozoans (figure 32.3).

Lophotrochozoans. Lophotrochozoan animals grow the same way you do, by adding additional mass to an existing body. Most live in water, and propel themselves through it using cilia. Many, although not all, lophotrochozoans have a type of free-living larva known as a **trochophore**.

Four major kinds of protostomes are assigned to the Lophotrochozoa:

The **lophophorate phyla,** which include bryozoans, phoronids, and brachiopods, are all coelomates with a horseshoe-shaped crown of ciliated tentacles called a **lophophore** around their mouths, used for feeding.

Flatworms, phylum Platyhelminthes, are acoelomates. They have a simple body plan with no enclosed body cavity and no organs for transporting oxygen to the body's interior.

Mollusks, phylum Mollusca, are unsegmented coelomates. They display a wide variety of body forms; included in this group are octopuses, snails, and clams.

Annelids, phylum Annelida, are segmented coelomate worms. This group includes both marine polychaetes and terrestrial earthworms.

Ecdysozoans. Ecdysozoans are the molting animals. They increase in size by molting their external skeletons, an ability that seems to have evolved only once. Molting animals have been successful in nearly all environments, and all move about by means other than ciliary action. Ecdysozoans all share a common set of homeobox genes, regulatory genes that govern the pattern of embryonic development.

Of the numerous phyla of protostomes assigned to the Ecdysozoa, two have been particularly successful:

Roundworms, phylum Nematoda, are pseudocoelomate worms that shed their hard cuticle four times as they grow to adult size. They are one of the most abundant and widely distributed of all animal phyla.

Arthropods, phylum Arthropoda, are coelomate animals with segmented external skeletons and jointed appendages. The most successful of all animal phyla, arthropods include insects, spiders, and crustaceans.

As more complete genomic comparisons become available, our picture of the evolutionary lineage of the protostomes will undoubtedly become clearer. For now, until a consensus emerges, this text will continue to adopt the traditional arrangement of the phyla.

> The protostomes are traditionally grouped according to the nature of their body cavity, but recent rRNA evidence suggests they might be better grouped according to whether or not they molt.

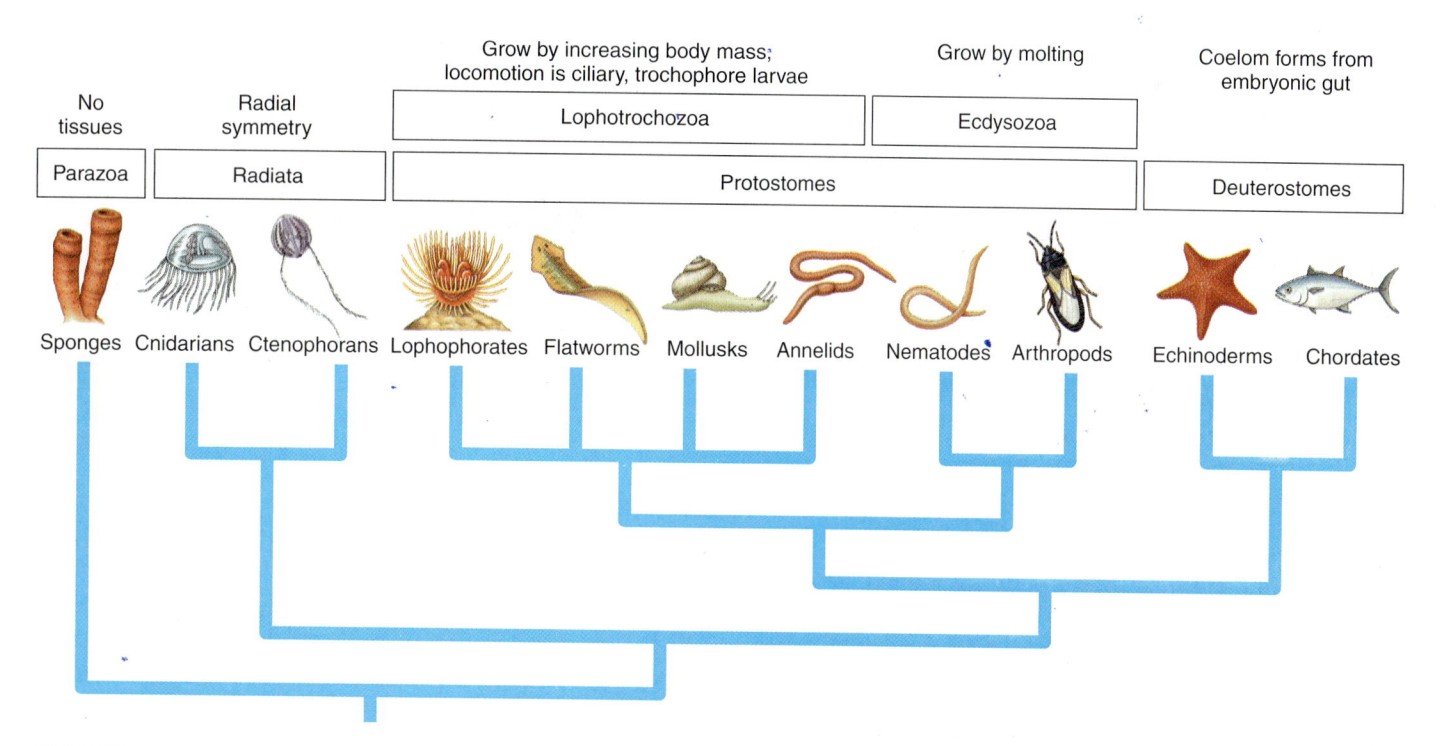

FIGURE 32.3
An rRNA protostome phylogeny. New phylogenies based on differences in ribosomal RNA genes suggest that protostomes might better be grouped according to whether they grow by molting (ecdysozoans) or by adding mass to an existing body (lophotrochozoans).

Chapter 32 Noncoelomate Invertebrates 635

32.2 The simplest animals are not bilaterally symmetrical.

Parazoa

The sponges are parazoans, animals that lack tissues and organs and a definite symmetry. However, sponges, like all animals, have true, complex *multicellularity*. The body of a sponge contains several distinctly different types of cells whose activities are loosely coordinated with one another. As we will see, the coordination between cell types in the eumetazoans increases and becomes quite complex.

The Sponges

There are perhaps 5000 species of marine sponges, phylum Porifera, and about 150 species that live in fresh water. In the sea, sponges are abundant at all depths. Although some sponges are tiny (no more than a few millimeters across), others, such as the loggerhead sponges, may reach 2 meters or more in diameter. A few small sponges are radially symmetrical, but most members of this phylum completely lack symmetry. Many sponges are colonial. Some have a low and encrusting form, while others may be erect and lobed, sometimes in complex patterns (figure 32.4). Although larval sponges are free-swimming, adults are **sessile**, or anchored in place to submerged objects.

Although sponges, like all animals, are composed of multiple cell types (figure 32.5), there is relatively little coordination among sponge cells. A sponge seems to be little more than a mass of cells embedded in a gelatinous matrix, but these cells are specialized for different functions and recognize one another with a high degree of fidelity.

The basic structure of a sponge can best be understood by examining the form of a young individual. A small, anatomically simple sponge first attaches to a substrate and then grows into a vaselike shape. The walls of the "vase" have three functional layers. First, facing into the internal cavity are specialized flagellated cells called **choanocytes**, or collar cells. These cells line either the entire body interior or, in many large and more complex sponges, specialized chambers. Second, the bodies of sponges are bounded by an outer epithelial layer consisting of flattened cells somewhat like those that make up the epithelia, or outer layers, of animals in other phyla. Some portions of this layer contract when touched or exposed to appropriate chemical stimuli, and this contraction may cause some of the pores to close. Third, between these two layers, sponges consist mainly of a gelatinous, protein-rich matrix called the **mesohyl**, within which various types of amoeboid cells occur. In addition, many kinds of sponges have minute needles of calcium carbonate or silica known as **spicules**, or fibers of a tough protein called **spongin**, or both, within this matrix. Spicules and spongin strengthen the bodies of the sponges in which they occur. A spongin skeleton is the model for the bathtub sponge, once the skeleton of a real animal, but now usually a cellulose or plastic mimic.

Sponges feed in a unique way. The beating of flagella that line the inside of the sponge draws water in through numerous small pores; the name of the phylum, Porifera, refers to this system of pores. Plankton and other small organisms are filtered from the water, which flows through passageways and eventually is forced out through an **osculum**, a specialized, larger pore.

Choanocytes. Each choanocyte closely resembles a protist with a single flagellum (figure 32.5), a similarity that reflects its evolutionary derivation. The beating of the flagella of the many choanocytes that line the body interior draws water in through the pores and through the sponge, thus bringing in food and oxygen and expelling wastes. Each choanocyte flagellum beats independently, and the pressure they create collectively in the cavity forces water out of the osculum. In some sponges, the inner wall of the body interior is highly convoluted, increasing the surface area and, therefore, the number of flagella that can drive the water. In such a sponge, 1 cubic centimeter of sponge can propel more than 20 liters of water per day.

Reproduction in Sponges. Some sponges will re-form themselves once they have passed through a silk mesh.

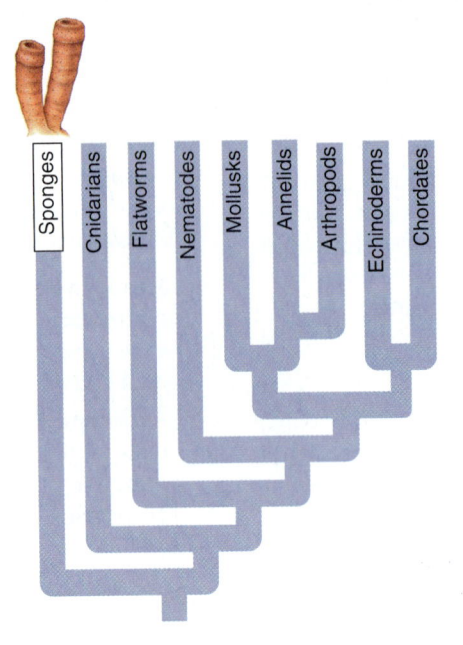

FIGURE 32.4
Aplysina longissima. This beautiful, bright orange and red elongated sponge is found on deep regions of coral reefs. The osculum is an opening at the top.

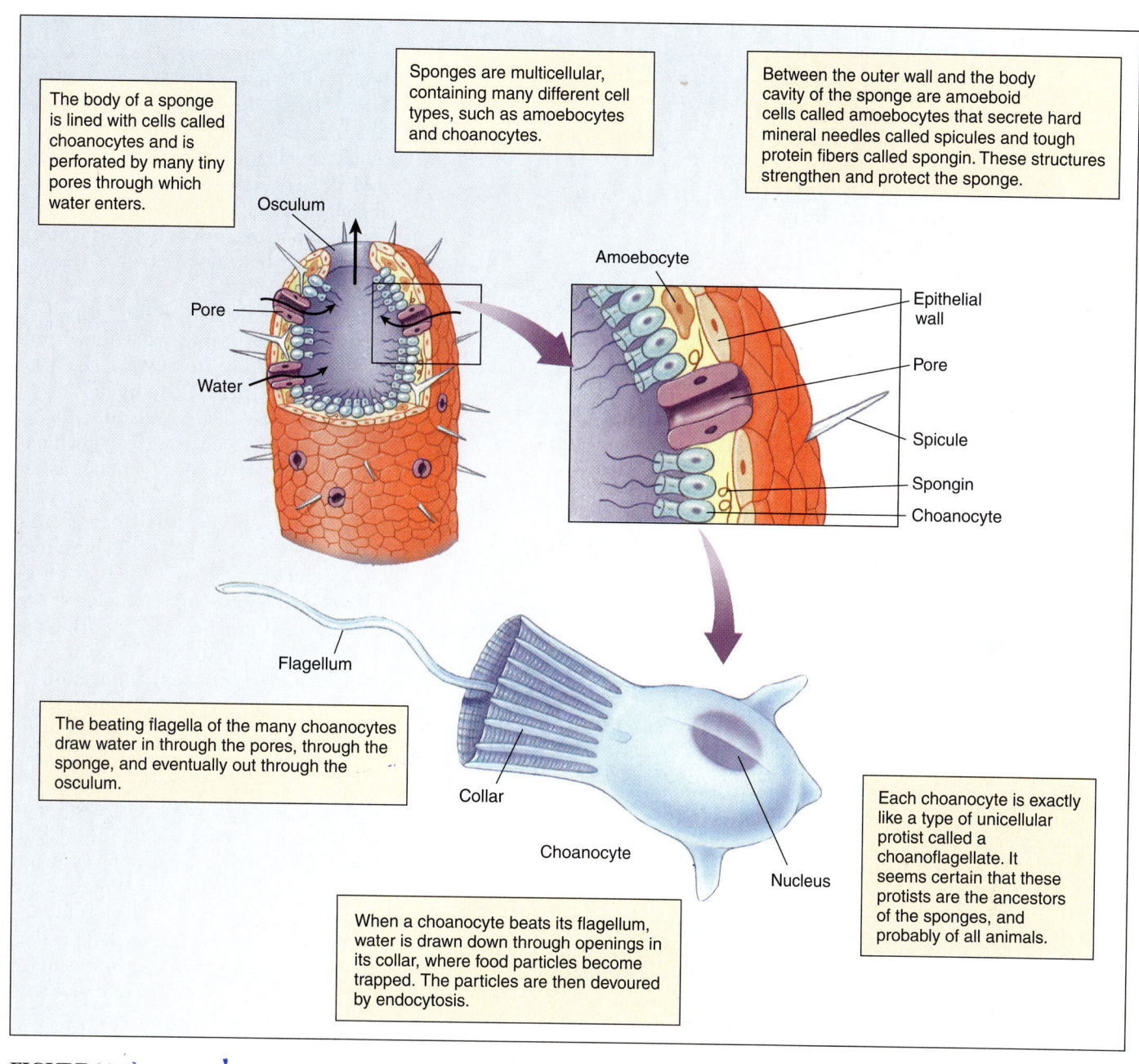

FIGURE 32.5

Phylum Porifera: Sponges. Sponges are composed of several distinctly different cell types, whose activities are coordinated with each other. The sponge body is not symmetrical and has no organized tissues.

Labels in figure:
- The body of a sponge is lined with cells called choanocytes and is perforated by many tiny pores through which water enters.
- Sponges are multicellular, containing many different cell types, such as amoebocytes and choanocytes.
- Between the outer wall and the body cavity of the sponge are amoeboid cells called amoebocytes that secrete hard mineral needles called spicules and tough protein fibers called spongin. These structures strengthen and protect the sponge.
- Osculum, Pore, Water, Amoebocyte, Epithelial wall, Pore, Spicule, Spongin, Choanocyte
- Flagellum, Collar, Choanocyte, Nucleus
- The beating flagella of the many choanocytes draw water in through the pores, through the sponge, and eventually out through the osculum.
- When a choanocyte beats its flagellum, water is drawn down through openings in its collar, where food particles become trapped. The particles are then devoured by endocytosis.
- Each choanocyte is exactly like a type of unicellular protist called a choanoflagellate. It seems certain that these protists are the ancestors of the sponges, and probably of all animals.

Thus, as you might suspect, sponges frequently reproduce by simply breaking into fragments. If a sponge breaks up, the resulting fragments usually are able to reconstitute whole new individuals. Sexual reproduction is also exhibited by sponges, with some mature individuals producing eggs and sperm. Larval sponges may undergo their initial stages of development within the parent. They have numerous external, flagellated cells and are free-swimming. After a short planktonic stage, they settle down on a suitable substrate, where they begin their transformation into adults.

> **Sponges probably represent the most primitive animals, possessing multicellularity but neither tissue-level development nor body symmetry. Their cellular organization hints at the evolutionary ties between the unicellular protists and the multicellular animals. Sponges are unique in the animal kingdom in possessing choanocytes, special flagellated cells whose beating drives water through the body interior.**

Chapter 32 Noncoelomate Invertebrates

Radiata

The subkingdom Eumetazoa contains animals that evolved the first key transition in the animal body plan: distinct *tissues*. Two distinct cell layers form in the embryos of these animals: an outer ectoderm and an inner endoderm. These embryonic tissues give rise to the basic body plan, differentiating into the many tissues of the adult body. Typically, the outer covering of the body (called the epidermis) and the nervous system develop from the ectoderm, and the layer of digestive tissue (called the **gastrodermis**) develops from the endoderm. A layer of gelatinous material, called the **mesoglea**, lies between the epidermis and gastrodermis and contains the muscles in most eumetazoans.

Eumetazoans also evolved true body symmetry and are divided into two major groups. The Radiata includes two phyla of radially symmetrical organisms: Phylum Cnidaria (pronounced ni-DAH-ree-ah), or the cnidarians, is composed of hydroids, jellyfish, sea anemones, and corals. Phylum Ctenophora (pronounced tea-NO-fo-rah) is composed of the comb jellies, also called the ctenophores. All other eumetazoans are in the Bilateria and exhibit a fundamental bilateral symmetry.

The Cnidarians

Cnidarians are nearly all marine, although a few live in fresh water. These fascinating and simply constructed animals are basically gelatinous in composition. They differ markedly from the sponges in organization; their bodies are made up of distinct tissues, although they have not evolved true organs. These animals are carnivores. For the most part, they do not actively move from place to place, but rather capture their prey (which includes fishes, crustaceans, and many other kinds of animals) with the tentacles that ring their mouths.

Cnidarians may have two basic body forms, polyps and medusae (figure 32.6). Polyps are cylindrical and are usually found attached to a firm substrate. They may be solitary or colonial. In a polyp, the mouth faces away from the substrate on which the animal is growing, and therefore often faces upward. Many polyps build up a chitinous or calcareous (made up of calcium carbonate) external or internal skeleton, or both. Only a few polyps are free-floating. In contrast, most medusae are free-floating and are often umbrella-shaped. Their mouths usually point downward, and the tentacles hang down around them. Medusae, particularly those of the class Scyphozoa, are commonly known as jellyfish because their mesoglea is thick and jellylike.

Many cnidarians occur only as polyps, while others exist only as medusae; still others alternate between these two phases during their life cycles. Both phases consist of diploid individuals. Polyps may reproduce asexually by budding; if they do, they may produce either new polyps or medusae. Medusae reproduce sexually. In most cnidarians, fertilized eggs give rise to free-swimming, multicellular, ciliated larvae known as **planulae**. Planulae are common in the plankton at times and may be dispersed widely in the currents.

A major evolutionary innovation in cnidarians, compared with sponges, is the internal extracellular digestion of food (figure 32.7). Digestion takes place within a gut cavity, rather than only within individual cells. Digestive enzymes, released from cells lining the walls of the cavity, partially break down food. Cells lining the gut subsequently engulf food fragments by phagocytosis.

The extracellular fragmentation that precedes phagocytosis and intracellular digestion allows cnidarians to digest animals larger than individual cells, an important improvement over the strictly intracellular digestion that occurs in sponges.

Nets of nerve cells coordinate contraction of cnidarian muscles, apparently with little central control. Cnidarians have no blood vessels, respiratory system, or excretory organs.

On their tentacles and sometimes on their body surface, cnidarians bear specialized cells called **cnidocytes**. The name of the phylum, Cnidaria, refers to these cells, which are highly distinctive and occur in no other group of organisms. Within each cnidocyte is a **nematocyst**, a small but powerful "harpoon." Each nematocyst features a coiled, threadlike tube. Lining the inner wall of the tube is a series of barbed spines. Cnidarians use the threadlike tube to spear their prey; then they draw the harpooned

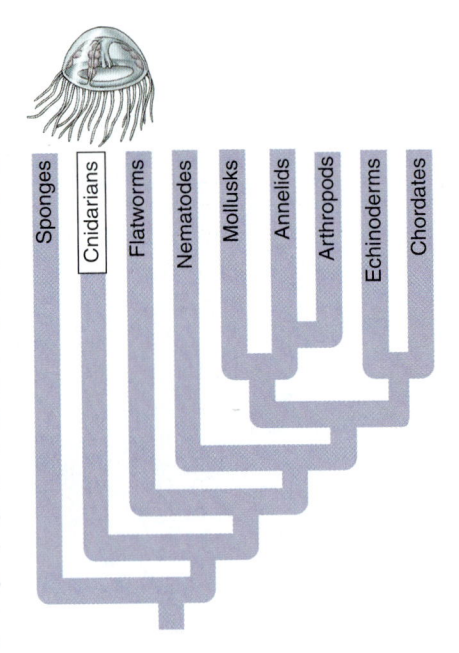

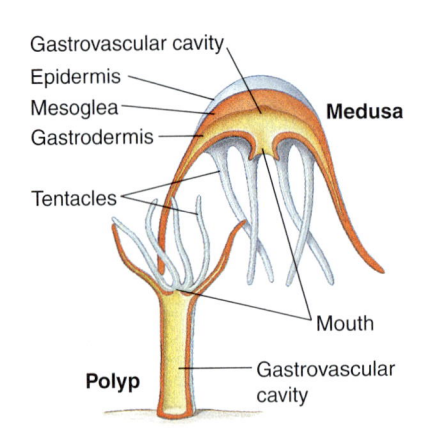

FIGURE 32.6
Two body forms of cnidarians, the medusa and the polyp. These two phases alternate in the life cycles of many cnidarians. But a number—including the corals and sea anemones—exist only as polyps, and some exist only as medusae. Both forms have two fundamental layers of cells, separated by a jellylike layer called the mesoglea.

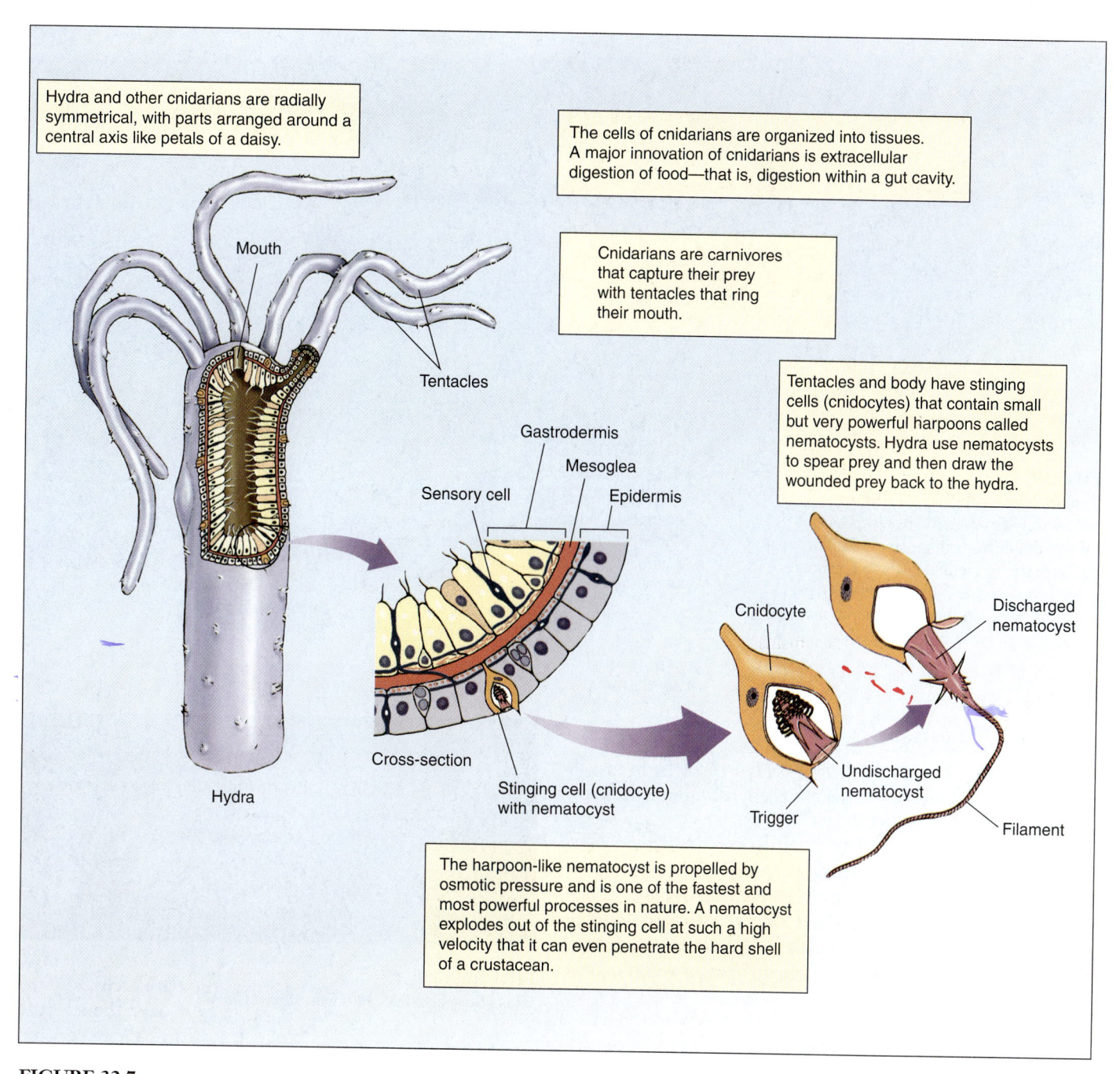

FIGURE 32.7

Phylum Cnidaria: cnidarians. The cells of a cnidarian such as this *Hydra* are organized into specialized tissues. The interior gut cavity is specialized for extracellular digestion—that is, digestion within a gut cavity rather than within individual cells. Cnidarians are radially symmetrical, with parts arranged around a central axis like the petals of a daisy.

prey back with the tentacle containing the cnidocyte. Nematocysts may also serve a defensive purpose. To propel the harpoon, the cnidocyte uses water pressure. Before firing, the cnidocyte builds up a very high internal osmotic pressure. This is done by using active transport to build a high concentration of ions inside, while keeping the cnidocyte's cell wall impermeable to water. Within the undischarged nematocyst, osmotic pressure reaches about 140 atmospheres.

When a flagellum-like trigger on the cnidocyte is stimulated to discharge, its walls become permeable to water, which rushes inside and violently pushes out the barbed filament. Nematocyst discharge is one of the fastest cellular processes in nature. The nematocyst is pushed outward so explosively that the barb can penetrate even the hard shell of a crab. A toxic protein often produces a stinging sensation, causing some cnidarians to be called "stinging nettles."

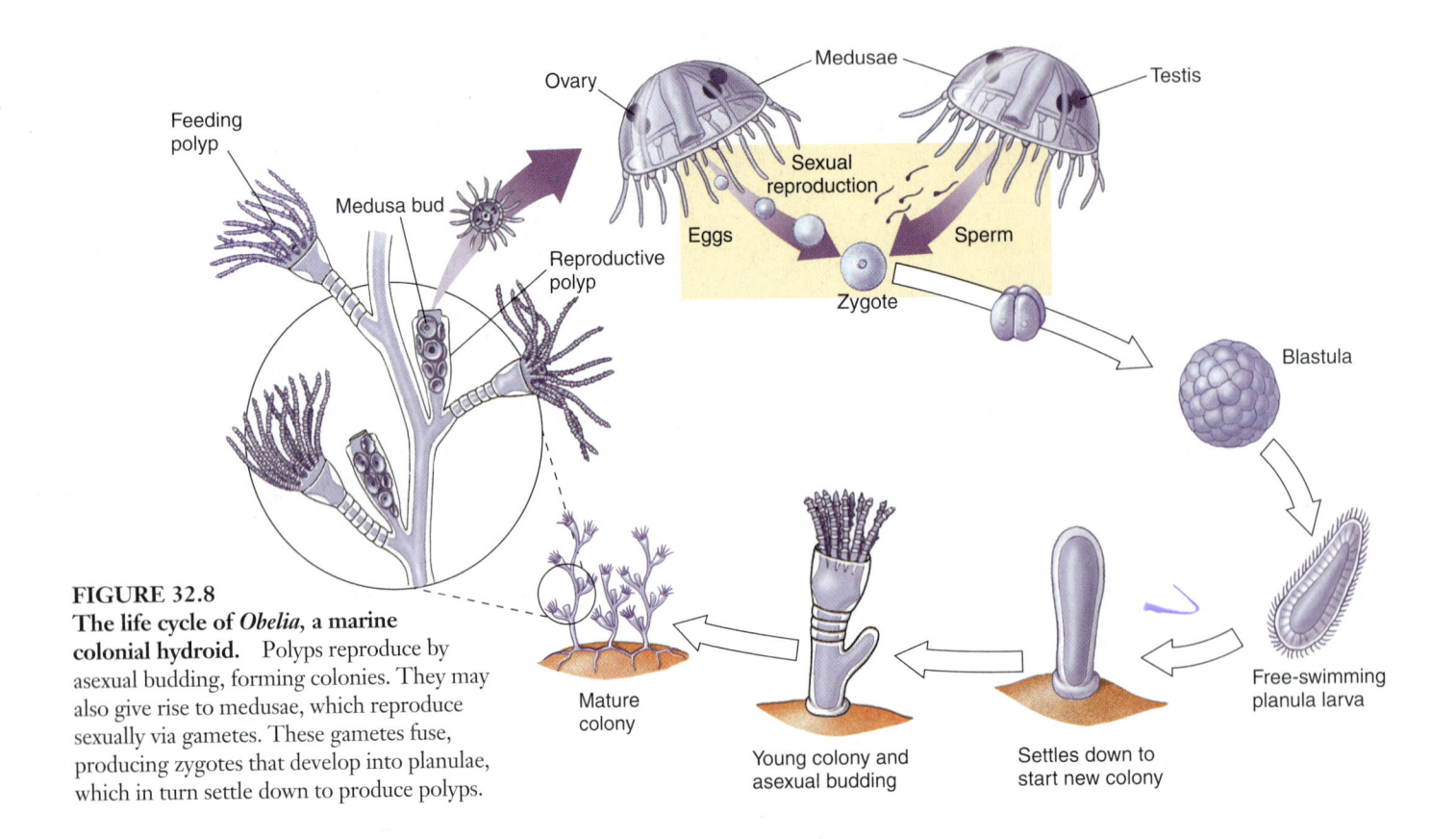

**FIGURE 32.8
The life cycle of *Obelia*, a marine colonial hydroid.** Polyps reproduce by asexual budding, forming colonies. They may also give rise to medusae, which reproduce sexually via gametes. These gametes fuse, producing zygotes that develop into planulae, which in turn settle down to produce polyps.

Classes of Cnidarians

There are four classes of cnidarians: Hydrozoa (hydroids), Scyphozoa (jellyfish), Cubozoa (box jellyfish), and Anthozoa (anemones and corals).

Class Hydrozoa: The Hydroids. Most of the approximately 2700 species of hydroids (class Hydrozoa) have both polyp and medusa stages in their life cycle (figure 32.8). Most of these animals are marine and colonial, such as *Obelia* and the very unusual Portuguese man-of-war. Some of the marine hydroids are bioluminescent.

A well-known hydroid is the abundant freshwater genus *Hydra*, which is exceptional in having no medusa stage and existing as a solitary polyp. Each polyp sits on a basal disk, which the hydra can use to glide around, aided by mucous secretions. It can also move by somersaulting—bending over and attaching itself to the substrate by its tentacles, and then looping over to a new location. If the polyp detaches itself from the substrate, it can float to the surface.

Class Scyphozoa: The Jellyfish. The approximately 200 species of jellyfish (class Scyphozoa) are transparent or translucent marine organisms, some of a striking orange, blue, or pink color (figure 32.9). These animals spend most of their time floating near the surface of the sea. In all of them, the medusa stage is dominant—much larger and more complex than the polyp stage. The medusae are bell-shaped, with hanging tentacles around

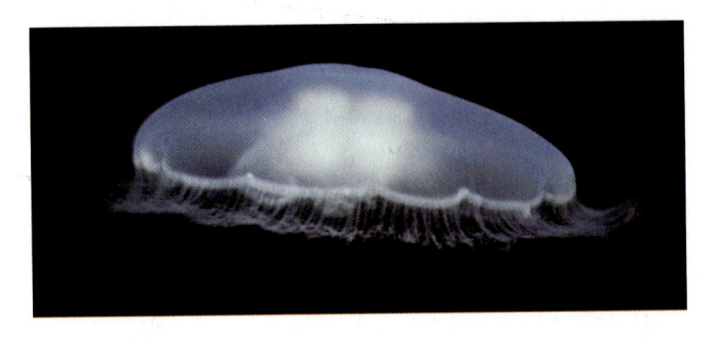

**FIGURE 32.9
Class Scyphozoa.** Jellyfish, *Aurelia aurita*.

their margins. The polyp stage is small, inconspicuous, and simple in structure.

The outer layer, or epithelium, of a jellyfish contains a number of specialized epitheliomuscular cells, each of which can contract individually. Together, the cells form a muscular ring around the margin of the bell that pulses rhythmically and propels the animal through the water. Jellyfish have separate male and female individuals. After fertilization, planulae form, which then attach and develop into polyps. The polyps can reproduce asexually as well as budding off medusae. In some jellyfish that live in the open ocean, the polyp stage is suppressed, and planulae develop directly into medusae.

Class Cubozoa: The Box Jellyfish. Until recently, the cubozoans were considered an order of Scyphozoa. As their name implies, they are box-shaped medusae; the polyp stage is inconspicuous and in many cases not known. Most are only a few centimeters in height, although some are 25 centimeters tall. A tentacle or group of tentacles is found at each corner of the box (figure 32.10). Box jellies are strong swimmers and voracious predators of fish. The stings of some species can be fatal to humans.

Class Anthozoa: The Sea Anemones and Corals. By far the largest class of cnidarians is Anthozoa, the "flower animals" (Greek *anthos*, "flower"). The approximately 6200 species of this group are solitary or colonial marine animals. They include stonelike corals, soft-bodied sea anemones, and other groups known by such fanciful names as sea pens, sea pansies, sea fans, and sea whips (figure 32.11). All of these names reflect a plantlike body topped by a tuft or crown of hollow tentacles. Like other cnidarians, anthozoans use these tentacles in feeding. Nearly all members of this class that live in shallow waters harbor symbiotic algae, which supplement the nutrition of their hosts through photosynthesis. Fertilized eggs of anthozoans usually develop into planulae that settle and develop into polyps; no medusae are formed.

Sea anemones are a large group of soft-bodied anthozoans that live in coastal waters all over the world and are especially abundant in the tropics. When touched, most sea anemones retract their tentacles into their bodies and fold up. Sea anemones are highly muscular and relatively complex organisms, with greatly divided internal cavities. These animals range from a few millimeters to about 10 centimeters in diameter and are perhaps twice that high.

Corals are another major group of anthozoans. Many of them secrete tough outer skeletons, or exoskeletons, of calcium carbonate and are thus stony in texture. Others, including the gorgonians, or soft corals, do not secrete exoskeletons. Some of the hard corals help form coral reefs, which are shallow-water limestone ridges that occur in warm seas. Although the waters where coral reefs develop are often nutrient-poor, the coral animals are able to grow actively because of the abundant algae found within them.

The Ctenophores (Comb Jellies)

The members of the small phylum Ctenophora range from spherical to ribbonlike and are known as comb jellies, or sea walnuts. Traditionally, the roughly 90 marine species of ctenophores were considered closely related to the cnidarians. However, ctenophores are structurally more complex than cnidarians. They have anal pores, so that water and other substances pass completely through the animal. Comb jellies, abundant in the open ocean, are transparent and usually only a few centimeters long. The members of one group have two long, retractable tentacles that they use to capture their prey.

Ctenophores propel themselves through water with eight comblike plates of fused cilia that beat in a coordinated fashion (figure 32.12). They are the largest animals that use cilia for locomotion. Many ctenophores are bioluminescent, giving off bright flashes of light particularly evident in the open ocean at night.

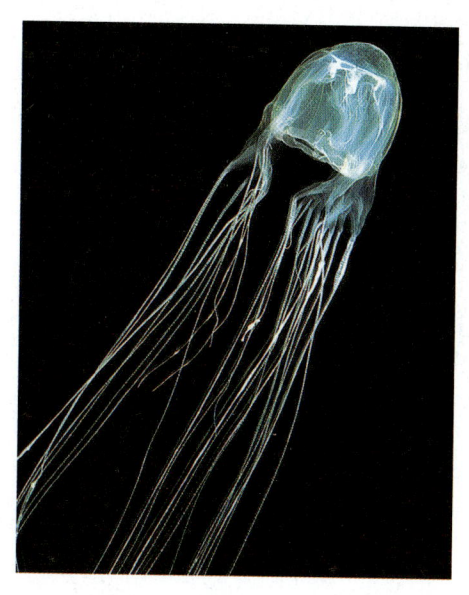

FIGURE 32.10
Class Cubozoa. Box jellyfish, *Chironex fleckeri*.

FIGURE 32.11
Class Anthozoa. The sessile, soft-bodied sea anemone.

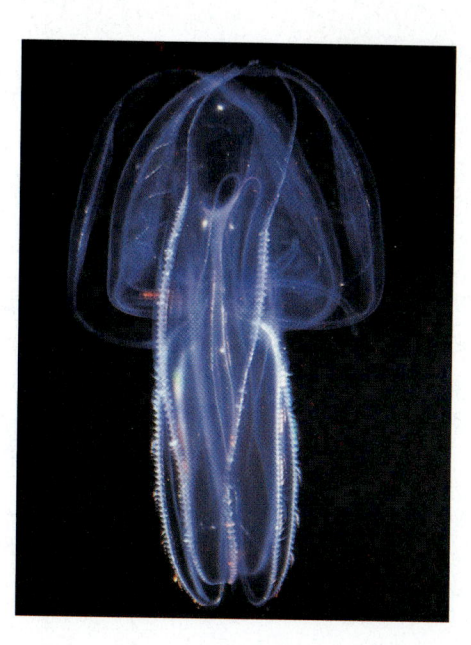

FIGURE 32.12
A comb jelly (phylum Ctenophora). Note the comblike plates along the ridges of the base.

> Cnidarians and ctenophores have tissues and radial symmetry. Cnidarians have a specialized kind of cell called a cnidocyte. Ctenophores propel themselves through the water by means of eight comblike plates of fused cilia.

Chapter 32 Noncoelomate Invertebrates **641**

32.3 Acoelomates are solid worms that lack a body cavity.

The Bilaterian Acoelomates

The Bilateria are characterized by the second key transition in the animal body plan, *bilateral symmetry*, which allowed animals to achieve high levels of specialization within parts of their bodies. The simplest bilaterians are the acoelomates; they lack any internal cavity other than the digestive tract. As discussed earlier, all bilaterians have three embryonic germ layers during development: ectoderm, endoderm, and mesoderm. We will focus our discussion of the acoelomates on the largest phylum of the group, the flatworms.

Phylum Platyhelminthes: The Flatworms

Phylum Platyhelminthes consists of some 20,000 species. These ribbon-shaped, soft-bodied animals are flattened dorsoventrally, from top to bottom. Flatworms are among the simplest of bilaterally symmetrical animals, but they do have a definite head at the anterior end and they do possess organs. Their bodies are solid; the only internal space consists of the digestive cavity (figure 32.13).

Flatworms range in length from a millimeter or less to many meters, as in some tapeworms. Most species of flatworms are parasitic, occurring within the bodies of many other kinds of animals (figure 32.14). Other flatworms are free-living, occurring in a wide variety of marine and freshwater habitats, as well as in moist places on land. Free-living flatworms are carnivores and scavengers; they eat various small animals and bits of organic debris. They move from place to place by means of ciliated epithelial cells, which are particularly concentrated on their ventral surfaces.

Those flatworms that have a digestive cavity have an incomplete gut, one with only one opening. As a result, they cannot feed, digest, and eliminate undigested particles of food simultaneously, and thus, flatworms cannot feed continuously, as more advanced animals can. Muscular contractions in the upper end of the gut cause a strong sucking force, allowing flatworms to ingest their food and tear it into small bits. The gut is branched and extends throughout the body, functioning in both digestion and transport of food. Cells that line the gut engulf most of the food particles by phagocytosis and digest them; but, as in the cnidarians, some of these particles are partly digested extracellularly. Tapeworms, which are parasitic flatworms, lack digestive systems. They absorb their food directly through their body walls.

Unlike cnidarians, flatworms have an excretory system, which consists of a network of fine tubules (little tubes) that

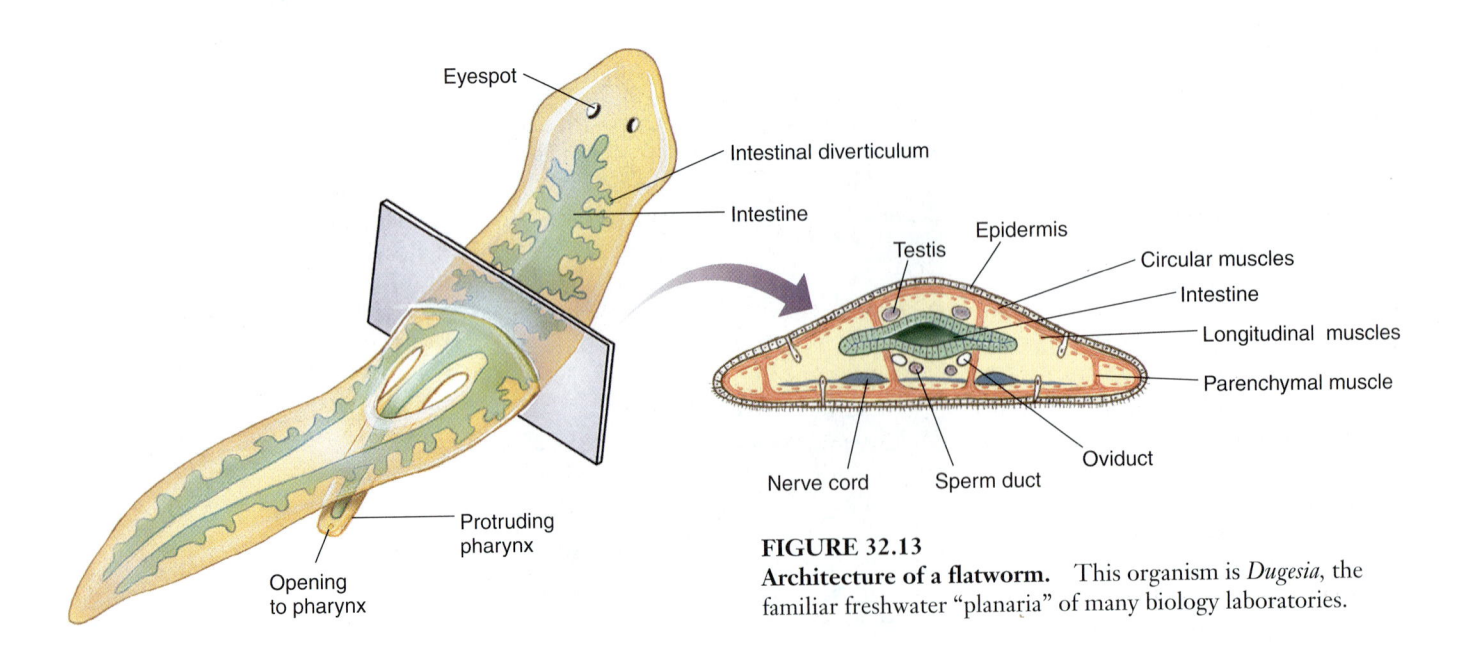

FIGURE 32.13
Architecture of a flatworm. This organism is *Dugesia*, the familiar freshwater "planaria" of many biology laboratories.

642   Part V   Diversity of Life on Earth

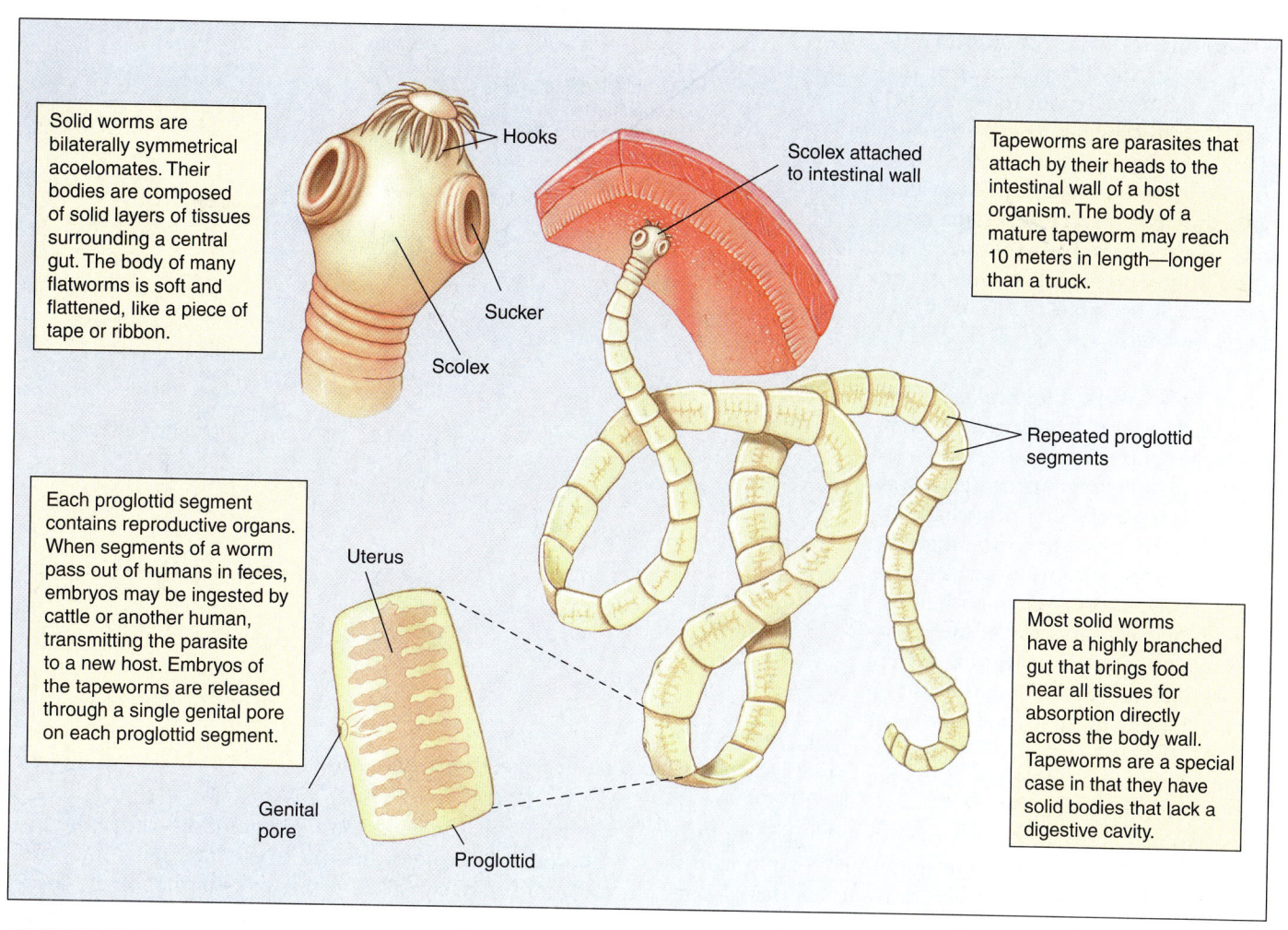

FIGURE 32.14

Phylum Platyhelminthes: Flatworms. Acoelomate flatworms such as this beef tapeworm, *Taenia saginata*, are bilaterally symmetrical solid worms. In addition, all bilaterians have three embryonic layers and a distinct head.

---

runs throughout the body. Cilia line the hollow centers of bulblike **flame cells** located on the side branches of the tubules. Cilia in the flame cells move water and excretory substances into the tubules and then to exit pores located between the epidermal cells. Flame cells were named because of the flickering movements of the tuft of cilia within them. They primarily regulate the water balance of the organism. The excretory function of flame cells appears to be secondary. A large proportion of the metabolic wastes excreted by flatworms diffuses directly into the gut and is eliminated through the mouth.

Like sponges, cnidarians, and ctenophores, flatworms lack circulatory systems for the transport of oxygen and food molecules. Consequently, all flatworm cells must be within diffusion distance of oxygen and food. Flatworms have thin bodies and highly branched digestive cavities that make such a relationship possible.

The nervous system of flatworms is very simple. Like cnidarians, some primitive flatworms have only a nerve net. However, most members of this phylum have longitudinal nerve cords that constitute a simple central nervous system.

Free-living members of this phylum have eyespots on their heads. These are inverted, pigmented cups containing light-sensitive cells connected to the nervous system. These eyespots enable the worms to distinguish light from dark; worms move away from strong light.

The reproductive systems of flatworms are complex. Most flatworms are **hermaphroditic,** with each individual containing both male and female sexual structures. In many of them, fertilization is internal. When they mate, each partner deposits sperm in the copulatory sac of the other. The sperm travel along special tubes to reach the eggs. In most free-living flatworms, fertilized eggs are laid in cocoons strung in ribbons and hatch into miniature adults. Some parasitic flatworms undergo a complex succession of distinct larval forms. Flatworms are also capable of asexual regeneration. In some genera, when a single individual is divided into two or more parts, each part can regenerate an entirely new flatworm.

Class Turbellaria: Turbellarians.
Only one of the three classes of flatworms, the turbellarians (class Turbellaria), are free-living. One of the most familiar is the freshwater genus *Dugesia*, the common planaria used in biology laboratory exercises. Other members of this class are widespread and often abundant in lakes, ponds, and the sea. Some also occur in moist places on land.

Class Trematoda: The Flukes. Two classes of parasitic flatworms live within the bodies of other animals: flukes (class Trematoda) and tapeworms (class Cestoda). Both groups of worms have epithelial layers resistant to the digestive enzymes and immune defenses produced by their hosts—an important feature in their parasitic way of life. However, they lack certain features of the free-living flatworms, such as cilia in the adult stage, eyespots, and other sensory organs that lack adaptive significance for an organism that lives within the body of another animal.

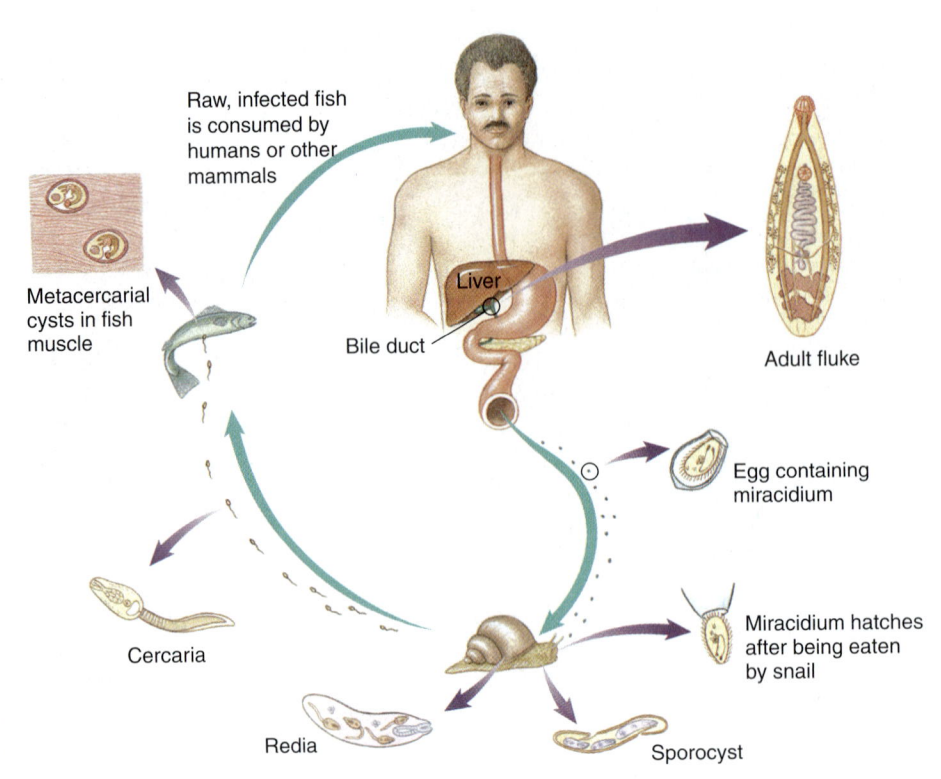

FIGURE 32.15
Life cycle of the human liver fluke, *Clonorchis sinensis*.

Flukes take in food through their mouth, just like their free-living relatives. There are more than 10,000 named species of flukes, ranging in length from less than 1 millimeter to more than 8 centimeters. Flukes attach themselves within the bodies of their hosts by means of suckers, anchors, or hooks. Some have a life cycle that involves only one host, usually a fish. Most have life cycles involving two or more hosts. Their larvae almost always occur in snails, and there may be other intermediate hosts. The final host of these flukes is almost always a vertebrate.

To human beings, one of the most important flatworms is the human liver fluke, *Clonorchis sinensis*. It lives in the bile passages of the liver of humans, cats, dogs, and pigs. It is especially common in Asia. The worms are 1 to 2 centimeters long and have a complex life cycle. Although they are hermaphroditic, cross-fertilization usually occurs between different individuals. Eggs, each containing a complete, ciliated first-stage larva, or **miracidium,** are passed in the feces (figure 32.15). If they reach water, they may be ingested by a snail. Within the snail, an egg transforms into a *sporocyst*, a baglike structure with embryonic germ cells. Within the sporocysts are produced **rediae,** which are elongated, nonciliated larvae. These larvae continue growing within the snail, giving rise to several individuals of the tadpole-like next larval stage, **cercariae.**

Cercariae escape into the water, where they swim about freely. If they encounter a fish of the family Cyprinidae—the family that includes carp and goldfish—they bore into the muscles or under the scales, lose their tails, and transform into **metacercariae** within cysts in the muscle tissue. If a human being or other mammal eats raw infected fish, the cysts dissolve in the intestine, and the young flukes migrate to the bile duct, where they mature. An individual fluke may live for 15 to 30 years in the liver. In humans, a heavy infestation of liver flukes may cause cirrhosis of the liver and death.

Other very important flukes are the blood flukes of genus *Schistosoma*. They afflict about 1 in 20 of the world's population, more than 200 million people throughout tropical Asia, Africa, Latin America, and the Middle East. Three species of *Schistosoma* cause the disease called schistosomiasis, or bilharzia. Some 800,000 people die each year from this disease.

Recently, a great deal of effort has gone into controlling schistosomiasis. The worms protect themselves from the body's immune system in part by coating themselves with a variety of the host's own antigens that effectively render the worm immunologically invisible (see chapter 48). Despite this difficulty, the search is on for a vaccine that would cause the host to develop antibodies to one of the antigens of the young worms before they protect themselves with host antigens. This vaccine would protect humans from infection. The disease can be cured with drugs after infection.

644   Part V   Diversity of Life on Earth

Class Cestoda: The Tapeworms. Class Cestoda is the third class of flatworms; like flukes, they live as parasites within the bodies of other animals. In contrast to flukes, tapeworms simply hang on to the inner walls of their hosts by means of specialized terminal attachment organs and absorb food through their skins. Tapeworms lack digestive cavities as well as digestive enzymes. They are extremely specialized in relation to their parasitic way of life. Most species of tapeworms occur in the intestines of vertebrates, about a dozen of them regularly in humans.

The long, flat bodies of tapeworms are divided into three zones: the **scolex,** or attachment organ; the unsegmented **neck;** and a series of repetitive segments, the **proglottids** (see figure 32.14). The scolex usually bears several suckers and may also have hooks. Each proglottid is a complete hermaphroditic unit, containing both male and female reproductive organs. Proglottids are formed continuously in an actively growing zone at the base of the neck, with maturing ones moving farther back as new ones are formed in front of them. Ultimately, the proglottids near the end of the body form mature eggs. As these eggs are fertilized, the zygotes in the very last segments begin to differentiate, and these segments fill with embryos, break off, and leave their host with the host's feces. Embryos, each surrounded by a shell, emerge from the proglottid through a pore or the ruptured body wall. They are deposited on leaves, in water, or in other places where they may be picked up by another animal.

The beef tapeworm *Taenia saginata* occurs as a juvenile in the intermuscular tissue of cattle but as an adult in the intestines of human beings. A mature adult beef tapeworm may reach a length of 10 meters or more. These worms attach themselves to the intestinal wall of their host by a scolex with four suckers. Segments shed from the end of the worm pass from the human in the feces and may crawl onto vegetation. These segments ultimately rupture and scatter the embryos. Embryos may remain viable for up to five months. If ingested by cattle, they burrow through the wall of the intestine and ultimately reach muscle tissues through the blood or lymph vessels. About 1% of the cattle in the United States are infected, and some 20% of the beef consumed is not federally inspected. Thus, when humans eat infected beef that is cooked "rare," infection by these tapeworms is likely. As a result, the beef tapeworm is a frequent parasite of humans.

Phylum Nemertea: The Ribbon Worms

The phylogenetic relationship of phylum Nemertea (figure 32.16) to other free-living flatworms is unclear. Nemerteans are often called ribbon worms, or proboscis worms. These aquatic worms have the body plan of a flatworm, but also possess a fluid-filled sac that may be a primitive coelom. This sac serves as a hydraulic power source for their proboscis, a long muscular tube that can be thrust out quickly from a sheath to capture prey. Shaped like a thread or a ribbon, ribbon worms are mostly marine and consist of about 900 species. Ribbon worms are large, often 10 to 20 centimeters, and sometimes many meters, in length. They are the simplest animals that possess a **complete digestive system,** one that has two separate openings, a mouth and an anus. Ribbon worms also exhibit a circulatory system in which blood flows in vessels. Many important evolutionary trends that become fully developed in more advanced animals make their first appearance in the Nemertea.

> The acoelomates, typified by flatworms, are the most primitive bilaterally symmetrical animals and the simplest animals in which organs occur.

FIGURE 32.16
A ribbon worm, *Lineus* (phylum Nemertea). This is the simplest animal with a complete digestive system.

Chapter 32 Noncoelomate Invertebrates 645

32.4 Pseudocoelomates have a simple body cavity.

The Pseudocoelomates

All bilaterians except solid worms possess an internal body cavity, the third key transition in the animal body plan. Seven phyla are characterized by their possession of a pseudocoel (see figure 31.5). Their evolutionary relationships remain unclear, with the possibility that the pseudocoelomate condition arose independently many times. The pseudocoel serves as a hydrostatic skeleton—one that gains its rigidity from being filled with fluid under pressure. The animals' muscles can work against this "skeleton," thus making the movement of the pseudocoelomates far more efficient than that of the acoelomates. Pseudocoelomates lack a defined circulatory system; this role is performed by fluids that move within the pseudocoel. In this section, we focus on three significant pseudocoelomate phyla.

Phylum Nematoda: The Roundworms

Nematodes, eelworms, and other roundworms constitute a large phylum, Nematoda, with some 20,000 recognized species. Scientists estimate that the actual number might approach 100 times that many. Members of this phylum are found everywhere. Nematodes are abundant and diverse in marine and freshwater habitats, and many members of this phylum are parasites of animals (figure 32.17) and plants. Many nematodes are microscopic and live in soil. A spadeful of fertile soil may contain, on the average, a million nematodes.

Nematodes are bilaterally symmetrical, unsegmented worms. They are covered by a flexible, thick cuticle, which is molted four times as they grow. Their muscles constitute a layer beneath the epidermis and extend along the length of the worm, rather than encircling its body. These longitudinal muscles pull against both the cuticle and the pseudocoel, which forms a hydrostatic skeleton. When nematodes move, their bodies whip about from side to side.

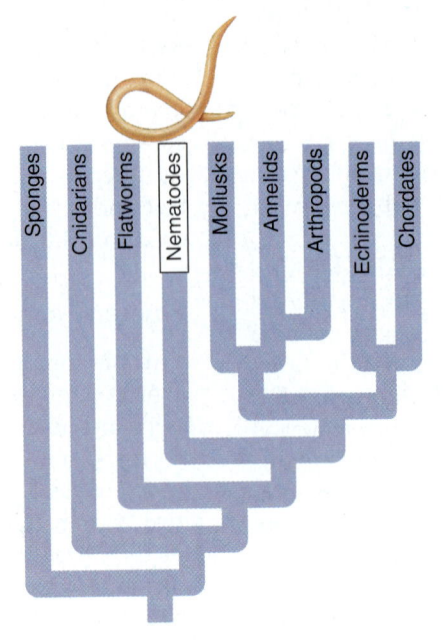

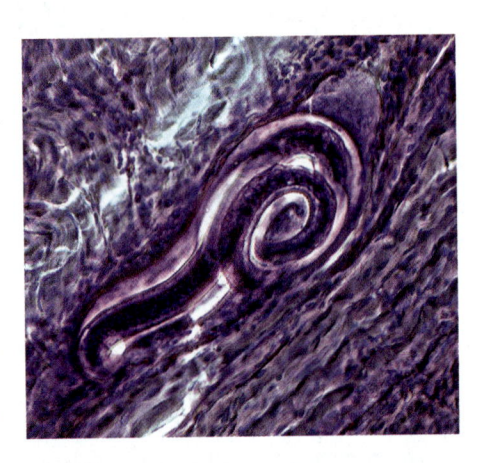

FIGURE 32.17
Trichinella **nematode encysted in pork.** The serious disease trichinosis can result from eating undercooked pork or bear meat containing such cysts.

Nematodes gather nutrients and exchange oxygen through their cuticle, but they also possess a well-developed digestive system. Near the mouth of a nematode, at its anterior end, are usually 16 raised, hairlike, sensory organs. The mouth is often equipped with piercing organs called **stylets.** Food passes through the mouth as a result of the sucking action produced by the rhythmic contraction of a muscular chamber called the **pharynx** at the worm's anterior end. After passing through a short corridor into the pharynx, food continues through the other portions of the digestive tract, where it is broken down and then digested. The wall of the intestine is only one cell layer thick. Some of the water with which the food has been mixed is reabsorbed near the end of the digestive tract, and material that has not been digested is eliminated through the anus (figure 32.18).

Nematodes completely lack flagella or cilia, even on sperm cells. Reproduction in nematodes is sexual, with the sexes usually separate. Their development is simple, and the adults consist of very few cells. For this reason, nematodes have become extremely important subjects for genetic and developmental studies (see chapter 19). The 1-millimeter-long *Caenorhabditis elegans* matures in only three days, its body is transparent, and it has only 959 cells. It is the only animal whose complete developmental cellular anatomy is known.

The diets of nematodes vary greatly. Many are active hunters, preying on protists and other small animals. Many species of nematodes are parasites, living within the bodies of larger animals. Almost every species of plant and animal that has been studied has been found to have at least one parasitic species of nematode living in it. The largest known nematode, reaching a length of 9 meters, is a parasite in the placenta of female sperm whales. About 50 species of nematodes, including several that are rather common in the United States, regularly parasitize human beings. For example, hookworms, mostly of the genus *Necator*, can be common in southern states. By sucking blood through the intestinal wall, they can produce anemia if untreated.

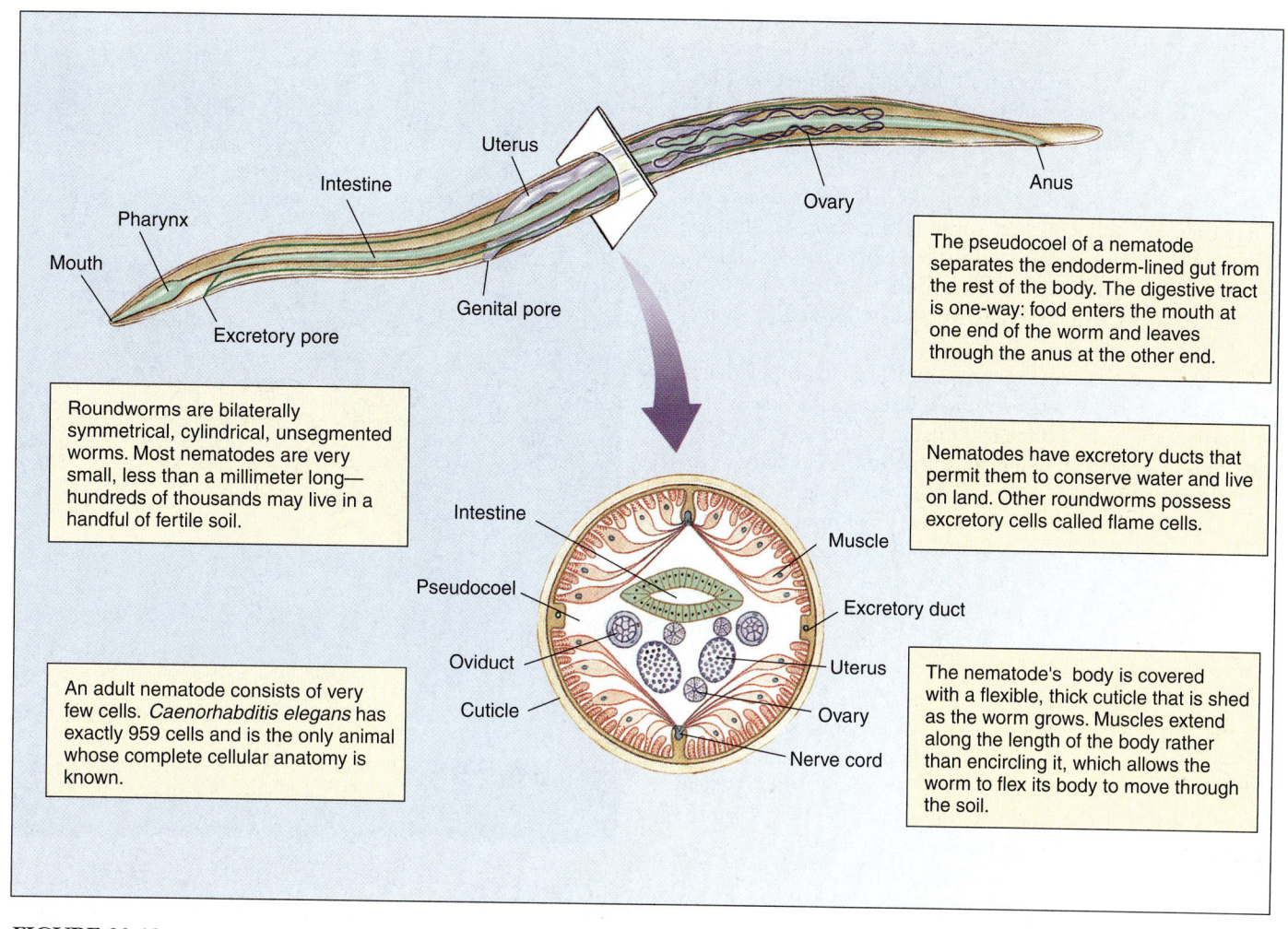

FIGURE 32.18
Phylum Nematoda: Roundworms. Roundworms such as this nematode possess a body cavity between the gut and the body wall called the pseudocoel. It allows nutrients to circulate throughout the body and prevents organs from being deformed by muscle movements.

Nematode-Caused Diseases

The most serious and common nematode-caused disease in temperate regions is trichinosis, caused by worms of the genus *Trichinella*. These worms live in the small intestine of pigs, where fertilized female worms burrow into the intestinal wall. Once it has penetrated these tissues, each female produces about 1500 live young. The young enter the lymph channels and travel to muscle tissue throughout the body, where they mature and form highly resistant, calcified cysts. Infection in human beings or other animals arises from eating undercooked or raw pork in which the cysts of *Trichinella* are present. If the worms are abundant, a fatal infection can result, but such infections are rare; in the United States, only about 20 deaths have been attributed to trichinosis during the past decade.

Pinworms, *Enterobius*, are abundant throughout the United States, where it is estimated they infect about 30% of all children and about 16% of adults. Adult pinworms live in the human large intestine. Fortunately, the symptoms they cause are not severe, and the worms can easily be controlled by drugs.

The intestinal roundworm, *Ascaris*, infects approximately one of six people worldwide but is rare in areas with modern plumbing. Like pinworms, these worms live in the intestine. Their fertilized eggs are spread in feces, and can remain viable for years in the soil. Adult females, which are up to 30 centimeters long, contain up to 30 million eggs, and can lay up to 20,000 of them each day!

Other nematode-caused diseases are extremely serious in the tropics. *Filaria* infects at least 250 million people worldwide. Up to 10 centimeters long, they live in the lymphatic system, which they may seriously obstruct, causing severe inflammation and swelling. Extreme filariasis produces the condition known as elephantiasis.

Phylum Rotifera: Rotifers

Rotifers (phylum Rotifera) are bilaterally symmetrical, unsegmented pseudocoelomates. Although they are pseudocoelomates, rotifers are very unlike nematodes. Several features suggest their ancestors may have resembled flatworms. Rotifers are very small—at 50–500 micrometers long, they are smaller than some ciliate protists. But rotifers have complex bodies with three cell layers and highly developed internal organs. A complete gut passes from mouth to anus. An extensive pseudocoel acts as a hydroskeleton to provide body rigidity.

Rotifers are basically aquatic animals that propel themselves through the water by rapidly beating cilia, like a boat with oars. There are about 1800 named species of this phylum. While a few rotifers live in soil or in the capillary water in cushions of mosses, most occur in fresh water, and they are common everywhere. Most live no longer than 1 or 2 weeks, but some species can survive in a desiccated inactive state on the leaves of plants; when rain falls, the rotifers regain activity and actively feed in the film of water that temporarily covers the leaf. Very few rotifers are marine.

Rotifers have a well-developed food-processing apparatus. A conspicuous organ on the tip of the head called the corona gathers food (figure 32.19). It is composed of a circle of cilia that sweeps food into the rotifer's mouth and is also used for locomotion. Rotifers are often called "wheel animals" because the cilia, when beating together, resemble the movement of spokes radiating from a wheel.

A Relatively New Phylum: Cycliophora

In December 1995, two Danish biologists reported the discovery of a strange new kind of creature, smaller than a period on a printed page. The tiny organism had a striking circular mouth surrounded by a ring of fine, hairlike cilia, and its life cycle is so unusual that they assigned it to an entirely new phylum, Cycliophora (Greek, "carrying a small wheel") (figure 32.20). There are 35 traditional animal phyla. A 36th, the loricifera, was discovered in 1983. Cycliophora is the 37th. Cycliophorans live on the mouthparts of lobsters. When the lobster to which it is attached starts to molt, the tiny symbiont begins a bizarre form of sexual reproduction. Dwarf males emerge, composed of nothing but brains and reproductive organs. Each dwarf male seeks out another female symbiont on the molting lobster and fertilizes its eggs, generating free-swimming individuals that can seek out another lobster and renew the life cycle.

> All pseudocoelomates have fluid-filled body cavities that separate endoderm from mesoderm. Nematodes are bilaterally symmetrical, unsegmented worms. Other pseudocoelomates have very different body plans. Rotifers, while tiny, are very complex, as are cycliophorans.

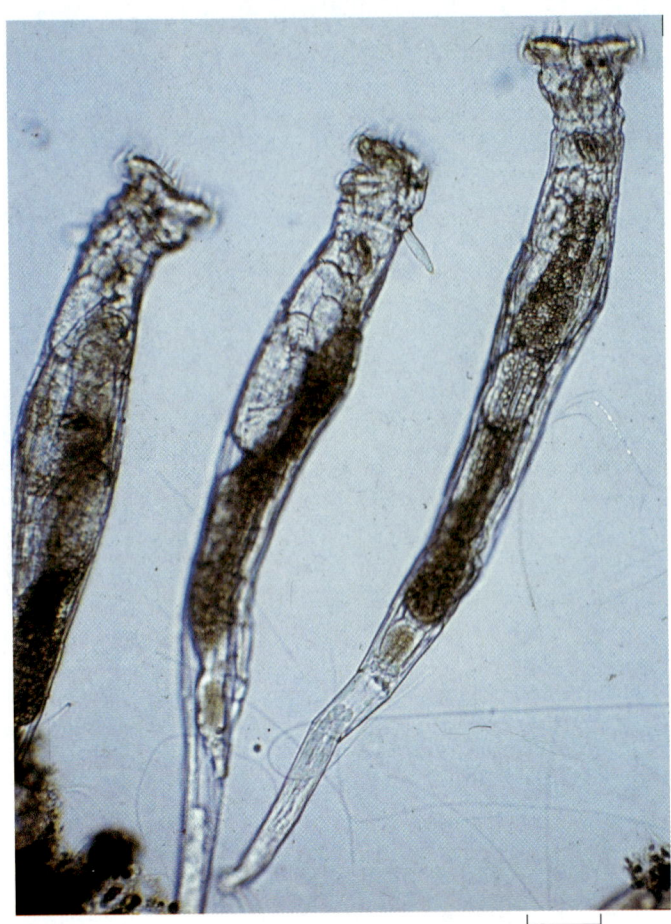

FIGURE 32.19
A rotifer. Microscopic in size, rotifers are smaller than some ciliate protists, and yet have complex internal organs.

FIGURE 32.20
Cycliophorans. Tinier than the period at the end of this sentence, these pseudocoelomates live on the mouthparts of lobsters. Many individuals of the species *Symbion pandora* are shown attached to a cilia-covered mouthpart of a lobster (50×).

648 Part V Diversity of Life on Earth

Concept Review

For interactive testing, visit the Online Learning Center with PowerWeb at www.mhhe.com/Raven7

32.1 The classification of invertebrates is currently being reevaluated.

An Uproar Over Invertebrate Phylogeny

- Great disagreement exists as to how the 35 animal phyla are related. (p. 634)
- Animal family trees have historically been based on lumping key aspects of body architecture, and more recently by focusing on differences in ribosomal RNA sequences. (p. 634)
- In the traditional animal family tree, bilaterally symmetrical animals are divided into acoelomates, pseudocoelomates, and coelomates. Phylogenies based on rRNA sequences recognize two clades: lophotrochozoans and ecdysozoans. (pp. 634–635)

32.2 The simplest animals are not bilaterally symmetrical.

Parazoa

- Sponges are multicellular animals that lack tissue-level development and body symmetry, and uniquely contain choanocytes with flagella, which line the internal cavity of the sponge. The beating of these flagella draws water in and through the sponge, thus bringing in food and oxygen and expelling wastes. (p. 636)

Radiata

- Distinct tissues are seen in the eumetazoans. Different cell layers in the embryo form different tissues and give rise to the basic body plan. (p. 638)
- Cnidarians are gelatinous organisms made of distinct tissues but lacking true organs. They have two basic body forms—polyps and medusae—and contain specialized cells called cnidocytes on their tentacles. (p. 638)
- There are four basic classes of cnidarians: Hydrozoa, Scyphozoa, Cubozoa, and Anthozoa. (pp. 640–641)
- Ctenophores propel themselves through the water with fused cilia. (p. 641)

32.3 Acoelomates are solid worms that lack a body cavity.

The Bilaterian Acoelomates

- Bilateria are characterized by bilateral symmetry, allowing high levels of specialization. (p. 642)
- Phylum Platyhelminthes, the flatworms, contains class Turbellaria, (Turbellarians), class Trematoda (flukes), and class Cestoda (tapeworms). (pp. 642–645)
- Phylum Nemertea contains the ribbon worms. (p. 645)

32.4 Pseudocoelomates have a simple body cavity.

The Pseudocoelomates

- In the pseudocoelomates, circulation occurs in a pseudocoel, not in a defined circulatory system. (p. 646)
- Phylum Nematoda contains roundworms, which are bilaterally symmetrical and unsegmented. (p. 646)
- Many diseases are caused by nematodes, such as *Trichinella*, *Enterobius*, *Ascaris*, and *Filaria*. (p. 647)
- Phylum Rotifera is composed of aquatic animals that propel themselves through the water by beating cilia. (p. 648)
- Phylum Cycliophora is composed of tiny organisms that are so unusual they were assigned a new phylum. (p. 648)

Test Your Understanding

Self Test

1. The traditional animal family tree places more significance on body plan (the presence of a coelom), while the rRNA tree places more significance on
   a. embryological development patterns.
   b. whether or not the animal molts.
   c. symmetry.
   d. all of these.
2. A key evolutionary development seen for the first time in the sponges is
   a. a complete digestive system.
   b. tissues.
   c. body symmetry.
   d. multicellularity.
3. All of the following are found in sponges except
   a. spicules.
   b. choanocytes.
   c. a digestive tract.
   d. sexual and/or asexual reproduction.
4. The first animal group to show extracellular digestion was the
   a. sponges.
   b. cnidarians.
   c. flatworms.
   d. roundworms.
5. Cnidarians project a nematocyst to capture their prey by
   a. building up a high internal osmotic pressure.
   b. ejecting it with a jet of water.
   c. using a springlike apparatus.
   d. muscle contractions that "throw" the nematocyst.
6. Which of the following is an example of an organism with the medusa body form?
   a. a hydra
   b. a coral
   c. an anemone
   d. a jellyfish
7. Key evolutionary advances of the flatworms are bilateral symmetry and
   a. a coelom.
   b. internal organs.
   c. a one-way digestive tract.
   d. a body cavity.
8. For excretion, flatworms use
   a. a miracidium.
   b. osmosis.
   c. flame cells.
   d. proglottids.
9. The type of body cavity seen in the roundworms is called a(n)
   a. coelom.
   b. acoelom.
   c. pseudocoel.
   d. gastrovascular cavity.
10. The type of pseudocoelomates found in soil, freshwater and marine environments, and as parasites, are
    a. nematodes.
    b. *Trichinella*.
    c. rotifers.
    d. Cycliophora.

Test Your Visual Understanding

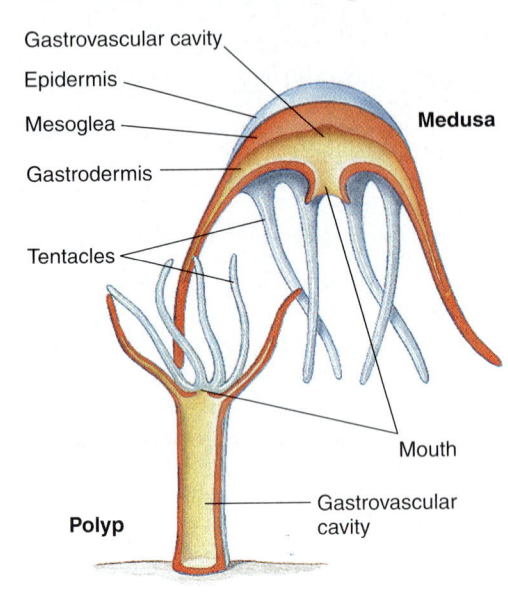

1. What phylum is represented by the two organisms pictured in this figure, and what are some general characteristics of this phylum?
2. What type of organism spends most of its life cycle in the medusa body form? What type of organism spends most of its life cycle in the polyp body form?

Apply Your Knowledge

1. A sponge is a filter feeder, which means it eats by filtering food from the water that passes through its pores and out its osculum. Assume that a sponge filters 1.8 milliliters of water per second. How much water is filtered in an hour? In a day?
2. Coral reefs are often found in nutrient-poor waters. What type of symbiotic relationship helps these coral animals grow actively? What are the advantages for each member in the relationship? What would happen to the coral reefs if the water they are growing in became cloudy with pollution?
3. Parasites, especially those that require two or more hosts to complete their life cycles, often produce very large numbers of offspring. What advantage would this present to them?

33

Coelomate Invertebrates

Concept Outline

33.1 Mollusks were among the first coelomates.

Mollusk Body Plan. The mollusk body plan is characterized by three distinct sections.

Major Classes of Mollusks. The three major classes of mollusks are gastropods, bivalves, and cephalopods.

33.2 Annelids were the first segmented animals.

Segmented Bodies. Annelids are segmented coelomate worms, most of which live in the sea.

Major Classes of Annelids. The three major classes of annelids are the polychaetes (marine worms), the oligochaetes (earthworms and related freshwater worms), and the hirudines (leeches).

33.3 Lophophorates appear to be a transitional group.

Lophophorates. The three phyla of lophophorates share a unique ciliated feeding structure.

33.4 Arthropods are the most diverse of all animal groups.

Arthropod Body Plan. Arthropods have segmented bodies and open circulatory systems.

A Major Group of Arthropods: Crustaceans. Crustaceans are a large, diverse, primarily marine subphylum of arthropods.

Major Classes of Arthropods: Arachnids. Spiders have eight legs.

Major Classes of Arthropods: Centipedes and Millipedes. Centipedes and millipedes have 30+ legs.

Major Classes of Arthropods: Insects. Insects, the most diverse class of animals, are the only arthropods that can fly.

33.5 Echinoderms are radially symmetrical as adults.

Deuterostome Development and an Endoskeleton. All echinoderms are marine deuterostomes with an endoskeleton.

Echinoderm Body Plan. Echinoderms are radially symmetrical as adults.

Major Classes of Echinoderms. All six classes of echinoderms exhibit a five-part body plan.

FIGURE 33.1
Coelomate invertebrates have evolved a very diverse array of body plans. This insect (*Polistes*, the common paper wasp) has a segmented body, jointed appendages, and an exoskeleton.

Although acoelomates and pseudocoelomates have proven very successful, a third way of organizing the animal body has also evolved, one that occurs in the bulk of the animal kingdom. We will begin our discussion of the coelomate invertebrate animals with mollusks, which include such animals as clams, snails, slugs, and octopuses. Annelids, such as earthworms, leeches, and seaworms, are also coelomates, but in addition, were the earliest group of animals to evolve segmented bodies. Arthropods, such as the paper wasp in figure 33.1, evolved jointed appendages, and have become the most successful of all animal groups. Echinoderms are exclusively marine animals that exhibit deuterostome development and endoskeletons, two evolutionary innovations that they share with the chordates, which are the subject of chapter 34.

651

33.1 Mollusks were among the first coelomates.

Mollusk Body Plan

The evolution of the *coelom* was a significant advance in the structure of the animal body. Coelomates have a new body design that repositions the fluid and allows complex tissues and organs to develop. This new body plan also made it possible for animals to evolve a wide variety of different body architectures and to grow to much larger sizes than acoelomate animals. Among the earliest groups of coelomates were the mollusks and the annelids.

Mollusks

Mollusks (phylum Mollusca) are an extremely diverse animal phylum, second only to the arthropods, with over 110,000 described species. Mollusks exhibit a variety of body forms and live in many different environments. Mollusks include snails, slugs, clams, scallops, oysters, cuttlefish, octopuses, and many other familiar animals (figure 33.2). The durable shells of some mollusks are often beautiful and elegant; they have long been favorite objects for professional scientists and amateurs alike to collect, preserve, and study. Chitons and nudibranchs are less familiar marine mollusks. Mollusks are characterized by a coelom, and while there is extraordinary diversity in this phylum, many of the basic components of the mollusk body plan can be seen in figure 33.3.

Mollusks evolved in the oceans, and most groups have remained there. Marine mollusks are widespread and often abundant. Some groups of mollusks that have invaded freshwater and terrestrial habitats include the snails and slugs that live in your garden. Terrestrial mollusks are often abundant in places that are at least seasonally moist. Some of these places, such as the crevices of desert rocks, may appear very dry, but even these habitats have at least a temporary supply of water at certain times.

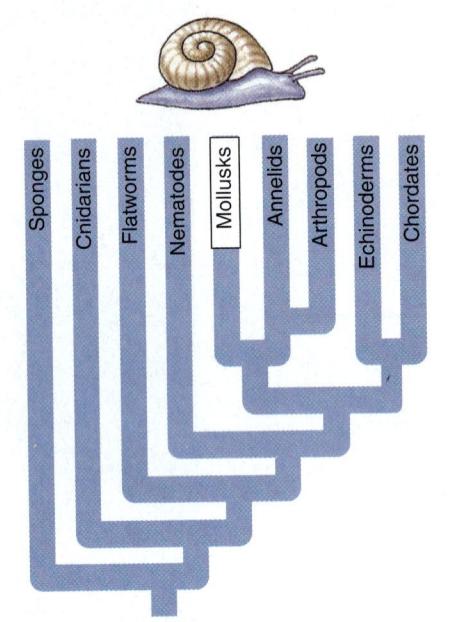

FIGURE 33.2
A mollusk. The blue-ringed octopus is one of the few mollusks dangerous to humans. Strikingly beautiful, it is equipped with a sharp beak and poison glands—divers give it a wide berth!

As a group, mollusks are an important source of food for humans. Oysters, clams, scallops, mussels, octopuses, and squids are among the culinary delicacies that belong to this large phylum. Mollusks are also economically significant in many other ways. For example, pearls are produced in oysters, and the material called mother-of-pearl, often used in jewelry and other decorative objects, is produced in the shells of a number of different mollusks, but most notably in the snail called abalone. Mollusks are not wholly beneficial to humans, however. Bivalve mollusks called shipworms burrow through wood submerged in the sea, damaging boats, docks, and pilings. The zebra mussel has recently invaded North American ecosystems via the ballast water of cargo ships from Europe, wreaking havoc in many aquatic ecosystems. Slugs and terrestrial snails often cause extensive damage to garden flowers, vegetables, and crops. Other mollusks serve as hosts to the intermediate stages of many serious parasites, including several nematodes and flatworms, discussed in chapter 32.

Mollusks range in size from almost microscopic to huge. Although most measure a few centimeters in their largest dimension, the giant squid, which is occasionally cast ashore but has rarely been observed in its natural environment, may grow up to 21 meters long! Weighing up to 250 kilograms, the giant squid is the largest invertebrate and one of the heaviest. Millions of giant squid probably inhabit the deep regions of the ocean, even though they are seldom caught. Another large mollusk is the bivalve *Tridacna maxima*, the giant clam, which may be as long as 1.5 meters and may weigh as much as 270 kilograms (figure 33.4).

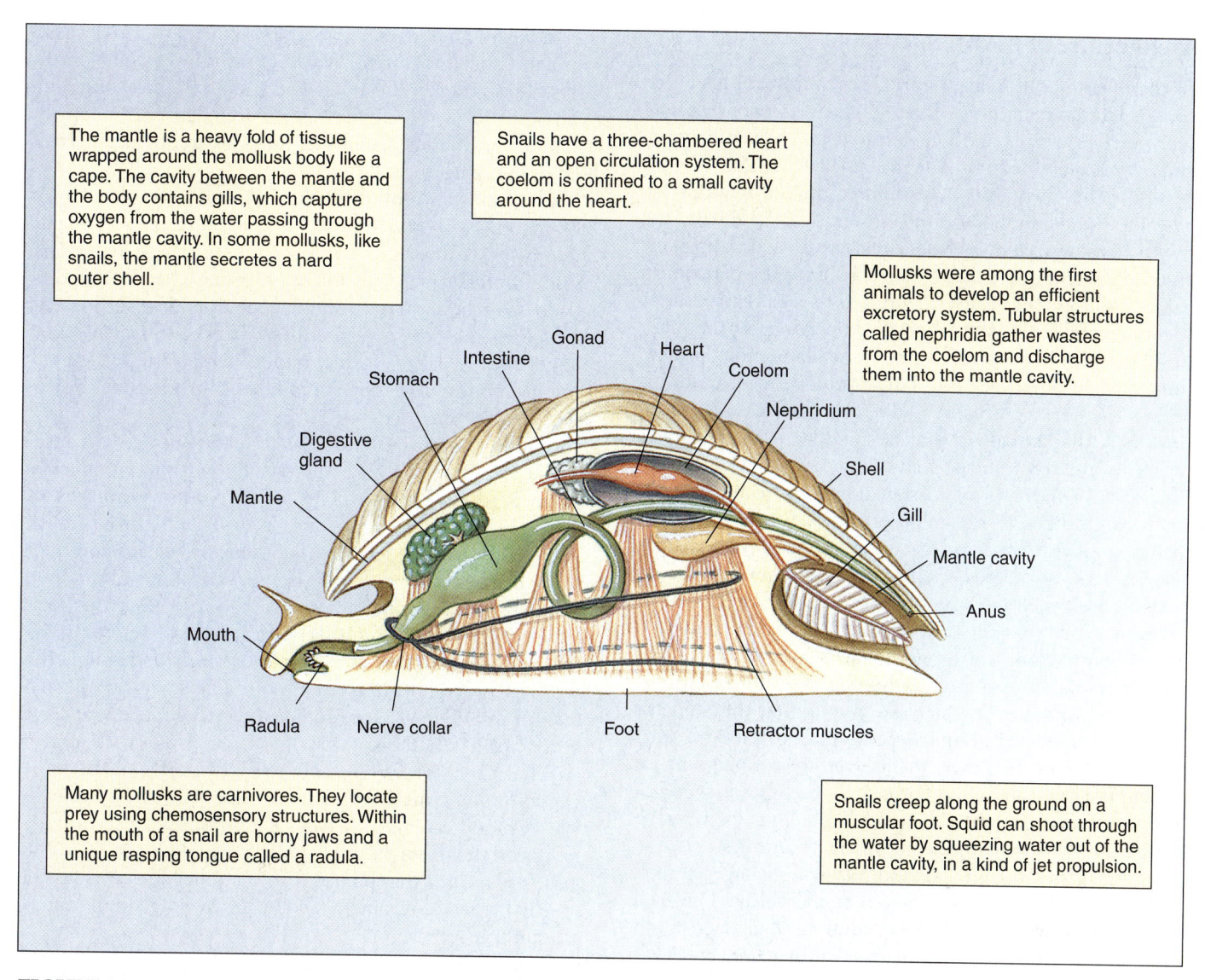

[Diagram of generalized molluscan body plan with labels and annotations:]

- The mantle is a heavy fold of tissue wrapped around the mollusk body like a cape. The cavity between the mantle and the body contains gills, which capture oxygen from the water passing through the mantle cavity. In some mollusks, like snails, the mantle secretes a hard outer shell.

- Snails have a three-chambered heart and an open circulation system. The coelom is confined to a small cavity around the heart.

- Mollusks were among the first animals to develop an efficient excretory system. Tubular structures called nephridia gather wastes from the coelom and discharge them into the mantle cavity.

- Many mollusks are carnivores. They locate prey using chemosensory structures. Within the mouth of a snail are horny jaws and a unique rasping tongue called a radula.

- Snails creep along the ground on a muscular foot. Squid can shoot through the water by squeezing water out of the mantle cavity, in a kind of jet propulsion.

Labels: Stomach, Intestine, Gonad, Heart, Coelom, Nephridium, Shell, Gill, Mantle cavity, Anus, Digestive gland, Mantle, Mouth, Radula, Nerve collar, Foot, Retractor muscles

FIGURE 33.3
Phylum Mollusca: mollusks. As shown in this generalized molluscan body plan, the body cavity of a mollusk is a coelom, which is completely enclosed within the mesoderm. This allows physical contact between the mesoderm and the endoderm, permitting interactions that lead to development of highly specialized organs such as a stomach.

FIGURE 33.4
Giant clam. Second only to the arthropods in number of described species, members of the phylum Mollusca occupy almost every habitat on earth. This giant clam, *Tridacna maxima*, has a green color caused by the presence of symbiotic dinoflagellates (zooxanthellae). Through photosynthesis, the dinoflagellates probably contribute most of the food supply of the clam, although it remains a filter feeder like most bivalves. Some individual giant clams may be nearly 1.5 meters long and weigh up to 270 kilograms.

Chapter 33 Coelomate Invertebrates

Architecture of the Mollusk Body

In their basic body plan (figure 33.5), mollusks have distinct bilateral symmetry. Their digestive, excretory, and reproductive organs are concentrated in a **visceral mass,** and a muscular **foot** is their primary mechanism of locomotion. They may also have a differentiated **head** at the anterior end of the body. Folds (often two) arise from the dorsal body wall and enclose a cavity between themselves and the visceral mass; these folds constitute the **mantle.** In some mollusks, the mantle cavity acts as a lung; in others, it contains gills. **Gills** are specialized portions of the mantle that usually consist of a system of filamentous projections rich in blood vessels. These projections greatly increase the surface area available for gas exchange and, therefore, the animal's overall respiratory potential. Mollusk gills are very efficient, and many gilled mollusks extract 50% or more of the dissolved oxygen from the water that passes through the mantle cavity. Finally, in most members of this phylum, the outer surface of the mantle also secretes a protective shell.

A mollusk shell consists of a horny outer layer, rich in protein, which protects the two underlying calcium-rich layers from erosion. The middle layer consists of densely packed crystals of calcium carbonate. The inner layer is pearly and increases in thickness throughout the animal's life. When it reaches a sufficient thickness, this layer is used as mother-of-pearl. Pearls themselves are formed when a foreign object, such as a grain of sand, becomes lodged between the mantle and the inner shell layer of **bivalve mollusks** (two-shelled), including clams and oysters. The mantle coats the foreign object with layer upon layer of shell material to reduce irritation caused by the object. The shell of mollusks serves primarily for protection. Many species can withdraw for protection into their shell if they have one.

In aquatic mollusks, a continuous stream of water passes into and out of the mantle cavity, drawn by the cilia on the gills. This water brings in oxygen and, in the case of the bivalves, food; it also carries out waste materials. When the gametes are being produced, they are frequently carried out in the same stream.

The foot of a mollusk is muscular and may be adapted for locomotion, attachment, food capture (in squids and octopuses), or various combinations of these functions. Some mollusks secrete mucus, forming a path that they glide along on their foot. In cephalopods—squids and octopuses–the foot is divided into arms, also called tentacles. In some *pelagic* forms, mollusks that are perpetually free-swimming, the foot is modified into winglike projections or thin fins.

One of the most characteristic features of all the mollusks except the bivalves is the **radula,** a rasping, tonguelike organ used for feeding. The radula consists primarily of dozens to thousands of microscopic, chitinous teeth arranged in rows (figure 33.6). Gastropods (snails and their relatives) use their radula to scrape algae and other food materials off their substrates and then to convey this food to the digestive tract. Other gastropods are active predators, some using a modified radula to drill through the shells of prey and extract the food. The small holes often seen in oyster shells are produced by gastropods that have bored holes to kill the oyster and extract its body for food.

The circulatory system of all mollusks except cephalopods consists of a heart and an open system in which blood circulates freely. The mollusk heart usually has three chambers, two that collect aerated blood from the gills, and a third that pumps it to the other body tissues. In mollusks, the coelom takes the form of a small cavity around the heart.

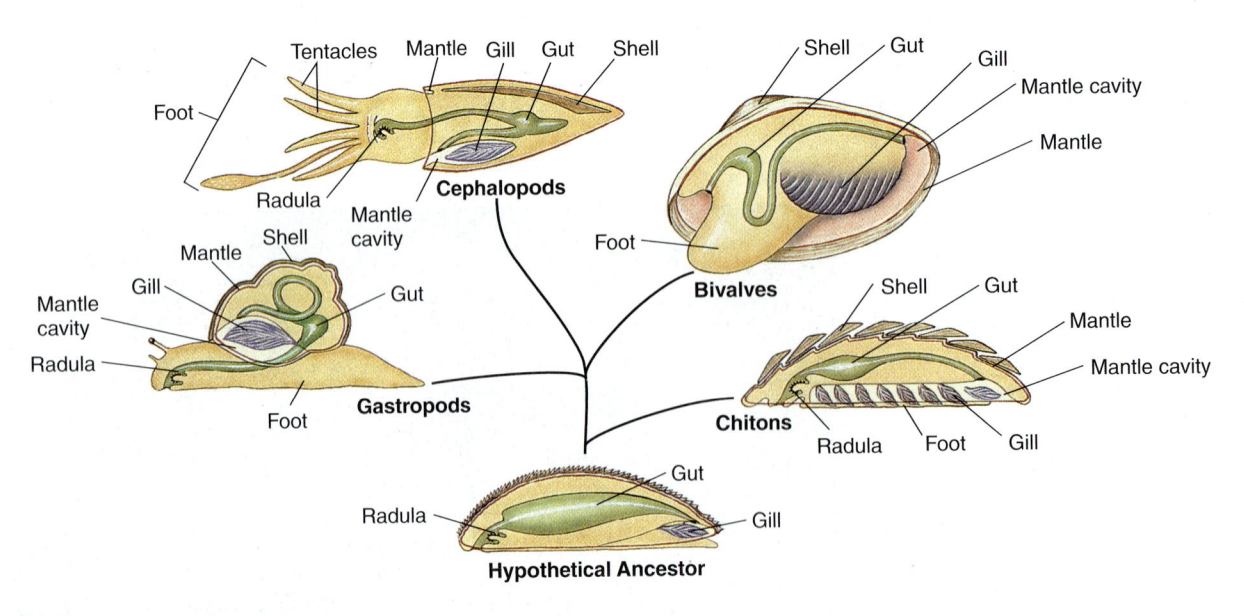

FIGURE 33.5
Body plans among the mollusks.

Nitrogenous wastes are removed from the mollusk by one or two tubular structures called **nephridia.** A typical nephridium has an open funnel, the **nephrostome,** which is lined with cilia. A coiled tubule runs from the nephrostome into a bladder, which in turn connects to an excretory pore. Wastes are gathered by the nephridia from the coelom and discharged into the mantle cavity. The wastes are then expelled from the mantle cavity by the continuous pumping of the gills. Sugars, salts, water, and other materials are reabsorbed by the walls of the nephridia and returned to the animal's body as needed to achieve an appropriate osmotic balance.

In animals with a closed circulatory system, such as annelids, cephalopod mollusks, and vertebrates, the coiled tubule of a nephridium is surrounded by a network of capillaries. Wastes are extracted from the circulatory system through these capillaries, transferred into the nephridium, and then subsequently discharged. Salts, water, and other associated materials may also be reabsorbed from the tubule of the nephridium back into the capillaries. For this reason, the excretory systems of these coelomates are much more efficient than the flame cells of the acoelomates, which pick up substances only from the body fluids. Mollusks were one of the earliest evolutionary lines to develop an efficient excretory system. Other than chordates, coelomates with closed circulation have similar excretory systems.

Reproduction in Mollusks

Most mollusks have distinct male and female individuals, although a few bivalves and many gastropods are hermaphroditic. Even in hermaphroditic mollusks, cross-fertilization is most common. Remarkably, some sea slugs and oysters are able to change from one sex to the other several times during a single season.

Most aquatic mollusks engage in external fertilization. The males and females release their gametes into the water, where they mix and fertilization occurs. However, gastropods more often have internal fertilization, with the male inserting sperm directly into the female's body. Internal fertilization is one of the key adaptations that allowed gastropods to colonize the land.

Many marine mollusks have free-swimming larvae called **trochophores** (figure 33.7a), which closely resemble the larval stage of many marine annelids. Trochophores swim by means of a row of cilia that encircles the middle of their body. In most marine snails and in bivalves, a second free-swimming stage, the veliger, follows

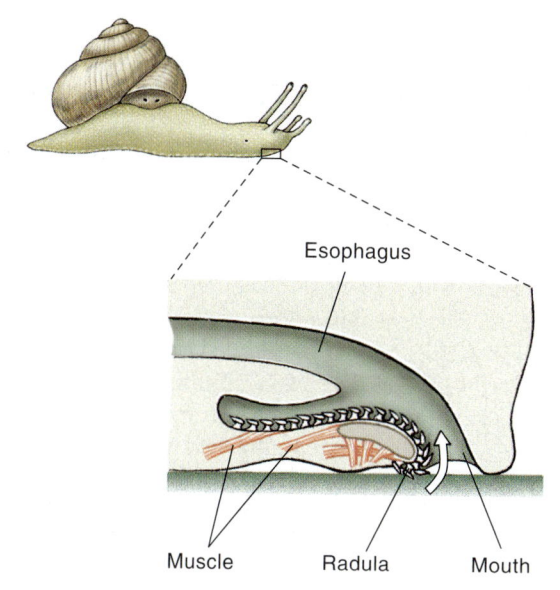

FIGURE 33.6
Structure of the radula in a snail. The radula consists of chitin and is covered with rows of teeth that extend backward. As the animal feeds, its mouth opens, and the radula brings food in by scraping backward.

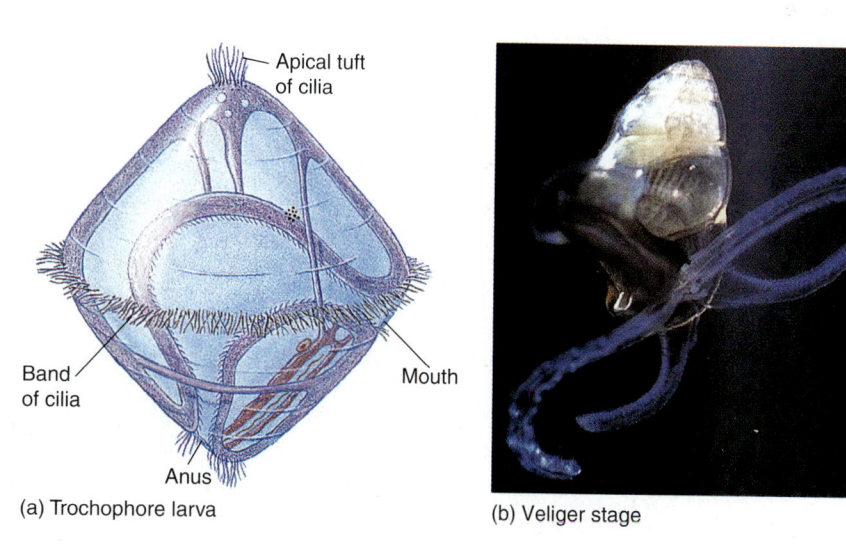

(a) Trochophore larva

(b) Veliger stage

FIGURE 33.7
Stages in the molluscan life cycle. (a) The trochophore larva of a mollusk. Similar larvae, as you will see, are characteristic of some annelid worms as well as a few other phyla. (b) The veliger stage of a mollusk.

the trochophore stage. This **veliger** stage has the beginnings of a foot, shell, and mantle (figure 33.7b). Trochophores and veligers drift widely in the ocean currents, dispersing mollusks to new areas.

> Mollusks were among the earliest animals to evolve an efficient excretory system. The mantle of mollusks not only secretes their protective shell, but also forms a cavity that is essential to respiration.

Chapter 33 Coelomate Invertebrates

Major Classes of Mollusks

Of the seven classes of mollusks, we will examine four as representatives of the phylum: (1) Polyplacophora—chitons; (2) Gastropoda—snails, slugs, limpets, and their relatives; (3) Bivalvia—clams, oysters, scallops, and their relatives; and (4) Cephalopoda—squids, octopuses, cuttlefishes, and nautilus. By studying living mollusks and the fossil record, some scientists have deduced that the ancestral mollusk was probably a dorsoventrally flattened, unsegmented, wormlike animal that glided on its ventral surface. This animal may also have had a chitinous cuticle and overlapping calcareous scales. Other scientists believe that mollusks arose from segmented ancestors and became unsegmented secondarily.

Class Polyplacophora: The Chitons

Chitons are marine mollusks that have oval bodies with eight overlapping calcareous plates. Underneath the plates, the body is not segmented. Chitons creep along using a broad, flat foot surrounded by a groove or mantle cavity in which the gills are arranged. Most chitons are grazing herbivores that live in shallow marine habitats, but some live at depths of more than 7000 meters.

Class Gastropoda: The Snails and Slugs

The class Gastropoda contains about 40,000 described species of snails, slugs, and similar animals. This class is primarily a marine group, but it also contains many freshwater and terrestrial mollusks (figure 33.8). Most gastropods have a shell, but some, such as slugs and nudibranchs, have lost their shells through the course of evolution. Gastropods generally creep along on a foot, which may be modified for swimming.

The heads of most gastropods have a pair of tentacles with eyes at the ends. These tentacles have been lost in some of the more advanced forms of the class. Within the mouth cavity of many members of this class are horny jaws and a radula.

During embryological development, gastropods undergo torsion. **Torsion** is the process by which the mantle cavity and anus are moved from a posterior location to the front of the body, where the mouth is located. Torsion is brought about by a disproportionate growth of the lateral muscles; that is, one side of the larva grows much more rapidly than the other. A 120° rotation of the visceral mass brings the mantle cavity above the head and twists many internal structures. In some groups of gastropods, varying degrees of detorsion have taken place. The **coiling,** or spiral winding, of the shell is a separate process. This process has led to the loss of the right gill and right nephridium in most gastropods. Thus, the visceral mass of gastropods has become bilaterally asymmetrical during the course of evolution.

FIGURE 33.8
A gastropod mollusk. The terrestrial snail *Allogona townsendiana*.

Gastropods display extremely varied feeding habits. Some are predatory, others scrape algae off rocks (or aquarium glass), and others are scavengers. Many are herbivores, and some terrestrial ones are serious garden and agricultural pests. The radula of oyster drills is used to bore holes in the shells of other mollusks, through which the contents of the prey can be removed. In cone shells, the radula has been modified into a kind of poisonous harpoon, which is shot with great speed into the prey.

Sea slugs, or nudibranchs, are active predators; a few species of nudibranchs have the extraordinary ability to extract the nematocysts from the cnidarian polyps they eat, transfer them through their digestive tract to the surface of their gills intact, and use them for their own protection. Nudibranchs get their name from their gills, which instead of being enclosed within the mantle cavity are exposed along the dorsal surface (*nudi,* "naked," + *branch,* "gill").

In terrestrial gastropods, the empty mantle cavity, which was occupied by gills in their aquatic ancestors, is extremely rich in blood vessels and serves as a lung, in effect. This structure evolved in animals living in environments with plentiful oxygen; it absorbs oxygen from the air much more effectively than a gill could, but is not as effective under water.

Class Bivalvia: The Bivalves

Members of the class Bivalvia include the clams, scallops, mussels, and oysters. Bivalves have two lateral (left and right) shells (valves) hinged together dorsally (figure 33.9). A ligament hinges the shells together and causes them to gape open. Pulling against this ligament are one or two large adductor muscles that can draw the shells together.

The mantle secretes the shells and ligament and envelops the internal organs within the pair of shells. The mantle is frequently drawn out to form two siphons, one for an incoming and one for an outgoing stream of water. The siphons often function as snorkels to allow bivalves to filter water through their body while remaining almost completely buried in sediments. A complex folded gill lies on each side of the visceral mass. These gills consist of pairs of filaments that contain many blood vessels. Rhythmic beating of cilia on the gills creates a pattern of water circulation. Most bivalves are sessile filter feeders. They extract small organisms from the water that passes through their mantle cavity.

Bivalves do not have distinct heads or radulas, differing from gastropods in this respect (see figure 33.5). However, most have a wedge-shaped foot that may be adapted, in different species, for creeping, burrowing, cleansing the animal, or anchoring it in its burrow. Some species of clams can dig into sand or mud very rapidly by means of muscular contractions of their foot.

Bivalves disperse from place to place largely as larvae. While most adults are adapted to a burrowing way of life, some genera of scallops can move swiftly through the water by using their large adductor muscles to clap their shells together. These muscles are what we usually eat as "scallops." The edge of a scallop's body is lined with tentacle-like projections tipped with complex eyes.

There are about 10,000 species of bivalves. Most species are marine, although many also live in fresh water. Over 500 species of pearly freshwater mussels, or naiads, occur in the rivers and lakes of North America.

Class Cephalopoda: The Octopuses, Squids, and Nautiluses

The more than 600 species of the class Cephalopoda—octopuses, squids, and nautiluses—are the most intelligent of the invertebrates. They are active marine predators that swim, often swiftly, and compete successfully with fish. The foot has evolved into a series of tentacles equipped with suction cups, adhesive structures, or hooks that seize prey efficiently. Squids have 10 tentacles; octopuses, as indicated by their name, have eight (figure 33.10); and the nautilus has about 80 to 90. After snaring prey with its tentacles, a cephalopod bites the prey with its strong, beaklike paired jaws and then pulls it into its mouth by the tonguelike action of the radula.

Cephalopods have highly developed nervous systems, and their brains are unique among mollusks. Their eyes are very elaborate, and have a structure much like that of vertebrate eyes, although they evolved separately (see chapter 46). Many cephalopods exhibit complex patterns of behavior and a high level of intelligence; octopuses can be easily trained to distinguish among classes of objects. Most members of this class have closed circulatory systems and are the only mollusks that do.

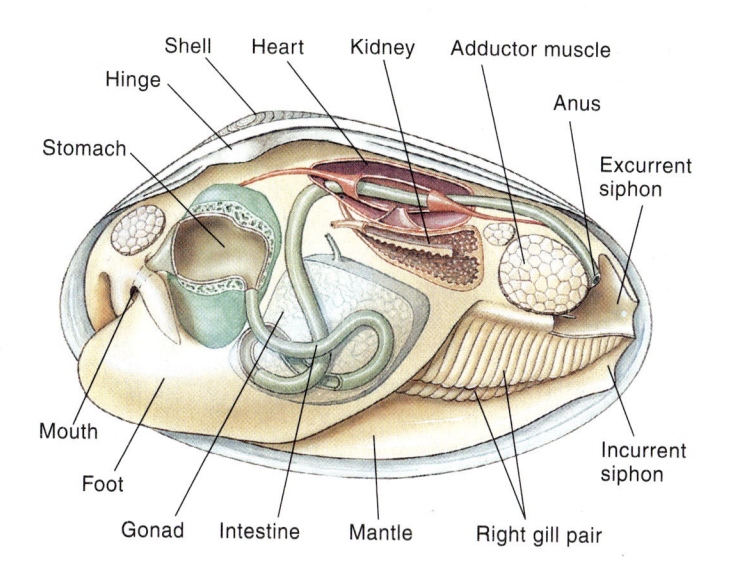

FIGURE 33.9
Diagram of a clam. Internal organs and the foot are shown. The left shell and mantle are removed. Bivalves such as this clam circulate water through their gills and filter out food particles.

FIGURE 33.10
A cephalopod. Pacific giant octopus (*Octopus defleini*).

Although they evolved from shelled ancestors, living cephalopods, except for the few species of nautilus, lack an external shell. Like other mollusks, cephalopods take water into the mantle cavity and expel it through a siphon. Cephalopods have modified this system into a means of jet propulsion. When threatened, they eject water violently and shoot themselves through the water.

Most octopuses and squids are capable of changing color to suit their background or display messages to one another. They accomplish this feat using their *chromatophores*, pouches of pigments embedded in the epithelium.

> Gastropods typically live in a hard shell. Bivalves have hinged shells but do not have a distinct head area. Cephalopods possess well-developed brains and are the most intelligent invertebrates.

Chapter 33 Coelomate Invertebrates

33.2 Annelids were the first segmented animals.

Segmented Bodies

A key transition in the animal body plan was *segmentation*, the building of a body from a series of similar segments. The first segmented animals to evolve were most likely **annelid worms,** phylum Annelida (figure 33.11). One advantage of having a body built from repeated units (segments) is that the development and function of these units can be more precisely controlled, at the level of individual segments or groups of segments. For example, different segments may possess different combinations of organs or perform different functions relating to reproduction, feeding, locomotion, respiration, or excretion.

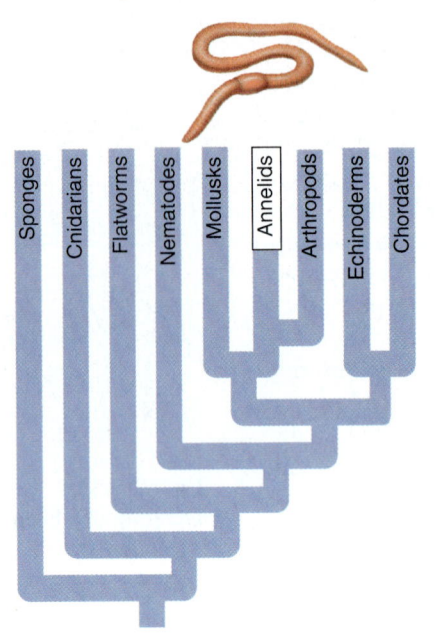

FIGURE 33.11
A polychaete annelid. *Nereis virens* is a wide-ranging, predatory, marine polychaete worm equipped with feathery parapodia for movement and respiration, as well as jaws for hunting. You may have purchased *Nereis* as fishing bait!

Annelids

Two-thirds of all annelids live in the sea (about 8000 species), and most of the rest, some 3100 species, are earthworms. Annelids are characterized by three principal features:

1. **Repeated segments.** The body of an annelid worm is composed of a series of ringlike segments running the length of the body, looking like a stack of doughnuts or a roll of coins (figure 33.12). Internally, the segments are divided from one another by partitions called **septa,** just as bulkheads separate the segments of a submarine. In each of the cylindrical segments, the excretory and locomotor organs are repeated. The fluid within the coelom of each segment creates a hydrostatic (liquid-supported) skeleton that gives the segment rigidity, like an inflated balloon. Muscles within each segment push against the fluid in the coelom. Because each segment is separate, each can expand or contract independently. This lets the worm move in complex ways.
2. **Specialized segments.** The anterior (front) segments of annelids have become modified to contain specialized sensory organs. Some are sensitive to light, and elaborate eyes with lenses and retinas have evolved in some annelids. A well-developed cerebral ganglion, or brain, is contained in one anterior segment.
3. **Connections.** Although partitions separate the segments, materials and information do pass between segments. Annelids have a closed circulatory system that carries blood from one segment to another. A ventral nerve cord connects the nerve centers or ganglia in each segment with one another and with the brain. These neural connections are critical features that allow the worm to function and behave as a unified and coordinated organism.

Body Plan of the Annelids

The basic annelid body plan is a tube within a tube, with the internal digestive tract—a tube running from mouth to anus—suspended within the coelom. The tube that makes up the digestive tract has several portions—the pharynx, esophagus, crop, gizzard, and intestine—that are specialized for different functions.

Annelids make use of their hydrostatic skeleton for locomotion. To move, annelids contract circular muscles running around each segment. Doing so squeezes the segment, causing the coelomic fluid to squirt outward, like a tube of toothpaste. Because the fluid is trapped in the segment by the septa, instead of escaping like toothpaste, the fluid causes the segment to elongate and get much thinner. By then contracting longitudinal muscles that run along the length of the worm, the segment is returned to its original shape. In most annelid groups, each segment typically possesses **setae,** bristles of chitin that help anchor the worms during locomotion. By extending the setae in some segments so that they anchor in the substrate and retracting them in other segments, the worm can squirt its body, section by section, in either direction.

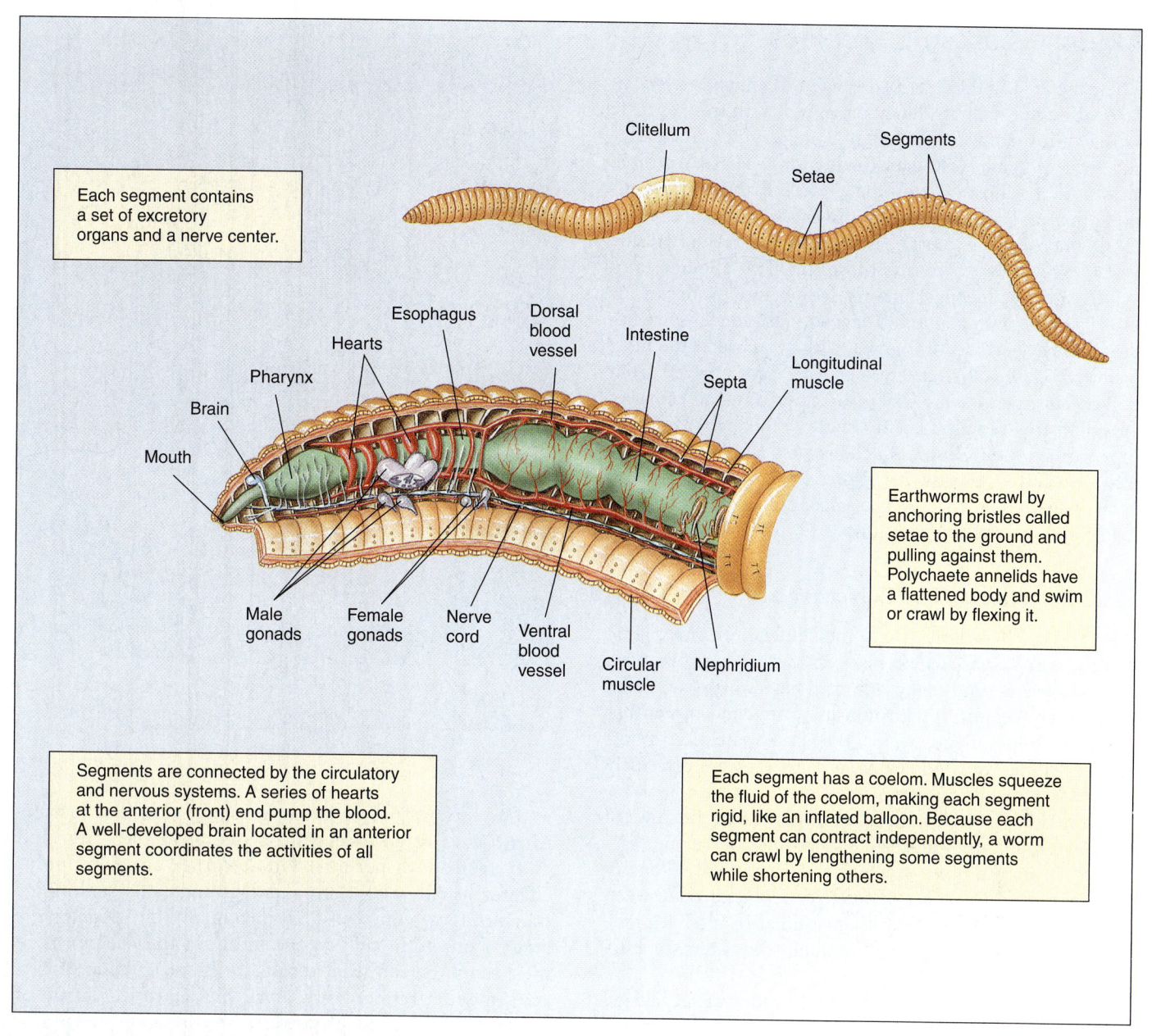

FIGURE 33.12
Phylum Annelida: annelids. Marine polychaetes and earthworms (phylum Annelida) were most likely the first organisms to evolve a body plan based on partly repeated body segments. Segments are separated internally from each other by septa.

Unlike the arthropods and most mollusks, most annelids have a closed circulatory system. Annelids exchange oxygen and carbon dioxide with the environment through their body surfaces; most lack gills or lungs. However, much of their oxygen supply reaches the different parts of their bodies through their blood vessels. Some of these vessels at the anterior end of the worm body are enlarged and heavily muscular, serving as hearts that pump the blood. Earthworms have five pulsating blood vessels on each side that serve as hearts, helping to pump blood from the main dorsal vessel, which is their major pumping structure, to the main ventral vessel.

The excretory system of annelids consists of ciliated, funnel-shaped nephridia generally similar to those of mollusks. These nephridia—each segment has a pair—collect waste products and transport them out of the body through the coelom by way of specialized excretory tubes.

Annelids are a diverse group of coelomate animals characterized by serial segmentation. Each segment in the annelid body has its own excretory and locomotor elements, and is connected to other segments by a common circulatory and neural system.

Chapter 33 Coelomate Invertebrates 659

Major Classes of Annelids

The roughly 12,000 described species of annelids occur in many different habitats. They range in length from as little as 0.5 millimeter to more than 3 meters in some polychaetes and giant Australian earthworms. There are three classes of annelids: (1) Polychaeta, free-living, almost entirely marine bristleworms, comprising some 8000 species; (2) Oligochaeta, terrestrial earthworms and related marine and freshwater worms, with some 3100 species; and (3) Hirudinea, leeches, mainly freshwater predators or bloodsuckers, with about 500 species. The annelids are believed to have evolved in the sea, with polychaetes the most primitive class. Oligochaetes seem to have evolved from polychaetes, perhaps by way of brackish water to estuaries and then to streams. Leeches share with oligochaetes an organ called a **clitellum**, which secretes a cocoon specialized to receive the eggs. It is generally agreed that leeches evolved from oligochaetes, specializing in their bloodsucking lifestyle as external parasites.

Class Polychaeta: The Polychaetes

Polychaetes (class Polychaeta) include clamworms, plume worms, scaleworms, lugworms, twin-fan worms, sea mice, peacock worms, and many others. These worms are often surprisingly beautiful, with unusual forms and sometimes iridescent colors (figure 33.13). Polychaetes are often a crucial part of marine food chains, because they are extremely abundant in certain habitats.

Some polychaetes live in tubes or permanent burrows of hardened mud, sand, mucuslike secretions, or calcium carbonate. From the tubes in which they live, these sedentary polychaetes project a set of feathery tentacles that sweep the water for food, making them primarily filter feeders. Other polychaetes are active swimmers, crawlers, or burrowers. Many are active predators.

Polychaetes have a well-developed head with specialized sense organs; they differ from other annelids in this respect. Their bodies are often highly organized into distinct regions formed by groups of segments related in function and structure. Their sense organs include eyes, which range from simple eyespots to quite large and conspicuous stalked eyes.

Another distinctive characteristic of polychaetes is the paired, fleshy, paddlelike flaps, called **parapodia**, on most of their segments. These parapodia, which bear bristlelike setae, are used in swimming, burrowing, or crawling. They also play an important role in gas exchange because they greatly increase the surface area of the body. Some polychaetes that live in burrows or tubes may have parapodia featuring hooks to help anchor the worm. Slow crawling is carried out by means of the parapodia. Rapid crawling and swimming are accomplished by undulating the body. In addition, the polychaete epidermis often includes ciliated cells that aid in respiration and food procurement.

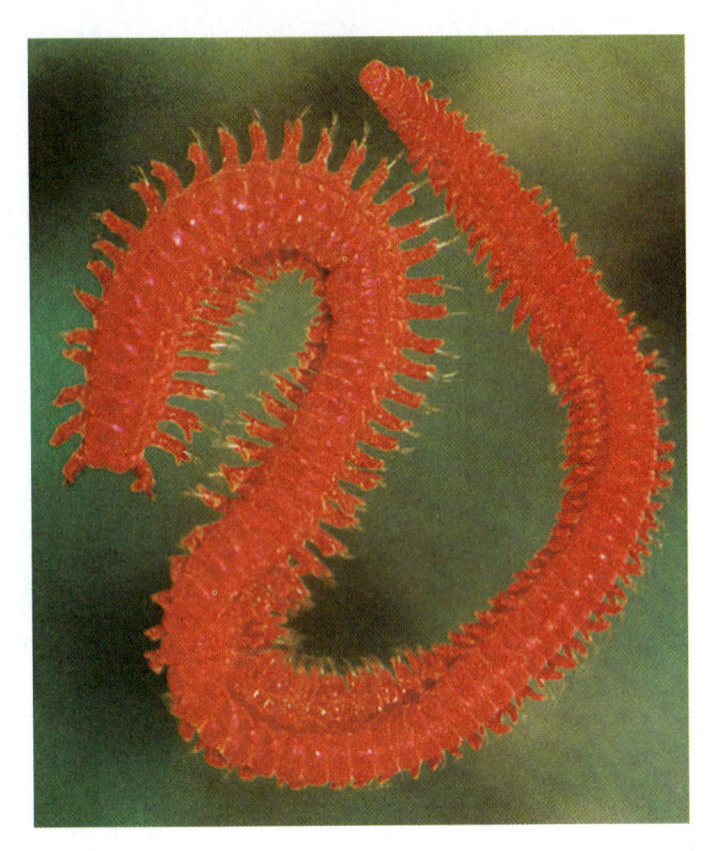

FIGURE 33.13
A polychaete. The shiny bristleworm, *Oenone fulgida*.

The sexes of polychaetes are usually separate, and fertilization is often external, occurring in the water and away from both parents. Unlike other annelids, polychaetes usually lack permanent **gonads,** the sex organs that produce gametes. They produce their gametes directly from germ cells in the lining of the coelom or in their septa. Fertilization results in the production of ciliated, mobile trochophore larvae similar to the larvae of mollusks. The trochophores develop for long periods in the plankton before beginning to add segments and thus changing to a juvenile form that more closely resembles the adult form.

Class Oligochaeta: The Earthworms

The body of an earthworm (class Oligochaeta) consists of 100 to 175 similar segments, with a mouth on the first and an anus on the last. Earthworms seem to eat their way through the soil because they suck in organic and other material by expanding their strong pharynx. Everything they ingest passes through their long, straight digestive tract. One region of this tract, the gizzard, grinds up the organic material with the help of soil particles.

The material that passes through an earthworm is deposited outside the opening of its burrow in the form of castings that consist of irregular mounds. In this way, earthworms aerate and enrich the soil. A worm can eat its own weight in soil every day.

In view of the underground lifestyle that earthworms have evolved, it is not surprising that they have no eyes. However, earthworms do have numerous light-, chemo-, and touch-sensitive cells, mostly concentrated in segments near each end of the body—those regions most likely to encounter light or other stimuli. Earthworms have fewer setae than polychaetes and no parapodia or head region.

Earthworms are hermaphroditic, another way in which they differ from most polychaetes. When they mate (figure 33.14), their anterior ends point in opposite directions, and their ventral surfaces touch. The *clitellum* is a thickened band on an earthworm's body; the mucus it secretes holds the worms together during copulation. Sperm cells are released from pores in specialized segments of one partner into the sperm receptacles of the other, the process going in both directions simultaneously.

Two or three days after the worms separate, the clitellum of each worm secretes a mucous cocoon, surrounded by a protective layer of chitin. As this sheath passes over the female pores of the body—a process that takes place as the worm moves—it receives eggs. As it subsequently passes along the body, it incorporates the sperm that were deposited during copulation. Fertilization of the eggs takes place within the cocoon. When the cocoon finally passes over the end of the worm, its ends pinch together. Within the cocoon, the fertilized eggs develop directly into young worms similar to adults.

Class Hirudinea: The Leeches

Leeches (class Hirudinea) occur mostly in fresh water, although a few are marine and some tropical leeches occupy terrestrial habitats. Most leeches are 2 to 6 centimeters long, but one tropical species reaches up to 30 centimeters. Leeches are usually flattened dorsoventrally, like flatworms. They are hermaphroditic, and develop a clitellum during the breeding season; cross-fertilization is obligatory because they are unable to self-fertilize.

A leech's coelom is reduced and continuous throughout the body, not divided into individual segments as in the polychaetes and oligochaetes. Leeches have evolved suckers at one or both ends of the body. Those that have suckers at both ends move by attaching first one and then the other end to the substrate, looping along. Many species are also capable of swimming. Except for one species, leeches have no setae.

Some leeches have evolved the ability to suck blood from animals. Many freshwater leeches live as external parasites. They remain on their hosts for long periods and suck their blood from time to time.

The best-known leech is the medicinal leech, *Hirudo medicinalis* (figure 33.15). Individuals of *Hirudo* are 10 to 12 centimeters long and have bladelike, chitinous jaws

FIGURE 33.14
Earthworms mating. The anterior ends are pointing in opposite directions.

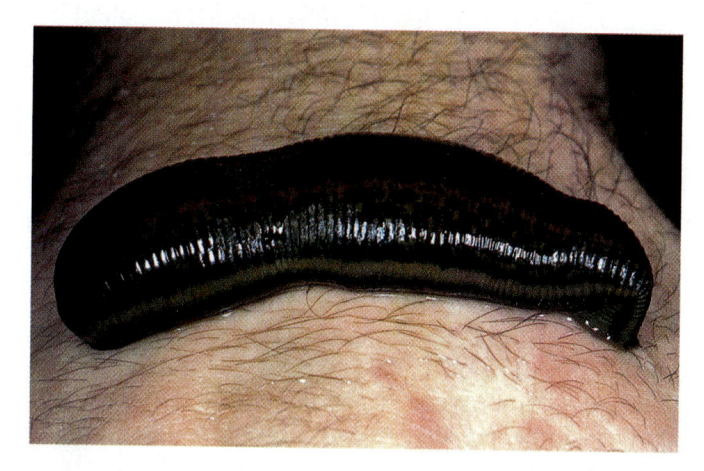

FIGURE 33.15
A leech. *Hirudo medicinalis*, the medicinal leech, is seen here feeding on a human arm. Leeches uses chitinous, bladelike jaws to make an incision to access blood and secrete an anticoagulant to keep the blood from clotting. Both the anticoagulant and the leech itself have made important contributions to modern medicine.

that rasp through the skin of the victim. The leech secretes an anticoagulant into the wound to prevent the blood from clotting as it flows out, and its powerful sucking muscles pump the blood out quickly once the hole has been opened. Leeches were used in medicine for hundreds of years to suck blood out of patients whose diseases were mistakenly believed to be caused by an excess of blood. Today, European pharmaceutical companies still raise and sell leeches, but they are used to remove excess blood after certain surgeries. Following the surgery, blood may accumulate because veins may function improperly and fail to circulate the blood. The accumulating blood "turns off" the arterial supply of fresh blood, and the tissue often dies. When leeches remove the excess blood, new capillaries form in about a week, and the tissues remain healthy.

Segmented annelids evolved in the sea. Earthworms are their descendants, as are parasitic leeches.

Chapter 33 Coelomate Invertebrates **661**

33.3 Lophophorates appear to be a transitional group.

Lophophorates

Three phyla of marine animals—Phoronida, Ectoprocta, and Brachiopoda—are characterized by a **lophophore,** a circular or U-shaped ridge around the mouth bearing one or two rows of ciliated, hollow tentacles. Because of this unusual feature, they are thought to be related to one another. The lophophore presumably arose in a common ancestor. The coelomic cavity of lophophorates extends into the lophophore and its tentacles. The lophophore functions as a surface for gas exchange and as a food-collection organ. Lophophorates use the cilia of their lophophore to capture the organic detritus and plankton on which they feed. Lophophorates are attached to their substrate or move slowly.

Lophophorates share some features with mollusks, annelids, and arthropods (all protostomes), and share others with deuterostomes. Cleavage in lophophorates is mostly radial, as in deuterostomes. The formation of the coelom varies; some lophophorates resemble protostomes in this respect, others deuterostomes. In the Phoronida, the mouth forms from the blastopore, while in the other two phyla, it forms from the end of the embryo opposite the blastopore. Molecular evidence shows that the ribosomes of all lophophorates are decidedly protostome-like, lending strength to the argument for placing them within the protostome phyla. Despite the differences among the three phyla, the unique structure of the lophophore seems to indicate that the members share a common ancestor. Their relationships continue to present a fascinating puzzle.

Phylum Phoronida: The Phoronids

Phoronids (phylum Phoronida) superficially resemble the common polychaete tube worms seen on dock pilings, but they have many important differences. Each phoronid secretes a chitinous tube and lives out its life within it (figure 33.16). They also extend tentacles to feed and quickly withdraw them when disturbed, but the resemblance to the tube worm ends there. Instead of a straight, tube-within-a-tube body plan, phoronids have a U-shaped gut. Only about 10 phoronid species are known, ranging in length from a few millimeters to 30 centimeters. Some species lie buried in sand; others are attached to rocks, either singly or in groups. Phoronids develop as protostomes, with radial cleavage and the anus developing secondarily.

Phylum Ectoprocta: The Bryozoans

Ectoprocts (phylum Ectoprocta) look like tiny, short versions of phoronids (figure 33.17). They are small—usually less than 0.5 millimeter long—and live in colonies that look

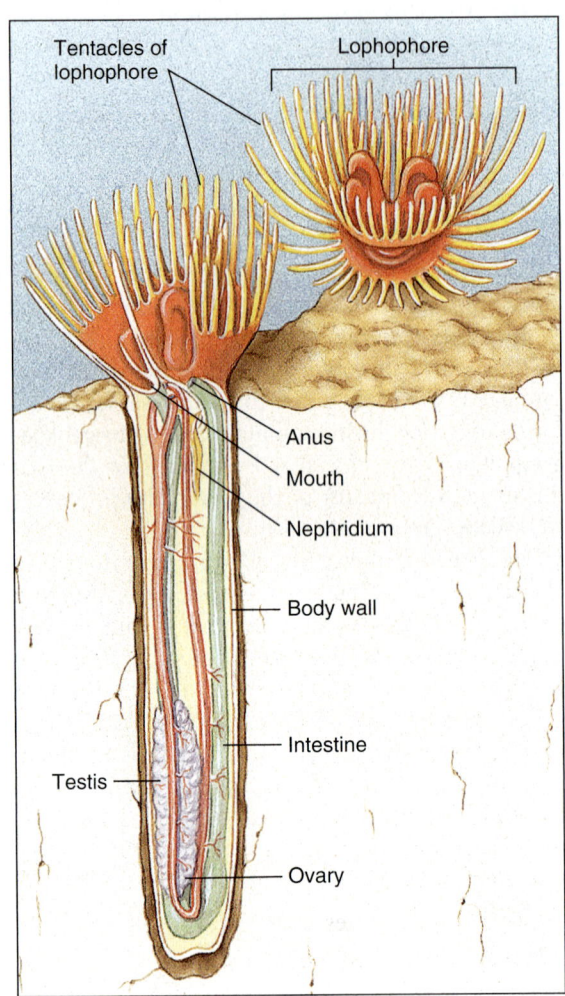

FIGURE 33.16
Phoronids (phylum Phoronida). A phoronid, such as *Phoronis*, lives in a chitinous tube that the animal secretes to form the outer wall of its body. The lophophore consists of two parallel, horseshoe-shaped ridges of tentacles and can be withdrawn into the tube when the animal is disturbed.

like patches of moss on the surfaces of rocks, seaweed, or other submerged objects (in fact, their common name, bryozoans, translates from Greek as "moss-animals"). The name Ectoprocta refers to the location of the anus (proct), which is external to the lophophore. The 4000 species include both marine and freshwater forms—the only nonmarine lophophorates. Individual ectoprocts secrete a tiny chitinous chamber called a **zoecium** that attaches to rocks and to other members of the colony. Individuals communicate chemically through pores between these chambers. Ectoprocts develop as deuterostomes, with the mouth developing secondarily; cleavage is radial.

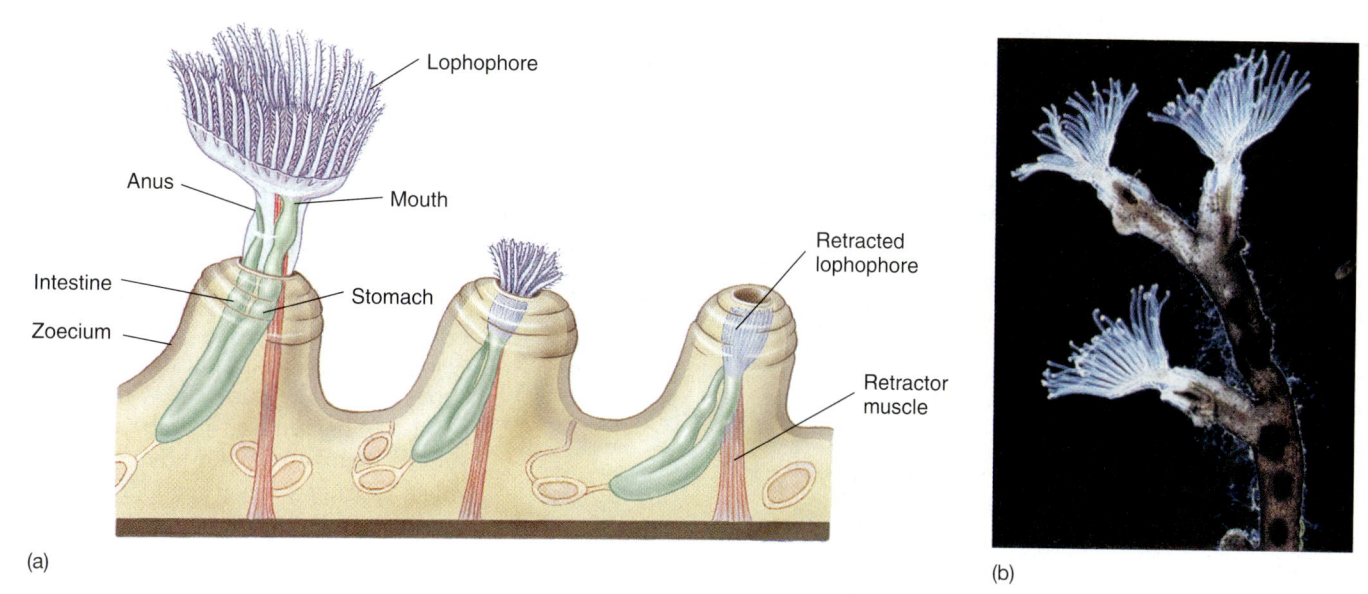

FIGURE 33.17
Ectoprocts (phylum Ectoprocta). (*a*) This drawing depicts a small portion of a colony of the freshwater ectoproct *Plumatella*, which grows on the underside of rocks. The individual at the left has a fully extended lophophore, the structure characteristic of the three lophophorate phyla. The tiny individuals of *Plumatella* disappear into their shells when disturbed. (*b*) *Plumatella repens*, a freshwater bryozoan.

Phylum Brachiopoda: The Brachiopods

Brachiopods, or lamp shells, superficially resemble clams, with two calcified shells (figure 33.18). Many species attach to rocks or sand by a stalk that protrudes through an opening in one shell. The lophophore lies within the shell and functions when the brachiopod's shells are opened slightly. Although a little more than 300 species of brachiopods (phylum Brachiopoda) exist today, more than 30,000 species of this phylum are known as fossils. Because brachiopods were common in the earth's oceans for millions of years and because their shells fossilize readily, they are often used as index fossils to define a particular time period or sediment type. Brachiopods develop as deuterostomes and show radial cleavage.

> The three phyla of lophophorates probably share a common ancestor, and they show a mixture of protostome and deuterostome characteristics.

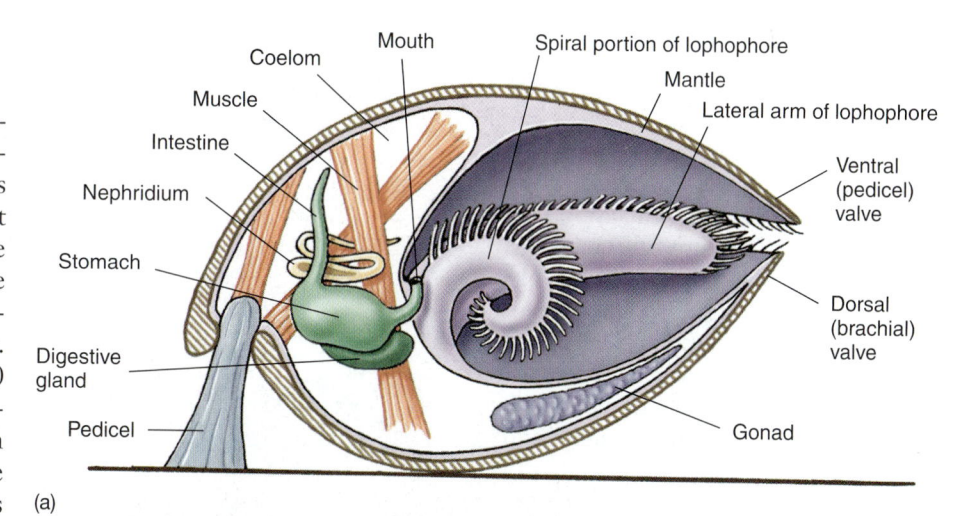

FIGURE 33.18
Brachiopods (phylum Brachiopoda). (*a*) The lophophore lies within two calcified shells, or valves. (*b*) The brachiopod *Terebratolina septentrionalis* is shown here slightly opened so that the lophophore is visible.

Chapter 33 Coelomate Invertebrates **663**

33.4 Arthropods are the most diverse of all animal groups.

Arthropod Body Plan

With the evolution of the first annelids, many of the major innovations of animal structure had already appeared: the division of tissues into three primary types (endoderm, mesoderm, and ectoderm), bilateral symmetry, a coelom, and segmentation. With arthropods, two more innovations arose—the development of *jointed appendages* and an *exoskeleton*. Jointed appendages and an exoskeleton have allowed arthropods (phylum Arthropoda) to become the most diverse phylum.

The Arthropods

Arthropods, especially the largest class—insects, are by far the most successful of all animals. Well over 1,000,000 species—about two-thirds of all the named species on earth—are members of this phylum (figure 33.19). One scientist recently estimated, based on the number and diversity of insects in tropical forests, that as many as 30 million species may comprise this one class alone. About 200 million insects are alive at any one time for each human! Insects and other arthropods (figure 33.20) abound in every habitat on the planet, but they especially dominate the land, along with flowering plants and vertebrates.

The majority of arthropod species consist of small animals, mostly about a millimeter in length. Members of the phylum range in adult size from about 80 micrometers long (some parasitic mites) to 3.6 meters across (a gigantic crab found in the sea off Japan).

Arthropods, especially insects, are of enormous economic importance and affect all aspects of human life. They compete with humans for food of every kind, play a key role in the pollination of certain crops, and cause billions of dollars of damage to crops, before and after harvest. They are by far the most important herbivores in all terrestrial ecosystems and are a valuable food source as well. Virtually every kind of plant is eaten by one or more species of insect. Diseases spread by insects cause enormous financial damage each year and strike every kind of domesticated animal and plant, as well as human beings.

Jointed Appendages

The name "arthropod" comes from two Greek words, *arthros*, "jointed," and *podes*, "feet." All arthropods have jointed appendages. The numbers of these appendages are reduced in the more advanced members of the phylum. Individual appendages may be modified into antennae, mouthparts of various kinds, or legs. Some appendages, such as the wings of certain insects, are not homologous to the other appendages; insect wings evolved separately.

To gain some idea of the importance of jointed appendages, imagine yourself without them—no hips, knees, ankles, shoulders, elbows, wrists, or knuckles. Without jointed appendages, you could not walk or grasp any object. Arthropods use jointed appendages such as legs for walking, antennae to sense their environment, and mouthparts for feeding.

An Exterior Skeleton

The arthropod body plan has a second major innovation: a rigid external skeleton, or **exoskeleton,** made of chitin and protein. In any animal, the skeleton functions to provide places for muscle attachment. In arthropods, the muscles attach to the interior surface of the hard exoskeleton, which also protects the animal from predators and impedes water loss. Chitin is chemically similar to cellulose, the dominant structural component of plants, and shares similar properties of toughness and flexibility. Together, the chitin and protein provide an external covering that is very strong but also capable of flexing in response to the contraction of muscles attached to it. In most crustaceans, the exoskeleton is made even tougher, although less flexible, with deposits of calcium salts. However, there is a limitation. The exoskeleton must be much thicker to bear the pull of the muscles in large insects than in small ones.

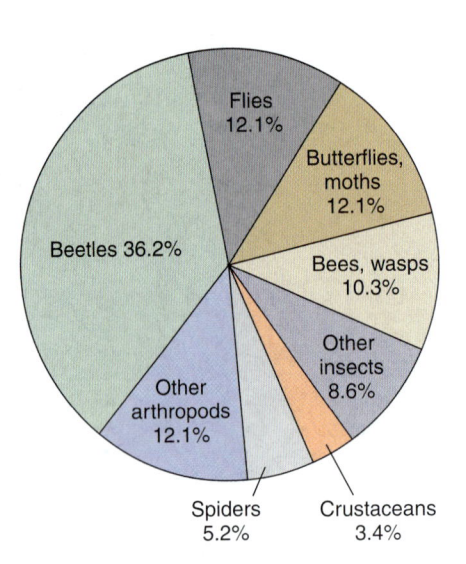

FIGURE 33.19
Arthropods are a successful group. About two-thirds of all named species are arthropods. About 80% of all arthropods are insects, and about half of the named species of insects are beetles.

- Beetles 36.2%
- Flies 12.1%
- Butterflies, moths 12.1%
- Bees, wasps 10.3%
- Other insects 8.6%
- Crustaceans 3.4%
- Spiders 5.2%
- Other arthropods 12.1%

664 Part V Diversity of Life on Earth

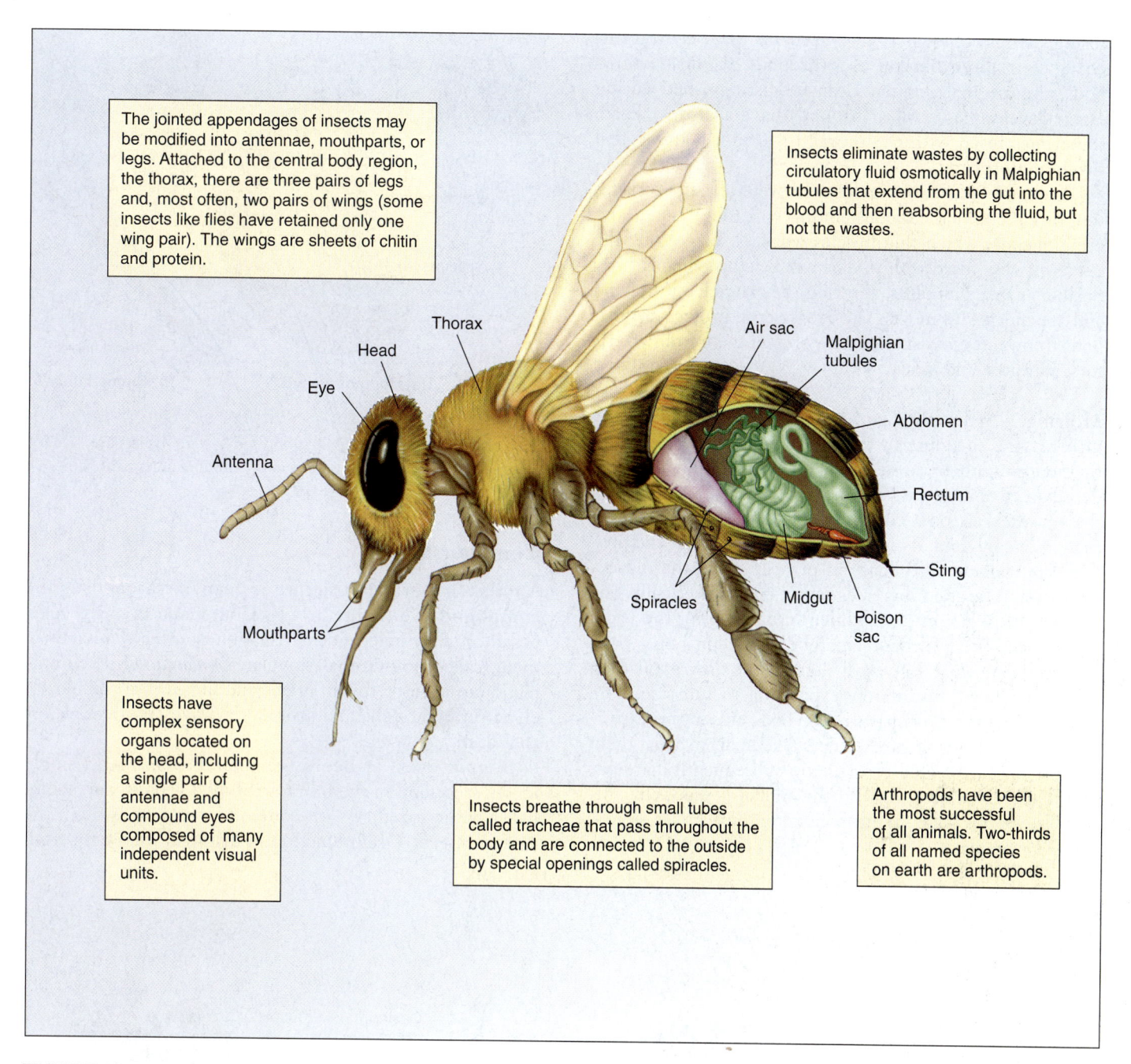

FIGURE 33.20
Phylum Arthropoda: arthropods. Insects and other arthropods (phylum Arthropoda) have a coelom, segmented bodies, and jointed appendages. The three body regions of an insect (head, thorax, and abdomen) are each actually composed of a number of segments that fuse during development. All arthropods have a strong exoskeleton made of chitin. One class, the insects, has evolved wings that permit them to fly rapidly through the air.

That is why you don't see beetles as big as birds, or crabs the size of a cow—the exoskeleton would be so thick the animal couldn't move its great weight. Because this size limitation is inherent in the body design of arthropods, there are practically no large arthropods—few are larger than your thumb.

Arthropod bodies are segmented like those of annelids, a phylum to which at least some arthropods are clearly related. Members of some classes of arthropods have many body segments. In others, the segments have become fused together into functional groups, or **tagmata** (singular, *tagma*), such as the head and thorax of an insect. This

Chapter 33 Coelomate Invertebrates 665

fusing process, known as *tagmatization*, is of central importance in the evolution of arthropods. In most arthropods, the original segments can be distinguished during larval development. All arthropods have a distinct head, sometimes fused with the thorax to form a tagma called the **cephalothorax**.

The bodies of all arthropods are covered by an exoskeleton, or cuticle, that contains chitin. This tough outer covering, against which the muscles work, is secreted by the epidermis and fused with it. The exoskeleton remains fairly flexible at specific points, allowing the exoskeleton to bend and appendages to move. The exoskeleton protects arthropods from water loss and helps to protect them from predators, parasites, and injury.

Molting. Arthropods periodically undergo **ecdysis**, or molting, the shedding of the outer cuticular layer. When they outgrow their exoskeleton, they form a new one underneath. This process is controlled by hormones. When the new exoskeleton is complete, it becomes separated from the old one by fluid. This fluid dissolves the chitin and protein (and calcium carbonate, if present) from the old exoskeleton. The fluid increases in volume until, finally, the original exoskeleton cracks open, usually along the back, and is shed. The arthropod emerges, clothed in a new, pale, and still somewhat soft exoskeleton. The arthropod then "puffs itself up," ultimately expanding to full size. The blood circulation to all parts of the body aids in this expansion, and many insects and spiders take in air to assist them as well. The expanded exoskeleton subsequently hardens. While the exoskeleton is soft, the animal is especially vulnerable. At this stage, arthropods often hide under stones, leaves, or branches.

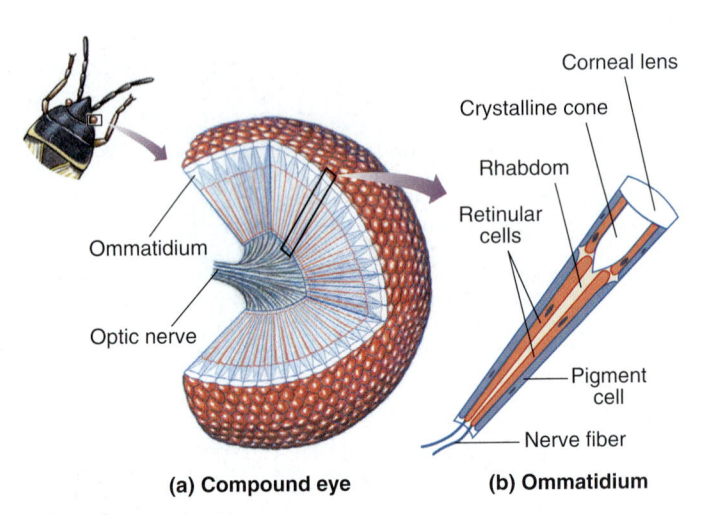

FIGURE 33.21
The compound eye. The compound eyes in insects are complex structures composed of many independent visual units called ommatidia.

Compound Eye

Another important structure in many arthropods is the **compound eye** (figure 33.21). Compound eyes are composed of many independent visual units, often thousands of them, called **ommatidia** (singular, *ommatidium*). Each ommatidium is covered with a lens and linked to a complex of eight retinular cells and a light-sensitive central core, or **rhabdom**.

Simple eyes, or **ocelli** (singular, *ocellum*) with single lenses are found in the other arthropod groups, and sometimes occur together with compound eyes, as is often the case in insects. Ocelli function in distinguishing light from

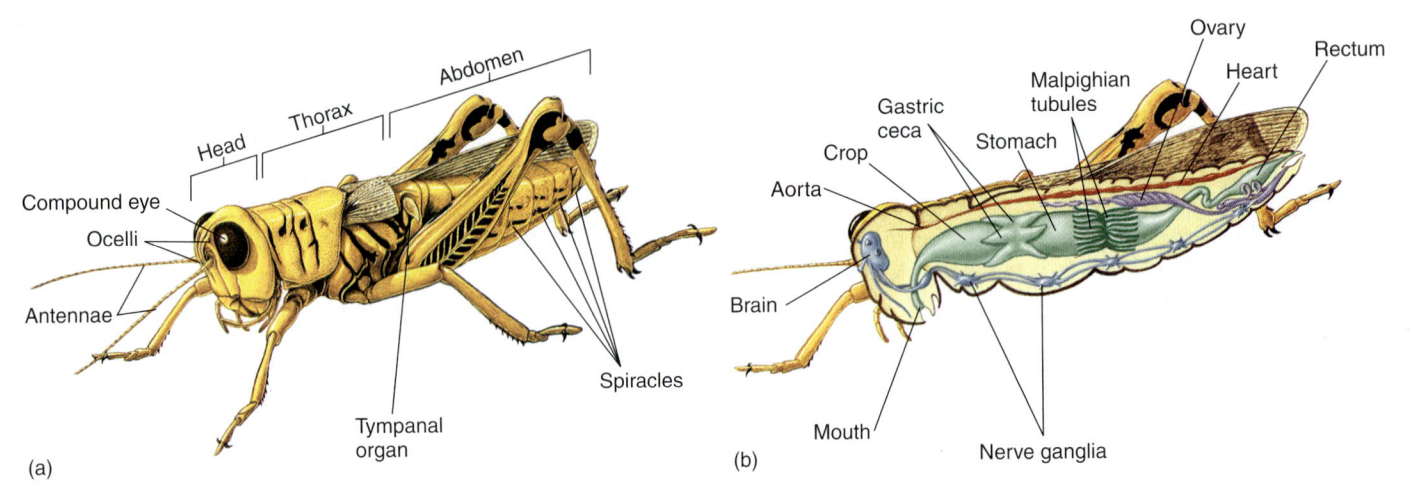

FIGURE 33.22
A grasshopper (order Orthoptera). This grasshopper illustrates the major structural features of the insects, the most numerous group of arthropods. (*a*) External anatomy. (*b*) Internal anatomy.

darkness. The ocelli of some flying insects, namely locusts and dragonflies, function as horizon detectors and help the insect visually stabilize its course in flight.

Circulatory System

In the course of arthropod evolution, the coelom has become greatly reduced, consisting only of cavities that house the reproductive organs and some glands. Arthropods completely lack cilia, both on the external surfaces of the body and on the internal organs. Like annelids, arthropods have a tubular gut that extends from mouth to anus.

The circulatory system of arthropods is open; their blood flows through cavities between the internal organs and not through closed vessels. The principal component of an insect's circulatory system is a longitudinal vessel called the heart. This vessel runs near the dorsal surface of the thorax and abdomen (figure 33.22b). When it contracts, blood flows into the head region of the insect.

When an insect's heart relaxes, blood returns to it through a series of valves. These valves are located in the posterior region of the heart and allow the blood to flow inward only. Thus, blood from the head and other anterior portions of the insect gradually flows through the spaces between the tissues toward the posterior end and then back through the one-way valves into the heart.

Nervous System

The central feature of the arthropod nervous system is a double chain of segmented ganglia running along the animal's ventral surface. At the anterior end of the animal are three fused pairs of dorsal ganglia, which constitute the brain. However, much of the control of an arthropod's activities is relegated to ventral ganglia. Therefore, the animal can carry out many functions, including eating, moving, and copulating, even if the brain has been removed. The brain of arthropods seems to be a control point, or inhibitor, for various actions, rather than a stimulator, as it is in vertebrates.

Respiratory System

Insects depend on their respiratory rather than their circulatory system to carry oxygen to their tissues. All parts of the body need to be near a respiratory passage to obtain oxygen. Along with the thickness of their chitin exoskeletons, this is another feature of arthropod design that places severe limitations on their size compared to that of vertebrates.

Unlike most animals, the arthropods have no single major respiratory organ. The respiratory system of most terrestrial arthropods consists of small, branched, cuticle-lined air ducts called **tracheae** (figure 33.23). These tra-

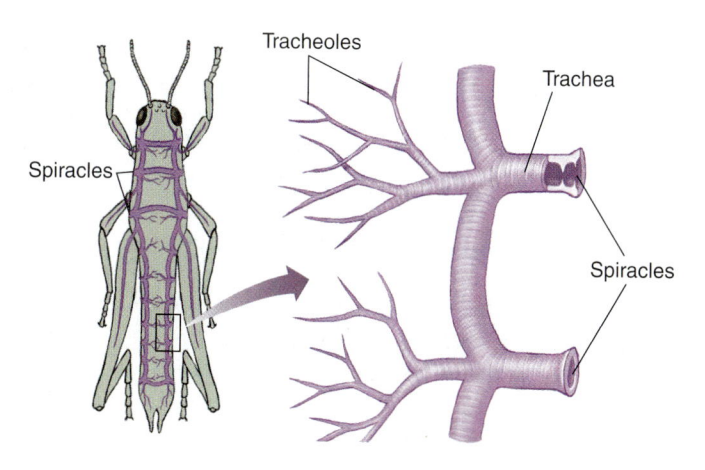

FIGURE 33.23
Tracheae and tracheoles. Tracheae and tracheoles are connected to the exterior by specialized openings called spiracles and carry oxygen to all parts of a terrestrial insect's body.

cheae, which ultimately branch into very small **tracheoles,** are a series of tubes that transmit oxygen throughout the body. Tracheoles are in direct contact with individual cells, and oxygen diffuses directly across the plasma membranes. Air passes into the tracheae by way of specialized openings in the exoskeleton called **spiracles,** which, in most insects, can be opened and closed by valves. The ability to prevent water loss by closing the spiracles was a key adaptation that facilitated the invasion of the land by arthropods.

Excretory System

Although various kinds of excretory systems are found in different groups of arthropods, we will focus here on the unique excretory system consisting of **Malpighian tubules,** which evolved in terrestrial insects. Malpighian tubules are slender projections from the digestive tract that are attached at the junction of the midgut and hindgut (see figure 49.11). Fluid passes through the walls of the Malpighian tubules to and from the blood in which the tubules are bathed. As this fluid passes through the tubules toward the hindgut, nitrogenous wastes are precipitated as concentrated uric acid or guanine. These substances are then emptied into the hindgut and eliminated. Most of the water and salts in the fluid are reabsorbed by the hindgut and rectum and returned to the arthropod's body. Malpighian tubules are an efficient mechanism for water conservation and were another key adaptation facilitating invasion of the land by arthropods.

> Arthropods are segmented animals with jointed appendages. All arthropods have a rigid chitin and protein exoskeleton, and an open circulatory system. Insects have many adaptations for terrestrial living.

A Major Group of Arthropods: Crustaceans

The crustaceans (subphylum Crustacea) are a large group of primarily aquatic organisms, consisting of some 35,000 species of crabs, shrimps, lobsters, crayfish, barnacles, water fleas, pillbugs, and related groups (table 33.1). Most crustaceans have two pairs of antennae, three types of chewing appendages, and various numbers of pairs of legs. All crustacean appendages, with the possible exception of the first pair of antennae, are basically biramous ("two-branched"). In some crustaceans, appendages appear to have only a single branch; in those cases, one of the branches has been lost during the course of evolutionary specialization. The **nauplius** larva stage through which all crustaceans pass (figure 33.24) provides evidence that all members of this diverse group are descended from a common ancestor. The nauplius hatches with three pairs of appendages and metamorphoses through several stages before reaching maturity. In many groups, this nauplius stage is passed in the egg, and development of the hatchling to the adult form is direct.

Crustaceans differ from insects but resemble centipedes and millipedes in that they have appendages on their abdomen as well as on their thorax. They are the only arthropods with two pairs of antennae. Their **mandibles** (biting jaws) likely originated from a pair of limbs that took on a chewing function during the course of evolution, a process that apparently occurred independently in the common ancestor of the terrestrial mandibulates. Many crustaceans have compound eyes. In addition, they have delicate tactile hairs that project from the cuticle all over the body. Larger crustaceans have feathery gills near the bases of their legs. In smaller members of this class, gas exchange takes place directly through the thinner areas of the cuticle or the entire body. Most crustaceans have separate sexes. Many different kinds of specialized copulation occur among the crustaceans, and the members of some orders carry their eggs with them, either singly or in egg pouches, until they hatch.

Decapod Crustaceans

Large, primarily marine crustaceans such as shrimps, lobsters, and crabs, along with their freshwater relatives, the crayfish, are collectively called *decapod crustaceans* (figure 33.25). The term *decapod* means "ten-footed." In these animals, the exoskeleton is usually reinforced with calcium carbonate. Most of their body segments are fused into a cephalothorax covered by a dorsal shield, or carapace, which arises from the head. The crushing pincers common in many decapod crustaceans are used in obtaining food—for example, by crushing mollusk shells.

In lobsters and crayfish, appendages called **swimmerets** occur in lines along the ventral surface of the abdomen and are used in reproduction and swimming. In addition, flattened appendages known as **uropods** form a kind of compound "paddle" at the end of the abdomen. These animals may also have a **telson,** or tail spine. By snapping its abdomen, the animal propels itself through the water rapidly and forcefully. Crabs differ from lobsters and crayfish in proportion; their carapace is much larger and broader, and the abdomen is tucked under it.

Table 33.1 Major Groups Within the Traditional Classification of the Phylum Arthropoda

Group	Characteristics	Members
Crustaceans	Mouthparts are mandibles (biting jaws); appendages are biramous ("two-branched")	Lobsters, crabs, shrimp, isopods, barnacles
Arachnids	Mouthparts are chelicerae (pincers or fangs)	Spiders, mites, ticks, scorpions, daddy longlegs
Centipedes and millipedes	Mouthparts are mandibles; bodies consist of a head and numerous body segments bearing paired uniramous ("single-branched") appendages	Centipedes, millipedes
Insects	Mouthparts are mandibles; appendages are uniramous	Beetles, bees, flies, fleas, true bugs, grasshoppers, butterflies, termites

FIGURE 33.24 The nauplius larva. Although crustaceans are diverse, they have fundamentally similar larvae. The nauplius larva of a crustacean is an important unifying feature found in all members of this group.

668 Part V Diversity of Life on Earth

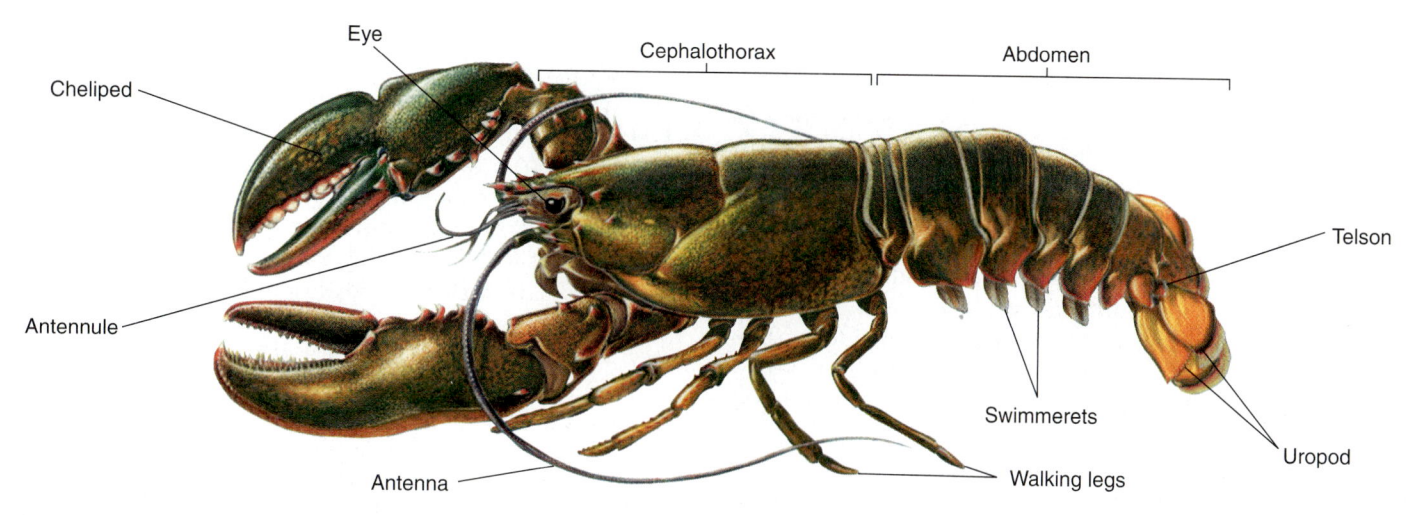

FIGURE 33.25
Decapod crustacean. A lobster, *Homarus americanus*, with its principal features labeled.

Terrestrial and Freshwater Crustaceans

Although most crustaceans are marine, many occur in fresh water, and a few have become terrestrial. These include pillbugs and sowbugs, the terrestrial members of a large order of crustaceans known as the isopods (order Isopoda). About half of the estimated 4500 species of this order are terrestrial and live primarily in places that are moist, at least seasonally. Sand fleas or beach fleas (order Amphipoda) are other familiar crustaceans, many of which are semiterrestrial (intertidal) species.

Along with the larvae of larger species, minute crustaceans are abundant in plankton. Especially significant are the tiny copepods (order Copepoda; figure 33.26), which are among the most abundant multicellular organisms on earth.

Sessile Crustaceans

Barnacles (order Cirripedia; figure 33.27) are a group of crustaceans that are sessile as adults. Barnacles have free-swimming larvae, which ultimately attach their heads to a piling, rock, or other submerged object and then stir food into their mouth with their feathery legs. Calcareous plates protect the barnacle's body, and these plates are usually attached directly and solidly to the substrate. Although most crustaceans have separate sexes, barnacles are hermaphroditic, but they generally cross-fertilize.

> Crustaceans include marine, freshwater, and terrestrial forms. All possess a nauplius larval stage and branched appendages.

FIGURE 33.26
Freshwater crustacean. A copepod with attached eggs, a member of an abundant group of marine and freshwater crustaceans (order Copepoda), most of which are a few millimeters long. Copepods are important components of plankton.

FIGURE 33.27
Gooseneck barnacles, *Lepas anatifera*, feeding. These are stalked barnacles; many others lack a stalk.

Major Classes of Arthropods: Arachnids

The largely terrestrial arthropod class Arachnida, with some 57,000 named species, occupies a distinct evolutionary line of arthropods in which the most anterior appendages, mouthparts called **chelicerae** (singular, *chelicera*), often function as fangs or pincers. Arachnids include spiders, ticks, mites, scorpions, and daddy longlegs. Arachnids have a pair of chelicerae, a pair of pedipalps, and four pairs of walking legs. The chelicerae are the foremost appendages; they consist of a stout basal portion and a movable fang often connected to a poison gland.

The next pair of appendages, **pedipalps,** resemble legs but have one less segment and are not used for locomotion. In male spiders, the pedipalps are specialized copulatory organs. In scorpions, they are large pincers.

Most arachnids are carnivorous. The main exception is mites, which are largely herbivorous. Most arachnids can ingest only preliquified food, which they often digest externally by secreting enzymes into their prey. They can then suck up the digested material with their muscular, pumping pharynx. Arachnids are primarily, but not exclusively, terrestrial. Some 4000 known species of mites and one species of spider live in fresh water, and a few mites live in the sea.

Many spiders have a unique respiratory system that involves **book lungs,** a series of leaflike plates within a chamber. Air is drawn into and expelled out of this chamber by muscular contraction. Book lungs may exist alongside tracheae, or they may function instead of tracheae.

Order Araneae: The Spiders

There are about 35,000 named species of spiders (order Araneae). These animals play a major role in virtually all terrestrial ecosystems. They are particularly important as predators of insects and other small animals. Spiders hunt their prey or catch it in silk webs of remarkable diversity. The silk is formed from a fluid protein that is forced out of spinnerets on the posterior portion of the spider's abdomen. The webs and habits of spiders are often distinctive.

Many kinds of spiders, including the familiar wolf spiders and tarantulas, do not spin webs but instead hunt their prey actively. Others, called trap-door spiders, construct silk-lined burrows with lids, seizing their prey as it passes by.

Spiders have poison glands leading through their chelicerae, which are pointed and used to bite and paralyze prey. The bites of some members of this order, such as the black widow and brown recluse spiders (figure 33.28), are poisonous to humans and other large mammals.

Order Acari: Mites and Ticks

The order Acari, the mites and ticks, is the largest in terms of number of species and the most diverse of the arachnids. Although only about 30,000 species of mites and ticks have been named, scientists that study the group estimate that a million or more members of this order may exist.

Most mites are small, less than 1 millimeter long, but adults of different species range from 100 nanometers to 2 centimeters. In most mites, the cephalothorax and abdomen are fused into an unsegmented, ovoid body. Respiration occurs either by means of tracheae or directly through the exoskeleton. Many mites pass through several distinct stages during their life cycle. In most, an inactive eight-legged prelarva gives rise to an active six-legged larva, which in turn produces a succession of three eight-legged stages and, finally, the adult males and females.

Mites and ticks are diverse in structure and habitat. They are found in virtually every terrestrial, freshwater, and marine habitat known, and feed on fungi, plants, and animals. They act as predators and as internal and external parasites of both invertebrates and vertebrates.

Ticks are blood-eating parasites that occur on the surface of their host. They are larger than most other mites and cause discomfort by sucking the blood of humans and other animals. Ticks can carry many diseases, including some caused by viruses, bacteria, and protozoa. The spotted fevers (for example, Rocky Mountain spotted fever) are caused by bacteria carried by ticks. Lyme disease is apparently caused by spirochaetes transmitted by ticks. Red-water fever, or Texas fever, is an important tick-borne protozoan disease of cattle, horses, sheep, and dogs.

FIGURE 33.28
Two common poisonous spiders. (*a*) The black widow spider, *Latrodectus mactans*. (*b*) The brown recluse spider, *Loxosceles reclusa*. Both species are common throughout temperate and subtropical North America, but bites are rare in humans.

Scorpions, spiders, and mites are all arachnids, and possess chelicerates.

Major Classes of Arthropods: Centipedes and Millipedes

Millipedes, centipedes, and insects all respire by means of tracheae and excrete their waste products through Malpighian tubules. These groups were certainly derived from annelids, probably those similar to the oligochaetes, which they resemble in their embryology.

The centipedes (class Chilopoda) and millipedes (class Diplopoda) both have bodies that consist of a head region followed by numerous segments, all more or less similar and nearly all bearing paired appendages. Although the name *centipede* would imply an animal with 100 legs and the name *millipede* one with 1000, adult centipedes usually have 30 or more legs, while adult millipedes have 60 or more. Centipedes have one pair of legs on each body segment (figure 33.29), while millipedes have two (figure 33.30). Each segment of a millipede is a tagma that originated during the group's evolution when two ancestral segments fused. This explains why millipedes have twice as many legs per segment as centipedes.

In both centipedes and millipedes, fertilization is internal and takes place by direct transfer of sperm. The sexes are separate, and all species lay eggs. Young millipedes usually hatch with three pairs of legs; they experience a number of growth stages, adding segments and legs as they mature, but do not change in general appearance.

Centipedes, of which some 2500 species are known, are all carnivorous and feed mainly on insects. The appendages of the first trunk segment are modified into a pair of poison fangs. The poison is often quite toxic to human beings, and many centipede bites are extremely painful, sometimes even dangerous.

In contrast, most millipedes are herbivores, feeding mainly on decaying vegetation. A few millipedes are carnivorous. Many millipedes can roll their bodies into a flat coil or sphere because the dorsal area of each of their body segments is much longer than the ventral one. More than 10,000 species of millipedes have been named, but this is estimated to be no more than one-sixth of the actual number of species that exists. In each segment of their body,

FIGURE 33.29
A centipede. Centipedes, such as this member of the genus *Scolopendra*, are active predators.

FIGURE 33.30
A millipede. Millipedes, such as this *Sigmoria* individual, are herbivores.

most millipedes have a pair of complex glands that produce a bad-smelling fluid. This fluid is exuded for defensive purposes out through openings along the sides of the body. The chemistry of the external secretions of different millipedes has become a subject of considerable interest because of the chemical diversity of the compounds involved and their effectiveness in protecting millipedes from attack. Some species produce cyanide gas from segments near their head end. Millipedes live primarily in damp, protected places, such as under leaf litter, in rotting logs, under bark or stones, or in the soil.

> Centipedes are segmented hunters with one pair of legs on each segment. Millipedes are segmented herbivores with two pairs of legs on each segment.

Major Classes of Arthropods: Insects

The insects, class Insecta, are by far the largest group of organisms on earth, whether measured in terms of numbers of species or numbers of individuals. Insects live in every conceivable habitat on land and in fresh water, and a few have even invaded the sea. More than half of all the named animal species are insects, and the actual proportion is doubtless much higher because millions of additional forms await detection, classification, and naming. Approximately 90,000 described species occur in the United States and Canada, and the actual number of species in this area probably approaches 125,000. A hectare of lowland tropical forest is estimated to be inhabited by as many as 41,000 species of insects, and many suburban gardens may have 1500 or more species. It has been estimated that approximately a billion billion (10^{18}) individual insects are alive at any one time. A glimpse at the enormous diversity of insects is presented in figure 33.31 and also later in table 33.2.

FIGURE 33.31
Insect diversity. (*a*) Luna moth, *Actias luna.* Luna moths and their relatives are among the most spectacular insects (order Lepidoptera). (*b*) Soldier fly, *Ptecticus trivittatus* (order Diptera). (*c*) Boll weevil, *Anthonomus grandis.* Weevils are one of the largest groups of beetles (order Coleoptera). (*d*) A thorn-shaped treehopper, *Umbonia crassicornis* (order Homoptera). (*e*) Grasshopper (order Orthoptera). (*f*) Termite, *Macrotermes bellicosus* (order Isoptera). The large, sausage-shaped individual is a queen, specialized for laying eggs; most of the smaller individuals around the queen are nonreproductive workers, but the larger individual at the lower left is a reproductive male.

Table 33.2 Major Orders of Insects

Order	Typical Examples	Key Characteristics	Approximate Number of Named Species
Coleoptera	Beetles	The most diverse animal order; two pairs of wings; front pair of wings is a hard cover that partially protects the transparent rear pair of flying wings; heavily armored exoskeleton; biting and chewing mouthparts; complete metamorphosis	350,000
Diptera	Flies	Some that bite people and other mammals are considered pests; front flying wings are transparent; hindwings are reduced to knobby balancing organs; sucking, piercing, and lapping mouthparts; complete metamorphosis	120,000
Lepidoptera	Butterflies, moths	Often collected for their beauty; two pairs of broad, scaly, flying wings, often brightly colored; hairy body; tubelike, sucking mouthparts; complete metamorphosis	120,000
Hymenoptera	Bees, wasps, ants	Often social, known to many by their sting; two pairs of transparent flying wings; mobile head and well-developed eyes; often possess stingers; chewing and sucking mouthparts; complete metamorphosis	100,000
Hemiptera and Homoptera	True bugs, bedbugs, leafhoppers, aphids, cicades	Often live on blood; some are plant-eaters; two pairs of wings, or wingless; piercing, sucking mouthparts; simple metamorphosis	60,000
Orthoptera	Grasshoppers, crickets, roaches	Known for their jumping; two pairs of wings or wingless; among the largest insects; biting and chewing mouthparts in adults; simple metamorphosis.	20,000
Odonata	Dragonflies	Among the most primitive of the insect order; two pairs of transparent flying wings; large, long, and slender body; chewing mouthparts; simple metamorphosis	5,000
Isoptera	Termites	One of the few types of animals able to eat wood; two pairs of wings, but some stages are wingless; social insects; there are several body types with division of labor; chewing mouthparts; simple metamorphosis	2,000
Siphonaptera	Fleas	Small, known for their irritating bites; wingless; small, flattened body with jumping legs; piercing and sucking mouthparts; complete metamorphosis	1,200

External Features

Insects are primarily a terrestrial group, and most, if not all, of the aquatic insects probably had terrestrial ancestors. Most insects are relatively small, ranging in size from 0.1 millimeter to about 30 centimeters in length or wingspan. Insects have three body sections, the head, thorax, and abdomen; three pairs of legs, all attached to the thorax; and one pair of antennae. In addition, they may have one or two pairs of wings. Insect mouthparts all have the same basic structure but are modified in different groups in relation to their feeding habits (figure 33.32). Most insects have compound eyes, and many have ocelli as well. The insect thorax consists of three segments, each with a pair of legs. Legs are completely absent in the larvae of certain groups—for example, in most flies (order Diptera) and mosquitoes (figure 33.33). If two pairs of wings are present, they attach to the middle and posterior segments of the thorax. If only one pair of wings is present, it usually attaches to the middle segment. The thorax is almost entirely filled with muscles that operate the legs and wings.

The wings of insects arise as saclike outgrowths of the body wall. In adult insects, the wings are solid except for the veins. Insect wings are not homologous to the other appendages. Basically, insects have two pairs of wings, but in some groups, such as flies, the second set has been reduced to a pair of balancing knobs called halteres during the course of evolution. Most insects can fold their wings over their abdomen when they are at rest; but a few, such as the dragonflies and damselflies (order Odonata), keep their wings erect or outstretched at all times.

Insect forewings may be tough and hard, as in beetles. If they are, they form a cover for the hindwings and usually open during flight. The tough forewings also serve a protective function in the order Orthoptera, which includes grasshoppers and crickets. The wings of insects are made of sheets of chitin and protein; their strengthening veins are tubules of chitin and protein. Moths and butterflies have wings covered with detachable scales that provide most of their bright colors (figure 33.34). In some wingless insects, such as the springtails or silverfish, wings never evolved. Other wingless groups, such as fleas and lice, are derived from ancestral groups of insects that had wings.

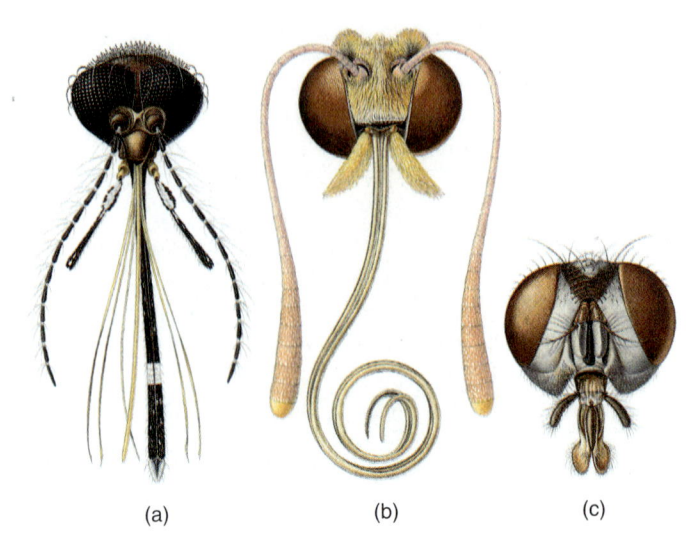

FIGURE 33.32
Modified mouthparts in three kinds of insects. Mouthparts are modified for (*a*) piercing in the mosquito, *Culex;* (*b*) sucking nectar from flowers in the alfalfa butterfly, *Colias;* and (*c*) sopping up liquids in the housefly, *Musca domestica.*

FIGURE 33.33
Larvae of a mosquito, *Culex pipiens.*
The aquatic larvae of mosquitoes are quite active. They breathe through tubes from the surface of the water, as shown here. Covering the water with a thin film of oil causes them to drown.

FIGURE 33.34
Scales on the wing of *Parnassius imperator*, a butterfly from China.
Scales of this sort account for most of the colored patterns on the wings of butterflies and moths.

Internal Organization

The internal features of insects resemble those of the other arthropods in many ways. The digestive tract is a tube, usually somewhat coiled. It is often about the same length as the body. However, in the orders Hemiptera and Homoptera, which contain the leafhoppers, cicadas, and related groups, and in many flies (order Diptera), the digestive tube may be greatly coiled and several times longer than the body. Such long digestive tracts are generally found in insects that have sucking mouthparts and feed on juices rather than on protein-rich solid foods because they offer a greater opportunity to absorb fluids and their dissolved nutrients. The digestive enzymes of the insect also are more dilute and thus less effective in a highly liquid medium than in a more solid one. Longer digestive tracts give these enzymes more time to work while food is passing through.

The anterior and posterior regions of an insect's digestive tract are lined with cuticle. Digestion takes place primarily in the stomach, or midgut, and excretion takes place through Malpighian tubules. Digestive enzymes are mainly secreted from the cells that line the midgut, although some are contributed by the salivary glands near the mouth.

The tracheae of insects extend throughout the body and permeate its different tissues. In many winged insects, the tracheae are dilated in various parts of the body, forming air sacs. These air sacs are surrounded by muscles and form a kind of bellows system to force air deep into the tracheal system. The spiracles, a maximum of 10 on each side of the insect, are paired and located on or between the segments along the sides of the thorax and abdomen. In most insects, the spiracles can be opened by muscular action. Closing the spiracles at times may be important in retarding water loss. In some parasitic and aquatic groups of insects, the spiracles are permanently closed. In these groups, the tracheae run just below the surface of the insect, and gas exchange takes place by diffusion.

The **fat body** is a group of cells located in the insect body cavity. This structure may be quite large in relation to the size of the insect, and it serves as a food-storage reservoir, also having some of the functions of a vertebrate liver. It is often more prominent in immature insects than in adults, and it may be completely depleted when metamorphosis is finished. Insects that do not feed as adults retain their fat bodies and live on the food stored in them throughout their adult lives (which may be very short).

Sense Receptors

In addition to their eyes, insects have several characteristic kinds of sense receptors. These include **sensory hairs,** which are usually widely distributed over their bodies. The sensory hairs are linked to nerve cells and are sensitive to mechanical and chemical stimulation. They are particularly abundant on the antennae and legs—the parts of the insect most likely to come into contact with other objects.

Sound, which is of vital importance to insects, is detected by tympanal organs in groups such as grasshoppers and crickets, cicadas, and some moths. These organs are paired structures composed of a thin membrane, the **tympanum,** associated with the tracheal air sacs. In many other groups of insects, sound waves are detected by sensory hairs. Male mosquitoes use thousands of sensory hairs on their antennae to detect the sounds made by the vibrating wings of female mosquitoes.

Sound detection in insects is important both for protection from enemies and for communication. Many insects communicate by making sounds, most of which are quite soft, very high-pitched, or both, and thus inaudible to humans. Only a few groups of insects, especially grasshoppers, crickets, and cicadas, make sounds that people can hear. Male crickets and longhorned grasshoppers produce sounds by rubbing their two front wings together. Short-horned grasshoppers do so by rubbing their hind legs over specialized areas on their wings. Male cicadas vibrate the membranes of air sacs located on the lower side of the most anterior abdominal segment.

In addition to using sound, nearly all insects communicate by means of chemicals or mixtures of chemicals known as **pheromones.** These compounds, extremely diverse in their chemical structure, are sent forth into the environment, where they are active in very small amounts and convey a variety of messages to other individuals. These messages not only convey the attraction and recognition of members of the same species for mating, but they also mark trails for members of the same species, as in the ants.

Insect Life Histories

During the course of their development, many insects undergo **metamorphosis.** In *simple metamorphosis,* seen in grasshoppers, immature stages are quite similar to adults, but gradually get bigger and more developed over a series of molts. In *complete metamorphosis,* immature stages called **larvae** are often wormlike and active in feeding. A resting stage, during which the insect is called a **pupa** or **chrysalis,** immediately precedes the final molt into the adult form.

> All insects possess three body segments (tagmata): the head, the thorax, and the abdomen. The three pairs of legs are attached to the thorax. Most insects have compound eyes, and many have one or two pairs of wings. Insects possess sophisticated means of sensing their environment, including sensory hairs, tympanal organs, and chemoreceptors. Many insects undergo simple or complete metamorphosis.

33.5 Echinoderms are radially symmetrical as adults.

Deuterostome Development and an Endoskeleton

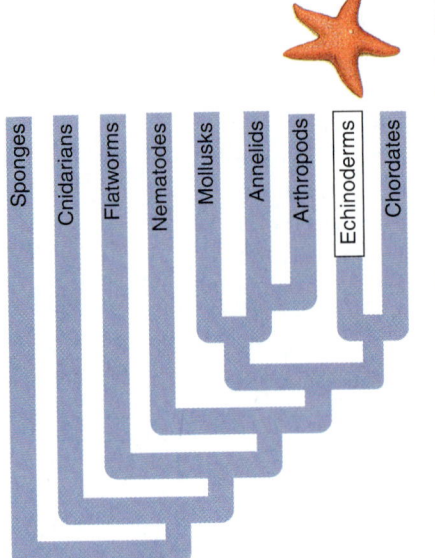

The mollusks, annelids, and arthropods discussed so far are protostomes. However, the echinoderms described in this section, are characterized by *deuterostome development*, a key evolutionary advance. The *endoskeleton* makes its first appearance in the echinoderms also.

The Echinoderms

Deuterostomate marine animals called **echinoderms** appeared nearly 600 million years ago. Echinoderms (phylum Echinodermata) are an ancient group of marine animals consisting of about 6000 living species (figure 33.35) and are also well represented in the fossil record. The term *echinoderm* means "spiny skin" and refers to an **endoskeleton** composed of hard, calcium-rich plates just beneath the delicate skin (figure 33.36). When they first form, the plates are enclosed in living tissue and so are truly an endoskeleton, although in adults they frequently fuse, forming a hard shell. Another innovation in echinoderms is the development of a hydraulic system to aid in movement or feeding. Called a **water-vascular system,** this fluid-filled system is composed of a central ring canal from which five radial canals extend out into the body and arms.

Many of the most familiar animals seen along the seashore, sea stars (starfish), brittle stars, sea urchins, sand dollars, and sea cucumbers, are echinoderms. All are radially symmetrical as adults. While some other kinds of animals are radially symmetrical, none have the complex organ systems of adult echinoderms. Echinoderms are well represented not only in the shallow waters of the sea but also in its abyssal depths. In the oceanic trenches, which are the deepest regions of the oceans, sea cucumbers account for more than 90% of the biomass! All of them are bottom-dwellers except for a few swimming sea cucumbers. The adults range from a few millimeters to more than a meter in diameter (for one species of sea star) or in length (for a species of sea cucumber).

There is an excellent fossil record of the echinoderms, extending back into the Cambrian period. However, despite this wealth of information, the origin of echinoderms remains unclear. They are thought to have evolved from bilaterally symmetrical ancestors because echinoderm larvae are bilateral. The radial symmetry that is the hallmark of echinoderms develops later, in the adult body. Many biologists believe that early echinoderms were sessile and evolved radiality as an adaptation to the sessile existence. Bilaterality is of adaptive value to an animal that travels through its environment, while radiality is of value to an animal whose environment meets it on all sides. Echinoderms attached to the sea bottom by a central stalk were once common, but only about 80 such species survive today.

Echinoderms are a unique, exclusively marine group of organisms in which deuterostome development and an endoskeleton are seen for the first time.

FIGURE 33.35
Diversity in echinoderms. (*a*) Sea star, *Oreaster occidentalis* (class Asteroidea), in the Gulf of California, Mexico. (*b*) Warty sea cucumber, *Parastichopus parvimensis* (class Holothuroidea), Philippines. (*c*) Sea urchin (class Echinoidea).

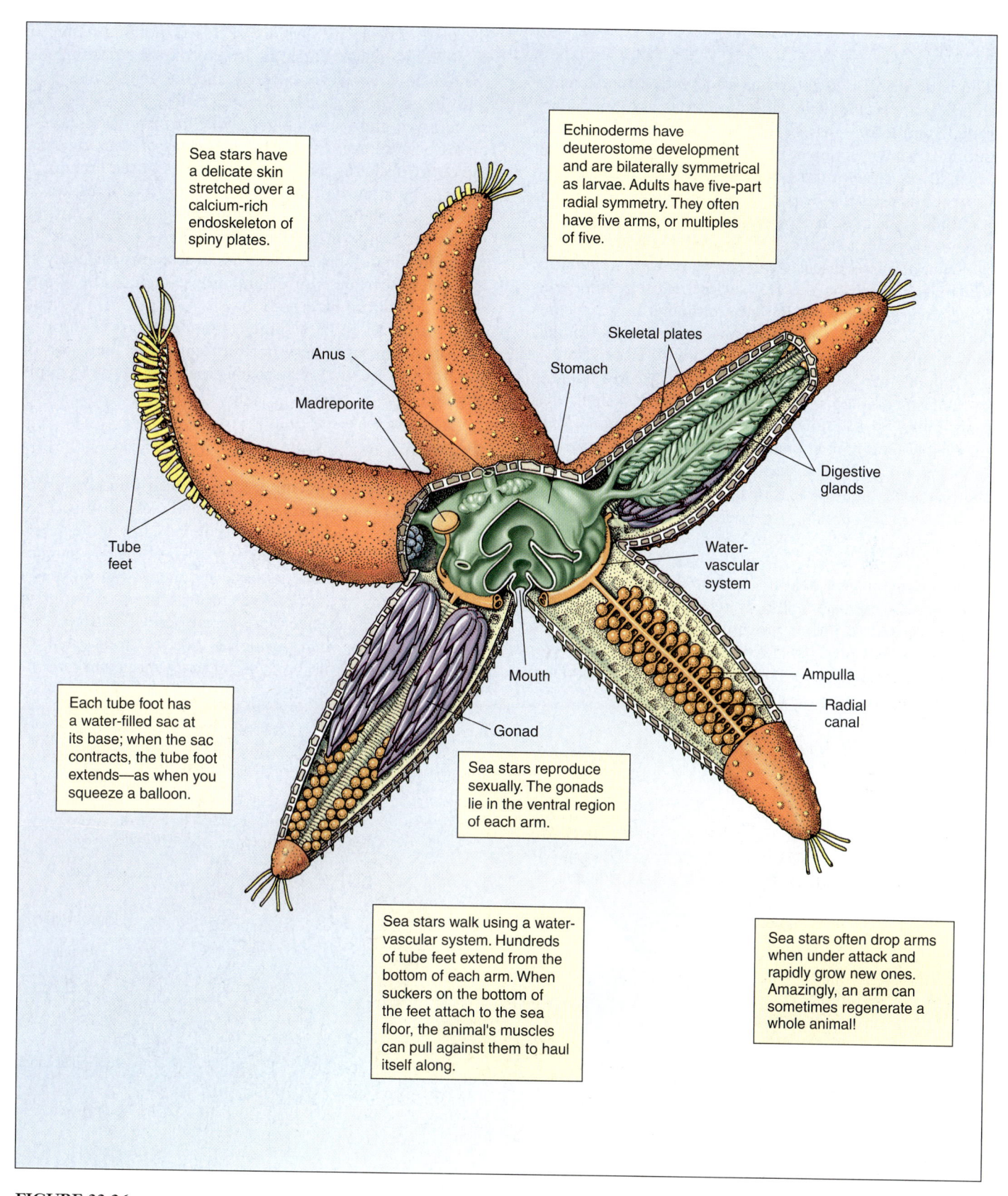

FIGURE 33.36
Phylum Echinodermata: echinoderms. Echinoderms, such as sea stars (phylum Echinodermata), are coelomates with a deuterostomate pattern of development. A delicate skin stretches over an endoskeleton made of calcium-rich plates, often fused into a continuous, tough, spiny layer.

Chapter 33 Coelomate Invertebrates 677

Echinoderm Body Plan

The body plan of echinoderms undergoes a fundamental shift during development. All echinoderms have **secondary radial symmetry**—that is, they are bilaterally symmetrical during larval development but become radially symmetrical as adults. Because of their radially symmetrical bodies, the usual terms used to describe an animal's body are not applicable: Dorsal, ventral, anterior, and posterior have no meaning without a head or tail. Instead, the body structure of echinoderms is discussed in reference to their mouths, which are located on the oral surface. Most echinoderms crawl along on their oral surfaces, although in sea cucumbers and some other echinoderms, the animal's axis lies horizontally, and they crawl with the oral surface in front.

Echinoderms have a five-part body plan corresponding to the arms of a sea star or the design on the "shell" of a sand dollar. These animals have no head or brain. Their nervous systems consist of a central **nerve ring** from which branches arise. The animals are capable of complex response patterns, but there is no centralization of function.

Endoskeleton

Echinoderms have a delicate epidermis, containing thousands of neurosensory cells, stretched over an endoskeleton composed of either movable or fixed calcium-rich (calcite) plates called ossicles. The animals grow more or less continuously, but their growth slows with age. When the plates first form, they are enclosed in living tissue. In some echinoderms, such as asteroids and holothuroids, the ossicles are widely scattered, and the body wall is flexible. In others, especially the echinoids, the ossicles become fused and form a rigid shell. In many cases, these plates bear spines. In nearly all species of echinoderms, the entire skeleton, even the long spines of sea urchins, is covered by a layer of skin. Another important feature of this phylum is the presence of mutable collagenous tissue, which can range in texture from tough and rubbery to weak and fluid. This amazing tissue accounts for many of the special attributes of echinoderms, such as the ability to rapidly autotomize body parts. The plates in certain portions of the body of some echinoderms are perforated by pores. Through these pores extend **tube feet,** part of the water-vascular system that is a unique feature of this phylum.

The Water-Vascular System

The water-vascular system of an echinoderm radiates from a ring canal that encircles the animal's esophagus. Five **radial canals,** their positions determined early in the development of the embryo, extend into each of the five parts of the body and determine its basic symmetry (figure 33.37). Water enters the water-vascular system through a **madreporite,** a sievelike plate on the animal's surface, and flows to the ring canal through a tube, or stone canal, so named because of the surrounding rings of calcium carbon-

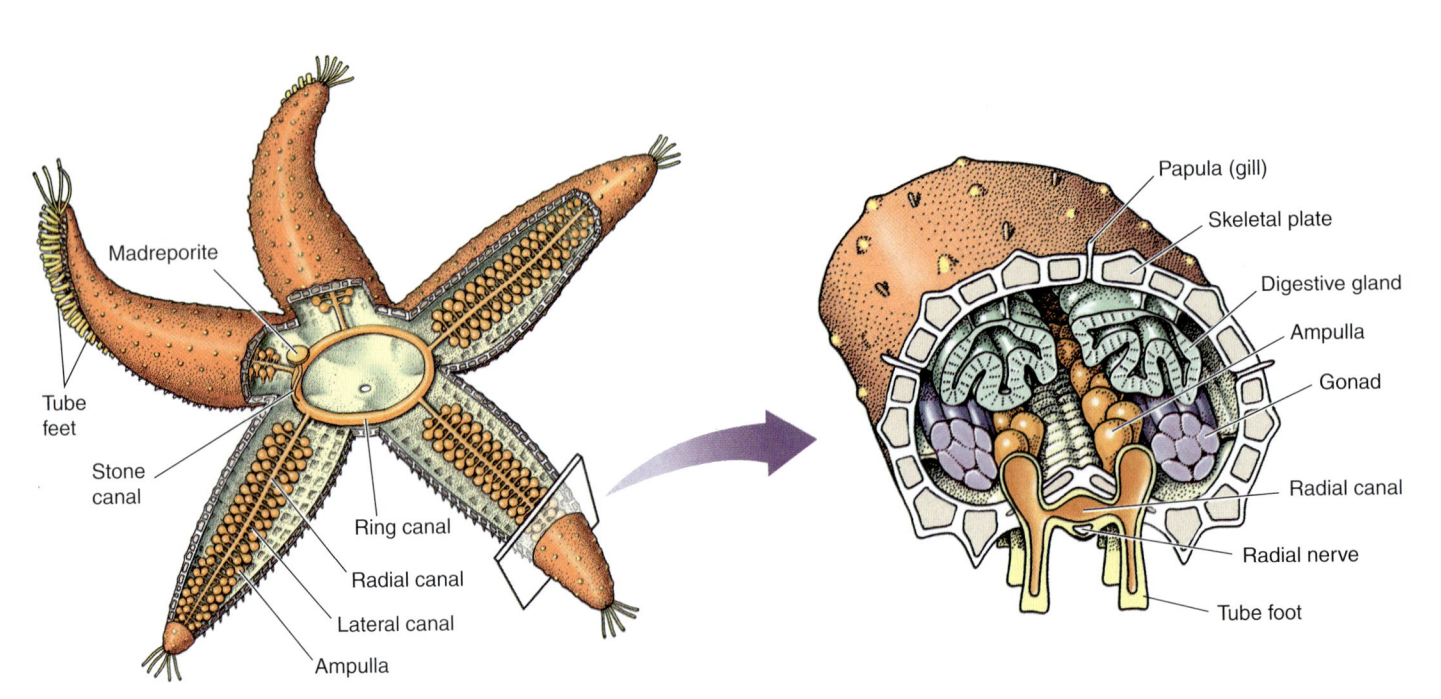

FIGURE 33.37
The water-vascular system of an echinoderm. Radial canals allow water to flow into the tube feet. As the ampulla in each tube foot contracts, the tube extends and can attach to the substrate. Subsequently, muscles in the tube feet contract, bending the tube foot and pulling the animal forward.

678   Part V   Diversity of Life on Earth

FIGURE 33.38
Tube feet. The nonsuckered tube feet of the sea star, *Ludia magnifica*, are extended.

ate. The five radial canals in turn extend out through short side branches into the hollow tube feet (figure 33.38). In some echinoderms, each tube foot has a sucker at its end; in others, suckers are absent. At the base of each tube foot is a muscular sac, the **ampulla,** which contains fluid. When the ampulla contracts, the fluid is prevented from entering the radial canal by a one-way valve and is forced into the tube foot, thus extending it. When extended, the foot can attach itself to the substrate. Longitudinal muscles in the tube foot wall then contract, causing the tube feet to bend. Relaxation of the muscles in the ampulla allows the fluid to flow back into the ampulla, which moves the foot. By the concerted action of a very large number of small, individually weak tube feet, the animal can move across the seafloor.

Sea cucumbers usually have five rows of tube feet on the body surface that are used in locomotion. They also have modified tube feet around their mouth cavity that are used in feeding. In sea lilies, the tube feet arise from the branches of the arms, which extend from the margins of an upward-directed cup. With these tube feet, the animals take food from the surrounding water. In brittle stars (see figure 33.41), the tube feet are pointed and specialized for feeding.

Body Cavity

In echinoderms, the coelom, which is proportionately large, connects with a complicated system of tubes and helps provide circulation and respiration. In many echinoderms, respiration and waste removal occur across the skin through small, fingerlike extensions of the coelom called papulae (see figure 33.37). They are covered with a thin layer of skin and protrude through the body wall to function as gills.

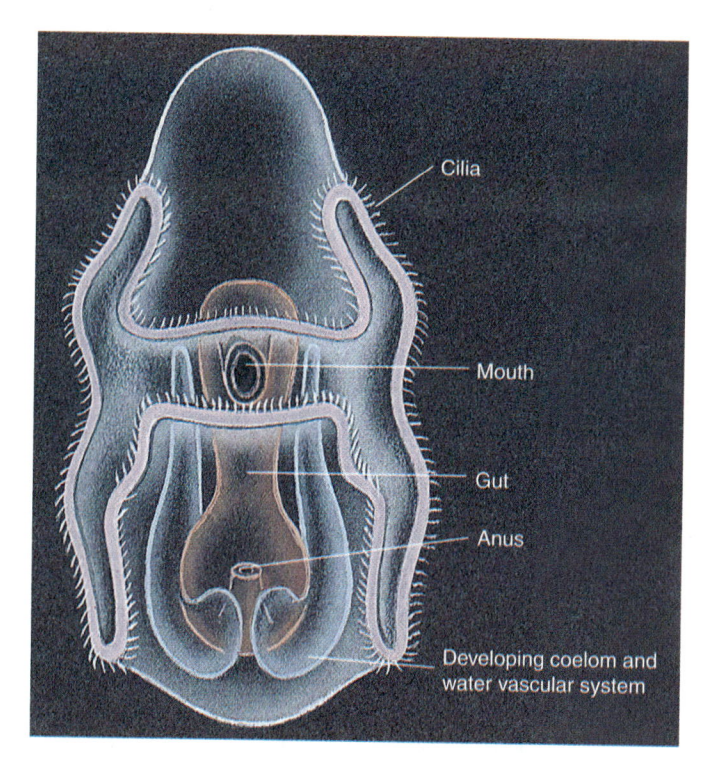

FIGURE 33.39
The free-swimming larva of an echinoderm. The bands of cilia by which the larva moves are prominent in this drawing. Such bilaterally symmetrical larvae suggest that the ancestors of the echinoderms were not radially symmetrical, like the living members of the phylum.

Reproduction

Many echinoderms are able to regenerate lost parts, and some, especially sea stars and brittle stars, drop various parts when under attack. In a few echinoderms, asexual reproduction takes place by splitting, and the broken parts of sea stars can sometimes regenerate whole animals. Some of the smaller brittle stars, especially tropical species, regularly reproduce by breaking into two equal parts; each half then regenerates a whole animal.

Despite the ability of many echinoderms to break into parts and regenerate new animals from them, most reproduction in the phylum is sexual and external. The sexes in most echinoderms are separate, although there are few external differences. Fertilized eggs of echinoderms usually develop into free-swimming, bilaterally symmetrical larvae (figure 33.39), which differ from the trochophore larvae of mollusks and annelids. These larvae form a part of the plankton until they metamorphose through a series of stages into the more sedentary adults.

> **Echinoderms are characterized by a secondary radial symmetry and a five-part body plan. They have characteristic calcium-rich plates called ossicles and a unique water-vascular system that includes hollow tube feet.**

Chapter 33 Coelomate Invertebrates 679

Major Classes of Echinoderms

There are more than 20 extinct classes of echinoderms and an additional 6 with living members: (1) Crinoidea, sea lilies and feather stars; (2) Asteroidea, sea stars, or starfish; (3) Ophiuroidea, brittle stars; (4) Echinoidea, sea urchins and sand dollars; (5) Holothuroidea, sea cucumbers; and (6) Concentricycloidea, sea daisies. All the classes of echinoderms exhibit a basic five-part body plan, although the basic symmetry of the phylum is not obvious in some. Even sea cucumbers, which are shaped like their plant namesake, have five radial grooves running along their soft, sluglike bodies. We describe three of the living classes here.

Class Asteroidea: The Sea Stars

Sea stars, or starfish (class Asteroidea; figure 33.40), are perhaps the most familiar echinoderms. Among the most important predators in many marine ecosystems, they range in size from a centimeter to a meter across. They are abundant in the intertidal zone, but they also occur at depths as great as 10,000 meters. Around 1500 species of sea stars occur throughout the world.

The body of a sea star is composed of a central disc that merges gradually with the arms. Although most sea stars have five arms, members of some families have many more, typically in multiples of five. The body is somewhat flattened, flexible, and covered with a pigmented epidermis.

Class Ophiuroidea: The Brittle Stars

Brittle stars (class Ophiuroidea; figure 33.41) are the largest class of echinoderms in numbers of species (about 2000), and they are probably the most abundant also. Secretive, they avoid light and are more active at night.

Brittle stars have slender, branched arms. The most mobile of echinoderms, brittle stars move by pulling themselves along, "rowing" over the substrate by moving their arms, often in pairs or groups, from side to side. Some brittle stars use their arms to swim, a very unusual habit among echinoderms.

Like sea stars, brittle stars have five arms. Closely related to the sea stars, on closer inspection they are surprisingly different. They have no pedicellariae, as sea stars have, and the groove running down the length of each arm is closed over and covered with ossicles. Their tube feet lack ampullae, have no suckers, and are used for feeding, not locomotion.

Class Echinoidea: The Sea Urchins and Sand Dollars

The members of the class Echinoidea, sand dollars and sea urchins, lack distinct arms but have the same five-part body plan as all other echinoderms. Five rows of tube feet pro-

FIGURE 33.40
Class Asteroidea. This class includes the familiar starfish, or sea stars.

FIGURE 33.41
Class Ophiuroidea. Brittle stars crawl actively across their marine substrates.

trude through the plates of the calcareous skeleton. About 950 living species constitute the class Echinoidea. Because of their calcareous plates, sea urchins and sand dollars are well preserved in the fossil record, with more than 5000 additional species described.

> **The six classes of echinoderms have five-part body symmetry, and are all radially symmetrical as adults.**

Part V Diversity of Life on Earth

Concept Review

For interactive testing, visit the Online Learning Center with PowerWeb at www.mhhe.com/Raven7

33.1 Mollusks were among the first coelomates.

Mollusk Body Plan

- The evolution of a coelom was a significant advance in animal body structure because it repositions fluid and allows complex tissues and organs to develop. (p. 652)
- Mollusks are extremely diverse; they exhibit a wide range of body forms and inhabit a large variety of habitats. (p. 652)
- The molluscan body plan is bilaterally symmetrical, with an efficient excretory system and a muscular foot for locomotion. (p. 654)
- Most mollusks have distinct sexes, and most aquatic mollusks engage in external fertilization. (p. 655)

Major Classes of Mollusks

- Major classes of mollusks include the class Polyplacophora (chitons), class Gastropoda (snails and slugs), class Bivalvia (bivalves), and class Cephalopoda (octopuses, squids, and nautiluses). (pp. 656–657)

33.2 Annelids were the first segmented animals.

Segmented Bodies

- Segmentation was a key transition in animal body plans, because similar segments can be individually controlled for different functions. (p. 658)
- Three principal features of annelids are repeated segments, specialized segments, and connections between segments. (p. 658)
- The annelid body plan is a tube within a tube. (p. 658)
- Locomotion is carried out by a squeezing motion as the hydrostatic skeleton forces fluid from one segment to the next. (p. 658)

Major Classes of Annelids

- The roughly 12,000 recognized annelid species are divided into three classes: Polychaeta (polychaetes), Oligochaeta (earthworms), and Hirudinea (leeches). (pp. 660–661)

33.3 Lophophorates appear to be a transitional group.

Lophophorates

- The marine phyla Phoronida (phoronids), Ectoprocta (bryozoans), and Brachiopoda (brachiopods) are characterized by a ridge around the mouth bearing ciliated, hollow tentacles (a lophophore) that functions in gas exchange and food collection. (pp. 662–663)

33.4 Arthropods are the most diverse of all animal groups.

Arthropod Body Plan

- Over two-thirds of all named species on the earth are arthropods. (p. 664)
- Jointed appendages used as legs, antennae, and mouthparts, and an exoskeleton of chitin and protein used for protection and water loss prevention, have allowed arthropods to become the most diverse phylum on the planet. (p. 664)
- Arthropods are segmented, with some segments becoming fused into tagmata (functional groups). (pp. 665–666)
- The exoskeleton is secreted by, and fused with, the epidermis, and provides a hard surface for muscle attachment. (p. 666)
- All arthropods have an open circulatory system; some have adaptations such as compound eyes, a respiratory system composed of tracheae, and an excretory system composed of Malpighian tubules. (pp. 666–667)

A Major Group of Arthropods: Crustaceans

- Most crustaceans (35,000 species) have two pairs of antennae, three types of chewing appendages, and various pairs of legs. (p. 668)
- Crustaceans are found in marine, freshwater, and terrestrial habitats. (pp. 668–669)

Major Classes of Arthropods: Arachnids

- Arachnids (57,000 species) have a pair of chelicerae, a pair of pedipalps, and four pairs of walking legs. (p. 670)
- Two major orders are Araneae (spiders) and Acari (mites and ticks). (p. 670)

Major Classes of Arthropods: Centipedes and Millipedes

- Centipedes (class Chilopoda) and millipedes (class Diplopoda) are made of a head region followed by numerous similar segments. Centipedes have one pair of legs per segment, and millipedes have two pairs per segment. (p. 671)

Major Classes of Arthropods: Insects

- Class Insecta is the largest group of organisms on the planet, living in nearly every possible habitat. (p. 672)
- Most are relatively small, and contain three body sections: head, thorax, and abdomen, with three pairs of legs attached to the thorax, and one pair of antennae. (p. 674)
- Most insects have compound eyes. (p. 674)
- Sensory hairs, tympanal organs, and chemoreceptors all act as sense receptors in insects. (p. 675)
- Many insects undergo either simple or complex metamorphosis. (p. 675)

33.5 Echinoderms are radially symmetrical as adults.

Deuterostome Development and an Endoskeleton

- Echinoderms are marine animals with hard calcium plates forming a true endoskeleton in young individuals. (p. 676)

Echinoderm Body Plan

- All echinoderms are bilaterally symmetrical during larval development, and become radially symmetrical as adults. (p. 678)
- Echinoderms have a five-part body plan with a central, branched nerve ring and an endoskeleton composed of calcium-rich plates (ossicles). (p. 678)
- Many echinoderms can regenerate lost parts, but in most of them, reproduction is sexual and external. (p. 679)

Major Classes of Echinoderms

- There are six living classes of Echinoderms: Crinoidea (sea lilies and feather stars), Asteroidea (sea stars or starfish), Ophiuroidea (brittle stars), Echinoidea (sea urchins and sand dollars), Holothuroidea (sea cucumbers), and Concentricycloidea (sea daisies). (p. 680)

Test Your Understanding

Self Test

1. Of the mollusks, snails are in the class of
   a. gastropods.
   b. bivalves.
   c. cephalopods.
   d. chitons.
2. A mantle is
   a. present only in bivalves.
   b. a structure that acts as a lung or contains gills.
   c. a rasping, tonguelike organ in mollusks.
   d. necessary for mollusks to be motile.
3. Segmentation was first apparent in the
   a. flatworms.
   b. annelids.
   c. mollusks.
   d. arthropods.
4. Which of the following is not present in polychaetes?
   a. a coelom
   b. parapodia
   c. permanent gonads
   d. setae
5. The lophophore, the structure characteristic of lophophorates,
   a. functions in gas exchange.
   b. functions in feeding.
   c. can be withdrawn when the animal is disturbed.
   d. All of these are correct.
6. The phylum that shows the greatest diversity, or the greatest number of species, is
   a. Arthropoda.
   b. Brachiopoda.
   c. Echinodermata.
   d. Mollusca.
7. Air for respiration enters the insect body through the
   a. tracheae.
   b. spiracles.
   c. tracheoles.
   d. Malpighian tubules.
8. Arthropods shed their old exoskeleton as they grow in a process known as
   a. tagmatization.
   b. metamorphosis.
   c. chrysalis.
   d. ecdysis.
9. Which animal group has radial symmetry, a water-vascular system, moves with tube feet, and has an endoskeleton?
   a. arachnids
   b. crustaceans
   c. echinoderms
   d. cnidarians
10. The echinoderms that lack distinct arms are the
    a. brittle stars.
    b. sea urchins.
    c. sea stars.
    d. Asteroidea.

Test Your Visual Understanding

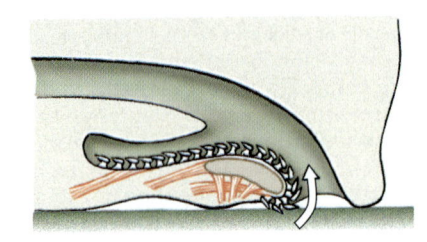

1. The radula shown in the figure above is a feeding structure found in individuals of the phylum Mollusca but is lacking in one group of mollusks. What group of mollusks does not have a radula? How do individuals in this group eat?

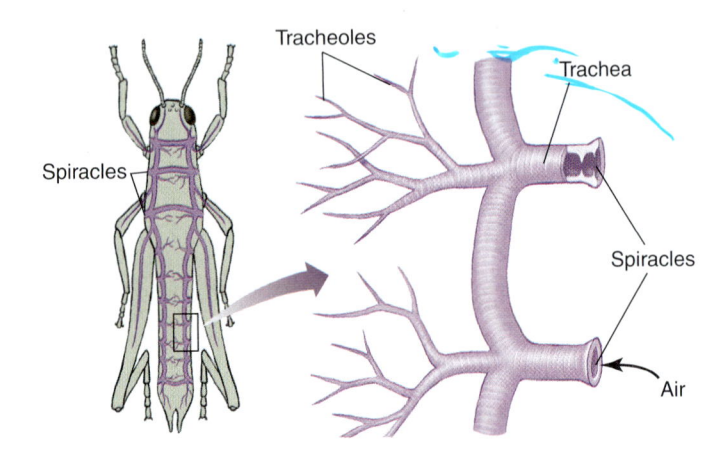

2. Horticultural oils are sometimes used as insecticides to eliminate insect pests from foliage by coating the insect with oil. Referring to the figure above, can you explain how this method of insecticidal control works?

Apply Your Knowledge

1. Freshwater bivalves are an important ecological resource because they filter freshwater systems. A population of freshwater bivalves located in a meter-squared area of substratum filters 10 m^3 of water a day. How many liters of water are filtered per day (1 m^3 equals 1000 liters)? How many liters of water are filtered per hour? How many liters of water would be filtered by a population filling a 5 m^2 area of substratum?
2. Scientists believe the ancestral mollusk had a very limited shell, consisting mainly of calcareous plates. The shell became more developed in some groups but was lost in others. What is the evolutionary advantage of having a shell? Of not having one?
3. Although arthropods are very successful in aquatic environments, what are the key adaptations that facilitated the invasion of the land by arthropods?
4. Why is it believed that echinoderms and chordates, which are so dissimilar, are members of the same evolutionary line?

34
Vertebrates

Concept Outline

34.1 Attaching muscles to an internal framework greatly improves movement.

The Chordates. Chordates have an internal flexible rod, the first stage in the evolution of a truly internal skeleton.

34.2 Nonvertebrate chordates have a notochord but no backbone.

The Nonvertebrate Chordates. Lancelets are thought to resemble the ancestors of vertebrates.

34.3 The evolution of vertebrates involved invasions of sea, land, and air.

Characteristics of Vertebrates. Vertebrates have a true, usually bony endoskeleton, with a backbone encasing the spinal column, and a skull-encased brain.
Fishes. Over half of all vertebrate species are fishes, which include the group from which all other vertebrates evolved.
Amphibians. The key innovation that made life on land possible for vertebrates was the pulmonary vein.
Reptiles. Reptiles were the first vertebrates to completely master the challenge of living on dry land.
The Rise and Fall of Dominant Reptile Groups. Now-extinct forms of reptiles dominated life on land for 250 million years. Four orders survive today.
Birds. Birds are much like reptiles, but with feathers.
History of the Birds. Birds are thought to have evolved from dinosaurs with adaptations of feathers and flight.
Mammals. Mammals are the only vertebrates that possess hair and mammary glands.
The Orders of Mammals. Mammals only became common when dinosaurs disappeared.

34.4 Evolution among the primates has focused on brain size and locomotion.

Primates. Only primates have binocular vision and grasping hands.
Australopithecines. The evolution of bipedalism was the first step on the hominid evolutionary path.
The Genus *Homo*. The genus of which humans are a member evolved in Africa.

FIGURE 34.1
A large vertebrate. Today, mammals, such as this elephant, dominate vertebrate life on land, but for over 200 million years, mammals were a minor group in a world dominated by reptiles.

Members of the phylum Chordata exhibit great improvements in the endoskeleton over what is seen in echinoderms. As we saw in chapter 33, the endoskeleton of echinoderms is functionally similar to the exoskeleton of arthropods; it is a hard shell that encases the body, with muscles attached to its inner surface. Chordates employ a very different kind of endoskeleton, one that is truly internal. Members of the phylum Chordata are characterized by a flexible rod that develops along the back of the embryo. Muscles attached to this rod allowed early chordates to swing their backs from side to side, swimming through the water. This key evolutionary advance, attaching muscles to an internal element, started chordates along an evolutionary path that led to the vertebrates—and, for the first time, to truly large animals (figure 34.1).

34.1 Attaching muscles to an internal framework greatly improves movement.

The Chordates

Chordates (phylum Chordata) are deuterostome coelomates whose nearest relatives in the animal kingdom are the echinoderms, the only other deuterostomes. However, unlike echinoderms, chordates are characterized by a *notochord, jointed appendages,* and *segmentation.* There are some 56,000 species of chordates, a phylum that includes birds, reptiles, amphibians, fishes, and mammals.

Four features characterize the chordates and have played an important role in the evolution of the phylum (figure 34.2):

1. A single, hollow **nerve cord** runs just beneath the dorsal surface of the animal. In vertebrates, the dorsal nerve cord differentiates into the brain and spinal cord.
2. A flexible rod, the **notochord,** forms on the dorsal side of the primitive gut in the early embryo and is present at some developmental stage in all chordates. The notochord is located just below the nerve cord. The notochord may persist throughout the life cycle of some chordates or be displaced during embryonic development, as in most vertebrates, by the vertebral column that forms around the nerve cord.
3. **Pharyngeal slits** connect the **pharynx,** a muscular tube that links the mouth cavity and the esophagus, with the outside. In terrestrial vertebrates, the slits do not actually connect to the outside and are better termed **pharyngeal pouches.** Pharyngeal pouches are present in the embryos of all vertebrates. They become slits, open to the outside in animals with gills, but disappear in those lacking gills. The presence of these structures in all vertebrate embryos provides evidence of their aquatic ancestry.
4. Chordates have a **postanal tail** that extends beyond the anus, at least during their embryonic development. Nearly all other animals have a terminal anus.

All chordates have all four of these characteristics at some time in their lives. For example, humans have pharyngeal pouches, a dorsal nerve cord, a postanal tail, and a notochord as embryos. As adults, the nerve cord remains, while the notochord is replaced by the vertebral column, and all but one pair of pharyngeal pouches are lost. This remaining pair forms the Eustachian tubes that connect the throat to the middle ear. The postanal tail regresses forming the tail bone.

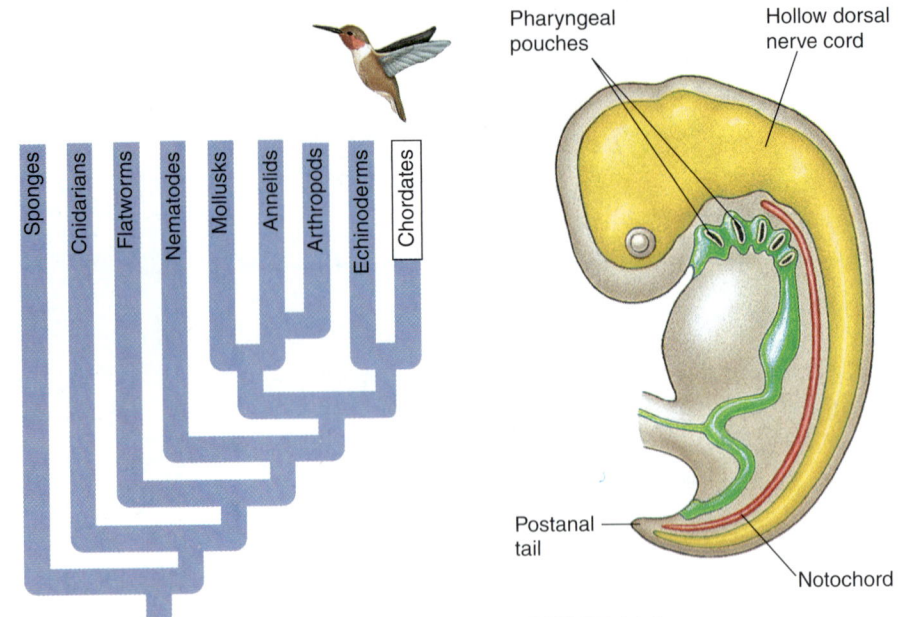

FIGURE 34.2
The four principal features of the chordates, as shown in a generalized embryo.

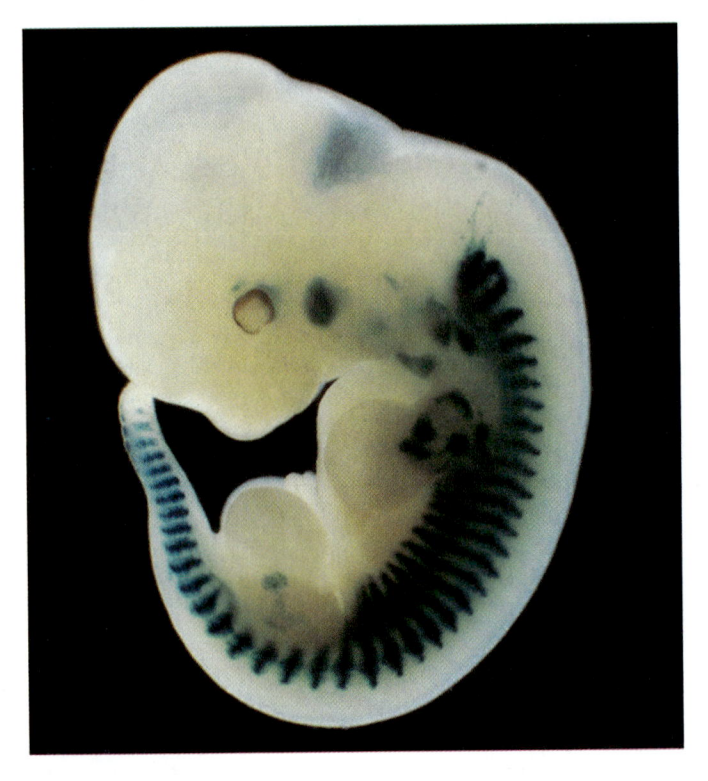

FIGURE 34.3
A mouse embryo. At 11.5 days of development, the mesoderm is already divided into segments called somites (stained dark in this photo), reflecting the fundamentally segmented nature of all chordates.

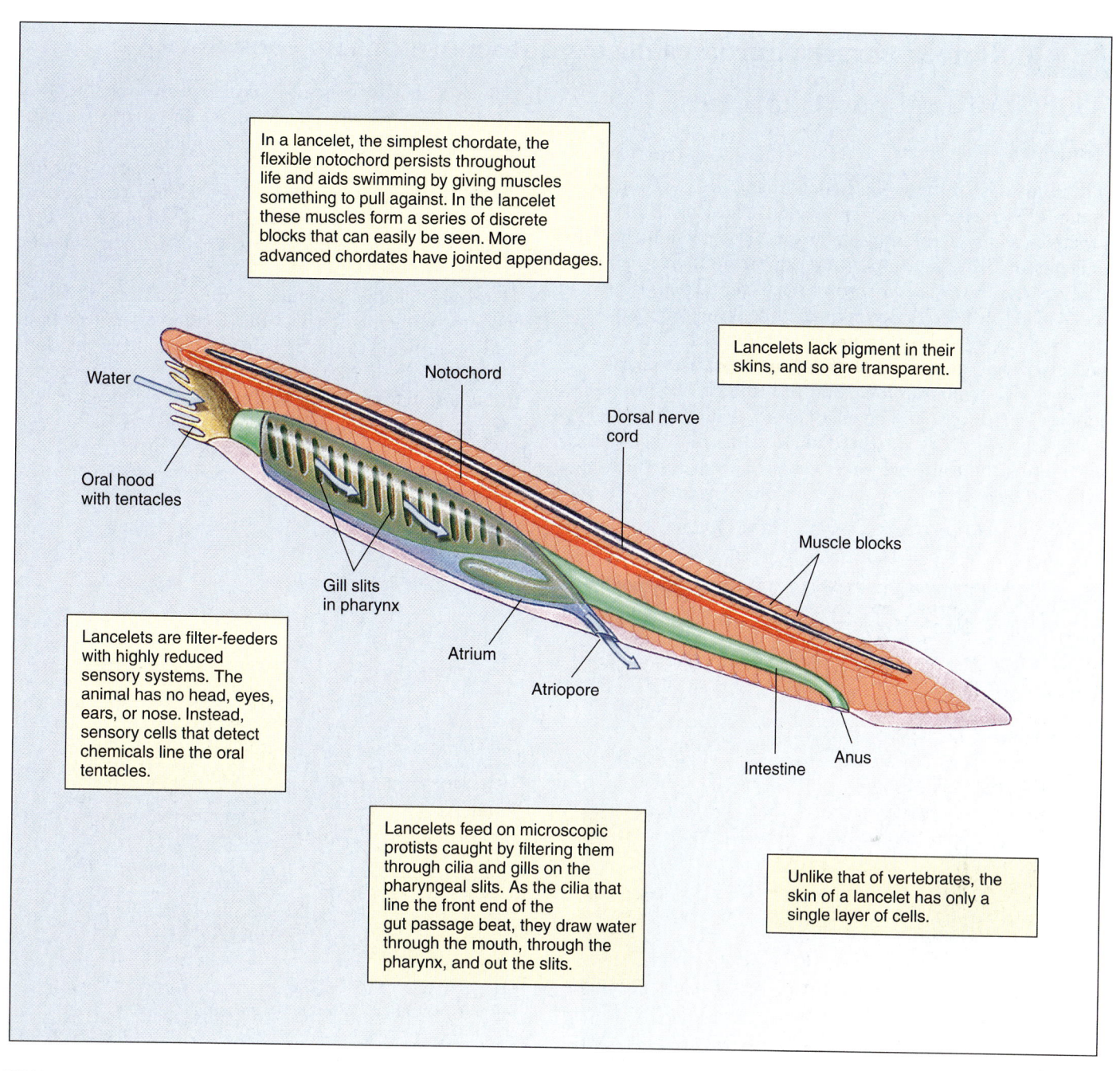

FIGURE 34.4
Phylum Chordata: chordates. Vertebrates, tunicates, and lancelets are chordates (phylum Chordata), coelomate animals with a flexible rod, the notochord, that provides resistance to muscle contraction and permits rapid lateral body movements. Chordates also possess pharyngeal pouches and slits (reflecting their aquatic ancestry and present habitat in some) and a hollow dorsal nerve cord. In vertebrates, the notochord is replaced during embryonic development by the vertebral column.

A number of other characteristics also distinguish the chordates fundamentally from other animals. Chordates' muscles are arranged in segmented blocks that affect the basic organization of the chordate body and can often be clearly seen in embryos of this phylum (figure 34.3). Most chordates have an internal skeleton against which the muscles work. Either this internal skeleton or the notochord (figure 34.4) makes possible the extraordinary powers of locomotion that characterize the members of this group.

Chordates are characterized by a hollow dorsal nerve cord, a notochord, pharyngeal pouches, and a postanal tail at some point in their development. The flexible notochord anchors internal muscles and allows rapid, versatile movement.

Chapter 34 Vertebrates 685

34.2 Nonvertebrate chordates have a notochord but no backbone.

The Nonvertebrate Chordates

Tunicates

The tunicates (subphylum Urochordata) are a group of about 1250 species of marine animals. Most of them are sessile as adults, with only the larvae having a notochord and nerve cord. As adults, they exhibit neither a major body cavity nor visible signs of segmentation (figure 34.5a,b). Most species occur in shallow waters, but some are found at great depths. In some tunicates, adults are colonial, living in masses on the ocean floor. The pharynx is lined with numerous cilia, and the animals obtain their food by ciliary action. The cilia beat, drawing a stream of water into the pharynx, where microscopic food particles are trapped in a mucous sheet secreted from a structure called an *endostyle*.

The tadpolelike larvae of tunicates plainly exhibit all of the basic characteristics of chordates and mark the tunicates as having the most primitive combination of features found in any chordate (figure 34.5c). The larvae do not feed and have a poorly developed gut. They remain free-swimming for only a few days before settling to the bottom and attaching themselves to a suitable substrate by means of a sucker.

Tunicates change so much as they mature and adjust developmentally to a sessile, filter-feeding existence that it would be difficult to discern their evolutionary relationships by examining an adult. Many adult tunicates secrete a **tunic,** a tough sac composed mainly of cellulose, a substance frequently found in the cell walls of plants and algae but rarely found in animals. The tunic surrounds the animal and gives the subphylum its name. Colonial tuni-

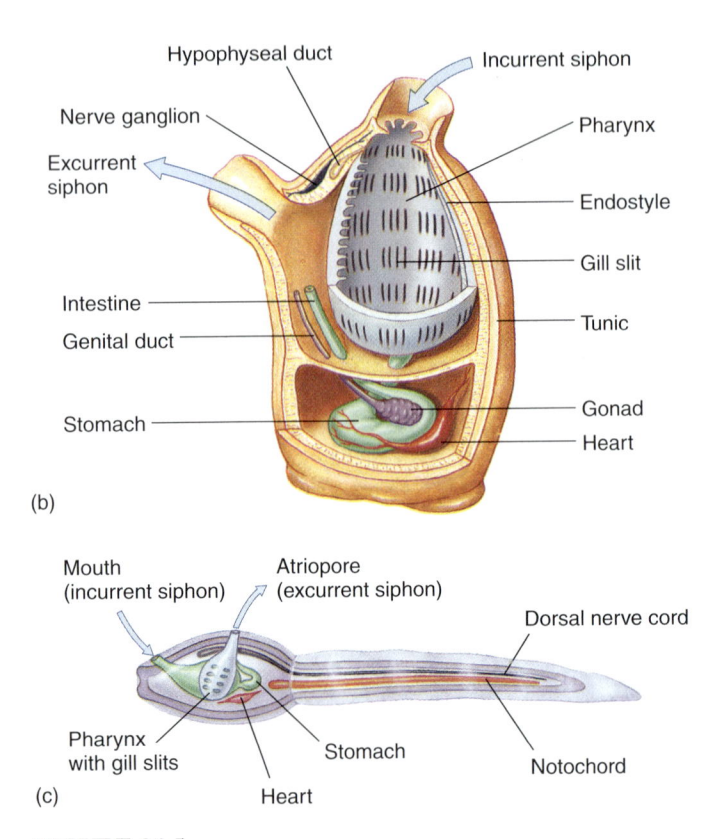

FIGURE 34.5
Tunicates (phylum Chordata, subphylum Urochordata). (*a*) The sea peach, *Halocynthia auranthium*. (*b*) Diagram of the structure of an adult tunicate. (*c*) Diagram of the structure of a larval tunicate, showing the characteristic tadpole-like form. Larval tunicates resemble the postulated common ancestor of the chordates.

686   Part V   Diversity of Life on Earth

cates may have a common sac and a common opening to the outside. A group of Urochordates, the Larvacea, retains the tail and notochord into adulthood. One theory of vertebrate origins involves a larval form, perhaps that of a tunicate, which acquires the ability to reproduce.

Lancelets

Lancelets are scaleless, fishlike marine chordates a few centimeters long that occur widely in shallow water throughout the oceans of the world. Lancelets (subphylum Cephalochordata) were given their English name because they resemble a lancet—a small, two-edged surgical knife. There are about 23 species of this subphylum. Most of them belong to the genus *Branchiostoma*, formerly called *Amphioxus*, a name still used widely. In lancelets, the notochord runs the entire length of the dorsal nerve cord and persists throughout the animal's life.

Lancelets spend most of their time partly buried in sandy or muddy substrates, with only their anterior ends protruding (figure 34.6). They can swim, although they rarely do so. Their muscles can easily be seen as a series of discrete blocks. Lancelets have many more pharyngeal gill slits than fishes, which they resemble in overall shape. They lack pigment in their skin, which has only a single layer of cells, unlike the multilayered skin of vertebrates. The lancelet body is pointed at both ends. There is no distinguishable head or sensory structures other than pigmented light receptors.

Lancelets feed on microscopic plankton, using a current created by beating cilia that line the oral hood, pharynx, and gill slits (figure 34.7). The gill slits provide an exit for the water and are an adaptation for filter feeding. The oral hood projects beyond the mouth and bears sensory tentacles, which also ring the mouth. Males and females are separate, but no obvious external differences exist between them.

Biologists are not sure whether lancelets are primitive or are actually degenerate fishes whose structural features have been reduced and simplified during the course of evolution. The fact that lancelets feed by means of cilia and have a single-layered skin, coupled with distinctive features of their excretory systems, suggests that this is an ancient group of chordates. The recent discovery of fossil forms similar to living lancelets in rocks 550 million years old—well before the appearance of any fishes—also argues for the antiquity of this group. Recent studies by molecular systematists further support the hypothesis that lancelets are vertebrates' closest ancestors.

FIGURE 34.6
Lancelets. Two lancelets, *Branchiostoma lanceolatum* (phylum Chordata, subphylum Cephalochordata), partly buried in shell gravel, with their anterior ends protruding. The muscle segments are clearly visible; the square objects along the side of the body are gonads.

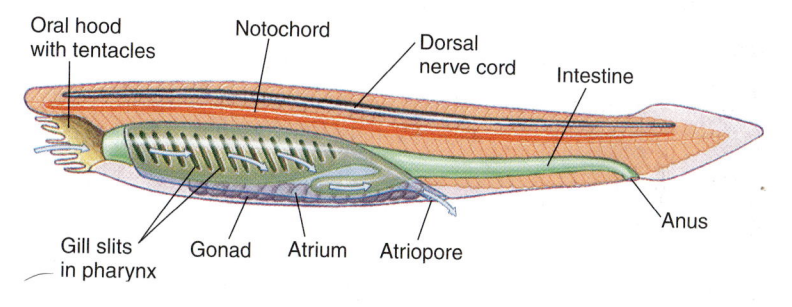

FIGURE 34.7
The structure of a lancelet. This diagram shows the path through which the lancelet's cilia pull water.

> Nonvertebrate chordates, including tunicates and lancelets, have notochords but no vertebrae. They are the closest relatives of vertebrates.

34.3 The evolution of vertebrates involved invasions of sea, land, and air.

Characteristics of Vertebrates

Vertebrates (subphylum Vertebrata) are chordates with a spinal column. The name *vertebrate* comes from the individual bony or cartilaginous segments called vertebrae that make up the spine. Vertebrates differ from the tunicates and lancelets in two important respects:

1. **Vertebral column.** In all vertebrates except the earliest fishes, the notochord is replaced during the course of embryonic development by a vertebral column (figure 34.8). The column is a series of bony or cartilaginous vertebrae that enclose and protect the dorsal nerve cord like a sleeve.
2. **Head.** Vertebrates have a distinct and well-differentiated head; the brain is fully encased within a protective box, the skull or cranium, made of bone or cartilage. For this reason, the vertebrates are sometimes called the **craniate chordates.**

In addition to these two key characteristics, vertebrates differ from other chordates in other important respects (figure 34.9):

1. **Neural crest.** A unique group of embryonic cells called the neural crest contributes to the development of many vertebrate structures. These cells develop on the crest of the neural tube as it forms by invagination and pinching together of the neural plate (see chapter 51 for a detailed account). Neural crest cells then migrate to various locations in the developing embryo, where they participate in the development of a variety of structures.
2. **Internal organs.** Internal organs characteristic of vertebrates include a liver, kidneys, and endocrine glands. The ductless endocrine glands secrete hormones that help regulate many of the body's functions. All vertebrates have a heart and a closed circulatory system. In both their circulatory and their excretory functions, vertebrates differ markedly from other animals.
3. **Endoskeleton.** The endoskeleton of most vertebrates is made of cartilage or bone. Cartilage and bone are specialized tissue containing fibers of the protein collagen compacted together. Bone also contains crystals of a calcium phosphate salt. Bone forms in two stages. First, collagen is laid down in a matrix of fibers along stress lines to provide flexibility, and then calcium minerals infiltrate the fibers, providing rigidity. The great advantage of bone over chitin as a structural material is that bone is strong without being brittle. The vertebrate endoskeleton makes possible the great size and extraordinary powers of movement that characterize this group.

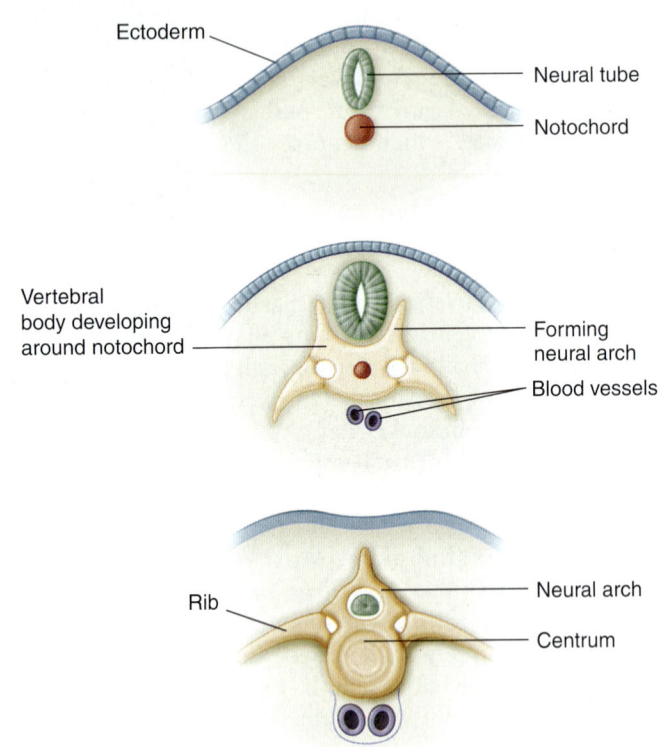

FIGURE 34.8
Embryonic development of a vertebra. During the evolution of animal development, the flexible notochord is surrounded and eventually replaced by a cartilaginous or bony covering, the centrum. The neural tube is protected by an arch above the centrum, and the vertebra may also have a hemal arch, which protects major blood vessels below the centrum. The vertebral column functions as a strong, flexible rod that the muscles pull against when the animal swims or moves.

Overview of the Evolution of Vertebrates

The first vertebrates evolved in the oceans about 470 million years ago. They were jawless fishes with a single caudal fin. Many of them looked like a flat hot dog, with a hole at one end and a fin at the other. The appearance of a hinged jaw was a major advancement, opening up new food options, and jawed fishes became the dominant creatures in the sea. Their descendants, the amphibians, invaded the land. Salamander-like amphibians and other, much larger, now-extinct amphibians were the first vertebrates to live successfully on land. Amphibians, in turn, gave rise to the first reptiles about 300 million years ago. Within 50 million years, reptiles, better suited than amphibians to living out of water, replaced them as the dominant land vertebrates.

With the success of reptiles, vertebrates truly came to dominate the surface of the earth. Many kinds of reptiles evolved, ranging in size from smaller than a chicken to

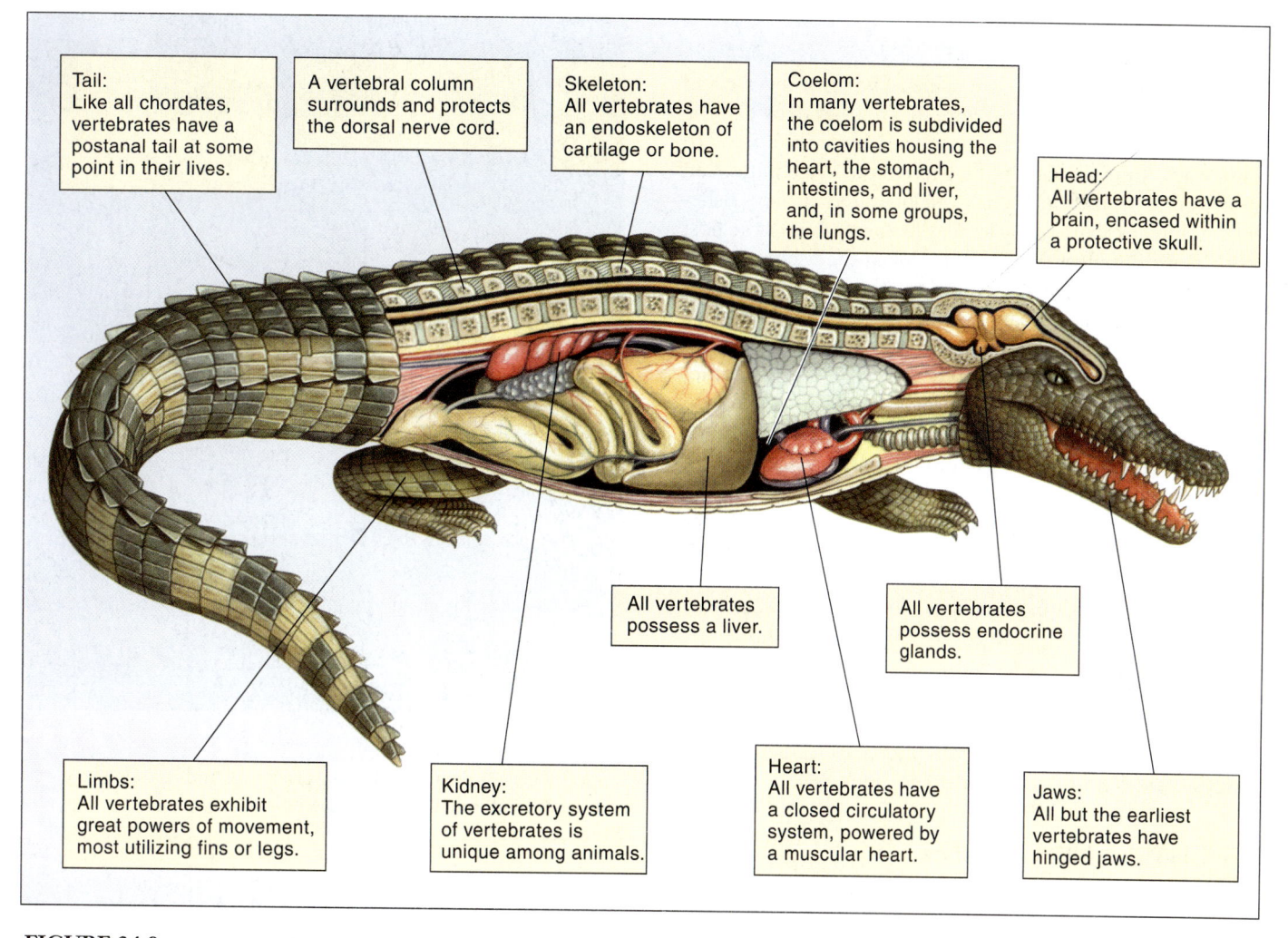

FIGURE 34.9
Major characteristics of vertebrates. Adult vertebrates are characterized by an internal skeleton of cartilage or bone, including a vertebral column and a skull. Several other internal and external features are characteristic of vertebrates.

- Tail: Like all chordates, vertebrates have a postanal tail at some point in their lives.
- A vertebral column surrounds and protects the dorsal nerve cord.
- Skeleton: All vertebrates have an endoskeleton of cartilage or bone.
- Coelom: In many vertebrates, the coelom is subdivided into cavities housing the heart, the stomach, intestines, and liver, and, in some groups, the lungs.
- Head: All vertebrates have a brain, encased within a protective skull.
- All vertebrates possess a liver.
- All vertebrates possess endocrine glands.
- Limbs: All vertebrates exhibit great powers of movement, most utilizing fins or legs.
- Kidney: The excretory system of vertebrates is unique among animals.
- Heart: All vertebrates have a closed circulatory system, powered by a muscular heart.
- Jaws: All but the earliest vertebrates have hinged jaws.

bigger than a truck. Some flew, and others swam. Among them evolved reptiles that gave rise to the two remaining great lines of terrestrial vertebrates, birds (descendants of the dinosaurs) and mammals. Dinosaurs and mammals appear at about the same time in the fossil record, 220 million years ago. For over 150 million years, dinosaurs dominated the face of the earth. Over all these centuries (think of it—over *a million centuries!*), the largest mammal was no bigger than a cat. Then, about 65 million years ago, the dinosaurs abruptly disappeared, for reasons that are still hotly debated. In their absence, mammals and birds quickly took their place, becoming in turn abundant and diverse.

The history of vertebrates has been a series of evolutionary advances that have allowed vertebrates to first invade the sea and then the land. In this chapter, we examine the key evolutionary advances that permitted vertebrates to invade the land successfully. As you will see, this invasion was a staggering evolutionary achievement, involving fundamental changes in many body systems.

Vertebrates are a diverse group, containing members adapted to life in aquatic habitats, on land, and in the air. There are nine principal classes of living vertebrates. Five of the classes are fishes that live in the water, and four are land-dwelling **tetrapods,** animals with four limbs. (The name tetrapod comes from two Greek words meaning "four-footed.") The extant classes of fishes are (1) the class Myxini, the hagfish, and (2) the class Cephalaspidomorphi, the lampreys, which together comprise the superclass Agnatha (the jawless fishes); (3) the class Chondrichthyes, the cartilaginous fishes, sharks, skates, and rays; and (4) class Actinopterygii and (5) class Sarcopterygii, comprising the bony fishes that are dominant today. The four classes of tetrapods are Amphibia, the amphibians; Reptilia, the reptiles; Aves, the birds; and Mammalia, the mammals.

Vertebrates, the principal chordate group, are characterized by a vertebral column and a distinct head.

Chapter 34 Vertebrates 689

Fishes

Over half of all vertebrates are fishes. The most diverse and successful vertebrate group (figure 34.10), they provided the evolutionary base for invasion of land by amphibians. In many ways, amphibians, the first terrestrial vertebrates, can be viewed as transitional—fish out of water. In fact, fishes and amphibians share many similar features, among the host of obvious differences. First, let us look at the fishes.

The story of vertebrate evolution started in the ancient seas of the Cambrian period (545 to 490 million years ago), when the first backboned animals appeared. Figure 34.11 shows the key vertebrate characteristics that evolved subsequently. Wriggling through the water, jawless and toothless, the first fishes sucked up small food particles from the ocean floor like miniature vacuum cleaners. Most were less than a foot long, respired with gills, and had no paired fins or vertebrae (although some had rudimentary vertebrae); they had a head and a primitive tail to push them through the water. For 50 million years, during the Ordovician period (490 to 438 million years ago), these simple fishes were the only vertebrates. By the end of this period, fish had developed primitive fins to help them swim and massive shields of bone for protection. Jawed fishes first appeared during the Silurian period (438 to 408 million years ago), and along with them came a new mode of feeding. Later, both the cartilaginous and bony fishes appeared.

FIGURE 34.10
Fish. Fish are the most diverse vertebrates and include more species than all other kinds of vertebrates combined.

Key Characteristics of Fishes

From whale sharks 18 meters long to tiny cichlids no larger than your fingernail, fishes vary considerably in size, shape, color, and appearance. Some live in freezing arctic seas, others in warm, freshwater lakes, and still others spend a lot of time out of water entirely. However varied, all fishes have important characteristics in common:

1. **Vertebral column.** Fish have an internal skeleton with a bony or cartilaginous spine surrounding the dorsal nerve cord, and a bony or cartilaginous skull encasing the brain. An exception is the agnathans, composed of the hagfish and lampreys. In hagfish, a cartilaginous skeleton and skull are present, but vertebrae are not; the notochord persists and provides support. In lampreys, a cartilaginous skeleton and notochord are present, but rudimentary cartilaginous vertebrae also surround the notochord in places.
2. **Jaws and paired appendages.** Fishes other than agnathans all have jaws and paired appendages, features that are seen in all tetrapods too (figure 34.11). Jaws allowed the predation of larger and active prey. Bony fishes and cartilaginous fishes have two pairs of fins—a pair of pectoral fins at the shoulder and a pair of pelvic fins at the hip. Paired appendages greatly improved maneuverability and are the precursors of arms and legs in tetrapods. In the lobe-finned fish, these pairs of fins became jointed.
3. **Gills.** Fishes are water-dwelling creatures and must extract oxygen dissolved in the water around them. They do this by directing a flow of water through their mouths and across their gills. The gills are composed of fine filaments of tissue that are rich in blood vessels. They are located at the back of the pharynx and supported by arches of cartilage. Blood moves through the gills in the opposite direction to the flow of water in order to maximize the efficiency of oxygen absorption (see chapter 44).
4. **Single-loop blood circulation.** Blood is pumped from the heart to the gills. From the gills, the oxygenated blood passes to the rest of the body, and then returns to the heart. The heart is a muscular tube-pump made of four chambers that contract in sequence.
5. **Nutritional deficiencies.** Fishes are unable to synthesize the aromatic amino acids and must consume them in their diet. This inability has been inherited by all their vertebrate descendants.

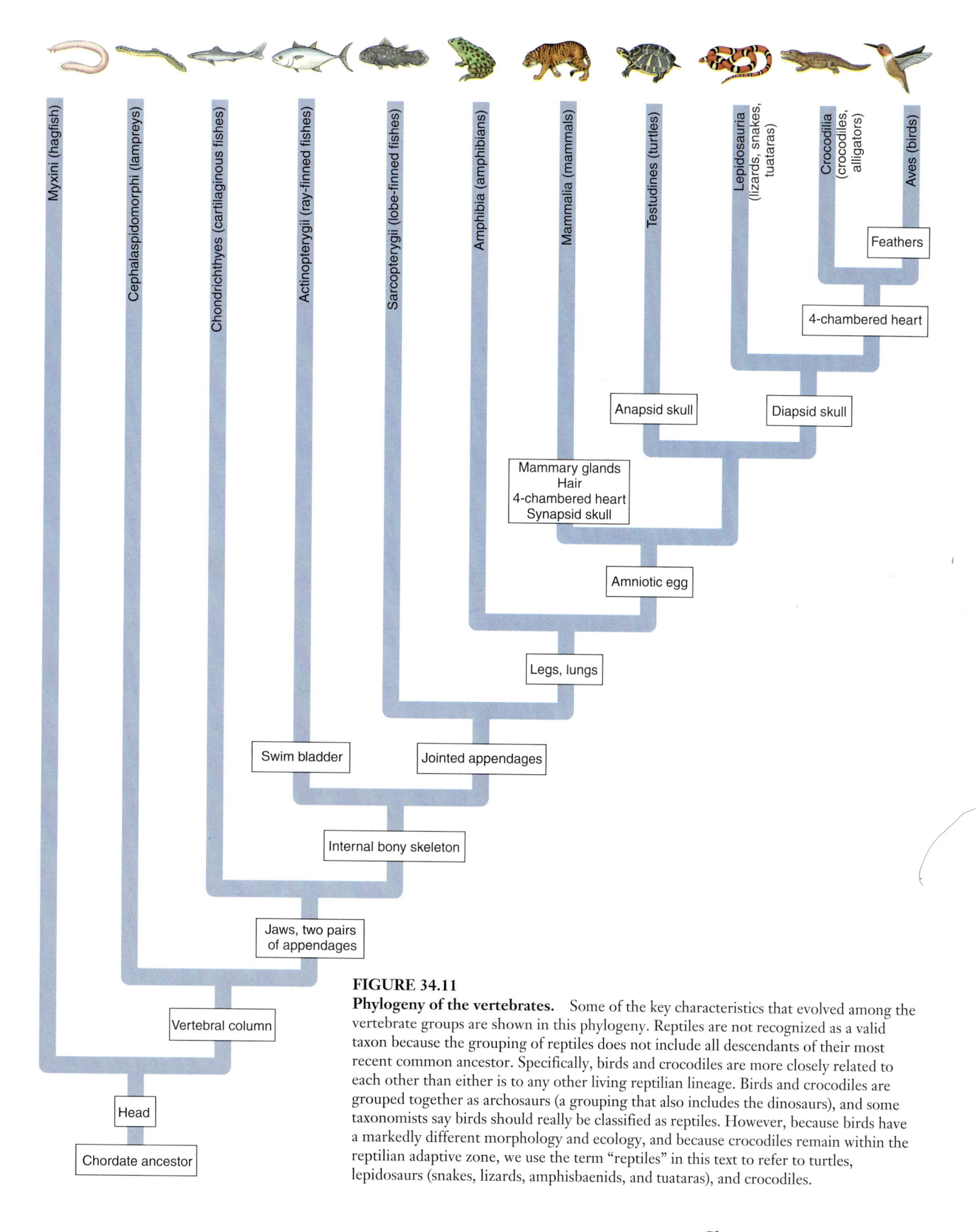

FIGURE 34.11
Phylogeny of the vertebrates. Some of the key characteristics that evolved among the vertebrate groups are shown in this phylogeny. Reptiles are not recognized as a valid taxon because the grouping of reptiles does not include all descendants of their most recent common ancestor. Specifically, birds and crocodiles are more closely related to each other than either is to any other living reptilian lineage. Birds and crocodiles are grouped together as archosaurs (a grouping that also includes the dinosaurs), and some taxonomists say birds should really be classified as reptiles. However, because birds have a markedly different morphology and ecology, and because crocodiles remain within the reptilian adaptive zone, we use the term "reptiles" in this text to refer to turtles, lepidosaurs (snakes, lizards, amphisbaenids, and tuataras), and crocodiles.

Chapter 34 Vertebrates 691

The First Fishes

The first fishes were members of the five orders of ostracoderms (a word meaning "shell-skinned"). Only their headshields were made of bone; their elaborate internal skeletons were constructed of cartilage. Many ostracoderms were bottom-dwellers, with a jawless mouth underneath a flat head, and eyes on the upper surface. Ostracoderms thrived in the Ordovician period (figure 34.12) and in the period that followed, the Silurian period (438–408 million years ago), only to become almost completely extinct at the close of the Devonian period (408–360 million years ago). One group, the jawless Agnatha, survive today as hagfish (class Myxini; table 34.1) and parasitic lampreys (class Cephalaspidomorphi).

Evolution of the Jaw

A fundamentally important evolutionary advance that occurred in the late Silurian period, 410 million years ago, was the development of jaws. Jaws evolved from the most anterior of a series of arch-supports made of cartilage that

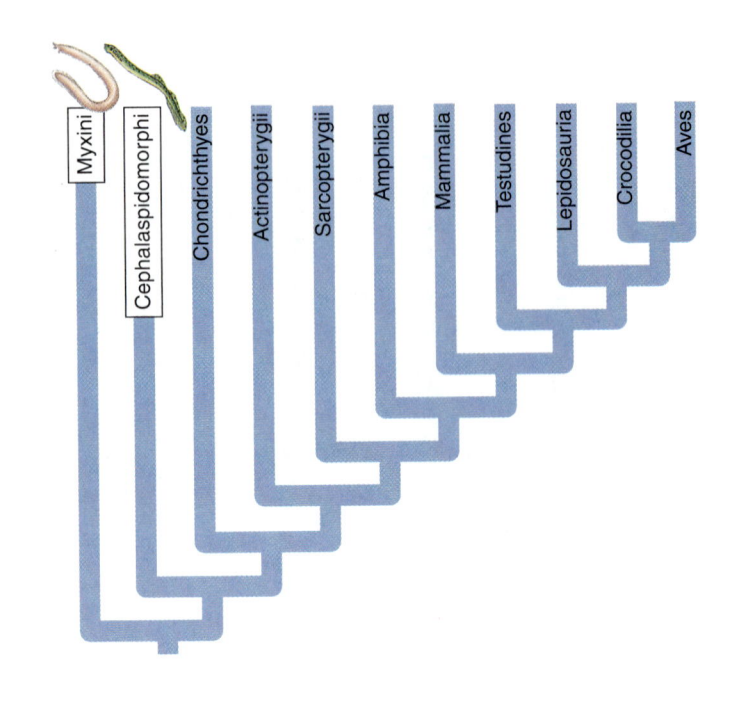

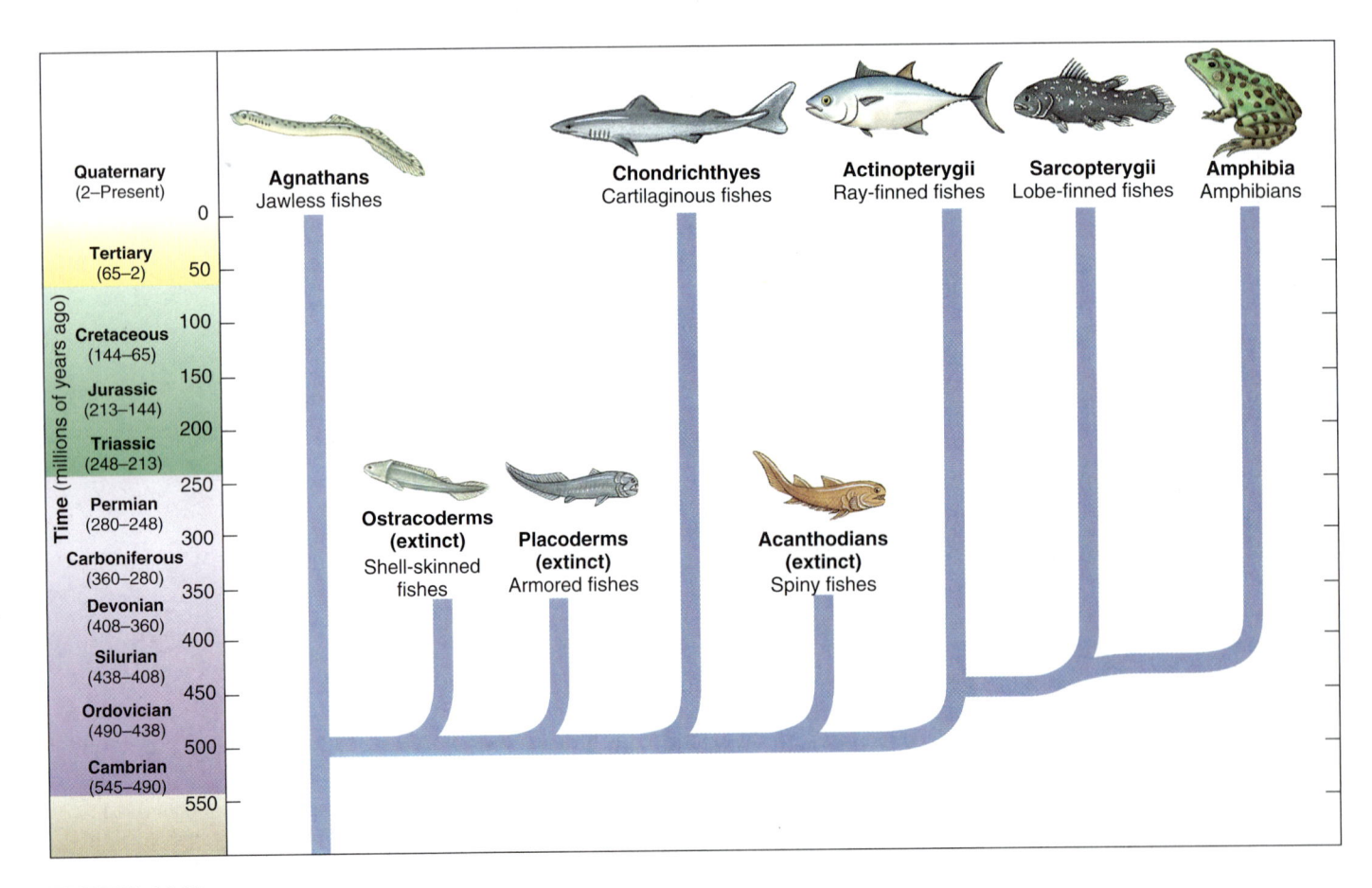

FIGURE 34.12
Evolution of the fishes. The evolutionary relationships among the different groups of fishes as well as between fishes and amphibians are shown. The spiny and armored fishes that dominated the early seas are now extinct.

692   Part V   Diversity of Life on Earth

Table 34.1 Major Classes of Fishes

Class	Typical Examples	Key Characteristics	Approximate Number of Living Species
Actinopterygii	Ray-finned fishes	Most diverse group of vertebrates; swim bladders and bony skeletons; paired fins supported by bony rays	30,000
Sarcopterygii	Lobe-finned fishes	Largely extinct group of bony fishes; ancestral to amphibians; paired lobed fins	7
Chondrichthyes	Sharks, skates, rays	Streamlined hunters; cartilaginous skeletons; no swim bladders; internal fertilization	750
Myxini	Hagfishes	Jawless fishes with no paired appendages; scavengers; mostly blind, but having a well-developed sense of smell	30
Cephalaspidomorphi	Lampreys	Largely extinct group of jawless fishes with no paired appendages; parasitic and nonparasitic types; all breed in fresh water	35
Placodermi	Armored fishes	Jawed fishes with heavily armored heads; many were quite large	Extinct
Acanthodii	Spiny fishes	Fishes with jaws; all now extinct; paired fins supported by sharp spines	Extinct

were used to reinforce the tissue between gill slits, holding the slits open (figure 34.13). This transformation was not as radical as it might at first appear. Each gill arch was formed by a series of several cartilages (later to become bones) arranged somewhat in the shape of a V turned on its side, with the point directed outward. Imagine the fusion of the front pair of arches at top and bottom, with hinges at the points, and you have the primitive vertebrate jaw. The top half of the jaw is not attached to the skull directly except at the rear. Teeth developed on the jaws from modified scales on the skin that lined the mouth.

Armored fishes called placoderms and spiny fishes called acanthodians both had jaws. Spiny fishes were very common during the early Devonian period, largely replacing ostracoderms, but became extinct themselves at the close of the Permian. Like ostracoderms, they had internal skeletons made of cartilage, but their scales contained small plates of bone, foreshadowing the much larger role bone would play in the future of vertebrates. Spiny fishes were predators and far better swimmers than ostracoderms, with as many as seven fins to aid their swimming. All of these fins were reinforced with strong spines, giving these fishes their name. No spiny fishes survive today.

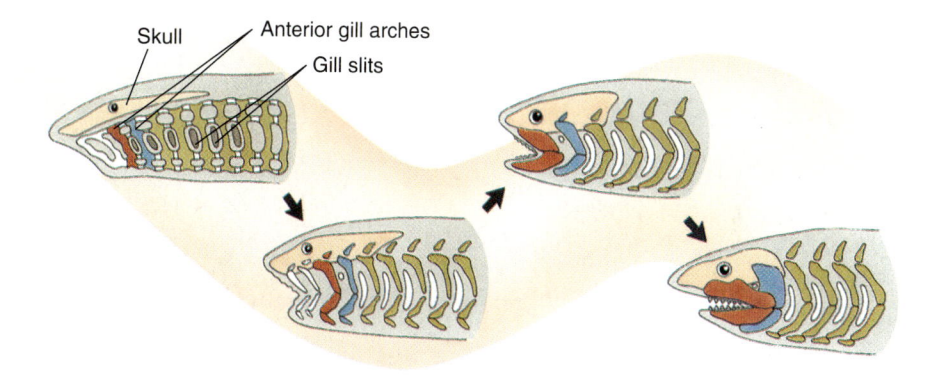

FIGURE 34.13
Evolution of the jaw. Jaws evolved from the anterior gill arches of ancient, jawless fishes.

By the mid-Devonian period, the heavily armored placoderms became common. A very diverse and successful group, seven orders of placoderms dominated the seas of the late Devonian, only to become extinct at the end of that period. The front of the placoderm body was more heavily armored than the rear. The placoderm jaw was much improved over the primitive jaw of spiny fishes, with the upper jaw fused to the skull and the skull hinged on the shoulder. Many of the placoderms grew to enormous sizes, some over 30 feet long, with two-foot skulls that had an enormous bite.

GEOLOGIC TIME SCALE	ERAS	PERIOD	EPOCH	MILLIONS OF YEARS AGO	BIOLOGICAL EVENTS
Cenozoic	Cenozoic	Quaternary	Recent	(.01-0)	Human civilization
Mesozoic			Pleistocene	(1.8-.01)	Ice ages; modern humans (genus *Homo*) appear
Paleozoic		Tertiary	Pliocene	(5-1.8)	Hominids appear; large carnivores
			Miocene	(23-5)	Apes arise; climate cooler; plains and grasslands
			Oligocene	(36-23)	Monkeys appear; mild climate
			Eocene	(57-36)	Radiation of flowering plants; most modern mammalian orders represented
			Paleocene	(65-57)	Radiation of mammals, birds, and insects; tropical conditions
		MASS EXTINCTION			
	Mesozoic	Cretaceous		(144-65)	Flowering plants appear; climax of dinosaurs followed by extinction
		Jurassic		(213-144)	First birds; dinosaurs radiate; lush forests
		MASS EXTINCTION			
		Triassic		(248-213)	Conifers dominant; dinosaurs arise; first mammals
	MASS EXTINCTION (96% OF SPECIES DISAPPEAR)				
Precambrian	Paleozoic	Permian		(280-248)	Radiation of reptiles; mammal-like reptiles arise
		Carboniferous		(360-280)	Forest of ferns and conifers; amphibians arise; insects radiate
		MASS EXTINCTION			
		Devonian		(408-360)	Sharks and bony fishes appear; lobe-finned fishes evolve
		Silurian		(438-408)	Jawless fishes diversify; first jawed fishes; colonization of land by plants and arthropods
		MASS EXTINCTION			
		Ordovician		(490-438)	First vertebrates (jawless fishes); first plants
		Cambrian		(545-490)	Cambrian explosion: nearly all the major animal body plans arise; earliest chordates
	Precambrian			600	Soft-bodied invertebrates and algae diversify
				900	Multicellular organisms appear
				1500	Oldest definite fossils of eukaryotes
				2500	Oldest definite fossils of prokaryotes
				4500	Formation of the Earth

FIGURE 34.14
The geological time scale. Vertebrates appear soon after the Cambrian period, and have played a major role in life's diversity ever since.

The Rise of Active Swimmers

At the end of the Devonian period, essentially all of these pioneer vertebrates disappeared, replaced by sharks and bony fishes in one of several mass extinctions that occurred during earth's geological and biological history (figure 34.14). Sharks and bony fishes first evolved in the early Devonian, 400 million years ago. In these fishes, the jaw was improved even further, with the first gill arch behind the jaws being transformed into a supporting strut or prop, joining the rear of the lower jaw to the rear of the skull. This allowed the mouth to open very wide, into almost a full circle. In a great white shark, this wide-open mouth can be a very efficient weapon.

The major factor responsible for the replacement of primitive fishes by sharks and bony fishes was that they had a superior design for swimming. The typical shark or bony fish is streamlined. The head of the fish acts as a wedge to cleave through the water, and the body tapers back to the tail, allowing the fish to slip through the water with a minimum amount of turbulence.

In addition, sharks and bony fishes have an array of mobile fins that greatly aid swimming. First, a propulsion fin, a large and efficient tail (caudal) fin, helps drive the fish through the water when it is swept side-to-side, pushing against the water and thrusting the fish forward. Second, stabilizing fins, one (or sometimes two) dorsal fins on the back, act as a stabilizer to prevent rolling as the fish swims through the water, while another ventral fin acts as a keel to prevent side-slip. Third, paired fins at shoulder and hip ("a fin at each corner") consist of a front (pectoral) pair and a rear (pelvic) pair. These fins act like the elevator flaps of an airplane to assist the fish in going up or down through the water, as rudders to help it turn sharply left or right, and as brakes to help it stop quickly.

Sharks Become Top Predators

In the period following the Devonian, the Carboniferous period (360 to 280 million years ago), sharks became the dominant predator in the sea. Sharks (class Chondrichythes) have a skeleton made of cartilage, like primitive fishes, but it is "calcified," strengthened by granules of calcium carbonate deposited in the outer layers of cartilage. The result is a very light and strong skeleton. Streamlined, with paired fins and a light, flexible skeleton, sharks are superior swimmers (figure 34.15). Their pectoral fins are particularly large, jutting out stiffly like airplane wings—and that is how they function, adding lift to compensate for the downward thrust of the tail fin. Sharks are very aggressive predators, and some early sharks reached enormous size.

Sharks were among the first vertebrates to develop teeth. These teeth evolved from rough scales on the skin and are not set into the jaw as human teeth are, but rather sit atop it. The teeth are not firmly anchored and are easily lost. In a shark's mouth, the teeth are arrayed in up to 20 rows; the teeth in front do the biting and cutting, while behind them other teeth grow and wait their turn. When a tooth breaks or is worn down, a replacement from the next row moves forward. One shark may eventually use more than 20,000 teeth. This programmed loss of teeth offers a great advantage: The teeth in use are always new and sharp. A shark's skin is covered with tiny, teethlike scales, giving it a rough "sandpaper" texture. Like the teeth, these scales are constantly replaced throughout the shark's life.

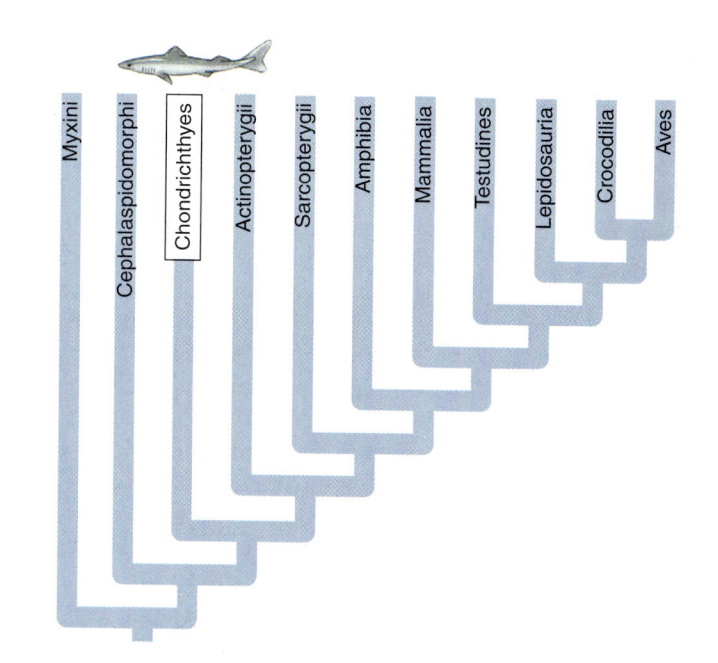

FIGURE 34.15 Chondrichthyes. Members of the class Chondrichthyes, such as this blue shark, are mainly predators or scavengers.

Reproduction among the class Chondrichythes is the most advanced of any fishes. Shark eggs are fertilized internally. During mating, the male grasps the female with modified fins called claspers. Sperm run from the male into the female through grooves in the claspers. Although a few species lay fertilized eggs, the eggs of most species develop within the female's body, and the pups are born alive.

Many of the early evolutionary lines of sharks died out during the great extinction at the end of the Permian period (280 to 248 million years ago). The survivors thrived and underwent a burst of diversification during the Mesozoic era, when most of the modern groups of sharks appeared. Skates and rays (flattened sharks that are bottom-dwellers) evolved at this time, some 200 million years after the sharks first appeared. Sharks competed successfully with the marine reptiles of that time and are still the dominant predators of the sea. Today, there are 275 species of sharks, more kinds than existed in the Carboniferous period.

Chapter 34 Vertebrates 695

Bony Fishes Dominate the Water

Bony fishes evolved at the same time as sharks, some 400 million years ago, but took quite a different evolutionary road. Instead of gaining speed through lightness, as sharks did, bony fishes adopted a heavy internal skeleton made completely of bone. Such an internal skeleton is very strong, providing a base against which very strong muscles can pull. The process of *ossification* (the evolutionary replacement of cartilage by bone) happened suddenly in evolutionary terms, completing a process started by sharks, who lay down a thin film of bone over their cartilage. Not only is the internal skeleton ossified, but so is the external skeleton, the outer covering of plates and scales. Many scientists believe bony fishes evolved from spiny sharks, which also had bony plates set in their skin. Bony fishes are the most successful of all fishes, indeed of all vertebrates. There are several dozen orders containing more than 30,000 living species.

Unlike sharks, bony fishes evolved in fresh water. The most ancient fossils of bony fishes are found in freshwater lake beds from the middle Devonian period. These first bony fishes were small and possessed paired air sacs connected to the back of the throat. These sacs could be inflated with air to buoy the fish up or deflated to sink it down.

Most bony fishes have highly mobile fins, very thin scales, and completely symmetrical tails (which keep the fish on a straight course as it swims through the water). This is a very successful design for a fish. Two great groups arose from these pioneers: the lobe-finned fishes (class Sarcopterygii), ancestors of the first tetrapods, and the ray-finned fishes (class Actinopterygii), which include the vast majority of today's fishes (figure 34.16).

The characteristic feature of all ray-finned fishes is an internal skeleton of parallel bony rays that support and stiffen each fin. There are no muscles within the fins; they are moved by muscles within the body. In ray-finned fishes, the primitive air sacs are transformed into an air pouch, which provides a remarkable degree of control over buoyancy.

Important Adaptations of Bony Fishes

The remarkable success of the bony fishes has resulted from a series of significant adaptations that have enabled them to dominate life in the water. These include the swim bladder, lateral line system, and gill cover.

Swim Bladder. Although bones are heavier than cartilaginous skeletons, bony fishes are still buoyant because they possess a swim bladder, a gas-filled sac that allows them to regulate their buoyant density and so remain suspended at any depth in the water effortlessly (figure 34.17). Sharks, by contrast, must move through the water or sink, because their bodies are denser than water. In primitive bony fishes, the swim bladder is a dorsal outpocketing of the pharynx behind the throat, and these species fill the swim bladder by simply gulping air at the surface of the water. In most of today's bony fishes, the swim bladder is an independent organ that is filled and drained of gases, mostly nitrogen and oxygen, internally. How do bony fishes manage this remarkable trick? It turns out that the gases are harvested from their blood by a unique gland that discharges the gases into the bladder when more buoyancy is required. To reduce buoyancy, gas is released by a muscular valve from the bladder back into the blood supply. A variety of physiological factors control the exchange of gases between the blood stream and the swim bladder.

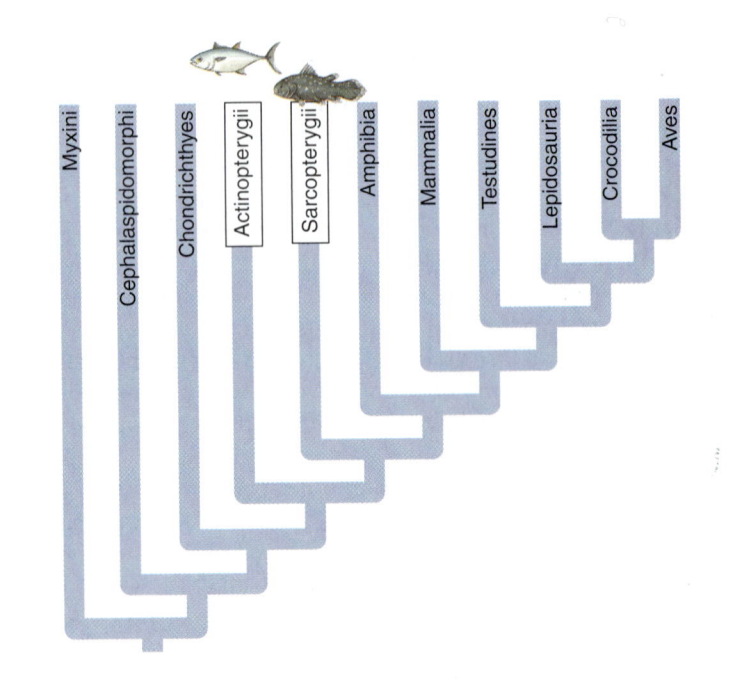

Lateral Line System. Although precursors are found in sharks, bony fishes possess a fully developed lateral line system. The lateral line system consists of a series of sensory organs that project into a canal beneath the surface of the skin. The canal runs the length of the fish's body and is open to the exterior through a series of sunken pits. Movement of water past the fish forces water through the canal. The sensory organs consist of clusters of cells with

FIGURE 34.16 Ray-finned fishes. The ray-finned fishes (class Actinopterygii) are extremely diverse. This Korean angelfish in Fiji is one of the many striking fishes that live around coral reefs in tropical seas.

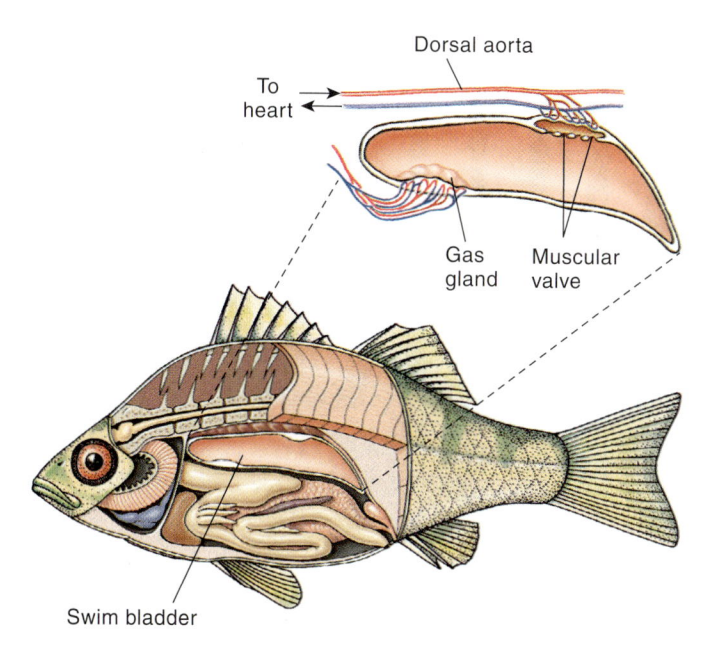

FIGURE 34.17
Diagram of a swim bladder. The bony fishes use this structure, which evolved as a dorsal outpocketing of the pharynx, to control their buoyancy in water. The swim bladder can be filled with or drained of gas to allow the fish to control buoyancy. Gases are taken from the blood, and the gas gland secretes the gases into the swim bladder; gas is released from the bladder by a muscular valve.

FIGURE 34.18
The living coelacanth, *Latimeria chalumnae*, a lobe-finned fish (class Sarcopterygii). Discovered in the western Indian Ocean in 1938, this coelacanth represents a group of fishes thought to have been extinct for about 70 million years. Scientists who studied living individuals in their natural habitat at depths of 100 to 200 meters observed them drifting in the current and hunting other fishes at night. Some individuals are nearly 3 meters long; they have a slender, fat-filled swim bladder. *Latimeria* is a strange animal, and its discovery was a complete surprise.

hairlike projections called cilia, embedded in a gelatinous cap. The hairs are deflected by the slightest movement of water over them. The pits are oriented so that some are stimulated no matter what direction the water moves (see chapter 46). Nerve impulses from these sensory organs permit the fish to assess its rate of movement through water, sensing the movement as pressure waves against its lateral line. This is how a trout orients itself with its head upstream.

The lateral line system also enables a fish to detect motionless objects at a distance by the movement of water reflected off the object. In a very real sense, this is the fish equivalent of hearing. The basic mechanism of cilia deflection by pressure waves is very similar to what happens in human ears (see chapter 46).

Gill Cover. Most bony fishes have a hard plate called the operculum that covers the gills on each side of the head. Flexing the operculum permits bony fishes to pump water over their gills. The gills are suspended in the pharyngeal slits that form a passageway between the pharynx and the outside of the fish's body. When the operculum is closed, it seals off the exit. When the mouth is open, closing the operculum increases the volume of the mouth cavity, so that water is drawn into the mouth. When the mouth is closed, opening the operculum decreases the volume of the mouth cavity, forcing water past the gills to the outside. Using this very efficient bellows, bony fishes can pass water over the gills while stationary in the water. That is what a goldfish is doing when it seems to be gulping in a fish tank.

The Path to Land

Lobe-finned fishes (class Sarcopterygii) evolved 390 million years ago, shortly after the first bony fishes appeared. Only seven species survive today, a single species of coelacanth (figure 34.18) and six species of lungfish. Lobe-finned fishes have paired fins that consist of a long fleshy muscular lobe (hence their name), supported by a central core of bones that form fully articulated joints with one another. There are bony rays only at the tips of each lobed fin. Muscles within each lobe can move the fin rays independently of one another, a feat no ray-finned fish could match. Although rare today, lobe-finned fishes played an important part in the evolutionary story of vertebrates. Amphibians almost certainly evolved from the lobe-finned fishes.

> Fishes were the first vertebrates. Fishes are characterized by gills and a simple, single-loop circulatory system. Cartilaginous fishes, such as sharks, are fast swimmers, while the very successful bony fishes have unique characteristics such as swim bladders and lateral line systems.

Amphibians

Frogs, salamanders, and caecilians, the damp-skinned vertebrates, are direct descendants of fishes. They are the sole survivors of a very successful group, the amphibians, the first vertebrates to walk on land. Most present-day amphibians are small and live largely unnoticed by humans. Amphibians are among the most numerous of terrestrial animals; there are more species of amphibians than of mammals. Throughout the world, amphibians play key roles in terrestrial food chains.

Characteristics of Living Amphibians

Biologists have classified living species of amphibians into three orders (table 34.2): 4200 species of frogs and toads in 22 families make up the order Anura ("without a tail"); 500 species of salamanders and newts in 9 families make up the order Urodela, or Caudata ("visible tail"); and 150 species (6 families) of wormlike, nearly blind organisms called caecilians that live in the tropics make up the order Apoda, or Gymnophiona ("without legs"). These amphibians have several key characteristics in common:

1. **Legs.** Frogs and salamanders have four legs and can move about on land quite well. Legs were one of the key adaptations to life on land. Caecilians have lost their legs during the course of adapting to a burrowing existence.
2. **Cutaneous respiration.** Frogs, salamanders, and caecilians all supplement the use of lungs by respiring directly across their skin, which is kept moist and provides an extensive surface area. This mode of respiration is only efficient for a high surface-to-volume ratio in an animal.
3. **Lungs.** Most amphibians possess a pair of lungs, although the internal surfaces are poorly developed, with much less surface area than reptilian or mammalian lungs have. Amphibians still breathe by lowering the floor of the mouth to suck in air, and then raising it back to force the air down into the lungs (see chapter 44).
4. **Pulmonary veins.** After blood is pumped through the lungs, two large veins called pulmonary veins return the aerated blood to the heart for repumping. This allows the aerated blood to be pumped to the tissues at a much higher pressure than when it leaves the lungs.
5. **Partially divided heart.** The initial chamber of the fish heart is absent in amphibians, and the second and last chambers are separated by a dividing wall that helps prevent aerated blood from the lungs from mixing with nonaerated blood being returned to the heart from the rest of the body. This separates the blood circulation into two separate paths, pulmonary and systemic. The separation is imperfect; the third chamber has no dividing wall (see Chapter 44).

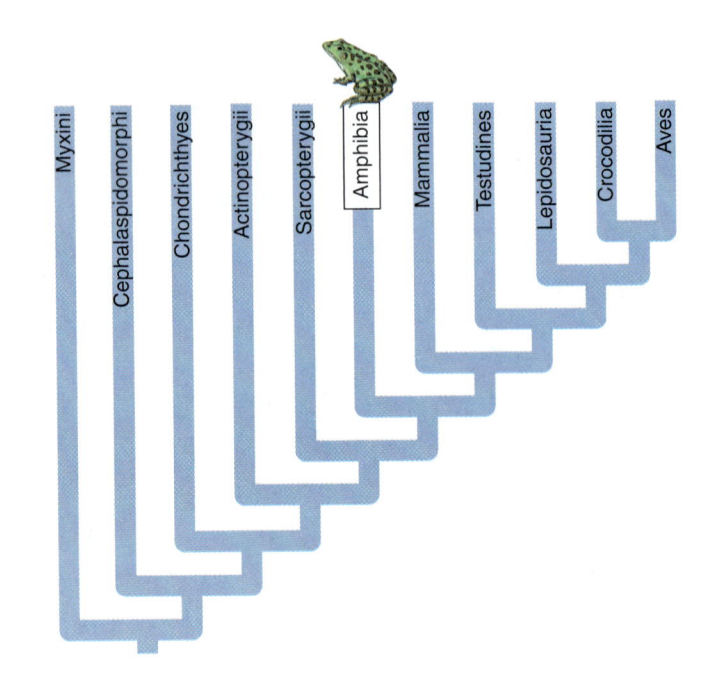

Several other specialized characteristics are shared by all present-day amphibians. In all three orders, there is a zone of weakness between the base and the crown of the teeth. They also have a peculiar type of sensory rod cell in the retina of the eye called a "green rod." The exact function of this rod is unknown.

Table 34.2 Orders of Amphibians

Order	Typical Examples	Key Characteristics	Approximate Number of Living Species
Anura	Frogs, toads	Compact, tailless body; large head fused to the trunk; rear limbs specialized for jumping	4200
Urodela (Caudata)	Salamanders, newts	Slender body; long tail and limbs set out at right angles to the body	500
Apoda (Gymnophiona)	Caecilians	Tropical group with a snakelike body; no limbs; little or no tail	150

698   Part V   Diversity of Life on Earth

Origin of Amphibians

The word *amphibia* (Greek, "both lives") nicely describes the essential quality of modern-day amphibians, reflecting their ability to live in two worlds—the aquatic world of their fish ancestors and the terrestrial world they first invaded. Here, we review the checkered history of this group, almost all of whose members have been extinct for the last 200 million years. Then we will examine in more detail what the few kinds of surviving amphibians are like.

Paleontologists (scientists who study fossils) agree that amphibians must have evolved from the lobe-finned fishes, although for some years these scientists have disagreed about whether the direct ancestors were coelacanths, lungfish, or the extinct rhipidistian fishes. Good arguments can be made for each. Many details of amphibian internal anatomy resemble those of the coelacanth. Lungfish and rhipidistians have openings in the tops of their mouths similar to the internal nostrils of amphibians. In addition, lungfish have paired lungs, like those of amphibians. Recent DNA analysis indicates lungfish are in fact far more closely related to amphibians than are coelacanths. Most paleontologists consider that amphibians evolved from rhipidistian fishes, rather than lungfish, because the patterns of bones in the early amphibian skull and limbs bear a remarkable resemblance to those seen in the rhipidistians. They also share a particular tooth structure.

The successful invasion of land by vertebrates involved a number of major adaptations:

1. Legs were necessary to support the body's weight as well as to allow movement from place to place (figure 34.19).
2. Lungs were necessary to extract oxygen from the air. Even though far more oxygen is available to gills in air than in water, the delicate structure of fish gills requires the buoyancy of water to support them, and they will not function in air.
3. The heart had to be redesigned to make full use of new respiratory systems and to deliver the greater amounts of oxygen required by walking muscles.
4. Reproduction had to be carried out in water until methods evolved to prevent eggs from drying out.
5. Most importantly, a system had to be developed to prevent the body itself from drying out.

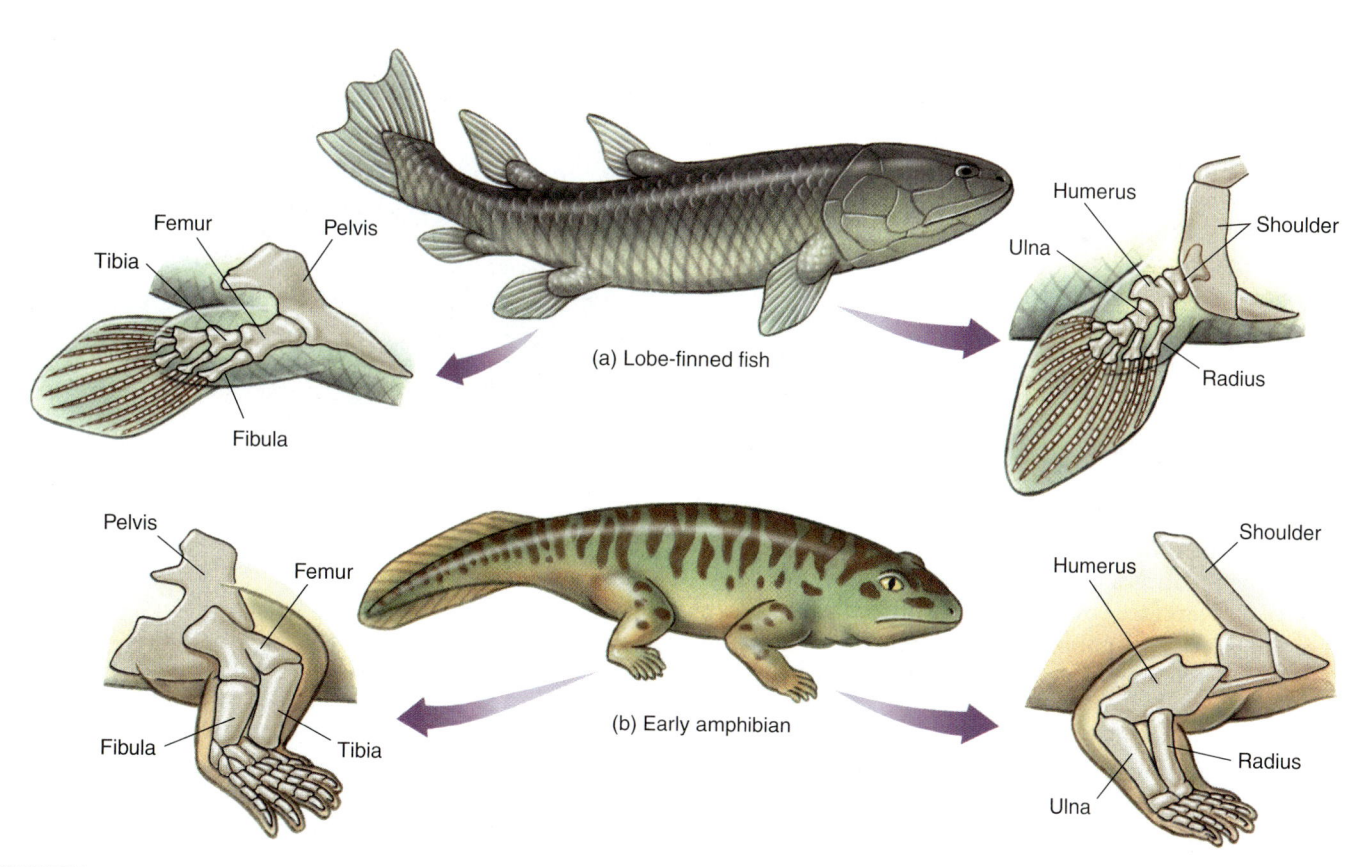

FIGURE 34.19
A comparison between the limbs of a lobe-finned fish and those of a primitive amphibian. (*a*) A lobe-finned fish. Some of these animals could probably move onto land. (*b*) A primitive amphibian. As illustrated by their skeletal structure, the legs of such an animal could clearly function on land much better than the fins of the lobe-finned fish.

Chapter 34 Vertebrates 699

The First Amphibian

Amphibians solved these problems only partially, but their solutions worked well enough that amphibians have survived for 350 million years. Evolution does not insist on perfect solutions, only workable ones.

Ichthyostega, the earliest amphibian fossil (figure 34.20), was found in a 370-million-year-old rock in Greenland. At that time, Greenland was part of the North American continent and lay near the equator. All amphibian fossils from the next 100 million years are found in North America. Only when Asia and the southern continents merged with North America to form the supercontinent Pangaea did amphibians spread throughout the world.

Ichthyostega was a strongly built animal, with four sturdy legs well supported by hip and shoulder bones. Unlike the bone structure of fish, the shoulder bones were no longer attached to the skull, and the hipbones were braced against the backbone, so the limbs could support the animal's weight. To strengthen the backbone further, long, broad ribs that overlap each other formed a solid cage for the lungs and heart. The rib cage was so solid that it probably couldn't expand and contract for breathing. Instead, *Ichthyostega* obtained oxygen somewhat as a frog does, by lowering the floor of the mouth to draw air in, and then raising it to push air down the windpipe into the lungs.

The Rise and Fall of Amphibians

Amphibians first became common during the Carboniferous period (360 to 280 million years ago). Fourteen families of amphibians are known from the early Carboniferous, nearly all of them aquatic or semi-aquatic, like *Ichthyostega*. By the late Carboniferous, much of North America was covered by low-lying tropical swamplands, and 34 families of amphibians thrived in this wet terrestrial environment, sharing it with pelycosaurs and other early reptiles. In the early Permian period that followed (280 to 248 million years ago), a remarkable change occurred among amphibians—they began to leave the marshes for dry uplands. Many of these terrestrial amphibians had bony plates and armor covering their bodies and grew to be very large, some as big as a pony (figure 34.21). Both their large size and the complete covering of their bodies indicate that these amphibians did not use the skin respiratory system of present-day amphibians, but rather had an impermeable leathery skin to prevent water loss. By the mid-Permian period, there were 40 families of amphibians. Only 25% of them were still semi-aquatic like *Ichthyostega*; 60% of the amphibians were fully terrestrial, and 15% were semiterrestrial. This was the peak of amphibian success, sometimes called the Age of Amphibians.

By the end of the Permian period, a reptile called a therapsid had become common, ousting the amphibians from their newly acquired niche on land. Following the mass extinction event at the end of the Permian, therapsids were the dominant land vertebrate, and most amphibians were aquatic. This trend continued in the following Triassic period (248 to 213 million years ago), which saw the virtual extinction of amphibians from land. By the end of the Triassic, there were only 15 families of amphibians (including the first frog), and almost without exception, they were aquatic. Some of these grew to great size; one was 3 meters long. Only two groups of amphibians are known from the subsequent Jurassic period (213 to 144 million years ago), the anurans (frogs and toads) and the urodeles (salamanders and newts). The Age of Amphibians was over.

FIGURE 34.20
Amphibians were the first vertebrates to walk on land. Reconstruction of *Ichthyostega*, one of the first amphibians with efficient limbs for crawling on land, had an improved olfactory sense associated with a lengthened snout and a relatively advanced ear structure for picking up airborne sounds. Despite these features, *Ichthyostega*, which lived about 350 million years ago, was still quite fishlike in overall appearance and represents a very early amphibian.

FIGURE 34.21
A terrestrial amphibian of the Permian period. *Cacops*, a large, extinct amphibian, had extensive body armor.

Amphibians Today

All of today's amphibians descended from the two families of amphibians that survived the Age of the Dinosaurs. During the Tertiary period (65 to 2 million years ago), these moist-skinned amphibians accomplished a highly successful invasion of wet habitats all over the world, and today there are over 4800 species of amphibians in 37 different families, comprising the orders Anura, Urodela, and Apoda.

Anura. Frogs and toads, amphibians without tails, live in a variety of environments, from deserts and mountains to ponds and puddles (figure 34.22a). Frogs have smooth, moist skin, a broad body, and long hind legs that make them excellent jumpers. Most frogs live in or near water, although some tropical species live in trees. Unlike frogs, toads have a dry, bumpy skin and short legs, and are well adapted to dry environments. All adult anurans are carnivores, eating a wide variety of invertebrates.

Most frogs and toads return to water to reproduce, laying their eggs directly in water. Their eggs lack watertight external membranes and would dry out quickly if out of the water. Eggs are fertilized externally and hatch into swimming larval forms called tadpoles. Tadpoles live in the water, where they generally feed on minute algae. After considerable growth, the body of the tadpole gradually changes into that of an adult frog. This process of abrupt change in body form is called **metamorphosis.**

Urodela (Caudata). Salamanders have elongated bodies, long tails, and smooth, moist skin (figure 34.22b). They typically range in length from a few inches to a foot, although giant Asiatic salamanders of the genus *Andrias* are as much as 1.5 meters long and weigh up to 33 kilograms. Most salamanders live in moist places, such as under stones or logs, or among the leaves of tropical plants. Some salamanders live entirely in water.

Salamanders lay their eggs in water or in moist places. Fertilization is usually external, although a few species practice a type of internal fertilization in which the female picks up sperm packets deposited by the male. Unlike anurans, the young that hatch from salamander eggs do not undergo profound metamorphosis, but are born looking like small adults and are carnivorous.

Apoda (Gymnophiona). Caecilians, members of the order Apoda (Gymnophiona), are a highly specialized group of tropical burrowing amphibians (figure 34.22c). These legless, wormlike creatures average about 30 centimeters long, but can be up to 1.3 meters long. They have very small eyes and are often blind. They resemble worms but have jaws with teeth. They eat worms and other soil invertebrates. The caecilian male deposits sperm directly into the female, and the female usually bears live young. Mud eels, small amphibians with tiny forelimbs and no hindlimbs that live in the eastern United States, are not apodans, but highly specialized urodelians.

FIGURE 34.22
Class Amphibia. (*a*) Red-eyed tree frog, *Agalychnis callidryas* (order Anura). (*b*) An adult barred tiger salamander, *Ambystoma tigrinum* (order Urodela). (*c*) A caecilian, *Caecilia tentaculata* (order Apoda).

Amphibians ventured onto land some 370 million years ago. They are characterized by moist skin, legs (secondarily lost in some species), lungs (usually), and a more complex and divided circulatory system. They are still tied to water for reproduction.

Reptiles

If we think of amphibians as a first draft of a manuscript about survival on land, then reptiles are the finished book. For each of the five key challenges of living on land, reptiles improved on the innovations first seen in amphibians. Legs were arranged to support the body's weight more effectively, allowing reptile bodies to be bigger and to run. Lungs and heart were altered to make them more efficient. The skin was covered with dry plates or scales to minimize water loss, and eggs were encased in watertight covers. Reptiles were the first truly *terrestrial* vertebrates.

Over 7000 species of reptiles (class Reptilia) now live on earth (table 34.3). They are a highly successful group in today's world; there are three reptile species for every two mammal species. While it is traditional to think of reptiles as more primitive than mammals, the great majority of reptiles living today evolved from lines that appeared after therapsids did (the line that leads directly to mammals).

Key Characteristics of Reptiles

All living reptiles share certain fundamental characteristics, features they retain from the time when they replaced amphibians as the dominant terrestrial vertebrates. Among the most important are:

1. **Amniotic egg.** Amphibians never succeeded in becoming fully terrestrial because amphibian eggs must be laid in water to avoid drying out. Most reptiles lay watertight eggs that contain a food source (the yolk) and a series of four membranes—the yolk sac, the amnion, the allantois, and the chorion (figure 34.23). Each membrane plays a role in making the egg an independent life-support system. The outermost membrane of the egg is the **chorion,** which lies just beneath the porous shell. It allows respiratory gases to pass through, but retains water within the egg. The **amnion** encases the developing embryo within a fluid-filled cavity. The **yolk sac** provides food from the yolk for the embryo via blood vessels connecting to the embryo's gut. The **allantois** surrounds a cavity into which waste products from the embryo are excreted. All modern reptiles (as well as birds and mammals) show exactly this same pattern of membranes within the egg. These three classes are called amniotes.

2. **Dry skin.** Living amphibians have moist skin and must remain in moist places to avoid drying out. Reptiles have dry, watertight skin. A layer of scales or armor covers their bodies, preventing water loss. These scales develop as surface cells fill with keratin, the same protein that forms claws, fingernails, hair, and bird feathers.

3. **Thoracic breathing.** Amphibians breathe by squeezing their throat to pump air into their lungs; this limits their breathing capacity to the volume of their mouth. Reptiles developed pulmonary breathing, expanding and contracting the rib cage to suck air into the lungs and then force it out. The capacity of this system is limited only by the volume of the lungs.

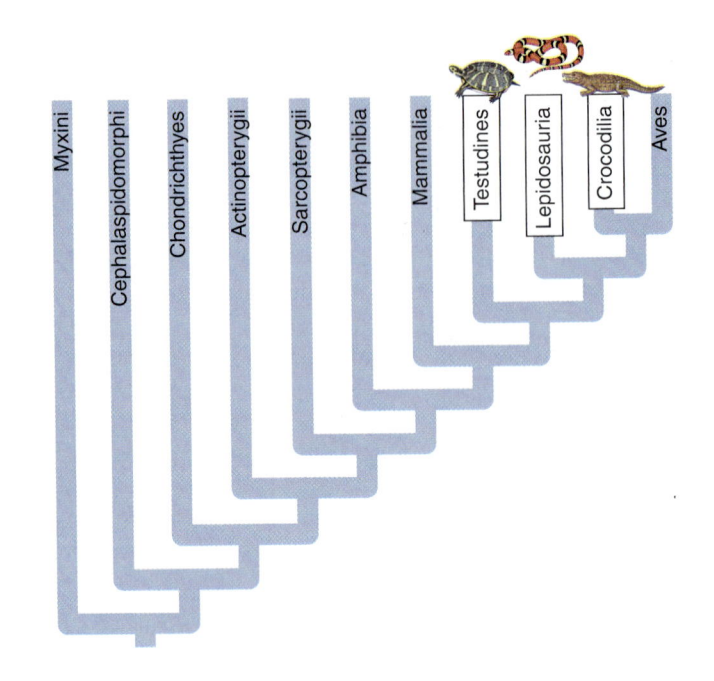

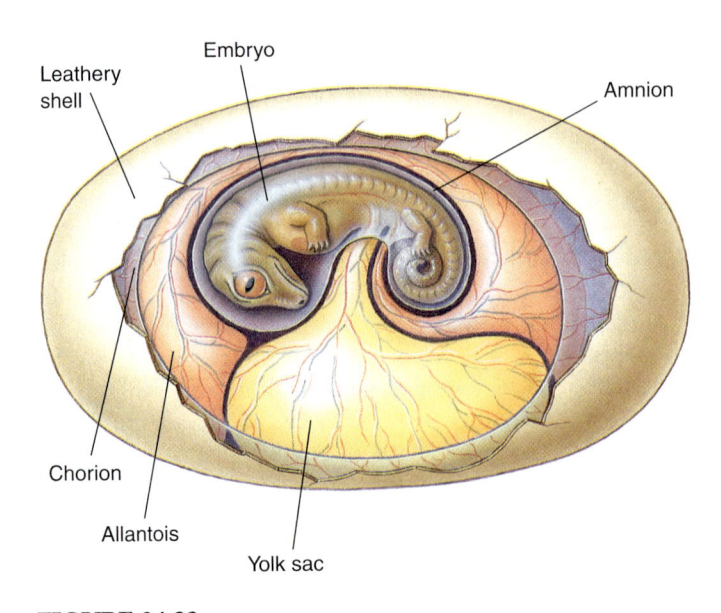

FIGURE 34.23
The watertight egg. The amniotic egg is perhaps the most important feature that allows reptiles to live in a wide variety of terrestrial habitats.

> Reptiles were the first vertebrates to completely master the challenge of living on dry land.

Table 34.3 Major Orders of Reptiles

Order	Typical Examples	Key Characteristics	Approximate Number of Living Species
Squamata, suborder **Sauria**	Lizards	Lizards; limbs set at right angles to body; anus is in transverse (sideways) slit; most are terrestrial	3800
Squamata, suborder **Serpentes**	Snakes	Snakes; no legs; move by slithering; scaly skin is shed periodically; most are terrestrial	3000
Chelonia	Turtles, tortoises, sea turtles	Ancient armored reptiles with shell of bony plates to which vertebrae and ribs are fused; sharp, horny beak without teeth	250
Crocodylia	Crocodiles, alligators, gavials, caimans	Advanced reptiles with four-chambered heart and socketed teeth; anus is a longitudinal (lengthwise) slit; closest living relatives to birds	25
Rhynchocephalia	Tuataras	Sole survivors of a once successful group that largely disappeared before dinosaurs; fused, wedgelike, socketless teeth; primitive third eye under skin of forehead	2
Ornithischia	Stegosaur	Dinosaurs with two pelvic bones facing backward, like a bird's pelvis; herbivores, with turtle-like upper beak; legs under body	Extinct
Saurischia	Tyrannosaur	Dinosaurs with one pelvic bone facing forward, the other back, like a lizard's pelvis; both plant- and flesh-eaters; legs under body	Extinct
Pterosauria	Pterosaur	Flying reptiles; wings were made of skin stretched between fourth fingers and body; wingspans of early forms typically 60 centimeters, later forms nearly 8 meters	Extinct
Plesiosaura	Plesiosaur	Barrel-shaped marine reptiles with sharp teeth and large, paddle-shaped fins; some had snakelike necks twice as long as their bodies	Extinct
Ichthyosauria	Ichthyosaur	Streamlined marine reptiles with many body similarities to sharks and modern fishes	Extinct

The Rise and Fall of Dominant Reptile Groups

During the 250 million years that reptiles were the dominant large terrestrial vertebrates, four major forms of reptiles took turns as the dominant type: pelycosaurs, therapsids, thecodonts, and dinosaurs.

Pelycosaurs: Becoming a Better Predator

Early reptiles such as *pelycosaurs* were better adapted to life on dry land than amphibians because they evolved watertight eggs. They had powerful jaws because of an innovation in skull design and muscle arrangement. Pelycosaurs were **synapsids,** meaning that their skulls had a pair of temporal holes behind the openings for the eyes. An important feature of reptile classification is the presence and number of openings behind the eyes (see figure 34.28). Their jaw muscles were anchored to these holes, which allowed them to bite more powerfully. An individual pelycosaur weighed about 200 kilograms. With long, sharp, "steak knife" teeth, pelycosaurs were the first land vertebrates to kill beasts their own size (figure 34.24). Dominant for 50 million years, pelycosaurs once made up 70% of all land vertebrates. They died out about 250 million years ago, replaced by their direct descendants—the therapsids.

Therapsids: Speeding Up Metabolism

Therapsids (figure 34.25) ate ten times more frequently than their pelycosaur ancestors. Evidence indicates that they may have been endotherms, able to regulate their own body temperature. The extra food consumption would have been necessary to produce body heat. This would have permitted therapsids to be far more active than other vertebrates of that time, when winters were cold and long. For 20 million years, therapsids (also called "mammal-like reptiles") were the dominant land vertebrate, until they were largely replaced 230 million years ago by a cold-blooded, or ectothermic, reptile line—the thecodonts. Therapsids became extinct 170 million years ago, but not before giving rise to their descendants—the mammals.

Thecodonts: Wasting Less Energy

Thecodonts were **diapsids,** their skulls having two pairs of temporal holes, and like amphibians and early reptiles, they were ectotherms (figure 34.26). Thecodonts largely replaced therapsids when the world's climate warmed 230 million years ago. In the warm climate, the therapsid's endothermy no longer offered a competitive advantage, and ectothermic thecodonts needed only a tenth as much food. Thecodonts were the first land vertebrates to be bipedal—to stand and walk on two feet. They were dominant through the Triassic period and survived for 15 million years, until they were replaced by their direct descendants—the dinosaurs.

FIGURE 34.24
A pelycosaur. *Dimetrodon,* a carnivorous pelycosaur, had a dorsal sail that is thought to have been used to dissipate body heat or gain it by basking.

FIGURE 34.25
A therapsid. This small, weasel-like cynodont therapsid, *Megazostrodon,* may have had fur. Living in the late Triassic period, this therapsid is so similar to modern mammals that some paleontologists consider it the first mammal.

FIGURE 34.26
A thecodont. *Euparkeria,* a thecodont, had rows of bony plates along the sides of the backbone, as seen in modern crocodiles and alligators.

Dinosaurs: Learning to Run Upright

Dinosaurs evolved from thecodonts about 220 million years ago. Unlike the thecodonts, their legs were positioned directly underneath their bodies, a significant improvement in body design (figure 34.27). This design placed the weight of the body directly over the legs, which allowed dinosaurs to run with great speed and agility. A dinosaur fossil can be distinguished from a thecodont fossil by the presence of a hole in the side of the hip socket. Because the dinosaur leg is positioned underneath the socket, the force is directed upward, not inward, thus requiring no bone on the side of the socket. Dinosaurs went on to become the most successful of all land vertebrates, dominating for 150 million years. All dinosaurs became extinct rather abruptly 65 million years ago, apparently as a result of an asteroid's impact.

Figures 34.28 and 34.29 summarize the evolutionary relationships among the extinct and living reptiles.

FIGURE 34.27
Mounted skeleton of *Afrovenator*. This bipedal carnivore was about 30 feet long and lived in Africa about 130 million years ago.

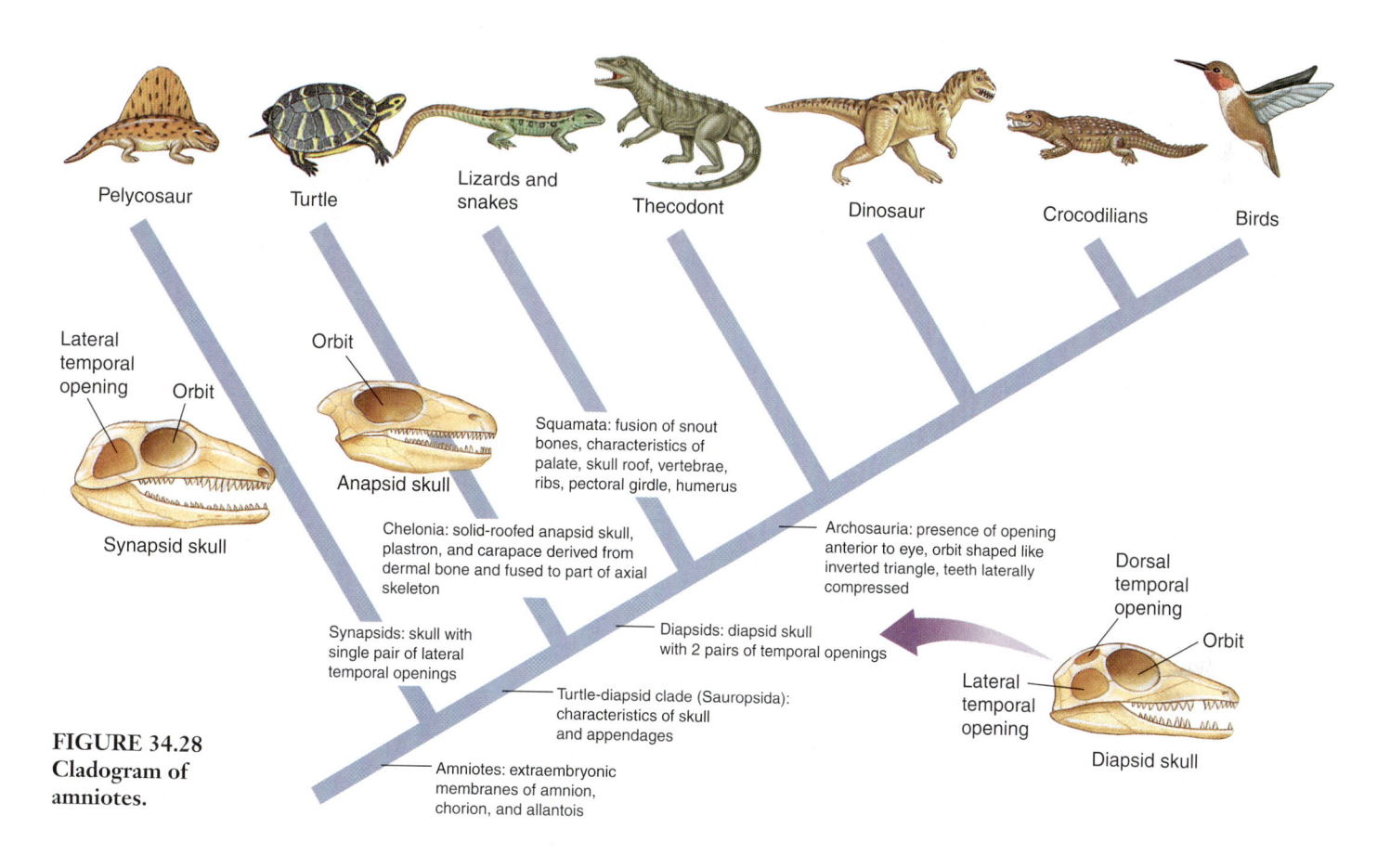

FIGURE 34.28 Cladogram of amniotes.

Pelycosaur — Lateral temporal opening, Orbit — Synapsid skull

Turtle — Orbit — Anapsid skull

Lizards and snakes

Thecodont

Dinosaur

Crocodilians

Birds — Dorsal temporal opening, Orbit, Lateral temporal opening — Diapsid skull

Squamata: fusion of snout bones, characteristics of palate, skull roof, vertebrae, ribs, pectoral girdle, humerus

Chelonia: solid-roofed anapsid skull, plastron, and carapace derived from dermal bone and fused to part of axial skeleton

Synapsids: skull with single pair of lateral temporal openings

Diapsids: diapsid skull with 2 pairs of temporal openings

Archosauria: presence of opening anterior to eye, orbit shaped like inverted triangle, teeth laterally compressed

Turtle-diapsid clade (Sauropsida): characteristics of skull and appendages

Amniotes: extraembryonic membranes of amnion, chorion, and allantois

Chapter 34 Vertebrates 705

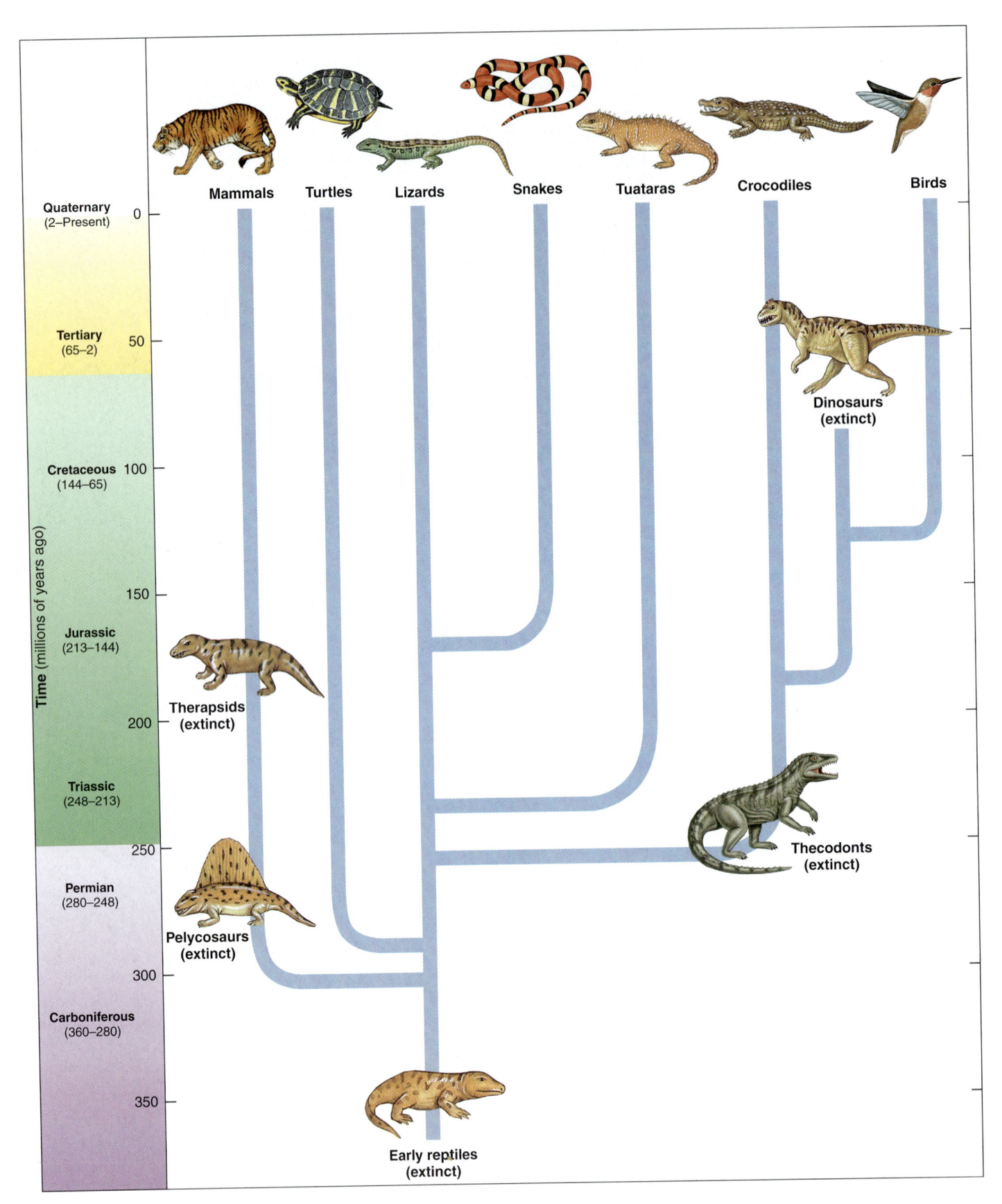

FIGURE 34.29
Evolutionary relationships among the reptiles. There are four orders of living reptiles: turtles, lizards and snakes, tuataras, and crocodiles. This phylogenetic tree shows how these four orders are related to one another and to dinosaurs, birds, and mammals.

706   Part V   Diversity of Life on Earth

FIGURE 34.30 A comparison of reptile and fish circulation. (*a*) In reptiles such as this turtle, blood is repumped after leaving the lungs, and circulation to the rest of the body remains vigorous. (*b*) The blood in fishes flows from the gills directly to the rest of the body, resulting in slower circulation.

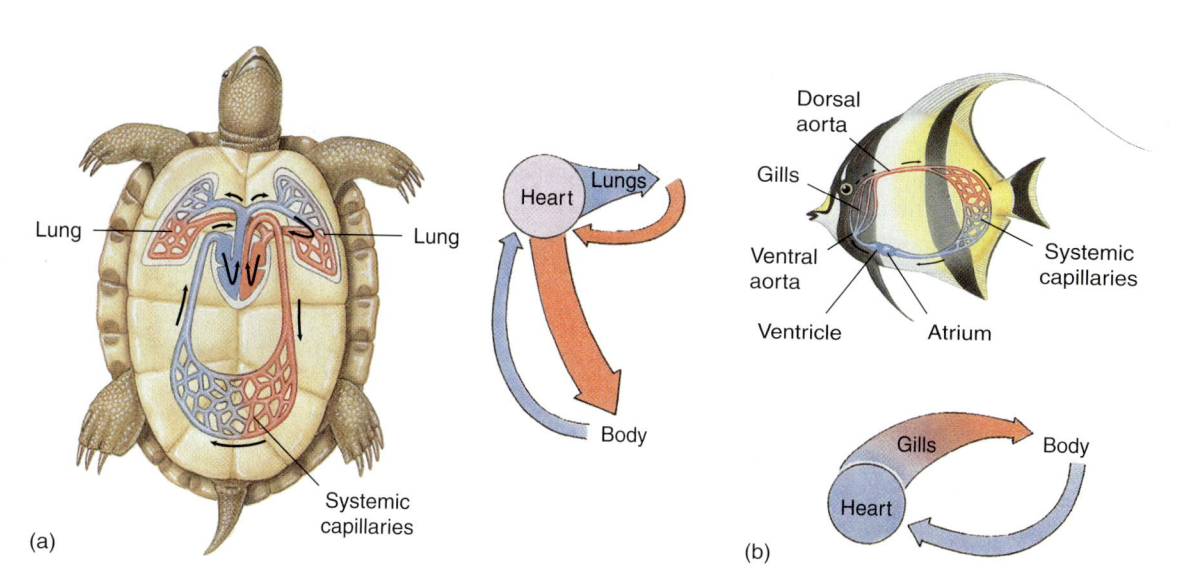

Today's Reptiles

Most of the major reptile orders are now extinct. Of the 16 orders of reptiles that have existed, only 4 survive:

1. **Turtles.** The most ancient surviving lineage of reptiles is that of turtles. Turtles have anapsid skulls much like those of the first reptiles. Turtles have changed little in the past 200 million years.
2. **Lizards and snakes.** Most reptiles living today belong to the second lineage to evolve, the lizards and snakes. Lizards and snakes are descended from an ancient lineage of lizardlike reptiles that branched off the main line of reptile evolution in the late Permian period, 250 million years ago, before the thecodonts appeared (see figure 34.29). Throughout the Mesozoic era, during the dominance of the dinosaurs, these reptiles survived as minor elements of the landscape, much as mammals did. Like mammals, lizards and snakes became diverse and common only after the dinosaurs disappeared.
3. **Tuataras.** The third lineage of surviving reptiles to evolve were the Rhynchocephalonts, small diapsid reptiles that appeared shortly before the dinosaurs. They lived throughout the time of the dinosaurs and were common in the Jurassic period. They began to decline in the Cretaceous period, apparently unable to compete with lizards, and were already rare by the time dinosaurs disappeared. Today, only two species of the order Rhynchocephalia survive, both tuataras living on small islands near New Zealand.
4. **Crocodiles.** The fourth lineage of living reptile, crocodiles, appeared on the evolutionary scene much later than other living reptiles. Crocodiles are descended from the same line of thecodonts that gave rise to the dinosaurs, and they resemble dinosaurs in many ways. They have changed very little in over 200 million years. Crocodiles, pterosaurs, thecodonts, and dinosaurs together make up a group called archosaurs ("ruling reptiles").

Other Important Characteristics

As you might imagine from the structure of the amniotic egg, reptiles and other amniotes do not practice external fertilization as most amphibians do. There would be no way for a sperm to penetrate the membrane barriers protecting the egg. Instead, the male places sperm inside the female, where the sperm fertilize the egg before the membranes are formed. This is called internal fertilization.

The circulatory system of reptiles is improved over that of fish and amphibians, providing oxygen to the body more efficiently (figure 34.30). The improvement is achieved by extending the septum within the heart from the atrium partway across the ventricle. This septum creates a partial wall that tends to lessen mixing of oxygen-poor blood with oxygen-rich blood within the ventricle. In crocodiles, the septum completely divides the ventricle, creating a four-chambered heart, just as it does in birds and mammals (and probably in dinosaurs).

All living reptiles are **ectothermic,** obtaining their heat from external sources. In contrast, **endothermic** animals are able to generate their heat internally. In addition, **homeothermic** animals have a constant body temperature, and **poikilothermic** animals have a body temperature that fluctuates with ambient temperature. Reptiles are largely ectothermic poikilotherms; their body temperature is largely determined by their surroundings, and therefore it fluctuates. Reptiles also regulate their temperature through behavior. They may bask in the sun to warm up or seek shade to prevent overheating. The thecodont ancestors of crocodiles were ectothermic, as crocodiles are today. The later dinosaurs from which birds evolved were endothermic. Crocodiles and birds differ in this one important respect. Ectothermy is a principal reason crocodiles have been grouped among the reptiles.

Chapter 34 Vertebrates 707

Kinds of Living Reptiles

The four surviving orders of reptiles contain about 7000 species. Reptiles occur worldwide except in the coldest regions, where it is impossible for ectotherms to survive. Reptiles are among the most numerous and diverse of terrestrial vertebrates. The four living orders of the class Reptilia are Chelonia, Rhynchocephalia, Squamata, and Crocodilia.

Order Chelonia: Turtles and Tortoises. The order Chelonia consists of about 250 species of turtles (most of which are aquatic; figure 34.31) and tortoises (which are terrestrial). They differ from all other reptiles because their bodies are encased within a protective shell. Many of them can pull their head and legs into the shell as well, for total protection from predators. Turtles and tortoises lack teeth but have sharp beaks.

Today's turtles and tortoises have changed very little since the first turtles appeared 200 million years ago. Turtles are anapsid, meaning that they lack the temporal openings in the skull characteristic of other living reptiles, which are diapsid. This evolutionary stability of turtles may reflect the continuous benefit of their basic design—a body covered with a shell. In some species, the shell is made of hard plates; in other species, it is a covering of tough, leathery skin. In either case, the shell consists of two basic parts. The carapace is the dorsal covering, while the plastron is the ventral portion. In a fundamental commitment to this shell architecture, the vertebrae and ribs of most turtle and tortoise species are fused to the inside of the carapace. All of the support for muscle attachment comes from the shell.

While most tortoises have a dome-shaped shell into which they can retract their head and limbs, water-dwelling turtles have a streamlined, disc-shaped shell that permits rapid turning in water. Freshwater turtles have webbed toes; in marine turtles, the forelimbs have evolved into flippers. Although marine turtles spend their lives at sea, they must return to land to lay their eggs. Many species migrate long distances to do this. Atlantic green turtles migrate from their feeding grounds off the coast of Brazil to Ascension Island in the middle of the South Atlantic—a distance of more than 2000 kilometers—to lay their eggs on the same beaches where they themselves hatched.

Order Rhynchocephalia: Tuataras. Today, the order Rhynchocephalia contains only two species of tuataras, large, lizardlike animals about half a meter long. The only place in the world where these endangered species are found is a cluster of small islands off the coast of New Zealand. The native Maoris of New Zealand named the tuatara for the conspicuous spiny crest running down its back.

An unusual feature of the tuatara (and some lizards) is the inconspicuous "third eye" on the top of its head, called a parietal eye. Concealed under a thin layer of scales, the eye has a lens and a retina and is connected by nerves to the brain. Why have an eye, if it is covered up? The parietal eye may function to alert the tuatara when it has been exposed to too much sun, protecting it against overheating. Unlike most reptiles, tuataras are most active at low temperatures. They burrow during the day and feed at night on insects, worms, and other small animals.

FIGURE 34.31
Red-bellied turtles, *Pseudemys rubriventris.* This turtle is common in the northeastern United States.

Order Squamata: Lizards and Snakes. The order Squamata (figure 34.32) consists of three suborders: Sauria, some 3800 species of lizards; Amphisbaenia, about 135 species of worm lizards; and Serpentes, about 3000 species of snakes. The distinguishing characteristics of this order are the presence of paired copulatory organs in the male and a lower jaw that is not joined directly to the skull. A movable hinge with five joints (a human jaw has only one) allows great flexibility in the movements of the jaw. In addition, the loss of the lower arch of bone below the lower opening in the skull of lizards makes room for large muscles to operate their jaws. Most lizards and snakes are carnivores, preying on insects and small animals, and these improvements in jaw design have made a major contribution to their evolutionary success.

The chief difference between lizards and snakes is that most lizards have limbs, and snakes do not. Snakes also lack movable eyelids and external ears. Lizards are a more ancient group than modern snakes, which evolved only 20 million years ago. Common lizards include iguanas, chameleons, geckos, and anoles. Most are small, measuring less than a foot in length. The largest lizards belong to the monitor family. The largest of all monitors is the Komodo dragon of Indonesia, which reaches 3 meters in length and weighs up to 100 kilograms. Snakes also vary in length from only a few inches to nearly 10 meters.

Lizards and snakes rely on agility and speed to catch prey and elude predators. Only two species of lizard are venomous, the Gila monster of the southwestern United States and the beaded lizard of western Mexico. Similarly, most species of snakes are nonvenomous. Of the 13 families of snakes, only 4 are venomous: the elapids

(a)

(b)

FIGURE 34.32
Representatives from the order Squamata. (*a*) An Australian skink, *Sphenomorophus*. Some burrowing lizards lack legs, and the snakes evolved from one line of legless lizards. (*b*) A smooth green snake, *Liochlorophis vernalis*.

FIGURE 34.33
Crocodile. Most crocodiles resemble birds and mammals in having four-chambered hearts; all other living reptiles have three-chambered hearts. Like birds, crocodiles are more closely related to dinosaurs than to any of the other living reptiles.

(cobras, kraits, and coral snakes); the sea snakes; the vipers (adders, bushmasters, rattlesnakes, water moccasins, and copperheads); and some colubrids (African boomslang and twig snake).

Many lizards, including skinks and geckos, have the ability to lose their tails and then regenerate a new one. This apparently allows these lizards to escape from predators.

Order Crocodilia: Crocodiles and Alligators. The order Crocodilia is composed of 25 species of large, primarily aquatic, primitive-looking reptiles (figure 34.33). In addition to crocodiles and alligators, the order includes two less familiar animals: the caimans and the gavials. Crocodilians have remained relatively unchanged since they first evolved.

Crocodiles are largely nocturnal animals that live in or near water in tropical or subtropical regions of Africa, Asia, and South America. The American crocodile is found in southern Florida, Cuba, and throughout tropical Central America. Nile crocodiles and estuarine crocodiles can grow to enormous size and are responsible for many human fatalities each year. There are only two species of alligators: one living in the southern United States and the other a rare endangered species living in China. Caimans, which resemble alligators, are native to Central America. Gavials are a group of fish-eating crocodilians with long, slender snouts that live only in India and Burma.

All crocodilians are carnivores. They generally hunt by stealth, waiting in ambush for prey, and then attacking ferociously. Their bodies are well adapted for this form of hunting: Their eyes are on top of their heads, and their nostrils are on top of their snouts, so they can see and breathe while lying quietly submerged in water. They have enormous mouths, studded with sharp teeth, and very strong necks. A valve in the back of the mouth prevents water from entering the air passage when a crocodilian feeds underwater.

In many ways, crocodiles resemble birds far more than they do other living reptiles. For example, alone among living reptiles, crocodiles care for their young (a trait they share with at least some dinosaurs) and have a four-chambered heart, as birds do. Why are crocodiles more similar to birds than to other living reptiles? Most biologists now believe that birds are in fact the direct descendants of dinosaurs. Both crocodiles and birds are more closely related to dinosaurs, and to each other, than to lizards and snakes.

Many major reptile groups that dominated life on land for 250 million years are now extinct. The four living orders of reptiles include the turtles, lizards and snakes, tuataras, and crocodiles.

Chapter 34 Vertebrates 709

Birds

Only four groups of animals have evolved the ability to fly—insects, pterosaurs, birds, and bats. Pterosaurs, flying reptiles, evolved from gliding reptiles and flew for 130 million years before becoming extinct with the dinosaurs. The ways these very different animals meet the challenges of flight are startingly similar. Like water running downhill through similar gullies, evolution tends to seek out similar adaptations, but major differences occur as well. The success of birds lies in the development of a structure unique in the animal world—the feather. Developed from reptilian scales, feathers are the ideal adaptation for flight, serving as lightweight airfoils that are easily replaced if damaged (unlike the vulnerable skin wings of pterosaurs and bats). Today, birds (class Aves) are the most successful and diverse of all terrestrial vertebrates, with 28 orders containing a total of 166 families and about 8600 species (table 34.4).

Key Characteristics of Birds

Modern birds lack teeth and have only vestigial tails, but they still retain many reptilian characteristics. For instance, birds lay amniotic eggs, although the shells of bird eggs are hard rather than leathery. Also, reptilian scales are present on the feet and lower legs of birds. What makes birds unique? Two primary characteristics distinguish them from living reptiles:

1. **Feathers.** Feathers are modified reptilian scales that serve two functions: providing lift for flight and conserving heat. The structure of feathers combines maximum flexibility and strength with minimum weight (figure 34.34). Feathers develop from tiny pits in the skin called follicles. In a typical flight feather, a shaft emerges from the follicle, and pairs of vanes develop from its opposite sides. At maturity, each vane has many branches called barbs. The barbs, in turn, have many projections called barbules that are equipped with microscopic hooks. These hooks link the barbs to one another, giving the feather a continuous surface and a sturdy but flexible shape. Like scales, feathers can be replaced. Among living animals, feathers are unique to birds. Recent fossil finds suggest that some dinosaurs may have had feathers.

2. **Flight skeleton.** The bones of birds are thin and hollow. Many of the bones are fused, making the bird skeleton more rigid than a reptilian skeleton. The fused sections of backbone and of the shoulder and hip girdles form a sturdy frame that anchors muscles during flight. The power for active flight comes from large breast muscles that can make up 30% of a bird's total body weight. They stretch down from the wing and attach to the breastbone, which is greatly enlarged and bears a prominent keel for muscle attachment. They also attach to the fused collarbones that form the so-called "wishbone." No other living vertebrates have a fused collarbone or a keeled breastbone.

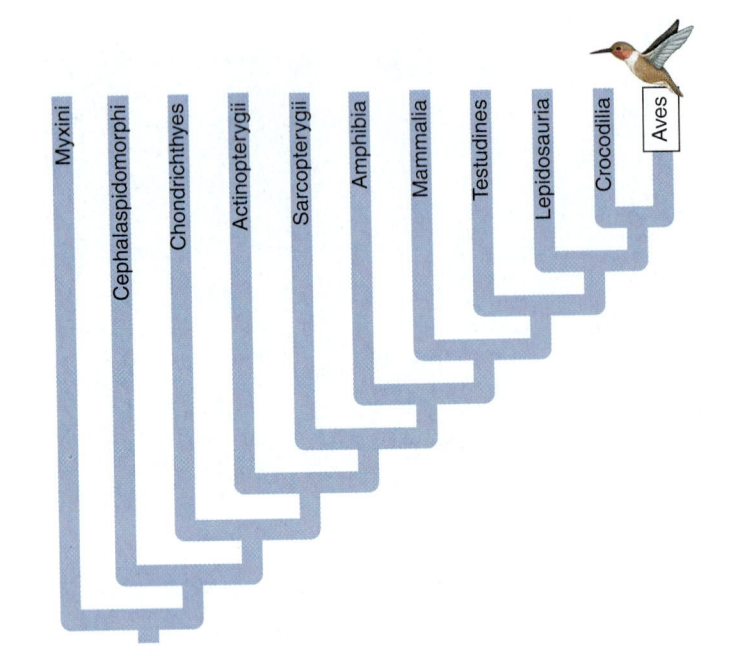

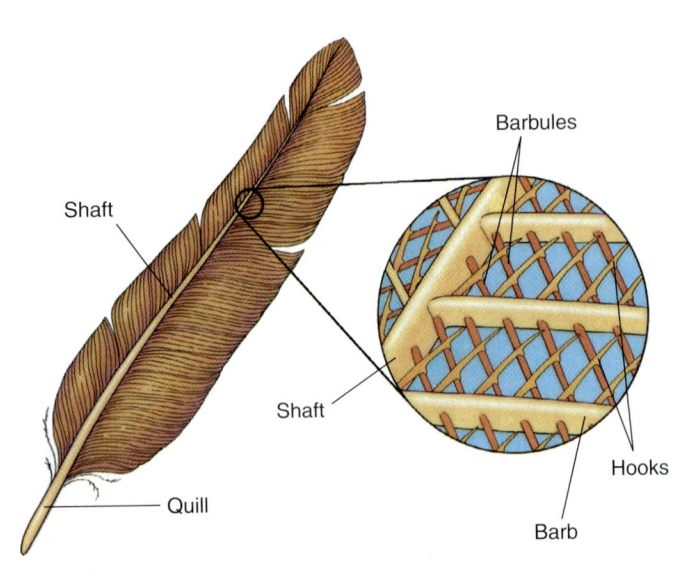

FIGURE 34.34
A feather. The enlargement shows how the secondary branches and barbs of the vanes are linked together by microscopic barbules.

Birds are the most diverse of all terrestrial vertebrates. They are closely related to reptiles, but unlike reptiles or any other animals, birds have feathers.

---

710   Part V   Diversity of Life on Earth

Table 34.4 Major Orders of Birds

Order	Typical Examples	Key Characteristics	Approximate Number of Living Species
Passeriformes	Crows, mockingbirds, robins, sparrows, starlings, warblers	*Songbirds* Well-developed vocal organs; perching feet; dependent young	5276 (largest of all bird orders; contains over 60% of all species)
Apodiformes	Hummingbirds, swifts	*Fast fliers* Short legs; small bodies; rapid wing beat	428
Piciformes	Honeyguides, toucans, woodpeckers	*Woodpeckers or toucans* Grasping feet; chisel-like, sharp bills can break down wood	383
Psittaciformes	Cockatoos, parrots	*Parrots* Large, powerful bills for crushing seeds; well-developed vocal organs	340
Charadriiformes	Auks, gulls, plovers, sandpipers, terns	*Shorebirds* Long, stiltlike legs; slender, probing bills	331
Columbiformes	Doves, pigeons	*Pigeons* Perching feet; rounded, stout bodies	303
Falconiformes	Eagles, falcons, hawks, vultures	*Birds of prey* Carnivorous; keen vision; sharp, pointed beaks for tearing flesh; active during the day	288
Galliformes	Chickens, grouse, pheasants, quail	*Gamebirds* Often limited flying ability; rounded bodies	268
Gruiformes	Bitterns, coots, cranes, rails	*Marsh birds* Long, stiltlike legs; diverse body shapes; marsh-dwellers	209
Anseriformes	Ducks, geese, swans	*Waterfowl* Webbed toes; broad bill with filtering ridges	150
Strigiformes	Barn owls, screech owls	*Owls* Nocturnal birds of prey; strong beaks; powerful feet	146
Ciconiiformes	Herons, ibises, storks	*Waders* Long-legged; large bodies	114
Procellariformes	Albatrosses, petrels	*Seabirds* Tube-shaped bills; capable of flying for long periods of time	104
Sphenisciformes	Emperor penguins, crested penguins	*Penguins* Marine; modified wings for swimming; flightless; found only in southern hemisphere; thick coats of insulating feathers	18
Dinornithiformes	Kiwis	*Kiwis* Flightless; small; primitive; confined to New Zealand	2
Struthioniformes	Ostriches	*Ostriches* Powerful running legs; flightless; only two toes; very large	1

History of the Birds

A 150-million-year-old fossil of the first known bird, *Archaeopteryx* (figure 34.35)—pronounced "archie-op-ter-ichs"—was found in 1862 in a limestone quarry in Bavaria, the impression of its feathers stamped clearly into the rocks.

Birds Are Descended from Dinosaurs

The skeleton of *Archaeopteryx* shares many features with small theropod dinosaurs. About the size of a crow, its skull has teeth, and very few of its bones are fused to one another, a feature that some paleontologists consider dinosaurian rather than avian. Its bones are thought to have been solid, not hollow like a bird's. Also, it has a long, reptilian tail and no enlarged breastbone such as modern birds use to anchor flight muscles. Finally, it has the forelimbs of a dinosaur. Because of its many dinosaur features, several *Archaeopteryx* fossils were originally classified as the coelurosaur *Compsognathus*, a small theropod dinosaur of similar size—until feathers were discovered on the fossils. What makes *Archaeopteryx* distinctly avian is the presence of feathers on its wings and tail. It also has other birdlike features, notably a wishbone. Dinosaurs lack a wishbone, although thecodonts had them.

The remarkable similarity of *Archaeopteryx* to *Compsognathus* has led almost all paleontologists to conclude that *Archaeopteryx* is the direct descendant of dinosaurs—indeed, that today's birds are "feathered dinosaurs." Some even speak flippantly of "carving the dinosaur" at Thanksgiving dinner. The recent discovery of feathered dinosaurs in China lends strong support to this inference. The dinosaur

FIGURE 34.35
Archaeopteryx. An artist's reconstruction of *Archaeopteryx*, an early bird about the size of a crow. Closely related to its ancestors among the bipedal dinosaurs, *Archaeopteryx* lived in the forests of central Europe 150 million years ago. The true feather colors of *Archaeopteryx* are not known.

Caudipteryx, for example, is clearly intermediate between *Archaeopteryx* and dinosaurs, having large feathers on its tail and arms but also many features of *Velociraptor* dinosaurs (figure 34.36). Because the arms of *Caudipteryx* were too short to use as wings, feathers probably didn't evolve for flight, but instead served as insulation, much as fur does for animals. Flight is an ability certain kinds of dinosaurs achieved as they evolved longer arms. We call these dinosaurs birds.

Despite their close affinity to dinosaurs, biologists continue to classify birds as Aves, a separate class, because of their unique adaptations and great diversity, including three key evolutionary novelties: feathers, hollow bones, and physiological mechanisms such as superefficient lungs

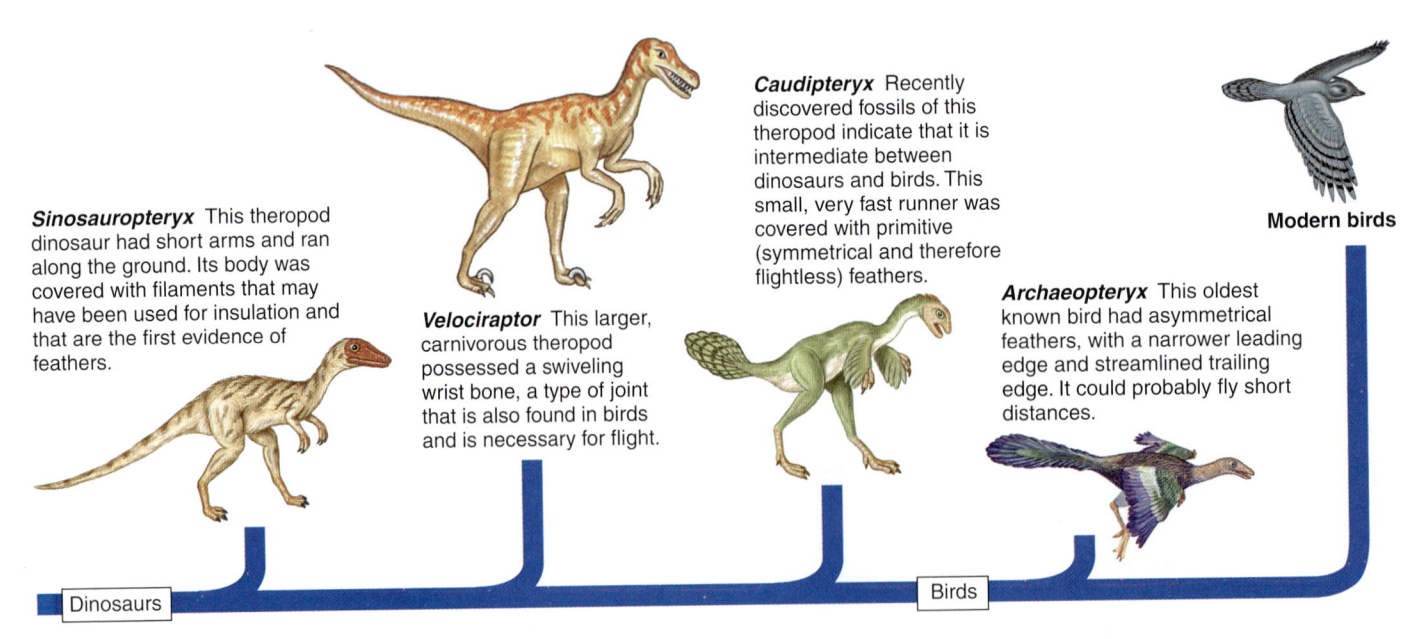

Sinosauropteryx This theropod dinosaur had short arms and ran along the ground. Its body was covered with filaments that may have been used for insulation and that are the first evidence of feathers.

Velociraptor This larger, carnivorous theropod possessed a swiveling wrist bone, a type of joint that is also found in birds and is necessary for flight.

Caudipteryx Recently discovered fossils of this theropod indicate that it is intermediate between dinosaurs and birds. This small, very fast runner was covered with primitive (symmetrical and therefore flightless) feathers.

Archaeopteryx This oldest known bird had asymmetrical feathers, with a narrower leading edge and streamlined trailing edge. It could probably fly short distances.

Modern birds

FIGURE 34.36
The evolutionary path to the birds. Almost all paleontologists now accept the theory that birds are the direct descendants of theropod dinosaurs.

712 Part V Diversity of Life on Earth

that permit sustained, powered flight. This practical judgment should not conceal the basic agreement among almost all biologists that birds are the direct descendants of theropod dinosaurs, as closely related to coelurosaurs as are other theropods (see figure 34.36).

By the early Cretaceous period, only a few million years after *Archaeopteryx* lived, a diverse array of birds had evolved, with many of the features of modern birds. Fossils in Mongolia, Spain, and China discovered within the last few years reveal a diverse collection of toothed birds with the hollow bones and breastbones necessary for sustained flight. Other fossils reveal highly specialized, flightless diving birds. The diverse birds of the Cretaceous shared the skies with pterosaurs for 70 million years.

Because the impression of feathers is rarely fossilized and modern birds have hollow, delicate bones, the fossil record of birds is incomplete. Relationships among the 166 families of modern birds are mostly inferred from studies of the degree of DNA similarity among living birds. These studies suggest that the most ancient living birds are the flightless birds, such as the ostrich. Ducks, geese, and other waterfowl evolved next, in the early Cretaceous, followed by a diverse group of woodpeckers, parrots, swifts, and owls. The largest of the bird orders, Passeriformes, or songbirds (60% of all species of birds today), evolved in the mid-Cretaceous. The more specialized orders of birds, such as shorebirds, birds of prey, flamingos, and penguins, did not appear until the late Cretaceous. All but a few of the modern orders of toothless birds are thought to have arisen before the disappearance of the pterosaurs and dinosaurs at the end of the Cretaceous period, 65 million years ago.

Birds Today

You can tell a great deal about the habits and food of a bird by examining its beak and feet. For instance, carnivorous birds such as owls have curved talons for seizing prey and sharp beaks for tearing apart their meal. The beaks of ducks are flat for shoveling through mud, while the beaks of finches are short, thick seed-crushers. There are 28 orders of birds, the largest consisting of over 5000 species (figure 34.37).

Many adaptations enabled birds to cope with the heavy energy demands of flight:

1. **Efficient respiration.** Flight muscles consume an enormous amount of oxygen during active flight. The reptilian lung has a limited internal surface area, not nearly enough to absorb all the oxygen needed. Mammalian lungs have a greater surface area, but as we will see in chapter 44, bird lungs satisfy this challenge with a radical redesign. When a bird inhales, the air goes past the lungs to a series of air sacs located near and within the hollow bones of the back; from there, the air travels to the lungs and then to a set of anterior air sacs before being exhaled. Because air always passes through the lungs in the same direction, and blood flows past the lung at right angles to the airflow, gas exchange is highly efficient.

2. **Efficient circulation.** The revved-up metabolism needed to power active flight also requires very efficient blood circulation, so that the oxygen captured by the lungs can be delivered to the flight muscles quickly. In the heart of most living reptiles, oxygen-rich blood coming from the lungs mixes with oxygen-poor blood returning from the body because the wall dividing the ventricle into two chambers is not complete. In birds, the wall dividing the ventricle is complete, and the two blood circulations do not mix, so flight muscles receive fully oxygenated blood.

   In comparison with reptiles and most other vertebrates, birds have a rapid heartbeat. A hummingbird's heart beats about 600 times a minute, and an active chickadee's heart beats 1000 times a minute. In contrast, the heart of the large, flightless ostrich averages 70 beats per minute—the same rate as the human heart.

3. **Endothermy.** Birds, like mammals, are endothermic. Many paleontologists believe the dinosaurs that birds evolved from were endothermic as well. Birds maintain body temperatures significantly higher than those of most mammals, ranging from 40° to 42°C (human body temperature is 37°C). Feathers provide excellent insulation, helping to conserve body heat. The high temperatures maintained by endothermy permit metabolism in the bird's flight muscles to proceed at a rapid pace, to provide the ATP necessary to drive rapid muscle contraction.

FIGURE 34.37
Class Aves. This summer tanager, *Piranga rubra*, is a member of the largest order of birds, the Passeriformes, with over 5000 species.

> The class Aves probably debuted 150 million years ago with *Archaeopteryx*. Modern birds are characterized by feathers, scales, a thin, hollow skeleton, auxiliary air sacs, and a four-chambered heart. Birds lay amniotic eggs and are endothermic.

Mammals

There are about 4500 living species of mammals (class Mammalia), the smallest number of species in any of the five classes of vertebrates. Most large, land-dwelling vertebrates are mammals (figure 34.38), and they tend to dominate terrestrial communities, as did the dinosaurs they replaced. When you look out over an African plain, you see the big mammals—the lions, zebras, gazelles, and antelope. Your eye does not as readily pick out the many birds, lizards, and frogs that live in the grassland community with them. But the typical mammal is not all that large. Of the 4500 species of mammals, 3200 are rodents, bats, shrews, or moles.

Key Mammalian Characteristics

Mammals are distinguished from all other classes of vertebrates by two fundamental characteristics—hair and mammary glands—and are marked by several other notable features:

1. **Hair.** All mammals have hair. Even apparently naked whales and dolphins grow sensitive bristles on their snouts. The evolution of fur and the ability to regulate body temperature enabled mammals to invade colder climates that ectothermic reptiles could not inhabit, and the insulation fur provided may have ensured the survival of mammals when the dinosaurs perished. Mammals are endothermic animals, and typically maintain body temperatures higher than the temperature of their surroundings. The dense undercoat of many mammals reduces the amount of body heat that escapes.

    Another function of hair is camouflage. The coloration and pattern of a mammal's coat usually matches its background. A little brown mouse is practically invisible against the brown leaf litter of a forest floor, while the orange and black stripes of a Bengal tiger disappear against the orange-brown color of the tall grass in which it hunts. Hairs also function as sensory structures. The whiskers of cats and dogs are stiff hairs that are very sensitive to touch. Mammals that are active at night or live underground often rely on their whiskers to locate prey or to avoid colliding with objects. Hair can also serve as a defense weapon. Porcupines and hedgehogs protect themselves with long, sharp, stiff hairs called quills.

    Unlike feathers, which evolved from modified reptilian scales, mammalian hair is a completely different form of skin structure. An individual mammalian hair is a long, protein-rich filament that extends like a stiff thread from a bulblike foundation beneath the skin known as a hair follicle. The filament is composed mainly of dead cells filled with the fibrous protein keratin.

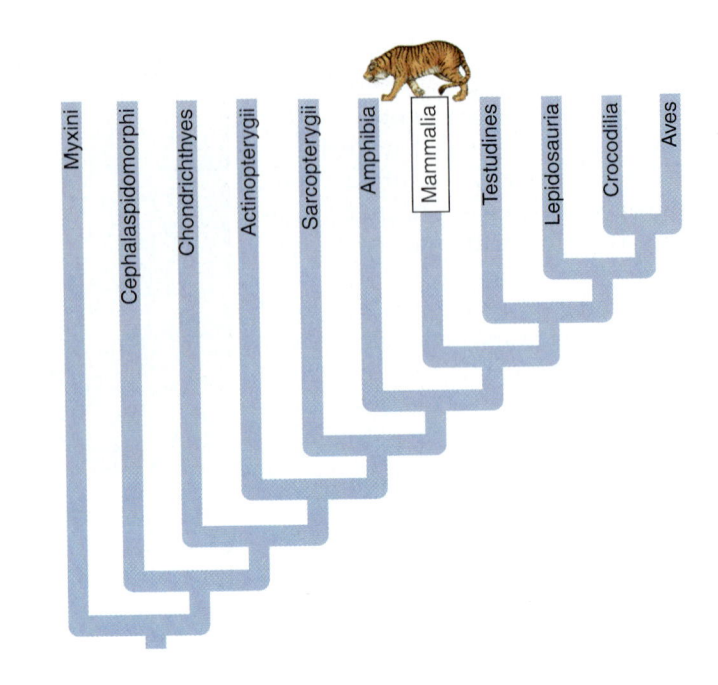

FIGURE 34.38
Mammals. African elephants, *Loxodonta africana* (order Proboscidea), at a water hole.

2. **Mammary glands.** All female mammals possess mammary glands that secrete milk. Newborn mammals, born without teeth, suckle this milk. Even baby whales are nursed by their mother's milk. Milk is a fluid rich in fat, sugar, and protein. A liter of human milk contains 11 grams of protein, 49 grams of fat, 70 grams of carbohydrate (chiefly the sugar lactose), and 2 grams of minerals critical to early growth, such as calcium. About 95% of the volume is water, critical to avoid dehydration. Milk is a very high calorie food (human milk has 750 kcal per liter), important because of the high energy needs of a rapidly growing

newborn mammal. About 50% of the energy in the milk comes from fat.

3. **Endothermy.** As stated previously, mammals are endothermic, a crucial adaptation that has allowed them to be active at any time of the day or night and to colonize severe environments, from deserts to ice fields. Many characteristics, including hair that provides insulation, played important roles in making endothermy possible. Also, the more efficient blood circulation provided by the four-chambered heart and the more efficient respiration provided by the *diaphragm* (a special sheet of muscles below the rib cage that aids breathing) make possible the higher metabolic rate upon which endothermy depends.

4. **Placenta.** In most mammal species, females carry their young in a uterus during development, nourishing them through a placenta, and give birth to live young. The placenta is a specialized organ within the uterus of the pregnant mother that brings the bloodstream of the fetus into close contact with the bloodstream of the mother (figure 34.39). Food, water, and oxygen can pass across from mother to child, and wastes can pass over to the mother's blood and be carried away.

5. **Teeth.** Reptiles have homodont dentition, meaning that their teeth are all the same. However, mammals have heterodont dentition, with different types of teeth that are highly specialized to match particular eating habits (figure 34.40). It is usually possible to determine a mammal's diet simply by examining its teeth. Compare the skull of a dog (a carnivore) and a deer (an herbivore). The dog's long canine teeth are well suited for biting and holding prey, and some of its premolar and molar teeth are triangular and sharp for ripping off chunks of flesh. In contrast, canine teeth are absent in deer; instead, the deer clips off mouthfuls of plants with flat, chisel-like incisors on its lower jaw. The deer's molars are large and covered with ridges to effectively grind and break up tough plant tissues. Rodents, such as beavers, are gnawers and have long incisors for chewing through branches or stems. These incisors are ever-growing; that is, the ends wear down, but new incisor growth maintains the length.

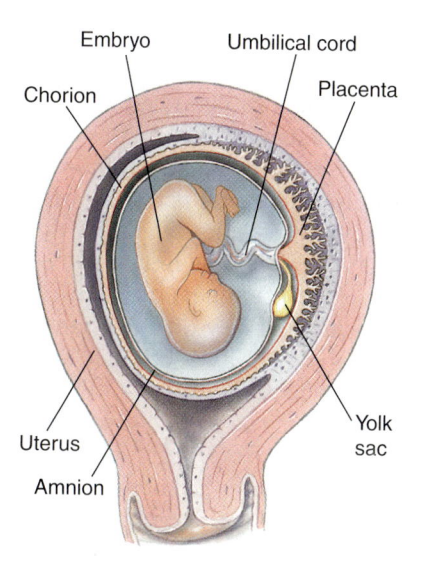

FIGURE 34.39
The placenta. The placenta is characteristic of the largest group of mammals, the placental mammals. It evolved from membranes in the amniotic egg. The umbilical cord evolved from the allantois. The chorion, or outermost part of the amniotic egg, forms most of the placenta itself. The placenta serves as the provisional lungs, intestine, and kidneys of the embryo, without ever mixing maternal and fetal blood.

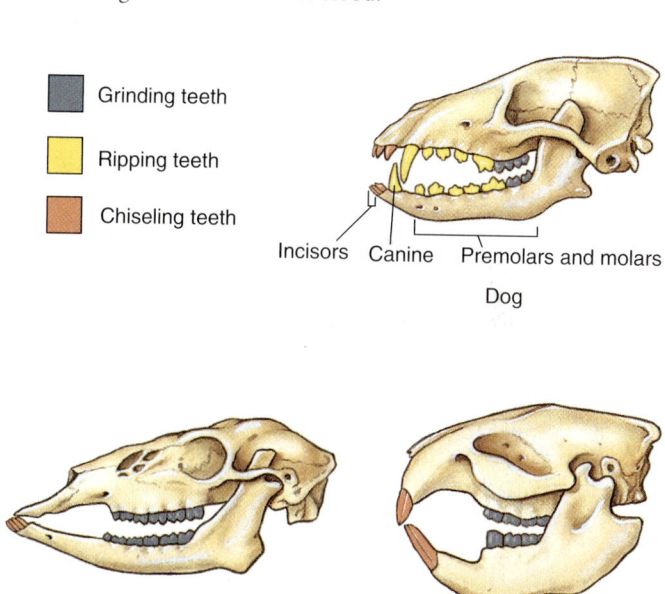

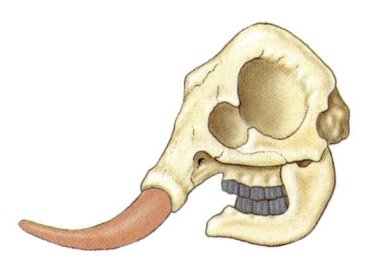

FIGURE 34.40
Mammals have different types of specialized teeth. While reptiles have all the same kind of teeth, mammals have different types of teeth specialized for different feeding habits. Carnivores, such as dogs, have canine teeth that are able to rip food; some of the premolars and molars in dogs are also ripping teeth. Herbivores, such as deer, have incisors to chisel off vegetation and molars designed to grind up the plant material. In the beaver, the chiseling incisors dominate. In the elephant, the incisors have become specialized weapons, and molars grind up vegetation. Humans are omnivores; we have ripping, chiseling, and grinding teeth.

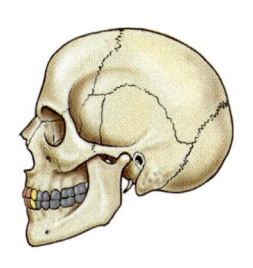

Chapter 34 Vertebrates 715

6. **Digestion of plants.** Most mammals are herbivores, eating mostly or only plants. Cellulose, the major component of plant cell walls, forms the bulk of a plant's body and is a major source of food for mammalian herbivores. The cellulose molecule has the structure of a pearl necklace, with each pearl a glucose sugar molecule. Mammals do not have enzymes that can break the links between the pearls to release the glucose elements for use as food. Herbivorous mammals rely on a mutualistic partnership with bacteria that have the necessary cellulose-splitting enzymes.

   Mammals such as cows, buffalo, antelopes, goats, deer, and giraffes have huge, four-chambered stomachs that function as fermentation vats. The first chamber is the largest and holds a dense population of cellulose-digesting bacteria. Chewed plant material passes into this chamber, where the bacteria attack the cellulose. The material is then digested further in the rest of the stomach.

   Rodents, horses, rabbits, and elephants are herbivores that employ mutualistic bacteria to digest cellulose in a different way. They have relatively small stomachs, and instead digest plant material in their large intestine, like a termite. The bacteria that actually carry out the digestion of the cellulose live in a pouch called the cecum that branches from the end of the small intestine.

   Even with these complex adaptations for digesting cellulose, a mouthful of plant is less nutritious than a mouthful of flesh. Herbivores must consume large amounts of plant material to gain sufficient nutrition. An elephant eats 135 to 150 kilograms of food (300 to 400 pounds) each day.

7. **Hooves and horns.** Keratin, the protein of hair, is also the structural building material in claws, fingernails, and hooves. Hooves are specialized keratin pads on the toes of horses, cows, sheep, antelopes, and other running mammals. The pads are hard and horny, protecting the toe and cushioning it from impact.

   The horns of cattle, sheep, and antelope are composed of a core of bone surrounded by a sheath of keratin. The bony core is attached to the skull, and the horn is not shed. The horn you see is the outer sheath, made of hairlike fibers of keratin compacted into a very hard structure. Deer antlers are made not of keratin but of bone. Male deer grow and shed a set of antlers each year. While growing during the summer, antlers are covered by a thin layer of skin known as velvet.

8. **Flight.** Bats are the only mammals capable of powered flight (figure 34.41). Like the wings of birds, bat wings are modified forelimbs. The bat wing is a leathery membrane of skin and muscle stretched over the bones of four fingers. The edges of the membrane attach to the side of the body and to the hind leg. When resting, most bats prefer to hang upside down by their toe claws. After rodents, bats are the second largest order of mammals. They have been a particularly successful group because many species have been able to utilize a food resource that most birds do not have access to—night-flying insects.

   How do bats navigate in the dark? Late in the eighteenth century, the Italian biologist Lazzaro Spallanzani showed that a blinded bat could fly without crashing into things and still capture insects. Clearly another sense other than vision was being used by bats to navigate in the dark. When Spallanzani plugged the ears of a bat, it was unable to navigate and collided with objects. Spallanzani concluded that bats "hear" their way through the night world.

   We now know that bats have evolved a sonar system that functions much like the sonar devices used by ships and submarines to locate underwater objects. As a bat flies, it emits a very rapid series of extremely high-pitched "clicking" sounds well above the range of human hearing. The high-frequency pulses are emitted either through the mouth or, in some cases, through the nose. The soundwaves bounce off obstacles or flying insects, and the bat hears the echo. Through sophisticated processing of this echo within its brain, a bat can determine not only the direction of an object but also the distance to the object (see chapter 46).

FIGURE 34.41
Greater horseshoe bat, *Rhinolophus ferrumequinum*. The bat is the only mammal capable of true flight.

> Mammals first appeared 220 million years ago, evolving to their present position of dominance in modern terrestrial ecosystems. Mammals are the only vertebrates that possess hair and mammary glands.

The Orders of Mammals

Mammals have been around since the time of the dinosaurs, although they were never common until the dinosaurs disappeared. We have learned a lot about the evolutionary history of mammals from their fossils. The first mammals arose from therapsids in the mid-Triassic period about 220 million years ago, just as the first dinosaurs evolved from thecodonts. Tiny, shrewlike creatures that lived in trees eating insects, the earliest mammals were only a minor element in a land that quickly came to be dominated by dinosaurs. Fossils reveal that these early mammals had large eye sockets, evidence that they may have been active at night. Early mammals had a single lower jawbone. Therapsid fossils show a change from the reptile lower jaw, having several bones, to a jaw closer to the mammalian type. Two of the bones forming the therapsid jaw joint retreated into the middle ear of mammals, linking with a bone already there and producing a three-bone structure that amplifies sound better than the reptilian ear.

Table 34.5 Some Groups of Extinct Mammals

Group	Description
Cave bears	Numerous in the ice ages; this enormous vegetarian bear slept through the winter in large groups.
Irish elk	Neither Irish nor an elk (it is a kind of deer), *Megaloceros* was the largest deer that ever lived, with horns spanning 12 feet. Seen in French cave paintings, they became extinct about 2500 years ago.
Mammoths	Although only two species of elephants survive today, the elephant family was far more diverse during the late Tertiary. Many were cold-adapted mammoths with fur.
Giant ground sloths	*Megatherium* was a giant, 20-foot ground sloth that weighed three tons and was as large as a modern elephant.
Sabertooth cats	The jaws of these large, lionlike cats opened an incredible 120° to allow the animal to drive its huge upper pair of saber teeth into prey.

Early Divergence in Mammals

For 155 million years, while the dinosaurs flourished, mammals were a minor group of small insectivores and herbivores. Only five orders of mammals arose in that time, and their fossils are scarce, indicating that mammals were not abundant. However, the two groups to which present-day mammals belong did appear. The most primitive mammals, direct descendants of therapsids, were members of the subclass Prototheria. Most prototherians were small and resembled modern shrews. All prototherians laid eggs, as did their therapsid ancestors. The only prototherians surviving today are the monotremes—the duckbilled platypus and the echidnas, or spiny anteaters. The other major mammalian group is the subclass Theria. All of the mammals you are familiar with, including humans, are therians. Therians are viviparous (that is, their young are born alive). The two major living therian groups are marsupials, or pouched mammals, and placental mammals. Kangaroos, opossums, and koalas are marsupials. Dogs, cats, humans, horses, and most other mammals are placentals.

The Age of Mammals

At the end of the Cretaceous period 65 million years ago, the dinosaurs and numerous other land and marine animals became extinct, but mammals survived, possibly because of the insulation their fur provided. In the Tertiary period (lasting from 65 million years to 2 million years ago), mammals rapidly diversified, taking over many of the ecological roles once dominated by dinosaurs (table 34.5). Mammals reached their maximum diversity late in the Tertiary period, about 15 million years ago. At that time, tropical conditions existed over much of the world. During the last 15 million years, world climates have deteriorated, and the area covered by tropical habitats has decreased, causing a decline in the total number of mammalian species. Seventeen of them (containing 94% of the species) are placental. The other two are the primitive monotremes and the marsupials.

Chapter 34 Vertebrates 717

Today's Mammals

Monotremes: Egg-laying Mammals. The duck-billed platypus and two species of echidna, or spiny anteater, are the only living monotremes (figure 34.42a). Among living mammals, only monotremes lay shelled eggs. The structure of their shoulder and pelvis is more similar to that of the early reptiles than to any other living mammal. Also like reptiles, monotremes have a cloaca, a single opening through which feces, urine, and reproductive products leave the body. Monotremes are more closely related to early mammals than is any other living mammal.

In addition to many reptilian features, monotremes have both of the defining mammalian features: fur and functioning mammary glands. Young monotremes drink their mother's milk after they hatch from eggs. Females lack well-developed nipples, so the babies cannot suckle. Instead, the milk oozes onto the mother's fur, and the babies lap it off with their tongues.

The platypus, found only in Australia, lives much of its life in the water and is a good swimmer. It uses its bill much as a duck does, rooting in the mud for worms and other soft-bodied animals. Echidnas of Australia and New Guinea have very strong, sharp claws, which they use for burrowing and digging. The echidna probes with its long, beaklike snout for insects, especially ants and termites.

Marsupials: Pouched Mammals. The major difference between marsupials (figure 34.42b) and other mammals is their pattern of embryonic development. In marsupials, a fertilized egg is surrounded by chorion and amniotic membranes, but no shell forms around the egg as it does in monotremes. During most of its early development, the marsupial embryo is nourished by an abundant yolk within the egg. Shortly before birth, a short-lived placenta forms from the chorion membrane. Soon after, sometimes within eight days of fertilization, the embryonic marsupial is born. It emerges tiny and hairless, and crawls into the marsupial pouch, where it latches onto a nipple and continues its development.

Marsupials evolved shortly before placental mammals, about 100 million years ago. Today, most species of marsupials live in Australia and South America, areas that have been historically isolated. Marsupials in Australia and New Guinea have diversified to fill ecological positions occupied by placental mammals elsewhere in the world. For example, kangaroos are the Australian grazers, playing the role antelope, horses, and buffalo perform elsewhere. The placental mammals in Australia and New Guinea today arrived relatively recently and include some introduced by humans. The only marsupial found in North America is the Virginia opossum.

Placental Mammals. As stated earlier in this section, mammals that produce a true placenta that nourishes the embryo throughout its entire development are called placental mammals (figure 34.42c). Most species of mammals living today, including humans, are in this group. Of the 19 orders of living mammals, 17 are placental mammals. Table 34.6 shows some of these orders, in addition to the marsupials. They are a very diverse group, ranging in size from 1.5-gram pygmy shrews to 100,000 kilogram whales.

Early in the course of embryonic development, the placenta forms. Both fetal and maternal blood vessels are abundant in the placenta, and substances can be exchanged efficiently between the bloodstreams of mother and offspring (see figure 34.39). The fetal placenta is formed from the membranes of the chorion and allantois. The maternal side of the placenta is part of the wall of the uterus, the organ in which the young develop. In placental mammals, unlike marsupials, the young undergo a considerable period of development before they are born.

FIGURE 34.42
Today's mammals. (a) This echidna, *Tachyglossus aculeatus*, is a monotreme. (b) Marsupials include kangaroos, such as this adult with young in its pouch. (c) This female African lion, *Panthera leo* (order Carnivora), is a placental mammal.

> Mammals were not a major group until the dinosaurs disappeared. Mammalian specializations include the placenta, a tooth design suited to diet, and specialized sensory systems.

Table 34.6 Major Orders of Mammals

Order	Typical Examples	Key Characteristics	Approximate Number of Living Species
Rodentia	Beavers, mice, porcupines, rats	*Small plant-eaters* Chisel-like incisor teeth	1814
Chiroptera	Bats	*Flying mammals* Primarily fruit- or insect-eaters; elongated fingers; thin wing membrane; nocturnal; navigate by sonar	986
Insectivora	Moles, shrews	*Small, burrowing mammals* Insect-eaters; the most primitive placental mammals; spend most of their time underground	390
Marsupialia	Kangaroos, koalas	*Pouched mammals* Young develop in abdominal pouch	280
Carnivora	Bears, cats, raccoons, weasels, dogs	*Carnivorous predators* Teeth adapted for shearing flesh; no native families in Australia	240
Primates	Apes, humans, lemurs, monkeys	*Tree-dwellers* Large brain size; binocular vision; opposable thumb; group that evolved from a line that branched off early from other mammals	233
Artiodactyla	Cattle, deer, giraffes, pigs	*Hoofed mammals* With two or four toes; mostly herbivores	211
Cetacea	Dolphins, porpoises, whales	*Fully marine mammals* Streamlined bodies; front limbs modified into flippers; no hindlimbs; blowholes on top of head; no hair except on muzzle	79
Lagomorpha	Rabbits, hares, pika	*Rodent-like jumpers* Four upper incisors (rather than the two seen in rodents); hind legs often longer than forelegs, an adaptation for jumping	69
Suborder Pinnipedia	Sea lions, seals, walruses	*Marine carnivores* Feed mainly on fish; limbs modified for swimming	34
Edentata	Anteaters, armadillos, sloths	*Toothless insect-eaters* Many are toothless, but some have degenerate, peglike teeth	30
Perissodactyla	Horses, rhinoceroses, zebras	*Hoofed mammals with one or three toes* Herbivorous teeth adapted for chewing	17
Proboscidea	Elephants	*Long-trunked herbivores* Two upper incisors elongated as tusks; largest living land animal	2

34.4 Evolution among the primates has focused on brain size and locomotion.

Primates

Primates are mammals with two distinct features that allowed them to succeed in the arboreal, insect-eating environment:

1. **Grasping fingers and toes.** Unlike the clawed feet of tree shrews and squirrels, primates have grasping hands and feet that let them grip limbs, hang from branches, seize food, and in some primates, use tools. The first digit in many primates is opposable and at least some, if not all, of the digits have nails.
2. **Binocular vision.** Unlike the eyes of shrews and squirrels, which sit on each side of the head so that the two fields of vision do not overlap, the eyes of primates are shifted forward to the front of the face. This produces overlapping binocular vision that lets the brain judge distance precisely—important to an animal moving through the trees.

Other mammals have binocular vision, but only primates have both binocular vision and grasping hands, making them particularly well adapted to their environment.

FIGURE 34.43
A prosimian. This tarsier, a prosimian native to tropical Asia, shows the characteristic features of primates: grasping fingers and toes and binocular vision.

The Evolution of Prosimians

About 40 million years ago, the earliest primates split into two groups: the prosimians and the anthropoids. The **prosimians** ("before monkeys") looked something like a cross between a squirrel and a cat and were common in North America, Europe, Asia, and Africa. Only a few prosimians survive today—lemurs, lorises, and tarsiers (figure 34.43). In addition to grasping digits and binocular vision, prosimians have large eyes with increased visual acuity. Most prosimians are nocturnal, feeding on fruits, leaves, and flowers, and many lemurs have long tails for balancing.

Anthropoids

Anthropoids include monkeys, apes, and humans. Anthropoids are almost all diurnal—that is, active during the day—feeding mainly on fruits and leaves. Evolution favored many changes in eye design, including color vision, that were adaptations to daytime foraging. An expanded brain governs the improved senses, with the braincase forming a larger portion of the head. Anthropoids, like the relatively few diurnal prosimians, live in groups with complex social interactions, and they tend to care for their young for prolonged periods, allowing for a long childhood of learning and brain development. About 30 million years ago, some anthropoids migrated to South America. Their descendants, known as the New World monkeys, are easy to identify: All are arboreal; they have flat, spreading noses; and many of them grasp objects with long, prehensile tails. Anthropoids that remained in Africa gave rise to two lineages: the Old World monkeys and the hominoids (apes and humans). Old World monkeys include ground-dwelling as well as arboreal species. None of them have prehensile tails, their nostrils are close together, their noses point downward, and some have toughened pads of skin for prolonged sitting.

The **hominoids** include the apes and the **hominids** (humans and their direct ancestors). The living apes consist of the gibbon (genus *Hylobates*), orangutan (*Pongo*), gorilla (*Gorilla*), and chimpanzee (*Pan*). Apes have larger brains than monkeys, and they lack tails. With the exception of the gibbon, which is small, all living apes are larger than any monkey. Apes exhibit the most adaptable behavior of any mammal except human beings. Once widespread in Africa and Asia, apes are rare today, living in relatively small areas. No apes ever occurred in North or South America.

Studies of ape DNA have explained a great deal about how the living apes evolved. The Asian apes evolved first. The line of apes leading to gibbons diverged from other apes about 15 million years ago, while orangutans split off about 10 million years ago (figure 34.44). Neither group is closely related to humans.

The African apes evolved more recently, between 6 and 10 million years ago. These apes are the closest living relatives to humans; some taxonomists have even advocated placing humans and the African apes in the same zoological family, the Hominidae. Fossils of the earliest hominids (humans and their direct ancestors), described later in this section, suggest that the common ancestor of the hominids was more like a chimpanzee than a gorilla. Based on

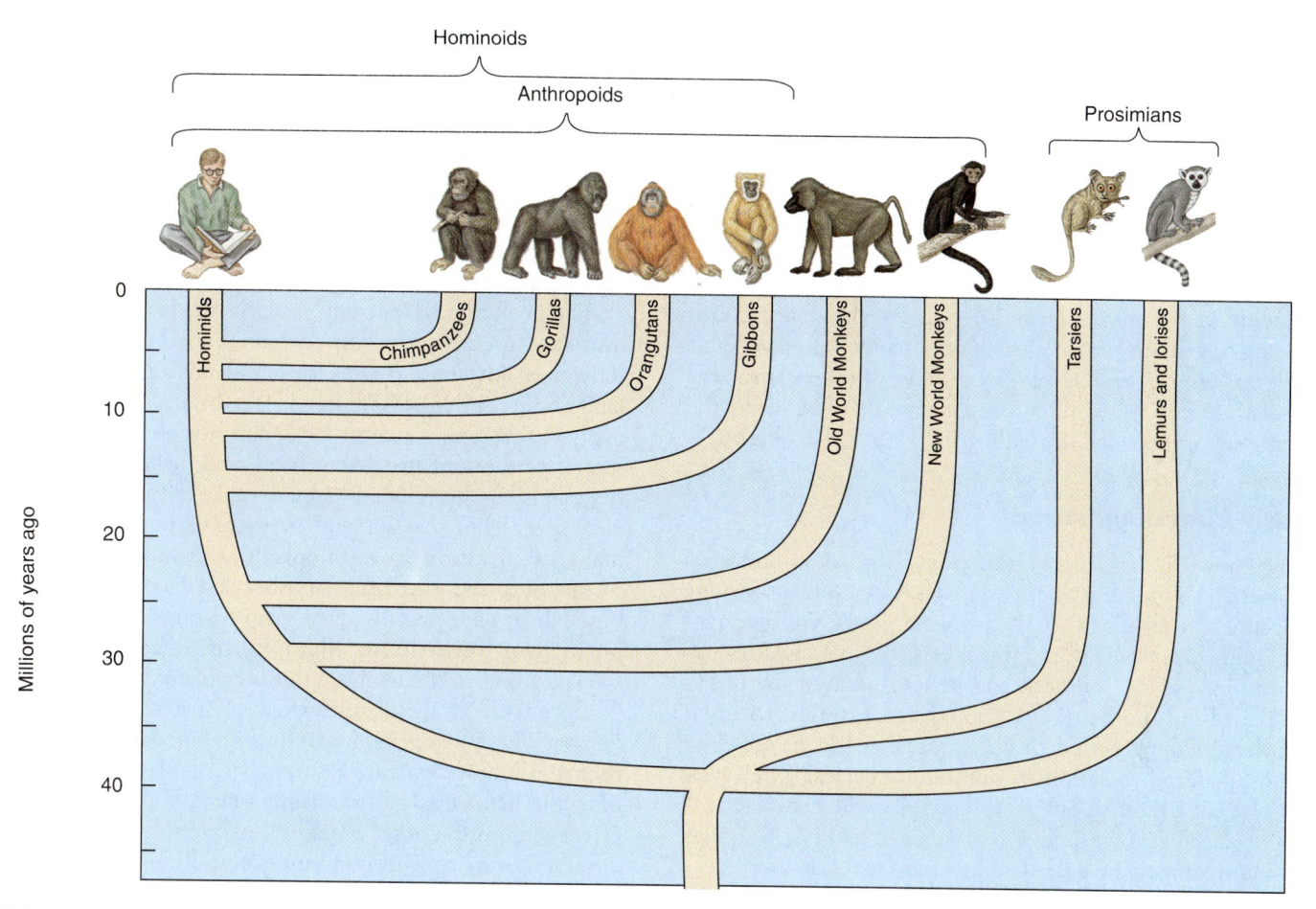

FIGURE 34.44
A primate evolutionary tree. The most ancient of the primates are the prosimians, while the hominids were the most recent to evolve.

genetic differences, scientists estimate that gorillas diverged from the line leading to chimpanzees and humans some 8 million years ago.

Soon after the gorilla lineage diverged, the common ancestor of all hominids split off from the chimpanzee line to begin the evolutionary journey leading to humans. Because this split was so recent, the genes of humans and chimpanzees have not had time to evolve many genetic differences. For example, a human hemoglobin molecule differs from its chimpanzee counterpart in only a single amino acid. In general, humans and chimpanzees exhibit a level of genetic similarity normally found between closely related sibling species of the same genus!

Comparing Apes to Hominids

The common ancestor of apes and hominids is thought to have been an arboreal climber. Much of the subsequent evolution of the hominoids reflected different approaches to locomotion. Hominids became **bipedal,** walking upright, while the apes evolved knuckle-walking, supporting their weight on the back sides of their fingers. (Monkeys, by contrast, use the palms of their hands.)

Humans depart from apes in several areas of anatomy related to bipedal locomotion. Because humans walk on two legs, their vertebral column is more curved than an ape's, and the human spinal cord exits from the bottom rather than the back of the skull. The human pelvis has become broader and more bowl-shaped, with the bones curving forward to center the weight of the body over the legs. The hip, knee, and foot (in which the human big toe no longer splays sideways) have all changed proportions. Being bipedal, humans carry much of the body's weight on the lower limbs, which comprise 32 to 38% of the body's weight and are longer than the upper limbs; human upper limbs do not bear the body's weight and make up only 7 to 9% of human body weight. African apes walk on all fours, with the upper and lower limbs both bearing the body's weight; in gorillas, the longer upper limbs account for 14 to 16% of body weight, the somewhat shorter lower limbs for about 18%.

Primates include the prosimians and anthropoids. The apes, monkeys, and hominids make up the anthropoids, and chimpanzees seem the most closely related to humans.

Chapter 34 Vertebrates 721

Australopithecines

Five to 10 million years ago, the world's climate began to get cooler, and the great forests of Africa were largely replaced with savannas and open woodland. In response to these changes, a new kind of hominoid was evolving, one that was bipedal. These new hominoids are classified as hominids—that is, of the human line.

The major groups of hominids include three to seven species of the genus *Homo* (depending how you count them), seven species of the older, smaller-brained genus *Australopithecus*, and several even older lineages. In every case where the fossils allow a determination to be made, the hominids are bipedal, the hallmark of hominid evolution.

Early Australopithecines

Our knowledge of australopithecines is based on hundreds of fossils, all found in South and East Africa (except for one specimen from Chad). It is probable, however, that australopithecines lived over a much broader area of Africa because only in the southern and eastern parts of the continent are sediments of the proper age exposed to fossil hunters. The evolution of hominids seems to have begun with an initial radiation of numerous species. The seven species identified so far provide ample evidence that australopithecines were a diverse group, and additional species will undoubtedly be described by future investigators.

These early hominids weighed about 18 kilograms and were about 1 meter tall. Their dentition was distinctly hominid, but their brains were no larger than those of apes, generally 500 cubic centimeters (cc) or less. *Homo* brains, by comparison, are usually larger than 600 cc; modern *H. sapiens* brains average 1350 cc.

The structure of australopithecine fossils clearly indicates that they walked upright. Evidence of bipedalism includes a set of some 69 hominid footprints found at Laetoli, East Africa. Two individuals, one larger than the other, walked upright side-by-side for 27 meters, their footprints preserved in a layer of 3.7-million-year-old volcanic ash. Importantly, the big toe is not splayed out to the side as in a monkey or ape, indicating that these footprints were clearly made by hominids.

The evolution of bipedalism marks the beginning of hominids. Bipedalism seems to have evolved as australopithecines left dense forests for grasslands and open woodland (figure 34.45). Whether larger brains or bipedalism evolved first was a matter of debate for some time. One school of thought proposed that hominid brains enlarged first, and then hominids became bipedal. Supporters of this view speculated that human intelligence was necessary to make the decision to walk upright and move out of the forests onto the grassland. Another school of thought saw bipedalism as a precursor to larger brains, arguing that bipedalism freed the forelimbs to manufacture and use tools, leading to the evolution of bigger brains. Recently, a treasure trove of fossils unearthed in Africa has settled the debate. These fossils demonstrate that bipedalism extended back 4 million years; knee joint, pelvis, and leg bones all exhibit the hallmarks of an upright stance. Substantial brain expansion, on the other hand, did not appear until roughly 2 million years ago. In hominid evolution, upright walking clearly preceded large brains.

The reason bipedalism evolved in hominids remains a matter of controversy. No tools appeared until 2.5 million years ago, so tool-making seems an unlikely cause. Alternative ideas suggest that walking upright is faster and uses less energy than walking on four legs; that an upright posture

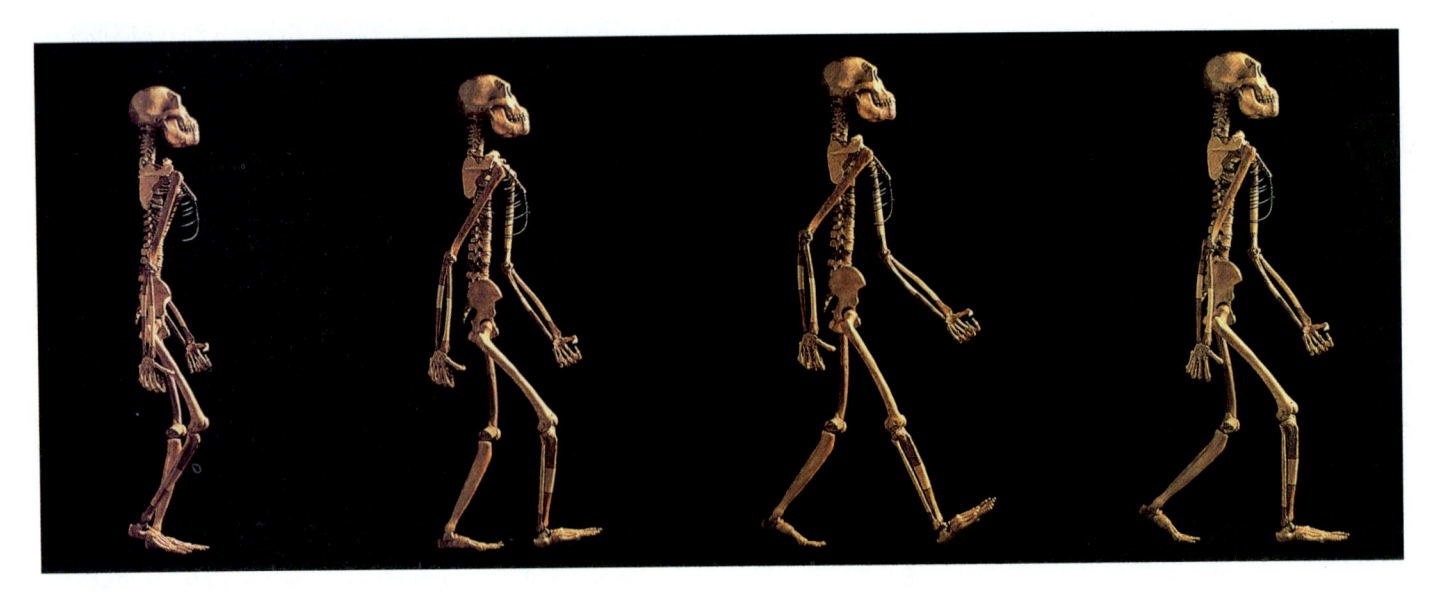

FIGURE 34.45
A reconstruction of an early hominid walking upright. These articulated plaster skeletons, made by Owen Lovejoy and his students at Kent State University, depict an early hominid (*Australopithecus afarensis*) walking upright.

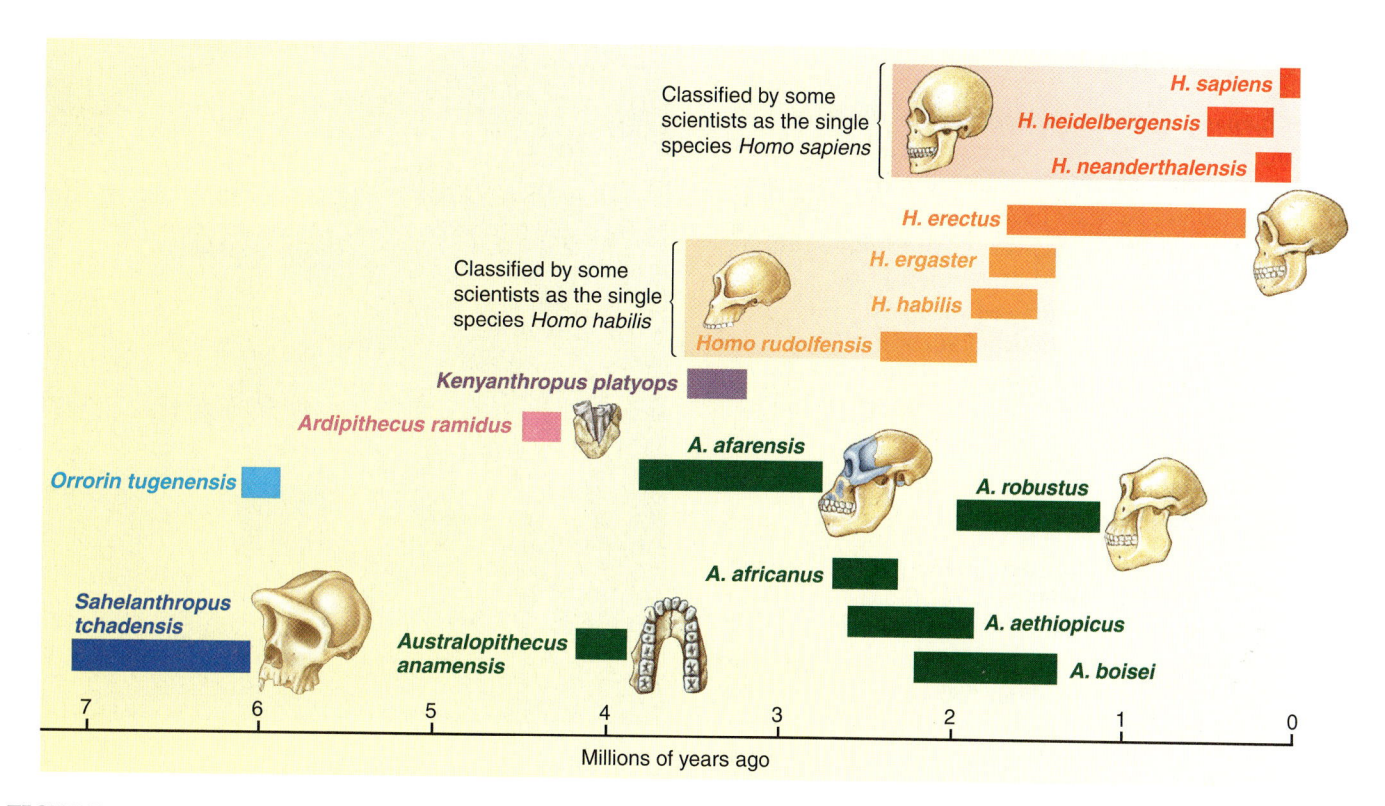

FIGURE 34.46
A hominid evolutionary tree. In this tree, the most widely accepted, the horizontal bars show the dates of first and last appearances of proposed species. Six species of *Australopithecus* and seven of *Homo* are included, as well as four other newly described early hominid genera. The oldest hominid, *Sahelanthropus tchadensis*, was discovered in Chad, Central Africa, in 2002. Some researchers question whether this 6–7 million-year-old fossil is a hominid, suggesting that it may instead be a common ancestor of chimpanzees and hominids.

permits hominids to pick fruit from trees and see over tall grass; that being upright reduces the body surface exposed to the sun's rays; that an upright stance aided the wading of aquatic hominids; and that bipedalism frees the forelimbs of males to carry food back to females, encouraging pair-bonding. All of these suggestions have their proponents, and none are universally accepted. The origin of bipedalism, the key event in the evolution of hominids, remains a mystery.

Early Hominids

In recent years, anthropologists have found a remarkable series of early hominid fossils extending as far back as 6–7 million years. Often displaying a mixture of primitive and modern traits, these fossils have thrown the study of early hominids into turmoil. While the inclusion of these fossils among the hominids seems warranted, only a few specimens of these early genera have been discovered, and they do not provide enough information to determine with any degree of certainty their relationships to australopithecines and humans. The search for additional early hominid fossils continues.

Differing Views of the Hominid Family Tree

Investigators take two different philosophical approaches to characterizing the diverse group of African hominid fossils. One group (the "lumpers") focuses on common elements in different fossils, and tends to lump together fossils that share key characters, attributing differences among the fossils to diversity within the group. Other investigators (the "splitters") are more inclined to assign fossils that exhibit differences to different species. For example, in the hominid evolutionary tree presented in figure 34.46, lumpers recognize three species of *Homo*, while splitters recognize no fewer than seven! At this point, it is not possible to decide which view is correct; more fossils are needed to determine how much the differences between fossils represent within-species variation and how much they characterize between-species differences.

> The evolution of bipedalism—walking upright—marks the beginning of hominid evolution, although no one is quite sure why bipedalism evolved. The root of the hominid evolutionary tree is only imperfectly known.

The Genus *Homo*

The first humans evolved from australopithecine ancestors about 2 million years ago. The exact ancestor has not been clearly defined, but is commonly thought to be *A. afarensis*. Only within the last 30 years have a significant number of fossils of early *Homo* been uncovered. An explosion of interest has fueled intensive field exploration in the last few years, and new finds are announced regularly; every year, our picture of the base of the human evolutionary tree grows clearer. The following account will undoubtedly be outdated by future discoveries, but it provides a good example of science at work.

The First Human: *Homo habilis*

In the early 1960s, stone tools were found scattered among hominid bones close to the site where *A. boisei* had been unearthed. Although the fossils were badly crushed, painstaking reconstruction of the many pieces suggested a skull with a brain volume of about 680 cubic centimeters, larger than the australopithecine range of 400 to 550 cubic centimeters. Because of its association with tools, this early human was called *Homo habilis*, meaning "handy man." Partial skeletons discovered in 1986 indicate that *H. habilis* was small in stature, with arms longer than its legs and a skeleton much like that of *Australopithecus*. Because of its general similarity to australopithecines, many researchers at first questioned whether this fossil was human.

How Diverse Was Early *Homo*?

Because so few fossils of early *Homo* have been found, lively debate has ensued concerning whether they should all be lumped into *H. habilis* or split into three species: *H. rudolfensis*, *H. habilis*, and *H. ergaster* (Greek *ergaster*, "workman"). If the three species designations are accepted, as increasing numbers of researchers are doing, it would appear that *Homo* underwent an adaptive radiation, with *H. rudolfensis* the most ancient species, followed by *H. habilis* and then *H. ergaster*. Because of its modern skeleton, *H. ergaster* (figure 34.47) is thought to be the most likely ancestor to later species of *Homo*.

Out of Africa: *Homo erectus*

Our picture of what early *Homo* was like lacks detail, because it is based on only a few specimens. We have much more information about the species that replaced it, *Homo erectus*.

Homo erectus was a lot larger than *Homo habilis*—about 1.5 meters tall. It had a large brain, about 1000 cubic centimeters, and walked erect. Its skull had prominent brow ridges and, like modern humans, a rounded jaw. Most interesting of all, the shape of the skull interior suggests that *H. erectus* was able to talk.

FIGURE 34.47
Early *Homo*. This skull of a boy, who apparently died in early adolescence, is 1.6 million years old and has been assigned to the species *Homo ergaster*. He was about 1.5 meters in height and weighed approximately 47 kilograms.

Where did *H. erectus* come from? It should be no surprise to you that it came out of Africa. In 1976, a complete *H. erectus* skull was discovered in East Africa. It was 1.5 million years old, a million years older than earlier Asian finds. Far more successful than *H. habilis*, *H. erectus* quickly became widespread and abundant in Africa, and within 1 million years had migrated into Asia and Europe. A social species, *H. erectus* lived in tribes of 20 to 50 people, often dwelling in caves. They successfully hunted large animals, butchered them using flint and bone tools, and cooked them over fires—a site in China contains the remains of horses, bears, elephants, and rhinoceroses.

Homo erectus survived for over a million years, longer than any other species of human. These very adaptable humans only disappeared in Africa about 500,000 years ago, as modern humans were emerging. Interestingly, they survived even longer in Asia, until 250,000 years ago.

The evolutionary journey entered its final phase when modern humans first appeared in Africa about 600,000 years ago. Investigators who focus on human diversity denote three species of modern humans: *Homo heidelbergensis*, *H. neanderthalensis*, and *H. sapiens*. Other investigators lump the three species into one, *H. sapiens* ("wise man"). The oldest modern human, *Homo heidelbergensis*, is known from a 600,000-year-old fossil from Ethiopia. Although it coexisted with *H. erectus* in Africa, *H. heidelbergensis* has more

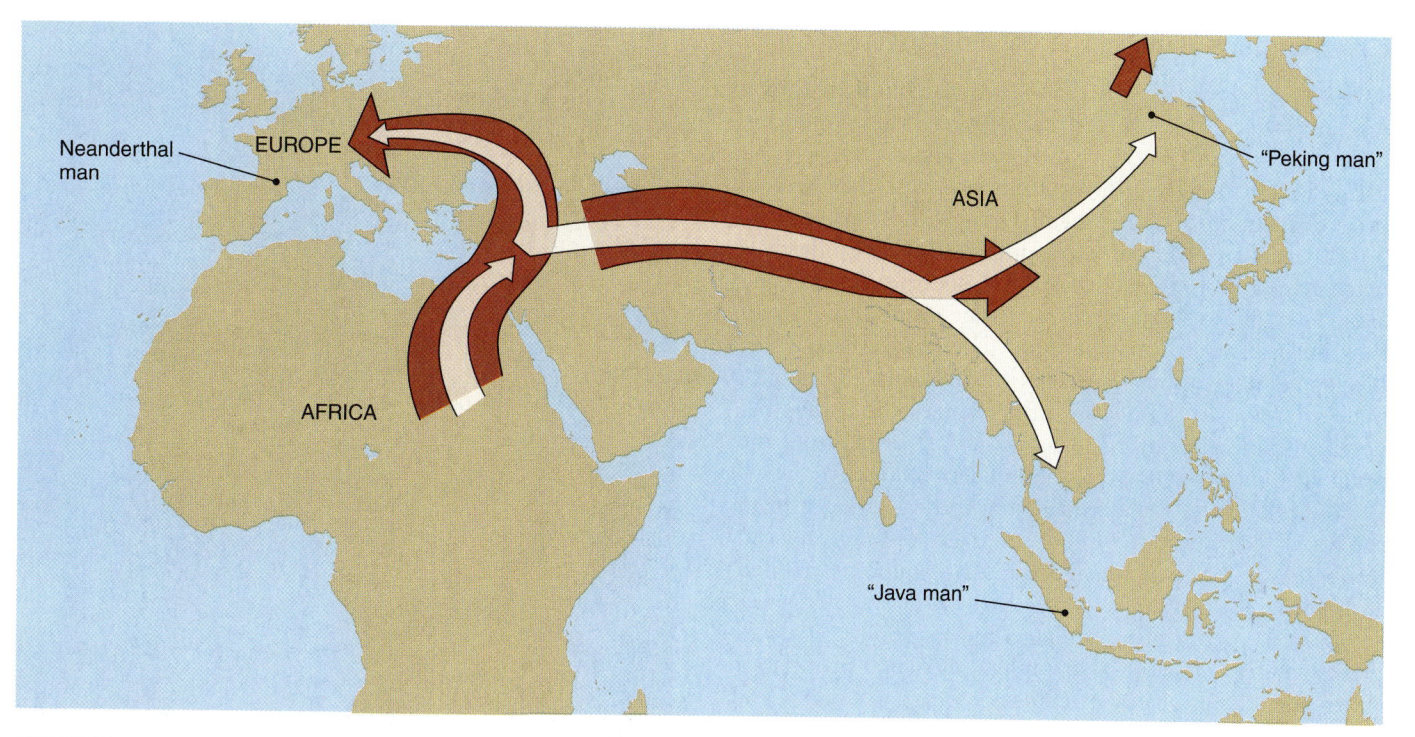

FIGURE 34.48
Out of Africa—many times. A still-controversial theory suggests that *Homo* spread from Africa to Europe and Asia repeatedly. First, *Homo erectus* (*white* arrow) spread as far as Java and China. Later, *H. erectus* was followed and replaced by *Homo neanderthalensis*, a pattern repeated again still later by *Homo sapiens* (*red* arrow).

advanced anatomical features, including a bony keel running along the midline of the skull, a thick ridge over the eye sockets, and a large brain. Also, its forehead and nasal bones are very like those of *H. sapiens*.

As *H. erectus* was becoming rarer, about 130,000 years ago, a new species of human arrived in Europe from Africa. *Homo neanderthalensis* likely branched off the ancestral line leading to modern humans as long as 500,000 years ago. Compared to modern humans, Neanderthals were short, stocky, and powerfully built; their skulls were massive, with protruding faces, heavy, bony ridges over the brows, and larger braincases.

Out of Africa—Again?

The oldest known fossil of *Homo sapiens*, our own species, is from Ethiopia and is about 130,000 years old. Other fossils from Israel appear to be between 100,000 and 120,000 years old. Outside of Africa and the Middle East, no clearly dated *H. sapiens* fossils older than roughly 40,000 years of age have been found. Proponents of a viewpoint known as the Recently-Out-of-Africa Hypothesis interpret this as evidence that *H. sapiens* evolved in Africa and then migrated to Europe and Asia. An opposing view, the Multiregional Hypothesis, argues that the human races independently evolved from *H. erectus* in different parts of the world.

Recently, scientists studying human mitochondrial DNA have added fuel to the fire of this controversy. Because DNA accumulates mutations over time, the oldest populations should show the greatest genetic diversity. Researchers sequencing the entire mDNA from 53 individuals of differing ethnic backgrounds found that modern humans shared a common ancestor 170,000 years ago, confirming that *H. sapiens* originated in Africa at about this time. The data also reveal a distinct branch on the human family tree 52,000 years ago, separating Africans from non-Africans. This is consistent with the hypothesis that humans originated in Africa, from there spreading to all parts of the world and retracing the path taken by *H. erectus* half a million years before (figure 34.48).

Another way to examine the human family tree is to look at genes on the Y chromosome, which does not undergo recombination. Y chromosomes pass down unchanged in males from one generation to the next. Any new changes that arise during evolution are easy to track on the family tree, because they too are passed down unchanged. The researchers in a large study looked at the pattern of gene variation among more than a thousand European males. While they identified many different patterns of variation, fully 80% of European males shared a single pattern, suggesting that modern Europeans have a common ancestor. The data indicate the pattern arose some 40,000 to 50,000 years ago. In other words, *H. sapiens* came to Europe recently.

By both these sets of evidence, the Multiregional Hypothesis is wrong. Our family tree has a single stem.

Chapter 34 Vertebrates 725

Cro-Magnons Replace the Neanderthals

The Neanderthals (classified by many paleontologists as a separate species, *Homo neanderthalensis*) were named after the Neander Valley of Germany where their fossils were first discovered in 1856. Rare at first outside of Africa, they became progressively more abundant in Europe and Asia, and by 70,000 years ago had become common. The Neanderthals made diverse tools, including scrapers, spearheads, and hand axes. They lived in huts or caves. Neanderthals took care of their injured and sick and commonly buried their dead, often placing food, weapons, and even flowers with the bodies. Such attention to the dead strongly suggests that they believed in a life after death. This is the first evidence of the symbolic thinking characteristic of modern humans.

Fossils of *H. neanderthalensis* abruptly disappear from the fossil record about 34,000 years ago and are replaced by fossils of *H. sapiens* called the Cro-Magnons (named after the valley in France where their fossils were first discovered). We can only speculate why this sudden replacement occurred, but it was complete all over Europe in a short period. A variety of evidence indicates that Cro-Magnons came from Africa—fossils of essentially modern aspect but as much as 100,000 years old have been found there. Cro-Magnons seem to have replaced the Neanderthals completely in the Middle East by 40,000 years ago, and then spread across Europe, coexisting with the Neanderthals for several thousand years. Recent analyses of Neanderthal DNA reveal it to be quite distinct from Cro-Magnon DNA, indicating the two species did not interbreed. Neanderthals are our cousins, not our ancestors. The Cro-Magnons that replaced the Neanderthals had a complex social organization and are thought to have had full language capabilities. Elaborate and often beautiful cave paintings made by Cro-Magnons can be seen throughout Europe (figure 34.49).

Humans of modern appearance eventually spread across Siberia to North America, where they arrived at least 13,000 years ago, after the ice had begun to retreat and a land bridge still connected Siberia and Alaska. By 10,000 years ago, about 5 million people inhabited the entire world (compared with more than 6 billion today).

Our Own Species: *Homo sapiens*

H. sapiens is the only surviving species of the genus *Homo*, and indeed the only surviving hominid. Some of the best fossils of *Homo sapiens* are 20 well-preserved skeletons with skulls found in a cave near Nazareth in Israel. Modern dating techniques estimate these humans to be between 90,000 and 100,000 years old. The skulls are modern in appearance and size, with high, short braincases, vertical foreheads with only slight brow ridges, and a cranial capacity of roughly 1550 cubic centimeters.

We humans are animals and the product of evolution. Our evolution has been marked by a progressive increase

FIGURE 34.49
Cro-Magnon art. Rhinoceroses are among the animals depicted in this remarkable cave painting found in 1995 near Vallon-Pont d'Arc, France.

in brain size, distinguishing us from other animals in several ways. First, humans are able to make and use tools effectively—a capability that, more than any other factor, has been responsible for our dominant position in the animal kingdom. Second, although not the only animal capable of conceptual thought, humans have refined and extended this ability until it has become the hallmark of our species. Finally, we use symbolic language and can, with words, shape concepts out of experience and transmit that accumulated experience from one generation to another. Thus, we have undergone what no other animal ever has: extensive cultural evolution. Through culture, we have found ways to change and mold our environment, rather than changing evolutionarily in response to the demands of the environment. We control our biological future in a way never before possible—an exciting potential and a frightening responsibility.

> Several species of *Homo* evolved in Africa, and some migrated from there to Europe and Asia. *Homo sapiens*, our species, seems to have evolved in Africa and then, like *H. erectus* before it, migrated to Europe and Asia. Our species, *Homo sapiens*, is proficient at conceptual thought and tool use, and is the only animal that uses symbolic language.

Concept Review

For interactive testing, visit the Online Learning Center with PowerWeb at www.mhhe.com/Raven7

34.1 Attaching muscles to an internal framework greatly improves movement.

The Chordates

- Four features characterize the chordates: (1) single, hollow nerve cord; (2) a flexible notochord present at some developmental stage; (3) pharyngeal pouches connecting the pharynx and the esophagus; (4) a postanal tail at least during embryonic development. (p. 684)

34.2 Nonvertebrate chordates have a notochord but no backbone.

The Nonvertebrate Chordates

- Tunicates possess a notochord and a nerve cord as larvae, but exhibit no major body cavity or segmentation as adults. (p. 686)
- Lancelets are fishlike marine chordates with a permanent notochord running the entire length of the dorsal nerve cord. (p. 687)

34.3 The evolution of vertebrates involved invasions of sea, land, and air.

Characteristics of Vertebrates

- Vertebrates differ from tunicates and lancelets in that they have a vertebral column instead of a notochord and a distinct, well-differentiated head. (p. 688)
- The history of the vertebrates includes a series of evolutionary advances that allowed them to invade the sea and then the land. (p. 689)

Fishes

- Fish were the first vertebrates and are the most diverse and successful vertebrate group. (p. 690)
- Key characteristics of fish include a vertebral column, jaws and paired appendages, gills, single-loop circulation, and nutritional deficiencies. (p. 690)
- The first fish had heads made of bone and internal skeletons made of cartilage. Jaws later evolved from cartilage arch supports. (pp. 692–693)
- Sharks eventually became dominant sea predators, partially due to a skeleton composed of calcified cartilage. Sharks were also among the first vertebrates to develop teeth. (p. 695)
- Bony fish evolved at the same time as sharks, but adopted a heavy internal skeleton made of bone. Such ossification provided a strong base for muscle attachment and evolved in fresh water. (p. 696)
- Bony fishes also evolved important adaptations such as a swim bladder for buoyancy, a lateral line sensory system, and a gill cover (operculum) to permit water to be pumped over the gills. (pp. 696–697)

Amphibians

- Key characteristics of living amphibians include legs, cutaneous respiration, lungs, pulmonary veins, and a partially divided heart. (p. 698)
- Paleontologists believe amphibians must have evolved from lobe-finned fishes. Amphibians today include frogs and toads, salamanders, and caecilians. (pp. 699–701)

Reptiles

- Key characteristics of reptiles include the amniotic egg, dry skin, and thoracic breathing. (p. 702)

The Rise and Fall of Dominant Reptile Groups

- Four major forms of reptiles took their respective turns as the dominant large terrestrial vertebrates: pelycosaurs, therapsids, theocodonts, and dinosaurs. (pp. 704–705)
- Other important characteristics of reptiles are that they practice internal fertilization, have an improved circulatory system, and are ectothermic. (p. 707)
- Only four reptilian orders survive today: turtles, lizards and snakes, tuataras, and crocodiles. (pp. 707–709)

Birds

- Modern birds retain many reptilian characteristics, but lack teeth and have vestigial tails. They are distinguished from living reptiles by feathers and the presence of a thin, hollow flight skeleton. (p. 710)

History of the Birds

- *Archaeopteryx* was probably the first bird, arising 150 MYA and likely a direct descendant of dinosaurs. (pp. 712–713)
- Aves continues to be listed as a separate class due to the evolutionary novelties of feathers and hollow bones and to physiological mechanisms such as efficient lungs capable of sustaining powered flight. (p. 713)

Mammals

- Key mammalian characteristics include hair, mammary glands, a placenta, heterodont dentition, the ability to digest plant material, keratinized hooves and horns, and flight capability (in bats). (pp. 714–716)

The Orders of Mammals

- Mammals were not common until dinosaurs disappeared. Modern mammals fall into one of three categories: monotremes, egg-laying mammals; marsupials, pouched mammals; and placentals. (pp. 717–718)

34.4 Evolution among the primates has focused on brain size and locomotion.

Primates

- Grasping fingers and toes and binocular vision are two features that allowed primates to flourish. (p. 720)
- Modern prosimians include lemurs, lorises, and tarsiers, while anthropoids include monkeys, apes, and humans. (pp. 720–721)

Australopithecines

- Bipedalism marked the beginning of hominid evolution, although the reason for such evolution remains controversial. (p. 722)
- Currently, two philosophical approaches, lumping and splitting, are used to characterize African hominid fossils. (p. 723)

The Genus Homo

- The first humans (*Homo habilis*) evolved from australopithecine ancestors about 2 MYA. (p. 724)
- *Homo erectus* replaced *H. habilis*, and is believed to have come out of Africa. (pp. 724–725)
- *Homo sapiens* is both the only surviving species of the genus *Homo* and the only surviving hominid. (p. 726)
- Humans are the only animals that can effectively make tools, that have refined and extended the ability to use conceptual thought, and that can use symbolic language and shape concepts and experiences with words. (p. 726)

Test Your Understanding

For interactive testing, visit the Online Learning Center with PowerWeb at www.mhhe.com/Raven7

Self Test

1. Which of the following is a characteristic of chordates but is *not* found in other animals?
   a. a notochord
   b. jointed appendages
   c. an exoskeleton
   d. all of these
2. In which animal(s) does the notochord persist in the adult?
   a. tunicates
   b. lampreys
   c. lancelets
   d. all of these
3. The very first vertebrates were
   a. cartilaginous fish.
   b. fishes with jaws.
   c. amphibians.
   d. jawless fish.
4. Which of the following is *not* a characteristic of fishes?
   a. gills
   b. lungs
   c. single-loop blood circulation
   d. nutritional deficiencies
5. What adaptation of bony fish allows them to detect and orient themselves in the upstream direction?
   a. the swim bladder
   b. lobed fins
   c. the operculum
   d. the lateral line system
6. In order for amphibians to be successful on land, they had to develop which of the following?
   a. a more efficient swim bladder
   b. cutaneous respiration and lungs
   c. more efficient gills
   d. shelled eggs
7. Amniotic eggs evolved as a means to
   a. protect the embryo while the parent sits on the egg.
   b. protect the embryo from predators.
   c. allow the parent to gather food, rather than sitting on the nest.
   d. prevent the embryo from drying out.
8. A group of early reptiles that may have been warm-blooded was the
   a. pelycosaurs.
   b. therapsids.
   c. thecodonts.
   d. all of these.
9. *Archaeopteryx* is believed to be the transition fossil between dinosaurs and birds because, like a bird, *Archaeopteryx* had feathers and
   a. a tail similar to that of modern birds.
   b. scales.
   c. a toothless, elongated mouth like a beak.
   d. a fused collarbone, indicating flying ability.
10. Mammals that have live births but incubate newborns in a pouch through the completion of development are
    a. monotremes.
    b. marsupials.
    c. duck-billed platypuses.
    d. placental mammals.

Test Your Visual Understanding

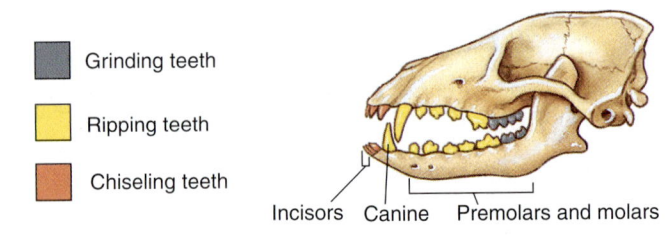

- Grinding teeth
- Ripping teeth
- Chiseling teeth

Dog — Incisors, Canine, Premolars and molars

Deer

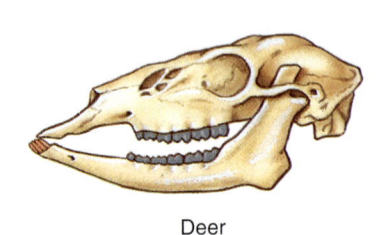

Beaver

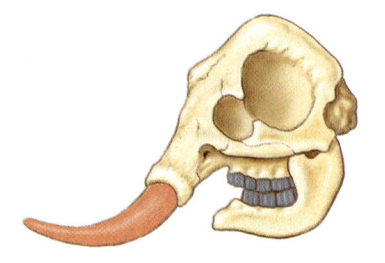

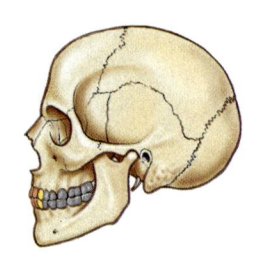

Elephant

Human

1. Based on this figure, predict whether each type of consumer is a carnivore, herbivore, or omnivore.
   a. dog
   b. deer
   c. beaver
   d. elephant
   e. human

Apply Your Knowledge

1. Homeothermic animals use 98% of cellular energy in metabolism and "store" 2% for growth. Poikilotherms have lower metabolisms and so are able to store 44% of cellular energy for growth. For homeotherms, 77.5% of chemical energy is converted into cellular energy with an efficiency of 77.5%. By comparison, poikilotherm efficiency is 41.9%. How much food must be consumed by each type of animal to gain one gram of weight?
2. What characteristics allowed vertebrates to attain great sizes?
3. What limits the ability of amphibians to occupy the full range of terrestrial habitats and allows other terrestrial vertebrates to occupy them successfully?
4. List some of the advantages that the early birds, in which flight was not nearly as efficient as it is in most of their modern descendants, might have had as a result of the presence of feathers.

52
Behavioral Biology

Chapter Outline

52.1 Many behavioral patterns are innate.

- **Approaches to the Study of Behavior.** Field biologists focus on evolutionary aspects of behavior.
- **Behavioral Genetics.** At least some behaviors are genetically determined.

52.2 Learning influences behavior.

- **Nonassociative Learning and Conditioning.** Conditioning requires association between two stimuli or between a stimulus and a response.
- **The Development of Behavior.** Parent-offspring interactions influence behavioral development.
- **Animal Cognition.** It is not clear to what degree animals "think."
- **Migratory Behavior.** Animals use many cues from the environment to navigate during migrations.

52.3 Communication is a key element of many animal behaviors.

- **Courtship.** Animals use signals to court one another.
- **Communication in Social Groups.** Bees and other social animals communicate in complex ways.

52.4 Evolutionary forces shape behavior.

- **Behavioral Ecology.** Behavior is shaped by natural selection.
- **Foraging Behavior.** Natural selection favors the most efficient foraging behavior.
- **Territorial Behavior.** Animals defend territories to increase reproductive advantage and foraging efficiency.
- **Reproductive Strategies.** The degree of parental investment influences other reproductive behaviors.
- **Reproductive Competition and Sexual Selection.** Mate choice affects reproductive success, and so is a target of natural selection.
- **Mating Systems.** Mating systems are reproductive solutions to particular ecological challenges.

52.5 There is considerable controversy about the evolution of social behavior.

- **Altruism and Group Living.** Many ideas have been put forward to explain the evolution of altruism.
- **Group Living and the Evolution of Social Systems.** Social organisms exhibit cooperation and altruism.

FIGURE 52.1
Rearing offspring involves complex behavior. Living in groups called prides makes lions better mothers. Females share the responsibilities of nursing and protecting the pride's young, increasing the probability that the youngsters will survive into adulthood.

Organisms interact with their environment in many ways. To understand these interactions, we need to appreciate the internal factors that shape the way an animal behaves, as well as aspects of the external environment that affect individual organisms. In this chapter, we explore the mechanisms that determine an animal's behavior (figure 52.1) and examine the field of behavioral ecology, which investigates how natural selection has molded behavior through evolutionary time.

Part VIII   Ecology and Behavior

52.1 Many behavioral patterns are innate.

Approaches to the Study of Behavior

During the past two decades, the study of animal behavior has emerged as an important and diverse science that bridges several disciplines within biology. Evolution, ecology, physiology, genetics, and psychology all have natural and logical linkages with the study of behavior, each discipline adding a different perspective and addressing different questions.

Research in animal behavior has made major contributions to our understanding of nervous system organization, child development, and human communication, as well as the speciation process, community organization, and the mechanism of natural selection itself. The study of the behavior of nonhuman animals has led to the identification of general principles of behavior, which have been applied, often controversially, to humans. This has changed the way we think about the origins of human behavior and the way we perceive ourselves.

Behavior can be defined as the way an animal responds to stimuli in its environment. A stimulus might be as simple as the odor of food. In this sense, a bacterial cell "behaves" by moving toward higher concentrations of the sugar. This behavior is very simple and well suited to the life of bacteria, allowing these organisms to live and reproduce. As animals evolved, they occupied different environments and faced diverse problems that affected their survival and reproduction. Their nervous systems and behavior concomitantly became more complex. Nervous systems perceive and process information concerning environmental stimuli and trigger adaptive motor responses, which we see as patterns of behavior.

When we observe animal behavior, we can explain it in two different ways. First, we might ask *how* it all works—that is, how the animal's senses, nerve networks, or internal state provide a physiological basis for the behavior. In this way, we would be asking a question about *proximate causation*. To analyze the proximate cause of behavior, we might measure hormone levels or record the impulse activity of neurons in the animal. We could also ask *why* the behavior evolved—that is, what is its adaptive value? This is a question concerning *ultimate causation*. To study the ultimate cause of a behavior, we would attempt to determine how it influenced the animal's survival or reproductive success. Thus, a male songbird may sing during the breeding season because it has a level of the steroid sex hormone testosterone, which binds to hormone receptors in the brain and triggers the production of song; this would be the proximate cause of the male bird's song. But the male sings to defend a territory from other males and to attract a female with which to reproduce; this is the ultimate, or evolutionary, explanation for the male's vocalization.

The study of behavior has had a long history of controversy. One source of controversy has been the question of whether behavior is determined more by an individual's genes or by its learning and experience. In other words, is behavior the result of nature (instinct) or nurture (experience)? In the past, this question has been considered an either/or proposition, but we now know that instinct and experience both play significant roles, often interacting in complex ways to produce the final behavior. The scientific study of instinct and learning, as well as their interrelationship, has led to the growth of several scientific disciplines, including ethology (the study of behavior), behavioral genetics, behavioral neuroscience, and comparative psychology.

Innate Behavior

Early research in the field of animal behavior—most notably by Nobel Prize winners Karl von Frisch, Konrad Lorenz, and Niko Tinbergen (figure 52.2)—focused on behavioral patterns that appeared to be instinctive or innate. Because behavior is often *stereotyped* (appearing in the same way in different individuals of a species), these early researchers argued that it must be based on preset paths in the nervous system. In their view, these paths are structured from genetic blueprints and cause animals to show essentially the same behavior from the first time it is produced throughout their lives.

These researchers based their opinions on behaviors such as egg retrieval by geese. Geese incubate their eggs in a nest. If a goose notices that an egg has been knocked out of the nest, it will extend its neck toward the egg, get up, and roll the egg back into the nest with a side-to-side motion of its neck while the egg is tucked beneath its bill (figure 52.3). Even if the egg is removed during retrieval, the goose completes the behavior, as if driven by a program released by the initial sight of the egg outside the nest. According to ethologists, egg retrieval behavior is triggered by a **sign stimulus** (also called a **key stimulus**), the appearance of an egg out of the nest; a component of the goose's nervous system, the **innate releasing mechanism**, provides the neural instructions for the motor program, or **fixed action pattern**. More generally, the sign stimulus is a "signal" in the environment that triggers a behavior. The innate releasing mechanism is the sensory mechanism that detects the signal, and the fixed action pattern is the stereotyped act.

One interesting aspect of sign stimuli is that they are often not very specific; in some situations, a wide variety of objects will trigger a fixed action pattern. For example, geese will attempt to roll baseballs and even beer cans back into their nests. Moreover, once the objects are in the nest, the goose recognizes that they are not eggs and removes them! A simi-

FIGURE 52.2
The founding fathers of ethology. Karl von Frisch, Konrad Lorenz, and Niko Tinbergen pioneered the study of behavioral science. In 1973, they received the Nobel Prize in physiology or medicine for their path-making contributions. Von Frisch led the study of honeybee communication and sensory biology. Lorenz focused on social development (imprinting) and the natural history of aggression. Tinbergen examined the functional significance of behavior and was the first behavioral ecologist.

lar example is provided by male stickleback fish. During the breeding season, males develop bright red coloration on their undersides. Territorial males react aggressively to the approach of other males, performing an aggressive display and even attacking. When Niko Tinbergen observed a male stickleback in a laboratory aquarium displaying aggressively when a red fire truck passed by the window, he realized that the red coloration was the sign stimulus. Subsequent experiments revealed that males would respond to many unfishlike models as long as the models had a red stripe.

This phenomenon is taken one step further by what are termed **supernormal stimuli.** Given a choice between two sign stimuli, one of normal size and the other much larger, many animals will respond to the larger of the two. Thus, geese given a choice of a normal goose egg and one the size of a volleyball will choose to roll the bigger one back to the nest. Why supernormal stimuli exist is not always clear. One aspect to keep in mind, however, is that in many cases, supernormal stimuli do not occur in nature. Thus, geese may prefer eggs the size of volleyballs, but they never encounter eggs of that size. It may be that geese have evolved to respond to the larger object so that they will attend to eggs, rather than smaller, circular rocks. As a result, natural selection may have favored the evolution of a preference for larger objects; this general response may lead to unexpected outcomes in experiments, but probably doesn't often lead to maladaptive behavior.

FIGURE 52.3
Innate egg-rolling response in geese. The series of movements used by a goose to retrieve an egg is a fixed action pattern. Once it detects the sign stimulus (in this case, an egg outside the nest), the goose goes through the entire set of movements: It will extend its neck toward the egg, get up, and roll the egg back into the nest with a side-to-side motion of its neck while the egg is tucked beneath its bill.

Early research in animal behavior emphasized innate behaviors that are the result of preset pathways in the nervous system and thus are likely to be genetically controlled.

Behavioral Genetics

In a famous experiment carried out in the 1940s, Robert Tryon studied the ability of rats to find their way through a maze with many blind alleys and only one exit, where a reward of food awaited. Some rats quickly learned to zip through the maze to the food, making few incorrect turns, while other rats took much longer to learn the correct path. Tryon bred the fast learners with one another to establish a "maze-bright" colony, and he bred the slow learners with one another to establish a "maze-dull" colony. He then tested the offspring in each colony to see how quickly they learned the maze. The offspring of maze-bright rats learned even more quickly than their parents had, while the offspring of maze-dull parents were even poorer at maze learning. After repeating this procedure over several generations, Tryon was able to produce two behaviorally distinct types of rat with very different maze-learning abilities (figure 52.4). Clearly, the ability to learn the maze was to some degree hereditary, governed by genes passed from parent to offspring. Furthermore, those genes appeared to be specific to this behavior, because the two groups of rats did not differ in their ability to perform other behavioral tasks, such as running a completely different kind of maze. Tryon's research demonstrates how a study can reveal that behavior has a heritable component.

Further support for the genetic basis of behavior has come from studies of hybrids. William Dilger of Cornell University examined two species of lovebird (genus *Agapornis*), which differ in the way they carry twigs, paper, and other materials used to build a nest. *Agapornis fischeri* holds nest material in its beak, while *A. roseicollis* carries material tucked under its flank feathers (figure 52.5). When Dilger crossed the two species to produce hybrids, he found that the hybrids carry nest material in a way that seems intermediate between that of the parents: They repeatedly shift material between the beak and the flank feathers. Other studies conducted on courtship songs in crickets and tree frogs also demonstrate the intermediate nature of hybrid behavior.

The role of genetics can also be seen in humans by comparing the behavior of identical twins. Identical twins are, as their name implies, genetically identical. However, most sets of identical twins are raised in the same environment, so it is not possible to determine whether similarities in behavior result from their genetic similarity or from experiences shared as they grew up (the classic nature-versus-nurture debate). However, in some cases, twins have been separated at birth. A recent study of 50 such sets of twins revealed many similarities in personality, temperament, and even leisure-time activities, even though the twins had often been raised in very different environments. These similarities indicate that genetics plays a role in determining behavior even in humans, although the relative importance of genetics versus environment is still hotly debated.

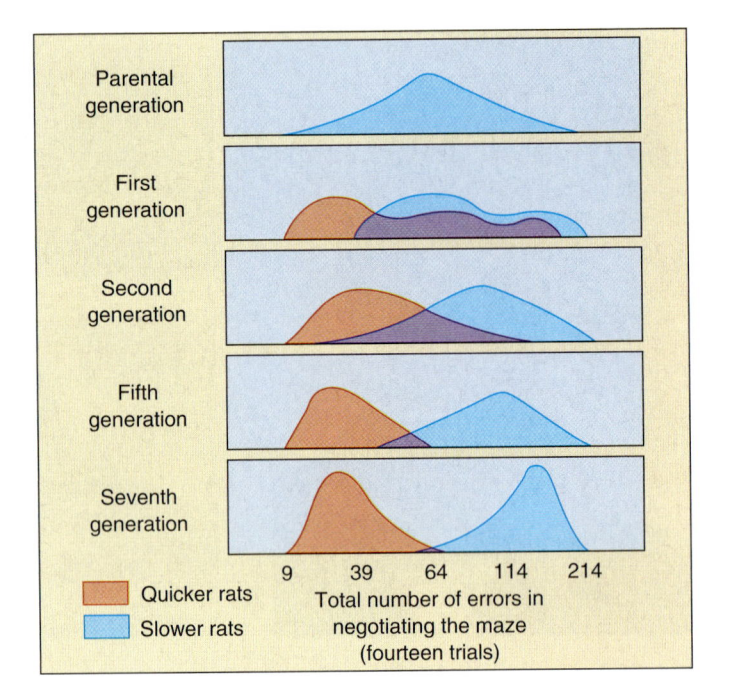

FIGURE 52.4
The genetics of learning. Selection experiments in the laboratory established a genetic basis for differences in the ability to learn to run through a maze.
What would happen if, after the seventh generation, rats were randomly assigned mates regardless of their ability to learn the maze?

FIGURE 52.5
Genetics of lovebird behavior. Some species of lovebirds carry nest material, such as these paper strips, under their flank feathers. When mated with species that carry material in their beaks, the hybrids show intermediate behavior, attempting to carry the material in both ways.

FIGURE 52.6
Genetically caused defect in maternal care. (*a*) In mice, normal mothers take very good care of their offspring, retrieving them if they move away and crouching over them. (*b*) Mothers with the mutant *fosB* allele perform neither of these behaviors, leaving their pups exposed. (*c*) Amount of time female mice were observed crouching in a nursing posture over offspring. (*d*) Proportion of pups retrieved when they were experimentally moved. **Why does the lack of *fosB* alleles lead to maternal inattentiveness?**

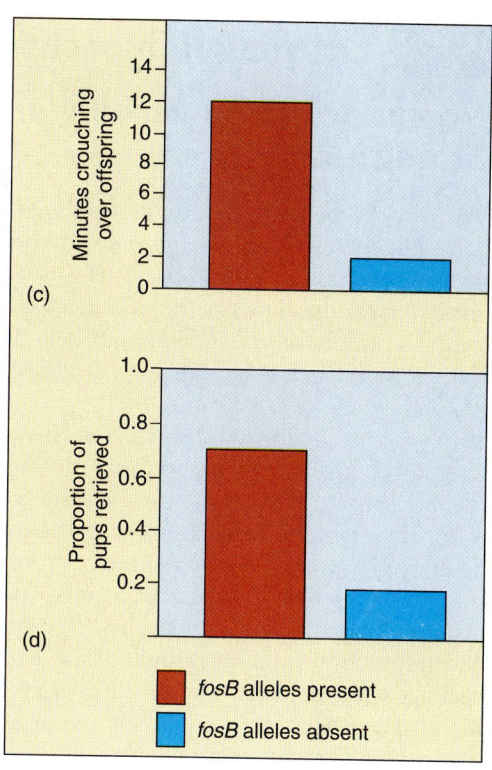

Single Gene Effects on Behavior

The maze-learning, hybrid, and identical twins studies just described suggest that genes play a role in behavior, but recent research has provided much greater detail on the genetic basis of behavior. In both *Drosophila* and mice, many mutations have been associated with particular behavioral abnormalities.

In fruit flies, for example, individuals that possess alternative alleles for a single gene differ greatly in their feeding behavior as larvae; larvae with one particular allele move around a great deal as they eat, whereas individuals with the alternative allele move hardly at all. A wide variety of mutations at other genes are now known in *Drosophila* that affect almost every aspect of courtship behavior.

The ways in which genetic differences affect behavior have been worked out for several mouse genes. For example, some mice with one mutation have trouble remembering information learned two days earlier about where objects are located. This difference appears to result because the mutant mice do not produce the enzyme α-calcium-calmodulin-dependent kinase II, which plays an important role in the functioning of the hippocampus, a part of the brain important for spatial learning.

Modern molecular biology techniques allow the role of genetics in behavior to be investigated with ever greater precision. For example, male mice genetically engineered to lack the ability to synthesize nitric oxide, a brain neurotransmitter, show increased aggressive behavior.

A particularly fascinating breakthrough occurred in 1996, when scientists discovered a new gene, *fosB*, that seems to determine whether or not female mice will nurture their young. Females with both *fosB* alleles disabled will initially investigate their newborn babies, but then ignore them, in stark contrast to the caring and protective maternal behavior displayed by normal females (figure 52.6).

The cause of this inattentiveness appears to result from a chain reaction. When mothers of new babies initially inspect them, information from their auditory, olfactory, and tactile senses is transmitted to the hypothalamus, where *fosB* alleles are activated, producing a particular protein, which in turn activates other enzymes and genes that affect the neural circuitry within the hypothalamus. These modifications within the brain cause the female to react maternally toward her offspring. In contrast, in mothers lacking the *fosB* alleles, this reaction is stopped midway. No protein is activated, the brain's neural circuitry is not rewired, and maternal behavior does not result.

As these genetic techniques are becoming used more widely, the next few years should see similar dramatic advances in our knowledge of how genes affect various human behaviors.

The genetic basis of behavior is supported by artificial selection experiments, hybridization studies, and studies on the behavior of mutants. Research has also identified specific genes that control behavior.

52.2 Learning influences behavior.

Nonassociative Learning and Conditioning

Many of the behavioral patterns displayed by animals are not solely the result of instinct. In many cases, animals alter their behavior as a result of previous experiences, a process termed **learning**. The role of learning was first studied intensively in laboratory rodents, but now researchers have learned much about the learning processes and capabilities of a wide range of organisms.

The simplest type of learning, **nonassociative learning**, does not require an animal to form an association between two stimuli or between a stimulus and a response. One form of nonassociative learning is **habituation**, which can be defined as a decrease in response to a repeated stimulus that has no positive or negative consequences. In many cases, the stimulus evokes a strong response when it is first encountered, but the magnitude of the response gradually declines with repeated exposure. For example, young birds see many types of objects moving overhead. At first, they may respond by crouching down and remaining still. Some of the objects, such as falling leaves or members of their own species flying by, are seen very frequently and have no positive or negative consequence to the nestlings. Over time, the young birds may habituate to such stimuli and stop responding. Thus, habituation can be thought of as learning not to respond to a stimulus. Being able to ignore unimportant stimuli is critical for an animal confronting a barrage of stimuli in a complex environment; animals that could not do so would fail to focus their attention on important activities, such as finding food and avoiding predators, and probably would leave few offspring in the next generation.

A change in behavior that involves an association between two stimuli or between a stimulus and a response is termed **associative learning** (figure 52.7). The behavior is modified, or **conditioned**, through the association. This form of learning is more complex than habituation. The two major types of associative learning are **classical conditioning** and **operant conditioning**; they differ in the way the associations are established.

Classical Conditioning

In classical conditioning, the paired presentation of two different kinds of stimuli causes the animal to form an association between the stimuli. Classical conditioning is also called **Pavlovian conditioning**, after Russian psychologist Ivan Pavlov, who first described it. Pavlov presented meat powder, an *unconditioned stimulus*, to a dog and noted that the dog responded by salivating, an *unconditioned response*. If an unrelated stimulus, such as the ring-

FIGURE 52.7
Learning what is edible. Associative learning is involved in predator-prey interactions. (*a*) A naive toad is offered a bumblebee as food. (*b*) The toad is stung, and (*c*) subsequently avoids feeding on bumblebees or any other insects having black-and-yellow coloration. The toad has associated the appearance of the insect with pain, and modifies its behavior.

1110   Part VIII   Ecology and Behavior

ing of a bell, was presented at the same time as the meat powder, over repeated trials the dog would salivate in response to the sound of the bell alone. The dog had learned to associate the unrelated sound stimulus with the meat powder stimulus. Its response to the sound stimulus was, therefore, conditioned, and the sound of the bell is referred to as a *conditioned stimulus*.

Operant Conditioning

In operant conditioning, an animal learns to associate its behavioral response with a reward or punishment. American psychologist B. F. Skinner studied operant conditioning in rats by placing them in an apparatus that came to be called a "Skinner box." As the rat explored the box, it would occasionally press a lever by accident, causing a pellet of food to appear. At first, the rat would ignore the lever, eat the food pellet, and continue to move about. Soon, however, it learned to associate pressing the lever (the behavioral response) with obtaining food (the reward). When it was hungry, it would spend all its time pressing the lever. This sort of trial-and-error learning is of major importance to most vertebrates.

Comparative psychologists used to believe that any two stimuli could be linked in classical conditioning and that animals could be conditioned to perform any learnable behavior in response to any stimulus by operant conditioning. As you will see in the following discussion, this view has changed. Today, it is thought that instinct guides learning by determining what type of information can be learned through conditioning.

Instinct and Learning

It is now clear that some animals have innate predispositions toward forming certain associations. For example, if a rat is offered a food pellet at the same time it is exposed to X rays (which later produce nausea), the rat will remember the taste of the food pellet but not its size. Similarly, pigeons can learn to associate *food* with colors but not with sounds, while they can associate *danger* with sounds but not with colors.

These examples of *learning preparedness* demonstrate that what an animal can learn is biologically influenced— that is, learning is possible only within the boundaries set by instinct. Innate programs have evolved because they underscore adaptive responses. In nature, food that is toxic to a rat is likely to have a particular taste; thus, it is adaptive to be able to associate a taste with a feeling of sickness that may develop hours later. The seed a pigeon eats may have a distinctive color that the pigeon can see, but it makes no sound the pigeon can hear. The study of learning has expanded to include its ecological significance, so that we are now able to consider the evolution of learning.

FIGURE 52.8
The Clark's nutcracker has an extraordinary memory. A Clark's nutcracker can remember the locations of up to 2000 seed caches months after hiding them. After conducting experiments, scientists have concluded that the birds use features of the landscape and other surrounding objects as spatial references to memorize the locations of the caches.

An animal's ecology, of course, is key to understanding its mental capabilities. Some species of birds, such as Clark's nutcracker, feed on seeds. Birds store seeds in caches they bury when seeds are abundant so they will have food during the winter. Thousands of seed caches may be buried and then later recovered, sometimes as much as nine months later. One would expect the birds to have an extraordinary spatial memory, and this is indeed what has been found (figure 52.8). Clark's nutcracker, and other seed-hoarding birds, have an unusually large hippocampus, the center for memory storage in the brain (see chapter 45).

> **Habituation is a simple form of learning in which there is no association between stimuli and responses. In contrast, associative learning (classical and operant conditioning) involves the formation of an association between two stimuli or between a stimulus and a response.**

The Development of Behavior

Behavioral biologists now recognize that behavior has both genetic and learned components. Thus far in this chapter, we have discussed the influence of genes and learning separately. But as we will see, these factors interact during development to shape behavior.

Parent-Offspring Interactions

As an animal matures, it may form social attachments to other individuals or develop preferences that will influence behavior later in life. This process, called **imprinting**, is sometimes considered a type of learning. In *filial imprinting*, social attachments form between parents and offspring. For example, young birds of some species begin to follow their mother within a few hours after hatching, and their following response results in a bond between mother and young. However, the young birds' initial experience determines how this imprint is established. The German ethologist Konrad Lorenz showed that birds will follow the first object they see after hatching and direct their social behavior toward that object. Lorenz raised geese from eggs, and when he offered himself as a model for imprinting, the goslings treated him as if he were their parent, following him dutifully (figure 52.9). Black boxes, flashing lights, and watering cans can also be effective imprinting objects (figure 52.10). The success of imprinting is highest during a **sensitive phase,** or a **critical period** (roughly 13 to 16 hours after hatching in geese).

Several studies demonstrate that the social interactions that occur between parents and offspring during the critical period are key to the normal development of behavior. The psychologist Harry Harlow gave orphaned rhesus monkey infants the opportunity to form social attachments with two surrogate "mothers," one made of soft cloth covering a wire frame and the other made only of wire. The infants chose to spend time with the cloth mother, even if only the wire mother provided food, indicating that texture and tactile contact, rather than provision of food, may be among the key qualities in a mother that promote infant social attachment. If infants are deprived of normal social contact, their development is abnormal. Greater degrees of deprivation lead to greater abnormalities in social behavior during childhood and adulthood. Studies of orphaned human infants suggest that a constant "mother figure" is required for normal growth and psychological development.

Recent research has revealed a biological need for the stimulation that occurs during parent-offspring interactions early in life. Female rats lick their pups after birth, and this stimulation inhibits the release of a hormonelike chemical that can block normal growth. Pups that receive normal tactile stimulation also have more brain receptors for glucocorticoid hormones, longer-lived brain neurons, and a greater tolerance for stress. Premature human infants who are massaged gain weight rapidly. These studies indicate that the need for normal social interaction is based in the brain and that touch and other aspects of contact between parents and offspring are important for physical as well as behavioral development.

Sexual imprinting, is a process in which an individual learns to direct its sexual behavior at members of its own species. **Cross-fostering** studies, in which individuals of one species are raised by parents of another species, reveal that this form of imprinting also occurs early in life. In most species of birds, these studies have shown that the fostered bird will attempt to mate with members of its foster species when it is sexually mature.

FIGURE 52.9
An unlikely parent. The eager goslings follow Konrad Lorenz as if he were their mother. He is the first object they saw when they hatched, and they have used him as a model for imprinting.

FIGURE 52.10
How imprinting is studied. Ducklings will imprint on the first object they see, even (*a*) a black box or (*b*) a white sphere.

Interaction Between Instinct and Learning

The work of Peter Marler and his colleagues on the acquisition of courtship song by white-crowned sparrows provides an excellent example of the interaction between instinct and learning in the development of behavior. Courtship songs are sung by mature males and are species-specific. By rearing male birds in soundproof incubators equipped with speakers and microphones, Marler could control what a bird heard as it matured and then record the song it produced as an adult. He found that white-crowned sparrows that heard no song at all during development or that heard only the song of a different species, the song sparrow, sang a poorly developed song as adults (figure 52.11). But birds that heard the song of their own species or that heard the songs of *both* the white-crowned sparrow and the song sparrow, sang a fully developed, white-crowned sparrow song as adults. These results suggest that these birds have a **genetic template**, or instinctive program, that guides them to learn the appropriate song. During a critical period in development, the template will accept the correct song as a model. Thus, song acquisition depends on learning, but only the song of the correct species can be learned. The genetic template for learning is *selective*. However, learning plays a prominent role as well. If a young white-crowned sparrow becomes deaf after it hears its species' song during the critical period, it will also sing a poorly developed song as an adult. Therefore, the bird must "practice" listening to himself sing, matching what he hears to the model his template has accepted.

Although this explanation of song development stood unchallenged for many years, recent research has shown that white-crowned sparrow males *can* learn another species' song under certain conditions. If a live male strawberry finch is placed in a cage next to a young male sparrow, the young sparrow will learn to sing the strawberry finch's song! This finding indicates that social stimuli may be more effective than a tape-recorded song in overriding the innate program that guides song development. Furthermore, the males of some bird species have no opportunity to hear the song of their own species. In such cases, it appears that the males, instinctively "know" their own species' song. For example, cuckoos are brood parasites; females lay their eggs in the nest of another species of bird, and the young that hatch are reared by the foster parents (figure 52.12). When the cuckoos become adults, they sing the song of their own species rather than that of their foster parents. Because male brood parasites would most likely hear the song of their host species during development, it is adaptive for them to ignore such "incorrect" stimuli. They hear no adult males of their own species singing, so no correct song models are available. In these species, natural selection has programmed the male with a genetically guided song.

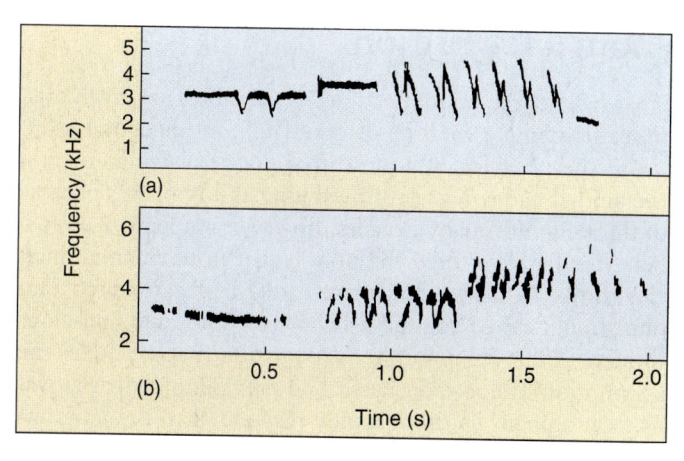

FIGURE 52.11
Song development in birds. (*a*) The sonograms of songs produced by male white-crowned sparrows that had been exposed to their own species' song during development are different from (*b*) those of male sparrows that heard no song during rearing. This difference indicates that the genetic program itself is insufficient to produce a normal song.

FIGURE 52.12
Brood parasite. Cuckoos lay their eggs in the nests of other species of birds. Because the young cuckoos (large bird to the right) are raised by a different species (such as this meadow pipit, smaller bird to the left), they have no opportunity to learn the cuckoo song; the cuckoo song they later sing is innate.

Interactions that occur during sensitive phases of imprinting are critical to normal behavioral development. Physical contact plays an important role in growth and in the development of psychological well-being.

Animal Cognition

To what degree animals "think" is a subject of lively dispute. It is likely each of us could tell an anecdotal story about the behavior of a pet cat or dog that would suggest the animal had a degree of reasoning ability or was capable of thinking. For many decades, however, students of animal behavior flatly rejected the notion that nonhuman animals can think. In fact, behaviorist Lloyd Morgan stated that one should never assume a behavior represents conscious thought if there is any other explanation that precludes the assumption of consciousness. The prevailing approach was to treat animals as though they responded to the environment through instinctive behaviors and simple, innately programmed learning.

In recent years, serious attention has been given to the topic of animal awareness. The central question is whether animals show **cognitive behavior**—that is, do they process information and respond in a manner that suggests thinking (figure 52.13)? What kinds of behavior would demonstrate cognition? Some birds in urban areas remove the foil caps from nonhomogenized milk bottles to get at the cream beneath, and this behavior is known to have spread within a population to other birds. Japanese macaques learn to wash potatoes and float grain to separate it from sand. A chimpanzee pulls the leaves off a tree branch and uses the branch to probe the entrance to a termite nest and gather termites. Vervet monkeys have a vocabulary that identifies specific predators (see figure 52.24).

Only a few experiments have tested the thinking ability of nonhuman animals. Some of these studies suggest that animals may deliberately give false information (that is, they "lie"). Currently, researchers are trying to determine if some primates deceive others to manipulate the behavior of the other members of their troop. Many anecdotal accounts appear to support the idea that deception occurs in some nonhuman primate species, such as baboons and chimpanzees, but it has been difficult to devise field-based experiments to test this idea. Much of this type of research on animal cognition is in its infancy, but it is sure to grow and to raise controversy. In any case, there is nothing to be gained by dogmatically denying the possibility of animal consciousness.

Some examples, particularly those involving problem solving by animals, are hard to explain in any way other

FIGURE 52.13

Animal thinking? (*a*) This chimpanzee is stripping the leaves from a twig, which it will then use to probe a termite nest. This behavior strongly suggests that the chimpanzee is consciously planning ahead, with full knowledge of what it intends to do. (*b*) This sea otter is using a rock as an "anvil," against which it bashes a clam to break it open. A sea otter will often keep a favorite rock for a long time, as though it has a clear idea of its future use of the rock. Behaviors such as these suggest that animals have cognitive abilities.

**FIGURE 52.14
Problem solving by a chimpanzee.** Unable to get the bananas by jumping, the chimpanzee devises a solution.

than as a result of some sort of mental process. For example, in a series of classic experiments conducted in the 1920s, a chimpanzee was left in a room with a banana hanging from the ceiling out of reach. Also in the room were several boxes, each lying on the floor. After some unsuccessful attempts to jump up and grab the bananas, the chimpanzee suddenly looked at the boxes and then immediately proceeded to move them underneath the banana, stack one on top of another, and climb up to claim its prize (figure 52.14).

Perhaps it is not surprising to find obvious intelligence in animals as closely related to us as chimpanzees. But recent studies have found that other animals also show evidence of cognition. Ravens have always been considered among the most intelligent of birds. Bernd Heinrich of the University of Vermont recently conducted an experiment using a group of hand-reared ravens that lived in an outdoor aviary. Heinrich placed a piece of meat on the end of a string and hung it from a branch in the aviary. The birds liked to eat meat, but had never seen string before and were unable to get at the meat. After several hours, during which time the birds periodically looked at the meat but did nothing else, one bird flew to the branch, reached down, grabbed the string, pulled it up, and placed it under his foot. He then reached down and grabbed another piece of the string, repeating this action over and over, each time bringing the meat closer (figure 52.15). Eventually, he brought the meat within reach and grasped it. The raven, presented with a completely novel problem, had devised a solution. Eventually, three of the other five ravens also figured out how to get the meat. Heinrich has conducted other similarly creative experiments that leave little doubt that ravens have advanced cognitive abilities.

> **Research on the cognitive behavior of animals is in its infancy, but some examples are compelling.**

**FIGURE 52.15
Problem solving by a raven.** Confronted with a problem it has never previously faced, the raven figures out how to get the meat at the end of the string by repeatedly pulling up a bit of string and stepping on it.

Chapter 52 Behavioral Biology 1115

Migratory Behavior

Orientation and Migration

Animals may travel to and from a nest to feed or move regularly from one place to another. To do so, they must orient themselves by tracking stimuli in the environment.

Movement toward or away from a stimulus is called **taxis**. The attraction of flying insects to outdoor lights is an example of *positive phototaxis*. Insects that avoid light, such as the common cockroach, exhibit *negative phototaxis*. Other stimuli may be used as orienting cues. For example, trout orient themselves in a stream so as to face against the current. However, not all responses involve a specific orientation. Some animals just become more or less active when stimulus intensity increases; such responses are called **kineses**.

Long-range, two-way movements are known as **migrations**. Ducks and geese migrate along flyways from Canada across the United States each fall and return each spring. Monarch butterflies migrate each fall from central and eastern North America to several small, geographically isolated areas of coniferous forest in the mountains of central Mexico (figure 52.16). Each August, the butterflies begin a flight southward to their overwintering sites. At the end of winter, the monarchs begin the return flight to their summer breeding ranges. What is amazing about the migration of the monarch, however, is that two to five generations may be produced as the butterflies fly north. The butterflies that migrate in the autumn to the precisely located overwintering grounds in Mexico have never been there before.

Recent geographic range expansions by some migrating birds have revealed how migratory patterns change. When colonies of bobolinks became established in the western United States, far from their normal range in the Midwest and East, they did not migrate directly to their winter range in South America. Instead, they migrated east to their ancestral range and then south along the original flyway (figure 52.17). Rather than changing the original migration pattern, they simply added a new pattern. Scientists continue to study the western bobolinks to learn whether, in time, a more efficient migration path will evolve or whether the birds will always follow their ancestral course.

How Migrating Animals Navigate

Biologists have studied migration with great interest, and we now have a good understanding of how these feats of navigation are achieved. It is important to understand the distinction between **orientation** (the ability to follow a bearing) and **navigation** (the ability to set or adjust a bearing, and then follow it). The former is analogous to using a compass, while the latter is like using a compass in conjunction with a map. Experiments on starlings indicate that inexperienced birds migrate by orientation, while older birds that have migrated previously use true navigation (figure 52.18).

Birds and other animals navigate by looking at the sun and the stars. The indigo bunting, which flies during the day and uses the sun as a guide, compensates for the movement of the sun in the sky as the day progresses by reference to the North Star, which does not move in the sky. Buntings also use the positions of the constellations

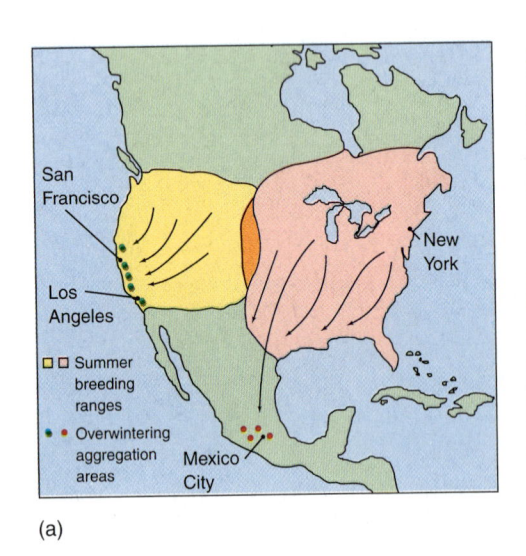

FIGURE 52.16
Migration of monarch butterflies. (*a*) Monarchs from western North America overwinter in areas of mild climate along the Pacific coast. Those from the eastern United States and southeastern Canada migrate to Mexico, a journey of over 3000 kilometers that takes from two to five generations to complete. (*b*) Monarch butterflies arrive at the remote fir forests of the overwintering grounds in Mexico, where they (*c*) form aggregations on the tree trunks.

FIGURE 52.17
Birds on the move. (*a*) The summer range of bobolinks recently extended to the far western United States from their more established range in the Midwest. When they migrate to South America in the winter, bobolinks that nested in the West do not fly directly to the winter range; instead, they fly to the Midwest first and then use the ancestral flyway. (*b*) The golden plover has an even longer migration route that is circular. These birds fly from arctic breeding grounds to wintering areas in southeastern South America, a distance of some 13,000 kilometers.

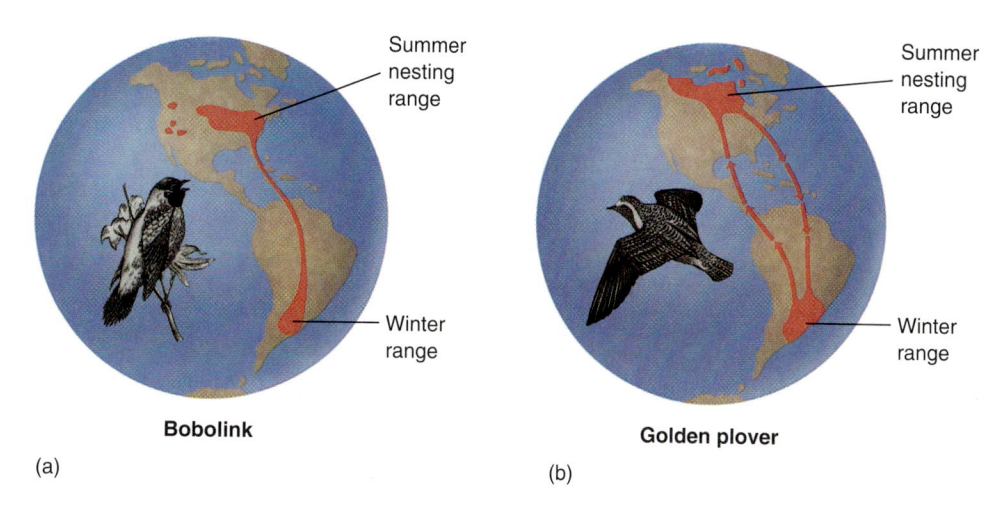

and the position of the pole star in the night sky, cues they learn as young birds. Starlings and certain other birds compensate for the sun's apparent movement in the sky by using an internal clock. If such birds are shown an experimental sun in a fixed position while in captivity, they will change their orientation to it at a constant rate of about 15° per hour.

Many migrating birds also have the ability to detect the earth's magnetic field and to orient themselves with respect to it. In a closed indoor cage, they will attempt to move in the correct geographic direction, even though there are no visible external cues. However, the placement of a powerful magnet near the cage can alter the direction in which the birds attempt to move. Researchers have found magnetite, a magnetized iron ore, in the heads of some birds, but have not been able to identify the sensory receptors birds employ to detect magnetic fields.

It appears that the first migration of a bird is innately guided by both celestial cues (the birds fly mainly at night) and the earth's magnetic field. These cues give the same information about the general direction of the migration, but when the two cues are experimentally manipulated to give conflicting directions, the information provided by the stars seems to override the magnetic information. Recent studies, however, indicate that celestial cues tell northern hemisphere birds to move south when they begin their migration, while magnetic cues give them the direction for the specific migratory path (perhaps a southeast turn the bird must make midroute). In short, these new data suggest that celestial and magnetic cues interact during development to fine-tune the bird's navigation.

We know relatively little about how other migrating animals navigate. For instance, green sea turtles migrate from Brazil halfway across the Atlantic Ocean to Ascension Island, where the females lay their eggs. How do they find this tiny island in the middle of the ocean, which they haven't seen for perhaps 30 years? How do the young that hatch on the island know how to find their way to Brazil? Recent studies suggest that wave action is an important cue.

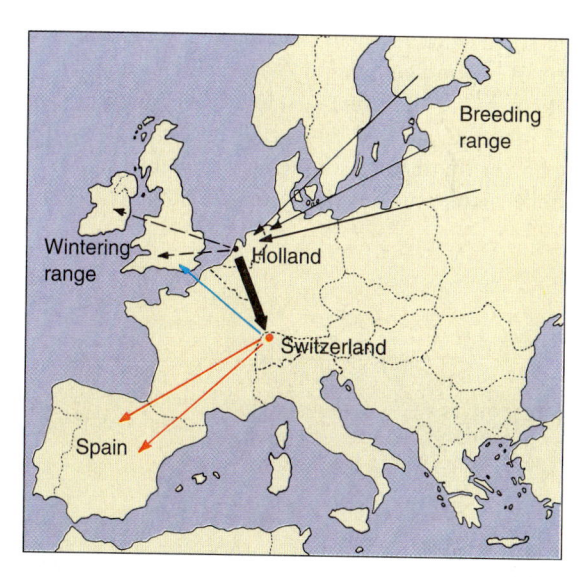

FIGURE 52.18
Migratory behavior of starlings. The navigational abilities of inexperienced birds differ from those of adults that have made the migratory journey before. Starlings were captured in Holland, halfway along their full migratory route from Baltic breeding grounds to wintering grounds in the British Isles; these birds were transported to Switzerland and released. Experienced older birds compensated for the displacement and flew toward the normal wintering grounds (*blue arrow*). Inexperienced young birds kept flying in the same direction, on a course that took them toward Spain (*red arrows*). These observations imply that inexperienced birds fly by orientation, while experienced birds learn true navigation.

> **Many animals migrate in predictable ways, navigating by looking at the sun and stars, and in some cases by detecting magnetic fields.**

52.3 Communication is a key element of many animal behaviors.

Much of the research in animal behavior is devoted to analyzing the nature of communication signals, determining how they are perceived, and identifying their ecological roles and their evolutionary origins.

Courtship

During courtship, animals produce signals to communicate with potential mates and with other members of their own sex. A *stimulus-response chain* sometimes occurs, in which the behavior of one individual in turn releases a behavior by another individual (figure 52.19).

Courtship Signaling

A male stickleback fish will defend the nest it builds on the bottom of a pond or stream by attacking *conspecific* males (that is, males of the same species) that approach the nest. Niko Tinbergen studied the social releasers responsible for this behavior by making simple clay models. He found that a model's shape and degree of resemblance to a fish were unimportant; any model with a red underside (like the underside of a male stickleback) could release the attack behavior. Tinbergen also used a series of clay models to demonstrate that a male stickleback recognizes a female by her abdomen when it is swollen with eggs.

Courtship signals are often *species-specific*, limiting communication to members of the same species and thus playing a key role in reproductive isolation (see chapter 23). The flashes of fireflies (which are actually beetles) are such species-specific signals. Females recognize conspecific males by their flash pattern (figure 52.20), and males recognize conspecific females by their flash response. This series of reciprocal responses provides a continuous "check" on the species identity of potential mates.

Long-Distance Communication

Chemical signals also mediate interactions between males and females. **Pheromones**, chemical messengers used for communication between individuals of the same species, serve as sex attractants, among other functions, in many animals. Even the human egg produces a chemical attractant to communicate with sperm! Female silk moths (*Bombyx mori*) produce a sex pheromone called *bombykol* in a gland

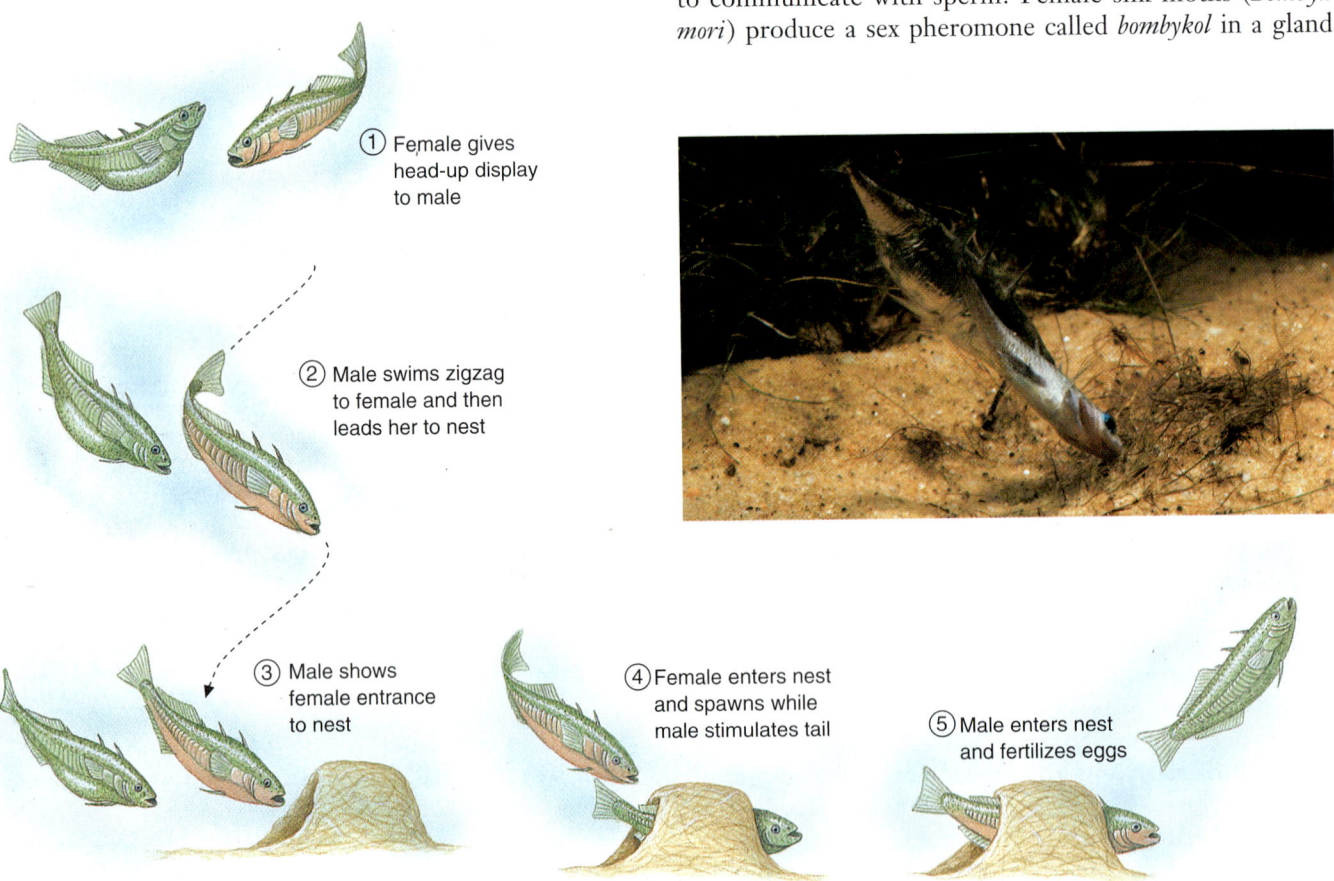

FIGURE 52.19
A stimulus-response chain. Stickleback courtship involves a sequence of behaviors leading to the fertilization of eggs.

1118 Part VIII Ecology and Behavior

associated with the reproductive system. Neurophysiological studies show that the male's antennae contain numerous sensory receptors specific for bombykol. These receptors are extraordinarily sensitive, enabling the male to respond behaviorally to concentrations of bombykol as low as one molecule in 10^{17} molecules of oxygen in the air.

Many insects, amphibians, and birds produce species-specific acoustic signals to attract mates. Bullfrog males call by inflating and discharging air from their vocal sacs, located beneath the lower jaw. Females can distinguish a conspecific male's call from the call of other frogs that may be in the same habitat and calling at the same time. As mentioned in section 52.2, male birds produce songs, complex sounds composed of notes and phrases, to advertise their presence and to attract females. In many species, variations in the males' songs identify *particular* males in a population. In these species, the song is individually specific as well as species-specific.

Level of Specificity

Different signals provide different levels of information about the sender. The **level of specificity** relates to the function of the signal. Many courtship signals are species-specific to help animals avoid making errors in mating that would produce inviable hybrids or otherwise waste reproductive effort. A male bird's song is individually specific because it allows his presence (as opposed to simply the presence of an unidentifiable member of the species) to be recognized by neighboring birds. When territories are being established, males may sing and aggressively confront neighboring conspecifics to defend their space. Aggression carries the risk of injury, and singing is energetically costly. After territorial borders have been established, intrusions by neighbors are few because the outcomes of the contests have already been determined. Each male then "knows" his neighbor by the song he sings, and also "knows" that male does not constitute a threat because they have already settled their territorial contests. So, all birds in the population can lower their energy costs by identifying their neighbors through their individual songs. Similarly, mammals mark their territories with pheromones that signal individual identity, encoded as a blend of a number of chemicals. Other signals, such as the mobbing and alarm calls of birds, are anonymous, conveying no information about the identity of the sender. These signals may permit communication about the presence of a predator common to several bird species.

Although in most cases signals are used for communication between members of the same species, sometimes two different species are involved. For example, fish with parasites adopt a specific posture in the presence of cleaner fish that signals to the cleaner fish that they are ready to be cleaned (figure 52.21). In a similar vein, some animals send signals to predators. White-tailed deer, for example, raise their tails to display their prominent white undersides while running away from a predator. These *pursuit deterrent* sig-

FIGURE 52.20
Firefly fireworks. The bioluminescent displays of these lampyrid beetles are species-specific and serve as behavioral mechanisms of reproductive isolation. Each number represents the flash pattern of a male of a different species.

FIGURE 52.21
Cleaner fish. This grouper has entered the cleaner fish's "station" and adopted a posture that allows the cleaner fish to enter the mouth and gills and feed on attached parasites.

nals are presumed to indicate to the predator that it has already been seen and thus should not waste its time trying to catch the deer.

Animal communications serve many purposes and are transmitted in many ways.

Communication in Social Groups

Many insects, fish, birds, and mammals live in social groups in which information is communicated between group members. For example, some individuals in mammalian societies serve as "guards." When a predator appears, the guards give an *alarm call*, and group members respond by seeking shelter. Social insects, such as ants and honeybees, produce *alarm pheromones* that trigger attack behavior. Ants also deposit *trail pheromones* between the nest and a food source to lead other colony members to food (figure 52.22). Honeybees have an extremely complex *dance language* that directs hivemates to rich nectar sources.

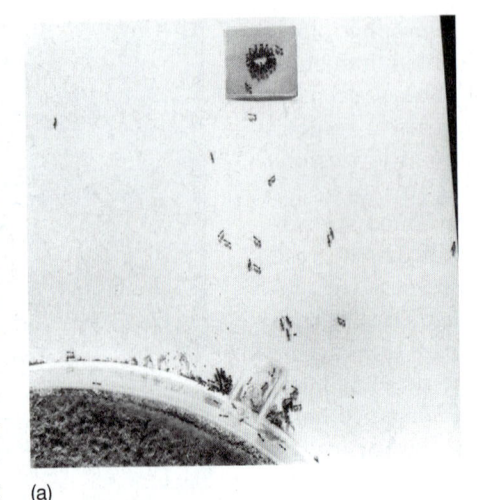

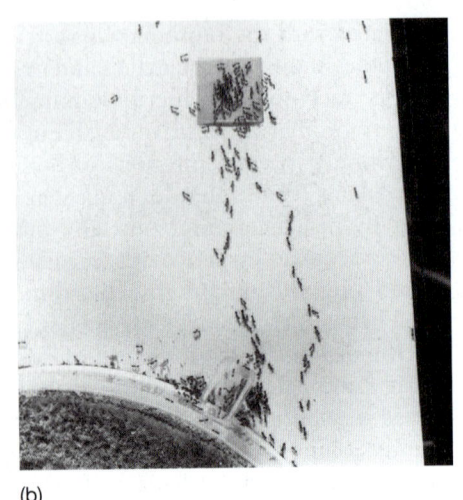

FIGURE 52.22
The chemical control of fire ant foraging. Trail pheromones, produced in an accessory gland near the fire ant's sting, organize cooperative foraging. The trails taken by the first ants to travel to a food source (*a*) are soon followed by most of the other ants (*b*).

The Dance Language of the Honeybee

The European honeybee, *Apis mellifera*, lives in hives consisting of 30,000 to 40,000 individuals whose behaviors are integrated into a complex colony. Worker bees may forage for miles from the hive, collecting nectar and pollen from a variety of plants and switching between plant species and populations on the basis of how energetically rewarding their food is. The food sources used by bees tend to occur in patches, and each patch offers much more food than a single bee can transport to the hive. A colony is able to exploit the resources of a patch because of the behavior of scout bees, which locate patches and communicate their location to hivemates through a dance language. Over many years, Nobel laureate Karl von Frisch was able to unravel the details of this communication system.

After a successful scout bee returns to the hive, she performs a remarkable behavior pattern called a *waggle dance* on a vertical comb (figure 52.23). The path of the bee during the dance resembles a figure-eight. on the straight part of the path, the bee vibrates or waggles her abdomen while producing bursts of sound. She may stop periodically to give her hivemates a sample of the nectar she has carried back to the hive in her crop. As she dances, she is followed closely by other bees, which soon appear as foragers at the new food source.

Von Frisch and his colleagues claimed that the other bees use information in the waggle dance to locate the food source. According to their explanation, the scout bee indicates the *direction* of the food source by representing the angle between the food source, the hive, and the sun as the deviation from vertical of the straight part of the dance performed on the hive wall (that is, if the bee moved straight, then the food source would be in the direction of the sun, but if the food were at a 30° angle relative to the sun's position, then the bee would move upward at a 30° angle from vertical). The *distance* to the food source is indicated by the tempo, or degree of vigor, of the dance.

Adrian Wenner, a scientist at the University of California, did not believe that the dance language communicated anything about the location of food, and he challenged von Frisch's explanation. Wenner maintained that flower odor was the most important cue leading bees to arrive at a new food source. A heated controversy ensued as the two groups of researchers published articles supporting their positions.

FIGURE 52.23
The waggle dance of honeybees. (*a*) The angle between the food source, the nest, and the sun is represented by a dancing bee as the angle between the straight part of the dance and vertical. The food is 20° to the right of the sun, and the straight part of the bee's dance on the hive is 20° to the right of vertical. (*b*) A scout bee dances on a comb in the hive.

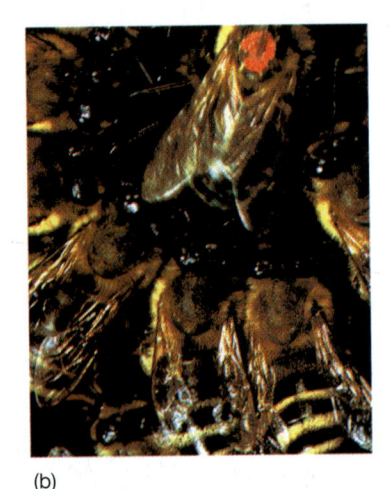

1120 Part VIII Ecology and Behavior

FIGURE 52.24

Primate semantics. (*a*) Predators, such as this leopard, attack and feed on vervet monkeys. (*b*) The monkeys give different alarm calls when troupe members sight an eagle, leopard, or snake. Each distinctive call elicits a different and adaptive escape behavior.

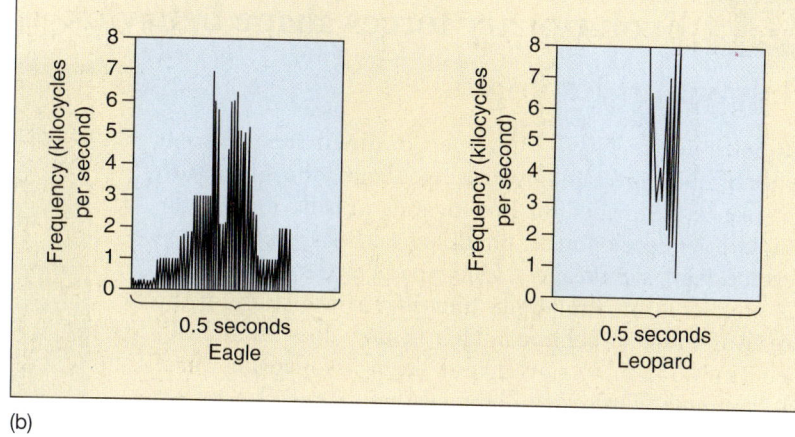

Such controversies can be very beneficial, because they often generate innovative experiments. In this case, the "dance language controversy" was resolved (in the minds of most scientists) in the mid-1970s by the creative research of James L. Gould. Gould devised an experiment in which hive members were tricked into misinterpreting the directions given by the scout bee's dance. As a result, Gould was able to manipulate where the hive members would go if they were using visual signals. If odor were the cue they were using, hive members would have appeared at the food source, but instead they appeared exactly where Gould had predicted. This confirmed von Frisch's ideas.

Recently, researchers have extended the study of the honeybee dance language by building robot bees whose dances can be completely controlled. Their dances are programmed by a computer and perfectly reproduce the natural honeybee dance; the robots even stop to give food samples! The use of robot bees has allowed scientists to determine precisely which cues direct hivemates to food sources.

Primate Language

Some primates have a "vocabulary" that allows individuals to communicate the identity of specific predators. The vocalizations of African vervet monkeys, for example, distinguish among eagles, leopards, and snakes (figure 52.24). Chimpanzees and gorillas can learn to recognize a large number of symbols and use them to communicate abstract concepts. The complexity of human language would at first appear to defy biological explanation, but closer examination suggests that the differences are in fact superficial—all languages share many basic structural similarities. All of the roughly 3000 languages draw from the same set of 40 consonant sounds (English uses two dozen of them), and any human can learn them. Researchers believe these similarities reflect the way our brains handle abstract information, a genetically determined characteristic of all humans.

Language develops at an early age in humans. Human infants are capable of recognizing the 40 consonant sounds characteristic of speech, including those not present in the particular language they will learn, while ignoring other sounds. In contrast, individuals who have not heard certain consonant sounds as infants can only rarely distinguish or produce them as adults. That is why English speakers have difficulty mastering the throaty French "r," French speakers typically replace the English "th" with "z," and native Japanese often substitute "r" for the unfamiliar English "l." Children go through a "babbling" phase, in which they learn by trial and error how to make the sounds of language. Even deaf children go through a babbling phase using sign language. Next, children quickly and easily learn a vocabulary of thousands of words. Like babbling, this phase of rapid learning seems to be genetically programmed. It is followed by a stage in which children form simple sentences that, though they may be grammatically incorrect, can convey information. Learning the rules of grammar constitutes the final step in language acquisition.

While language is the primary channel of human communication, odor and other nonverbal signals (such as "body language") may also convey information. However, it is difficult to determine the relative importance of these other communication channels in humans.

The study of animal communication involves analysis of the specificity of signals, their information content, and the methods used to produce and receive them.

Chapter 52 Behavioral Biology **1121**

52.4 Evolutionary forces shape behavior.

Behavioral Ecology

Nobel laureate Niko Tinbergen outlined the different types of questions biologists can ask about animal behavior. In essence, he divided the investigation of behavior into the study of its development, physiological basis, and function (evolutionary significance). One type of evolutionary analysis pioneered by Tinbergen himself was the study of the **survival value** of behavior. That is, how does an animal's behavior allow it to stay alive or keep its offspring alive? For example, Tinbergen observed that after gull nestlings hatch, the parents remove the eggshells from the nest. To understand why this behavior occurs, he camouflaged chicken eggs by painting them to resemble the natural background where they would lie and distributed them throughout the area in which the gulls were nesting (figure 52.25). He placed broken eggshells next to some of the eggs, and as a control, he left other camouflaged eggs alone without eggshells. He then noted which eggs were found more easily by crows. Because the crows could use the white interior of a broken eggshell as a cue, they ate more of the camouflaged eggs that were near eggshells. Thus, Tinbergen concluded that eggshell removal behavior is *adaptive*: It reduces predation and thus increases the offspring's chances of survival.

Tinbergen is credited with being one of the founders of **behavioral ecology,** the study of how natural selection shapes behavior. This branch of ecology examines the **adaptive significance** of behavior, or how behavior may increase survival and reproduction. Current research in behavioral ecology focuses on how behavior contributes to an animal's reproductive success, or **fitness.** As we saw in section 52.1, differences in behavior among individuals often result from genetic differences. Thus, natural selection operating on behavior has the potential to produce evolutionary change. To study the relation between behavior and fitness, then, is to study the process of adaptation itself.

Consequently, the field of behavioral ecology is concerned with two questions. First, is behavior adaptive? Although it is tempting to assume that the behavior produced by individuals must in some way represent an adaptive response to the environment, this need not be the case. As discussed in chapter 21, traits can evolve for many reasons other than natural selection, such as genetic drift, gene flow, or the correlated consequences of selection on other traits. Moreover, traits may be present in a population because they evolved as adaptations in the past, but are no longer useful. These possibilities hold true for behavioral traits as much as for any other kind of trait.

If a trait is adaptive, the question then becomes: How is it adaptive? Although the ultimate criterion is reproductive success, behavioral ecologists are interested in *how* a trait can lead to greater reproductive success. By enhancing energy intake, thus increasing the number of offspring produced? By improving success in getting more matings? By decreasing the chance of predation? The job of a behavioral ecologist is to determine the effect of a behavioral trait on each of these activities and then to discover whether increases in, for example, foraging efficiency, translate into increased fitness.

FIGURE 52.25
The adaptive value of egg coloration. Niko Tinbergen painted chicken eggs to resemble the mottled brown camouflage of gull eggs. The eggs were used to test the hypothesis that camouflaged eggs are more difficult for predators to find and thus increase the young's chances of survival.

> **Behavioral ecology is the study of how natural selection shapes behavior.**

Foraging Behavior

The best way to introduce behavioral ecology is by examining one well-defined behavior in detail. While many behaviors might be chosen, we will focus on foraging behavior. For many animals, food comes in a variety of sizes. Larger foods may contain more energy but may be harder to capture and less abundant. In addition, some types of food may be farther away than other types. Hence, for these animals foraging involves a trade-off between a food's energy content and the cost of obtaining it. The *net energy* (in calories or Joules) gained by feeding on prey of each size is simply the energy content of the prey minus the energy costs of pursuing and handling it. According to **optimal foraging theory,** natural selection favors individuals whose foraging behavior is as energetically efficient as possible. In other words, animals tend to feed on prey that maximize their net energy intake per unit of foraging time.

A number of studies have demonstrated that foragers do preferentially utilize prey that maximize the energy return. Shore crabs, for example, tend to feed primarily on intermediate-sized mussels, which provide the greatest energetic return; larger mussels yield more energy, but also take considerably more energy to crack open (figure 52.26).

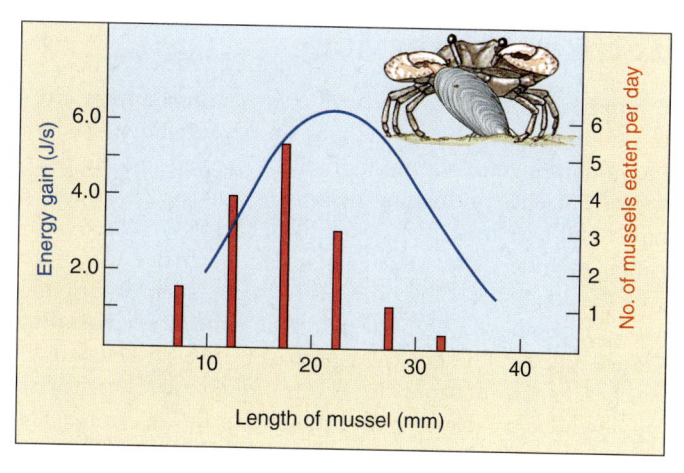

This optimal foraging approach makes two assumptions. First, natural selection will only favor behavior that maximizes energy acquisition if increased energy reserves lead to increases in reproductive success. In some cases, this is true. For example, in both Columbian ground squirrels and captive zebra finches, a direct relationship exists between net energy intake and the number of offspring raised; similarly, the reproductive success of orb-weaving spiders is related to how much food they can capture.

However, animals have other needs besides energy, and sometimes these needs conflict. One obvious alternative is avoiding predators: Often, the behavior that maximizes energy intake is not the one that minimizes predation risk. Thus, the behavior that maximizes fitness often may reflect a trade-off between obtaining the most energy at the least risk of being eaten. Not surprisingly, many studies have shown that a wide variety of animal species alter their foraging behavior—becoming less active, spending more time watching for predators, or staying nearer to cover—when predators are present. Still another alternative is finding mates: Males of many species, for example, will greatly reduce their feeding rate to enhance their ability to attract and defend females. Even within foraging, trade-offs must be made, because maximizing energy is not the sole goal of foraging; particular nutrients are needed as well. Moose, for example, will feed on energy-poor aquatic vegetation to ensure that they get an adequate supply of calcium.

The second assumption of optimal foraging theory is that it has resulted from natural selection. As we have seen, natural selection can lead to evolutionary change only when differences among individuals have a genetic basis. Few studies have investigated whether differences among individuals in their ability to maximize energy intake are the result of genetic differences, but there are some exceptions. For example, one study found that female zebra finches that were particularly successful in maximizing net energy intake tended to have similarly successful offspring. Because birds were removed from their mothers before they left the nest, this similarity indicated that foraging behavior probably has a genetic basis rather than being a result of young birds learning to forage from their mothers.

Differences in foraging behavior among individuals may also be a function of age. Inexperienced yellow-eyed juncos (a small North American bird), for example, have not learned how to handle large prey items efficiently. As a result, the energetic costs of eating such prey are higher than the benefits, and these birds tend to focus on smaller prey. Only when the birds are older and more experienced do they learn to easily dispatch these prey, which are then included in the diet.

FIGURE 52.26
Optimal diet. The shore crab selects a diet of energetically profitable prey. The curve describes the net energy gain (equal to energy gained minus energy expended) derived from feeding on different sizes of mussels. The bar graph shows the numbers of mussels of each size in the diet. Shore crabs tend to feed on those mussels that provide the most energy.
What factors might be responsible for the slight difference in peak prey length relative to the length optimal for maximum energy gain?

> **Natural selection may favor the evolution of foraging behaviors that maximize the amount of energy gained per unit time spent foraging. Animals that acquire energy efficiently during foraging may increase their fitness by having more energy available for reproduction, but other considerations, such as avoiding predators, are also important in determining reproductive success.**

Territorial Behavior

Animals often move over a large area, or **home range**, during their daily course of activity. In many animal species, the home range of several individuals overlap in time or in space, but each individual defends a portion of its home range and uses it exclusively. This behavior, in which individual members of a species maintain exclusive use of an area that contains some limiting resource, such as foraging ground, food, or potential mates, is called **territoriality** (figure 52.27). The critical aspect of territorial behavior is *defense* against intrusion by other individuals. Territories are defended by displays advertising that the territories are occupied and by overt aggression. A bird sings from its perch within a territory to prevent takeover by a neighboring bird. If an intruder is not deterred by the song, the territory owner may attack and try to drive it away. However, territorial defense has its costs. Singing is energetically expensive, and attacks can lead to injury. In addition, advertisement through song or visual display can reveal one's position to a predator.

Why does an animal bear the costs of territorial defense? Over the past two decades, it has become increasingly clear that an economic approach can be useful in answering this question. Although there are costs to defending a territory, there are also benefits; these benefits may take the form of increased food intake, exclusive access to mates, or access to refuges from predators. Studies of nectar-feeding birds such as hummingbirds and sunbirds provide an example (figure 52.28). A bird benefits from having the exclusive use of a patch of flowers because it can efficiently harvest the nectar they produce. To maintain exclusive use, however, the bird must actively defend the flowers. The benefits of exclusive use outweigh the costs of defense only under certain conditions. Sunbirds, for example, expend 3000 calories per hour chasing intruders from a territory. Whether or not the benefit of defending a territory will exceed this cost depends upon the amount of nectar in the flowers and how efficiently the bird can collect it. For example, if flowers are very scarce or nectar levels are very low, a nectar-feeding bird may not gain enough energy to balance the energy used in defense. Under this circumstance, it is not advantageous to be territorial. Similarly, if flowers are very abundant, a bird can efficiently meet its daily energy requirements without behaving territorially and adding the costs of defense. From an energetic standpoint, defending abundant resources isn't worth the cost. Territoriality thus only occurs at intermediate levels of flower availability and nectar production, where the benefits of defense outweigh the costs.

In many species, exclusive access to females is a more important determinant of territory size for males than is food availability. In some lizards, for example, males maintain enormous territories during the breeding season. These territories, which encompass the territories of several females,

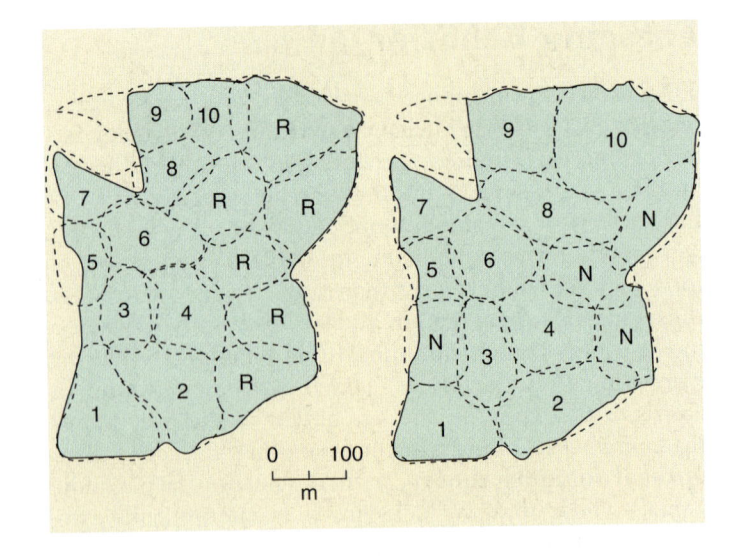

FIGURE 52.27
Competition for space. Territory size in birds is adjusted according to the number of competitors. When six pairs of great tits (*Parus major*) were removed from their territories (indicated by R in the left figure), their territories were taken over by other birds in the area and by four new pairs (indicated by N in the right figure). Numbers correspond to the birds present before and after.

FIGURE 52.28
The benefit of territoriality. Sunbirds, found in Africa and ecologically similar to New World hummingbirds, increase nectar availability by defending flowers.

are much larger than what would be required to supply enough food, and they are defended vigorously. In the nonbreeding season, by contrast, male territory size decreases dramatically, as does aggressive territorial behavior.

> An economic approach can be used to explain the evolution and ecology of behaviors such as territoriality. This approach assumes that animals that gain more energy from a behavior than they expend will have an advantage in survival and reproduction over animals that behave in less efficient ways.

Reproductive Strategies

During the breeding season, animals make several important "decisions" concerning their choice of mates, how many mates to have, and how much time and energy to devote to rearing offspring. These decisions are all aspects of an animal's **reproductive strategy,** a set of behaviors that presumably have evolved to maximize reproductive success. Reproductive strategies have evolved partly in response to the energetic costs of reproduction and the way food resources, nest sites, and members of the opposite sex are spatially distributed in the environment.

Parental Investment and Mate Choice

Males and females usually differ in their reproductive strategies. Darwin was the first to observe that females often do not simply mate with the first male they encounter, but instead seem to evaluate a male's quality and then decide whether to mate. This behavior, called **mate choice,** has since been described in many invertebrate and vertebrate species.

By contrast, mate choice by males is much less common. Why should this be? Many of the differences in reproductive strategies between the sexes can be understood by comparing the parental investment made by males and females. **Parental investment** refers to the contributions each sex makes in producing and rearing offspring; it is, in effect, an estimate of the energy expended by males and females in each reproductive event.

Many studies have shown that parental investment is high in females. One reason is that eggs are much larger than sperm—195,000 times larger in humans! Eggs contain proteins and lipids in the yolk and other nutrients for the developing embryo, but sperm are little more than mobile DNA. Furthermore, in some groups of animals, females are responsible for gestation and lactation, costly reproductive functions only they can carry out.

The consequence of such great disparities in reproductive investment is that the sexes should face very different selective pressures. Because any single reproductive event is relatively cheap for males, they can best increase their fitness by mating with as many females as possible—male fitness is rarely limited by the amount of sperm they can produce. By contrast, each reproductive event for females is much more costly, and the number of eggs that can be produced often does limit reproductive success. For this reason, a female has an incentive to be choosy, trying to pick the male that can provide the greatest benefit to her offspring. As we shall see, this benefit can take a number of different forms.

These conclusions only hold when female reproductive investment is much greater than that of males. In species with biparental care, males may contribute equally to the cost of raising young; in this case, the degree of mate choice should be equal between the sexes.

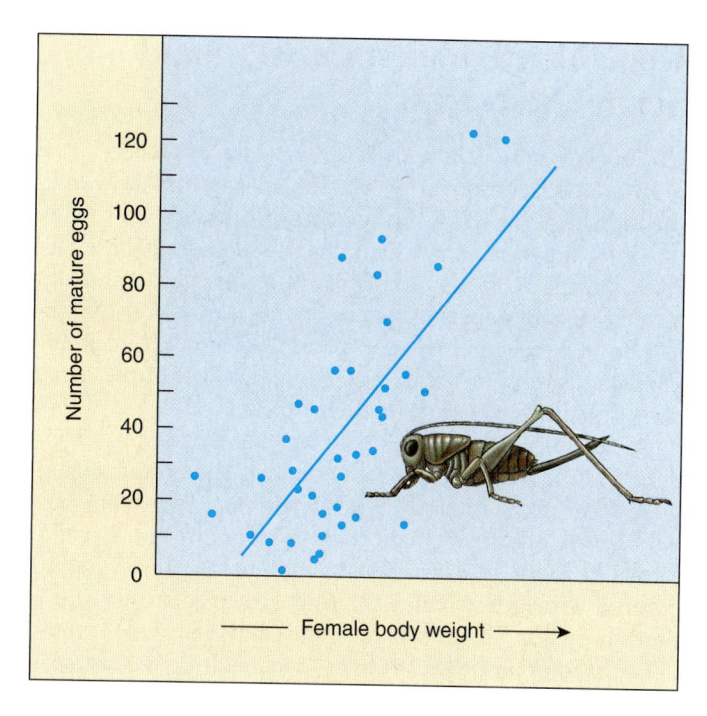

FIGURE 52.29
The advantage of male mate choice. Male mormon crickets choose heavier females as mates, and larger females have more eggs. Thus, male mate selection increases fitness.
Is there a benefit to females for mating with large males?

Furthermore, in some cases, male investment exceeds that of females. For example, male mormon crickets transfer a protein-containing spermatophore to females during mating. Almost 30% of a male's body weight is made up by the spermatophore, which provides nutrition for the female and helps her develop her eggs. As we might expect, in this case it is the females that compete with each other for access to males that are the choosy sex. Indeed, males are quite selective, favoring heavier females. The selective advantage of this strategy results because heavier females have more eggs; thus, males that choose larger females leave more offspring (figure 52.29).

Another example comes from species in which the males take care of the eggs and developing young. This reversal of the normal sex roles occurs in a wide variety of species, including seahorses and a number of bird and insect species. In such species, as with mormon crickets, males are often choosy, and females must compete for mates.

The relative reproductive investment of the sexes determines reproductive strategies.

Chapter 52 Behavioral Biology 1125

Reproductive Competition and Sexual Selection

As discussed in chapter 21, the reproductive success of an individual is determined by a number of factors: how long the individual lives, how frequently it mates, and how many offspring it produces per mating. The second of these factors, competition for mating opportunities, has been termed **sexual selection.** Some people consider sexual selection to be distinctive from natural selection, but others see it as a subset of natural selection, just one of the many ways organisms can increase their fitness.

Sexual selection involves both *intrasexual selection*, or interactions between members of one sex ("the power to conquer other males in battle," as Darwin put it), and *intersexual selection*, essentially mate choice ("the power to charm"). Sexual selection leads to the evolution of structures used in combat with other males, such as a deer's antlers and a ram's horns, as well as ornaments used to "persuade" members of the opposite sex to mate, such as long tail feathers and bright plumage (figure 52.30a,b). These traits are called **secondary sexual characteristics.**

Intrasexual Selection

In many species, individuals of one sex—usually males—compete with each other for the opportunity to mate with individuals of the other sex. These competitions may take place over ownership of a territory in which females reside or direct control of the females themselves. The latter case is exemplified by many species, such as the impala, in which females travel in large groups with a single male that gets exclusive access to mate with the females and thus strives vigorously to defend this access against other males that would like to supplant him.

In mating systems such as these, a few males may get an inordinate number of matings, and most males do not mate at all. In elephant seals, males control territories on the breeding beaches, and a few dominant males do most of the breeding (see figure 52.32). On one beach, for example, eight males impregnated 348 females, while the remaining males mated rarely, if at all.

For this reason, selection strongly favors any trait that confers greater ability to outcompete other males. In many cases, larger males are able to dominate smaller ones. As a result, in many territorial species, males have evolved to be considerably larger than females, for the simple reason that the largest males are the ones that get to mate. Such differences between the sexes are referred to as **sexual dimorphism.** In other species, males have evolved structures used for fighting, such as horns, antlers, and large canine teeth. These traits are also often sexually dimorphic and may have evolved because of the advantage they give in intrasexual conflicts.

Sometimes competition is not between the males themselves, but between their sperm, a phenomenon called **sperm competition.** In species whose females mate with multiple males, many features have evolved to maximize success at sperm competition. For example, in such species, the testes are often large to produce many sperm per mating, and the sperm themselves are also often larger so that they swim more rapidly and thus enhance the likelihood of fertilizing an egg.

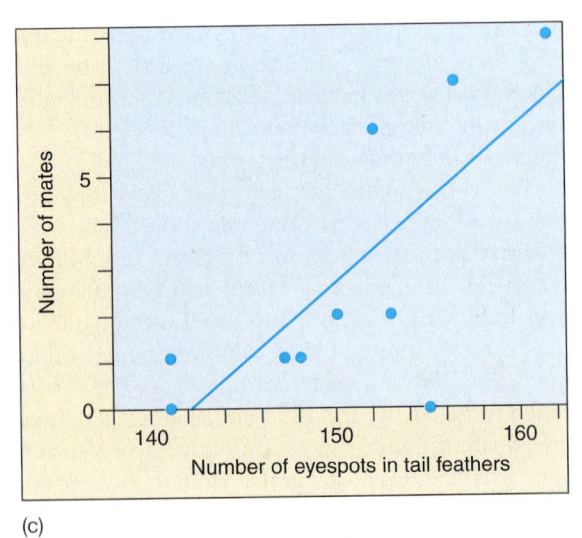

FIGURE 52.30
Products of sexual selection. Attracting mates with long feathers is common in bird species such as (*a*) the African paradise whydah and (*b*) the peacock, which show pronounced sexual dimorphism. (*c*) Female peahens prefer to mate with males having greater numbers of eyespots in their tail feathers.
Why do females prefer males with more spots?

Intersexual Selection

Peahens prefer to mate with peacocks that have more spots in their long tail feathers (figure 52.30*b,c*). Similarly, female frogs prefer to mate with males having more complex calls. Why did such mating preferences evolve?

Direct Benefits of Mate Choice. In some cases, the benefits of mate choice are obvious. In many species of birds and mammals and in some species of other types of animals, males help raise the offspring. In these cases, females would benefit by choosing the male that can provide the best care—the better the parent, the more offspring she is likely to rear.

In other species, males provide no care, but maintain territories that provide food, nesting sites, and predator refuges. In such species, females that choose males with the best territories will maximize their reproductive success.

Indirect Benefits. In other species, however, males provide no direct benefits of any kind to females. In such cases, it is not intuitively obvious what females have to gain by being choosy. Moreover, what could be the possible benefit of choosing a male with an extremely long tail or a complex song?

A number of theories have been proposed to explain the evolution of such preferences. One idea is that females choose the male that is the healthiest or oldest. Large males, for example, have probably been successful at living long, acquiring a lot of food, and resisting parasites and disease. Similarly, in guppies and some birds, the brightness of a male's color reflects the quality of his diet and overall health. Females may gain two benefits from mating with large or colorful males. First, to the extent that the males' success in living long and prospering is the result of genetic makeup, the female will be ensuring that her offspring receive good genes from their father. Indeed, according to several studies, males that are preferred by females produce offspring that are more vigorous and survive better than offspring of males that are not preferred. Second, healthy males are less likely to be carrying diseases, which might be transmitted to the female during mating.

A variant of this theory goes one step further. In some cases, females prefer mates with traits that are detrimental to survival (figure 52.30). The long tail of the peacock is a hindrance in flying and makes males more vulnerable to predators. Why should females prefer males with such traits? The **handicap hypothesis** states that only genetically superior mates can survive with such a handicap. By choosing a male with the largest handicap, the female is ensuring that her offspring will receive these quality genes. Of course, the male offspring will also inherit the genes for the handicap. For this reason, evolutionary biologists are still debating this hypothesis.

FIGURE 52.31
Male Túngara frog calling. Female frogs of the genus *Physalaemus* prefer males that include a "chuck" in their call. However, only males of the species *Physalaemus pustulosus* produce such calls.

Other courtship displays appear to have evolved from a predisposition in the female's sensory system toward a certain type of stimulus. For example, females may be better able to detect particular colors or sounds at a certain frequency. **Sensory exploitation** involves the evolution in males of an attractive signal that "exploits" these preexisting biases—for example, if females are particularly adept at detecting red objects, then males will often evolve red coloration. Consider the vocalizations of the Túngara frog (*Physalaemus pustulosus*) (figure 52.31). Unlike related species, males include a "chuck" in their calls. Recent research suggests that not only are females of this species particularly attracted to calls of this sort, but so are females of related species, even though males of these species do not produce "chucks." Why this preference evolved is unknown, but males of the Túngara frog have evolved to take advantage of it.

A great variety of other hypotheses have been proposed to explain the evolution of mating preferences. Many of these hypotheses may be correct in some circumstances, but none seems capable of explaining all of the variation in mating behavior in the animal world. This is an area of vibrant research, with new discoveries appearing regularly.

> Natural selection has favored the evolution of behaviors that maximize the reproductive success of males and females. By evaluating and selecting mates with superior qualities, an animal can increase its reproductive success.

Mating Systems

The number of individuals with which an animal mates during the breeding season varies throughout the animal kingdom. Mating systems include monogamy (one male mates with one female), polygyny (one male mates with more than one female; figure 52.32), and polyandry (one female mates with more than one male). Like mate choice, mating systems have evolved to maximize reproductive fitness. Much research has shown that mating systems are strongly influenced by ecology. For instance, a male may defend a territory that holds nest sites or food sources necessary for a female to reproduce, and the territory may have resources sufficient for more than one female. If males differ in the quality of the territories they hold, a female's fitness is maximized if she mates with a male holding a high-quality territory. Such a male may already have a mate, but it is still more advantageous for the female to breed with that male than with an unmated male that defends a low-quality territory. In this case, natural selection would favor the evolution of polygyny.

Mating systems are also constrained by the needs of offspring. If the presence of both parents is necessary for young to be reared successfully, then monogamy may be favored. This is generally the case in birds, in which over 90% of all species are monogamous. A male may either remain with his mate and provide care for the offspring or desert that mate to search for others; both strategies may increase his fitness. The strategy that natural selection will favor depends upon the requirement for male assistance in feeding or defending the offspring. In some species, offspring are **altricial**—they require prolonged and extensive care. In these species, the need for care by two parents reduces the tendency for the male to desert his mate and seek other matings. In species where the young are **precocial** (requiring little parental care), males may be more likely to be polygynous.

Although polygyny is much more common, polyandrous systems—in which one female mates with several males—are known in a variety of animals. For example, in spotted sandpipers, males take care of all incubation and parenting, and females mate and leave eggs with two or more males.

In recent years, researchers have uncovered many unexpected aspects of animal mating. Some of these discoveries have resulted from the application of new technologies, whereas others have come from detailed and intensive field studies.

Extra-Pair Copulations

Chapter 16 describes how DNA fingerprinting can be used to identify blood samples. Another common use of this technology is to establish paternity. DNA fingerprints are so variable that each individual's pattern is unique. Thus, by comparing the DNA of a man and a child, experts can establish with a relatively high degree of confidence whether the man is the child's father.

This approach is now commonly used in paternity lawsuits, but it has also become a standard weapon in the arsenal of behavioral ecologists. By establishing paternity, researchers can precisely quantify the reproductive success of individual males and thus assess how successful their particular reproductive strategies have been (figure 52.33a). In one classic study of red-winged blackbirds (figure 52.33b), researchers established that half of all nests contained at least one bird fertilized by a male other than the territory owner; overall, 20% of the offspring were the result of such **extra-pair copulations (EPCs).**

Studies such as this have established that EPCs—"cheating"—are much more pervasive in the bird world than originally suspected. Even in some species that were believed to be monogamous on the basis of behavioral observations, the incidence of offspring being fathered by a male other than the female's mate is sometimes surprisingly high.

Why do individuals have extra-pair copulations? For males, the answer is obvious: increased reproductive success. For females, it is less clear, because in most cases, it does not result in an increased number of offspring. One possibility is that females mate with genetically superior individuals, thus enhancing the genes passed on to their off-

FIGURE 52.32
Female defense polygyny in elephant seals. Male elephant seals fight with each other for possession of territories. Only the largest males can hold territories, which contain many females.

spring. Another possibility is that females can increase the amount of help they get in raising their offspring. If a female mates with more than one male, each male may help raise the offspring. This is exactly what happens in a common English bird, the dunnock. Females mate not only with the territory owner, but also with subordinate males that hang around the edge of the territory. If these subordinates mate enough with a female, they will help raise her young, presumably because they may have fathered some of these young.

Alternative Mating Tactics

Natural selection has led to the evolution of a variety of other means of increasing reproductive success. For example, in many species of fish, there are two genetic classes of males. One group is large and defends territories to obtain matings. The other type of male is small and adopts a completely different strategy. They do not maintain territories, but loiter at the edge of the territories of large males. Just at the end of a male's courtship, when the female is laying her eggs and the territorial male is depositing sperm, the smaller male will dart in and release its own sperm into the water, thus fertilizing some of the eggs. If this strategy is successful, natural selection will favor the evolution of these two different male reproductive strategies.

Similar patterns are seen in other organisms. In some dung beetles, territorial males have large horns that they use to guard the chambers in which females reside, whereas genetically small males don't have horns; instead, they dig side tunnels and attempt to intercept the female inside her chamber. In isopods, there are three genetic size classes. The medium-sized males pass for females and enter a large male's territory in this way; the smallest class are so tiny, they are able to sneak in completely undetected.

This is just a glimpse of the rich diversity in mating systems that have evolved. The bottom line is: If there is a way of increasing reproductive success, natural selection will favor its evolution.

> **Mating systems represent reproductive adaptations to ecological conditions.** The need for parental care and the ability of both sexes to provide it are important influences on the evolution of monogamy, polygyny, and polyandry. Detailed studies of animal mating systems, along with the use of modern molecular techniques, are revealing many surprises. This diversity is a testament to the power of natural selection to favor any trait that increases an animal's fitness.

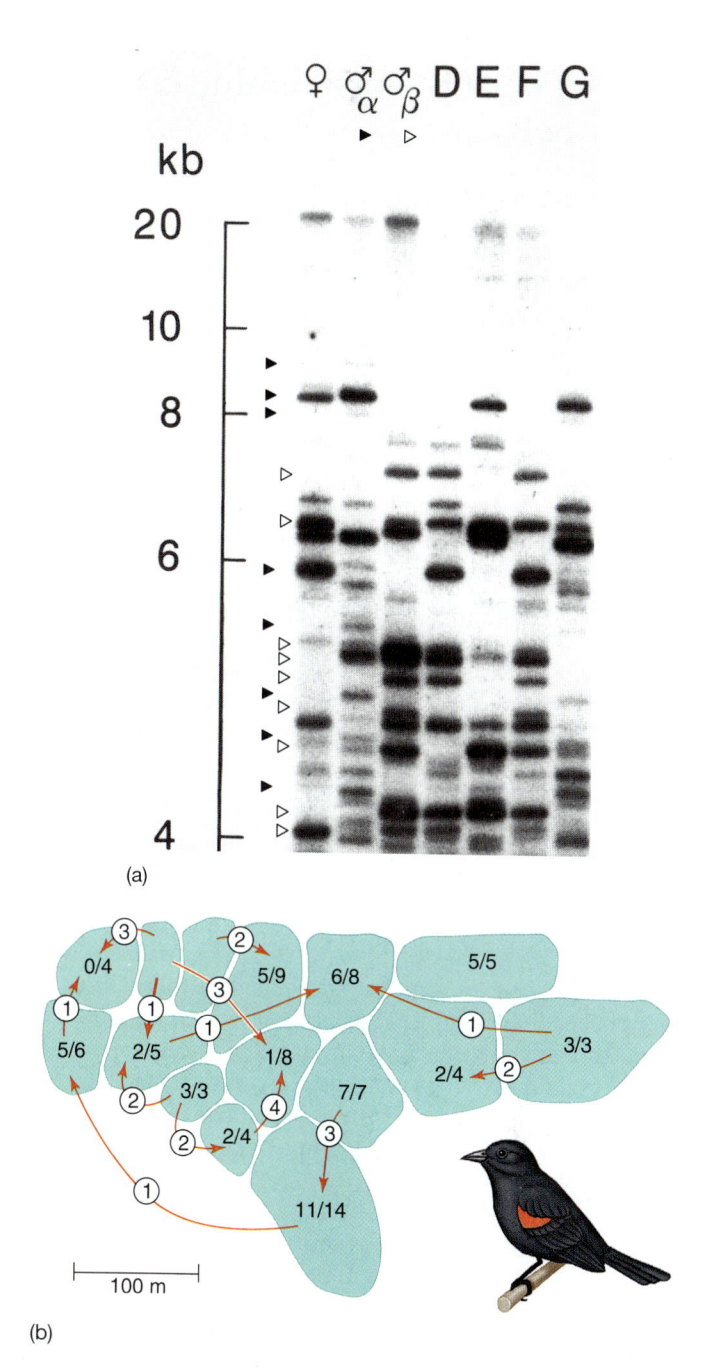

FIGURE 52.33
The study of paternity. (*a*) A DNA fingerprinting gel from the dunnock. The bands represent fragments of DNA of different lengths. The four nestlings (D–G) were in the nest of the female. By comparing the bands present in the two males, we can determine which male fathered which offspring. The triangles point to the bands that are diagnostic for one male and not the other. In this case, the β male fathered three (D, E, F, but not G) of the four offspring. (*b*) Results of a DNA fingerprinting study in red-winged blackbirds. Fractions indicate the proportion of offspring fathered by the male in whose territory the nest occurred. Arrows indicate how many offspring were fathered by particular males outside of each territory. Nests on some territories were not sampled.

52.5 There is considerable controversy about the evolution of social behavior.

Altruism and Group Living

Altruism—the performance of an action that benefits another individual at a cost to the actor—occurs in many guises in the animal world. In many bird species, for example, parents are assisted in raising their young by other birds, which are called *helpers at the nest*. In species of both mammals and birds, individuals that spy a predator will give an alarm call, alerting other members of their group, even though such an act would seem to call the predator's attention to the caller. Finally, lionesses with cubs will allow all cubs in the pride to nurse, including cubs of other females.

The existence of altruism has long perplexed evolutionary biologists. If altruism imposes a cost to an individual, how could an allele for altruism be favored by natural selection? One would expect such alleles to be at a disadvantage, and thus their frequency in the gene pool should decrease through time.

A number of explanations have been put forward to explain the evolution of altruism. One suggestion often heard on television documentaries is that such traits evolve for the good of the species. The problem with such explanations is that natural selection operates on individuals within species, not on species themselves. Thus, natural selection will not favor alleles that lead an individual to act in ways that benefit others at a cost to itself; it is even possible for traits to evolve that are detrimental to the species as a whole, as long as they benefit the individual. In some cases, selection can operate on groups of individuals, but such **group selection** is rare. For example, if an allele for cannibalism evolved within a population, individuals with that allele would be favored because they would have more to eat; however, the group might eventually eat itself to extinction, and the allele would be removed from the species. Although group selection can occur, the necessary conditions are rarely met in nature. In most cases, consequently, the "good of the species" cannot explain the evolution of altruistic traits.

Another possibility is that seemingly altruistic acts aren't altruistic after all. For example, helpers at the nest are often young and gain valuable parenting experience by assisting established breeders. Moreover, by hanging around an area, such individuals may inherit the territory when the established breeders die. Similarly, alarm callers may actually benefit by causing other animals to panic. In the ensuing confusion, the caller may be able to slip off undetected. Detailed field studies in recent years have demonstrated that some acts truly are altruistic, but others are not.

Reciprocity

Robert Trivers, now of Rutgers University, proposed that individuals may form "partnerships" in which mutual exchanges of altruistic acts occur because they benefit both participants. In the evolution of such **reciprocal altruism**, "cheaters" (nonreciprocators) are discriminated against and are cut off from receiving future aid. According to Trivers, if the altruistic act is relatively inexpensive, the small benefit a cheater receives by not reciprocating is far outweighed by the potential cost of not receiving future aid. Under these conditions, cheating should not occur.

Vampire bats roost in hollow trees in groups of 8 to 12 individuals. Because these bats have a high metabolic rate, individuals that have not fed recently may die. Bats that have found a host imbibe a great deal of blood, so giving up a small amount to keep a roostmate from starvation presents no great energy cost to the donor. Vampire bats tend to share blood with past reciprocators. If an individual fails to give blood to a bat from which it received blood in the past, it will be excluded from future bloodsharing.

Kin Selection

The most influential explanation for the origin of altruism was presented by William D. Hamilton in 1964. It is perhaps best introduced by quoting a passing remark made in a pub in 1932 by the great population geneticist J. B. S. Haldane. Haldane said that he would willingly lay down his life for *two brothers* or *eight first cousins*. Evolutionarily speaking, Haldane's statement makes sense, because for each allele Haldane received from his parents, his brothers each had a 50% chance of receiving the same allele (figure 52.34). Consequently, it is statistically expected that two of his brothers would pass on as many of Haldane's particular combination of alleles to the next generation as Haldane himself would. Similarly, Haldane and a first cousin would share an eighth of their alleles (see figure 52.34). Their parents, who are siblings, would each share half their alleles, and each of their children would receive half of these, of which half on the average would be in common: $½ × ½ × ½ = ⅛$. Eight first cousins would therefore pass on as many of those alleles to the next generation as Haldane himself would. Hamilton saw Haldane's point clearly: Natural selection will favor any strategy that increases the net flow of an individual's alleles to the next generation.

Hamilton showed that by directing aid toward kin, or close genetic relatives, an altruist may increase the reproductive success of its relatives enough to compensate for the reduction in its own fitness. Because the altruist's

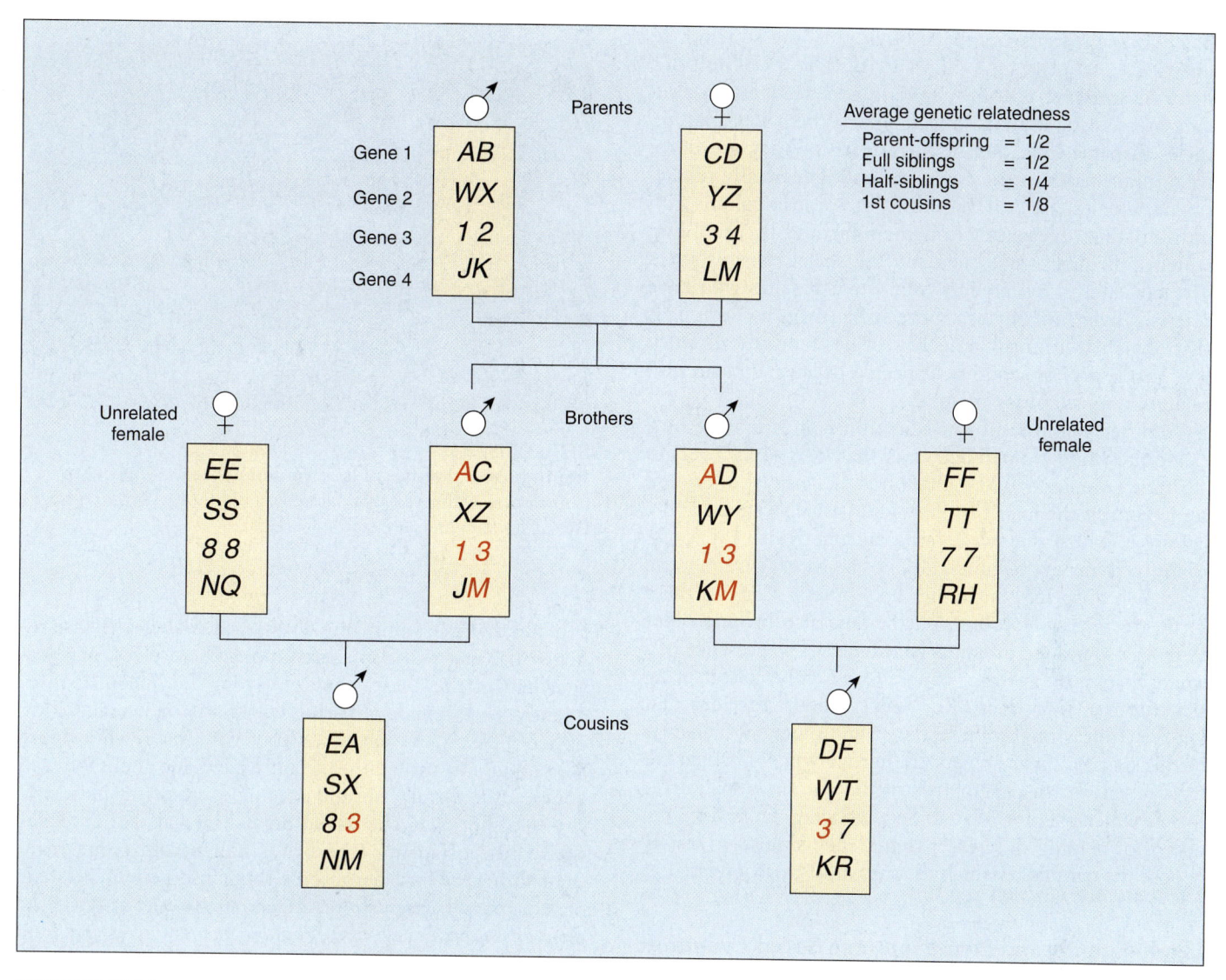

FIGURE 52.34
Hypothetical example of genetic relationships. On average, full siblings share half of their alleles. By contrast, cousins only share one-eighth of their alleles on average.

behavior increases the propagation of alleles in relatives, it will be favored by natural selection. Selection that favors altruism directed toward relatives is called **kin selection.** Although the behaviors being favored are cooperative, the genes are actually "behaving selfishly," because they encourage the organism to support copies of themselves in other individuals. In other words, if an individual has a dominant allele that causes altruism, any action that increases the frequency of this allele in future generations will be favored, even if that action is detrimental to the particular individual taking the action.

Hamilton's kin selection model predicts that altruism is likely to be directed toward close relatives. The more closely related two individuals are, the greater the potential genetic payoff. This relationship is described by **Hamilton's rule,** which states that altruistic acts are favored when $rb > c$. In this expression, b and c are the benefits and costs of the altruistic act, respectively, and r is the coefficient of relatedness, the proportion of alleles shared by two individuals through common descent. For example, an individual should be willing to have one less child ($c = 1$) if such actions allow a half-sibling, which shares one-quarter of its genes ($r = 0.25$), to have five or more additional offspring ($b = 5$).

Examples of Kin Selection

Many examples of kin selection are known from the animal world. For example, Belding's ground squirrels give alarm calls when they spot a predator such as a coyote or a

Chapter 52 Behavioral Biology 1131

badger. Such predators may attack a calling squirrel, so giving a signal places the caller at risk. The social unit of a ground squirrel colony consists of a female and her daughters, sisters, aunts, and nieces. When they mature, males disperse long distances from where they are born; thus, adult males in the colony are not genetically related to the females. By marking all squirrels in a colony with an individual dye pattern on their fur and by recording which individuals gave calls and the social circumstances of their calling, researchers found that females who have relatives living nearby are more likely to give alarm calls than females with no kin nearby. Males tend to call much less frequently, as would be expected because they are not related to most colony members.

Another example of kin selection comes from a bird called the white-fronted bee-eater that lives along rivers in Africa in colonies of 100 to 200 birds. In contrast to ground squirrels, it is the male bee-eaters that usually remain in the colony in which they were born, and the females that disperse to join new colonies. Many bee-eaters do not raise their own offspring, but rather help others. Many of these birds are relatively young, but helpers also include older birds whose nesting attempts have failed. The presence of a single helper, on average, doubles the number of offspring that survive. Two lines of evidence support the idea that kin selection is important in determining helping behavior in this species. First, helpers are normally males, which are usually related to other birds in the colony, and not females, which are not related. Second, when birds have the choice of helping different parents, they almost invariably choose the parents to which they are most closely related.

Haplodiploidy and Hymenopteran Social Evolution

Probably the most famous application of kin selection theory has been to social insects. A hive of honeybees consists of a single queen, who is the sole egg-layer, and up to 50,000 of her offspring, nearly all of whom are female workers with nonfunctional ovaries (figure 52.35). In addition to this reproductive division of labor, honeybees exhibit cooperative care of the brood and overlap of generations, such that queens live alongside their offspring. These are the hallmarks of a **eusocial** system.

The evolutionary origin of eusociality was long a mystery. How could natural selection favor the evolution of sterile workers that provided no offspring? Hamilton explained the origin of eusociality in hymenopterans (that is, bees, wasps, and ants) with his kin selection model. In these insects, males are haploid and females are diploid. This unusual system of sex determination, called *haplodiploidy*, leads to an unusual situation. If the queen is fertilized by a single male, then all female offspring will inherit exactly the same alleles from their father (because he

FIGURE 52.35
Reproductive division of labor in honeybees. The queen (shown here with a red spot painted on her thorax) is the sole egg-layer. Her daughters are sterile workers.

is haploid and has only one copy of each allele). These female offspring will also share among themselves, on average, half of the alleles they get from the queen. Consequently, each female offspring will share, on average, 75% of her alleles with each sister (to verify this, rework figure 52.34, but allow the father to only have one allele for each gene). By contrast, should a female offspring have offspring of her own, she would only share half of her alleles with these offspring (the other half would come from their father). Thus, because of this close genetic relatedness, *workers propagate more of their own alleles by giving up their own reproduction to assist their mother in rearing their sisters, some of whom will be new queens and start new colonies and reproduce.* Thus, this unusual haplodiploid system may have set the stage for the evolution of eusociality in hymenopterans, and indeed, such systems have evolved as many as 12 or more times in the Hymenoptera.

One wrinkle in this theory, however, is that eusocial systems have evolved in several other groups, including thrips, termites, and naked mole rats. Although thrips are also haplodiploid, both termites and naked mole rats are not. Thus, although haplodiploidy may have facilitated the evolution of eusociality, it is not a necessary prerequisite.

> **Many factors could be responsible for the evolution of altruistic behaviors. Individuals may benefit directly if altruistic acts are reciprocated; kin selection explains how alleles for altruism can increase in frequency if altruistic acts are directed toward relatives. Kin selection is a potent force favoring, in some situations, the evolution of altruism and even complex social systems.**

Group Living and the Evolution of Social Systems

Organisms as diverse as prokaryotes, cnidarians, insects, fish, birds, prairie dogs, lions, whales, and chimpanzees exist in social groups. To encompass the wide variety of social phenomena, we can broadly define a **society** as a group of organisms of the same species that are organized in a cooperative manner.

Why have individuals in some species given up a solitary existence to become members of a group? We have just seen that one explanation is kin selection: Groups may be composed of close relatives. In other cases, individuals may benefit directly from social living. For example, a bird that joins a flock may receive greater protection from predators. As flock size increases, the risk of predation decreases because there are more individuals to scan the environment for predators (figure 52.36). A member of a flock may also increase its feeding success if it can acquire information from other flock members about the location of new, rich food sources. In some predators, hunting in groups can increase success and allow the group to tackle prey too large for any one individual.

Insect Societies

In insects, sociality has chiefly evolved in two orders, the Hymenoptera (ants, bees, and wasps) and the Isoptera (termites), although a few other insect groups include social species. As we have just discussed, a number of types of insects have evolved eusocial systems. These social insect colonies are composed of different **castes**, groups of individuals that differ in size and morphology and perform different tasks, such as workers and soldiers (figure 52.37).

In honeybees, the queen maintains her dominance in the hive by secreting a pheromone, called "queen substance," that suppresses development of the ovaries in other females, turning them into sterile workers. Drones (male bees) are produced only for purposes of mating. When the colony grows larger in the spring, some members do not receive a sufficient quantity of queen substance, and the colony begins preparations for swarming. Workers make several new queen chambers, in which new queens begin to develop. Scout workers look for a new nest site and communicate its location to the colony. The old queen and a swarm of female workers then move to the new site. Left behind, a new queen emerges, kills the other potential queens, flies out to mate, and returns to assume "rule" of the hive.

Leafcutter ants provide another fascinating example of the remarkable lifestyles of social insects. Leafcutters live in colonies of up to several million individuals, growing crops of fungi beneath the ground. Their moundlike nests are underground "cities" covering more than 100 square meters, with hundreds of entrances and chambers as deep as 5 meters beneath the ground. The division of labor among the worker ants is related to their size. Every day, workers travel along trails from the nest to a tree or a bush, cut its leaves into small pieces, and carry the pieces back to the nest (see figure 52.37). Smaller workers chew the leaf fragments into a mulch, which they spread like a carpet in the underground fungus chambers. Even smaller workers implant fungal hyphae in the mulch; recent molecular studies suggest that ants have been cultivating these fungi for more than 50 million years! Soon a luxuriant garden of fungi is growing. While other workers weed out undesirable kinds of fungi, nurse ants carry the larvae of the nest to choice spots in the garden, where the larvae graze. Some of these larvae grow into reproductive queens that will disperse from the parent nest and start new colonies, repeating the cycle.

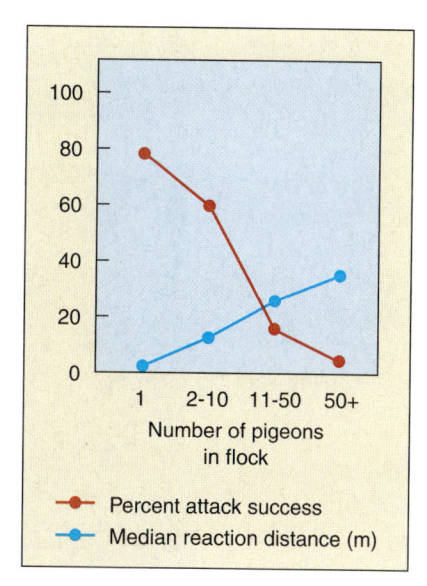

FIGURE 52.36
Flocking behavior decreases predation. As the size of a pigeon flock increases, hawks are less successful at capturing pigeons. Also, when more pigeons are present in the flock, they can detect hawks at greater distances, thus allowing more time for the pigeons to escape.
Would living in a flock affect the time available for foraging in pigeons?

FIGURE 52.37
Castes of ants. These two leaf-cutter ants are members of different castes. The large ant is a worker carrying leaves to the nest, whereas the smaller ants are protecting the worker from attack.

Chapter 52 Behavioral Biology 1133

Vertebrate Societies

In contrast to the highly structured and integrated insect societies and their remarkable forms of altruism, vertebrate social groups are usually less rigidly organized and cohesive. It seems paradoxical that vertebrates, which have larger brains and are capable of more complex behaviors, are generally less altruistic than insects. Nevertheless, in some complex vertebrate social systems, individuals may be exhibiting both reciprocity and kin-selected altruism. But vertebrate societies also generally display more conflict and aggression among group members than do insect societies. Conflict in vertebrate societies generally centers on access to food and mates.

Like insect societies, vertebrate societies have particular types of organization. Each social group of vertebrates has a certain size, stability of members, number of breeding males and females, and type of mating system. Behavioral ecologists have learned that the way a group is organized is influenced most often by ecological factors such as food type and predation. For example, meerkats take turns watching for predators while other group members forage for food (figure 52.38).

African weaver birds, which construct nests from vegetation, provide an excellent example of the relationship between ecology and social organization. Their roughly 90 species can be divided according to the type of social group they form. One set of species lives in the forest and builds camouflaged, solitary nests. Males and females are monogamous; they forage for insects to feed their young. The second group of species nests in colonies in trees on the savanna. They are polygynous and feed in flocks on seeds. The feeding and nesting habits of these two sets of species are correlated with their mating systems. In the forest, insects are hard to find, and both parents must cooperate in feeding the young. The camouflaged nests do not call the attention of predators to their brood. On the open savanna, building a hidden nest is not an option. Rather, savanna-dwelling weaver birds protect their young from predators by nesting in trees, which are not very abundant. This shortage of safe nest sites means that birds must nest together in colonies. Because seeds occur abundantly, a female can acquire all the food needed to rear young without a male's help. The male, free from the duties of parenting, spends his time courting many females—a polygynous mating system.

One exception to the general rule that vertebrate societies are not organized like those of insects is the naked mole rat, a small, hairless rodent that lives in and near East Africa. Unlike other kinds of mole rats, which live alone or in small family groups, naked mole rats form large underground colonies with a far-ranging system of tunnels and a central nesting area. It is not unusual for a colony to contain 80 individuals.

Naked mole rats feed on bulbs, roots, and tubers, which they locate by constant tunneling. As in insect societies, there is a division of labor among the colony members, with some mole rats working as tunnelers while others perform different tasks, depending upon the size of their bodies. Large mole rats defend the colony and dig tunnels.

Naked mole rat colonies have a reproductive division of labor similar to the one normally associated with the eusocial insects. All of the breeding is done by a single female, or "queen," who has one or two male consorts. The workers, consisting of both sexes, keep the tunnels clear and forage for food.

FIGURE 52.38
Foraging and predator avoidance. A meerkat sentinel on duty. Meerkats, *Suricata suricata*, are a species of highly social mongoose living in the semiarid sands of the Kalahari Desert in southern Africa. This meerkat is taking its turn to act as a lookout for predators. Under the security of its vigilance, the other group members can focus their attention on foraging.

> Eusocial insects exhibit an advanced social structure that includes reproductive division of labor and workers with different tasks. Social behavior in vertebrates is often characterized by kin-selected altruism. Altruistic behavior is involved in cooperative breeding in birds and alarm-calling in mammals.

Concept Review

For interactive testing, visit the Online Learning Center with PowerWeb at www.mhhe.com/Raven7

52.1 Many behavioral patterns are innate.

Approaches to the Study of Behavior
- Behavior can be defined as the way an animal responds to stimuli in its environment. (p. 1106)
- Proximate causation defines how something works, while ultimate causation addresses why something works. (p. 1106)
- Innate (instinctive) behaviors were argued to be the result of preprogrammed nervous system pathways under genetic control. (p. 1106)

Behavioral Genetics
- Experimental evidence suggests that some behaviors are hereditary, governed by genes passed from one generation to the next. (p. 1108)
- Additional experimentation has identified specific genes that may be associated with behavioral control. (p. 1109)

52.2 Learning influences behavior.

Nonassociative Learning and Conditioning
- Learning occurs when animals alter their behavior as a result of previous experiments. (p. 1110)
- Nonassociative learning, such as habituation, does not require the animal to form an association, while associative learning (classical or operant) requires development of an association between stimuli or between a stimulus and a response. (p. 1110)
- Many experiments have suggested that learning is possible only within the boundaries set by instinct. (p. 1111)

The Development of Behavior
- Behavioral biologists now recognize that behavior has both genetic and learned components. (p. 1112)
- Imprinting occurs when a maturing animal forms social attachments to other individuals or forms preferences that will influence behavior later in life; it is most often successful during a sensitive phase. (p. 1112)
- Experiments with song-learning in birds have suggested that in some cases the genetic template is selective, and learning must also play a prominent role. (p. 1113)

Animal Cognition
- Recent studies have shown compelling evidence of cognition in several species of animals including chimpanzees and ravens. (pp. 1114–1115)

Migratory Behavior
- Birds and other animals navigate (setting a bearing and following it) by looking at the sun and the stars, and potentially by detecting magnetic fields as well. (pp. 1116–1117)

52.3 Communication is a key element of many animal behaviors.

Courtship
- Courtship signals are often species-specific, limiting communication to members of the same species. (p. 1118)
- Different signals provide different levels of information about the sender, and the level of specificity relates to the function of the signal. (p. 1119)

Communication in Social Groups
- Social organisms utilize many modes of communication, including alarm calls, alarm pheromones, trail pheromones, and a dance language. (p. 1120)
- Upon close examination, the complexity of human languages is in fact a superficial difference, and all languages share many basic structural similarities. (p. 1121)

52.4 Evolutionary forces shape behavior.

Behavioral Ecology
- Behavioral ecology is the study of how natural selection shapes behavior. (p. 1122)

Foraging Behavior
- Optimal foraging theory states that natural selection produces animals that tend to feed on prey that maximize their net energy intake per unit foraging time. (p. 1123)

Territorial Behavior
- Territoriality occurs when individuals of a species maintain exclusive use of an area containing some limiting resource. (p. 1124)

Reproductive Strategies
- Because a reproductive event for females is often more costly than it is for males, females tend to be more choosy in mate selection. (p. 1125)

Reproductive Competition and Sexual Selection
- Sexual selection involves both intrasexual and intersexual selection. (pp. 1126–1127)

Mating Systems
- In mating systems such as monogamy, polygyny, and polyandry the number of mates an individual will have during the breeding season varies. (p. 1128)
- Several studies have established that extra-pair copulations are more pervasive in the animal kingdom, especially in birds, than originally suspected. (p. 1128)

52.5 There is considerable controversy about the evolution of social behavior.

Altruism and Group Living
- Altruistic behaviors may have evolved because they impart some benefit to the individual or the individual's relatives. (pp. 1130–1131)
- Kin selection suggests that aiding kin may increase the reproductive success of relatives enough to compensate for the reduction in a helper's own fitness. (pp. 1131–1132)

Group Living and the Evolution of Social Systems
- A society can be broadly defined as a group of organisms of the same species that are organized in a cooperative manner. (p. 1133)
- Social insect colonies are composed of different castes of workers that differ in size and morphology and have different tasks to perform. (p. 1133)
- Vertebrate social behavior is often characterized by kin-selected altruism such as cooperative feeding and rearing of young. (p. 1134)

Test Your Understanding

For interactive testing, visit the Online Learning Center with PowerWeb at www.mhhe.com/Raven7

Self Test

1. A male stickleback's display of aggression at a red object is an example of
   a. innate behavior.
   b. operant conditioning.
   c. associative learning.
   d. habituation.
2. What type of behavior is associated with a "critical period" in which a stimulus must be detected to trigger the response?
   a. cognitive behavior
   b. taxis
   c. imprinting
   d. associative learning
3. The study of song development in sparrows showed that
   a. the acquisition of a species-specific song is innate.
   b. there are two components to this behavior—a genetic template and learning.
   c. song acquisition is an example of associative learning.
   d. All of these are correct.
4. The level of specificity in courtship signaling is often
   a. individual-specific.
   b. anonymous.
   c. species-specific.
   d. any of these.
5. Which of the following is not a function of pheromones?
   a. sex attractants
   b. trigger alarm behaviors
   c. trail markers
   d. All of these are functions of pheromones.
6. Behavioral ecology is the study of
   a. how natural selection shapes behaviors.
   b. how environmental conditions affect animal behaviors.
   c. how feeding opportunities affect animal behaviors.
   d. how interactions with other individuals of the same species affects behaviors.
7. Which of the following would be most likely to occur according to the optimal foraging theory?
   a. Shore crabs eat the largest mussels even though they're harder to crack open.
   b. Columbian ground squirrels eat more because larger squirrels produce more offspring.
   c. Smaller yellow-eyed juncos eat larger prey, although they are harder to manage.
   d. An antelope will venture out on the open plain to feed even though lions are around.
8. Which of the following statements about territories is true?
   a. Territories and home ranges describe the same area of land.
   b. The defense of a territory is always beneficial to the animal.
   c. A territory contains resources exclusively for the animal that defends it.
   d. Territories overlap in time or space with other territories.
9. In the haplodiploidy system of sex determination, males are
   a. haploid.
   b. diploid.
   c. sterile.
   d. not present because bees exist as single-sex populations.
10. According to kin selection, saving the life of your _____ would do the least for increasing your inclusive fitness.
    a. mother
    b. brother
    c. sister-in-law
    d. niece

Test Your Visual Understanding

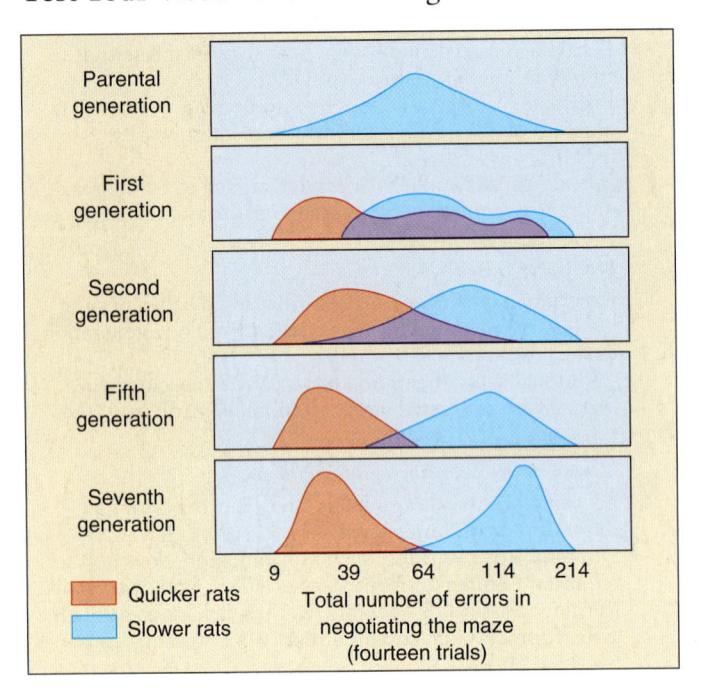

1. Can you predict what the graph in the seventh generation would look like if instead of mating the maze-bright rats with each other and the maze-dull rats with each other, you always mated maze-bright rats with maze-dull rats?

Apply Your Knowledge

1. Imagine an animal society in which the individuals share an average of one-third of their alleles with one another. If the benefits to an individual of performing an altruistic act are twice the costs, does the kin selection model predict that these individuals are likely to be altruistic? Explain your reasoning.
2. If a young animal exhibits a behavior immediately after being born or hatched, is it reasonable to conclude that the behavior is instinctive rather than learned? If not, what sorts of experiments might be performed to distinguish between these two possibilities?
3. Swallows often hunt in groups, while hawks and other predatory birds are usually solitary hunters. What do you think is the basis for this difference?

53

Population Ecology

Concept Outline

53.1 Organisms must cope with a varied environment.

The Environmental Challenge. Organisms cope with environmental variation with physiological, morphological, and behavioral adaptations.

53.2 Populations are groups of individuals of the same species that live in the same place.

Population Ranges. Population borders are determined by areas in which individuals cannot survive and reproduce.

Pattern of Spacing of Individuals in a Population. The distribution of individuals in a population can be random, uniform, or clumped.

Metapopulations. Sometimes populations are arranged in networks connected by the exchange of individuals.

53.3 Population dynamics depend critically upon age distribution.

Demography. A population's growth rate is a function of its age structure and age-specific mortality rates.

53.4 Life histories often reflect trade-offs between reproduction and survival.

The Cost of Reproduction. Evolutionary success is a trade-off between investing in current reproduction and investing in growth that promotes future reproduction.

53.5 Population growth is limited by the environment.

Population Growth. Populations grow if births exceed deaths until they reach the carrying capacity of their environment.

Factors that Regulate Populations. Some of the factors that regulate a population's growth depend upon the size of the population; others do not.

53.6 The human population has grown explosively in the last three centuries.

The Advent of Exponential Growth. Human populations will continue to grow in developing countries because of the number of young people entering their reproductive years.

FIGURE 53.1
Life takes place in populations. This population of gannets is subject to the rigorous effects of reproductive strategy, competition, predation, and other limiting factors.

Ecology, the study of how organisms relate to one another and to their environments, is a complex and fascinating area of biology that has important implications for each of us. In our exploration of ecological principles, we first consider how organisms respond to the abiotic environment in which they occur and how these responses affect the properties of populations, emphasizing population dynamics (figure 53.1). In chapter 54, we discuss communities of coexisting species and the interactions that occur among them. In subsequent chapters, we discuss the functioning of entire ecosystems and of the biosphere, concluding with a consideration of the problems facing our planet and our fellow species.

53.1 Organisms must cope with a varied environment.

The Environmental Challenge

The nature of the physical environment in large measure determines what organisms live there. Key elements include:

Temperature. Most organisms are adapted to live within a relatively narrow range of temperatures and will not thrive if temperatures are colder or warmer. The growing season of plants, for example, is importantly influenced by temperature.

Water. Plants and all other organisms require water. On land, water is often scarce, so patterns of rainfall have a major influence on life.

Sunlight. Almost all ecosystems rely on energy captured by photosynthesis; the availability of sunlight influences the amount of life an ecosystem can support, particularly below the surface in marine communities.

Soil. The physical consistency, pH, and mineral composition of soil often severely limit plant growth, particularly the availability of nitrogen and phosphorus.

Approaches to Coping with Environmental Variation

An individual encountering environmental variation may maintain a "steady-state" internal environment, a condition known as **homeostasis**. Many animals and plants actively employ physiological, morphological, or behavioral mechanisms to maintain homeostasis. The beetle in figure 53.2 is using a behavioral mechanism to cope with drastic changes in water availability. Other animals and plants are known as **conformers** because they conform to the environment in which they find themselves, their bodies adopting the temperature, salinity, and other physical aspects of their surroundings.

Responses to environmental variation can be seen over both the short and the long term. In the short term, spanning periods of a few minutes to an individual's lifetime, organisms have a variety of ways of coping with environmental change. Over longer periods, natural selection can operate to make a population better adapted to the environment.

Individual Responses to Environmental Change.

Physiology. Many organisms are able to adapt to environmental change by making physiological adjustments. Thus, your body constricts the blood vessels on the surface of your face on a cold day, reducing heat loss (and also giving your face a "flush"). Similarly, humans who visit high altitudes initially experience altitude sickness—the symptoms of which include heart palpitations, nausea, fatigue, headache, mental impairment, and in serious cases, pulmonary edema—because of the lower atmospheric pressure and consequent lower oxygen availability in the air. After several days, however, the same people feel fine, because a number of physiological changes have increased the delivery of oxygen to their body tissues (table 53.1).

Some insects avoid freezing in the winter by adding glycerol "antifreeze" to their blood; others tolerate freezing by converting much of their glycogen reserves into alcohols that protect their cell membranes from freeze damage.

Morphology. Animals that maintain a constant internal temperature (endotherms) in a cold environment have adaptations that tend to minimize energy expenditure. Many mammals grow thicker coats during the winter, utilizing their fur as insulation to retain body heat. In general, the thicker the fur, the greater the insulation (figure 53.3). Thus, a wolf's fur is about three times as thick in winter as in summer and insulates more than twice as well. Other mammals escape some of the costs of maintaining a constant body temperature during winter by hibernating during the coldest season, behaving, in effect, like conformers.

Behavior. Many animals deal with variation in the environment by moving from one patch of habitat to another, avoiding areas that are unsuitable. The tropical lizard in figure 53.4 manages to maintain a fairly uniform body temperature in an open habitat by basking in patches of sun and then retreating to the shade when it becomes too hot.

FIGURE 53.2
Meeting the challenge of obtaining moisture. On the dry sand dunes of the Namib Desert in southwestern Africa, the beetle *Onymacris unguicularis* collects moisture from the fog by holding its abdomen up at the crest of a dune to gather condensed water.

Table 53.1 Physiological Changes at High Altitude
Increased rate of breathing
Increased erythrocyte production, raising the amount of hemoglobin in the blood
Decreased binding capacity of hemoglobin, increasing the rate at which oxygen is unloaded in body tissues
Increased density of mitochondria, capillaries, and muscle myoglobin

By contrast, in shaded forests, the same lizard does not have the opportunity to regulate its body temperature through behavioral means. Thus, it becomes a conformer and adopts the temperature of its surroundings.

Behavioral adaptations can be extreme. Spadefoot toads (genus *Scaphiophus*), which live in the deserts of North America, can burrow nearly a meter below the surface and remain there for as long as nine months of each year, their metabolic rates greatly reduced as they live on fat reserves. When moist, cool conditions return, the toads emerge and breed. The young toads mature rapidly and burrow back underground.

Evolutionary Responses to Environmental Variation. These examples represent different ways in which organisms may adjust to changing environmental conditions. The ability of an individual to alter its physiology, morphology, or behavior is itself an evolutionary adaptation, the result of natural selection. The results of natural selection can also be detected by comparing closely related species that live in different environments. In such cases, species often have evolved striking adaptations to the particular environment in which they live.

For example, animals that live in different climates show many differences. Mammals from colder climates tend to have shorter ears and limbs—a phenomenon termed "Allen's Rule"—which reduces the surface area across which animals lose heat. Lizards that live in different climates exhibit physiological adaptations for coping with life at different temperatures. Desert lizards are unaffected by high temperatures that would kill a lizard from northern Europe, but the northern lizards are capable of running, capturing prey, and digesting food at cooler temperatures at which desert lizards would be completely immobilized.

Many species also exhibit adaptations to living in areas where water is scarce. Everyone knows of the camel and other desert animals that can go extended periods without drinking water. Another example of desert adaptation is seen in frogs. Most frogs have moist skins through which water permeates readily. Such animals could not survive in arid climates because they would rapidly dehydrate and die. However, some frogs have solved this problem by evolving a greatly reduced rate of water loss through the skin. One species, for example, secretes a waxy substance from specialized glands that waterproofs its skin and reduces rates of water loss by 95%.

Adaptation to different environments can also be studied experimentally. For example, when strains of *E. coli* were grown at high temperatures (42°C), the speed at which resources are utilized improved through time. After 2000 generations, this ability increased 30% over what it had been when the experiment started. The mechanism by which efficiency of resource use was increased is still unknown and is the focus of current research.

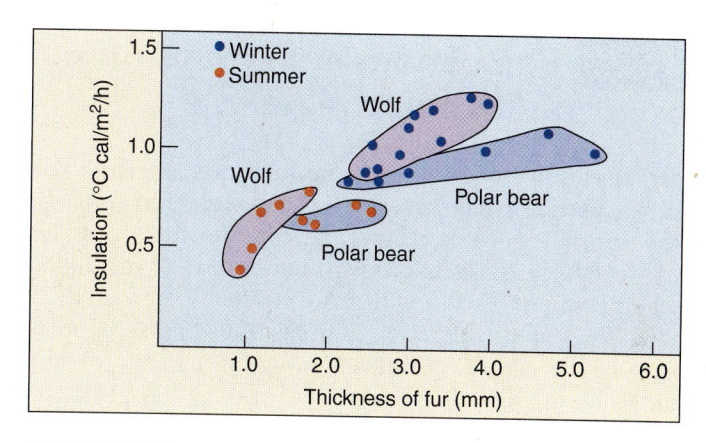

FIGURE 53.3
Morphological adaptation. Fur thickness in North American mammals has a major impact on the degree of insulation the fur provides.
Polar bears are able to live in zoos in warm climates. How thick would you expect the hair of a polar bear to be in a zoo in Miami, Florida?

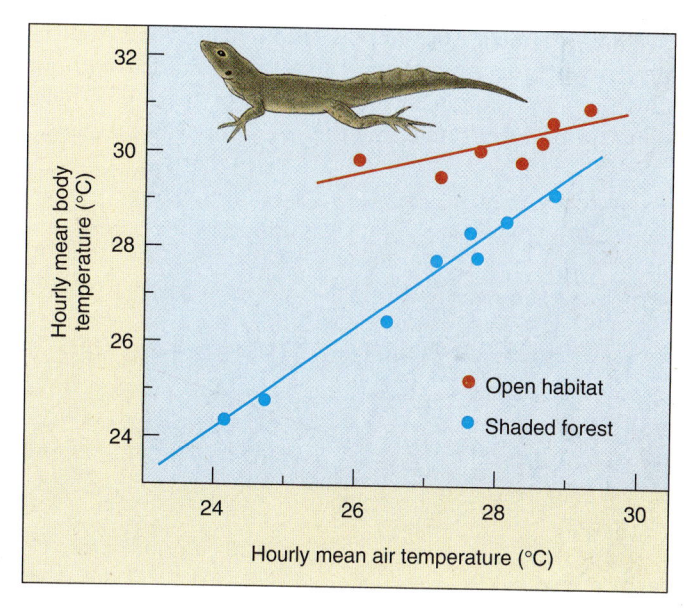

FIGURE 53.4
Behavioral adaptation. The Puerto Rican lizard *Anolis cristatellus* maintains a relatively constant temperature by seeking out and basking in patches of sunlight; in shaded forests, this behavior is not possible, and body temperature conforms to the surroundings.
When given the opportunity, lizards regulate their body temperature to maintain a temperature optimal for physiological functioning. Would lizards in open habitats exhibit different escape behaviors than lizards in shaded forest?

Organisms use a variety of physiological, morphological, and behavioral mechanisms to adjust to environmental variation. Over time, species evolve adaptations to living in different environments.

53.2 Populations are groups of individuals of the same species that live in the same place.

Organisms live as members of **populations**, groups of individuals that occur together at one place and time. In the remainder of this chapter, we consider the properties of populations, focusing on elements that influence whether a population will grow or shrink, and at what rate. The explosive growth of the world's human population in the last few centuries provides a focus for our inquiry.

The term "population" can be defined narrowly or broadly. This flexibility allows us to speak in similar terms of the world's human population, the population of protists in the gut of a termite, or the population of deer that inhabit a forest. Sometimes the boundaries defining a population are sharp, such as the edge of an isolated mountain lake for trout, and sometimes they are fuzzier, as when individual deer readily move back and forth between two forests separated by a cornfield.

Three characteristics of population ecology are particularly important: population range, the area throughout which a population occurs; the pattern of spacing of individuals within that range; and the size a population attains.

Population Ranges

No population, not even one composed of humans, occurs in all habitats throughout the world. Most species, in fact, have relatively limited geographic ranges, and the range of some species is miniscule. The Devil's Hole pupfish, for example, lives in a single hot-water spring in southern Nevada (figure 53.5), and the Socorro isopod is known from a single spring system in New Mexico. At the other extreme, some species are widely distributed. The common dolphin (*Delphinus delphis*), for example, is found throughout all the world's oceans.

As we just saw in section 53.1, organisms must be adapted for the environment in which they occur. Polar bears are exquisitely adapted to survive the cold of the Arctic, but you won't find them in the tropical rain forest. Certain prokaryotes can live in the near-boiling waters of Yellowstone's geysers, but they do not occur in cooler streams nearby. Each population has its own requirements—temperature, humidity, certain types of food, and a host of other factors—that determine where it can live and reproduce and where it can't. In addition, in places that are otherwise suitable, the presence of predators, competitors, or parasites may prevent a population from occupying an area, a topic we will take up in chapter 54.

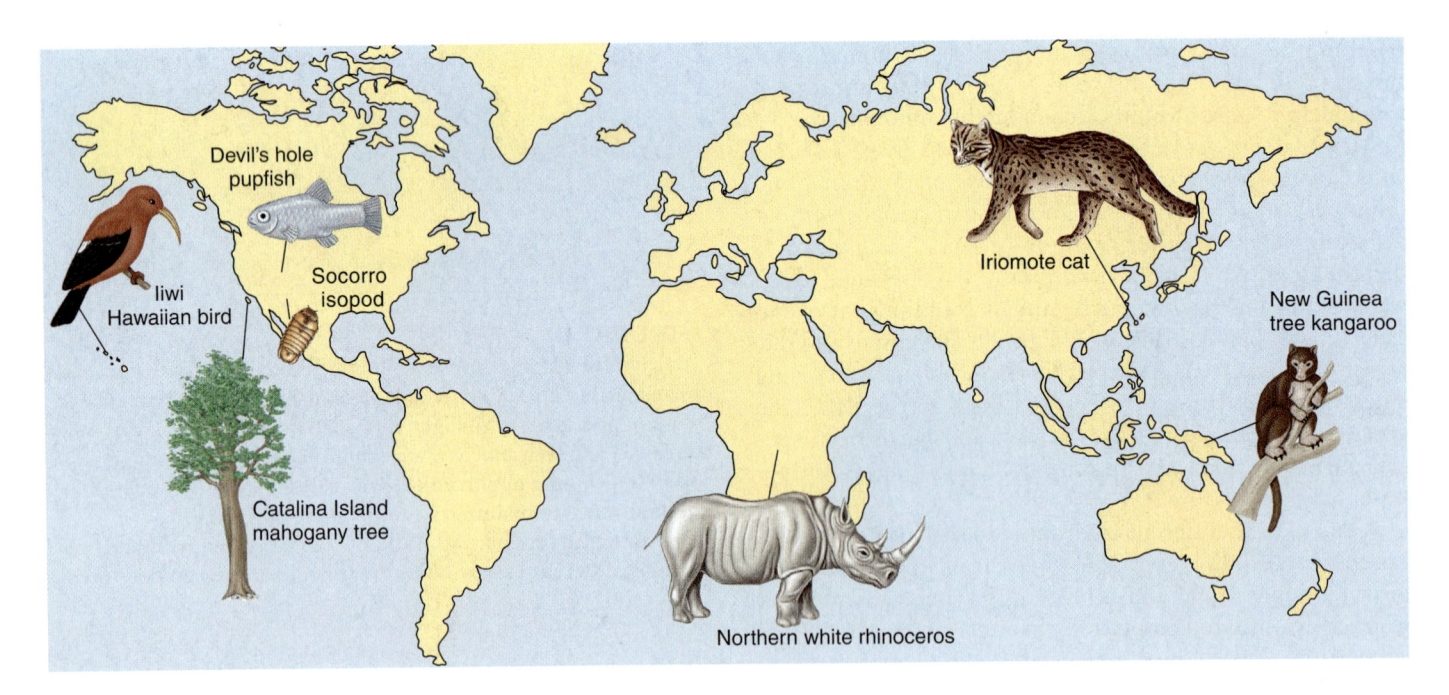

FIGURE 53.5
Species that occur in only one place. These species, and many others, are only found in a single population. All are endangered species, and should anything happen to their single habitat, the population—and the species—would become extinct.

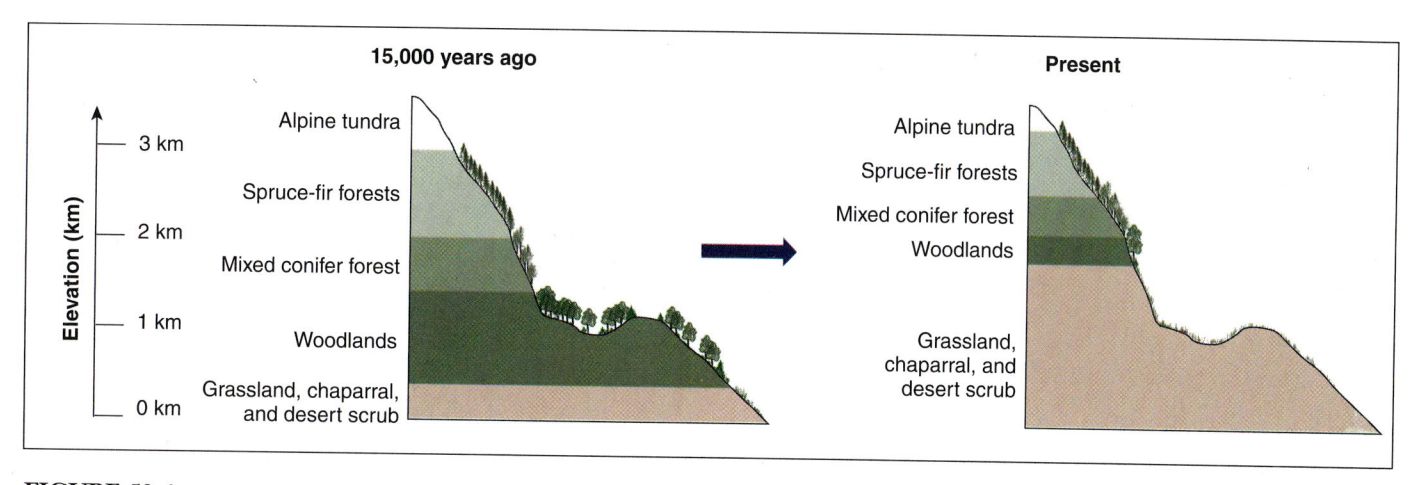

FIGURE 53.6
Altitudinal shifts in population ranges in the mountains of southwestern North America. During the glacial period 15,000 years ago, conditions were cooler than they are now. As the climate has warmed, tree species that require colder temperatures have shifted their range upward in altitude so that they live in the climatic conditions to which they are adapted.

Range Expansions and Contractions

Population ranges are not static; rather, they change through time. These changes occur for two reasons. In some cases, the environment changes. For example, as the glaciers retreated at the end of the last ice age, approximately 10,000 years ago, many North American plant and animal populations expanded northward. At the same time, as climates warmed, species experienced shifts in the elevation at which they could live (figure 53.6).

In addition, populations can expand their ranges when they are able to circumvent inhospitable habitat to colonize suitable, previously unoccupied areas. For example, the cattle egret is native to Africa. Some time in the late 1800s, these birds appeared in northern South America, having made the nearly 2000-mile transatlantic crossing, perhaps aided by strong winds. Since then, they have steadily expanded their range and now can be found throughout most of the United States (figure 53.7).

> A population is a group of individuals of the same species existing together in an area. Its range, the area a population occupies, changes over time.

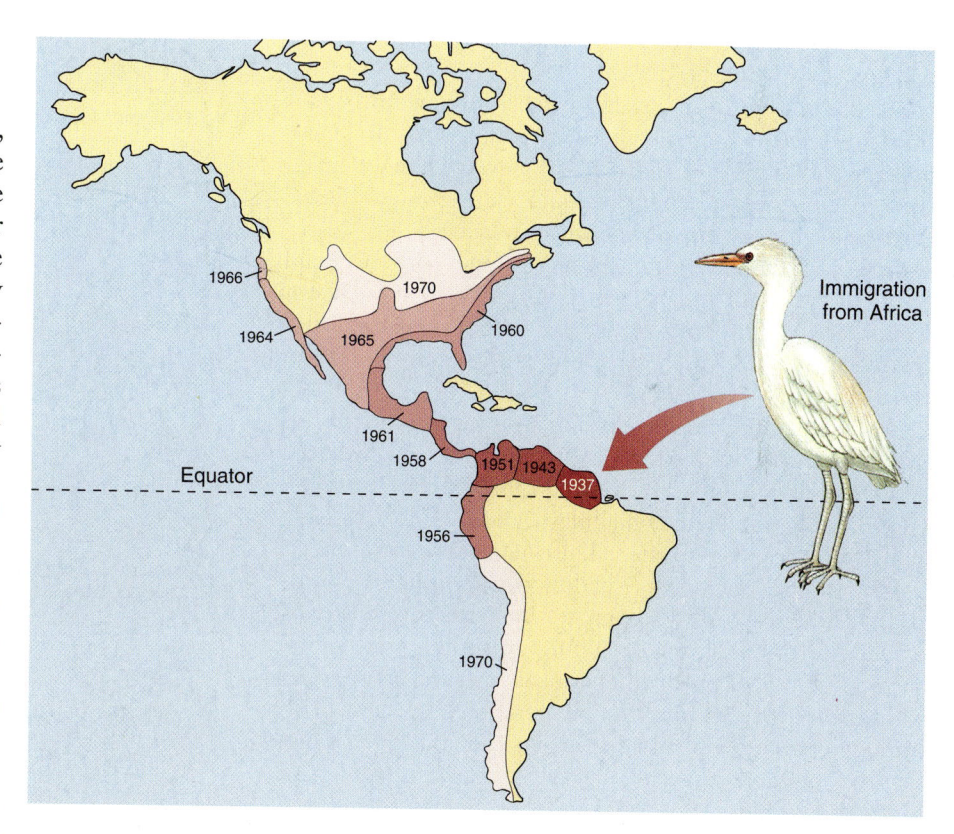

FIGURE 53.7
Range expansion of the cattle egret. The cattle egret—so named because it follows cattle and other hoofed animals, catching any insects or small vertebrates that they disturb—first arrived in South America in the late 1800s. Since the 1930s, the range expansion of this species has been well documented, as it has moved westward and up into much of North America, as well as down the western side of the Andes to near the southern tip of South America.

Chapter 53 Population Ecology 1141

Pattern of Spacing of Individuals in a Population

Another key characteristic of population structure is the way in which individuals of a population are distributed. They may be randomly spaced, uniformly spaced, or clumped (figure 53.8).

Randomly Spaced

Individuals are randomly spaced within populations when they do not interact strongly with one another or with nonuniform aspects of their environment. Random distributions are not common in nature. Some species of trees, however, appear to exhibit random distributions in Panamanian rain forests (figure 53.8b–d).

Uniformly Spaced

Uniform spacing within a population may often, but not always, result from competition for resources. The means by which it is accomplished, however, varies.

In animals, uniform spacing often results from behavioral interactions, as discussed in chapter 52. In many species, individuals of one or both sexes defend a territory from which other individuals are excluded. These territories provide the owner with exclusive access to resources such as food, water, hiding refuges, or mates and tend to space individuals evenly across the habitat. Even in nonterritorial species, individuals often maintain a defended space into which other animals are not allowed to intrude.

Among plants, uniform spacing is also a common result of competition for resources. In this case, however, the spacing results from direct competition for the resources. Closely spaced individual plants will compete for available sunlight, nutrients, or water. These contests can be direct, as when one plant casts a shadow over another, or indirect, as when two plants compete by extracting nutrients or water from a shared area. In addition, some plants, such as creosote, produce chemicals in the surrounding soil that are toxic to other members of their species. In all of these cases, only plants that are spaced an adequate distance from each other will be able to coexist, leading to uniform spacing.

FIGURE 53.8

Population dispersion. The different patterns of dispersion are exhibited by (*a*) different arrangements of bacterial colonies and (*b-d*) three different species of trees from the same locality in Panama.

Source: Data from Elizabeth Losos, Center for Tropical Forest Science, Smithsonian Tropical Research Institute.

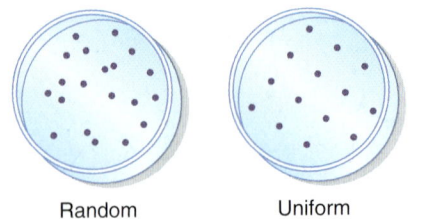

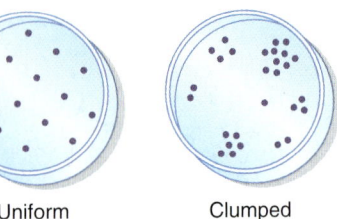

(a) Bacterial colonies — Random, Uniform, Clumped

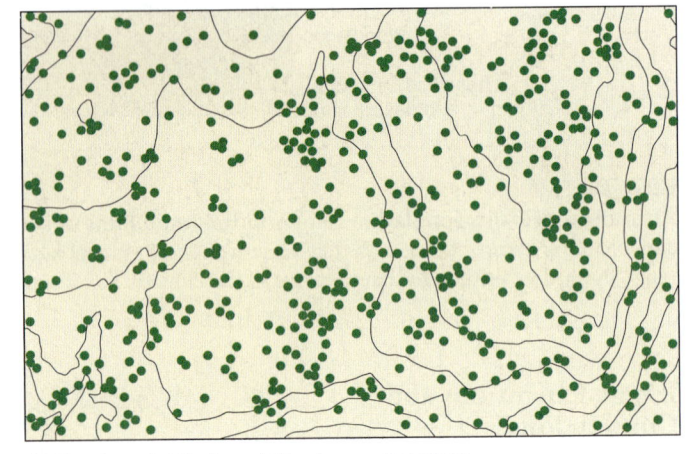

(b) Random distribution of *Brosimum alicastrum*

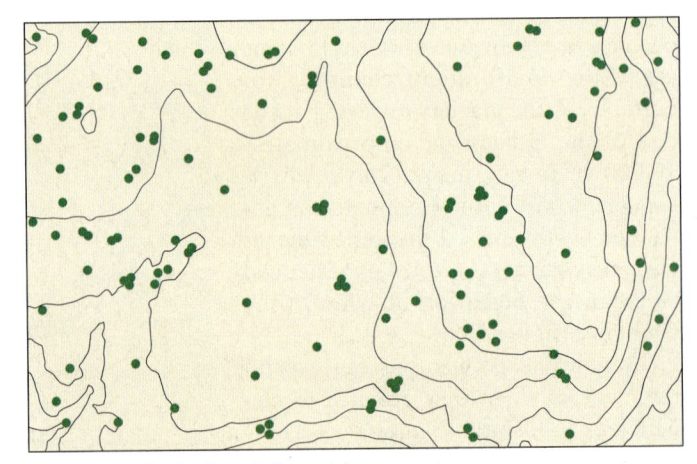

(c) Uniform distribution of *Coccoloba coronata*

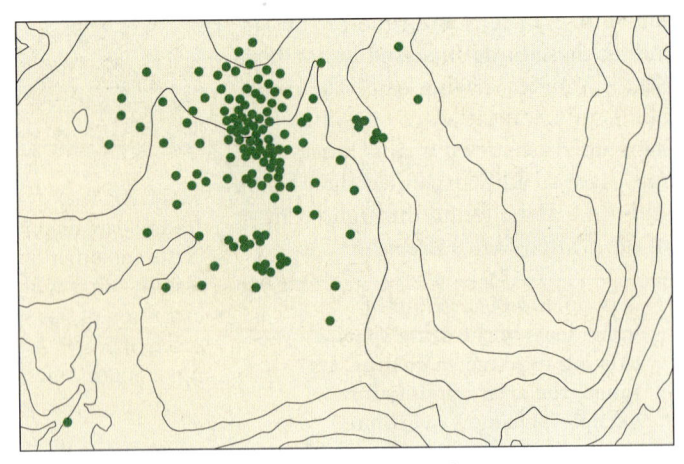

(d) Clumped distribution of *Chamguava schippii*

1142 Part VIII Ecology and Behavior

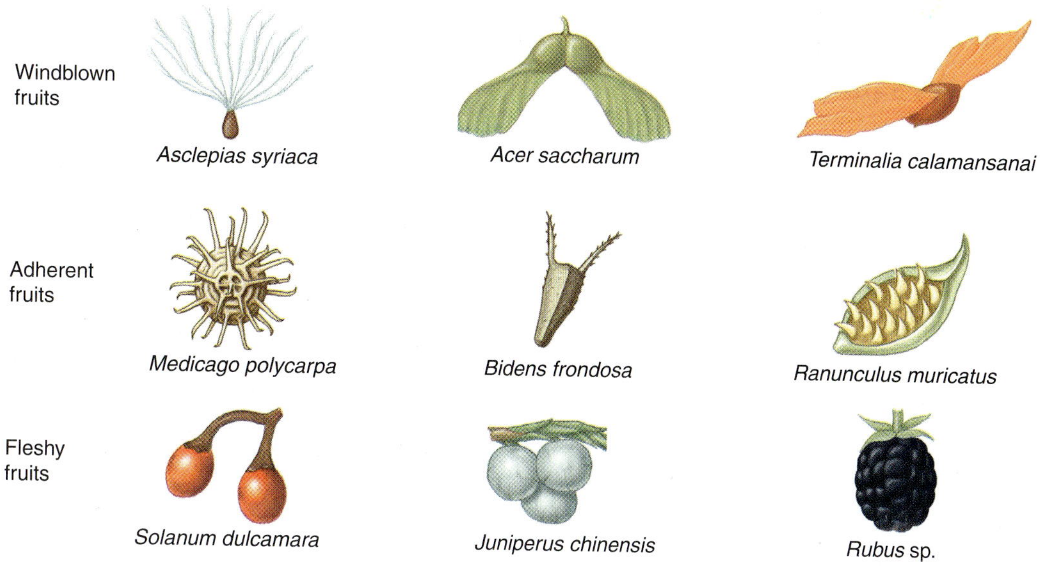

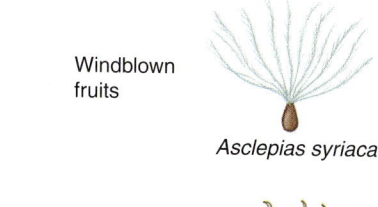

FIGURE 53.9 Some of the many adaptations of seeds. Seeds have evolved a number of different means of facilitating dispersal from their maternal plant. Some seeds can be transported great distances by the wind, whereas seeds enclosed in adherent and fleshy fruits can be transported by animals.

Windblown fruits: *Asclepias syriaca*, *Acer saccharum*, *Terminalia calamansanai*
Adherent fruits: *Medicago polycarpa*, *Bidens frondosa*, *Ranunculus muricatus*
Fleshy fruits: *Solanum dulcamara*, *Juniperus chinensis*, *Rubus* sp.

Clumped Spacing

Individuals clump into groups or clusters in response to uneven distribution of resources in their immediate environments. Clumped distributions are common in nature because individual animals, plants, and microorganisms tend to prefer habitats defined by soil type, moisture, or other aspects of the environment to which they are best adapted.

Social interactions also can lead to clumped distributions. Many species live and move around in large groups, which go by a variety of names (for example, flock, herd, pride). Such groupings can provide many advantages, including increased awareness of and defense against predators, decreased energetic cost of moving through air and water, and access to the knowledge of all group members.

At a broader scale, populations are often most densely populated in the interior of their range and less densely distributed toward the edges. Such patterns usually result from the manner in which the environment changes in different areas. Populations are often best adapted to the conditions in the interior of their distribution. As environmental conditions change, individuals are less well adapted, and thus densities decrease. Ultimately, the point is reached at which individuals cannot persist at all; this marks the edge of a population's range.

The Human Effect

By altering the environment, humans have allowed some species, such as coyotes, to expand their ranges and move into areas they previously did not occupy. Moreover, humans have served as an agent of dispersal for many species. Some of these transplants have been widely successful as is discussed in greater detail in chapter 57. For example, 100 starlings were introduced into New York City in 1896 in a misguided attempt to establish every species of bird mentioned by Shakespeare. Their population steadily spread so that by 1980, they occurred throughout the United States. Similar stories could be told for countless plants and animals, and the list increases every year. Unfortunately, the success of these invaders often comes at the expense of native species.

Dispersal Mechanisms

Dispersal to new areas can occur in many ways. Lizards, for example, have colonized many distant islands, probably due to individuals or their eggs floating or drifting on vegetation. Bats are often the only mammals on distant islands because they can fly to them. Seeds of plants are designed to disperse in many ways (figure 53.9). Some seeds are aerodynamically designed to be blown long distances by the wind. Others have structures that stick to the fur or feathers of animals, so that they are carried long distances before falling to the ground. Still others are enclosed in fleshy fruits. These seeds can pass through the digestive systems of mammals or birds and then germinate at the spot upon which they are defecated. Finally, seeds of *Arceuthobium* are violently propelled from the base of the fruit in an explosive discharge. Although the probability of long-distance dispersal events leading to successful establishment of new populations is slim, over millions of years, many such dispersals have occurred.

> The distribution of individuals within a population can be random, uniform, or clumped and is determined in part by the availability of resources.

Metapopulations

Species often exist as a network of distinct populations that interact with each other by exchanging individuals. Such networks, termed **metapopulations,** usually occur in areas in which suitable habitat is patchily distributed and separated by intervening stretches of unsuitable habitat.

To what degree populations within a metapopulation interact depends on the amount of dispersal and is often not symmetrical: Populations increasing in size tend to send out many dispersers, whereas populations at low levels tend to receive more immigrants than they send off. In addition, relatively isolated populations tend to receive relatively few arrivals.

Not all suitable habitats within a metapopulation's area may be occupied at any one time. For various reasons, some individual populations may become extinct, perhaps as a result of an epidemic disease, a catastrophic fire, or inbreeding depression. However, because of dispersal from other populations, such areas may eventually be recolonized. In some cases, the number of habitats occupied in a metapopulation may represent an equilibrium in which the rate of extinction of existing populations is balanced by the rate of colonization of empty habitats.

A species may also exhibit a metapopulation structure in areas in which some habitats are suitable for long-term population maintenance, whereas others are not. In these situations, termed **source-sink metapopulations,** the populations in the better areas (the sources) continually send out dispersers that bolster the populations in the poorer habitats (the sinks). In the absence of such continual replenishment, sink populations would have a negative growth rate and would eventually become extinct.

Metapopulations of butterflies have been studied particularly intensively (figure 53.10). In one study, researchers sampled populations of the glanville fritillary butterfly at 1600 meadows in southwestern Finland. On average, every year, 200 populations became extinct, but 114 empty meadows were colonized. A variety of factors seemed to increase the likelihood of a population's extinction, including small population size, isolation from sources of immigrants, low resource availability (as indicated by the number of flowers on a meadow), and lack of genetic variation within the population. The researchers attribute the greater number of extinctions than colonizations to a string of very dry summers. Because none of the populations is large enough to survive on its own, continued survival of the species in southwestern Finland would appear to require the continued existence of a metapopulation network in which new populations are continually created and existing populations are supplemented by emigrants. Continued bad weather thus may doom the species, at least in this part of its range.

Metapopulations, where they occur, can have two important implications for the range of a species. First, by continually colonizing empty patches, they prevent long-term extinction. If no such dispersal existed, then each population might eventually perish, leading to disappearance of the species from the entire area. Moreover, in source-sink metapopulations, the species occupies a larger area than it otherwise might occupy, including marginal areas that could not support a population without a continual influx of immigrants. For these reasons, the study of metapopulations has become very important in conservation biology as natural habitats become increasingly fragmented.

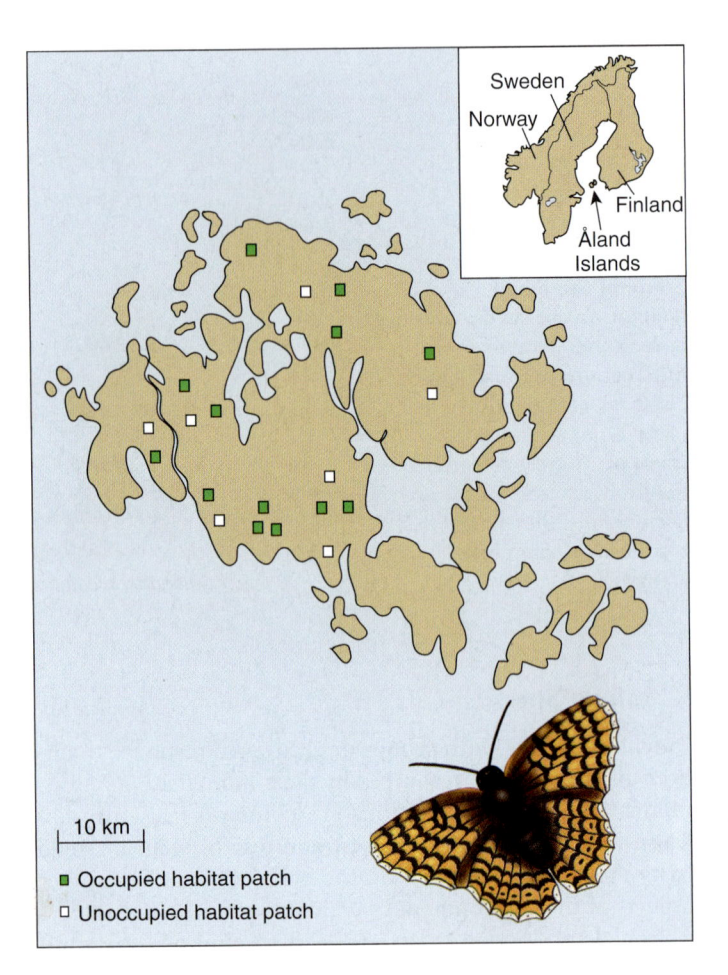

FIGURE 53.10
Metapopulations of butterflies. The glanville fritillary butterfly occurs in metapopulations in southwestern Finland on the Åland Islands. None of the populations is large enough to survive for long on its own, but continual emigration of individuals from other populations allows some populations to survive. In addition, continual establishment of new populations tends to offset extinction of established populations, although in recent years, extinctions have outnumbered colonizations.

> **Across broader areas, individuals may occur in populations that are loosely interconnected, termed metapopulations.**

1144 Part VIII Ecology and Behavior

53.3 Population dynamics depend critically upon age distribution.

The dynamics of a population—how it changes through time—are affected by many factors. One important factor is the age distribution of individuals—that is, what proportion of individuals are adults, juveniles, and babies.

Demography

Demography (from the Greek *demos*, "the people," + *graphos*, "measurement") is the statistical study of populations. How the size of a population changes through time can be studied at two levels: as a whole or broken down into parts. At the most inclusive level, we can study the whole population to determine whether it is increasing, decreasing, or remaining constant. Populations grow if births outnumber deaths and shrink if deaths outnumber births. Understanding these trends is often easier, however, if we break the population into smaller units composed of individuals of the same age (for example, 1-year-olds) and study the factors affecting birth and death rates for each unit separately.

Factors Affecting Population Growth Rates

Growth rates can be influenced by the population's **sex ratio**. The number of births in a population is usually directly related to the number of females, but may not be as closely related to the number of males in species in which a single male can mate with several females. In many species, males compete for the opportunity to mate with females (a situation discussed in chapter 52); consequently, a few males get many matings, whereas many males do not mate at all. In such species, a female-biased sex ratio would not affect population growth rates; reduction in the number of males simply changes the identities of the reproductive males without reducing the number of births. By contrast, among monogamous species, such as many birds, in which pairs form long-lasting reproductive relationships, a reduction in the number of males can directly reduce the number of births.

Generation time, the average interval between the birth of an individual and the birth of its offspring, can also affect population growth rates. Species differ greatly in generation time. Differences in body size can explain much of this variation—mice go through approximately 100 generations during the course of one elephant generation—but not all of it (figure 53.11). Newts, for example, are smaller than mice, but have considerably longer generation times. Everything else equal, populations with short generations can increase in size more quickly than populations with long generations. Conversely, because generation time and life span are usually closely correlated, populations with short generation times may also diminish in size more rapidly if birthrates suddenly decrease.

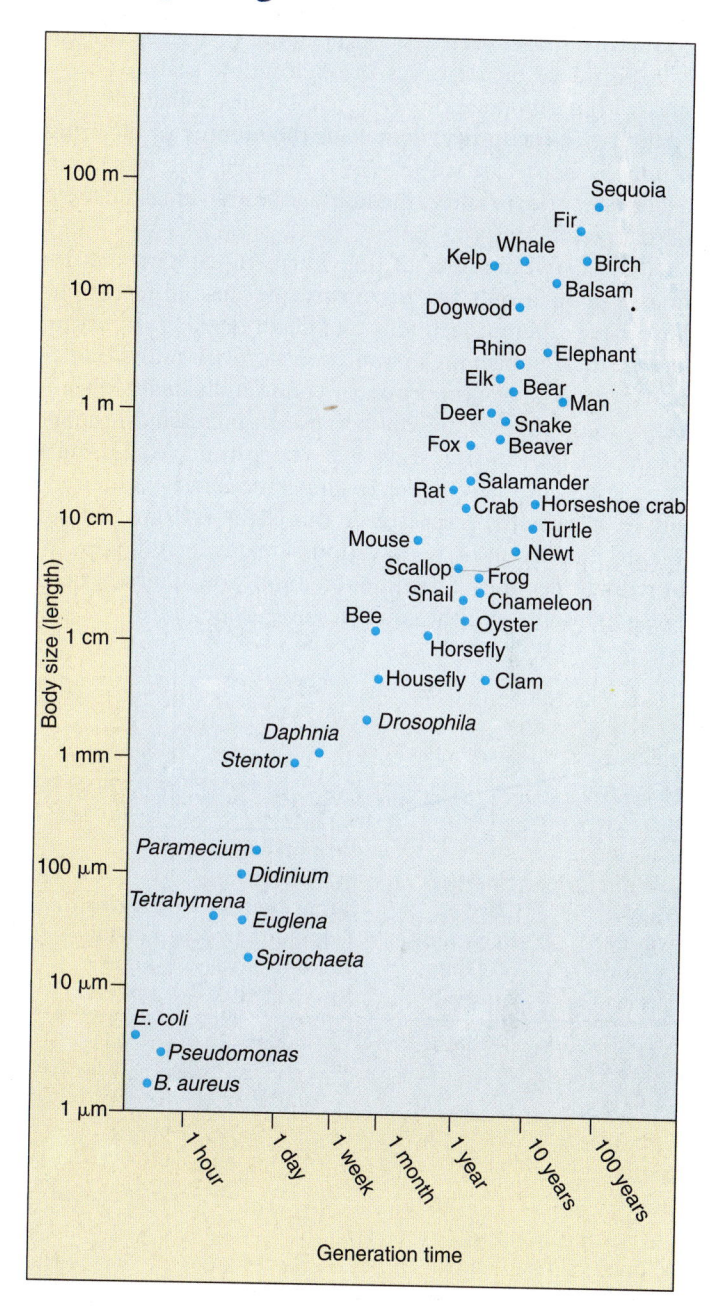

FIGURE 53.11
The relationship between body size and generation time. In general, larger animals have longer generation times, although there are exceptions.
If resources became more abundant, would you expect smaller or larger species to increase in population size more quickly?

Age Structure

A group of individuals of the same age is referred to as a **cohort**. In most species, the probability that an individual will reproduce or die varies through its life span. As a result, within a population, every cohort has a characteristic birthrate, or **fecundity**, defined as the number of offspring produced in a standard time (for example, per year), and death rate, or **mortality**, the number of individuals that die in that period.

The relative number of individuals in each cohort defines a population's **age structure**. Because different cohorts have different fecundity and death rates, age structure has a critical impact on a population's growth rate. Populations with a large proportion of young individuals, for example, tend to grow rapidly because an increasing proportion of their individuals are reproductive. Human populations in many underdeveloped countries are an example, as we will discuss later in this chapter. Conversely, if a large proportion of a population is relatively old, populations may decline. This phenomenon now characterizes Japan and some wealthy countries in Europe.

Life Tables and Population Change Through Time

To assess how populations in nature are changing, ecologists use a **life table**, which tabulates the fate of a cohort from birth until death, showing the number of offspring produced and the number of individuals that die each year. A very nice life table analysis is exhibited in a study of the meadow grass *Poa annua* (table 53.2). This study follows the fate of 843 individuals through time, charting how many survive in each interval and how many offspring each survivor produces.

In table 53.2, the first column indicates the age of the cohort (that is, the number of 3-month intervals from the start of the study). The second and third columns indicate the number of survivors and the proportion of the original cohort still alive at the beginning of that interval. The fifth column presents the **mortality rate,** the proportion of individuals that started that interval alive but died by the end of it. The seventh column indicates the average number of seeds produced by each surviving individual in that interval, and the last column presents the number of seeds produced relative to the size of the original cohort.

Table 53.2 Life Table for a Cohort of the Grass *Poa annua*, Containing 843 Seedlings

Age (in 3-month intervals)	Number Alive at Beginning of Time Interval	Proportion of Cohort Surviving to Beginning of Time Interval (survivorship)	Deaths During Time Interval	Mortality Rate During Time Interval	Seeds Produced During Time Interval	Seeds Produced per Surviving Individual (fecundity)	Seeds Produced per Member of Cohort (fecundity × survivorship)
0	843	1.000	121	0.143	0	0.00	0.00
1	722	0.857	195	0.271	303	0.42	0.36
2	527	0.625	211	0.400	622	1.18	0.74
3	316	0.375	172	0.544	430	1.36	0.51
4	144	0.171	90	0.626	210	1.46	0.25
5	54	0.064	39	0.722	60	1.11	0.07
6	15	0.018	12	0.800	30	2.00	0.04
7	3	0.004	3	1.000	10	3.33	0.01
8	0	0.000	—		Total = 1665		Total = 1.98

Much can be learned by examining life tables. In the case of *P. annua*, we see that both the probability of dying and the number of offspring per surviving individual produced steadily increases with age. By adding up the numbers in the last column, we get the total number of offspring produced per individual in the initial cohort. This number is almost 2, which means that for every original member of the cohort, on average two individuals have been produced. A figure of 1.0 would be the break-even number, the point at which the population was neither growing nor shrinking. In this case, the population appears to be growing rapidly.

In most cases, life table analysis is more complicated than this. First, except for organisms with short life spans, it is difficult to track the fate of a cohort until the death of the last individual. An alternative approach is to construct a cross-sectional study, examining the fate of all cohorts over a single year. In addition, many factors—such as offspring reproducing before all members of their parental generation's cohort have died—complicate the interpretation of whether populations are growing or shrinking.

Survivorship Curves

The percentage of an original population that survives to a given age is called its **survivorship**. One way to express some aspects of the age distribution of populations is through a *survivorship curve*. Examples of different survivorship curves are shown in figure 53.12. In hydra, animals related to jellyfish, individuals are equally likely to die at any age, as indicated by the straight survivorship curve (type II). Oysters produce vast numbers of offspring, only a few of which live to reproduce. However, once they become established and grow into reproductive individuals, their mortality rate is extremely low (type III survivorship curve). Finally, even though human babies are susceptible to death at relatively high rates, mortality rates in humans, as in many other animals and in protists, rise steeply in the postreproductive years (type I survivorship curve). Of course, these are just generalizations, and many organisms show more complicated patterns. Examination of the data for *P. annua*, for example, reveals that it is most similar to a type II survivorship curve (figure 53.13).

> The growth rate of a population is a sensitive function of its age structure. The age structure of a population and the manner in which mortality and birthrates vary among different age cohorts determine whether a population will increase or decrease in size.

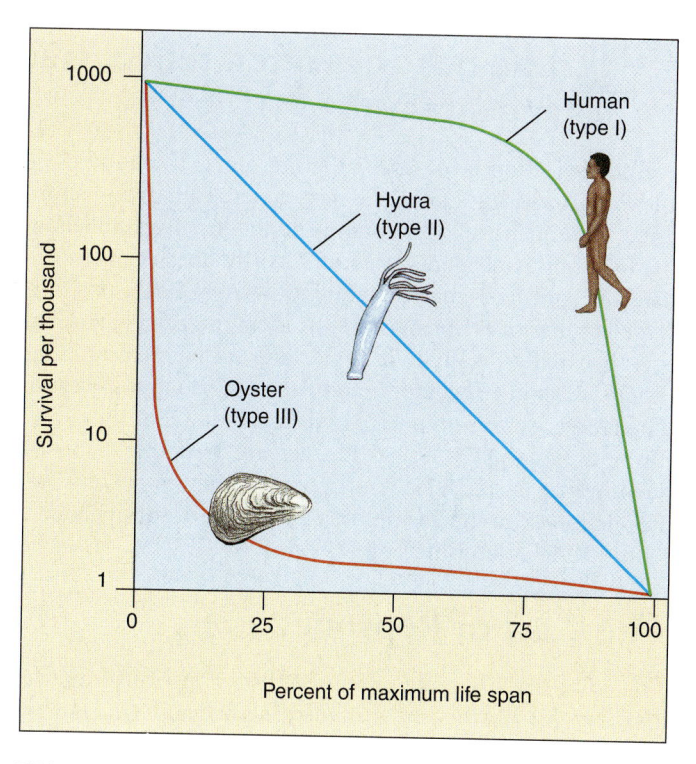

FIGURE 53.12
Survivorship curves. By convention, survival (the vertical axis) is plotted on a log scale. Humans have a type I life cycle, the hydra (an animal related to jellyfish) type II, and oysters type III.

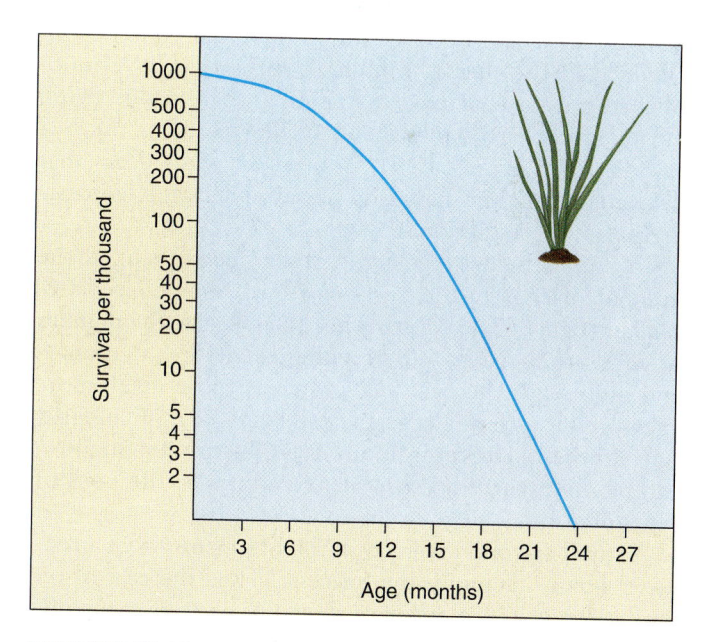

FIGURE 53.13
Survivorship curve for a cohort of the meadow grass *Poa annua*. After several months of age, mortality increases at a constant rate through time.
Suppose you wanted to keep meadow grass in your room as a houseplant. Suppose, too, that you wanted to buy a plant that was likely to live as long as possible. What age plant would you buy?

Chapter 53 Population Ecology 1147

53.4 Life histories often reflect trade-offs between reproduction and survival.

Natural selection favors traits that maximize the number of surviving offspring left in the next generation. Two factors affect this quantity: how long an individual lives and how many young it produces each year. Why doesn't every organism reproduce immediately after its own birth, produce large families of offspring, care for them intensively, and do this repeatedly throughout a long life, while outcompeting others, escaping predators, and capturing food with ease? The answer is that no one organism can do all of this, simply because not enough resources are available. Consequently, organisms allocate resources either to current reproduction or to increasing their prospects of surviving and reproducing at later life stages.

The Cost of Reproduction

The complete life cycle of an organism constitutes its **life history**. All life histories involve significant trade-offs. Because resources are limited, a change that increases reproduction may decrease survival and reduce future reproduction. Thus, a Douglas fir tree that produces more cones increases its current reproductive success, but it also grows more slowly; because the number of cones produced is a function of how large a tree is, this diminished growth will decrease the number of cones it can produce in the future. Similarly, birds that have more offspring each year have a higher probability of dying during that year or producing smaller clutches the following year (figure 53.14). Conversely, individuals that delay reproduction may grow faster and larger, enhancing future reproduction.

In one elegant experiment, researchers changed the number of eggs in the nests of a bird, the collared flycatcher (figure 53.15). Birds whose clutch size (the number of eggs produced in one breeding event) was decreased laid more eggs the next year, whereas those given more eggs produced fewer eggs the following year. Ecologists refer to the reduction in future reproductive potential resulting from current reproductive efforts as the **cost of reproduction**.

Natural selection will favor the life history that maximizes lifetime reproductive success. When the cost of reproduction is low, individuals should produce as many offspring as possible because there is little cost. Low costs of reproduction may occur when resources are abundant, such that producing offspring does not impair survival or the ability to produce many offspring in subsequent years. Costs of reproduction are also low when overall mortality rates are high. In such cases, individuals may be unlikely to survive to the next breeding season anyway, so the incremental effect of increased reproductive efforts may not make a difference in future survival.

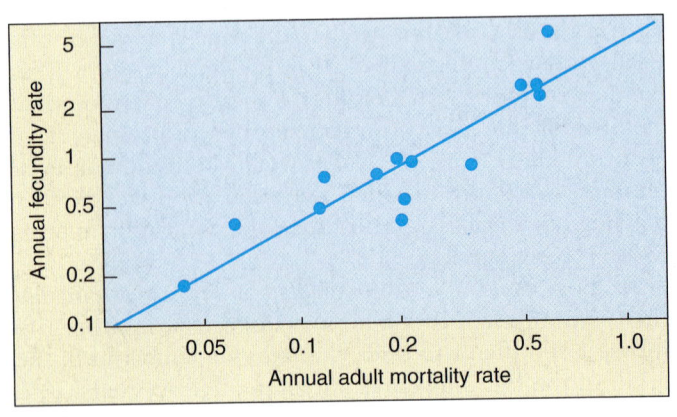

FIGURE 53.14
Reproduction has a price. Increased fecundity in birds correlates with higher mortality in several populations of birds, ranging from the albatross (lowest) to the sparrow (highest). Birds that raise more offspring per year have a higher probability of dying during that year.
Do you think that species in this graph are ordered by body size?

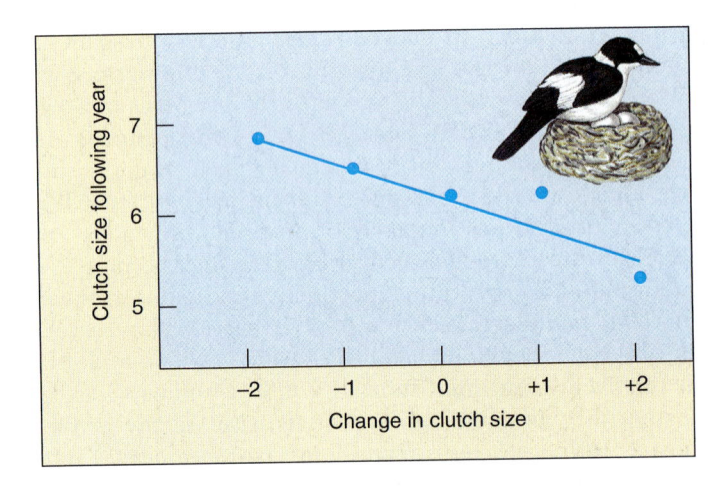

FIGURE 53.15
Reproductive events per lifetime. Adding eggs to nests of collared flycatchers (which increases the reproductive efforts of the female rearing the young) decreases clutch size the following year; removing eggs from the nest increases the next year's clutch size. This experiment demonstrates the trade-off between current reproductive effort and future reproductive success.
Why does this relationship exist?

Alternatively, when costs of reproduction are high, lifetime reproductive success may be maximized by deferring or minimizing current reproduction to enhance growth and survival rates. This may occur when costs of reproduction significantly affect the ability of an individual to survive or decrease the number of offspring that can be produced in the future.

1148 Part VIII Ecology and Behavior

Investment per Offspring

In terms of natural selection, the number of offspring produced is not as important as how many of those offspring themselves survive to reproduce.

A key reproductive trade-off concerns how many resources to invest in producing any single offspring. Assuming that the amount of energy to be invested in offspring is limited, a trade-off must occur between the number of offspring produced and the size of each offspring (figure 53.16). This trade-off has been experimentally demonstrated in the side-blotched lizard, *Uta stansburiana*, which normally lays on average four and a half eggs at a time. When some of the eggs are removed surgically early in the reproductive cycle, the female lizard produces only 1 to 3 eggs, but supplies each of these eggs with greater amounts of yolk, producing eggs and, subsequently, hatchlings that are much larger than normal (figure 53.17).

In the side-blotched lizard and many other species, the size of offspring critically affects their survival prospects—larger offspring have a greater chance of survival. Producing many offspring with little chance of survival might not be the best strategy, but producing only a single, extraordinarily robust offspring also would not maximize the number of surviving offspring. Rather, an intermediate situation, in which several fairly large offspring are produced, should maximize the number of surviving offspring.

Reproductive Events per Lifetime

The trade-off between age and fecundity plays a key role in many life histories. Annual plants and most insects focus all their reproductive resources on a single large event and then die. This life history adaptation is called **semelparity** (from the Latin *semel*, "once," + *parito*, "to beget"). Organisms that produce offspring several times over many seasons exhibit a life history adaptation called **iteroparity** (from the Latin *itero*, "to repeat"). Species that reproduce yearly must avoid overtaxing themselves in any one reproductive episode so that they will be able to survive and reproduce in the future. Semelparity, or "big bang" reproduction, is usually found in short-lived species that have a low probability of staying alive between broods, such as plants growing in harsh climates. Semelparity is also favored when fecundity entails large reproductive cost, as when Pacific salmon migrate upriver to their spawning grounds. In these species, rather than investing some resources in an unlikely bid to survive until the next breeding season, individuals place all their resources into reproduction.

Age at First Reproduction

Among mammals and many other animals, longer-lived species put off reproduction longer than short-lived species

FIGURE 53.16
The relationship between clutch size and offspring size. In great tits, the size of the nestlings is inversely related to the number of eggs laid. The more mouths they have to feed, the less the parents can provide to any one nestling.
Would natural selection favor producing many small young or a few large ones?

FIGURE 53.17
Variation in baby lizard size produced by experimental manipulations. In clutches in which some developing eggs were surgically removed, the remaining offspring were larger (*center*) than lizards produced in control clutches in which all the eggs were allowed to develop (*right*). In experiments in which some of the yolk was removed from the eggs, smaller lizards hatched (*left*).

(relative to expected life span). The advantage of delayed reproduction is that juveniles gain experience before expending the high costs of reproduction. In long-lived animals, this advantage outweighs the energy that is invested in survival and growth rather than reproduction. In shorter-lived animals, on the other hand, time is of the essence; thus, quick reproduction is more critical than juvenile training, and reproduction tends to occur earlier.

> Life history adaptations involve many trade-offs between reproductive cost and investment in survival. Different kinds of animals and plants employ quite different approaches.

53.5 Population growth is limited by the environment.

Population Growth

Populations often remain at a relatively constant size, regardless of how many offspring are born. As you saw in chapter 1, Darwin based his theory of natural selection partly on this seeming contradiction. Natural selection occurs because of checks on reproduction, with some individuals producing fewer surviving offspring than others. To understand populations, we must consider how they grow and what factors in nature limit population growth.

The Exponential Growth Model

The rate of population increase, r, is defined as the difference between the birthrate (b) and the death rate (d) corrected for any movement of individuals in or out of the population, whether net emigration (e, movement out of the area) or net immigration (i, movement into the area). Thus,

$$r = (b - d) + (i - e)$$

Movements of individuals can have a major impact on population growth rates. For example, the increase in human population in the United States during the closing decades of the twentieth century was mostly due to immigration. Less than half of the increase came from the reproduction of the people already living there.

The simplest model of population growth assumes that a population grows without limits at its maximal rate and also that rates of immigration and emigration are equal. This rate, called the **biotic potential**, is the rate at which a population of a given species will increase when no limits are placed on its rate of growth. In mathematical terms, this is defined by the following formula:

$$\frac{dN}{dt} = r_i N$$

where N is the number of individuals in the population, dN/dt is the rate of change in its numbers over time, and r_i is the intrinsic rate of natural increase for that population—its innate capacity for growth.

The biotic potential of any population is exponential (red line in figure 53.18). Even when the *rate* of increase remains constant, the actual *number* of individuals accelerates rapidly as the size of the population grows. The result of unchecked exponential growth is a population explosion. A single pair of houseflies, laying 120 eggs per generation, could produce more than 5 trillion descendants in a year. In 10 years, their descendants would form a swarm more than 2 meters thick over the entire surface of the earth! In practice, such patterns of unrestrained growth prevail only for short periods, usually when an organism reaches a new habitat with abundant resources.

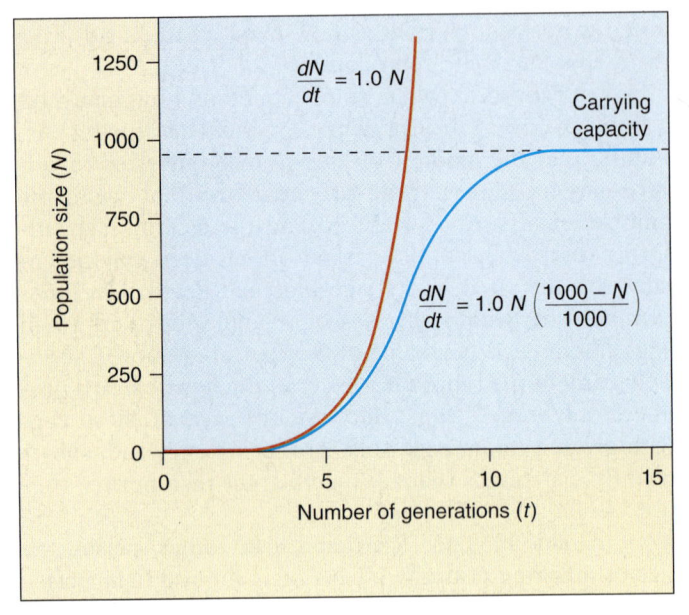

FIGURE 53.18
Two models of population growth. The red line illustrates the exponential growth model for a population with an r of 1.0. The blue line illustrates the logistic growth model in a population with $r = 1.0$ and $K = 1000$ individuals. At first, logistic growth accelerates exponentially; then, as resources become limiting, the death rate increases and growth slows. Growth ceases when the death rate equals the birthrate. The carrying capacity (K) ultimately depends on the resources available in the environment.

Natural examples include dandelions arriving in the fields, lawns, and meadows of North America from Europe for the first time; algae colonizing a newly formed pond; or the first terrestrial immigrants landing on an island recently thrust up from the sea.

Carrying Capacity

No matter how rapidly populations grow, they eventually reach a limit imposed by shortages of important environmental factors, such as space, light, water, or nutrients. A population ultimately may stabilize at a certain size, called the **carrying capacity** of the particular place where it lives. The carrying capacity, symbolized by K, is the maximum number of individuals that the environment can support.

The Logistic Growth Model

As a population approaches its carrying capacity, its rate of growth slows greatly, because fewer resources remain for each new individual to use. The growth curve of such a

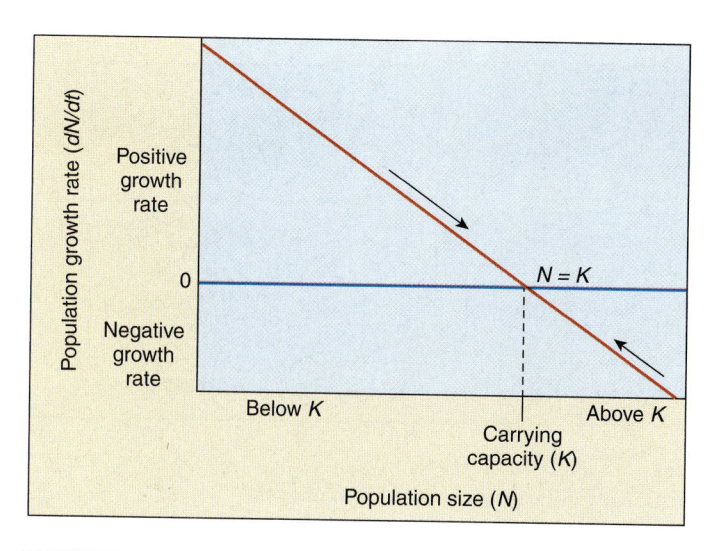

FIGURE 53.19

Relationship between population growth rate and population size. Populations far from the carrying capacity (K) will have high growth rates—positive if the population is below K, and negative if it is above K. As the population approaches K, growth rates approach zero.

Why does the growth rate converge on zero?

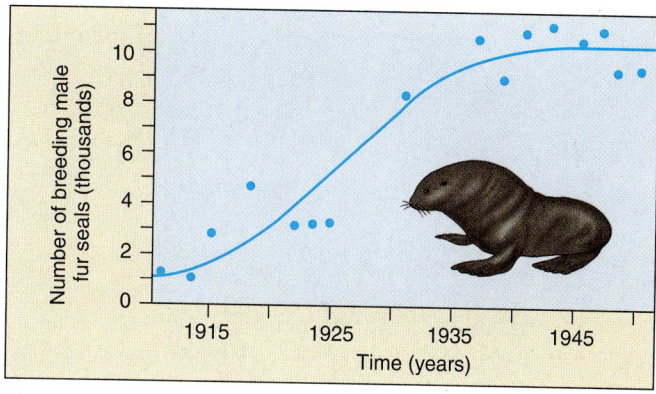

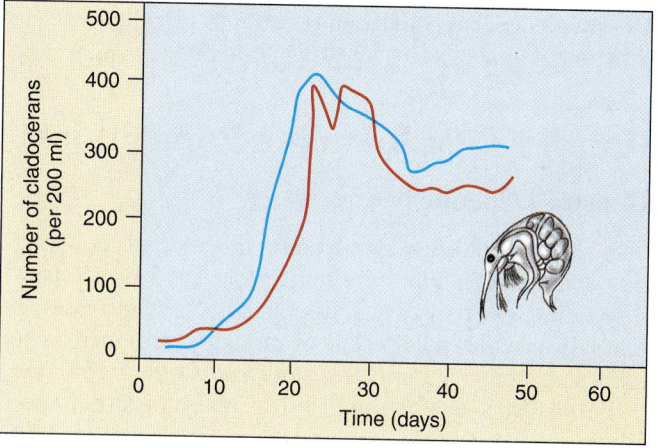

FIGURE 53.20

Many populations exhibit logistic growth. (*a*) A fur seal (*Callorhinus ursinus*) population on St. Paul Island, Alaska. (*b*) Two laboratory populations of the cladoceran *Bosmina longirostris*. Note that the populations first exceeded the carrying capacity, before decreasing to a size that was then maintained.

Why is there a hump in the population growth curve in (*b*), followed by a decline in the population?

population, which is always limited by one or more factors in the environment, can be approximated by the following logistic growth equation:

$$\frac{dN}{dt} = rN\left(\frac{K - N}{K}\right)$$

In this model of population growth, the growth rate of the population (*dN/dt*) equals its intrinsic rate of natural increase (*r* multiplied by *N*, the number of individuals present at any one time), adjusted for the amount of resources available. The adjustment is made by multiplying *rN* by the fraction of *K* still unused (*K* minus *N*, divided by *K*). As *N* increases (the population grows in size), the fraction by which *r* is multiplied (the remaining resources) becomes smaller and smaller, and the rate of increase of the population declines.

Graphically, if you plot *N* versus *t* (time), you obtain a **sigmoidal growth curve** characteristic of many biological populations. The curve is called "sigmoidal" because its shape has a double curve like the letter S. As the size of a population stabilizes at the carrying capacity, its rate of growth slows down, eventually coming to a halt (blue line in figure 53.18).

In mathematical terms, as *N* approaches *K*, the *rate of population growth* (*dN/dt*) begins to slow, reaching 0 when *N = K* (figure 53.19). Conversely, if the population size exceeds the carrying capacity, then *K − N* will be negative, and the population will experience a negative growth rate. As the population size declines toward the carrying capacity, the magnitude of this negative growth rate will decrease until it reaches 0 when *N = K*. Notice that the population will tend to move toward the carrying capacity regardless of whether it is initially above or below it. For this reason, logistic growth tends to return a population to the same size. In this sense, such populations are considered to be in equilibrium because they would be expected to be at or near the carrying capacity at most times.

In many cases, real populations display trends corresponding to a logistic growth curve. This is true not only in the laboratory, but also in natural populations (figure 53.20*a*). In some cases, however, the fit is not perfect (figure 53.20*b*), and as we shall see shortly, many populations exhibit other patterns.

> The size at which a population stabilizes in a particular place is defined as the carrying capacity of that place for that species. Populations often grow to the carrying capacity of their environment.

Chapter 53 Population Ecology **1151**

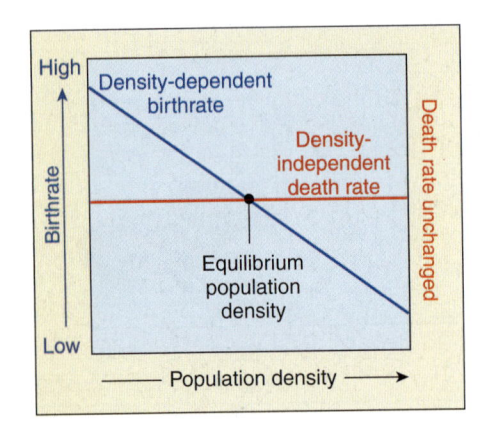

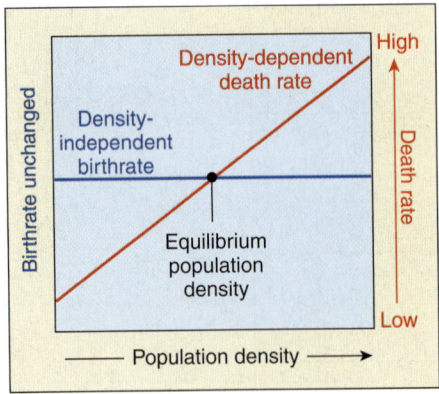

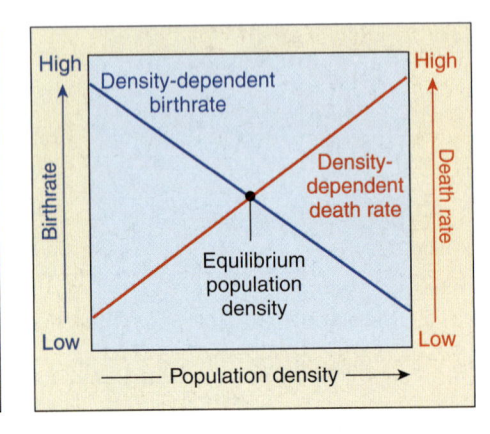

FIGURE 53.21

Density-dependent population regulation. Density-dependent factors can affect birthrates, death rates, or both. Why might birthrates be density-dependent?

Factors That Regulate Populations

Density-Dependent Effects

The reason population growth rates are affected by population size is that many important processes have **density-dependent effects.** That is, as population size increases, either reproductive rates decline or mortality rates increase, or both, a phenomenon termed *negative feedback* (figure 53.21).

Populations can be regulated in many different ways. When populations approach their carrying capacity, competition for resources can be severe, leading both to a decreased birthrate and an increased risk of death (figure 53.22). In addition, predators often focus their attention on particularly common prey, which also results in increasing rates of mortality as populations increase. High population densities can also lead to an accumulation of toxic wastes in the environment.

Behavioral changes may also affect population growth rates. Some species of rodents, for example, become antisocial, fighting more, breeding less, and generally acting stressed-out. These behavioral changes result from hormonal actions, but their ultimate cause is not yet clear; most likely, they have evolved as adaptive responses to situations in which resources are scarce. In addition, in crowded populations, the population growth rate may decrease because of an increased rate of emigration of individuals attempting to find better conditions elsewhere (figure 53.23).

FIGURE 53.22

Density dependence in the song sparrow (*Melospiza melodia*) on Mandarte Island. Reproductive success decreases and mortality rates increase as population size increases.
What would happen if researchers supplemented the food available to the birds?

FIGURE 53.23

Density-dependent effects. Migratory locusts, *Locusta migratoria*, are a legendary plague of large areas of Africa and Eurasia. At high population densities, the locusts have different hormonal and physical characteristics and take off as a swarm. The most serious infestation of locusts in 30 years occurred in North Africa in 1988.

1152   Part VIII   Ecology and Behavior

However, not all density-dependent factors are negatively related to population size. In some cases, growth rates increase with population size. This phenomenon is referred to as the **Allee effect** (after Warder Allee, who first described it), and is an example of *positive feedback*. The Allee effect can take several forms. Most obviously, in populations that are too sparsely distributed, individuals may have difficulty finding mates. Moreover, some species may rely on large groups to deter predators or to provide the necessary stimulation for breeding activities.

Density-Independent Effects

Growth rates in populations sometimes do not correspond to the logistic growth equation. In many cases, such patterns result because growth is under the control of **density-independent effects**. In other words, the rate of growth of a population at any instant is limited by something unrelated to the size of the population.

A variety of factors may affect populations in a density-independent manner. Most of these are aspects of the external environment, such as extremely cold winters, droughts, storms, or volcanic eruptions. Individuals often will be affected by these activities regardless of the size of the population. Populations in areas where such events occur relatively frequently will display erratic growth patterns in which the populations increase rapidly when conditions are benign, but exhibit large reductions whenever the environment turns hostile (figure 53.24). Needless to say, such populations do not produce the sigmoidal growth curves characteristic of the logistic equation.

Population Cycles

In some populations, density-dependent effects lead not to an equilibrium population size but to cyclic patterns of increase and decrease. Ecologists have studied cycles in hare populations since the 1820s. They have found that the North American snowshoe hare (*Lepus americanus*) follows a "10-year cycle" (in reality, it varies from 8 to 11 years). Its numbers fall 10-fold to 30-fold in a typical cycle, and 100-fold changes can occur (figure 53.25). Two factors appear to be generating the cycle: food plants and predators.

Food plants. The preferred foods of snowshoe hares are willow and birch twigs. As hare density increases, the quantity of these twigs decreases, forcing the hares to feed on high-fiber (low-quality) food. Lower birthrates, low juvenile survivorship, and low growth rates follow.

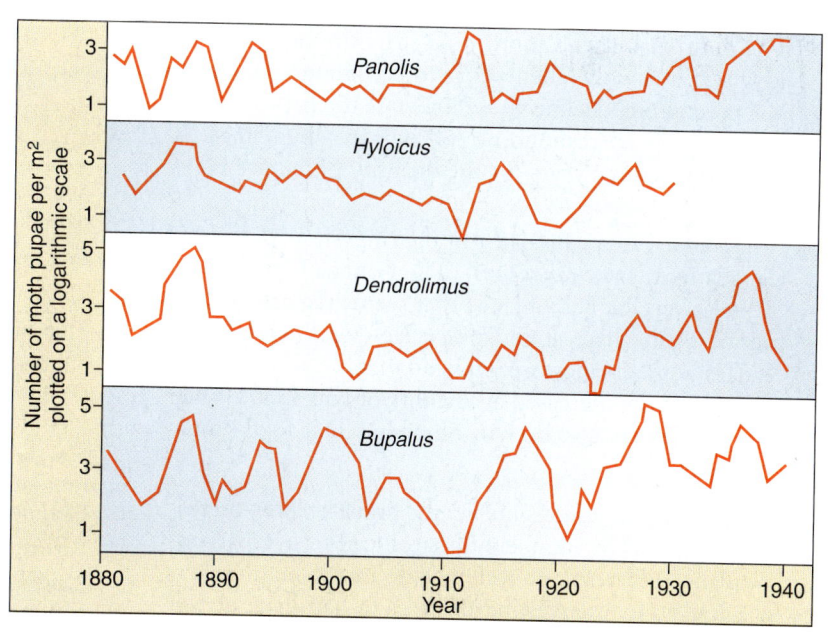

FIGURE 53.24
Fluctuations in the number of pupae of four moth species in Germany. The population fluctuations suggest that density-independent factors are regulating population size. The species concordance in trends through time suggests that the same factors are regulating population size in all species.
What might those factors be?

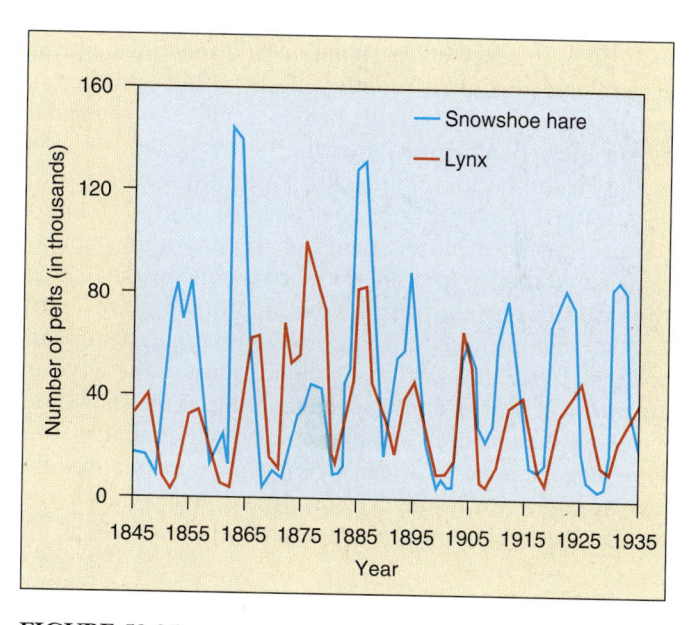

FIGURE 53.25
Linked population cycles of the snowshoe hare and the northern lynx. These data are based on records of fur returns from trappers in the Hudson Bay region of Canada. The lynx population carefully tracks that of the snowshoe hare, but lags behind it slightly.
Suppose experimenters artificially kept the hare population at a high and constant level; what would happen to the lynx population? Conversely, if experimenters artificially kept the lynx population at a high and constant level, what would happen to the hare population?

Chapter 53 Population Ecology 1153

The hares also spend more time searching for food, an activity that exposes them more to predation. The result is a precipitous decline in willow and birch twig abundance, and a corresponding fall in hare abundance. It takes two to three years for the quantity of mature twigs to recover.

Predators. A key predator of the snowshoe hare is the Canada lynx (*Lynx canadensis*). The Canada lynx shows a "10-year" cycle of abundance that seems remarkably entrained to the hare abundance cycle (see figure 53.25). As hare numbers increase, lynx numbers do too, rising in response to the increased availability of lynx food. When hare numbers fall, so do lynx numbers, their food supply depleted.

Which factor is responsible for the predator-prey oscillations? Do increasing numbers of hares lead to overharvesting of plants (a hare-plant cycle), or do increasing numbers of lynx lead to overharvesting of hares (a hare-lynx cycle)? Field experiments carried out by C. Krebs and coworkers in 1992 provide an answer. In Canada's Yukon, Krebs set up experimental plots that contained hare populations. If food is added (no food shortage effect) and predators are excluded (no predator effect) in an experimental area, hare numbers increase tenfold and stay there—the cycle is lost. However, the cycle is retained if either of the factors is allowed to operate alone: exclude predators but don't add food (food shortage effect alone), or add food in the presence of predators (predator effect alone). Thus, both factors can affect the cycle, which in practice seems to be generated by the interaction between the two.

Population cycles traditionally have been considered to occur rarely. However, a recent review of nearly 700 long-term (25 years or more) studies of trends within populations found that cycles were not uncommon; nearly 30% of the studies—including birds, mammals, fish, and crustaceans—provided evidence of some cyclic pattern in population size through time, although most of these cycles are nowhere near as dramatic in amplitude as the snowshoe hare and lynx cycles. In some cases, such as that of the snowshoe hare and lynx, density-dependent factors may be involved, whereas in other cases, density-independent factors, such as cyclic climatic patterns, may be responsible.

Population Growth Rates and Life History Models

As we have seen, some species usually maintain stable population sizes near the carrying capacity, whereas the population sizes of other species fluctuate markedly and are often far below carrying capacity. As we saw in our discussion of life histories, the selective factors affecting such species differ markedly. Populations near their carrying capacity may face stiff competition for limited resources. By contrast, resources are abundant in populations far below carrying capacity.

Table 53.3 *r*-Selected and *K*-Selected Life History Adaptations

Adaptation	*r*-Selected Populations	*K*-Selected Populations
Age at first reproduction	Early	Late
Life span	Short	Long
Maturation time	Short	Long
Mortality rate	Often high	Usually low
Number of offspring produced per reproductive episode	Many	Few
Number of reproductions per lifetime	Usually one	Often several
Parental care	None	Often extensive
Size of offspring or eggs	Small	Large

We have already seen the consequences of such differences. When resources are limited, the cost of reproduction often will be very high. Consequently, selection will favor individuals that can compete effectively and utilize resources efficiently. Such adaptations often come at the cost of lowered reproductive rates. Such populations are termed **K-selected** because they are adapted to thrive when the population is near its carrying capacity (K). Table 53.3 lists some of the typical features of *K*-selected populations. Examples of *K*-selected species include coconut palms, whooping cranes, whales, and humans.

By contrast, in populations far below the carrying capacity, resources may be abundant. Costs of reproduction will be low, and selection will favor those individuals that can produce the maximum number of offspring. Selection here favors individuals with the highest reproductive rates; such populations are termed **r-selected.** Examples of organisms displaying *r*-selected life history adaptations include dandelions, aphids, mice, and cockroaches.

Most natural populations show life history adaptations that exist along a continuum ranging from completely *r*-selected traits to completely *K*-selected traits. Although these tendencies hold true as generalities, few populations are purely *r*- or *K*-selected and show all of the traits listed in table 53.3. These attributes should be treated as generalities, with the recognition that many exceptions exist.

> Density-dependent effects are caused by factors that come into play particularly when the population size is larger; density-independent effects result from factors that operate regardless of population size. Some life history adaptations favor near-exponential growth; others favor the more competitive logistic growth. Most natural populations exhibit a combination of the two.

53.6 The human population has grown explosively in the last three centuries.

The Advent of Exponential Growth

Humans exhibit many *K*-selected life history traits, including small brood size, late reproduction, and a high degree of parental care. These life history traits evolved during the early history of hominids, when the limited resources available from the environment controlled population size. Throughout most of human history, our populations have been regulated by food availability, disease, and predators. Although unusual disturbances, including floods, plagues, and droughts, no doubt affected the pattern of human population growth, the overall size of the human population grew slowly during our early history. Two thousand years ago, perhaps 130 million people populated the earth. It took a thousand years for that number to double, and it was 1650 before it had doubled again, to about 500 million. In other words, for over 16 centuries, the human population was characterized by very slow growth. In this respect, human populations resembled many other species with predominantly *K*-selected life history adaptations.

Starting in the early 1700s, changes in technology gave humans more control over their food supply, enabled them to develop superior weapons to ward off predators, and led to the development of cures for many diseases. At the same time, improvements in shelter and storage capabilities made humans less vulnerable to climatic uncertainties. These changes allowed humans to expand the carrying capacity of the habitats in which they lived, and thus to escape the confines of logistic growth and reenter the exponential phase of the sigmoidal growth curve.

Responding to the lack of environmental constraints, the human population has grown explosively over the last 300 years. While the birthrate has remained unchanged at about 30 per 1000 per year over this period, the death rate has fallen dramatically, from 20 per 1000 per year to its present level of 13 per 1000 per year. The difference between birth and death rates meant that the population grew as much as 2% per year, although the rate has now declined to 1.3% per year.

A 1.3% annual growth rate may not seem large, but it has produced a current human population of 6.3 billion people (figure 53.26)! At this growth rate, 82 million people are added to the world population annually, and the human population will double in 53 years. As we shall see, both the current human population level and the projected growth rate have potentially grave consequences for our future.

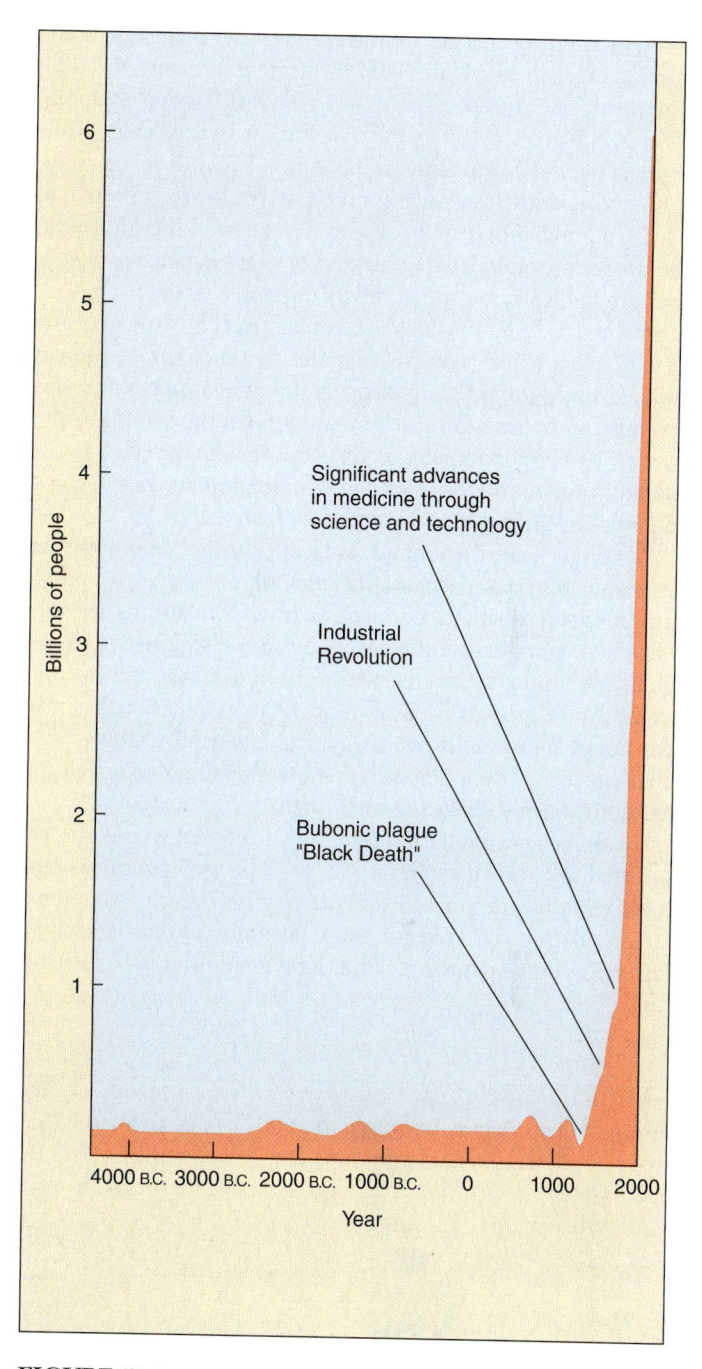

FIGURE 53.26

History of human population size. Temporary increases in death rate, even a severe one such as that occurring during the Black Death of the 1300s, have little lasting impact. Explosive growth began with the industrial revolution in the 1800s, which produced a significant, long-term lowering of the death rate. The current world population is 6.3 billion, and at the present rate, it will double in 53 years.

Based on what we have learned about population growth, what do you predict will happen to human population size?

Chapter 53 Population Ecology 1155

Population Pyramids

While the human population as a whole continues to grow rapidly at the beginning of the twenty-first century, this growth is not occurring uniformly over the planet. In some countries, such as Mexico, the birthrate is greatly exceeding the death rate (figure 53.27). As a result, if Mexico's population keeps growing at its current rate, it is projected to increase from 102 million in 2002 to 151 million in 2050. Other countries are growing much more slowly. The rate at which a population can be expected to grow in the future can be assessed graphically by means of a *population pyramid*, a bar graph displaying the numbers of people in each age category (figure 53.28). Males are conventionally shown to the left of the vertical age axis, females to the right. A human population pyramid thus displays the age composition of a population by sex. In most human population pyramids, the number of older females is disproportionately large compared to the number of older males, because females in most regions have a longer life expectancy than males.

Viewing such a pyramid, we can predict demographic trends in births and deaths. In general, a rectangular pyramid is characteristic of countries whose populations are stable, their numbers neither growing nor shrinking. A triangular pyramid is characteristic of a country that will exhibit rapid future growth because most of its population has not yet entered the child-bearing years. Inverted triangles are characteristic of populations that are shrinking, usually as a result of sharply declining birthrates.

Examples of population pyramids for Sweden and Kenya in 2000 are shown in figure 53.28. The two countries exhibit very different age distributions. The nearly rectangular population pyramid for Sweden indicates that its population is not expanding. The very triangular pyramid of Kenya, by contrast, predicts explosive future growth. The difference is most apparent when we consider that only 20% of Sweden's population is less than 15 years old, compared to nearly half of all Kenyans. Moreover, the fertility rate (offspring per woman) in Sweden is 1.6; in Kenya, it is 4.4. As a result, Kenya's population could double in less than 35 years, whereas Sweden's will remain stable.

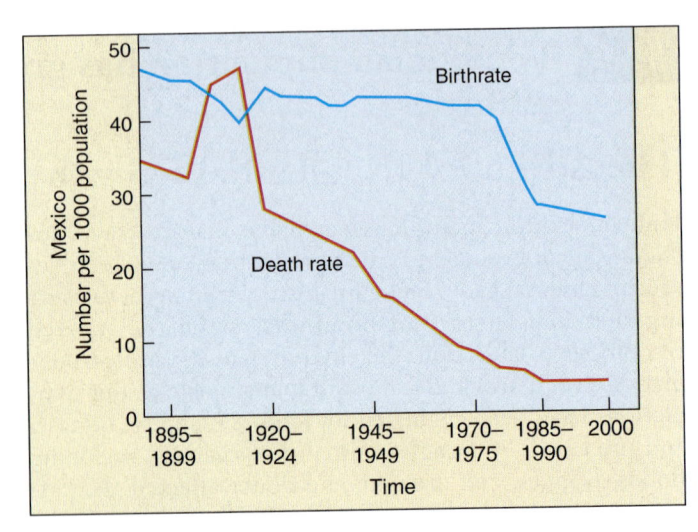

FIGURE 53.27
Why the population of Mexico is growing. The death rate (*red line*) in Mexico fell steadily throughout the last century, while the birthrate (*blue line*) remained fairly steady until 1970. The difference between birth and death rates has fueled a high growth rate. Efforts begun in 1970 to reduce the birthrate have been quite successful, but the growth rate remains high.
Is population growth rate increasing?

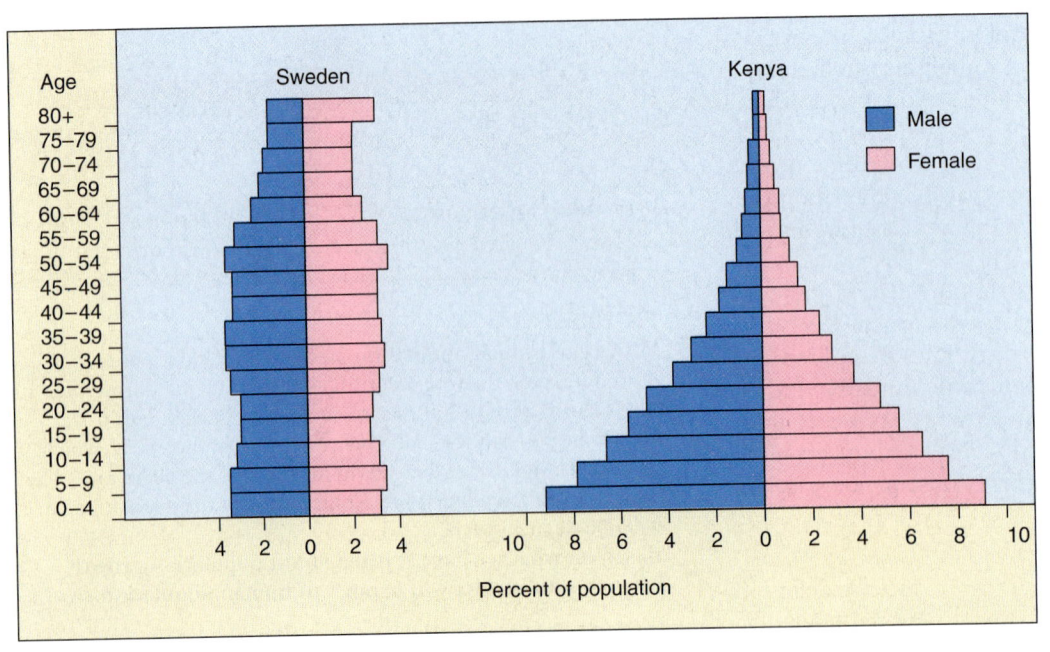

FIGURE 53.28
Population pyramids from 2000. Population pyramids are graphed according to a population's age distribution. Kenya's pyramid has a broad base because of the great number of individuals below childbearing age. When the young people begin to bear children, the population will experience rapid growth. The Swedish pyramid exhibits a slight bulge among middle-aged Swedes, the result of the "baby boom" that occurred in the middle of the twentieth century.
What will the population distributions look like in 20 years?

1156 Part VIII Ecology and Behavior

Table 53.4 A Comparison of 2002 Population Data in Developed and Developing Countries

	United States (highly developed)	Brazil (moderately developed)	Ethiopia (poorly developed)
Fertility rate	2.1	2.2	5.9
Doubling time at current rate (yr)	115	53	28
Infant mortality rate (per 1000 births)	6.6	33	97
Life expectancy at birth (yrs)	77	69	52
Per capita GNP (U.S. $)	$34,100	$7300	$660
Population < 15 years old (%)	21	30	44

An Uncertain Future

The earth's rapidly growing human population constitutes perhaps the greatest challenge to the future of the **biosphere,** the world's interacting community of living things. Humanity is adding 82 million people a year to the earth's population—over a million every five days, 150 every minute! In more rapidly growing countries, the resulting population increase is staggering (table 53.4). India, for example, had a population of 1.05 billion in 2002; by 2050, its population will exceed 1.6 billion.

A key element in the world's population growth is its uneven distribution among countries. Of the billion people added to the world's population in the 1990s, 90% live in developing countries (figure 53.29). This is leading to a major reduction in the fraction of the world's population that lives in industrialized countries. In 1950, fully one-third of the world's population lived in industrialized countries; by 1996, that proportion had fallen to one-quarter; and in 2020, the proportion will have fallen to one-sixth. Thus, the world's population growth will be centered in the parts of the world least equipped to deal with the pressures of rapid growth.

Rapid population growth in developing countries has the harsh consequence of increasing the gap between rich and poor. Today, the 19% of the world's population that lives in the industrialized world have a per capita income of $22,060, while 81% of the world's population lives in developing countries and has a per capita income of only $3,580. Further, of the people in the developing world, about one-quarter of the population gets by on $1 per day. Eighty percent of all the energy used today is consumed by the industrialized world, while only 20% is used by developing countries. Perhaps most worrisome for the future, fully 94% of all scientists and engineers reside in the industrialized world, and only 6% live in developing countries. Thus, the problems created by the future's explosive population growth will be faced by countries with little of the world's scientific or technological expertise.

No one knows whether the world can sustain today's population of 6 billion people, much less the far greater numbers expected in the future. As chapter 56 outlines, the world ecosystem is already under considerable stress. We cannot reasonably expect to expand its carrying capacity indefinitely, and indeed we already seem to be stretching the limits. De-

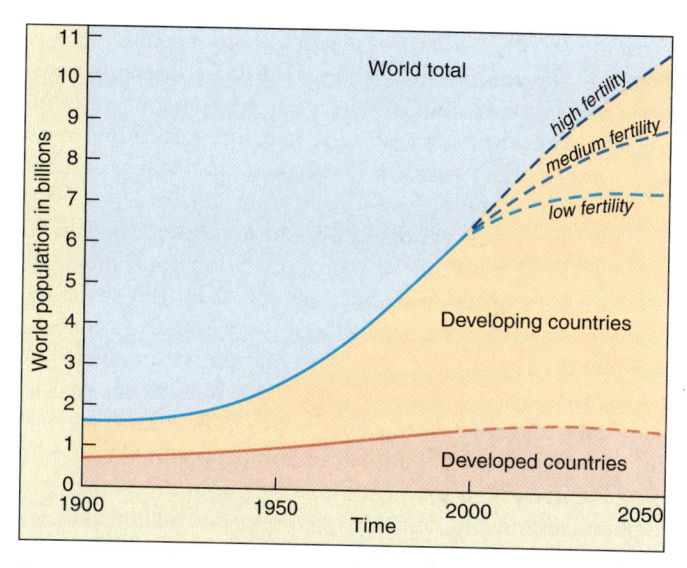

FIGURE 53.29
Distribution of population growth. Most of the worldwide increase in population since 1950 has occurred in developing countries. The age structures of developing countries indicate that this trend will increase in the near future. World population in 2050 likely will be between 7.3 and 10.7 billion, according to a recent United Nations study. Depending on fertility rates, the population at that time will either be increasing rapidly or slightly, or in the best case, declining slightly.
Is this an example of density-dependent population regulation? If so, what factors are regulating population size?

spite using an estimated 45% of the total biological productivity of the earth's landmasses and more than one-half of all renewable sources of fresh water, between one-fourth and one-eighth of all people in the world are malnourished. Moreover, as anticipated by Thomas Malthus in his famous 1798 *Essay on the Principle of Population*, death rates are beginning to rise in some areas. In sub-Saharan Africa, for example, population projections for the year 2025 have been scaled back from 1.33 billion to 1.05 billion (21%) because of the impact of AIDS. Similar decreases are projected for Russia as a result of higher death rates due to disease. If we are to avoid catastrophic increases in the death rate, birthrates must fall dramatically. Faced with this grim dichotomy, significant efforts are underway worldwide to lower birthrates.

Population Growth Rate on the Decline

The world population growth rate is declining, from a high of 2.0% in the period 1965–1970 to 1.3% in 2002. Nonetheless, because of the larger population, this amounts to an increase of 82 million people per year to the world population, compared to 53 million per year in the 1960s.

The United Nations attributes the growth rate decline to increased family planning efforts and the increased economic power and social status of women. The United States has led the world in funding family planning programs abroad, but some groups oppose spending money on international family planning. The opposition states that money is better spent on improving education and the economy in other countries, leading to an increased awareness and lowered fertility rates. The U.N. certainly supports the improvement of education programs in developing countries, but interestingly, it has reported increased education levels *following* a decrease in family size as a result of family planning.

Most countries are devoting considerable attention to slowing the growth rate of their populations, and there are genuine signs of progress. For example, from 1984 to 2000, family planning programs in Kenya succeeded in reducing the fertility rate from 8.0 to 4.4 children per couple, thus lowering the population growth rate from 4.0% per year to 2.1% per year. Because of these efforts, the global population may stabilize at about 8.9 billion people by the middle of the current century. How many people the planet can support sustainably depends on the quality of life that we want to achieve; there are already more people than can be sustainably supported with current technologies.

Consumption in the Developed World Is Also a Problem

Population size is not the only factor that determines resource use; per capita consumption is also important. In this respect, we in the industrialized world need to pay more attention to lessening the impact each of us makes because, even though the vast majority of the world's population is in developing countries, the vast majority of resource consumption occurs in the industrialized countries. Indeed, the wealthiest 20% of the world's population accounts for 86% of the world's consumption of resources and produces 53% of the world's carbon dioxide emissions, whereas the poorest 20% of the world is responsible for only 1.3% of consumption and 3% of carbon dioxide emissions. Looked at another way, in terms of resource use, a child born today in the industrialized world will consume many more resources over the course of his or her life than a child born in the developing world. One way of quantifying this disparity is by calculating what has been termed the **ecological footprint**, which is the amount of productive land required to support an individual at the standard of living of a particular population through the course of his

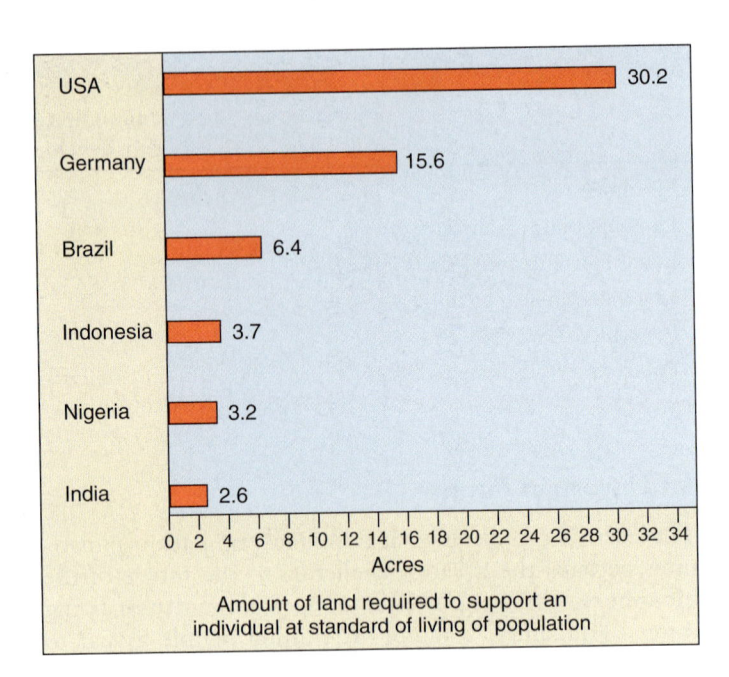

FIGURE 53.30
Ecological footprints of individuals in different countries. An ecological footprint calculates how much land is required to support a person through his or her life, including the acreage used for production of food, forest products, and housing, in addition to the forest required to absorb the carbon dioxide produced by the combustion of fossil fuels.
Which is a more important cause of resource depletion, overpopulation or overconsumption?

or her life. This figure estimates the acreage used for the production of food (both plant and animal), forest products, and housing, as well as the area of forest required to absorb carbon dioxide produced by the combustion of fossil fuels. As figure 53.30 illustrates, the ecological footprint of an individual in the United States is more than 10 times greater than that of someone in India. Based on these measurements, researchers have calculated that resource use by humans is now one-third greater than the amount that nature can sustainably replace. Moreover, consumption is increasing rapidly in parts of the developing world; if all humans lived at the standard of living in the industrialized world, two additional planet earths, would be needed!

Building a sustainable world is the most important task facing humanity's future. The quality of life available to our children will depend to a large extent on our success in limiting both population growth and the amount of per capita resource consumption.

> In 2002, the global human population of 6.3 billion people was growing at a rate of approximately 1.3% annually. At that rate, the population would double in 53 years. Growth rates, however, are declining, but consumption per capita in the developed world is also a significant drain on resources.

Concept Review

For interactive testing visit the Online Learning Center with PowerWeb at www.mhhe.com/Raven7

53.1 Organisms must cope with a varied environment.

The Environmental Challenge

- Many plants and animals actively employ physiological, morphological, or behavioral mechanisms to maintain homeostasis. Such mechanisms are the result of natural selection. (pp. 1138–1139)

53.2 Populations are groups of individuals of the same species that live in the same space.

Population Ranges

- Boundary edges between populations can range from sharp and impermeable to indistinct and easily permeable. (p. 1140)
- Three aspects of populations are especially important: the range in which a population occurs, the pattern of spacing of individuals throughout their range, and the size a population eventually attains. (p. 1140)
- Population ranges are not static and can change through time, usually as a consequence of environmental change or physiological adaptation. (p. 1141)

Pattern of Spacing of Individuals in a Population

- Individuals are randomly spaced when interactions between individuals are weak or when microenvironmental aspects of the habitat are nonuniform. (p. 1142)
- Uniform spacing often results from resource competition. (p. 1142)
- Clumping or clustering usually occurs as a response to uneven distribution of resources in the immediate environment. (p. 1143)
- Natural dispersal can occur through many avenues, including individuals or their gametes traveling via wind, water, or other organisms. (p. 1143)

Metapopulations

- Metapopulations are a network of distinct populations interacting and exchanging gametes and/or individuals. Such networks usually occur in habitat patches separated by inhospitable habitat. (p. 1144)
- At any given time, not all suitable habitat patches are inhabited, and periodically an inhabited patch will go extinct, only to be colonized later by individuals from a nearby patch. (p. 1144)
- Source-sink metapopulations occur when only some habitats are suitable for long-term population maintenance. In these cases, more permanent populations (sources) continually send individuals to less permanent, poorer habitat patches (sinks), which are periodically purged, often by environmental conditions, and then recolonized from the source population. (p. 1144)

53.3 Population dynamics depend critically upon age distribution.

Demography

- Demography is the statistical study of populations and their changes through time. Populations can be studied as a whole or broken down into constituent parts. (p. 1145)
- Many factors affect population growth, including sex ratio, generation time, and age structure of the populations. (pp. 1145–1146)
- Life tables are used to assess how populations change through time, and are often constructed by following a cohort of individuals from birth until death. (p. 1146)
- Survivorship curves can be constructed by graphing the percentage of an original population, or cohort, that survives to a given age. These curves can be divided into three basic categories based on the period of life in which individuals are most likely to die. (p. 1147)

53.4 Life histories often reflect trade-offs between reproduction and survival.

The Cost of Reproduction

- The complete life cycle of an organism constitutes its life history, and all life histories involve significant trade-offs. (p. 1148)
- Natural selection will favor life histories that maximize lifetime reproductive success. (p. 1148)
- A key reproductive trade-off concerns how many resources to invest in producing any single offspring; thus, a trade-off must exist between the number of offspring produced and the size of each offspring. (p. 1149)
- The trade-off between age and fecundity also plays a key role in many life histories. (p. 1149)
- Organisms that invest all their energy into a single large event and die afterwards are termed semelparous, while iteroparous organisms produce offspring several times over many seasons. (p. 1149)

53.5 Population growth is limited by the environment.

Population Growth

- The actual rate of population increase is defined as the difference between the birthrate and the death rate corrected for movement of animals into and out of a population. (p. 1150)
- The innate capacity of growth for any population is exponential. (p. 1150)
- In logistic growth, as a population approaches carrying capacity, its rate of growth slows as fewer resources remain available for use, resulting in a sigmoidal growth curve. (pp. 1150–1151)

Factors That Regulate Populations

- Density-dependent effects are factors that exert increasing pressure on a population as the population increases in size. (p. 1152)
- Density-independent effects are factors that affect the rate of population growth independent of the size of the population. (p. 1153)

53.6 The human population has grown explosively in the last three centuries.

The Advent of Exponential Growth

- Starting in the early 1700s, changes in technology have given humans more control over their food supply and death rate and made them less vulnerable to climatic uncertainties. The human population has thus been able to escape the confines of logistic growth and has grown explosively over the past 300 years. (p. 1155)
- Population pyramids graphically depict the number of people in each age category. (p. 1156)
- Population growth has been more rapid in developing countries, and the age structures of developing countries indicate that this trend will increase in the near future. (p. 1157)
- Per capita resource consumption is a significant factor determining resource consumption, which is highest in the industrialized countries. (p. 1158)

Test Your Understanding

For interactive testing, visit the Online Learning Center with PowerWeb at www.mhhe.com/Raven7

Self Test

1. The term *homeostasis* refers to
   a. the maintenance of a consistent internal environment.
   b. the ability to conform internal temperature to environmental temperature.
   c. an organism's biotic potential.
   d. the carrying capacity of a population.
2. Which of the following is *not* considered a population?
   a. the ginkgo trees (*Ginkgo biloba*) in New York City
   b. the birds in your hometown
   c. the human inhabitants of Pennsylvania
   d. the grizzly bears (*Ursus arctos*) of Alaska.
3. A clumped population may be due to
   a. weak interactions between the members of a population.
   b. intense competition for uniformly distributed resources.
   c. uneven distribution of resources in the environment.
   d. intense territoriality.
4. The tidewater goby (*Eucyclogobius newberryi*) is an endangered species of fish that occurs as metapopulations in isolated coastal wetlands of California. Large wetlands serve as sources of individuals for populations in small wetlands, which function as sinks due to inferior habitat quality. What effect would most likely be seen on the population of gobies if a barrier to migration between wetlands developed?
   a. The populations in the small and large wetlands would evolve independently of each other.
   b. The populations in the large wetlands would most likely go extinct.
   c. The populations in the small wetlands would most likely go extinct.
   d. The populations in the small and large wetlands would most likely go extinct.
5. Which of the following factors does *not* determine the growth rate of a population?
   a. the population's sex ratio
   b. the species' generation time
   c. the age structure of the population
   d. the optimal temperature at which an organism can reproduce.
6. A population with a larger proportion of older individuals than younger individuals will likely
   a. grow larger and then decline rapidly.
   b. continue to grow larger indefinitely.
   c. grow smaller and may stabilize at a smaller population size.
   d. not experience a change in population size.
7. Humans are an example of an organism with a type I survivorship curve. This means
   a. mortality rates are highest for younger individuals.
   b. mortality rates are highest for older individuals.
   c. mortality rates are constant over the life span of individuals.
   d. the population growth rate is high.
8. According to the Population Reference Bureau (2002), the worldwide intrinsic rate of human population growth (*r*) is currently 1.3%. In the United States, $r = 0.6$%. How will the U.S. population change relative to the world population?
   a. The world population will grow, while the population of the United States will decline.
   b. The world population will grow, while the population of the United States will remain the same.
   c. Both the world and the U.S. populations will grow, but the world population will grow more rapidly.
   d. The world population will decline, while the U.S. population will increase.
9. The logistic population growth model, $dN/dt = rN[(K - N)/K]$, describes a population's growth when an upper limit to growth is assumed. This upper limit to growth is known as the population's _____, and as *N* gets larger, dN/dt _____.
   a. biotic potential/increases
   b. biotic potential/decreases
   c. carrying capacity/increases
   d. carrying capacity/decreases
10. Which of the following is *not* an example of a density-dependent effect on population growth?
    a. an extremely cold winter
    b. competition for food resources
    c. stress-related illness associated with overcrowding
    d. competition for nesting sites

Test Your Visual Understanding

1. The song sparrow *Melospiza melodia* exhibits density-dependent population growth on Mandarte Island. List three reasons that the number of young per female may be sensitive to population size.

Apply Your Knowledge

1. Throughout most of North America, large carnivores have either been extirpated (driven to local extinction) or their populations are at extremely low levels. Explain how the loss of carnivore populations may have changed the vegetation of North America.
2. Do humans show more *r*-selected life-history traits or *K*-selected traits? How does this correlate with current global human population growth?
3. Give your opinion: What constitutes the greatest threat to the future of the planet, the rapidly growing population in developing parts of the world or high resource consumption in the developed world?

54

Community Ecology

Concept Outline

54.1 Biological communities are composed of species that occur together.

Concepts of Communities. Whether a community is more than the sum of its parts is debated.

54.2 Interactions among competing species shape ecological niches.

Fundamental and Realized Niches. Interspecific interactions can restrict niche use.
Gause and the Principle of Competitive Exclusion. If resources are limited, no two species can occupy the same niche indefinitely.
Resource Partitioning. Species that live together often partition the available resources, reducing competition.
Detecting Interspecific Competition. Experiments often can detect competition, but they have limitations.

54.3 Predation has ecological and evolutionary effects.

Predation and Prey Populations. Predators can limit the size of populations.
Plant Defenses Against Herbivores. Plants use chemicals to defend themselves against animals trying to eat them.
Animal Defenses Against Predators. Animals defend themselves with camouflage, chemicals, and stings.
Mimicry. Species can copy the appearance of others.

54.4 Species within a community interact in many ways.

Coevolution and Symbiosis. Organisms have evolved many adjustments to living together.
Commensalism. Some organisms use others, neither hurting nor helping their benefactors.
Mutualism. Often species interactions benefit both.
Parasitism. Sometimes one organism serves as the food supply of another, much smaller one.
Interactions Among Ecological Processes. Multiple processes may occur simultaneously.

54.5 Ecological succession may increase the species richness of communities.

Succession. Species replacements often occur in a regular pattern.
The Role of Disturbance. Disturbances can disrupt successional change. In some cases, moderate amounts of disturbance increase species diversity.

FIGURE 54.1
Communities involve interactions between disparate groups. This clownfish is one of the few species that can nestle safely among the stinging tentacles of the sea anemone. In addition to safe harbor, the clownfish further benefits by eating scraps of food left over from the anemone's meals. In turn, the sea anemone benefits by ingesting scraps of food dropped by the fish. The sea anemone–clownfish relationship is thus a classic case of symbiosis.

All the organisms that live together in a place are called a community. The myriad of species that inhabit a tropical rain forest are a community. Indeed, every inhabited place on earth supports its own particular array of organisms. Over time, the different species have made many complex adjustments to community living (figure 54.1), evolving together and forging relationships that give the community its character and stability. Both competition and cooperation have played key roles; in this chapter, we look at these and other factors in community ecology.

1161

54.1 Biological communities are composed of species that occur together.

Almost any place on earth is occupied by species, sometimes by many of them, as in the rain forests of the Amazon, and sometimes by only a few, as in the near-boiling waters of Yellowstone's geysers (where a number of microbial species live). The term **community** refers to the species that occur at any particular locality (figure 54.2). Communities can be characterized either by their constituent species or by their properties, such as species richness or primary productivity.

Interactions among community members govern many ecological and evolutionary processes. These interactions, such as predation and mutualism, affect the population biology of particular species—whether a population increases or decreases in abundance, for example—as well as the ways in which energy and nutrients cycle through the ecosystem. Moreover, the community context in many ways affects the patterns of natural selection faced by a species, and thus the evolutionary course it takes.

Scientists study biological communities in many ways, ranging from detailed observations to elaborate, large-scale experiments. In some cases, such studies focus on the entire community, whereas in other cases only a subset of species that are likely to interact with each other are studied. Although scientists sometimes refer to such subsets as communities (for example, the "spider community"), the term **assemblage** is probably more appropriate to connote that the species included are only a portion of those present within the entire community.

FIGURE 54.2
A Tanzanian savanna community. A community consists of all the species—plants, animals, fungi, protists, and prokaryotes—that occur at a locality, in this case Lake Manyara National Park in Tanzania.

Concepts of Communities

Two views exist on the makeup and functioning of communities. The **individualistic concept** of communities, first championed by H. A. Gleason of the University of Chicago early in the twentieth century, holds that a community is nothing more than an aggregation of species that happen to co-occur at one place. By contrast, the **holistic concept** of communities, which can be traced to the work of F. E. Clements, also about a century ago, views communities as an integrated unit. In this sense, the community could be viewed as a superorganism whose constituent species have coevolved to the extent that they function as part of a greater whole, just as the kidneys, heart, and lungs all function together within an animal's body. In this view, then, a community would amount to more than the sum of its parts.

These two views make differing predictions about the integrity of communities across space and time. If, as the individualistic view implies, communities are nothing more than a combination of species that occur together, then moving geographically across the landscape or back through time, we would not expect to see the same community. That is, species should appear and disappear independently, as a function of each species' own unique ecological requirements. By contrast, if a community is an integrated whole, then we would make the opposite prediction: Communities should stay the same through space or time, until being replaced by completely different communities when environmental differences are sufficiently great.

Communities Across Space and Time

Most ecologists today favor the individualistic concept. For the most part, species seem to respond independently to changing environmental conditions. As a result, community composition changes gradually across landscapes as some species appear and become more abundant, while others decrease in abundance and eventually disappear. A famous example of this pattern is the abundance of tree species in the Santa Catalina Mountains of Arizona along a

1162 Part VIII Ecology and Behavior

geographic gradient running from very dry to very moist. Figure 54.3 shows that species can change abundance in patterns that are for the most part independent of each other. As a result, tree communities at different localities in these mountains are a continuum, one merging into the next, rather than representing discretely different sets of species.

Similar patterns through time are seen in paleontological studies. For example, a very good fossil record exists for the trees and small mammals that occurred in North America over the past 20,000 years. Examination of prehistoric communities shows little similarity to those that occur today. Many species that occur together today were never found together in the past. Conversely, species that used to occur in the same communities often do not overlap in their geographic ranges today. These findings suggest that as climate has changed during the waxing and waning of the Ice Ages, species have responded independently, rather than shifting their distributions together, as would be expected if the community were an integrated unit.

Nonetheless, in some cases the abundance of species in a community does change geographically in a synchronous pattern. Often, this occurs at **ecotones,** places where the environment changes abruptly. For example, in the western United States, certain patches of habitat have serpentine soils. Such soil differs from normal soil in many ways (for example, high concentrations of nickel, chromium, and iron; low concentrations of copper and calcium). Comparison of the plant species that occur on different soils shows that distinct communities exist on each type, with an abrupt transition from one to the other over a short distance (figure 54.4). Similar transitions are seen wherever greatly different habitats come into contact, such as at the interface between terrestrial and aquatic habitats or where grassland and forest meet.

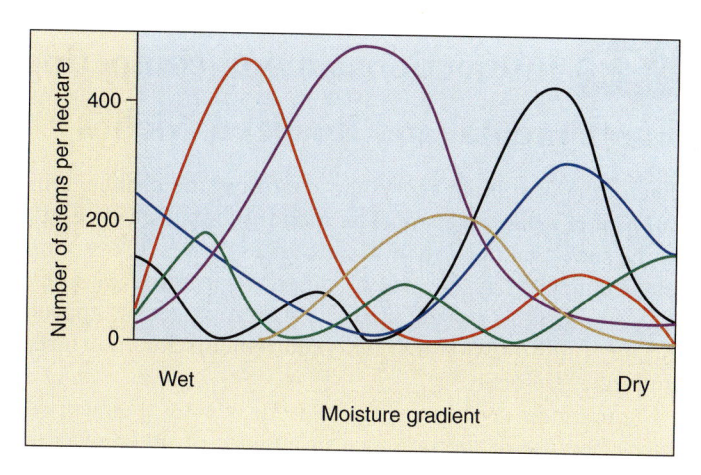

FIGURE 54.3
Abundance of tree species along a moisture gradient in the Santa Catalina Mountains of southeastern Arizona. The species' patterns of abundance are independent of each other. Thus, community composition changes continuously along the gradient.
Why do species exhibit different patterns of response to change in moisture?

> A community comprises all species that occur at one site. In most cases, community members vary independently of each other in abundance across space and through time.

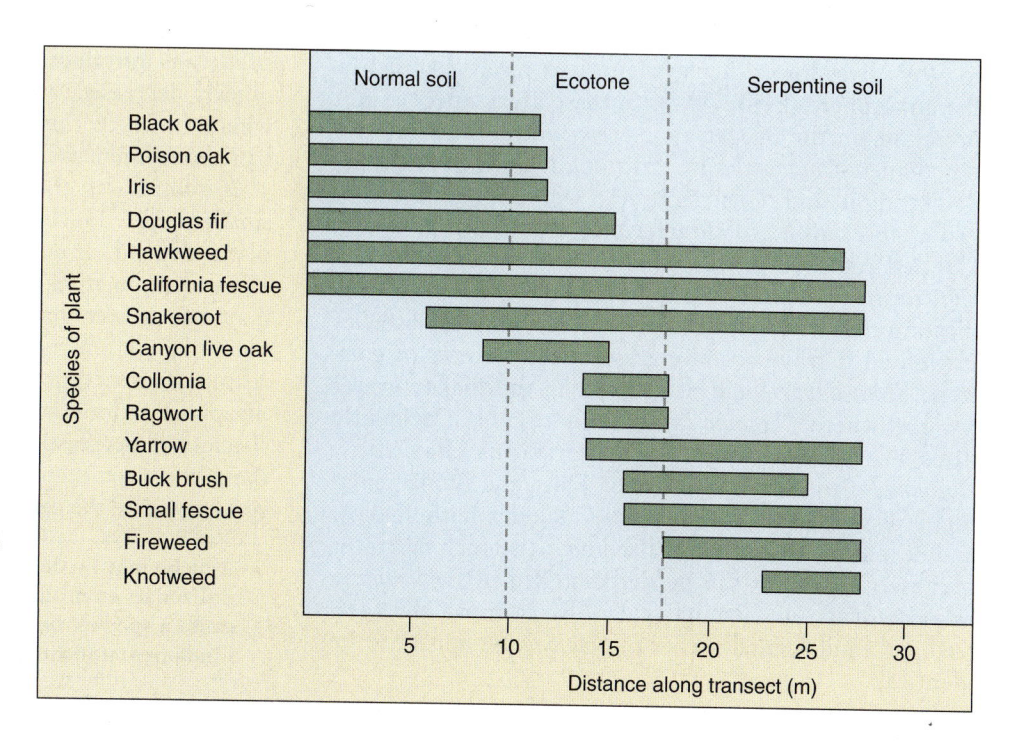

FIGURE 54.4
Change in community composition across an ecotone. The plant communities on normal and serpentine soils are greatly different, and the transition from one community to another occurs over a short distance.
Why is there a sharp transition between the two community types?

Chapter 54 Community Ecology 1163

54.2 Interactions among competing species shape ecological niches.

Fundamental and Realized Niches

Each organism in an ecosystem confronts the challenge of survival in a different way. The **niche** an organism occupies is the sum total of all the ways it utilizes the resources of its environment. A niche may be described in terms of space utilization, food consumption, temperature range, appropriate conditions for mating, requirements for moisture, and other factors.

Sometimes species are not able to occupy their entire niche because of the presence or absence of other species. Species can interact with each other in a number of ways, and these interactions can either have positive or negative effects. One type of interaction is **interspecific competition**, which occurs when two species attempt to utilize the same resource and there is not enough of the resource to satisfy both. Fighting over resources is referred to as **interference competition**; consuming shared resources is called **exploitative competition**.

The entire niche that a species is capable of using, based on its physiological tolerance limits and resource needs, is called the **fundamental niche**. The actual niche the species occupies is its **realized niche**. Because of interspecific interactions, the realized niche of a species may be considerably smaller than its fundamental niche.

In a classic study, J. H. Connell of the University of California, Santa Barbara, investigated competitive interactions between two species of barnacles that grow together on rocks along the coast of Scotland. Of the two species Connell studied, *Chthamalus stellatus* lives in shallower water, where tidal action often exposes it to air, and *Semibalanus balanoides* (called *Balanus balanoides* prior to 1995) lives lower down, where it is rarely exposed to the atmosphere (figure 54.5). In these areas, space is at a premium. In the deeper zone, *Semibalanus* could always outcompete *Chthamalus* by crowding it off the rocks, undercutting it, and replacing it even where it had begun to grow, an example of interference competition. When Connell removed *Semibalanus* from the area, however, *Chthamalus* was easily able to occupy the deeper zone, indicating that no physiological or other general obstacles prevented it from becoming established there. In contrast, *Semibalanus* could not survive in the shallow-water habitats where *Chthamalus* normally occurs; it evidently does not have the special adaptations that allow *Chthamalus* to occupy this zone. Thus, the fundamental niche of the barnacle *Chthamalus* includes both shallow and deeper zones, but its realized niche is much narrower because *Chthamalus* can be outcompeted by *Semibalanus* in parts of its fundamental niche. By contrast, the realized and fundamental niches of *Semibalanus* appear to be identical.

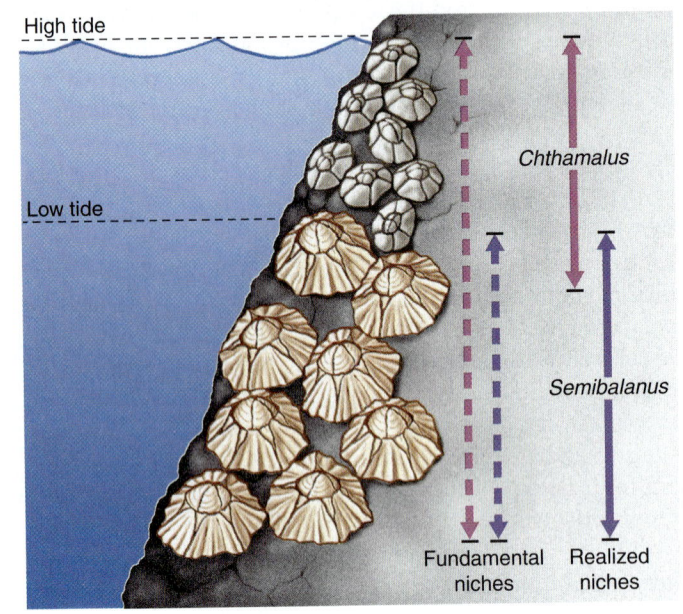

FIGURE 54.5
Competition among two species of barnacles. *Chthamalus* can live in both deep and shallow zones (its fundamental niche), but *Semibalanus* forces *Chthamalus* out of the part of its fundamental niche that overlaps the realized niche of *Semibalanus*.

Processes other than competition can also restrict the realized niche of a species. For example, the plant St. John's-wort was introduced and became widespread in open rangeland habitats in California until a specialized beetle was introduced to control it. Populations of the plant quickly decreased, and it is now only found in shady sites where the beetle cannot thrive. In this case, the presence of a predator limits the realized niche of a plant.

In some cases, the absence of another species leads to a smaller realized niche. For example, many North American plants depend on insects for pollination; indeed, the value of insect pollination for American agriculture has been estimated at greater than $2 billion per year. However, pollinator populations are currently declining for a variety of reasons. Conservationists are concerned that if these insects disappear from some habitats, the niche of many plant species will decrease or even disappear entirely. In this case, then, the absence—rather than the presence—of another species will be the cause of a relatively small realized niche.

> A niche may be defined as the way in which an organism utilizes its environment. Interspecific interactions may cause a species' realized niche to be smaller than its fundamental niche.

Gause and the Principle of Competitive Exclusion

In classic experiments carried out in 1934 and 1935, Russian ecologist G. F. Gause studied competition among three species of *Paramecium*, a tiny protist. All three species grew well alone in culture tubes, preying on bacteria and yeasts that fed on oatmeal suspended in the culture fluid (figure 54.6*a*). However, when Gause grew *P. aurelia* together with *P. caudatum* in the same culture tube, the numbers of *P. caudatum* always declined to extinction, leaving *P. aurelia* the only survivor (figure 54.6*b*). Why? Gause found that *P. aurelia* could grow six times faster than its competitor *P. caudatum* because it was able to better utilize the limited available resources, an example of exploitative competition.

From experiments such as this, Gause formulated what is now called the principle of **competitive exclusion**. This principle states that if two species are competing for a limited resource, the species that uses the resource more efficiently will eventually eliminate the other locally. In other words, no two species with the same niche can coexist when resources are limiting.

Niche Overlap

In a revealing experiment, Gause challenged *Paramecium caudatum*—the defeated species in his earlier experiments—with a third species, *P. bursaria*. Because he expected these two species to also compete for the limited bacterial food supply, Gause thought one would win out, as had happened in his previous experiments. But that's not what happened. Instead, both species survived in the culture tubes, dividing the food resources. How did they do it? In the upper part of the culture tubes, where the oxygen concentration and bacterial density were high, *P. caudatum* dominated because it was better able to feed on bacteria. However, in the lower part of the tubes, the lower oxygen concentration favored the growth of a different potential food, yeast, and *P. bursaria* was better able to eat this food. The fundamental niche of each species was the whole culture tube, but the realized niche of each species was only a portion of the tube. Because the niches of the two species did not overlap too much, both species were able to survive. However, competition did have a negative effect on the participants (figure 54.6*c*). When grown without a competitor, both species reached densities three times greater than when they were grown with a competitor.

Competitive Exclusion

Gause's principle of competitive exclusion can be restated to say that *no two species can occupy the same niche indefinitely when resources are limiting*. Certainly species can and do coexist while competing for some of the same resources. Nevertheless, Gause's hypothesis predicts that when two species coexist on a long-term basis, either resources must not be limited or their niches will always differ in one or more features; otherwise, one species will outcompete the other, and the extinction of the second species will inevitably result.

> If resources are limiting, no two species can occupy the same niche indefinitely without competition driving one to local extinction.

**FIGURE 54.6
Competitive exclusion among three species of *Paramecium*.** In the microscopic world, *Paramecium* is a ferocious predator. (*a*) In his experiments, Gause found that three species of *Paramecium* grew well alone in culture tubes. (*b*) However, *P. caudatum* declined to extinction when grown with *P. aurelia* because they shared the same realized niche, and *P. aurelia* outcompeted *P. caudatum* for food resources. (*c*) *P. caudatum* and *P. bursaria* were able to coexist because the two have different realized niches and thus avoided competition.

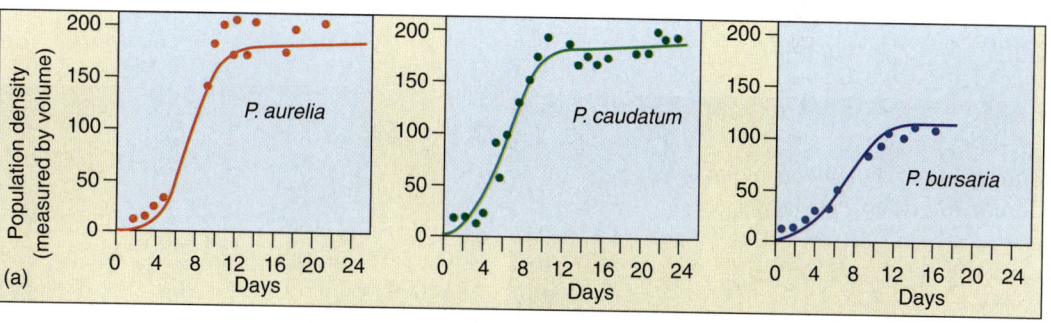

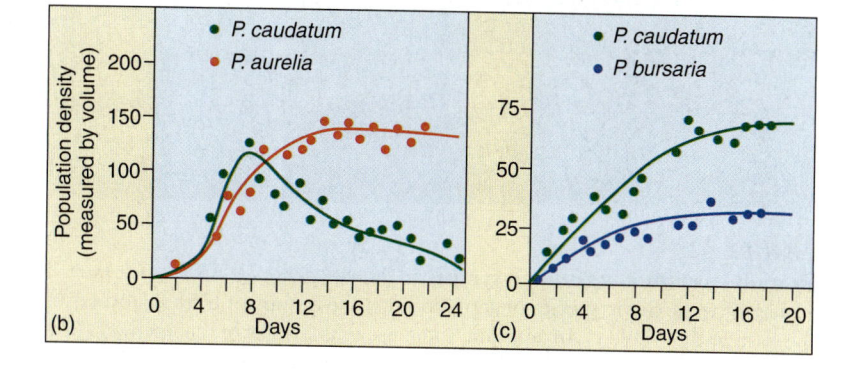

Resource Partitioning

Gause's exclusion principle has a very important consequence: If competition for a limited resource is intense, then either one species will drive the other to extinction, or natural selection will reduce the competition between them. When the late Princeton ecologist Robert MacArthur studied five species of warblers, small insect-eating forest songbirds, he found that they all appeared to be competing for the same resources. However, when he studied them more carefully, he found that each species actually fed in a different part of spruce trees and so ate different subsets of insects. One species fed on insects near the tips of branches, a second within the dense foliage, a third on the lower branches, a fourth high on the trees, and a fifth at the very apex of the trees. Thus, each species of warbler had evolved so as to utilize a different portion of the spruce tree resource. They had *subdivided the niche*, partitioning the available resource to avoid direct competition with one another.

Resource partitioning is often seen in similar species that occupy the same geographic area. Such **sympatric species** often avoid competition by living in different portions of the habitat or by utilizing different food or other resources (figure 54.7). This pattern of resource partitioning is thought to result from the process of natural selection causing initially similar species to diverge in resource use in order to reduce competitive pressures.

Evidence for the role of evolution comes forth by comparing species whose ranges are only partially overlapping. Where the two species co-occur, they tend to exhibit greater differences in morphology (the form and structure of an organism) and resource use than do allopatric populations of the same species. Called **character displacement**, the differences evident between sympatric species are thought to have been favored by natural selection as a mechanism to facilitate habitat partitioning and thus reduce competition. Thus, the two Darwin's finches in figure 54.8 have bills of similar size where the finches are allopatric, each living on an island where the other does not occur. On islands where they are sympatric, the two species have evolved beaks of different sizes, one adapted to larger seeds and the other to smaller ones.

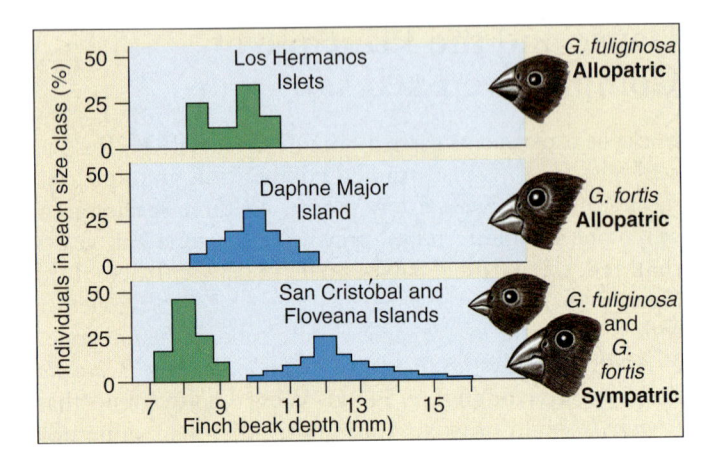

FIGURE 54.8
Character displacement in Darwin's finches. These two species of finches (genus *Geospiza*) have bills of similar size when allopatric, but different size when sympatric.
Why do populations have different bill sizes when they occur with other species?

> Sympatric species often partition available resources, reducing competition between them.

FIGURE 54.7
Resource partitioning among sympatric lizard species. Species of *Anolis* lizards on Caribbean islands partition their tree habitats in a variety of ways. Some species of anoles occupy the canopy of trees (*a*), others use twigs on the periphery (*b*), and still others are found at the base of the trunk (*c*). In addition, some use grassy areas in the open (*d*). When two species occupy the same part of the tree, they either utilize different-sized insects as food or partition the thermal microhabitat; for example, one might only be found in the shade, whereas the other would only bask in the sun.

1166   Part VIII   Ecology and Behavior

Detecting Interspecific Competition

It is not simple to determine when two species are competing. The fact that two species use the same resources need not imply competition if that resource is not in limited supply. If the population sizes of two species are negatively correlated, such that where one species has a large population, the other species has a small population and vice versa, the two species need not be competing for the same limiting resource. Instead, the two species might be independently responding to the same feature of the environment—perhaps one species thrives best in warm conditions and the other in cool conditions.

Experimental Studies of Competition

Some of the best evidence for the existence of competition comes from experimental field studies. By setting up experiments in which two species occur either alone or together, scientists can determine whether the presence of one species has a negative effect on a population of the second species. For example, a variety of seed-eating rodents occur in North American deserts. In 1988, researchers set up a series of 50 meter × 50 meter enclosures to investigate the effect of kangaroo rats on smaller, seed-eating rodents. Kangaroo rats were removed from half of the enclosures, but not from the other enclosures. The walls of all of the enclosures had holes that allowed rodents to come and go, but in the kangaroo rat removal plots, the holes were too small to allow the kangaroo rats to enter. Over the course of the next three years, the researchers monitored the number of the smaller rodents present in the plots. As figure 54.9 illustrates, the number of other rodents was substantially higher in the absence of kangaroo rats, indicating that kangaroo rats compete with the other rodents and limit their population sizes.

A great number of similar experiments have indicated that interspecific competition occurs between many species of plants and animals. The effects of competition can be seen in aspects of population biology other than population size, such as behavior and individual growth rates. For example, two species of *Anolis* lizards occur on the Caribbean island of St. Maarten. When one of the species, *A. gingivinus*, is placed in 12 meter × 12 meter enclosures without the other species, individual lizards grow faster and perch lower than do lizards of the same species when placed in enclosures in which *A. pogus*, a species normally found near the ground, is also present.

Caution is Necessary

Although experimental studies can be a powerful means of understanding interactions between coexisting species, they have their limitations.

First, care is necessary in interpreting the results of field experiments. Negative effects of one species on another do not automatically indicate the existence of competition. For example, many similar-sized fish have a negative effect on each other, but it results not from competition, but from the fact that adults of each species prey on juveniles of the other species. In addition, the presence of one species may attract predators, which then also prey on the second species. In this case, the second species may have a lower population size in the presence of the first species due to the predators, even if they are not competing at all. Thus, experimental studies are most effective when combined with detailed examination of the ecological mechanism causing the negative effect of one species on another species.

Second, experimental studies are not always feasible. For example, the coyote population has increased in the United States in recent years simultaneously with the decline of the grey wolf. Is this trend an indication that the species compete? Because of the size of the animals and the large geographic areas occupied by each individual, manipulative experiments involving fenced areas with only one or both species—with each experimental treatment replicated several times for statistical analysis—are not practical. Similarly, studies of slow-growing trees might require many centuries to detect competition between adult trees. In such cases, detailed studies of the ecological requirements of the species are our best bet for understanding interspecific interactions.

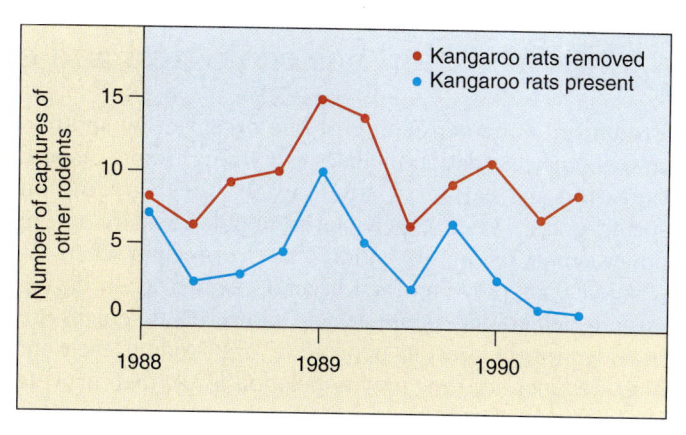

FIGURE 54.9

Detecting interspecific competition. This experiment tests how removal of kangaroo rats affects the population size of other rodents. Immediately after kangaroo rats were removed, the number of other rodents increased relative to the enclosures that still contained kangaroo rats. Notice that population sizes (as estimated by number of captures) changed in synchrony in the two treatments, probably reflecting changes in the weather. **Why are there more individuals of other rodent species when kangaroo rats are excluded?**

Experimental studies can provide strong tests of the hypothesis that interspecific competition occurs, but such studies have limitations. Detailed ecological studies are important regardless of whether experiments are conducted.

54.3 Predation has ecological and evolutionary effects.

Predation is the consuming of one organism by another. In this sense, predation includes everything from a leopard capturing and eating an antelope, to a deer grazing on spring grass. When experimental populations are set up under simple laboratory conditions, the predator often exterminates its prey and then becomes extinct itself, having nothing left to eat (figure 54.10). However, if refuges are provided for the prey, its population drops to low levels but not to extinction. Low prey population levels then provide inadequate food for the predators, causing the predator population to decrease. When this occurs, the prey population can recover.

Predation and Prey Populations

In nature, predators often have large effects on prey populations. Some of the most dramatic examples involve situations in which humans have either added or eliminated predators from an area. For example, the elimination of large carnivores from much of the eastern United States has led to population explosions of white-tailed deer, which strip the habitat of all edible plant life within their reach. Similarly, when sea otters were hunted to near extinction on the western coast of the United States, populations of sea urchins, a principal prey item of the otters, exploded.

Conversely, the introduction of rats, dogs, and cats to many islands around the world has led to the decimation of native faunas. For example, populations of Galápagos tortoises on several islands are endangered by introduced rats, dogs, and cats, which eat the eggs and the young tortoises. Similarly, in New Zealand, several species of birds and reptiles have been eradicated by rat predation and now only occur on a few offshore islands that the rats have not reached. In addition, on Stephens Island, near New Zealand, every individual of the now-extinct Stephens Island wren was killed by a single lighthouse keeper's cat!

A classic example of the role predation can play in a community involves the introduction of prickly pear cactus to Australia in the nineteenth century. In the absence of predators, the cactus spread rapidly, so that by 1925 it occupied 12 million hectares of rangeland in an impenetrable morass of spines that made cattle ranching difficult. To control the cactus, a predator from its natural habitat in Argentina, the moth *Cactoblastis cactorum*, was introduced, beginning in 1926. By 1940, cactus populations had been decimated, and it now generally occurs in small populations.

Predation and Evolution

Predation provides strong selective pressures on prey populations. Any feature that would decrease the probability of capture should be strongly favored. In the next three pages, we discuss a number of defense mechanisms in plants and animals. In turn, the evolution of such features causes natural selection to favor counteradaptations in predator populations. In this way, a *coevolutionary arms race* may ensue in which predators and prey are constantly evolving better defenses and better means of circumventing these defenses.

One example comes from the fossil record of mollusks and their predators. During the Mesozoic period (approximately 65 to 225 million years ago), new forms of predatory fish and crustaceans evolved that were able to crush or tear open shells. As a result, a variety of defensive measures evolved in mollusks, including spines, thicker shells, and shells too smooth for predators to grasp. In turn, these adaptations may have pressured predators to evolve more effective predatory adaptations and tactics, such as bigger and stronger claws and the ability to drill into shells.

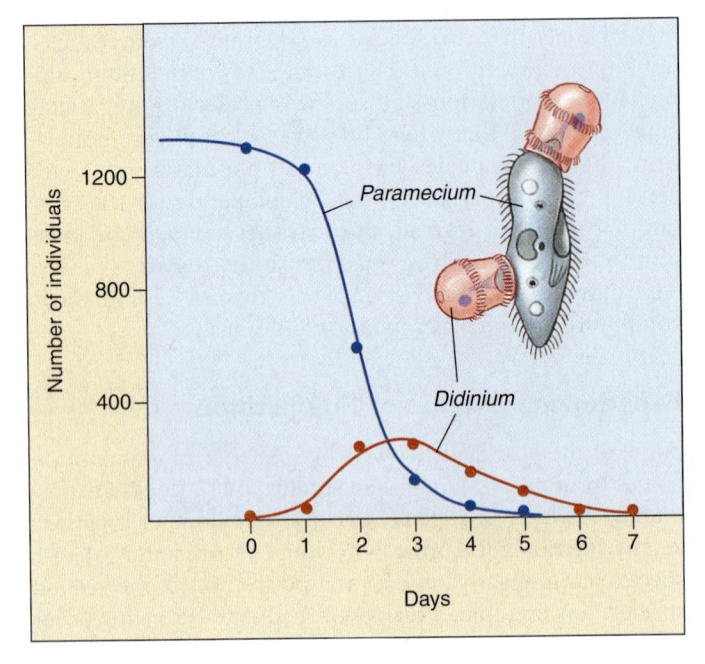

FIGURE 54.10
Predator-prey in the microscopic world. When the predatory *Didinium* is added to a *Paramecium* population, the numbers of *Didinium* initially rise, while the numbers of *Paramecium* steadily fall. When the *Paramecium* population is depleted, however, the *Didinium* individuals also die.
Can you think of any ways this experiment could be changed so that *Paramecium* might not go extinct?

> Predation can have substantial effects on prey populations. As a result, prey species often evolve defensive adaptations.

FIGURE 54.11
Insect herbivores well suited to their hosts. (*a*) The green caterpillars of the cabbage butterfly, *Pieris rapae*, are camouflaged on the leaves of cabbage and other plants on which they feed. Although mustard oils protect these plants against most herbivores, the cabbage butterfly caterpillars are able to break down the mustard oil compounds. (*b*) An adult cabbage butterfly.

Plant Defenses Against Herbivores

Plants have evolved many mechanisms to defend themselves from herbivores. The most obvious are morphological defenses: Thorns, spines, and prickles play an important role in discouraging browsers, and plant hairs, especially those that have a glandular, sticky tip, deter insect herbivores. Some plants, such as grasses, deposit silica in their leaves, both strengthening and protecting themselves. If enough silica is present, these plants are simply too tough to eat.

Chemical Defenses

As significant as morphological adaptations are, the chemical defenses that occur so widely in plants are even more crucial. Best known and perhaps most important in the defenses of plants against herbivores are **secondary chemical compounds.** These are distinguished from primary compounds, which are the components of a major metabolic pathway, such as respiration. Many plants, and apparently many algae as well, contain structurally diverse secondary compounds that are either toxic to most herbivores or disturb their metabolism greatly, preventing, for example, the normal development of larval insects. Consequently, most herbivores tend to avoid the plants that possess these compounds.

The mustard family (Brassicaceae) produces a group of chemicals known as mustard oils. These are the substances that give the pungent aromas and tastes to such plants as mustard, cabbage, watercress, radish, and horseradish. The same tastes we enjoy signal the presence of chemicals that are toxic to many groups of insects. Similarly, plants of the milkweed family (Asclepiadaceae) and the related dogbane family (Apocynaceae) produce a milky sap that deters herbivores from eating them. In addition, these plants usually contain cardiac glycosides, molecules that can produce drastic deleterious effects on the heart function of vertebrates.

The Evolutionary Response of Herbivores

Certain groups of herbivores are associated with each family or group of plants protected by a particular kind of secondary compound. These herbivores are able to feed on these plants without harm, often as their exclusive food source. For example, cabbage butterfly caterpillars (subfamily Pierinae) feed almost exclusively on plants of the mustard and caper families, as well as on a few other small families of plants that also contain mustard oils (figure 54.11). Similarly, caterpillars of monarch butterflies and their relatives (subfamily Danainae) feed on plants of the milkweed and dogbane families. How do these animals manage to avoid the chemical defenses of the plants, and what are the evolutionary precursors and ecological consequences of such patterns of specialization?

We can offer a potential explanation for the evolution of these particular patterns. Once the ability to manufacture mustard oils evolved in the ancestors of the caper and mustard families, the plants were protected for a time against most or all herbivores that were feeding on other plants in their area. At some point, certain groups of insects—for example, the cabbage butterflies—evolved the ability to break down mustard oils and thus feed on these plants without harming themselves. Having developed this ability, the butterflies were able to use a new resource without competing with other herbivores for it. Often, in groups of insects such as cabbage butterflies, sense organs have evolved that are able to detect the secondary compounds their food plants produce. Clearly, the relationship that has formed between cabbage butterflies and the plants of the mustard and caper families is an example of long-term, reciprocal evolutionary adjustment of the characteristics of the members of a biological community, a process termed **coevolution.**

> The members of many groups of plants are protected from most herbivores by their secondary compounds. Once the members of a particular herbivore group evolve the ability to feed on one of these types of plants, these herbivores gain access to a new resource, which they can exploit without competition from other herbivores.

Animal Defenses Against Predators

Some animals that feed on plants rich in secondary compounds receive an extra benefit. For example, when the caterpillars of monarch butterflies feed on plants of the milkweed family, they do not break down the cardiac glycosides that protect these plants from herbivores. Instead, the caterpillars concentrate and store the cardiac glycosides in fat bodies; they then pass them through the chrysalis stage to the adult and even to the eggs of the next generation. The incorporation of cardiac glycosides thus protects all stages of the monarch life cycle from predators. A bird that eats a monarch butterfly quickly regurgitates it (figure 54.12) and in the future avoids the conspicuous orange-and-black pattern that characterizes the adult monarch. Some bird species have evolved the ability to tolerate the protective chemicals. These birds eat the monarchs.

Chemical Defenses

Animals also manufacture and use a startling array of substances to perform a variety of defensive functions. Bees, wasps, predatory bugs, scorpions, spiders, and many other arthropods use chemicals to defend themselves and to kill their prey. In addition, various chemical defenses have evolved among marine animals and the vertebrates, including venomous snakes, lizards, fishes, and some birds. The poison-dart frogs of the family Dendrobatidae produce toxic alkaloids in the mucus that covers their brightly colored skin; these alkaloids are deadly to animals that try to eat the frogs (figure 54.13). Some of these toxins are so powerful that a few micrograms will kill a person if injected into the bloodstream. More than 200 different alkaloids have been isolated from these frogs, and some are playing important roles in neuromuscular research. An intensive investigation of marine animals, algae, and flowering plants is under way in search of new drugs to fight cancer and other diseases, or to use as sources of antibiotics.

Defensive Coloration

Many insects that feed on milkweed plants are brightly colored; they advertise their poisonous nature using an ecological strategy known as *warning coloration*, or aposematic coloration. Showy coloration is characteristic of animals that use poisons and stings to repel predators, while organisms that lack specific chemical defenses are seldom brightly colored. In fact, many have *cryptic coloration*—color that blends with the surroundings and thus hides the individual from predators (figure 54.14). Camouflaged animals usually do not live together in groups because a predator that discovers one individual gains a valuable clue to the presence of others.

Animals defend themselves against predators with chemical defenses, warning coloration, and camouflage.

FIGURE 54.12
A blue jay learns that monarch butterflies taste bad. (*a*) This cage-reared jay had never seen a monarch butterfly before it tried eating one. (*b*) The same jay regurgitated the butterfly a few minutes later. This bird will probably avoid trying to capture all orange-and-black insects in the future.

FIGURE 54.13
Vertebrate chemical defenses. Frogs of the family Dendrobatidae, abundant in the forests of Latin America, are extremely poisonous to vertebrates. Dendrobatids advertise their toxicity with bright coloration. As a result of either instinct or learning, predators avoid such brightly colored species that might otherwise be suitable prey.

FIGURE 54.14
Cryptic coloration. An inchworm caterpillar (*Necophora quernaria*) (hanging from the upper twig) closely resembles a twig.

Mimicry

During the course of their evolution, many species have come to resemble distasteful ones that exhibit warning coloration. The mimic gains an advantage by looking like the distasteful model. Two types of mimicry have been identified: Batesian and Müllerian mimicry.

Batesian Mimicry

Batesian mimicry is named for Henry Bates, the British naturalist who first brought this type of mimicry to general attention in 1857. In his journeys to the Amazon region of South America, Bates discovered many instances of palatable insects that resembled brightly colored, distasteful species. He reasoned that the mimics would be avoided by predators, who would be fooled by the disguise into thinking the mimic was the distasteful species.

Many of the best-known examples of Batesian mimicry occur among butterflies and moths. Obviously, predators in systems of this kind must use visual cues to hunt for their prey; otherwise, similar color patterns would not matter to potential predators. Increasing evidence indicates that Batesian mimicry can involve nonvisual cues, such as olfaction, although such examples are less obvious to humans.

The kinds of butterflies that provide the models in Batesian mimicry are, not surprisingly, members of groups whose caterpillars feed on only one or a few closely related plant families. The plant families on which they feed are strongly protected by toxic chemicals. The model butterflies incorporate the poisonous molecules from these plants into their bodies. The mimic butterflies, in contrast, belong to groups in which the feeding habits of the caterpillars are not so restricted. As caterpillars, these butterflies feed on a number of different plant families unprotected by toxic chemicals.

One often-studied mimic among North American butterflies is the viceroy, *Limenitis archippus* (figure 54.15*a*). This butterfly, which resembles the poisonous monarch, ranges from central Canada through much of the United States and into Mexico. The caterpillars feed on willows and cottonwoods, and neither caterpillars nor adults were originally thought to be distasteful to birds, although recent findings may dispute this. Interestingly, the Batesian mimicry seen in the adult viceroy butterfly does not extend to the caterpillars: Viceroy caterpillars are camouflaged on leaves, resembling bird droppings, while the monarch's distasteful caterpillars are very conspicuous.

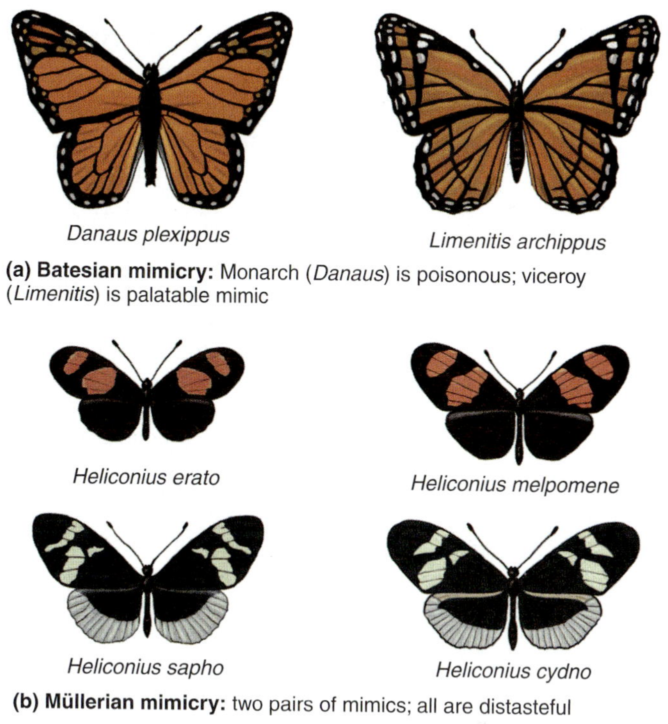

Müllerian Mimicry

Another kind of mimicry, **Müllerian mimicry,** was named for the German biologist Fritz Müller, who first described it in 1878. In Müllerian mimicry, several unrelated but protected animal species come to resemble one another

(a) Batesian mimicry: Monarch (*Danaus*) is poisonous; viceroy (*Limenitis*) is palatable mimic

(b) Müllerian mimicry: two pairs of mimics; all are distasteful

FIGURE 54.15
Mimicry. (*a*) Batesian mimicry. Monarch butterflies (*Danaus plexippus*) are protected from birds and other predators by the cardiac glycosides they incorporate from the milkweeds and dogbanes they feed on as larvae. Adult monarch butterflies advertise their poisonous nature with warning coloration. Viceroy butterflies (*Limenitis archippus*) are Batesian mimics of the poisonous monarch. (*b*) Pairs of Müllerian mimics. *Heliconius erato* and *H. melpomene* are sympatric, and *H. sapho* and *H. cydno* are sympatric. All of these butterflies are distasteful. They have evolved similar coloration patterns in sympatry to minimize predation; predators need only learn one pattern to avoid.

(figure 54.15*b*). If animals that resemble one another are all poisonous or dangerous, they gain an advantage because a predator will learn more quickly to avoid them. In some cases, predator populations even evolve an innate avoidance of species; such evolution may occur more quickly when multiple dangerous prey look alike.

In both Batesian and Müllerian mimicry, mimic and model must not only look alike but also act alike if predators are to be deceived. For example, the members of several families of insects that closely resemble wasps behave surprisingly like the wasps they mimic, flying often and actively from place to place.

> In Batesian mimicry, unprotected species resemble others that are distasteful. Both species exhibit warning coloration. In Müllerian mimicry, two or more unrelated but protected species resemble one another, thus achieving a kind of group defense.

Chapter 54 Community Ecology 1171

54.4 Species within a community interact in many ways.

Coevolution and Symbiosis

The plants, animals, protists, fungi, and prokaryotes that live together in communities have changed and adjusted to one another continually over a period of millions of years. For example, many features of flowering plants have evolved in relation to the dispersal of the plant's gametes by animals (figure 54.16). These animals, in turn, have evolved a number of special traits that enable them to obtain food or other resources efficiently from the plants they visit, often from their flowers. While doing so, the animals pick up pollen, which they may deposit on the next plant they visit, or seeds, which may be left elsewhere in the environment, sometimes a great distance from the parent plant.

Such interactions are examples of coevolution, a phenomenon we have already seen in predator-prey interactions.

Examples of Symbiosis

Another type of coevolution involves **symbiosis,** in which two or more kinds of organisms interact in often elaborate and more-or-less permanent relationships. All symbiotic relationships carry the potential for coevolution between the organisms involved, and in many instances the results of this coevolution are fascinating. Examples of symbiosis include *lichens*, which are associations of certain fungi with green algae or cyanobacteria. Lichens are discussed in more detail in chapter 30. Another important example are *mycorrhizae*, associations between fungi and the roots of most kinds of plants. The fungi expedite the plant's absorption of certain nutrients, and the plants in turn provide the fungi with carbohydrates. Similarly, root nodules that occur in legumes and certain other kinds of plants contain bacteria that fix atmospheric nitrogen and make it available to their host plants.

In the tropics, leafcutter ants are often so abundant that they can remove a quarter or more of the total leaf surface of the plants in a given area. They do not eat these leaves directly; rather, they take them to underground nests, where they chew them up and inoculate them with the spores of particular fungi. These fungi are cultivated by the ants and brought from one specially prepared bed to another, where they grow and reproduce. In turn, the fungi constitute the primary food of the ants and their larvae. The relationship between leafcutter ants and these fungi is an excellent example of symbiosis. Recent phylogenetic studies using DNA and assuming a molecular clock (see chapter 22) suggest that these symbioses are ancient, perhaps evolving more than 50 million years ago.

FIGURE 54.16
Pollination by a bat. Many flowers have coevolved with other species to facilitate pollen transfer. Insects are widely known as pollinators, but they're not the only ones: birds, bats, and even opossums and lizards serve as pollinators for some species. Notice the cargo of pollen on the bat's snout.

Kinds of Symbiosis

The major kinds of symbiotic relationships include (1) **commensalism,** in which one species benefits while the other neither benefits nor is harmed; (2) **mutualism,** in which both participating species benefit; and (3) **parasitism,** in which one species benefits but the other is harmed. Parasitism can also be viewed as a form of predation, although the organism that is preyed upon does not necessarily die.

Coevolution is a term that describes the long-term evolutionary adjustments of species to one another. In symbiosis, two or more species interact closely, with at least one species benefitting.

Commensalism

Commensalism is a symbiotic relationship that benefits one species and neither hurts nor helps the other. In nature, individuals of one species are often physically attached to members of another. For example, epiphytes are plants that grow on the branches of other plants. In general, the host plant is unharmed, while the epiphyte that grows on it benefits. Similarly, various marine animals, such as barnacles, grow on other, often actively moving sea animals, such as whales, and thus are carried passively from place to place. These "passengers" presumably gain more protection from predation than they would if they were fixed in one place, and they also reach new sources of food. The increased water circulation that such animals receive as their host moves around may be of great importance, particularly if the passengers are filter feeders. The gametes of the passenger are also more widely dispersed than would be the case otherwise.

When Is Commensalism Not Commensalism?

One of the best-known examples of symbiosis involves the relationships between certain small tropical fishes and sea anemones (see chapter 32). The fish have evolved the ability to live among the stinging tentacles of sea anemones, even though these tentacles would quickly paralyze other fishes that touched them (see figure 54.1). The anemone fishes feed on the detritus left from the meals of the host anemone, remaining uninjured under remarkable circumstances.

On land, an analogous relationship exists between birds called oxpeckers and grazing animals such as cattle or rhinoceroses (figure 54.17). The birds spend most of their time clinging to the animals, picking off parasites and other insects, carrying out their entire life cycles in close association with the host animals.

In each of these instances, it is difficult to be certain whether the second partner receives a benefit or not; there is no clear-cut boundary between commensalism and mutualism. For instance, it may be advantageous to the sea anemone to have particles of food removed from its tentacles because it may then be better able to catch other prey. Similarly, while often thought of as commensalism, the association of grazing mammals and gleaning birds is actually an example of mutualism. The mammal benefits by having parasites and other insects removed from its body, but the birds also benefit by gaining a dependable source of food.

On the other hand, commensalism can easily transform itself into parasitism. For example, oxpeckers are also known to pick not only parasites, but also scabs off their grazing hosts. Once the scab is picked, the birds drink the blood that flows from the wound. Occasionally, the cumulative effect of persistent attacks can greatly weaken the herbivore, particularly when conditions are not favorable, such as during droughts.

FIGURE 54.17
Commensalism, mutualism, or parasitism? In this symbiotic relationship, oxpeckers definitely receive a benefit in the form of nutrition from the ticks and other parasites they pick off their host (in this case, an impala) and eat. But the effect on the host is not always clear. If the ticks are harmful, their removal benefits the host, and the relationship is mutually beneficial. If the oxpeckers also pick at scabs, causing blood loss and possible infection, the relationship may be parasitic. If the hosts are unharmed by either the ticks or the oxpeckers, the relationship may be an example of commensalism.

Commensalism is the benign use of one organism by another.

Chapter 54 Community Ecology 1173

Mutualism

Mutualism is a symbiotic relationship among organisms in which both species benefit. Mutualistic relationships are of fundamental importance in determining the structure of biological communities. Some of the most spectacular examples of mutualism occur among flowering plants and their animal visitors, including insects, birds, and bats. As discussed in chapter 29, during the course of their evolution, the characteristics of flowers have evolved in large part in relation to the characteristics of the animals that visit them for food and, in the process, spread their pollen from individual to individual. At the same time, characteristics of the animals have changed, increasing their specialization for obtaining food or other substances from particular kinds of flowers.

Another example of mutualism involves ants and aphids. Aphids, also called greenflies, are small insects that suck fluids from the phloem of living plants with their piercing mouthparts. They extract a certain amount of the sucrose and other nutrients from this fluid, but they excrete much of it in an altered form through their anus. Certain ants have taken advantage of this—in effect, domesticating the aphids. The ants carry the aphids to new plants, where they come into contact with new sources of food, and then consume as food the "honeydew" that the aphids excrete.

Ants and Acacias

A particularly striking example of mutualism involves ants and certain Latin American species of the plant genus *Acacia*. In these species, certain leaf parts, called stipules, are modified as paired, hollow thorns. The thorns are inhabited by stinging ants of the genus *Pseudomyrmex*, which do not nest anywhere else (figure 54.18). Like all thorns that occur on plants, the acacia thorns serve to deter herbivores.

At the tip of the leaflets of these acacias are unique, protein-rich bodies called Beltian bodies, named after the nineteenth-century British naturalist Thomas Belt. Beltian bodies do not occur in species of *Acacia* that are not inhabited by ants, and their role is clear: They serve as a primary food for the ants. In addition, the plants secrete nectar from glands near the bases of their leaves. The ants consume this nectar as well, feeding it and the Beltian bodies to their larvae.

Obviously, this association is beneficial to the ants, and one can readily see why they inhabit acacias of this group. The ants and their larvae are protected within the swollen thorns, and the trees provide a balanced diet, including the sugar-rich nectar and the protein-rich Beltian bodies. What, if anything, do the ants do for the plants?

Whenever any herbivore lands on the branches or leaves of an acacia inhabited by ants, the ants, which continually patrol the acacia's branches, immediately attack and devour the herbivore. The ants that live in the acacias also help their hosts compete with other plants. The ants cut away any branches of other plants that touch the acacia in which they are living. They create, in effect, a tunnel of light through which the acacia can grow, even in the lush tropical rain forests of lowland Central America. In fact, when an ant colony is experimentally removed from a tree, the acacia is unable to compete successfully in this habitat. Finally, the ants bring organic material into their nests. The parts they do not consume, together with their excretions, provide the acacias with an abundant source of nitrogen.

As with commensalism, however, things are not always as they seem. Ant-acacia mutualisms also occur in Africa. In Kenya, several species of acacia ants occur, but only a single species occurs on any tree. One species, *Crematogaster nigriceps*, is competitively inferior to two of the other species. To prevent invasion by other ant species, *C. nigriceps* prunes the branches of the acacia, preventing it from coming into contact with branches of other trees, which would serve as a bridge for invaders. Although this behavior is beneficial to the ant, it is detrimental to the tree because it destroys the tissue from which flowers are produced, essentially sterilizing the tree. In this case, what initially evolved as a mutualistic interaction has instead become a parasitic one.

FIGURE 54.18
Mutualism: ants and acacias. Ants of the genus *Pseudomyrmex* live within the hollow thorns of certain species of acacia trees in Latin America. The nectaries at the bases of the leaves and the Beltian bodies at the ends of the leaflets provide food for the ants. The ants, in turn, supply the acacias with organic nutrients and protect the acacia from herbivores and shading from other plants.

Mutualism involves interactions between species that are mutually beneficial.

Parasitism

Parasitism may be regarded as a special form of symbiosis in which the parasite usually is much smaller than the prey and remains closely associated with it. Parasitism is harmful to the prey organism and beneficial to the parasite. In many cases, the parasite kills its host, and thus the ecological effects of parasitism can be similar to those of predation.

External Parasites

Parasites that feed on the exterior surface of an organism are external parasites, or **ectoparasites** (figure 54.19). Many instances of external parasitism are known in both plants and animals. **Parasitoids** are insects that lay eggs on living hosts. This behavior is common among wasps, whose larvae feed on the body of the unfortunate host, often killing it.

Internal Parasites

Vertebrates are parasitized internally by **endoparasites,** members of many different phyla of animals and protists. Invertebrates also have many kinds of parasites that live within their bodies. Internal parasitism is generally marked by much more extreme specialization than external parasitism, as shown by the many protist and invertebrate parasites that infect humans. The more closely the life of the parasite is linked with that of its host, the more its morphology and behavior are likely to have been modified during the course of its evolution. The same is true of symbiotic relationships of all sorts. Conditions within the body of an organism are different from those encountered outside and are apt to be much more constant. Consequently, the structure of an internal parasite is often simplified, and unnecessary armaments and structures are lost as it evolves (for example, see descriptions of tapeworms in chapter 32).

Parasites Can Manipulate Host Behavior

Many parasites have complex life cycles that require several different hosts for growth to adulthood and reproduction. Recent research has revealed the remarkable adaptations of certain parasites that alter the behavior of the host and thus facilitate transmission from one host to the next. For example, many parasites cause their hosts to behave in ways that make them more vulnerable to their predators; when the host is ingested, the parasite is able to infect the predator. One of the most famous examples involves a parasitic flatworm, *Dicrocoelium dendriticum*, which lives in ants as an intermediate host, but reaches adulthood in large herbivorous mammals such as cattle and deer. Transmission from an ant to a cow might seem difficult, because cows do not normally eat insects. The flatworm, however, has evolved a remarkable adaptation. When an ant is infected, one of the

FIGURE 54.19
An external parasite. The flowering plant dodder (*Cuscuta*) is a parasite that has lost its chlorophyll and its leaves in the course of its evolution. Because it is heterotrophic (unable to manufacture its own food), dodder obtains its food from the host plants it grows on.

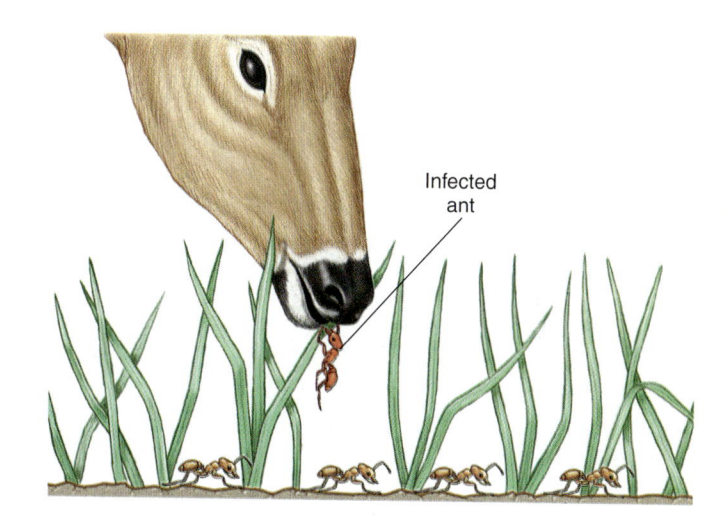

FIGURE 54.20
Parasitic manipulation of host behavior. Due to a parasite in its brain, an ant climbs to the top of a grass blade, where it may be eaten by a grazing herbivore, thus passing the parasite from insect to mammal.

flatworms migrates into the brain and causes the ant to climb to the top of vegetation and lock its mandibles onto a grass blade at the end of the day, just when herbivores are grazing (figure 54.20). The result is that the ant is eaten along with the grass, leading to infection of the grazer.

> Parasitism is a form of symbiosis that is beneficial to the parasite, but harmful to the host.

Chapter 54 Community Ecology 1175

Interactions Among Ecological Processes

We have seen the different ways in which species can interact with each other. In nature, however, more than one type of interaction usually occurs at the same time. In many cases, the outcome of one type of interaction is modified or even reversed when another type of interaction is also occurring.

Predation Reduces Competition

When resources are limiting, a superior competitor can eliminate other species from a community. However, predators can prevent or greatly reduce competitive exclusion by reducing the numbers of individuals of competing species. A given predator may often feed on two, three, or more kinds of plants or animals in a given community. The predator's choice depends partly on the relative abundance of the prey options. In other words, a predator may feed on species *A* when it is abundant and then switch to species *B* when *A* is rare. Similarly, a given prey species may be a primary source of food for increasing numbers of species as it becomes more abundant. In this way, superior competitors may be prevented from outcompeting other species.

Such patterns are often characteristic of biological communities in marine intertidal habitats. For example, in preying selectively on bivalves, starfish prevent bivalves from monopolizing such habitats, opening up space for many other organisms (figure 54.21). When starfish are removed from a habitat, species diversity falls precipitously, and the seafloor community comes to be dominated by a few species of bivalves. Predation tends to reduce competition in natural communities, so it is usually a mistake to attempt to eliminate a major predator, such as wolves or mountain lions, from a community because the result may be a decrease in the biological diversity of the community.

Parasitism May Counter Competition

Parasites may affect sympatric species differently and thus influence the outcome of interspecific interactions. In a classic experiment, Thomas Park of the University of Chicago investigated interactions between two flour beetles, *Tribolium castaneum* and *T. confusum*, with a parasite, *Adelina*. In the absence of the parasite, *T. castaneum* is dominant and *T. confusum* normally goes extinct. When the parasite is present, however, the outcome is reversed and *T. castaneum* perishes. Similar effects of parasites in natural systems have been observed in many species. For example, in the *Anolis* lizards of St. Maarten mentioned previously, the competitively inferior species is resistant to lizard malaria (a form of the disease related to human malaria), whereas the other species is highly susceptible.

FIGURE 54.21
Predation reduces competition. (*a*) In a controlled experiment in a coastal ecosystem, Robert Payne of the University of Washington removed a key predator, starfishes (*Pisaster*). (*b*) In response, fiercely competitive mussels exploded in growth, effectively crowding out seven other indigenous species.

Only in areas in which the malaria parasite occurs are the two species capable of coexisting.

Indirect Effects

In some cases, species may not directly interact, yet the presence of one species may affect a second species by way of interactions with a third species. Such effects are termed **indirect effects.** For example, the desert rodents described in section 54.2 (see page 1167) eat seeds, and so do the ants in their community; thus, we might expect them to compete with each other. However, when all rodents were completely

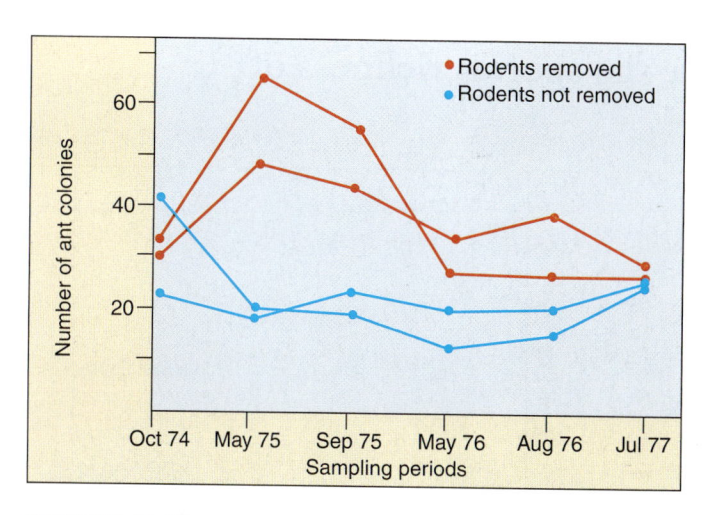

FIGURE 54.22
Change in ant population size after the removal of rodents. Ants initially increased in population size relative to ants in the two enclosures from which rodents weren't removed, but then these ant populations declined.
Why do ant populations increase and then decrease in the absence of rodents?

removed from large experimental enclosures and not allowed back in (unlike the previous experiment, no holes were placed in the walls), ant populations first increased but then declined (figure 54.22). The initial increase was the expected result of removing a competitor; why did it reverse? The answer reveals the intricacies of natural ecosystems (figure 54.23). Rodents prefer large seeds, whereas ants prefer smaller seeds. Further, in this system, plants with large seeds are competitively superior to plants with small seeds. Thus, the removal of rodents leads to an increase in the number of plants with large seeds, which reduces the number of small seeds available to ants, which leads to a decline in ant populations. Thus, the effect of rodents on ants is complicated: a direct, negative effect of resource competition and an indirect, positive effect mediated by plant competition.

Keystone Species

Species that have particularly strong effects on the composition of communities are termed **keystone species**. Predators, such as the starfish, can often serve as keystone species by preventing one species from outcompeting others, thus maintaining high levels of species richness in a community.

There are, however, a wide variety of other types of keystone species. Some species manipulate the environment in ways that create new habitats for other species. Beavers, for example, change running streams into small impoundments, altering the flow of water and flooding areas (figure 54.24). Similarly, alligators excavate deep holes at the bottoms of lakes. In times of drought, these holes are the only areas where water remains, thus allowing aquatic species that otherwise would perish to persist until the drought ends and the lake refills.

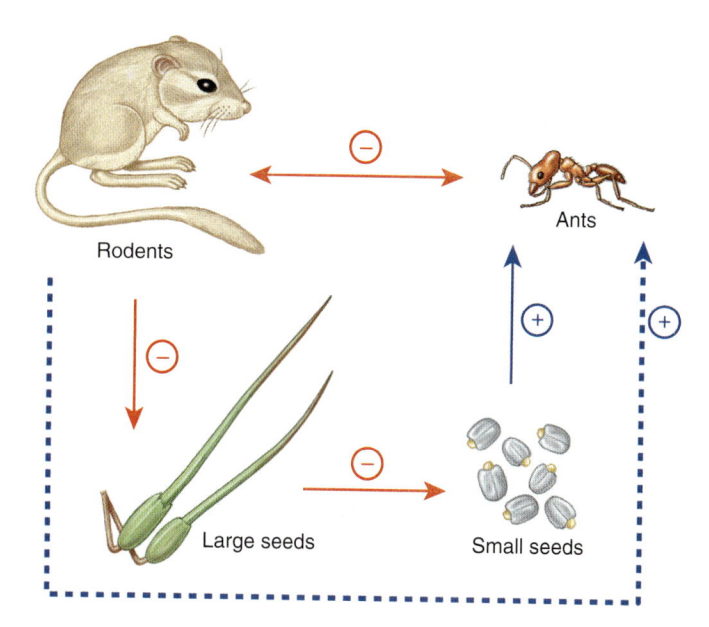

FIGURE 54.23
Rodent-ant interactions. Rodents and ants both eat seeds, so the presence of rodents has a negative effect on ants, and vice versa. However, the presence of rodents has a negative effect on large seeds. In turn, the number of plants with large seeds has a negative effect on plants that produce small seeds. Hence, the presence of rodents should increase the number of small seeds. In turn, the number of small seeds has a positive effect on ant populations. Thus, indirectly, the presence of rodents has a positive effect on ant population size.

FIGURE 54.24
Example of a keystone species. Beavers, by constructing dams and transforming flowing streams into ponds, create new habitats for many plant and animal species.

Many different processes are likely to be occurring simultaneously within communities. Only by understanding how these processes interact will we be able to understand how communities function.

Chapter 54 Community Ecology 1177

54.5 Ecological succession may increase the species richness of communities.

Even when the climate of an area remains stable year after year, communities have a tendency to change from simple to complex in a process known as **succession.** This process is familiar to anyone who has seen a vacant lot or cleared woods slowly become occupied by an increasing number of plants, or a pond become dry land as vegetation encroaches from the sides.

Succession

If a wooded area is cleared and left alone, plants will slowly reclaim the area. Eventually, all traces of the clearing will disappear, and the area will again be woods. This kind of succession, which occurs in areas where an existing community has been disturbed but soil still remains, is called **secondary succession.**

In contrast, **primary succession** occurs on bare, lifeless substrate, such as rocks, or in open water, where organisms gradually move into an area and change its nature. Primary succession occurs in lakes left behind after the retreat of glaciers, on volcanic islands that rise above the sea, and on land exposed by retreating glaciers (figure 54.25). Primary succession on glacial moraines provides an example (figure 54.26). On the bare, mineral-poor ground exposed when glaciers recede, soil pH is basic as a result of carbonates in the rocks, and nitrogen levels are low. Lichens are the first vegetation able to grow under such conditions. Acidic secretions from the lichens help break down the substrate and reduce the pH, as well as adding to the accumulation of soil. Mosses then colonize these pockets of soil, eventually building up enough nutrients in the soil for alder shrubs to take hold. Over a hundred years, the alders, which have symbiotic bacteria that fix atmospheric nitrogen (see chapter 27), increase soil nitrogen levels, and their acidic leaves further lower soil pH. At this point, spruce are able to become established, and they eventually crowd out the alder and form a dense spruce forest.

In a similar example, an **oligotrophic** lake—one poor in nutrients—may gradually, by the accumulation of organic matter, become **eutrophic**—rich in nutrients. As this occurs, the composition of communities will change, first increasing in species richness and then declining.

Primary succession in different habitats often eventually arrives at the same kinds of vegetation—vegetation characteristic of the region as a whole. This relationship led American ecologist F. E. Clements, as part of his holistic concept of communities, to propose the concept of a final *climax community*. With an increasing realization that (1) the climate keeps changing, (2) the process of succession is often very slow, and (3) the nature of a region's vegetation is being determined to an increasing extent by human activities, ecologists today do not consider the concept of "climax community" as useful as they once did.

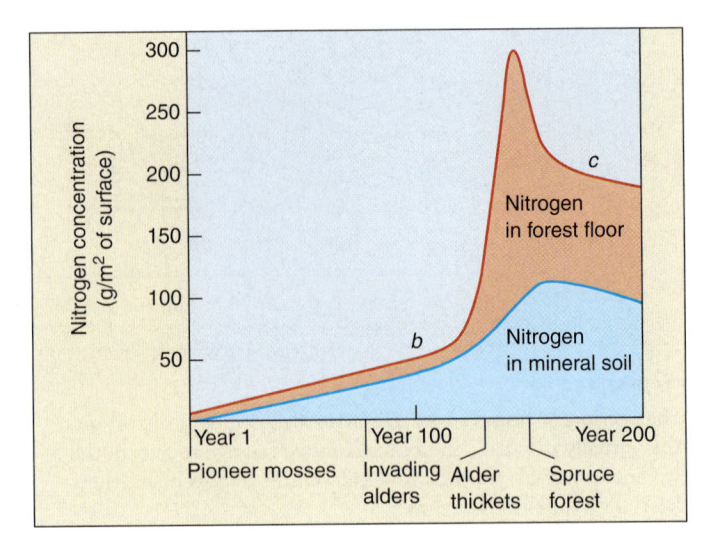

FIGURE 54.25
Plant succession produces progressive changes in the soil. Initially, the glacial moraine at Glacier Bay, Alaska (portrayed in figure 54.26), had little soil nitrogen, but nitrogen-fixing alders led to a buildup of nitrogen in the soil, encouraging the subsequent growth of the coniferous forest. Letters in the graph correspond to photographs in parts *b* and *c* of figure 54.26.

Why Succession Happens

Succession happens because species alter the habitat and the resources available in it in ways that favor other species. Three dynamic concepts are of critical importance in the process: tolerance, facilitation, and inhibition.

1. **Tolerance.** Early successional stages are characterized by weedy, *r*-selected species that are tolerant of the harsh, abiotic conditions in barren areas.
2. **Facilitation.** The weedy early successional stages introduce local changes in the habitat that favor other, less weedy species. Thus, the mosses in the Glacier Bay succession convert nitrogen to a form that allows alders to invade. The alders in turn lower soil pH as their fallen leaves decompose, and spruce and hemlock, which require acidic soil, are able to invade.
3. **Inhibition.** Sometimes the changes in the habitat caused by one species, while favoring other species, inhibit the growth of the species that caused them. Alders, for example, do not grow as well in acidic soil as the spruce and hemlock that replace them.

Over the course of succession, the number of species typically increases as the environment becomes more hospitable. In some cases, however, as ecosystems mature,

1178 Part VIII Ecology and Behavior

FIGURE 54.26
Primary succession at Alaska's Glacier Bay. (*a*) The sides of the glacier have been retreating at a rate of about 8 meters per year, leaving behind exposed soil from which nitrogen and other minerals have been leached. The first invaders of these exposed sites are pioneer moss species with nitrogen-fixing, mutualistic microbes. (*b*) Within 20 years, young alder shrubs take hold. Rapidly fixing nitrogen, they soon form dense thickets. (*c*) As soil nitrogen levels rise, spruce crowd out the mature alders, forming a forest.

more *K*-selected species replace *r*-selected ones, and superior competitors force out other species, leading ultimately to a decline in species richness.

Succession in Animal Communities

The species of animals present in a community also change through time in a successional pattern. Usually this results because species have different habitat requirements, and as the vegetation changes during succession, habitat disappears for some species and appears for others. A particularly striking example occurred on the Krakatau islands, which were devastated by an enormous volcanic eruption in 1883. Initially composed of nothing but barren ash-fields, the vegetation of the three islands of the group experienced rapid successional change: A few blades of grass appeared the next year, and within 15 years the coastal vegetation was well established and the interior was covered with dense grasslands. By 1930, the islands were almost entirely forested (figure 54.27).

The fauna of Krakatau changed in synchrony with the vegetation. Nine months after the eruption, the only animal found was a single spider, but by 1908, 200 animal species were found in a three-day exploration. For the most part, the first animals were grassland inhabitants, but as trees came to dominate, some of these early colonists, such as the zebra dove and the long-tailed shrike, disappeared and were replaced by forest-inhabiting species, such as fruit bats and frugivorous birds.

Although patterns of succession of animal species have been caused by vegetational succession, changes in the composition of the animal community in turn have affected plant occurrences. In particular, many plant species that are animal-dispersed or pollinated could not colonize Krakatau until their dispersers or pollinators had become established. For example, fruit bats were slow to colonize Krakatau, and until they appeared, few bat-dispersed plant species were present.

FIGURE 54.27
Succession after a volcanic eruption. A major volcanic explosion in 1883 on the island of Krakatau destroyed all life on the island. (*a*) This photo shows a later, much less destructive eruption of the volcano. (*b*) Krakatau, forested and populated by animals.

> Communities change through time by a process termed succession. Species richness tends to increase through time, though ultimately it may decline.

Chapter 54 Community Ecology 1179

The Role of Disturbance

Traditionally, many ecologists considered biological communities to be in a state of **equilibrium,** a stable condition that resisted change and fairly quickly returned to its original state if disturbed by humans or natural events. Such stability was usually attributed to the process of interspecific competition.

In recent years, this viewpoint has been reevaluated. Increasingly, scientists are recognizing that communities are constantly changing as a result of climatic changes, species invasions, and disturbance events. As a result, many ecologists now invoke **nonequilibrium** models that emphasize change, rather than stability. A particular focus of ecological research concerns the role that disturbances play in determining the structure of communities.

Disturbances can be widespread or local. Severe disturbances, such as forest fires, drought, and floods, may affect large areas. Animals may also cause severe disruptions. Gypsy moths can devastate a forest by consuming its trees. Unregulated deer populations may grow explosively, the deer overgrazing and so destroying the forest in which they live. On the other hand, local disturbances may affect only a small area, as when a tree falls in a forest or an animal digs a hole and uproots vegetation.

Intermediate Disturbance Hypothesis

In some cases, disturbance may act to increase the species richness of an area. According to the *intermediate disturbance hypothesis,* communities experiencing moderate amounts of disturbance will have higher levels of species richness than communities experiencing either little or great amounts of disturbance. Two factors could account for this pattern. First, in communities where moderate amounts of disturbance occur, patches of habitat will exist at different successional stages. Thus, within the area as a whole, species diversity will be greatest because the full range of species—those characteristic of all stages of succession—will be present. For example, a pattern of intermittent episodic disturbance that produces gaps in the rain forest (as when a tree falls) allows invasion of the gap by other species (figure 54.28). Eventually, the species inhabiting the gap will go through a successional sequence, one tree replacing another, until a canopy tree species comes again to occupy the gap. But if there are many gaps of different ages in the forest, many different species will coexist, some in young gaps and others in older ones.

Second, moderate levels of disturbance may prevent communities from reaching the final stages of succession, in which a few dominant competitors eliminate most of the other species. On the other hand, too much disturbance might leave the community continually in the earliest stages of succession, when species richness is relatively low.

FIGURE 54.28
Intermediate disturbance. A single fallen tree created a small light gap in the tropical rain forest of Malawi, Africa. Such gaps play a key role in maintaining the high species diversity of the rain forest. In this case, a banana plant is able to sprout up among the dense foliage of trees in the forest.

Ecologists are increasingly realizing that disturbance is common, rather than exceptional, in many communities. As a result, the idea that communities inexorably move along a successional trajectory culminating in the development of a climax community is no longer widely accepted. Rather, predicting the state of a community in the future may be difficult because the unpredictable occurrence of disturbances will often counter successional changes. Understanding the role that disturbances play in structuring communities is currently an important area of investigation in ecology.

> **Change is often common in ecological communities. In some cases, intermediate levels of disturbance may maximize species richness.**

Concept Review

For interactive testing, visit the Online Learning Center with PowerWeb at www.mhhe.com/Raven7

54.1 Biological communities are composed of species that occur together.

Concepts of Communities

- The term community refers to the collection of species interacting in a particular area. (p. 1162)
- The individualistic concept of communities holds that a community is nothing more than an aggregation of species co-occurring in one area. (p. 1162)
- The holistic concept of communities views communities as an integrated unit. (p. 1162)
- Recent studies have found that species changed independently, not synchronously, as climate changed, suggesting the individualistic community view over the holistic view. (pp. 1162–1163)

54.2 Interactions among competing species shape ecological niches.

Fundamental and Realized Niches

- An organism's niche is the sum total of all the ways it utilizes environmental resources. (p. 1164)
- Interspecific competition occurs when two species attempt to utilize the same limited resource. (p. 1164)
- A fundamental niche is the entire niche a species is capable of using, while the realized niche is the actual portion a species occupies. (p. 1164)

Gause and the Principle of Competitive Exclusion

- The principle of competitive exclusion states that if two species are competing for a limited resource, the species that uses the resource more efficiently will eventually limit the other, at least locally. (p. 1165)
- Niche overlap can occur between two or more species, as long as the degree of overlap does not lead to significant negative effects on one or more of the species. (p. 1165)
- Gause's theory predicts when two species coexist on a long-term basis, resources must not be limiting, or else the niches differ by one or more resources. (p. 1165)

Resource Partitioning

- Sympatric species often reduce, or avoid, competition by living in different portions of the habitat or by utilizing different resources. (p. 1166)

Detecting Interspecific Competition

- Although experimental studies can reveal the existence of interspecific competition, and field experiments can often overcome some of the limitations inherent in laboratory conditions, both methods have their limitations that must be overcome. (p. 1167)

54.3 Predation has ecological and evolutionary effects.

Predation and Prey Populations

- Predation provides strong selective pressures on prey populations. (p. 1168)
- Prey species often evolve better defenses against predators, and predators often then evolve better means of circumventing such defenses. (p. 1168)

Plant Defenses Against Herbivores

- Plants have developed many mechanisms for protection against herbivory, including morphological defenses, such as structures to deter herbivory, and chemical defenses, such as secondary chemical compounds. (p. 1169)

Animal Defenses Against Predators

- Defensive coloration can be used either as aposematic (warning) coloring or cryptic coloring (camouflage). (p. 1170)

Mimicry

- Batesian mimicry occurs when palatable insects resemble distasteful or toxic species and thus gain some protection from predation as predators learn to avoid them. (p. 1171)
- Müllerian mimicry occurs when unrelated but distasteful species come to resemble one another. (p. 1171)

54.4 Species within a community interact in many ways.

Coevolution and Symbiosis

- Symbiosis is a form of coevolution in which two or more kinds of organisms live together in a complex, long-term relationship. (p. 1172)

Commensalism

- Commensalism occurs when the relationship between two species benefits one species and neither harms nor benefits the other species. (p. 1173)

Mutualism

- Mutualism occurs when both species involved in the relationship benefit. (p. 1174)

Parasitism

- Parasitism is a special form of symbiosis in which the parasite is usually much smaller than the prey, and the two remain closely associated. Parasitism is harmful to the prey and beneficial to the predator. (p. 1175)
- Parasites can either live outside the host (ectoparasites, such as mosquitoes) or inside the host (endoparasites, such as tapeworms). (p. 1175)
- Recent studies have shown that some parasites can alter the behavior of their host in ways that facilitate transmission between hosts. (p. 1175)

Interactions Among Ecological Processes

- Predators can prevent or greatly reduce competitive exclusion by reducing the numbers of individuals of competing species. (p. 1176)
- Parasites may affect sympatric species differently and thus influence the outcome of interspecific interactions. (p. 1176)
- Keystone species are those that have particularly strong effects on community composition. (p. 1177)

54.5 Ecological succession may increase the species richness of communities.

Succession

- Primary succession occurs when a life-less substrate is gradually inhabited by organisms, which changes the environment. (p. 1178)
- Secondary succession occurs in an area where an existing community has been disturbed. (p. 1178)

The Role of Disturbance

- Increasingly, scientists are recognizing communities as dynamic entities that change due to climatic shifts, species invasions, and disturbance events. (p. 1180)
- The intermediate disturbance hypothesis predicts that communities experiencing moderate levels of disturbance will have higher species richness than communities experiencing either smaller or greater amounts of disturbance. (p. 1180)

Test Your Understanding

Self Test

1. The entire range of factors an organism is able to exploit in its environment is its
   a. community.
   b. realized niche.
   c. fundamental niche.
   d. habitat.
2. The phenomenon known as character displacement is associated with
   a. sympatric species.
   b. allopatric species.
   c. competitive exclusion.
   d. primary succession.
3. As the number of individuals of the predatory species increases, the prey population
   a. increases.
   b. decreases.
   c. stabilizes.
   d. first increases and then begins to decrease.
4. In order for mimicry to be effective in protecting a species from predation, it must
   a. occur in a palatable species that looks like a distasteful species.
   b. have cryptic coloration.
   c. occur such that mimics look and act like models.
   d. occur in only poisonous or dangerous species.
5. Brightly colored poison-dart frogs are an example of
   a. Batesian mimicry.
   b. Müllerian mimicry.
   c. cryptic coloration.
   d. aposematic coloration.
6. Which of the following is an example of commensalism?
   a. a tapeworm living in the gut of its host
   b. a clownfish living among the tentacles of a sea anemone
   c. a lichen
   d. bees feeding on nectar from a flower
7. A parasite that feeds on its host from inside is a(n)
   a. endoparasite.
   b. ectoparasite.
   c. parasitoid.
   d. predator.
8. A species that interacts with many other organisms in its community in critical ways can be called a
   a. predator.
   b. primary succession species.
   c. secondary succession species.
   d. keystone species.
9. Succession that occurs on abandoned agricultural fields is best described as
   a. coevolution.
   b. primary succession.
   c. secondary succession.
   d. prairie succession.
10. Lichen growing on the surface of rocks provides an example of
    a. facilitation.
    b. tolerance.
    c. inhibition.
    d. secondary succession.

Test Your Visual Understanding

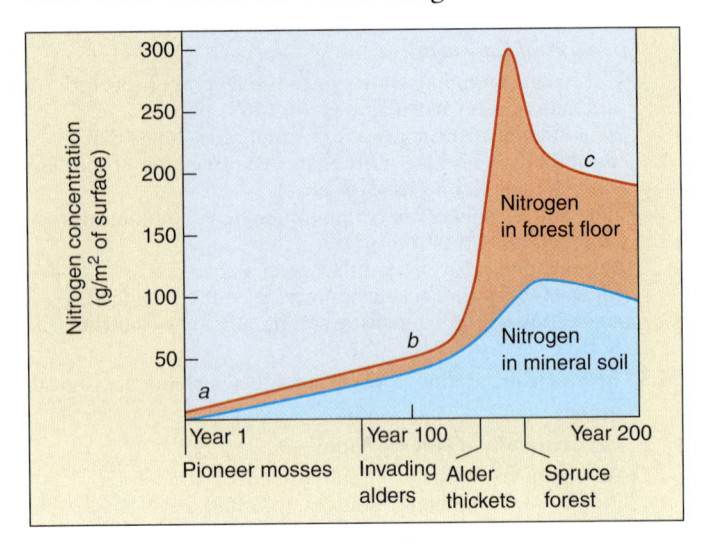

1. The graph above shows three general stages of succession indicated by the labels *a*, *b*, and *c* that occurred after the receding of a glacier at Glacier Bay, Alaska. Using the concepts of tolerance, facilitation, and inhibition, explain what is happening at each stage of succession.

Apply Your Knowledge

1. Through genetic engineering, bacterial toxin genes from a soil bacterium, *Bacillus thuringienis* (Bt), have been inserted into corn plants, making them resistant to certain insects. The corn plants produce a toxin that kills insects but doesn't harm the plant or people. However, because the corn pollen, which also contains the toxins, could spread to neighboring milkweed plants, there was a concern that monarch butterflies, which are killed by the toxin, could be at risk. The lethal dose of toxin consists of 2500 pollen grains/in^2 of milkweed leaf. Milkweed plants within a field of corn contain on average 500 pollen grains/in^2 of milkweed leaf. Assume that the level of pollen decreases by 60% with every 10 feet. How much pollen will reach milkweed plants that are 10 feet from the cornfield? 20 feet from the cornfield?

55

Dynamics of Ecosystems

Concept Outline

55.1 Chemicals cycle within ecosystems.

The Water Cycle. Water cycles between the atmosphere and the oceans, although deforestation has broken the cycle in some ecosystems.

The Carbon Cycle. Photosynthesis captures carbon from the atmosphere; respiration returns it.

The Nitrogen Cycle. Nitrogen is captured from the atmosphere by the metabolic activities of bacteria; other bacteria degrade organic nitrogen, returning it to the atmosphere.

The Phosphorus Cycle. Of all the nutrients that plants require, phosphorus tends to be the most limiting.

Biogeochemical Cycles Illustrated: Recycling in a Forest Ecosystem. In a classic experiment, the role of forests in retaining nutrients is assessed.

55.2 Energy flows through ecosystems.

Trophic Levels. Energy passes through ecosystems in a limited number of steps, typically three or four. Plants produce biomass by photosynthesis, while animals produce biomass by consuming plants or other animals.

Energy in Food Chains. As energy passes through an ecosystem, a good deal is converted to heat at each step.

Ecological Pyramids. The biomass of a trophic level is lower, the farther it is from the primary production of photosynthesizers.

55.3 Interactions occur among different trophic levels.

Trophic Cascades. Processes at one trophic level can have effects on higher or lower levels of the food chain.

55.4 Biodiversity promotes ecosystem stability.

Effects of Species Richness. Species-rich communities are more productive and resistant to disturbance than communities that have fewer kinds of species.

Causes of Species Richness. Ecosystem productivity, spatial heterogeneity, and climate all affect the number of species in an ecosystem.

Biogeographic Patterns of Species Diversity. Many more species occur in the tropics than in temperate regions.

Island Biogeography. Species richness on islands may be a dynamic equilibrium between extinction and colonization.

FIGURE 55.1
Pond in a rain forest. Interactions among species determine the rate at which chemicals and energy move between living elements of the ecosystem and how they reenter the abiotic realm, such as the atmosphere, the ground, or water.

The earth is a closed system with respect to chemicals, but, thanks to input from the sun, an open system in terms of energy. Collectively, the organisms in ecosystems regulate the capture and expenditure of energy and the cycling of chemicals (figure 55.1). As we will see in this chapter, all organisms, including humans, depend on the ability of other organisms—plants, algae, fungi, and some prokaryotes—to recycle the basic components of life. In chapters 56 and 57, we consider the many different types of ecosystems that constitute the biosphere and discuss the many threats to the biosphere and the species it contains.

1183

55.1 Chemicals cycle within ecosystems.

An **ecosystem** includes all the organisms that live in a particular place, plus the abiotic environment in which they live and with which they interact. Two processes occur within ecosystems: (1) Energy enters ecosystems—usually from the sun, but sometimes from other sources—and is converted to chemical energy by photosynthesis or other mechanisms. Subsequently, this energy is passed on in organic compounds to other organisms or is dissipated as heat. (2) Chemical elements move through ecosystems in **biogeochemical cycles**. On a global scale, only a very small portion of these substances is contained within the bodies of organisms; almost all exist in nonliving reservoirs: the atmosphere, water, or rocks. Biogeochemical cycles occur at all levels, from molecular (within an organism) to planetary (within the atmosphere or lithosphere), and at all time scales, from seconds (cellular biochemical reactions) to millennia (weathering of rocks, sedimentation). Most cycles occur simultaneously; we consider them separately only for convenience.

Physical processes such as evaporation and precipitation move these elements through the abiotic portion of the environment. Organisms capture these elements through photosynthesis and transform them into organic compounds, which then may be passed through the biotic portion of the environment. Eventually, they are returned to the abiotic realm through respiration, decomposition, or other pathways.

Both energy and chemical elements move through the biotic environment along the same pathways, primarily photosynthesis, feeding, and decomposition. Nonetheless, important differences exist. In contrast to matter, energy cannot be recycled and is eventually converted to heat. Consequently, continual input from the sun is needed, whereas matter continuously recirculates through the environment and is used and reused.

The Water Cycle

The water cycle (figure 55.2) is the most familiar of all biogeochemical cycles. All life depends directly on the presence of water; the bodies of most organisms consist mainly of this substance. Water is the source of hydrogen ions, whose movements generate ATP in organisms. For that reason alone, water is indispensable to their functioning.

The Path of Free Water

The oceans cover three-fourths of the earth's surface. From the oceans, water evaporates into the atmosphere, a process powered by energy from the sun. Over land, approximately 90% of the water that reaches the atmosphere is moisture that evaporates from the surface of plants through a process called transpiration (see chapter 37). Most precipitation falls directly into the oceans, but some falls on land, where it passes into surface and subsurface bodies of fresh water. Only about 2% of all the water on earth is captured in any form—frozen, held in the soil, or incorporated into the bodies of organisms. All of the rest is free water, circulating between the atmosphere, the land, and the oceans.

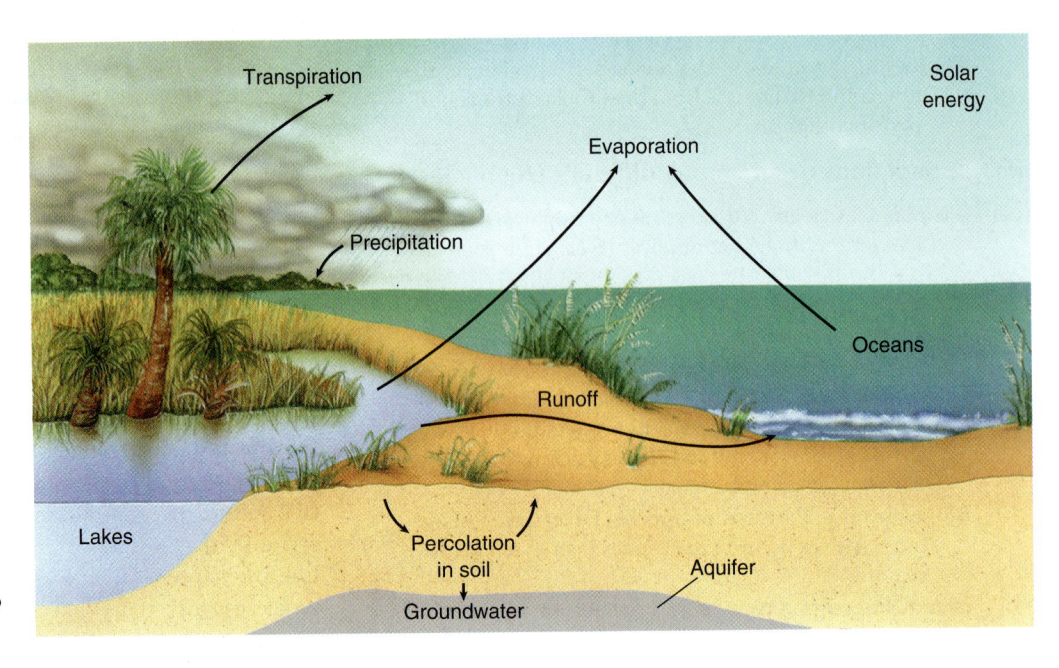

FIGURE 55.2
The water cycle. Water circulates from the atmosphere to the earth and back again.

The Importance of Water to Organisms

Organisms live or die on the basis of their ability to capture water and incorporate it into their bodies. Plants take up water from the earth in a continuous stream. Crop plants require about 1000 kilograms of water to produce one kilogram of food, and the ratio in natural communities is similar. Animals obtain water directly or from the plants or other animals they eat. The amount of free water available at a particular place often determines the nature and abundance of the living organisms present there.

Groundwater

Much less obvious than surface water, which we see in streams, lakes, and ponds, is groundwater, which occurs in aquifers—permeable, saturated, underground layers of rock, sand, and gravel. In many areas, groundwater is the most important reservoir of water. It amounts to more than 96% of all fresh water in the United States. The upper, unconfined portion of the groundwater constitutes the *water table*, which flows into streams and is partly accessible to plants; the lower, confined layers are generally out of reach, although they can be "mined" by humans. The water table is recharged by water that percolates through the soil from precipitation as well as by water that seeps downward from ponds, lakes, and streams. The deep aquifers (underground material reservoirs of water) are recharged very slowly from the water table.

Groundwater flows much more slowly than surface water, anywhere from a few millimeters to a meter or so per day. In the United States, groundwater provides about 25% of the water used for all purposes and supplies about 50% of the population with drinking water. Rural areas tend to depend almost exclusively on wells to access groundwater, and its use is growing at about twice the rate of surface water use. In the Great Plains of the central United States, the extensive use of the Ogallala Aquifer as a source of water for agricultural needs as well as for drinking water is depleting it faster than it can be naturally recharged. This seriously threatens the agricultural production of the area; similar problems are appearing throughout the drier portions of the globe.

Because of the greater rate of groundwater use and because it flows so slowly, the increasing chemical pollution of groundwater is also a very serious problem. It is estimated that about 2% of the groundwater in the United States is already polluted, and the situation is worsening. Pesticides, herbicides, and fertilizers have become a serious threat to water purity. Other key sources of groundwater pollution are the roughly 200,000 surface pits, ponds, and lagoons that are actively used for chemical waste disposal in the United States alone. Because of the large volume of water, its slow rate of turnover, and its inaccessibility, removing pollutants from aquifers is virtually impossible.

FIGURE 55.3
Deforestation breaks the water cycle. As time goes by, the consequences of tropical deforestation may become even more severe, as the extensive erosion in this deforested area on the southeast Asian island of Borneo shows.

Breaking the Water Cycle

In dense forest ecosystems, such as tropical rain forests, more than 90% of the moisture is taken up by plants and then transpired back into the air. Because so many plants in a rain forest are doing this, the vegetation is the primary source of local rainfall. In a very real sense, these plants create their own rain: The moisture that travels up from the plants into the atmosphere falls back to earth as rain.

Where forests are cut down, the water cycle is broken, and moisture is not returned to the atmosphere. Water drains away from the area to the sea instead of rising to the clouds and falling again on the forest. As early as the late 1700s, the great German explorer Alexander von Humbolt reported that stripping the trees from a tropical rain forest in Colombia had prevented water from returning to the atmosphere and created a semiarid desert. It is a tragedy of our time that just such a transformation is occurring in many tropical areas, as tropical and temperate rain forests are being clear-cut or burned (figure 55.3). In the twentieth century, much of Madagascar, a California-sized island off the east coast of Africa that contains many species found nowhere else in the world, was transformed from lush tropical forest into semiarid desert by deforestation. Because the rain no longer falls, there is no practical way to reforest this land. The water cycle, once broken, cannot be reestablished easily.

Water cycles between oceans and the atmosphere. Some 96% of the fresh water in the United States consists of groundwater, which provides 25% of all the water used in this country.

The Carbon Cycle

Carbon dioxide makes up only about 0.03% of the atmosphere. Worldwide, the synthesis of organic compounds from carbon dioxide and water through photosynthesis (see chapter 10) utilizes about 10% of the carbon dioxide in the atmosphere each year. This enormous amount of biological activity takes place as a result of the combined activities of photosynthetic bacteria, protists, and plants. All terrestrial heterotrophic organisms obtain their carbon indirectly from photosynthetic organisms. When the bodies of dead organisms decompose, microorganisms release carbon dioxide, which goes back into the atmosphere or, if in water, settles into the sediment (in fact, marine sediments are by far the largest store of carbon on earth). From there, it can be reincorporated into the bodies of other organisms. Figure 55.4 illustrates the carbon cycle.

Most of the organic compounds formed as a result of carbon dioxide fixation in the bodies of photosynthetic organisms are ultimately broken down and released back into the atmosphere or water. Certain carbon-containing compounds, such as cellulose, are more resistant to breakdown than others, but certain prokaryotes and fungi, as well as a few kinds of insects, are able to accomplish this feat. Some cellulose, however, accumulates as undecomposed organic matter such as peat. The carbon in this cellulose may eventually be incorporated into fossil fuels such as oil or coal.

In addition to the roughly 700 billion metric tons of carbon dioxide in the atmosphere, approximately 1 trillion metric tons are dissolved in the ocean. More than half of this quantity is in the upper layers, where photosynthesis takes place. The fossil fuels, primarily oil and coal, contain more than 5 trillion additional metric tons of carbon, and between 600 million and 1 trillion metric tons are locked up in living organisms at any one time. In global terms, respiration and photosynthesis (see chapters 9 and 10) are approximately balanced, but the balance has been shifted recently because of the consumption of fossil fuels. The combustion of coal, oil, and natural gas has released large stores of carbon into the atmosphere as carbon dioxide. The increased amount of carbon dioxide in the atmosphere appears to be changing global climates, and may do so even more rapidly in the future, as is discussed in chapter 56.

> About 10% of the estimated 700 billion metric tons of carbon dioxide in the atmosphere is fixed annually by the process of photosynthesis.

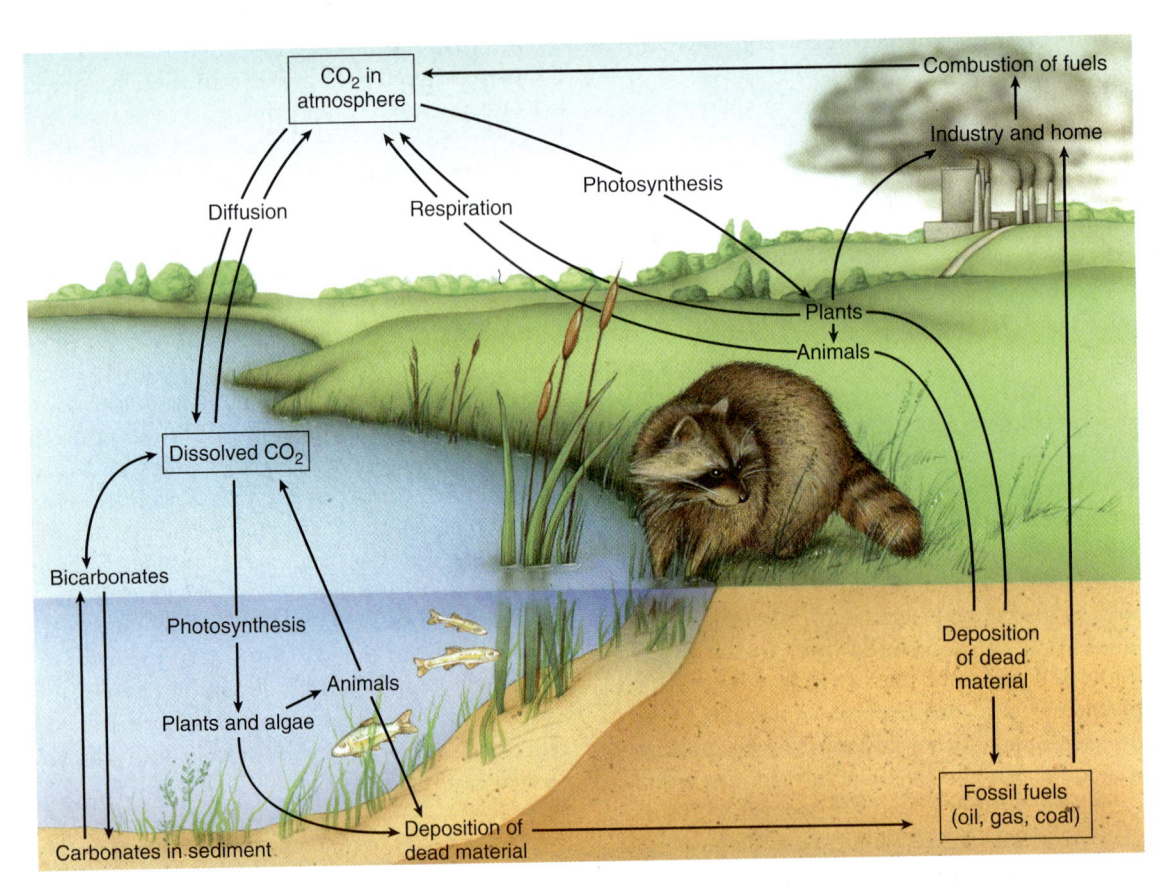

FIGURE 55.4
The carbon cycle. Photosynthesis captures carbon; respiration returns it to the atmosphere. Dissolved carbon dioxide in aquatic ecosystems mostly reacts with water to form bicarbonates.

The Nitrogen Cycle

Nitrogen is an essential component of organic compounds such as proteins and amino acids. Via the nitrogen cycle, a few kinds of organisms—all of them prokaryotes—can convert, or "fix" (see chapter 27), atmospheric nitrogen (78% of the earth's atmosphere) into forms that can be used for biological processes (figure 55.5). The triple bond that links together the two atoms that make up diatomic atmospheric nitrogen (N_2) makes it a very stable molecule. In living systems, the cleavage of atmospheric nitrogen is catalyzed by a complex of three proteins—ferredoxin, nitrogen reductase, and nitrogenase. This process uses ATP as a source of energy, electrons derived from photosynthesis or respiration, and a powerful reducing agent. The overall reaction of nitrogen fixation is written:

$$N_2 + 3H_2 \rightarrow 2NH_3$$

Some genera of prokaryotes have the ability to fix atmospheric nitrogen. Most are free-living, but some form symbiotic relationships with the roots of legumes (plants of the pea family, Fabaceae) and other plants. Only the symbiotic bacteria fix enough nitrogen to be of major significance in nitrogen production. Because of the activities of such organisms in the past, a large reservoir of ammonia and nitrates now exists in most ecosystems. This reservoir is the immediate source of much of the nitrogen used by organisms. In the oceans, however, no such reservoir exists, and nitrogen is the major limiting nutrient.

Nitrogen-containing compounds, such as proteins in plant and animal bodies, are decomposed rapidly by certain prokaryotes and fungi. These prokaryotes and fungi use the amino acids they obtain through decomposition to synthesize their own proteins and to release excess nitrogen in the form of ammonium ions (NH_4^+), a process known as **ammonification.** The ammonium ions, which also are excreted by many animals, can be converted to soil nitrites and nitrates by certain kinds of organisms and can then be absorbed by plants.

A certain proportion of the fixed nitrogen in the soil is steadily lost. Under anaerobic conditions, nitrate is often converted to nitrogen gas (N_2) and nitrous oxide (N_2O), both of which return to the atmosphere. This process, which several genera of bacteria carry out, is called **denitrification.**

Large amounts of nitrogen are added to the environment in the form of fertilizers applied to agricultural lands. As a result, humans now play an important role in the global nitrogen cycle.

> **Nitrogen becomes available to organisms almost entirely through the metabolic activities of prokaryotes, some free-living and others that live symbiotically in the roots of legumes and other plants.**

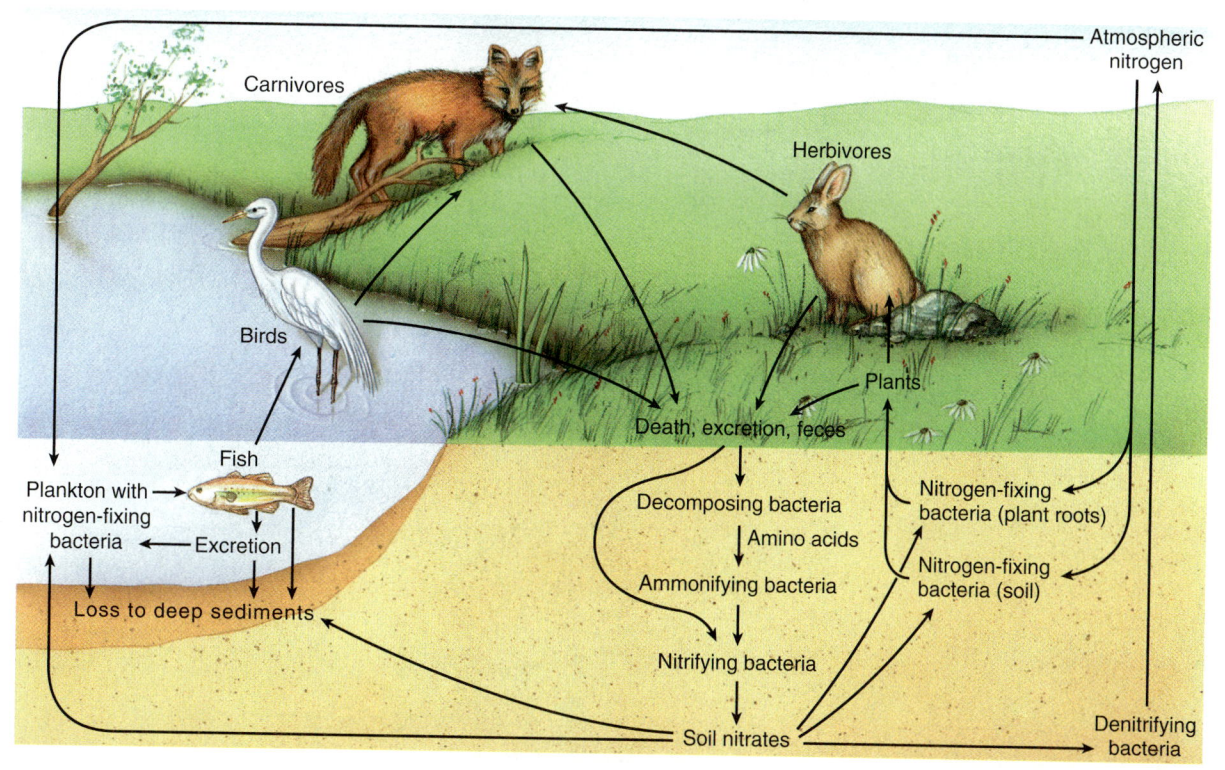

FIGURE 55.5

The nitrogen cycle. Certain bacteria fix atmospheric nitrogen, converting it to a form living organisms can use. Other bacteria decompose nitrogen-containing compounds from plant and animal materials, returning them to the atmosphere.

The Phosphorus Cycle

In all biogeochemical cycles other than those involving water, carbon, oxygen, and nitrogen, the reservoir of the nutrient exists in mineral form (that is, in rocks), rather than in the atmosphere. The phosphorus cycle (figure 55.6) is presented as a representative example of all other mineral cycles. Phosphorus, a component of ATP, phospholipids, and nucleic acids, plays a critical role in plant nutrition.

Of all the required nutrients other than nitrogen, phosphorus is the most likely to be scarce enough to limit plant growth. Phosphates, in the form of phosphorus anions, exist in soil only in small amounts. This is because they are relatively insoluble and are present only in certain kinds of rocks. As phosphates weather out of soils, they are transported by rivers and streams to the oceans, where they accumulate in sediments. They are naturally brought back up again only by upwelling ocean currents, as occur along the Pacific coast of North and South America. Phosphates brought to the surface are assimilated by algae, which are eaten by fish, which are in turn eaten by birds. Seabirds deposit enormous amounts of guano (feces) rich in phosphorus along certain coasts. Guano deposits have traditionally been used for fertilizer. Crushed phosphate-rich rocks, found in certain regions, are also used for fertilizer. The seas are the only inexhaustible source of phosphorus, making deep-seabed mining increasingly commercially attractive.

Every year, millions of tons of phosphate are added to agricultural lands in the belief that it becomes fixed to and enriches the soil. In general, three times more phosphate than a crop requires is added each year. This is usually in the form of **superphosphate,** which is soluble calcium dihydrogen phosphate, $Ca(H_2PO_4)_2$, derived by treating bones or apatite, the mineral form of calcium phosphate, with sulfuric acid. But the enormous quantities of phosphates that are being added annually to the world's agricultural lands are not leading to proportionate gains in crops. Plants can apparently use only so much of the phosphorus that is added to the soil.

> **Phosphates are relatively insoluble and are present in most soils only in small amounts. They often are so scarce that their absence limits plant growth.**

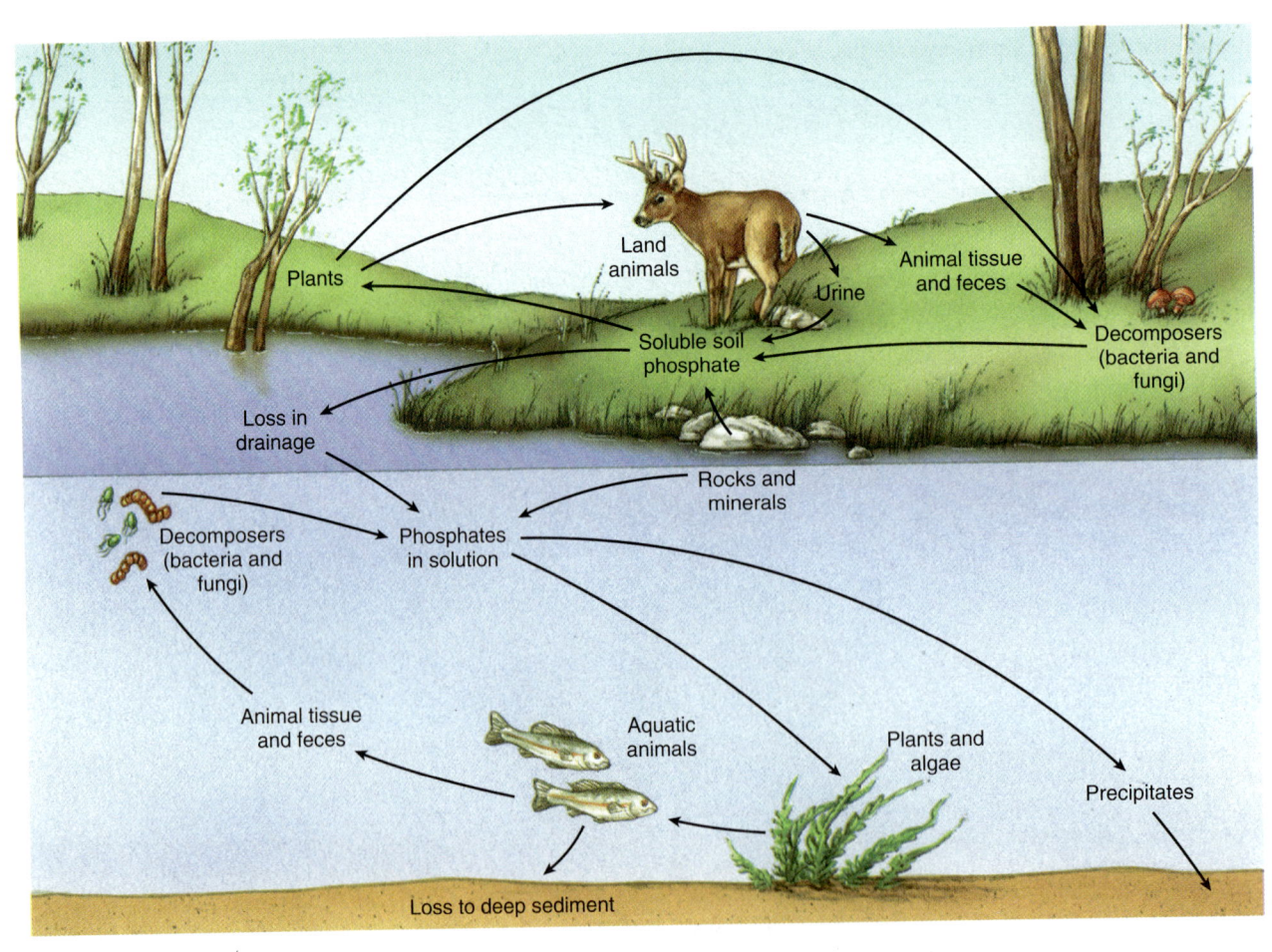

FIGURE 55.6
The phosphorus cycle. Phosphates weather from soils into water, enter plants and animals, and are redeposited in the soil when plants and animals decompose.

1188 Part VIII Ecology and Behavior

Biogeochemical Cycles Illustrated: Recycling in a Forest Ecosystem

An ongoing series of studies conducted at the Hubbard Brook Experimental Forest in New Hampshire has revealed in impressive detail the overall recycling pattern of nutrients in an ecosystem. The way this particular ecosystem functions, and especially the way nutrients cycle within it, has been studied since 1963 by Herbert Bormann of the Yale School of Forestry and Environmental Studies, Gene Likens of the Institute of Ecosystem Studies, and their colleagues. These studies have yielded much of the available information about the cycling of nutrients in forest ecosystems. They have also provided the basis for the development of much of the experimental methodology that is being applied successfully to the study of other ecosystems.

Hubbard Brook is the central stream of a large watershed that drains a region of temperate deciduous forest (figure 55.7a). To measure the flow of water and nutrients within the Hubbard Brook ecosystem, concrete dams with V-shaped notches were built across six tributary streams. All of the water that flowed out of the valleys had to pass through the notches. The researchers measured the precipitation that fell in the six valleys, and determined the amounts of nutrients that were present in the water flowing in the six streams. By these methods, they demonstrated that the undisturbed forests in this area were very efficient at retaining nutrients; the small amounts of nutrients that precipitated from the atmosphere with rain and snow were approximately equal to the amounts of nutrients that ran out of the valleys. These quantities were very low in relation to the total amount of nutrients in the system. There was a small net loss of calcium—about 0.3% of the total calcium in the system per year—and small net gains of nitrogen and potassium.

In 1965 and 1966, the investigators felled all the trees and shrubs in one of the six watersheds and then prevented regrowth by spraying the area with herbicides. The effects were dramatic. The amount of water running out of that valley increased by 40%. This indicated that water that previously would have been taken up by vegetation and ultimately evaporated into the atmosphere was now running off. For the four-month period from June to September 1966, the runoff was four times higher than it had been during comparable periods in the preceding years. The amounts of nutrients running out of the system also greatly increased; for example, the loss of calcium was 10 times higher than it had been previously. Phosphorus, on the other hand, did not increase in the stream water; it apparently was locked up in the soil.

The change in the status of nitrogen in the disturbed valley was especially striking (figure 55.7b). The undisturbed ecosystem in this valley had been accumulating nitrogen at a rate of about 2 kilograms per hectare per year, but the deforested ecosystem *lost* nitrogen at a rate of about 120 kilograms per hectare per year. The nitrate level of the water rapidly increased to a level exceeding that judged safe for human consumption, and the stream that drained the area generated massive blooms of cyanobacteria and algae. In other words, the fertility of this logged-over valley decreased rapidly, while at the same time the danger of flooding greatly increased. This experiment is particularly instructive at the start of the twenty-first century, as large areas of tropical rain forest are being destroyed to make way for cropland, a topic that is discussed further in chapter 56.

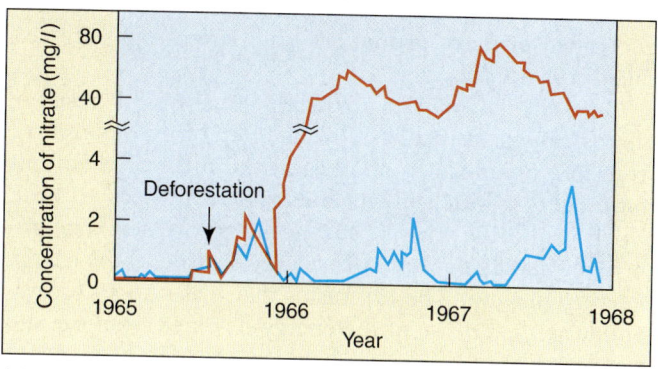

FIGURE 55.7

The Hubbard Brook experiment. (*a*) A 38-acre watershed was completely deforested, and the runoff monitored for several years. (*b*) Deforestation greatly increased the loss of minerals in runoff water from the ecosystem. The red curve represents nitrate in the runoff water from the deforested watershed; the blue curve, nitrate in runoff water from an undisturbed neighboring watershed.

When the trees and shrubs in one of the valleys in the Hubbard Brook watershed were cut down and the area was sprayed with herbicide, water runoff and the loss of nutrients from that valley increased. Nitrogen, which had been accumulating at a rate of about 2 kilograms per hectare per year, was lost at a rate of 120 kilograms per hectare per year.

55.2 Energy flows through ecosystems.

Trophic Levels

An ecosystem includes autotrophs and heterotrophs. **Autotrophs** manufacture their own food. Plants, algae, and some prokaryotes use light energy, but other organisms, such as chemoautotrophic prokaryotes that live in deepwater hydrothermal vents and derive energy by oxidizing sulfur compounds, transform inorganic molecules to manufacture their food. To support themselves, **heterotrophs,** which include animals, fungi, most protists and prokaryotes, and nongreen plants, must obtain organic molecules that have been synthesized by autotrophs. Autotrophs are also called **primary producers,** and heterotrophs are also called **consumers.**

Once energy enters an ecosystem, usually as the result of photosynthesis, it is passed, by ingestion or decomposition, from one organism to another in the form of chemical-bond energy. Much of this energy is converted to heat, which cannot be used to sustain life processes. The autotrophs that first acquire this energy provide all the energy heterotrophs use. The organisms that make up an ecosystem delay the release of the energy obtained from the sun back into space.

Green plants, the primary producers of a terrestrial ecosystem, generally capture about 1% of the energy that falls on their leaves, converting it to food energy. In especially productive systems, this percentage may be a little higher. When these plants are consumed by other organisms, only a portion of the plants' accumulated energy is actually converted into the bodies of the organisms that consume them, because most of it is dissipated as heat. The proportion of the ingested energy that is stored as chemical-bond energy varies among organisms. In an invertebrate, for example, about 10% of the food it eats is turned into its own body and thus into potential food for its predators. Although the comparable figure varies from approximately 5% in carnivores to nearly 20% in herbivores, 10% is a good average value for the amount of organic material that reaches the next trophic level.

Several different levels of consumers exist. **Primary consumers,** or herbivores, feed directly on green plants. **Secondary consumers,** carnivores and the parasites of animals, feed in turn on the herbivores. **Decomposers** break down the organic matter accumulated in the bodies of other organisms. Another more general term that includes decomposers is **detritivore.** Detritivores live on the refuse of an ecosystem. They include large scavengers, such as crabs, vultures, and jackals, as well as decomposers.

All of these categories occur in any ecosystem. They represent different **trophic levels,** from the Greek word *trophos,* which means "feeder." Organisms from each

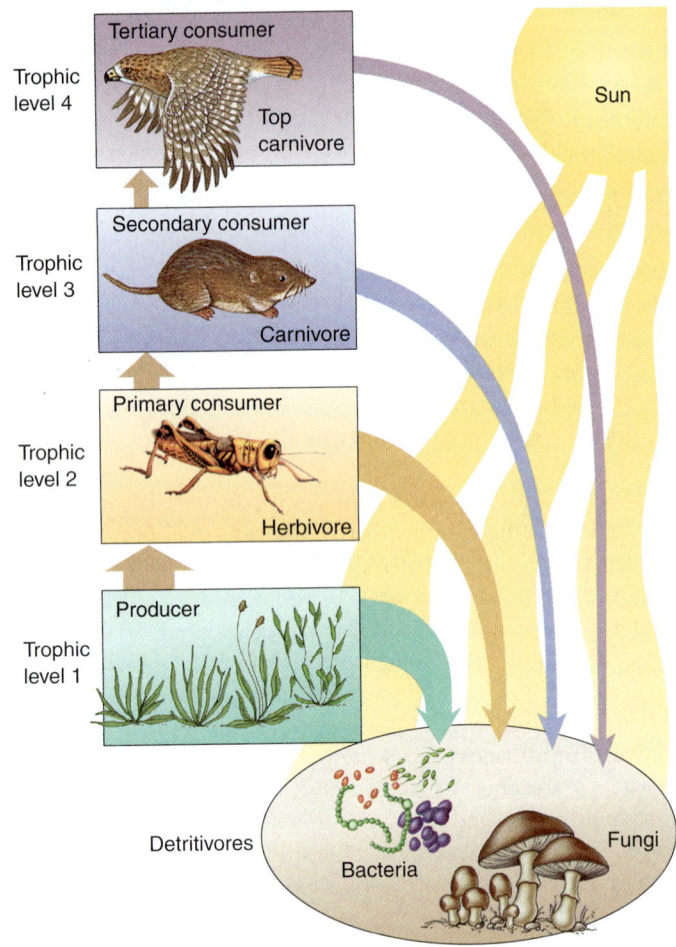

FIGURE 55.8
Trophic levels within a food chain. Plants obtain their energy directly from the sun, placing them at trophic level 1. Animals that eat plants, such as grasshoppers, are primary consumers, or herbivores, and are at trophic level 2. Animals that eat plant-eating animals, such as shrews, are carnivores and are at trophic level 3 (secondary consumers). Animals that eat carnivorous animals, such as hawks, are tertiary consumers at trophic level 4. Detritivores use all trophic levels for food.

trophic level, feeding on one another, make up a series called a **food chain** (figure 55.8). The length and complexity of food chains vary greatly. In real life, it is rather rare for a given organism to feed only on one other type of organism. Usually, each organism feeds on two or more kinds and in turn is eaten by several other kinds of organisms. When diagrammed, the relationship appears as a series of branching lines, rather than a straight line, and is called a **food web.**

1190 Part VIII Ecology and Behavior

Primary Productivity

Approximately 1% of the solar energy that falls on a plant is converted to the chemical bonds of organic material. **Primary production** and **primary productivity** are terms used to describe the amount of energy produced by photosynthetic organisms in a community. *Gross primary productivity* is the total organic matter produced, including that used by the photosynthetic organisms for respiration. *Net primary productivity* (NPP), therefore, is a measure of the amount of organic matter produced in a community in a given time that is available for heterotrophs. It equals the gross primary productivity minus the amount of energy expended by the metabolic activities of the photosynthetic organisms. The total mass of all of the organisms living in an ecosystem, called its **biomass,** increases as a result of its net productivity.

Productive Biological Communities. Some ecosystems have a high net primary productivity. For example, tropical forests and wetlands normally produce between 1500 and 3000 grams of organic material per square meter per year. By contrast, some corresponding figures for other communities are 1200 to 1300 grams for temperate forests, 900 grams for savanna, and 90 grams for deserts (table 55.1).

Secondary Productivity

The rate of biomass production by heterotrophs is called **secondary productivity.** Because herbivores and carnivores cannot carry out photosynthesis, they do not manufacture biomolecules directly from CO_2. Instead, they obtain them by eating plants or other heterotrophs. Secondary productivity by herbivores is approximately an order of magnitude less than the primary productivity upon which it is based. Where does all the energy in plants go? First, much of the biomass is not consumed by herbivores and instead supports the decomposer community (prokaryotes, fungi, and detritivorous animals). Second, some energy is not assimilated by the herbivore's body but is passed on as feces to the decomposers (figure 55.9). Third, not all the chemical-bond energy that herbivores assimilate is retained as chemical-bond energy in the organic molecules of their tissues. Much of it is lost as heat produced by work.

> Energy passes through ecosystems, a good deal being converted to heat at each step. Primary productivity occurs as a result of photosynthesis, which is carried out by green plants, algae, and some prokaryotes. Secondary productivity is the production of new biomass by heterotrophs.

Table 55.1 Ecosystem Productivity Per Year

Ecosystem Type	NPP per Unit Area (g/m²)	World NPP (10¹² kg)
Algal beds and reefs	2500	1.6
Tropical rain forest	2200	37.4
Wetlands	2000	4.0
Tropical seasonal forest	1600	12.0
Estuaries	1500	2.1
Temperate evergreen forest	1300	6.5
Temperate deciduous forest	1200	8.4
Savanna	900	13.5
Boreal forest	800	9.6
Woodland and shrubland	700	6.0
Cultivated land	650	9.1
Temperate grassland	600	5.4
Continental shelf	360	9.6
Lake and stream	250	0.5
Tundra and alpine	140	1.1
Open ocean	125	41.5
Desert and semidesert shrub	90	1.6
Extreme desert, rock, sand, and ice	3	0.07

Source: Data in: Begon, Harper, and Townsend, *Ecology* 3/e, Blackwell Science 1996, page 715. Original source: Whittaker, R. H. *Communities and Ecosystems,* 2/e Macmillan, London, 1975.

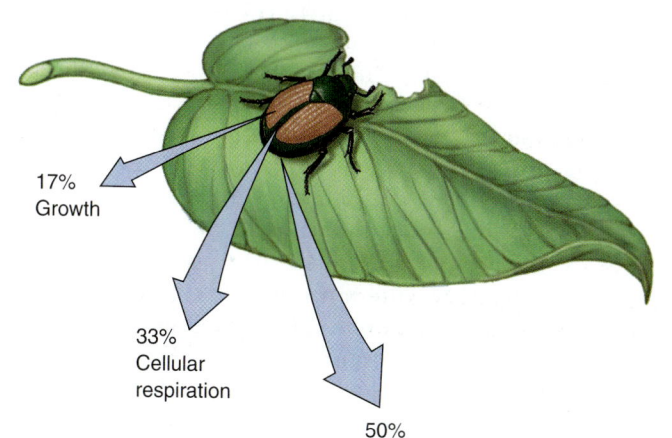

FIGURE 55.9
How heterotrophs utilize food energy. A heterotroph assimilates only a fraction of the energy it consumes. For example, if the bite of an herbivorous insect comprises 500 Joules of energy (1 Joule = 0.239 calories), about 50%, 250 J, is lost in feces, about 33%, 165 J, is used to fuel cellular respiration, and about 17%, 85 J, is converted into insect biomass. Only this 85 J is available to the next trophic level.

Chapter 55 Dynamics of Ecosystems 1191

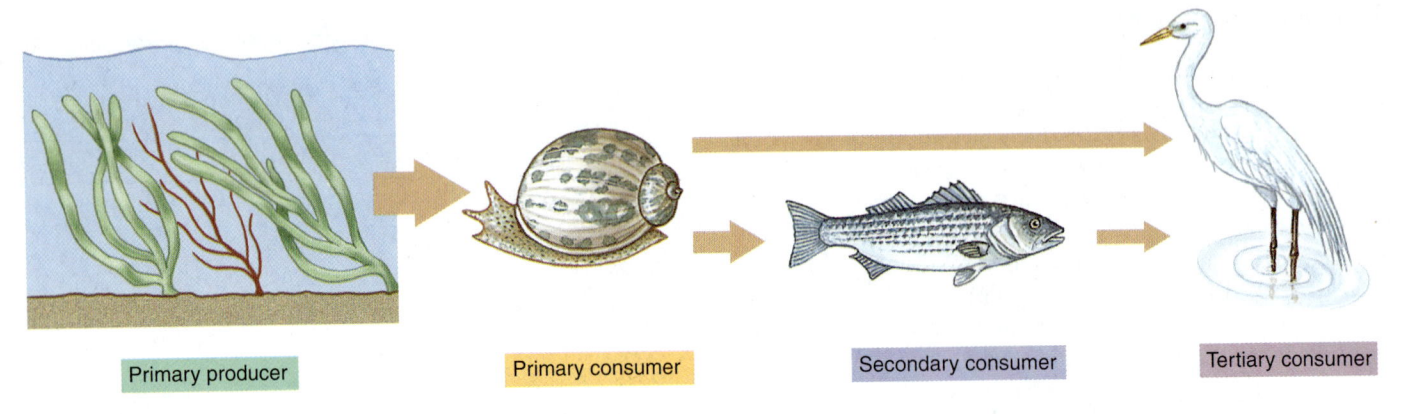

FIGURE 55.10
A food chain. Because so much energy is converted to heat at each step, food chains usually consist of just three or four steps.

Energy in Food Chains

Food chains generally consist of only three or four steps (figure 55.10). So much energy is converted to heat at each step that very little usable energy remains in the system after it has been incorporated into the bodies of organisms at four successive trophic levels.

Community Energy Budgets

Lamont Cole of Cornell University studied the flow of energy in a freshwater ecosystem in Cayuga Lake in upstate New York. He calculated that about 150 of each 1000 calories of potential energy fixed by algae and cyanobacteria are transferred into the bodies of small heterotrophs (figure 55.11). Of these, about 30 calories are incorporated into the bodies of smelt, small fish that are the principal secondary consumers of the system. If humans eat the smelt, they gain about 6 of the 1000 calories that originally entered the system. If trout eat the smelt and humans eat the trout, humans gain only about 1.2 calories.

Factors Limiting Community Productivity

Communities with higher productivity can in theory support longer food chains. The limit on a community's productivity is determined ultimately by the amount of sunlight it receives, for this determines how much photosynthesis can occur. This is why in the deciduous forests of North America the net primary productivity increases as the growing season lengthens. NPP is higher in warm climates than cold ones not only because of the longer growing seasons, but also because more nitrogen tends to be available in warm climates, where nitrogen-fixing bacteria are more active.

> Considerable energy is converted to heat at each stage in food chains, which limits their length. In general, more productive communities can support longer food chains.

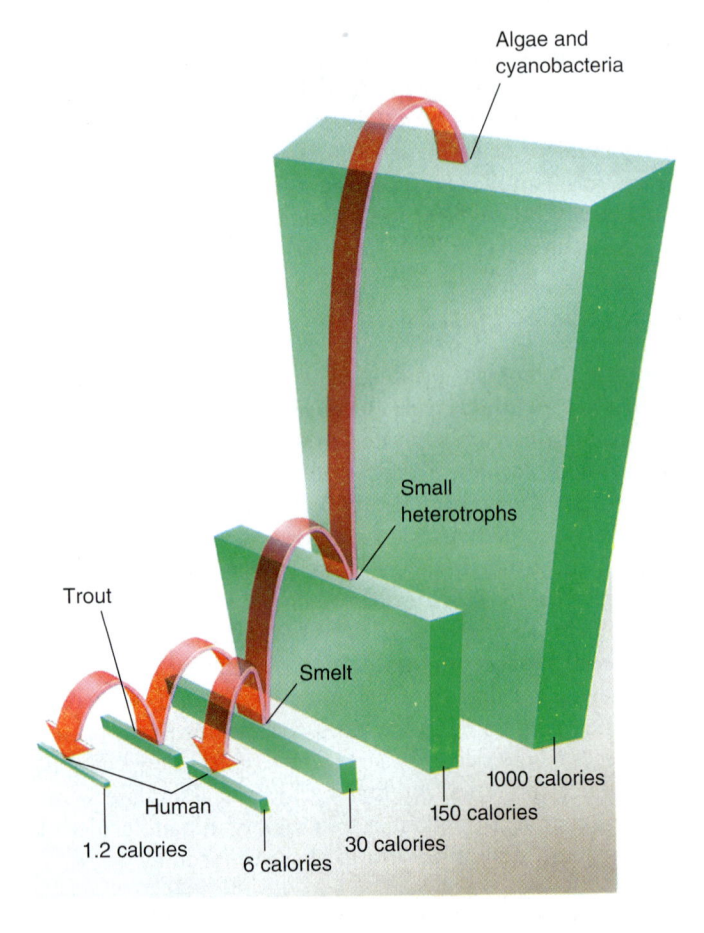

FIGURE 55.11
The food chain in Cayuga Lake. Autotrophic plankton (algae and cyanobacteria) fix the energy of the sun, heterotrophic plankton feed on them, and both are consumed by smelt. The smelt are eaten by trout, with about a fivefold loss in fixed energy. For humans, the amount of smelt biomass is at least five times greater than that available in trout, but humans prefer to eat trout.
Why does it take so many calories of algae to support so few calories of humans?

1192   Part VIII   Ecology and Behavior

Ecological Pyramids

As previously mentioned, a plant fixes about 1% of the sun's energy that falls on its green parts. The successive members of a food chain, in turn, process into their own bodies about 10% of the energy available in the organisms on which they feed. For this reason, there are generally far more individuals at the lower trophic levels of any ecosystem than at the higher levels. Similarly, the biomass of the primary producers present in a given ecosystem is greater than the biomass of the primary consumers, with successive trophic levels having less biomass and correspondingly less potential energy.

These relationships, if shown diagrammatically, appear as pyramids (figure 55.12). We can speak of pyramids of biomass, pyramids of energy, pyramids of number, and so forth, as characteristic of ecosystems.

Inverted Pyramids

Some aquatic ecosystems have inverted biomass pyramids. For example, in a planktonic ecosystem—dominated by small organisms floating in water—the turnover of photosynthetic phytoplankton at the lowest level is very rapid, with zooplankton consuming phytoplankton so quickly that the phytoplankton (the producers at the base of the food chain) can never develop a large population size. Because the phytoplankton reproduce very rapidly, the community can support a population of heterotrophs that is larger in biomass and more numerous than the phytoplankton (see figure 55.12b). Although in such cases the pyramid of biomass is inverted, it is important to remember that the pyramid of energy cannot be inverted. Because so little energy in one trophic level can be converted into body mass at the next, higher trophic level, the distribution of energy production in trophic levels will always be pyramidal.

Top Carnivores

The loss of energy that occurs at each trophic level places a limit on how many top-level carnivores a community can support. As we have seen, only about one thousandth of the energy captured by photosynthesis passes all the way through a three-stage food chain to a tertiary consumer. This explains why no predators subsist on lions or eagles—the biomass of these animals is simply insufficient to support another trophic level.

In the pyramid of numbers, top-level predators tend to be fairly large animals. Thus, the small residual biomass available at the top of the pyramid is concentrated in a relatively small number of individuals.

> Because energy is converted to heat at every step of a food chain, the biomass of primary producers (photosynthesizers) tends to be greater than that of the herbivores that consume them, and herbivore biomass tends to be greater than the biomass of the predators that consume them.

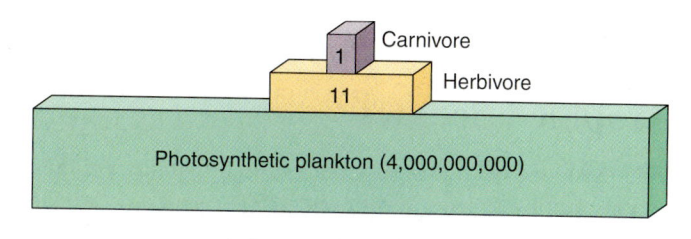

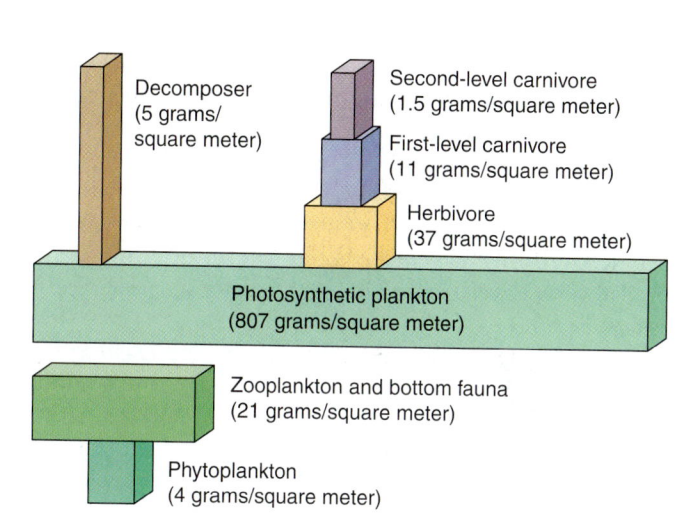

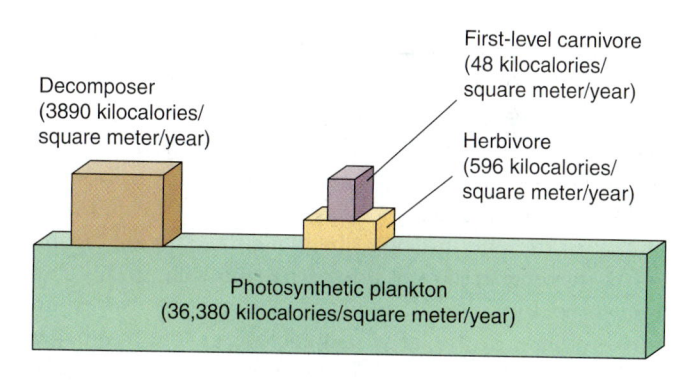

FIGURE 55.12
Ecological pyramids. Ecological pyramids measure different characteristics of each trophic level. (*a*) Pyramid of numbers. (*b*) Pyramids of biomass, both normal (*top*) and inverted (*bottom*). (*c*) Pyramid of energy.
How can the existence of inverted pyramids of biomass be explained?

55.3 Interactions occur among different trophic levels.

Trophic Cascades

The existence of food webs creates the possibility of interactions among species at different trophic levels. Predators not only have effects on the species upon which they prey, but also, indirectly, on the plants eaten by these prey. Conversely, increases in primary productivity not only provide more food for herbivores but, indirectly, lead also to more food for carnivores.

When we look at the world around us, we see a profusion of plant life. Why is this? Why don't herbivore populations increase to the extent that all available vegetation is consumed? The answer, of course, is that predators keep the herbivore populations in check, thus allowing plant populations to thrive. This phenomenon, in which the effect of one trophic level flows up or down to other levels, is called a **trophic cascade**. When the effect flows from predators to lower trophic levels, the cascade is referred to as *top-down*.

Experimental studies have confirmed the existence of trophic top-down cascades. For example, in one study in New Zealand, sections of a stream were isolated with a mesh that prevented fish from entering. In some of the enclosures, brown trout were added, whereas other enclosures were left without large fish. After 10 days, the number of invertebrates in the trout enclosures was one-half the number in the controls (figure 55.13). In turn, the biomass of algae, which invertebrates feed upon, was five times greater in the trout enclosures than in the controls.

The logic of trophic cascades leads to the prediction that a fourth trophic level, carnivores that prey on other carnivores, would also lead to cascading effects. In this case, the top predators would keep lower-level predator populations in check, which should lead to a profusion of herbivores and a paucity of vegetation. In an experiment similar to the one just described, enclosures were created in free-flowing streams in northern California. In this case, large predatory fish were added to some enclosures and not to others. In the large fish enclosures, the number of smaller predators, such as damselfly nymphs, was greatly reduced, leading to an increase in their prey, including algae-eating insects, which led, in turn, to decreases in the biomass of algae (figure 55.14).

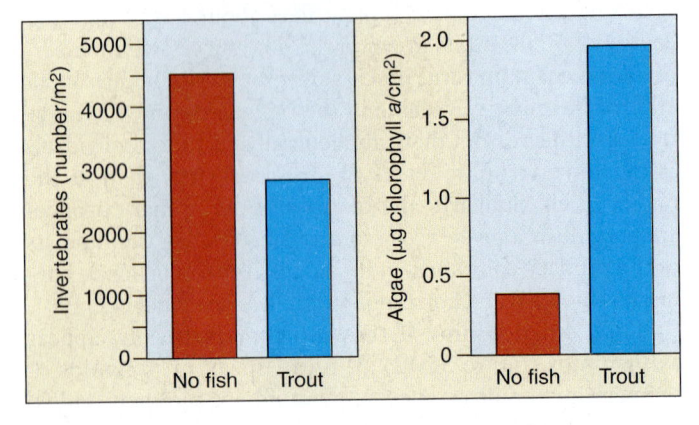

FIGURE 55.13
Trophic cascades. Streams with trout have fewer herbivorous invertebrates and more algae than streams without trout.
Why do streams with trout have more algae?

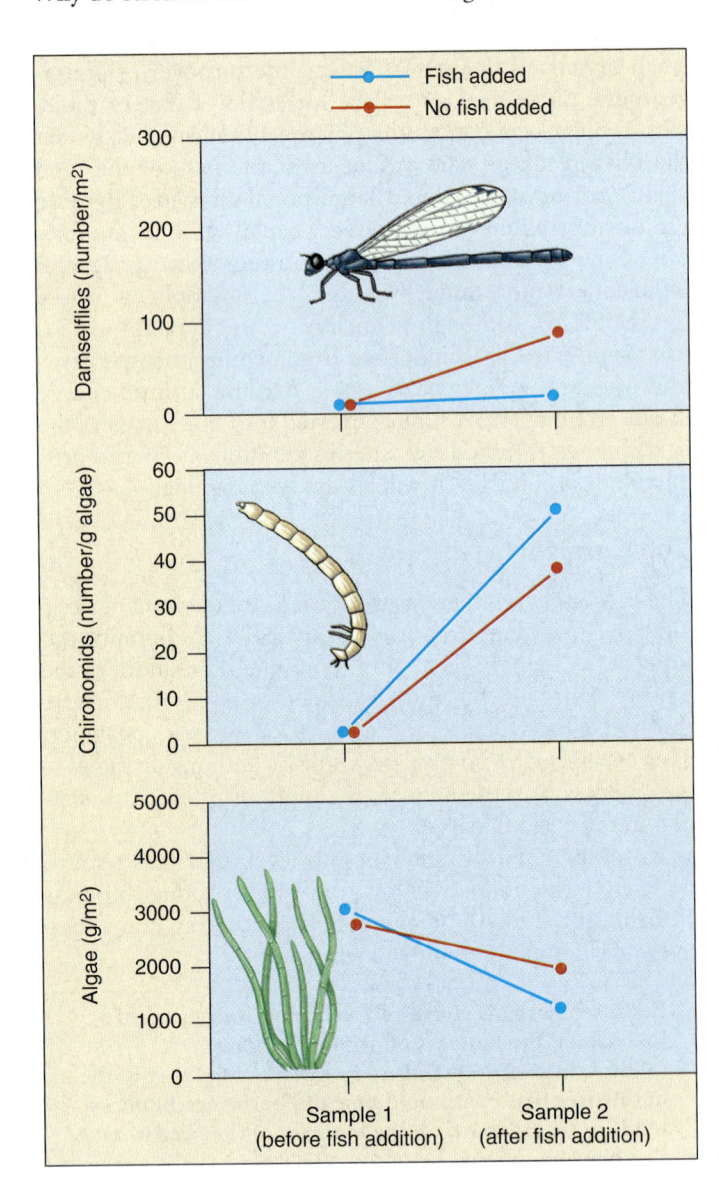

FIGURE 55.14
Four-level trophic cascades. Streams with large, carnivorous fish have fewer lower-level predators, such as damselflies, more herbivorous insects (exemplified by the number of chironomids, a type of aquatic insect), and lower levels of algae.
What might be the effect if snakes that prey on fish were added to the enclosures?

1194   Part VIII   Ecology and Behavior

Human Effects on Trophic Cascades

Humans have inadvertently created a test of the trophic cascade hypothesis by removing top predators from ecosystems. The great naturalist Aldo Leopold captured the results long before the trophic cascade hypothesis had ever been scientifically articulated when he wrote in the *Sand County Almanac*:

"I have lived to see state after state extirpate its wolves. I have watched the face of many a new wolfless mountain, and seen the south-facing slopes wrinkle with a maze of new deer trails. I have seen every edible bush and seedling browsed, first to anemic desuetude, and then to death. I have seen every edible tree defoliated to the height of a saddle horn."

Many similar examples exist in which the removal of predators has led to cascading effects on lower trophic levels. On Barro Colorado Island, a hilltop turned into an island by the construction of the Panama Canal at the beginning of the last century, large predators such as jaguars and mountain lions are absent. As a result, smaller predators whose populations are normally held in check—including monkeys, peccaries (a relative of the pig), coatimundis and armadillos—have become extraordinarily abundant. These animals eat almost anything they find. Ground-nesting birds are particularly vulnerable, and many species have declined; at least 15 bird species have vanished from the island entirely. Similarly, in woodlots in the midwestern United States, raccoons, opossums, and foxes have become abundant due to the elimination of larger predators, and populations of ground-nesting birds have declined greatly.

Bottom-Up Effects

In a manner opposite to top-down effects, factors acting at the bottom of food webs may have consequences that ramify to higher trophic levels, leading to what are termed *bottom-up* effects. The basic idea is that when the productivity of an ecosystem is low, herbivore populations will be too small to support any predators. Increases in productivity will be entirely devoured by the herbivores, whose populations will increase in size. At some point, herbivore populations will become large enough that predators can be supported. Thus, further increases in productivity will not lead to increases in herbivore populations, but rather, to increases in predator populations. Again, at some level, top predators will become established that can prey on lower-level predators. With the lower-level predator populations in check, herbivore populations will again increase as productivity increases (figure 55.15).

Experimental evidence for the role of bottom-up effects was provided in an elegant study conducted on the Eel River in northern California. Enclosures were constructed that excluded large fish. A roof was placed above each enclosure. Some roofs were clear and let light pass

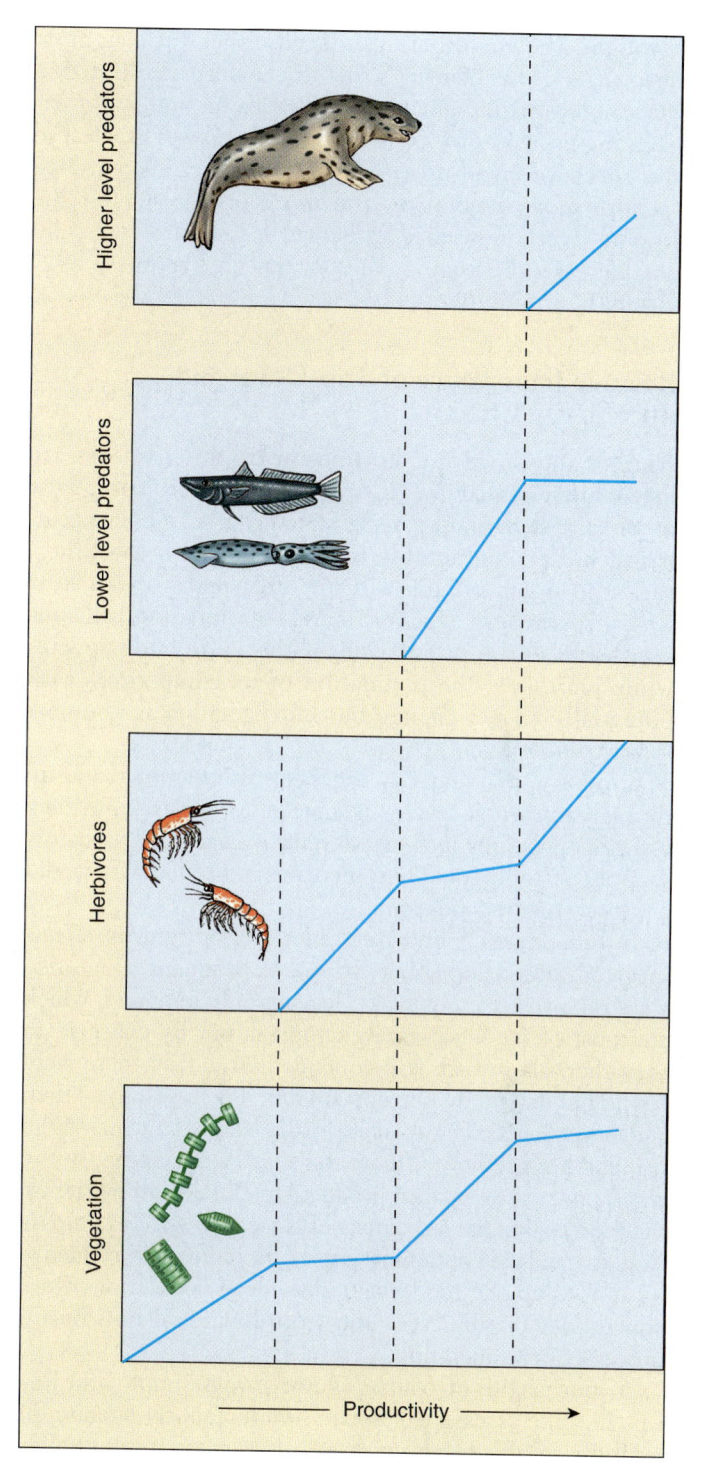

FIGURE 55.15
Bottom-up effects. At low levels of productivity, herbivore populations cannot be maintained. Above some threshold, increases in productivity lead to increases in herbivore biomass; vegetation biomass no longer increases with productivity because it is converted into herbivore biomass. Similarly, above another threshold, herbivore biomass gets converted to carnivore biomass. At this point, vegetation biomass is no longer constrained by herbivores, and so again increases with increasing productivity. **Why are there portions of the curves where vegetation biomass does not increase as productivity increases?**

through, whereas others produced either little light or deep shade. The result was that the enclosures differed in the amount of sunlight reaching them. As one might expect, the primary productivity differed and was greatest in the unshaded enclosures. This increased productivity led to both more vegetation and more predators, but the trophic level sandwiched in between, the herbivores, did not increase, precisely as the bottom-up hypothesis predicted (figure 55.16).

Relative Importance of Top-Down and Bottom-Up Effects

Neither top-down nor bottom-up trophic cascades are inevitable. For example, if two species of herbivores exist in an ecosystem and compete strongly, and if one species is much more vulnerable to predation than the other, then top-down effects will not propagate to the next lower trophic level. Rather, increased predation will simply decrease the population of the vulnerable species while increasing the population of its competitor, with potentially no net change on the vegetation in the next lower trophic level.

Similarly, productivity increases might not move up through all trophic levels. In some cases, for example, herbivore populations increase so quickly that their predators cannot control them. Thus, increases in productivity would not move up the food chain.

In other cases, trophic cascades and bottom-up effects may reinforce each other. In one experiment, large fish were removed from one lake, leaving only minnows, which ate most of the algae-eating zooplankton. By contrast, in the other lake, which still contained large fish, there were few minnows and much zooplankton. The researchers then added nutrients to both lakes. In the lake from which large fish had been removed, there were few zooplankton, so the resulting increase in algal productivity did not propagate up the food chain, and large mats of algae formed. By comparison, in the lake containing large fish, increased productivity moved up the food chain, and algae populations were controlled. In this case, both top-down and bottom-up processes were operating.

A food chain, of course, is not always simple and linear. In some cases, species may simultaneously operate on multiple trophic levels, as when a jaguar eats both smaller carnivores and herbivores, or a bear eats both fish and berries. Ecologists are currently working to apply theories of food chain interactions to these more complicated situations.

> Because of the linked nature of food webs, species at different trophic levels affect each other, and these effects can promulgate both up and down the food web.

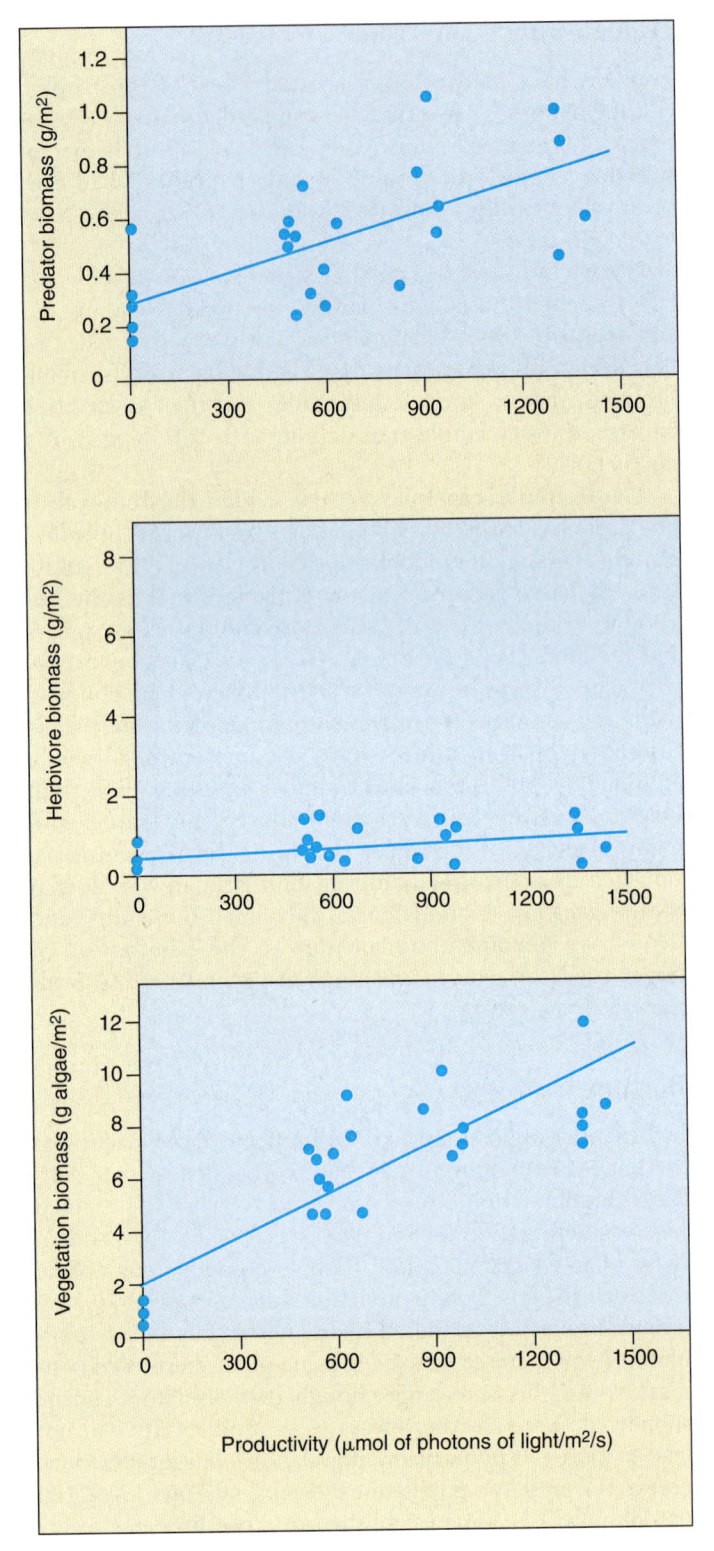

FIGURE 55.16
Bottom-up effects on a stream ecosystem. Increases in the amount of light hitting the stream lead to increases in the amount of vegetation. However, herbivore biomass does not increase with increased productivity because it is converted into predator biomass. **Why is the amount of light an important determinant of predator biomass?**

55.4 Biodiversity promotes ecosystem stability.

Effects of Species Richness

Ecologists have long debated the consequences of differences in species richness among communities. One theory is that species-rich communities are more stable—that is, more constant in composition and better able to resist disturbance. This hypothesis has been elegantly studied by David Tilman and colleagues at the University of Minnesota's Cedar Creek Natural History Area. These workers monitored 207 small rectangular plots of land (8 to 16 m^2) for 11 years (figure 55.17a). In each plot, they counted the number of prairie plant species and measured the total amount of plant biomass (that is, the mass of all plants on the plot). Over the course of the study, plant species richness was related to community stability—plots with more species showed less year-to-year variation in biomass. Moreover, in two drought years, the decline in biomass was negatively related to species richness; in other words, plots with more species were less affected. In a related experiment, when seeds of other plant species were added to different plots, the ability of these species to become established was negatively related to species richness (figure 55.17b). More-diverse communities, in other words, are more resistant to invasion by new species, another measure of community stability.

Species richness may also affect other ecosystem processes. Tilman and colleagues monitored 147 experimental plots that varied in number of species to estimate how much growth was occurring and how much nitrogen the growing plants were taking up from the soil. Tilman found that the more species a plot had, the greater the nitrogen uptake and total amount of biomass produced. In his study, increased biodiversity clearly leads to greater productivity.

Laboratory studies on artificial ecosystems have provided similar results. In one elaborate study, ecosystems covering 1 square meter were constructed in growth chambers that controlled temperature, light levels, air currents, and atmospheric gas concentrations. A variety of plants, insects, and other animals were introduced to construct ecosystems composed of 9, 15, or 31 species, with the lower diversity treatments containing a subset of the species in the higher diversity enclosures. As with Tilman's experiments, the amount of biomass produced was related to species richness, as was the amount of carbon dioxide consumed, another measure of the productivity of the ecosystem.

Tilman's conclusion that healthy ecosystems depend on diversity is not accepted by all ecologists. Critics question the validity and relevance of these biodiversity studies, arguing that the more species are added to a plot, the greater the probability that one will be highly productive. To show that high productivity results from high species richness per se, rather than from the presence of particular highly productive species, experimental plots would have to exhibit "overyielding"; in other words, plot productivity would have to be greater than that of the single most productive species grown in isolation. Although this point is still debated, recent work at Cedar Creek and elsewhere has provided evidence of overyielding, supporting the claim that species richness of communities enhances community productivity and stability.

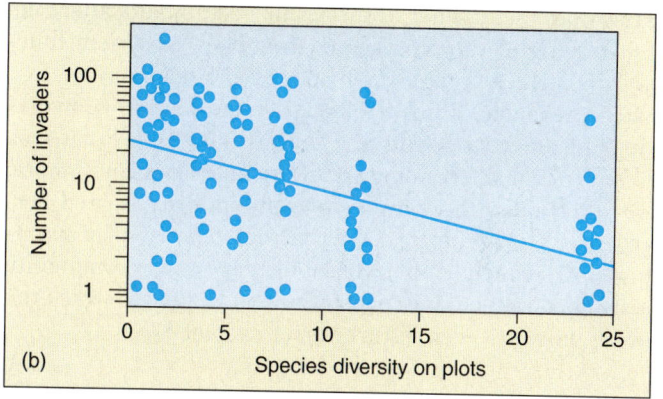

FIGURE 55.17
Effect of species richness on ecosystem stability and productivity. (a) One of the Cedar Creek experimental plots. (b) Community stability can be assessed by looking at the effect of species richness on community invasibility. Each dot represents data from one experimental plot in the Cedar Creek experimental fields. Plots with more species are harder to invade by non-native species, which occur in surrounding agricultural fields.
How could you devise an experiment on invasibility that didn't rely on species from surrounding areas?

Experimental field studies support the hypothesis that species-rich communities are more stable and productive.

Causes of Species Richness

While ecologists still argue about why some ecosystems are more stable than others—that is, better able to avoid permanent change and return to normal after disturbances such as land clearing, fire, invasion by insects, or severe storm damage—most ecologists now accept as a working hypothesis that biologically diverse ecosystems are generally more stable than simple ones. Ecosystems with many different kinds of organisms support a more complex web of interactions, and an alternative species is thus more likely to exist to compensate for the effect of a disruption.

Factors Promoting Species Richness

How does the number of species in a community affect the functioning of the ecosystem? How does ecosystem functioning affect the number of species in a community? It is often extremely difficult to sort out the relative contributions of different factors. Of the many variables that may play a role in determining species richness in a community, we will discuss three: ecosystem productivity, spatial heterogeneity, and climate. Other factors that may play an important role include the evolutionary age of the community (examined later in this chapter) and the degree to which the community has been disturbed (see chapter 54).

Ecosystem Productivity. Ecosystems differ in productivity, which is a measure of how much new growth they can produce. Surprisingly, the relationship between productivity and species richness is not linear. Rather, ecosystems with intermediate levels of productivity tend to have the most species (figure 55.18a). Why this is so is a topic of considerable current debate. One possibility is that levels of productivity are linked to numbers of predators. At low productivity, there are few predators, and superior competitors eliminate most species, whereas at high productivity, there are so many predators that only the most predation-resistant species survive. At intermediate levels, however, predators may act as keystone species, maintaining species richness (see chapter 54).

Spatial Heterogeneity. Environments that contain more soil types, topographies, and other habitat variations can be expected to accommodate more species because they provide a greater variety of habitats, climates, places to hide from predators, and so on. In general, the species richness of animals tends to reflect the species richness of the plants in their community—that is, the more plant species there are, the greater the variety of microhabitats. In turn, plant species richness reflects the spatial heterogeneity of the ecosystem. Thus, the number of lizard species in the American Southwest mirrors the structural diversity of the plants there (figure 55.18b).

Climate. The role of climate is more difficult to assess. On one hand, more species might be expected to coexist in a seasonal environment than in a constant one, because a changing climate may favor different species at different times of the year. On the other hand, stable environments are able to support specialized species that would be unable to survive where conditions fluctuate. Although other explanations are possible, data for mammal species along the west coast of North America supports the latter prediction (figure 55.18c).

> Species richness promotes ecosystem productivity and is fostered by spatial heterogeneity and stable climate.

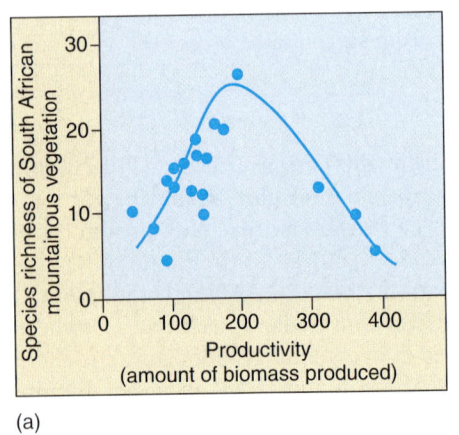

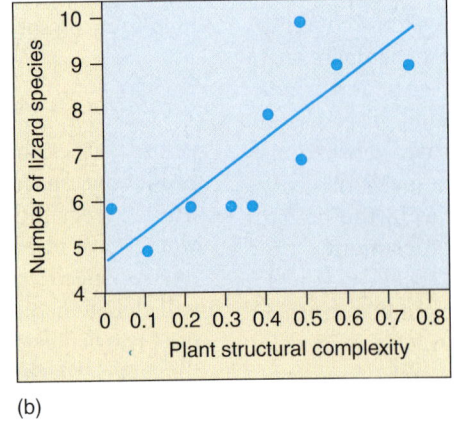

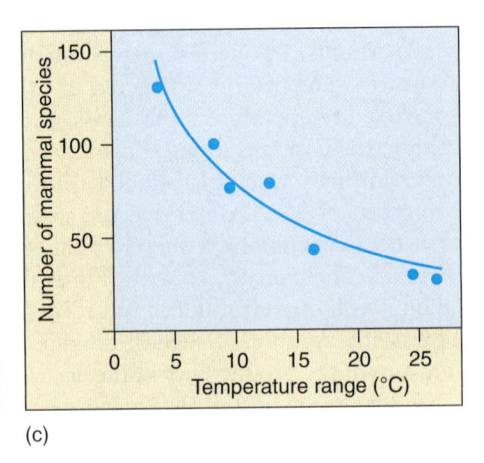

FIGURE 55.18
Factors that affect species richness. (a) *Productivity:* In fynbos plant communities of mountainous areas of South Africa, species richness of plants peaks at intermediate levels of productivity (biomass). (b) *Spatial heterogeneity:* The species richness of desert lizards is positively correlated with the structural complexity of the plant cover in desert sites in the American Southwest. (c) *Climate:* The species richness of mammals is inversely correlated with monthly mean temperature range along the west coast of North America.
(*a*) Why is species richness greatest at intermediate levels of productivity? (*b*) Why do more structurally complex areas have more species? (*c*) Why do areas with less variation in temperature have more species?

Biogeographic Patterns of Species Diversity

Since before Darwin, biologists have recognized that more different kinds of animals and plants inhabit the tropics than the temperate regions. For many species, there is a steady increase in species richness from the arctic to the tropics. Called a **species diversity cline,** such a biogeographic gradient in numbers of species correlated with latitude has been reported for plants and animals, including birds (figure 55.19), mammals, and reptiles.

Why Are There More Species in the Tropics?

For the better part of a century, ecologists have puzzled over the cline in species diversity from the arctic to the tropics. The difficulty has not been in forming a reasonable hypothesis of why there are more species in the tropics, but rather in sorting through the many reasonable hypotheses that suggest themselves. Here we will consider five of the most commonly discussed suggestions:

Evolutionary Age. It has often been proposed that the tropics have more species than temperate regions because the tropics have existed over long, uninterrupted periods of evolutionary time, while temperate regions have been subject to repeated glaciations. The greater age of tropical communities would have allowed complex population interactions to coevolve within them, fostering a greater variety of plants and animals in the tropics.

However, recent work suggests that the long-term stability of tropical communities has been greatly exaggerated. An examination of pollen within undisturbed soil cores reveals that during glaciations the tropical forests contracted to a few small refuges surrounded by grassland. This suggests that the tropics have not had a continuous record of species richness over long periods of evolutionary time.

Higher Productivity. A second often-advanced hypothesis is that the tropics contain more species because this part of the earth receives more solar radiation than temperate regions do. The argument is that more solar energy, coupled to a year-round growing season, greatly increases the overall photosynthetic activity of plants in the tropics. If we visualize the tropical forest as a pie (total resources) being cut into slices (species niches), we can see that a larger pie accommodates more slices. However, as noted previously, many field studies have indicated that species richness is highest at intermediate levels of productivity. Accordingly, increasing productivity would be expected to lead to lower, not higher, species richness. Perhaps the long column of vegetation down through which light passes in a tropical forest produces a wide range of light frequencies and intensities, creating a greater variety of light environments and so promoting species diversity.

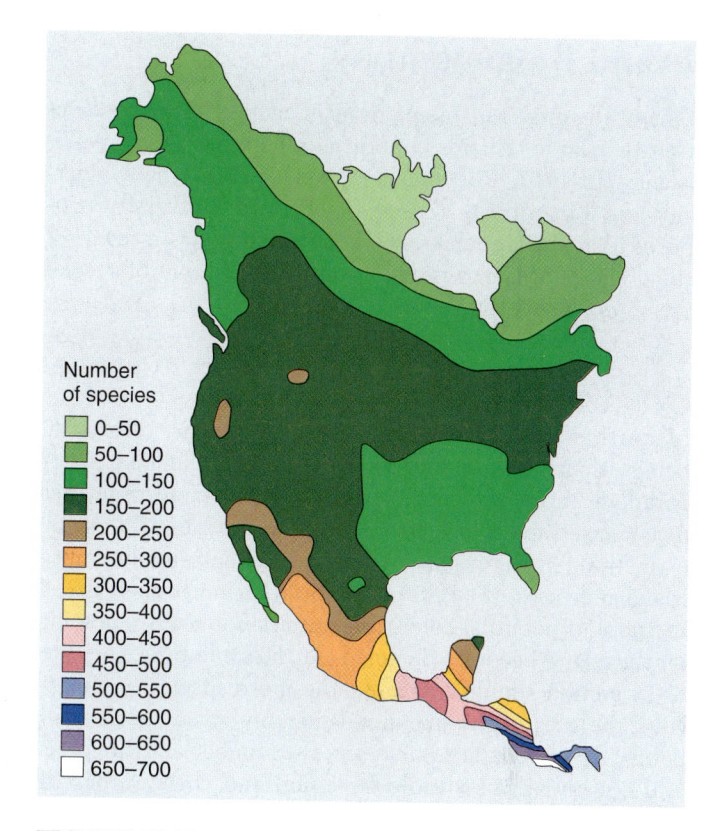

FIGURE 55.19
A latitudinal cline in species richness. Among North and Central American birds, a marked increase in the number of species occurs moving toward the tropics. Fewer than 100 species are found at arctic latitudes, while more than 600 species live in southern Central America.

Predictability. Seasonal variation, though it does exist in the tropics, is generally substantially less than in temperate areas. This reduced seasonality might encourage specialization, with niches subdivided to partition resources and so avoid competition. The expected result would be a larger number of more specialized species in the tropics, which is what we see. Many field tests of this hypothesis have been carried out, and almost all support it, reporting larger numbers of narrower niches in tropical communities than in temperate areas.

Predation. Many reports indicate that predation may be more intense in the tropics. In theory, more intense predation could reduce the importance of competition, permitting greater niche overlap and thus promoting greater species richness.

Spatial Heterogeneity. As noted earlier, spatial heterogeneity promotes species richness. Tropical forests, by virtue of their complexity, create a variety of microhabitats and so may foster larger numbers of species.

No one really knows why more species are present in the tropics, but there are plenty of hypotheses.

Chapter 55 Dynamics of Ecosystems 1199

Island Biogeography

One of the most reliable patterns in ecology is the observation that larger islands contain more species than smaller islands. In 1967, Robert MacArthur of Princeton University and Edward O. Wilson of Harvard University proposed that this **species-area relationship** was a result of the effect of area and isolation on the likelihood of species extinction and colonization.

The Equilibrium Model

MacArthur and Wilson reasoned that species are constantly being dispersed to islands, so islands have a tendency to accumulate more and more species. At the same time that new species are added, however, other species are lost by extinction. As the number of species on an initially empty island increases, the rate of colonization must decrease as the pool of potential colonizing species not already present on the island becomes depleted. At the same time, the rate of extinction should increase—the more species on an island, the greater the likelihood that any given species will perish. As a result, at some point, the number of extinctions and colonizations should be equal and the number of species should then remain constant. Every island of a given size, then, has a characteristic equilibrium number of species that tends to persist through time (the intersection point in figure 55.20a), although the species composition will change as some species become extinct and new species colonize.

MacArthur and Wilson's equilibrium model proposes that island species richness is a dynamic equilibrium between colonization and extinction. Both island size and distance from the mainland would play important roles. We would expect smaller islands to have higher rates of extinction because their population sizes would, on average, be smaller. Also, we would expect fewer colonizers to reach islands that lie farther from the mainland. Thus, small islands far from the mainland would have the fewest species; large islands near the mainland would have the most (figure 55.20b).

The predictions of this simple model bear out well in field data. Asian Pacific bird species (figure 55.20c) exhibit a positive correlation of species richness with island size, but a negative correlation of species richness with distance from the source of colonists.

Testing the Equilibrium Model

Field studies in which small islands have been censused, cleared, and allowed to recolonize tend to support the equilibrium model. However, long-term experimental field studies are suggesting that the situation is more complicated than MacArthur and Wilson envisioned. Their model predicts a high level of **species turnover** as some species perish and others arrive. However, studies of island birds and spiders indicate that very little turnover occurs from year to year. Moreover, those species that do come and go comprise a subset of species that never attain high populations. A substantial proportion of the species appears to maintain high populations and rarely go extinct. These studies, of course, have only been going on for a relatively short period of time. It is possible that over periods of centuries, the equilibrium model is a good description of what determines island species richness.

Species richness on islands is a dynamic equilibrium between colonization and extinction.

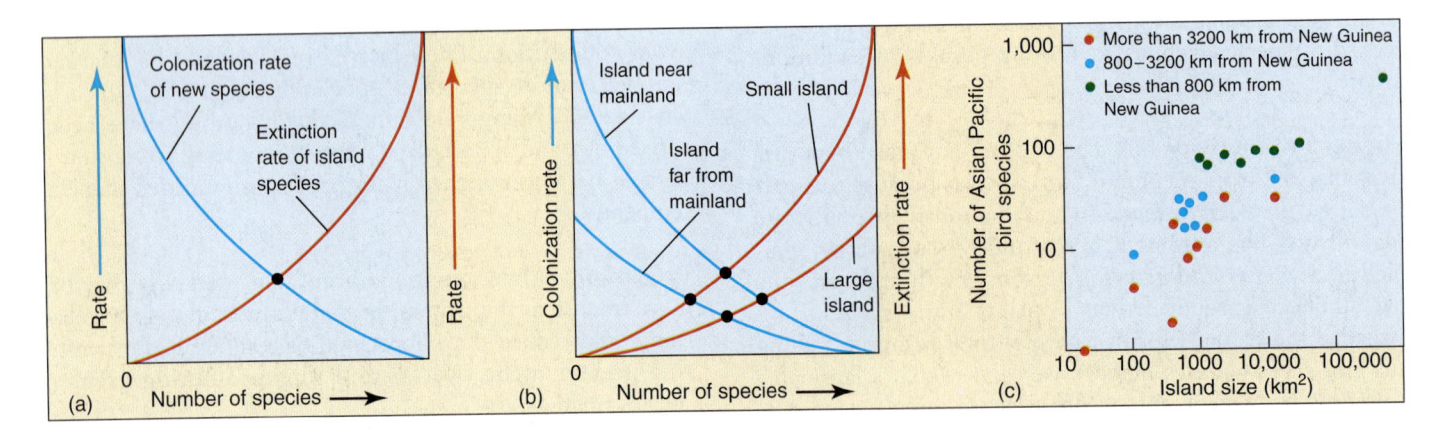

FIGURE 55.20

The equilibrium model of island biogeography. (*a*) Island species richness reaches an equilibrium (*black dot*) when the colonization rate of new species equals the extinction rate of species on the island. (*b*) The equilibrium shifts depending on the rate of colonization, the size of an island, and its distance to sources of colonists. Species richness is positively correlated with island size and inversely correlated with distance from the mainland. Smaller islands have higher extinction rates, shifting the equilibrium point to the left. Similarly, more distant islands have lower colonization rates, again shifting the equilibrium point leftward. (*c*) The effect of distance from a larger island, which can be the source of colonizing species, is readily apparent. More distant islands have fewer species compared to nearer islands of the same size.
Do islands with the same number of species necessarily have the same extinction rate?

Concept Review

55.1 Chemicals cycle within ecosystems.

- An ecosystem includes all of the organisms living in a particular place, plus the abiotic environment in which they live. (p. 1184)

The Water Cycle

- Water continuously cycles between the bodies of water, land, organisms, and atmosphere. (p. 1184)
- More than 96% of all fresh water in the United States is groundwater. (p. 1185)
- An estimated 2% of the groundwater in the Untied States is already polluted. (p. 1185)
- Cutting down forests disrupts the water cycle as lower amounts of moisture are returned to the atmosphere. (p. 1185)

The Carbon Cycle

- About 10% of the carbon dioxide in the atmosphere is fixed annually by photosynthesis. (p. 1186)
- When the bodies of dead organisms decompose, microorganisms release carbon dioxide into the atmosphere, freeing it for incorporation into other organisms. (p. 1186)
- Large amounts of carbon dioxide are also dissolved in the world's oceans. (p. 1186)

The Nitrogen Cycle

- Certain bacteria and fungi can fix atmospheric nitrogen and decompose nitrogen-containing compounds into forms useful in biological processes. (p. 1187)
- Some nitrogen-fixing bacteria form symbiotic relationships with the roots of legumes and other plants. (p. 1187)

The Phosphorus Cycle

- Phosphates are relatively insoluble and are present only in certain types of rocks. But, other than nitrogen, phosphorus is the required nutrient most likely to be scarce enough to limit plant growth. (p. 1188)

Biogeochemical Cycles Illustrated: Recycling in a Forest Ecosystem

- Deforestation in the Hubbard Brook Experimental Forest led to increased loss of minerals in runoff water from the ecosystem. (p. 1189)

55.2 Energy flows through ecosystems.

Trophic Levels

- Autotrophs (producers) manufacture their own food, while heterotrophs (consumers) must obtain organic molecules that have been synthesized by autotrophs. (p. 1190)
- Once energy enters an ecosystem, it is passed from one organism to another in the form of chemical bonds, and much of the energy is converted to heat. (p. 1190)
- Overall, an average of 10% of organic matter is transferred from one trophic level to the next. (p. 1190)
- Organisms from different trophic levels (feeding levels) make up a food chain, and the length and complexity of food chains vary greatly. Most often, trophic levels exist and are discussed in terms of food webs, or interactions of multiple food chains. (p. 1190)
- Primary productivity describes the amount of organic matter produced from solar energy in a given area during a given period of time, and is usually divided into gross primary productivity and net primary productivity. (p. 1191)
- Secondary productivity refers to the rate of biomass production by heterotrophs. (p. 1191)

Energy in Food Chains

- Food chains are limited in the number of trophic levels that can be supported because so much energy is lost at each step that very little energy usually remains after three or four steps. (p. 1192)
- In theory, communities with higher productivity can support longer food chains. (p. 1192)

Ecological Pyramids

- Due to the loss of useful energy at each trophic level, higher trophic levels generally have fewer individuals than lower levels. (p. 1193)
- Some aquatic ecosystems have inverted biomass pyramids due to extremely rapid turnover of biomass. (p. 1193)
- The loss of energy also places a limit on the number of top-level carnivores that can be supported in a community. (p. 1193)

55.3 Interactions occur among different trophic levels.

Trophic Cascades

- Top-down effects are said to occur when the effect of one trophic level flows down to the next level. (p. 1194)
- Humans have inadvertently tested the trophic cascade hypothesis by removing top predators from the ecosystem. (p. 1195)
- Bottom-up effects are said to occur when factors at the bottom of food webs cause ramifications at higher trophic levels. (p. 1195)
- Neither top-down nor bottom-up effects are inevitable, for many reasons; for example, in some cases, species may simultaneously operate on multiple trophic levels. (p. 1196)

55.4 Biodiversity promotes ecosystem stability.

Effects of Species Richness

- One theory is that species-rich communities are more stable, more constant in composition, and thus better able to resist disturbance. (p. 1197)

Causes of Species Richness

- Three of the main variables determining species richness are ecosystem productivity, spatial heterogeneity, and climate. (p. 1198)

Biogeographic Patterns of Species Diversity

- Five of the most common hypotheses for the cline of low species diversity in the arctic to high species diversity in the tropics are evolutionary age, higher productivity, predictability, predation, and spatial heterogeneity. (p. 1199)

Island Biogeography

- The equilibrium theory of island biogeography proposes that island species richness is a dynamic equilibrium between colonization and extinction. (p. 1200)

Test Your Understanding

For interactive testing, visit the Online Learning Center with PowerWeb at www.mhhe.com/Raven7

Self Test

1. Over land, most of the water in the atmosphere results from _____, whereas over the oceans, most of the water in the atmosphere results from _____.
   a. evaporation/transpiration
   b. evaporation/evaporation
   c. transpiration/evaporation
   d. transpiration/transpiration
2. Which of the statements about groundwater is *not* accurate?
   a. In the U.S., groundwater provides 50% of the population with drinking water.
   b. Groundwaters are being depleted faster than they can be recharged.
   c. Groundwaters are becoming increasingly polluted.
   d. Removal of pollutants from groundwaters is easily achieved.
3. The largest store of carbon molecules on earth is in
   a. the atmosphere.
   b. fossil fuels.
   c. marine sediments.
   d. living organisms.
4. Some bacteria have the ability to "fix" nitrogen. This means
   a. they convert ammonia into nitrites and nitrates.
   b. they convert atmospheric nitrogen gas into biologically useful forms of nitrogen.
   c. they break down nitrogen-rich compounds and release ammonium ions.
   d. they convert nitrate into nitrogen gas.
5. The phosphorus cycle differs from the water, carbon, and nitrogen cycles in that
   a. the reservoir for phosphorus exists in mineral form in rocks rather than in the atmosphere.
   b. phosphorus is far more abundant than water, carbon, or nitrogen.
   c. phosphorus is less important to biological systems than water, carbon, or nitrogen.
   d. phosphorus, once used by an organism, does not cycle back to the environment.
6. In the Hubbard Brook experiments, which of the following situations did *not* occur?
   a. The undisturbed forests were effective at retaining nutrients.
   b. The deforested areas gained nitrogen.
   c. The deforested areas lost more water to runoff than the undisturbed forests.
   d. The deforested areas lost more minerals to runoff than the undisturbed forests.
7. What percentage of the energy that hits a plant's leaves is converted into chemical (food) energy?
   a. 1%
   b. 5%
   c. 10%
   d. 20%
8. According to the trophic cascade hypothesis, the removal of carnivores from an ecosystem may result in
   a. a decline in the number of herbivores and a decline in the amount of vegetation.
   b. a decline in the number of herbivores and an increase in the amount of vegetation.
   c. an increase in the number of herbivores and an increase in the amount of vegetation.
   d. an increase in the number of herbivores and a decrease in the amount of vegetation.
9. Experimental evidence from the Cedar Creek Natural History Area supports the hypothesis that
   a. species richness is related to community stability.
   b. human activities are upsetting the trophic cascade.
   c. bottom-up effects can regulate the number of top carnivores in a system.
   d. some aquatic ecosystems have inverted biomass pyramids.
10. The equilibrium model of island biogeography suggests all of the following except
    a. larger islands have more species than smaller islands.
    b. the species richness of an island is determined by colonization and extinction.
    c. smaller islands have lower rates of extinction.
    d. islands closer to the mainland will have higher colonization rates.

Test Your Visual Understanding

Table 55.1 Ecosystem Productivity Per Year

Ecosystem Type	NPP per Unit Area (g/m^2)	World NPP (10^{12} kg)
Algal beds and reefs	2500	1.6
Tropical rain forest	2200	37.4
Wetlands	2000	4.0
Tropical seasonal forest	1600	12.0
Estuaries	1500	2.1
Temperate evergreen forest	1300	6.5
Temperate deciduous forest	1200	8.4
Savanna	900	13.5
Boreal forest	800	9.6
Woodland and shrubland	700	6.0
Cultivated land	650	9.1
Temperate grassland	600	5.4
Continental shelf	360	9.6
Lake and stream	250	0.5
Tundra and alpine	140	1.1
Open ocean	125	41.5
Desert and semidesert shrub	90	1.6
Extreme desert, rock, sand, and ice	3	0.07

1. Using this table relate productivity to species richness.

Apply Your Knowledge

1. Nitrogen is a limiting nutrient to plant growth in most ecosystems. To increase plant production, fertilizers that contain nitrogen (and phosphorus) are used. What effects might this influx of nitrogen have on ecosystems?
2. Many U.S. communities struggle with issues of deer overpopulation. Explain how human activities have created this situation.

56

The Biosphere

Concept Outline

56.1 Climate shapes the character of ecosystems.

Effects of the Sun and Atmospheric Circulation. The sun powers major movements in atmospheric circulation.
Atmospheric Circulation, Precipitation, and Climate. Latitude and elevation have important effects on climate, although other factors affect regional climate.

56.2 Biomes are widespread terrestrial ecosystems.

The Major Biomes. Characteristic communities called biomes occur in different climatic regions. Variations in temperature and precipitation are good predictors of which biomes will occur where.

56.3 Aquatic ecosystems cover much of the earth.

Patterns of Circulation in the Oceans. The world's oceans circulate in huge circles deflected by the continents.
Marine Ecosystems. The communities of the ocean are delineated primarily by depth.
Freshwater Habitats. Like miniature oceans, ponds and lakes support different communities at different depths. Freshwater ecosystems are often highly productive.

56.4 Human activity is placing the biosphere under increasing stress.

Pollution. Human industrial and agricultural activity introduces significant levels of many harmful chemicals into ecosystems.
Acid Precipitation. Burning of cheap, high-sulfur coal has introduced sulfur to the upper atmosphere, where it combines with water to form sulfuric acid that falls back to earth, harming ecosystems.
Destruction of the Tropical Forests. Many of the world's tropical forests are being destroyed by human activity.
The Ozone Hole. Industrial chemicals called CFCs are destroying the atmosphere's ozone layer, removing an essential shield from the sun's UV radiation.
Carbon Dioxide and Global Warming. The world's industrialization has led to a marked increase in the atmosphere's level of carbon dioxide, with resulting warming of climates.

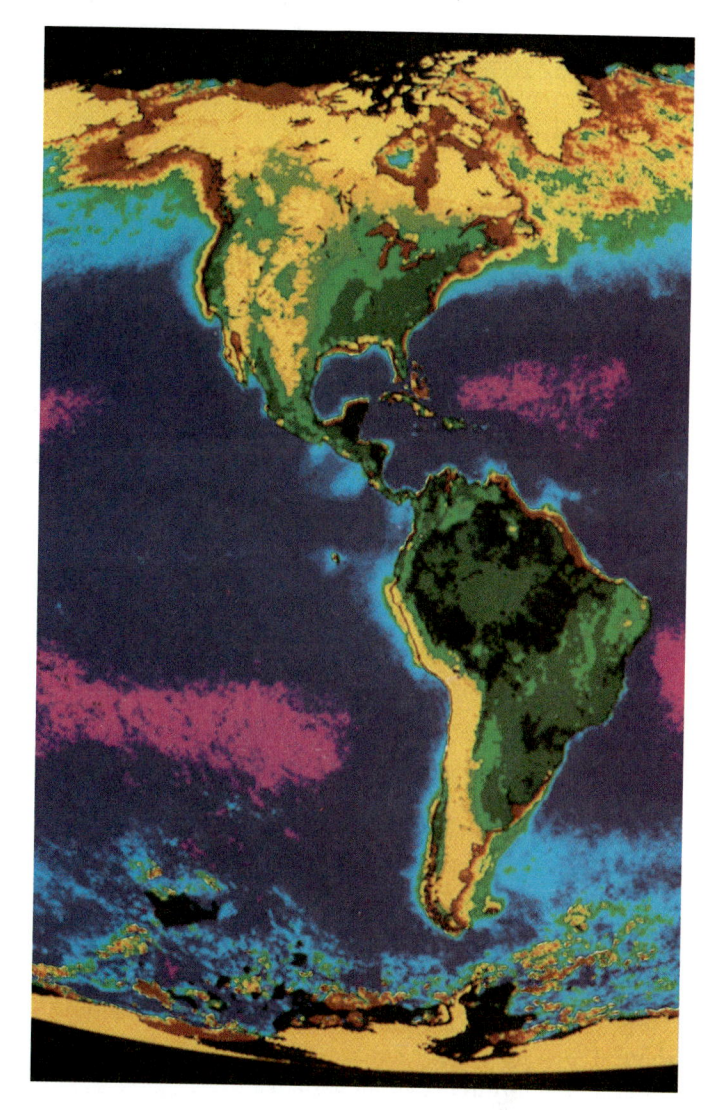

FIGURE 56.1
Life in the biosphere. In this satellite image, yellow zones are largely arid and have relatively little green plant cover, whereas the opposite is true of green zones. Almost every environment on earth can be described in terms of temperature and moisture. These physical parameters have great bearing on the forms of life that are able to inhabit a particular region.

The biosphere includes all living communities on earth, from the profusion of life in the tropical rain forests to the photosynthetic phytoplankton in the world's oceans. In a very general sense, the distribution of life on earth reflects variations in the world's environments, principally in temperature and the availability of water. Figure 56.1 is a satellite image of North and South America, collected over eight years. The colors are keyed to the relative abundance of chlorophyll, a good indicator of rich biological communities. Phytoplankton and algae produce the dark red zones in the oceans and along the seacoasts. Green and dark green areas on land are dense forests, while yellow areas include the deserts of western South America and the tundra of the far north, which have much lower productivity.

56.1 Climate shapes the character of ecosystems.

The distribution of biomes (see section 56.2) results from the interaction of the features of the earth itself, such as different soil types or the occurrence of mountains and valleys, with two key physical factors: (1) the amount of solar heat that reaches different parts of the earth and seasonal variations in that heat; and (2) global atmospheric circulation and the resulting patterns of oceanic circulation. Together, these factors dictate local climate, and so determine the amounts and distribution of precipitation.

Effects of the Sun and Atmospheric Circulation

The earth receives an enormous quantity of heat from the sun in the form of shortwave radiation, and it radiates an equal amount of heat back to space in the form of longwave radiation. About 10^{24} calories arrive at the upper surface of the earth's atmosphere each year, or about 1.94 calories per square centimeter per minute. About half of this energy reaches the earth's surface. The wavelengths that reach the earth's surface are not identical to those that reach the outer atmosphere. Most of the ultraviolet radiation is absorbed by the oxygen (O_2) and ozone (O_3) in the atmosphere. As we will see in section 56.4, the depletion of the ozone layer, apparently as a result of human activities, poses serious ecological problems.

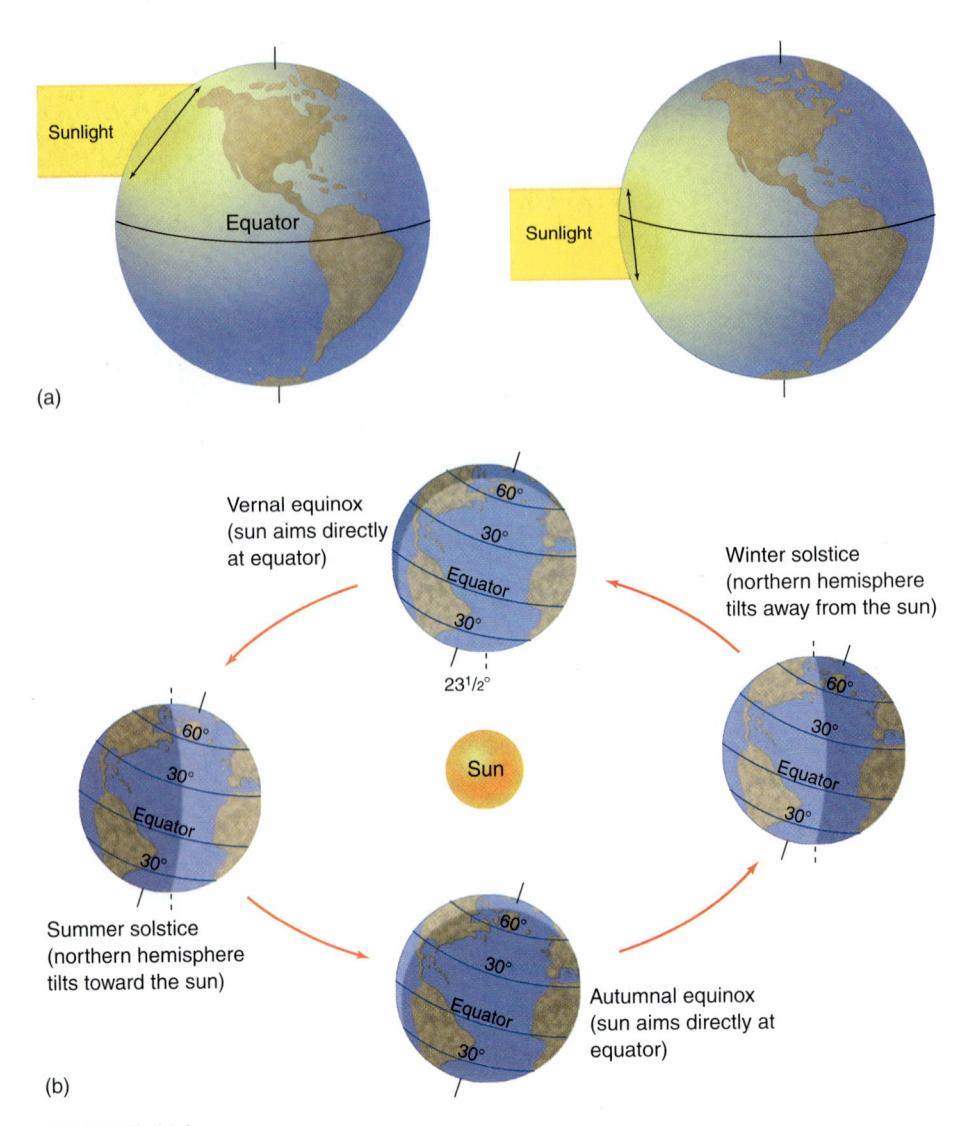

FIGURE 56.2

Relationships between the earth and the sun are critical in determining the nature and distribution of life on earth. (*a*) A beam of solar energy striking the earth in the middle latitudes spreads over a wider area of the earth's surface than a similar beam striking the earth near the equator. (*b*) The rotation of the earth around the sun has a profound effect on climate. In the northern and southern hemispheres, temperatures change in an annual cycle because the earth tilts slightly on its axis in relation to the path around the sun.

How the Sun Affects Climate

The world contains a great diversity of biomes because its climate varies so much from place to place. On a given day, Miami, Florida, and Bangor, Maine, often have very different weather. There is no mystery about this. Because the earth is a sphere, some parts of it receive more energy from the sun than others. This variation is responsible for many of the major climatic differences that occur over the earth's surface, and indirectly, for much of the diversity of biomes. The tropics are warmer than temperate regions because the sun's rays arrive almost perpendicular to regions near the equator. Near the poles, the angle of incidence of the sun's rays spreads them out over a much greater area, providing less energy per unit area (figure 56.2*a*).

1204   Part VIII   Ecology and Behavior

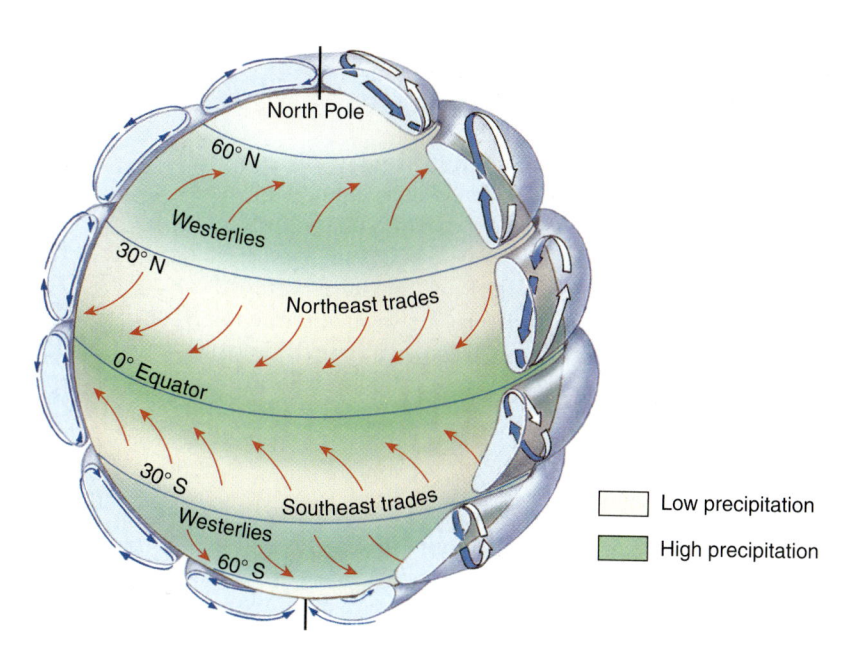

FIGURE 56.3
General patterns of atmospheric circulation. The pattern of air movement toward and away from the earth's surface is shown. Rising air that is cooled creates a band of high precipitation at the equator and at 60°N and 60°S. Reheated, falling air can hold more moisture and creates a band of low precipitation at 30°N and 30°S and near the poles. As air masses change latitudes, they are also deflected by the earth's rotational velocity, producing the trade winds and the westerlies.

The earth's annual orbit around the sun and its daily rotation on its own axis are also important in determining world climates (figure 56.2b). Because of the annual cycle, and the inclination of the earth's axis at approximately 23.5° from its plane of revolution around the sun, a progression of seasons occurs in all parts of the earth away from the equator. One pole or the other is tilted closer to the sun at all times, except during the spring and autumn equinoxes.

Major Atmospheric Circulation Patterns

The moisture-holding capacity of air increases when it warms and decreases when it cools. High temperatures near the equator encourage evaporation and create warm, moist air. As this air rises, it is replaced by air that is drawn toward the equator from the north and south. In turn, the movement of this air pulls the heated tropical air away from the equator. As it moves away from the equator, it cools and loses most of its moisture (figure 56.3). Consequently, the greatest amounts of precipitation on earth fall near the equator, which is where earth's tropical rain forests are found. When the air masses that have risen reach about 30° north and south latitude, the dry air, now cooler, sinks and becomes reheated. As the air reheats, its evaporative capacity increases, creating a zone of decreased precipitation and hot deserts. Some of this air mass, still warmer than in the polar regions, continues to flow toward the poles. It rises again at about 60° north and south latitude, producing another zone of high precipitation occupied by relatively wet and cool coniferous forests. At this latitude, there is another low-pressure area, the polar front. Some of this rising air flows back to the equator, and some continues north and south, descending near the poles and producing another zone of low precipitation before it returns to the equator.

Air Currents Generated by the Earth's Rotation

The earth's rotational velocity is greatest at the equator and decreases toward the poles. Air masses move at the same velocity as the earth does at the same latitude. As an air mass moves toward the equator, its velocity is less than that of the underlying earth. As a result, the air mass is deflected to the northwest south of the equator and to the southwest north of the equator. Conversely, air moving away from the equator meets a slower rotational spin on the earth, and thus is deflected in the opposite direction, to the northeast north of the equator and to the southeast south of the equator.

These deflected air masses between 30° and the equator form the trade winds, which blow all year long and are the steadiest winds found anywhere on earth. They are stronger in winter and weaker in summer. Between 30° and 60° north and south latitude, strong prevailing westerlies blow from west to east and dominate climatic patterns in these latitudes, particularly along the western edges of the continents. Weaker winds, blowing from east to west, occur farther north and south in their respective hemispheres.

> Warm air rises near the equator, descends and produces arid zones at about 30° north and south latitude, flows toward the poles, then rises again at about 60° north and south latitude, and moves back toward the equator. Part of this air, however, moves toward the poles, where it produces zones of low precipitation.

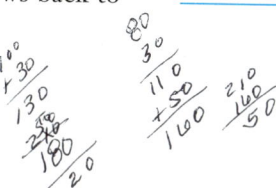

Chapter 56 The Biosphere **1205**

Atmospheric Circulation, Precipitation, and Climate

As we have discussed, precipitation is generally low near 30° north and south latitude, where air is falling and warming, and relatively high near 60° north and south latitude, where it is rising and cooling. Partly as a result of these factors, all the great deserts of the world lie near 30° north or south latitude. Some major deserts are formed in the interiors of large continents. These areas have limited precipitation because of their distance from the sea, the ultimate source of most moisture.

Rain Shadows

Other deserts occur because mountain ranges intercept moisture-laden winds from the sea. When this occurs, the air rises and the moisture-holding capacity of the air decreases, resulting in increased precipitation on the windward side of the mountains—the side from which the wind is blowing. As the air descends the other side of the mountains, the leeward side, it is warmed, and its moisture-holding capacity increases, tending to block precipitation. In California, for example, the eastern sides of the Sierra Nevada Mountains are much drier than the western sides, and the vegetation is often very different. This phenomenon is called the **rain shadow effect** (figure 56.4).

Regional Climates

Four relatively small areas, each located on a different continent, share a climate that resembles that of the Mediterranean region. So-called Mediterranean climates are found in portions of southern California and Oregon; in central Chile; in southwestern Australia; and in the Cape region of South Africa. In all of these areas, the prevailing westerlies blow during the summer from a cool ocean onto warm land. As a result, the air's moisture-holding capacity increases, the air absorbing moisture and creating hot, rainless summers. Such climates are unusual on a world scale. In the regions where they occur, many unique kinds of plants and animals, often local in distribution, have evolved. Because of the prevailing westerlies, the great deserts of the world (other than those in the interiors of continents) and the areas of Mediterranean climate lie on the western sides of the continents.

Another kind of regional climate occurs in southern Asia. The monsoon climatic conditions characteristic of India and southern Asia appear during the summer months. During the winter, the trade winds blow from the east-northeast off the cool land onto the warm sea. From June to October, though, when the land is heated, the direction of the air flow reverses, and the winds veer around to blow

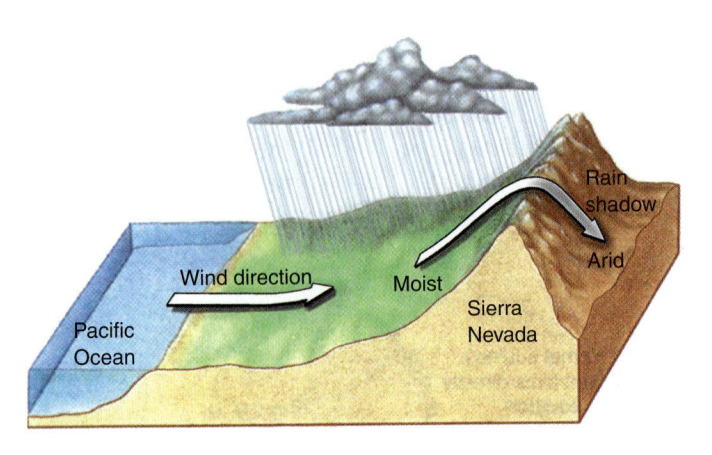

FIGURE 56.4
The rain shadow effect. Moisture-laden winds from the Pacific Ocean rise and are cooled when they encounter the Sierra Nevada Mountains. As their moisture-holding capacity decreases, precipitation occurs, making the middle elevation of the range one of the snowiest regions on earth; it supports tall forests, including those that contain the famous giant sequoias (*Sequoiadendron giganteum*). As the air descends on the east side of the range, its moisture-holding capacity increases again, and the air picks up moisture from its surroundings. As a result, desert conditions prevail on the east side of the mountains.

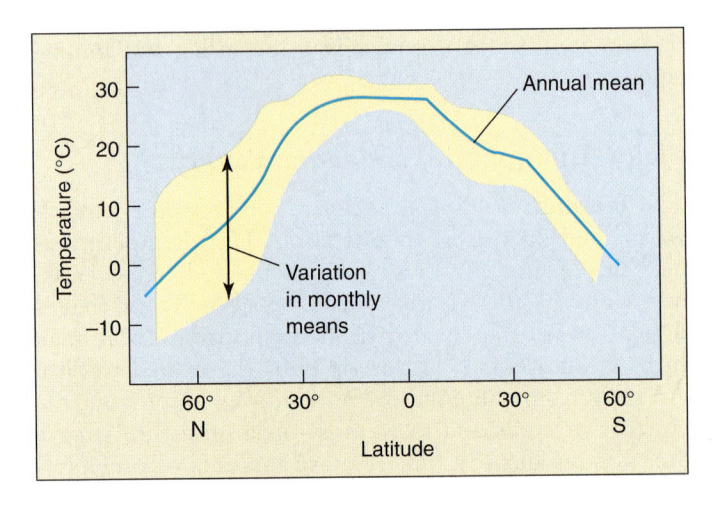

FIGURE 56.5
Temperature varies with latitude. The blue line represents the annual mean temperature at latitudes from the North Pole to Antarctica.
Why is it hotter at low latitudes?

onto the Indian subcontinent and adjacent areas from the southwest, bringing rain. The duration and strength of the monsoon winds spell the difference between food sufficiency and starvation for hundreds of millions of people in this region each year.

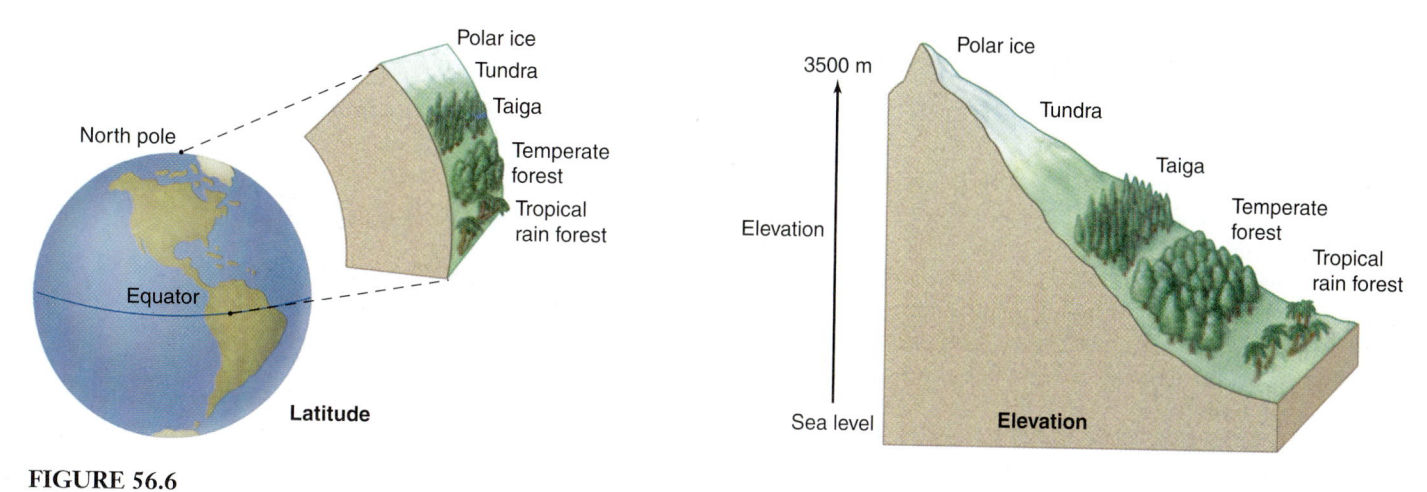

FIGURE 56.6
Elevation affects the distribution of biomes much as latitude does. Biomes that normally occur far north and far south of the equator at sea level also occur in the tropics at high mountain elevations. Thus, on a tall mountain in the tropics, one might see a sequence of biomes like the one illustrated at right.

Seasonal changes in wind circulation also produce changes in ocean currents, sometimes causing nutrient-rich cold water to well up from ocean depths. This produces "blooms" among the plankton and other organisms living near the surface. Similar turnover occurs seasonally in freshwater lakes and ponds, bringing nutrients from the bottom to the surface in the fall and again in the spring.

Latitude

As noted earlier, temperatures are higher in tropical ecosystems for a simple reason: More sunlight per unit area falls on tropical latitudes. Solar radiation is most intense when the sun is directly overhead, and this only occurs in the tropics, where sunlight strikes the equator perpendicularly (see figure 56.2*a*). As figure 56.5 shows, the highest mean global temperatures occur near the equator (that is, 0 latitude). Because the tropics lie near the equator, there is little variation in mean monthly temperature in tropical ecosystems. As you move from the equator into temperate latitudes, sunlight strikes the earth at a more oblique angle, so that less falls on a given area. As a result, mean temperatures are lower. At temperate latitudes, temperature variation increases because of the increasingly marked seasons.

Elevation

Temperature also varies with elevation, with higher altitudes becoming progressively colder. At any given latitude, air temperature falls about 6°C for every 1000-meter increase in elevation. The ecological consequences of temperature varying with elevation are the same as those for temperature varying with latitude (figure 56.6). Thus, in North America, a 1000-meter increase in elevation results in a temperature drop equal to that of an 880-kilometer increase in latitude. This is one reason "timberline" (the elevation above which trees do not grow) occurs at progressively lower elevations the further the area is from the equator.

Microclimate

Climate also varies on a very fine scale within ecosystems. Within the litter on a forest floor, there is considerable variation in shading, local temperatures, and rates of evaporation from the soil. Called **microclimate,** these very localized climatic conditions can be very different from those of the overhead atmosphere. Gardeners spread straw over newly seeded lawns to create such a moisture-retaining microclimate.

Animals are able to take advantage of these microclimates to meet their needs. For example, many ectothermic animals, such as butterflies and lizards, shuttle between warm and cool microclimates to precisely regulate their body temperature. In some cool, mountainous areas, lizards are able to maintain their body temperature 20°C above air temperature by basking in spots that are bathed by the sun but sheltered from the wind.

The great deserts and associated arid areas of the world mostly lie along the western sides of continents at about 30° north and south latitude. Mountain ranges tend to intercept rain, creating deserts in their shadow. In general, temperatures are warmer in the tropics and at lower elevations.

Chapter 56   The Biosphere   1207

56.2 Biomes are widespread terrestrial ecosystems.

The Major Biomes

Biomes are major communities of organisms that have a characteristic appearance and are distributed over a wide land area defined largely by regional variations in climate. As you might imagine from such a broad definition, there are many ways to classify biomes, and different ecologists may assign the same community to different biomes. Few disagree, however, about the reality of biomes as major biological communities.

Terrestrial biomes are defined by their characteristic vegetational structure, which in turn, is affected by the climate of a region. Because areas within similar climates impose similar selective conditions, the plants that occur in these regions tend to evolve similar adaptive responses, even if they are not closely related. This is the phenomenon of convergent evolution discussed in chapter 22 and is the reason that deserts around the world are dominated by cactuslike plants, whereas tropical rain forests are composed of tall, lushly vegetated trees.

Distribution of the Major Biomes

Eight major biome categories are presented in this text: tropical rain forest, savanna, desert, temperate grassland, temperate deciduous forest, temperate evergreen forest, taiga, and tundra. These biomes occur worldwide, occupying large regions that can be defined by rainfall and temperature.

Six additional biomes are considered by some ecologists to be subsets of the eight major ones: polar ice, mountain zone, chaparral, warm moist evergreen forest, tropical monsoon forest, and semidesert. They vary remarkably from one another because they have evolved in regions with very different climates. Distributions of these 14 biomes are mapped in figure 56.7

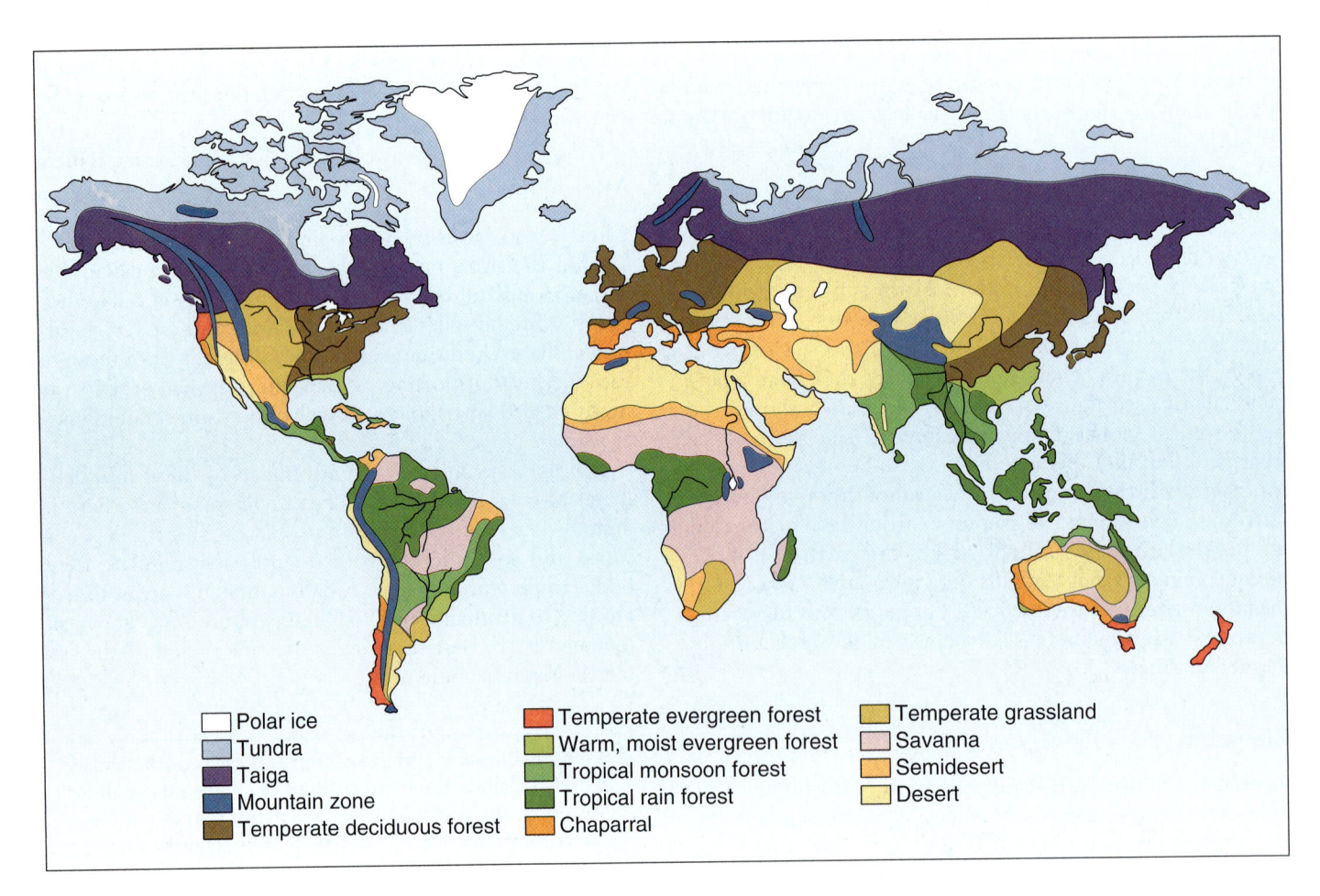

FIGURE 56.7
The distribution of biomes. Each biome is similar in vegetational structure and appearance wherever it occurs on earth.

Part VIII   Ecology and Behavior

Biomes and Climate

Many different environmental factors play a role in determining which biomes are found where. Two key parameters are moisture and temperature, both strong influences on ecosystem productivity. Figure 56.8 presents data on primary productivity as a function of annual precipitation and annual mean temperature. However, different places with the same annual precipitation and temperature sometimes support different biomes, so other factors must also be important. These factors include soil structure and its mineral composition (discussed in detail in chapter 38) and seasonal versus constant climate. Nevertheless, precipitation and temperature do a fine job of predicting which biomes will occur in most places, as figure 56.9 illustrates.

If there were no mountains and no climatic effects caused by the irregular outlines of continents and by different sea temperatures, each biome would form an even belt around the globe, defined largely by latitude. But in truth, these other factors also greatly affect the distribution of biomes. Distance from the ocean has a major impact on rainfall, and elevation affects temperature—for example, the summits of the Rocky Mountains are covered with a vegetation type that resembles the tundra, which normally occurs at a much higher latitude.

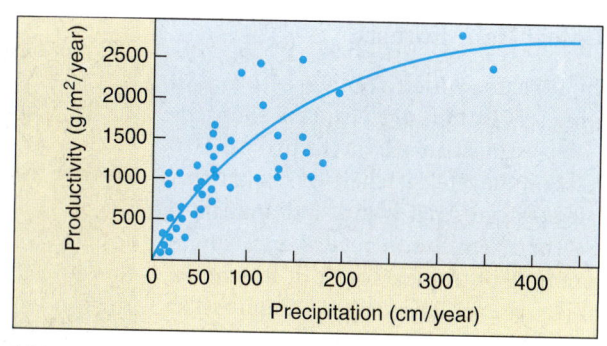

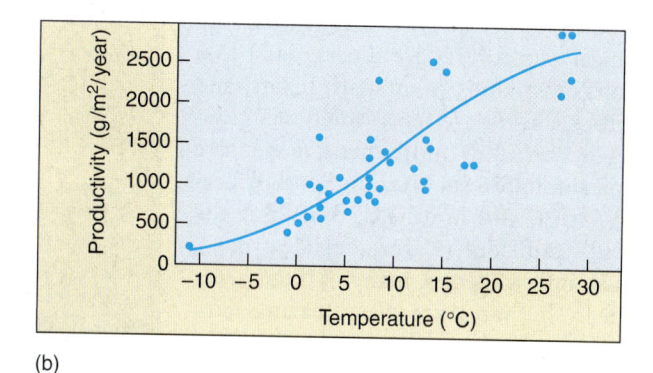

FIGURE 56.8

The effects of precipitation and temperature on primary productivity. The net primary productivity of ecosystems at 52 locations around the globe depends significantly upon (*a*) mean annual precipitation and (*b*) mean annual temperature.

Why does productivity increase with increasing precipitation and temperature?

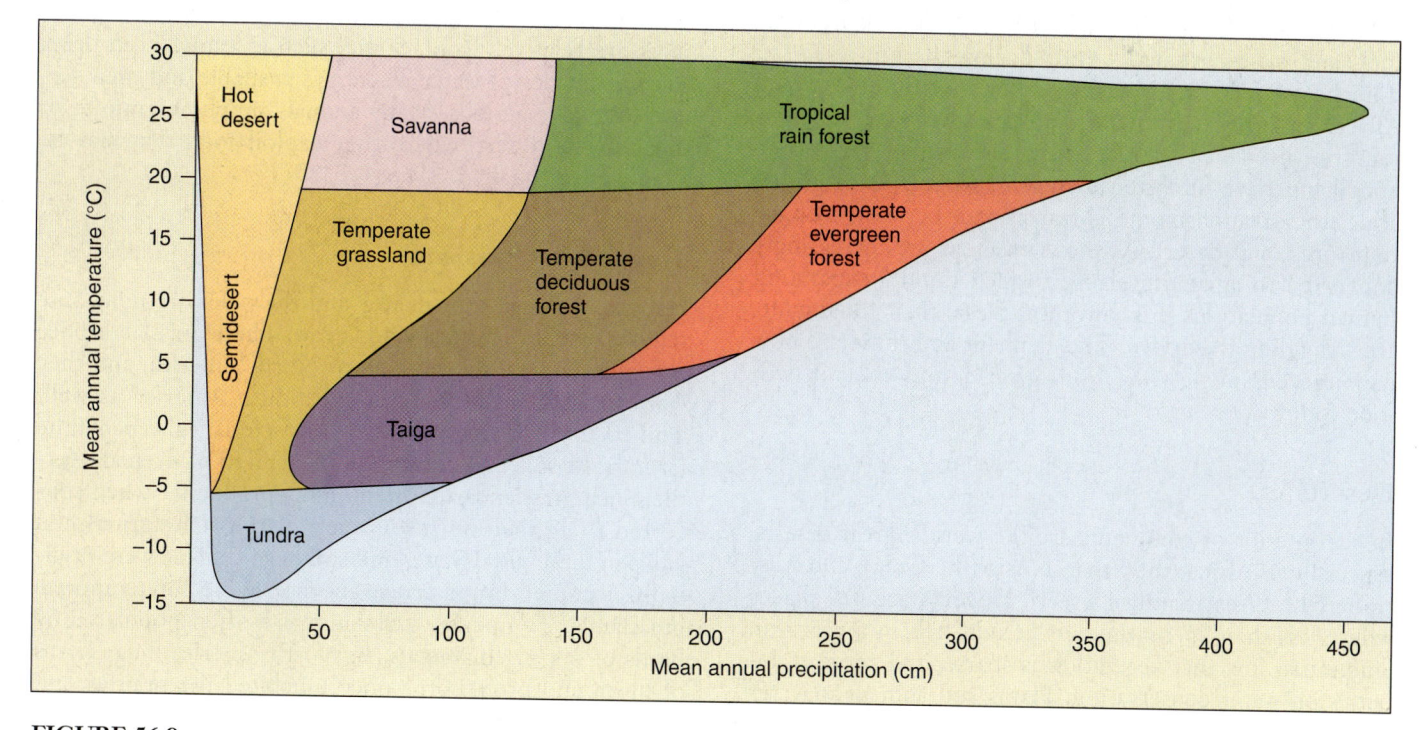

FIGURE 56.9

Predictors of biome distribution. Temperature and precipitation are excellent predictors of biome distribution. At mean annual precipitations between 50 and 150 cm, other factors—such as seasonal drought, fire, and grazing—also have a major influence.

Chapter 56 The Biosphere 1209

Tropical Rain Forests

Rain forests, which receive 140 to 450 centimeters of rain per year, are the richest ecosystems on earth (figure 56.10). They contain at least half of the earth's species of terrestrial plants and animals—more than 2 million species! In a single square mile of tropical forest in Rondonia, Brazil, there are 1200 species of butterflies—twice the total number found in the United States and Canada combined. The communities that make up tropical rain forests are diverse, and as a result, most kinds of animals, plants, and microorganisms are represented in a given area by very few individuals. Extensive tropical rain forests occur in South America, Africa, and Southeast Asia. But the world's rain forests are being destroyed, and countless species, many of them never seen by humans, are disappearing with them. A quarter of the world's species will disappear with the rain forests during the lifetime of many of us.

FIGURE 56.10
Tropical rain forest.

Savannas

In the dry climates that border the tropics are the world's great grasslands, called **savannas.** Savanna landscapes are open, often with widely spaced trees, and rainfall (75 to 125 centimeters annually) is seasonal. Many of the animals and plants are active only during the rainy season. The huge herds of grazing animals that inhabit the African savanna are familiar to all of us. Such animal communities lived in North America during the Pleistocene epoch but have persisted mainly in Africa. On a global scale, the savanna biome is transitional between tropical rain forest and desert. As these savannas are increasingly converted to agricultural use to feed rapidly expanding human populations in subtropical areas, their inhabitants are struggling to survive. The elephant and rhino are now endangered species; the lion, giraffe, and cheetah will soon follow.

Deserts

In the interior of continents are the world's great deserts, especially in Africa (the Sahara), Asia (the Gobi), and Australia (the Great Sandy Desert). **Deserts** are dry places where less than 25 centimeters of rain falls in a year—an amount so low that vegetation is sparse, and survival depends on water conservation. Plants and animals may restrict their activity to favorable times of the year, when water is present. To avoid high temperatures, most desert vertebrates live in deep, cool, and sometimes even somewhat moist burrows. Those that are active over a greater portion of the year emerge only at night, when temperatures are relatively cool. Some, such as camels, can drink large quantities of water when it is available and then survive long, dry periods. Many animals simply migrate to or through the desert, where they exploit food that may be abundant seasonally.

Temperate Grasslands

Halfway between the equator and the poles are temperate regions where rich **grasslands** grow. These grasslands once covered much of the interior of North America, and they were widespread in Eurasia and South America as well. The roots of perennial grasses characteristically penetrate far into the soil, and grassland soils tend to be deep and fertile. Such grasslands are often highly productive when converted to agricultural use. Many of the rich agricultural lands in the United States and southern Canada were originally occupied by **prairies,** another name for temperate grasslands. Temperate grasslands are often populated by herds of grazing mammals. In North America, huge herds of bison and pronghorns once inhabited the prairies, but these herds are now almost all gone.

Temperate Deciduous Forests

Mild climates (warm summers and cool winters) and plentiful rains promote the growth of **deciduous** (hardwood) **forests** in Eurasia, the northeastern United States, and eastern Canada. A deciduous tree is one that drops its leaves in the winter (figure 56.11). Deer, bears, beavers, and raccoons are familiar animals of the temperate regions. Because the temperate deciduous forests represent the remnants of more extensive forests that stretched across North America and Eurasia several million years ago, the remaining areas in eastern Asia and eastern North America share animals and plants that were once more widespread. The deciduous forest in eastern Asia is rich in species because climatic conditions have historically remained constant. Many perennial herbs live in areas of temperate deciduous forest.

FIGURE 56.11
Temperate deciduous forest.

Temperate Evergreen Forests

Temperate evergreen forests occur in regions where winters are cold and there is a strong, seasonal dry period. The pine forests and oak woodlands of the western United States and California are typical temperate evergreen forests. Temperate evergreen forests are characteristic of regions with nutrient-poor soils. Temperate-mixed evergreen forests represent a broad transitional zone between temperate deciduous forests to the south and taiga to the north. Many of these forests are endangered by overlogging, particularly in the western United States.

Taiga

A great ring of northern forests of coniferous trees (spruce, hemlock, and fir) extends across vast areas of Asia and North America. Coniferous trees have leaves like needles that are kept all year long. This ecosystem, called **taiga**, is one of the largest on earth. The winters are long and cold, and most of the limited precipitation falls in the summer. Because the taiga has too short a growing season for farming, few people live there. Many large herbivores, including elk, moose, deer, and such carnivores as wolves, bears, lynx, and wolverines, live in the taiga. Traditionally, fur trapping has been extensive in this region, as has lumber production. Marshes, lakes, and ponds are common, and are often fringed by willows or birches. Most of the trees occur in dense stands of one or a few species.

Tundra

In the far north, above the great coniferous forests and south of the polar ice, few trees grow. The grassland, called **tundra,** is open, windswept, and often boggy. Enormous in extent, this ecosystem covers one-fifth of the earth's land surface. Very little rain or snow falls. When rain does fall during the brief arctic summer, it sits on frozen ground, creating a sea of boggy ground. **Permafrost,** or permanent ice, usually exists within a meter of the surface. Trees are small and are mostly confined to the margins of streams and lakes. As in taiga, herbs of the tundra are perennials that grow rapidly during the brief summers. Large grazing mammals, including musk-oxen, caribou and reindeer, and carnivores, such as wolves, foxes, and lynx, live in the tundra. Lemming populations rise and fall on a long-term cycle, with important effects on the animals that prey on them.

> Major biological communities called biomes can be distinguished in different climatic regions. These communities, which occur in regions of similar climate, are much the same wherever they are found. Variation in annual mean temperature and precipitation are good predictors of which biome will occur where.

56.3 Aquatic ecosystems cover much of the earth.

More than 75% of the world's surface is covered by water. Most of this area is ocean, and the remainder is freshwater area on land. This section considers each in turn.

Patterns of Circulation in the Oceans

Patterns of ocean circulation are determined by the patterns of atmospheric circulation, but they are modified by the locations of landmasses. Ocean circulation is dominated by huge surface gyres (figure 56.12), which move around the subtropical zones of high pressure between approximately 30° north and 30° south latitude. The direction of these gyres is affected by the earth's rotational velocity, in the same way as are the trade winds. Thus, these gyres move clockwise in the northern hemisphere and counterclockwise in the southern hemisphere. The ways they redistribute heat profoundly affect life not only in the oceans but also on coastal lands. For example, the Gulf Stream, in the North Atlantic, swings away from North America near Cape Hatteras, North Carolina, and brings warm water to Europe near the southern British Isles. Because of the Gulf Stream, western Europe is much warmer and more temperate than eastern North America at similar latitudes. As a general principle, western sides of continents in temperate zones of the northern hemisphere are warmer than their eastern sides; the opposite is true of the southern hemisphere.

In South America, the Humboldt Current carries phosphorus-rich cold water northward up the west coast. Phosphorus is brought up from the ocean depths by the upwelling of cool water that occurs as offshore winds blow from the mountainous slopes that border the Pacific Ocean. This nutrient-rich current helps make possible the abundance of marine life that supports the fisheries of Peru and northern Chile. Marine birds, which feed on these organisms, are responsible for the commercially important, phosphorus-rich, guano deposits on the seacoasts of these countries.

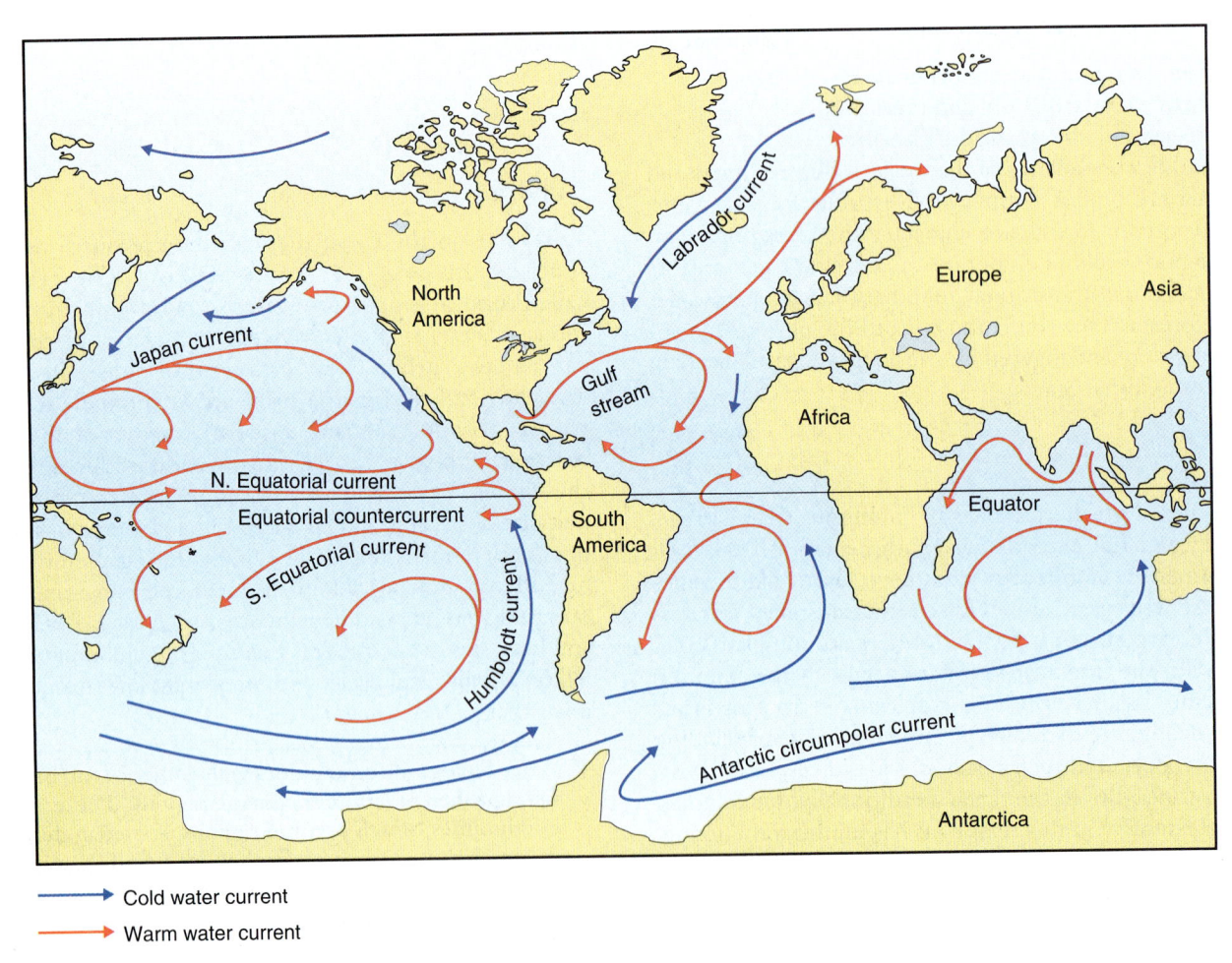

FIGURE 56.12
Ocean circulation. Water moving in great surface spiral patterns called gyres profoundly affects the climate on adjacent land masses.

1212 Part VIII Ecology and Behavior

El Niño and Ocean Ecology

Every Christmas, a tepid current sweeps down the coast of Peru and Ecuador from the tropics, reducing the fish population slightly and giving local fishermen some time off. The fishermen named this Christmas current *El Niño* (literally, "the child," after the Christ Child). Now, though, the term is reserved for a catastrophic version of the same phenomenon, one that occurs every two to seven years and is felt not only locally but globally.

Scientists now have a pretty good idea of what goes on in an El Niño. Normally, the Pacific Ocean is fanned by constantly blowing, east-to-west trade winds that push warm surface water away from the ocean's eastern side (Peru, Ecuador, and Chile) and allow cold water to well up from the depths in its place, carrying nutrients that feed plankton and hence fish. This surface water piles up in the west, around Australia and the Philippines, making that area several degrees warmer and a meter or so higher than the eastern side of the ocean. But if the winds slacken briefly, warm water begins to slosh back across the ocean.

Once this happens, ocean and atmosphere conspire to ensure that it keeps happening. The warmer the eastern ocean gets, the warmer and lighter the air above it becomes, and hence the more similar to the air on the western side. This reduces the difference in pressure across the ocean. Because a pressure difference is what makes winds blow, the easterly trade winds weaken further, letting the warm water continue its eastward advance.

The result is to shift the weather systems of the western Pacific Ocean 6000 km eastward. The tropical rainstorms that usually drench Indonesia and the Philippines occur when warm seawater abutting these islands causes the air above it to rise, cool, and condense its moisture into clouds. When the warm water moves east, so do the clouds, leaving the previously rainy areas in drought. Conversely, the western edge of South America, where coastal waters are usually too cold to trigger much rain, gets a soaking, while the upwelling slows down. During an El Niño, commercial fish stocks virtually disappear from the waters of Peru and northern Chile as the fish move south to find colder water, and plankton drop to one-twentieth of their normal abundance.

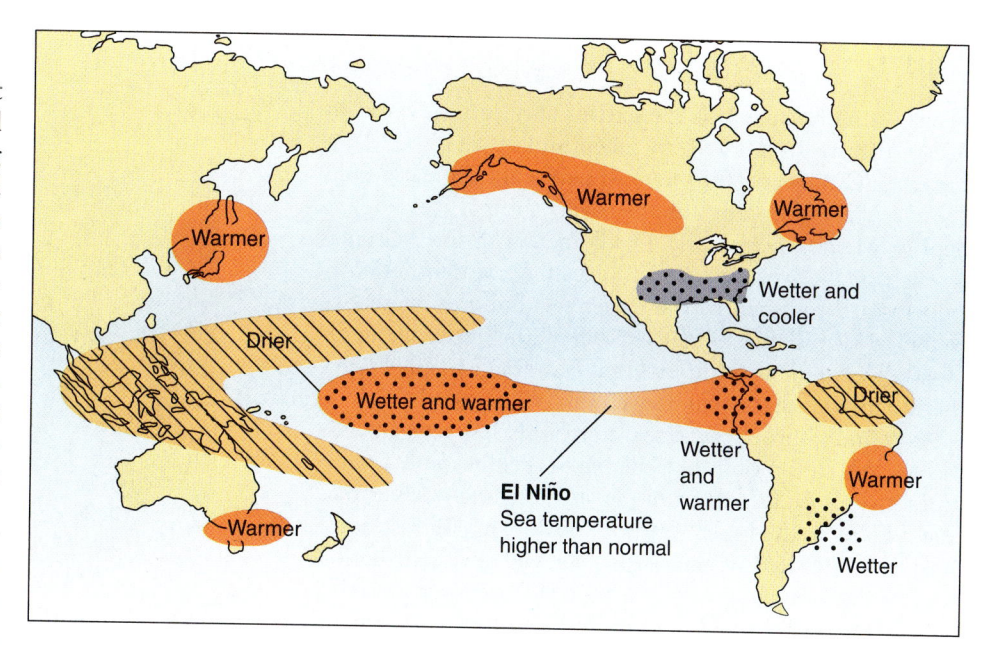

FIGURE 56.13
An El Niño winter. El Niño currents produce unusual weather patterns all over the world as warm waters from the western Pacific move eastward.

That is just the beginning. El Niño's effects are propagated across the world's weather systems (figure 56.13). Violent winter storms lash the coast of California, accompanied by flooding, and El Niño produces colder and wetter winters than normal in Florida and along the Gulf Coast. The American Midwest experiences heavier-than-normal rains, as do Israel and its neighbors.

El Niño can wreak havoc on ecosystems. In the Galápagos Islands, for example, seabird and sea lion populations crash as animals starve due to the lack of fish. By contrast, on land, the heavy rains produce a bumper crop of seeds, and land birds flourish. In Chile, similar effects on seed abundance propagate up the food chain, leading first to increased rodent populations and then to increased predator populations, a nice example of a bottom-up trophic cascade, discussed in chapter 55.

Although the effects of El Niños are now fairly clear, what triggers them remains a mystery. Models of weather patterns suggest that the climatic change that triggers El Niño is "chaotic." Wind and ocean currents return again and again to the same condition, but never in a regular pattern, and small nudges can send them off in many different directions—including an El Niño.

> The world's oceans circulate in huge gyres deflected by continental landmasses. Circulation of ocean water redistributes heat, warming the western side of continents. Disturbances in ocean currents, such as El Niño, can have profound influences on world climate.

Chapter 56 The Biosphere 1213

Marine Ecosystems

Nearly three-quarters of the earth's surface is covered by ocean. Oceans have an average depth of more than 3 kilometers, and they are, for the most part, cold and dark. Heterotrophic organisms inhabit even the greatest ocean depths, which reach nearly 11 kilometers in the Marianas Trench of the western Pacific Ocean. Photosynthetic organisms are confined to the upper few hundred meters of water. Most organisms that live below this level obtain almost all of their food indirectly, as a result of photosynthetic activities that occur above.

As temperature increases, water holds less oxygen. For this reason, the amount of available oxygen becomes an important limiting factor for organisms in some warmer marine regions of the globe. Carbon dioxide, in contrast, is almost never limited in the oceans. Indeed, the distribution of minerals is much more uniform in the ocean than it is on land, where individual soils reflect the composition of the parent rocks from which they have weathered. Nonetheless, this is not true for all minerals, including several—iron, nitrogen, and phosphorus—that affect productivity and are responsible for the differences in plankton density illustrated in figure 56.1 (see also text page 1203).

Some marine habitats, such as coral reefs and estuaries, are remarkably productive (see table 55.1). Overall, approximately 50% of the world's primary production comes from aquatic systems. Because much of this productivity is rapidly grazed and replenished, aquatic systems support about three times more animal production than terrestrial systems. The marine environment consists of three major habitats: (1) the **neritic zone**, the coastal shallow waters; (2) the **pelagic zone**, the area of water above the ocean floor; and (3) the **benthic zone**, the actual ocean floor (figure 56.14). The part of the ocean floor that drops to depths where light does not penetrate is called the **abyssal zone**.

The Neritic Zone

The neritic zone of the ocean is the area less than 300 meters below the surface along the coasts of continents and islands. The zone is small in area, but it is inhabited by large numbers of species (figure 56.15). Intense and sometimes violent interaction between sea and land gives a selective advantage to well-secured organisms that can withstand being washed away by the continual beating of the waves. Part of this zone, the **intertidal (or littoral) region**, is exposed to the air whenever the tides recede.

The world's great fisheries are in shallow waters over continental shelves, either near the continents themselves or in the open ocean, where the seabed is near the surface. Nutrients, derived from land, are much more abundant in coastal and other shallow regions, where upwelling from the depths occurs, than in the open ocean. This accounts for the great productivity of the continental shelf fisheries.

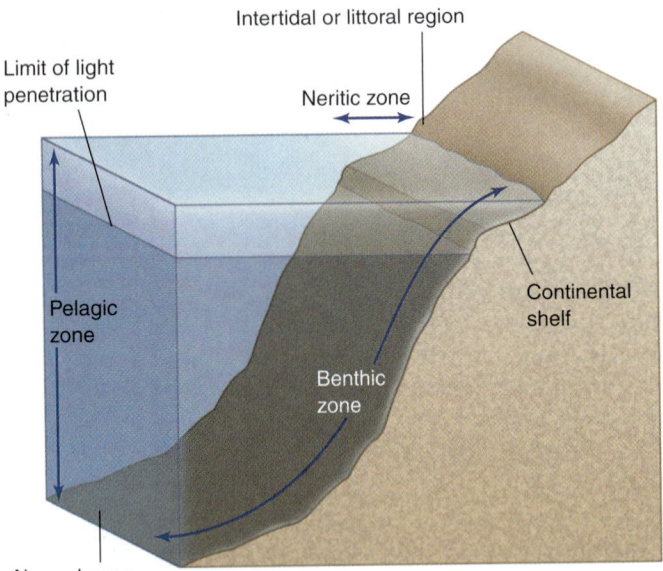

FIGURE 56.14
Marine ecosystems. Ecologists classify marine communities into neritic, pelagic, and benthic zones, according to depth (which affects how much light penetrates) and distance from shore. The abyssal zone is that portion of the benthic zone that lies beyond the limit of light penetration.

FIGURE 56.15
Diversity in coastal regions. Fishes and many other kinds of animals find food and shelter among the coral in coastal waters.

The preservation of these fisheries, a source of high-quality protein exploited throughout the world, has become a growing concern. For example, in Chesapeake Bay, environmental stresses have become so severe that they not only threaten the continued existence of formerly highly productive fisheries, but also diminish the quality of human life in these regions.

The Pelagic Zone

Drifting freely in the upper waters of the pelagic zone, a diverse biological community exists, primarily consisting of

1214   Part VIII   Ecology and Behavior

microscopic organisms called **plankton.** Fish and other larger organisms that swim in these waters constitute the *nekton*, whose members feed on plankton and one another. Some members of the plankton, including protists and some bacteria, are photosynthetic. Collectively, these organisms account for about 40% of all photosynthesis that takes place on earth. Most plankton live in the top 100 meters of the sea, the zone into which light from the surface penetrates freely. Perhaps half of the total photosynthesis in this zone is carried out by organisms less than 10 micrometers in diameter—at the lower limits of size for organisms. These include cyanobacteria and algae, organisms so small that their abundance and ecological importance have been unappreciated until relatively recently.

Many heterotrophic protists and animals live in the plankton and feed directly on photosynthetic organisms and on one another. Gelatinous animals, especially jellyfish and ctenophores, are abundant in the plankton. The largest animals that have ever existed on earth, baleen whales, graze on plankton and nekton, as do a number of other organisms, such as fishes and crustaceans.

Populations of organisms that make up plankton can increase rapidly, and the turnover of nutrients in the sea is great, although the productivity in these systems is quite low. Because nitrogen and phosphorus are often present in only small amounts and organisms may be relatively scarce, the productivity that does occur is more a result of rapid use and recycling rather than an abundance of these nutrients.

The Benthic Zone

The seafloor at depths below 1000 meters has about twice the area of all the land on earth. The seafloor itself is a thick blanket of mud, consisting of fine particles that have settled from the overlying water and accumulated over millions of years. Because of high pressures (an additional atmosphere of pressure for every 10 meters of depth), cold temperatures (2° to 3°C), darkness, and lack of food, biologists originally thought that nothing could live on the seafloor. In fact, recent work has shown that the number of species living at great depth is quite high. Rough estimates of deep-sea diversity have soared to millions of species. Many appear endemic (local). The diversity of species is so high it may rival that of tropical rain forests. Most of these animals are only a few millimeters in size, although larger ones also occur in these regions. Some of the larger ones are bioluminescent (figure 56.16a) and thus are able to communicate with one another or attract their prey.

FIGURE 56.16
Life in the benthic zone. (*a*) The luminous spot below the eye of this deep-sea fish results from the presence of a symbiotic colony of luminous bacteria. Similar luminous signals are a common feature of mobile deep-sea animals. (*b*) These giant beardworms live along vents where water jets from fissures at 350°C and then cools to the 2°C of the surrounding water.

Animals on the sea bottom depend on the meager leftovers from organisms living kilometers overhead. The low densities and small size of most deep-sea animals is in part a consequence of this limited food supply. In 1977, oceanographers diving in a research submarine were surprised to find dense clusters of large animals living on geothermal energy at a depth of 2500 meters. These deep-sea oases occur where seawater circulates through porous rock at sites where molten material from beneath the earth's crust comes close to the rocky surface. A series of these areas occur on the Mid-Ocean Ridge, where basalt erupts through the ocean floor.

This water is heated to temperatures in excess of 350°C and, in the process, becomes rich in reduced compounds. These compounds, such as hydrogen sulfide, provide energy for bacterial primary production through chemosynthesis instead of photosynthesis. Mussels, clams, and large red-plumed worms in a phylum unrelated to any shallow-water invertebrates cluster around the vents (figure 56.16*b*). Prokaryotes live symbiotically within the tissues of these animals. The animal supplies a place for the prokaryotes to live and transfers CO_2, H_2S, and O_2 to them for their growth; the bacteria supply the animal with organic compounds to use as food. Polychaete worms (see chapter 33), anemones, and limpets live on free-living chemosynthetic bacteria. Crabs act as scavengers and predators, and some of the fish are also predators. This is one of the few ecosystems on earth that does not depend on the sun's energy.

> **About 40% of the world's photosynthetic productivity is estimated to occur in the oceans. The turnover of nutrients in the plankton is much more rapid than in most other ecosystems, and the total amounts of nutrients are very low.**

Chapter 56 The Biosphere 1215

Freshwater Habitats

Freshwater habitats are distinct from both marine and terrestrial ones, but they are limited in area. Inland lakes cover about 1.8% of the earth's surface, and running water (streams and rivers) covers about 0.3%. All freshwater habitats are strongly connected with terrestrial ones; marshes and swamps constitute intermediate habitats. In addition, a large amount of organic and inorganic material continuously enters bodies of fresh water from communities growing on the land nearby (figure 56.17). Many kinds of organisms are restricted to freshwater habitats. When organisms live in rivers and streams, they must be able to swim against the current or attach themselves in such a way as to resist the effects of the current and avoid being swept away.

Ponds and Lakes

Small bodies of fresh water are called ponds, and larger ones lakes. Because water absorbs light passing through it at wavelengths critical to photosynthesis (every meter absorbs 40% of the red and about 2% of the blue), the distribution of photosynthetic organisms is limited to the upper **photic zone**; only heterotrophic organisms occur in the lower **disphotic** and **aphotic zones**, where very little or no light penetrates.

Like the ocean, ponds and lakes have three zones where organisms occur, distributed according to the depth of the water and its distance from shore (figure 56.18). The *littoral zone* is the shallow area along the shore. The *limnetic zone* is the well-illuminated surface water away from the shore, inhabited by plankton and other organisms that live in open water. The *profundal zone* is the area below the limits where light can effectively penetrate.

Thermal Stratification

Thermal stratification is characteristic of larger lakes in temperate regions (figure 56.19). In summer, warmer water forms a layer at the surface known as the *epilimnion*. Cooler water, called the *hypolimnion* (about 4°C), lies below. An abrupt change in temperature, the thermocline, separates these two layers. Depending on the climate of the particular area, the epilimnion may become as much as 20 meters thick during the summer.

In autumn, the temperature of the epilimnion drops until it is the same as that of the hypolimnion, 4°C. When this occurs, epilimnion and hypolimnion mix—a process called fall overturn. Because water is densest at about 4°C, further cooling of the water as winter progresses creates a layer of cooler, lighter water, which freezes to form a layer of ice at the surface. Below the ice, the water temperature remains between 0° and 4°C, and plants and animals can survive. In spring, the ice melts, and the surface water warms up. When it warms back to 4°C, it again mixes with

FIGURE 56.17
A nutrient-rich stream. Much organic material falls or seeps into streams from communities along the edges. This input increases the stream's biological productivity.

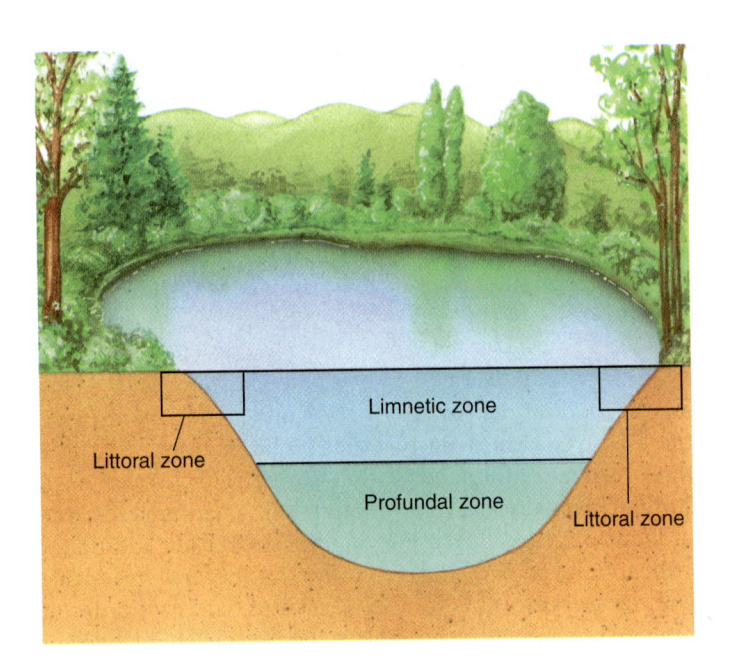

FIGURE 56.18
The three zones in ponds and lakes. A shallow "edge" (littoral) zone lines the periphery of the lake, where attached algae and their insect herbivores live. An open-water surface (limnetic) zone lies across the entire lake and is inhabited by floating algae, plankton, and fish. A dark, deep-water (profundal) zone overlies the sediments at the bottom of the lake and contains numerous prokaryotes and wormlike organisms that consume dead debris settling at the bottom of the lake.

the water below. This process is known as spring overturn. When lake waters mix in the spring and fall, nutrients formerly held in the depths of the lake are returned to the surface, and oxygen from surface waters is carried to the depths. Without such mixing, survival and growth would be impossible for many organisms at all levels of the ecosystem.

Productivity of Freshwater Ecosystems

Some aquatic communities, such as fast-moving streams, are not highly productive. Because the moving water washes away plankton, the photosynthesis that supports the community is limited to algae attached to the surface and to rooted plants.

Productivity of Lakes. Lakes can be divided into two categories based on their production of organic matter. **Eutrophic lakes** contain an abundant supply of minerals and organic matter. As the plentiful organic material drifts below the thermocline from the well-illuminated surface waters of the lake, it provides a source of energy for other organisms. Most of these are oxygen-requiring organisms that can easily deplete the oxygen supply below the thermocline during the summer months. The oxygen supply of the deeper waters cannot be replenished until the layers mix in the fall. This lack of oxygen in the deeper waters of some lakes may have profound effects, as when it allows relatively harmless materials such as sulfates and nitrates to convert into toxic materials such as hydrogen sulfide and ammonia.

In **oligotrophic lakes,** organic matter and nutrients are relatively scarce. Such lakes are often deeper than eutrophic lakes and have very clear, blue water. Their hypolimnetic water is always rich in oxygen.

Human activities can transform oligotrophic lakes into eutrophic ones. In many lakes, phosphorus is in short supply and is the nutrient that limits growth. When excess phosphorus from sources such as fertilizer runoff, sewage, and detergents enters lakes, it can quickly lead to harmful effects. In many cases, this creates perfect conditions for the growth of blue-green algae, which proliferate immensely. Soon, larger plants are outcompeted and disappear, along with the animals that live on them. In addition, as these phytoplankton die and decompose, oxygen in the water is used up, killing the natural fish and invertebrate populations. This situation can be remedied if the continual input of phosphorus is diminished. Given time, lakes can recover and return to prepollution states.

Productivity of Wetlands. Swamps, marshes, bogs, and other **wetlands** covered with water support a wide variety of water-tolerant plants, called hydrophytes ("water plants"), and a rich diversity of invertebrates, birds, and other animals. Wetlands are among the most productive ecosystems on earth (see table 55.1). They also play a key

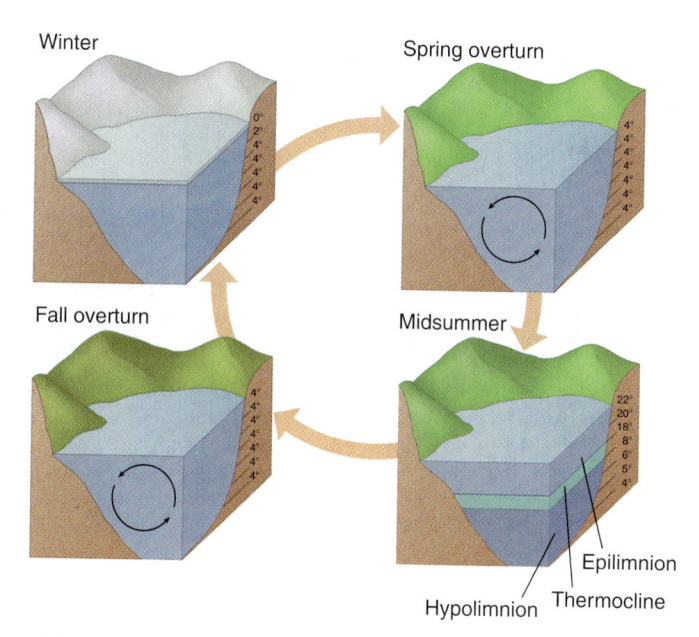

FIGURE 56.19
Stratification in fresh water. The pattern of stratification in a large pond or lake in temperate regions is upset in the spring and fall overturns. Of the three layers of water shown, the hypolimnion consists of the densest water, at 4°C; the epilimnion consists of warmer water that is less dense; and the thermocline is the zone of abrupt change in temperature that lies between them.

ecological role by providing water storage basins that moderate flooding. Many wetlands are being disrupted as humans "develop" what is sometimes perceived as otherwise useless land, but government efforts are now under way to protect the remaining wetlands.

Differences Between Aquatic and Terrestrial Ecosystems

Overall, a number of differences characterize aquatic ecosystems in comparison to terrestrial ecosystems:

1. Almost 100 times more inhabitable space (a conservative limit including only the upper 4000 meters of ocean).
2. A more stable temperature regime.
3. No shortage of water, but light and nutrients can be limiting.
4. Primary producers are generally microscopic, with high rates of turnover, mostly as the result of consumption by grazers.
5. Most aquatic grazers are ectotherms and may have low metabolic costs which allows larger population sizes for a given amount of energy.

The most productive freshwater ecosystems are wetlands. Most lakes are far less productive, limited by lack of nutrients.

Chapter 56 The Biosphere 1217

56.4 Human activity is placing the biosphere under increasing stress.

Pollution

The Rhine is one of the most beautiful rivers on earth. On the first day of November in 1986, the Rhine almost died. The blow that struck the Rhine did not at first seem deadly. That morning, firefighters were battling a blaze in Basel, Switzerland. The fire was gutting a huge warehouse, into which firefighters shot streams of water to dampen the flames. The warehouse belonged to Sandoz, a giant chemical company. In the rush to contain the fire, no one thought to ask what chemicals were stored in the warehouse. By the time the fire was out, streams of water had washed 30 tons of mercury and pesticides into the Rhine.

Flowing downriver, the deadly wall of poison killed everything it passed. For hundreds of kilometers, dead fish blanketed the surface of the river. Many cities that use the water of the Rhine for drinking had little time to make other arrangements. Even the plants in the river began to die. From Switzerland, all across Germany, to the sea, the river reeked of rotting fish, and not one drop of water was safe to drink.

Six months later, Swiss and German environmental scientists monitoring the effects of the accident were able to report that the blow to the Rhine had not been mortal. Enough small aquatic invertebrates and plants had survived to provide a basis for the eventual return of fish and other water life, and the river was rapidly washing out the remaining residues from the spill. As a lesson difficult to ignore, the spill on the Rhine caused the governments of Germany and Switzerland to intensify efforts to protect the river from future industrial accidents and to regulate the establishment and growth of chemical and industrial plants on its shores.

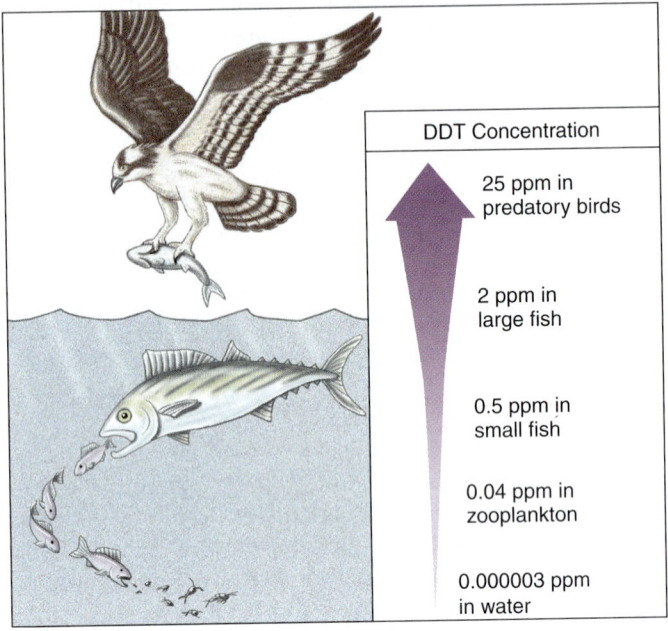

The Threat of Pollution

Stories similar to the one about the pollution of the Rhine can be told countless times in different places in the industrial world, from the Love Canal in New York to the James River in Virginia to the town of Times Beach in Missouri. But not all pollutants that threaten the sustainability of life are as immediately toxic as those that affected the Rhine. Many forms of pollution arise gradually as by-products of industry. For example, the polymers known as plastics, which we produce in abundance, break down slowly in nature. Much is burned or otherwise degraded to smaller vinyl chloride units. Virtually all the plastic that has ever been produced is still with us, in one form or another. Collectively, it constitutes a new form of pollution.

Widespread agriculture, carried out increasingly by modern methods, introduces large amounts of many new kinds of chemicals into the global ecosystem, including pesticides, herbicides, and fertilizers. Industrialized countries such as the United States now attempt to carefully monitor the side effects of these chemicals. Unfortunately, large quantities of many toxic chemicals no longer manufactured still circulate in the ecosystem.

For example, chlorinated hydrocarbons, a class of compounds that includes DDT, chlordane, lindane, and dieldrin, have been banned from normal use in the United States. They are still manufactured in the United States, however, and exported to other countries, where their use continues. Chlorinated hydrocarbon molecules break down slowly and accumulate in animal fat. Furthermore, as they pass through a food chain, they become increasingly concentrated in a process called **biological magnification** (figure 56.20). DDT caused serious problems by leading to the production of thin, fragile eggshells in many predatory bird species in the United States and elsewhere until the late 1960s, when it was banned in time to save the birds from extinction. Chlorinated compounds have other undesirable side effects and exhibit hormonelike activities in the bodies of animals, disrupting normal hormonal cycles, with sometimes potentially serious consequences.

FIGURE 56.20
Biological magnification of DDT. Because DDT accumulates in animal fat, the compound becomes increasingly concentrated in higher levels of the food chain. Before DDT was banned in the United States, predatory bird populations drastically declined because DDT made their eggshells thin and fragile enough to break during incubation.

DDT Concentration
- 25 ppm in predatory birds
- 2 ppm in large fish
- 0.5 ppm in small fish
- 0.04 ppm in zooplankton
- 0.000003 ppm in water

> Pollution is causing ecosystems to accumulate many harmful substances as the result of spills and runoff from agricultural or urban use.

1218 Part VIII Ecology and Behavior

Acid Precipitation

The Four Corners power plant in New Mexico burns coal, sending smoke high into the atmosphere through its smokestacks, each over 65 meters tall. The smoke the stacks belch out contains high concentrations of sulfur dioxide and other sulfates, which produce acid when they combine with water vapor in the air. The intent of those who designed the plant was to release the sulfur-rich smoke high enough that winds would disperse and dilute it, carrying the acids far away. They failed to foresee that the environmental effects of this acidity would be serious.

Sulfur introduced into the upper atmosphere combines with water vapor to produce sulfuric acid, and when the water later falls as rain or snow, the precipitation is acidic. Natural rainwater rarely has a pH lower than 5.6; in the northeastern United States, however, rain and snow now have a pH of 4.2 or below, with occasional storms bringing precipitation with pH as low as 3.0 (figure 56.21).

Acid precipitation destroys life. At pH levels below 5.0, many fish species and other aquatic animals die, unable to reproduce. Thousands of lakes in southern Sweden and Norway no longer support fish; these lakes are now eerily clear. In the northeastern United States and eastern Canada, tens of thousands of lakes are also dying biologically as a result of acid precipitation. In southern Sweden and elsewhere, groundwater now has a pH between 4.0 and 6.0, as acid precipitation slowly filters down into the underground reservoirs.

Enormous forest damage has occurred in the Black Forest in Germany and in the forests of the eastern United States and Canada. It has been estimated that at least 3.5 million hectares of forest in the northern hemisphere are being affected by acid precipitation (figure 56.22), and the problem is clearly growing.

The solution seems obvious: capture and remove the emissions instead of releasing them into the atmosphere. However, there are serious difficulties with this plan. First, it is expensive. The costs of installing and maintaining the necessary industrial equipment in the United States are estimated to be $4 to $5 billion per year. An additional difficulty is that the polluter and the recipient of the pollution are far from each other, and neither wants to pay for what they view as someone else's problem. Nonetheless, the Clean Air Act revisions of 1990 significantly addressed this problem in the United States for the first time, and innovative programs provided incentives to reduce emissions. As a result, the costs have been much less than expected, and the acidity of precipitation is decreasing, at least in some parts of the United States.

> Industrial pollutants such as nitric and sulfuric acids, introduced into the upper atmosphere by factory smokestacks, are spread over wide areas by prevailing winds and fall to earth as "acid rain," lowering the pH of water on the ground and killing life.

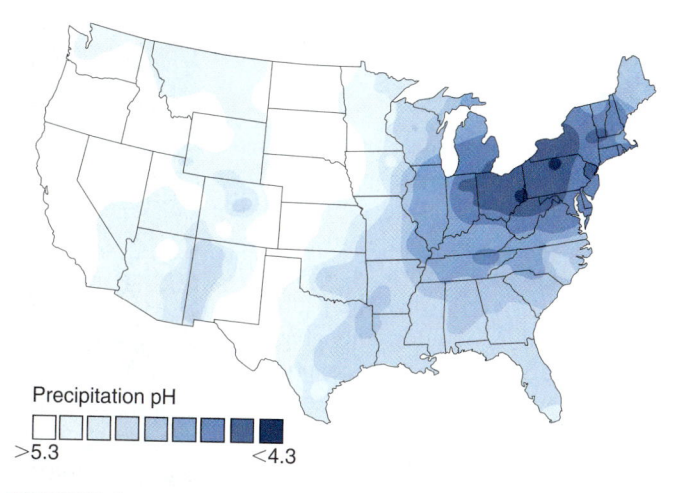

FIGURE 56.21
pH values of rainwater in the United States. Precipitation in the United States, especially in the Northeast, is more acidic than natural rainwater, which has a pH of about 5.6.

FIGURE 56.22
Damage to trees at Clingman's Dome, Tennessee. Acid precipitation weakens trees and makes them more susceptible to pests and predators.

Chapter 56 The Biosphere 1219

Destruction of the Tropical Forests

More than half of the world's human population lives in the tropics, and this percentage is increasing rapidly. For global stability and for the sustainable management of the world ecosystem, it will be necessary to solve the problems of food production and regional stability in these areas. World trade, political and economic stability, and the future of most species of plants, animals, fungi, and microorganisms depend on our addressing these problems.

Loss of Rain Forests

Tropical rain forests are biologically the richest of the world's biomes. Most other kinds of tropical forest, such as seasonally dry forests and savanna forests, have already been largely destroyed, because they tend to grow on more fertile soils and thus were exploited by humans a long time ago. Now the rain forests, which grow on poor soils, are being destroyed. In the mid-1990s, only about 5.5 million square kilometers of tropical rain forest were estimated to still exist in a relatively undisturbed form. This area, about two-thirds the size of the United States (excluding Alaska), represents about half of the original extent of the rain forest. From it, about 160,000 square kilometers are being clear-cut every year, with perhaps an equivalent amount severely disturbed by shifting cultivation, firewood gathering, and the clearing of land for cattle ranching. The total area of tropical rain forest destroyed—and therefore permanently removed from the world total—amounts to an area greater than the size of Indiana each year. At this rate, all of the tropical rain forests in the world would be gone in about 30 years—but in fact, the rate of destruction is even more rapid in many regions. As a result of this overexploitation, experts predict there will be little undisturbed tropical forest left anywhere in the world within a few decades. Many areas now occupied by dense, species-rich forests may still be tree-covered, but the stands will be sparse and species-poor.

Loss of Productivity

Not only does the disappearance of tropical forests lead to a tragic loss of largely unknown biodiversity, but the loss of the forests themselves is ecologically serious.

(a)   (b)

FIGURE 56.23
Destroying the tropical forests. (*a*) When tropical forests are cleared, the ecological consequences can be disastrous. These fires are destroying rain forest in Brazil and clearing it for cattle pasture. (*b*) The consequences of deforestation can be seen on these middle-elevation slopes in Ecuador, which now support only low-grade pastures and permit topsoil to erode into the rivers (note the color of the water, stained brown by high levels of soil erosion). In the 1970s, these areas supported highly productive forest, which protected the watersheds of the area.

Tropical forests are complex, productive ecosystems that function well in the areas where they have evolved. When people cut a forest or open a prairie in the north temperate zone, they provide farmland that we know can be worked for generations. But in most areas of the tropics, people are unable to engage in continuous agriculture. When they clear a tropical forest, they consume natural resources that will never be available again (figure 56.23). The complex ecosystems built up over millions of years are now being dismantled, in almost complete ignorance, by humans.

Biologists must learn more about the construction of sustainable agricultural ecosystems that can meet human needs in tropical and subtropical regions. The ecological concepts we have been reviewing in this chapter and in chapter 55 are universal principles. The undisturbed tropical rain forest has one of the highest rates of net primary productivity of any plant community on earth, and it is therefore imperative to develop ways to harvest it for human purposes in a sustainable, intelligent way.

More than half of the tropical rain forests have been destroyed by human activity, and the rate of loss is accelerating.

The Ozone Hole

The swirling colors of the satellite photo in figure 56.24 represent different concentrations of **ozone** (O_3), a form of oxygen gas that makes up a layer of the stratosphere, where it filters out ultraviolet radiation from the sun. As you can see, over Antarctica there is an "ozone hole" three times the size of the United States, an area where the ozone concentration is much less than elsewhere. The ozone thinning appeared for the first time in 1975, and has reappeared each year since for a few months during the Antarctic winter. Each year, the layer of ozone is thinner and the hole is larger.

The major cause of the ozone depletion had already been suggested in 1974 by Sherwood Roland and Mario Molina, who were awarded the Nobel Prize for their work in 1995. They proposed that chlorofluorocarbons (CFCs), relatively inert chemicals used in cooling systems, fire extinguishers, and Styrofoam containers, were percolating up through the atmosphere and reducing O_3 molecules to O_2. They predicted that one chlorine atom from a CFC molecule could destroy 100,000 ozone molecules in the following mechanism:

UV radiation causes CFCs to release Cl atoms:
$$CCl_3F \xrightarrow{UV} Cl + CCl_2F$$

UV creates oxygen free radicals:
$$O_2 \longrightarrow 2O$$

Cl atoms and O free radicals interact with ozone:
$$2Cl + 2O_3 \longrightarrow 2ClO + 2O_2$$
$$2ClO + 2O \longrightarrow 2Cl + 2O_2$$

Net reaction: $2O_3 \longrightarrow 3O_2$

Although other factors have also been implicated in ozone depletion, the role of CFCs is so predominant that worldwide agreements have been signed to phase out their production. The United States and most other countries banned the production of CFCs and other ozone-destroying chemicals after 1995, and concentrations of ozone-depleting chemicals in the upper atmosphere are decreasing. Nonetheless, the CFCs that were manufactured earlier are moving slowly upward and will remain in the atmosphere. As a result, the ozone layer will not fully recover until the latter half of this century.

Thinning of the ozone layer in the stratosphere, 25 to 40 kilometers above the surface of the earth, is a matter of serious concern. This layer protects life from the harmful ultraviolet rays of the sun that bombard the earth continuously. Life appeared on land only after the oxygen layer was sufficiently thick to generate enough ozone to shield the surface of the earth from these destructive rays.

Ultraviolet radiation is a serious human health concern. Every 1% drop in atmospheric ozone is estimated to lead to a 6% increase in the incidence of skin cancers. At middle latitudes, the approximately 3% ozone drop that has already occurred worldwide is estimated to have increased skin cancer incidence by as much as 20%. A type of skin cancer called melanoma is one of the more lethal human diseases.

Industrial CFCs released into the atmosphere react with ozone, converting it to oxygen gas. This has the effect of destroying the earth's ozone shield and exposing the earth's surface to increased levels of harmful UV radiation.

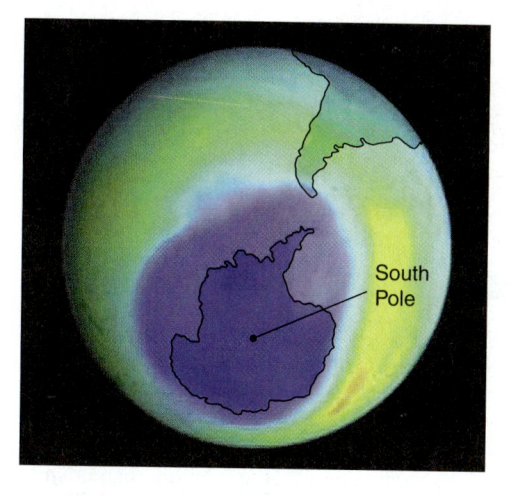

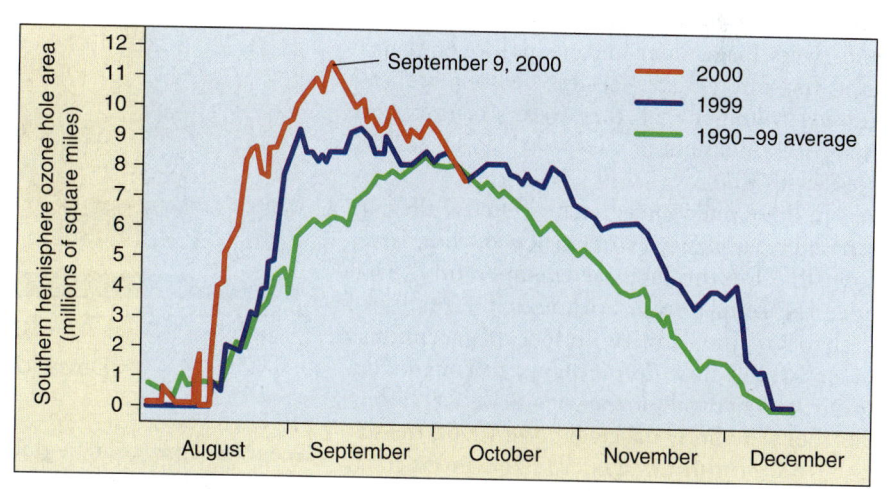

FIGURE 56.24
The growing ozone hole over Antarctica. For decades, NASA satellites have tracked the extent of ozone depletion over Antarctica. Every year since 1975, an ozone "hole" has appeared in August when sunlight triggers chemical reactions in cold air trapped over the South Pole during the Antarctic winter. The hole intensifies during September before tailing off as temperatures rise in November–December. In 2000, the 11.4 million-square-mile hole (*dark purple* in the satellite image) covered an area larger than the United States, Canada, and Mexico combined, the largest hole ever recorded. In September 2000, the hole extended over Punta Arenas, a city of about 120,000 people in southern Chile, exposing residents to very high levels of UV radiation.

Carbon Dioxide and Global Warming

By studying the earth's history and making comparisons with other planets, scientists have determined that concentrations of gases in the atmosphere, particularly carbon dioxide, maintain the average temperature on earth about 25°C higher than it would be if these gases were absent. The particles of carbon dioxide and other gases reflect the longer wavelengths of infrared light, or heat, radiating from the surface of the earth, keeping them in the atmosphere and creating what is known as a **greenhouse effect** (figure 56.25).

Roughly seven times as much carbon dioxide is locked up in fossil fuels as exists in the atmosphere today. Before industrialization, the concentration of carbon dioxide in the atmosphere was approximately 260 to 280 parts per million (ppm). Since the extensive use of fossil fuels began, the amount of carbon dioxide in the atmosphere has been increasing rapidly. During a 25-year period starting in 1958, the concentration of carbon dioxide increased from 315 ppm to more than 340 ppm, and it continues to rise. Climatologists have calculated that the actual mean global temperature has increased about 0.6°C since 1900, a change known as **global warming**. In some areas, however, the increase has been much greater; on Alaska's North Slope, for example, temperatures have increased 2.2° to 3.2°C (figure 56.26). Moreover, the years 1998, 2001, and 2002 were the three warmest on record since instruments started collecting data more than a century ago. Evidence for this warming can be seen in many ways. For example, ice on lakes and rivers forms later and melts sooner than it used to—on average, ice-free seasons are now 2.5 weeks longer than they were a century ago. Also, ice thickness at the North Pole has decreased 40%.

Are these phenomena related? And if the current increase in levels of carbon dioxide is responsible for the increased temperature, what does the future hold? Although the reality of global warming and the effect of greenhouse gases (which include methane, nitrous oxide, ozone, and hydrofluorocarbons as well as carbon dioxide) were hotly debated in the 1990s, the scientific community has, for the most part, reached a consensus that temperatures are increasing and that human-released greenhouse gases are a major cause of this increase. Given that carbon dioxide concentrations in the year 2100 are predicted to be between 490 and 1250 ppm (with somewhere in the middle probably most likely), the question becomes: How much will temperatures rise?

In a recent study, the U.N.-sponsored Intergovernmental Panel on Climate Change, composed of an international team of climate experts, predicted that global temperatures will rise between 1.4° and 5.8°C by the end of the twenty-first century. Such an increase in temperature will have drastic effects—some beneficial, but most detrimental—on both natural ecosystems and human populations.

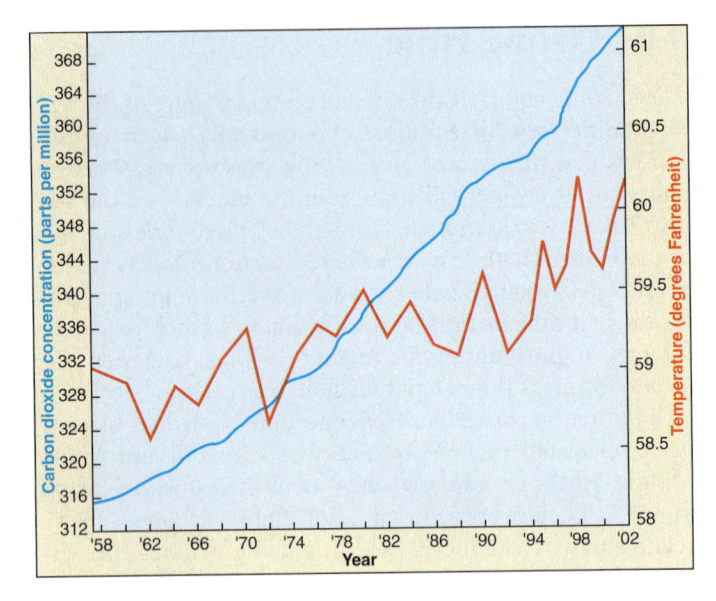

FIGURE 56.25
The greenhouse effect. The concentration of carbon dioxide in the atmosphere has steadily increased since the 1950s (*blue line*). The red line shows the general increase in average global temperature for the same period of time.
Why has temperature change been erratic, even though carbon dioxide levels have risen steadily?

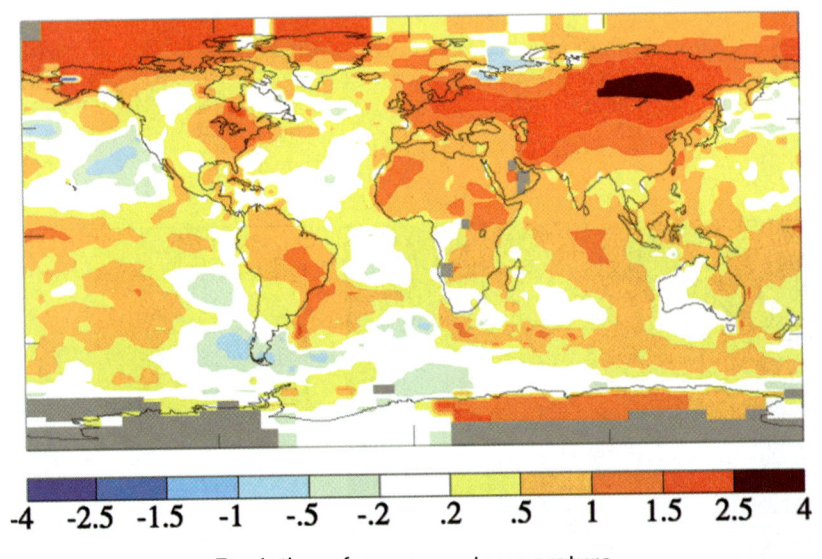

FIGURE 56.26
Geographic variation in global warming. The years 2001 and 2002 were the second and third warmest on record, but some areas heated up more than others. Colors indicate how much warming has occurred in different areas relative to the mean temperature from 1951 to 1980.

Effects of Global Warming on Natural Ecosystems

Prehistoric Climate Change. Global warming—and cooling—have occurred in the past, most recently during the ice ages and intervening warm periods that punctuated the last 1.8 million years. During these times, global temperatures changed as much as 10°C from one extreme to the other; sometimes these changes may have occurred rapidly, over periods of only a few decades.

Species responded to climate change by shifting their geographic distributions, tracking their environments. For example, a number of cold-adapted North American tree species that are now found only in the far north or at high elevations lived much further south or at substantially lower elevations 10,000 to 20,000 years ago, when conditions were much cooler.

Range Shifts in Contemporary Species. Present-day global warming is having a similar effect. Examples of range shifts include the expansion of shrubs into areas of the Arctic that were previously shrub-free; northward shifts of at least 39 butterfly species, some as much as 200 kilometers in 27 years (figure 56.27); and northward range changes of 12 British bird species, averaging 400 kilometers over 20 years. Global warming has also led to elevational changes because higher altitudes are now warmer. Such changes include shifts in alpine plants in the European Alps, lowland birds in Costa Rica, and butterflies in California.

Life Cycle Changes. In addition, a changing climate changes the temperature-dependent aspects of the life cycle of many plants and animals. Many migratory birds arrive earlier and depart later from their summer breeding grounds. Many insects and amphibians breed earlier in the year, and many plants flower earlier. Leaves now change color later, and the growing season in Europe has extended 3.6 days later over the past 50 years.

Effects of Global Warming on Species

As many species change their geographic ranges and life cycles, conservation biologists are concerned for several reasons. First, sedentary species, such as some plants and invertebrates, may not be able to disperse rapidly enough to keep up with changing temperatures and environments. Perhaps more importantly, many species may simply not have the option. Although plant and animal species could move freely across the landscape in prehistoric times, that is no longer the case. Now, habitats have been fragmented, and many populations and entire species are isolated in "islands" of habitat surrounded by unsuitable areas that have been altered by humans. If environmental changes render their current habitats uninhabitable, species stuck in these islands may not be able to move and may thus become extinct. Moreover, high-altitude species may run out of mountain to move up. Indeed, a number of high montane tropical species of frogs have recently disappeared in Costa Rica, and many scientists believe that climate change rendered their habitat unsuitable. Because they already occurred at the highest elevations, they had nowhere to go.

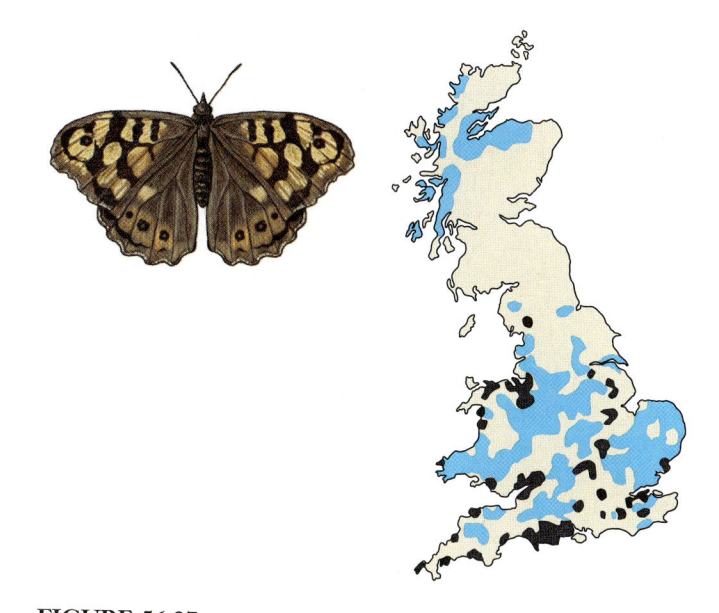

FIGURE 56.27
Butterfly range shift. Distribution of the speckled wood butterfly, *Pararge aegeria*, in Great Britain in 1915–1939 (*black*) and 1970–1997 (*blue*).

Species also may have trouble adapting to new conditions. One example involves spring breeding in birds. Prior to recent climate changes, the timing of reproduction of many species was finely tuned so that they would produce offspring when food resources were flush. However, the cue that triggers breeding in some species is day length, which of course has not changed. By contrast, as a result of climate change, insects, the food of these birds, appear and reproduce earlier in the year; consequently, the birds are now breeding too late in the season, and not enough food is available to feed their young. Perhaps as a result, populations of many insectivorous migratory birds are declining.

Another problem with global warming concerns species having **temperature-sensitive sex determination.** In other words, the sex of the offspring is determined by the temperature they experience during development. As a result of rising temperatures, sex ratios may be skewed. In turtles, for example, higher temperatures usually produce males, and there is concern that for some species, if temperatures rise enough, no females may be produced.

In many other ways, changing temperatures may affect how species interact with each other and with their environment. Some changes may not be detrimental, or some species may be able to adapt to the new selective pressures they experience. But in other cases, the result may be disruption of ecosystems, population declines, and extinction.

Effects of Global Warming on Humans

Global warming may affect human health and welfare in a variety of ways. Some of these changes may be beneficial, but even if they are detrimental, some countries—particularly the wealthier ones—will be able to adjust. On the other hand, some changes will require extremely costly countermeasures that even wealthy countries will be hardpressed to afford, while poorer countries may find even less detrimental consequences to be very expensive.

Rising Sea Levels. Much of the earth's supply of water is locked up in ice at either high latitudes or high altitudes. As the earth warms, it is likely that some—perhaps much—of this ice will melt. Already, glaciers are retreating in some locations around the world, sometimes at an alarming rate. Similarly, polar ice is melting, and the recent collapse of some Antarctic ice shelves is an ominous indication of what is to come. Most of this ice melt will end up in the oceans, and as a result, sea levels may rise by as much as a meter by the end of this century. Such an increase would cause increased erosion of low-lying land and coastal marshes, and other habitats would also be imperiled. As many as 200 million people would be affected by increased flooding. Should sea levels continue to rise, coastal cities and some entire islands, such as the Maldives in the Indian Ocean, would be endangered.

Other Climatic Effects. Global warming is predicted to have a variety of effects besides increased temperatures. In particular, the frequency of extreme events—such as heat waves, droughts, severe storms, and hurricanes—is expected to increase, and El Niño events, with their attendant climatic effects, may become more common. In addition, rainfall patterns are likely to shift, and those areas that are already water stressed, which are currently home to nearly two billion people, will likely face even graver water shortage problems in the years to come. Some evidence suggests that these effects are already evident in an increase in storms, hurricanes, and the frequency of El Niño events over the past few years.

Effects on Agriculture. Global warming will have both positive and negative effects on agriculture. On the positive side, warmer temperatures and increased atmospheric carbon dioxide tend to increase growth of some crops and thus may increase agricultural yields: other crops, however, will be negatively affected. Further, most crops will be affected by increased frequencies of droughts. Also, on the negative side, changes in rainfall patterns, temperature, pest distributions, and various other factors will require many adjustments. Such changes may come relatively easily for farmers in the developed world, but the associated costs may be devastating for those in the developing countries. Moreover, although crops in north temperate regions may flourish with higher temperatures, many tropical crops are already growing at their maximal temperatures, so increased temperatures may lead to reduced crop yields.

Effects on Human Health. Increasingly frequent storms, flooding, and drought will have adverse consequences on human health. Aside from their direct impact, such events often disrupt the fragile infrastructure of developing countries, leading to the loss of safe drinking water and other problems. As a result, epidemics of cholera and other diseases may be expected to occur more often as a consequence of these events.

In addition, as temperatures rise, areas suitable for tropical organisms will expand northward. Of particular concern are those organisms that cause human diseases. Many diseases currently limited to tropical areas may expand their range and become problematic in non-tropical countries. Diseases transmitted by mosquitoes, such as malaria, dengue fever, and several types of encephalitis, are examples. The distribution of mosquitoes is limited by cold; winter freezes kill many mosquitoes and their eggs. As a result, malaria only occurs in areas where temperatures are usually above 16°C, and yellow fever and dengue fever occur in areas where temperatures are normally above 10°C. (The difference results because the diseases are transmitted by different mosquito species.) Moreover, at higher temperatures, the malaria pathogen matures more rapidly. Malaria already kills one million people every year; some projections suggest that the percentage of the human population at risk for malaria may increase by 33% by the end of the twenty-first century. Moreover, as predicted, malaria already appears to be on the move. By 1980, malaria had been eradicated from all of the United States except California, but in recent years it has appeared in a variety of southern, and even a few northern, states.

Dengue fever (sometimes called "breakbone fever" because of the pain it causes) is also spreading. Previously a disease restricted to the tropics and subtropics, where it infects 50–100 million people, it now occurs in the United States, southern South America, and northern Australia.

One of the most alarming aspects of these diseases is that no vaccines are available. Drug treatment is available (for malaria), but the parasites are rapidly evolving resistance and rendering the drugs ineffective. There is no drug treatment for dengue fever.

Solving the global warming problem will not be easy. It will require large reductions in the amount of carbon dioxide injected into the atmosphere. Some nations are taking steps to reduce their emissions, but others are not. Coordinated international effort is required to slow down the increase in global temperatures. Although the predicted effects of global warming are uncertain, most scientists agree that the impact will be severe.

Global temperatures are rising, at least in part as a result of human activities. Increased temperature will have many detrimental effects on natural ecosystems and human welfare.

Concept Review

56.1 Climate shapes the character of ecosystems.
Effects of the Sun and Atmospheric Circulation
- The tropics are warmer than temperate regions because the sun's rays arrive almost perpendicular to equatorial regions. (p. 1204)
- High temperatures near the equator encourage evaporation and create warm, moist air, causing large amounts of precipitation near the equator. (p. 1205)
- When the air masses reach about 30° north and south latitude, the cool, dry air sinks, forming areas of low precipitation (deserts). (p. 1205)

Atmospheric Circulation, Precipitation, and Climate
- The rain shadow effect helps create deserts at places other than 30° north and south latitude, as mountain ranges intercept moisture-laden winds and cause most of the precipitation to fall on the windward side of the range, blocking moisture from the leeward side. (p. 1206)
- Unique regional climates are also formed due to prevailing winds such as the westerlies and the trade winds. (p. 1206)
- Temperature tends to vary according to latitude—warm in the tropics, and cooler as you move away—and according to elevation, with temperature decreasing as elevation increases. (p. 1207)

56.2 Biomes are widespread terrestrial ecosystems.
The Major Biomes
- Biomes are major communities of organisms with a characteristic appearance that are distributed over areas defined by temperature and precipitation differences. (p. 1208)
- Eight of the major biomes are tropical rain forest, savanna, desert, temperate grassland, temperate deciduous forest, temperate evergreen forest, taiga, and tundra. (p. 1208)

56.3 Aquatic ecosystems cover much of the earth.
Patterns of Circulation in the Oceans
- Ocean circulation patterns are determined by patterns of atmospheric circulation, but are also modified by the location of landmasses. (p. 1212)
- When the east-west trade winds in the Pacific Ocean slacken, warm water begins to move back across the ocean from the coast of South America, causing a phenomenon called El Niño, which widely influences the world's weather systems. (p. 1213)

Marine Ecosystems
- The marine environment consists of three major habitats: the neritic zone (shallow water containing most of the world's major fisheries), the pelagic zone (water above the ocean floor), and the benthic zone (ocean floor). (p. 1214)
- Approximately 40% of all photosynthesis on earth takes place in the oceans. (p. 1215)

Freshwater Habitats
- Freshwater habitats are much more limited in area than marine habitats. (p. 1216)
- Photosynthetic organisms are limited to the upper photic zone of ponds and lakes. (p. 1216)
- Ponds and lakes have three zones where organisms are found, which are distributed according to the distance from shore: the littoral, limnetic, and profundal zones. (p. 1216)
- Thermal stratification is characteristic of larger lakes in temperate regions. (p. 1216)
- Eutrophic lakes contain an abundant supply of organic matter, while organic matter and nutrients are relatively scarce in oligotrophic lakes. (p. 1217)
- Human activities can lead to the eutrophication of oligotrophic waters. (p. 1217)
- Wetlands are often very productive ecosystems that also serve as water storage basins helping to moderate flooding. (p. 1217)

56.4 Human activity is placing the biosphere under increasing stress.
Pollution
- Widespread modern agriculture introduces large amounts of new chemicals into the global ecosystem, including pesticides, herbicides, and fertilizers. (p. 1218)
- Stable, long-lasting chlorinated hydrocarbons such as DDT can become increasingly concentrated in an ecosystem due to biological magnification. (p. 1218)

Acid Precipitation
- Industrial pollution contains many chemicals. For example, sulfur, when introduced into the upper atmosphere, can combine with water vapor to produce sulfuric acid, which can fall to the ground in many forms, including snow and rain. (p. 1219)
- Precipitation with an acidic pH can cause many environmental problems, including lake acidification, groundwater contamination, and forest damage. (p. 1219)

Destruction of the Tropical Forests
- More than half of the world's human population lives in the tropics, and this percentage is increasing. (p. 1220)
- Rain forests grow on poor soil, thus they can become hard to regenerate once they are cleared. (p. 1220)
- Increasingly, larger numbers of people moving into tropical areas are clearing larger areas of rain forest. (p. 1220)

The Ozone Hole
- Industrial chlorofluorocarbons (CFCs) released into the atmosphere for many decades have led to a thinning of the earth's stratospheric ozone layer, which shields the planet from harmful ultraviolet radiation. (p. 1221)

Carbon Dioxide and Global Warming
- The greenhouse effect is caused when carbon dioxide and other gases allow short-wavelength solar radiation into the atmosphere, but trap longer-wavelength heat radiation from escaping, thus warming the earth's atmosphere. (p. 1222)
- In the 1990s, the scientific community reached a consensus that the earth's average temperature is increasing, and human-related greenhouse gases are the major cause of the increase. (p. 1222)
- The effects of global warming on natural ecosystems may include the shifting of the geographic distribution of organisms tracking environmental conditions, as well as life cycle changes to adapt to changing environmental conditions. (p. 1223)
- Continued global warming may also influence human conditions due to rising sea levels, effects on agriculture, and human health issues. (p. 1224)

Test Your Understanding

For interactive testing, visit the Online Learning Center with PowerWeb at www.mhhe.com/Raven7

Self Test

1. A rain shadow results in
   a. extremely wet conditions due to loss of moisture from winds rising over a mountain range.
   b. dry air moving toward the poles that cools and sinks in regions 15° to 30° north/south latitude.
   c. global polar regions that rarely receive moisture from the warmer, tropical regions, and are therefore dryer.
   d. desert conditions on the down-wind side of a mountain due to increased moisture-holding capacity of the winds coming from the seas.
2. What two factors are most important in biome distribution?
   a. temperature and latitude
   b. rainfall and temperature
   c. latitude and rainfall
   d. temperature and soil type
3. Savannas are best described as areas with
   a. extremely dry conditions and sparse vegetation.
   b. cold, dry conditions with herbs and few trees.
   c. warm summers, cool winters, and abundant rainfall that promotes tree growth.
   d. seasonal rainfall, few trees, and abundant grasses.
4. The cacti found in the deserts of North and South America look very much like the euphorbs found in the deserts of Africa. However, these plants are not closely related. The similarities in these plants are due to
   a. convergent evolution as a result of similar environmental pressures.
   b. artificial selection for these similar traits.
   c. differences in rainfall between the two deserts.
   d. differences in pollinator species in the two deserts.
5. Which of the following is not a result of an El Niño event?
   a. The trade winds relax in the central and western Pacific.
   b. The sea surface is about a meter higher at the Philippines than at Ecuador.
   c. A rise in sea surface temperature and a decline in primary productivity adversely affect fisheries in Ecuador and Peru.
   d. Flooding and strong winter storms occur in California.
6. The neritic zone is best described as the
   a. area of water above the ocean floor where a diversity of plankton species are concentrated.
   b. ocean floor that is made up of mud and other fine particles that have settled from the water.
   c. area less than 300 meters below the surface of the oceans along the coasts of continents and islands.
   d. part of the ocean floor that drops to depths where light does not penetrate.
7. The limnetic zone of a lake is best described as the
   a. shallow area along the shore.
   b. area below the limits where light can penetrate.
   c. zone where photosynthesis cannot occur.
   d. well-illuminated surface waters away from the shore.
8. In temperate regions, lakes are thermally stratified, with warm waters at the top and cooler waters at the bottom during the summer. The region of abrupt change between these layers is known as the
   a. thermocline.
   b. hypolimnion.
   c. epilimnion.
   d. fall overturn.
9. Oligotrophic lakes can be turned into eutrophic lakes as a result of human activities such as
   a. overfishing of sensitive species, which disrupts fish communities.
   b. introducing nutrients into the water, which stimulates plant and algal growth.
   c. disrupting terrestrial vegetation near the shore, which causes soil to run into the lake.
   d. spraying pesticides into the water to control aquatic insect populations.
10. The loss of the ozone layer has serious implications for the quality of the environment because
    a. ozone (O_3) protects organisms from ultraviolet radiation that can cause cancer.
    b. a depleted ozone layer causes rainwater to have a lower pH that kills plant life.
    c. loss of the ozone layer causes the sun's rays to get trapped in the atmosphere and increase global temperatures.
    d. a depleted ozone layer can interact with toxic chemicals to increase their effect on organismal health.

Test Your Visual Understanding

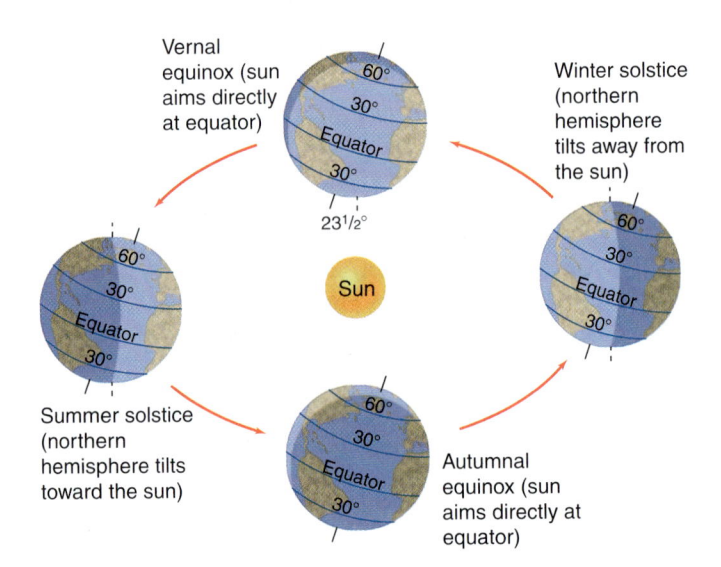

1. Predict what changes to global climate would occur if the earth's axis were not tilted.

Apply Your Knowledge

1. Some vegetarians argue that it is more ethical to eat "lower on the food chain" (eat more grains and vegetables) than to consume meat. Explain this argument in terms of energy conversions in ecosystems.
2. The Clean Water Act protects wetlands. Under the act, there is to be no net loss of wetlands and any wetland losses should be offset through restoration projects or creation of new wetlands. Why are wetlands given this status?
3. The concentration of CO_2 in the atmosphere has steadily increased since the 1950s. A general increase in average global temperature has also occurred. How is global climate change predicted to affect the area where you live?

57

Conservation Biology

Concept Outline

57.1 The new science of conservation biology is focused on conserving biodiversity.

Overview of the Biodiversity Crisis. In prehistoric times, humans decimated the faunas of many areas. Today, worldwide extinction rates are accelerating.

Species Endemism and Hotspots. Some geographic areas are particularly rich in species that occur nowhere else.

What's So Bad About Losing Biodiversity? Biodiversity has considerable direct economic value and provides key support to the biosphere.

57.2 The extinction crisis is a result of many factors.

Factors Responsible for Extinction. Most recorded extinctions can be attributed to a few causes. Sometimes more than one factor can affect a species at the same time.

Habitat Loss. Without a place to live, species cannot survive.

Overexploitation. Species cannot persist if too many individuals are removed by humans.

Detrimental Effects of Introduced Species. Introduced species can wreak havoc on native species and ecosystems.

Disruption of Ecosystems. Extinction of one species can have a cascading effect throughout the food web, making other species vulnerable as well.

The Perils of Small Population Size. Small populations are vulnerable to genetic and demographic problems.

57.3 Successful recovery efforts need to be multidimensional.

Approaches for Preserving Endangered Species. Species preservation efforts take many forms and must be tailored toward the particular threats that species face.

Conservation of Ecosystems. Maintaining large preserves and focusing on the health of the entire ecosystem may be the best means of preserving biodiversity.

FIGURE 57.1
Endangered. The Siberian tiger is in grave danger of extinction, being hunted for its pelt and having its natural habitat greatly reduced. A concerted effort to save it is using many of the approaches discussed in this chapter.

Among the greatest challenges facing the biosphere is the accelerating pace of species extinctions. Not since the Cretaceous period have so many species become extinct in so short a time span (figure 57.1). This challenge has led to the emergence in the last decade of the discipline of conservation biology. Conservation biology is an applied science that seeks to learn how to preserve species, communities, and ecosystems. It studies the causes of declines in species richness and attempts to develop methods for preventing such declines. In this chapter, we first examine the biodiversity crisis and its importance. Then, using case histories, we identify and study factors that have played key roles in many extinctions. We finish with a review of recovery efforts at the species and community levels.

57.1 The new science of conservation biology is focused on conserving biodiversity.

Overview of the Biodiversity Crisis

Extinction is a fact of life. Most species—probably all—become extinct eventually. More than 99% of species known to science (most from the fossil record) are now extinct. However, current rates are alarmingly high. Taking into account the rapid and accelerating loss of habitat that is occurring, especially in the tropics, it has been calculated that as much as 20% of the world's biodiversity may be lost by the middle of this century. In addition, many of these species may be lost before we are even aware of their existence. Scientists estimate that no more than 15% of the world's eukaryotic organisms have been discovered and given scientific names, and this proportion is probably much lower for tropical species.

These losses will not only affect poorly known groups. As many as 50,000 species of the world's total of 250,000 species of plants, 4000 of the world's 20,000 species of butterflies, and nearly 2000 of the world's 9000 species of birds could be lost during this time period. Considering that the human species has been in existence for only 600,000 years of the world's 4.5-billion-year history, and that our ancestors developed agriculture only about 10,000 years ago, this is an astonishing—and dubious—accomplishment.

FIGURE 57.2
North America before human inhabitants. Animals found in North America prior to the migration of humans included birds and large mammals, such as the ancient North American camel, saber-toothed cat, giant ground sloth, and teratorn vulture.

Extinctions Due to Prehistoric Humans

A great deal can be learned about current rates of extinction by studying the past, and in particular the impact of human-caused extinctions. In prehistoric times, members of *Homo sapiens* wreaked havoc whenever they entered a new area. For example, at the end of the last Ice Age, approximately 12,000 years ago, the fauna of North America was composed of a diversity of large mammals similar to those living in Africa today: mammoths and mastodons, horses, camels, giant ground sloths, saber-toothed cats, and lions, among others (figure 57.2). Shortly after humans arrived, 74 to 86% of the *megafauna* (that is, animals weighing more than 100 pounds) became extinct. These extinctions are thought to have been caused by hunting and, indirectly, by burning and clearing of forests. (Some scientists attribute these extinctions to climate change, but that hypothesis doesn't explain why the end of earlier ice ages was not associated with mass extinctions, nor does it explain why extinctions occurred primarily among larger animals, with smaller species relatively unaffected.)

Around the globe, similar results have followed the arrival of humans. Forty thousand years ago, Australia was occupied by a wide variety of large animals, including marsupials similar in size and ecology to hippos and leopards, a kangaroo nine feet tall, and a 20-foot-long monitor lizard. These all disappeared, at approximately the same time as humans arrived. Smaller islands have also been devastated. Madagascar has seen the extinction of at least 15 species of lemurs, including one the size of a gorilla; a pygmy hippopotamus; and the flightless elephant bird, *Aepyornis*, the largest bird to ever live (more than 3 meters tall and weighing 450 kilograms). On New Zealand, 30 species of birds went extinct, including all 13 species of moas, another group of large, flightless birds. Interestingly, one continent that seems to have been spared these megafaunal extinctions is Africa. Scientists speculate that this lack of extinction in prehistoric Africa may have resulted because much of human evolution occurred in Africa. Consequently, other African species had been coevolving with humans for several million years and thus had evolved counteradaptations to human predation.

1228 Part VIII Ecology and Behavior

Extinctions in Historical Time

Historical extinction rates are best known for birds and mammals because these species are conspicuous—that is, relatively large and well studied. Estimates of extinction rates for other species are much rougher. The data presented in table 57.1, based on the best available evidence, show recorded extinctions from 1600 to the present. These estimates indicate that about 85 species of mammals and 113 species of birds have become extinct since the year 1600. That is about 2.1% of known mammal species and 1.3% of known birds. The majority of extinctions have occurred in the last 150 years. The extinction rate for birds and mammals was about one species every decade from 1600 to 1700, but it rose to one species every year during the period from 1850 to 1950, and to four species per year between 1986 and 1990 (figure 57.3). This increase in the rate of extinction is the heart of the biodiversity crisis.

Unfortunately, the biodiversity crisis seems to be worsening. For example, the number of bird species recognized as "critically endangered" increased 8% from 1996 to 2000, and a 2002 report suggested that as many as half of the earth's plant species may be threatened with extinction. Some researchers predict that two-thirds of all vertebrate species could perish by the end of this century.

The majority of historic extinctions—though by no means all of them—have occurred on islands. For example, of the 90 species of mammals that have gone extinct in the last 500 years, 73% lived on islands (and another 19% in Australia). The particular vulnerability of island species probably results from a number of factors: Such species have often evolved in the absence of predators, and so have lost their ability to escape both humans and introduced predators such as rats and cats. In addition, humans have introduced competitors and diseases; avian malaria, for example has devastated the bird fauna of the Hawaiian Islands. Finally, island populations are often relatively small, and thus particularly vulnerable to extinction, as we shall see later in this chapter.

In recent years, the extinction crisis has moved from islands to continents. Most species now threatened with extinction occur on continents, and these areas will bear the brunt of the extinction crisis in this century.

Some people have argued that we should not be concerned, because extinctions are a natural event and mass extinctions (the extinction of large numbers of species within a geologically short period of time) have occurred in the past. Indeed, mass extinctions have taken place several times over the past half-billion years. However, the current mass extinction event is notable in several respects. First, it is the only such event triggered by a single species. Moreover, although species diversity usually recovers after a few million years, this is a long time to deny our descendants the benefits and joys of biodiversity. In addition, it is not clear that biodiversity will rebound this time. After previous mass extinctions, new species have evolved to utilize resources available due to extinctions of the species that previously used them. Today, however, such resources are unlikely to be available, because humans are destroying the habitats and taking the resources for their own use.

Table 57.1 Recorded Extinctions Since 1600 A.D.

Taxon	Mainland	Island	Ocean	Total	Approximate Number of Species	Percent of Taxon Extinct
Mammals	30	51	4	85	4,000	2.1
Birds	21	92	0	113	9,000	1.3
Reptiles	1	20	0	21	6,300	0.3
Amphibians*	2	0	0	2	4,200	0.05
Fish	22	1	0	23	19,100	0.1
Invertebrates	49	48	1	98	1,000,000+	0.01
Flowering plants	245	139	0	384	250,000	0.2

*An alarming decline in amphibian populations has occurred recently, and many species may be on the verge of extinction.

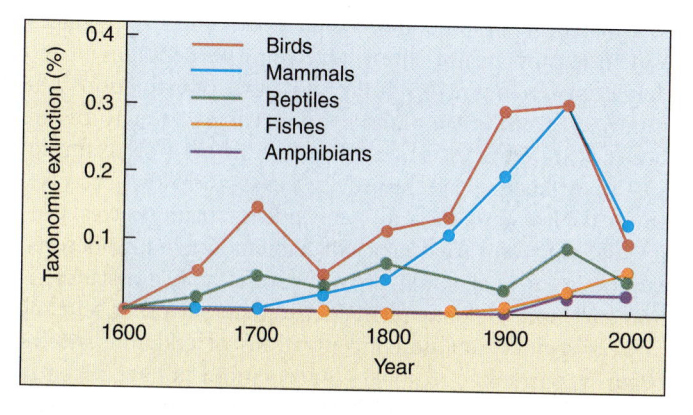

FIGURE 57.3
Trends in species loss. These graphs present data on recorded animal extinctions since 1600. The majority of extinctions have occurred on islands, with birds and mammals particularly affected (although this may reflect to some degree our more limited knowledge of other groups).
Why are extinction rates highest for birds and mammals?

> Since prehistoric times, humans have had a devastating effect on biodiversity almost everywhere in the world. Most historical extinctions have occurred on islands, but most future extinctions will occur on continents.

Chapter 57 Conservation Biology 1229

Species Endemism and Hotspots

A species found naturally in only one geographic area and no place else is said to be **endemic** to that area. The area over which an endemic species is found may be very large. For example, the black cherry tree (*Prunus serotina*) is endemic to all of temperate North America. More typically, however, endemic species occupy restricted ranges. The Komodo dragon (*Varanus komodoensis*) lives only on a few small islands in the Indonesian archipelago, while the Mauna Kea silversword (*Argyroxiphium sandwicense*) lives in a single volcano crater on the island of Hawaii.

Isolated geographic areas, such as oceanic islands, lakes, and mountain peaks, often have high percentages of endemic species, many in significant danger of extinction. The number of endemic plant species varies greatly in the United States from one state to another. For example, 379 plant species are found in Texas and nowhere else, whereas New York has only one endemic plant species. California, with its varied array of habitats, including deserts, mountains, seacoast, old-growth forests, and grasslands, is home to more endemic plant species than any other state.

Worldwide, notable concentrations of endemic species occur in particular regions. Conservationists have recently identified areas, termed **hotspots,** that have high endemism and are disappearing at a rapid rate. Such hotspots include Madagascar, a variety of tropical rain forests, the eastern Himalayas, areas with Mediterranean climates such as California, South Africa, and Australia, and in several other climatic areas (figure 57.4 and table 57.2). Overall, 25 such hotspots have been identified, which in total contain nearly half of all the terrestrial species in the world.

Why these areas contain so many endemic species is a topic of active scientific research. Some of these hotspots occur in areas of high species diversity, and the explanations for high species diversity in general (see chapter 55), such as high productivity, probably apply to these hotspots as well. In addition, some hotspots occur on isolated islands, such as New Zealand, New Caledonia, and Polynesia (including the Hawaiian Islands), where evolutionary diversification over long periods of time has resulted in rich biotas composed of plant and animal species found nowhere else in the world.

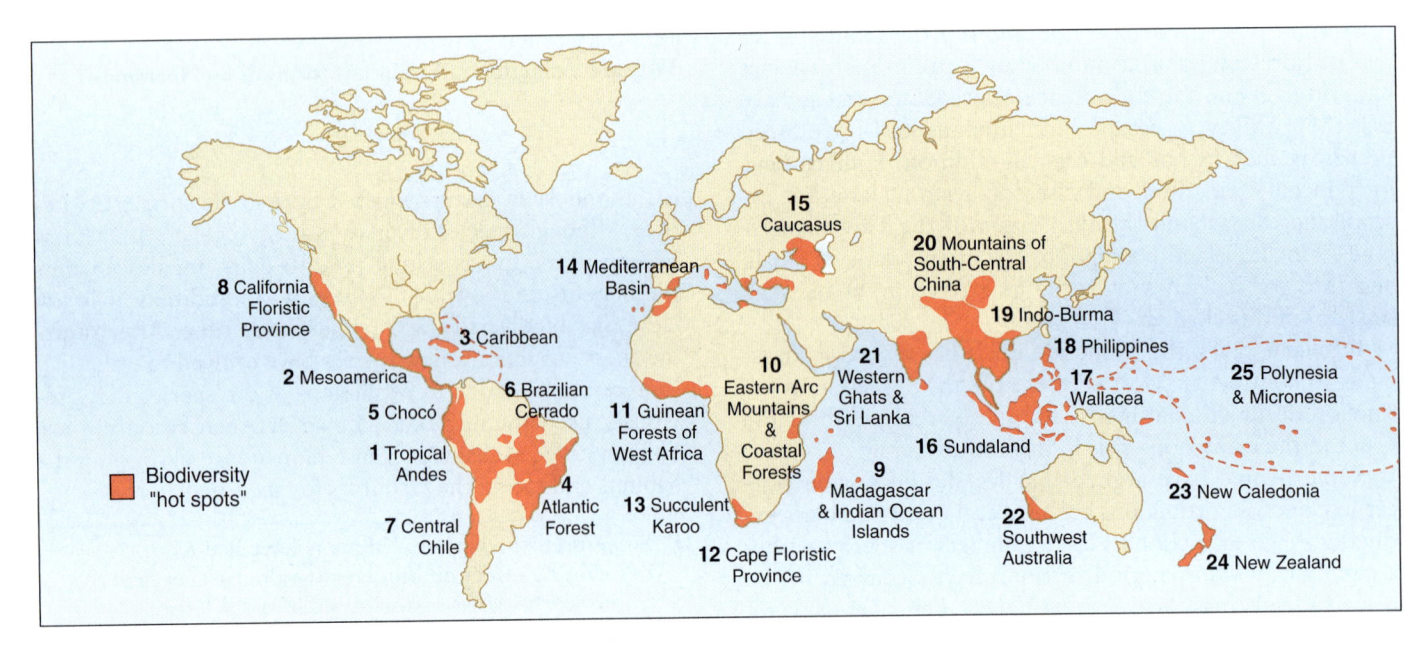

Table 57.2 Numbers of Endemic Species in Some Hotspot Areas

Region	Mammals	Reptiles	Amphibians	Plants
Atlantic coastal Brazil	160	60	253	6,000
South American Chocó	60	63	210	2,250
Philippines	115	159	65	5,832
Tropical Andes	68	218	604	20,000
Southwestern Australia	7	50	24	4,331
Madagascar	84	301	187	9,704
Cape region (South Africa)	9	19	19	5,682
California Floristic Province	30	16	17	2,125
New Caledonia	6	56	0	2,551
South-central China	75	16	51	3,500

FIGURE 57.4
Hotspots of high endemism. These areas are rich in endemic species under threat of imminent extinction.

1230 Part VIII Ecology and Behavior

Population Growth in Hotspots

Because of the great number of endemic species that hotspots contain, conserving their biological diversity must be an important component of efforts to safeguard the world's biological heritage. Or, to look at it another way, by protecting just 1.4% of the world's land surface, 44% of the world's vascular plants and 35% of its terrestrial vertebrates can be preserved.

Unfortunately, hotspots contain not only many endemic species, but also growing human populations. In 1995, these areas contained 1.1 billion people—20% of the world's population—sometimes at high densities (figure 57.5a). More importantly, human populations were growing in all but one of these hotspots as the result of both high birth and immigration rates; overall, the rate of growth exceeded the global average in 19 hotspots (figure 57.5b). Indeed, in some hotspots, the rate of growth is nearly twice that of the rest of the world.

Not surprisingly, many of these areas are experiencing high rates of habitat destruction as land is cleared for agriculture, housing, and economic development. More than 70% of the original area of each hotspot has already disappeared, and in 14 hotspots, 15% or less of the original habitat remains. In Madagascar, it is estimated that 90% of the original forest has already been lost—this on an island where 85% of the species are found nowhere else in the world. In the forests of the Atlantic coast of Brazil, the extent of deforestation is even higher: 95% of the original forest is gone.

Of course, population pressure is not the only cause of habitat destruction in hotspots. Commercial exploitation to meet the demands of more affluent people in the developed world also plays an important role. For example, large-scale logging of tropical rain forests occurs in countries around the world to provide lumber, most of which ends up in the United States, western Europe, and Japan. Similarly, many forests in Central and South America are cleared to make way for cattle ranches that produce cheap meat for fast-food restaurants. Hotspots in more affluent countries are often at risk because they occur in areas where land has great value for real estate and commercial purposes.

Regardless of its cause, decimation of hotspots will take a great toll on the world's biological diversity. As we shall see shortly, such massive rates of habitat destruction result in extremely high rates of species extinction.

Some areas of the earth have particularly high levels of species endemism. Unfortunately, many of these areas are currently in great jeopardy due to human population growth and habitat destruction, with correspondingly high rates of species extinction.

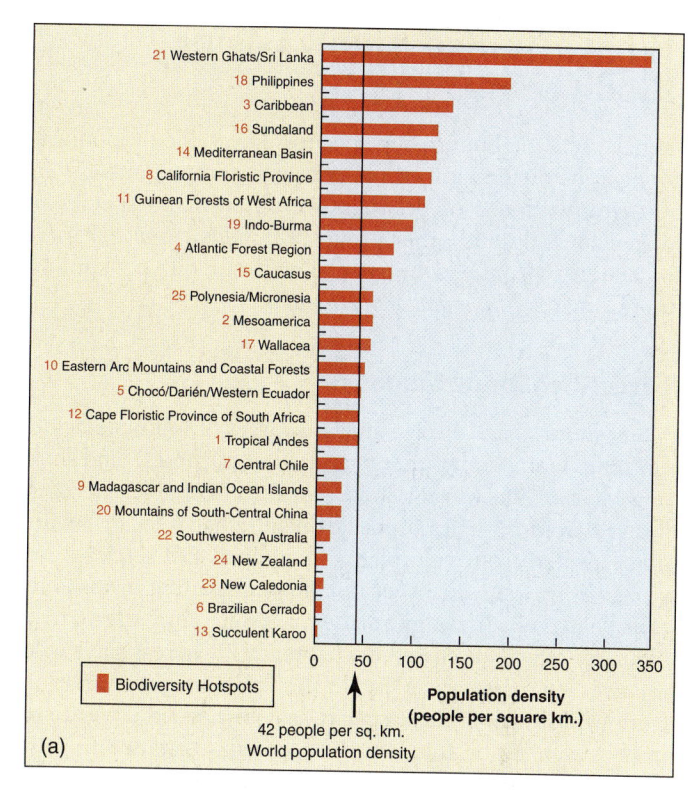

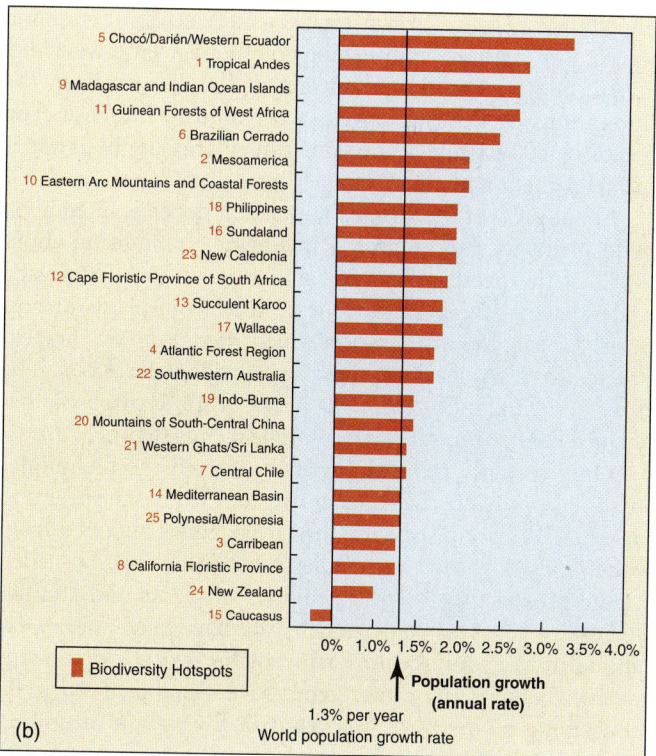

FIGURE 57.5

Human populations in hotspots. The rich biodiversity in many hotspots is under pressure from (a) dense and (b) rapidly growing human populations.

Why do population density and growth rates differ among hotspots?

Chapter 57 Conservation Biology **1231**

What's So Bad About Losing Biodiversity?

What's so bad about losing species? The value of biodiversity can be divided into three principal components: (1) *direct economic value* of products we obtain from species of plants, animals, and other groups; (2) *indirect economic value* of benefits produced by species without our consuming them; and (3) *ethical and aesthetic values*.

Direct Economic Value

Many species have direct value as sources of food, medicine, clothing, biomass (for energy and other purposes), and shelter. Most of the world's food crops, for example, are each derived from a small number of plants that were originally domesticated from wild plants in tropical and semiarid regions. As a result, many of our most important crops contain relatively little genetic variation (equivalent to a "founder effect" discussed in chapter 21), whereas their wild relatives have great diversity. In the future, genetic variation from wild strains of these species may be needed if we are to improve yields or find a way to breed resistance to new pests. In fact, recent agricultural breeding experiments have illustrated the value of conserving wild relatives of common crops. For example, by breeding commercial varieties with a small, oddly colored wild species of the tomato from the mountains of Peru, scientists were able to increase crop yields by 50%, while at the same time increasing both nutritional content and color.

About 70% of the world's population depends directly on wild plants as their source of medicine. In addition about 40% of the prescription and nonprescription drugs used today have active ingredients extracted from plants or animals. Aspirin, the world's most widely used drug, was first extracted from the leaves of the tropical willow, *Salix alba*. The rosy periwinkle, *Catharanthus roseus*, from Madagascar has yielded potent drugs for combating leukemia (figure 57.6).

Only recently have biologists perfected the techniques that make possible the transfer of genes from one species to another. We are just beginning to use genes obtained from other species to our advantage (see chapter 16). So-called "gene prospecting" of the genomes of plants and animals for useful genes has only begun. We have been able to examine only a minute proportion of the world's organisms to see whether any of their genes have useful properties. By conserving biodiversity, we maintain the option of finding useful benefits in the future; unfortunately, many of the most promising species occur in habitats that are being destroyed at an alarming rate, such as tropical rain forests.

Indirect Economic Value

Diverse biological communities are of vital importance to healthy ecosystems. They help maintain the chemical

FIGURE 57.6
The rosy periwinkle. Two drugs extracted from the Madagascar periwinkle *Catharanthus roseus*, vinblastine and vincristine, effectively treat common forms of childhood leukemia, increasing chances of survival from 20% to over 95%.

quality of natural water, buffer ecosystems against floods and drought, preserve soils and prevent loss of minerals and nutrients, moderate local and regional climate, absorb pollution, and promote the breakdown of organic wastes and the cycling of minerals. By destroying biodiversity, we are creating conditions of instability and lessened productivity and promoting desertification, waterlogging, mineralization, and many other undesirable outcomes throughout the world.

Economists have recently been able to compare the societal value, in monetary terms, of intact habitats compared with the value of destroying those habitats. Surprisingly, in most studies conducted so far, intact ecosystems are more valuable than the products derived by destroying them. For example, in Thailand, coastal mangrove habitats are commonly cleared so that shrimp farms can be established. Although the shrimp produced are valuable, their value is vastly outweighed by the benefits in timber, charcoal production, offshore fisheries, and storm protection provided by the mangroves (figure 57.7a). Similarly, intact tropical rain forest in Cameroon, West Africa, provides fruit and other forest materials. Clearing the forest for agriculture or palm plantations leads to stream-polluting erosion as well as increased flooding. Combining all the costs and benefits of the three options, maintaining intact forests has the highest economic value (figure 57.7b).

Probably the most famous example of the value of intact ecosystems is provided by the watersheds of New York City. Ninety percent of the water for the New York area's nine million residents comes from the Catskill Mountains and the nearby headwaters of the Delaware River (figure 57.8). Water that runs off from over 1600 square miles of rural, mountainous areas is collected into reservoirs and then transported by aqueduct more than 85 miles to New York City at a rate of 1.3 billion gallons per day.

In the 1990s, New York City faced a dilemma. New federal water regulations were requiring ever cleaner water, even as development and pollution in the source areas of the water were threatening to compromise water quality. The city had two choices: either work to protect the functioning ecosystem so that it could produce clean water, or construct filtration plants to clean it upon arrival. Economic analysis made the choice clear: Building the plants would cost $6 billion, with annual operating costs of $300 million, whereas spending a billion dollars over ten years could preserve the ecosystem and maintain water purity. The decision was easy.

These examples provide some idea of the value of the services that ecosystems provide. But maintaining ecosystems is not always more valuable than converting them to other uses. Certainly, when the United States was being settled and land was plentiful, ecosystem conversion was beneficial. Even today, habitat destruction may sometimes be economically desirable. Nonetheless, we still have only a rudimentary knowledge of the many ways intact ecosystems provide services. In many cases, it is not until they are lost that the value becomes clear, as unexpected negative effects, such as increased flooding and pollution or decreased rainfall, become apparent.

The same argument can be made for preserving particular species within ecosystems. Given how little we know about the biology of most species, particularly in the tropics, it is impossible to predict all the consequences of removing a species. Imagine taking a parts list for an airplane and randomly changing a digit in one of the part numbers: You might change a cushion to a roll of toilet paper—but you might just as easily change a key bolt holding up the wing to a pencil. By removing biodiversity, we are gambling with the future of the ecosystems upon which we depend and whose functioning we understand very little.

Ethical and Aesthetic Values

Many people believe that preserving biodiversity is an ethical issue because every species is of value in its own right, even if humans are not able to exploit or benefit from it. These people feel that along with the power to exploit and destroy other species comes responsibility: As the only organisms capable of eliminating large numbers of species and entire ecosystems, and as the only organisms capable of reflecting upon what we are doing, humans should act as guardians or stewards for the diversity of life around us.

Almost no one would deny the aesthetic value of biodiversity—a beautiful flower or a noble elephant—but how do we place a value on beauty? Perhaps the best we can do is to appreciate the deep sense of loss we would feel if it no longer existed.

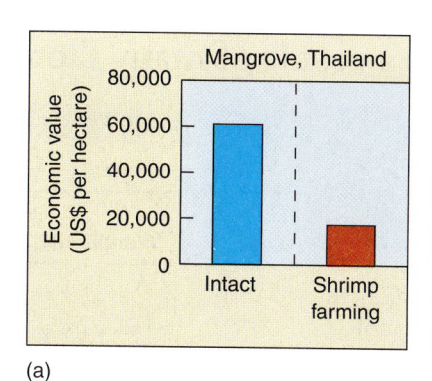

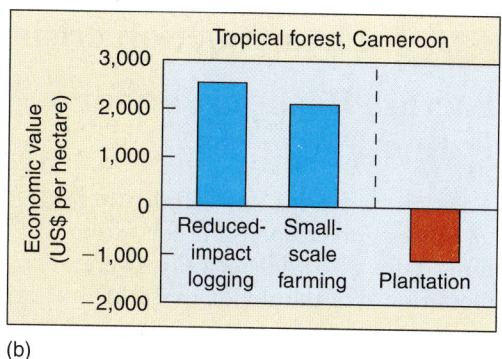

FIGURE 57.7

The economic value of maintaining habitats. (*a*) Mangroves in Thailand and (*b*) rain forests in Cameroon provide more economic benefits if they are left standing than if they are destroyed and the land used for other purposes.

If shrimp farms established on cleared mangrove habitats make money, how can clearing mangroves not be an economic plus?

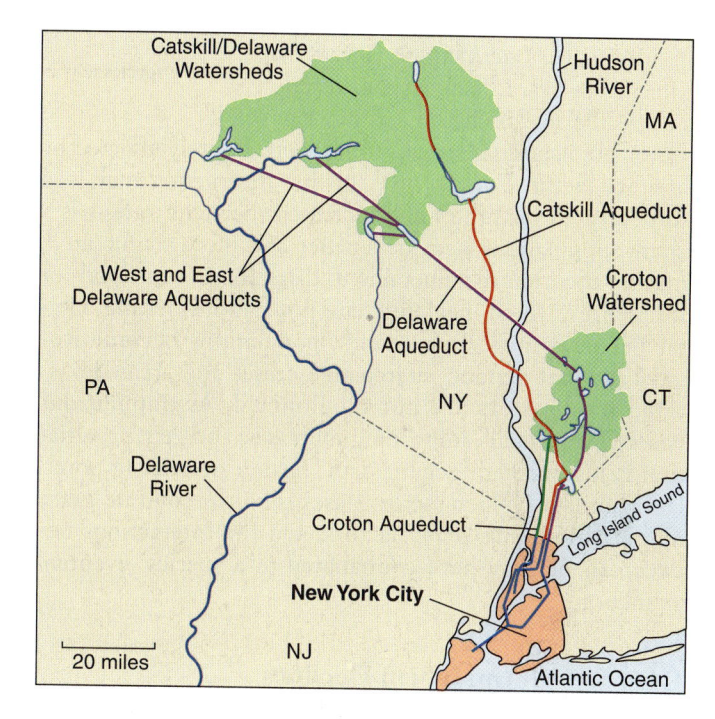

FIGURE 57.8

New York City's water source. New York gets its water from distant rain catchments. Preserving the ecological integrity of these areas is cheaper than building new water treatment plants.

> Biodiversity is of great value in its own right, as well as for the products it provides, its contributions to the health of the ecosystems we depend on, and the beauty it offers.

57.2 The extinction crisis is a result of many factors.

Factors Responsible for Extinction

A variety of causes, independently or in concert, are responsible for extinctions (table 57.3). Historically, overexploitation was the major cause of extinction; although overexploitation is still a factor, habitat loss is the major problem for most groups today, while introduced species rank second. Many other factors can contribute to species extinctions as well, including disruption of ecosystem interactions, pollution, loss of genetic variation, and catastrophic disturbances, either natural or man-made.

More than one of these factors may affect a species. In fact, a chain reaction is possible in which the action of one factor predisposes a species to be more severely affected by another factor. For example, habitat destruction may lead to decreased birthrates and increased mortality rates. As a result, populations become smaller and more fragmented, making them more vulnerable to disasters such as floods or forest fires, which may eliminate populations. As the habitat becomes more fragmented, populations become isolated, so that genetic interchange ceases and areas devastated by disasters are not recolonized. As populations become very small, inbreeding increases, and genetic variation is lost through genetic drift, further decreasing population fitness. Which factor causes the final coup de grace may be irrelevant; many factors, and the interactions between them, may have contributed to a species' eventual extinction.

Table 57.3 Causes of Extinctions

	Percentage of Species Influenced by the Given Factor*				
Group	Habitat Loss	Overexploitation	Species Introduction	Other	Unknown
EXTINCTIONS					
Mammals	19	23	20	2	36
Birds	20	11	22	2	45
Reptiles	5	32	42	0	21
Fish	35	4	30	4	48
THREATENED EXTINCTIONS					
Mammals	68	54	6	20	—
Birds	58	30	28	2	—
Reptiles	53	63	17	9	—
Fish	78	12	28	2	—

*Some species may be influenced by more than one factor; thus, some rows may exceed 100%.

Case Study: Amphibian Declines

In 1963, herpetologist Jay Savage was hiking through pristine cloud forest in Costa Rica. Reaching a wind-swept ridge, he couldn't believe his eyes. Before him was a huge aggregation of breeding toads. What was so amazing was the color of the toads: bright, eye-dazzling orange, unlike anything he had ever seen before (figure 57.9). The color of the toads was so amazing and unexpected that Savage briefly considered the possibility that colleagues had played a practical joke, getting to the clearing before him and somehow coloring normal toads orange. Realizing that this could not be, he went on to study the toads, eventually describing a species new to science, the golden toad, *Bufo periglenes*.

For the next 24 years, large numbers of toads were seen during the breeding season each spring. Their home

FIGURE 57.9
An extinct species. The golden toad was last seen in the wild in 1989.

was legally recognized as the Monteverde Cloud Forest Reserve, a well-protected, intact, and functioning ecosystem, seemingly a successful model of conservation. Then, in 1988, few toads were seen, and in 1989, only a single male was observed. Since then, despite exhaustive efforts, no more golden toads have been found. Despite living in a well-protected ecosystem, with no obvious threats from pollution, introduced species, overexploitation, or any other factor, the species appears to have gone extinct,

FIGURE 57.10 Amphibian extinction crisis. Circles indicate the number of extinct (*red*) and endangered (*yellow*) species around the world. These numbers are rapidly being revised upward as scientists focus their attention on little-known species, many of which turn out to be in grave danger.

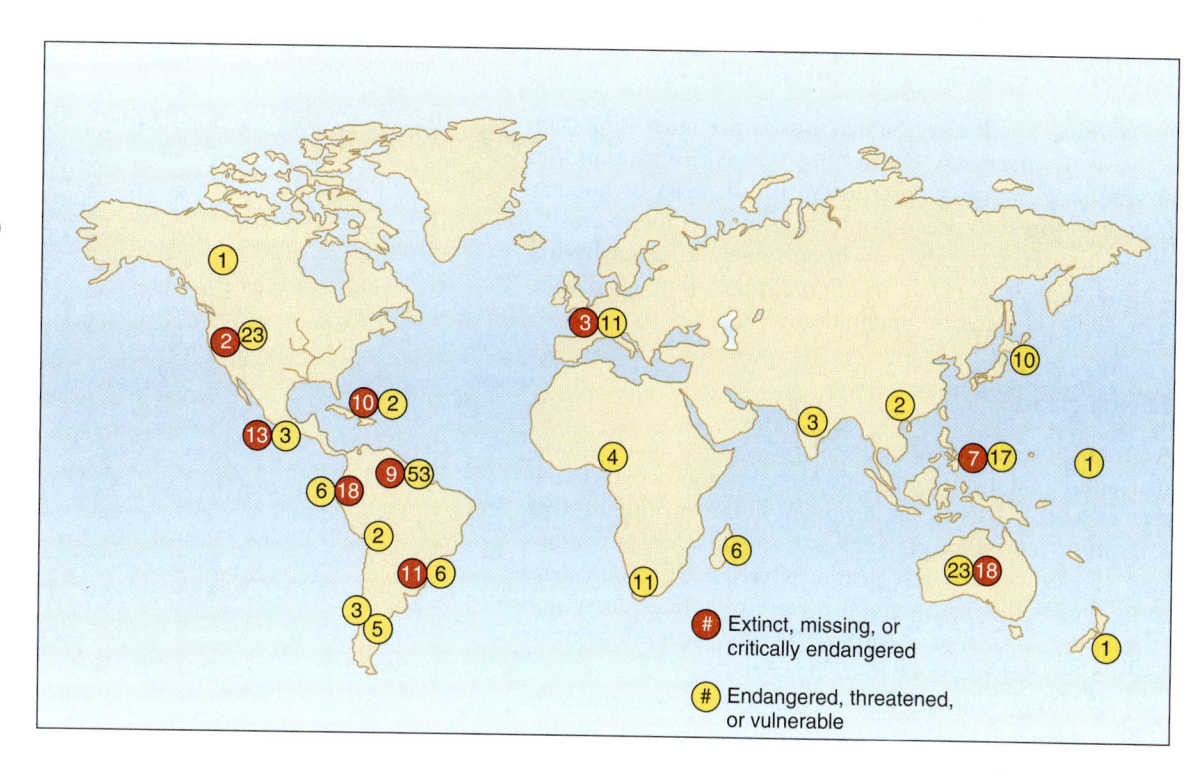

right under the eyes of watchful scientists and conservationists. How could this happen?

Frogs in Trouble. At the first World Herpetological Congress in 1989 in Canterbury, England, frog experts from around the world met to discuss conservation issues relating to frogs and toads. At this meeting, it became clear that the golden toad story was not unique. Experts reported case after case of similar stories: Frog populations that had once been abundant were now decreasing or entirely gone.

Since then, scientists have devoted a great deal of time and effort to determining whether frogs and other amphibian species truly are in trouble and, if so, why. Unfortunately, the situation appears to be even worse than originally suspected. In 2002, experts associated with the University of California at Berkeley reported that at least 32 amphibian species have gone extinct in recent years; another 26 are "missing in action," not having been seen for many years and possibly extinct; and 91 species, in countries as different as Ecuador, Venezuela, Australia, and the United States, are critically endangered (figure 57.10). Moreover, these numbers are probably underestimates; little information exists from many areas of the world, such as Southeast Asia and central Africa. Indeed, in that same year, researchers suggested that as many as 100 species from the island nation of Sri Lanka have recently gone extinct, news that is perhaps not surprising given that 95% of the nation's rain forests have also disappeared in recent times.

Cause for Concern. Amphibian declines are worrisome for several reasons. First, many of the species—including the golden toad—have declined in pristine, well-protected habitats. If species are going extinct in such areas, it bodes ill for our ability to preserve global biological diversity. Second, many amphibian species are particularly sensitive to the state of the environment because of their moist skin, which allows chemicals from the environment to pass into the body, and their use of aquatic habitats for larval stages (such as the tadpoles of frogs), which requires unpolluted water. In other words, amphibians may be analogous to the canaries formerly used in coal mines to detect problems with air quality: If the canaries keeled over, the miners knew they had to get out.

Third, no single cause for amphibian declines is apparent. Although a single cause would be of concern, it would also suggest that a coordinated global effort could reverse the trend, as happened with chlorofluorocarbons and decreasing ozone levels (see chapter 56). However, different species are afflicted by different problems, including habitat destruction, global warming–induced environmental changes, pollution, decreased ozone levels, parasite epidemics, and introduced species. This is an area of active scientific research, and the implication is that the global environment is deteriorating in many different ways. Could amphibians be global "canaries," serving as indicators that the world's environment is in serious trouble?

> **Many factors are responsible for extinction.** Many amphibian species are in trouble around the world. No single cause is responsible, which raises worry about the overall state of the global environment.

Chapter 57 Conservation Biology **1235**

Habitat Loss

As table 57.3 indicates, habitat loss is the most important cause of modern-day extinction. Given the tremendous amounts of ongoing destruction of all types of habitat, from rain forest to ocean floor, this should come as no surprise. Natural habitats may be adversely affected by humans in four ways: (1) destruction, (2) pollution, (3) disruption, and (4) habitat fragmentation.

Destruction

A proportion of the habitat available to a particular species may simply be destroyed. This is a common occurrence in the "clear-cut" harvesting of timber, in the burning of tropical forest to produce grazing land, and in urban and industrial development. Forest clearance has been, and continues to be, by far the most pervasive form of habitat disruption (figure 57.11). Many tropical forests are being cut or burned at a rate of 1% or more per year.

To estimate the effect of reductions in habitat available to a species, biologists often use the well-established observation that larger areas support more species (see figure 55.20). Although this relationship varies according to geographic area, type of organism, and type of area (for example, oceanic islands or patches of habitat on the mainland), in general a tenfold increase in area leads to approximately a doubling in the number of species. This relationship suggests, conversely, that if the area of a habitat is reduced by 90%, so that only 10% remains, then half of all species will be lost. Evidence for this theory comes from a study of extinction rates of birds on habitat islands (that is, islands of a particular type of habitat surrounded by unsuitable habitat) in Finland where the extinction rate was found to be inversely proportional to island size (figure 57.12).

Pollution

Habitat may be degraded by pollution to the extent that some species can no longer survive there. Degradation occurs as a result of many forms of pollution, from acid rain to pesticides. Aquatic environments are particularly vulnerable; for example, many northern lakes in both Europe and North America have been essentially sterilized by acid rain.

Disruption

Human activities may disrupt a habitat enough to make it untenable for some species. For example, visitors to caves in Alabama and Tennessee caused significant population declines in bats over an eight-year period, some as great as 100%. When visits were fewer than one per month, less

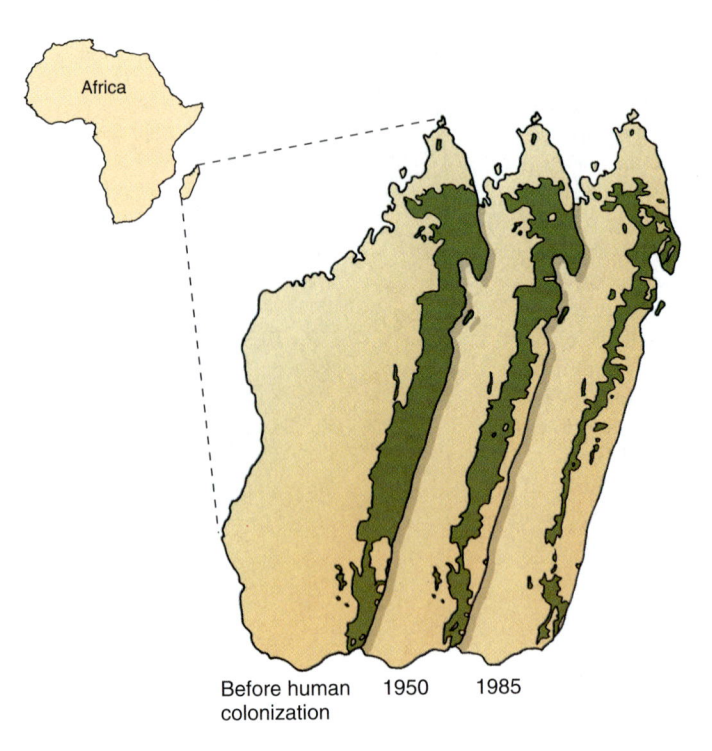

FIGURE 57.11
Extinction and habitat destruction. The rain forest covering the eastern coast of Madagascar, an island off the coast of East Africa, has been progressively destroyed as the island's human population has grown. Ninety percent of the original forest cover is now gone. Many species have become extinct, and many others are threatened, including 16 of Madagascar's 31 primate species.

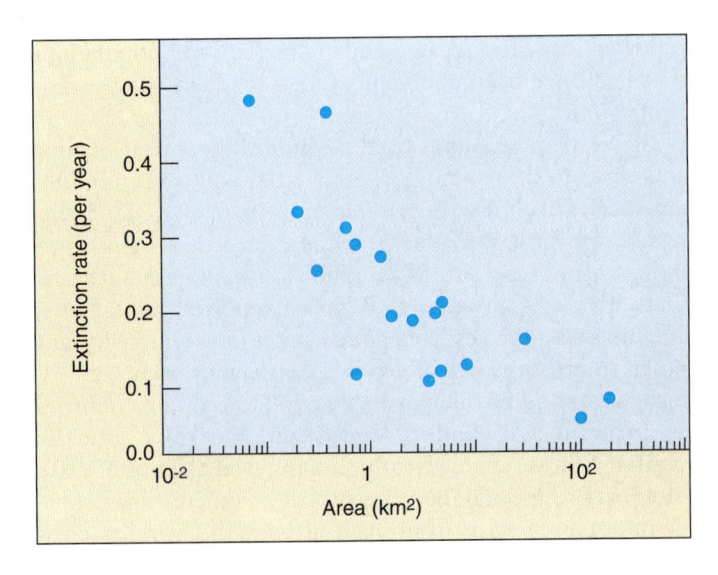

FIGURE 57.12
Extinction and the species-area relationship. The data present percent extinction rates as a function of habitat area for birds on a series of Finnish islands. Smaller islands experience far greater local extinction rates.
Why does extinction rate increase with decreasing island size?

1236   Part VIII   Ecology and Behavior

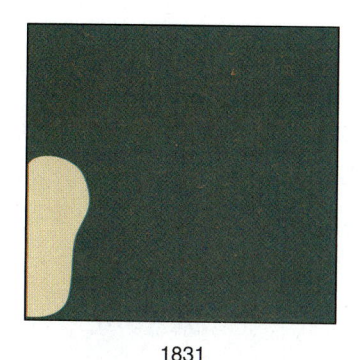

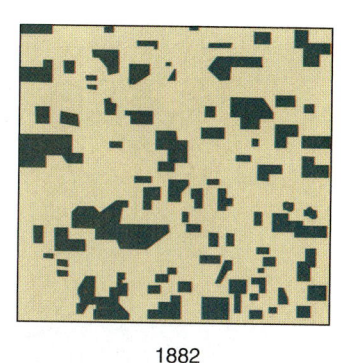

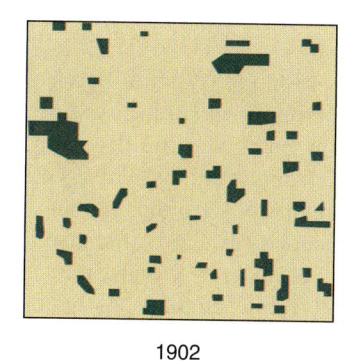

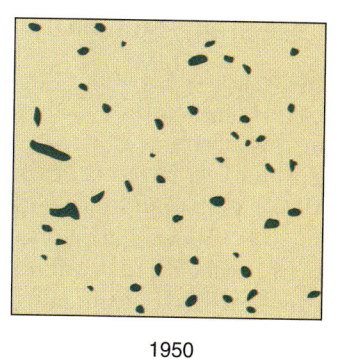

FIGURE 57.13
Fragmentation of woodland habitat. From the time of settlement of Cadiz Township, Wisconsin, the forest has been progressively reduced from a nearly continuous cover to isolated woodlots covering less than 1% of the original area.

than 20% of bats were lost, but caves having more than four visits per month suffered population declines of 86–95%.

Habitat Fragmentation

Loss of habitat by a species frequently results not only in lowered population numbers, but also in fragmentation of the population into unconnected patches (figure 57.13).

A habitat may become fragmented in nonobvious ways, as when roads and habitation intrude into forest. The effect is to carve the populations living in the habitat into a series of smaller populations, often with disastrous consequences because of the relationship between range size and extinction rate. Although detailed data are not available, fragmentation of wildlife habitat in developed temperate areas is thought to be very substantial.

As habitats become fragmented and shrink in size, the relative proportion of the habitat that occurs on the boundary, or edge, increases. **Edge effects** can significantly degrade a population's chances of survival. Changes in microclimate (temperature, wind, humidity, etc.) near the edge may reduce appropriate habitat for many species more than the physical fragmentation suggests. In isolated fragments of rain forest, for example, trees on the edge are exposed to direct sunlight and, consequently, hotter and drier conditions than they are accustomed to in the cool, moist forest interior. As a result, in one study, the biomass of trees within 100 meters of the forest edge decreased by 36% in the first 17 years after fragment isolation.

Also, increasing habitat edges opens up opportunities for parasites and predators, which are both more effective at edges. As fragments decrease in size, the proportion of habitat that is distant from any edge decreases, and consequently, more and more of the habitat is within the range of these predators. Habitat fragmentation is blamed for local extinctions in a wide range of species.

FIGURE 57.14
A study of habitat fragmentation. Biodiversity was monitored in the isolated patches of rain forest in Manaus, Brazil, before and after logging. Fragmentation led to significant species loss within patches.

The impact of habitat fragmentation can be seen clearly in a major study done in Manaus, Brazil, where the rain forest was commercially logged. Landowners agreed to preserve patches of rain forest of various sizes, and censuses of these patches were taken before the logging started, while they were still part of a continuous forest. After logging, species began to disappear from the now-isolated patches (figure 57.14). First to go were the monkeys, which have large home ranges. Birds that prey on ant colonies followed, disappearing from patches too small to maintain enough ant colonies to support them.

Because some species, such as monkeys, require large patches, large fragments are indispensable if we wish to preserve high levels of biodiversity. The take-home lesson is that preservation programs will need to provide suitably large habitat fragments to avoid this impact.

Chapter 57 Conservation Biology 1237

Case Study: Songbirds

Every year since 1966, the U.S. Fish and Wildlife Service has organized thousands of amateur ornithologists and bird-watchers in an annual bird count called the Breeding Bird Survey. In recent years, a shocking trend has emerged. While year-round residents that prosper around humans, such as robins, starlings, and blackbirds, have increased their numbers and distribution over the last 30 years, forest songbirds have declined severely. The decline has been greatest among long-distance migrants such as thrushes, orioles, tanagers, catbirds, vireos, buntings, and warblers. These birds nest in northern forests in the summer, but spend their winters in South or Central America or the Caribbean Islands.

In many areas of the eastern United States, more than three-quarters of the tropical migrant bird species have declined significantly. Rock Creek Park in Washington, D.C., for example, has lost 90% of its long-distance migrants in the past 20 years. Nationwide, American redstarts declined about 50% in the single decade of the 1970s. Studies of radar images from National Weather Service stations in Texas and Louisiana indicate that only about half as many birds fly over the Gulf of Mexico each spring compared to the numbers in the 1960s. This suggests a total loss of about half a billion birds.

The culprit responsible for this widespread decline appears to be habitat fragmentation and loss. Fragmentation of breeding habitat and nesting failures in the summer nesting grounds of the United States and Canada have had a major negative impact on the breeding of woodland songbirds. Many of the most threatened species are adapted to deep woods and need an area of 25 acres or more per pair to breed and raise their young. As woodlands are broken up by roads and developments, it is becoming increasingly difficult for them to find enough contiguous woods to nest successfully.

A second and perhaps even more important factor is the availability of critical winter habitat in Central and South America. Studies of the American redstart clearly indicate that birds with better winter habitat have a superior chance of successfully migrating back to their breeding grounds in the spring. Peter Marra and Richard Holmes of Dartmouth College and Keith Hobson of the Canadian Wildlife Service captured birds, took blood samples, and measured the levels of the stable carbon isotope ^{13}C. Plants growing in the best overwintering habitats in Jamaica and Honduras (mangroves and wetland forests) have low levels of ^{13}C, and so do the redstarts that feed on them. Of these wet-forest birds, 65% maintained or gained weight over the winter. By contrast, plants growing in substandard dry scrub have high levels of ^{13}C, and so do the redstarts that feed on them. Scrub-dwelling birds lost up to 11% of their body mass over the winter. Now here's the key: Birds that winter in the substandard scrub leave later in the spring on the long flight to northern breeding grounds, arrive later at their summer homes, and have fewer young (figure 57.15). The proportion of ^{13}C in birds arriving in New Hampshire breeding grounds increases as spring wears on and scrub-overwintering stragglers belatedly arrive. Thus, loss of mangrove habitat in the neotropics is having a real negative impact. As the best habitat disappears, overwintering birds fare poorly, and this leads to population declines. Unfortunately, the Caribbean lost about 10% of its mangroves in the 1980s, and continues to lose about 1% per year. This loss of key habitat appears to be a driving force in the looming extinction of songbirds.

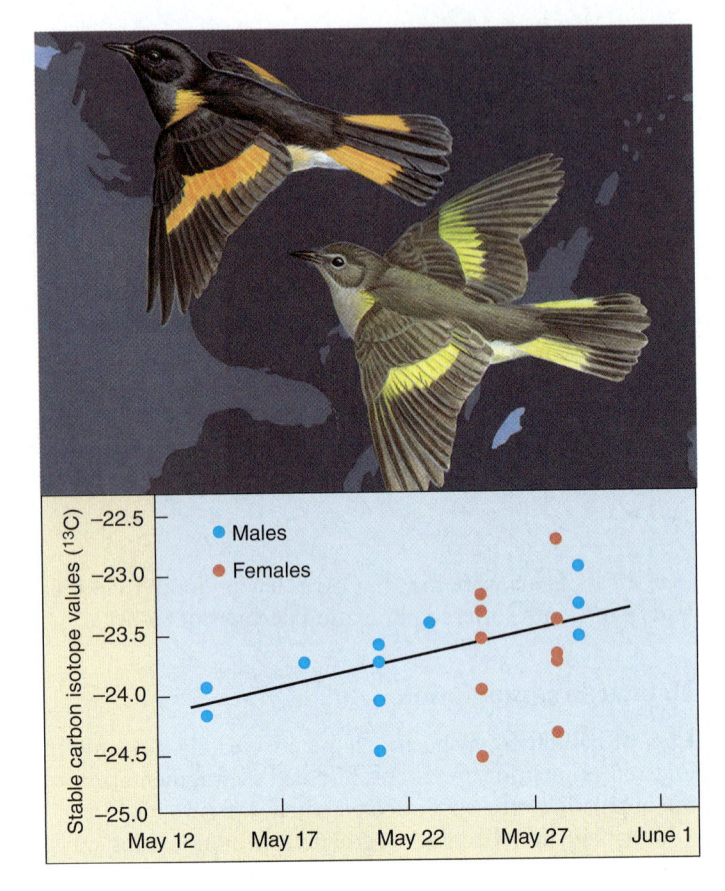

FIGURE 57.15

The American redstart, a migratory songbird. The numbers of this species are in serious decline. The graph presents data on the level of ^{13}C in redstarts arriving at summer breeding grounds. Early arrivals, with the best shot at reproductive success, have lower levels of ^{13}C, indicating they wintered in more favorable mangrove-wetland forest habitats.

> As habitats are destroyed, remaining habitat becomes fragmented, increasing the threat to many species. Fragmentation of summer breeding grounds and loss of high-quality overwintering habitat seem to be contributing to a marked decline in migratory songbird species.

Overexploitation

Species that are hunted or harvested by humans have historically been at grave risk of extinction, even when the species is initially very abundant. A century ago, the skies of North America were darkened by huge flocks of passenger pigeons, but after being hunted as free and tasty food, they were driven to extinction. The bison that used to migrate in enormous herds across the central plains of North America only narrowly escaped the same fate.

The existence of a commercial market often leads to overexploitation of a species. The international trade in furs, for example, has severely reduced the numbers of chinchilla, vicuña, otter, and many cat species. The harvesting of commercially valuable trees provides another example: Almost all West Indies mahogany trees (*Swietenia mahogani*) have been logged, and the extensive cedar forests of Lebanon, once widespread at high elevations, now survive in only a few isolated groves.

A particularly telling example of overexploitation is the commercial harvesting of fish in the North Atlantic. During the 1980s, fishing fleets continued to harvest large amounts of cod off Newfoundland, even as the population numbers declined precipitously. By 1992, the cod population had dropped to less than 1% of its original numbers. The American and Canadian governments have closed the fishery, but no one can predict whether the fish populations will recover. The Atlantic bluefin tuna has experienced a 90% population decline in the past 10 years. The swordfish has declined even further. In both cases, the drop has led to even more intense fishing of the remaining populations.

Case Study: Whales

Whales, the largest living animals that ever evolved, are rare in the world's oceans today, their numbers driven down by commercial whaling. Before the advent of cheap, high-grade oils manufactured from petroleum in the early twentieth century, oil made from whale blubber was an important commercial product in the worldwide marketplace. In addition, the fine, latticelike structure termed "baleen" used by baleen whales to filter-feed plankton from seawater was used in undergarments. Because a whale is such a large animal, each individual captured is of significant commercial value.

Right whales were the first to bear the brunt of commercial whaling. They were called "right" whales because they were slow and easy to capture, and they provided up to 150 barrels of blubber oil and abundant baleen, making them the right whale for a commercial whaler to hunt.

As the right whale declined in the eighteenth century, whalers turned to the gray, humpback, and bowhead whales. As their numbers declined, whalers turned to the blue, the largest of all whales, and when those were decimated, to the fin, then the Sei, and then the sperm whales.

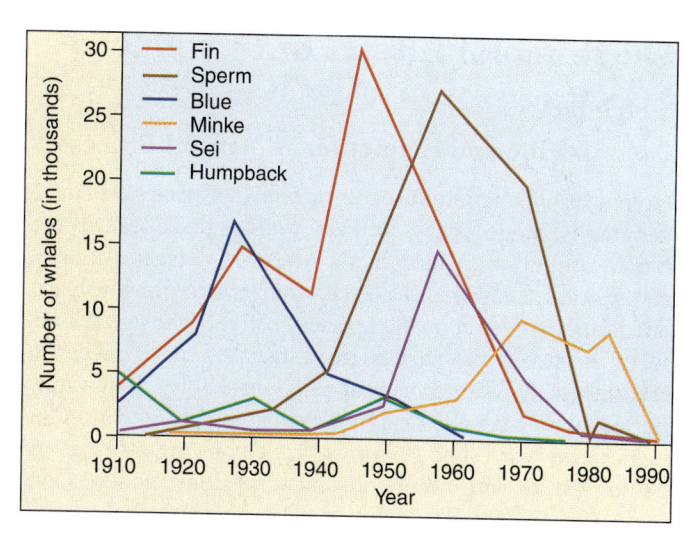

FIGURE 57.16
World catch of whales in the twentieth century. Each species is hunted in turn until its numbers fall so low that hunting it becomes commercially unprofitable.
Why might whale populations fail to recover once hunting is stopped?

As each species of whale became the focus of commercial whaling, its numbers began a steep decline (figure 57.16).

Hunting of right whales was made illegal in 1935. By then, they had been driven to the brink of extinction, their numbers less than 5% of what they had been. Although protected ever since, their numbers have not recovered in either the North Atlantic or the North Pacific. By 1946, several other whale species faced imminent extinction, and whaling nations formed the International Whaling Commission (IWC) to regulate commercial whale hunting. Like a fox guarding the henhouse, the IWC for decades did little to limit whale harvests, and whale numbers continued to decline steeply. Finally, in 1974, when the numbers of all but the small minke whales had been driven down, the IWC banned hunting of blue, gray, and humpback whales, and instituted partial bans on other species. The rule was violated so often, however, that the IWC in 1986 instituted a worldwide moratorium on all commercial killing of whales. While some commercial whaling continues, often under the guise of harvesting for scientific studies, annual whale harvests have dropped dramatically in the last 20 years.

Some species appear to be recovering, while others do not. Humpback numbers have more than doubled since the early 1960s, increasing nearly 10% annually, and Pacific gray whales have fully recovered to their previous numbers of about 20,000 animals after being hunted to less than 1000. Right, sperm, fin, and blue whales have not recovered, and no one knows whether they will.

> Overharvesting has driven most large whale species to the brink of extinction. Stopping the harvest has allowed recovery of some, but not all, species.

Chapter 57 Conservation Biology 1239

Detrimental Effects of Introduced Species

Colonization and Extinction as Natural Processes

Colonization, a natural process by which a species expands it geographic range, occurs in many ways: A flock of birds gets blown off course, a bird eats a fruit and defecates its seed miles away, or lowered sea levels connect two previously isolated landmasses, allowing species to freely move back and forth. Such events—particularly those leading to successful establishment of a new population—probably occur rarely, but when they do, the resulting change to natural communities can be large. The reason is that colonization brings together species with no previous history of interaction. Consequently, ecological interactions may be particularly strong because the species have not evolved ways of adjusting to the presence of each other, such as adaptations to avoid predation or minimize competitive effects.

The paleontological record documents many cases in which geologic changes brought previously isolated species together, such as when the Isthmus of Panama emerged above the sea approximately three million years ago, connecting the previously isolated faunas and floras of North and South America. In some cases, the result has been an increase in species diversity, but in other cases, invading species have led to the extinction of natives.

Human Influence on the Process

Unfortunately, what was naturally a rare process has become all too common in recent years, thanks to the actions of humans. Species introductions due to human activities occur in many ways, sometimes intentionally, but usually not. Plants and animals can be transported in the ballast of large ocean vessels, in nursery plants, as stowaways in boats, cars, and planes, as beetle larvae within wood products—even as seeds in the mud stuck to the bottom of a shoe! Overall, some researchers estimate that as many as 50,000 species have been introduced into the United States. The results of such introductions are occasionally catastrophic.

The effects of introductions on humans have been enormous. In the United States alone, nonnative species cost the economy an estimated $140 billion per year. For example, dozens of foreign weeds in Colorado have covered more than a million acres. Just three of these species cost wheat farmers tens of millions of dollars. At the same time, leafy spurge, a plant from Europe, outcompetes native grasses, ruining rangeland for cattle at a price tag of $144 million per year. The zebra mussel, a mollusk native to the Black Sea, is a huge problem throughout much of the eastern and central United States, where it can attain densities as high as 700,000/m^2, clogging pipes, including those for water and power plants, and causing an estimated $3–5 billion damage a year.

Introduced species can have impacts on human health as well. For example, West Nile fever, the death rate of which

FIGURE 57.17
The akiapolaau, an endangered Hawaiian bird. More than two-thirds of Hawaii's native bird species are now extinct or have been greatly reduced in population size. Bird faunas on islands around the world have experienced similar declines after human arrival.

is steadily mounting, was probably introduced from Africa or the Middle East in the late 1990s.

The effect of species introductions on native ecosystems is equally dramatic. Islands have been particularly affected. For example, a lighthouse keeper's cat singlehandedly (singlepawedly?) wiped out an entire species, the Stephens Island wren. Rats had a devastating effect throughout the South Pacific where bird species nested on the ground and had no defense against the voracious predators to which they were evolutionarily naive. More recently, the brown tree snake, introduced to the island of Guam, essentially eliminated all species of forest birds. In Hawaii, the problem has been slightly different: Introduced mosquitoes brought with them avian malaria, to which the native species had evolved no resistance. The result is that more than 100 species (more than 70% of the native fauna) either went extinct or are now restricted to higher and cooler elevations where the mosquitoes don't occur (figure 57.17).

The effects of introduced species are not always direct, but instead may reverberate throughout an ecosystem. For example, the Argentine ant has spread through much of the southern United States, eliminating most native ant species with which it comes in contact. The extinction of these ant species has had a dramatic negative effect on the coastal horned lizard, which specializes on the larger native species. In their absence, the lizards have shifted to less-preferred prey species. In addition, the native species consume seeds, and in the process, play an important role in seed dispersal. Argentine ants, by contrast, do not eat seeds. In South Africa, where the Argentine ant has also appeared, at least one plant species has experienced decreased reproductive success due to the loss of its dispersal agent.

The most dramatic effects of introduced species, however, occur when entire ecosystems are transformed. Some

1240 Part VIII Ecology and Behavior

plant species can completely overrun a habitat, displacing all native species and turning the area into a monoculture (that is, an area occupied by a single species). In California, the yellow star thistle now covers 4 million hectares of what was once highly productive grassland. In Hawaii, a small tree native to the Canary Islands, *Myrica faya*, has spread widely. Because it is able to fix nitrogen at high rates, it has caused a 90-fold increase in the nitrogen content of the soil, thus allowing other, nitrogen-requiring species to invade. In South Africa, several introduced tree species have such high water uptake rates that nearby rivers are now dry.

Efforts to Combat Introduced Species

Once an introduced species becomes established, eradicating it is often extremely difficult, expensive, and time-consuming. Some efforts—such as the removal of goats and rabbits from certain small islands—have been successful, but many other efforts have failed. The best hope for stopping the ravages of introduced species is to prevent species from being introduced in the first place. Although easier said than done, government agencies are now working strenuously to put into place procedures that can intercept species in transit, before they have the opportunity to become established.

Case Study: Lake Victoria Cichlids

Lake Victoria, an immense, shallow, freshwater sea about the size of Switzerland in the heart of equatorial East Africa, had until 1954 been home to an incredibly diverse collection of over 300 species of cichlid fishes (see figure 23.14). These small, perchlike fish range from 2 to 10 inches in length, with males having endless varieties of color. Today, most of these cichlid species are threatened, endangered, or extinct.

What happened to bring about the abrupt loss of so many endemic cichlid species? In 1954, the Nile perch, *Lates niloticus*, a commercial fish with a voracious appetite, was introduced on the Ugandan shore of Lake Victoria. Nile perch, which grow to over 4 feet in length, were to form the basis of a new fishing industry (figure 57.18). For decades, these perch did not seem to have a significant impact; over 30 years later, in 1978, Nile perch still made up less than 2% of the fish harvested from the lake.

Then something happened to cause the Nile perch population to explode and to spread rapidly through the lake, eating their way through the cichlids. By 1986, Nile perch constituted nearly 80% of the total catch of fish from the lake and the endemic cichlid species were virtually gone. Over 70% of cichlid species disappeared, including all open-water species.

So what happened to kick-start the mass extinction of the cichlids? The trigger seems to have been eutrophication. Before 1978, Lake Victoria had high oxygen levels at

FIGURE 57.18
Victor and vanquished. The Nile perch (larger fishes in foreground), a commercial fish introduced into Lake Victoria as a potential food source, is responsible for the virtual extinction of hundreds of species of cichlid fishes (smaller fishes in tub).

all depths, down to the bottom layers more than 60 meters deep. However, by 1989 high inputs of nutrients from agricultural runoff and sewage from towns and villages had led to algal blooms that severely depleted oxygen levels in deeper parts of the lake. Cichlids feed on algae, and initially their population numbers are thought to have risen in response to this increase in their food supply, but unlike the conditions during similar algal blooms of the past, the Nile perch was present to take advantage of the situation. With a sudden increase in its food supply (cichlids), the numbers of Nile perch exploded, and they simply ate all available cichlids.

Since 1990, the situation has been compounded by the introduction into Lake Victoria of a floating water weed from South America, the water hyacinth *Eichornia crassipes*. Extremely fecund under eutrophic conditions, thick mats of water hyacinth soon covered entire bays and inlets, choking off the coastal habitats of non-open-water cichlids.

Human activities are increasingly introducing many species all over the world. Such introductions often have catastrophic effects on the native fauna and flora, and can cause enormous economic damage.

Chapter 57 Conservation Biology **1241**

Disruption of Ecosystems

Species often become vulnerable to extinction when their web of ecological interactions becomes seriously disrupted. Because of the many relationships linking species in an ecosystem (see chapter 55), human activities that affect one species can ramify throughout an ecosystem, ultimately affecting many other species.

A recent case in point are the sea otters that live in the cold waters off Alaska and the Aleutian Islands. Otter populations have declined sharply in recent years. In a 500-mile stretch of coastline, otter numbers have dropped from 53,000 in the 1970s to an estimated 6000, a plunge of nearly 90%. Investigating this catastrophic decline, marine ecologists uncovered a chain of interactions among the species of the ocean and kelp forest ecosystems, a falling-domino series of lethal effects that illustrates the concepts of both top-down and bottom-up trophic cascades discussed in chapter 55.

Case Study: Alaskan Near-Shore Habitat

The first in a series of events leading to the sea otter's decline seems to have been the heavy commercial harvesting of whales (see the case study earlier in this section, page 1239). Without whales to keep their numbers in check, ocean zooplankton thrived, leading in turn to proliferation of a species of fish called pollock that feeds on the abundant zooplankton. Given this ample food supply, the pollock proved to be very successful competitors of other northern Pacific fish, such as herring and ocean perch, so that levels of these other fish fell steeply in the 1970s.

Then the falling chain of dominoes began to accelerate. The decline in the nutritious forage fish led to an ensuing crash in Alaskan populations of Steller's sea lions and harbor seals, for which pollock did not provide sufficient nourishment. Numbers of these pinniped species have fallen precipitously since the 1970s.

Pinnipeds are the major food of orcas, also called killer whales. Faced with a food shortage, some orcas turned to the next best thing: sea otters. In one bay where the entrance from the sea was too narrow and shallow for orcas to enter, only 12% of the sea otters have disappeared, while in a similar bay that orcas could enter easily, two-thirds of the otters disappeared in a year's time.

Without otters to eat them, the population of sea urchins exploded, eating the kelp and thus "deforesting" the kelp forests and denuding the ecosystem (figure 57.19). As a result, fish species that live in the kelp forest, such as sculpins and greenlings (a cod relative), are declining.

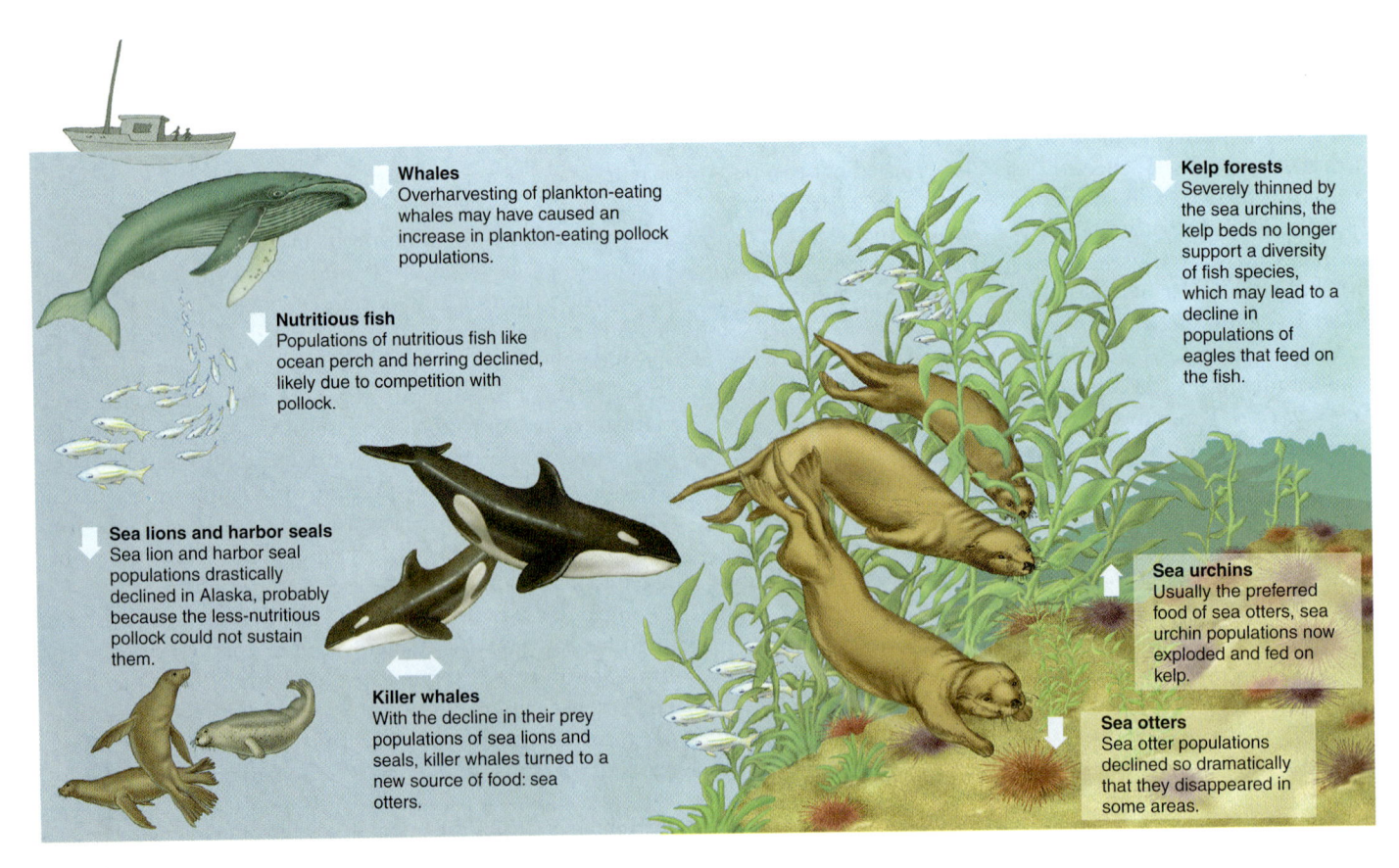

FIGURE 57.19
Disruption of the kelp forest ecosystem. Overharvesting by commercial whalers altered the balance of fish in the ocean ecosystem, inducing killer whales to feed on sea otters, a keystone species of the kelp forest ecosystem.

Preserving Keystone Species

As discussed in chapter 54, a keystone species is a species that exerts a particularly strong influence on the structure and functioning of a certain ecosystem. The sea otters of figure 57.19 are a keystone species of the kelp forest ecosystem, and their removal can have disastrous consequences. There is no hard-and-fast line that allows us to clearly identify keystone species. Rather, it is a qualitative concept, a statement that a species plays a particularly important role in its community. Keystone species are usually characterized by the strength of their impact on their community. *Community importance* measures the change in some quantitative aspect of the ecosystem (for example, species richness, productivity, or nutrient cycling) per unit of change in the abundance of a species.

Case Study: Flying Foxes. The severe decline of many species of pteropodid bats, or "flying foxes," in the Old World tropics is an example of how the loss of a keystone species can dramatically affect the other species living within an ecosystem, sometimes even leading to a cascade of further extinctions (figure 57.20). These bats have very close relationships with important plant species on the islands of the Pacific and Indian Oceans. The family Pteropodidae contains nearly 200 species, approximately one-quarter of them in the genus *Pteropus*, and is widespread on the islands of the South Pacific, where they are the most important—and often the only—pollinators and seed dispersers. A study in Samoa found that 80 to 100% of the seeds landing on the ground during the dry season were deposited by flying foxes. Many species are entirely dependent on these bats for pollination. Some have evolved features such as night-blooming flowers that prevent any other potential pollinators from taking over the role of the fruit bats.

In Guam, where the two local species of flying fox have recently been driven extinct or nearly so, the impact on the ecosystem appears to be substantial. Botanists have found that some plant species are not fruiting or are doing so only marginally, producing fewer fruits than normal. Fruits are not being dispersed away from parent plants, so offspring shoots are being crowded out by the adults.

Flying foxes are being driven to extinction by human hunters who kill them for food and for sport, and by orchard farmers who consider them pests. Flying foxes are particularly vulnerable because they live in large, easily seen groups of up to a million individuals. Because they move in regular and predictable patterns and can be easily tracked to their home roost, hunters can easily bag thousands at a time.

Programs aimed at preserving particular species of flying foxes are only just beginning. One particularly successful example is the program to save the Rodrigues fruit bat, *Pteropus rodricensis*, which occurs only on Rodrigues Island in the Indian Ocean near Madagascar. The population dropped from about 1000 individuals in 1955 to fewer than 100 by 1974, largely due to the loss of the fruit bat's forest habitat to farming. Since 1974, the species has been legally protected, and the forest area of the island is being increased through a tree-planting program. Eleven captive-breeding colonies have been established, and the bat population is now increasing rapidly. The combination of legal protection, habitat restoration, and captive breeding has in this instance produced a very effective preservation program.

FIGURE 57.20
Preserving keystone species. The flying fox is a keystone species on many Old World tropical islands. It pollinates many of the plants, and is a key disperser of seeds. Its elimination due to hunting and habitat loss is having a devastating effect on the ecosystems of many South Pacific Islands.

> Because of the interrelationships among species within an ecosystem, activities that directly harm one species may indirectly have large impacts on many other species as well.

Chapter 57 Conservation Biology **1243**

The Perils of Small Population Size

Because of the factors just discussed, populations of many species are fragmented and reduced in size. Such populations are particularly prone to extinction.

Demographic Factors

Small populations are vulnerable to a variety of demographic factors. By nature of their small size, they are ill-equipped to withstand a catastrophic event, such as a flood, forest fire, or disease epidemic. One example is the heath hen. Although the species was once common throughout the eastern United States, hunting pressure in the eighteenth and nineteenth centuries eventually eliminated all but one population, on the island of Martha's Vineyard near Cape Cod, Massachusetts. Protected in a nature preserve, the population was increasing in number until a fire destroyed most of the preserve's habitat. The small surviving population was then ravaged the next year by an unusual congregation of predatory birds, followed shortly thereafter by a disease epidemic.

When populations become extremely small, bad luck can spell the end. For example, the dusky seaside sparrow (figure 57.21), a now-extinct subspecies that was found on the east coast of Florida, dwindled to a population of five individuals, all of which happened to be males. In a large population, the probability that all individuals will be of one sex is infinitesimal. But in small populations, just by the luck of the draw, it is possible that five or ten or even 20 consecutive births will all be individuals of one sex, and that can be enough to send a species to extinction. In addition, when populations are small, individuals may have trouble finding each other (the Allee effect discussed in chapter 53), thus leading the population into a downward spiral toward extinction.

Lack of Genetic Variability

Small populations face a second dilemma. Because of their low numbers, such populations are prone to the loss of genetic variation as a result of genetic drift (figure 57.22). Indeed, many small populations contain little or no genetic variability. The result of such genetic homogeneity can be catastrophic. Genetic variation is beneficial to a population both because of heterozygote advantage (see chapter 21) and because genetically variable individuals tend not to have two copies of deleteriously recessive alleles. As a result, populations lacking variation are often composed of sickly, unfit, or sterile individuals. Indeed, in the laboratory, groups of rodents and fruit flies that are maintained at small population sizes often perish after a few generations as each generation becomes less robust and fertile than the preceding one. Although it is difficult to demonstrate that a species has gone

FIGURE 57.21
Alive no more. This male was one of the last dusky seaside sparrows, a subspecies that no longer exists.

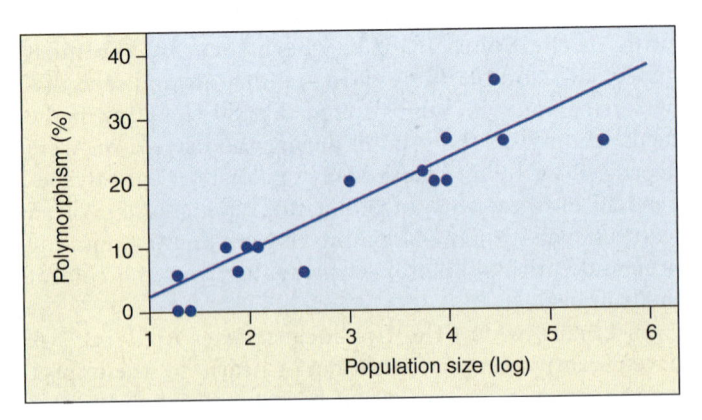

FIGURE 57.22
Loss of genetic variability in small populations. The percentage of polymorphic genes in isolated populations of the tree *Halocarpus bidwillii* in the mountains of New Zealand is a sensitive function of population size.
Why do small populations lose genetic variation?

extinct because of lack of genetic variation, studies of both zoo and natural populations clearly reveal that more genetically variable individuals have greater fitness. Furthermore, in the longer term, populations with limited genetic variation have diminished ability to adapt to changing environments.

Demographic and Genetic Factors Interact

As populations decrease in size, demographic and genetic factors feed on each other, causing what has been termed an "extinction vortex." That is, as a population gets smaller, it becomes more vulnerable to demographic catastrophes. In turn, genetic variation starts to be lost, causing reproductive rates to decline and population numbers to decline even further, and so on. Eventually, the population disappears entirely, but attributing its demise to one particular factor would be misleading.

Case Study: Prairie Chickens

The greater prairie chicken (*Tympanuchus cupido pinnatus*) is a showy, two-pound wild bird renowned for its flamboyant mating rituals (figure 57.23). Abundant in many midwestern states, the prairie chickens in Illinois have in the past six decades undergone a population collapse. Once, enormous numbers of birds occurred throughout the state, but with the 1837 introduction of the steel plow, the first that could slice through the deep, dense root systems of prairie grasses, the Illinois prairie began to be replaced by farmland, and by the turn of the century, the prairie had vanished. By 1931, a related subspecies known as the heath hen (*Tympanuchus cupido cupido*) had become extinct in Illinois. The greater prairie chicken fared little better, its numbers falling to 25,000 statewide in 1933 and then to 2000 in 1962. In surrounding states with less intensive agriculture, it continued to prosper.

In 1962 and 1967, sanctuaries were established in Illinois to attempt to preserve the prairie chicken. But privately owned grasslands kept disappearing, along with their prairie chickens, and by the 1980s the birds were extinct in Illinois except for the two preserves. And even there, their numbers kept falling. By 1990, the egg hatching rate, which at one time had averaged between 91 and 100%, had dropped to an extremely low 38%. By the mid-1990s, the count of males had dropped to as low as six in each sanctuary.

What was wrong with the sanctuary populations? One suggestion was that because of very small population sizes and a mating ritual whereby one male may dominate a flock, the Illinois prairie chickens had lost so much genetic variability as to create serious inbreeding problems. To test this idea, biologists at the University of Illinois compared DNA from frozen tissue samples of birds that had died in Illinois between 1974 and 1993, and found that in recent years Illinois birds had indeed become genetically less diverse. After extracting DNA from tissue in the roots of feathers from stuffed birds collected in the 1930s from the same population, the researchers found that Illinois birds had lost fully one-third of the genetic diversity of birds living in the same place before the population collapse of the 1970s. By contrast, prairie chicken populations in other states still contained much of the genetic variation that had disappeared from Illinois populations.

Now the stage was set to halt the Illinois prairie chicken's race toward extinction. Wildlife managers began to transplant birds from genetically diverse populations of Minnesota, Kansas, and Nebraska to Illinois. Between 1992 and 1996, a total of 518 out-of-state prairie chickens were brought in to interbreed with the Illinois birds, and hatching rates were back up to 94% by 1998. It looks as though the Illinois prairie chickens have been saved from extinction.

The key lesson to be learned is the importance of not allowing things to go too far—not to drop down to a single isolated population. Without the outlying genetically different populations, the prairie chickens in Illinois could not have been saved. For example, when the last population of the dusky seaside sparrow lost its last female, there was no other source of females and the subspecies went extinct.

FIGURE 57.23
A mating ritual. The male prairie chicken inflates bright orange air sacs, part of his esophagus, into balloons on each side of his head. As air is drawn into the sacs, it creates a three-syllable "boom-boom-boom" that can be heard for miles.

Small populations face a variety of perils: They have less ability to rebound from catastrophes and are vulnerable to loss of genetic variation. Worst of all, these problems build on each other, magnifying their impact.

57.3 Successful recovery efforts need to be multidimensional.

Approaches for Preserving Endangered Species

Once the cause of a species' endangerment is known, it becomes possible to design a recovery plan. If the cause is commercial overharvesting, regulations can be issued to lessen the impact and protect the threatened species. If the cause is habitat loss, plans can be instituted to restore the habitat. Loss of genetic variability in isolated subpopulations can be countered by transplanting individuals from genetically different populations. Populations in immediate danger of extinction can be captured, introduced into a captive-breeding program, and later reintroduced to other suitable habitat.

Of course, all of these solutions are extremely expensive. As Bruce Babbitt, Secretary of the Interior in the Clinton administration, noted, it is much more economical to prevent "environmental trainwrecks" from occurring than to clean them up afterwards. Preserving ecosystems and monitoring species before they are threatened is the most effective means of protecting the environment and preventing extinctions.

Habitat Restoration

Conservation biology typically concerns itself with preserving populations and species in danger of decline or extinction. Conservation, however, requires that there be something left to preserve, while in many situations, conservation is no longer an option. Species, and in some cases whole communities, have disappeared or been irretrievably modified. The clear-cutting of the temperate forests of Washington State leaves little behind to conserve, as does converting a piece of land into a wheat field or an asphalt parking lot. Redeeming these situations requires restoration rather than conservation.

Three quite different sorts of habitat restoration programs might be undertaken, depending on the cause of the habitat loss.

Pristine Restoration.
In ecosystems where all species have been effectively removed, conservationists might attempt to restore the plants and animals that are the natural inhabitants of the area, if these species can be identified. When abandoned farmland is to be restored to prairie, as in figure 57.24, how would you know what to plant? Although it is in principle possible to reestablish each of the original species in their original proportions, rebuilding a community requires knowing the identities of all the original inhabitants and the ecologies of each of the species. We rarely have this much information, so no restoration is ever truly pristine.

FIGURE 57.24
Habitat restoration. The University of Wisconsin–Madison Arboretum has pioneered restoration ecology. (*a*) The restoration of the prairie was at an early stage in November 1935. (*b*) The prairie as it looks today. This picture was taken at approximately the same location as the 1935 photograph.

Removing Introduced Species.
Sometimes the habitat of a species has been destroyed by a single introduced species. In such a case, habitat restoration involves removing the introduced species. Restoration of the once-diverse cichlid fishes to Lake Victoria will require more than breeding and restocking the endangered species. The introduced water hyacinth and Nile perch populations will have to be brought under control or removed, and eutrophication will have to be reversed.

It is important to act quickly if an introduced species is to be removed. When aggressive African bees (the so-called "killer bees") were inadvertently released in Brazil, they remained confined to the local area only one season. Now they occupy much of the western hemisphere.

Cleanup and Rehabilitation.
Habitats seriously degraded by chemical pollution cannot be restored until the pollution is cleaned up. The successful restoration of the Nashua River in New England is one example of how a concerted effort can succeed in restoring a heavily polluted habitat to a relatively pristine condition.

Captive Breeding

Recovery programs, particularly those focused on one or a few species, must sometimes involve direct intervention in natural populations to avoid an immediate threat of extinction.

The Peregrine Falcon. American populations of birds of prey, such as the peregrine falcon (*Falco peregrinus*), began an abrupt decline shortly after World War II. Of the approximately 350 breeding pairs east of the Mississippi River in 1942, all had disappeared by 1960. The culprit proved to be the chemical pesticide DDT (dichlorodiphenyltrichloroethane) and related organochlorine pesticides. Birds of prey are particularly vulnerable to DDT because they feed at the top of the food chain, where DDT becomes concentrated. DDT interferes with the deposition of calcium in the bird's eggshells, causing most of the eggs to break before they are ready to hatch.

The use of DDT was banned by federal law in 1972, causing levels in the eastern United States to fall quickly. However, there were no peregrine falcons left in the eastern United States to reestablish a natural population. Falcons from other parts of the country were used to establish a captive-breeding program at Cornell University in 1970, with the intent of reestablishing the peregrine falcon in the eastern United States by releasing offspring of these birds. By the end of 1986, over 850 birds had been released in 13 eastern states, producing an astonishingly strong recovery (figure 57.25).

The California Condor. Numbers of the California condor (*Gymnogyps californianus*), a large, vulturelike bird with a wingspan of nearly 3 meters, have been declining gradually for the past 200 years. By 1985, condor numbers had dropped so low that the bird was on the verge of extinction. Six of the remaining 15 wild birds disappeared that year alone. The entire breeding population of the species consisted of the birds remaining in the wild and an additional 21 birds in captivity. In a last-ditch attempt to save the condor from extinction, the remaining birds were captured and placed in a captive-breeding population. The breeding program was set up in zoos, with the goal of releasing offspring on a large, 5300-ha ranch in prime condor habitat. Birds were isolated from human contact as much as possible, and closely related individuals were prevented from breeding. By early 2002, the captive population of California condors had reached over 120 individuals. Thirty-two captive-reared condors have been released successfully in California at two sites in the mountains north of Los Angeles, after extensive prerelease training to avoid power poles and people. All of the released birds seem to be doing well, and another 21 birds released into the Grand Canyon have adapted successfully. Biologists are particularly excited by breeding activities that resulted in the first-ever offspring produced in the wild by captive-reared parents.

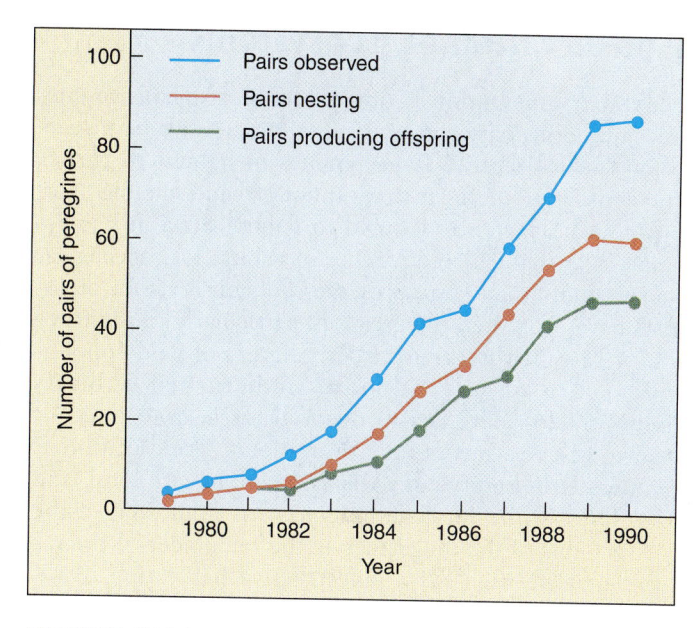

FIGURE 57.25
Captive breeding. The peregrine falcon has been reestablished in the eastern United States by releasing captive-bred birds over a period of 10 years.

Yellowstone Wolves. The ultimate goal of captive-breeding programs is not simply to preserve interesting species, but rather to restore ecosystems to a balanced, functional state. Yellowstone Park has been an ecosystem out of balance, due in large part to the systematic extermination of the gray wolf (*Canis lupus*) in the park early in this century. Without these predators to keep their numbers in check, herds of elk and deer expanded rapidly, damaging vegetation so that the elk themselves starve in time of scarcity. In an attempt to restore the park's natural balance, two complete wolf packs from Canada were released into the park in 1995 and 1996. The wolves adapted well, breeding so successfully that by 2002 the park contained 16 free-ranging packs and more than 200 wolves.

While ranchers near the park have been unhappy about the return of the wolves, little damage to livestock has been noted, and the ecological equilibrium of Yellowstone Park seems well on the way to being regained. Elk are congregating in larger herds, and their populations are not growing as rapidly as in years past. Importantly, wolves are killing coyotes and their pups, driving them out of some areas. Coyotes, the top predators in the absence of wolves, are known to attack cattle on surrounding ranches, so reintroduction of wolves to the park may actually benefit the cattle ranchers who are opposed to it.

> **Efforts to preserve endangered species are as diverse as the causes of endangerment. Although captive breeding is not a solution in all, or even most, cases, it has helped restore several vertebrate species.**

Conservation of Ecosystems

Habitat fragmentation is one of the most pervasive enemies of biodiversity conservation efforts. As we have seen, some species require large patches of habitat to thrive, and conservation efforts that cannot provide suitable habitat of such a size are doomed to failure. Since it has become clear that isolated patches of habitat lose species far more rapidly than large preserves do, conservation biologists have promoted the creation, particularly in the tropics, of so-called *megareserves*, large areas of land containing a core of one or more undisturbed habitats (figure 57.26). The key to devoting such large tracts of land to reserves successfully over a long period of time is to operate the reserve in a way compatible with local land use. Thus, while no economic activity is allowed in the core regions of the megareserve, the remainder of the reserve may be used for nondestructive harvesting of resources. Linking preserved areas to carefully managed land zones creates a much larger total "patch" of habitat than would otherwise be economically practical, and thus addresses the key problem created by habitat fragmentation. Pioneering these efforts, eight such megareserves have been created in Costa Rica (figure 57.27) to jointly manage biodiversity and economic activity.

In addition to this focus on maintaining large enough reserves, in recent years, conservation biologists also have recognized that the best way to preserve biodiversity is to focus on preserving intact ecosystems, rather than particular species. For this reason, attention in many cases is turning to identifying those ecosystems most in need of preservation and devising the means to protect not only the species within the ecosystem, but the functioning of the ecosystem itself.

> Efforts are being undertaken worldwide to preserve biodiversity in megareserves designed to counter the influences of habitat fragmentation. Focusing on the health of entire ecosystems, rather than particular species, can often be a more effective means of preserving biodiversity.

FIGURE 57.26
Design of a megareserve. A megareserve, or biosphere reserve, recognizes the need for people to have access to resources. Critical ecosystems are preserved in the core zone. Research and tourism are allowed in the buffer zone. Sustainable resource harvesting and permanent habitation are allowed in the multiple-use areas surrounding the buffer.

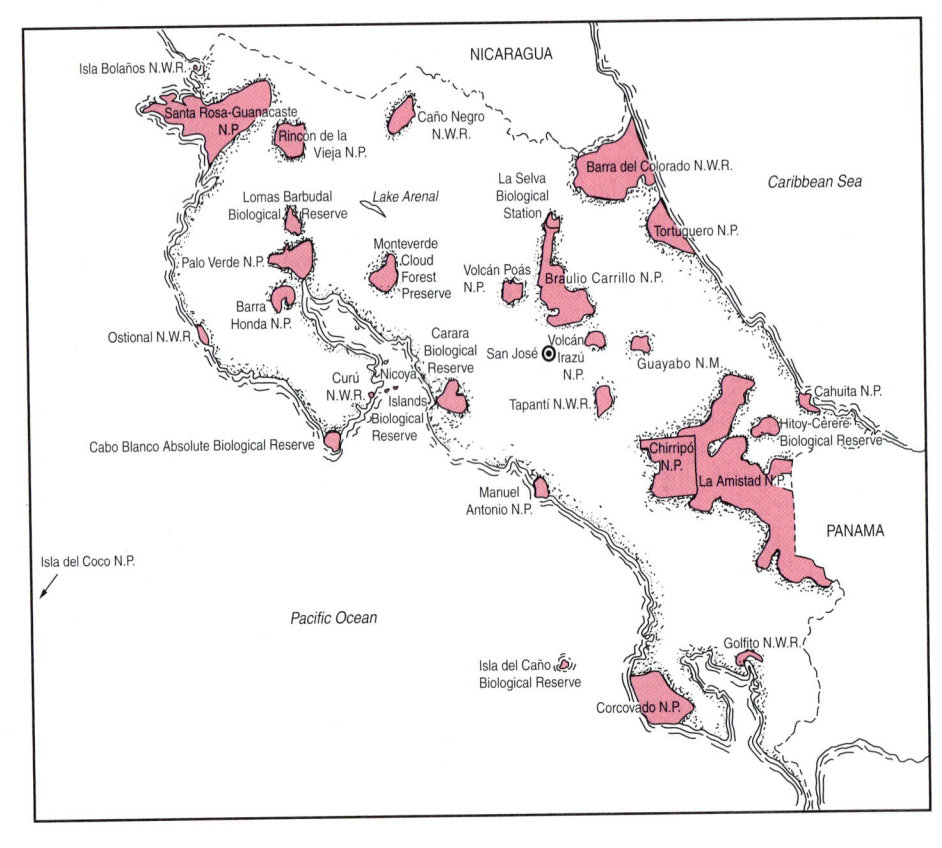

FIGURE 57.27
Biosphere reserves in Costa Rica. Costa Rica, a country the size of West Virginia, has placed about 12% of its land into national parks and eight megareserves.

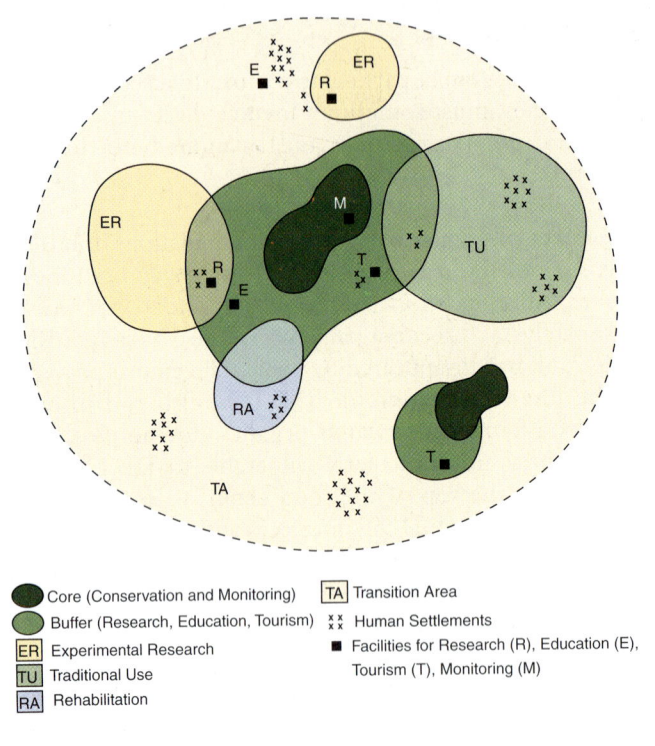

Concept Review

57.1 The new science of conservation biology is focused on conserving biodiversity.

Overview of the Biodiversity Crisis
- More than 99% of species known to science are now extinct. (p. 1228)
- It is estimated that as much as 20% of the world's biodiversity may be lost during the next 50 years. (p. 1228)
- Shortly after humans arrived, 74–86% of the earth's megafauna became extinct. (p. 1228)
- Estimates indicate that about 85 species of mammals and 113 species of birds have become extinct in the past 400 years. (p. 1229)
- The majority of historic extinctions have occurred on islands, although most species now threatened by extinction live on continents. (p. 1229)

Species Endemism and Hotspots
- An endemic species is found naturally in only one geographic area. (p. 1230)
- Isolated, geographic areas often have high percentages of endemic species. (p. 1230)
- Worldwide, concentrations of endemic species occur in several hotspots that must be the focus of conservation efforts. Unfortunately, these areas currently also contain 20% of the world's human population. (pp. 1230–1231)
- Population pressure as well as habitat loss due to commercial exploitation play an important role in biodiversity loss. (p. 1231)

What's So Bad About Losing Biodiversity?
- The value of biodiversity can be divided into direct economic value (food, clothing), indirect economic value (ecosystem maintenance), and ethical and aesthetic value (beauty). (pp. 1232–1233)

57.2 The extinction crisis is a result of many factors.

Factors Responsible for Extinction
- Historically, overexploitation was the major factor responsible for extinction, but habitat loss is the most significant problem for most groups today. (p. 1234)
- Other factors, such as ecosystem disruption, pollution, loss of genetic diversity, catastrophic disturbance, and habitat fragmentation, can also contribute to species extinction. (p. 1234)
- Many amphibian species around the world are in decline, and because no single cause can be determined, fears are rising as to the overall health of the global environment. (p. 1235)

Habitat Loss
- Natural habitats may be adversely affected by humans in four ways: destruction, pollution, human disruption, and fragmentation. (pp. 1236–1237)
- Fragmentation of summer breeding grounds and loss of overwintering habitats appear to be contributing to an overall decline in the numbers of migratory songbirds. (p. 1238)

Overexploitation
- Species that are harvested by people have historically been at grave risk of extinction. (p. 1239)
- The existence of a commercial market often leads to overexploitation of a species. (p. 1239)
- Overharvesting has brought most species of large whales to the brink of extinction. (p. 1239)

Detrimental Effects of Introduced Species
- Although colonization and range expansion is a natural process, humans are increasingly causing the introduction of species to new, unintended areas. (p. 1240)
- Some species introductions can lead to the transformation of an ecosystem, the displacement of native species, and the production of a monoculture. (pp. 1240–1241)
- The introduction of Nile perch to Lake Victoria, along with eutrophication, caused a large reduction in the number of endemic cichlid species in Lake Victoria. (p. 1241)

Disruption of Ecosystems
- Species often become vulnerable to extinction when their web of ecological interactions becomes seriously disrupted. (p. 1242)
- Activities that detrimentally affect one species may indirectly affect other species as well. (pp. 1242–1243)

The Perils of Small Population Size
- Small populations are ill-equipped to withstand a catastrophic event, and single, natural events can lead to the extinction of a population or a species. (p. 1244)
- Small populations are also prone to the loss of genetic variation due to genetic drift. (p. 1244)
- As a population gets smaller, it becomes more vulnerable to demographic catastrophes, in turn causing a decline in genetic variation and a subsequent decline in reproductive rates, which set in motion a cascading effect. (p. 1245)

57.3 Successful recovery efforts need to be multidimensional.

Approaches for Preserving Endangered Species
- Many approaches exist for preserving and fostering endangered species, and these include habitat restoration and captive propagation. (p. 1246)
- Habitat restoration can involve many actions, including pristine restoration, removal of introduced species, and cleanup and rehabilitation. (p. 1246)
- Captive propagation has produced many success stories, including the peregrine falcon, the california condor, and wolves in Yellowstone National Park. (p. 1247)

Conservation of Ecosystems
- Conservation efforts are increasingly being turned toward preserving large tracts of land over long periods of time in order to preserve intact ecosystems rather than individual or particular species. (p. 1248)

Test Your Understanding

Self Test

1. Historically, island species have tended to become extinct faster than species living on a mainland. Which of the following reasons can not be used to explain this phenomenon?
   a. Island species have often evolved in the absence of predators and have no natural avoidance strategies.
   b. Humans have introduced diseases and competitors to islands, which negatively impacts island populations.
   c. Island populations are usually smaller than mainland populations.
   d. Island populations are usually less fit than mainland populations.
2. An endemic species is
   a. one found in many different geographic areas.
   b. one found naturally in just one geographic area.
   c. one found only on islands.
   d. one that has been introduced to a new geographic area.
3. Conservation hotspots are best described as
   a. areas with large numbers of endemic species that are disappearing rapidly.
   b. areas where people are particularly active supporters of biological diversity.
   c. islands that are experiencing high rates of extinction.
   d. areas where native species are being replaced with introduced species.
4. Which of the following does *not* distinguish the current mass extinction event from past events?
   a. This is the only extinction event to be triggered by a single species.
   b. This is the only extinction event from which biodiversity may not recover.
   c. This is the only extinction event in which the majority of losses are occurring in mainland areas.
   d. This is the only extinction event in which the majority of losses are occurring on islands.
5. The ability of an intact ecosystem, such as a wetland, to buffer against flooding and filter pollutants from water is a(n) _____ value of biodiversity.
   a. direct economic
   b. indirect economic
   c. ethical
   d. aesthetic
6. Which of the following is currently considered the leading cause of extinction?
   a. overexploitation of species
   b. competition from introduced species
   c. habitat loss
   d. pollution
7. What percentage of Madagascar's original forests have been lost due to human activity?
   a. 10%
   b. 30%
   c. 60%
   d. 90%
8. Numbers of migratory songbirds are declining in North America. Which of the following factors is important in this decline?
   a. pollution
   b. human disruption of breeding behavior
   c. habitat fragmentation in the United States
   d. global climate change
9. Most large whale species have been driven to the brink of extinction. Which of the following is the most accepted explanation for this situation?
   a. overexploitation
   b. habitat loss
   c. pollution
   d. competition from introduced species
10. A keystone species is one that
    a. has a higher likelihood of extinction than a nonkeystone species.
    b. exerts a strong influence on an ecosystem.
    c. causes other species to become extinct.
    d. has a weak influence on an ecosystem.

Test Your Visual Understanding

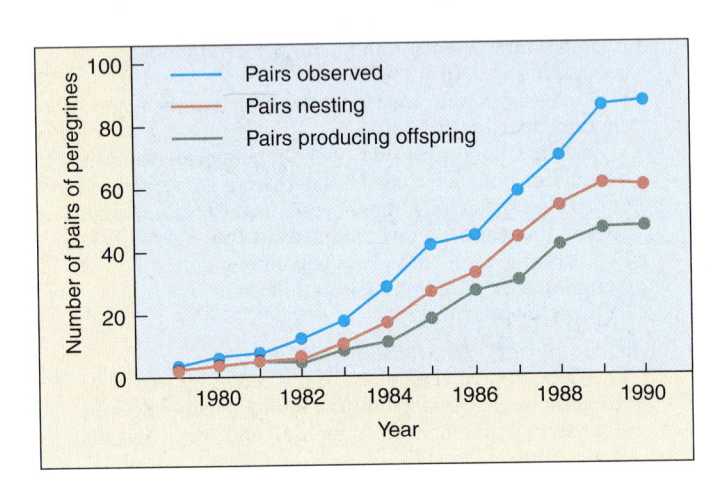

1. List the possible factors that explain the trend in this graph.

Apply Your Knowledge

1. Do you know of any endangered or threatened species that live near you? What factors have led to their decline? What efforts are being made to preserve any endangered species in your area?
2. Exotic species can have detrimental impacts on native species. Are all exotics bad? Can you name any instances in which exotics have not had detrimental effects?
3. Recent advances in genetic technologies have allowed scientists to develop microbes that can metabolize toxins in the environment, such as pesticides and oil slicks. What are the dangers of releasing these genetically modified organisms into the wild?

Glossary

A

ABO blood group A set of four phenotypes produced by different combinations of three alleles at a single locus; blood types are A, B, AB, and O, depending on which alleles are expressed as antigens on the red blood cell surface.

abscission (L. *ab*, away, off, + *scissio*, dividing) In vascular plants, the dropping of leaves, flowers, fruits, or stems at the end of the growing season, as the result of the formation of a layer of specialized cells (the abscission zone) and the action of a hormone (ethylene).

acquired immune deficiency syndrome (AIDS) An infectious and usually fatal human disease caused by a retrovirus, HIV (human immunodeficiency virus), which attacks T cells. The affected individual is helpless in the face of microbial infections.

actin (Gr. *actis*, a ray) One of the two major proteins that make up vertebrate muscle; the other is myosin.

action potential A transient, all-or-none reversal of the electric potential across a membrane; in neurons, an action potential initiates transmission of a nerve impulse.

activation energy The energy that must be processed by a molecule in order for it to undergo a specific chemical reaction.

active site The region of an enzyme surface to which a specific set of substrates binds, lowering the activation energy required for a particular chemical reaction and so facilitating it.

active transport The pumping of individual ions or other molecules across a cellular membrane from a region of lower concentration to one of higher concentration (that is, against a concentration gradient); this transport process requires energy, which is typically supplied by the expenditure of ATP.

adaptation (L. *adaptare*, to fit) A peculiarity of structure, physiology, or behavior that promotes the likelihood of an organism's survival and reproduction in a particular environment.

adaptive radiation The evolution of several divergent forms from a primitive and unspecialized ancestor.

adenosine diphosphate (ADP) A nucleotide consisting of adenine, ribose sugar, and two phosphate groups; formed by the removal of one phosphate from an ATP molecule.

adenosine triphosphate (ATP) A nucleotide consisting of adenine, ribose sugar, and three phosphate groups; ATP is the energy currency of cellular metabolism in all organisms.

adenylyl cyclase An enzyme that produces large amounts of cAMP from ATP; the cAMP acts as a second messenger in a target cell.

adipose cells Fat cells, found in loose connective tissue, usually in large groups that form adipose tissue. Each adipose cell can store a droplet of fat (triacylglyceride).

adventitious (L. *adventicius*, not properly belonging to) Referring to a structure arising from an unusual place, such as stems from roots or roots from stems.

aerobic (Gr. *aer*, air, + *bios*, life) Requiring free oxygen; any biological process that can occur in the presence of gaseous oxygen (O_2).

alga, pl. **algae** A unicellular or simple multicellular photosynthetic organism lacking multicellular sex organs.

allantois (Gr. *allas*, sausage, + *eidos*, form) A membrane of the amniotic egg that functions in respiration and excretion in birds and reptiles and plays an important role in the development of the placenta in most mammals.

allele (Gr. *allelon*, of one another) One of two or more alternative states of a gene.

allometric growth (Gr. *allos*, other, + *ergon*, action) A pattern of growth in which different components grow at different rates.

allopatric speciation The differentiation of geographically isolated populations into distinct species.

allosteric site A part of an enzyme, away from its active site, that serves as an on/off switch for the function of the enzyme.

alternation of generations A reproductive cycle in which a haploid (n) phase, the gametophyte, gives rise to gametes, which, after fusion to form a zygote, germinate to produce a diploid ($2n$) phase, the sporophyte. Spores produced by meiotic division from the sporophyte give rise to new gametophytes, completing the cycle.

altruism Self-sacrifice for the benefit of others; in formal terms, the behavior that increases the fitness of the recipient while reducing the fitness of the altruistic individual.

alveolus, pl. **alveoli** (L., a small cavity) One of many small, thin-walled air sacs within the lungs in which the bronchioles terminate.

amino acid The subunit structure from which proteins are produced, consisting of a central carbon atom with a carboxyl group (—COOH), an amino group (—NH_2), a hydrogen, and a side group (*R* group); only the side group differs from one amino acid to another.

amniocentesis (Gr. *amnion*, membrane around the fetus, + *centes*, puncture) Examination of a fetus indirectly, by tests on cell cultures grown from fetal cells obtained from a sample of the amniotic fluid surrounding the developing embryo or tests on the fluid itself.

amnion (Gr., membrane around the fetus) The innermost of the extraembryonic membranes; the amnion forms a fluid-filled sac around the embryo in amniotic eggs.

amniotic egg An egg that is isolated and protected from the environment by a more or less impervious shell during the period of its development and that is completely self-sufficient, requiring only oxygen.

amyloplast (Gr. *amylon*, starch, + *plastos*, formed) A plant organelle called a plastid that specializes in storing starch.

anabolism (Gr. *ana*, up, + *bolein*, to throw) The biosynthetic or constructive part of metabolism; those chemical reactions involved in biosynthesis.

anaerobic (Gr. *an*, without, + *aer*, air, + *bios*, life) Any process that can occur without oxygen, such as anaerobic fermentation or H_2S photosynthesis.

analogous (Gr. *analogos*, proportionate) Structures that are similar in function but different in evolutionary origin, such as the wing of a bat and the wing of a butterfly.

anaphase In mitosis and meiosis II, the stage initiated by the separation of sister chromatids, during which the daughter chromosomes move to opposite poles of the cell; in meiosis I, marked by separation of replicated homologous chromosomes.

androecium (Gr. *andros*, man, + *oilos*, house) The floral whorl that comprises the stamens.

aneuploidy (Gr. *an*, without, + *eu*, good, + *ploid*, multiple of) The condition in an organism whose cells have lost or gained a chromosome; Down syndrome, which results from an extra copy of human chromosome 21, is an example of aneuploidy in humans.

angiosperms The flowering plants, one of five phyla of seed plants. In angiosperms, the ovules at the time of pollination are completely enclosed by tissues.

animal pole In fish and other aquatic vertebrates with asymmetrical yolk distribution in their eggs, the hemisphere of the blastula comprising cells relatively poor in yolk.

anion (Gr. *anienae*, to go up) A negatively charged ion.

anther (Gr. *anthos*, flower) In angiosperm flowers, the pollen-bearing portion of a stamen.

antheridium, pl. **antheridia** A sperm-producing organ.

antibody (Gr. *anti*, against) A protein called immunoglobulin that is produced by lymphocytes in response to a foreign substance (antigen) and released into the bloodstream.

anticodon The three-nucleotide sequence at the end of a transfer RNA molecule that is complementary to, and base-pairs with, an amino-acid–specifying codon in messenger RNA.

antigen (Gr. *anti*, against, + *genos*, origin) A foreign substance, usually a protein or polysaccharide, that stimulates an immune response.

anus The terminal opening of the gut; the solid residues of digestion are eliminated through the anus.

aorta (Gr. *aeirein*, to lift) The major artery of vertebrate systemic blood circulation; in mammals, carries oxygenated blood away from the heart to all regions of the body except the lungs.

apical meristem (L. *apex*, top, + Gr. *meristos*, divided) In vascular plants, the growing point at the tip of the root or stem.

apoptosis A process of programmed cell death, in which dying cells shrivel and shrink; used in all animal cell development to produce planned and orderly death of cells not destined to be present in the final tissue.

aposematic coloration An ecological strategy of some organisms that "advertise" their poisonous nature by the use of bright colors.

aquifers Permeable, saturated, underground layers of rock, sand, and gravel, which serve as reservoirs for groundwater.

archaebacteria A group of bacteria that are among the most primitive still in existence; characterized by the absence of peptidoglycan in their cell walls, a feature that distinguishes them from all other bacteria.

archegonium, pl. **archegonia** (Gr. *archegonos*, first of a race) The multicellular egg-producing organ in bryophytes and some vascular plants.

archenteron (Gr. *arche*, beginning, + *enteron*, gut) The principal cavity of a vertebrate embryo in the gastrula stage; lined with endoderm, it opens up to the outside and represents the future digestive cavity.

arteriole A smaller artery, leading from the arteries to the capillaries.

arteriosclerosis Hardening and thickening of the wall of an artery.

ascomycetes A large group comprising part of the "true fungi." They are characterized by separate hyphae, asexually produced conidiospores, and sexually produced ascospores within asci.

ascospore A fungal spore produced within an ascus.

ascus, pl. **asci** (Gr. *askos*, wineskin, bladder) A specialized cell, characteristic of the ascomycetes, in which two haploid nuclei fuse to produce a zygote that divides immediately by meiosis; at maturity, an ascus contains ascospores.

asexual reproduction The process by which an individual inherits all of its chromosomes from a single parent, thus being genetically identical to that parent; cell division is by mitosis only.

aster In animal cell mitosis, a radial array of microtubules extending from the centrioles toward the plasma membrane, possibly serving to brace the centrioles for retraction of the spindle.

atrioventricular (AV) node A slender connection of cardiac muscle cells that receives the heartbeat impulses from the sinoatrial node and conducts them by way of the bundle of His.

atrium (L., vestibule or courtyard) An antechamber; in the heart, a thin-walled chamber that receives venous blood and passes it on to the thick-walled ventricle; in the ear, the tympanic cavity.

autonomic nervous system (Gr. *autos*, self, + *nomos*, law) The involuntary neurons and ganglia of the peripheral nervous system of vertebrates; regulates the heart, glands, visceral organs, and smooth muscle.

autosome (Gr. *autos*, self, + *soma*, body) Any eukaryotic chromosome that is not a sex chromosome; autosomes are present in the same number and kind in both males and females of the species.

autotroph (Gr. *autos*, self, + *trophos*, feeder) An organism able to build all the complex organic molecules that it requires as its own food source, using only simple inorganic compounds.

auxin (Gr. *auxein*, to increase) A plant hormone that controls cell elongation, among other effects.

axon (Gr., axle) A process extending out from a neuron that conducts impulses away from the cell body.

B

bacteriophage (Gr. *bakterion*, little rod, + *phagein*, to eat) A virus that infects bacterial cells; also called a *phage*.

Barr body A deeply staining structure, seen in the interphase nucleus of a cell of an individual with more than one X chromosome, that is a condensed and inactivated X. Only one X remains active in each cell after early embryogenesis.

basal body A self-reproducing, cylindrical, cytoplasmic organelle composed of nine triplets of microtubules from which the flagella or cilia arise.

base-pair A complementary pair of nucleotide bases, consisting of a purine and a pyrimidine.

basidiospore A spore of the basidiomycetes, produced within and borne on a basidium after nuclear fusion and meiosis.

basidium, pl. **basidia** (L., a little pedestal) A specialized reproductive cell of the basidiomycetes, often club shaped, in which nuclear fusion and meiosis occur.

basophil A leukocyte containing granules that rupture and release chemicals that enhance the inflammatory response. Important in causing allergic responses.

Batesian mimicry A situation in which a palatable or nontoxic organism resembles another kind of organism that is distasteful or toxic. Both species exhibit warning coloration.

B cell A type of lymphocyte that, when confronted with a suitable antigen, is capable of secreting a specific antibody protein.

behavioral ecology The study of how natural selection shapes behavior.

biennial A plant that normally requires two growing seasons to complete its life cycle. Biennials flower in the second year of their lives.

bile salts A solution of organic salts that is secreted by the vertebrate liver and temporarily stored in the gallbladder; emulsifies fats in the small intestine.

binary fission (L. *binarius*, consisting of two things or parts, + *fissus*, split) Asexual reproduction by division of one cell or body into two equal or nearly equal parts.

binomial distribution The distribution of phenotypes seen among the progeny of a cross in which there are only two alternative alleles.

biodiversity The number of species and their range of behavioral, ecological, physiological, and other adaptations, in an area.

biomass (Gr. *bios*, life, + *maza*, lump or mass) The total mass of all the living organisms in a given population, area, or other unit being measured.

biome One of the major terrestrial ecosystems, characterized by climatic and soil conditions; the largest ecological unit.

bipedal Able to walk upright on two feet.

bipolar cell A specialized type of neuron connecting cone cells to ganglion cells in the visual system. Bipolar cells receive a hyperpolarized stimulus from the cone cell and then transmit a depolarization stimulus to the ganglion cell.

blade The broad, expanded part of a leaf; also called the lamina.

blastocoel (Gr. *blastos*, sprout, + *koilos*, hollow) The central cavity of the blastula stage of vertebrate embryos.

blastodisc (Gr. *blastos*, sprout, + *discos*, a round plate) In the development of birds, a disclike area on the surface of a large, yolky egg that undergoes cleavage and gives rise to the embryo.

blastomere (Gr. *blastos*, sprout, + *meros*, part) One of the cells of a blastula.

blastopore (Gr. *blastos*, sprout, + *poros*, a path or passage) In vertebrate development, the opening that connects the archenteron cavity of a gastrula stage embryo with the outside.

blastula (Gr., a little sprout) In vertebrates, an early embryonic stage consisting of a hollow, fluid-filled ball of cells one layer thick; a vertebrate embryo after cleavage and before gastrulation.

Bohr effect The release of oxygen by hemoglobin molecules in response to elevated ambient levels of CO_2.

Bowman's capsule In the vertebrate kidney, the bulbous unit of the nephron, which surrounds the glomerulus.

β-oxidation In the cellular respiration of fats, the process by which two-carbon acetyl groups are removed from a fatty acid and combined with coenzyme A to form acetyl-CoA, until the entire fatty acid has been broken down.

bronchus, pl. **bronchi** (Gr. *bronchos*, windpipe) One of a pair of respiratory tubes branching from the lower end of the trachea (windpipe) into either lung.

bryophytes Nonvascular plants, including mosses, hornworts, and liverworts.

bud An asexually produced outgrowth that develops into a new individual. In plants, an embryonic shoot, often protected by young leaves; buds may give rise to branch shoots.

C

C₃ photosynthesis The main cycle of the dark reactions of photosynthesis, in which CO_2 binds to ribulose 1,5-bisphosphate (RuBP) to form two three-carbon phosphoglycerate (PGA) molecules.

C₄ photosynthesis A process of CO_2 fixation in photosynthesis by which the first product is the four-carbon oxaloacetate molecule.

callus (L. *callos*, hard skin) Undifferentiated tissue; a term used in tissue culture, grafting, and wound healing.

Calvin cycle The dark reactions of C₃ photosynthesis; also called the Calvin-Benson cycle.

calyx (Gr. *kalyx*, a husk, cup) The sepals collectively; the outermost flower whorl.

capillary (L. *capillaris*, hairlike) The smallest of the blood vessels;

the very thin walls of capillaries are permeable to many molecules, and exchanges between blood and the tissues occur across them; the vessels that connect arteries with veins.

capsid The outermost protein covering of a virus.

carapace (Fr. from Sp. *carapacho*, shell) Shieldlike plate covering the cephalothorax of decapod crustaceans; the dorsal part of the shell of a turtle.

carbohydrate (L. *carbo*, charcoal, + *hydro*, water) An organic compound consisting of a chain or ring of carbon atoms to which hydrogen and oxygen atoms are attached in a ratio of approximately 2:1; having the generalized formula $(CH_2O)_n$; carbohydrates include sugars, starch, glycogen, and cellulose.

carbon fixation The conversion of CO_2 into organic compounds during photosynthesis; the first stage of the dark reactions of photosynthesis, in which carbon dioxide from the air is combined with ribulose 1,5-bisphosphate.

carpel (Gr. *karpos*, fruit) A leaflike organ in angiosperms that encloses one or more ovules.

carrying capacity The maximum population size that a habitat can support.

cartilage (L. *cartilago*, gristle) A connective tissue in skeletons of vertebrates. Cartilage forms much of the skeleton of embryos, very young vertebrates, and some adult vertebrates, such as sharks and their relatives.

catabolism (Gr. *ketabole*, throwing down) In a cell, those metabolic reactions that result in the breakdown of complex molecules into simpler compounds, often with the release of energy.

catalysis The process by which chemical subunits of larger organic molecules are held and positioned by enzymes that stress their chemical bonds, leading to the disassembly of the larger molecule into its subunits, often with the release of energy.

cation (Gr. *katienai*, to go down) A positively charged ion.

cecum (L. *caecus*, blind) In vertebrates, a blind pouch at the beginning of the large intestine.

cell cycle The repeating sequence of growth and division through which cells pass each generation.

cell plate The structure that forms at the equator of the spindle during early telophase in the dividing cells of plants and a few green algae.

cell surface receptor A cell surface protein that binds a signal molecule and converts the extracellular signal into an intracellular one.

cellular respiration The metabolic harvesting of energy by oxidation, ultimately dependent on molecular oxygen; carried out by the Krebs cycle and oxidative phosphorylation.

cellulose (L. *cellula*, a little cell) The chief constituent of the cell wall in all green plants, some algae, and a few other organisms; an insoluble complex carbohydrate formed of microfibrils of glucose molecules.

cell wall The rigid, outermost layer of the cells of plants, some protists, and most bacteria; the cell wall surrounds the plasma membrane.

central nervous system (CNS) That portion of the nervous system where most association occurs; in vertebrates, it is composed of the brain and spinal cord; in invertebrates, it usually consists of one or more cords of nervous tissue, together with their associated ganglia.

centriole (Gr. *kentron*, center, + L. *olus*, little one) A cytoplasmic organelle located outside the nuclear membrane, identical in structure to a basal body; found in animal cells and in the flagellated cells of other groups; divides and organizes spindle fibers during mitosis and meiosis.

centromere (Gr. *kentron*, center, + *meros*, a part) Condensed region on a eukaryotic chromosome where sister chromatids are attached to each other after replication. See kinetochore.

cerebellum (L., little brain) The hindbrain region of the vertebrate brain that lies above the medulla (brain stem) and behind the forebrain; it integrates information about body position and motion, coordinates muscular activities, and maintains equilibrium.

cerebral cortex The thin surface layer of neurons and glial cells covering the cerebrum; well developed only in mammals, and particularly prominent in humans. The cerebral cortex is the seat of conscious sensations and voluntary muscular activity.

cerebrum (L., brain) The portion of the vertebrate brain (the forebrain) that occupies the upper part of the skull, consisting of two cerebral hemispheres united by the corpus callosum. It is the primary association center of the brain. It coordinates and processes sensory input and coordinates motor responses.

chelicera, pl. **chelicerae** (Gr. *chele*, claw, + *keras*, horn) The first pair of appendages in horseshoe crabs, sea spiders, and arachnids—the chelicerates, a group of arthropods. Chelicerae usually take the form of pincers or fangs.

chemiosmosis The mechanism by which ATP is generated in mitochondria and chloroplasts; energetic electrons excited by light (in chloroplasts) or extracted by oxidation in the Krebs cycle (in mitochondria) are used to drive proton pumps, creating a proton concentration gradient; when protons subsequently flow back across the membrane, they pass through channels that couple their movement to the synthesis of ATP.

chiasma An X-shaped figure that can be seen in the light microscope during meiosis; evidence of crossing over, where two chromatids have exchanged parts; chiasmata move to the ends of the chromosome arms as the homologues separate.

chitin (Gr. *chiton*, tunic) A tough, resistant, nitrogen-containing polysaccharide that forms the cell walls of certain fungi, the exoskeleton of arthropods, and the epidermal cuticle of other surface structures of certain other invertebrates.

chloroplast (Gr. *chloros*, green, + *plastos*, molded) A cell-like organelle present in algae and plants that contains chlorophyll (and usually other pigments) and carries out photosynthesis.

chorion (Gr., skin) The outer member of the double membrane that surrounds the embryo of reptiles, birds, and mammals; in placental mammals, it contributes to the structure of the placenta.

chromatid (Gr. *chroma*, color, + L. -*id*, daughters of) One of the two daughter strands of a duplicated chromosome that is joined by a single centromere.

chromatin (Gr. *chroma*, color) The complex of DNA and proteins of which eukaryotic chromosomes are composed; chromatin is highly uncoiled and diffuse in interphase nuclei, condensing to form the visible chromosomes in prophase.

chromosome (Gr. *chroma*, color, + *soma*, body) The vehicle by which hereditary information is physically transmitted from one generation to the next; in a bacterium, the chromosome consists of a single naked circle of DNA; in eukaryotes, each chromosome consists of a single linear DNA molecule and associated proteins.

cilium A short cellular projection from the surface of a eukaryotic cell, having the same internal structure of microtubules in a 9 + 2 arrangement as seen in a flagellum.

circadian rhythm An endogenous cyclical rhythm that oscillates on a daily (24-hour) basis.

cisterna A small collecting vessel that pinches off from the end of a Golgi body to form a transport vesicle that moves materials through the cytoplasm.

cladistics A taxonomic technique used for creating hierarchies of organisms that represent true phylogenetic relationship and descent.

class A taxonomic category between phyla and orders. A class contains one or more orders, and belongs to a particular phylum.

classical conditioning The repeated presentation of a stimulus in association with a response that causes the brain to form an association between the stimulus and the response, even if they have never been associated before.

clathrin A protein located just inside the plasma membrane in eukaryotic cells, in indentations called clathrin-coated pits.

cleavage In vertebrates, a rapid series of successive cell divisions of a fertilized egg, forming a hollow sphere of cells, the blastula.

climax vegetation Vegetation encountered in a self-perpetuating community of plants that has proceeded through all the stages of succession and stabilized.

cloaca (L., sewer) In some animals, the common exit chamber from the digestive, reproductive, and urinary system; in others, the cloaca may also serve as a respiratory duct.

cloning Producing a cell line or culture all of whose members contain identical copies of a particular nucleotide sequence; an essential element in genetic engineering.

coacervate (L. *coacervatus*, heaped up) A spherical aggregation of lipid molecules in water, held together by hydrophobic forces.

cochlea (Gr. *kochlios*, a snail) In terrestrial vertebrates, a tubular cavity of the inner ear containing the essential organs for hearing.

codon (L., code) The basic unit of the genetic code; a sequence of three adjacent nucleotides in DNA or mRNA that codes for one amino acid.

coenzyme (L. *co-*, together, + Gr. *en*, in, + *zyme*, leaven) A nonprotein organic molecule such as NAD^+ that plays an accessory role in enzyme-catalyzed processes, often by acting as a donor or acceptor of electrons.

G-3

coevolution (L. *co-*, together, + *e-*, out, + *volvere*, to fill) The simultaneous development of adaptations in two or more populations, species, or other categories that interact so closely that each is a strong selective force on the other.

cofactor One or more nonprotein components required by enzymes in order to function; many cofactors are metal ions, others are organic coenzymes.

collenchyma cell (Gr. *kolla*, glue, + *en*, in, + *chymein*, to pour) In plants, the cells that form a supporting tissue called collenchyma; often found in regions of primary growth in stems and in some leaves.

commensalism (L. *cum*, together with, + *mensa*, table) A relationship in which one individual lives close to or on another and benefits, and the host is unaffected; a kind of symbiosis.

community (L. *communitas*, community, fellowship) All of the species inhabiting a common environment and interacting with one another.

companion cell A specialized parenchyma cell that is associated with each sieve-tube member in the phloem of a plant.

competitive exclusion The hypothesis that two species with identical ecological requirements cannot exist in the same locality indefinitely, and that the more efficient of the two in utilizing the available scarce resources will exclude the other; also known as Gause's principle.

complementary Describes genetic information in which each nucleotide base has a complementary partner with which it forms a base-pair.

complement system The chemical defense of a vertebrate body that consists of a battery of proteins that become activated by the walls of bacteria and fungi.

concentration gradient The concentration difference of a substance across a distance; in a cell, a greater concentration of its molecules in one region than in another.

cone (1) In plants, the reproductive structure of a conifer. (2) In vertebrates, a type of light-sensitive neuron in the retina, concerned with the perception of color and with the most acute discrimination of detail.

conidia An asexually produced fungal spore.

conjugation (L. *conjugare*, to yolk together) Temporary union of two unicellular organisms, during which genetic material is transferred from one cell to the other; occurs in bacteria, protists, and certain algae and fungi.

contractile vacuole In protists and some animals, a clear fluid-filled vacuole that takes up water from within the cell and then contracts, releasing it to the outside through a pore in a cyclical manner; functions primarily in osmoregulation and excretion.

conus arteriosus The anteriormost chamber of the embryonic heart in vertebrate animals.

convergent evolution The independent development of similar structures in organisms that are not directly related; often found in organisms living in similar environments.

cork cambium The lateral meristem that forms the periderm, producing cork (phellem) toward the surface (outside) of the plant and phelloderm toward the inside.

cornea (L. *corneus*, horny) The transparent outer layer of the vertebrate eye.

corolla (L., a little crown) The petals, collectively; usually the conspicuously colored flower whorl.

corpus callosum The band of nerve fibers that connects the two hemispheres of the cerebrum in humans and other primates.

corpus luteum (L., yellowish body) A structure that develops from a ruptured follicle in the ovary after ovulation.

cortex (L., bark) The outer layer of a structure; in animals, the outer, as opposed to the inner, part of an organ; in vascular plants, the primary ground tissue of a stem or root.

cotyledon (Gr. *kotyledon*, a cup-shaped hollow) A seed leaf that generally stores food in dicots or absorbs it in monocots, providing nourishment used during seed germination.

crassulacean acid metabolism (CAM) A mode of carbon dioxide fixation by which CO_2 enters open leaf stomata at night and is used in photosynthesis during the day, when stomata are closed to prevent water loss.

cross-current flow In bird lungs, the latticework of capillaries arranged across the air flow, at a 90° angle.

crossing over In meiosis, the exchange of corresponding chromatid segments between homologous chromosomes; responsible for genetic recombination between homologous chromosomes.

cuticle (L. *cuticula*, little skin) A waxy or fatty, noncellular layer (formed of a substance called cutin) on the outer wall of epidermal cells.

cyanobacteria A group of photosynthetic bacteria, sometimes called the "blue-green algae," that contain the chlorophyll pigments most abundant in plants and algae, as well as other pigments.

cyclic AMP (cAMP) A form of adenosine monophosphate (AMP) in which the atoms of the phosphate group form a ring; found in almost all organisms, cAMP functions as an intracellular second messenger that regulates a diverse array of metabolic activities.

cystic fibrosis An autosomal disorder that produces the most common fatal genetic disease in Caucasians, characterized by secretion of thick mucus that clogs passageways in the lungs, liver, and pancreas.

cytochrome (Gr. *kytos*, hollow vessel, + *chroma*, color) Any of several iron-containing protein pigments that serve as electron carriers in transport chains of photosynthesis and cellular respiration.

cytokinesis (Gr. *kytos*, hollow vessel, + *kinesis*, movement) Division of the cytoplasm of a cell after nuclear division.

cytoplasm (Gr. *kytos*, hollow vessel, + *plasma*, anything molded) The material within a cell, excluding the nucleus; the protoplasm.

cytoskeleton A network of protein microfilaments and microtubules within the cytoplasm of a eukaryotic cell that maintains the shape of the cell, anchors its organelles, and is involved in animal cell motility.

cytotoxic T cell (Gr. *kytos*, hollow vessel, + toxin) A special T cell activated during cell-mediated immune response that recognizes and destroys infected body cells.

D

dehydration synthesis (L. *co-*, together, + *densare*, to make dense) A type of chemical reaction in which two molecules join to form one larger molecule, simultaneously splitting out a molecule of water; one molecule is stripped of a hydrogen atom, and another is stripped of a hydroxyl group (—OH), resulting in the joining of the two molecules, while the H^+ and —OH released may combine to form a water molecule.

demography (Gr. *demos*, people, + *graphein*, to draw) The properties of the rate of growth and the age structure of populations.

denaturation The loss of the native configuration of a protein or nucleic acid as a result of excessive heat, extremes of pH, chemical modification, or changes in solvent ionic strength or polarity that disrupt hydrophobic interactions; usually accompanied by loss of biological activity.

dendrite (Gr. *dendron*, tree) A process extending from the cell body of a neuron, typically branched, that conducts impulses toward the cell body.

deoxyribonucleic acid (DNA) The genetic material of all organisms; composed of two complementary chains of nucleotides wound in a double helix.

depolarization The movement of ions across a plasma membrane that locally wipes out an electrical potential difference.

derived character A characteristic used in taxonomic analysis representing a departure from the primitive form.

desmosome A type of anchoring junction that links adjacent cells by connecting their cytoskeletons with cadherin proteins.

diaphragm (Gr. *diaphrassein*, to barricade) (1) In mammals, a sheet of muscle tissue that separates the abdominal and thoracic cavities and functions in breathing. (2) A contraceptive device used to block the entrance to the uterus temporarily and thus prevent sperm from entering during sexual intercourse.

diastolic pressure In the measurement of human blood pressure, the minimum pressure between heartbeats (repolarization of the ventricles). *Compare with* systolic pressure.

dicot Short for dicotyledon; a class of flowering plants generally characterized as having two cotyledons, net-veined leaves, and flower parts usually in fours or fives.

differentiation A developmental process by which a relatively unspecialized cell undergoes a progressive change to a more specialized form or function.

diffusion (L. *diffundere*, to pour out) The net movement of dissolved molecules or other particles from a region where they are more concentrated to a region where they are less concentrated.

digestion (L. *digestio*, separating out, dividing) The breakdown of complex, usually insoluble foods into molecules that can be absorbed into cells, where they are degraded to yield energy and the raw materials for synthetic processes.

dihybrid (Gr. *dis*, twice, + *hibridia*, mixed offspring) An individual heterozygous at two different loci; for example *A/a B/b*.

dikaryotic (Gr. *di*, two, + *karyon*, kernel) In fungi, having pairs of nuclei within each cell.

dioecious (Gr. *di*, two, + *oikos*, house) Having the male and female elements on different individuals.

diploid (Gr. *diploos*, double, + *eidos*, form) Having two sets of chromosomes (2*n*); in animals, twice the number characteristic of gametes; in plants, the chromosome number characteristic of the sporophyte generation; in contrast to haploid (*n*).

directional selection A form of selection in which selection acts to eliminate one extreme from an array of phenotypes.

disaccharide A carbohydrate formed of two simple sugar molecules bonded covalently.

disruptive selection A form of selection in which selection acts to eliminate rather than favor the intermediate type.

diurnal (L. *diurnalis*, day) Active during the day.

DNA ligase The enzyme that links together Okazaki fragments in DNA replication of the lagging strand; it also links other broken areas of the DNA backbone.

domain (1) A distinct modular region of a protein that serves a particular function in the action of the protein, such as a regulatory domain or a DNA-binding domain. (2) In taxonomy, the level higher than kingdom. The three domains currently recognized are Bacteria, Archaea, and Eukarya.

dominant An allele that is expressed when present in either the heterozygous or the homozygous condition.

double fertilization The fusion of the egg and sperm (resulting in a 2*n* fertilized egg, the zygote) and the simultaneous fusion of the second male gamete with the polar nuclei (resulting in a primary endosperm nucleus, which is often triploid, 3*n*); a unique characteristic of all angiosperms.

Down syndrome A congenital syndrome caused by the presence of an extra copy of chromosome 21.

duodenum (L. *duodeni*, 12 each—from its length, about 12 fingers' breadth) In vertebrates, the upper portion of the small intestine.

E

ecdysis (Gr. *ekdysis*, stripping off) Shedding of outer, cuticular layer; molting, as in insects or crustaceans.

ecdysone (Gr. *ekdysis*, stripping off) Molting hormone of arthropods, which triggers when ecdysis occurs.

ecology (Gr. *oikos*, house, + *logos*, word) The study of interactions of organisms with one another and with their physical environment.

ecosystem (Gr. *oikos*, house, + *systems*, that which is put together) A major interacting system that includes organisms and their nonliving environment.

ecotype (Gr. *oikos*, house, + L. *typus*, image) A locally adapted variant of an organism; differing genetically from other ecotypes.

ectoderm (Gr. *ecto*, outside, + *derma*, skin) One of the three embryonic germ layers of early vertebrate embryos; ectoderm gives rise to the outer epithelium of the body (skin, hair, nails) and to the nerve tissue, including the sense organs, brain, and spinal cord.

ectomycorrhizae Externally developing mycorrhizae that do not penetrate the cells they surround.

ectotherms Animals such as reptiles, fish, or amphibians, whose body temperature is regulated by their behavior or by their surroundings.

electronegativity A property of atomic nuclei that refers to the affinity of the nuclei for valence electrons; a nucleus that is more electronegative has a greater pull on electrons than one that is less electronegative.

electron transport chain The passage of energetic electrons through a series of membrane-associated electron-carrier molecules to proton pumps embedded within mitochondrial or chloroplast membranes. *See* chemiosmosis.

endergonic Describes a chemical reaction in which the products contain more energy than the reactants, so that free energy must be put into the reaction from an outside source to allow it to proceed.

endocrine gland (Gr. *endon*, within, + *krinein*, to separate) Ductless gland that secretes hormones into the extracellular spaces, from which they diffuse into the circulatory system.

endocytosis (Gr. *endon*, within, + *kytos*, hollow vessel) The uptake of material into cells by inclusion within an invagination of the plasma membrane; the uptake of solid material is phagocytosis, while that of dissolved material is pinocytosis.

endoderm (Gr. *endon*, within, + *derma*, skin) One of the three embryonic germ layers of early vertebrate embryos, destined to give rise to the epithelium that lines internal structures and most of the digestive and respiratory tracts.

endodermis (Gr. *endon*, within, + *derma*, skin) In vascular plants, a layer of cells forming the innermost layer of the cortex in roots and some stems.

endometrium (Gr. *endon*, within, + *metrios*, of the womb) The lining of the uterus in mammals; thickens in response to secretion of estrogens and progesterone and is sloughed off in menstruation.

endomycorrhizae Mycorrhizae that develop within cells.

endoplasmic reticulum (ER) An internal membrane system that forms a netlike array of channels and interconnections of organelles within the cytoplasm of eukaryotic cells.

endorphin One of a group of small neuropeptides produced by the vertebrate brain; like morphine, endorphins modulate pain perception.

endosperm (Gr. *endon*, within, + *sperma*, seed) A storage tissue characteristic of the seeds of angiosperms, which develops from the union of a male nucleus and the polar nuclei of the embryo sac. The endosperm is digested by the growing sporophyte either before maturation of the seed or during its germination.

endosymbiosis Theory that proposes that eukaryotic cells evolved from a symbiosis between different species of prokaryotes.

endotherm An animal capable of maintaining a constant body temperature. *See* homeotherm.

energy The capacity to do work.

enhancer A site of regulatory protein binding on the DNA molecule distant from the promoter and start site for a gene's transcription.

enthalpy In a chemical reaction, the energy contained in the chemical bonds of the molecule, symbolized as H; in a cellular reaction, the free energy is equal to the enthalpy of the reactant molecules in the reaction.

entropy (Gr. *en*, in, + *tropos*, change in manner) A measure of the randomness or disorder of a system; a measure of how much energy in a system has become so dispersed (usually as evenly distributed heat) that it is no longer available to do work.

enzyme (Gr. *enzymes*, leavened, from *en*, in, + *zyme*, leaven) A protein that is capable of speeding up chemical reactions by lowering the required activation energy.

epicotyl The region just above where the cotyledons are attached.

epidermis (Gr. *epi*, on or over, + *derma*, skin) The outermost layers of cells; in plants, the exterior primary tissue of leaves, young stems, and roots; in vertebrates, the nonvascular external layer of skin, of ectodermal origin; in invertebrates, a single layer of ectodermal epithelium.

epididymis (Gr. *epi*, on, + *didymos*, testicle) A sperm storage vessel; a coiled part of the sperm duct that lies near the testis.

epistasis (Gr. *epi*, on, + *stasis*, standing still) Interaction between two nonallelic genes in which one of them modifies the phenotypic expression of the other.

epithelium (Gr. *epi*, on, + *thele*, nipple) In animals, a type of tissue that covers an exposed surface or lines a tube or cavity.

equilibrium (L. *aequus*, equal, + *libra*, balance) A stable condition; the point at which a chemical reaction proceeds as rapidly in the reverse direction as it does in the forward direction, so that there is no further net change in the concentrations of products or reactants. In ecology, a stable condition that resists change and fairly quickly returns to its original state if disturbed by humans or natural events.

erythrocyte (Gr. *erythros*, red, + *kytos*, hollow vessel) Red blood cell, the carrier of hemoglobin.

erythropoiesis The manufacture of blood cells in the bone marrow.

estrous cycle The periodic cycle in which periods of estrus correspond to ovulation events.

estrus (L. *oestrus*, frenzy) The period of maximum female sexual receptivity, associated with ovulation of the egg.

ethology (Gr. *ethos*, habit or custom, + *logos*, discourse) The study of patterns of animal behavior in nature.

euchromatin (Gr. *eu*, good, + *chroma*, color) That portion of a eukaryotic chromosome that is transcribed into mRNA; contains active genes that are not tightly condensed during interphase.

eukaryote (Gr. *eu*, good, + *karyon*, kernel) A cell characterized by membrane-bounded organelles, most notably the nucleus, and one that possesses chromosomes whose DNA is associated with proteins; an organism composed of such cells.

eutrophic (Gr. *eutrophos*, thriving) Refers to a lake in which an abundant supply of minerals and organic matter exists.

evolution (L. *evolvere*, to unfold) Genetic change in a population of organisms; in general, evolution leads to progressive change from simple to complex.

exergonic Describes a chemical reaction in which the products contain less free energy than the reactants, so that free energy is released in the reaction.

exocrine gland (Gr. *ex*, out of, + *krinein*, to separate) A type of gland that releases its secretion through a duct, such as a digestive gland or a sweat gland.

exocytosis (Gr. *ex*, out of, + *kytos*, hollow vessel) A type of bulk transport out of cells in which a vacuole fuses with the plasma membrane, discharging the vacuole's contents to the outside.

exon (Gr. *exo*, outside) A segment of DNA that is both transcribed into RNA and translated into protein.

exoskeleton (Gr. *exo*, outside, + *skeletos*, hand) An external skeleton, as in arthropods.

exteroceptor (L. *exter*, outward, + *capere*, to take) A receptor that is excited by stimuli from the external world.

F

facilitated diffusion Carrier-assisted diffusion of molecules across a cellular membrane through specific channels from a region of higher concentration to one of lower concentration; the process is driven by the concentration gradient and does not require cellular energy from ATP.

fat A molecule composed of glycerol and three fatty acid molecules.

feedback inhibition Control mechanism whereby an increase in the concentration of some molecules inhibits the synthesis of that molecule.

fermentation (L. *fermentum*, ferment) The enzyme-catalyzed extraction of energy from organic compounds without the involvement of oxygen.

fertilization The fusion of two haploid gamete nuclei to form a diploid zygote nucleus.

fibroblast (L. *fibra*, fiber, + Gr. *blastos*, sprout) A flat, irregularly branching cell of connective tissue that secretes structurally strong proteins into the matrix between the cells.

First Law of Thermodynamics Energy cannot be created or destroyed, but can only undergo conversion from one form to another; thus, the amount of energy in the universe is unchangeable.

fitness The genetic contribution of an individual to succeeding generations. relative fitness refers to the fitness of an individual relative to other individuals in a population.

fixed action pattern A stereotyped animal behavior response, thought by ethologists to be based on programmed neural circuits.

foraging behavior A collective term for the many complex, evolved behaviors that influence what an animal eats and how the food is obtained.

founder effect The effect by which rare alleles and combinations of alleles may be enhanced in new populations.

fovea (L., a small pit) A small depression in the center of the retina with a high concentration of cones; the area of sharpest vision.

free energy Energy available to do work.

free radical An ionized atom with one or more unpaired electrons, resulting from electrons that have been energized by ionizing radiation being ejected from the atom; free radicals react violently with other molecules, such as DNA, causing damage by mutation.

fruit In angiosperms, a mature, ripened ovary (or group of ovaries), containing the seeds.

functional genomics Research into the function of specific genes in an organism.

fundamental niche Also referred to as the hypothetical niche, this is the entire niche an organism could fill if there were no other interacting factors (such as competition or predation).

G

gametangium, pl. **gametangia** (Gr. *gamein*, to marry, + L. *tangere*, to touch) A cell or organ in which gametes are formed.

gamete (Gr., wife) A haploid reproductive cell.

gametocytes Cells in the malarial sporozoite life cycle capable of giving rise to gametes when in the correct host.

gametophyte In plants, the haploid (*n*), gamete-producing generation, which alternates with the diploid (*2n*) sporophyte.

ganglion, pl. **ganglia** (Gr., a swelling) An aggregation of nerve cell bodies; in invertebrates, ganglia are the integrative centers; in vertebrates, the term is restricted to aggregations of nerve cell bodies located outside the central nervous system.

gap junction A junction between adjacent animal cells that allows the passage of materials between the cells.

gastrula (Gr., little stomach) In vertebrates, the embryonic stage in which the blastula with its single layer of cells turns into a three-layered embryo made up of ectoderm, mesoderm, and endoderm.

gene (Gr. *genos*, birth, race) The basic unit of heredity; a sequence of DNA nucleotides on a chromosome that encodes a protein, tRNA, or rRNA molecule, or regulates the transcription of such a sequence.

gene conversion Alteration of one homologous chromosome by the cell's error-detection and repair system to make it resemble the sequence on the other homologue.

genetic drift Random fluctuation in allele frequencies over time by chance.

genome The entire DNA sequence of an organism.

genomics The study of genomes as opposed to individual genes.

genotype (Gr. *genos*, birth, race, + *typos*, form) The genetic constitution underlying a single trait or set of traits.

genus, pl. **genera** (L., race) A taxonomic group that ranks below a family and above a species.

germination (L. *germinare*, to sprout) The resumption of growth and development by a spore or seed.

germ layers The three cell layers formed at gastrulation of the embryo that foreshadow the future organization of tissues; the layers, from the outside inward, are the ectoderm, the mesoderm, and the endoderm.

gill (1) In aquatic animals, a respiratory organ, usually a thin-walled projection from some part of the external body surface, endowed with a rich capillary bed and having a large surface area. (2) In basidiomycete fungi, the plates on the underside of the cap.

glomerular filtrate The fluid that passes out of the capillaries of each glomerulus.

glomerulus (L., a little ball) A cluster of capillaries enclosed by Bowman's capsule.

glucagon (Gr. *glykys*, sweet, + *ago*, to lead forward) A vertebrate hormone produced in the pancreas that acts to initiate the breakdown of glycogen to glucose subunits.

gluconeogenesis The synthesis of glucose from noncarbohydrates (such as proteins or fats).

glucose A common six-carbon sugar ($C_6H_{12}O_6$); the most common monosaccharide in most organisms.

glycocalyx A "sugar coating" on the surface of a cell resulting from the presence of polysaccharides on glycolipids and glycoproteins embedded in the outer layer of the plasma membrane.

glycogen (Gr. *glykys*, sweet, + *gen*, of a kind) Animal starch; a complex branched polysaccharide that serves as a food reserve in animals, bacteria, and fungi.

glycolipid Lipid molecule modified within the Golgi complex by having a short sugar chain (polysaccharide) attached.

glycolysis (Gr. *glykys*, sweet, + *lyein*, to loosen) The anaerobic breakdown of glucose; this enzyme-catalyzed process yields two molecules of pyruvate with a net of two molecules of ATP.

glycoprotein Protein molecule modified within the Golgi complex by having a short sugar chain (polysaccharide) attached.

glyoxysome A small cellular organelle or microbody containing enzymes necessary for conversion of fats into carbohydrates.

Golgi apparatus A collection of flattened stacks of membranes (each called a Golgi body) in the cytoplasm of eukaryotic cells; functions in collection, packaging, and distribution of molecules synthesized in the cell.

G protein A protein that binds guanosine triphosphate (GTP) and assists in the function of cell surface receptors. When the receptor binds its signal molecule, the G protein binds GTP and is activated to start a chain of events within the cell.

gravitropism (L. *gravis*, heavy, + *tropes*, turning) Growth response to gravity in plants; formerly called geotropism.

ground meristem (Gr. *meristos*, divisible) The primary meristem, or meristematic tissue, that gives rise to the plant body (except for the epidermis and vascular tissues).

guttation (L. *gutta*, a drop) The exudation of liquid water from leaves due to root pressure.

gymnosperm (Gr. *gymnos*, naked, + *sperma*, seed) A seed plant with seeds not enclosed in an ovary; conifers are gymnosperms.

gynoecium (Gr. *gyne*, woman, + *oikos*, house) The aggregate of carpels in the flower of a seed plant.

H

habitat (L. *hibitare*, to inhabit) The environment of an organism; the place where it is usually found.

habituation (L. *habitus*, condition) A form of learning; a diminishing response to a repeated stimulus.

haplodiploidy A phenomenon occurring in certain organisms such as wasps, wherein both haploid (male) and diploid (female) individuals are encountered.

haploid (Gr. *haploos*, single, + *ploion*, vessel) Having only one set of chromosomes (*n*), in contrast to diploid (*2n*).

Hardy-Weinberg equilibrium A mathematical description of the fact that allele and genotype frequencies remain constant in a random-mating population in the absence of inbreeding, selection, or other evolutionary forces; usually stated: if the frequency of allele *a* is *p* and the frequency of allele *b* is *q*, then the genotype frequencies after one generation of random mating will always be $p^2 + 2pq + q^2 = 1$.

Haversian canal After Clopton Havers, English anatomist. Narrow channels that run parallel to the length of a bone and contain blood vessels and nerve cells.

helper T cell A class of white blood cells that initiates both the cell-mediated immune response and the humoral immune response; helper T cells are the targets of the AIDS virus (HIV).

hemoglobin (Gr. *haima*, blood, + L. *globus*, a ball) A globular protein in vertebrate red blood cells and in the plasma of many invertebrates that carries oxygen and carbon dioxide.

hemophilia (Gr. *haima*, blood, + *philios*, friendly) A group of hereditary diseases characterized by failure of the blood to clot.

hemopoietic stem cell (Gr. *haima*, blood, + *poiesis*, a making) The cells in bone marrow where blood cells are formed.

hermaphroditism (Gr. *hermaphroditos*, containing both sexes) Condition in which an organism has both male and female functional reproductive organs.

heterochromatin (Gr. *heteros*, other, + *chroma*, color) The portion of the eukaryotic chromosome that is not transcribed into RNA; remains condensed in interphase and stains intensely in histological preparations.

heterokaryotic (Gr. *heteros*, other, + *karyon*, kernel) In fungi, having two or more genetically distinct types of nuclei within the same mycelium.

heterosporous (Gr. *heteros*, other, + *sporos*, seed) In vascular plants, having spores of two kinds, namely, microspores and megaspores.

heterotroph (Gr. *heteros*, other, + *trophos*, feeder) An organism that cannot derive energy from photosynthesis or inorganic chemicals, and so must feed on other plants and animals, obtaining chemical energy by degrading their organic molecules.

heterozygous Having two different alleles of the same gene; the term is usually applied to one or more specific loci, as in "heterozygous with respect to the *W* locus" (that is, the genotype is *W/w*).

histone (Gr. *histos*, tissue) One of a group of relatively small, very basic polypeptides, rich in arginine and lysine, forming the core of nucleosomes around which DNA is wrapped in the first stage of chromosome condensation.

holoblastic cleavage (Gr. *holos*, whole, + (*blastos*, germ) Process in vertebrate embryos in which the cleavage divisions all occur at the same rate, yielding a uniform cell size in the blastula.

homeobox A sequence of 180 nucleotides located in homeotic genes that produces a 60-amino-acid peptide sequence (the homeodomain) active in transcription factors.

homeostasis (Gr. *homos*, similar, + *stasis*, standing) The maintenance of a relatively stable internal physiological environment in an organism; usually involves some form of feedback self-regulation.

homeotherm (Gr. *homos*, same or similar, + *therme*, heat) An organism, such as a bird or mammal, capable of maintaining a stable body temperature independent of the environmental temperature. *See* endotherm.

homeotic gene One of a series of "master switch" genes that determine the form of segments developing in the embryo.

hominid (L. *homo*, man) Any primate in the human family, Hominidae. *Homo sapiens* is the only living representative.

hominoid (L. *homo*, man) Collectively, hominids and apes; the monkeys and hominoids constitute the anthropoid primates.

homokaryotic (Gr. *homos*, same, + *karyon*, kernel) In fungi, having nuclei with the same genetic makeup within a mycelium.

homologue One of a pair of chromosomes of the same kind located in a diploid cell; one copy of each pair of homologues comes from each gamete that formed the zygote.

homologous structure (Gr. *homologia*, agreement) A condition in which the similarity between two structures or functions is indicative of a common evolutionary origin.

homosporous (Gr. *homos*, same or similar, + *sporos*, seed) In some plants, production of only one type of spore rather than differentiated types. *Compare with* heterosporous.

homozygous Being a homozygote, having two identical alleles of the same gene; the term is usually applied to one or more specific loci, as in "homozygous with respect to the *W* locus" (that is, the genotype is *W/W* or *w/w*).

hormone (Gr. *hormaein*, to excite) A molecule, usually a peptide or steroid, that is produced in one part of an organism and triggers a specific cellular reaction in target tissues and organs some distance away.

human immunodeficiency virus (HIV) The virus responsible for AIDS, a deadly disease that destroys the human immune system. HIV is a retrovirus (its genetic material is RNA) that is thought to have been introduced to humans from chimpanzees.

hybridization The mating of unlike parents.

hydrostatic skeleton (Gr. *hydro*, water, + *skeletos*, hard) The skeleton of most soft-bodied invertebrates that have neither an internal nor an external skeleton. They use the relative incompressibility of the water within their bodies as a kind of skeleton.

hyperosmotic The condition in which a (hyperosmotic) solution has a higher osmotic concentration than that of a second solution. *Compare with* hypoosmotic.

hyperpolarization Above-normal negativity of a cell membrane during its resting potential.

hypersensitive response Plants respond to pathogens by selectively killing plant cells to block the spread of the pathogen.

hypha, pl. **hyphae** (Gr. *hyphe*, web) A filament of a fungus or oomycete; collectively, the hyphae comprise the mycelium.

hypocotyl The region immediately below where the cotyledons are attached.

hypoosmotic The condition in which a (hypoosmotic) solution has a lower osmotic concentration than that of a second solution. *Compare with* hyperosmotic.

hypothalamus (Gr. *hypo*, under, + *thalamos*, inner room) A region of the vertebrate brain just below the cerebral hemispheres, under the thalamus; a center of the autonomic nervous system, responsible for the integration and correlation of many neural and endocrine functions.

I

imaginal disk One of about a dozen groups of cells set aside in the abdomen of a larval insect and committed to forming key parts of the adult insect's body.

immune response In vertebrates, a defensive reaction of the body to invasion by a foreign substance or organism. *See* antibody and B cell.

immunoglobulin (L. *immunis*, free, + *globus*, globe) An antibody.

inbreeding The breeding of genetically related plants or animals; inbreeding tends to increase homozygosity.

inclusive fitness Describes the sum of the number of genes directly passed on in an individual's offspring and those genes passed on indirectly by kin (other than offspring) whose existence results from the benefit of the individual's altruism.

induction Process by which one group of embryonic cells affects an adjacent group, thereby inducing those cells to differentiate in a manner they otherwise would not have.

industrial melanism (Gr. *melas*, black) Phrase used to describe the evolutionary process in which initially light-colored organisms become dark as a result of natural selection.

inflammatory response (L. *inflammare*, to flame) A generalized nonspecific response to infection that acts to clear an infected area of infecting microbes and dead tissue cells so that tissue repair can begin.

initiation factor One of several proteins involved in the formation of an initiation complex in prokaryote polypeptide synthesis.

insertional inactivation Destruction of a gene's function by the insertion of a transposon.

instar A larval developmental stage in insects.

interferon In vertebrates, a protein produced in virus-infected cells that inhibits viral multiplication.

interneuron (association neuron) A nerve cell found only in the middle of the spinal cord that acts as a functional link between sensory neurons and motor neurons.

internode In plants, the region of a stem between two successive nodes.

interoceptor A receptor that senses information related to the body itself, its internal condition, and its position.

interphase The period between two mitotic or meiotic divisions in which a cell grows and its DNA replicates; includes G_1, S, and G_2 phases.

G-7

intron (L. *intra*, within) Portion of mRNA as transcribed from eukaryotic DNA that is removed by enzymes before the mature mRNA is translated into protein. See exon.

inversion (L. *invertere*, to turn upside down) A reversal in order of a segment of a chromosome; also, to turn inside out, as in embryogenesis of sponges or discharge of a nematocyst.

ion Any atom or molecule containing an unequal number of electrons and protons and therefore carrying a net positive or net negative charge.

ionizing radiation High-energy radiation that is highly mutagenic, producing free radicals that react with DNA; includes X rays and gamma rays.

isomer (Gr. *isos*, equal, + *meros*, part) One of a group of molecules identical in atomic composition but differing in structural arrangement; for example, glucose and fructose.

isosmotic The condition in which the osmotic concentrations of two solutions are equal, so that no net water movement occurs between them by osmosis.

K

karyotype (Gr. *karyon*, kernel, + *typos*, stamp or print) The morphology of the chromosomes of an organism as viewed with a light microscope.

keratin (Gr. *kera*, horn, + *in*, suffix used for proteins) A tough, fibrous protein formed in epidermal tissues and modified into skin, feathers, hair, and hard structures such as horns and nails.

kidney In vertebrates, the organ that filters the blood to remove nitrogenous wastes and regulates the balance of water and solutes in blood plasma.

kilocalorie Unit describing the amount of heat required to raise the temperature of a kilogram of water by one degree Celsius (°C); sometimes called a Calorie, equivalent to 1000 calories.

kinesis (Gr. *kinesis*, motion) Changes in activity level in an animal that are dependent on stimulus intensity. See kinetic energy.

kinetic energy The energy of motion.

kinetochore (Gr. *kinetikos*, putting in motion, + *choros*, chorus) Disc-shaped protein structure within the centromere to which the spindle fibers attach during mitosis or meiosis. See centromere.

kingdom The second highest commonly used taxonomic category.

kin selection Selection favoring relatives; an increase in the frequency of related individuals (kin) in a population, leading to an increase in the relative frequency in the population of those alleles shared by members of the kin group.

Krebs cycle Another name for the citric acid cycle; also called the tricarboxylic acid (TCA) cycle.

L

labrum (L., a lip) The upper lip of insects and crustaceans situated above or in front of the mandibles.

larynx The voice box; a cartilaginous organ that lies between the pharynx and trachea and is responsible for sound production in vertebrates.

lateral line system A sensory system encountered in fish, through which mechanoreceptors in a line down the side of the fish are sensitive to motion.

lateral meristems (L. *latus*, side, + Gr. *meristos*, divided) In vascular plants, the meristems that give rise to secondary tissue; the vascular cambium and cork cambium.

Law of Independent Assortment Mendel's second law of heredity, stating that genes located on nonhomologous chromosomes assort independently of one another.

Law of Segregation Mendel's first law of heredity, stating that alternative alleles for the same gene segregate from each other in production of gametes.

leaf primordium, pl. **primordia** (L. *primordium*, beginning) A lateral outgrowth from the apical meristem that will eventually become a leaf.

lenticels (L. *lenticella*, a small window) Spongy areas in the cork surfaces of stem, roots, and other plant parts that allow interchange of gases between internal tissues and the atmosphere through the periderm.

leucoplast (Gr. *leukos*, white, + *plasein*, to form) In plant cells, a colorless plastid in which starch grains are stored; usually found in cells not exposed to light.

leukocyte (Gr. *leukos*, white, + *kytos*, hollow vessel) A white blood cell; a diverse array of nonhemoglobin-containing blood cells, including phagocytic macrophages and antibody-producing lymphocytes.

lichen Symbiotic association between a fungus and a photosynthetic organism such as a green alga or cyanobacterium.

limbic system The hypothalamus, together with the network of neurons that link the hypothalamus to some areas of the cerebral cortex. Responsible for many of the most deep-seated drives and emotions of vertebrates, including pain, anger, sex, hunger, thirst, and pleasure.

lipase (Gr. *lipos*, fat, + *-ase*, enzyme suffix) An enzyme that catalyzes the hydrolysis of fats.

lipid (Gr. *lipos*, fat) A nonpolar hydrophobic organic molecule that is insoluble in water (which is polar) but dissolves readily in nonpolar organic solvents; includes fats, oils, waxes, steroids, phospholipids, and carotenoids.

lipid bilayer The structure of a cellular membrane, in which two layers of phospholipids spontaneously align so that the hydrophilic head groups are exposed to water, while the hydrophobic fatty acid tails are pointed toward the center of the membrane.

locus The position on a chromosome where a gene is located.

loop of Henle In the kidney of birds and mammals, a hairpin-shaped portion of the renal tubule in which water and salt are reabsorbed from the glomerular filtrate by diffusion.

lophophore A ring or U-shaped arrangement of tentacles, often ciliated, into which the coelom of the animal extends. The organ functions in feeding and gas exchange and is encountered in only a few phyla of small marine invertebrates.

luteal phase The second phase of the female reproductive cycle, during which the mature eggs are released into the fallopian tubes, a process called ovulation.

lymph (L. *lympha*, clear water) In animals, a colorless fluid derived from blood by filtration through capillary walls in the tissues.

lymphatic system In animals, an open vascular system that reclaims water that has entered interstitial regions from the bloodstream (lymph); includes the lymph nodes, spleen, thymus, and tonsils.

lymphocyte (L. *lympha*, water, + Gr. *kytos*, hollow vessel) A type of white blood cell. Lymphocytes are responsible for the immune response; there are two principal classes—B cells and T cells.

lymphokine A regulatory molecule that is secreted by lymphocytes. In the immune response, lymphokines secreted by helper T cells unleash the cell-mediated immune response.

lysis (Gr., a loosening) Disintegration of a cell by rupture of its plasma membrane.

M

macroevolution (Gr. *makros*, large, + L. *evolvere*, to unfold) The creation of new species and the extinction of old ones.

macromolecule (Gr. *makros*, large, + L. *moleculus*, a little mass) An extremely large biological molecule; refers specifically to proteins, nucleic acids, polysaccharides, lipids, and complexes of these.

macronutrients (Gr. *makros*, large, + L. *nutrire*, to nourish) Inorganic chemical elements required in large amounts for plant growth, such as nitrogen, potassium, calcium, phosphorus, magnesium, and sulfur.

macrophage A large phagocytic cell that is able to engulf and digest cellular debris and invading bacteria.

major histocompatibility complex (MHC) A set of protein cell-surface markers anchored in the plasma membrane, which the immune system uses to identify "self." All the cells of a given individual have the same "self" marker, called an MHC protein.

Malpighian tubules Blind tubules opening into the hindgut of terrestrial arthropods; they function as excretory organs.

mandibles (L. *mandibula*, jaw) In crustaceans, insects, and myriapods, the appendages immediately posterior to the antennae; used to seize, hold, bite, or chew food.

mantle The soft, outermost layer of the body wall in mollusks; the mantle secretes the shell.

marsupial (L. *marsupium*, pouch) A mammal in which the young are born early in their development, sometimes as soon as eight days after fertilization, and are retained in a pouch.

mass flow hypothesis The overall process by which materials move in the phloem of plants.

matrix (L. *mater*, mother) In mitochondria, the solution in the interior space surrounded by the cristae that contains the enzymes and other molecules involved in oxidative respiration; more generally, that part of a tissue within which an organ or process is embedded.

menstrual cycle (L. *mens*, month) In humans and higher primates, the cyclic changes in the ovaries and uterine endometrium; lasts about a month in humans.

menstruation Periodic sloughing off of the blood-enriched lining of the uterus when pregnancy does not occur.

meristem (Gr. *merizein*, to divide) Undifferentiated plant tissue from which new cells arise.

meroblastic cleavage (Gr. *meros*, part, + *blastos*, sprout) A type of cleavage in the eggs of reptiles, birds, and some fish. Occurs only on the blastodisc.

mesoderm (Gr. *mesos*, middle, + *derma*, skin) One of the three embryonic germ layers that form in the gastrula; gives rise to muscle, bone and other connective tissue, the peritoneum, the circulatory system, and most of the excretory and reproductive systems.

mesophyll (Gr. *mesos*, middle, + *phyllon*, leaf) The photosynthetic parenchyma of a leaf, located within the epidermis.

messenger RNA (mRNA) The RNA transcribed from structural genes; RNA molecules complementary to a portion of one strand of DNA, which are translated by the ribosomes to form protein.

metabolism (Gr. *metabole*, change) The sum of all chemical processes occurring within a living cell or organism.

metamorphosis (Gr. *meta*, after, + *morphe*, form, + *osis*, state of) Process in which a marked change in form takes place during postembryonic development as, for example, from tadpole to frog.

metaphase (Gr. *meta*, after, + *phasis*, form) The stage of mitosis or meiosis during which microtubules become organized into a spindle and the chromosomes come to lie in the spindle's equatorial plane.

metastasis The process by which cancer cells move from their point of origin to other locations in the body; also, a population of cancer cells in a secondary location, the result of movement from the primary tumor.

methanogens Obligate, anaerobic archaebacteria that produce methane.

microarray DNA sequences are placed on a microscope slide or chip with a robot. The microarray can then be probed with RNA from specific tissues to identify expressed DNA.

microbody A cellular organelle bounded by a single membrane and containing a variety of enzymes; generally derived from endoplasmic reticulum; includes peroxisomes and glyoxysomes.

microevolution (Gr. *mikros*, small, + L. *evolvere*, to unfold) Refers to the evolutionary process itself. Evolution within a species. Also called adaptation.

micronutrient (Gr. *mikros*, small, + L. *nutrire*, to nourish) A mineral required in only minute amounts for plant growth, such as iron, chlorine, copper, manganese, zinc, molybdenum, and boron.

micropyle In the ovules of seed plants, an opening in the integuments through which the pollen tube usually enters.

microtubule (Gr. *mikros*, small, + L. *tubulus*, little pipe) In eukaryotic cells, a long, hollow protein cylinder, composed of the protein tubulin; these influence cell shape, move the chromosomes in cell division, and provide the functional internal structure of cilia and flagella.

microvillus (Gr. *mikros*, small, + L. *villus*, shaggy hair) Cytoplasmic projection from epithelial cells; microvilli greatly increase the surface area of the small intestine.

middle lamella The layer of intercellular material, rich in pectic compounds, that cements together the primary walls of adjacent plant cells.

mimicry (Gr. *mimos*, mime) The resemblance in form, color, or behavior of certain organisms (mimics) to other more powerful or more protected ones (models).

mitosis (Gr. *mitos*, thread) Somatic cell division; nuclear division in which the duplicated chromosomes separate to form two genetically identical daughter nuclei.

monocot Short for monocotyledon; flowering plant in which the embryos have only one cotyledon, the floral parts are generally in threes, and the leaves typically are parallel-veined.

monocyte (Gr. *monos*, single, + *kytos*, hollow vessel) A type of leukocyte that becomes a phagocytic cell (macrophage) after moving into tissues.

monoecious (Gr. *monos*, single, + *oecos*, house) A plant in which the staminate and pistillate flowers are separate, but borne on the same individual.

monophyletic In phylogenetic classification, a group that includes the most recent common ancestor of the group and all its descendants. A clade is a monophyletic group.

monosaccharide (Gr. *monos*, single, + L. *saccharum*, sugar) A simple sugar that cannot be decomposed into smaller sugar molecules.

monotreme (Gr. *monos*, single, + *treme*, hole) An egg-laying mammal.

morphogen A signal molecule produced by an embryonic organizer region that informs surrounding cells of their distance from the organizer, thus determining relative positions of cells during development.

morphology The form and structure of an organism.

morula (L., a little mulberry) Solid ball of cells in the early stage of embryonic development.

mosaic development A pattern of embryonic development in which initial cells produced by cleavage divisions contain different developmental signals (determinants) from the egg, setting the individual cells on different developmental paths.

motor (efferent) neuron Neuron that transmits nerve impulses from the central nervous system to an effector, which is typically a muscle or gland.

Müllerian mimicry After Fritz Müller, German biologist. A phenomenon in which two or more unrelated but protected species resemble one another, thus achieving a kind of group defense.

multigene family A collection of related genes on a single chromosome or on different chromosomes.

muscle fiber A long, cylindrical, multinucleated cell containing numerous myofibrils, which is capable of contraction when stimulated.

mutagen (L. *mutare*, to change, + Gr. *genaio*, to produce) An agent that induces changes in DNA (mutations); includes physical agents that damage DNA and chemicals that alter DNA bases.

mutant (L. *mutare*, to change) A mutated gene; alternatively, an organism carrying a gene that has undergone a mutation.

mutation A permanent change in a cell's DNA; includes changes in nucleotide sequence, alteration of gene position, gene loss or duplication, and insertion of foreign sequences.

mutualism (L. *mutuus*, lent, borrowed) A symbiotic association in which two (or more) organisms live together, and both members benefit.

mycelium, pl. **mycelia** (Gr. *mykes*, fungus) In fungi, a mass of hyphae.

mycorrhiza, pl. **mycorrhizae** (Gr. *mykes*, fungus, + *rhiza*, root) A symbiotic association between fungi and the roots of a plant.

myelin sheath (Gr. *myelinos*, full of marrow) A fatty layer surrounding the long axons of motor neurons in the peripheral nervous system of vertebrates.

myofilament (Gr. *myos*, muscle, + L. *filare*, to spin) A contractile microfilament, composed largely of actin and myosin, within muscle.

myosin (Gr. *myos*, muscle, + *in*, belonging to) One of the two protein components of microfilaments (the other is actin); a principal component of vertebrate muscle.

N

natural killer cell A cell that does not kill invading microbes, but rather, the cells infected by them.

natural selection The differential reproduction of genotypes; caused by factors in the environment; leads to evolutionary change.

negative feedback A homeostatic control mechanism whereby an increase in some substance or activity inhibits the process leading to the increase; also known as feedback inhibition.

nephridium, pl. **nephridia** (Gr. *nephros*, kidney) In invertebrates, a tubular excretory structure.

nephrid organ A filtration system of many freshwater invertebrates in which water and waste pass from the body across the membrane into a collecting organ, from which they are expelled to the outside through a pore.

nephron (Gr. *nephros*, kidney) Functional unit of the vertebrate kidney; one of numerous tubules involved in filtration and selective reabsorption of blood; each nephron consists of a Bowman's capsule, an enclosed glomerulus, and a long attached tubule; in humans, called a renal tubule.

nerve A group or bundle of nerve fibers (axons) with accompanying neurological cells, held together by connective tissue; located in the peripheral nervous system.

neural crest A special strip of cells that develops just before the neural groove closes over to form the neural tube in embryonic development.

neural groove The long groove formed along the long axis of the embryo by a layer of ectodermal cells.

neural tube The dorsal tube, formed from the neural plate, that differentiates into the brain and spinal cord.

neuroglia (Gr. *neuron*, nerve, + *glia*, glue) Nonconducting nerve cells that are intimately associated with neurons and appear to provide nutritional support.

neuromuscular junction The structure formed when the tips of axons contact (innervate) a muscle fiber.

neuron (Gr., nerve) A nerve cell specialized for signal transmission; includes cell body, dendrites, and axon.

neurotransmitter (Gr. *neuron*, nerve, + L. *trans*, across, + *mittere*, to send) A chemical released at the axon terminal of a neuron that travels across the synaptic cleft, binds a specific receptor on the far side, and depending on the nature of the receptor, depolarizes or hyperpolarizes a second neuron or a muscle or gland cell.

neurulation A process in early embryonic development by which a dorsal band of ectoderm thickens and rolls into the neural tube.

neutrophil An abundant type of granulocyte capable of engulfing microorganisms and other foreign particles; neutrophils comprise about 50 to 70% of the total number of white blood cells.

niche The role played by a particular species in its environment.

nicotinamide adenine dinucleotide (NAD⁺) A molecule that becomes reduced (to NADH) as it carries high-energy electrons from oxidized molecules and delivers them to ATP-producing pathways in the cell.

nociceptor A naked dendrite that acts as a receptor in response to a pain stimulus.

nocturnal (L. *nocturnus*, night) Active primarily at night.

node (L. *nodus*, knot) The part of a plant stem where one or more leaves are attached. *See* internode.

node of Ranvier After L.A. Ranvier, French histologist. A gap formed at the point where two Schwann cells meet and where the axon is in direct contact with the surrounding intercellular fluid.

nonassociative learning A learned behavior that does not require an animal to form an association between two stimuli, or between a stimulus and a response.

nonsense codon One of three codons (UAA, UAG, and UGA) that are not recognized by tRNAs, thus serving as "stop" signals in the mRNA message and terminating translation.

notochord (Gr. *noto*, back, + L. *chorda*, cord) In chordates, a dorsal rod of cartilage that runs the length of the body and forms the primitive axial skeleton in the embryos of all chordates.

nucellus (L. *nucella*, a small nut) Tissue composing the chief pair of young ovules, in which the embryo sac develops; equivalent to a megasporangium.

nucleic acid A nucleotide polymer; chief types are deoxyribonucleic acid (DNA), which is double-stranded, and ribonucleic acid (RNA), which is typically single-stranded.

nucleolus (L., a small nucleus) In eukaryotes, the site of rRNA synthesis; a spherical body composed chiefly of rRNA in the process of being transcribed from multiple copies of rRNA genes.

nucleosome (L. *nucleus*, kernel, + *soma*, body) The fundamental packaging unit of eukaryotic chromosomes; a complex of DNA and histone proteins in which the double-helical DNA winds around eight molecules of histone; chromatin is composed of long sequences of nucleosomes.

nucleotide A single unit of nucleic acid, composed of a phosphate, a five-carbon sugar (either ribose or deoxyribose), and a purine or a pyrimidine.

nucleus In atoms, the central core, containing positively charged protons and (in all but hydrogen) electrically neutral neutrons; in eukaryotic cells, the membranous organelle that houses the chromosomal DNA; in the central nervous system, a cluster of nerve cell bodies.

O

ocellus, pl. **ocelli** (L., little eye) A simple light receptor common among invertebrates.

Okazaki fragment A short segment of DNA produced by discontinuous replication elongating in the 5′→3′ direction away from the replication.

olfaction (L. *olfactum*, smell) The process or function of smelling.

ommatidium, pl. **ommatidia** (Gr., little eye) The visual unit in the compound eye of arthropods; contains light-sensitive cells and a lens able to form an image.

oncogene (Gr. *oncos*, cancer, + *genos*, birth) A mutant form of a growth-regulating gene that is inappropriately "on," causing unrestrained cell growth and division.

oocyst The zygote in a sporozoan life cycle. It is surrounded by a tough cyst to prevent dehydration or other damage.

operant conditioning A learning mechanism in which the reward follows only after the correct behavioral response.

operator A site of gene regulation; a sequence of nucleotides overlapping the promoter site and recognized by a repressor protein; binding of the repressor prevents binding of the polymerase to the promoter site and so blocks transcription of the structural gene.

operculum A flat, bony, external protective covering over the gill chamber in fish.

operon (L. *operis*, work) A cluster of adjacent structural genes transcribed as a unit into a single mRNA molecule.

order A category of classification above the level of family and below that of class.

organ (Gr. *organon*, tool) A body structure composed of several different tissues grouped in a structural and functional unit.

organelle (Gr. *organella*, little tool) Specialized part of a cell; literally, a small cytoplasmic organ.

orthologs Genes that reflect the conservation of a single gene found in an ancestor.

osmoconformer An animal that maintains the osmotic concentration of its body fluids at about the same level as that of the medium in which it is living.

osmosis (Gr. *osmos*, act of pushing, thrust) The diffusion of water across a selectively permeable membrane (a membrane that permits the free passage of water but prevents or retards the passage of a solute); in the absence of differences in pressure or volume, the net movement of water is from the side containing a lower concentration of solute to the side containing a higher concentration.

osmotic pressure The potential pressure developed by a solution separated from pure water by a differentially permeable membrane. The higher the solute concentration, the greater the osmotic potential of the solution; also called *osmotic potential.*

osteoblast (Gr. *osteon*, bone, + *blastos*, bud) A bone-forming cell.

osteocyte (Gr. *osteon*, bone, + *kytos*, hollow vessel) A mature osteoblast.

outcrossing Breeding with individuals other than oneself or one's close relatives.

ovary (L. *ovum*, egg) (1) In animals, the organ in which eggs are produced. (2) In flowering plants, the enlarged basal portion of a carpel, which contains the ovule(s); the ovary matures to become the fruit.

oviduct (L. *ovum*, egg, + *ductus*, duct) In vertebrates, the passageway through which ova (eggs) travel from the ovary to the uterus.

oviparity (L. *ovum*, egg, + *parere*, to bring forth) Refers to a type of reproduction in which the eggs are developed after leaving the body of the mother, as in reptiles.

ovoviviparity Refers to a type of reproduction in which young hatch from eggs that are retained in the mother's uterus.

ovulation In animals, the release of an egg or eggs from the ovary.

ovum, pl. **ova** (L., egg) The egg cell; female gamete.

oxidation (Fr. *oxider*, to oxidize) Loss of an electron by an atom or molecule; in metabolism, often associated with a gain of oxygen or a loss of hydrogen.

oxidative respiration Process of cellular activity in which glucose or other molecules are broken down to water and carbon dioxide with the release of energy.

oxygen debt The amount of oxygen required to convert the lactic acid generated in the muscles during exercise back into glucose.

oxytocin (Gr. *oxys*, sharp, + *tokos*, birth) A hormone of the posterior pituitary gland that affects uterine contractions during childbirth and stimulates lactation.

ozone O_3, a stratospheric layer of the earth's atmosphere responsible for filtering out ultraviolet radiation supplied by the sun.

P

pacemaker A patch of excitatory tissue in the vertebrate heart that initiates the heartbeat.

palisade parenchyma (L. *palus*, stake) In plant leaves, the columnar, chloroplast-containing parenchyma cells of the mesophyll. Also called *palisade cells.*

papilla A small projection of tissue.

paracrine A type of chemical signaling between cells in which the effects are local and short-lived.

paralogs Two genes within an organism that arose from the duplication of one gene in an ancestor.

paraphyletic In phylogenetic classification, a group that includes the most recent common ancestor of the group, but not all its descendants.

parapodia (Gr. *para*, beside, + *pous*, foot) One of the paired lateral processes on each side of most segments in polychaete annelids.

parasexuality In certain fungi, the fusion and segregation of heterokaryotic haploid nuclei to produce recombinant nuclei.

parasitism (Gr. *para*, beside, + *sitos*, food) A living arrangement in which an organism lives on or in an organism of a different species and derives nutrients from it.

parenchyma cell The most common type of plant cell; characterized by large vacuoles, thin walls, and functional nuclei.

parthenogenesis (Gr. *parthenos*, virgin, + *genesis*, birth) The development of an egg without fertilization, as in aphids, bees, ants, and some lizards.

partial pressure The components of each individual gas—such as nitrogen, oxygen, and carbon dioxide—that together comprise the total air pressure.

pelagic Free-swimming, usually in open water.

pellicle A tough, flexible covering in ciliates and euglenoids.

peptide bond (Gr. *peptein*, to soften, digest) The type of bond that links amino acids together in proteins through a dehydration reaction.

perianth (Gr. *peri*, around, + *anthos*, flower) In flowering plants, the petals and sepals taken together.

pericycle (Gr. *peri*, around, + *kykos*, circle) In vascular plants, one or more cell layers surrounding the vascular tissues of the root, bounded externally by the endodermis and internally by the phloem.

periderm (Gr. *peri*, around, + *derma*, skin) Outer protective tissue in vascular plants that is produced by the cork cambium and functionally replaces epidermis when it is destroyed during secondary growth; the periderm includes the cork, cork cambium, and phelloderm.

peristalsis (Gr. *peristaltikos*, compressing around) In animals, a series of alternating contracting and relaxing muscle movements along the length of a tube such as the oviduct or alimentary canal that tend to force material such as an egg cell or food through the tube.

peroxisome A microbody that plays an important role in the breakdown of highly oxidative hydrogen peroxide by catalase.

petal A flower part, usually conspicuously colored; one of the units of the corolla.

petiole (L. *petiolus*, a little foot) The stalk of a leaf.

phagocyte (Gr. *phagein*, to eat, + *kytos*, hollow vessel) Any cell that engulfs and devours microorganisms or other particles.

phagocytosis (Gr., cell-eating) Endocytosis of a solid particle; the plasma membrane folds inward around the particle (which may be another cell) and engulfs it to form a vacuole.

pharynx (Gr., gullet) In vertebrates, a muscular tube that connects the mouth cavity and the esophagus; it serves as the gateway to the digestive tract and to the trachea.

phenotype (Gr. *phainein*, to show, + *typos*, stamp or print) The realized expression of the genotype; the physical appearance or functional expression of a trait.

pheromone (Gr. *pherein*, to carry, + *hormonos*, exciting, stirring up) Chemical substance released by one organism that influences the behavior or physiological processes of another organism of the same species. Some pheromones serve as sex attractants, as trail markers, and as alarm signals.

phloem (Gr. *phloos*, bark) In vascular plants, a food-conducting tissue basically composed of sieve elements, various kinds of parenchyma cells, fibers, and sclereids.

phosphodiester bond The type of bond that links nucleotides in a nucleic acid; formed when the phosphate group of one nucleotide binds to the 3′ hydroxyl group of the sugar of another.

phospholipid Similar in structure to a fat, but having only two fatty acids attached to the glycerol backbone, with the third space linked to a phosphorylated molecule; contains a polar hydrophilic "head" end (phosphate group) and a nonpolar hydrophobic "tail" end (fatty acids).

photoperiodism (Gr. *photos*, light, + *periodos*, a period) The tendency of biological reactions to respond to the duration and timing of day and night; a mechanism for measuring seasonal time.

photoreceptor (Gr. *photos*, light) A light-sensitive sensory cell.

photosystem An organized collection of chlorophyll and other pigment molecules embedded in the thylakoid of chloroplasts; traps photon energy and channels it as energetic electrons to the thylakoid membrane.

phototropism (Gr. *photos*, light, + *trope*, turning) In plants, a growth response to a light stimulus.

phycologist (Gr. *phykos*, seaweed) The scientist who studies algae.

phylogeny The evolutionary history of an organism, including which species are closely related and in what order related species evolved; often represented in the form of an evolutionary tree.

phylum, pl. phyla (Gr. *phylon*, race, tribe) A major category, between kingdom and class, of taxonomic classifications.

phytochrome (Gr. *phyton*, plant, + *chroma*, color) A plant pigment that is associated with the absorption of light; photoreceptor for red to far-red light.

phytoremediation The process of removing contamination from soil or water using plants.

pigment (L. *pigmentum*, paint) A molecule that absorbs light.

pilus, pl. pili Extensions of a bacterial cell enabling it to transfer genetic materials from one individual to another or to adhere to substrates.

pinocytosis (Gr. *pinein*, to drink, + *kytos*, hollow vessel, + *osis*, condition) The process of fluid uptake by endocytosis in a cell.

pistil (L. *pistillum*, pestle) Central organ of flowers, typically consisting of ovary, style, and stigma; a pistil may consist of one or more fused carpels and is more technically and better known as the gynoecium.

pith The ground tissue occupying the center of the stem or root within the vascular cylinder.

placenta, pl. placentae (L., a flat cake) (1) In flowering plants, the part of the ovary wall to which the ovules or seeds are attached. (2) In mammals, a tissue formed in part from the inner lining of the uterus and in part from other membranes, through which the embryo (later the fetus) is nourished while in the uterus and through which wastes are carried away.

plankton (Gr. *planktos*, wandering) Free-floating, mostly microscopic, aquatic organisms.

plasma (Gr., form) The fluid of vertebrate blood; contains dissolved salts, metabolic wastes, hormones, and a variety of proteins, including antibodies and albumin; blood minus the blood cells.

plasma cell An antibody-producing cell resulting from the multiplication and differentiation of a B lymphocyte that has interacted with an antigen.

plasma membrane The membrane surrounding the cytoplasm of a cell; consists of a single phospholipid bilayer with embedded proteins.

plasmid (Gr. *plasma*, a form or mold) A small fragment of extrachromosomal DNA, usually circular, that replicates independently of the main chromosome, although it may have been derived from it.

plasmodesmata In plants, cytoplasmic connections between adjacent cells.

plasmodium (Gr. *plasma*, a form or mold, + *eidos*, form) Stage in the life cycle of myxomycetes (plasmodial slime molds); a multinucleate mass of protoplasm surrounded by a membrane.

plastid (Gr. *plastos*, formed or molded) An organelle in the cells of photosynthetic eukaryotes that is the site of photosynthesis and, in plants and green algae, of starch storage.

platelet (Gr. dim. of *plattus*, flat) In mammals, a fragment of a white blood cell that circulates in the blood and functions in the formation of blood clots at sites of injury.

pleiotropy Condition in which an individual allele has more than one effect on production of the phenotype.

plumule The epicotyl of a plant with its two young leaves.

point mutation An alteration of one nucleotide in a chromosomal DNA molecule.

polar body Minute, nonfunctioning cell produced during the meiotic divisions leading to gamete formation in vertebrates.

pollen tube A tube formed after germination of the pollen grain; carries the male gametes into the ovule.

pollination The transfer of pollen from an anther to a stigma.

polyandry The condition in which a female mates with more than one male.

polyclonal antibody An antibody response in which an antigen elicits many different antibodies, each fitting a different portion of the antigen surface.

polygyny (Gr. *poly*, many, + *gyne*, woman, wife) A mating choice in which a male mates with more than one female.

polymer (Gr. *poly*, many, + *meris*, part) A molecule composed of many similar or identical molecular subunits; starch is a polymer of glucose.

polymerase chain reaction (PCR) A process by which DNA polymerase is used to copy a sequence of interest repeatedly, making millions of copies of the same DNA.

polymorphism (Gr. *poly*, many, + *morphe*, form) The presence in a population of more than one allele of a gene at a frequency greater than that of newly arising mutations.

polypeptide (Gr. *poly*, many, + *peptein*, to digest) A molecule consisting of many joined amino acids; not usually as complex as a protein.

polyphyletic In phylogenetic classification, a group that does not include the most recent common ancestor of all members of the group.

polyploidy Condition in which one or more entire sets of chromosomes is added to the diploid genome.

polysaccharide (Gr. *poly*, many, + *sakcharon*, sugar, from Latin *sakara*, gravel, sugar) A carbohydrate composed of many monosaccharide sugar subunits linked together in a long chain; examples are glycogen, starch, and cellulose.

polyunsaturated fat A fat molecule having at least two double bonds between adjacent carbons in one or more of the fatty acid chains.

population (L. *populus*, the people) Any group of individuals, usually of a single species, occupying a given area at the same time.

posttranscriptional control A mechanism of control over gene expression that operates after the transcription of mRNA is complete.

potential energy Energy that is not being used, but could be; energy in a potentially usable form; often called "energy of position."

precapillary sphincter A ring of muscle that guards each capillary loop and that, when closed, blocks flow through the capillary.

primary endosperm nucleus In flowering plants, the result of the fusion of a sperm nucleus and the (usually) two polar nuclei.

primary growth In vascular plants, growth originating in the apical meristems of shoots and roots; results in an increase in length.

primary immune response The first response of an immune system to a foreign antigen. If the system is challenged again with the same antigen, the memory cells created during the primary response will respond more quickly.

primary induction Inductions between the three primary tissue types—ectoderm, mesoderm, and endoderm.

primary nondisjunction Failure of chromosomes to separate properly at meiosis I.

primary phloem In plant phloem, the cells involved in food conduction.

primary productivity The amount of energy produced by photosynthetic organisms in a community.

primary structure The specific amino acid sequence of a protein.

primary tissues Tissues that comprise the primary plant body.

primary transcript The initial mRNA molecule copied from a gene by RNA polymerase, containing a faithful copy of the entire gene, including introns as well as exons.

primary wall In plants, the wall layer deposited during the period of cell expansion.

primate Monkeys and apes (including humans).

primitive streak (L. *primus*, first) In the early embryos of birds, reptiles, and mammals, a dorsal, longitudinal strip of ectoderm and mesoderm that is equivalent to the blastopore in other forms.

principle of parsimony Principle stating that scientists should favor the hypothesis that requires the fewest assumptions.

prions Infectious proteinaceous particles.

procambium (Gr. *pro*, before, + *cambiare*, to exchange) In vascular plants, a primary meristematic tissue that gives rise to primary vascular tissues.

prokaryote (Gr. *pro*, before, + *karyon*, kernel) A bacterium; a cell lacking a membrane-bounded nucleus or membrane-bounded organelles.

promoter A specific nucleotide sequence to which RNA polymerase attaches to initiate transcription of mRNA from a gene.

prophase (Gr. *pro*, before, + *phasis*, form) An early stage in nuclear division, characterized by the formation of a microtubule spindle along the future axis of division, the shortening and thickening of the chromosomes, and their movement toward the equator of the spindle (the "metaphase plate").

proprioceptor (L. *proprius*, one's own) In vertebrates, a sensory receptor that senses the body's position and movements.

prostaglandins (Gr. *prostas*, a porch or vestibule, + *glans*, acorn) A group of modified fatty acids that function as chemical messengers.

prostate gland (Gr. *prostas*, a porch or vestibule) In male mammals, a mass of glandular tissue at the base of the urethra that secretes an alkaline fluid that has a stimulating effect on the sperm as they are released.

protein (Gr. *proteios*, primary) A chain of amino acids joined by peptide bonds.

protein kinase An enzyme that adds phosphate groups to proteins, changing their activity.

proteomics The study of all proteins in an organism.

proton pump A protein channel in a membrane of the cell that expends energy to transport protons against a concentration gradient; involved in the chemiosmotic generation of ATP.

proto-oncogene A normal gene that promotes cell division, so called because mutations that cause these genes to become overexpressed convert them into oncogenes that produce excessive cellular proliferation (i.e., cancer).

protozoa (Gr. *protos*, first, + *zoon*, animal) The traditional name given to heterotrophic protists.

pseudogene (Gr. *pseudos*, false, + *genos*, birth) A copy of a gene that is not transcribed.

pseudopod (Gr. *pseudes*, false, + *pous*, foot) A nonpermanent cytoplasmic extension of the cell body. Also called a *pseudopodium*.

punctuated equilibrium A hypothesis about the mechanism of evolutionary change proposing that long periods of little or no change are punctuated by periods of rapid evolution.

pupa (L., girl, doll) A developmental stage of some insects in which the organism is nonfeeding, immotile, and sometimes encapsulated or in a cocoon; the pupal stage occurs between the larval and adult phases.

purine (Gr. *purinos*, fiery, sparkling) The larger of the two general kinds of nucleotide base found in DNA and RNA; a nitrogenous base with a double-ring structure, such as adenine or guanine.

pyrimidine (alt. of pyridine, from G. *pyr*, fire) The smaller of two general kinds of nucleotide base found in DNA and RNA; a nitrogenous base with a single-ring structure, such as cytosine, thymine, or uracil.

Q

quaternary structure The structural level of a protein composed of more than one polypeptide chain, each of which has its own tertiary structure; the individual chains are called subunits.

R

radicle (L. *radicula*, root) The part of the plant embryo that develops into the root.

radioactivity The emission of nuclear particles and rays by unstable atoms as they decay into more stable forms.

radula (L., scraper) Rasping tongue found in most mollusks.

realized niche The actual niche occupied by an organism when all biotic and abiotic interactions are taken into account.

receptor-mediated endocytosis Process by which specific macromolecules are transported into eukaryotic cells at clathrin-coated pits, after binding to specific cell-surface receptors.

receptor protein A highly specific cell-surface receptor embedded in a cell membrane that responds only to a specific messenger molecule.

recessive An allele that is only expressed when present in the homozygous condition, while being "hidden" by the expression of a dominant allele in the heterozygous condition.

reciprocal altruism Performance of an altruistic act with the expectation that the favor will be returned. A key and very controversial assumption of many theories dealing with the evolution of social behavior. *See* altruism.

reciprocal recombination A mechanism of genetic recombination that occurs only in eukaryotic organisms, in which two chromosomes trade segments; can occur between nonhomologous chromosomes as well as the more usual exchange between homologous chromosomes in meiosis.

recombinant DNA Fragments of DNA from two different species, such as a bacterium and a mammal, spliced together in the laboratory into a single molecule.

reduction (L. *reductio*, a bringing back; originally "bringing back" a metal from its oxide) The gain of an electron by an atom, often with an associated proton.

reflex (L. *reflectere*, to bend back) In the nervous system, a motor response subject to little associative modification; a reflex is among the simplest neural pathways, involving only a sensory neuron, sometimes (but not always) an interneuron, and one or more motor neurons.

reflex arc The nerve path in the body that leads from stimulus to reflex action.

refractory period The recovery period after membrane depolarization during which the membrane is unable to respond to additional stimulation.

replication fork The Y-shaped end of a growing replication bubble in a DNA molecule undergoing replication.

repolarization Return of the ions in a nerve to their resting potential distribution following depolarization.

repressor (L. *reprimere*, to press back, keep back) A protein that regulates DNA transcription by preventing RNA polymerase from attaching to the promoter and transcribing the structural gene. *See* operator.

residual volume The amount of air remaining in the lungs after the maximum amount of air has been exhaled.

resting membrane potential The charge difference (difference in electric potential) that exists across a neuron at rest (about 70 millivolts).

restriction endonuclease An enzyme that cleaves a DNA duplex molecule at a particular base sequence, usually within or near a palindromic sequence; also called a restriction enzyme.

restriction fragment length polymorphism (RFLP) Restriction enzymes recognize very specific DNA sequences. Alleles of the same gene or surrounding sequences may have base-pair differences, so that DNA near one allele is cut into a different length fragment than DNA near the other allele. These different fragments separate based on size on gels.

retina (L., a small net) The photosensitive layer of the vertebrate eye; contains several layers of neurons and light receptors (rods and cones); receives the image formed by the lens and transmits it to the brain via the optic nerve.

retrovirus (L. *retro*, turning back) An RNA virus. When a retrovirus enters a cell, a viral enzyme (reverse transcriptase) transcribes viral RNA into duplex DNA, which the cell's machinery then replicates and transcribes as if it were its own.

Rh blood group A set of cell surface markers (antigens) on the surface of red blood cells in humans and rhesus monkeys (for which it is named); although there are several alleles, they are grouped into two main types, called Rh-positive and Rh-negative.

rhizome (Gr. *rhizoma*, mass of roots) In vascular plants, a usually more or less horizontal underground stem; may be enlarged for storage or may function in vegetative reproduction.

ribonucleic acid (RNA) A class of nucleic acids characterized by the presence of the sugar ribose and the pyrimidine uracil; includes mRNA, tRNA, and rRNA.

ribosomal RNA (rRNA) A class of RNA molecules found, together with characteristic proteins, in ribosomes; transcribed from the DNA of the nucleolus.

ribosome The molecular machine that carries out protein synthesis; the most complicated aggregation of proteins in a cell, also containing three different rRNA molecules.

ribozyme An RNA molecule that can behave as an enzyme, sometimes catalyzing its own assembly; rRNA also acts as a ribozyme in the polymerization of amino acids to form protein.

RNA polymerase An enzyme that catalyzes the assembly of an mRNA molecule, the sequence of which is complementary to a DNA molecule used as a template. *See* transcription.

RNA primer In DNA replication, a sequence of about 10 RNA nucleotides complementary to unwound DNA that attaches at a replication fork; the DNA polymerase uses the RNA primer as a starting point for addition of DNA nucleotides to form the new DNA strand; the RNA primer is later removed and replaced by DNA nucleotides.

RNA splicing A nuclear process by which intron sequences of a primary mRNA transcript are cut out and the exon sequences spliced together to give the correct linkages of genetic information that will be used in protein construction.

rod Light-sensitive nerve cell found in the vertebrate retina; sensitive to very dim light; responsible for "night vision."

root The usually descending axis of a plant, normally below ground, which anchors the plant and serves as the major point of entry for water and minerals.

rumen An "extra stomach" in cows and related mammals wherein digestion of cellulose occurs and from which partially digested material can be ejected back into the mouth.

S

saltatory conduction A very fast form of nerve impulse conduction in which the impulses leap from node to node over insulated portions.

saprobes Heterotrophic organisms that digest their food externally (such as most fungi).

sarcolemma The specialized cell membrane in a muscle cell.

sarcoma A cancerous tumor arising from cells of connective tissue, bone, or muscle.

sarcomere (Gr. *sarx*, flesh, + *meris*, part of) Fundamental unit of contraction in skeletal muscle; repeating bands of actin and myosin that appear between two Z lines.

sarcoplasmic reticulum (Gr. *sarx*, flesh, + *plassein*, to form, mold; L. *reticulum*, network) The endoplasmic reticulum of a muscle cell. A sleeve of membrane that wraps around each myofilament.

satellite DNA A nontranscribed region of the chromosome with a distinctive base composition; a short nucleotide sequence repeated tandemly many thousands of times.

saturated fat A fat composed of fatty acids in which all the internal carbon atoms contain the maximum possible number of hydrogen atoms.

Schwann cells The supporting cells associated with projecting axons, along with all the other nerve cells that make up the peripheral nervous system.

sclereid (Gr. *skleros*, hard) In vascular plants, a sclerenchyma cell with a thick, lignified, secondary wall having many pits; not elongate like a fiber.

sclerenchyma cell (Gr. *skleros*, hard, + *en*, in, + *chymein*, to pour) A type of tissue made up of sclerenchyma cells.

scrotum (L., bag) The pouch that contains the testes in most mammals.

scuttellum The modified cotyledon in cereal grains.

secondary cell wall In plants, the innermost layer of the cell wall. Secondary walls have a highly organized microfibrillar structure and are often impregnated with lignin.

secondary growth In vascular plants, an increase in stem and root diameter made possible by cell division of the lateral meristems.

secondary immune response The swifter response of the body the second time it is invaded by the same pathogen because of the presence of memory cells, which quickly become antibody-producing plasma cells.

secondary induction An induction between tissues that have already differentiated.

secondary structure In a protein, hydrogen-bonding interactions between CO and NH groups of the primary structure.

Second Law of Thermodynamics A statement concerning the transformation of potential energy into heat; it says that disorder (entropy) is continually increasing in the universe as energy changes occur, so disorder is more likely than order.

second messenger A small molecule or ion that carries the message from a receptor on the target cell surface into the cytoplasm.

segregation The process by which alternative forms of traits are expressed in offspring rather than blending each trait of the parents in the offspring.

selection The process by which some organisms leave more offspring than competing ones, and their genetic traits tend to appear in greater proportions among members of succeeding generations than the traits of those individuals that leave fewer offspring.

selectively permeable Condition in which a membrane is permeable to some substances but not to others.

self-fertilization The union of egg and sperm produced by a single hermaphroditic organism.

semen (L., seed) In reptiles and mammals, sperm-bearing fluid expelled from the penis during male orgasm.

semicircular canal Any of three fluid-filled canals in the inner ear that help to maintain balance.

semiconservative replication DNA replication in which each strand of the original duplex serves as the template for construction of a totally new complementary strand, so the original duplex is partially conserved in each of the two new DNA molecules.

senescent Aged, or in the process of aging.

sensory (afferent) neuron A neuron that transmits nerve impulses from a sensory receptor to the central nervous system or central ganglion.

sepal (L. *sepalum*, a covering) A member of the outermost floral whorl of a flowering plant.

septum, pl. **septa** (L., fence) A wall between two cavities.

seta, pl. **setae** (L., bristle) In an annelid, bristles of chitin that help anchor the worm during locomotion or when it is in its burrow.

sex-linked A trait determined by a gene carried on the X chromosome and absent on the Y chromosome.

sexual reproduction The process of producing offspring through an alternation of fertilization (producing diploid cells) and meiotic reduction in chromosome number (producing haploid cells).

sexual selection A type of differential reproduction that results from variable success in obtaining mates.

shoot In vascular plants, the aboveground portions, such as the stem and leaves.

sieve cell In the phloem of vascular plants, a long, slender sieve element with relatively unspecialized sieve areas and with tapering end walls that lack sieve plates.

sinoatrial (SA) node *See* pacemaker.

sinus (L., curve) A cavity or space in tissues or in bone.

G-13

sister chromatid One of two identical copies of each chromosome, still linked at the centromere, produced as the chromosomes duplicate for mitotic division; similarly, one of two identical copies of each homologous chromosome present in a tetrad at meiosis.

sodium-potassium pump Transmembrane channels engaged in the active (ATP-driven) transport of sodium ions, exchanging them for potassium ions, where both ions are being moved against their respective concentration gradients; maintains the resting membrane potential of neurons and other cells.

solute A molecule dissolved in some solution; as a general rule, solutes dissolve only in solutions of similar polarity; for example, glucose (polar) dissolves in (forms hydrogen bonds with) water (also polar), but not in vegetable oil (nonpolar).

solvent The medium in which one or more solutes is dissolved.

somatic cell Any of the cells of a multicellular organism except those that are destined to form gametes (germ-line cells).

somatic mutation A change in genetic information (mutation) occurring in one of the somatic cells of a multicellular organism, not passed from one generation to the next.

somatic nervous system (Gr. *soma*, body) In vertebrates, the neurons of the peripheral nervous system that control skeletal muscle.

somite (Gr. *soma*, body) One of the blocks, or segments, of tissue into which the mesoderm is divided during differentiation of the vertebrate embryo.

Southern blot A procedure used for identifying a specific gene, in which DNA from the source being tested is cut into fragments with restriction enzymes and separated by gel electrophoresis, then blotted onto a sheet of nitrocellulose and probed with purified, labeled, single-stranded DNA corresponding to a specific gene; if the DNA matching the specific probe is present in the source DNA, it is visible as a band of radioactive label on the sheet.

species, pl. species (L., kind, sort) A kind of organism; species are designated by binomial names written in italics.

spectrin A scaffold of proteins that links plasma membrane proteins to actin filaments in the cytoplasm of red blood cells, producing their characteristic biconcave shape.

sperm (Gr. *sperma*, seed) A mature male gamete, usually motile and smaller than the female gamete.

spermatid (Gr. *sperma*, seed) In animals, each of four haploid (*n*) cells that result from the meiotic divisions of a spermatocyte; each spermatid differentiates into a sperm cell.

spermatozoa The male gamete, usually smaller than the female gamete, and usually motile.

sphincter (Gr. *sphinkter*, band, from *sphingein*, to bind tightly) In vertebrate animals, a ring-shaped muscle capable of closing a tubular opening by constriction (such as between stomach and small intestine or between anus and exterior).

spindle apparatus The assembly that carries out the separation of chromosomes during cell division; composed of microtubules (spindle fibers) and assembled during prophase at the equator of the dividing cell.

spiracle (L. *spiraculum*, from *spirare*, to breathe) External opening of a trachea in arthropods.

spongy parenchyma A leaf tissue composed of loosely arranged, chloroplast-bearing cells. *See* palisade parenchyma.

sporangium, pl. sporangia (Gr. *spora*, seed, + *angeion*, a vessel) A structure in which spores are produced.

spore A haploid reproductive cell, usually unicellular, capable of developing into an adult without fusion with another cell.

sporophyte (Gr. *spora*, seed, + *phyton*, plant) The spore-producing, diploid (2*n*) phase in the life cycle of a plant having alternation of generations.

stabilizing selection A form of selection in which selection acts to eliminate both extremes from a range of phenotypes.

stamen (L., thread) The organ of a flower that produces the pollen; usually consists of anther and filament; collectively, the stamens make up the androecium.

starch (Mid. Eng. *sterchen*, to stiffen) An insoluble polymer of glucose; the chief food storage substance of plants.

statocyst (Gr. *statos*, standing, + *kystis*, sac) A sensory receptor sensitive to gravity and motion.

stele The central vascular cylinder of stems and roots.

stem cell A relatively undifferentiated cell in animal tissue that can divide to produce more differentiated tissue cells.

stereoscopic vision (Gr. *stereos*, solid, + *opitkos*, pertaining to the eye) Ability to perceive a single, three-dimensional image from the simultaneous but slightly divergent two-dimensional images delivered to the brain by each eye.

stigma (Gr., mark, tattoo mark) (1) In angiosperm flowers, the region of a carpel that serves as a receptive surface for pollen grains. (2) Light-sensitive eyespot of some algae.

stipules Leaflike appendages that occur at the base of some flowering plant leaves or stems.

stolon (L. *stolo*, shoot) A stem that grows horizontally along the ground surface and may form adventitious roots, such as runners of the strawberry plant.

stoma, pl. stomata (Gr., mouth) In plants, a minute opening bordered by guard cells in the epidermis of leaves and stems; water passes out of a plant mainly through the stomata.

stratum corneum The outer layer of the epidermis of the skin of the vertebrate body.

striated muscle (L. *striare*, to groove) Skeletal voluntary muscle and cardiac muscle.

stromatolite A fossilized mat of ancient bacteria formed as long as 2 billion years ago, in which the bacterial remains individually resemble some modern-day bacteria.

style (Gr. *stylos*, column) In flowers, the slender column of tissue that arises from the top of the ovary and through which the pollen tube grows.

substrate (L. *substratus*, strewn under) (1) The foundation to which an organism is attached. (2) A molecule upon which an enzyme acts.

succession In ecology, the slow, orderly progression of changes in community composition that takes place through time.

summation Repetitive activation of the motor neuron resulting in maximum sustained contraction of a muscle.

surface tension A tautness of the surface of a liquid, caused by the cohesion of the molecules of liquid. Water has an extremely high surface tension.

surface-area-to-volume ratio Relationship of the surface area of a structure, such as a cell, to the volume it contains.

swim bladder An organ encountered only in the bony fish that helps the fish regulate its buoyancy by increasing or decreasing the amount of gas in the bladder via the esophagus or a specialized network of capillaries.

symbiosis (Gr. *syn*, together with, + *bios*, life) The condition in which two or more dissimilar organisms live together in close association; includes parasitism (harmful to one of the organisms), commensalism (beneficial to one, of no significance to the other), and mutualism (advantageous to both).

sympatric speciation The differentiation of populations within a common geographic area into species.

synapomorphy In systematics, a derived character that is shared by clade members.

synapse (Gr. *synapsis*, a union) A junction between a neuron and another neuron or muscle cell; the two cells do not touch, the gap being bridged by neurotransmitter molecules.

synapsis (Gr., contact, union) The point-by-point alignment (pairing) of homologous chromosomes that occurs before the first meiotic division; crossing over takes place during synapsis.

synaptic cleft The space between two adjacent neurons.

synaptic vesicle A vesicle of a neurotransmitter produced by the axon terminal of a nerve. The filled vesicle migrates to the presynaptic membrane, fuses with it, and releases the neurotransmitter into the synaptic cleft.

synaptonemal complex A protein lattice that forms between two homologous chromosomes in prophase I of meiosis, holding the replicated chromosomes in precise register with each other so that base-pairs can form between nonsister chromatids for crossing over that is usually exact within a gene sequence.

syncytial blastoderm A structure composed of a single large cytoplasm containing about 4000 nuclei in embryonic development of insects such as *Drosophila*.

syngamy (Gr. *syn*, together with, + *gamos*, marriage) The process by which two haploid cells (gametes) fuse to form a diploid zygote; fertilization.

synteny Extensive conserved regions of DNA among different species.

systematics The reconstruction and study of evolutionary relationships.

systolic pressure A measurement of how hard the heart is contracting. When measured during a blood pressure reading, ventricular systole (contraction) is what is being monitored.

T

tagma, pl. **tagmata** (Gr., arrangement, order, row) A compound body section of an arthropod resulting from embryonic fusion of two or more segments; for example, head, thorax, abdomen.

taxis, pl. **taxes** (Gr., arrangement) An orientation movement by a (usually) simple organism in response to an environmental stimulus.

taxonomy The science of classifying living things. By agreement among taxonomists, no two organisms can have the same name, and all names are expressed in Latin.

T cell A type of lymphocyte involved in cell-mediated immunity and interactions with B cells; the "T" refers to the fact that T cells are produced in the thymus.

telencephalon (Gr. *telos*, end, + *encephalon*, brain) The most anterior portion of the brain, including the cerebrum and associated structures.

telomere A specialized nontranscribed structure that caps each end of a chromosome.

tendon (Gr. *tendon*, stretch) A strap of cartilage that attaches muscle to bone.

tertiary structure The folded shape of a protein, produced by hydrophobic interactions with water, ionic and covalent bonding between side chains of different amino acids, and van der Waal's forces; may be changed by denaturation so that the protein becomes inactive.

testcross A mating between a phenotypically dominant individual of unknown genotype and a homozygous "tester," done to determine whether the phenotypically dominant individual is homozygous or heterozygous for the relevant gene.

testis, pl. **testes** (L., witness) In mammals, the sperm-producing organ.

tetanus Sustained forceful muscle contraction with no relaxation.

thalamus (Gr. *thalamos*, chamber) That part of the vertebrate forebrain just posterior to the cerebrum; governs the flow of information from all other parts of the nervous system to the cerebrum.

thermodynamics (Gr. *therme*, heat, + *dynamis*, power) The study of transformations of energy, using heat as the most convenient form of measurement of energy.

thigmotropism In plants, unequal growth in some structure that comes about as a result of physical contact with an object.

threshold The minimum amount of stimulus required for a nerve to fire (depolarize).

thylakoid (Gr. *thylakos*, sac, + *-oides*, like) A saclike membranous structure containing chlorophyll in cyanobacteria and the chloroplasts of eukaryotic organisms.

tight junction Region of actual fusion of plasma membranes between two adjacent animal cells that prevents materials from leaking through the tissue.

tissue (L. *texere*, to weave) A group of similar cells organized into a structural and functional unit.

totipotent A cell that possesses the full genetic potential of the organism.

trachea, pl. **tracheae** (L., windpipe) A tube for breathing; in terrestrial vertebrates, the windpipe that carries air between the larynx and bronchi (which leads to the lungs); in insects and some other terrestrial arthropods, a system of chitin-lined air ducts.

tracheids In plant xylem, dead cells that taper at the ends and overlap one another.

transcription (L. *trans*, across, + *scribere*, to write) The enzyme-catalyzed assembly of an RNA molecule complementary to a strand of DNA.

transcription factor One of a set of proteins required for RNA polymerase to bind to a eukaryotic promoter region, become stabilized, and begin the transcription process.

transfection The transformation of eukaryotic cells in culture.

transfer RNA (tRNA) (L. *trans*, across, + *ferre*, to bear or carry) A class of small RNAs (about 80 nucleotides) with two functional sites; at one site, an "activating enzyme" adds a specific amino acid, while the other site carries the nucleotide triplet (anticodon) specific for that amino acid.

translation (L. *trans*, across, + *latus*, that which is carried) The assembly of a protein on the ribosomes, using mRNA to specify the order of amino acids.

translocation (L. *trans*, across, + *locare*, to put or place) (1) In plants, the long-distance transport of soluble food molecules (mostly sucrose), which occurs primarily in the sieve tubes of phloem tissue. (2) In genetics, the interchange of chromosome segments between nonhomologous chromosomes.

transpiration (L. *trans*, across, + *spirare*, to breathe) The loss of water vapor by plant parts; most transpiration occurs through the stomata.

transposition Type of genetic recombination in which transposable elements (transposons) move from one site in the DNA sequence to another, apparently randomly.

transposon (L. *transponere*, to change the position of) A DNA sequence capable of transposition.

triglyceride (triacylglycerol) An individual fat molecule, composed of a glycerol and three fatty acids.

triploid Possessing three sets of chromosomes.

trochophore A free-swimming larval stage unique to the mollusks and annelids.

trophic level (Gr. *trophos*, feeder) A step in the movement of energy through an ecosystem.

trophoblast (Gr. *trephein*, to nourish, + *blastos*, germ) In vertebrate embryos, the outer ectodermal layer of the blastodermic vesicle; in mammals, it is part of the chorion and attaches to the uterine wall.

tropism (Gr. *trope*, a turning) A response to an external stimulus.

tropomyosin (Gr. *tropos*, turn, + *myos*, muscle) Low-molecular-weight protein surrounding the actin filaments of striated muscle.

troponin Complex of globular proteins positioned at intervals along the actin filament of skeletal muscle; thought to serve as a calcium-dependent "switch" in muscle contraction.

tubulin (L. *tubulus*, small tube, + *in*, belonging to) Globular protein subunit forming the hollow cylinder of microtubules.

tumor-suppressor gene A gene that normally functions to inhibit cell division; mutated forms can lead to the unrestrained cell division of cancer, but only when both copies of the gene are mutant.

turgor (L. *turgor*, a swelling) The pressure within a cell resulting from the movement of water into the cell; a cell with high turgor pressure is said to be turgid. *See* osmotic pressure.

U

unequal crossing over A process by which a crossover in a small region of misalignment at synapsis causes two homologous chromosomes to exchange segments of unequal length.

unsaturated fat A fat molecule in which one or more of the fatty acids contain fewer than the maximum number of hydrogens attached to their carbons.

urea (Gr. *ouron*, urine) An organic molecule formed in the vertebrate liver; the principal form of disposal of nitrogenous wastes by mammals.

urethra (Gr. from *ourein*, to urinate) The tube carrying urine from the bladder to the exterior of mammals.

uric acid Insoluble nitrogenous waste products produced largely by reptiles, birds, and insects.

urine (Gr. *ouron*, urine) The liquid waste filtered from the blood by the kidney and stored in the bladder pending elimination through the urethra.

uterus (L., womb) In mammals, a chamber in which the developing embryo is contained and nurtured during pregnancy.

V

vacuole A membrane-bounded sac in the cytoplasm of some cells, used for storage or digestion purposes in different kinds of cells; plant cells often contain a large central vacuole that stores water, proteins, and waste materials.

vascular cambium In vascular plants, a cylindrical sheath of meristematic cells, the division of which produces secondary phloem outwardly and secondary xylem inwardly; the activity of the vascular cambium increases stem or root diameter.

vascular tissue (L. *vasculum*, a small vessel) Containing or concerning vessels that conduct fluid.

vas deferens (L. *vas*, a vessel, + *ferre*, to carry down) In mammals, the tube carrying sperm from the testes to the urethra.

vasopressin A posterior pituitary hormone that regulates the kidney's retention of water.

vegetal pole The hemisphere of the zygote comprising cells rich in yolk.

vein (L. *vena*, a blood vessel) (1) In plants, a vascular bundle forming a part of the framework of the conducting and supporting tissue of a stem or leaf. (2) In animals, a blood vessel carrying blood from the tissues to the heart.

veliger The second larval stage of mollusks following the trochophore stage, during which the beginning of a foot, shell, and mantle can be seen.

ventricle A muscular chamber of the heart that receives blood from an atrium and pumps blood out to either the lungs or the body tissues.

vesicle (L. *vesicula*, a little bladder) A small intracellular, membrane-bounded sac in which various substances are transported or stored.

vessel element In vascular plants, a typically elongated cell, dead at maturity, which conducts water and solutes in the xylem.

vestibular apparatus The complicated sensory apparatus of the inner ear that provides for balance and orientation of the head in vertebrates.

villus, pl. villi (L., a tuft of hair) In vertebrates, one of the minute, fingerlike projections lining the small intestine that serve to increase the absorptive surface area of the intestine.

visceral mass (L., internal organs) Internal organs in the body cavity of an animal.

vitamin (L. *vita*, life, + *amine*, of chemical origin) An organic substance that cannot be synthesized by a particular organism but is required in small amounts for normal metabolic function.

viviparity (L. *vivus*, alive, + *parere*, to bring forth) Refers to reproduction in which eggs develop within the mother's body and young are born free-living.

voltage-gated ion channel A transmembrane pathway for an ion that is opened or closed by a change in the voltage, or charge difference, across the plasma membrane.

W

water potential The potential energy of water molecules. Regardless of the reason (e.g., gravity, pressure, concentration of solute particles) for the water potential, water moves from a region where water potential is greater to a region where water potential is lower.

wild type In genetics, the phenotype or genotype that is characteristic of the majority of individuals of a species in a natural environment.

X

xylem (Gr. *xylon*, wood) In vascular plants, a specialized tissue, composed primarily of elongate, thick-walled conducting cells, which transports water and solutes through the plant body.

Y

yolk plug A plug occurring in the blastopore of amphibians during formation of the archenteron in embryological development.

Z

zona pellucida An outer membrane that encases a mammalian egg.

zoospore A motile spore.

zooxanthellae—Symbiotic photosynthetic protists in the tissues of corals.

zygomycetes (Gr. *zygon*, yoke, + *mykes*, fungus) A type of fungus whose chief characteristic is the production of sexual structures called zygosporangia, which result from the fusion of two of its simple reproductive organs.

zygote (Gr. *zygotos*, paired together) The diploid ($2n$) cell resulting from the fusion of male and female gametes (fertilization).

Credits

Photo's

Chapter 1
Figure 1.1: © Christopher Ralling; **1.2 (top left photo):** From C.P. Morgan & R.A. Jersid, *Anatomical Record*, 166:575-586, 1970 © John Wiley & Sons; **1.2 (bottom left photo):** © Lennart Nilsson; **1.2 (top center):** © Ed Reschke; **1.2 (bottom center):** © PhotoDisc/Volume 4; **1.2 (top right):** © PhotoDisc/Volume 44; **1.2 (second from top, both):** © PhotoDisc/Volume 44; **1.2 (right 3rd from top):** © John D. Cunningham/Visuals Unlimited; **1.2 (bottom right):** © Robert & Jean Pollock; **1.A:** © N.H. (Dan) Cheatham/National Audubon Society Collection/Photo Researchers; **1.5:** Huntington Library/Superstock; **1.7:** From DARWIN by Adrian Desmond, © 1991 by Adrian Desmond and James Moore, by permission of Warner Books, Inc.; **1.12:** © Mary Evans Picture Library/Photo Researchers; **1.16:** © Dennis Kunkel/Phototake

Chapter 2
Figure 2.1: © Irving Geis/Photo Researchers; **2.10b:** © Hulton/Archive; **2.11:** © PhotoDisc/Volume 6; **2.11 (right)** © Corbis/Volume 98; **2.14:** © Hermann Eisenbeiss/National Audubon Society Collection/Photo Researchers

Chapter 3
Figure 3.1: Courtesy University of California Lawrence Livermore National Library and the U.S. Department of Energy; **3.4 (top left):** © PhotoDisc/Volume 6; **3.4 (top right):** © Manfred Kage/Peter Arnold, Inc.; **3.4 (bottom left):** © PhotoDisc/Volume 9; **3.4 (bottom center, right):** © PhotoDisc/Volume 6; **3.4 (bottom right):** © Scott Blackman/Tom Stack & Associates; **3.12:** Courtesy of Lawrence Berkeley National Laboratory; **3.13:** © Driscoll, Youngquist, & Baldeschwieler, Caltech/SPL/Photo Researchers, Inc.; **3.28:** © J.D. Litvay/Visuals Unlimited; **3.29:** © Scott Johnson/Animals Animals/Earth Scenes

Chapter 4
Figure 4.1: © PhotoDisc/BS Volume 15; **4.2:** © Edward S. Ross; **4.3:** © Y. Arthur Bertrand/Peter Arnold, Inc.; **4.4:** © T.E. Adams/Visuals Unlimited; **4.5:** © Bob McKeever/Tom Stack & Associates; **4.9:** Courtesy of J. William Schopf, UCLA; **4.10:** © R. Robinson/Visuals Unlimited; **4.11:** © Dwight R. Kuhn; **4.12:** © Andrew H. Knoll, Harvard University; **4.15 (top left):** © Alfred Pasieka/Science Photo Library/Photo Researchers; **4.15 (top center):** © Kari Lounatman/Photo Researchers; **4.15 (top right):** © Corbis/Volume 64; **4.15 (bottom left):** © PhotoDisc/BS Volume 15; **4.15 (bottom center):** © Corbis/Volume 46; **4.15 (bottom right):** © PhotoDisc/Volume 44; **4.17:** NASA

Chapter 5
Figure 5.1: © Dr. Gopal Murti/Science Photo Library/Photo Researchers; **p. 83 (top left):** © David M. Phillips/Visuals Unlimited; **p. 83 (left 2nd from top):** © Mike Abbey/Visuals Unlimited; **p. 83 (left 3rd from top):** © David M. Phillips/Visuals Unlimited; **p. 83 (left 4th from top):** © Mike Abbey/Visuals Unlimited; **p. 83 (top right):** © K.G. Murti/Visuals Unlimited; **p. 83 (right 2nd from top):** © David Becker/Science Photo Library/Photo Researchers; **p.83 (right 3rd from top):** © Microworks/Phototake; **p. 83 (right 4th from top):** © Stanley Flegler/Visuals Unlimited; **5.6:** © I. M. Pope/Tom Stack & Associates; **5.7:** Courtesy of E.H. Newcomb & T.D. Pugh, University of Wisconsin; **5.10b:** Courtesy of Dr. Thomas Tillack; **5.10c:** Photo J. David Robertson, from Charles Flickinger, *Medical Cellular Biology*, W.B. Saunders, 1979; **5.12:** © Ed Reschke; **5.13:** © R. Bolender & D. Fawcett/Visuals Unlimited; **5.15:** Courtesy of Dr. Charles Flickinger, *Medical Cellular Biology*, W.B. Saunders, 1979; **5.18:** Courtesy of E.H. Newcomb & S.E. Frederick, U. of Wisconsin, Reprinted with permission from *Science*, Vol. 163, 1353-1355 © 1969 American Association for the Advancement of Science; **5.20:** From C.P. Morgan & R.A. Jersid, *Anatomical Record* 166: 575-586, © John Wiley & Sons; **5.21b:** © Don W. Fawcett/Visuals Unlimited; **5.22:** Courtesy of Dr. Kenneth Miller, Brown University; **5.27 (both):** Courtesy of William Dentler; **5.28:** © Biophoto Associates/Photo Researchers; **5.29:** © Biophoto Associates/Photo Researchers

Chapter 6
Figure 6.1: © K.R. Porter/Photo Researchers; **6.6:** Courtesy of Dr. Roger C. Wagner; **6.15 (all):** © David M. Phillips/Visuals Unlimited; **6.16 (top):** Centers for Disease Control/Dr. Edwin P. Ewing Jr.; **6.16 (2nd from top):** © BCC Microimaging Inc.; **6.16 (3rd from top, & bottom):** Courtesy of M.M. Perry & A.B. Gilbert, *Cell Science*, 39-257, 1979; **6.17:** Courtesy of Dr. Birgit H. Satir

Chapter 7
Figure 7.1: © Cabisco/Visuals Unlimited; **7.15b:** © Don Fawcett/Visuals Unlimited

Chapter 8
Figure 8.1: © Robert A. Caputo/Aurora & Quanta Productions, Inc.; **8.5 (both):** © Spencer Grant/Photo Edit; **8.10b:** Courtesy of Dr. Lester J. Reed, University of Texas, Austin

Chapter 9
Figure 9.1: © Jane Buron/Bruce Coleman Inc.; **9.9:** © PhotoDisc/Vol. 19; **p. 180 (top left):** © Corbis/Vol. 145; **p. 180 (top right):** © PhotoDisc/Vol. 44; **p. 180 (bottom left):** © PhotoDisc/Volume 44; **p. 180 (bottom center):** © Edward S. Ross; **p. 180 (bottom right):** © Edward S. Ross

Chapter 10
Figure 10.1: © Corbis/ Volume 102; **10.2 (left):** © Manfred Kage/Peter Arnold; **10.2 (right):** Courtesy of Dr. Kenneth Miller, Brown University; **10.7:** © Eric V. Grave/Photo Researchers Inc.; **10.8 (left):** © Eric Soder/Tom Stack & Associates; **10.8 (right):** © Eric Soder/Tom Stack & Associates; **10.21 (left):** © Joseph Nettis/National Audubon Society Collection/Photo Researchers; **10.21 (right):** © Clyde H. Smith/Peter Arnold

Chapter 11
Figure 11.1: © George Musil/Visuals Unlimited; **11.3a,b:** Courtesy of William Margolin; **11.5:** © Biophoto Associates/Photo Researchers Inc.; **11.7:** © Science Photo Library/Photo Researchers; **11.11:** © Dr. Andrew S. Bajer; **11.12:** © Andrew S. Bajer; **11.13 (both):** Dr. Jeremy Pickett-Heaps; **11.14 (left):** © David M. Phillips/Visuals Unlimited; **11.14 (right):** © Dr. Guenter Albrecht-Buehler; **11.15a:** © B.A. Palevits & E.H. Newcomb/BPS/Tom Stack & Associates

Chapter 12
Figure 12.1: © L. Maziarski/Visuals Unlimited; **12.7:** Diter von Wettstein; Reproduced with permission from *Annual Review of Genetics*, v. 6, 1972, by Annual Review, Inc.; **12.11 (all):** © C.A. Hasenkampf/Biological Photo Service; **12.12 (left):** Courtesy of Sheldon Wolff & Judy Bodycote; **12.12 (right):** Courtesy of Prof. William F. Morgan

Chapter 13
Figure 13.1: © Corbis; **13.2:** George Johnson; **13.3:** © Visuals Unlimited; **13.4:** © Richard Gross/Biological Photography; **13.5, 13.6, 13.9:** Courtesy of V. Orel, Mendelianum Musei Moraviae, Brno; **13.10:** Courtesy of R.W. Van Norman; **13.16:** From Albert & Blakeslee Corn and Man *Journal of Heredity*, Vol. 5, pg. 511, 1914, Oxford University Press; **13.18 (both):** © Fred Bruemmer; **13.20 (far left):** © Richard Hutchings/Photo Researchers; **13.20 (left):** © Cheryl A. Ertelt/Visuals Unlimited; **13.20 (right):** © William H. Mullins/Photo Researchers; **13.20 (far right):** © Gerard Lacz/Peter Arnold Inc.; **13.23:** © Corbis-

C-1

Bettmann; **13.24:** © Science Photo Library/Photo Researchers; **13.26:** © Alfred Paseika/Science Photo Library/Photo Researchers; **13.28 (both):** © Cabisco/Phototake; **13.34:** © Leonard Lessin/Peter Arnold; **13.36:** © Hans Reinhard/Okapia/Photo Researchers; **13.37 (left):** Courtesy of Loris McGavaran, Denver Children's Hospital; **13.37 (right):** © Richard Hutchings/Photo Researchers Inc.

Chapter 14
Figure 14.1: © PhotoDisc/Volume 29; **14.9 (both):** From "The Double Helix," by J.D. Watson, Atheneum Press, N.Y. 1968; **14.10a:** © Barrington Brown/Photo Researchers Inc.; **14.11:** From M. Meselson and F.W. Stahl/ *Proceedings of the Nat. Acad. of Sci.* 44 (1958):671; **14.17 (both):** From *Biochemistry* 4/e by Stryer © 1995 by Lubert Stryer. Used with permission of W.H. Freeman and Company; **14.19:** © Prof. Ulrich Laemmli/Photo Researchers Inc.; **14.20a:** Courtesy of Dr. David Wolstenholme

Chapter 15
Figure 15.1: © K.G. Murti/Visuals Unlimited; **15.3:** N. Ban, P. Nissen, J. Hansen, P.B. Moore & T.A. Steitz, "The Complete Atomic Structure of the Large Ribosomal subunit at 2.4A Resolution," Reprinted with permission from *Science*, v. 289 #5481, p917, © 2000 American Association for the Advancement of Science; **15.7:** From R.C. Williams; *Proc. Nat. Acad. of Sci.* 74 (1977):2313; **15.13:** Courtesy of Dr. Oscar L. Miller; **15.18b:** Courtesy of Dr. Bert O'Malley, Baylor College of Medicine

Chapter 16
Figure 16.1: © Stanley Cohen/Science Photo Library/Photo Researchers Inc.; **16.7b:** Courtesy of Bio-Rad Laboratories; **16.15:** Courtesy of Lifecodes Corp., Stamford, CT; **16.16:** AP/Wide World Photos; **16.17:** R.L. Brinster, U. of Pennsylvania Sch. of Vet. Med.; **16.20:** Courtesy of Monsanto

Chapter 17
Figure 17.1: Courtesy of Robert D. Fleischmann, The Institute for Genomic Research; **17.4:** Courtesy of Celera Genomics; **17.11 (all):** Reproduced with permission from Altpeter et al, *Plant Cell Reports* 16:12-17, 1996, photos provided by Indra Vasil; **17.12:** Courtesy of Research Collaboratory for Structural Bioinformatics; **17.13:** © Corbis/R-F Website; **17.14:** © Grant Heilman/Grant Heilman Photography

Chapter 18
Figure 18.1: Courtesy of Dr. Claus Pelling; **18.15 (both):** Courtesy of Dr. Harrison Echols; **18.18a:** Courtesy of Dr. Victoria Foe

Chapter 19
Figure 19.1: © Cabisco/Visuals Unlimited; **19.3:** Photo Lennart Nilsson/Albert Bonniers Forlag AB, *A Child is Born*, Dell Publishing Company; **19.4 (all):** © Cabisco/Phototake; **19.6:** © Ed Lewis; **19.15 (top left):** Dr. Christiane Nusslein-Volhard/Max Planck Institute as published in From Egg to Adult, © 1992 HHMI; **19.15b-d:** James Langeland, Stephen Paddock & Sean Carroll as published in *From Egg to Adult*, © 1992 HHMI; **19.16 (all):** Courtesy of Manfred Frasch; **19.17:** Courtesy of E.B. Lewis

Chapter 20
Figure 20.1: © University of Wisconsin-Madison News & Public Affairs, Photo by Jeff Miller; **20.3:** Courtesy of Dr. Charles Brinton; **20.10:** © Custom Medical Stock Photo; **20.18:** Courtesy of American Cancer Society; **20.23:** AP/Wide World Photos; **20.25:** © University of Wisconsin-Madison News & Public Affairs

Chapter 21
Figure 21.1: © Corbis/R-F Website; **21.3:** Biological Photo Service; **21.19:** Courtesy of H. Rodd

Chapter 22
Figure 22.1: © PhotoDisc Website; **22.4 (both):** © Breck P. Kent/Animals Animals/Earth Scenes; **22.9 (both):** Courtesy of Lyudmilla N. Trut, Institute of Cytology & Genetics, Siberian Dept. of the Russian Academy of Sciences

Chapter 23
Figure 23.1: Jonathan Losos; **23.3:** © Porterfield/Chickering/Photo Researchers Inc.; **23.4:** © Barbara Gerlach/Visuals Unlimited; **23.6 (top left):** © John Shaw/Tom Stack & Associates; **23.6 (bottom left):** © Rob & Ann Simpson/Visuals Unlimited; **23.6 (top right):** © Suzanne L. Collins & Joseph T. Collins/National Audubon Society Collection/Photo Researchers; **23.6 (bottom right):** © Phil A. Dotson/National Audubon Society Collection/Photo Researchers; **23.8 (left):** © Chas. McRae/Visuals Unlimited; **23.8 (right):** Jonathan Losos; **23.12 (left):** © Jeffrey Taylor; **23.12 (right):** © William P. Mull; **23.15:** © G.R. Roberts

Chapter 24
Figure 24.1: Courtesy of Richard P. Elinson; **p. 492a:** © Dr. R. Clark & M. Goff/Science Photo Library/Photo Researchers; **p. 492b:** © PhotoDisc Green/Getty Images; **p. 492c:** © BIOS(C. Ruoso)/Peter Arnold, Inc.; **p. 492d:** Darwin Dale/Photo Researchers; **p. 492e:** Centers for Disease Control; **p. 492f:** © Paul G. Young/The Institute for Genomic Research; **p. 492g:** Dr. Jeremy Burgess/Science Photo Library/Photo Researchers; **p. 492h:** © Science Photo Library/Photo Researchers; **p. 492I:** © PhotoDisc/Getty Images; **p. 492j:** © CNRI/Science Photo Library/Photo Researchers; **24.4:** © PhotoDisc/Getty Images; **24.10:** Courtesy of Anna Di Gregorio; **24.13 (top left):** © Image Bank/Getty Images; **24.13 (top right):** © Darwin Dale/Photo Researchers; **24.13 (bottom left):** © Aldo Brando/Peter Arnold, Inc.; **24.13 (bottom right):** © Tom E. Adams/Peter Arnold, Inc.; **24.14 (both):** Courtesy of Walter Gehring, reprinted with permission from Induction of Ectopic Eyes by Targeted Expression of the Eyeless Gene in Drosophila, G. Halder, P. Callaerts, Walter J. Gehring, *Science* Vol. 267, © 24 March 1995 American Association for the Advancement of Science; **24.15 (both):** Courtesy of Dr. William Jeffery

Chapter 25
Figure 25.1: © Corbis/Volume 8; **25.2 (top left):** © Corbis/Volume 102; **25.2 (top right):** © PhotoDisc/Volume 56; **25.2 (bottom left):** © PhotoDisc/Volume 1; **25.2 (bottom right):** © Corbis/Volume 8; **25.6:** Image # 5789, photo by D. Finnin/American Museum of Natural History

Chapter 26
Figure 26.1: © K.G. Murti/Visuals Unlimited; **26.3a:** © Dept. of Microbiology, Biozentrum/Science Photo Library/Photo Researchers; **26.7:** Courtesy of Katherine Sutliff, from *Science* 300: 1377, May 30, 2003, p. 1377, © American Association for the Advancement of Science, Reprinted with permission; **26.8:** © Corbis/Volume 40

Chapter 27
Figure 27.1: © David M. Phillips/Visuals Unlimited; **p. 547 (above):** © Abraham & Beachey/BPS/Tom Stack & Associates; **p. 547 (below):** © F. Widell/Visuals Unlimited; **27.3a:** © Science Photo Library/Photo Researchers; **27.3b:** © University of Regensburg, Courtesy of Reinhard Rachel; **27.3c:** © Andrew Syred/Science Photo Library/Photo Researchers; **27.3d:** © Microfield Scientific Ltd./Science Photo Library/Photo Researchers; **27.3e:** © Alfred Paseika/Science Photo Library/Photo Researchers; **27.3g:** © S.W. Watson/ Visuals Unlimited; **27.3h:** © Dennis Kunkel Microscopy, Inc., **27.3I:** Courtesy of Dr. Hans Reichenbach; **27.4:** © G.W. Willis; **27.5:** © Julius Adler/Visuals Unlimited; **27.6 (left):** © W. Watson/ Visuals Unlimited; **27.6 (right):** © Norma J. Lang/ Biological Photo Service; **27.8:** © CNRI/SPL/Photo Researchers; **27.10:** © R. Calentine/ Visuals Unlimited; **27.11:** © Science/Visuals Unlimited

Chapter 28
Figure 28.1: © John D. Cunningham/Visuals Unlimited; **28.2:** Courtesy of Dr. Edward W. Daniels, Argonne National Lab, the University of Illinois College of Medicine at Chicago; **28.5:** © John D. Cunningham/Visuals Unlimited; **28.6a:** © Michael Abby/ Visuals Unlimited; **28.7 (left):** © Manfred Kage/Peter Arnold Inc.; **28.7 (right):** Edward S. Ross; **28.11a:** © Brian Parker/ Tom Stack & Associates; **28.12:** © Gregory Ochocki/Photo Researchers; **28.13:** © D.P. Wilson/Photo Researchers; **28.15:** © John D. Cunningham/ Visuals Unlimited; **28.16:** © Phil A. Harrington/Peter Arnold, Inc.; **28.18:** © Richard Rowan/Photo Researchers; **28.17:** © Manfred Kage/ Peter Arnold; **28.19:** © Edward S. Ross; **28.20a:** © John Shaw/Tom Stack & Associates; **28.20b:** © John Shaw/Tom Stack & Associates; **28.21:** © Genichiro Higuchi,

C-2   Credits

Higuchi Science Laboratory; **p.578:** © D.P. Wilson/Photo Researchers

Chapter 29

Figure 29.1: © Stephen J. Krasemann/DRK Photo; **29.4:** © Edward S. Ross; **29.6:** © Kirtley Perkins/Visuals Unlimited; **29.7:** © Kingsley R. Stern; **29.8:** Courtesy of Hans Steur, The Netherlands; **29.9:** Darrell Vodopich; **29.10:** © Kingsley R. Stern; **29.11, 29.12, 29.14:** © Edward S. Ross; **29.16 (left):** © Walter H. Hodge/Peter Arnold Inc.; **29.16 (center):** © Kjell Sandved/ Butterfly Alphabet; **29.16c:** © Runk/Schoenberger/Grant Heilman Photography; **29.17:** Courtesy of David Dilcher & Ge Sun; **29.18:** Courtesy of Sandra Floyd

Chapter 30

Figure 30.1: © Bill Keogh/Visuals Unlimited; **30.2a:** Courtesy of Dr. Peter Daszak; **30.2b:** © Cabisco/Visuals Unlimited; **30.2c:** © Robert Simpson/Tom Stack & Associates; **30.2d:** © Michael & Patricia Fogden; **30.3:** Courtesy of E.C. Setliff & W. L. MacDonald; **30.4:** © Kjell Sandved/ Butterfly Alphabet; **30.5:** © Microfield Scientific Ltd/Photo Researchers; **30.6:** © L. West/Photo Researchers; **30.7:** Ralph Williams/USDA Forest Service; **30.9a:** © 1997 Regents of the University of Michigan; **30.10:** © Cabisco/Phototake; **30.11a:** © Alexandra Lowry/The National Audubon Society Collection/ Photo Researchers; **30.12a:** © Ed Pembleton; **30.12b:** © Kjell Sandved/Butterfly Alphabet; **30.13:** © David M. Phillips/Visuals Unlimited; **30.14a:** © James Castner; **30.14b:** © Edward S. Ross; **30.15:** © Ed Reschke; **30.16b:** © D.H. Marx/ Visuals Unlimited; **30.17:** © Scott Camazine/Photo Researchers; **30.18:** Courtesy of Zoology Department/University of Canterbury, New Zealand; **30.19a:** Courtesy of Laura E. Sweets, University of Missouri; **30.19b:** © Manfred Kage/Peter Arnold, Inc.

Chapter 31

Figure 31.1: © Corbis/Volume 53; **p. 619a:** © Corbis/Volume 86; **p. 619b:** © Corbis/Volume 65; **p. 619c:** © David M. Phillips/Visuals Unlimited; **p. 619d:** © Royalty-Free/Corbis **p. 619e:** © Edward S. Ross; **p. 619f:** © Corbis; **p. 619g:** © Cleveland P. Hickman Jr.; **p. 619h:** © Cabisco/Phototake; p. 619I: © Ed Reschke; **31.2 (top left):** © Corbis/Volume 53; **31.2 (top right):** © Corbis; **31.2 (bottom left):** © Corbis; **31.2d:** © Corbis; **31.8:** Courtesy of Dr. Igor Eeckhaut

Chapter 32

Figure 32.1: © Denise Tackett/Tom Stack & Associates **32.4:** © Andrew J. Martinez/Photo Researchers; **32.9:** © Gwen Fidler/Tom Stack & Associates; **32.10:** © Kelvin Aitken/Peter Arnold Inc.; **32.11:** © Daniel Gotshall; **32.12:** © David Wrobel/Visuals Unlimited; **32.16:** © Kjell Sandved/Butterfly Alphabet; **32.17:** © Biology Media/R. Knauf/Photo Researchers; **32.19:** © T.E. Adams/Visuals Unlimited; **32.20:** Courtesy of Matthias Obst

Chapter 33

Figure 33.1: © James H. Robinson/Animals Animals/Earth Scenes; **33.2:** © Alex Kerstich/Visuals Unlimited; **33.4:** © A. Flowers & L. Newman/Photo Researchers; **33.7b:** © Kjell Sandved/Butterfly Alphabet; **33.8:** © Milton Rand/Tom Stack & Associates; **33.10:** © Fred Bavendam/Peter Arnold Inc.; **33.11:** © Visuals Unlimited; **33.13:** © Kjell Sandved/Butterfly Alphabet; **33.14:** © David Dennis/Tom Stack & Associates; **33.15:** © Cleveland P. Hickman; **33.17b:** © Robert Brons/Biological PhotoService; **33.18b:** © Fred Bavendam/Peter Arnold Inc.; **33.26:** © T.E. Adams/Visuals Unlimited; **33.27:** © Kjell Sandved/Butterfly Alphabet; **33.28a:** © Rod Planck/Tom Stack & Associates; **33.28b:** © Ann Moreton/Tom Stack & Associates; **33.29:** © Alex Kerstich/Visuals Unlimited; **33.30:** © Edward S. Ross; **33.31a:** © Cleveland P. Hickman; **33.31b:** © Kjell Sandved/Butterfly Alphabet; **33.31c:** © Norm Thomas/Photo Researchers; **33.31d:** © Valorie Hodgson/Visuals Unlimited; **33.31e:** © Corbis/Volume 6; **33.31f:** © Kjell Sandved/Butterfly Alphabet; **33.33:** © John Shaw/Tom Stack & Associates; **33.34:** © Kjell Sandved/Butterfly Alphabet; **33.35a:** © Alex Kerstitch/Visuals Unlimited; **33.35b:** © Randy Morse/Tom Stack & Associates; **33.35c:** © Daniel W. Gotshall/Visuals Unlimited; **33.38:** © Kjell Sandved/Butterfly Alphabet; **33.40:** © Daniel W. Gotshall/Visuals Unlimited; **33.41:** © Kjell Sandved/Visuals Unlimited

Chapter 34

Figure 34.1: © Corbis; **34.3:** © Eric N. Olson, PhD/The University of Texas MD Anderson Cancer Center; **34.5a:** Rick Harbo Marine Images; **34.6:** © Heather Angel; **34.10:** © Corbis; **34.15:** © Corbis/Volume 53; **34.16:** © John D. Cunningham/Visuals Unlimited; **34.18:** © 1979 Peter Scoones/Contact Press Images/Woodfin Camp & Associates; **34.21:** © Natural History Museum/J. Sibbick; **34.22a:** © John Shaw/Tom Stack & Associates; **34.22b:** © Suzanne L. Collins & Joseph T. Collins/Photo Researchers; **34.22c:** © Jany Sauvanet/Photo Researchers; **34.24:** © Natural History Museum/J. Sibbick; **34.27:** Photo by Paul Sareno/University of Chicago; **34.31:** © William J. Weber/Visuals Unlimited; **34.32a:** © John Cancalosi/Tom Stack & Associates; **34.32b:** © Rod Planck/Tom Stack & Associates; **34.33:** © Corbis/Volume 6; **34.37:** © Corbis; **34.38:** © PhotoDisc/Volume 44; **34.41:** © Stephen Dalton/National Audubon Society Collection/Photo Researchers; **34.42:** © B.J. Alcock/Visuals Unlimited; **34.42b:** © Corbis/Volume 6; **34.42c:** © Stephen J. Krasemann/DRK Photo; **34.43:** © Alan Nelson/Animals Animals/Earth Scenes; **34.45 (all):** © David L. Brill; **34.47:** © 1985 David L. Brill/Fossil credit: National Museums of Kenya, Nairobi; **34.49:** AP/Wide World Photos

Chapter 35

Figure 35.1: Photo by Susan Singer; **35.2:** © Terry Ashley/Tom Stack & Associates; **35.4a:** © Hart-Davis/Science Photo Library/Photo Researchers; **35.4b:** R. Calentine/Visuals Unlimited; **35.8 (both):** © Heidi Mullen; **35.9:** Courtesy of Liming Zhao & Fred Sack, Ohio State University; **35.10:** © Andrew Syred/Science Photo Library/Photo Researchers; **35.11 (both):** Courtesy of Alan Lloyd; **35.12a:** © Biophoto Associates/Photo Researchers Inc.; **35.12b:** © John D. Cunningham/Visuals Unlimited; **35.12c:** © Lawrence Mellinchamp/Visuals Unlimited; **35.13c:** Courtesy of Wilfred Cote, Suny College of Environmental Forestry; **35.14:** © Randy Moore/Visuals Unlimited; **35.15:** © E.J. Cable/Tom Stack & Associates; **35.16a:** Courtesy John Schiefelbein, from Myeong Min Lee & John Schiefelbein, "WEREWolf, MYB-related Protein in Arabidopsis," *Cell* V. 99:473-483, Nov. 24, 1999; **35.16b:** Courtesy of Dr. Philip Benfey, from Wysocka-Diller, J.W., Helariutta, Y., Fukaki, H., Malamy, J.E. and P.N. Benfey (2000) Molecular analysis of SCARECROW function reveals a radial patterning mechanism common to root and shoot Development, *Cell* 127, 595-603; **35.17:** Photomicrograph by G.S. Ellmore; **35.19a:** © Kingsley R. Stern; **35.19b:** Photomicrograph by G.S. Ellmore; **35.21a:** © Walter H. Hodge/Peter Arnold, Inc.; **35.21b:** Courtesy of Robert A. Schlising; **35.21c:** © Kingsley R. Stern; **35.22:** Courtesy of J.H. Troughton and L. Donaldson and Industrial Research Ltd.; **35.24:** © John D. cunningham/Visuals Unlimited; **35.25 (both):** © Ed Reschke; **35.26:** © Ed Reschke/Peter Arnold Inc.; **35.27a:** © Ed Reschke; **35.27b:** © Jack M. Bostrack/Visuals Unlimited; **35.29:** Richard Waites & Andrew Hudson, from Phantastica: a gene required for dorsoventrality of leaves in Antirrhinum majus, *Development* 121, 2143-2154 (1995) © The Company of Biologists Limited 1995; **35.30a:** © Kjell Sandved/Butterfly Alphabet; **35.30b:** © Pat Anderson/Visuals Unlimited; **35.31a:** © Edward S. Ross; **35.31b:** © Glenn M. Oliver/Visuals Unlimited; **35.31c:** © Joel Arrington/Visuals Unlimited; **35.34:** © Ed Reschke; **35.35 (above):** © Michael P. Godomski/Photo Researchers; **35.35 (below):** © Ed Reschke/Peter Arnold, Inc.; **p.753:** © Kingsley R. Stern

Chapter 36

Figure 36.1: © Norm Thomas/The National Audubon Society Collection/Photo Researchers; **36.4:** Courtesy of E.C. Yeung & D.W. Meinke; **36.5:** Courtesy of Dr. Chun-Ming Liu; **36.6:** Courtesy of Kathy Barton & Jeff Lang; **36.7a(above):** © Runk/Schoenberger/Grant Heilman Photography; **36.7b:** © Kevin & Betty Collins/Visuals Unlimited; **36.8:** © Jack M. Bostrack/Visuals Unlimited; **36.10a:** © Ed Reschke/Peter Arnold, Inc.

36.10b: © David Sieren/Visuals Unlimited; **36.11 (top left):** © Kingsley R. Stern; **36.11 (top center):** © James Richardson/Visuals Unlimited; **36.11 (top right):** © Kingsley R. Stern; **36.11 (middle left):** © Kingsley R. Stern; **36.11 (center):** © Kingsley R. Stern; **36.11 (middle right):** © Barry L. Runk/Grant Heilman Photography; **36.11 (bottom left):** Courtesy of Robert A. Schlising; **36.11 (bottom right):** © Charles D.Winters/Photo Researchers; **36.12a:** © Edward S. Ross; **36.12b, 36.13, 36.14:** © James Castner; **36.16:** Courtesy of Prof. Tuan-hua David Ho

Chapter 37
Figure 37.1: © Richard Rowan's Collection, Inc./Photo Researchers; **37.7:** © John D. Cunningham/Visuals Unlimited; **37.8:** © Terry Ashley/Tom Stack & Associates; **37.9:** © Grant Heilman/Grant Heilman Photography; **37.10:** © Bruce Iverson Photomicrography; **37.12a:** © Andrew Syred/Science Photo Library/Photo Researchers; **37.12b:** © Bruce Iverson/Science Photo Library/Photo Researchers

Chapter 38
Figure 38.1: © Scott T. Smith/Corbis **38.2 (all):** Courtesy of Dr. Emmanuel Epstein; **38.4:** © Michael P. Gadomski/Photo Researchers; **38.6 (both):** Courtesy of Nicholas School of the Environment and Earth Sciences, Duke University; **38.7:** Courtesy of Sharon Long; **38.9:** © Kjell Sandved/Butterfly Alphabet; **38.10:** © Runk/Schoenberger/Grant Heilman Photography; **38.11:** © Barry Rice; **38.12:** © Don Albert; **38.14:** U.S. EPA National Research Lab, Cincinnati; **38.15 (all):** © AP/Wide World Photos; **p.793:** U.S. EPA National Research Lab, Cincinnati

Chapter 39
Figure 39.1: © Holt Studios International (Nigel Cattlin)/Photo Researchers; **39.2:** © Jane Grushow/Grant Heilman Photography; **39.3:** USDA/Agricultural Research Service; **39.4 (both):** USDA/Agricultural Research Service; **39.6:** © C. Allan Morgan/Peter Arnold, Inc.; **39.7:** © Runk/Schoenberger/Grant Heilman Photography

Chapter 40
Figure 40.1: © John D. Cunningham/Visuals Unlimited; **40.5:** © Runk/Schoenberger/Grant Heilman Photography; **40.6:** © John D. Cunningham/Visuals Unlimited; **40.7:** © S.J. Krasemann/Peter Arnold Inc.; **40.8:** Courtesy of Frank B. Salisbury; **40.9a:** © T. Walker; **40.9b:** © R.J. Delorit, Agronomy Publications; **40.10a:** © Jim Zipp/The National Audubon Society Collection/Photo Researchers; **40.10b:** © Runk/Schoenberger/Grant Heilman Photography; **40.11:** © Prof. Malcolm B. Wilkins, Botany Dept., Glasgow University; **40.18 (all):** © Prof. Malcolm B. Wilkins, Botany Dept., Glasgow University; **40.20:** © Robert Calentine/Visuals Unlimited; **40.21:** ©Runk/Schoenberger/Grant Heilman Photography; **40.22:** © Sylvan H. Wittwer/Visuals Unlimited; **40.25a:** © John Solden/Visuals Unlimited; **40.25b:** Courtesy of Donald R. McCarty, from "Molecular Analysis of viviparous-1: An Abscisic Acid-Insensitive Mutant of Maize", *The Plant Cell*, v.1, 523-532, © 1989 American Society of Plant Physiologists; **40.25c:** © David M. Phillips/Visuals Unlimited;

Chapter 41
Figure 41.1: © Richard La Val/Animals Animals; **41.3a:** © Stephen G. Maka/DRK Photo; **41.3b:** © Heidi Mullen **41.4** Courtesy of Lingjing Chen & Renee Sung; **41.5 (both):** Detlef Weigel & Ove Nilsson, The Salk Institute for Biological Studies; **41.7:** © Jim Strawser/Grant Heilman Photography; **41.12 (all):** Courtesy of John L. Bowman; **41.14:** © John Bishop/Visuals Unlimited; **41.15:** © Paul Gier/Visuals Unlimited; **41.16:** Courtesy of Enrico Coen; **41.18a:** Courtesy of William F. Chissoe, Noble Microscopy Lab, U. of Oklahoma; **41.18b:** Courtesy of Dr. Joan Nowicke, Smithsonian Institution; **41.19:** © Kingsley R. Stern; **41.20:** © Edward S. Ross; **41.21:** © Michael & Patricia Fogden; **41.22 (both):** © Thomas Eisner; **41.23** © John D. Cunningham/Visuals Unlimited; **41.24:** © Edward S. Ross; **41.25a:** © David Sieren/Visuals Unlimited; **41.25b:** © Barbara Gerlach/Visuals Unlimited; **41.28:** © Jerome Wexler/Photo Researchers; **41.29 (all):** Courtesy of Dr. Hans Ulrich Koop, from *Plant Cell Reports*, 17:601-604; **41.30 (both):** © Edward S. Ross

Chapter 42
Figure 42.1: Photo Lennart Nilsson/Albert Bonniers Forlag AB, *Behold Man*, Little Brown & Co.; **p. 861 (top 3):** © Ed Reschke; **p. 861 (4th from top):** © Fred Hossler/visuals Unlimited; **p. 861 (bottom):** © Ed Reschke; **42.6:** © J. Gross/Science Photo Library/Photo Researchers; **42.7:** © Biophoto Associates/Photo Researchers; **p. 863 (top):** © Biophoto Associates/Photo Researchers; **p. 863 (2nd from top):** © Cleveland P. Hickman; **p. 863 (3rd from top):** © Chuck Brown/Photo Researchers; **p. 863 (4th from top):** © Ed Reschke; **p. 863 (bottom):** © Ken Edward/Science Source/Photo Researchers; **42.8b:** © Ed Reschke; **42.9:** © Ed Reschke; **42.10:** © David M. Phillips/Visuals Unlimited; **p. 867 (all):** © Ed Reschke; **42.13:** © Anthony Bannister/Animals Animals/Earth Scenes; **42.19, 42.20:** © Dr. H.E. Huxley; **42.31:** © Treat Davidson/Photo Researchers; **p. 886:** © Dr. H.E. Huxley

Chapter 43
Figure 43.1: © John Gerlach/Animals Animals/Earth Scenes; **43.20:** From O.T. Avery, C.M. Macleod & M. McCarty, "Studies on the chemical nature of the substance inducing transformation of pneumococcal types," reproduced from the *Journal of Experimental Medicine* 79 (1944): 137-158, fig. 1 by copyright permission of the Rockefeller University Press, reproduced by permission. Photograph made by Mr. Joseph B. Haulenbeek

Chapter 44
Figure 44.1: © Professors P.M. Motta & S. Correr/Science Photo Library/Photo Researchers; **44.11:** © Ed Reschke; **44.17 (all):** Courtesy of Frank P. Sloop, Jr.; **44.18:** © Frans Lanting/Minden Pictures

Chapter 45
Figure 45.1: Courtesy of David I. Vaney, University of Queensland, Australia; **45.4:** © C.S. Raines/Visuals Unlimited; **45.14:** © John Heuser, Washington University School of Medicine, St. Louis, MO; **45.16:** © Ed Reschke; **45.18b:** © E.R. Lewis, YY Zeevi, T.E. Everhart, U. of California/Biological Photo Service; **45.27:** Dr. Marcus E. Rachle, Washington University, McDonnell Center for High Brain Function; **45.28:** Photo Lennart Nilsson/Albert Bonniers Forlag AB, *Behold Man*, Little Brown & Co.; **45.31:** © E.R. Lewis/Biological Photo Service

Chapter 46
Figure 46.1: © Omikron/Photo Researchers; **46.6:** © Ed Reschke; **46.23:** © Leonard L. Rue, III

Chapter 47
Figure 47.1: © Francois Gohier/Science Source/Photo Researchers; **47.11:** © Corbis/Bettmann; **47.15:** © John Paul Kay/Peter Arnold, Inc.; **47.20:** © Robert & Linda Mitchell

Chapter 48
Figure 48.1: National Library of Medicine; **48.3:** © Manfred Kage/Peter Arnold, Inc.; **48.7:** © Visuals Unlimited; **48.9 (both):** © Dr.Andrejs Liepins Science Photo Library/Photo Researchers; **48.18:** © Stuart Fox; **48.23:** © CDC/Science Source/Photo Researchers

Chapter 49
Figure 49.1: © Belinda Wright/DRK Photo

Chapter 50
Figure 50.1: © Michael Fogden/DRK Photo; **50.2a:** © Chuck Wise/Animals Animals/Earth Scenes; **50.2b:** © Fred McConnaughey/The National Audubon Society Collection/Photo Researchers; **50.4:** © David Doubilet; **50.5:** © Hans Pfletschinger/Peter Arnold Inc.; **50.7:** © Cleveland P. Hickman Jr.; **50.8:** © Frans Lanting/Minden Pictures; **50.9a:** © Jean Phillippe Varin/Jacana/Photo Researchers; **50.9b:** © Tom McHugh/The National Audubon Society Collection/Photo Researchers; **50.9c:** © Corbis/Volume 86; **50.12a:** © David M. Phillips/Photo Researchers; **50.13:** Photo Lennart Nilsson/Bonniers Forlag AB, *A Child is Born*, Dell Publishing Co.; **50.18:** © Ed Reschke; **50.22 (all):** © McGraw-Hill Higher Education/Bob Coyle, photographer

Chapter 51
Figure 51.1: Photo Lennart Nilsson/Bonniers Forlag AB, *A*

C-4   Credits

Child is Born, Dell Publishing Co.; **51.2b:** © P. Bagavandoss/Science Source/Photo Researchers; **51.2c:** © David M. Phillips/Visuals Unlimited; **51.3b:** Courtesy of Dr. Everett Anderson; **51.6:** © David M. Phillips/Visuals Unlimited; **51.7:** © Cabisco/Phototake; **51.8:** © David M. Phillips/Visuals Unlimited; **51.21:** Photo Lennart Nilsson/Albert Bonniers Forlag AB, *A Child is Born*, Dell Publishing Company; **51.22 (all):** Photo Lennart Nilsson/Albert Bonniers Forlag AB, *A Child is Born*, Dell Publishing Company

Chapter 52

Figure 52.1: © K. Ammann/Bruce Coleman Inc.; **52.2 (left):** © UPI/Corbis Bettmann; **52.2 (center):** © Corbis/ Bettmann; **52.2 (right):** © UPI/ Corbis-Bettmann; **52.3:** From J.L. Gould, Ehology, Norton 1982; **52.5:** © William C. Dilger, Cornell University; **52.6:** From J.R. Brown et al, "A defect in nurturing mice lacking . . . gene for fosB" *Cell* v. 86, 1996 pp 297-308, © Cell Press; **52.7 (all):** © Lee Boltin Picture Library; **52.8:** © William Grenfell/Visuals Unlimited; **52.9a:** Thomas McAvoy, Life Magazine/© Time, Inc.; **52.10 (both):** Grzimek's *Encyclopedia of Ethology* Van Nostrand-Reinhold Co.; **52.12:** © Roger Wilmshurst/The National Audubon Society Collection/ Photo Researchers; **52.13a:** © Linda Koebner/Bruce Coleman; **52.13b:** © Jeff Foott/Tom Stack & Associates; **52.14 (all):** Superstock; **52.15:** Courtesy of Bernd Heinrich; **52.16b:** © Fred Bruenner/Peter Arnold Inc.; **52.16c:** © James L. Amos/Peter Arnold Inc.; **52.19:** © Dwight R. Kuhn/DRK Photo; **52.21** © Corbis/ Volume 53; **52.22:** © Sol Mednick; **52.23b:** © Dr. Mark Moffett/Minden Pictures; **52.24a:** © S. Osolinski /OSF/Animals Animals/Earth Scenes; **52.25:** Nina Leen, Life Magazine, © Time Inc.; **52.28:** © Bios(C.Thouvenin)/Peter Arnold, Inc.; **52.30b:** © B. Chudleigh/Vireo; **52.32:** © George D. Lepp/Corbis; **52.33a:** Courtesy of T.A. Burke, Reprinted by permission from *Nature*, "Parental care and mating behavior of polyandrous dunnocks," 338: 247-251, 1989; **52.35:** © Edward S. Ross; **52.37:** © Mark Moffett/Minden Pictures; **52.38:** © Nigel Dennis/National Audubon Society Collection/Photo Researchers

Chapter 53

Figure 53.1: © PhotoDisc/Volume 44; **53.2:** Courtesy of William J. Hamilton III; **53.17:** Courtesy of Barry Sinervo; **53.23 (left):** © Jean Vie/Gamma; **53.23 (right):** © Gianni Tortole/National Audubon Society Collection/Photo Researchers

Chapter 54

Figure 54.1: © Corbis; **54.2:** © Tim Davis/Photo Researchers; **54.7 (all):** J.B. Losos; **54.11 (both):** © Edward S. Ross; **54.12 (both):** © Lincoln P. Brower; **54.13:** © Michael & Patricia Fogden/Corbis; **54.14:** © James L. Castner; **54.16:** © Merlin D. Tuttle/Bat Conservation International; **54.17:** PhotoDisc/Volume 44; **54.18:** © Michael Fogden/DRK Photo; **54.19:** © Edward S. Ross; **54.21a:** F. Stuart Westmorland/Photo Researchers; **54.21b:** © Anne Wertheim/ Animals Animals/Earth Scenes; **54.24:** © David Hosking/National Audubon Society Collection/Photo Researchers; **54.26 (all):** © Tom Bean; **54.27 (both):** Courtesy of Robert Whittaker; **54.28:** © Edward S. Ross

Chapter 55

Figure 55.1: © Corbis/Volume 46; **55.3:** © Martin Harvey, Gallo Images/Corbis; **55.7a:** U.S. Forest Service; **55.17a:** © Layne Kennedy/Corbis

Chapter 56

Figure 56.1: NASA; **56.10:** © Michael Graybill & Jan Hodder/Biological Photo Service; **56.11:** © IFA/Peter Arnold, Inc.; **56.15:** © Digital Vision/Picture Quest; **56.16a:** © Jim Church; **56.16b:** Courtesy of J. Frederick Grassel, Woods Hole Oceanographic Institution; **56.17:** © Edward S. Ross; **56.22:** © Gilbert S. Grant/National Audubon Society Collection/Photo Researchers; **56.23a:** © Peter May/Peter Arnold Inc.; **56.23b:** © Frans Lanting/Minden Pictures; **56.24a:** NASA

Chapter 57

Figure 57.1: © Tom & Pat Leeson/Photo Researchers; **57.2:** Photo by Tom McHugh, © Natural History Museum of Los Angeles County/Photo Researchers; **57.6:** © Edward S. Ross; **57.9:** © Michael Fogden/DRK Photo; **57.14:** © 1990 R.O. Bierregaard; **57.15:** Reprinted with permission from *Science*, Vol 282 Dec. © 1998 American Association for the Advancement of Science; **57.17:** © Jack Jeffrey; **57.18:** Mark Chandler; **57.20:** Merlin D. Tuttle/Bat Conservation International; **57.21:** U.S. Fish & Wildlife Service; **57.23:** © Wm. J. Weber/Visuals Unlimited; **57.24 (both):** University of Wisconsin-Madison Arboretum

Text

Chapter 1

Box 1.1: From Howard Neverov, "The Consent" in *The Collected Works of Howard Nemerov*, 7th Edition, 1981. Reprinted by permission.

Chapter 5

Figure 5.6: Copyright © 2002 from *Molecular Biology of the Cell* by Bruce Alberts, et al. Reproduced by permission of Routledge/Taylor & Francis Books, Inc.

Chapter 6

Figure 6.10: Modified from Alberts, et al., *Molecular Biology of the Cell*, 3rd Edition, 1994 Garland Publishing, New York, NY.

Chapter 10

Figure 10.7: From Raven, et al., *Biology of Plants*, 5th edition. Reprinted by permission of Worth Publishers. **Figure 10.13:** From Lincoln Taiz and Eduardo Zeiger, *Plant Physiology*, 1991 Benjamin-Cummings Publishing. Reprinted with permission of the authors.

Chapter 11

Figure 11.4: Copyright © 2002 from *Molecular Biology of the Cell* by Bruce Alberts, et al. Reproduced by permission of Routledge/Taylor & Francis Books, Inc.

Chapter 14

Table 14.1: Data from E. Chargaff and J. Davidson (editors), *The Nucleic Acids*, 1955, Academic Press, New York, NY.

Chapter 16

TA 16.1: CALVIN AND HOBBES © 1995 Watterson. Reprinted with permission of Universal Press Syndicate. All rights reserved.

Chapter 17

Table 17.1: Reprinted with permission from *Nature*. Copyright Macmillan Magazines Limited. **Figure 17.9:** G. More, K.M. Devos, Z. Wang, and M.D. Gale: "Grasses, line up and form a circle," *Current Biology*, 1995, vol. 5, pp. 737-739. **Figure 17.11:** Modified from Keho Villiard and Sommerville, "DNA Microarrays for studies of Photosynthetic Organisms," *Trends in Plant Science*, 1999.

Chapter 18

Figure 18.22: From an *Introduction of Genetic Analysis* 5/e by Anthony J.F. Griffiths, et al., Copyright © 1976, 1981, 1986, 1989, 1993 by W.H. Freeman and Company. Used with permission.

Chapter 19

Figure 19.5: Copyright © John Kochik for Howard Hughes Medical Institute. **Figure 19.9 (text):** From H. Robert Horvitz, *From Egg to Adult*, published by Howard Hughes Medical Institute. Copyright © 1992. Reprinted by permission. **Figure 19.9 (top):** M.E. Challinor illustration. From Howard Hughes Medical Institute © as published in *From Egg to Adult*, 1992. Reprinted by permission. **Figure 19.9 (bottom):** Illustration by: The studio of Wood Ronsaville Harlin, Inc. **Figure 19.20:** Modified from John Kochik for Howard Hughes Medical Institute.

Chapter 20

Table 20.2: Data from the American Cancer Society, Inc., 2002.

Chapter 21

Figure 21.8: Data from P.A. Powers, et al., "A Multidisciplinary Approach to the Selectionist/Neutralist Controversy." *Oxford Surveys in Evolutionary Biology*. Oxford University Press, 1993. **Figure 21.10:** From R.F. Preziosi and D.J. Fairbairn, "Sexual Size Dimorphism and Selection in the Wild in the Waterstrider *Aquarius remigis*: Lifetime Fecundity Selection on Female Total Length and Its Components," *Evolution, International Journal of Organic Evolution* 51:467-474, 1997. **Figure 21.11:** Data from M.R. MacNair in J.M. Bishops & L.M. Cook, *Genetic Consequences of Man-Made Change*, Academic Press, 1981, p. 177-207. **Figure 21.12:** Adapted from Clark, B. "Balanced Polymorphism and the Diversity of Sympatric Species." Syst. Assoc. Publ., Vol. 4, 1962.

Chapter 22

Figure 22.3A: Data from Grant, "Natural Selection and Darwin's Finches" in *Scientific American*, October 1991. **Figure 22.3b:** Data from Grant, "Natural Selection and Darwin's Finches" in *Scientific American*, October 1991. **Figure 22.5:** Data from Grant, et al., "Parallel Rise and Fall of Melanic Peppered Moths" in *Journal of*

Credits   C-5

Heredity, vol. 87, 1996, Oxford University Press. **Figure 22.6:** Data from G. Dayton and A. Roberson, *Journal of Genetics*, Vol. 55, p. 154, 1957.

Chapter 23

Figure 23.2: Data from R. Conant & J.T. Collins, *Reptiles & Amphibians of Eastern/Central North America*, 3rd edition, 1991. Houghton Mifflin Company. **Figure 23.10:** Data from B.M. Bechler, et al., *Birds of New Guinea*, 1986, Princeton University Press. **Figure 23.18:** Data from D. Futuyma, *Evolutionary Biology*, 1998, Sinauer.

Chapter 25

Figure 25.8: From Richard O. Prum and Alan H. Brush, "The Evolutionary Origin and Diversification of Feathers," *Quarterly Review of Biology* 77 (3): 261-295, September 2002. Reprinted with permission of The University of Chicago Press.

Chapter 27

Figure 27.9: Data from U.S. Centers for Disease Control and Prevention, Atlanta, GA.

Chapter 31

Figure 31.09: Courtesy of Joel Cracraft.

Chapter 36

Figure 36.3: From Ralph Quantrano, Washington University. **Figure 36.16:** From G.B. Fincher, "Molecular and Cellular Biology Associated with Endosperm Mobilization in Germinating Cereal Grains." Reprinted with permission from the *Annual Review of Plant Physiology and Plant Molecular Biology*, Volume 40. Copyright © 1989 by Annual Reviews www.annualreviews.org.

Chapter 40

Figure 40.10: Data from Hong et al., Arabidopsis *not* Mutants Define Multiple Functions Required for Acclimation to High Temperatures, *Plant Physiology*, Vol. 132, pp. 757–767, 2003.

Chapter 41

Figure 41.8: After McDaniel, 1996. **Figure 41.9:** After McDaniel, 1996. **Figure 41.10:** After McDaniel, 1996.

Chapter 42

Figure 42.32: From *New York Times*, December 15, 1998. Copyright © 1998 *New York Times*. Reprinted with permission.

Chapter 48

Figure 48.21: From Beck & Habicht, "Immunity and the Invertebrates" in *Scientific American*, November 1996. Reprinted by permission of Roberto Osti Illustrations. **Figure 48.25:** Data from U.S. Centers for Disease Control and Prevention, Atlanta, GA.

Chapter 49

Figure 49.4: Data from B. Heinrich, *Science*, American Association for the Advancement of Science.

Chapter 50

Table 50.2: Data from American College of Obstetricians and Gynecologists: Contraception, Patient Education Pamphlet No. AP005.ACOG, Washington, D.C., 1990.

Chapter 52

Figure 52.6: Data from J.R. Brown et al, "A Defect in Nurturing in Mice Lacking the Immediate Early Gene for fosB", Cell, 1996. **Figure 52.11:** Reprinted from *Animal Behaviour*, Vol. 51, M.D. Beecher, P.K. Stoddard, S.E. Campbell, and C.L. Horning, "Repertoire Matching Between Neighbouring Song Sparrows," pp. 917-923. Copyright © 1996, with permission from Elsevier. **Figure 52.20:** From John Alcock, *Animal Behavior*, 1989. **Figure 52.24b:** From John Alcock, *Journal of Animal Behavior*, 1988. Reprinted by permission of Academic Press, Ltd., London. **Figure 52.30C:** Data from M. Petrie, et al. "Peahens Prefer Peacocks with Elaborate Trains, *Animal Behavior*, 1991. **Figure 52.33B:** Data from H.L. Gibbs et al., Realized Reproductive Excess of Polygynous Red-Winged Blackbirds Revealed by NDA Markers," *Science*, 1990.

Chapter 53

Figure 53.24: From G.C. Varley, "Population Changes in German Forest Pests," *Journal of Animal Ecology*, Vol. 18, May 1949. Reprinted with permission of British Ecology Society. **Table 53.1:** Based on Table 14-11 in A.J. Vander, J.H. Sherman, and D.S. Luciano, *Human Physiology*, 5th ed. Copyright © 1997 McGraw-Hill Companies, Inc., Dubuque, Iowa. All Rights Reserved. **Figure 53.5:** After E.R. Pianka, *Evolutionary Ecology*., 4th edition, New York, Harper & Row, 1987. **Figure 53.6:** Data from Brown & Lomolino, *Biogeography*, 3rd edition, 1998, Sinauer Associates, Inc. **Figure 53.7:** Data from Brown & Lomolino, *Biogeography*, 3rd edition, 1998, Sinauer Associates, Inc. After A.T. Smith *Ecology*, 1974. **Figure 53.8:** Data from Elizabeth Losos, Center for Tropical Forest Science, Smithsonian Tropical Research Institute. **Figure 53.10:** Data from *Patch Occupation and Population Size of the Glanville Fritillary in the Anland Islands*, Metapopulation Research Group, Helsinki, Finland. **Figure 53.11:** Data from Bonner, 1965. **Figure 53.2:** Modified from Ricklefs, 1997. **Figure 53.13:** Modified from Ricklefs, 1997. **Figure 53.16:** Data from C.M. Perrins, *Animal Ecology*, 1995. **Figure 53.20B:** Data from C.E. Goulden, L.L. Henry, and A.J. Tessier, *Ecology*, 1982. **Figure 53.22:** Data from Arcese & Smith J. *Animal Biology*, vol. 57, pp. 119-136, 1989 and Smith et al. 1991. **Table 53.3:** After E.R. Pianka, *Evolutionary Ecology*, 4th edition, 1987, New York, Harper & Row, 1987. **Figure 53.30:** Data from National Geographic, July 2001.

Chapter 54

Figure 54.3: From *The Economy of Nature* 4/e, by Robert E. Ricklefs. Copyright © 1973, 1979 by Chiron Press Inc.; Copyright © 1990, 2000 by W.H. Freeman and Company. Used with permission. **Figure 54.4:** From *The Economy of Nature* 4/e, by Robert E. Ricklefs. Copyright © 1973, 1979 by Chiron Press Inc.; Copyright © 1990, 2000 by W.H. Freeman and Company. Used with permission. **Figure 54.6:** Data from Begon et al., *Ecology*, 1996. After: W.B. Clapham, *Natural Ecosystems*, Clover, Macmillan. **Figure 54.8:** Data from E.J. Heske, et al., *Ecology*, 1994. **Figure 54.9:** Data from E.J. Heske, et al., *Ecology*, 1994. **Figure 54.22:** Data from D.W. Davidson et al., "Granivory in a Desert Ecosystem." *Ecology*, 1984.

Chapter 55

Figure 55.18: Data from F. Morrin, *Community Ecology*, Blackwell, 1999. **Figure 55.16:** Data from J.T. Wooten & M.E. Power, "Productivity, Consumers & the Structure of a River Food Chain," *Proceedings National Academic Sciences*, 1993. **Figure 55.13:** Data from Flecker, A.S. and Townsend, C.R., "Community-Wide Consequences of Trout Introduction in New Zealand Streams." In *Ecosystem Management: Selected Readings*, F.B. Samson and F.L. Knopf eds., Springer-Verlag, New York, 1996. **Figure 55.14:** Data from M. Power "Habitat Heterogenieity and the Functional Significance of Fish," *Ecology*, 1997. **Table 55.1:** After Whittaker, 1975.

Chapter 56

Figure 56.13: Map from "An Act of God," July 19, 1997, *The Economist*. Copyright © 1997 *The Economist* Newspaper Ltd. All rights reserved. Reprinted with permission. Further reproduction prohibited. www.economist.com. **Figure 56.25:** Data from Geophysical Monograph, American Geophysical Union, National Academy of Sciences, and National Center for Atmospheric Research.

Chapter 57

Figure 57.3: After Smith et al., 1993. **Figure 57.10:** IUCN 2000, AmphibiaWeb, Hero J.M. & L. Shoo, 2003. Chapter 7 in *Amphibian Conservation*, Smithsonian Press. Background biodiversity hotspots map from Myers et al, 2000. *Nature* 403:853-858 c/o Conservation International. Prepared by J.M. Hero, April 2002. **Figure 57.11:** After Green and Sussman, 1990. **Figure 57.12:** Data assembled by Pima, 1991. **Figure 57.15:** Data from Marra, Hobson, Holmes, "Linking Winter & Summer Events," in *Science*, Dec. 1998. **Figure 57.16:** Data from UNEP, *Environmental Data Report*, 1993, 1994. **Figure 57.22:** Data from H.L. Billington, "Effects of Population Size on a Genetic Variation in a Dioecious Conifer" in *Conservation Biology*, Blackwell Scientific Publication, Inc., 1991. **Figure 57.25:** Data from The Peregrine Fund. **Figure 57.27:** From "Quetzal and The Macaw," The Story of Costa Rica's National Parks, by David Rains Wallace, 1992 Sierra Club. Reprinted courtesy of David Rains Wallce. **Figure 57.4:** From *Nature's Place: Human Population and the Future of Biological Diversity*, by Richard P. Cincotta and Robert Engelman, 2000. Reprinted with permission of Population Action International. **Figure 57.5:** From *Nature's Place: Human Population and the Future of Biological Diversity*, by Richard P. Cincotta and Robert Engelman, 2000. Reprinted with permission of Population Action International. **Figure 57.26:** From *Nature and Resources*, Vol. XXII, 1986. Copyright © 1986 UNESCO Press. Reproduced by permission of UNESCO.

Index

Boldface page numbers correspond with **boldface terms** in the text. Page numbers followed by an "f" indicate figures; page numbers followed by a "t" indicate tabular material.

A

Aardvark, 528, 528f
Aarskog-Scott syndrome, 269f
Abalone, 652
A band, 874–77, 874–77f
ABC model, of floral organ specification, 838, 839f
Abiotic realm, 807, 1184
ABO blood group, 109t, 136, **260**, 260f, 435, **1029**
Abomasum, 898, 898–99f
Abortion, spontaneous, 272, 1098
Abscisic acid, 763, 773, 816, 817t, **827**, 827f
Abscission, 816f, 821, **852**, 852f
Abscission zone, 852
Absolute dating, 460
Absorption, in digestive tract, 888–89, 896
Absorption maximum, 985
Absorption spectrum, **191**
  of photosynthetic pigments, 191, 191f
Abstinence, 1076
Abyssal zone, **1214**, 1214f
Acacia, mutualism with ants, 801, 801f, 1174, 1174f
Acanthodian, 692f, 693t, 694
Acari (order), 670
Accessory digestive organs, 889, 889f, 895, 895f, 902
Accessory pigment, 191, 193f, 194, 198–99
Accessory sex organs
  female, 1072
  male, 1070, 1070f
Accumulated mutation hypothesis, of aging, 401
Acer saccharum, 1143f
Acetabularia, 80
  Hammerling's experiments with, 280, 280f
Acetaldehyde, 36f, 167f, 181
Acetate, in muscle, 181, 181f
Acetic acid, 36f
Acetone, commercial production of, 558
Acetylation, of histones, 373
Acetylcholine (ACh), 129, **879**, 948f, **949**–50, 963t, 965–66, 966f
Acetylcholine (ACh) receptor, 944f, 952, 966, 966f, 1026
Acetylcholinesterase (AChE), **950**
Acetyl-CoA, **168**
  from fat catabolism, 179, 179f
  from glycolysis, 162, 163f
  oxidation in Krebs cycle, 169–70, 169f, 171f
  from protein catabolism, 178, 178f
  from pyruvate, 167–68f, 168
  uses of, 168
ACh. *See* Acetylcholine
AChE. *See* Acetylcholinesterase
Acid, 31, 31f
Acid growth hypothesis, **821**, 821f
Acid precipitation, 1219, 1219f, 1236
Acinar cells, 901f
Acini, 895
Acoelomate, 621f, 623t, **625**, 625f, 628, 634, 634f, 642–45
Aconitase, 171f, 378
Acorn worm, 623t
Acquired characteristics, inheritance of, **434**, 434f
Acquired immunity. *See* Active immunity
Acquired immunodeficiency syndrome. *See* AIDS
Acromegaly, 1002
Acrosome, **1068**, 1069f, 1070, 1082, 1082–83f
ACTH. *See* Adrenocorticotropic hormone
Actin, 39t, 40, **98, 876**, 876f
Actin filament, 86–87f, **98**–100, 98f, 139, 139f, 218, 574, 868, 868f, 1087. *See also* Thin myofilament
Actinobacteria, 549f
Actinomyces, 549f, 554
Actinopoda (phylum), 574
Actinopterygii (class), 689, 691–92, 693t, 696, 696f
Actinosphaerium, 574f
Action potential, **944**–47
  all-or-none law of, 946
  falling phase of, 945f, 946
  generation of, 944–47, 945–47f
  propagation of, 946, 946f
  refractory period after, 946
  rising phase of, 945–46, 945f
  undershoot of, 945f, 946
Action spectrum, **192**
  of chlorophyll, 192, 193f
Activating enzyme, 310
Activation energy, **148**, 148f, 156
Activator, **152, 368**–69, 368–69f, **370**–71, 371–72f, 373
Active immunity, **1018**, 1026–27, 1027f
Active site, 149, **150**, 150f
Active transport, **120**–22, 122t
Adaptation, 237, 1122
  as characteristic of life, 2
  to different environments, 1139, 1139f
  speciation and, 479
Adaptive radiation, **482**, 482f, 485
Adaptive significance, **1122**
Adder, 709
Adder's tongue fern, 210t
Addiction. *See* Drug addiction
Adelina, 1176
Adenine, 49–50, 49–50f, 154, 284, 284f, 287, 287f, 376, 822, 822f
Adeno-associated virus, vector for gene therapy, 264, 264f
Adenosine deaminase deficiency, 263, 333
Adenosine diphosphate. *See* ADP
Adenosine triphosphate. *See* ATP
Adenovirus, 263f, 524f
  in cancer treatment, 422
  E1B-deficient, 423
  vector for gene therapy, 263
Adenylyl cyclase, **132**, 133–34f, **998**, 998f
ADH. *See* Antidiuretic hormone
Adherens junction, 128t, 137f, **139**
Adherent fruit, 1143f
Adhesion, **29**, 768, 772
ADH gene, 435
Adipose cells, **862**, 863t, 1042
Adipose tissue, 862, 862f
ADP, **154**
Adrenal cortex, 995t, 1007, 1007f, 1101
Adrenal gland, 858f, 965, 993t, 1007, 1007f
Adrenal hypoplasia, 269f
Adrenaline. *See* Epinephrine
Adrenal medulla, 993, 995t, 1007, 1007f
Adrenergic receptor
  alpha-adrenergic receptor, 998
  beta-adrenergic receptor, 132, 132f, 998
Adrenocorticotropic hormone (ACTH), 994t, 1001–2, 1001f, 1003f, 1007
Adventitious plantlet, 849, 849f
Adventitious root, 743, 747, 764f, 774, 832–33, 832f, 836, 837f
Aerenchyma, **774**, 775f
Aerial root, 743
Aerobic capacity, 881
Aerobic respiration, **160, 162**, 163f, 166, 177
  ATP yield from, 176, 176f
  evolution of, 182
  regulation of, 177, 177f
Aerotropism, 811
Aesthetic value, of biodiversity, 1233
Afferent arteriole, 1052
Afferent neuron. *See* Sensory neuron
Aflatoxin, 614
Africa, human migration out of, 725, 725f
African boomslang, 709
African clawed frog (*Xenopus laevis*)
  development in, 392, 392f, 397
  nuclear transplantation in, 393
  rRNA genes of, 323
African elephant (*Loxodonta africana*), 714f
African sleeping sickness. *See* Trypanosomiasis
African violet, 748f
Afrovenator, 705f
Afterbirth, 1101
Agammaglobulinemia, 269f
Agave, 594
Age, at first reproduction, 1149
A gene, 366f
Agent Orange, 821
Age of Amphibians, 700
Age of Mammals, 717
Age structure
  of population, **1146**
  population pyramids, 1156, 1156f
Agglutination reaction, 1024, **1029**
Aggregate fruit, 761f
Aggressive behavior, 1107, 1109, 1119, 1134
Aging
  cancer death rate and, 419, 419f
  premature, 402
  theories of, 401–2
Agonist (hormone), 1010
Agnatha (superclass), 689–90, 692, 692f, 864, 1010
Agriculture
  applications of genetic engineering to, 335–38, 335–38f
  applications of genomics to, 357–58, 357–58f
  effect of global warming on, 785, 785f, 1224
  pollution due to, 1218
  reproductive cloning of animals, 424–25
  sustainable, 1220
Agrobacterium tumefaciens, 335, 335f, 822, 822f
AIDS, 533t, 536, 540, 1157
  deaths in United States, 538
  fatality rate of, 1033
  gene therapy for, 333t
  in United States, 1034f
  vaccine against, 538, 539f, 1034
Air pollution
  industrial melanism and, 456–57
  monitoring with lichens, 611
Air sac, 929, 929f
Akiapolaau, 1240f
Alanine, 36f, 38t, 41, 42f, 178
Alarm call, 1119–20, 1121f, 1130–32
Alarm pheromone, 1120
Alaskan near-shore habitat, 1242–43, 1242f
Albatross, 711t
Albinism, 249t, 253
Albinism-deafness syndrome, 269f
Albumin, 39t, 866, **910**
Alcardi syndrome, 269f
Alcohol abuse, 902
Alcoholic beverage, 181
Alder, 1178–79f
Aldolase, 165
Aldose reductase, 356f
Aldosterone, 920, 993, 995t, 1007, 1056–58, 1058f
Aleurone, **763**, 764f
Alfalfa, 787
Alfalfa butterfly (*Colias*), 673f
Alfalfa plant bug, 796, 796f
Algae
  multicellular, 72f
  sexual life cycle in, 228, 229f
Algal bed, 1191t
Alkaloid, 613, 798, 799t, 800, 1170
Alkaptonuria, 249t, 296
Allantoin, **1051**
Allantois, **702**, 702f, 715f, 718, 1066, **1096**, 1096f
Allee, Warder, 1153
Allee effect, **1153**, 1244
Allele, **249**
  multiple, 260, 260f
  temperature-sensitive, 256, 256f

I-1

Allele frequency, 433–34, **436**
　changes in populations, 436–45
Allelopathy, **798**, 798f
Allen's Rule, 1139
Allergen, 1036
Allergy, 1025, 1036, 1036f
　to genetically modified food, 339
　to mold, 614
Alligator, 703t, 709, 1177
Allolactose, 368f
Allometric growth, 1102, 1102f
Allomyces, 605t, 606f
Allopatric speciation, **480**, 481f
Allophycocyanin, 572
Allopolyploidy, **480**–81, 498f
Allosteric inhibitor, **152**, 153f
Allosteric site, **152**, 156, 156f
Alper, T., 542
Alpha-1-antitrypsin deficiency, 333t
α cells, 895f, 902, 902f, 1008
Alpha-glucose, **58**, 58f
Alpha helix, 44f, 45
α turn α motif, 44f, 45
Alpha wave, 959
Alpine tundra, 1141f
Alport syndrome, 269f
Alternate leaf, 749, 749f
Alternation of generations, 606f, 842
Alternative splicing, **314**–15, **349**, 349f, 351, 356, 376, 376f
Altitude
　air pressure and, 924, 924f
　physiological changes at high altitude, 1138t
Altitude sickness, 1138
Alton giant, 1002f
Altricial young, **1128**
Altruism, 1130–32, 1134
　reciprocal, **1130**, 1134
ALU element, 350, 411
Aluminum, in plants, 783
Alveolar duct, 1101–2, 1101f
Alveolar sac, 928f
Alveolata, 564f, 568–70, 568–70f
Alveoli, 923f, **928**, 928f, 930
Alzheimer disease, 45, 272, 358, 960
Amacrine cells, 986, 986f
Amanita muscaria, 600f
Amborella, 594f
Amborella trichopoda, 593, 593f
Ambulocetus natans, 461, 461f
American basswood, 827f
American redstart, 1238, 1238f
American woodcock, 987
Ames test, 414, 414f
Amine, biogenic, 950
Amine hormone, 993
Amino acid, 37–38, **41**
　abbreviations for, 42f
　absorption in small intestine, 896, 896f
　catabolism of, 178, 178f
　chemical classes of, 41, 42f
　genetic code, 304–5, 304f, 305t
　prebiotic chemistry, 66, 67f
　in proteins, 41
　reabsorption in kidney, 1048f, 1054
　structure of, 41, 41–42f
　twenty common, 42f
Aminoacyl-tRNA synthetase, **310**, 311f
Amino group, 36f
Amish population, 439
Amiskwia, 630f
Ammonia, 67f, 902, **1051**, 1051f, 1217
Ammonification, **1187**, 1187f
Amniocentesis, **274**, 274f
Amnion, **702**, 702f, 715f, 1066, **1096**, 1097f, 1098
Amniote, 702, 707
　cladogram of, 705f

Amniotic egg, 702, 710, 715f, 1066
Amniotic fluid, 1096
Amniotic membrane, **1096**
Amoeba, 565, 574, 574f
　slime mold, 576, 576f
Amoeba proteus, 574f
Amoebocyte, 637f
Amphibia (class), 689, 691f
Amphibian, 622t, 692f, 698–701
　brain of, 956, 956f
　characteristics of, 698, 698t
　chytridiomycosis in, 614, 614f
　circulation in, 698, 916, 916f
　cleavage in, 1084, 1084f
　development in, 1065, 1065f
　evolution of, 688, 700, 700f
　extinctions, 1229f, 1229t
　first, 700
　gastrulation in, 1088, 1088f
　heart of, 698–99, 916, 916f
　invasion of land by, 699, 699f
　kidney of, 1050
　larva of, 977
　legs of, 699, 699–700f
　lungs of, 698–99
　metamorphosis in, 1005, 1005f, 1065
　nitrogenous wastes of, 1051
　orders of, 701
　population declines in, 614, 1234–35, 1234–35f
　present day, 701
　reproduction in, 699, 1065, 1065f
　respiration in, 698, 916, 923f, 927, 927f
　swimming in, 882
Amphioxus. See *Branchiostoma*
Amphipoda (order), 669
Amphisbaenia (suborder), 708
Ampulla (inner ear), 979, 979f
Ampulla (tube feet), 678f, **679**
Ampullae of Lorenzini, 971t, 988
Amygdala, **958**, 960
Amylase
　pancreatic, 895
　salivary, 890, 900t
α-Amylase, 764f, 824
Amyloid plaque, 45, 960
β-Amyloid protein, 960
Amylopectin, 57
Amyloplast, **97**, 739, **763**, 810
Amylose, 57
Anabaena, 549f, 557, 557f
Anabolism, **155**
Anaerobic growth, **70**
Anaerobic respiration, **160**, 163, 177, 881
Analogous structures, 13, 504, 516
Anaphase
　meiosis I, 230, 231f, 233–34f, 236
　meiosis II, 231f, 235f, 236
　mitotic, 213, 213f, **216**, 216–17f, 231f, 233f
Anaphase-promoting complex (APC), 217, 221, 221f
Anaphylactic shock, **1036**
Anapsid, 707–8
Anatomical dead space, 931
Ancestral characters, **512**–13
Anchoring junction, 137–38f, 138–39
Anchoring protein, 112, 113f
Andrews, Tommie Lee, 332, 332f
Andrias, 701
Androecium, 589, **594**, **840**
Androgen, 1009–10
Androgen insensitivity, 269f
Anemia, 264
Anesthetic, 959
Aneuploidy, **272**, **411**
Angelfish, 696f

Angina pectoris, **921**
Angiogenesis, 423
Angiogenesis inhibitor, 423, 423f
Angiosperm. See Flowering plant
Angiostatin, 423
Angiotensin, 920
Angiotensin I, 1058, 1058f
Angiotensin II, 1007, 1058, 1058f
Angiotensinogen, 1058f
Angular acceleration, detection of, 978–79, 978–79f
Anhidrotic ectodermal dysplasia, 269f
Animal(s)
　asexual reproduction in, 237
　body plan of, evolution of, 624–27
　classification of, 527–28, 527–28f, 620, 621f, 622–23t, 628–30
　coevolution of animals and plants, 801, 831, 831f, 1172, 1172f, 1174
　development in, 382, 382f, 619t
　diversity in, 617–30
　evolution of, 628–30, 629f
　fruit dispersal by, 762, 762f
　general features of, 618, 618–19t
　habitats of, 619t
　invasion of land by, 460f, 699, 699f
　major phyla of, 622–23t
　movement in, 618t, 870–73
　multicellularity in, 618t
　obtaining nutrients, 618t
　phylogeny of, 620, 621f, 628, 629f
　pollination by, 840, 844–45
　reproductive cloning of, 424–25
　sexual life cycle in, 228, 229f
　sexual reproduction in, 619t
　tissues of, **619**
　transgenic, 338, 338f
Animal breeding, 12, 238, 450, 450f, 459, 459f
　thoroughbred horses, 238, 450, 450f
Animal cells
　cell division in, 209f, 213
　cytokinesis in, 218, 218f
　lack of cell walls, 618t
　mitosis in, 215
　structure of, 86f, 102t
Animal cognition, 1114–15, 1114–15f
Animalcules, 81
Animal disease
　fungal, 604, 614
　prion, 542
　viral, 540–41
Animal fat, 53–54, 54f
Animalia (domain), 16, 16f
Animalia (kingdom), 74–75f, 511f, **518**, 518f, 521, 521f, 523t
Animal pole, **383**, 392, 392f, 1084, 1084f, 1088, 1088f
Anion, **21**, 114
Annelid, 634–35, **635**, 658–61
　body plan of, 658–59, 659f
　circulation in, 908
　classes of, 660–61
　connections between segments of, 658
　excretory organs of, 1046f
　nervous system of, 954, 954f
　segmentation in, 527, 527f, 627, 627f, 658, 659f
Annelida (phylum), 621f, 622t, 635, 658–61
Annotation, 351
Annual growth layers, 733f
Annual plant, **594**, 814, **851**, 851f
Anole, 708
Anolis lizard
　courtship display of, 479, 479f
　dewlap of, 479, 479f
　interspecific competition in, 1167

　malaria in, 1176
　resource partitioning among species of, 1166f
　thermoregulation in, 1139f
Anomalocaris canadensis, 630f
Anopheles mosquito, 348f, 493t, 496
Anorexia nervosa, 903
Anseriformes (order), 711t
Ant, 675t, 1120, 1133, 1240
　ant farmer-fungi symbiosis, 613, 613f
　ant-rodent interactions, 1176–77, 1177f
　flatworm parasites of, 1175, 1175f
　mutualism with acacias, 801, 801f, 1174, 1174f
　mutualism with aphids, 1174
Antagonist (hormone), 1010
Antagonist (muscle), **873**, 873f
Antagonistic effector, 1043, 1043f
Antarctic circumpolar current, 1212f
Anteater, 467f, 528, 528f, 719t
Antelope, 716
Antenna, 665–66f, 668, 669f
Antenna complex (photosynthesis), 195–96, 195f, 198–99
Antennal gland, 1046
Antennapedia complex, **398**–99, 398f
Antennapedia gene, 398
Antennule, 669f
Anterior end, 624, 624f
Anterior pituitary, 993, 994t, **1000**–1004, 1003f, 1071, 1074–75
Anther, 245f, 589, **594**, 595, 595–96f, 838, **840**, 840f, 842f, 843, 847f
Antheridium, **582**, 583f, 586, 588–89, 588f, 609f
Anthocerotophyta (phylum), 583
Anthocyanin, 257, 852
Anthophyta (phylum), 585t
Anthozoa (class), 641, 641f
Anthrax, 357, 357f, 555t, 558
Anthropoid, **720**, 721f
Antibacterial soap, 552
Antibiotic, 898
　sources of, 603
　susceptibility of bacteria to, 84
Antibiotic resistance, 327, 327f, 408, 552
Antibody, 39t, 866, **1018**, 1023, 1025–27, 1028f, 1032. See also Immunoglobulin
　diversity of, 1025–26
　immunocytochemistry, 83
　in medical diagnosis, 1029–30
　monoclonal. See Monoclonal antibody
　polyclonal, 1030
　production of, 1019
　structure of, 1025, 1025f
Anticoagulant, of medicinal leech, 661, 661f
Anticodon, 303f, 310, 311f
Antidepressant drug, 950
Antidiuretic hormone (ADH), 39t, 920, 994t, **1000**, 1000–1001f, **1056**–57, 1057f
Antifreeze, 815, 1138
Antigen, **1018**, 1019
Antigenic determinant site, **1018**
Antigen-presenting cells, **1020**–21, 1021f
Antigen shifting, **1035**
Antioxidant, **400**
Antiparallel strands, in DNA, 287, 287f
Antipodal, **595**, 842f, 843, 843f, 848f
Antiport, **122**
Antisense RNA, 422, 826, 826f
Antisense strand, 306

Antisocial behavior, 273
Antler, 716, 1126
Anura (order), 698t, 700–701, 701f
Anus, 888, 888–89f, 899f
Aorta, 916, 917f, **918**, 1007
Aortic arch, 974, 1091
Aortic body, **932**
Aortic valve, 917f, **918**
APC. *See* Anaphase-promoting complex
APC gene, 417t, 419f
Ape, 719t, 720, 904, 1067
  chromosome number in, 495f
  compared to hominids, 721
  evolution of, 720
Aperture (pollen grain), 595
APETALA1 gene, 837–38
Aphasia, 960
Aphid
  feeding on phloem, 776, 776f
  mutualism with ants, 1174
Aphotic zone, **1216**, 1216f
Apical bud, 822f
Apical complex, 568
Apical dominance, 836, 836f
Apical meristem, 388, 389f, **730–31**, 730–33f, 737, 739f, 740, 742f, 744–45, 744f, 748, 756, 756f, 758, 822f
Apicomplexes, **568–69**, 569f
Apicoplast, 496
Aplysina longissima, 636f
apoB gene, 376–77
Apocynaceae (family), 1169
Apoda (order), 698t, 701, 701f
Apodiformes (order), 711t
Apolipoprotein B, 376–77
  APOB100, 378
  APOB48, 378
Apomixis, 849
Apoptosis, 223, **400**, 400f, 1016
Appendicitis, 465
Appendicular locomotion, 882
Appendicular skeleton, 871, 871f
Appendix, 465, 889f, 897, 897f
Apple, 736, 821, 849
Applied research, 6–7
Apyrimidinic (AP) site, **412**
Aquaporin, **116**, 122t, **768**, 769f
Aquatic ecosystem, 1212–18
  compared to terrestrial ecosystems, 1217
Aqueous solution, 116
Aquifer, 1184f, 1185
Aquifex, 519, 520f, 549f
Aquificae, 549f
Arabidopsis, 355, 838
  aquaporins of, 768
  auxin transport in, 820
  CONSTANS gene in, 835
  det2 mutant in, 808
  development in, 389f, 758f
  EMBRYONIC FLOWER gene in, 833, 833f
  genome of, 348f, 352, 493t, 496
  homeodomain proteins of, 16
  hot mutants in, 815, 815f
  LEAFY COTYLEDON gene in, 759
  LEAFY gene in, 833, 833f, 835–36
  MONOPTEROUS gene in, 758
  response to touch, 811
  scarecrow mutant in, 740, 740f, 810
  shootmeristemless mutant in, 758, 758f
  short root mutant in, 810
  small RNAs in, 374
  suspensor mutant in, 757, 757f
  tissue-specific gene expression in, 740, 740f
  too many mouths mutant in, 734, 734f
  trichome mutation in, 735f
  vernalization in, 836
Arachidonic acid, 994
Arachnid, 668t, 670
Arachnida (class), 670
Araneae (order), 670, 670f
Arapsid, 705f
Arbuscular mycorrhizae, 607, **612**, 612f
Arceuthobium, 1143
Archaea (domain), 16, 16f, 74f, 518–19, 518–20f, 520t, 548
Archaebacteria, **70**, 70f, 84, 519, 546, 548, 549f. *See also* Prokaryote
  bacteria versus, 548
  cell wall of, 71, 548
  DNA of, 71
  gene architecture in, 548
  membrane lipids of, 71
  nonextreme, 519
  plasma membrane of, 548
  thermophilic, 71
Archaebacteria (kingdom), 74–75f, **518**, 518f, 521f, 523t
Archaefructaceae (family), 593
Archaefructus, 583f, 593, 594f
Archaeopteryx, 461, 517f, 712, 712f
Archegonium, **582**, 583f, 586, 588–89, 588f, 591
Archenteron, **626**, 627, 1087–88, 1087f, 1091f
Archosaur, 515f, 691f, 705f, 707
Arctic, species richness in, 1199, 1199f
Arctic fox, coat color in, 256f
Arcyria, 576f
Ardipithecus ramidus, 723f
ARF. *See* Auxin response factor
Argentine ant, 1240
Argentine hemorrhagic fever, 357t
Arginine, 42f
  biosynthesis in *Neurospora*, 297, 297f
Arithmetic progression, 11, 11f
Armadillo, 10f, 528, 528f, 719t, 1195
  feral, 555f
Armillaria, 599, 604f
Armored fish, 692f, 693
Arnold, William, 194, 194f
Arousal
  sexual, 1000
  state of consciousness, 959
Arrow worm, 623t
Arsenic, 415, 415t, 791–92
Arteriole, **912**
Arteriosclerosis, 921
Artery, 858f, 908, **912**, 912f
Arthropod, 622t, 634–35f, **635**, 665f, 871
  body plan of, 664–67
  circulatory system of, 665f, 667
  economic importance of, 664
  excretory system of, 665–66f, 667
  groups of, 668–74, 668t
  jointed appendages of, 664, 665f
  locomotion in, 883–84
  molting in, 666
  nervous system of, 665–66f, 667, 954, 954f
  reproduction in, 1062
  respiratory system of, 665–67f, 667
  segmentation in, 527, 527f, 627, 627f, 665, 665f
  taste receptors in, 975, 975f
Arthropoda (phylum), 621f, 622t, **635**, 664–75
Articular cartilage, 865, 872f
Artificial selection, **12**, **440**, 458–59, 458–59f, 841
  domestication, 459, 459f
  laboratory experiments, 458, 458f

Artiodactyla (order), 719t
Arylamide, 415
Asbestos, 415, 415t
Ascaris, 228, 622t
Ascidian, 501, 502f
Asclepiadaceae (family), 1169
Asclepias syriaca, 1143f
Ascocarp, **609**, 609f
Ascogonium, 609
Ascomycetes, 382, 600, 600f, 602, 609–11, 609–10f, 613–14
Ascomycota (phylum), 605, 605f, 605t, 609–10, 609–10f
Ascospore, **609**, 609f, 610
Ascus, **609**, 609f, 610
Asexual reproduction, **237**, 1062–63
  in cnidarians, 1062
  in plants, 831, 849, 849f, 851
  in protists, 1062
Ash, 749, 761f, 762
Ashkenazi Jews, 259, 264t
Asian flu, 540
Asparagine, 42f
Aspartic acid, 42f, 178
Aspen, 851
  transgenic, 833, 833f
Aspergillus flavus, 614, 614f
Aspirin, 995, 1232
Assemblage, **1162**
Association cortex, 958
Association neuron. *See* Interneuron
Associative activity, 956f
Associative learning, **1110**, 1110f
Assortative mating, **438**
Aster (mitosis), **215**, 215–16f
Asteroidea (class), 676f, 678, 680, 680f
Atherosclerosis, 54, **921**
Athlete's foot, 604
Atmosphere
  of early earth, 65–66
  reducing, 65–66
Atmosphere (pressure unit), **924**
Atmospheric circulation, 1204–7, 1205f, 1212
Atmospheric pressure, 924, 924f
Atom, 2, 3f, **20–21**
  chemical behavior of, 22–23, 22–23f
  energy within, 22–23, 23f
  isotopes of, 21–22, 21f
  kinds of, 24, 24f
  neutral, **21**
  structure of, 20, 20f
Atomic mass, **20**
Atomic number, **20**, 24
ATP, 51, **154**, 159–60
  energy storage molecule, 154
  production of, 161. *See also* ATP synthase
    in electron transport chain, 162, 172–73, 172f, 175–76, 175–76f
    in fat catabolism, 179, 179f
    in glucose catabolism, 162, 162–63f
    in glycolysis, 163–65f, 164–67, 167f, 176, 176f
    in Krebs cycle, 162, 170, 175f, 176, 176f
    in photosynthesis, 187, 187f, 189, 194, 196–99, 196–99f
  regulation of aerobic respiration, 177, 177f
  structure of, 51f, 154, 154f, 160, 161f
  theoretical and actual yields from aerobic respiration, 176, 176f
  uses of
    in active transport, 120, 121f
    in Calvin cycle, 200–201, 200–201f, 204f
    in cell movement, 160

  in endergonic reactions, 160–61
  in energy-requiring reactions, 154
  in glycolysis, 164, 164–65f
  in muscle contraction, 876, 877f, 881
  in phloem transport, 776–77
  in sodium-potassium pump, 942f
ATP synthase, **161**, 161f, 175, 175f, 194, 199
Atrial natriuretic hormone, 920, 1010, 1057–58, 1058f
Atrial peptide, **333**
  genetically engineered, 333
Atriopore, 685–87
Atrioventricular (AV) bundle. *See* Bundle of His
Atrioventricular (AV) node, 919, 919f
Atrioventricular (AV) valve, **918**
Atrium, 685f, **915**, 915f
  left, 916–18, 916–17f
  right, 916–18, 916–17f
Auditory cortex, 957
Auditory nerve, 981f
Auditory receptor, 980–82
Auk, 711t
Aurelia aurita, 640f
Australopithecine, 722–24, 722f
  early, 722–23
Australopithecus, 722, 722f
Australopithecus aethiopicus, 723f
Australopithecus afarensis, 722–23, 723f
Australopithecus africanus, 723f
Australopithecus anamensis, 723f
Australopithecus boisei, 723f
Australopithecus robustus, 723f
Autoimmune disease, 1026, 1036
Automated DNA sequencer, 346, 346f
Autonomic motor neuron, **940**
Autonomic nervous system, 940f, 963–66, 963–66f, 965t
Autophosphorylation, 809, 809f
Autopolyploidy, **480**
Autoradiography, 328, 328f
Autosome, **270**
  nondisjunction involving, 272
Autotroph, **160**, 180, **553**, **1190**
Autumnal equinox, 1204
Auxin, 739, 809–11, 816, 817t, **818**, 822–23, 822–23f, 827
  discovery of, 818–19, 818–19f
  effects of, 819–21, 820–21f
  phototropism and, 819–20, 820f
  in root-shoot axis formation, 758–59
  synthetic, 820f, 821
Auxin receptor, 818
Auxin response factor (ARF), 758
Avery, Oswald, 283
Aves (class), 689, 691f, 710–13
AV node. *See* Atrioventricular node
avr gene, 803, 803f
AV valve. *See* Atrioventricular valve
Axial locomotion, 882
Axial skeleton, 871, 871f
Axil, 747
Axillary bud, 732, 732f, **744**, 744f, 796, 796f, 836, 836f
Axolotl, 924
Axon, 869, 869f, 940–41f, **941**
  conduction velocities of, 947t
  myelinated, 941, 946–47f, 947t, 963f
  unmyelinated, 941, 946–47, 947f, 947t
Axopodia, 565
Aysheaia, 630f
Aznalcóllar mine spill (Spain), 792, 792f
Azolla, 587
AZT, 538, 539f, 1034

B

Babbling, 1121
Baboon, 1114
BAC. *See* Bacterial artificial chromosome
Bacillary dysentery, 554
Bacillus, **546**, 549f
Bacillus anthracis, 357t, 549f, 555t
Bacillus subtilis, 208
Bacillus thuringiensis insecticidal protein, 336, 339, 558
Back-mutation, 414
Bacon, Francis, 4
Bacteria, **71**, 71f, 84, 520, 545–58, 549f. *See also* Prokaryote
  ancient, 70, 70f
  antibiotic sensitivity of, 84
  archaebacteria versus, 548
  as biofactories, 558
  cell wall of, 548
  colonial, 522
  flagella of, 84, 84–85f
  gene architecture in, 548
  genetically engineered, 558
  Gram staining of, 84
  intestinal, 557, 893, 897–98
  mutant hunt in, 552f
  photosynthetic, 71, 71f, 85, 85f, 112, 113f, 196, 196f, 546, 551f, 557
  plasma membrane of, 548
Bacteria (domain), 16, 16f, 74f, 518, 518–20f, 520, 520t, 548
Bacteria (kingdom), 74–75f, **518**, 518f, 521f, 523t
Bacterial artificial chromosome (BAC), **347**, 351
Bacterial disease
  in humans, 554–56, 555t
  in plants, 553, 803f
Bacteriochlorophyll, 553
Bacteriophage, **283**, 321, 532f, **534**
  cloning vector, 322, 322f
  Hershey-Chase experiment with, 283, 283f
  lysogenic cycle of, **534**, 535f
  lytic cycle of, 283, 534, 535f
  structure of, 534f
Bacteriophage lambda, 534
  cloning vector, 322, 322f
  lambda repressor, 364f
Bacteriophage T2, 283, 283f
Bacteriophage T4, 524f, 534, 535f
Bacteriorhodopsin, 112, 131
Bacteroid, 787f
Bakanae, 824
Balance, 978–79
Bald cypress (*Taxodium*), 774f
Baleen whale, 465, 465f, 1215, 1239
Ballast water, 652, 1240
Balsam poplar, 477
Bamboo, 852
Banana, 497, 497f, 826, 849, 1180
Bankivia fasciata, 435f
Barb (feather), 710, 710f
Barbiturate, 959
Barbule (feather), 710, 710f
Bark, 731, 732f, 735, 742, 796
  outer, 746
Barnacle, 668–69, 669f, 1173
  competition among species of, 1164, 1164f
Barn owl, 711t
Baroreceptor, **920**, 971t, **974**
Baroreceptor reflex, 920
Barr body, **271**, 271f, 273
Barred tiger salamander (*Ambystoma tigrinum*), 701f

Barrel sponge, 622t
Barro Colorado Island, 1195
Bar-shaped eye, in fruit flies, 267, 267f
Basal body, **100**, 100f
Basal ganglia, 955t, 958
Basal lamina, 137f
Basal metabolic rate (BMR), 903, 1005
Basal transcription factor, **370**, 370f, 372f
Base, 31, 31f
Base-pairs, 50f, 344–45
Base substitution, 410, 410t
Basic research, 6
Basidiocarp, **608**
Basidiomycetes, 382, 600, 600–1f, 602, 608, 608f, 613
Basidiomycota (phylum), 605, 605f, 605t, 608, 608f
Basidiospore, **608**, 608f
Basidium, **608**, 608f
Basilar membrane, 980–81, 981–82f
Basket sponge, 622t
Basking, 707, 1042, 1138, 1139f, 1207
Basophils, 866, 910f, **911**
Bat, 515f, 528f, 716, 716f, 719t, 1143
  echolocation in, 716, 982
  pollination by, 595, 845, 1172, 1172f, 1243, 1243f
  wings of, 884, 884f
Bates, Henry, 1171
Batesian mimicry, **1171**, 1171f
Bateson, William, 296
Bathtub sponge, 636
Batrachochytrium dendrobatidis, 600f, 614
bax gene, 400, 400f
bc₁ complex, 174, 174f
B cell(s), 910f, 1018–19, 1019–20t, 1023–24, 1023–24f, 1026–27, 1028f, 1032f
B cell receptor, 136f, 1023–26, 1023f, 1025f
B cell-stimulating factor, 1021
bcl genes, 400, 400f, 417t
Bdellovibrio, 549f
Beach flea, 669
Beaded lizard, 708
Beadle, George, 296–97
Beadle and Tatum experiment, 296–97, 296–97f
Beak
  of bird, 446, 447f, 713
    Darwin's finches, 10, 10f, 444, 454–55, 454–55f, 483, 483f, 1166, 1166f
  of turtle, 708
Bean, 759, 759f, 761f, 763, 764f, 791, 812, 851
Bear, 510f, 646f, 719t, 888, 1211
Beardworm, 1215f
Beaver, 715, 715f, 719t, 1177, 1177f, 1211
Becker muscular dystrophy, 269f
Bedbug, 675
Bee, 270t, 664f, 675t, 884, 1133, 1170
  African, 1246
  chromosome number in, 210t
  pollination by, 844, 844–45f, 847f
Beef tapeworm (*Taenia saginata*), 643f, 645
Beer-making, 558, 603, 610
"Bee's purple," 845, 845f
Beeswax, 53
Beet, 743, 851
Beetle, 622t, 664f, 675t, 884, 1138f
Behavior, 1105, **1106**–34. *See also specific types*
  adaptation to environmental change, 1138
  cognitive, **1114**–15, 1114–15f

communication and, 1118–21
development of, 1112–13
evolution and, 1122–29
foraging, 1123, 1123f
innate, 1106–7, 1107f
learning and, 1110–17
migratory, 1116–17, 1116–17f
reproductive strategies, **1125**, 1125f
stereotyped, 1106
study of, 1106–7, 1107f
territorial, 1124, 1124f
Behavioral ecology, **1122**, 1122f
Behavioral genetics, 1106, 1108–9, 1108–9f
Behavioral genomics, 358
Behavioral isolation, 473t, 474–75, 475f, 1119f
Belding's ground squirrel, 1131–32
Bell-shaped curve, 255, 255f
Belt, Thomas, 1174
Beltian body, 1174
Bent grass (*Agrostis tenuis*), metal tolerance in, 443, 443f
Benthic zone, **1214**–15, 1214–15f
Benzene, 415, 415f
Benzo[*a*]pyrene, 421
Benzo[*a*]pyrene-diolepoxide, 421
6-Benzylamino purine, 822f
Bergey's Manual of Systematic Bacteriology, 548
Beriberi, 904t
Berry, true, 761f
Beta-adrenergic receptor, 132, 132f, 998
βαβ motif, 44f, 45
β barrel, 45, 112, 113f
Beta-carotene, 193, 337, 337f
β cells, 895f, 902, 902f, 1008
Betacyanin, 852
Beta-glucose, **58**, 58f
Beta-pleated sheet, 44f, 45, 112, 113f
Beta wave, 959
b₆-f complex, 196–98f, 198
Bicarbonate, 32, 32f, 149, 910, 935–36, 936f, 1008
  in carbon cycle, 1186
  in pancreatic juice, 889, 894–95, 900, 900t, 901f
  reabsorption in kidney, 1056, 1056f
Biceps muscle, 859f, 873
bicoid gene, 396, 399
Bicoid protein, 395f, 396
Bicuspid (mitral) valve, 917f
Biennial plant, **851**
Bilateral symmetry, 620, **624**, 624f, 634, 642, 677f, 679f
Bilateria, **620**, 620–21f, 624, 642–45
Bile, 889, 894–95, 895f, 900, 902
Bile pigment, 895, 895f
Bile salt, 895, 895f
Bilharzia, 644
Binary fission, 207–9f, **208**, 214, 237, 547, **565**
Bindweed, 811
Binocular vision, 720, **987**
Binomial distribution, **253**, 253t
Binomial expansion, 436
Binomial name, 510
Biochemical pathway, **155**, 155f
  evolution of, 155
  regulation of, 156, 156f
Biodiversity, 509, 1197–1200. *See also* Species richness
  biodiversity crisis, 1228–33, 1228–33f, 1229t
  conservation biology, 1227–48
  economic value of, 1232–33, 1232f
  ethical and aesthetic values of, 1233

in rain forests, 1210
speciation and extinction through time, 487, 487f
Bioenergetics, 143
Biogenic amine, 950
Biogenic law, 1095
Biogeochemical cycle, **1184**–89, 1184–89f
  in forest ecosystem, 1189, 1189f
Biogeography
  island, 1200, 1200f
  patterns of species diversity, 1199, 1199f
Bioinformatics, **349**, 354, 503
Biological clock, 1009–10
Biological magnification, **1218**, 1218f
Biological species concept, **473**, 477
Biological weapons, 357
Biology, properties of life, 2
Bioluminescence, 1119f, 1215, 1215f
Biomass, **1191**
Biome, **1208**–11
  climate and, 1209, 1209f
  distribution of, 1208, 1208f
  predictors of biome distribution, 1209, 1209f
Bioremediation, 558, 558f, **603**
Biosphere, **1157**, 1203–24
  influence of human activity on, 1218–24
Biosphere reserve, 1248, 1248f
Biotechnology, 333–34
Bioterrorism, 357, 357t, 558, 610, 614
Biotic potential, **1150**
Biotin, 904t
Bioweapons, 558
Bipedalism, **721**–23, 722f
  in dinosaurs, 705, 705f
Bipolar cells, 986–87, 986f
Biramous appendage, 527–28, 528f, 668, 668t
Birch (*Betula*), 749, 845–46, 846f, 1211
Bird, 622t, 691f, 710–13
  altruism in, 1130
  bones of, 710
  brain of, 956, 956f
  characteristics of, 710
  circulation in, 713, 916–17, 917f
  cleavage in, 1086
  cognitive behavior in, 1114–15, 1115f
  digestive tract of, 890f
  eggs of, 710
  evolution of, 517f, 689, 706f, 712–13, 712f
  extinctions, 1229, 1229f, 1229t, 1234f
  fecundity in, 1148, 1148f
  flocking behavior in, 1133, 1133f
  gastrulation in, 1089, 1089f
  habituation in, 1110
  heart of, 916–17, 917f
  kidney of, 1050
  mating systems in, 1128–29, 1129f
  migration of, 988, 1116–17, 1117f
  nitrogenous wastes of, 1051, 1051f
  pollination by, 845, 845f
  present day, 713
  respiration in, 713, 928–29, 929f
  sex chromosomes of, 270, 270t
  swimming in, 882
  territorial behavior in, 1124, 1124f
  thermoregulation in, 713
  vitamin K requirement of, 898
  wings of, 884, 884f
Birds of prey, 711t, 713
Bird song, 1106, 1113, 1113f, 1119, 1124
Birth control, 1076–78, 1076f, 1077t, 1078f

I-4 Index

Birth control pill. *See* Oral contraceptives
Birthrate, 1150, 1157
　human, 1155
Birth weight, in humans, 447, 447f
Bison, 1210, 1239
1,3-Bisphosphoglycerate, 165f, 201f
Bithorax complex, **398**–99, 398f
bithorax gene, 398, 398f
Bittern, 711t
Bitter taste, 975
Bivalve mollusk, 652, **654**, 654f, 656–57, 657f
Bivalvia (class), 656–57, 657f
Black-and-white vision, 985
Blackberry, 761f, 849
Blackbird, 1238
Black cherry (*Prunus serotina*), 1230
Black Death. *See* Bubonic plague
Black Forest (Germany), 1219
Black locust, 748
Blackman, F.F., 200
Black walnut (*Juglans nigra*), 798, 798f
Black widow spider (*Latrodectus mactans*), 670f
Bladder
　swim. *See* Swim bladder
　urinary. *See* Urinary bladder
Bladder cancer, 414–15t, 417t
Bladderwort (*Utricularia*), 789
Blade, of leaf, 732f, 744, 744f, 748–49
"-Blast," 866
Blastocoel, 1084, 1086–88, 1086f, 1088f
Blastocyst, 384f, **385**, 393f, 426, 430, 1086, 1086f, 1097–98
Blastoderm, syncytial, **386**, 387f
Blastodisc, 1086, 1086f, 1089, 1089f
Blastomere, **383**, 383–84f, 385, 392, 397, **1084**, 1086
Blastopore, **619**, **626**, 1087–88, 1087–88f, 1090
　fate of, 626f, 627
Blastula, 384f, **385**, **619**, **626**, **1084**, 1086–87, 1086f, 1094f, 1097
Bleaching reaction, 986
Blending inheritance, 243, 248, **436**
Blights (plant disease), 553
Blind spot, 464–65, 465f
Blinking, 961
Blood, 856f, 863t, 866, 907, 910–11, 910–11f
　pH of, 32, 32f, 47, 932, 933f, 935, 935f
Blood cells, 910–11, 910f
Blood clotting, 40, 40f, 468, 898, 904t, 909, 910f, 911, 911f, 1043
Blood flow, 920–21
　resistance to, 912, 920
Blood fluke (*Schistosoma*), 644
Blood group, 128t
　ABO, **260**, 260f, 435, **1029**
　genetic variation in, 435
　Rh, **260**, 1029
Blood pressure, 914–15, 920–21, 1000f, 1007, 1048, 1054, 1057–58, 1058f
　baroreceptor reflex and, 920
　measurement of, 918, 918f
　sensing of, 971t, 974
Blood stem cells, 427
Blood transfusion, 260
Blood typing, 1029, 1029f
Blood vessel, 907–8
　characteristics of, 912–14, 912–13f
　innervation of, 965t
　walls of, 868, 912, 912f
Blood volume, 920, 1000f, 1007, 1054, 1057–58, 1058f
　regulation of, 920–21
Blowfly, 975f

Blueberry, 761f
Bluefin tuna, 1239
Blue-footed booby, 475f
Bluehead wrasse (*Thalassoma bifasciatum*), 1062f
Blue jay, 1170f
Blue-light receptor, in plants, 809, 809f
Blue-ringed octopus, 652f
Blue shark, 695
Blue whale, 1239, 1239f
BMR. *See* Basal metabolic rate
Boa constrictor, 465
Bobolink, 1116, 1117f
Bodybuilder, 1010
Body cavity
　of echinoderms, 679
　evolution of, 625, 625f
　kinds of, 625
Body color, in fruit fly, 268f
"Body language," 1121
Body plan
　animal, evolution of, 624–27
　of vertebrates, 856–57, 856–59f
Body position, 955
　sensing of, 974, 978–79, 978–79f
Body size
　circulatory and respiratory adaptations to, 915–17
　generation time and, 1145, 1145f
　in horses, 462, 462–63f
Bohr effect, 935
Bollgard gene, 336
Boll weevil (*Anthonomus grandis*), 336, 672f
Bombykol, 1118–19
Bone, 688, 855–56f, 863t, 864, 871
　of birds, 710
　compact, 865, 865f
　formation of, 688, 865
　remodeling of, 865
　spongy, 865, 865f
　structure of, 864–65, 865f
Bone cancer, 417t
Bone marrow, 910
Bone marrow stem cells, 429
Bony fish, 690, 695–97
Book lungs, 670, 670f
Boreal forest, 1191f
Boring sponge, 622t
Bormann, Herbert, 1189
Boron, in plants, 782, 782t
Borrelia burgdorferi, 549t, 555t
Borthwick, Harry A., 808
Bosmina longirostris, 1151f
Bottleneck effect, 439, 439f
Bottom-up effect, 1195–96, 1195–96f, 1213, 1242
Botulism, 357, 555t
Bovine somatotropin (BST), recombinant, 338, 338f
Bowerbankia, 623t
Bowhead whale, 1239
Bowman's capsule, 1053, 1053–55f
Box elder, 749
Box jellyfish, 641, 641f
Boyer, Herbert, 323
Boyle's Law, 931
Boysen-Jensen, Peter, 818
Brachial artery, 918, 918f
Brachiopoda (phylum), 623t, 635, 663, 663f
Brachydactyly, 249t
Brachyury gene, 501, 502f
Bradykinin, 994
Brain, 856f, 858f, 869t, 940f, 955t
　of amphibians, 956, 956f
　of birds, 956, 956f
　evolution of, 954–56
　of mammals, 956, 956f

　of reptiles, 956, 956f
　size of, 722, 726, 956, 956f
　of vertebrates, 954–56, 955f
Brain hormone, **1009**, 1009f
Branchial chamber, 924
Branchiostoma, 687, 687f
　Hox genes in, 399
Brassicaceae (family), 1169
Brassica juncea, 791
Brassica rapa, 824
Brassinolide, 825f
Brassinosteroid, 808, 816, 817t, **825**, 825f
BRCA genes, 414, 417t
Bread-making, 181, 603, 610
Bread mold, 607
Breastbone, keeled, 710
Breast cancer, 414, 414t, 417t, 422, 800, 1022
　hereditary, 414
Breast-feeding, 1000
Breath holding, 932
Breathing, 922
　in amphibians, 927, 927f
　in birds, 929, 929f
　in mammals, 930–36
　measurements of, 931
　mechanics of, 930–31, 931f
　negative pressure, 927
　positive pressure, 927, 927f
　regulation of, 932, 932–33f, 976
Breeding Bird Survey, 1238
Breeding season, 475, 1010, 1106, 1125, 1128, 1223
Briggs, Robert, 281
Briggs, Winslow, 820, 820f
Bright-field microscope, 83t
Bristlecone pine (*Pinus longaeva*), 590, 851
Bristle number, in *Drosophila*, 458, 458f
Bristleworm (*Oenone fulgida*), 660
Brittle star, 618f, 676, 679–80, 680f
Broca's area, 957f, 959–60
Broccoli, 847
Bronchi, **928**, 928f
Bronchiole, **928**, 928f
Brood parasite, 1113, 1113f
Brosimum alicastrum, 1142f
Brown, Robert, 88
Brown algae, 521f, 522, 563, **571**, 571f, 580, 757, 757f
Brown recluse spider (*Loxosceles reclusa*), 670f
Brown tree snake, 1240
Brusca, Richard, 528
Bryophyta (phylum), 582, 582f
Bryophyte, **582**
Bryozoa (phylum), 623t, 635, 662, 663f
BST. *See* Bovine somatotropin
Bt corn, 339
Bubble hypothesis, 68–69, 69f
Bubonic plague, 1155f
Buccal cavity, 925, 925f, 927f
Buckeye (*Aesculus*), 749
Bud, 747
　apical, 822f
　axillary, 732, 732f, **744**, 744f, 796, 796f, 836, 836f
　lateral, 822f, 827
　terminal, 732f, **744**, 744f
Budding
　asexual reproduction in animals, 237, 1062
　in protists, **565**
　virus release from cells, 536
　in yeast, 610, 610f
Bud primordium, 742f
Bud scale, 744, 827f
Bud scale scar, 744

Budworm, 336
Buffalo, 716
Buffer, 32, 32f
　in human blood, 32, 32f
Bug, true, 675t
Bulb (plant), 594, 743, 747, 747f, 849
Bulbourethral gland, 1068f, 1070
Bulimia, 903
Bullfrog, 963f, 1119
Bumblebee (*Bombus*), 844, 844f, 1110f
Bundle of His, **919**, 919f
Bundle scar, 744, 744f
Bundle-sheath cells, 203, 204f
Bunting, 1238
Burgrass, 762, 762f
Burial ritual, 726
Bushmaster, 709
Butter, 37t
Buttercup (*Ranunculus*), 742f, 840
　alpine, New Zealand, 485, 485f
Butterfly, 664f, 673, 673f, 675t, 798, 844, 1042, 1207, 1210
　Batesian mimicry in, 1171, 1171f
　eyespot on wings of, 503, 503f
　metapopulations of, 1144, 1144f
　range shifts in response to global warming, 1223
Buttress root, 743, 743f

C

CAAT box, 308, 370f
Cabbage, 851, 1169
Cabbage butterfly (*Pieris rapae*), 1169, 1169f
Cabbage palmetto, 748f
c-ABL gene, 416
Cacops, 700f
Cactoblastis cactorum, 1168
Cactus, 736, 751, 774
Cactus finch (*Geospiza scandens*), 10f, 454–55, 454f, 477, 483, 483f
Cadherin, **138**–39, 139f, 390–91
Cadherin domain, 391
Cadmium, 791–92
Caecilian, 698, 698t, 701, 701f, 883
Caecilia tentaculata, 701f
Caenorhabditis elegans, 646, 647f
　development in, 382f, 390, 390f, 400, 400f
　genome of, 348f, 352
　small RNAs in, 374
CAF, 538, 539f
Caffeine, 798
Caiman, 703t, 709
Calciferol. *See* Vitamin D
Calcitonin, 376, 376f, 994t, 1005
Calcitonin gene-related peptide (CGRP), 376, 376f
Calcium
　blood, 1005–6, 1006f
　homeostasis, 1005–6, 1006f
　intestinal absorption of, 1006, 1006f
　IP_3/calcium second-messenger system, 998–99, 999f
　in muscle contraction, 132, 878–79, 878f, 881
　in plants, 772, 782, 782t
　reabsorption in kidneys, 1006, 1006f
　release from bone, 1005–6, 1006f
　as second messenger, 132, 133f, 135
　in synapse, 948, 948f
α-Calcium-calmodulin-dependent kinase II, 1109
Calcium carbonate, 641, 654, 660, 978
Calcium channel, 114, 132
Calcium phosphate, 864
Calico cat, 271, 271f

I-5

California condor (*Gymnogyps californianus*), 1247
Callus (plant), 822, 850, 850f
Callus (skin), 860
Calmodulin, 39t, 132, 133f, 356, 999, 999f
Caloric intake, 903
  life span and, 402
Calorie, 903
Calorimeter, 903
Calvin, Melvin, 200
Calvin cycle, **187**, 187f, 189, **200**–202, 200–204f, 785
  carbon fixation in, 200–201
  discovery of, 200
  output of, 202
  reactions of, 201–2, 201f
Calyx, **840**
Cambium
  cork, **731**, 731f, 735, 742, 742f, 746, 746f, 822
  vascular, 731, 731f, 737, 741, 745, 745f
Cambrian explosion, **630**, 630f
Camel, 1050, 1210, 1228, 1228f
Camouflage, 440, 714, 1122, 1122f, 1169f, 1170
cAMP. *See* Cyclic AMP
Campbell, Keith, 394, 424
CAM plants, **204**, 204f, 774
Camptodactyly, 249f
Canada lynx (*Lynx canadensis*), population cycles of, 1153f, 1154
Canadia, 630f
Canaliculi, 864, 865f
Canary grass (*Phalaris canariensis*), 818
Cancer, 135
  of bladder, 414–15f, 417t
  of bone, 417t
  of breast, 414, 414t, 417t, 422, 800, 1022
  causes of, 414–15
  cell cycle and, 416–19, 416f, 417t
  cell cycle control in, 223–24, 223–24f
  of cervix, 414t, 541
  of colon, 414, 414t, 417t, 419f, 422–23, 897
  death rate from, 419, 419f
  definition of, 413
  epigenetic events and, 418–19
  growth factors and, 224
  of head and neck region, 415t, 417t
  incidence in United States, 414t
  invasion of surrounding tissue, 413f
  of kidney, 414t
  of liver, 414–15t, 541
  of lung, 413f, 414–15, 414–15t, 417t, 420–21, 420–21f, 423
  multistep nature of, 419, 419f
  mutations and, 413–23, 419f
  of nervous system, 414t
  of oral cavity, 414t
  of ovary, 414t, 417t
  of pancreas, 414t, 417t
  preventing spread of, 423
  prevention of start of, 422–23
  of prostate, 414t, 800
  renal cell, 417t
  of salivary gland, 417t
  of scrotum, 415
  of skin, 191, 415t, 1221
  smoking and, 223, 420–21, 420–21f
  of stomach, 414t
  T cells in surveillance against, 1022, 1022f
  of thyroid, 417t
  treatment of, 422–23, 423f, 1022
    gene therapy, 333t
  viruses and, 540–41

Candida, 614
Candida milleri, 610
Canine teeth. *See* Cuspid
Cannabalism, ritual, 542
CAP. *See* Catabolite activator protein
5′ Cap, mRNA, **309**, 309f, 316
CAP-binding site, 369f
Capillary, 908, **912**, 912f
Capillary action, 29, 29f
Capsid, viral, **532**, 532f
Capsule
  of bacteria, 84, 84f, 550
  surrounding organs, 862
Captive breeding, 1243, 1247, 1247f
Carapace
  of crustaceans, 668
  of turtle shell, 708
Carbohydrates, 37, 55–58
  catabolism of, 160
  disassembly of, 37, 37f
  functions of, 37t
  kilocalories per gram, 54
  structural, 58, 58f
  structure of, 37t
Carbon
  chemistry of, 36–37
  isotopes of, 21, 21f
  in plants, 782, 782t
Carbon-13, 21f
Carbon-14, 21f
Carbon cycle, 557, 1186, 1186f
Carbon dioxide
  atmospheric, 785, 924, 1222–24, 1222–23f
  in carbon cycle, 1186, 1186f
  diffusion from tissue, 936, 936f
  diffusion into alveoli, 936, 936f
  as electron acceptor, 163
  entry into plants, 772–73
  from ethanol fermentation, 181
  from Krebs cycle, 169–70, 171f
  partial pressure in blood, 930, 932
  from photorespiration, 203
  prebiotic chemistry, 67f
  from pyruvate oxidation, 168
  transport in blood, 115, 149, 909–10, 910f, 922, 935–36, 936f
  use in photosynthesis, 187–89, 187f, 189f, 200–201, 200–201f, 782, 785
Carbon fixation, 187, **189**, 200–**201**
Carbonic acid, 32, 32f, 149, 932, 935–36, 936f
Carbonic anhydrase, 149, 936f
Carbonyl group, 36f
Carboxyl group, 36f
Carboxypeptidase, 153
Carcinogen, **414**, 414f, 415, 415t
  in workplace, 415t
Carcinoma, 413f, **414**
Cardiac cycle, **918**–19, 918–19f
Cardiac glycoside, 1169–70, 1170–71f
Cardiac muscle, 856f, 867t, 868
Cardiac output, **920**
  exercise and, 920
Cardiovascular control center, 920
Cardiovascular disease, 921f, 921f
Caribou, 1211
Carnivora (order), 719t
Carnivore, 180f, 528, 528f, 618t, 719t, **888**, 889–90, 1190f, 1193f
  digestive system of, 899f
  saber-toothed, 516, 516f
  teeth of, 715, 715f
  top, 1193
  in trophic cascade, 1194, 1194–95f
Carnivorous fungus, 603f
Carnivorous plant, 788–89, 788–89f

Carotene, 37t, 985
Carotenoid, 191–**93**, 191f, 193f, 566, 568, 571, 573, 845, 852
Carotid body, 932
Carotid sinus, 974
Carpel, 245f, 589, 593, 593f, **594**, 595, 595f, 761, 838, 839f, **840**, 845, 848f
Carrier (gene disorder), 261, 261f, 274
Carrier protein, 112, **115**, 115f, 122t
Carroll, Sean, 395f
Carrot, 743, 851
Carrying capacity, **1150**–52, 1150f, 1154–55, 1157
Cartilage, 688, 863t, **864**–65, 864f, 871
  articular, 865, 872f
Cartilaginous fish, 689–90, 691–92f, 864, 1049
Cartilaginous joint, 872
Casein, 39t
Casparian strip, **741**, 741f, 768, 770, 770f
Cassava (*Manihot esculenta*), 799t
Caste, insect, **1133**
Castor bean (*Ricinus*), 745f
Cat, 717, 719t, 871f, 888, 1067, 1072
  coat color in, 256, 271, 271f
  predation by, 1168
Catabolism, **155**, **160**
  evolution of, 182
Catabolite activator protein (CAP), 364f, 366f, 368–69f, **369**
Catalase, 94
Catalysis, 37, **148**, 148f, 156
Catalyst, 26, 148, 148f
Catbird, 1238
Catecholamine, 950, **993**, 998
Caterpillar, 1169, 1169f
Catfish, 975, 980
Cation, **21**, 114, 943, 943f
Cattail, 594
Cattle, 716, 719t, 888–89, 898, 898f, 1072, 1173
  transgenic, 338
Cattle egret, 1141, 1141f
Caudal fin, 695
Caudata. *See* Urodela
Caudipteryx, 517f, 712, 712f
Causation
  proximate, 1106
  ultimate, 1106
Cave bear, 717t
Cavefish, 504, 505f, 977
Cave painting, 726, 726f
Cayuga Lake food web, 1192, 1192f
CCK. *See* Cholecystokinin
CCR5 receptor, 536, 537f, 538–39, 539f
CD4 coreceptor, 536, 1020, 1020t, 1033, 1033f
CD8 coreceptor, 1020, 1020t, 1033f
Cdc2 kinase, 220
Cdk. *See* Cyclin-dependent protein kinase
cDNA library, **324**, 324f
Cech, Thomas, 66, 151
Cecum, 465, **889**, 889f, 897–98, 897f, 899f
Cedar, 590, 1239
Cedar Creek experimental fields, 1197, 1197f
ced genes, 400, 400f
Celecoxib, 995
Celery, 37t, 736
Cell(s), 857f
  characteristics of, 80–83
  earliest, 70–71, 70–71f
  in hierarchical organization of living things, 2, 3f

origin of, 68–69, 69f
shape of, 99–100
size of, 80–81, 82f
  in prokaryotes, 546–47, 547t
  visualizing structure of, 82–83
Cell adhesion, 137–40, 390
Cell adhesion protein, 111, 111f
Cell body, of neuron, 869, 869f, **940**–41, 940–41f
Cell-cell interactions, 125–40
  in development, 391–92, 1092–93, 1092–93f
Cell cycle, **213**–17
  cancer and, 416–19, 416f, 417t
  duration of, 213, 222
  growth factors and, 222
Cell cycle control
  architecture of control system, 219
  in cancer cells, 223–24, 223–24f
  checkpoints, 219–21, 219–21f, 416
  general strategies of, 219, 219f
  molecular mechanisms of, 220–22, 220–22f
  in multicellular eukaryotes, 221
Cell death program, 400
Cell division, 207–24
  in animal cells, 209f, 213
  in cancer cells, 413
  in eukaryotes, 547
  Hayflick limit, 401
  in plant cells, 213
  in prokaryotes, 207–9f, 208–9, 547, 547t
  protein assemblies in different organisms, 209f
  in protists, 209f
  in yeast, 209f
Cell division amplification cascade, 135
Cell identity, 136
Cell junction, **137**–38, 137f
Cell-mediated immune response, **1018**, 1021–22, 1021–22f, **1031**–32
Cell membrane, 102t. *See also* Plasma membrane
  of archaebacteria, 519
Cell migration, in development, 384f, 385, 391, 1087
Cell plate, 217f, **218**, 218f
Cell signaling
  between cells
    by direct contact, 127, 127f
    endocrine signaling, 127, 127f
    paracrine signaling, 127, 127f
    synaptic signaling, 127, 127f
  intracellular, **128**–29, 128–29f, 128t
    amplification of signal, 134–35, 134–35f
  in cancer, 416, 417t
  growth factors in, 222, 222f
  initiation of, 132, 132–33f
  receptor proteins and, 126
Cell surface
  of prokaryotes, 550, 550f
  of protists, 565
Cell surface marker, 80, 108f, 109, 109f, 111, 111f, 128t, 136
Cell surface receptor, 40, 126, 128t, **130**–31, 130f, 134f. *See also* Receptor protein
  chemically gated ion channels, 130, 130f
  enzymic, 130, 130f
  G-protein-linked receptor, 130f, 131
Cell theory, **15**, 80–**81**
Cell-to-substrate interactions, 391
Cellular counterattack, **1014**–16
Cellular immune response, **1031**–32

I-6   Index

Cellular movement, 98–100
  crawling, 100
  energy for, 160
  swimming, 100
Cellular organization, as characteristic of life, 63, 63f, 80–81
Cellular respiration, 160, 177, 182, 922
Cellular slime mold, 575–76, 576f
Cellulase, 557
Cellulose, 56, 58, 58f, 87, 101, 218, 388, 716
  breakdown of, 557, 603, 889, 898, 1186
Cell wall, **84**, 87, **523f**, 618t
  of archaebacteria, 71, 519, 548
  of bacteria, 548
  of eukaryotes, 87, 87f, 90t, 101, 101f, 102t
  of fungi, 600–601
  of plant cells, 58, 58f, 117, 117f, 119, 818, 820–21, 825
  primary, **101**, 101f
  of prokaryotes, 84, 84f, 102t, 546f
  secondary, **101**, 101f
Cementum, 891f
Centimorgan, **268**, 344
Centipede, 527, 668t, 671, 672f
Central chemoreceptor, **932**, 933f, 976
Central Dogma, 51, 302–**3**, 303f
Central nervous system, 856f, 869, **940**, 940f, 954–62
Central sulcus, 957, 957f
Central vacuole, **86**, **101**, 101f
Centriole, 86f, **99**, 99f, 102t, 209f, **214**, 215, 215–16f, 562, 601
Centromere, 210f, **212**, **214**, 214–16f, 215–17, 232–33, 233–35f, 236, 350
Centrosome, **99**
Centrum, 688f
Cephalaspidomorphi (class), 689, 691f, 692, 693t
Cephalization, **624**, 954f
Cephalochordata (subphylum), 687, 687f
Cephalopoda (class), 654f, 657, 657f
Cephalothorax, **666**
Ceratium, 568f
Cercariae, **644**, 644f
Cereal grains, genome analysis of, 352, 353f
Cerebellum, 954–55f, **955**, 955t, 957f
Cerebral cortex, 955t, **957**–58, 957–58f
Cerebral hemisphere, 957, 957f
  dominant hemisphere, 959–60
Cerebrospinal fluid, pH of, 932, 932–33f, 976
Cerebrum, 954–57f, 955t, **956**
Cervical cancer, 414t, 541
Cervical cap, 1076, 1077t
Cervical nerves, 954f
Cervix, 1072, 1072–73f, 1101f
Cestoda (class), 645
Cetacea (order), 719t
cf gene, 263
cGMP. See Cyclic GMP
CGRP. See Calcitonin gene-related peptide
Chaetognatha (phylum), 623t
Chagas disease, 567
Chameleon, 708
Chamguava schippii, 1142f
Channel protein, 112, 113f, 122t
Chaparral, 1141f, 1208, 1208f
Chaperone protein, 45–**46**, 46f
Character, **243**
Character displacement, **482**, **1166**, 1166f

Charales, 526, 526f, 580
Charcot-Marie-Tooth disease, 269f, 411
Chargaff, Erwin, 285, 285t
Chargaff's rules, **285**
Checkpoint, cell cycle, 219–21, 219–21f
Cheese, 603
Cheetah, 1210
Chelicerae, 668t, **670**
Cheliped, 669f
Chelonia (order), 703t, 708
Chemical bond, **25**. *See also specific types of bonds*
Chemical carcinogenesis theory, **414**
Chemical defenses
  of animals, 1170
  of plants, 1169
Chemical digestion, 888
Chemical evolution, 19, 66–67
Chemically gated ion channel, 130
Chemical messenger, 992–95
Chemical reaction, **26**
  activation energy, 148, 148f
  energy changes in, 147, 147f
Chemical synapse, **127**, 948
Chemiosmosis, **175**, 175f, 177, 194, 199, 199f
Chemoautotroph, 553
Chemoheterotroph, 553
Chemokine, **538**
  HIV-inhibiting, 538, 539f
Chemoreceptor, **970**, **975**
  central, **932**, 933f, 976
  internal, 976
  peripheral, 932, 933f, 976
Chemosynthesis, 1215
Chemotropism, 811
Cherry (*Prunus cerasifera*), 746f, 748, 761f, 849
Chesapeake Bay, 1214
Chestnut blight (*Cryphonectria parasitica*), 609
Chest pain, 921
Chewing, 888, 890
Chewing appendage, 668
"Chewing the cud," 898
Chiasmata, 230, **232**, 233–34f, 236, 267
  terminal, 233
Chickadee, 713
Chicken, 210t, 711t, 1084f
Chicken pox, 533t
Chief cells, 892, 893f, 901f
Childbirth, 1000, 1101, 1101f. *See also Uterine contractions*
Chilopoda (class), 671
Chimera, **323**, **393**, 393f
Chimney sweep, 415, 415f
Chimpanzee (*Pan*), 14, 466, 720–21, 721f, 1102, 1102f
  chromosome number in, 210t
  cognitive behavior in, 1114–15, 1114–15f
  genome of, 495, 495f
  language in, 1121
Chinchilla, 1239
Chipmunk, 887f
Chironex fleckeri, 641f
Chiroptera (order), 719t
Chitin, 58, 58f, 87, 101, 600–601, 664, 665f, 666, 673, 871, 871f
Chitinase, 58
Chiton, 652, 654f, 656
Chlamydia, 549f, 555t
  heart disease and, 556
  sexually transmitted disease, **556**, 556f
Chlamydia trachomatis, 555t
Chlamydomonas, 573, 573f
Chloramphenicol, 520t, 551
Chlordane, 1218

Chlorella, 194, 573
Chlorenchyma, 736, 750
Chloride, 25, 25f
  in cytoplasm and extracellular fluid, 943t, 1044
  reabsorption in kidney, 1048f, 1054–56, 1055f, 1058f
Chloride channel, 114, 950
Chloride shift, 936
Chlorinated hydrocarbons, 1218
Chlorine, in plants, 782, 782f
  deficiency of, 783f
Chloroflexi, 549f
Chlorofluorocarbons, 1221
Chlorokybales, 526f
Chlorophyll, 53, 53f, 187, 191, **192**–93, 195–96, 195f, 1203, 1203f
  action spectrum of, 192, 193f
  structure of, 192, 192f
Chlorophyll *a*, 191–92, 191–92f, 194, 196, 198–99, 458, 553, 566, 571, 573
Chlorophyll *b*, 191, 191–92f, 566, 573
Chlorophyll *c*, 568, 571
Chlorophyta (phylum), 526, 526f, 564f, 573, 573f
Chloroplast, 87f, 90t, **96**, 97f, 102t, 186–87, 186f, 189
  diversity of, 563
  DNA of, 96–97, 563, 572
  energy cycle in, 202, 202f
  of euglenoids, 566, 566f
  genetic code in, 305
  genome of, 352
  origin of, 72–73, 73f, 97, 97f, 521, 521f, 563
  structure of, 187, 187f
Choanocyte, 622t, 636, **636**, 637f
Choanoflagellate, 525, 525f, 574, 574f, 637f
Cholecystokinin (CCK), 900, 900t, 901f
Choleochaetales, 526f
Cholera, 85f, 535, 554, 555t, 1224
Cholesterol, 37t, 53, 904
  blood, 921
  in cardiovascular disease, 921
  dietary, 921
  hormones derived from, 996f
  in membranes, 80f, 107, 108f
  structure of, 53f
  uptake by cells, 119
Cholesterol oxidase, 336
Cholesterol receptor, 264t
Chondrichthyes (class), 689, 691–92f, 693t, 695, 695f, 1044, 1049
Chondrocytes, 863t, **864**, 864f
Chondrodysplasia punctata, 269f
Chondroitin, 864
Chondromcyes, 549f
Chordata (phylum), 511f, 621f, 622t, 629f, 683–85, 685f
Chordate, 622t, 634–35f, **684**–85, 684–85f
  characteristics of, 684
  cleavage in, 1084
  craniate, **688**
  gastrulation in, 1087
  jointed appendages of, 684
  nonvertebrate, 686–87, 686–87f
  segmentation in, 527, 527f, 627, 627f, 684
Chorioderemia, 269f
Chorion, **702**, 702f, 715f, 718, 1066–67, **1096**, 1096–97f, 1098
Chorionic frondosum, 1096–97f, 1098
Chorionic membrane, 1096
Chorionic villi sampling, **274**

Chromatid, **212**, 214. *See also* Sister chromatid(s)
Chromatin, **89**, **211**, 214, 373, 373f
Chromatin remodeling complex, 373
Chromatophore, 657
Chromosomal rearrangement, 410t, 411, 497
  rate of, 495
Chromosomal theory of inheritance, **265**
Chromosome, 37–38t, **87**, **89**, 90t, 102t, 214, 373. *See also Karyotype*
  artificial. *See Bacterial artificial chromosome; Yeast artificial chromosome*
  of bacteria, 301f
  coiling of, 211, 211f
  composition of, 211
  condensation of, 89f
  discovery of, 210
  Drosophila, 361f
  of eukaryotes, 209–12, 211f, 547
  genes on, 265–74
  harlequin, 236f
  homologous, **212**, 212f, 214, 216f, 230, 230–31f, 232, 233–34f, 267, 409, 1068
  human. *See Human chromosomes*
  Philadelphia, 416
  of prokaryotes, 547, 547t
  structure of, 211–12, 211f
Chromosome abnormality, activation of proto-oncogene, 416
Chromosome number, 210, 210t, 228, 230f, 497, 497f
  human, alterations in, 272–73, 272–73f
Chromosome puff, 361f
Chronic granulomatous disease, 269f, 333t
Chronic myelogenous leukemia, 416
Chrysalis, **674**
Chrysanthemum, 835
Chrysolaminarin, 571
Chrysophyta (phylum), 571
Chthamalus stellatus, 1164, 1164f
Chylomicron, 896, 896f
Chyme, 893–95
Chymotrypsin, 895, 900t
Chytrid, 600, 600f, 602, 606, 606f, 614
Chytridiomycosis, 614, 614f
Chytridiomycota (phylum), 605–6, 605–6f, 605f
Cicada, 674, 1009f
Cichlid fish
  Lake Barombi Mbo, 481
  Lake Victoria, 484, 484f, 1241, 1241f, 1246
  pike cichlid, 448–49, 448–49f
Ciconiiformes (order), 711t
Cigarette smoking. *See Smoking*
Cilia, 86f, **100**, 100f, 102t, 547. *See also Ciliate*
  of ctenophores, 641
Ciliary muscle, 984
Ciliate, 305, 520f, 565f, 569–70, 569–70f
Cinchona officinalis, 799t, 800
Circadian clock, in plants, 813, 813f
Circulation, 907–21
Circulatory system, **625**, 857t, 858f, 907–21
  of amphibians, 698, 916, 916f
  of annelids, 659, 659f, 908
  of arthropods, 665f, 667
  of birds, 713, 916–17, 917f
  closed, **625**, 655, 657, 659, 689f, **908**–9, 908f

I-7

Circulatory system—*Cont.*
  of fish, 690, 707f, 915, 915f
  functions of, 908–9, 909f
  of invertebrates, 908, 908f
  of mammals, 916–17, 917f
  of mollusks, 654–55, 657
  open, **625**, 667, **908–9**, 908f
  of reptiles, 707, 707f, 916
  of vertebrates, 908–9, 909f, 915–17, 915–17f
Cirripedia (order), 669
Cisternae, of Golgi body, **92**
Cisternal space, **90**, 91, 91f
Citrate, 170, 171f
  stimulation of phosphofructokinase, 177, 177f
Citrate synthetase, 171f, 177, 177f
Citric acid, commercial production of, 603
Citric acid cycle. *See* Krebs cycle
Clade, **513**, 525, 628
Cladina evansii, 611f
Cladistics, **512–14**, 513f
  molecular, 525
Cladoceran, 1151f
Cladogram, **513–14**, 513f
Cladophyll, **747**, 747f
Clam, 635, 652, 654, 656, 657f, 1215
Clamworm, 660
Clark's nutcracker, 1111, 1111f
Class (taxonomic), **511**
Classical conditioning, **1110–11**
Classification
  of animals, 527–28, 527–28f, 620, 621f, 622–23t, 628–30
  impact of molecular data on, 525–28
  of organisms, 510–11, 511f
  of plants, 526, 526f
  of prokaryotes, 548
  of protists, 525, 564, 564f
  systematics and, 514
  of viruses, 524
Clathrin, 109t, 119
Claw, 716
Clay, origin of life within, 65
Clean Air Act, 457, 1219
Cleaner fish, 1119, 1119f
Clear-cut harvesting of timber, 1236
Cleavage, **383**, 383–84f, 619t, 626, 626f, 1074f, 1084–86, 1084–86f, 1085t
  in amphibians, 1084, 1084f
  in birds, 1086
  in chordates, 1084
  in fish, 1084
  holoblastic, **1084**, 1084f, 1086
  in humans, 1097
  in mammals, 1086, 1086f
  meroblastic, **1086**, 1086f
  radial, **626**, 626f
  in reptiles, 1086
  spiral, **626**, 626f
Cleavage furrow, 217–18f, 218
Cleft palate, 269f
Clematis, 811
Clements, F. E., 1162, 1178
Climate, 1206–7. *See also* Global climate change; Global warming
  biomes and, 1209, 1209f
  effects on ecosystems, 1204–7
  elevation and, 1207, 1207f
  El Niño and, 1213, 1213f
  latitude and, 1206–7f, 1207
  microclimate, **1207**, 1237
  regional, 1206–7
  selection to match climatic conditions, 441, 441f
  solar energy and, 1204, 1204f
  species richness and, 1198, 1198f

Climax community, 1178, 1180
Clingman's Dome, Tennessee, 1219f
Clitellum, **660–61**
Clitoris, 1072, 1072f
Cloaca, 888f, 889, 897, 1065
Clonal deletion, 1026
Clonal selection, 1026–**27**, 1027f
Clonal suppression, 1026
Clone, 1027. *See also* Gene clone
Clone-by-clone sequencing, 347, 347f, 351
Cloning, 320–24, 326, 326f
  DNA libraries, **324**, 324f
  host/vector systems, 321–22
  of humans, 425
  reproductive, 424–25, 424–25f, 429f, 430
  of sheep, 394, 394f, 424–25, 424–25f
  therapeutic, **428–30**, 428–29f
  tools required for, 320
Cloning vector, **321**–22
  phages, 322, 322f
  plasmids, 321–26, 322f, 326f
  using to transfer genes, 323, 323f
  virus, 325, 326f
  yeast artificial chromosome, 323
Closed circulatory system, **625**, 655, 657, 659, 689f, **908**–9, 908f
Clostridium, 549f, 557
Clostridium botulinum, 357t, 549f, 555t
Clotting factors, 910
Clover, 835
Clownfish, 1161f
Club moss, 584, 585t, 586, 586f
CMT gene, 411
c-myc gene, 417t
Cnidaria (phylum), 619t, 621f, 622t, 624
Cnidarian, 622t, 634–35f, 638–41, 638–41f, 908
  classes of, 640–41
  digestive cavity of, 888, 888f
  nervous system of, 954, 954f
  reproduction in, 1062
Cnidocyte, 622t, **638–39**, 639f
Coacervate, **68–69**
Coactivator, 371, 372f, 373
Coal, 1186, 1186f
Coal tar, 415
Coastal redwood (*Sequoia sempervirens*), 590
Coat color, 714
  in arctic fox, 256f
  in cats, 256, 271, 271f
  in dogs, 258, 258f
  in mice, 255, 440f
  in rabbits, 256
Coated pit, 109t, 118f, 119
Coatimundi, 1195
Cobra, 709
Cocaine, 798, 952, 952f
Coccidioides posadasii, 610
Coccoloba coronata, 1142f
Coccus, **546**
Coccyx, 1098
Cochlea, 980, 981f
  frequency localization in, 980–82, 982f
  transduction in, 980
Cochlear duct, 978f, 980, 981f
Cochlear nerve, 978f
Cockatoo, 711t
Cockroach, 898
Coconut (*Cocos nucifera*), 759, 762, 762f
Coconut milk, 822–23
Coconut oil, 53–54
Cocoon, earthworm, 661
Coding strand, **306**, 306–7f
Codominance, 256, **260**, 260f
Codon, **304**, 305t, 310, 315f
  nonsense, **310**, 312f

Coelacanth (*Latimeria chalumnae*), 697, 697f, 699
Coelom, 620, **625**, 625f, 633, 653f, 689f, 1090, 1091f
  extraembryonic, 1096
  formation of, 626f, 627
Coelomate, 621f, 622t, **625**, 625f, 627, 634, 634f, 651–80, 684
Coelophysis, 517f
Coelurosaur, 712
Coenzyme, **153**, 156
Coevolution, **1169**
  of insects and plants, 844
  of plants and animals, 801, 831, 831f, 1169, 1172, 1172f, 1174
  symbiosis and, 1172
Cofactor, 145f, **153**, 153f
Cognitive behavior, 1114–15, 1114–15f
Cohen, Stanley, 319, 323
Cohesin, 216–17
Cohesion, 28t, **29**, 768, 772
Coiling, of gastropod shell, **656**
Cold receptor, 971v
Cold tolerance, in plants, 815
Cole, Lamont, 1192, 1192f
Coleochaetales, 526, 526f
Coleoptera (order), 672f, 675t
Coleoptile, 730, 731f, 764f
Coleorhiza, 764f
Coleus, 730f
Collagen, 39t, 40, 40f, 102, 102f, 391, 402, 688, 862, 862f, 864
Collar cell. *See* Choanocyte
Collared flycatcher, 478, 478f, 1148, 1148f
Collecting duct, 1048f, 1052–53f, **1053**, 1055f, 1056–57
Collenchyma, 736, 745–46f
Collenchyma cells, **736**, 736f
Colon. *See* Large intestine
Colon cancer, 414, 414f, 417t, 419f, 422–23, 897
Colonial flagellate hypothesis, for origin of metazoans, **630**
Colonization, 1240
  of empty habitats, 1144
  of island, 1200, 1200f
Colony, bacterial, 546
Colorado potato beetle, 336
Coloration
  cryptic, 1170, 1170f
  warning, 1170, 1171f
Color blindness, 249t, 269f, **987**
Colorectal cancer. *See* Colon cancer
Color vision, 720, 969f, 985–86, 985f
Colostrum, 1102
Colubrid, 709
Columbiformes (order), 711t
Columella root cap, 739, 739f
Columnar epithelium, 856f
  pseudostratified, 861t
  simple, 860, 861t
Combination therapy, for HIV, **538**, 539f
Comb jelly, 623t, 624, 638, 641, 641f
Commensalism, 557, 604, **1172–73**, 1173f
Commitment (development), **1093**
Common ancestor, 512–14, 515f, 520f, 525
Common bile duct, 895, 895f
Common name, 510f
Communicating junction, 137–38f, **140**, 140f
Communication
  behavior and, 1119–21
  level of specificity in, **1118–19**
  long-distance, 1118–19
  in social groups, 1120–21, 1120–21f

Community, **2**, 3f, 1161–62, 1162f
  across space and time, 1162–63, 1163f
  climax, 1178, 1180
  concepts of, 1162–63
  energy budget for, 1192, 1192f
  fossil records of, 1163
Community ecology, 1161–80
Community importance, 1243
Compact bone, 865, 865f
Companion cells, 738, 738f, 776
Comparative anatomy, 13, 13f
Comparative genomics, 352, 353f, 491–96, 492–93t
  origins of genomic differences, 497–501
Compartmentalization
  in eukaryotes, 522, 547
  in prokaryotes, 547, 547t
Competence, in plants, **832–33**
Competition, 1142, 1161
  among barnacle species, 1164, 1164f
  effect of parasitism on, 1176
  experimental studies of, 1167, 1167f
  exploitative, **1164**
  interference, **1164**
  interspecific, **1164**, 1164f, 1167, 1167f
  reduction by predation, 1176, 1176f
  reproductive, 1126–27
Competitive exclusion, **1165**, 1165f, 1176
Competitive inhibitor, **152**, 153f
Complementary base pairing, **50**, 50f, 285, 286f, **288**, 374, 374f. *See also* Base-pairs
Complement system, 1016, 1016f, 1023–25
Complete digestive system, **645**, 645f
Complete flower, 840
Complete metamorphosis, 674
Complexity, as characteristic of life, 62–63
Composite gene, 1026, 1026f
Compound, **25**
Compound eye, 504, 505f, 665–66f, 666–67
Compound leaf, **749**, 749f
Compound microscope, **82**
Compsognathus, 712
Concentration gradient, 115–16
Concurrent flow, 926, 926f
Condensation, of chromosomes, **214**, 216f
Conditioned stimulus, 1111
Conditioning, 1110
  classical, **1110–11**
  operant, **1110–11**
Condom, 556, 1076, 1076f, 1077t
Cone (eye), 869t, 969f, 971t, 985–87, 985–86f
Cone (plant), 590–92, 591f, 760, 760f
Cone shell, 656
Confocal microscope, 83t
Conformer, **1138**
Congestive heart failure, 917
Conidia, **609**, 609f, 614f
Conidiophore, **609**
Conifer, 585t, 590–91, 590–91f, 594f, 844, 852
Coniferophyta (phylum), 585t, 590–91, 590–91f
Coniferous forest, 590, 1178f
Conjugation
  in bacteria, **407**, 552
  gene transfer by, 407, 407f
  in ciliates, 570, 570f
Conjugation bridge, 407, 407f

Conjugation map, of *Escherichia coli*, 408f
Connective tissue, 856, 856f, 862–66
  dense, **862**, 863t
  dense irregular, **862**
  dense regular, **862**
  loose, 856f, **862**, 863t
  special, **862**, 864–66
Connective tissue proper, **862**
Connell, J. H., 1164, 1164f
Connexon, 140, 140f
Consciousness, 959
Consensus sequence, **347**
Conservation biology, 1227–48
Conservation of synteny, **495**
Conservative replication, 288
Conspecific male, 1118
CONSTANS gene, of *Arabidopsis*, 835
Constitutive heterochromatin, 350
Consumer, 618, **1190**
  primary, **1190**, 1190f, 1192f, 1193
  secondary, **1190**, 1190f, 1192f
  tertiary, 1190f, 1192f
Contact dermatitis, **1036**
Contagion hypothesis, for origin of sex, 237
Contig, **344**, 347, 351
Continental shelf, 1191t, 1214, 1214f
Continuous variation, **255**, 255f
Contraception. *See* Birth control
Contraceptive implant, 1077t
Contractile protein, 40
Contractile root, 743
Contractile vacuole, 117, 119, 569f, 570, 1046
Control experiment, **6**
Conus arteriosus, 915–16, 915–16f
Convergent evolution, **467**, 467f, 504, **512**, 1208
Cooke, Howard, 401
Cookeina tricholoma, 600f
Cooksonia, 584, 584f
Coot, 711t
Copepoda (order), 669, 669f, 898
Copper, 934
  in plants, 782, 782t
    deficiency of, 783f
Copperhead, 709
Coprophagy, **898**
Copulation, 1066
  extra-pair, **1128**–29, 1129f
Copulatory organ, 475
Coral, 509f, 617, 622t, 624, 638, 641
Coral reef, 641, 1214
Coral snake, 709
Corepressor, 373
Cork, 746f
Cork cambium, **731**, 731f, 735, 742, 742f, 746, 746f, 822
Cork cells, 746
Corm, 747, 849
Corn (*Zea mays*), 203, 358f, 510f, 614, 614f, 731, 731f, 734f, 743, 745f, 756, 759, 759f, 763, 764f, 774, 841, 846, 851
  artificial selection in, 458, 458f
  chromosome number in, 210t
  genome of, 353f
  grain color in, 257, 257f
  oil content of kernels, 458, 458f
  transgenic, 336–37, 339
Cornea, 983, 984f
Corn oil, 37t, 53
Corolla, **840**
Corona, 648
Coronary artery, **918**, 920, 921f
Coronavirus, 533t
Corpora allata, 1009, 1009f

Corpora cavernosa, 1070, 1070f, 1072
Corpus callosum, 955t, 957, 957f
Corpus luteum, 1074f, 1075, 1075f, 1098, 1100
Corpus spongiosum, 1070, 1070f
Corrective lenses, 984f
Correns, Karl, 265
Cortex (plant), 730f, **741**, 742f, 745
Cortical nephron, 1052–53
Corticosteroid, **993**, 1007, 1036, 1101
Corticotropin-releasing hormone (CRH), 1003, 1003f
Cortisol, 129, 825f, 993, 995t, 996f, 1007
Corynebacterium diphtheriae, 555t
Costa Rica, biosphere reserves in, 1248, 1248f
Cost of reproduction, **1148**–49, 1148–49f, 1154
Cotton, 614
  transgenic, 336–37
Cotton bollworm, 336
Cottonwood (*Populus*), 477, 746f, 762, 845
Cotyledon, 388, 389f, 596f, 756f, 759–60f, 763, 764f
Countercurrent exchange, 923f
Countercurrent flow, **925**–26, 925–26f
Countercurrent heat exchange, **909**, 909f
Countercurrent multiplier system, 1056
Countertransport, **122**
Coupled transport, 120–22, 121f, 122t
Courtship behavior/signaling, 474, 475f, 479, **1118**–19, 1118–19f, 1126f
  of *Anolis* lizards, 479, 479f
Courtship song, 1113, 1113f
Covalent bond, 25, **26**–27, 26f
Cow(s). *See* Cattle
Cowper's gland. *See* Bulbourethral gland
Cowpox, 1018, 1018f
Cox. *See* Cyclooxygenase
Coyote, 1143, 1167, 1247
CpG island, 419
CpG repeat site, 419
C₃ photosynthesis, **200**, 203–4
C₄ photosynthesis, **203**–4, 204f
Crab, 37t, 622t, 668, 871f, 882, 1123, 1123f, 1190
Crane, 711t
Cranial cavity, 856f
Craniate chordate, **688**
Crassulaceae (family), 204
Crassulacean acid pathway. *See* CAM plants
Crawling, cellular, 100
Crayfish, 668
Creatine phosphate, 881
Creationism, 8, 64
Crematogaster nigriceps, 1174
Crenarchaeota, 549f
Creosote bush, 851
Crested penguin, 1066f
Cretaceous period, 487
Cretinism, 1005
Creutzfeldt-Jakob disease, 542
CRH. *See* Corticotropin-releasing hormone
Crick, Francis, 286–87, 286–87f, 304
Cricket, 673–74, 675f
Critical period, **1112**
Crocodile, 515f, 691f, 703, 705–6f, 707, 709, 709f, 916, 1050
Crocodilia (order), 703t, 709
Crocus, 747
Cro-Magnons, 726, 726f

Crop
  of annelids, 658
  of birds, 890f
  of earthworm, 888f
  of insects, 666f
"Crop milk," 1002
Crop plant, 851. *See also* specific crops
  artificial selection in, 458, 458f
  breeding of, 357–58
  effect of global warming on, 785, 785f, 1224
  genetic variation in, 1232
  increasing nutrient content of, 783
  transgenic, 335–38. *See also* Transgenic plants
  wild relatives of, 1232
Crop productivity, 358f
Crop rotation, **784**
Cross-bridge, **876**, 877f, 878, 878f
Cross-bridge cycle, **876**, 877f
Cross-current flow, 929, 929f
Cross-fertilization, **244**, 245f
Cross-fostering, **1112**
Crossing over, **230**, 230–32f, 232, 234f, 236f, 238, **267**, 267f, 409, 522
  within inverted segment, 411, 411f
  reciprocal, 232
  unequal, 409, 409f
Cross-pollination, 244
Cross-sectional study, 1147
Crow, 711t, 888, 1122
Crowded population, 1152, 1152f
Crown gall, 822, 822f
Crustacea (subphylum), 668–69, 668t
Crustacean, 635, 664f, 668–69, 668t, 871
  decapod, 668, 669f
  freshwater, 669, 669f
  locomotion in, 883
  respiration in, 924–25
  sessile, 669, 669f
  terrestrial, 669
Cryptic coloration, 1170, 1170f
Cryptochrome, 835, 838f
Crystal, 25
Crystal violet, 550f
Ctenophora (phylum), 621f, 623t, 624, 634–35f, 638, 641, 641f
Ctenophore, 1215
Cuboidal epithelium, 856f
  simple, 860, 861t
Cubozoa (class), 641, 641f
Cuckoo, 1113, 1113f
Cud, 898
Cuenot, Lucien, 255
Culex, 673f
Cultivation, 784
Cultural evolution, 726
Cup fungus, 232f, 600f, 609, 609f
Cupula, 971t, 977, 977f, 979, 979f
Cuspid, 715, 715f, 890, 890–91f
Cusp of tooth, 891f
Cutaneous receptor, **973**, 973f
Cutaneous respiration, 698, **916**, 922, 923f, 927
Cutaneous spinal reflex, 962f
Cuticle
  of arthropods, 666
  of nematodes, 647f
  of plant, **580**, 584, **732**, 750, 750f, 772
Cutin, **732**
Cuttlefish, 652
CXCR4 receptor, 536–38, 537f, 539f
Cyanobacteria, 71, 71f, 85, 182, 520f, 546, 549f, 551f, 553, 557. *See also* Lichen
Cyanocobalamin. *See* Vitamin B₆
Cyanogenic glycoside, 798, 799f

Cycad, 585t, 590, 592, 592f, 594f
Cycadophyta (phylum), 585t, 592, 592f
Cyclic AMP (cAMP), 576, **998**
  in glucose repression, 369, 369f
  as second messenger, 132, 132–34f, 998, 998f
Cyclic GMP (cGMP), 129, 135, 135f, 986
Cyclic ovulator, 1067
Cyclic photophosphorylation, **196**, 199
Cyclin, 219–22, 220–21f, 224, 416–17, 417t, 418f
Cyclin-dependent protein kinase (Cdk), **219**–22, 220–22f, 224, 416–17, 417t, 418f
Cycliophora (phylum), 648, 648f
CYCLOIDIA gene, of snapdragons, 841, 841f
Cyclooxygenase-1 (cox-1), 995
Cyclooxygenase-2 (cox-2), 995
Cyclosporin, 1022
Cyclostome, 1010
Cynodont, 704f
Cypress, 590
Cyst, of protists, 565
Cysteine, 41, 42f
Cystic fibrosis, 45, 255, 264t, 274, 494
  gene therapy for, 263, 333, 333t
"-Cyte," 866
Cytochrome, 39t
Cytochrome *b*, 484
Cytochrome *c*, 14f, 174, 174f
Cytochrome oxidase, 174, 174f
Cytokine, **994**, **1021**
Cytokinesis, **213**, 213f, 214, 217f, **218**, 231f, 234–35f
  in animal cells, 218, 218f
  in fungi, 218
  in plant cells, 218, 218f
  in protists, 218
Cytokinin, 774, 816, 817t, **822**–23, 822–23f
  synthetic, 822f
Cytoplasm, **80**, 80f, 86–87f
  ion composition of, 943t
Cytoplasmic streaming, 574, 601, 601f
Cytoproct, 569f, **570**
Cytosine, 49–50, 49–50f, 284, 284f, 287, 287f, 376
Cytoskeleton, 79f, 86f, **87**, 87f, 90t, **98**–100, 523t
  attachments to, 111, 111f
Cytosol, **90**
Cytotoxic T cells, 1019–20, 1019t, 1021–22f, 1022, 1028f

D

2,4-D, 820f, 821
Dachshund, 459, 459f
Daddy longlegs, 670
Dalton (unit of mass), 20
dam mutation, 412
Damselfly, 673
Danainae (subfamily), 1169
Dance language, of honeybees, 1120, 1120f
Dandelion, 743, 762, 849, 1150
Darevsky, Ilya, 1062
Dark current, 986
Dark-field microscope, 83t
"Dark meat," 880
Darwin, Charles, 242–43, 454–55, 454–55f. *See also* Galápagos entries
  critics of, 468
Descent of Man, The, 12
evidence for evolution, 10, 10f

I-9

Darwin, Charles—*Cont.*
 invention of theory of natural
  selection, 11–12, 11–12f
 Malthus and, 11
 On the Origin of Species, 8, 11f, 12,
  434
 photograph of, 8f
 pigeon breeding, 12
 Power of Movement of Plants, The,
  818, 818f
 theory of evolution, 8–9
 voyage on *Beagle*, 1, 1f, 8–9, 9f,
  454–55
 Wallace and, 12
Darwin, Francis, 818, 818f
Darwin's frog, 1065f
Dating, of fossils, 460, 460f
Daughter cells, 208, 231f
Davson-Danielli model, 108
Day-neutral plant, **835**, 836
DCC gene, 419f
DDT, 569, 1218, 1218f, 1247
Dead space, anatomical, 931
Deafness with stapes fixation, 269f
Deamination, **178**, 178f
 of adenine, 376
 of cytosine, 376
Death, as characteristic of life, 62
Death cap mushroom (*Amanita
  phalloides*), 608f
Death rate, 1150, 1157
 human, 1155
Decapod crustacean, 668, 669f
Decidua basalis, 1086, 1097f, 1098
Deciduous forest, **1211**
 productivity of, 1192
 temperate. *See* Temperate deciduous
  forest
Deciduous plant, 774, 808, 814, 852
Deciduous teeth, 890
Decomposer, 553, 604, 784, **1190**, 1193f
Decomposition, 557, 1184
 in carbon cycle, 1186, 1186f
 in nitrogen cycle, 1187, 1187f
 in phosphorus cycle, 1188f
Decrement, 972
Deductive reasoning, **4**, 4f
Deep-sea hydrothermal vent, 519
 extreme thermophiles from, 71
 origin of life in, 65
Deer, 528f, 715–16, 719f, 898, 1057,
  1067f, 1126, 1180, 1211
Deer mouse, 541
De-etiolated mutant, of *Arabidopsis*, 808
Defecation, 888, 897
Defensive coloration, 1170
Defoliant, 821
Deforestation, 1185, 1185f, 1189, 1189f,
  1220, 1220f, 1232, 1236, 1236f
Dehydration, 860, 920, 1000f, 1044,
  1057, 1057f
Dehydration synthesis, **37**, 37f, 284
Delayed hypersensitivity, **1036**
Deletion, 410–**11**, 410t
Delta wave, 959
De Mairan, Jean, 813
Demography, **1145**–47
Denaturation, **47**
 of enzymes, 152
 of proteins, 45, 47, 47f
Dendrite, 869, 869f, **940**, 940–41f
Dendritic spine, 941
Dendrobatidae (family), 1170, 1170f
Dengue fever, 1224
Denitrification, **1187**, 1187f
Denitrifier, 557
Dense connective tissue, **862**, 863t
Dense irregular connective tissue, **862**
Dense regular connective tissue, **862**

Density-gradient centrifugation, 288,
  288–89f
Dental caries, 458, **554**, 555t
Dental plaque, 554
Dentin, 891f
Deoxyhemoglobin, **934**
Deoxyribonucleic acid. *See* DNA
Deoxyribose, 49, 49f, 55f
Depolarization, **919**, 944–46, 945f,
  949–50, 972, 972f
Depo-Provera, 1076f, 1077t
Depression, clinical, 950
Derived characters, **512**–14, 513f, 628
Dermal tissue, of plants, **732**, 734–35,
  734–35f, **758**, 796
Dermatitis, contact, **1036**
Dermis, 1014
Descent of Man, The (Darwin), 12
"Descent with modification," 434
Desert, 1206, 1206f, 1208, 1208f, **1210**
 animals of, 1139
 productivity of, 1191f
Desert scrub, 1141f
Desiccation, **580**, 1064, 1066
Desmosome, 128t, 137f, **139**, 139f
Detergent, 1217
Determinant, **391**
Determinate development, 626f, **627**
Determination, **393**–94, 393–94f, **1093**
 floral, 836–37, 837f
 irreversibility of, 393–94, 394f
 mechanism of, 393
Detritivore, 618t, **1190**, 1190f
Deuterostome, 396, 527, 621f, 622t,
  626–27, 626f, 634, 634–35f,
  676, 677f, 684
 evolution of deuterostome
  development, 626–27, 626f
Developed countries, 1157–58,
  1157–58f, 1157t, 1231
Developing countries, 1157–58,
  1157–58f, 1157t
Development, 1064
 aging, 401–2
 in amphibians, 1065, 1065f
 in animals, 382, 382f, 619t
 apoptosis in, 400, 400f
 of behavior, 1112–13
 in *Caenorhabditis elegans*, 382f, 390,
  390f, 400, 400f
 cell migration in, 384f, 385, 391, 1087
 cellular mechanisms of, 381–402
 as characteristic of life, 2, 63
 determinate, 626f, **627**
 determination, 393–94, 393–94f
 deuterostome, evolution of,
  626–27, 626f
 in *Drosophila*, 386
 evidence for evolution, 464
 evolution of, 501–3, 501–3f, 1094–95
 of eye, 392f, 501, 504–6, 505f,
  1093, 1093f
 in frogs, 491, 491f
 in fungi, 382
 gene expression in, 362
 homeotic genes and, 398–400
 in humans, 383f, 1097–1101,
  1097–1101f
 indeterminate, 626f, **627**
 induction, 391–**92**, 392f
 in insects, 382f, 386, 386–87f
 of limbs, 502–3, 502f, 528
 mosaic, **391**
 overview of, 382, 382f
 pattern formation, 395–97, 395–97f
 in plants, 382, 382f, 388, 389f
  embryonic, 756–57, 756–57f
  establishment of tissue systems,
   758–59, 758–59f

food storage, 759, 759f
 fruit formation, 761–62
 morphogenesis, 759
 seed formation, 760, 760f
 vegetative, 755–63
postnatal, 1102, 1102f
regulative, **391**, 397
in sea urchins, 501, 501f
in vertebrates, 382–85, 382–85f,
  1081–1102, 1085t, 1095f
of wings, 502f
in *Xenopus*, 392, 392f, 397
Developmental decisions, 1093
Devil's Hole pupfish, 1140, 1140f
de Vries, Hugo, 267
Dewlap, of *Anolis* lizards, 479, 479f
Diabetes insipidus, 269f, 1057
Diabetes mellitus, 902, 1008, 1054
 treatment of, 426, 428
 type I (insulin-dependent), 1008
 type II (non-insulin-dependent), 1008
Diagnostics, 357, 1029–30
Diaphragm (birth control), 1076,
  1076f, 1077t
Diaphragm (muscle), 715, 856, 856f,
  930–32, 931f
Diapsid, **704**, 705f, 707
Diastole, **918**
Diastolic pressure, **918**, 918f
Diatom, 565, **571**, 572f
Diazepam, 950
Dicarboxylate cotransporter, 402
Dicer, 374, 375f
Dichlorophenoxyacetic acid. *See* 2,4-D
Dichogamous plant, 846, 847f
Dichromat, 987
Dicot, 593–94, 593f
 broad-leaved, 821
 development of seedling, 731f
 leaves of, 734f, 748, 748f, 750
 root of, 742, 742f
 shoot development in, 764f
 stem of, 745f
Dicrocoelium denditicum, 1175, 1175f
Dictyostelium discoideum, 348f, 576, 576f
Didinium, 1168f
Dieldrin, 1218
Diencephalon, 955t
Dienococcus, 549f
Diesel exhaust, 415t
Differential-interference-contrast
  microscope, 83t
Differentiation, **393**
Diffusion, **114**, 114f, 122t, 908
 facilitated, **115**, 115f, 122t
 Fick's Law of, **923**, 928
 through ion channels, 114
Digestion, 160, 887–904
 chemical, 888
 in cnidarians, 638
 external, 603
 extracellular, 638, 639f, 888
 intracellular, 94, 94f, 638, 888
 of plant material, 716
 in small intestine, 894, 894f
 in stomach, 892–93
Digestive enzymes, 94, 889, 894–95,
  900t, 1008
Digestive system, 857t, 859f
 of annelids, 658
 of birds, 890f
 of carnivores, 899f
 complete, **645**, 645f
 of flatworms, 642, 642f
 of herbivores, 899f
 of insectivores, 899f
 of insects, 674
 of invertebrates, 888, 888f
 of nematodes, 646, 647f

of ruminants, 898, 899f
 types of, 888, 888f
 of vertebrates, 888–89, 888–89f
 variations in, 898, 898–99f
Digestive tract, 889, 889f
 as barrier to infection, 860, 893, 1014
 cells lining, 138
 innervation of, 965t
 layers of, 889, 889f
 neural and hormonal regulation of,
  900, 900t, 901f
Digestive vacuole, 63f
Dihybrid, **253**
Dihybrid cross, 253–54, 254f, 257
Dihydroxyacetone phosphate, 165f
1,25-Dihydroxyvitamin D, 1006
Dikaryon, 600
Dikaryotic hyphae, **602**
Dilger, William, 1108
Dimetrodon, 704f
Dinoflagellate, 568, 568f, 653f
Dinomischus, 630f
Dinornithiformes (order), 711t
Dinosaur, 460f, 487, 689, 691f, 703t,
  705, 705–6f, 709, 712–13, 712f
 eggs of, 516, 516f
 feathered, 712
 parental care in, 516, 516f
Dioecious plant, **846**
Dioxin, 821
Diphtheria, 554, 555t
Diploblastic animal, 620
Diploid (2*n*), **212**, **228**–29, 228–29f,
  249, 497f, 522, 1062
Diplomonad, 520f
Diplontic life cycle, **581**
Diplopoda (class), 671
Diptera (order), 672f, 674, 675t
Dipterocarp, 613
Direct contact, cell signaling by,
  127, 127f
Directional selection, 446–47f, **447**
Dirigible, 26f
Disaccharide, **56**, 56f, 57, 57f
Disassortative mating, **438**
Disorder, 146, 146f
Dispersive replication, 288
Disphotic zone, **1216**, 1216f
Disruptive selection, **446**, 446–47f, 481
Dissociation, **47**
 of proteins, 47
Distal convoluted tubule, 1048f, **1053**,
  1053f, 1055f, 1056–57
Distal-less gene, 528
Disturbance, interruption of succession
  by, 1180, 1180f
Disulfide bridge, 43f, 47f
Diving, by elephant seals, 922f
DNA, **15**, 37–38f, **48**, 79f, 80, 279–98.
  See also Gene
 antiparallel strands, 287, 287f
 of archaebacteria, 71
 base composition of, 285, 285t
 central dogma, 51, 302–3, 303f
 of centromeres, 212, 350
 of chloroplasts, 96–97, 563, 572
 in chromosomes. *See* Chromosome
 cloning of. *See* Cloning
 coding strand, **306**, 306–7f
 complementary. *See* cDNA library
 damage to, 223
 directionality of, 285
 double helix, 15f, 48f, 50, 50f,
  286–87, 286–87f
 double-strand breaks in, 410
 evolution from RNA, 51
 functions of, 48, 50
 gel electrophoresis of, 325, 325f,
  330, 330f

I-10 Index

genetic engineering. *See* Genetic engineering
hormone response elements, 996f, 997
hydrogen bonds in, 50, 50f, 287, 287f
junk. *See* DNA, noncoding
major groove of, 286f, **363**–64, 363f, 367, 367–68f
methylation of, 373, 373f, 412, 418–19, 425
minor groove of, 286f
of mitochondria, 96–97, 402, 562, 725
mutations in. *See* Mutation
nicked, 320
noncoding, 350–51, 466, 494, 496
polymorphisms in, 435
of prokaryotes, 85, 208–9
proof that it is genetic material, 280–83
protein-coding, 298, 298f
recombinant, 320–22, 321–22f, **323**, 323f
repetitive, 351
replication of. *See* Replication
RNA versus, 51, 51f
scanning-tunneling micrograph of, 48, 48f
segmental duplications, 349–50, 350t
sequencing of. *See* Genome sequencing
simple sequence repeats, 350, 350t
with sticky ends, 320, 321f, 326f
structural, 350, 350t
structure of, 15f, 49–50, 49–50f, 279f, 284–87
of telomeres, 350
template strand, 290–91f, 306, 306–7f
three-dimensional structure of, 286–87, 286–87f
in transformation. *See* Transformation
X-ray diffraction pattern of, 286, 286f
DNA-binding motifs, in regulatory proteins, **363**–65, 364–65f
DNA fingerprint, 329, **331**, 332, 332f, 1128, 1129f
DNA gyrase, 292, 293t, 294
DNA helicase, 291, 292f, 293, 293t, 402
DNA library, 322, **324**, 324f
DNA ligase, 292, 293t, 294, **320**, 321f, 326f
DNA microarray, 354–55, 354f
DNA polymerase, 290, 295
proofreading function of, 290
Taq polymerase, 329, 329f
DNA polymerase I, 290, 292, 292f, 293t
DNA polymerase II, 290, 292
DNA polymerase III, 290, 291–94f, 292–94, 293t
DNA primase, 291, 292f, 293, 293t
DNA repair, 223, 412, 422–23
DNA repair hypothesis, for origin of sex, 237
DNA vaccine, **334**, 1035
DNA virus, 532, 532f, 533t
Dodder (*Cuscuta*), 743, 789, 811, 1175f
Doering, William, 800
Dog, 715, 715f, 717, 719t, 1072
breeds of, 459, 459f
chromosome number in, 210t
coat color in, 258, 258f
hearing in, 981
Pavlovian conditioning in, 1110
predation by, 1168
retinal degeneration in, 264, 264f
Dogbane, 1169
Dogwood, 751, 751f
Doherty, Peter, 334
Doll, Richard, 415

"Dolly" (cloned sheep), 394, 394f, 424–25, 424–25f
Dolphin (*Delphinus delphis*), 719t, 982, 1140
Domain (protein), 43, 44f, **45**, 502
Domain (taxonomic), 16, 16f, 74, 511, **518**, 518f
Domestication, 459, 459f
Dominant hemisphere, 959–60
Dominant:recessive ratio, 246t, 247–48, 248f, 251f
Dominant trait, 246t, **247**–51, 248f, 251f
in humans, 249t, 259
incomplete dominance, 256, 256f
L-Dopa, 950
Dopamine, **950**, 952, 952f
Dormancy
in plants, 774, 814, 814f, 827, 827f
in seeds, 814, 814f, 827, 827f
dorsal (transcription factor), 396
Dorsal lip transplant experiment, 1092–93, 1092f
Dorsal portion, 624, 624f
Dorsal root, **963**
Dorsal root ganglia, 961f, **963**, 1091
Double bond, **26**
Double fertilization, 589, 596, 596f, 840, **848**, 848f
Double helix, 15f, 48f, **50**, 50f, 286–87, 286–87f
Douche, 1076
Dove, 711t
Down, J. Langdon, 272
Down feather, 517f
Down syndrome, 210, **272**, 282f
maternal age and, 272, 273f, 274
translocation, 272
DPC4 gene, 417t
Draft sequence, 351–**52**, 494
Dragonfly, 673, 675t
hybridization between species, 474
Dreaming, 959
Drone (insect), 1062, 1133
Drought tolerance, in plants, 814, 814f
Drug abuse, 902
Drug addiction, 951–53, 952–53f
Drupe, 761f
Duchenne muscular dystrophy, 249t, 264t, 269f
Duck, 711t, 882
Duck-billed platypus, 717–18, 882, 988, 1067, 1067f
Dugesia, 642f, 644
Duke Experimental Forest, 785, 786f
Dung beetle, 1129
Dunnock, 1129, 1129f
Duodenal ulcer, 893
Duodenum, 893f, **894**, 895f, 900t
Duplication (mutation), 409, 409f, **411**, 497
Dusky seaside sparrow, 1244, 1244f
Dutch elm disease (*Ophiostoma ulmi*), 609
Dwarfism, pituitary, 1002
Dyad symmetry, 320
Dye, fungal, 611
Dynactin, 99, 99f
Dynein, 98, **99**–100, 99–100f, 396
Dyskeratosis congenita, 269f

E

Eagle, 711t, 888
Ear, 980–82, 981–82f
sensing gravity and angular acceleration, 978–79, 978–79f
structure of, 980, 981f

Ear canal, 980
Eardrum, 980, 981f
Ear popping, 980
Earth
age of, 13, 61
atmosphere of early earth, 65–66
formation of, 62
orbit around sun, 1204–5
origin of life within crust of, 65
rotation of, 1204–5f, 1205
Earthworm, 622t, 635, 659f, 660–61, 661f, 871, 908, 908f, 1062
digestive tract of, 888, 888f
locomotion in, 871f
Earwax, 53
East Coast fever, 567
Easter lily (*Lilium candidum*), 843f
Eastern bluebird, 884f
Eastern gray squirrel (*Sciurus carolinensis*), 511
Eastern milk snake (*Lampropeltis triangulum triangulum*), 472f
Eating disorder, 903
Ebola hemorrhagic fever, 357, 357t, 533t
Ebola virus, 524f, **541**, 541f
Ecdysis, **666**
Ecdysone. *See* Molting hormone
Ecdysozoan, 527, 628, 629f, **635**, 635f
ECG. *See* Electrocardiogram
Echidna, 717–18, 718f, 1067
Echinoderm, 622t, 634–35f, **676**, 677f
body cavity of, 679
body plan of, 678–79, 678–79f
classes of, 680
endoskeleton of, 676, 677f, 678
evolution of, 676
nervous system of, 954f
regeneration in, 679
reproduction in, 679, 679f
respiration in, 923f, 924
water-vascular system of, **676**, 677–79f, 678–79
Echinodermata (phylum), 621f, 622t, 629f, 676, 677f
Echinoidea (class), 676f, 678, 680
Echolocation, 716, 982
Ecological footprint, **1158**, 1158f
Ecological isolation, 473t, 474
Ecological processes, interactions among, 1176–77, 1176–77f
Ecological pyramid, 1193, 1193f
inverted, 1193
Ecological species concept, 477
Ecology, 1137
behavioral, **1122**, 1122f
community, 1161–80
of fungi, 604, 604f
population, 1137–58
Economic value, of biodiversity, 1232–33, 1232f
*Eco*RI, 321f, 323
Ecosystem, 2, 3f, **1184**. *See also specific types*
biogeochemical cycles in, 1184–89, 1184–89f
climate effects on, 1204–7
conservation of, 1248, 1248f
disruption of, 1242–43, 1242–43f
dynamics of, 1183–1200
effect of human activity on, 1218–24
energy flow through, 1184, 1190–93
interactions among trophic levels, 1194–96

stability of, 1197–1200, 1197f
trophic levels in, **1190**–91, 1190f
Ecotone, **1163**, 1163f
Ectoderm, 384f, 385, 392, 620, **624**, 625f, 638, 642, **856**, 860, 1085t, **1087**–90, 1087–90f, 1087t, 1093, 1093–94f, 1098
Ectomycorrhizae, **612**–13, 612f
Ectoparasite, **1175**, 1175f
Ectoprocta (phylum), 662, 663f
Ectotherm, **707**
Edema, 902, 914, 917, 1017
Edentata (order), 719t
Edge effect, **1237**
EEG. *See* Electroencephalogram
Eel, 882, 882f
Eel River (California), 1195–96
Eelworm, 646
Effector, 367, 369, 962f, 963t, **1040**, 1040–41f
antagonistic, 1043, 1043f
Efferent arteriole, 1053
Efferent neuron. *See* Motor neuron
EGF. *See* Epidermal growth factor
Egg, 228f, **1062**
amniotic. *See* Amniotic egg
of birds, 710, 1084f
of dinosaurs, 516, 516f
fertilization, 1082–83, 1082–83f
of fish, 381f
of frogs, 1084f
incubation by birds, 1066, 1066f
of lancelets, 1084f
of monotremes, 718, 1067, 1067f
of reptiles, 702, 1066
yolk distribution in, 1084, 1084f
Egg activation, 1082–83
Egg cells
"nursing of," 119
plant, **595**
Egg-rolling response, in geese, 1106, 1107f
Egg shell, 1066
coloration of, 1122, 1122f
effect of DDT on, 1218, 1218f
of reptilian egg, 702f
Egg white, 1066
Ehrlich, Paul, 1031
Einstein, Albert, 190
Ejaculation, 1070
Ejaculatory duct, 1068f, 1070
EKG. *See* Electrocardiogram
Elapid, 708–9
Elasmobranch, 988, 1049
Elasticity, of thorax and lungs, 931
Elastin, 102, 102f, 391, 402, 862
Elderberry (*Sambucus canadensis*), 736f, 746f
Eldonia, 630f
Eldredge, Niles, 486
Electrical fish, 475, 988
Electrical synapse, 948
Electricity, detection of, 988
Electrocardiogram (ECG, EKG), **919**, 919f
Electrochemical gradient, 174–75
Electroencephalogram (EEG), 959
Electromagnetic spectrum, 190, 190f
Electron, **20**–21, 20f
in chemical behavior of atoms, 22–23, 22–23f
energy level of, **23**, 23f
valence, **24**
Electron acceptor, 144–45, 153, 160
Electronegativity, 27f, **28**, 36, 172
Electron microscope, 82, 82f, 83t
microscopy of plasma membrane, 110, 110f
scanning, **82**–83, 83t, 110

I-11

Electron acceptor—*Cont.*
  specimen preparation for, 110, 110f
  transmission, **82**, 83t, 110
Electron orbital, 20–21, 20f, **22**–**23**, 22–23f
Electron transport chain, 162, 163f, 172–**73**, 172f, **174**–**75**, 174–75f
  ATP production in, 176, 176f
  electrochemical gradient, 174–75
  moving electrons through, 174, 174–75f
  photosynthetic, **194**–**97**, 195–97f
  production of ATP by chemiosmosis, 175, 175f
Electroreceptor, 475, 988
Electrotropism, 811
Element, **21**
  inert, **24**
  periodic table, 24, 24f
Elephant, 528, 528f, 683f, 714–15f, 716, 719t, 1039f, 1210
Elephant bird (*Aepyornis*), 1228
Elephantiasis, 647
Elephant seal, 439, 922f, 1126, 1128f
Elevation, climate and, 1207, 1207f
Elk, 1247
Elm, 761f, 762
El Niño, 1213, 1213f, 1224
Elongation factor, 311, 312f
Embryonic development
  human, 1097–1100, 1097–1100f
  in plants, 756–57, 756–57f
EMBRYONIC FLOWER gene, of *Arabidopsis*, 833, 833f
Embryonic stem cells, 405f, **426**, 426–27f, 430, 1086
Embryo sac, 589, **595**, 838, 840, **842**, 842f, 843
Emergent properties, **2**
Emerging viruses, **541**
Emerson, R. A., 194, 257
Emery-Dreifuss muscular dystrophy, 269f
Emigration, 437, 1150
Emotional state, 958
Emphysema, **931**–32
Emulsification, 895
Enamel, 554, 891f
Encephalartos transvenosus, 592f
Encephalitozoon cuniculi, 348f
Endangered species, 1140f, 1210
  conservation biology, 1227–48
  preservation of, 1246–48
Endemic species, **1230**
Endergonic reaction, **147**, 147f, 154, 156, 160–61
Endler, John, 449
Endocrine-disrupting chemical, 1010
Endocrine gland, 689f, **860**, **992**–93, 992–93f, 994–95t. *See also specific glands*
Endocrine signaling, 127, 127f
Endocrine system, 857t, 858f, 991–1010
  neural and endocrine interactions, 993
Endocytosis, 92, **118**–**19**, 118f, 122t
  receptor-mediated, 118–**19**, 118f, 122t
Endoderm, 384f, 385, 392, 620, **624**, 625, 625f, 638, 642, 653f, **856**, 860, 1085t, **1087**–**90**, 1087–90f, 1087t, 1093, 1094f, 1098
Endodermis, 730f, **741**, 742f, 770, 770f
Endogenous opiate, 951
Endolymph, 978, 979f
Endomembrane system, **86**, **90**–**94**, 99
Endometrium, 1072, 1074–75, 1086

Endonuclease, **290**
Endoparasite, **1175**
Endophyte, **613**
Endoplasmic reticulum (ER), **90**–92, 90t, 91f, 99, 102t, 105f
  calcium release from, 999
  origin of, 72, 72f
  rough, 86–87f, **91**, 91f, 93f, 95
  smooth, 86–87f, 91f, **92**
Endopodite, 528f
Endorphin, **950**
Endoskeleton, **676**, 683, **871**, 871f
  of echinoderms, 676, 677f, 678
  of vertebrates, 688
Endosperm, 589, 594, 596, 596f, 756, 756f, 759, 759–60f, 763, 764f, 840, 848, 848f
Endosperm nucleus, 596
Endosperm tissue, **596**
Endospore, **550**
Endostatin, 423
Endostyle, 686, 686f
Endosymbiont theory, 97, 97f
Endosymbiosis, **72**–**73**, 73f, **97**, 97f, **521**, **562**–**63**, 562–63f
Endothelium, 912, 912f
Endotherm, **707**, **713**–**15**, **909**, 909f, 917, 1042, 1138
Energy, 143, **144**–**56**. *See also specific types of energy*
  in atom, 22–23, 23f
  flow in living things, 144–45
  flow through ecosystem, 1184, 1190–93
  in food, 160
  in food chains, 1192
  forms of, 144
  harvesting by cells, 159–82
  laws of thermodynamics, 146
  for movement, 870, 876, 877f
  for phloem transport, 776–77
Energy budget, community, 1192, 1192f
Energy cycle, in chloroplasts and mitochondria, 202, 202f
Energy expenditure, 903
Energy intake, 1123
  maximizing, 1123, 1123f
Energy level, **23**, 23f
Englemann, T. W., 192, 193f
engrailed gene, 395f, 396, 399
Enhancement effect, **197**, 197f
Enhancer, **371**, 371–72f, 416
Enkephalin, **950**
Enolase, 165f
Entamoebae, 520f
Enteric bacteria, 549f
Enterobacteriaceae (family), 552
Enterogastrone, **900**
Enterovirus, 533t
Enthalpy, 147
Entropy, **146**–47, 146f, 156
Envelope, viral, **532**, 532f
Environment
  effect on gene expression, 256, 256f
  endocrine-disrupting chemicals in, 1010
  human effect on, 1143
  impact of genetically modified crops on, 339
  individual responses to changes in, 1138, 1138t
  limitations on population growth, 1150–54
  variations in individuals due to, 440
Environmental variation
  coping with, 1138
  evolutionary response to, 1139
Enzyme, 26, 37, 39, 39t, **149**
  attached to membranes, 111, 111f

catalytic cycle of, 150f
cofactors, **153**, 153f
computer-generated model of, 356f
defects in gene disorders, 296
denaturation of, 47, 152
electrophoresis of, 435
genetic variation in, 435
inhibitors and activators of, 152
mechanism of action of, 150
multienzyme complex, 151, 151f, 168
one-gene/one-enzyme hypothesis, **297**
pH effect on, 47
polymorphic, 435
receptors that act as, 129
RNA, 151
temperature effect on, 47, 152, 152f
Enzyme-substrate complex, **150**, 150f
Enzymic receptor, 128t, 129–30, 130f, 134
Eosinophils, 866, 910f, **911**
Ephedra, 585t, 592
Ephedrine, 592
Epicotyl, 730, 731f, 764f
Epidermal cells, of plants, **388**, 389f, **734**, 741, 742f
Epidermal growth factor (EGF), 222, 994
Epidermal growth factor (EGF) receptor, 224, 416, 422
Epidermis
  of animals, 638, 638–39f
  of plant, **732**, 733–34f, 745, 745–46f, 750, 750f
  of skin, 860, 1014
Epididymis, 1068f, 1070, 1070f
Epigenetic events, 374
  cancer and, 418–19
Epiglottis, 891, 891–92f
Epilimnion, 1216–17, 1217f
Epimysium, 862
Epinephrine, 132, 132f, 965–66, 965f, 993, 995t, 998–99, 998–99f, 1007, 1091
Epiparasite, 612–13
Epiphyte, **1173**
Epistasis, **257**–**58**, 257–58f, **450**
Epithelial membrane, **860**
Epithelial tissue, 856, 856f, 860, 861f
  columnar, 856f
  cuboidal, 856f
  keratinized, 860
  simple, **860**, 861t
  stratified, 856f, **860**
Epitheliomuscular cells, 640
Epithelium, **860**
EPSP. *See* Excitatory postsynaptic potential
EPSP synthetase, 336
Epstein-Barr virus, 533t
Equatorial countercurrent, 1212f
Equilibrium (balance), 978–79
Equilibrium, in biological community, **1180**
Equilibrium model, of island biogeography, 1200, 1200f
Equilibrium potential, **943**, 943t
Equisetum, 586, 587f
Equus, 462–63, 462–63f
ER. *See* Endoplasmic reticulum
Eratothenes, 4, 4f
erb-B gene, 417t
Erection, 951, 1070
Erythroblastosis fetalis, 260, 1029
Erythrocytes, 82f, 88, 863t, 866, 866f, 907f, 909, **910**–**11**, 910f
  cell surface antigens of, 260, 260f
  facilitated diffusion in, 115, 117f
  membrane of, 109, 109t, 113f

Erythropoiesis, 911
Erythropoietin, 222, 264, 911, 1010
  genetically engineered, 333
Escherichia coli, 549f, 552–53
  cell division in, 207f, 208, 209f
  chromosome of, 301
  cloning in, 321, 322f
  conjugation map of, 408f
  DNA of, 285t
  at high temperature, 1139
  lac operon of, 366f
  mutations in, 552
  replication in, 290–94
  translation in, 310f
Esophagus, 658, 859f, 888–89f, 889, 891–93f, 892, 899f
Essay on the Principle of Population (Malthus), 11, 1157
Essential mineral, **904**
Essential nutrient, **904**, 904t
  in plants, 782t
EST. *See* Expressed sequence tag
Estradiol, 825f, 995t, 996f, 1071t, 1072, 1074–75, 1075f, 1077, 1100–1102, 1100f
Estrogen, 37t, 53, 129, 800, 1009, 1078
Estrogen receptor, 800
Estrous cycle, **1067**, 1073, 1075
Estrus, **1067**
Estuary, 1191f, 1214
Ethanol, 36f, 167f
Ethanol fermentation, 181, 181f
Ethics
  of stem cell research, 430, 430t
  value of biodiversity, 1233
Ethidium bromide, 325f
Ethology, 1106
Ethylene, 335, 774, 811, 816, 817t, **826**, 826f
Ethylene receptor, 826
Etiolated seedling, 808
Eucalyptus, 613, 750
Euchromatin, **211**, 214
Eudicot, 594f
Euglena, 566, 566f
Euglenoid, 566, 566f
Euglenozoa, 564f, 566–67, 566–67f
Eukarya (domain), 16, 16f, 74f, 511f, 518, 518–20f, 520t, 521–22
Eukaryote, **70**, **72**, **80**, 80f, 521–22
  cell division in, 547
  cell wall of, 87, 87f, 90t, 101, 101f, 102t
  chromosomes of, 209–12, 211f, 547
  compartmentalization in, 522, 547
  cytoskeleton of, 98–100, 98f
  endomembrane system of, 90–94
  evolution of, 70, 72–74, 72–75f, 209, 562–63, 562–63f
  flagella of, 86f, 90t, 100, 100f, 102t, 547
  gene expression in, 315–16f, 316, 377f
  genome of, 348f
    gene organization in, 349, 350t
    noncoding DNA in, 350–51
  interior organization of, 86–101
  key characteristics of, 522, 523t
  kingdoms of, 521
  multicellularity in, 522, 523t
  organelles containing DNA, 96–97
  origin of, 521, 521f
  posttranscriptional control in, 374–78
  prokaryotes versus, 102t, 546–47, 547f
  promoters of, 370f
  recombination in, 547
  replication in, 295, 295f
  ribosomes of, 95

I-12 Index

sexuality in, 522
transcriptional control in, 362, 370–72, 370–72f
transcription in, 308–9, 308–9f, 315–16f, 316
translation in, 311, 315–16f, 316
unicellular, 520
vacuoles of, 101, 101f, 102t
Eumelanin, 258
Eumetazoa (subkingdom), **620,** 620–21f, 634, 638
Euparkeria, 704f
Europa, 64–65, 76, 76f
European corn borer, 336
Euryarchaeota, 549f
Eusocial system, **1132,** 1132f
Eustachian tube, 980, 981f
Eutherian, 528, 528f
Eutrophic lake, **1178, 1217,** 1241
Evaporation, 1184, 1184f
from leaves, 772
Evaporative cooling, 29
Evening primrose, 812, 844, 850f, 851
even-skipped gene, 395f
Evergreen forest
temperate. *See* Temperate evergreen forest
warm moist, 1208
Evergreen plant, 852
"Evo-devo," 630
Evolution, 6, **8,** 75f. *See also* Coevolution
of aerobic respiration, 182
agents of, 437t, 438, 438f
interactions among, 443, 443f
of amphibians, 688, 700, 700f
of angiosperms, 593, 593–94f
of animal body plan, 624–27
of animals, 628–30, 629f
behavior and, 1122–29
of bilateral symmetry, 624
of biochemical pathways, 155
of birds, 517f, 689, 706f, 712–13, 712f
of body cavity, 625, 625f
of brain, 954–56
of catabolism, 182
of cellular respiration, 182
chemical, 19, 66–67
of complex characters, 517, 517f
controversial nature of theory of, 12, 12f, 468
convergent, 504, **512,** 1208
cultural, 726
Darwin's theory of, 8–9
of development, 501–3, 501–3f, 1094–95
of echinoderms, 676
of endothermy, 917
of eukaryotes, 70, 72–74, 72–75f, 209, 562–63, 562–63f
evidence for, 453–68
age of earth, 13
anatomical record, 464–65
comparative anatomy, 13, 13f
convergence, 467, 467f
development, 464
experimental tests, 448–49, 448–49f, 458, 458f
fossil record, 10, 10f, 13, 64, 460–63, 460–63f, 468
homologous structures, 464, 464f
imperfect structures, 464–65, 465f
mechanism of heredity, 13
molecular biology, 14, 14f
molecular clocks, 14, 14f
molecular record, 466, 466f
phylogenetic trees, 14
vestigial structures, 465, 465f
evidence for Darwin's, 10, 10f

of eye, 450, 450f, 504–6, 504–6f, 983–84, 983–84f
of eyespot on butterfly wings, 503, 503f
of fish, 688, 692f
of flight, 517, 517f, 884
of flowers, 831, 840–41, 1174
future of, 488
genetic variation and, 434–35, 434f, 450, 450f
of genomes, 492–500
of glycolysis, 166, 182
of G-protein-linked receptors, 131
of hemoglobin, 466, 466f
of herbivores, 1159
of heterotrophy, 176
of homeobox genes, 399, 399f
of hominids, 722–23, 723f
of horses, 462–63, 462–63f
human impact on, 487–88
of humans. *See* Human evolution
of immune system, 1031–32, 1032f
of insects, 527–28
on islands, 467, 480, 481–82f, 482–83, 1168
of jaws, 692–93, 692–93f
of kidney, 1048–50
of land plants, 526, 526f
of mammals, 528, 528f, 689, 717, 717t
marsupial-placental convergence, 467, 467f
of meiosis, 522
of metabolism, 182
of multicellularity, 73, 522
mutation and, 437–38, 437f, 438f
natural selection and. *See* Natural selection
of nitrogen fixation, 182
of nuclear membrane, 522
of photosynthesis, 182, 196
of plants, 579–80, 733
predation and, 1168
of primates, 720–26, 721f
of prosimians, 720
rate of, 482–86, 486f, 512, 514
of reproduction, 1064–67
of reptiles, 688–89, 704–5, 705f, 706t
responses to environmental variation, 1139
of segmentation in animals, 527, 527f, 627, 627f
of sexual reproduction, 73, 237–38
of social behavior, 1130–35
in spurts, 486
of tissues, 624
use of comparative genomics, 352
of vertebrates, 461, 688–718, 691f, 694f
of wheat, 498f
of wings, 503
Evolutionary age, species richness and, 1199
Excision repair, **412**
Excitation-contraction coupling, **879**
Excitatory postsynaptic potential (EPSP), **949**–51, 949f, 951f
Excretion, by kidney, 1054
Excretory duct, 647f
Excretory organs
of annelids, 1046f
of insects, 1046–47, 1047f
Excretory pore, 647f, 1046, 1046f
Excretory system
of annelids, 659, 659f
of arthropods, 665–66f, 667
of flatworms, 642–43
of mollusks, 653f, 655
of vertebrates, 689f

Excurrent siphon, 657f, 686f
Exercise
cardiac output and, 920
effect on metabolic rate, 903
muscle metabolism during, 880–81
oxygen requirement for, 934
Exergonic reaction, **147**–48, 147f, 156
Exhalation. *See* Expiration
Exocrine gland, **860**
Exocytosis, **119,** 119f, 122t
Exon, 45, **313,** 313f, 315f, 349–50, 376
Exon shuffling, 314
Exonuclease, **290,** 292, 294
Exopodite, 528f
Exoskeleton, 58, 58f, 387f, 664–66, 665f, **871,** 871f, 873f, 1009
Experiment, **5**–6, 5f
control, 6
Expiration, 929, 931–32, 931f
Exploitative competition, **1164**
Exponential growth model, 1150, 1150f
Expressed sequence tag (EST), **349,** 351
External digestion, 603
External environment, sensing of, 970–71, 970f
External fertilization, **1064**–65, 1065f, 1082
External genitalia, 1068, 1068f
External gills, 924f
External respiration, 922
Exteroreceptor, **970,** 971t
Extinction, 487, 487f, 1144, 1227f
conservation biology, 1227–48
disruption of ecosystems and, 1242–43, 1242–43f
due to human activities, 487
due to prehistoric humans, 1228, 1228f
factors responsible for, 1234–46
genetic variation and, 1244–45, 1244f
habitat loss and, 1234t, 1236–38, 1236–38f
in historical time, 1229, 1229f, 1229t
introduced species and, 1234t, 1240–41
on island, 1200, 1200f
of Lake Victoria cichlid fish, 484, 1241, 1241f, 1246
overexploitation and, 1234t, 1239, 1239f
population size and, 1244–45, 1245f
Extinction vortex, 1245
Extracellular compartment, 1045f
Extracellular digestion, 888
Extracellular fluid, 1044–51
ion composition of, 943f
Extracellular matrix, 102, 102f, 391, 423, **862,** 866
Extraembryonic coelom, 1096
Extraembryonic membranes, 1096, 1096f
Extra-pair copulation, **1128**–29, 1129f
Extraterrestrial life, 64, 76, 76f
Extreme halophile, **71,** 73f
Extreme thermophile, **71,** 73f
Extremophile, 519
Eye, 983–87
of cephalopods, 657
compound, 504, 505f, 665–66f, 666–67
development of, 392f, 501, 504–6, 505f, 1093, 1093f
evolution of, 450, 450f, 504–6, 504–6f, 983–84, 983–84f
focusing of, 984f
image-forming, 983
innervation of, 965t

of insects, 450, 450f, 504, 504f, 666–67, 666f, 673
of mollusks, 464–65, 465f, 504, 504f
simple, 504, 505f, **666**
structure of, 983–84, 983–84f
of vertebrates, 464–65, 465f, 504–5, 504–5f, 969f, 983–84, 983–84f
Eye color, in fruit fly, 265–66, 265–68f
eyeless gene, 504, 505f
Eye muscles, 881f, 983f
Eye of potato, 747, 849
Eyespot
on butterfly wings, 503, 503f
of flatworms, 642f, 643, 983, 983f
of green algae, 573
on peacock feathers, 1126f, 1127
of planarian, 504–6, 504f, 506f
regeneration in ribbon worm, 505, 506f

F

Fabaceae (family), 814
Fabry disease, 269f
Facial recognition, 960
Facilitated diffusion, **115,** 115f, 122t
Factor(s), Mendelian, 249
Factor VIII, 264t
FAD, 51, 904t
FADH$_2$
contributing electrons to electron transport chain, 163f, 173–76, 174–76f
from fatty acid catabolism, 179, 179f
from Krebs cycle, 163f, 169f, 170, 171f
Falcon, 711t
Falconiformes (order), 711t
Fallopian tube, 859f, 1072, 1072f, 1074, 1074f, 1078, 1078f
Fall overturn, 1216, 1217f
Fallow, 784
False dandelion (*Pyrrhopappus carolinianus*), 762f
Family (taxonomic), **511**
Family planning program, 1158
Family resemblance, 242, 242f
Fanconi anemia, 333t
Farm animals
reproductive cloning of, 424–25
transgenic, 338, 338f
Farnesyl transferase, 422
Farnesyl transferase inhibitor, 422
Farsightedness, 984, 984f
Fasting, 902
Fast-twitch muscle fiber, **880**–81, 881f
Fat(s), 37–38t, **53,** 903
caloric content of, 179
catabolism of, 178–79, 178–79f
digestion of, 895–96, 900, 900t
as energy-storage molecules, 54
kilocalories per gram, 54
structure of, 53
Fat body, **674**
Fat globule, 862f
Fatty acid(s), 37–38t, 52–53, 52f, 106, 106f, 862
absorption in small intestine, 896, 896f
catabolism of, 179, 179f
polyunsaturated, **53**–54, 54f
saturated, **53**–54, 54f, 815
unsaturated, **53**–54, 54f, 904
Fatty acid synthetase, 151
Feather, 40f, 517, 517f, 710, 710f, 712, 712f, 1126f, 1127
Feces, 893, 897–98

Index I-13

Fecundity, **1146**, 1146t, 1148f, 1149
Feedback inhibition, 152, **156**, 1003–4, 1003f
Feeding phase, 575
Female reproduction, hormonal control of, 1071t
Female reproductive system, 859f, 1072–76, 1072–76f
Femoral nerve, 954f
Femur, 699f, 858f, 871–72f
Fermentation, **160**, 167, 167f, 177, **181**, 181f, 603
　ethanol, 181, 181f
　lactic acid, 181, 181f
Fermented foods, 603
Fern, 581, 584, 585t, 587–88, 587–88f, 747, 824
Ferredoxin, 196f, 198f, 199, 1187
Ferritin, 39t, 378
Ferritin gene, 337, 337f
Fertility, soil, 784
Fertilization, **228**, 229f, 238, 383, 1064, 1074, **1082**–83, 1082–83f, 1085t
　double, 589, **596**, 596f, 840, **848**, 848f
　external, 655, **1064**–65, 1065f, 1082
　internal, 655, 707, **1064**, 1066, 1082
　in plants, 595, 848, 848f
　site of sperm entry, 397
Fertilizer, 1185
　chemical, 784
　nitrogen, 335–36, 784, 1187
　organic, 784
　phosphorus, 784, 1188
　pollution from, 1217–18
　potassium, 784
Fetal development, human, 1100
Fetal membranes, **1096**
Fetal zone, 1101
Fetus, 383f, 1081f
Fever, 1017
Fever blister, 533t
F₁ generation. *See* First filial generation
F₂ generation. *See* Second filial generation
F₃ generation. *See* Third filial generation
Fiber (dietary), 897
Fiber (sclerenchyma), 736
Fibrin, 39t, 40, 40f, 910, **911**, 911f
Fibrinogen, 866, 910, 911f
Fibroblast(s), 79f, 222, 863t
Fibroblast growth factor, 962
Fibroclasts, 862
Fibrocytes, 862
Fibronectin, **102**, 102f, 391, 423
Fibrous root system, 743
Fibula, 699f, 871f
Fick's Law of Diffusion, **923**, 928
Fiddlehead, 587
Fig, 743, 743f
"Fight or flight" response, 965, 965f, 1007, 1091
Filament (flower), 589, **594**, 595f, **840**, 840f
Filaria, 622t, 647
Filariasis, 647
Filial imprinting, 1112
Filopodia, 565
Filovirus, 357t, 541
Filter feeder, 657
Filtration, 1046
　in kidney, 1052, 1053f
Fimbriae, 1074, 1074f
Fin(s), 690, 695–96

Finch, Darwin's, 10, 10f, 467, 477
　beaks of, 444, 454–55, 454–55f, 483, 483f, 1166, 1166f
Finger, grasping, 720
Fingernail, 716, 859f
Finished sequence, 351
Fin whale, 1239, 1239f
Fir, 585t, 590
Fire ant, 1120f
Fire-bellied seedcracker finch (*Pyronestes ostrinus*), 446, 447f
Fire blight, 553
Firefly, 1118, 1119f
Fireweed (*Epilobium angustifolium*), 847f
First, Neil, 424
First filial generation, **246**–47, 246t, 248f, 250, 251f
First Law of Thermodynamics, **146**
First trimester, 1097–1100, 1097–1100f
Fish, 622t, 690–97, 690f
　armored, 692f, 693
　bony, 690, 695–97
　cartilaginous, 689–90, 691–92f, 864, 1049
　characteristics of, 690
　chemical defenses of, 1170
　circulation in, 690, 707f, 915, 915f
　cleavage in, 1084
　eggs of, 381f
　electrical, 988
　evolution of, 688, 692f
　extinctions, 1229f, 1229t, 1234t
　heart in, 915, 915f
　jawed, 688, 690, 692–93, 693f
　jawless, 688, 692, 692f
　kidney of
　　cartilaginous fish, 1049
　　freshwater fish, 1048, 1049f
　　marine bony fish, 1048, 1049f
　lateral line system of, 977, 977f
　lobe-finned, 690, 692f, 693t, 696–97, 697f, 699, 699f
　nervous system of, 954–55, 955f
　nitrogenous wastes of, 1051, 1051f
　path to land, 697
　ray-finned, 689, 691–92f, 693t, 696, 696f
　reproduction in, 1065
　respiration in, 915, 923f, 924–25, 925f
　shell-skinned, 692, 692f
　spiny, 692f, 693
　swimming by, 695, 882, 882f
　taste buds in, 975
　thermoregulation in, 1042
　viviparous, 1064f, 1065
Fisheries/commercial fishing, 1214, 1239
Fission, **1062**
Fitness, **442**, 442f, **1122**
Fixed action pattern, **1106**, 1107f
Fixed anions, 942, 943f
Flagella, **84**
　of bacteria, 84, 84–85f
　of chytrids, 606
　of dinoflagellates, 568
　of eukaryotes, 86f, 90t, 100, 102t, 547
　of prokaryotes, 102t, 547, 547t, 550, 551f
　of protists, 565–66, 566f
　swimming with, 84, 85f
Flagellar motor, 551f
Flagellate, 520f
Flagellin, 85f, 550, 551f
Flame cells, **643**, **1046**, 1046f
Flamingo, 713
Flatworm, 621f, 622t, 633–35f, **635**, 642–45, 882, 908, 1046, 1046f
　digestive cavity of, 888
　eyespot of, 983, 983f

nervous system of, 954, 954f
　parasitic, 1175, 1175f
Flavin adenine dinucleotide. *See* FAD
Flavivirus, 533t
Flavobacteria, 520f
Flavonoid, 787f
Flavr Savr tomato, 335
Flax, 736
Flea, 673, 675t
Fleming, Walther, 210
Fleshy fruit, 1143f
Flight, 618t, 710, 884, 884f
　evolution of, 517, 517f, 884
　in mammals, 716, 716f
Flight feather, 710
Flight muscles, 884
Flight skeleton, 710
Flipper, 882
Flocking behavior, in birds, 1133, 1133f
Flooding, plant responses to, 774–75, 774–75f
Floral determination, **836**–37, 837f
Floral leaf, 751
Floral meristem, 838, 838–39f
Floral organ identity genes, 838, 839f
Flour beetle (*Tribolium*), 1176
Flower, 584, 732
　complete, 840, 840f
　evolution of, 840–41, 1174
　floral specialization, 841, 841f
　floral symmetry, 841, 841f
　incomplete, 840
　initiation of flowering, 832, 832f
　pistillate, 846, 847f
　production of, 834–39, 834–39f
　　autonomous pathway of, 836–37, 836–38f
　　flowering hormone, 835
　　formation of floral meristems and floral organs, 838, 838–39f
　　light-dependent pathway, 834–35, 834–35f, 838f
　　temperature-dependent pathway of, 836, 838f
　staminate, 846, 847f
　structure of, 245f, 475, 594–95, 595f, 840f
Flower color, 838, 845
　in four o'clocks, 256
　in garden pea, 243–47, 246t, 248f, 250–52, 251f
Flowering hormone, 835
Flowering plant, 580–81, 580f, 585t, 593–96, 593–96f, 831f
　asymmetric development in, 757
　comparative genomics, 496
　dichogamous, **846**, 847f
　dioecious, **846**
　diversity of, 840
　evolution of, 593, 593–94f, 831
　extinctions, 1229t
　fertilization in, 848, 848f
　gamete formation in, 838, 842–43, 842–43f
　life cycle of, 595–96, 596f
　metamorphosis in, 832–33, 832–33f
　monoecious, **846**
　pollination in. *See* Pollination
　reproduction in, 831–52
Fluidity, membrane, 107, 107f
Fluid mosaic model, **108**–9, 108f
Fluke, 644, 644f
Fluorescence microscopy, 79f, 83t
Fluoride, 554
Fluoxetine, 950
Fly, 664f, 674, 675t, 884, 975
Fly agaric, 600f
Flying fox, 884f
　declining populations of, 1243, 1243f

Flying phalanger, 467f
Flying squirrel, 467f
FMN, 904t
Folic acid, 904t
Foliose lichen, 611f
Follicle (feather), 710
Follicle (fruit), 761f
Follicle-stimulating hormone (FSH), 994t, 1001f, 1002, 1003f, 1067, 1071, 1071f, 1071t, 1075, 1075f
Food
　caloric content of, 903
　fortified, 783
　genetically modified, 339
　labeling of, 340
Food chain, **1190**, 1193
　energy in, 1192, 1192f
　length of, 180
　trophic levels within, 1190–91, 1190f
Food energy, 903
Food intake, regulation of, 903
Food poisoning, 549f
Food spoilage, 47, 604
Food storage, in plants, 759, 759f
Food storage root, 743
Food supply, 1155
　population cycles and, 1153–54
　worldwide, 357–58, 357–58f
Food vacuole, 565, 569–70, 569f
Food web, **1190**
Foolish seedling disease, 824
Foot
　of birds, 713
　of mollusks, 653–54f, **654**, 656–57, 882–83
Footprints, fossil, 722
Foraging behavior, 1123, 1123f, 1134f
Foraminifera (phylum), 565, 575, 575f
Forebrain, 955–56, 955f, 955t
　human, 957–60, 957–59f
Forelimb, of vertebrates, 13f, 464, 464f
Forensic science, 331–32, 332f
Forest ecosystem
　biogeochemical cycles in, 1189, 1189f
　effect of acid precipitation on, 1219, 1219f
　water cycle in, 1185, 1185f
Formaldehyde, 415t
N-Formylmethionine, 310, 311f, 520t
fosB gene, 1109
Fos protein, 224
Fossil
　creation of, 460
　dating of, 460, 460f
Fossil fuel, 785, 1186, 1186f, 1222
Fossil record, 64, 453f, 460–63, 460–63f
　community, 1163
　evidence for evolution, 10, 10f, 13
　gaps in, 461, 461f, 468
　history of evolutionary change, 460–61, 460f, 468
Founder effect, **439**, 479
Four o'clock, 812
　flower color in, 256
Fovea, 984f, 987
Fowl cholera, 1018
Fox, Sidney, 69
Fox, 1195, 1211
F plasmid, 407, 407f
Fragile-X syndrome, 269f, 358
Frameshift mutation, 304f, 410
Francisella tularensis, 357t
Frankia, 557
Franklin, Rosalind, 286, 286f
Free energy, **147**, 147f, 156

I-14　Index

Freely movable joint, 872, 872f
Free nerve ending, 973, 973f
Free radicals, 400, 402
Freeze-fracture microscopy, 110, 110f
Freezing tolerance, in plants, 815
Frequency-dependent selection, **444**, 444f
Freshwater ecosystem, 1216–17, 1216–17f
   plant adaptations to, 774, 774–75f
   productivity of, 1217
Freshwater vertebrate, 1044
Frictional drag, 882
Frog (*Rana*), 475, 698, 698t, 701, 701f, 888, 1065
   chromosome number in, 210t
   declining populations of, 1223, 1234–35, 1234–35f
   desert adaptations in, 1139
   development in, 491, 491f, 1088f
   eggs of, 1084f
      transplantation of nucleus of, 281, 281f
   gastrulation in, 1088f
   locomotion by, 882–83, 883f
   mating calls of, 1127, 1127f
Froglet, 1065f
Frond, 587
Frontal lobe, 957, 957f
Fructose, 55f, 56, 56f
Fructose 1,6-bisphosphate, 165f
Fructose 6-phosphate, 165f, 201
Fruit, 584, 593, **595**, 732, 736, 840, 848, 1143
   development of, 819, 822, 826
   dispersal of, 762, 762f, 845
   formation of, 816
   kinds of, 761f
   ripening of, 335, 826, 826f
Fruit bat, 1179
Fruit drop, 821
Fruit fly (*Drosophila*)
   ADH genes of, 435
   bar-shaped eye in, 267, 267f
   body color in, 268f
   bristle number in, 458, 458f
   chromosome number in, 210t
   chromosome puffs, 361f
   crossing over in, 267, 267f
   development in, 386
      pattern formation, 395–98, 395f, 397f
   dorsal view of, 386f
   eye color in, 265–66, 265–68f
   genetic map of, 344
   genome of, 348f, 352, 492t, 496
   Hawaiian, 479, 482, 482f
   heterozygosity in, 435
   homeotic genes in, 365, 398–99, 398–99f
   life span of, 402
   maternal genes in, 386
   metamorphosis in, 386, 387f
   Morgan's experiments with, 265–66, 265–66f
   pattern formation in, 395–98, 395f, 397f
      forming the axis, 395f, 396
      producing the body plan, 395f, 396–97
   selection for negative phototropism in, 447, 447f
   sex chromosomes of, 270, 270t
   sex-linked genes in, 266, 266f, 268f
   single gene effects on behavior, 1109
   transposons in, 411, 500
   wing traits in, 268f
   X chromosome of, 268f
Fruticose lichen, 611f

FSH. *See* Follicle-stimulating hormone
FtsZ protein, 208–9, 208–9f
Fucus, cell division in, 757, 757f
Fumarase, 171f
Fumarate, 170, 171f
Functional genomics, **354**–58, 503
Functional group, **36**, 36f
Fundamental niche, **1164**–65, 1164f
Fungal disease, 604
   in animals, 604, 614
   in humans, 604
   in plants, 604, 604f, 608–9, 614, 614f, 796, 797f, 803f
Fungal garden, of leafcutter ants, 613, 613f, 1133, 1172
Fungi, 599–614. *See also* Lichen; Mycorrhizae
   in bioremediation, 603
   body of, 601, 601f
   carnivorous, 603f
   cell types in, 600
   cytokinesis in, 218
   development in, 382
   ecology of, 604, 604f
   endophytic, **613**
   fungal-animal mutualisms, 614
   key characteristics of, 600–604
   major groups of, 605, 605f, 605t
   mating type in, 602, 608
   metabolic pathways in, 603
   mitosis in, 600–601
   obtaining nutrients, 500, 603, 603f
   phylogeny of, 605, 605f
   reproduction in, 600, 602, 602f
   in rumen, 613
   in symbioses, 604, 611–14
Fungi (kingdom), 16, 16f, 74–75f, **518**, 518f, 521, 521f, 523t
Funiculus, 595f
Fur, 714. *See also* Coat color
   thickness of, 1138, 1139f
Fur seal (*Callorhinus ursinus*), 882, 1151f
Fur trade, 1239
Fusarium, 614
fushi-tarazu gene, 395f

G

GABA, **950**
GABA receptor, 950
Galactose, 55f, 56, 56f
Galápagos finch, 10, 10f, 444, 454–55, 454–55f, 467, 477, 483, 483f, 1166, 1166f
Galápagos tortoise, 467
Gallbladder, 889, 889f, 894–95, 895f, 900, 900t, 901f, 965t
Galliformes (order), 711t
Galloping, 883
Gallstone, 895
Gamebird, 811t
Gametangium, **581**, 583, 588, 606–7f, 607
Gamete, 227, **228**, 228–29f, 265, **1062**
   plant, 581, 581f, 838, 842–43, 842–43f
   prevention of fusion of, 473t, 475
Gametic meiosis, 565
Gametocyte, 569f
Gametophyte, **581**, 581f, 582–84, 583f, 586–87, 588f, 589, 591, 595, 596f, 606f, 838, 842
Gametophytic self-incompatibility, 847, 847f
Ganglia, 869, 963
Ganglion cells, 986–87, 986f
Ganglioside, 259

Gannet, 1137f
Gap, **745**
Gap genes, 395f, 396
Gap junction, 128t, 137f, **140**, 140f, 868, 919
Garden pea (*Pisum sativum*)
   chromosome number in, 210t
   flower color in, 243–47, 246t, 248t, 250–52, 251f
   flower structure in, 245f
   genetic map of, 268f
   Knight's experiments with, 243
   Mendel's experiments with, 244–54
      choice of garden pea, 244
      experimental design, 245, 245f
      Mendel's interpretation of his results, 250–54
      page from notebook, 247f
      results of experiments, 246–49, 246t, 247–48f
   photoperiod in, 835
   plant height in, 246t
   seed traits in, 244–45, 246t, 247f, 253–54, 254f
Garrold, Archibald, 296
Gas exchange, 922
   in animals, 922–23, 923f
   in aquatic plants, 774–75, 775f
   in leaves, 750
   in lungs, 930, 930f
   in tissues, 930, 930f
Gastric gland, 893f, 900
Gastric inhibitory peptide (GIP), 900, 900t, 901f
Gastric juice, 892–93, 893f, 898, 898f, 900
Gastric pit, 893f
Gastric ulcer, 893
Gastrin, 900, 900t, 901f
Gastrocnemius muscle, 859f, 881f
Gastrodermis, **638**, 638–39f
Gastrointestinal tract. *See* Digestive tract
Gastropod, 654f, 655
Gastropoda (class), 656, 656f
Gastrovascular cavity, 638f, 888, 888f, 908, 908f
Gastrula, 383f, **385**, **619**, 627, 1087, 1094f
Gastrulation, 384f, **385**, 1085t, **1087**–89, 1087–89f
   in amphibians, 1088, 1088f
   in aquatic vertebrates, 1088, 1088f
   in birds, 1089, 1089f
   in chordates, 1087
   in humans, 1098
   in lancelet, 1087f
   in mammals, 1089, 1089f
   in reptiles, 1089
Gated ion channel, **944**
Gaucher disease, 333t
Gavial, 703t, 709
GC hairpin, 307, 307f
G_1 checkpoint, 223, 418f
GDP, 966
Gecko, 708–9
Gehring, Walter, 504
Gel electrophoresis
   of DNA, 325, 325f, 330, 330f
   of enzymes, 435
Gemmule (hereditary material), 243
Gene, **15**, 249t, **282**, **298**
   on chromosomes, 265–74
   composite, 1026, 1026f
   co-option of existing gene for new function, 501–3
   inactivation of, 499, 499f
   minimal number to support life, 352

   one-gene/one-polypeptide hypothesis, 296–97
   pleiotropic effects of, **255**, 450, 459
   in populations, 433–50
   protein-coding, 298, 298f, 350t
   segmental duplication, 349, 350t
   single-copy, 349
   tandem clusters of, 349
Genealogical species concept, 477
Gene clock hypothesis, of aging, 402
Gene clone, working with
   distinguishing differences in DNA, 331, 331f
   DNA fingerprinting, 332, 332f
   getting sufficient amount of DNA, 329, 329f
   identification of DNA, 330, 330f
Gene conversion, **409**
Gene disorder, 259–64
   due to alterations of proteins, 262, 262f
   enzyme deficiency in, 296
   gene therapy for. *See* Gene therapy
   genetic counseling in, 274
   important disorders, 249t, 264t
   pleiotropic effects in, 255
   protein changes in, 298, 298f
   rareness of, 259
   screening for, 358
Gene duplication, 497–99, 499f
Gene expression, 89, 301–16, **303**
   Central Dogma, 302–3, 303f
   chromatin structure and, 373, 373f
   control of, 361–78
   in development, 381–402
   developmental decisions, 1093
   environmental effects on, 256, 256f
   in eukaryotes, 315–16f, 316, 377f
   genetic code, 304–5, 304f, 305t
   intracellular receptors as regulators, 129, 129f
   microarray technology, 354–55, 354f
   posttranscriptional control, **362**, 374–78
   in prokaryotes, 316, 316f, 366–69, 366–69f
   proteomics. *See* Proteomics
   regulatory proteins, 363–65
   tissue-specific, 136
   in plants, 740, 740f
   transcriptional control, **362**, 366–72, 366–72f
   translational control, 377f, 378
Gene flow, 437–**38**, 437t, 438f, 473, 477–79, 488
   from genetically modified crops, 339
   interactions among evolutionary forces, 443, 443f
Gene-for-gene response, 803, 803–4f
Gene interactions, 450
Gene mobilization, **408**
Gene pool, **473**
Gene prospecting, 1232
Generation time, 494, **1145**
   body size and, 1145, 1145f
   of prokaryotes, 552
Generative cell, 591, 595, 596f, 842f, 843, 848, 848f
Gene-related patent, 358
Gene technology, 319–41
Gene therapy, 263–64, 333, 333t
   vectors for, 263–64, 264f
Genetic code, 304–5, 304f, 305t
   in chloroplasts, 305
   in ciliates, 305
   deciphering of, 304, 304f
   in mitochondria, 305
   triplet nature of, 304
   universality of, 305

Index   I-15

Genetic counseling, 274
Genetic disease. *See* Gene disorder
Genetic drift, 437–39, 437t, 438–39f, 443, 443f, 479, 488, 1244
Genetic engineering, 325–33
 agricultural applications of, 335–38, 335–38f
 bacteria and, 558
 medical applications of, 333–34, 333–34f
 risk and regulation of, 339–40, 340f
 stages of
  cloning, 326, 326f
  DNA cleavage, 325, 326f
  finding gene of interest, 326f, 328, 328f
  preliminary screening of clones, 326–37f, 327
  production of recombinant DNA, 325, 326f
 working with gene clones, 329–32
Genetic map, 268, **344**, 351
 of *Drosophila*, 344
 of garden pea, 268f
 of humans, 269, 269f
 using recombination to make maps, 268, 268f
Genetic marker, anonymous, 269
Genetic material. *See* Hereditary material
Genetic privacy, 358
Genetic recombination. *See* Recombination
Genetics
 behavioral, 1106, 1108–9, 1108–9f
 population, 433–34, **435**–50
 symbols used in, 250
Genetic screen, 327, 327f, 358
Genetic similarity, 466
Genetic system, 63
Genetic template, **1113**
Genetic testing, 259
Genetic variation, 13, 227, 237–38, 238f, 241f, 259, 351
 conditions for natural selection, 440
 in crop plants, 1232
 evolution and, 434–35, 434f, 450, 450f
 genes within populations, 433–50
 in human genome, 351
 loss of, 1234, 1244–45, 1244f
 measuring levels of, 435
 in nature, 435, 435f
 in prokaryotes, 552
Gene transfer, 406–8, 406–8f
 by conjugation, **407**, 407f
 lateral, **499**–500, 500f, 519
 by transposition, 407–8, 408f
 using vectors, 323, 323f
 vertical, **499**
Gene transfer therapy, **263**
Genistein, 799t, 800
Genital pore, 643f, 647f
Genome, **15**, 343–58. *See also specific organisms*
 of chloroplasts, 352
 eukaryotic, 348f
  gene organization in, 349, 350t
  noncoding DNA in, 350–51
 evolution of, 492–500
 finding genes in, 349, 349f
 human. *See* Human genome
 minimal size to support life, 352
 of mitochondria, 352
 origins of genomic differences, 497–501
 prokaryotic, 348f

size and complexity of, 348, 348f, 494, 497
 of virus, 532–33
Genome map, 344–45, 344–45f
 genetic. *See* Genetic map
 physical. *See* Physical map
Genome sequencing, 346–47, 346–47f
 clone-by-clone method, 347, 347f, 351
 databases, 349
 DNA preparation for, 346
 draft sequence, 351–**52**, 494
 evolutionary relationships from, 519
 finished sequence, 351
 shotgun method, 347, 347f, 351
 using artificial chromosomes, 347
Genomic imprinting, 428
Genomic library, **324**
Genomics, **348**, 352
 applications of, 357–58, 357–58f
 behavioral, 358
 comparative, 352, 353f, 491–96, 492–93f
 functional, **354**–58, 503
 vocabulary of, 351
Genotype, 249–50
 testcross to determine, 252, 252f
Genotype frequency, **436**–37
Genus, **510**–11
Geographic distribution, variation within species, 472–73, 472f
Geographic isolation, 473t, 488, 488f
Geography, of speciation, 480–81, 480–81f
Geological timescale, 75f, 694f
Geomagnetotropism, 811
Geometric progression, 11, 11f
Geranium, 744
Gerbil, 1050
German measles. *See* Rubella
Germ cells, 1068, 1069f
Germinal center, 914
Germinal epithelium, 1068
Germination, of seeds, 388, 389f, 596f, 756, 760, 763, **763**, 764f, 808, 814, 824, 827, 840
Germ layer, 384f, **385**, 624, 856, **1087**
 developmental fates of, 1087f, 1094f
 formation of, 1087–89, 1087–89f
Germ-line cells, 229, 229f, 410
GH. *See* Growth hormone
GHIH. *See* Growth hormone-inhibiting hormone
GHRH. *See* Growth hormone-releasing hormone
Giant clam (*Tridacna maxima*), 652, 653f
Giant ground sloth (*Megatherium*), 717, 1228, 1228f
Giant ragweed, 851
Giant sequoia (*Sequoiadendron giganteum*), 851f, 1206f
Giant squid, 652
Giant tube worm, 623t
Giardia, 562
Gibberellic acid, 763, 764f
Gibberellin, 774, 816, 817t, **824**, 824f, 826, 833–34, 836, 838f
Gibbon (*Hylobates*), 720, 721f
Gibbs' free energy. *See* Free energy
Gigantism, 1002, 1002f
Gila monster, 708
Gill(s), 922
 of animals, 684
 of bivalves, 657
 of crustaceans, 668
 external, **924**
 of fish, 590, 915, 923f, 924–25, 925f
 of mollusks, 653f, **654**, 657f, 924
 of mushroom, 600f, 608, 608f

Gill arch, 693, 693f, **925**, 925f
Gill chamber, 1091
Gill cover, 697
Gill filament, 925, 925f
Gill raker, 925f
Gill slits, 464, 685f, 687, 687f, 890
Gilman, Alfred, 131
Gingiva, 891f
Gingko, 7, 7f, 585t, 590, 592, 592f, 594f
Gingko biloba. *See* Maidenhair tree
Ginkgophyta (phylum), 585t, 592, 592f
GIP. *See* Gastric inhibitory peptide
Giraffe (*Giraffa camelopardalis*), 62f, 242, 434, 434f, 464, 515f, 716, 719t, 1210
Girdling of tree, 738
Gizzard
 of annelids, 658
 of birds, 888, 890, 890f
 of earthworm, 660, 888, 888f
Glaciation, 485, 485f
Glacier, 1141, 1178, 1178–79f, 1224
Glacier Bay, Alaska, succession at, 1178–79f
Gladiolus, 747
Gland (animal)
 endocrine, **860**
 exocrine, **860**
Gland (plant), 748
Glanville fritillary butterfly, 1144, 1144f
Gleaning bird, 1173, 1173f
Gleason, H. A., 1162
Glenn, John, 810
Gliding bacteria, 546, 549f
Glioblastoma, 417t
Global climate change, 358, 488, 1163, 1186
 crop production and, 785–86
 prehistoric, 1223
Global warming, 785–86, **1222**–24, 1222–23f
 effect on humans, 1224
 effect on natural ecosystems, 1223
 effect on species, 1223, 1223f
 geographic variation in, 1222, 1222f
β-Globin gene, 262, 298, 298f, 378
Globular protein, 44f
Globulin, **910**
Glomales, 605, 607, 612
Glomeromycota (phylum), 605
Glomerular filtrate, 1048, **1053**, 1054–55
Glomerulus, 1048, **1052**, 1053–55f
Glomus, 606
Glottis, 891, 891f, 927, 927–28f
Glucagon, 895f, 902, 902f, 995t, 1008, 1008f
Glucocorticoid, 1007
Gluconate metabolism, 450
Gluconeogenesis, **902**, 1007
Glucose, 37t, 55
 alpha form of, **58**, 58f
 beta form of, **58**, 58f
 blood, 880, 993, 1008, 1054
  regulation of, 902, 902f, 1042–43, 1042f
 catabolism of, 160, 162–63
  regulation of, 177, 177f
 homeostasis, 1007
 metabolism of, 1007
 priming of, 164
 production in photosynthesis, 187, 187f, 189, 201f, 203
 reabsorption in kidney, 1054
 structure of, 55f
 urine, 1008, 1054
Glucose 1-phosphate, 202

Glucose 6-phosphate, 165f, 902
Glucose-6-phosphate dehydrogenase deficiency, 269f
Glucose repression, 368–69, 369f
Glucose transporter, 39t, 115, 120–21, 121f
Glutamate, 178, 178f, **950**
Glutamic acid, 41, 42f
Glutamine, 42f
Glycation, of proteins, 402
Glyceraldehyde, 55f
Glyceraldehyde 3-phosphate, 164, 165f, 166, 200–201f, 201–2
Glycerol, 37t, 52, 52f, 106, 106f, 179, 1138
Glycerol kinase deficiency, 269f
Glycerol phosphate, 36f
Glycine, 42f, **950**
Glycocalyx, 109
Glycogen, **54**, 57
 breakdown of, 998, 998f
 liver, 902, 1008
 muscle, 880–81
 synthesis of, 903, 999, 1042
Glycogenolysis, 902
Glycolipid, 92, 109, 109t, 136
Glycolysis, 162, 164–67, 164–65f, 167f
 ATP production in, 176, 176f
 evolution of, 166, 182
 regulation of, 177, 177f
Glycophorin, 109t
Glycoprotein, 92, **102**, 108f, 109, 109t, 536
Glycoprotein hormone, 993, 994t, 998
Glyoxysome, **94**
Glyphosate, **336**, 336f
Glyptodont, 10f
G₂/M checkpoint, 219–20, 219f, 221f
Gnetophyta (phylum), 585t, 590, 592, 592f, 594f
Gnetum, 585t, 592
GnRH. *See* Gonadotropin-releasing hormone
Goat, 716
Goblet cells, 860
Goiter, 1004, 1004f
Golden mean, 744
Golden plover, 1117f
Golden rice, 357
Goldenrod, 834f, 835
Golden toad (*Bufo periglenes*), 1061, 1061f, 1234–35, 1234f
Golgi, Camillo, 92
Golgi apparatus, 86–87f, 90t, 91–**92**, 92–94f, 99, 102t
Golgi body, **92**, 739
Golgi tendon organ, 974
Gompertz number, 401
Gompertz plot, 401, 401f
Gonad
 indifferent, 1063, 1063f
 of polychaetes, **660**
Gonadotropin, 1002
Gonadotropin-releasing hormone (GnRH), 1003, 1003f, 1071, 1071f
Gonorrhea, 555t, 556, 556f
Gonos, 242
Gonyaulax, 568f
Goose, 464, 711t, 882, 1066
 egg retrieval behavior in, 1106, 1107f
 imprinting in, 1112, 1112f
Gooseneck barnacle (*Lepas anatifera*), 669f
Gorgonian coral, 641
Gorilla (*Gorilla*), 14, 466, 495, 495f, 513, 720–21, 721f, 1121
Gould, James L., 1121
Gould, John, 454

I-16 Index

Gould, Stephen Jay, 486
Gout, 269f, 1051
gp120 glycoprotein, 536, 537f
G$_0$ phase, **213**, 222
G$_1$ phase, **213**–14, 213f
G$_2$ phase, **213**–14, 213f
G protein, 131–32, **131**, 133–34f, 135, 416f, **966**, 966f, 975, 986, 998–99, 998–99f
G-protein-linked receptor, 128t, 130–31f, 131–32, 134
Graafian follicle, 1073, 1073f
Graded potential, 944–47, 944f
Gradualism, **486**, 486f
Graft rejection, 1022
Grain color, in corn, 257, 257f
Gram, Hans Christian, 84, 550f
Grammar, 1121
Gram-negative bacteria, **84, 550,** 550–51f
Gram-positive bacteria, **84,** 520f, 549f, **550,** 550f
Gram stain, 84, **550,** 550f
Grana, **96,** 97f, 186–87f, **187,** 192f
Grant, Peter, 455, 477
Grant, Rosemary, 455, 477
Granular leukocytes, **911**
Granulosa cells, 1072, 1073f, 1075, 1082, 1082f
Granzyme, 1016
Grape, 747, 747f, 761f, 795f, 824, 824f
Grasping fingers/toes, 720
Grass, 594, 747–48, 845, 846f, 849
Grasshopper, 270t, 666f, 672f, 673–74, 675t, 883, 1190f
  jumping of, 873f
Grasshopper mouse, 467f
Grassland, 1141f, **1210**
  temperate. *See* Temperate grassland
Gravitropism, **810,** 810f
  negative, 810, 810f
  positive, 810
Gravity, pollination by, 595
Gravity sensing, 971t
  in animals, 978–79, 978–79f
  in plants, 739
Gray crescent, 1083, 1083f, 1092
Gray matter, **941,** 961, 961–62f
Gray whale, 1239
Gray wolf (*Canis lupus*), release into Yellowstone Park, 1167, 1247
Grazing mammals, gleaning birds and, 1173, 1173f
Great Barrier Reef, 509f
Greater horseshoe bat (*Rhinolophus ferrumequinum*), 716f
Great tit (*Parus major*), 1124f, 1149f
Great white shark, 695
Green algae, 521f, 522, 525, 525f, 563, 573, 573f, 580, 808. *See also* Lichen
Greenbrier (*Smilax*), 741f
Green fluorescent protein, 740f
Greenhouse effect, **1222,** 1222f
Greenling, 1242
Green rod, 698
Green sea turtle, 708, 1117
Green snake (*Liochlorophis vernalis*), 709f
Gregarine, 569
Greyhound dog, 459, 459f
Griffith, Frederick, 282
Griffith, J., 542
Gross primary productivity, 1191
Ground finch, 483, 483f
  large ground finch (*Geospiza magnirostris*), 10f, 454–55, 454f, 483, 483f

medium ground finch (*Geospiza fortis*), 444, 454–55, 454–55f, 477, 1166, 1166f
small ground finch (*Geospiza fuliginosa*), 454–55, 454f, 477, 483, 483f, 1166, 1166f
Ground meristem, 730f, **731,** 731f, 739f, 741, 745, 756f
Ground squirrel, 1123
Ground substance, **862**
Ground tissue, **388,** 389f, **732,** 736, 736f, 739–40f, 745f, **758**
Groundwater, 791, 1184f, 1185, 1219
Grouper, 1119f
Group living, 1130–34
Group selection, **1130**
Grouse, 711t
Growth, **385,** 385f
  allometric, 1102, 1102f
  as characteristic of life, 2, 63
  gene expression and, 362
  in plants
    primary, 730, 732–33f, 733
    secondary, **731,** 732–33f, 733, **742**
Growth-deficient mutants, isolation of, 297
Growth factor, **221,** 221f, 378, 416, 416f, 422, **994**
  cancer and, 224
  cell cycle and, 222
  characteristics of, 222, 222f
Growth factor receptor, 222, 222f, 224, 224f, 416, 416f, 417t, 422
Growth hormone (GH), 994t, 1001–2, 1001–2f
  genetically engineered, 333, 333f
Growth hormone-inhibiting hormone (GHIH), 1003
Growth hormone-releasing hormone (GHRH), 1003
Gruiformes (order), 711t
G$_1$/S checkpoint, 219–20, 219f, 221f
GTP, 170, 171f, 309, 309f, 416, 966, 998f
GTP-binding protein. *See* G protein
Guanine, 49–50, 49–50f, 284, 284f, 287, 287f
Guano, 1051, 1188, 1212
Guanylyl cyclase, 129
Guard cells, **734,** 750, 750f, 772–73, 773f, 827f
Guide RNA, 567
Guinea pig, 904
Gulf Stream, 1212, 1212f
Gull, 711t, 1122f
Gullet, of ciliates, 569, 569f
Guppy, 1064, 1127
  selection on color in, 448–49, 448–49f
Gurdon, John, 281, 393
GUS gene, 355f
Gut, 888
Guthrie family, 259
Guttation, **771,** 771f
Gymnophiona. *See* Apoda
Gymnosperm, 580–81, 580f, 590–92, 590–592f, 594f
Gynoecium, 589, **594, 840**
Gypsy moth, 1180
Gyre, 1212
Gyrus, 957

H

Haberlandt, Gottlieb, 822
Habitat destruction, 1231–33, 1236, 1236f

Habitat fragmentation, 1237–38, 1237f, 1248
Habitat loss, 1228, 1234t, 1236–38, 1236–38f
Habitat restoration, 1246
  cleanup and rehabilitation, 1246
  pristine, 1246
  removal of introduced species, 1246
Habituation, 951, **1110**
Haeckel, Ernst, 1095
Haemophilus influenzae, 343f
Hagfish, 689–90, 691f, 692, 693t, 1044
Hair, 37f, 40, 714, 859f
  red, 242f
Hair cells, 977–82, 977–79f, 981f
Hair-cup moss (*Polytrichum*), 582f
Hair dye, 415t
Hair follicle receptor, 971t, 973, 973f
hairy gene, 395f, 396
Haldane, J.B.S., 68, 1130
Half-life, **21,** 460
Halichondrites, 630f
Hallucigenia, 630f
Hallucinogenic fungi, 603
Halobacterium, 520f, 549f
Halobacterium halobium, 113f
Halocarpus bidwillii, 1244f
Halophile, 519, 521f
  extreme, **71,** 73f
Haltere, 673
Hamilton, William D., 1130–31
Hamilton's rule, **1131**
Hamlet bass (*Hypoplectrus*), 1062f
Hammerling, Joachim, 280, 280f
Hamstring muscles, 873, 873f
Handicap hypothesis, **1127**
Hansen disease, 555t
Hantavirus, **541**
Haplodiploidy, 1132, 1132f
Haplodiplontic life cycle, 581, 581f
Haploid (*n*), **212, 228**–29, 228–29f, 236, 249, 497f, 522, 1062, 1068
Haplopappus gracilis, 210t
Haplotype, genomic, 351
Hapsburg family, 242
Harbor seal, 1242, 1242f
Hard coral, 641
Hard fat, 54f
Hard palate, 891f
Hard wood, 590
Hardy, G. H., 436
Hardy-Weinberg equilibrium, **436**–37, 436f
Hardy-Weinberg principle, 436–37
Hare, 719t, 898
Harlequin chromosome, 236f
Harlow, Harry, 1112
Harris tweed, 611
Hartl, Dan, 450
Harvest mouse, 159f
Hashimoto thyroiditis, 1036
Haustoria, 743, 797
Haversian canal, 865, 865f
Haversian system, 865, 865f
Hawaiian *Drosophila,* 479, 482, 482f
Hawaiian Islands, 762
Hawk, 515f, 711t, 1190f
Hay fever, 1036
Hayflick limit, 401, 401f
H band, 875–77, 875–77f
hCG. *See* Human chorionic gonadotropin
Head
  of mollusks, **654**
  of vertebrates, 689f
Head and neck cancer, 415t, 417t
Health effect, of global warming, 1224
Hearing, 971t, 980–82, 981–82f, 1091

Heart, 858f, **908**
  of amphibians, 698–99, 916, 916f
  of annelids, 659, 659f
  of birds, 916–17, 917f
  cardiac cycle, 918–19, 918–19f
  contraction of, 919
  electrical excitation in, 919, 919f
  embryonic, 1098
  of fish, 915, 915f
  of humans, 918–19, 918–19f
  innervation of, 965t
  of insects, 666f, 667, 908, 908f
  of mammals, 715, 916–17, 917f
  of mollusks, 654
  of reptiles, 702, 707, 709, 916
  three-chambered, 653f
  of vertebrates, 689f
Heart attack, 921
  stem cell therapy in, 426, 428
Heartbeat, 915
  fetal, 1100
Heart disease, 54
  chlamydia and, 556
Heart rate, 920
Heat, **146**
  sensing by pit vipers, 988, 988f
Heath hen (*Tympanuchus cupido cupido*), 1244–45
Heat loss, 909, 909f, 1191
Heat of vaporization, **29**
  of water, 28t, 29
Heat receptor, 971t
Heat sensing, 971t
Heat shock protein (HSP), 46, 46f, **815**
Heavy chain, **1025**–26, 1025–26f
Heavy metal
  phytoremediation for, 791–92, 792f
  tolerance in bent grass, 443, 443f
Hedgehog, 714
Height, 255, 255f
Heinrich, Bernd, 1115
Helical virus, **532**
Helicase. *See* DNA helicase
Helicobacter, 549f
Helicobacter pylori, 555t, 893
Heliconia imbricata, 845f
Helioconius sara, 798
Helium, 23f
Helix-turn-helix motif, **364,** 364–65f
Helpers at the nest, 1130
Helper T cells, 1019–23, 1019t, 1021f, 1023f, 1028f, 1033
Hemagglutinin, 540–41
Hemal arch, 688f
Hematocrit, 910
Heme group, 934, 934f
Hemichordata (phylum), 623t, 629f
Hemidesmosome, 137f, 138
Hemiptera (order), 674, 675t
Hemlock, 590, 800
Hemocyanin, 909
Hemoglobin, 37t, 39, 39t, 43, 262, 262f, 264t, 866, 911, **934**
  affinity for carbon dioxide, 935–36, 936f
  affinity for nitric oxide, 936
  affinity for oxygen, 935
  evolution of, 14, 14f, 466, 466f
  structure of, 45, 934, 934f
Hemolymph, **908,** 908f
Hemolytic disease of the newborn, 260, 1029
Hemophilia, 249t, **261,** 261f
  gene therapy for, 333t
  Royal pedigree, 261, 261f
  X-linked, 411
Hemopoietic stem cells, 1026
Hemorrhage, 920
Hepadnavirus, 533t

Index I-17

Heparin, 862
Hepaticophyta (phylum), 583
Hepatic portal vein, 896
Hepatitis B, 533t, 541
Hepatitis virus, 540
   vaccine for, 334, 334f
HER2 protein, 422
Herb, 585t, 1211
Herbal remedies, 800
Herbicide, 820f, 821, 1185, 1218
Herbicide resistance, in transgenic plants, 336, 339, 355f
Herbivore, 180f, 618t, 716, **888**, 889–90, 1190f, 1191, 1193f
   digestive system of, 899f
   evolution of, 1159
   global climate change and, 786
   plant defenses against, 1169, 1169f
   teeth of, 715, 715f
   in trophic cascade, 1194, 1195f, 1196
Herceptin, 422
Hereditary material, 48, 80–83, 279–98
   in nucleus, 280–81, 280–81f
   passage between organisms, 282, 282f
Heredity, **63**, 242–59. *See also* Gene entries
   early ideas about, 241–43
   first law of, 251
   mechanism as evidence for evolution, 13
   Mendel's model of, 248–49
   molecular basis of, 15
   second law of, 253–54, 254f
Hermaphrodite, **643**, 655, 661
Hermaphroditism, **1062**–63, 1062f
   sequential, 1062–63
Hermit hummingbird, 845f
Heroin, 951
Heron, 711t
Herpes simplex virus, 524f, 533t
   vaccine for, 334, 334f
Herring, 285t, 1242, 1242f
Hershey-Chase experiment, 283, 283f
Hertz, Heinrich, 190
Hesperidia, 761f
Heterochromatin, **211**, 214, 232
   constitutive, 350
Heterocyst, 557, 557f
Heterodont dentition, 715
Heterokaryon, 608
Heterosporous plant, **584**, 585t, 589, 591
Heterotroph, **160**, 180, **553**, 600, 887, 887f, **1190**, 1191f
   evolution of, 176
   infection of host organisms, 553
Heterozygosity, **435**
Heterozygote, 249t, 251–52f, 252, 274
Heterozygote advantage, **445**, 445f, 1244
Hexaploid, 498f
Hexokinase, 165f
Hexosaminidase A, 259, 264t
Hfr cells, 407
Hibernation, 1138
Hierarchical organization of living things, 2, 3f
Hill, John, 415
Hill, Robin, 189
Himalayan rabbit, 256
Hindbrain, 955, 955–96f, 955t
Hindenburg (dirigible), 26f
Hip, 872f
Hippocampus, 955, 955t, 958, 960, 1109
Hippocrates, 242
Hippopotamus, 528, 528f
Hirudinea (class), 660–61, 661f
Histamine, 862, 1016–17, 1017f, 1025, 1036

Histidine, 42f
Histogram, 255, 255f
Histone, **89**, 89f, 211, 211f, 373, 373f, 378, 523t, 568
HIV. *See* Human immunodeficiency virus
HLA. *See* Human leukocyte antigen
H.M.S. *Beagle* (Darwin's ship), 1, 1f, 8–9, 9f, 454–55
Hobson, Keith, 1238
Holistic concept, of communities, **1162**
Holly, 821
Hollyhock, 835
Holmes, Richard, 1238
Holoblastic cleavage, **1084**, 1084f, 1086
Holothuroidea (class), 676f, 678
Homeobox, **399**
Homeobox genes, 399, 399f, 635
Homeodomain, 398f, 758
Homeodomain motif, **365**, 365f
Homeodomain protein, 16, 16f
Homeostasis, **362**, 991, **1040**-58, **1138**
   as characteristic of life, 2, 63
   need to maintain, 1040–43
Homeotherm, 1066
Homeotic genes, 398–400, 398–99f
   in *Drosophila*, 365, 398–99, 398–99f
   role of, 398, 398f
Home range, **1124**, 1237
Hominid, **720**, 722
   compared to apes, 721
   early, 723
   evolution of, 722–23, 723f
   first, 460f
Hominoid, **720**, 721f
Homo erectus, 723f, 724–25, 725f
Homo ergaster, 723–24f, 724
Homogentisic acid, 296
Homo habilis, 723f, 724
Homo heidelbergensis, 723f, 724–25
Homologous chromosomes, **212**, 212f, 214, 216f, 230, 230–31f, 232, 233–34f, 267, 409, 1068
Homologous structures, **13**, 13f, **464**, 464f, 502, 516, 1072
Homologues, **212**
Homo neanderthalensis, 723f, 724–25, 725f
Homoplasy, **514**
Homoptera (order), 672f
Homo rudolfensis, 723f, 724
Homo sapiens, 723f, 724–26, 725f
Homosporous plant, **584**, 585t, 589
Homozygote, 249t, 251–52f, 252
Honeybee (*Apis mellifera*), 510–11, 986, 1062, 1120, 1120f, 1132–33, 1132f
   dance language of, 1120, 1120f
   pollination by, 844
Honeydew, 1174
Honeyguide, 711t, 898
Honey locust (*Gleditsia triacanthos*), 751
Honeysuckle (*Lonicera hispidula*), 762f
Hong Kong flu, 540–41
Hoof, 462, 716, 719t
Hoof-and-mouth disease, 524
Hooke, Robert, 15, 81
Hookworm, 622t, 646
Hopping, 883f
Horizontal cells, 986, 986f
Hormonal control
   of digestive tract, 900, 901f
   of osmoregulatory functions, 1057–58

Hormone, **127**, 860, **992**–93, 992f, 994–95t. *See also specific hormones*
   lipophilic, **993**, 996–99, 996f
   lipophobic, 996–99
   plant. *See* Plant hormone
   protein, 39t, 40
   that do not enter cells, 998–99, 998–99f
   that enter cells, 996–97, 996–97f
   transport in blood, 909–10
Hormone-receptor complex, 996–97f, 997
Hormone response element, 996f, 997
Horn (animal), 716, 1126
Horned lizard, 1240
Hornwort, 583, 583f
Horse, 528, 528f, 716–17, 719t, 883, 888–89, 892, 898, 1228
   chromosome number in, 210t
   evolution of, 462–63, 462–63f
   thoroughbred, 238, 450, 450f
Horseradish, 1169
Horsetail, 210t, 584, 585t, 586–87, 587f, 731
Host range, of virus, **532**
hot mutant, in *Arabidopsis*, 815, 815f
Hotspot, **1230**, 1230f
   population growth in, 1231, 1231f
Hot springs, 546
Hot sulfur springs, 519
Housefly (*Musca domestica*), 673f, 1150
   pesticide resistance in, 441, 441f
Hox genes, **399**, 527–28, 527f, 630, 758
HSP. *See* Heat shock protein
Hubbard Brook Experimental Forest, 1189, 1189f
Human, 719t
   birth weight in, 447, 447f
   cleavage in, 1097
   cloning of, 425
   development in, 383f, 1097–1101, 1097–1101f
   dispersal of animals and plants by, 1143
   DNA of, base composition, 285, 285t
   effect of global warming on, 1224
   effect on biosphere, 1218–24
   effect on ecosystems, 1218–24
   environmental problems caused by, 1195, 1236–37
   essential nutrients for, 904, 904t
   extinctions due to
     in historical time, 1229, 1229f, 1229t
     in prehistoric times, 1228, 1228f
   forebrain of, 957–60, 957–59f
   gastrulation in, 1098
   gene disorders in. *See* Gene disorder
   genetic map of, 269, 269f
   genetic variation in, 241f
   language development in, 1121
   migration from Africa, 725, 725f
   migration of, 357
   milk of, 714
   species introductions, 1240–41
   survivorship curve for, 1147, 1147f
   teeth of, 890
Human chorionic gonadotropin (hCG), 1075, 1098, 1100, 1100f
Human chromosomes, 210, 210f, 210t, 270–71, 295f
   alterations in chromosome number, 272–73, 272–73f
   chromosome 2, 495
   chromosome number, 210t
   composition of, 211
   karyotype, 212f, 270f
   sex chromosomes, 270, 270t
Human disease
   bacterial, 554–56, 555t
   effect of global warming on, 1224

   fungal, 604, 614
   nematodes, 646f, 647
Human evolution, 357, 439, 720–21
   future of, 488
Human Gene Mutation Database, 259, 262
Human genetics, 259–64. *See also* Gene disorder
Human genome, 348, 348f, 352, 492t
   comparative genomics, 494–95, 500
   foreign DNA in, 500
   genetic privacy, 358
   noncoding DNA in, 350
   segmental duplication in, 497–99, 499f
   single nucleotide polymorphisms in, 351, 355, 357
   transposable elements in, 350–51
Human Genome Project, 274, 348, 351
Human immunodeficiency virus (HIV), 109, 524f, 532f, 533t, 1033–34, 1033–34f
   effect on immune system, 536
   infection cycle of, 536–37, 537f
   latency period in humans, 536
   transmission of, 1033–34, 1076
   treatment of
     blocking or disabling receptors, 539, 539f
     chemokines and CAF, 538, 539f
     combination therapy, **538**
     vaccine therapy, 538
   tuberculosis and, 554
Human leukocyte antigen (HLA), **1020**
Human population
   in developing and developed countries, 1157–58, 1157–58f, 1157t
   growth of, 1146, 1155–58
     decline in growth rate, 1158
     exponential, 1155–56, 1155f
     future situation, 1157, 1157f, 1157t
     in hotspots, 1231, 1231f
     population pyramids, 1156, 1156f
Human remains, identification of, 357
Humboldt Current, 1212, 1212f
Humerus, 699f, 871f
Hummingbird, 711t, 713, 845, 845f, 1124, 1124f
Humoral immune response, 1023–24, 1023–24f, **1031**–32
Humoral immunity, **1018**
Humpback whale, 1239, 1239f
Humus, **784**
hunchback gene, 395f, 396
Hunger, 958
Hunter syndrome, 269f, 333t
Hunting, 1228, 1239
Huntingtin, 259
Huntington disease, 249t, 259, 264t, 950
Hurricane, 1224
Hybrid inviability, 473t
Hybridization (between species), **243**, 474–78, 474f, 497
Hybridization (nucleic acid), **328**, 328f, 330, 330–31f
Hybridoma, 1030, 1030f
Hybrid sterility, 473t
Hydra, 622t, 639f, 640, 888f, 908, 1147
Hydration shell, **30**, 116, 116f
Hydraulic propulsion, 882
Hydrocarbon, **36**
Hydrochloric acid, 31
   gastric, 892–93, 900, 900t, 901f
Hydrocortisone. *See* Cortisol
Hydrofluorocarbon, 1222

I-18   Index

Hydrogen, 26, 26f
  atomic structure of, 20f
  in plants, 782, 782f
  prebiotic chemistry, 67f
Hydrogenated oils, 54
Hydrogen bond, **28**
  in DNA, 50, 50f, 287, 287f
  in proteins, 43–44f, 45
  in water, 28, 28f, 30, 30f
Hydrogen cyanide, prebiotic chemistry, 66, 67f
Hydrogen ion, **31**–32
  excretion into urine, 1056, 1056f
Hydrogen peroxide, 94, 94f, 803
Hydrogen sulfide, 163, 182, 189, 196, 546, 553, 1217
Hydroid, 638, 640, 640f
Hydrolysis, **37**, 37f
Hydronium ion, 31
Hydrophilic molecule, **30**, 106, 106f
Hydrophobic exclusion, **30**, 45
Hydrophobic molecule, **30**, 106, 106f
Hydrophyte, 1217
Hydroponic culture, 782–83, 783f
Hydrostatic pressure, **117**, 117f
Hydrostatic skeleton, 646, 658, **871**, 871f
Hydrotropism, 811
Hydroxide ion, **31**–32
Hydroxyapatite, 864
Hydroxyl group, 36, 36f
Hydrozoa (class), 640
Hyena, 180f
Hymenoptera (order), 675t
  social systems in, 1132–33, 1132–33f
Hyoseris longiloba, 843f
Hyperaccumulating plant, 791–92
Hypercholesterolemia, 119, 249t, 264t, 333t, 411
Hyperosmotic solution, **116**, 117f, 1044
Hyperpolarization, 944–45, 945f, 950
Hypersensitive response, in plants, **803**, 803–4f
Hypersensitivity, 1036
  delayed, **1036**
  immediate, **1036**, 1036f
Hypertension, 918, 921
Hypertonic solution, **1044**
Hypertrophy, **881**
Hyperventilation, **931**–32
Hyphae, 600, 600f, **601**–3, 797, 797f
Hypocotyl, 730, 756f, 764f
Hypolimnion, 1216–17, 1217f
Hypoosmotic solution, **116**, 117f, 1044
Hypophosphatemia, 269f
Hypophyseal duct, 686f
Hypothalamohypophyseal portal system, 1003–4, 1003f
Hypothalamus, 955f, 955t, **956**, 958, 973, 993, 1000, 1000–1001f, 1010, 1042, 1109
  control of anterior pituitary by, 1003, 1003f
Hypothesis, **5**–7, 6f
Hypothyroidism, 1005
Hypotonic solution, **1044**
Hypoventilation, **931**–32
Hyracotherium, 462–63, 462–63f

I

IAA. See Indoleacetic acid
IBA. See Indolebutyric acid
I band, 874–77, 874–77f
Ibis, 711t
Ice, 27f, 28t, 29, 30f
Ice Age, 1163, 1228
Ichthyosauria (order), 703t
Ichthyosis, 269f
Ichthyostega, 700, 700f
Icosahedron, **532**
Identical twins, 1108
Idiomorph, 237
Ig. See Immunoglobulin
I gene (ABO blood group), 260, 260f
I gene (*lac* operon), 366f
Ig fold, 1032
Iguana, 708
  Galápagos, 471f
Iiwi, 1140f
Ileocecal valve, 897f
Ileum, **894**
Illicium, 594f
Imaginal disc, 387f
Immediate hypersensitivity, **1036**, 1036f
Immigration, 437, 1150
Immovable joint, 872, 872f
Immune regulation, in vertebrates, 1009–10
Immune response, 909, 1018, 1028f
  cell-mediated, **1018**, **1031**–32
  cells involved in, 1019, 1019t
  concepts of specific immunity, 1018
  discovery of, 1018
  humoral, **1018**, 1023–24, 1023–24f, **1031**–32
  initiation of, 1020, 1020t
  primary, **1027**, 1027f
  secondary, **1027**, 1027f
Immune surveillance, 910f, **1022**
Immune system, 857t, 858f, 909, 1013–36, **1014**
  cells of, 910f
  defeat of, 1033–36
  effect of HIV on, 536
  evolution of, 1031–32, 1032f
  of invertebrates, 1031, 1031f
  of vertebrates, 1032
Immunity
  active, **1018**, 1026–27, 1027f
  passive, **1018**
Immunization, passive, 1029
Immunochemistry, characterization of receptor proteins, 126
Immunocytochemistry, 83
Immunodeficiency, X-linked with hyper IgM, 269f
Immunoglobulin (Ig), 39t, 136, **1023**–24. See also Antibody
  classes of, 1024t
  structure of, 136f
Immunoglobulin A (IgA), 1024–25, 1024f
Immunoglobulin D (IgD), 1024–25, 1024f
Immunoglobulin E (IgE), 1024–25, 1024f, 1036
Immunoglobulin G (IgG), 1024–25, 1024f, 1024t
Immunoglobulin M (IgM), 1024–25, 1024f, 1024t
Immunoglobulin superfamily, 1032
Immunological tolerance, **1026**
Impala, 1126, 1173f
Implantation, 1074, 1074f, 1098
  prevention of, 1078
Imprinting (behavior), **1112**, 1112f
  filial, 1112
  sexual, **1112**
Imprinting, genomic, 428
Inchworm caterpillar (*Necophora quernaria*), 1170f
Incisor, 715, 715f, 890, 890–91f
Incomplete flower, 840
Incontinentia pigmenti, 269f
Incurrent siphon, 657f, 686f

Incus, 980, 981f
Independent assortment, **236**, 238, 238f, 267
  law of, 253–**54**, 254f
Indeterminate development, 626f, **627**
Indian grass (*Sorghastrum nutans*), 835
Indian pipe (*Hypopitys uniflora*), 789, 789f
Indifferent gonad, 1063, 1063f
Indigo bunting, 1116–17
Indirect effect, **1176**
Individualistic concept, of communities, **1162**
Indoleacetic acid (IAA), 820, 820f
Indolebutyric acid (IBA), 821
Induced fit, 150, 150f
Induced ovulator, 1067
Inducer T cells, 1019, 1019t, 1033
Induction (development), 391–92, 392f, 1090, **1093**, 1093f
  primary, **1093**
  secondary, **1093**
Induction of protein, 366–67
Inductive reasoning, **4**
Industrialized countries. See Developed countries
Industrial melanism, **456**
  agent of selection, 457
  in peppered moth, 456–57, 456–57f
  selection against melanism, 457, 457f
  selection for melanism, 456, 456f
Industrial pollution, 457, 1218
Industrial Revolution, 1155f
Indy gene, 402
Inert element, **24**
Infant, growth of, 1102
Infection thread, 787f
Inferior vena cava, 917f, **918**
Inflammatory response, 538, 910f, 1017, 1017f, 1025
Influenza, 533t, 540
  pandemic of, 1013
Influenza virus, 524f, 531f, 533t
  antigen shifting in, 1035
  subtypes of, 540
  types of, 540
  vaccine for, 334
Information molecules, 48–51
Infrared radiation, sensing of, 988, 988f
Ingenhousz, Jan, 188
Ingram, Vernon, 298
Inguinal canal, 1068, 1070
Inhalation. See Inspiration
"In heat," 1067
Inheritance
  of acquired characteristics, **434**, 434f
  blending, 243, 248, **436**
  chromosomal theory of, **265**
  patterns of, 241–74
Inhibin, 1071, 1071f
Inhibitor, **152**
  allosteric, **152**, 153f
  competitive, **152**, 153f
  noncompetitive, **152**, 153f
Inhibitory postsynaptic potential (IPSP), 949f, **950**–51, 951f, 963
Initiation complex, 308–9, 308f, **310**–11, 311f, 370, 370f, 372f, 373
Initiation factor, **310**, 311f
Injectable contraceptive, 1077t
Innate behavior, 1106–7, 1107f
Innate releasing mechanism, **1106**
Inner cell mass, 1086, 1086f, 1089, 1096–97
Inner ear, 978, 980, 981f
Inner membrane
  of chloroplasts, 96, 97f, 186f

  of mitochondria, 96, 96f, 163f, 176, 562
Inonotus tomentosus, 601f
Inositol triphosphate (IP$_3$), 132, 133f, 135
Inositol triphosphate (IP$_3$)/calcium second-messenger system, 998–99, 999f
Insect, 635, 651f, 664, 664–65f, 668t, 672–74, 675t, 871
  chromosome number in, 210t
  coevolution of insects and plants, 844
  development in, 382f, 386, 386–87f
  digestive system of, 674
  diversity among, 672f
  evolution of, 527–28
  excretory organs in, 1046–47, 1047f
  external features of, 673, 673f
  eyes of, 450, 450f, 504–5, 504–5f, 666–67, 666f, 673
  heart of, 908, 908f
  internal organization of, 674
  locomotion in, 883–84
  metamorphosis in, **674**, 1009, 1009f
  molting in, 871, 1009, 1009f
  orders of, 675t
  pheromones of, 674
  pollination by, 595, 844, 1164
  respiration in, 923f, 926–27
  segmentation in, 665f
  selection for pesticide resistance in, 441, 441f
  sense receptors of, 674
  sex chromosomes of, 270, 270t
  social, 1120, 1120f, 1132–33, 1132–33f
  thermoregulation in, 1042, 1042f
  wings of, 503, 664, 665f, 673, 673f, 884
Insecticide, 336
Insectivora (order), 719t
Insectivore, digestive system of, 899f
Insectivorous leaf, 751
Insect resistance, in transgenic plants, 336
Insertional inactivation, **408**, 410t, 411
Insertion of muscle, **873**
Inspiration, 929–32, 931f
Instar, **386**, 387f
Instinct, 1106, 1111–13, 1113f
Insulin, 39t, 298, 402, 895f, 902, 902f, 993, 995f, 999, 1008, 1008f, 1042, 1042f
  genetically engineered, 333, 558, 1008
Insulin-like growth factor, 994, **1002**
Insulin-like receptor, 402
Integrating center, **1040**, 1040–41f, 1057
Integrin, 102, 102f, 139, 391
Integument (flower), 589, **595**–96, 595f, 760, 838
Integumentary system, 857t, 859f
Intelligent design theory, against theory of evolution, 468
Interarterial pathway, 919
Intercalary meristem, **731**, 756
Intercalated disc, 867f, 868
Intercostal muscles, 930–32, 931f, 1101f
Interference competition, **1164**
Interferon, 1022
  alpha-interferon, 1016
  beta-interferon, 1016
  gamma-interferon, 1016, 1022
Intergovernmental Panel on Climate Change, 785
Interleukin, 1021
Interleukin-1, 1017, 1020, **1021**, 1021f, 1023f, 1028f

Index I-19

Interleukin-2, 1021–**22**, 1021f, 1023, 1023f, 1028f
  genetically engineered, 1022
Interleukin-4, 1021
Intermediate disturbance hypothesis, 1180, 1180f
Intermediate filament, 38t, 86–87f, **98**–100, 98f, 137f, 138–39
Intermembrane space, of mitochondria, **96**, 96f, 163f
Internal environment
  maintenance of, 1039–58
  sensing of, 970, 970f
Internal fertilization, 707, **1064**, 1066, 1082
Internal membranes, of prokaryotes, 551, 551f
Internal respiration, 922
International Human Genome Sequencing Consortium, 348
International Whaling Commission, 1239
Interneuron, 869t, **940**, 940f, 954, 962, 962f
Internodal pathway, 919, 919f
Internode, 732, 732f, **744**, 744f, 747
Interoceptor, **971**, 971f
Interphase, **213**, 214, 216f
Intersexual selection, 1126–27, 1126–27f
Interspecific competition, **1164**, 1164f, 1167, 1167f
Interstitial fluid, **908**, 910, 913–14, 914f
Intertidal region, 1176, **1214**, 1214f
Intervertebral disc, 864, 872, 872f
Intestinal roundworm (*Ascaris*), 647
Intestine. *See also* Large intestine; Small intestine
  bacteria in, 557, 893, 897–98
Intracellular compartment, 1045f
Intracellular digestion, 888
Intracellular receptor, **128**–29, 128–29f, 128t
  enzymic, 129
  gene regulators, 129, 129f
Intracellular receptor superfamily, 129
Intrasexual selection, 1126
Intrauterine device (IUD), 1077t, 1078
Intrinsic factor, 892
Introduced species, 1234t, 1240–41, 1246
Intron, 313–14, **313**, 313f, 315f, 316, 349–50, 350t, 376, 494, 520t, 542, 548
Invagination, **1087**
Inversion, 411, 411f
Invertebrate, **619**, 651–80
  circulation in, 908, 908f
  extinctions, 1229t
  immune system of, 1031, 1031f
  locomotion in, 883
  noncoelomate, 633–48
  osmoregulatory organs of, 1046, 1046f
  phylogeny of, 634–35, 634–35f
  respiration in, 922–23
Involution, **1087**, 1089
Iodine, 997, 997f
  deficiency of, 1004, 1004f
Ion(s), **21**
Ion channel, 109t, **114**, 128t, **130**
  gated, **944**
  hormones acting on, 999
  ligand-gated, 130, 130f, **944**, 944f, 949
  stimulus-gated, 972, 972f
  voltage-gated, 944–45, 945f, 972f
Ionic bond, **25**, 25f, 43f

Ionic compound, **25**
Ionization, **31**
  of water, 31–32, 31–32f
IP₃. *See* Inositol triphosphate
IPSP. *See* Inhibitory postsynaptic potential
Iriomote cat, 1140f
Iris (eye), 868, 983, 984f
Iris (plant), 594, 747, 747f, 834f, 835, 849
Irish elk (*Megaloceros*), 717t
Irish potato famine, 572
Iron
  absorption by intestine, 337
  deficiency of, 337, 337f
  in hemoglobin, 934, 934f
  in plants, 772, 782, 782t
Irreducible complexity argument, against theory of evolution, 468
Island
  biogeography of, 1200, 1200f
  endemic species, 1230
  evolution on, 439, 467, 480, 481–82f, 482–83, 1168
  extinctions on, 1229, 1229f, 1236, 1236f
  species introductions, 1240
Islets of Langerhans, **895**, 895f, 902, 902f, 1008, 1008f, 1042, 1042f
Isocitrate, 170, 171f
Isocitrate dehydrogenase, 171f
Isolating mechanism, 473–74, 473t
  as by-product of evolutionary change, 478–79
  incomplete, 478–79
  postzygotic, 473–**74**, 473t, 476, 476f
  prezygotic, 473, 473t, **474**–75, 478
  reinforcement by selection, 478–79, 478f
Isoleucine, 42f
Isomer, **56**, 56f
  of sugars, 56, 56f
Isomerase, 165f
Isometric contraction, **873**
Isometric virus, **532**
Isopod, 669, 1129, 1140
Isopoda (order), 669
Isoptera (order), 672f, 675t, 1133
Isosmotic solution, **116**–17, 117f, 1044
Isotonic contraction, **873**
Isotonic solution, **1044**
Isotope, 21–22, 21f
  radioactive, 21
Italian ryegrass (*Lolium multiflorum*), 613
Iteroparity, **1149**
IUD. *See* Intrauterine device
Ivy, 743, 747, 749f, 832–33, 832f, 835

J

Jackal, 1190
Jack pine, 760f
Jacob syndrome, 273
Jaguar, 1195
James River, 1218
Janssens, F. A., 267
Japan current, 1212f
Jasmonic acid, 802, 802f, 804
Jaundice, 895
Java man, 725f
Jaw(s)
  of cichlid fish, 484, 484f
  evolution of, 692–93, 692–93f
  of fish, 690
  of mammals, 717
  of vertebrates, 689f
Jawed fish, 690, 692–93, 693f

Jawless fish, 688, 692, 692f
Jejunum, **894**
Jellyfish, 619t, 622t, 624, 638, 640, 640f, 871, 1215
Jelly fungi, 608
Jenner, Edward, 334, 1018, 1018f, 1027f
Jet propulsion, 653f
Jimsonweed, 844
Joint
  movement at, 872–73
  types of, 872
Jointed appendages, 883
  of arthropods, 664, 665f
  of chordates, 684
Joule, **144**
Juniperus chinensis, 1143f
Jun protein, 224
Juvenile hormone, **1009**, 1009f
Juxtaglomerular apparatus, 1058f
Juxtamedullary nephron, 1052–53, 1052f

K

Kalanchoë, 751, 849, 849f
Kallmann syndrome, 269f
Kangaroo, 528, 528f, 717, 718f, 719t, 883, 1067, 1067f
Kangaroo rat, effect on smaller, seed-eating rodents, 1167, 1167f
Kaposi's sarcoma, 1022
Karyogamy, 609–10
Karyotype, **212**, 212f, 274
  human, 212f, 270f
Katydid, 801f
Kaufmann, Thomas, 398
kdr gene, 441, 441f
Kelp, giant, 571f
Kelp forest, 1242–43, 1242f
Kennedy disease, 269f
Kenyanthropus platyops, 723f
Keratin, 39t, 40, 40f, 99, 702, 714, 716, 860, 1014
Keratinized epithelium, 860
α-Ketoglutarate, 170, 171f, 178, 178f
α-Ketoglutarate dehydrogenase, 171f
Kettlewell, Bernard, 456–57
Key stimulus. *See* Sign stimulus
Keystone species, **1177**, 1177f, 1242f, 1243
  preservation of, 1243
Khorana, Har Gobind, 304
Kidney, 859f
  of amphibians, 1050
  of birds, 1050
  evolution of, 1048–50
  excretion in, 1054
  filtration in, 1052, 1053f
  of fish
    cartilaginous fish, 1049
    freshwater fish, 1048, 1049f
    marine bony fish, 1048, 1049f
  hormonal regulation of, 1057–58
  of mammals, 1050, 1052–54, 1052–54f
  reabsorption in, 1054–56, 1054–56f
  of reptiles, 1050
  secretion in, 1054, 1054f
  of vertebrates, 1047–50
Kidney cancer, 414t
Kidney transplant, 1022
Killer strain, *Paramecium*, 570
Killer T cells, 1035
Killer whale, 27f, 909f, 1242, 1242f
Killifish (*Rivulus hartii*), 448–49, 448–49f
Kilobase, 344

Kilocalorie, **144**, 156, 903
Kinase, 39t, 220
α-Kinase, 132
Kinectin, 99
Kineses, **1116**
Kinesin, 98, **99**, 396
Kinetic energy, **144**, 144f
Kinetin, 822f
Kinetochore, 209f, **214**, 214f, 215, 216–17f, 217, 230, 233, 233f
Kinetochore microtubules, 209f, 214f, 215, 216f, 234–35f
Kinetoplastid, 567, 567f
King, Thomas, 281
Kingdom (taxonomy), 16, 16f, 74, **511**, 518, 518f
  evolutionary relationships among kingdoms, 521f
Kinocilium, 977–79, 977f
Kin selection, 1130–**31**, 1131f, 1132, 1134
Kipukas, 482
Kiwi (bird), 711t
Klebsomidiales, 526f
Klinefelter syndrome, 273, 273f
Knee-jerk reflex, 961f, 962
Knight, T.A., 243
knirps gene, 395f
Koala, 717, 719t
Koelreuter, Josef, 243
Komodo dragon (*Varanus komodoensis*), 708, 1230
Kornberg, Arthur, 263
Krait, 709
Krakatau islands, succession after volcanic eruption, 1179, 1179f
K-ras gene, 416f, 417t, 419f
Krebs, C., 1154
Krebs cycle, 162, **169**, 171f, 175f
  ATP production in, 176, 176f
  products of, 170, 171f
  reactions of, 170
  regulation of, 177, 177f
Kreitman, Martin, 435
Krüppel gene, 395f
K-selected population, **1154**–55, 1154t, 1179
Kuru, 542
Kwashiorkor, 914

L

Labia majora, 1072, 1072f
Labor (childbirth), 1101
Labrador Current, 1212f
Labrador retriever, coat color in, 258, 258f
Lacerta, 1062
Lacewing (*Chrysoperla*), courtship song of, 475
lac operator, 366f, 368f
lac operon, 366, 366f, 369, 369f
lac promoter, 366f, 368f, 369
lac repressor, 39t, 368f, 369, 369f
Lactase, 57, 894, 900t
Lactate, 167f, 554
  blood, 881
Lactate dehydrogenase, 181, 441, 441f
Lactation, 1101–2, 1101f
Lacteal, 894
Lactic acid fermentation, 181, 181f
Lactiferous duct, 1101f
Lactose, 56f, 57, 369, 369f, 894
Lactose intolerance, 894
Lacunae
  within bone, 865, 865f
  within cartilage, 864, 864f
LacZ' gene, 321–22, 322f, 327, 327f

I-20 Index

Laetoli footprints, 722
Lagena, 980
Lagging strand, **290**–91, 291–92f, 294, 294f
Lagomorpha (order), 719t
Lake, 1184f, 1185, 1216, 1216f
   eutrophic, **1178**, **1217**, 1241
   oligotrophic, **1178**, **1217**
   productivity of, 1191f, 1217
   thermal stratification of, 1216–17, 1217f
Lake Barombi Mbo cichlid fish, 481
Lake Victoria cichlid fish, 484, 484f, 1241, 1241f, 1246
Lamarck, Jean-Baptiste, 434
Lamellae
   of bone, 865, 865f
   of gills, 925–26, 925f
Lamellipodia, 385
Lamprey, 689–90, 691f, 692, 693t, 1032, 1032f
Lamp shell, 623t, 663
Lancelet, 685f, 687, 687f, 915, 1087
   eggs of, 1084f
   gastrulation in, 1087f
Landmarks, on physical map, **344**–45, 344–45f
Land plants, evolution of, 526, 526f
Land snail (*Cepaea nemoralis*), 440
Langerhans, Paul, 1008
Language, 726, 959–60, 959f
   development in humans, 1121
   of primates, 1121, 1121f
Lanugo, 464, **1100**
Larch, 590
Large intestine, 859f, 889, 889f, 897, 897f
Large offspring syndrome, 425
Lariat structure, 314, 314f
Larkspur, 761f
Larva, **386**, **674**
   of amphibians, 977
   of bivalves, 657
   of echinoderms, 679, 679f
   of insects, 673f, 674
   of sponges, 617f, 637
   of tunicates, 686, 686f
Larvacea, 687
Larynx, 864, 864f, 891, 891–92f, **928**, 928f
Lassa fever, 357t
Late blight of potatoes, 572
Lateral bud, 822f, 827
Lateral bud primordium, 730f
Lateral gene transfer, **499**–500, 500f, 519
Lateral geniculate nucleus, 987, 987f
Lateral line organ, 971t
Lateral line system, 696–97, 977, 977f, 980
Lateral meristem, **730**, 731, 756
Lateral root, 741
Lateral root cap, 739, 739f
Lateral sulcus, 957f
Lateral ventricle, 957f
Latex, 736
Latitude, climate and, 1206–7f, 1207, 1209
Latitudinal cline, in species richness, 1199f
LDL. *See* Low-density lipoprotein
Lead, 791–92
Leader sequence, 310, 311f
Leading strand, **290**–91, 291–92f, 294, 294f
Leaf, 580, 732, 732f
   abscission of, 816f, 821, **852**, 852f
   alternate, 749, 749f
   compound, **749**, 749f

   external structure of, 748–49, 748–49f
   fall colors, 193f, 852
   internal structure of, 750, 750f
   modified, 751, 751f
   opposite, 749, 749f
   organization of, 186, 186f
   palmately compound, 749, 749f
   pinnately compound, 749, 749f
   simple, **749**, 749f
   transpiration of water from, 772
   whorled, 749, 749f
Leafcutter ant, 614, 1133, 1172
Leafhopper, 674, 675t
Leaf primordium, 730f, 742f
Leaf scar, 744, 744f, 852
LEAFY COTYLEDON gene, of *Arabidopsis*, 759
LEAFY gene, 837–38
   of *Arabidopsis*, 833, 833f, 835–36
Leafy spurge, 1240
Leak channel, **942**–43
Learning, 951, 960, **1110**
   associative, **1110**, 1110f
   behavior and, 1110–17
   genetics of, 1108, 1108f
   instinct and, 1111–13, 1113f
   nonassociative, **1110**
Learning preparedness, 1111
Leather, 1014
Lecithin, 37t
Lectin, 1031–32
Leder, Philip, 304
Lederberg, Joshua, 263, 406
Leech, 622t, 660–61, 661f, 882
Leeuwenhoek, Antonie van, 15, 81
Leg(s)
   of amphibians, 698–99, 699–700f
   of reptiles, 702
Leghemoglobin, 787f
Legionella, 549f
Legume, 761f, 787, 812, 814
   caesalpinoid, 613
Leishmaniasis, 567
Lemming, 1211
Lemon, 761f, 826
Lemon shark, 1064f
Lemur, 467, 719t, 720, 721f, 1228
Lens, 504, 983–84, 983–84f, 987f, 1093, 1093f
Lenticel, 746, 746f, 774–75, 775f
Leopard frog (*Rana*), postzygotic isolation in, 476, 476f
Leopold, Aldo, 1195
Lepidoptera (order), 672f, 675t, 844
Lepidosaur, 691f
Leprosy, 555t
Leptin, 903, 903f
Lesch-Nyhan syndrome, 269f
Lettuce, 835
Leucine, 41, 42f
Leucine zipper motif, **365**, 365f
Leucoplast, 97
Leukemia, 414–15, 417t, 1232f
   chronic myelogenous, 416
   myeloid, 417t
Leukocytes, 94, 536, 863t, 866, 866f, 909, 910f, **911**, 1015, 1019
   endocytosis in, 119
   granular, **911**
   movement of, 100
   nongranular, **911**
Level of specificity, **1119**
Levene, P. A., 284
Lewis, Edward, 398
Leydig cells, 1068, 1071, 1071f
LH. *See* Luteinizing hormone
Lichen, **604**, 611, 611f, 1172, 1178
   as air quality indicators, 611

   foliose, 611f
   fruticose, 611f
Life
   characteristics of, 62–63, 62f
   chemical building blocks of, 35–58
   diversity of, 74
   hierarchical organization of living things, 2, 3f
   origin of. *See* Origin of life
   properties of, 2
   unity of, 16
Life cycle
   of *Allomyces*, 606f
   of ascomycetes, 609f
   of basidiomycetes, 608, 608f
   of brown algae, 571
   changes due to global warming, 1223
   of *Chlamydomonas*, 573f
   of fern, 588f
   of foraminiferans, 575
   of moss, 583f
   of *Obelia*, 640f
   of *Paramecium*, 570f
   of pine, 591f
   of plants, 581, 581f, 595–96, 596f
   of *Plasmodium*, 569f
   of *Rhizopus*, 607f
Life expectancy, 1156
Life history, **1148**–49
Life history model, 1154
Life span, 402, 1145
   of *Drosophila*, 402
   of plants, 851–52
Life table, **1146**–47, 1146t
Lift, 884
Ligament, 862, 872
Ligand-gated channel, **944**, 944f, 949
Light. *See also* Sunlight
   cue to flowering in plants, 834–35, 834–35f, 838f
   wavelength of, 190, 190f
Light chain, **1025**–26, 1025–26f
Light-dependent reactions, of photosynthesis, 186–87f, **187**, 189, 196–97f, 196–99f
Light-independent reactions, of photosynthesis, **187**–89
Light microscope, 82, 82f, 83t
Lignification, 736
Lignin, 603–4, **736**, 737
Likens, Gene, 1189
Lilac, 749
Lily, 594, 743, 747, 843f
Limb, development of, 502–3, 502f, 528
Limb bud, 1098
Limbic system, 952, **958**
Lime, 761f
Limestone, 71, 575, 575f, 641
Limnetic zone, 1216, 1216f
Limpet, 1215
Lindane, 1218
LINE. *See* Long interspersed element
Lineage map, 390, 390f
Linen, 736
Lineus, 645f
Lingula, 623t
Linkage disequilibrium, **351**
Linkage map. *See* Genetic map
Linnaeus, Carolus, 510
Lion (*Panthera leo*), 62f, 143f, 180f, 474, 474f, 718f, 1105f, 1130, 1210, 1228
Lipase, 895, 900t
Lipid(s), 36, 52–53. *See also* Fat(s)
   functions of, 37t
   membrane, 520t
   of archaebacteria, 71
   structure of, 37t

Lipid bilayer, 52, 53f, 80, **106**, 107f, 108, 108f, 109t
   anchoring proteins in, 112, 112f
   fluidity of, 107, 107f
Lipid raft, 109
Lipophilic hormone, **993**, 996–99, 996f
Lipophobic hormone, 996–99
Liposome, 68
Littoral zone, 1216, 1216f
Liver, 689f, 859f, 860, 888f, 889, 889f, 895
   cell cycle in liver cells, 213, 222
   innervation of, 965t
   regulatory functions of, 902
Liver cancer, 414–15, 541
Liver fluke (*Clonorchis sinensis*), 622t, 644, 644f
Liverwort, 583
Lizard, 703t, 705–6f, 707–8, 1050, 1170, 1198f, 1207
   parthenogenesis in, 1062
   population dispersion, 1143
   territories of, 1124
   thermoregulation in, 1017, 1042, 1138–39, 1139f
Lobe-finned fish, 690, 692f, 693t, 696–97, 697f, 699, 699f
Lobopodia, 565
Lobster, 58f, 668, 669f
   cycliophorans on mouthparts of, 648, 648f
Locomotion, 870–73
   in air, 884, 884f
   appendicular, 882
   axial, 882
   on land, 883, 883f
   in water, 882, 882f
Locomotor organelles, of protists, 565
Locus, **249**
LOD (log of odds ratio), 269
Loggerhead sponge, 636
Logging, 1236–37
Logistic growth model, 1150–51, 1150–51f
Logarithmic scale, 31
"Lollipops," 408f
Long bone, 865
Long-day plant, **834**–35, 834f
   facultative, 835
Longhorned grasshopper, 674
Long interspersed element (LINE), 350
Long-tailed shrike, 1179
Long terminal repeat (LTR), 350–51
Long-term memory, 955, 960
Long-term potentiation (LTP), 960
Loop of Henle, 1048f, 1050, **1053**, 1053f, 1055–56, 1055f
Loose connective tissue, 856f, **862**, 863f
Lophophorate, 621f, 628, 634–35f, **635**, 662–63, 662–63f
Lophophore, 623t, **635**, **662**–63, 662–63f
Lophotrochozoan, 527, 628, 629f, 635, 635f
Lorenz, Konrad, 1106, 1107f, 1112, 1112f
Loricifera (phylum), 623t
Loris, 720, 721f
Louse, 555t, 673
Lovebird (*Agapornis*), nesting behavior in, 1108, 1108f
Love Canal, 1218
Low-density lipoprotein (LDL), 119, 378
Low-density lipoprotein (LDL) receptor, 411

Index I-21

Lowe syndrome, 269f
LSD, 950
LTP. *See* Long-term potentiation
LTR. *See* Long terminal repeat
Ludia magnifica, 679f
Lugworm, 660
Lumbar nerves, 954f
Luna moth (*Actias luna*), 672f
Lung(s), 858f, 923f, **927**
 of amphibians, 698–99, 927, 927f
 of birds, 928–29, 929f
 book, 670, 670f
 innervation of, 965t
 of mammals, 928, 928f
 of reptiles, 702, 927–28
Lung cancer, 413f, 414, 414–15t, 417t, 420–21, 420–21f, 423
 smoking and, 415
Lungfish, 697, 699
Luteinizing hormone (LH), 994t, 1001–2, 1001f, 1003f, 1067, 1071, 1071f, 1071t, 1074–75, 1075f
Luteolysin, 1075
Lycophyta (phylum), 584, 585t, 586, 586f, 748
Lycopodium lucidulum, 586f
Lyell, Charles, 10
Lyme disease, 554, 555t, 670
Lymph, 908, **914**
Lymphatic capillary, 896, 896f, 914, 914f
Lymphatic duct, 894f
Lymphatic system, 857t, 858f, 908, 913–14, 914f, 1015, 1015f
Lymphatic vessel, 858f, **908**, 914, 914f, 1015f
Lymph heart, **914**
Lymph node, 858f, 914, 1015f
Lymphocyte(s), 866, **911**, 914, 1019
Lymphocyte-like cells, of invertebrates, 1031
Lymphokine, **1021**
Lymphoma, 414t, 417t, 1022
Lymphoproliferative syndrome, 269f
Lynx, 1211
Lysenko, 836
Lysine, 42f
Lysogenic cycle, of bacteriophage, **534**, 535f
Lysogeny, **534**
Lysosome, 86–87f, 90t, **94**, 94f, 102t, 565, 1015
 primary, 94
 secondary, 94
Lysozyme, 150f, **1014**
Lytic cycle, of bacteriophage, **534**, 535f

M

Macaque, 466, 991f, 1114
MacArthur, Robert, 1166, 1200
McCarty, Maclyn, 283
McClintock, Barbara, 350, 406
MacLeod, Colin, 283
Macromolecule, 3f, 35, 35f, **36**
 polymer, 38t
 production of, 36–37, 37f
 structure of, 37t
Macronucleus, 569–70, 569–70f
Macronutrients, in plants, 782, 782t
Macrophage(s), 862, 863f, **1015**, 1015f, 1019t, 1020, 1020t, 1021f, 1023f, 1024, 1028f
 HIV entry into, 536, 537f
Macrophage colony-stimulating factor, 1022

Macrosiphon rosae, 776f
Macrotermes bellicosus, 672f
Mad cow disease, 542
Madreporite, 677f, **678**, 678f
Magnesium, in plants, 782, 782t
Magnetic field
 migration of birds and, 1117
 sensing of, 971v, 988
Magnetic receptor, 988
Magnetite, 1117
Magnolia, 594f
Mahogany tree (*Swietenia mahogani*), 1140f, 1239
Maidenhair tree (*Gingko biloba*), 592, 592f
Major groove, 286f, **363**–64, 363f, 367, 367–68f
Major histocompatibility complex (MHC), 128t, **1020**, 1033
 MHC proteins, 39f, 109t, 136, 136f, **1020**–21, 1020t, 1021f, 1023f
Malaria, 496, 568–69, 800, 1035, 1224
 in *Anolis* lizards, 1176
 avian, 1229, 1240
 eradication of, 568–69
 sickle cell anemia and, 262, 262f, 445, 445f
Malate, 170, 171f, 203, 204f
Malate dehydrogenase, 171f
Male reproduction, hormonal control of, 1071, 1071f, 1071t
Male reproductive system, 859f, 1068–71, 1068–71f
Mallard duck, 474
Malleus, 980, 981f
Malpighian tubule, 665–66f, **667**, 671, 674, 1046–47, 1047f
Maltase, 900t
Malthus, Thomas, 11, 1157
Maltose, 56f, 57, 57f, 890
Mammal, 622t, 714–18
 brain of, 956, 956f
 breathing in, 930–36
 characteristics of, 714–16
 circulation in, 916–17, 917f
 cleavage in, 1086, 1086f
 digestion of plants by, 716
 egg-laying. *See* Monotreme
 evolution of, 528, 528f, 689, 717, 717f
 extinctions, 717t, 1229, 1229f, 1229t, 1234t
 flying, 716, 716f, 719t
 gastrulation in, 1089, 1089f
 heart of, 715, 916–17, 917f
 kidney of, 1050, 1052–54, 1052–54f
 marine, 719t
 nitrogenous wastes of, 1051, 1051f
 orders of, 717–18, 719t
 placental. *See* Placental mammal
 pouched. *See* Marsupial
 present day, 718, 718f
 respiration in, 923f, 928, 928f
 saber-toothed, 516f
 sex determination in, 1063, 1063f
 teeth of, 715
 thermoregulation in, 714–15
Mammalia (class), 511f, 689, 691f, 714–18
Mammary gland, 714–15, 1101–2, 1101f
Mammoth, 717t, 1228
Manatee, 465
Manaus, Brazil, 1237, 1237f
Mandible, of crustaceans, **668**, 668t
Manganese, 153
 in plants, 782, 782t
Mangrove, 775, 775f, 1232, 1233f, 1238
Manic-depressive illness, 269f

Manihotoxin, 798, 799t
Mantle, 653–54f, **654**
Mantle cavity, 654f, 656, 924
Manual dexterity, 957
Map unit, 268, 344
Maple, 749, 761f, 762
Marburg hemorrhagic fever, 357t
Marchantia, 583f
Margulis, Lynn, 72
Marianas Trench, 1214
Marine bird, 1050f
Marine ecosystem, 1214–15, 1214–15f
 plant adaptations to, 775, 775f
Marine invertebrate, 1044
Mariner transposon, 411
Marine vertebrate, 1044
Marler, Peter, 1113
Marra, Peter, 1238
Marrella splendens, 630f
Mars, life on, 64, 76
Marsh, 1216
Marsilea, 587
Marsupial, 528, 528f, 717–18, 718f, **1067**, 1067f, 1072
 marsupial-placental convergence, 467, 467f
 saber-toothed, 516f
Marsupialia (order), 719t
Mass, 20
Mass extinction, **487**, 487f, 694f, 695, 700, 1229
Mass-flow hypothesis, of phloem transport, 776–77, 777f
Mast cells, 862, 863f, 1019t, **1025**
Mastication. *See* Chewing
Mastodon, 1228
Mate choice, **1125**, 1125f, 1127
Maternal age, Down syndrome and, 272, 273f, 274
Maternal care, in mice, 1109, 1109f
Maternal genes, 386, 387f
Maternity plant, 849, 849f
Mating. *See also* Courtship *entries*
 assortative, **438**
 disassortative, **438**
 random, 436–38, 437t, 438f
Mating behavior, 1061, 1061f
 selection acting on, 479
Mating call, 1119, 1127f
Mating factor, of yeast, 131
Mating ritual, 474
 of prairie chickens, 1245, 1245f
Mating success, 442
Mating system, 1128–29, 1128–29f, 1134
Mating type
 in ciliates, 570
 in fungi, 602, 607f, 608
Matrix
 extracellular. *See* Extracellular matrix
 of mitochondria, **96**, 96f, 174–75
Matter, **20**
Mauna Kea silversword (*Argyroxiphium sandwicense*), 1230
Maximal oxygen uptake, 881
Maximum likelihood method, 514
Maxwell, James Clerk, 190
Mayr, Ernst, 480
Maze-learning ability, in rats, 1108, 1108f
MCS. *See* Multiple cloning site
MDM2 gene, 417t
Measles, 533t
Mechanical isolation, 473t, 475
Mechanoreceptor, **970**, **973**
Mediator, 371
Medicago polycarpa, 1143f
Medicinal leech (*Hirudo medicinalis*), 661, 661f

Medicine, applications of genomics to, 357, 357f
Mediterranean climate, 1206, 1230
Medulla oblongata, 932, 933f, 955f, 955t, 957f, 964, 976
Medusa, 638, 638f, 640–41, 640f
Meerkat (*Suricata suricata*), 1134, 1134f
Megafauna, 1228
Megagametophyte, 596f, 756, 842, 842f
Megaphyll, 748
Megareserve, 1248, 1248f
Megasporangium, 591
Megaspore, 591, 591f, 595, 596f, 843, 843f
Megaspore mother cell, 591, 591f, 595, 595–96f, 838, 842f, 843
Megazostrodon, 704f
Meiosis, 227, **228**–38, 229f, **522**, 1062
 compared to mitosis, 230, 231f
 discovery of reduction division, 228–29
 errors in, 480
 evolution of, 522
 gametic, 565
 in plant cells, 227f
 sequence of events during, 232–36, 232–36f
 unique features of, 230–31, 230–31f
Meiosis I, **230**, 230f, 232–33, 233–34f, 236, 1069f, 1073, 1074f
Meiosis II, **230**, 230f, 235f, 236, 1069f, 1073–74, 1074f, 1083
Meissner's corpuscle, 971t, 973, 973f
Melanin, 256, 1002, 1010
Melanocyte-stimulating hormone (MSH), 994t, 1001f, 1002
Melanoma, 414t, 1022
Melanosome, 258
Melanotropin-inhibiting hormone (MIH), 1003
Melatonin, 995t, 1010
Membrane(s), 105–22. *See also specific membranes*
Membrane attack complex, 1016, 1016f
Membrane potential, 942–47
Membranous labyrinth, 978–79, 978–79f
Memory, 951, 958, 960
 long-term, 955, 960
 short-term, 960
Memory cells, 1019, 1023, 1023f, 1027, 1027–28f
Menarche, 1072
Mendel, Gregor Johann, 244, 244f
 experiments with garden pea, 244–54
 choice of garden pea, 244
 experimental design, 245, 245f
 Mendel's interpretation of results, 250–54
 page from notebook, 247f
 results of experiments, 246–49, 246t, 247–48f
 first law of heredity, 251
 model of heredity, 248–49
 second law of heredity, 253–54, 254f
Mendeleev, Dmitri, 24
Mendelian ratio, **248**
 modified, 255–58, **257**
Meninges, 961
Meningioma, 417t
Menkes syndrome, 269f
Menstrual cycle, 1010, 1067, 1072–73
 follicular phase of, 1073, 1073f, 1075f
 luteal phase of, **1075**, 1075f
 ovulation, 1074, 1074–75f
Menstrual phase, of endometrium, **1075**

Menstruation, 1067, 1075
β-Mercaptoethanol, 36f
Mercury, 1218
Meristem, **388**, 730–31, 733, 756
 apical, 388, 389f, **730**–31, 730–33f, 737, 739f, 740, 742f, 744–45, 744f, 748, 756, 756f, 758, 822f
 floral, 838, 838–39f
 ground, 730f, **731**, 731f, 739f, 741, 745, 756f
 intercalary, **731**, 756
 lateral, **730**, 731, 756
 primary, **730**, 740, 745
Meristematic development, 388, 389f
Merkel cells, 971t, 973, 973f
Meroblastic cleavage, **1086**, 1086f
Merozoite, 569f
Merychippus, 462–63f
Meselson-Stahl experiment, 288, 288–89f
Mesencephalon. *See* Midbrain
Mesoderm, 384f, 385, 392, 620, **624**, 625, 625f, 627, 642, 653f, **856**, 860, 1085t, **1087**–90, 1087–90f, 1087t, 1092–93, 1094f, 1098
Mesoglea, **638**, 638–39f
Mesohippus, 462–63f
Mesohyl, **636**
Mesophyll, 203, 204f, **750**, 772
 palisade, **750**, 750f, 771f
 spongy, **750**, 750f, 771f
Mesostigma, 526
Mesostigmatales, 526f
Mesothelioma, 415t
Messenger RNA (mRNA), 37t, 95, **302**. *See also* Primary transcript
 artificial, 304
 5′ cap, **309**, 309f, 316
 degradation of, 374, 377f, 378
 leader sequence on, 310, 311f
 making cDNA library, 324, 324f
 monocistronic, 310
 poly-A tail of, **309**, 309f, 316, 378
 polycistronic, 310, 316
 posttranscriptional control in eukaryotes, 374–78
 produced by maternal genes, 386, 387f
 stability of, 309
 translation of. *See* Translation
 transport from nucleus, 315–16, 316
 transport in phloem, 776
Metabolic efficiency, 180
Metabolic rate, 903
Metabolism, **155**, 156
 biochemical pathways, 155, 155f
 evolution of, 182
 using chemical energy to drive, 160–61
Metabolite, secondary, 798, 799t, 800
Metacercaria, **644**, 644f
Metal, phytoremediation for, 791–92, 792f
Metallothionin, 337, 337f
Metamerism. *See* Segmentation (animals)
Metamorphosis, **674**, **701**
 in amphibians, 701, 1005, 1005f, 1065
 complete, 674
 in *Drosophila*, 386, 387f
 in insects, **386**, 674, 1009, 1009f
 in plants, 832–33, 832–33f
 simple, 674
Metaphase
 meiosis I, 231f, 233, 233–34f
 meiosis II, 231f, 235f, 236
 mitotic, 213, 213f, **215**, 215–16f, 231f, 233

Metaphase plate, 215, 215f, 219, 230, 231f, 233, 233–35f
Metapopulation, **1144**, 1144f
 source-sink, **1144**
Metastases, **413**, 413f
 prevention of, 423
Metazoan, origin of, 630
Metchnikoff, Elie, 1031, 1031f
Meteorite, from Mars, 76
Methane, 546, 1222
 prebiotic chemistry, 66, 67f
Methanobacterium, 520f
Methanococcus, 71, 520f, 549f
Methanogen, **70**, 70f, 74f, 163, **519**, 521f
Methanopyrus, 520f
Methanosarcina barkeri, 70f
Methionine, 41, 42f, 311, 520t
Methylation of DNA, 373, 373f, 412, 418–19, 425
5-Methylcytosine, 373, 373f
Methyl group, 36f
MHC. *See* Major histocompatibility complex
Micelle, 52, 53f, 68
Microarray
 DNA. *See* DNA microarray
 protein, 356
Microartophil, 568
Microbody, 90t, **94**
Microclimate, **1207**, 1237
Microfossil, **70**, 70f, 72, 72f
Microgametophyte, 596f, 842, 842f
Micronucleus, 569–70, 569–70f
Micronutrients, 337
 in plants, 782, 782t
Microorganisms, in soil, 784
Microphyll, 748
Microphyll leaf, 585t
Micropyle, 489, **591**, 595, 595f
MicroRNA (miRNA), **374**, 375f
Microscope, 82–83, 82–83f. *See also specific types of microscopes*
 invention of, 81
Microsphere, 63, 68–69
Microsporangium, 591, 840
Microspore, 591, 591f, 595, 596f, 842f, 843
Microspore mother cell, 591, 591f, 596f, 842f, 843
Microsporidia, 520f, 605
Microtome, 110
Microtubule(s), 86–87f, **98**–99, 98–99f, 102t, 209f, 214
 kinetochore, 209f, 214f, 215, 216f, 234–35f
 spindle, 209f, 217, 233, 233–34f, 236
Microtubule-organizing center, 99, 214
Microvilli, 86f, 137f, 860, 894, 894f
Midbrain, 955, 955f, 955t
Mid-digital hair, 249t
Middle ear, 980, 981f
Middle lamella, **101**, 101f, **218**
Mid-Ocean Ridge, 1215
Midrib, 749
Miescher, Friedrich, 284
Migration, **1116**–17, 1116–17f
 of birds, 1116–17, 1117f
 of human populations, 357
 of monarch butterfly, 1116, 1116f
 of sea turtles, 708, 1117
Migratory locust (*Locusta migratoria*), 1152f
MIH. *See* Melanotropin-inhibiting hormone
Milk, 714–15, 1101
 antibodies in, 1025
 containing BST, 338

Milk-ejection reflex, 1000, 1102
Milk snake (*Lampropeltis triangulum*), geographic variation in, 472f
Milk sugar. *See* Lactose
Milkweed, 761f, 762, 1169
Miller, Stanley L., 66, 68
Miller-Urey experiment, 66–68, 66–67f
Millipede, 527, 668t, 671, 671f
Mimicry
 Batesian, **1171**, 1171f
 Müllerian, **1171**, 1171f
Mineral(s), **784**
 absorption by plants, 770–71, 770–71f
 essential, **904**
 in plants, 782t, 783f
 in soil, 784
 transport in plants, 768–76, 768–76f
Mineral fibers, synthetic, 415t
Mineralocorticoid, 1007
Mineral oil, 415t
Mine spill, 792, 792f
Minimal medium, 296f, 297
Minke whale, 1239, 1239f
Minnow, 980, 1196
Minor groove, 286f
Minotaur, 242
Mint, 593, 841
-10 sequence, **306**, 306f
-35 sequence, **306**, 306f
Miracidium, **644**, 644f
miRNA. *See* MicroRNA
Mismatch pair, **409**
Mismatch repair system, **412**
Miso, 603
Mispairing, slipped, 409–10
Mite, 670
Mitochondria, 86–87f, 90t, **96**, 96f, 102t, 105f, 162, 163f, 174–75, 523t
 division of, 96
 DNA of, 96–97, 562, 725
  C150T mutation, 402
 energy cycle in, 202, 202f
 genetic code in, 305
 genome of, 352
 of kinetoplastids, 567
 origin of, 72–73, 73f, 97, 97f, 521, 521f, 562
 ribosomes of, 562
Mitosis, 209, **210**, **213**–18, 213f, 215–17f
 in animal cells, 215
 compared to meiosis, 230, 231f
 in dinoflagellates, 568
 in fungi, 600–601
 in plant cells, 215
 in protists, 562
Mitral valve, 917f
Mixed conifer forest, 1141f
Moa, 1228
Mobbing behavior (birds), 1119
Mockingbird, 711t
Molar (teeth), 715, 715f, 890, 890–91f
Mold, 609
Mole, **31**, 467f, 719t
Molecular cladistics, 525
Molecular clock, **14**, 14f, **514**
Molecular formula, 26
Molecular motor, 99, 99f
Molecular phylogeny, 628, 629f
Molecular record, evidence for evolution, 14, 14f, 466, 466f
Molecular systematics, 525, **628**, 630
Molecule, **2**, 3f, 19–32, 19f, **25**
 polar, **28**, 36
Molina, Mario, 1221

Mollusca (phylum), 621f, 622t, 635, 652–57
Mollusk, 622t, 634–35f, **635**
 body plan of, 652–55, 653–54f
 classes of, 656–57
 economic significance of, 652
 eye of, 464–65, 465f, 504–5, 504–5f
 locomotion in, 882–83
 marine, 652
 pelagic forms, 654
 predation on, 1168
 reproduction in, 655, 655f
 respiration in, 924
 terrestrial, 652
Molly, 1064
Molting
 in arthropods, 666
 in insects, 871, 1009, 1009f
Molting hormone, 1009, 1009f
Molybdenum, 153, 904
 in plants, 782, 782t
Monarch butterfly (*Danaus plexippus*), 339, 1169
 defenses against predators, 1170–71, 1170–71f
 migration of, 1116, 1116f
Monitor lizard, 708, 1228
Monkey, 719t, 720, 845, 904, 1195
 New World, 720, 721f
 Old World, 720, 721f
Monocistronic mRNA, 310
Monoclonal antibody, 126, 1030
 in cancer treatment, 422
 production of, 1030, 1030f
 using to detect antigens, 1030, 1030f
Monocot, 593–94, 594f, 730
 development of seedling, 731f
 leaves of, 734f, 748, 748f, 750
 root of, 741f, 742
 stem of, 745f, 764f
Monocytes, 866, 910–11f, 1015, 1017, 1019f
Monoecious plant, **846**
Monogamy, 1128
Monokaryotic hyphae, **602**
Monomer, 38t
Mononucleosis, 533t
Monophyletic group, **514**, 515f
MONOPTEROUS gene, in *Arabidopsis*, 758
Monosaccharide, 38t, **55**, 55f, 896, 896f
Monosomy, 210, **272**
Monotreme, 528, 528f, 717–18, 718f, **1067**, 1067f
Monsoon climate, 1206
Monsoon forest, tropical, 1208, 1208f
Monteverde Cloud Forest Reserve (Costa Rica), 1234–35, 1234f
Moose, 1123
Morel (*Morchella esculenta*), 601, 605t, 609, 609f
Morgan, Lloyd, 1114
Morgan, Thomas Hunt, experiments with fruit flies, 265–66, 265–66f
Mormon cricket, 1125, 1125f
Morning-after pill, 1078
Morphine, 798, 799t, 951
Morphogen, 392, 392f, 397
Morphogenesis
 in human embryos, **1098**
 in plants, 388, 389f, 759
Morphology, adaptation to environmental change, 1138
Mortality, **1146**
Mortality rate, 401, **1146**–48, 1146t
Morula, **619**, 1074f, **1084**
Mosaic development, 391

Index   I-23

Mosquito, 210t, 568, 569f, 673f, 884, 1224
Mosquito fish, 1064
Moss, 581–82, 582–83f, 1178, 1178–79f
Moss animal, 623t
Moth, 664f, 672f, 673, 675t, 844
"Mother figure," 1112
Mother-of-pearl, 652, 654
Motif, protein, 43, 44f, 45
Motility, 523t
Motion sensing, 971t
Motor cortex, primary, 958f
Motor effectors, 940
Motor neuron, 869t, 879, **940**, 940f, 961–62f, 963–66
   autonomic, **940**
   somatic, 879, **940**
Motor protein, 98–99, 99f, 214, 396, 428
Motor unit, **879**, 879f
Mountain ash (*Sorbus*), 749f
Mountain lion, 1195
Mountain zone, 1208, 1208f
Mouse (*Mus musculus*), 719t
   chimeric, 393f
   coat color in, 255
   development in, 382–85f, 383–85, 397, 684f
   genome of, 348f, 352, 492t, 494–95
   homeodomain proteins of, 16f
   homeotic genes in, 399f
   marsupial, 467f
   maternal care in, 1109, 1109f
   ob gene in, 903, 903f
   single gene effects on behavior, 1109
   transgenic, 333, 333f
Mouth, 888–91, 888f, 891f
Mouthparts
   of arthropods, 668t
   of insects, 665f, 673, 673f
Movement
   in animals, 870–73
   as characteristic of life, 62, 62f
MPF. *See* M-phase-promoting factor
M phase. *See* Mitosis
M-phase-promoting factor (MPF), 219–20
mRNA. *See* Messenger RNA
MSH. *See* Melanocyte-stimulating hormone
MTS1 gene, 417t
Mucilage, 751, 788
Mucilaginous lubricant, of roots, 739
Mucosa, of gastrointestinal tract, 889, 889f, 894f
Mucous cells, 893f
Mucus, 654, 860, 1014
Mud eel, 701
Mulberry, 761f, 846
Mule, 476
Mullein, 851
Müller, Fritz, 1171
Müllerian mimicry, **1171**, 1171f
Muller's ratchet, for origin of sex, 238
Mullis, Kary, 329
Multicellularity, **73**
   in animals, 618t
   evolution of, 73
   in sponges, 636
Multicellular organism, 522, 523t
   cell cycle control in, 221
   development in, 381–402
Multienzyme complex, 151, 151f, 168
Multigene family, 349
Multinucleate hypothesis, for origin of metazoans, **630**
Multiple alleles, 260, 260f
Multiple cloning site (MCS), **321**
Multiple fruit, 761f

Multiple sclerosis, 427
Multiregional Hypothesis, 725
Mummichog (*Fundulus heteroclitus*), 441, 441f
Muscle
   insertion of, **873**
   lactic acid accumulation in, 181, 181f
   length and tension of, 974
   metabolism during rest and exercise, 880–81
   origin of, **873**
Muscle cells, 81, 99
   cell cycle in, 222
Muscle contraction, 132, 868, 872–73, 949, 963
   control of, 878–79
   isometric, **873**
   isotonic, **873**
   sensing of, 971t, 974, 974f
   sliding filament model of, 874, 874–77f, **875–77**
Muscle fascicle, 874f
Muscle fatigue, **881**
Muscle fiber, 868, 868f, **874**, 874f, 879
   fast-twitch (type II), **880**–81, 881f
   slow-twitch (type I), **880**, 881f
   types of, 879–80
Muscle spindle, 961f, 962, 974, 974f
Muscle stretch reflex, 962
Muscle tissue, 856, 856f, 867–68, 867t, 868f
Muscular dystrophy
   Becker, 269f
   Duchenne, 249t, 264t, 269f
   Emery-Dreifuss, 269f
   gene therapy for, 264, 333
Muscularis, of gastrointestinal tract, 889, 889f, 894f
Muscular system, 857t, 859f
Mushroom, 599–601, 605t, 608
Musical ability, 960
Musk-oxen, 1211
Mussel, 652, 656–57, 1123, 1123f, 1215
   starfish predation on, 1176, 1176f
Mustard, 845, 1169
Mustard oil, 1169, 1169f
Mutagen, 410, 414f, 415, 420–21
Mutant, **265**
Mutation, 410, 488, 497
   aging and, 401
   altering DNA sequence, 410, 410t
   arising from changes in gene position, 410t, 411
   cancer and, 224f, 413–23, 419f
   disorder-causing, 259
   evolution and, 437–38, 437t, 438f
   in germ-line tissue, 410
   interactions among evolutionary forces, 443, 443f
   kinds of, 410–11, 410t
   Muller's ratchet, 238
   in prokaryotes, 552, 552f
   somatic, 410, **1026**
Mutation rate, 438, 443, 494
Mut genes, 412
Mutualism, 557, 604, 611, **1172**, 1174, 1174f
   fungal-animal, 613, 613f
Myasthenia gravis, 1026
myb gene, 764f
Mycelium, **601**, 601f
   primary, 608, 608f
   secondary, 608, 608f
myc gene, 416, 417t
Mycobacterium leprae, 555t
Mycobacterium tuberculosis, 285t, 554, 554f, 555t
Mycologist, **600**

Mycoplasma, 555t
Mycorrhizae, **604**, **612**–13, 612f, 789, 1172
   arbuscular, **612**, 612f
   ectomycorrhizae, **612**–13, 612f
Myc protein, 224
Myelin sheath, 869, 869f, **941**, 941f, 946–47, 946–47f, 947f, 963f
Myeloma cells, 1030, 1030f
Myocardium, 868
Myofibril, 868, 868f, **874**, 874–75f, 878f
Myofilament, 874, 874f
Myoglobin, 39, 39f, 43, **880**
Myosin, 39t, 40, 100, **876**, 876f. *See also* Thick myofilament
Myotubular myopathy, 269f
Myrica faya, 1241
Myxini (class), 689, 691f, 692, 693t
Myxobacteria, 549t
Myxomycota (phylum), 576f
Myzostoma mortenensis, 628f
Myzostomid, 628, 628f

N

NAA. *See* Naphthalene acetic acid
NAD+, 51, **153**, 904t
   as electron acceptor, 153
   in oxidation-reduction reactions, 145f
   regeneration of, 166–67, 167f, 173, 173f, 181, 181f
   structure of, 153, 153f, 173, 173f
NADH
   contributing electrons to electron transport chain, 162, 163f, 173–76, 174–76f
   from fatty acid catabolism, 179, 179f
   from glycolysis, 163–65f, 166
   inhibition of pyruvate dehydrogenase, 177, 177f
   from Krebs cycle, 163f, 169f, 170, 171f
   from pyruvate oxidation, 162, 163f, 168, 168f
   recycling into NAD+, 166–67, 167f, 173, 173f, 181, 181f
   structure of, 173, 173f
NADH dehydrogenase, **174**, 174f
NADP+, 904t
NADPH
   production in photosynthesis, 187, 187f, 189, 197–99, 197–99f
   use in Calvin cycle, 200–201, 200–201f
NADP reductase, 199
Naiad, 657
Naked mole rat, 1132, 1134
Naked-seeded plant. *See* Gymnosperm
Nanaloricus mysticus, 623t
Nannippus, 463f
Nanoarchaeum equitans, 519
Nanos protein, **396**
Naphthalene acetic acid (NAA), 821
Nasal cavity, 928f
Nasal passage, 976, 976f
Natural killer cells, 1015–16, 1016f, 1019t, 1022
Natural selection, 8, **12**, 433, **434**, 434f, **440**, 1123, 1130
   conditions needed for, 440
   ecological species concept and, 477
   evolution and, 454–59, 468, 486
   invention of theory of, 11–12, 11–12f
   maintenance of variation in populations, 444–45, 444–45f
   in speciation, 479
Nature-versus-nurture debate, 1106, 1108

Nauplius larva, **668**, 668f
Nautilus, 657
Navigation, 988, **1116**–17, 1117f
Neandertals, 725–26, 725f
Near-shore habitat, Alaskan, 1242–43, 1242f
Nearsightedness, 984, 984f
Neck, of tapeworms, **645**
Necrosis, **400**, 796
Nectar, 595, 736, 844–45, 845f, 1120, 1124
Nectary, 595, 801, 844, 1174f
nef gene, 538, 539f
Negative feedback loop, 900, 1003, 1003f, 1040, 1040–41f, 1152
   blood glucose, 1042, 1042f
   thermoregulation, 1040, 1041f, 1042
Negative gravitropism, 810, 810f
Negative phototaxis, 1116
Negative pressure breathing, 927
Neisseria gonorrhoeae, 555t, 556, 556f
Nekton, 1215
Nematocyst, 622t, **638**–39, 639f, 656
Nematoda (phylum), 621f, 622t, 635, 646, 646–47f
Nematode, 634–35f, 646. *See also Caenorhabditis elegans*
   digestive tract of, 888, 888f
   disease-causing, 646, 647
   eaten by fungi, 603, 603f
   plant parasites, 796, 797f
   root-knot, 796, 797f
Nemertea (phylum), 623t, 645, 645f
Neonate, 1101
Neotenic larva, 924
Neotiella rutilans, 232f
Neotyphodium, 613
Nephridia, 653f, **655**, 659, 659f, 1046f
Nephron, **1048**, 1048f, 1052–58
   organization of, 1048f
   structure and filtration, 1052–53
   transport processes in, 1055–56, 1055–56f
Nephrostome, **655**, 1046, 1046f
Nereis virens, 658f
Neritic zone, **1214**, 1214f
Nerve, 858f, 869, **941**, 963, 963f
   stimulation of muscle contraction, 879
Nerve cord, 647f, 954, 954f
   dorsal, **684**, 684–85f, 688, 689f, 1094f
Nerve gas, 950
Nerve growth factor (NGF), 222, 994
Nerve impulse, 869, 869f, 942–47
Nerve net, 954, 954f
Nerve ring, **678**
Nervous system, 523t, 857t, 858f, 939–66
   of annelids, 659f, 954, 954f
   of arthropods, 665–66f, 667, 954, 954f
   central. *See* Central nervous system
   of cephalopods, 657
   of cnidarians, 954, 954f
   of echinoderms, 954f
   of fish, 954–55, 955f
   of flatworms, 643, 954, 954f
   neural and endocrine interactions, 993
   neurons and supporting cells, 940–41, 940–41f
   peripheral. *See* Peripheral nervous system
   regulation of digestion, 900
Nervous tissue, 856, 856f, 869, 869f, 869t
Nest, of weaver bird, 1134
Nesting behavior, 1108, 1108f

Net energy, 1123
Net primary productivity, 1191
Nettle, 845
Neural cavity, 1093f
Neural crest, 384f, **385**, 688, 1085t, **1090**–91, 1091f, 1094, 1094f
Neural fold, 1088f, 1090f
Neural groove, 384f, 1085t, **1090**, 1090f
Neural plate, 384f, 1088f, 1090f
Neural stem cells, 427
Neural tube, 384f, **385**, 688, 688f, 1085t, **1090**–91, 1090–91f, 1098
Neuraminidase, 540–51
Neuroblastoma, 417t
Neuroendocrine reflex, **1000**
Neurofibroma, 417t
Neurofilament, 99
Neuroglia, 869, **941**
Neurohormone, **992**
Neuromodulator, **950**
Neuromuscular junction, 619t, 949, 949f
Neuron, 81, 869, 869f, 939, 939f, 941f.
See also specific types of neurons
cell cycle in, 222
organization of, 940–41, 940–41f
Neuropeptide, **950**
Neurospora
Beadle and Tatum's experiment with, 296–97, 296–97f
chromosome number in, 210t
nutritional mutants in, 296–97, 296–97f
Neurotransmitter, **127**, 879, 944f, **948**–53, 948f, 953f, 963t, 992, 992f
drug addiction and, 951–53, 952–53f
Neurotrophin, **994**
Neurula, 383f
Neurulation, 384f, **385**, 391, 1085t, **1090**–91, 1090–91f, 1098
Neutral atom, **21**
Neutron, **20**–21, 20f
Neutrophils, 866, 910f, **911**, **1015**, 1017, 1019f
Newt, 698, 698t
Newton, Isaac, 4
New World monkey, 720, 721f
New York City, watersheds of, 1232–33, 1233f
New Zealand alpine buttercup, 485, 485f
NF-1 gene, 417t
NF-2 gene, 417t
NGF. *See* Nerve growth factor
Niacin. *See* Vitamin B$_3$
Niche, **1164**
fundamental, **1164**–65, 1164f
realized, **1164**–65
Niche overlap, 1165
Nicolson, G. J., 108
Nicotinamide adenine dinucleotide. *See* NAD$^+$
Nicotine, 798, 952–53
Nicotine patch, 953
Nicotine receptor, 952–53
nif genes, **335**–36, 557
Night blindness, 904t
Nile perch (*Lates niloticus*), 484, 1241, 1241f, 1246
Nimravid, saber-toothed, 516f
Nine + two structure, **100**, 100f, 547
Nipple, 1101
Nirenberg, Marshall, 304
Nitrate, 1187, 1187f, 1189, 1189f
Nitric oxide, 126, 128t, 129, 1010, 1109
as neurotransmitter, 951
as paracrine regulator, 994

penile erection and, 1070
in plants, 803
regulation of blood pressure and flow by, 920–21
super nitric oxide, 936
transport in blood, 936
Nitrification, 553, 557, 1187f
Nitrifier, 553
Nitrogen
atmospheric, 924, 1187, 1187f
electron energy levels for, 23f
fertilizer, 335–36, 784, 1187
in plants, 772, 782, 782t
prebiotic chemistry, 67f
Nitrogenase, 336, 557, 787f, 1187
Nitrogen cycle, 1187, 1187f, 1189
Nitrogen fixation, 182, 546, 557, 557f, 741, 787, 787f, 787f, 1172, 1178, 1178–79f, 1187, 1187f, 1241
evolution of, 182
in transgenic plants, 335–36
Nitrogenous base, 49, 49f, 154, 284, 284f
Nitrogenous wastes, **1051**, 1051f, 1054
Nitrogen reductase, 1187
Nitroglycerin, 921
Nitrosomonas, 549f
Nitrous oxide, 1222
N-myc gene, 417t
Nociceptor, 971t, **973**
Noctiluca, 568f
Node (plant stem), 732, 732f, **744**, 744f, 747
Nodes of Ranvier, 869, **941**, 941f, 947, 947f
Nod factor, 787f
Nonassociative learning, **1110**
Noncoelomate, 633–48
Noncompetitive inhibitor, **152**, 153f
Noncyclic photophosphorylation, **198**
Nondisjunction, **272**
involving autosomes, 272
involving sex chromosomes, 273, 273f
Nonequilibrium model, of biological community, **1180**
Nonextreme archaebacteria, **519**
Nongranular leukocytes, **911**
Nonsense codon, **310**, 312f
Nonspecific immune defenses, 860, 1014–17
Nonsteroidal anti-inflammatory drug (NSAID), **995**, 1036
Nonsulfur purple bacteria, 562
Nonvascular plant, **580**, 580f, 582–83, 582–83f
Nonvertebrate chordate, 686–87, 686–87f
Noradrenaline. *See* Norepinephrine
Norepinephrine, **950**, 963t, 965–66, 965f, 965f, 992–93, 995t, 998, 1007, 1091
Nori, 572
Norplant, 1077t
North equatorial current, 1212f
Nose, 864
Nostril, 928f
Notochord, 384f, 385, 501, 502f, 622t, **684**–87, 684–87f, 1085t, **1090**, 1090–91f, 1092
N-ras gene, 416f, 417t
NSAID. *See* Nonsteroidal anti-inflammatory drug
NtrC protein (activator), 372f
Nucellus, 589, **591**, 595f
Nuclear envelope, **80**, 80f, 86–88f, **89**, 209f, 215, 216–17f, 522, 523t
Nuclear pore, 88f, **89**, 377f, 378
Nuclease, 900t
Nucleation center, 98

Nucleic acids, 36, **49**, **284**. *See also* DNA; RNA
catabolism of, 178f
functions of, 37t, 48
structure of, 37t, 49, 49f
Nuclein, 284
Nucleoid, **80**, 208, **547**
Nucleoid region, **551**
Nucleolus, 86–88f, **88**, 90t, **95**, 95f, 217
Nucleosome, **89**, 89f, **211**, 211f, 214, 373–74, 373f
Nucleotide, 15, 15f, 37–38t, **49**, 49f, **284**, 284f, 285
numbering carbon atoms in, 284, 284f
Nucleus, atomic, 20, 20f
Nucleus, cellular, **70**, 79–80f, **80**, 86–88f, 88–89, 90t, 102t
hereditary information in, 280–81, 280–81f
origin of, 72, 72f
transplantation of, 281, 281f
cloning of animals, 424, 424f
in *Xenopus*, 393–94, 394f
transport of RNA out of, 89
Nudibranch, 622t, 652, 656
Numbat, 467f
Number of offspring per mating, 442
Nurse cells, 386, 387f
Nursing, 1101–2, 1101f
Nüsslein-Volhard, Christiane, 395f
Nutrient
essential, **904**, 904t
exchange in capillaries, 913
plant, 782–83, 782f, 783f
transport in blood, 909–10
Nutrition, 523t
Nutritional deficiencies, in fish, 690
Nutritional mutants, in *Neurospora*, 296–97, 296–97f
Nymphaeales, 594f

O

Oak (*Quercus*), 613, 737, 749, 832, 832f, 845–46, 1211
Oak, hybridization between species, 474
Oat (*Avena sativa*), 818–19, 819f
Obelia, 640, 640f
Obesity, **903**
ob gene, in mice, 903, 903f
Observation, 4–5, 5f
Occipital lobe, 957, 957f, 987, 987f
Ocean, 1184f
origin of life at ocean's edge, 62, 65, 68
productivity of, 1191f
Ocean circulation, 1204, 1207, 1212–13, 1212–13f
Ocean perch, 1242, 1242f
Ocelli, **666**, 666f, 673
Ocelot, 467f
Octet rule, **24**, 26
Octopus, 622t, 635, 652, 652f, 657, 657f, 882
Ocular albinism, 269f
Odonata (order), 673, 675t
Odontogriphus, 630f
Offspring
number of, 1149, 1149f
parental investment per offspring, 1149
parent-offspring interactions, 1105f, 1112, 1112f
size of each, 1149, 1149f
Ogallala Aquifer, 1185

Oil (fossil fuel), 1186, 1186f
clean up of oil spill, 558f
oil-degrading bacteria, 558f
Oil gland. *See* Sebaceous gland
Oils (plant), 53, 798
in corn kernels, 458, 458f
Okazaki fragment, **290**–91, 291–92f, 294f
Old World monkey, 720, 721f
Olfaction. *See* Smell
Olfactory bulb, 955f
Olfactory nerve, 976f
Olfactory receptor genes, 499, 499f
Oligochaeta (class), 660–61, 661f
Oligodendrocytes, **941**
Oligosaccharin, 816, 817t, **825**
Oligotrophic lake, **1178**, **1217**
Olive oil, 53
Omasum, 898, 898–99f
Ommatidia, **666**, 666f
phenotypic variation in, 450, 450f
Omnivore, **888**, 890
Oncogene, 224, **416**, 416f, 419f
Oncotic pressure, 914
One-gene/one-enzyme hypothesis, **297**
One-gene/one-polypeptide hypothesis, 296–97
Onion, 747, 747f, 763
Onion root tip, 89f
On the Origin of Species (Darwin), 8, 11f, 12, 434
"Ontogeny recapitulates phylogeny," 1094–95, 1094f
Onychophora (phylum), 623t
Onymacris unguicularis, 1138f
Oocyst, 569f
Oocyte
primary, 1073, 1074f
secondary, 1073–74, 1073–74f, 1082
Oogenesis, 1072, 1074f
Oomycete, **571**
Opabinia, 630f
Oparin, Alexander, 68
Open circulatory system, **625**, 667, **908**–9, 908f
Open reading frame (ORF), **349**
Operant conditioning, **1110**–11
Opercular cavity, 925, 925f
Operculum, **925**, 925f
Operon, **366**
Ophiuroidea (class), 680, 680f
Opiate, endogenous, 951
Opiate receptor, 378, 951
Opium, 950–51
Opium poppy (*Papaver somniferum*), 799t
Opossum, 210t, 717, 1067, 1072, 1195
Opposite leaf, 749, 749f
Opsin, 985–86
Optic chiasm, 955f, 957f, 987f
Optic lobe, **955**
Optic nerve, 464, 666f, 983–84f, 986, 986–87f
Optic stalk, 1093, 1093f
Optic tectum, 955–56f
Optimal foraging theory, **1123**
Optimum pH, 152
Optimum temperature, 152, 152f
Oral cavity, 927
Oral cavity cancer, 414t
Oral contraceptives, 1076–78, 1076f, 1077t
risk involved with, 1078
Orange, 761f, 826, 840
Orangutan (*Pongo*), 466, 495, 495f, 720, 721f
Orbital of electron, 20–21, 20f, **22**–23, 22–23f
Orca. *See* Killer whale

Index I-25

Orchid, 594, 613, 743, 762, 841, 841f
Orchidaceae (family), 841f
Order, as characteristic of life, 2
Order (taxonomic), **511**
Oreaster occidentalis, 676f
ORF. *See* Open reading frame
Organ, 2, 3f, **857**, 857f
Organelle, **2**, 3f, 63f, **80**, 80f, **86**, 90t, 102t, 520t
Organism, 2, 3f
Organizer, **392**, 392f, **1092**–93
Organochlorine pesticide, 1247
Organ of Corti, 971t, **980**, 981f
Organogenesis, 384f, **385**, 397, 397f, 1085t, 1098
Organ system, 2, 3f, **857**, 857–59f, 857t
Orgasm, 1000
oriC site, 290, 294
Orientation, 988, **1116**
Origin of life, 19, 27, 61–62, 61f
  deep in earth's crust, 65
  at deep-sea vents, 65
  extraterrestrial, 64, 76, 76f
  Miller-Urey experiment, 66–67, 66–67f
  at ocean's edge, 62, 65, 68
  place where life began, 65
  scientific viewpoint, 64
  special creation, 8, 64
  spontaneous origin, 64
  within clay, 65
Origin of muscle, **873**
Origin of replication, 208, 208f, 290, 290f, 294–95, 321
Oriole, 1238
Ornithine transcarbamylase deficiency, 269f
Ornithischia (order), 703t
Orrorin tugenensis, 723f
Ortholog, **495**, 503
Orthoptera (order), 666f, 672f, 673, 675t
Oscillating selection, **444**, 486, 512
Osculum, 636, 636f
Oskar protein, **396**
Osmoconformer, **1044**
Osmolality, **1044**, 1045f
  plasma, 1000, 1000f
Osmoreceptor, 920, 1000f, 1057, 1057f
Osmoregulator, 1044
Osmoregulatory functions, of hormones, 1057–58
Osmoregulatory organs, 857t, 1046–47, 1046–47f
Osmosis, 116–17, 116f, 122t, 768–69, 914
Osmotic balance, 117, 1044, 1045f
Osmotic concentration, **116**
Osmotic potential. *See* Solute potential
Osmotic pressure, **117**, 117f, **1044**
Osmotic protein, 39t
Osmotroph, **565**
Ossicle, 678, 980
Ossification, 696
Osteoblasts, 864–65, 865f
Osteoclasts, 865, 1006, 1006f
Osteocytes, 863t, 864–65, 865f
Ostracoderm, 692, 692f
Ostrich, 711t, 713, 1066
Otolith, 978f, 980
Otolith membrane, 978
Otter, 1239
Ottoia, 630f
Outcrossing, 846–47, 847f
Outer bark, 746
Outer ear, 980, 981f
Outer membrane
  of chloroplasts, 96, 97f, 186f
  of mitochondria, 96, 96f, 562

Outgroup, **512**
Outgroup comparison, **512**
Ovalbumin gene, 313f
Oval window, 980, 981f
Ovarian cancer, 414t, 417t
Ovarian cycle, 1010
Ovarian follicle, 1072
Ovary, 858–59f, 993f, 995t, 1072–73f
Ovary (plant), 589, **595**, 595–96f, **840**, 840f, 848f
Overexploitation, 1234t, 1239, 1239f
Overyielding, 1197
Oviduct. *See* Fallopian tube
Oviparity, 1064, 1066–67
Oviraptor, 516f
Ovoviviparity, 1064, 1066
Ovulation, 1067, 1074, 1074–75f
  prevention of, 1076–78
Ovulator
  cyclic, 1067
  induced, 1067
Ovule, 589, **590**, 591, **595**, 595f, 756, 838, **840**, 840f, 842f, 843, 848
Owl, 711t, 713, 982
Oxaloacetate, 169–70, 171f, 178, 204f
Oxidation, **23**, 23f, **144**–45, 145f, 156, 177
β-Oxidation, 178–79f, 179
Oxidation-reduction (redox) reaction, 144–**45**, 145f, 153, 172
Oxpecker, 1173, 1173f
Oxygen
  in air, 924, 934
  atomic structure of, 20f
  diffusion from environment into cells, 923
  diffusion into tissues, 934
  as electron acceptor, 173–74, 174f
  exchange in capillaries, 913
  in freshwater ecosystems, 1217
  in marine ecosystems, 1214
  partial pressure in blood, 930, 932, 934
  from photosynthesis, 65, 71, 182, 187–89, 187f, 196, 198, 557
  in plants, 782, 782t
  transport in blood, 909–11, 910f, 922, 934–35, 934–35f
Oxygen free radicals, 1015
Oxyhemoglobin, **934**
Oxyhemoglobin dissociation curve, 934, 935f
Oxytocin, 39t, 994t, **1000**, 1001f, 1004, 1070t, 1101–2
Oyster, 622t, 652, 654–56, 1147
  shell shape in, 461, 461f
Oyster drill, 656
Oyster mushroom (*Pleurotus osteatus*), 603, 603f
Ozone, 826, **1221**
Ozone hole, 1221, 1221f
Ozone layer, 65, 71, 191, 1204

P

P$_{680}$, 196, 198
P$_{700}$, 196
P$_{870}$, 196
p16 gene, 418f
p21 gene, 418f
p53 gene, 223, 417t, 418, 418–19f, 421
p53 protein, 220, 223, 223–24f, 422–23
Paal, Arpad, 818
Pacemaker, cardiac, 917, 919, 919f
Pacific giant octopus (*Octopus defleini*), 657f
Pacific yew (*Taxus brevifolia*), 799t, 800
Pacinian corpuscle, 971t, **973**, 973f
Pain, perception of, 950, 971t, 973

Pair-bonding, 1076, 1102
Paired appendages, of fish, 690
Pair-rule genes, 395f, **396**
Pakicetus attocki, 461, 461f
Paleontologist, 461
Palisade mesophyll, **750**, 750f, 771f
Palm, 594
Palmately compound leaf, 749, 749f
Palm oil, 54
Palo Verde tree (*Cercidium floridum*), 814f
Pancreas, 858f, 888–89f, 894, 900t, 901–2f, 993f
  as endocrine organ, 995t, 1008, 1008f
  secretions of, 895, 895f
Pancreatic cancer, 414t, 417t
Pancreatic duct, 895, 895f, 1008
Pancreatic juice, 889, 895, 895f
Pangaea, 700
Panspermia, **64**
Pantothenic acid. *See* Vitamin B$_5$
Paper, 37f
Papermaking, 737
Paper wasp (*Polistes*), 651f
Papillae, of tongue, 975, 975f
Papillomavirus, human, 541
Papuan kingfisher (*Tanysiptera hydrocharis*), 480, 481f
Parabronchi, 929, 929f
Paracrine regulation, **992**, 992f, 994–95, 1093
Paracrine signaling, 127, 127f
Paradise whydah, 1126f
Parallax, 987
Paralog, **499**, 503
Paramecium, 63f, 82f, 117, 569–70, 569f, 1168f
  competitive exclusion among species of, 1165, 1165f
  killer strains of, 570
  life cycle of, 570f
Paramylon granule, 566f
Paramyxovirus, 533t
Paraphyletic group, **514**, 515f
Parapodia, 658f, **660**
Parasite, 604
  effect on competition, 1176
  external, 1175, 1175f
  internal, 1175
  manipulation of host behavior, 1175, 1175f
Parasitic plant, 789, 789f
Parasitic root, 743
Parasitism, 557, **1172**
  brood, 1113, 1113f
Parasitoid, **1175**
Parasitoid wasp, 801, 801f
Parastichopus parvimensis, 676f
Parasympathetic division, **940**, 940f, 964, 964f
Parathion, 950
Parathyroid gland, 993f, 994t, 1005–6, 1006f
Parathyroid hormone (PTH), 994t, 1006, 1006f
Paratyphoid fever, 554
Parazoa, **620**, 620–21f, 634, 634–37f, 636–37
Parenchyma, **732**, 736, 745–46, 775f
Parenchyma cells, **736**, 736f, 737, 741, 768, 776
Parental care, in dinosaurs, 516, 516f
Parental investment, **1125**
Parent-offspring interactions, 1105f, 1112, 1112f
Parietal cells, 892, 893f, 901f
Parietal eye, 708
Parietal lobe, 957, 957f
Parietal pleural membrane, 930

Park, Thomas, 1176
Parkinson disease, 426, 950, 958
Parmotrema gardneri, 611f
Parnassius imperator, 673f
Parrot, 711t, 713
Parrot feather (*Myriophyllum spicatum*), 791
Parsimony, principle of, **514**
Parsnip, 743
Parthenogenesis, **237**, **1062**, 1083
Partial pressure, **924**
Passenger pigeon, 1239
Passeriformes (order), 711t, 713
Passion vine, 798
Passive immunity, **1018**
Passive immunization, 1029
Pasteur, Louis, 1018
Patella, 961f
Patellar ligament, 961f
Patent, gene-related, 358
Paternity testing, 1128, 1129f
Pathogen, 604, 860, 1014, 1018
Pattern formation, 395–97, 395–97f
  in *Drosophila*, 395–98, 395f, 397f
Pavlov, Ivan, 1110
Pavlovian conditioning. *See* Classical conditioning
Pax6 gene, 501, 503–6, 505f
Payne, Robert, 1176, 1176f
PCB, 415t
PCNA. *See* Proliferating cell nuclear antigen
PCR. *See* Polymerase chain reaction
PDGF. *See* Platelet-derived growth factor
Pea, 593, 734f, 747, 759, 761f, 763, 787f, 836, 841
Peach, 761f
Peacock, 1126f, 1127
Peacock worm, 660
Peanut, 614
Peanut allergy, 1036
Pear, 736, 736f
Pearl, 652, 654
Peat, 1186
Peat moss (*Sphagnum*), 582
Peccary, 1195
Pectin, 57
Pectoral fin, 690, 695
Pectoralis major muscle, 859f, 1101
Pectoralis minor muscle, 1101
Pedicel, **594**, 595f, 663f
Pedigree analysis, **261**, 261f, 269, 274
Pedipalp, **670**
Peer review, 7
Peking man, 725f
Pelizaeus-Merzbacher disease, 269f
Pellagra, 904t
Pellicle, 566, 566f, 569, 569f
Pelomyxa palustris, 562, 562f
Pelvic bones, of baleen whale, 465, 465f
Pelvic fin, 690, 695
Pelvic girdle, 872f
Pelvic inflammatory disease (PID), 556
Pelvis, 721, 858f, 871f
Pelycosaur, 704, 704–6f
pen gene, 441, 441f
Penguin, 711t, 713, 882, 1066
Penicillin, 84
Penicillin allergy, 1036
Penicillin resistance, 552
Penicillium, 603, 609
Penis, 859f, 1068, 1068f, 1070f
Pepper, 761f
Peppered moth (*Biston betularia*), industrial melanism and, 456–57, 456–57f

I-26 Index

Peppermint, 798
Pepsin, 152, 892–93, 900t
Pepsinogen, 892, 893f, 900, 901f
Peptic ulcer, 555t
Peptidase, 900t
Peptide bond, 41, 41f, 311
Peptide hormone, 993, 994–95t, 998
Peptide-nucleic acid (PNA) world, 67
Peptidoglycan, **71**, 84, 520t, 548, 550–51f
Per capita income, 1157
Per capita resource consumption, 1158
Peregrine falcon (*Falco peregrinus*), captive breeding of, 1247, 1247f
Perennial plant, 814, **851**–52, 851f
Perforin, 1016, 1016f
Pericardial cavity, 856, 856f
Perichondrium, 864
Pericycle, **741**, 742, 742f, 770f, 787f
Periderm, 746f
Perilymph, 978
Perineurium, 862
Periodic isolation, 485
Periodic table, 24, 24f
Periodontal ligament, 891f
Periosteum, 862
Peripheral chemoreceptor, 932, 933f, 976
Peripheral nervous system, 869, 940, 940f, 963–66, 963–66f
Peripheral vascular disease, 333t
Perissodactyla (order), 719t
Peristalsis, 892, 892f
Peritoneal cavity, **856**, 856f
Peritubular capillary, 1053, 1053f
Periwinkle, 749f
Permafrost, **1211**
Pernicious anemia, 892, 904t
Peroxisome, 86–87f, **94**, 94f
Persimmon cells, 125f
Pesticide, 1185, 1218, 1247
Pesticide resistance, in insects, 441, 441f
Petal, **594**, 595f, 839f, 840, 840f
Petiole, 744f, 748, 811, 852, 852f
Petrel, 711t
Petunia, 847
 transgenic, 336f, 338
PGK deficiency, 269f
pH, 31–32
 of blood, 932, 933f, 935, 935f
 of cerebrospinal fluid, 932, 932–33f, 976
 effect on enzymes, 47
 pH scale, **31**, 31f
 of urine, 1054
Phage. See Bacteriophage
Phage conversion, **535**
 in *Vibrio cholerae*, 535
Phagocytes, of invertebrates, 1031, 1031f
Phagocytosis, 94, 94f, 118–**19**, 118f, 122t, 1015, 1015f, 1024, 1031
Phagosome, 565
Phagotroph, 565
phantastica mutant, in snapdragon, 748, 748f
Pharmaceuticals
 applications of genetic engineering, 333, 333f
 from plants, 799f, 800, 1232
Pharyngeal pouch, 464, **684**
Pharyngeal slits, **684**, 684–85f, 1091, 1095
Pharynx, 646, 647f, 658, **684**, 889–91, 889f, 891f, 928f
Phase change, in plants, **832**–33, 832–33f
Phase-contrast microscope, 83t
Pheasant, 711t
Phelloderm, 746

Phenotype, **249**–50
Phenotype frequency, 437
Phenotypic ratio, 254f
Phenylalanine, 41, 42f
Phenylalanine hydroxylase, 264t
Phenylketonuria (PKU), 264t, 274
Phenylthiocarbamide (PTC) sensitivity, 249t
Pheromone, **474**–75, 674, **674**, 992, **1118**–19, 1133
 alarm, 1120
 in ferns, 824
 trail, 1120, 1120f
Philadelphia chromosome, 416
Phloem, 584, **732**, **738**, 738f, 745f, 768f, 796
 mass-flow hypothesis, **776**–77, 777f
 primary, 730f, 733f, **742**, 742f, 745, 745f
 secondary, 731, 733f, 745f, 746
 sugar and hormone transport in, 776–77, 776–77f
Phloem loading, 776
Phlox, 844
Phoronida (phylum), 623t, 635, 662, 662f
Phoronis, 623t, 662f
Phosphatase, 220
Phosphate group, 36f, 49, 49f, 52, 106, 154, 154f, 160, 284
Phosphodiesterase, 135f, 986
Phosphodiester bond, 49, 49f, **285**, 285f, 287, 287f
Phosphoenolpyruvate, 162f, 165f, 203
Phosphoenolpyruvate carboxylase, 203
Phosphofructokinase, 165f, 177, 177f
Phosphoglucoisomerase, 165f
2-Phosphoglycerate, 165f
3-Phosphoglycerate, 165f, 200–201f, 201
Phosphoglycerokinase, 165f
Phosphoglyceromutase, 165f
Phospholipase C, 132, 133f, 998, 999f
Phospholipid, **52**, **106**
 in membranes, 52, 52f, 80, 80f, 106–7, 107–8f
 structure of, 52, 52f, 106, 106f
Phosphorus
 fertilizer, 784, 1188
 in plants, 772, 782–83, 782t, 789
 polluted lakes, 1217
Phosphorus cycle, 1188–89, 1188f, 1212
Phosphorylase, 998, 998f
Phosphorylation, of proteins, 120, 121f, 132, 134, 220, 220f, 224, 377f, 998–99, 998f
Photic zone, **1216**, 1216f
Photoautotroph, **553**
Photoefficiency, 192
Photoelectric effect, 190
Photoheterotroph, 553
Photomorphogenesis, **808**, 808f
Photon, 190–91
Photoperiod, **834**–35
Photophosphorylation
 cyclic, **196**, 199
 noncyclic, **198**
Photopigment, 985–86, 985f
Photopsin, 985
Photoreceptor, 968f, **970**, 983
 sensory transduction in, 986
 in vertebrates, 985–86, 985–86f
Photorespiration, **203**–4, 203–4f, 785, 785f
Photosynthesis, 144, 177, 185–204, 1184, 1190
 anaerobic, 182
 in bacteria, 71, 71f, 85, 85f, 112, 113f, 196, 196f, 546, 551f, 557

C_3, **200**, 203–4
C_4, **203**–4, 204f
Calvin cycle, **187**, 187f, 189, 200–202, 200–204f, 785
 in carbon cycle, 1186, 1186f
 electron transport system in, **194**–97, 195–97f
 evolution of, 182, 196
 in food chain, 180f
 global climate change and, 785, 785f
 light-dependent reactions of, 186–87f, **187**, 189, 196–99, 196–99f
 light-independent reactions of, **187**–89
 overall reaction for, 188–89
 oxygen from, 65, 71, 182
 in pelagic zone, 1215
 soil and water in, 188
 stages of, 187
 summary of, 186–87
 sunlight in, 185–86, 188–90
 temperature dependence of, 188–89, 189f
Photosynthetic pigments, 187, 190–95
 absorption spectra of, 191
 organizing into photosystem, 194–95, 194–95f
Photosystem, **187**, 194–95, **194**, 194–95f
 architecture of, 194–95, 195f
 of bacteria, 196, 196f
 conversion of light to chemical energy, 196–97, 196–97f
 discovery of, 194, 194f
 of plants, 196–99, 197–99f
Photosystem I, **196**, 197–99f, 198–99
Photosystem II, **196**, 197–99f, 198–99
Phototaxis
 negative, 1116
 positive, 1116
Phototroph, **565**
Phototropin, 809, 809f
Phototropism, **808**–9, 809f, 818
 auxin and, 819–20, 820f
 negative, in *Drosophila*, 447, 447f
pH-tolerant archaebacteria, 519
Phycobilisome, 572
Phycocyanin, 572
Phycoerythrin, 572
Phyllotaxy, **744**
Phylogenetics, 516–17, 516–17f
Phylogenetic tree, **14**, 519
Phylogeny, **512**–14
 of animals, 620, 621f, 628, 629f
 based on rRNA sequences, 499–500, 500f
 of fungi, 605, 605f
 of invertebrates, 634–35, 634–35f
 molecular, 628, 629f
 "ontogeny recapitulates phylogeny," 1094–95, 1095f
 protostome
   rRNA, 635, 635f
   traditional, 634, 634f
 systematic, 525
Phylum, **511**
Physical map, 268, 320, **344**
 landmarks on, 344–45, 344–45f
Physical training, 881
Physiology, adaptation to environmental change, 1138, 1138t
Phytase, 337f
Phytate, 337
Phytoaccumulation, 790f
Phytoalexin, 803

Phytochrome, **808**, 808f, 835, 838f
Phytodegradation, 790f
Phytoestrogen, 799f, **800**
Phytophthora infestans, 572
Phytoplankton, 1193, 1193f
Phytoremediation, **790**–92, 790–92f
 for heavy metals, 791–92, 792f
 for trichloroethylene, 790–91, 790–91f
 for trinitrotoluene, 791
Phytovolatilization, 790f
Piciformes (order), 711t
Pickling, 47
PID. See Pelvic inflammatory disease
Pied flycatcher, 478, 478f
Pierinae (subfamily), 1169
PIF. See Prolactin-inhibiting factor
Pig, 719t, 888
 transgenic, 338
Pigeon, 711t, 1111
Pigeon breeding, 12
Pigment, **191**
 bile. See Bile pigment
 photosynthetic. See Photosynthetic pigments
Pigweed (*Amaranthus hybridus*), 441
Pika, 719t
Pikaia, 630f
Pike cichlid (*Crenicichla alta*), 448–49, 448–49f
Pillbug, 668–69
Pilobolus, 600f, 605t
Pilus, 84f, **407**, 407f, 535, 546f, 550
Pine, 585f, 590–91, 590–91f, 613, 1211
Pineal gland, 957f, 993, 993f, 995t, 1010
Pineapple, 761f, 821
Pine cone, 760f
Pine needle, 590
Pink bollworm, 336
Pinna, 864, 981f
Pinnately compound leaf, 749f, 749f
Pinnipedia (suborder), 719t
Pinocytosis, 118–**19**, 118f, 122t
Pintail duck, 474
Pinworm (*Enterobius*), 622t, 647
Pisolithus, 612f
Pistil, 840, 846
Pistillate flower, 846, 847f
Pit(s) (tracheids), 737, 737f, 772
Pitch (sound), 981–82f
Pitcher plant (*Nepenthes*), 751, 788, 788f
Pith, 732f, **742**, 745, 745f
Pit organ, 971t, 988, 988f
Pituitary dwarfism, 1002
Pituitary gland, 858f, 955f, 957f, 958, 993
 anterior, 993, 994t, **1000**–1004, 1003f, 1071, 1074–75
 posterior, 993, 994t, **1000**, 1001f, 1057
Pituitary stalk, 1003, 1003f
Pit viper, 988, 988f
PKU. See Phenylketonuria
Placenta, 715, 715f, 718, 1010, 1067, 1086, 1098f
 formation of, 1098
 functions of, 1098
 hormonal secretion by, 1100, 1100f
 structure of, 1097f
Placental mammal, 528, 528f, 717–18, 718f, **1067**, 1067f
 marsupial-placental convergence, 467, 467f
Placode, **1091**
Placoderm, 692f, 693, 693t
Plague, 357t

Index I-27

Planarian, 622t, 642–45, 642f, 908, 908f
  eyespot of, 504–6, 504f, 506f
Plankton, 669, 669f, 1193f, 1214, **1215**
Plant(s). *See also* Flowering plant
  annual, 814, **851**, 851f
  asexual reproduction in, 237, 849, 849f, 851
  biennial, **851**
  body plan in, 730–33
  carnivorous, 788–89, 788–89f
  circadian clocks in, 813, 813f
  classification of, 526, 526f
  coevolution of animals and plants, 801, 831, 831f, 1169, 1172, 1172f, 1174
  coevolution of insects and plants, 844
  coevolution with herbivores, 1169
  defenses against herbivores, 1169, 1169f
  development in, 382, 382f, 388, 389f
    embryonic, 756–57, 756–57f
    establishment of tissue systems, 758–59, 758–59f
    food storage, 759, 759f
    fruit formation, 761–62
    morphogenesis, 759
    seed formation, 760, 760f
    vegetative, 755–63
  diversity among, 579–96
  dormancy in, 814, 814f, 827, 827f
  evolution of, 579–80, 733
    land plants, 526, 526f, 580
  global climate change and, 785, 785f, 1224
  gravitropism in, **810**, 810f
  growth of, 807f
  heterozygosity in, 435
  hyperaccumulating, 791–92
  leaves of, 748–51. *See also* Leaf
  life cycles of, 581, 581f
  life span of, 851–52
  major groups of, 580f
  nonvascular, **580**, 580f, 582–83, 582–83f
  nutrient requirements of, 782–83, 782t, 783f
  nutritional adaptations in, 788–89, 788–89f
  organization of plant body, 732, 732f
  parasitic, 789, 789f
  perennial, 814, **851**–52, 851f
  photomorphogenesis in, 808, 808f
  photosystems of, 196–97, 197–99f
  phototropism in, **808**–9, 809f
  phytoremediation, **790**–92, 790–92f
  polyploidy in, 480–81, 496
  primary growth in, 730, 732–33f, 733
  primary plant body, **730**
  primary tissues of, **730**
  reproduction in, 831–52
  responses to flooding, 774–75, 774–75f
  roots of, 739–43. *See also* Root
  secondary growth in, **731**, 732–33f, 733, **742**
  secondary metabolites of, 798, 799t, 800
  secondary tissues of, **731**, **746**
  seed, 589–96
  sensory systems in, 807–27
  sexual life cycle in, 228, 229f
  stem of, 744–47. *See also* Stem
  thermotolerance in, 815, 815f
  tissue culture, 850, 850f
  tissues of, 734–38
    formation of, 388, 389f
  totipotency in, 281
  transgenic. *See* Transgenic plants
  transport in, 767–77
  turgor movement in, 812–13, 812–13f
  vascular. *See* Vascular plant
  vegetative propagation of, 747
  wound response in, 802, 802f
Plantae (kingdom), 16, 16f, 74–75f, **518**, 518f, 520, 521f, 523t
Plant breeding, 841
Plant cells
  cell division in, 213
  cell wall of, 87f, 101, 101f, 117, 117f, 119
  cytokinesis in, 218, 218f
  meiosis in, 227f
  mitosis in, 215
  structure of, 87f, 102t
Plant defenses, 795–804
  animals that protect plants, 801, 801f
  morphological and physiological, 796, 796–97f
  pathogen-specific, 803–4, 803–4f
  toxins, 798, 798f, 799t
Plant disease
  bacterial, 553, 803f
  fungal, 604, 604f, 608–9, 614, 614f, 796, 797f, 803f
  nematodes, 796, 797f
  viral, 803f
Plant fat, 54, 54f
Plant height, in garden pea, 246t
Plant hormone, 388, 816–27
  functions of, 817t
  production and location of, 817t
  transport in phloem, 776–77, 776–77f
Plantlet, 751
  adventitious, 849, 849f
Planula larva, **638**, 640, 640f
Plasma, 866, **910**, 914f
Plasma cells, 1019, 1019t, 1023–24, 1023f, 1027, 1027f
Plasma membrane, **80**, 87f, 90t, **105**, 546f. *See also* Lipid bilayer
  active transport across, 120–22
  of archaebacteria, 548
  of bacteria, 548
  bulk transport across, 118–19
  components of, 108–9, 108f, 109t
  electron microscopy of, 110, 110f
  of eukaryotes, 86f
  fluid mosaic model, 108–9, 108f
  passive transport across, 114–17
  of prokaryotes, 85
  structure of, 80f
  water movement across, 768
Plasma proteins, 866, 902, 910, 914
Plasmid, 319f, 321, **406**, 551–52
  antibiotic resistance genes on, 408
  cloning vector, 321–26, 322f, 326f
  creation of, 406, 406f
  gene transfer by, 406, 406f
  integration into genome, 406
Plasmid pSC101, 319f
  construction of, 323
  using to make recombinant DNA, 323, 323f
Plasmodesmata, 87f, 128t, **140**, 140f, **738**, 768, 770f
Plasmodial slime mold, 575, 576f
Plasmodium, 568–69, 569f
Plasmodium (slime mold), **575**, 576f
Plasmodium falciparum, 445, 445f, 800
  antigen shifting in, 1035
  genome of, 348f, 352, 493t, 496
Plasmodium malaria, 800
Plasmodium ovale, 800
Plasmodium vivax, 800
Plastic pollution, 1218
Plastid, **97**
Plastocyanin, 196–98f, 198–99
Plastoquinone, 197f, 198, 198f

Plastron, 708
Platelet(s), 222, 866, 909–10, 910–11f, **911**
Platelet-derived growth factor (PDGF), 222, 416f, 417t, 994
Platelet-derived growth factor (PDGF) receptor, 224, 416f
Platelet plug, 911f
Platyhelminthes (phylum), 621f, 622t, 625, 642–45
Platypus, 528, 528f
Pleasure messages, 952, 952f
Pleiotropic effect, **255**, 450, 459
Plesiomorphy, **513**
Plesiosaura (order), 464, 703t
Pleural cavity, 856, 856f, **930**
Pleural membrane
  parietal, **930**
  visceral, **930**
Plexus, 889, 889f
Pliohippus, 463f
Plover, 711t
Plum, 761f
Plumatella, 663f
Plume worm, 660
Plumule, 764f
Pluripotent stem cells, 429, 429f
Pluteus larva, 501
PNA world. *See* Peptide-nucleic acid world
Pneumatophore, 743, 743f, 775, 775f
Pneumocystis carinii. *See Pneumocystis jiroveci*
Pneumocystis jiroveci, 604
Pneumonia
  bacterial, 554, 555t
  viral, 533t
Poa annua, 1146–47, 1146t, 1147f
Pocket mouse
  coat color in, 440f
  urine of, 1050
"Pocket protein," 224
Podia, 575, 575f
Pogonophora (phylum), 623t
Poikilotherm, **707**, 1066
Point mutation, 410
Poinsettia, 751, 835, 835f, 845
Poison-dart frog, 1065f, 1170, 1170f
Poison ivy, 1036
Poison oak, 1036
Poisonous spider, 670, 670f
Poison sumac, 1036
Polar bear, 882, 1139f, 1140
Polar body, 1073, 1074f, 1082f
Polar front, 1205
Polar ice, 1207f, 1208, 1208f, 1224
Polarity, of water, 27f
Polar molecule, **28**, 36
Polar nucleus, 595, 843, 843f, 848f
Polio, 533t, 540
Poliovirus, 524f
Pollen, 591f
  dispersal of, 438, 443, 443f
Pollen cells, 227f
Pollen grain, 589, 591, **595**, 596f, **842**, 843f, 848f
  formation of, 838, 842f, 843
Pollen sac, 843
Pollen tube, 589–91, 591f, 595–96, 596f, 840, 843, 847, 847–48f, **848**
Pollination, 589, 591f, **595**, 844
  by animals, 840, 844–45
  by bats, 595, 845, 1172, 1172f, 1243, 1243f
  by bees, 844, 844–45f, 847f
  by birds, 845, 845f
  in early seed plants, 844
  by gravity, 595

by insects, 595, 844, 1164
by wind, 595, 844–45, 846f
Pollinator, 840
Pollution, 1218
  of groundwater, 1185
  habitat loss and, 1236
  phytoremediation, **790**–92, 790–92f
  threat of, 1218
Polyandry, 1128
Poly-A polymerase, 309
Poly-A tail, of mRNA, **309**, 309f, 316, 378
Polychaeta (class), 660, 660f
Polychaete, 622t, 635, 658–59f, 660, 660f, 1215
Polycistronic mRNA, 310, 316
Polyclonal antibody, 1030
Polydactyly, 249t, 439
Polygalacturonidase, 335
Polygyny, 1128, 1128f
Polymer, **36**, 38t
Polymerase, 39t
Polymerase chain reaction (PCR), 269, **329**, 329f, 345, 345f, 351
  stages of
    annealing of primers, 329, 329f
    denaturation of DNA, 329, 329f
    primer extension, 329, 329f
Polymerization, **98**
Polymorphism, **435**
  in DNA sequence, 435
  in enzymes, 435
Polyp
  of cnidarians, 638, 638f, 640–41, 640f, 656
  colorectal, 419f
Polypeptide, 38t, **41**
Polyphyletic group, **514**, 515f
Polyphyletic origin hypothesis, for origin of metazoans, **630**
Polyplacophora (class), 656
Polyploidization, 497
Polyploidy, **411**, **480**, 496, 497f
  speciation through, 480–81
Polyribosome, 310f
Polysaccharide, **56**
  digestion of, 895, 900t
  storage, 57
PolyU, 304
Polyunsaturated fatty acid, **53**–54, 54f
Pond, 1216, 1216f
Pondweed, 594
Pons, 955t, 957f
Popcorn, 759
Poplar (*Populus*), 786, 791
Popper, Karl, 6
Population, 2, 3f, **1140**
  age structure of, **1146**
  change through time, 1146–47
  crowded, 1152, 1152f
  human. *See* Human population
  metapopulations, **1144**, 1144f
  survivorship curves for, 1147, 1147f
Population cycle, 1153–54, 1153f
Population dispersion, 1142–43, 1142f
  human effect on, 1143
  mechanisms of, 1143
  randomly spaced, 1142, 1142f
  uniformly spaced, 1142, 1142f
Population ecology, 1137–58
Population genetics, 433–34, **435**–50
Population growth
  factors affecting growth rate, 1145, 1145f
  in hotspots, 1231, 1231f
  limitations by environment, 1150–54
  rate of, 1154
Population pyramid, 1156, 1156f

I-28 Index

Population range, 1140–41, 1140f, 1223
　expansion and contraction of,
　　1141, 1141f
Population size
　density-dependent effects on,
　　1152–53, 1152f
　density-independent effects on,
　　1153, 1153f
　extinction of small populations,
　　1244–45, 1245f
p orbital, 22, 22f
Porcupine, 714, 719t
Pore protein, 112, 113f
Porifera (phylum), 621f, 622t
Porin, 112, 113f
Porphyrin ring, 192, 192f
Porpoise, 719t, 956
"Portal," 896
Portal system, 1003
Portuguese man-of-war, 640
Positional label, **393,** 394
Positive feedback loop, 1004, 1043,
　1043f, 1153
Positive gravitropism, 810
Positive phototaxis, 1116
Positive pressure breathing, 927, 927f
Postanal tail, **684,** 684f, 689f
Posterior end, 624, 624f
Posterior pituitary, 993, 994t, **1000,**
　1001f, 1057
Postganglionic neuron, 964–65, 964f
Postnatal development, 1102, 1102f
Post-replication repair, **412**
Postsynaptic cells, 944f, 948, 948f, 952,
　953f, 992
Posttranscriptional control, **362,** 374–78
　alternative splicing of primary
　　transcript, 376, 376f
　RNA editing, 376–78
　small RNAs, 374, 374–75f
Posttranscriptional modification,
　309, 309f
Posttranslational modification, 377f
Postzygotic isolating mechanisms,
　473–**74,** 473t, 476, 476f
Potassium
　in action potential, 944–47, 945–47f
　blood, 993, 1058
　in cytoplasm and extracellular
　　fluid, 943t
　fertilizer, 784
　in guard cells, 773, 773f
　in plants, 772, 782, 782t
　in resting membrane potential,
　　942–43, 942–43f
　secretion in kidney, 1007, 1056,
　　1056f, 1058
Potassium channel, 109t, 114
Potato, 37t, 747, 785, 822, 849
　eye of, 747, 849
　Irish potato famine, 572
　transgenic, 336
Potato starch, 57
Potential difference, 942
Potential energy, 22–23, **144,** 144f
Potrykus, Ingo, 337, 337f
Pott, Sir Percivall, 415
Power of Movement of Plants, The
　(Darwin), 818, 818f
Power stroke, 876
Prairie, **1210**
Prairie chicken (*Tympanuchus cupido
　pinnatus*), 1245, 1245f
Prebiotic chemistry, 66, 67f
Precapillary sphincter, **912,** 913f
Precipitation (rain), 1184, 1184f, 1189,
　1206–7, 1209, 1209f
　acid. *See* Acid precipitation
　effect of global warming on, 1224

Precocial young, **1128**
Predation, **1168**
　evolution and, 1168
　evolution of prey population,
　　448–49, 448–49f
　population cycles and, 1154
　prey populations and, 1168, 1168f
　reduction of competition by,
　　1176, 1176f
　species richness and, 1199
Predator, 1168
　animal defenses against, 1170, 1170f
　predator-prey oscillations,
　　1153–54, 1153f
　search image for prey, 444, 444f
　selection to avoid, 440, 440f
　in trophic cascade, 1194–96, 1194f
Predator avoidance, 440, 440f,
　1133, 1134f
Prediction, 5f, 6
Preganglionic neuron, 964–65, 964f
Pregnancy, 1097–1101, 1097–1101f
　high-risk, 274
　Rh incompatibilities in, 260, 1029
Pregnancy test, 1030, 1075, 1098
Prehensile tail, 720
Premature aging, 402
Premolar, 715f, 890, 890–91f
Prenatal diagnosis, 274
Pressure, detection of, 973
Pressure potential, 769, 769f
Pressure-tolerant archaebacteria, 519
Presynaptic cells, 948
Prey, 1168
　minimizing predation risk, 1123
　predator-prey oscillations,
　　1153–54, 1153f
Prezygotic isolating mechanisms, 473,
　473t, **474**–75, 478
PR gene, 803
Prickle, 1169
Prickly pear, 747f, 1168
Pride (group), 1105f
Priestly, Joseph, 188
Primary abiogenesis, **68**
Primary cell wall, **101,** 101f
Primary consumer, **1190,** 1190f,
　1192f, 1193
Primary endosperm nucleus, 589
Primary growth, in plants, 730,
　732–33f, 733
Primary immune response,
　1027, 1027f
Primary induction, **1093**
Primary lysosome, 94
Primary meristem, **730,** 740, 745
Primary motor cortex, 957, 958f
Primary mycelium, 608, 608f
Primary oocyte, 1073, 1074f
Primary phloem, 730f, 733f, **742,** 742f,
　745, 745f
Primary plant body, **730**
Primary producer, **1190,** 1192f, 1193
Primary productivity, **1191,** 1194,
　1209, 1209f, 1214, 1220
　gross, 1191
　net, 1191
Primary somatosensory cortex,
　957, 958f
Primary spermatocyte, 1068, 1069f
Primary structure, of proteins, 43,
　44f, 47f
Primary succession, **1178**
Primary tissue, 856
　of plant, **730**
Primary transcript, 313f, 315f, **376.** *See
　also* Messenger RNA
　passage through nuclear membrane,
　　377f, 378

posttranscriptional control in
　eukaryotes, 374–78
processing of, 313–15
Primary xylem, 730f, 733f, 737, 742,
　742f, 745, 745f
Primate, 719t, 720–21
　evolution of, 720–26, 721f
　language of, 1121, 1121f
Primer, for replication, **290**
Primitive gut, 627
Primitive streak, **1089,** 1089f, 1097
Primordium, **594, 744,** 748
Primosome, 293–95
Principles of Geology (Lyell), 10
Prion, **542,** 542f
Prion protein, 542, 542f
Probability, 251, 253
Probe, **328,** 328f, 331f
Proboscidea (order), 714f, 719t
Proboscis (insect), 975, 975f
Procambium, 730f, **731,** 731f, 739f,
　745, 756f, 758, 760f
Procellariformes (order), 711t
Processivity, of DNA polymerase III,
　292–93
Prochloron, 85, 85f, 563
Producer, 1190f
　primary, **1190,** 1192f, 1193
Productivity
　factors limiting, 1192
　of freshwater ecosystems, 1217
　marine, 1214
　primary, **1191,** 1194, 1209, 1209f,
　　1214, 1220
　secondary, **1191**
　species richness and, 1197–99, 1198f
Product of reaction, **26,** 150, 150f
Profundal zone, 1216, 1216f
Progesterone, 129, 995t, 1009, 1071t,
　1072, 1075–78, 1075f, 1100,
　1100f, 1102
Proglottid, 643f, **645**
Programmed cell death. *See* Apoptosis
Progymnosperm, 589
Prokaryote, **70, 80,** 81, 84–85, 545–58.
　See also Bacteria
　asexual reproduction in, 237
　benefits of, 557–58
　cell division in, 207–9f, 208–9,
　　547, 547t
　cell organization of, 84–85, 84f
　cell surface of, 550, 550f
　cell structure in, 546f
　cell walls of, 84, 84f, 102t
　characteristics of, 523
　chromosomes of, 301f, 547, 547t
　classification of, 548
　colony of, 545f
　compartmentalization in, 547, 547t
　disease-causing, 553–56, 555t
　diversity in structure and
　　metabolism, 548, 548f
　DNA of, 85, 208–9
　eukaryotes versus, 102t, 546–47, 547f
　flagella of, 84, 84–85f, 102t, 547,
　　547t, 550, 551f
　forms of, 546
　gene expression in, 316, 316f,
　　366–69, 366–69f
　generation time of, 552
　genetic variation in, 552
　genome of, 348f
　interior organization of, 85, 85f
　internal membranes of, 551, 551f
　kinds of, 548
　metabolic diversity in, 547, 547f
　metabolism in, 553
　mutations in, 552, 552f
　prevalence of, 546–47

promoters of, 306, 306f
recombination in, 547, 552
replication in, 208–9, 208f, 290–94
ribosomes of, 551
size of, 546–47, 547t
symbiotic, 557
temperature dependence of, 1140
transcriptional control in, 362
transcription in, 306–7, 306–7f,
　316, 316f
translation in, 310, 316, 316f, 548
unicellularity of, 546, 547f
Prolactin, 994t, 1001f, 1002,
　1070t, 1102
Prolactin-inhibiting factor
　(PIF), 1003
Proliferating cell nuclear antigen
　(PCNA), 295
Proliferative phase, of
　endometrium, **1075**
Proline, 41, 42f
Promoter, **303, 362**
　efficiency of, 306
　in eukaryotes, 308, 308f, 370f
　in prokaryotes, 306, 306f
　regulation of access to, 362
Pronghorn, 1210
Proofreading function, of DNA
　polymerase, 290
Propane, 36
Prophage, **534,** 535f
Prophase
　meiosis I, 230, 231f, 232, 234f, 412
　meiosis II, 231f, 235f, 236
　mitotic, 213, 213f, **215,** 216f, 231f
Prophenyloxidase system, 1031
Proprioceptor, **974**
Prop root, 743
Prosencephalon. *See* Forebrain
Prosimian, **720,** 721f
Prostaglandin, 37t, 53, **994**–95, 1017,
　1017f, 1101
Prostate cancer, 414t, 800
Prostate gland, 1068f, 1070
Protandry, **1063**
Protanopia, 987
Protease, 39t
Protease inhibitor, 538, 539f, 801–2,
　802f, 1034
Protective coloring, in guppies,
　448–49, 448–49f
Protective layer, 852
Protein, 36
　catabolism of, 178–79, 178f, 1187
　central dogma, 51, 302–3, 303f
　changes in gene disorders, 262, 262f,
　　298, 298f
　denaturation of, 45, 47, 47f
　digestion of, 893, 895, 900t
　dissociation of, 47
　domains of, 43, 44f, 45
　folding of, 46, 46f
　functions of, 37t, 39–40, 39t, 40f
　glycation of, 402
　induced, 366–67
　in membranes, 80, 80f, 106,
　　108–9, 108f
　kinds of, 111, 111f
　movement of, 107, 107f
　structure of, 112, 112–13f
　motifs of, 43, 44f, 45
　nonpolar regions of, 43, 43f, 45,
　　112, 112f
　one-gene/one-polypeptide
　　hypothesis, 296–97
　phosphorylation of, 120, 121f, 132,
　　134, 220, 220f, 224, 377f,
　　998–99, 998f
　plasma. *See* Plasma proteins

I-29

Protein—Cont.
  polar regions of, 43, 43f, 112, 112f
  primary structure of, 43, 44f, 47f, 298, 298f
  quaternary structure of, 43, 44f, 45, 47
  repressed, 366–67
  secondary structure of, 43, 44f, 45
  structure of, 37t, 41, 43–45, 43–44f
  synthesis of. *See* Translation
  tertiary structure of, 43, 44f, 45, 47f
  transport within cells, 92, 93f
Protein-encoding gene, 350t
Protein kinase, **130**, 134f, 135, 222f, 422
  cyclin-dependent. *See* Cyclin-dependent protein kinase
  membrane/cytoskeleton, 416f
  serine/threonine-specific, 416f
  tyrosine-specific, 416f
Protein kinase-A, **998**, 998f
Protein microarray, 356
Protein microsphere, 63
Protein world, 66–67
Proteobacteria, 549f
Proteoglycan, 102, 102f, 391
Proteome, 351, 356
Proteomics, 315, **356**, 356f
Prothoracic gland, 1009, 1009f
Prothrombin, 911f
Protist, 561–76
  asexual reproduction in, 237, 1062
  cell division in, 209f
  cell surface of, 565
  classification of, 525, 564, 564f
  colonial, 561f
  cysts of, 565
  cytokinesis in, 218
  flagella of, 566, 566f
  general biology of, 565
  groups of, 566–76
  locomotor organelles of, 565
  mitosis in, 562
  predatory, 1168f
  reproduction in, 237, 565
Protista (kingdom), 16, 16f, 74–75f, **518**, 518f, 520, 521f, 523t, 525, 525f, 564
Protobiont, **68**
Protocell, 68
Protoderm, 730f, **731**, 731f, 739f, 745, 756f
Protogyny, 1062f, **1063**
Proton, **20**–21, 20f
Protonephridia, 1046, 1046f
Proton gradient, 198f, 773
Proton pump, 39f, 94, 161, 161f, 169, 174–75, 174f, 196, 551f, 770
Proto-oncogene, **224**, 224f, **416**, 416f
Protoplast, plant, 850, 850f
Protoplast fusion, 850
Protostome, 396, 527, 621f, **626**–28, 626f, 634, 634–35f, 662
  phylogeny of
    rRNA, 635, 635f
    traditional, 634, 634f
Prototheria (subclass), 717
Proximal convoluted tubule, 1048f, **1053**, 1053f, 1055, 1055f
Proximate causation, 1106
Prusiner, Stanley, 542
Pseudocoel, 646, 647f, 648
Pseudocoelomate, 621f, 622–23f, **625**, 625f, 628, 634–35, 634f, 646–48
Pseudogene, 349–50, 350t, **499**, 503
Pseudomonad, 553
Pseudomyrmex, 1174
Pseudopod, 565, 574, 574f
Pseudostratified columnar epithelium, 861t

Psilotum, 586
Psittaciformes (order), 711t
Psoriasis, 1014
PTC sensitivity. *See* Phenylthiocarbamide sensitivity
Pterophyta (phylum), 584, 585t, 586–88, 586–88f
Pteropodidae (family), 1243
Pterosaur, 703t, 707, 884, 884f
Pterosauria (order), 703t
PTH. *See* Parathyroid hormone
Ptychodiscus, 568f
Puberty
  in females, 1010, 1072
  in males, 1010
Puffball, 599f, 601, 608
Pufferfish (*Fugu rubripes*), 348f, 352, 492t, 494
Pulmonary arteriole, 928f
Pulmonary artery, 916, 917f, **918**
Pulmonary circulation, 698, **916**–17, 916–17f
Pulmonary valve, 917f, **918**
Pulmonary vein, 698, 916, 916–17f, **918**
Pulmonary venule, 928f
Pulp of tooth, 891f
Pulse, 918
Pulvini, **812**, 812f
Pumpkin, 743, 747, 846
Punctuated equilibrium, **486**, 486f
Punnett, Reginald Crundall, 250
Punnett square, **250**–51, 250f, 253
Pupa, **386**, 387f, **674**
Pupil, 983, 984f
Pure-breeding variety, **245**
Purine, 49, 49f, **284**, 284f, 285
Purine nucleoside phosphorylase deficiency, 333t
Purkinje fibers, **919**, 919f
Purple bacteria, 520–21f
Purple nonsulfur bacteria, 182, 521f, 553
Purple sulfur bacteria, 189
Pursuit deterrent signal, 1119–20
Pus, 1017
Pygmy hippopotamus, 1228
Pygmy marsupial frog, 1065f
Pyloric sphincter, 893, 893f
Pyramid of biomass, 1193, 1193f
  inverted, 1193
Pyramid of energy, 1193, 1193f
Pyramid of numbers, 1193, 1193f
Pyrimidine, 49, 49f, **284**, 284f, 285
Pyrodictium, 520f
Pyrolobus fumarii, 519
Pyrogen, 1017
Pyruvate, 36f
  conversion to acetyl-CoA, 167–68f, 168
  conversion to ethanol, 167f, 181, 181f
  conversion to lactate, 167f, 181
  from glycolysis, 162, 165f, 166
  from malate, 203, 204f
  oxidation of, 162, 163f, 168, 168f, 177, 177f
  from protein and fat catabolism, 178, 178f
Pyruvate dehydrogenase, 151, 151f, 168, 177, 177f
Pyruvate kinase, 165f

Q

Quadriceps muscle, 859f, 873, 873f, 961f
Quail, 711t
Quantitative traits, 255

Quaternary structure, of proteins, 43, 44f, 45, 47
Queen Anne's lace, 851
Queen bee, 1062, 1132–33, 1132f
Queen substance, 1133
Quiescent center, 740
Quill (feather), 710f
Quill (porcupine), 714
Quinine, 569, 799t
Quinone, 195, 198

R

Rabbit, 528f, 716, 719t, 883, 888–89, 898, 1067
  coat color in, 256
Rabies, 533t
Raccoon, 719t, 1195, 1211
Rachis, 749
Radial canal, 677–78f, **678**–79
Radial cleavage, **626**, 626f
Radial nerve (echinoderms), 954f
Radial symmetry, 620, **624**, 624f, 676, 677f
  secondary, **678**
Radiata, **620**, 620–21f, 624, 634–35f, 638–41, 638–41f
Radiation-sensitive badge, 21
Radicle, 763, 764f
Radioactive decay, dating of fossils using, 460, 460f
Radioactive isotope, **21**
Radioactivity, 21
Radiolarian, 574
Radio waves, 190, 190f
Radish, 743, 763, 1169
Radius, 699f, 871f
Radula, 622f, 653–55f, **654**, 656–57
raf gene, 416, 416f
Ragweed (*Ambrosia*), 1036
Rail, 711t
Rain forest, 1183f, 1185, **1210**, 1236f, 1237, 1237f
  biodiversity within, 1210
  loss of, 1220, 1220f
  tropical. *See* Tropical rain forest
Rain shadow effect, **1206**, 1206f
Ram ventilation, **925**
Random mating, 436–38, 437t, 438f
Ranunculus muricatus, 1143f
Rape case, 332, 332f
Raphe, 571
ras gene/protein, 135, 224, 224f, 416, 422
Raspberry, 751, 849
Rat, 719t
  DNA of, 285t
  genome of, 495
  introduced species, 1240
  maze learning behavior in, 1108, 1108f
  operant conditioning in, 1111
  predation by, 1168
  resistance to tooth decay, 458
  warfarin resistance in, 441
Rat flea, 555t
Rattlesnake, 709, 870, 870f, 988
Raven, cognitive behavior in, 1115, 1115f
Ray (fish), 693t, 695, 988, 1049
Ray (parenchyma cells), 737
Ray-finned fish, 689, 691–92f, 693t, 696, 696f
Ray initial, 737
Rb gene, 224, 224f, 417–18, 417t, 418f
Rb protein. *See* Retinoblastoma protein
Reabsorption, 1046
  in kidney, 1054–56, 1054–56f
Reactant, **26**
Reaction center, **194**–96, 195–97f, 198

Reading, 960
Reading frame, **304**, 310
Realized niche, **1164**–65
Reasoning
  deductive, **4**, 4f
  inductive, **4**
Rebek, Julius, 66–67
Recently-Out-of-Africa Hypothesis, 725
Receptacle (flower), **594**, 595f, 840f
Receptor. *See* Cell surface receptor
Receptor-mediated endocytosis, 118–**19**, 118f, 122t
Receptor potential, **972**, 972f, 973
Receptor protein, 80, 108f, 109t, 111–12, 111f, **126**, **948**, 948f. *See also* specific types of receptors
  characterization of, 126, 128t
  intracellular, **128**–29, 128–29f, 128t
Receptor protein family, 126
Recessive trait, 246t, **247**–51, 248f, 251f
  in humans, 249t, 259
Reciprocal altruism, **1130**, 1134
Reciprocal exchange, 232, **406**, 406f
Reciprocal recombination, 406, 409
Recognition helix, 364, 364f
Recognition site, **406**
Recombinant DNA, 320–21, 321f, **323**, 323f
  production of large quantities of, 321–22, 321–22f
Recombination, 230, **267**–68, 406–9, 523t
  crossing over and, **267**, 267f
  disadvantages of, 237
  in eukaryotes, 547
  gene transfer, 406–8, 407–8f
  in prokaryotes, 547, 552
  reciprocal, 406, 409
  using to make genetic maps, 268, 268f
  in viruses, 540–41
Recombinational repair, **412**
Recombination nodule, 232
Recruitment, 503, **879**
Rectum, 889f
Rectus abdominis muscle, 859f
Red algae, 521f, 522, 526f, 563, 572, 580
Red-bellied turtle (*Pseudemys rubriventris*), 708f
Red blood cell(s). *See* Erythrocytes
Red blood cell antigens, 1029, 1029f
Red-eyed tree frog (*Agalychnis callidryas*), 701f
Red fiber, 880
Red-green color blindness, 249t
Red hair, 242f
Rediae, **644**, 644f
Red maple (*Acer rubrum*), 737f
Red marrow, 865f
Red milk snake (*Lampropeltis triangulum syspila*), 472f
Redox reaction. *See* Oxidation-reduction reaction
Red-plumed worm, 1215
Red Queen hypothesis, for origin of sex, 238
Red tide, 568
Reducing atmosphere, 65–66
Reducing power, 145
Reduction, **23**, 23f, **145**, 145f, 156
Reduction division, 228–30
Red-water fever, 670
Red-winged blackbird, 1128, 1129f
Redwood, 585t
Reflex, 891, 961
Reflex arc, 961f, 962
Refrigeration, 47
Refuse utilizer, 180f

Regeneration
 in echinoderms, 679
 in lizards, 709
 in plants, 751
 of ribbon worm eyespot, 505, 506f
 in sea stars, 677f
 of spinal cord, 962
Regulation, as characteristic of life, 63
Regulative development, **391**, 397
Regulatory molecules, 992–95, 992f
Regulatory proteins, 363–65, 378
 DNA-binding motifs in, 363–65, 364–65f
Reindeer, 1211
Reinforcement, **478**–79, 478f
Relative dating, 460
Release factor, **312**, 312f
Remora, 925
REM sleep, 959
Renal carcinoma, 417t, 1022
Renal cortex, **1052**, 1052–53f, 1055f
Renal medulla, **1052**, 1052–53f, 1055f, 1056
Renal pelvis, 1052f
Renewable resources, 1157
Renin, 920, 1058, 1058f
Renin-angiotensin-aldosterone system, 1007, 1058, 1058f
Renner, Otto, 772
Repetitive DNA, 351
Replica plating, 552f
Replication, 214, 288–95
 conservative, 288
 direction of, 208, 290, 294
 dispersive, 288
 elongation stage of, 294
 enzymes needed for, 291–93, 293t
 errors in, 350, 412
 in eukaryotes, 295, 295f
 initiation stage of, 294
 lagging strand, **290**–91, 291–92f, 294, 294f
 leading strand, **290**–91, 291–92f, 294, 294f
 Meselson-Stahl experiment on, 288, 288–89f
 Okazaki fragments, **290**–91, 291–92f, 294f
 in prokaryotes, 208–9, 208f, 290–94
 rolling-circle, **407**
 semiconservative, **288**, 289f
 termination stage of, 294
 of virus, 532
Replication fork, **292**, 292f, 294–95, 295f
Replication origin, 208, 208f, 290, 290f, 295, 321
Replicon, 290, 295, 295f
Replisome, **293**–94
Repolarization, 945f, 946
Repression, 366–67
Repressor, **367**
Reproduction, 1061–78
 age at first reproduction, 1149
 in amphibians, 699, 1065, 1065f
 in arthropods, 1062
 asexual. *See* Asexual reproduction
 as characteristic of life, 2, 63
 cost of, **1148**–49, 1148–49f, 1154
 in echinoderms, 679, 679f
 evolution of, 1064–67
 in fish, 1065
 in flatworms, 643, 643f
 in fungi, 600, 602, 602f
 investment per offspring, 1149, 1149f
 in mammals, 1067, 1067f
 in mollusks, 655, 655f
 in nematodes, 646
 in plants, 831–52
 in polychaetes, 660

 in protists, 565
 reproductive events per lifetime, 1149
 in reptiles, 1066, 1066f
 sexual. *See* Sexual reproduction
 in sponges, 636–37
Reproductive cloning, 424–25, 424–25f, 429f, 430
 problems with, 425
 reasons for failure of, 425
Reproductive competition, 1126–27
Reproductive isolation, **473**, 477–79, 481, 1119f
 as by-product of evolutionary change, 478–79
 random changes may cause, 479
Reproductive leaf, 751
Reproductive strategy, **1125**, 1125f
Reproductive success, 440, 442, 1122–23, 1128–29, 1148
Reproductive system
 female, 857t, 859f, 1072–76, 1072–76f
 male, 857t, 859f, 1068–71, 1068–71f
Reprogramming, 425
Reptile, 622t, 702–9
 brain of, 956, 956f
 characteristics of, 702, 703t, 707
 circulation in, 707, 707f, 916
 cleavage in, 1086
 eggs of, 702
 evolution of, 688–89, 704–5, 705f, 706t
 extinctions, 1229f, 1229t, 1234t
 gastrulation in, 1089
 heart of, 702, 707, 709, 916
 kidney of, 1050
 legs of, 702
 lungs of, 702
 nitrogenous wastes of, 1051, 1051f
 present day, 707–9, 708–9f
 reproduction in, 1066, 1066f
 respiration in, 702, 916, 927–28
 skin of, 702
 swimming in, 882
 thermoregulation in, 704, 707, 1042, 1207
Reptilia (class), 689, 702
Research, 6–7
Resin, 590, 736
Resistance transfer factor (RTF), **408**
Resolution (microscope), **82**
Resource competition, 444
Resource partitioning, 1166, 1166f
Respiration, 922–29, 1184
 aerobic. *See* Aerobic respiration
 in air-breathing animals, 926–27
 in amphibians, 698, 916, 923f, 927, 927f
 anaerobic, **160**, 163, 177, 881
 in aquatic vertebrates, 924–25
 in birds, 713, 928–29, 929f
 in carbon cycle, 1186, 1186f
 cellular, 160, 177, 182, 922
 in crustaceans, 924–25
 cutaneous, 698, **916**, 922, 923f, 926
 in echinoderms, 923f, 924
 efficiency of, 922–29
 external, 922
 in fish, 915, 923f, 924–25, 925f
 in insects, 923f, 926–27
 internal, 922
 in invertebrates, 922–23
 in mammals, 923f, 928, 928f
 in mollusks, 924
 in plants, effect of global warming on, 786
 in reptiles, 702, 916, 927–28
 sulfate, 163

Respiratory control center, 932, 933f
Respiratory system, 857t, 858f
 of arthropods, 665–67f, 667
 as barrier to infection, 1014
 of vertebrates, 915–17, 915–17f
Resting membrane potential, 942–43, 942–43f, 943t, 945f
Restoration ecology, 1246, 1246f
Restriction endonuclease, **320**, 321f, 325, 326f, 330–31, 330–31f, 344, 344f, 351, 1031
Restriction fragment length polymorphism (RFLP) analysis, **331**, 331f, 344–45, 344f, 351
Restriction map, 320
Restriction site, **320**, 321f
RET gene, 416f, 417t
Reticular activating system, 959
Reticular formation, 959
Reticulin, 862
Reticulum, 898, 899f
Retina, 939f, 983–84f, 985–86, 986f, 1093f
 visual processing in, 987, 987f
Retinal, 53, 53f, 192–93, 985–86, 985f
Retinal degeneration, canine, 264, 264f
Retinitis pigmentosa, 269f
Retinoblastoma, 224, 417t, 418f
Retinoblastoma (Rb) protein, 220, 221f, 422
Retinoid, 996
Retinoschisis, 269f
Retinular cells, 666f, 983f
Retrotransposon, 494, 496
Retrovirus, 324, **536**
Reverse transcriptase, **324**, 324f, 350, **536**, 537f, 800, 1034
Reznick, David, 449
RFLP analysis. *See* Restriction fragment length polymorphism analysis
R gene, 803, 803f
R group, 41, 42f
Rhabdom, **666**, 666f
Rhabdovirus, 533f
Rh blood group, **260**, 1029
Rhesus monkey, 1112
Rheumatic fever, 554
Rheumatoid arthritis, 333t, 1007
Rhine River, 1218
Rhinoceros, 719t, 1140f, 1173, 1210
Rhinovirus, 524f
Rhipidistian, 699
Rhizobium, 557, 787, 787f
Rhizoid, **582**–83, 583f, 587, 588f, 757, 757f
Rhizome, 587, 588f, 747, 747f, 849
Rhizopogon, 612f
Rhizopus, 605f, 607f
Rh-negative individual, **1029**
RhoC enzyme, 423
Rhodophyta (phylum), 564f, **572**
Rhodopsin, 131, 135, 135f, 573
Rhombencephalon. *See* Hindbrain
Rh-positive individual, **1029**
Rhynchocephalia (order), 703t, 707–8
Rhyniophyta (phylum), 584, 584f
Rib(s), 871f, 930
Ribbon worm, 645, 645f
 regeneration of eyespot, 505, 506f
Riboflavin. *See* Vitamin B$_2$
Ribonuclease, 47f
Ribonucleic acid. *See* RNA
Ribose, 49–50, 49f, 55f, 154
Ribosomal proteins, 302, 302f
Ribosomal RNA (rRNA), 95, **302**, 302f
 of archaebacteria, 519
 of bacteria, 520
 evolutionary relationships from, 519–20, 520f

 phylogenic relationships from, 499–500, 500f
 protostome phylogeny based on, 635, 635f
 synthesis of, 215
Ribosomal RNA (rRNA) genes, 95, 215, 217, 308, 323, 323f, 349
Ribosome, 82f, 84f, 86–87f, 90t, **91**, 91f, **95**, 95f, 102f, **302**
 A site on, 302, 302f, 310, 311–12f, 312, 314f
 E site on, 302, 302f, 310, 311–12f, 312, 314f
 of eukaryotes, 95
 free, 95
 of mitochondria, 562
 of prokaryotes, 551
 P site on, 302, 302f, 310, 311–12f, 312, 314f
 structure of, 302, 302f
 in translation, 310–12, 310–12f
Ribozyme, 66, 151, 302f
Ribulose 1,5-bisphosphate, 200f, 201, 203, 785, 785f
Ribulose bisphosphate carboxylase/oxygenase. *See* Rubisco
Rice (*Oryza sativa*), 357f
 foolish seedling disease of, 824
 genome of, 348, 348f, 352, 353f, 493t, 496–97
 golden, 357
 transgenic, 337–38, 337f
 world demand for, 357
Rickets, 904t, 1006
Rickettsia, 549f
Rickettsia tsutsugamushi, 118f
Rickettsia typhi, 555t
Right whale, 1239
Rigor mortis, 877
Ring canal, 678f
Ringworm, 604
Ripening, of fruit, 335
RISC (enzyme complex), 374, 375f
River, 1218
RNA, 37t, **48**
 antisense. *See* Antisense RNA
 catalytic activity of, 66, 151
 Central Dogma, 51, 302–3, 303f
 chromosome-associated, 211
 directionality of, 285
 DNA versus, 51, 51f
 double-stranded loops in, 374, 374–75f
 evolution of DNA from, 51
 functions of, 48
 guide, 567
 kinds of, 302
 messenger. *See* Messenger RNA
 microRNA, **374**, 375f
 ribosomal. *See* Ribosomal RNA
 small, 374, 374–75f
 structure of, 49–51, 49f, 284
 synthesis of. *See* Transcription
 transfer. *See* Transfer RNA
 translation of. *See* Translation
RNA editing, 376–78
RNA interference, **374**, 375f
RNA polymerase, **303**, 307, 307f, 310f, 315f, 520t
 binding to promoter, 362, 366–67
 in eukaryotes, 308
 in prokaryotes, 306, 306f
RNA polymerase I, 308
RNA polymerase II, 308, 308f, 370, 371f, 372, 372f
RNA polymerase III, 308

Index I-31

RNA splicing, 313–14, 314f, **376**, 377f
  alternative splicing, 314–15, **349**, 349f, 351, 356, 376, 376f
RNA transcript, 307
  transport across nuclear membrane, 522
RNA virus, 524, 532, 532f, 533t
RNA world, 66, 151
Roach, 675t
Robin, 711t, 1238
Rock fossil, 460
Rocky Mountain spotted fever, 670
Rod(s), 135, 869t, 971t, 985–87, 985–86f
  green rod, 698
Rodbell, Martin, 131
Rodent, 528, 528f, 716, 892, 898
  desert, ant-rodent interactions, 1176–77, 1177f
Rodentia (order), 511f, 719t
Rodhocetus kasrani, 461, 461f
Rodrigues fruit bat (*Pteropus rodricensis*), 1243
Roland, Sherwood, 1221
Rolling-circle replication, **407**
Root, 732f
  absorption by plants, 770–71, 770–71f
  adventitious, 743, 747, 764f, 774, 832–33, 832f, 836, 837f
  arbuscular mycorrhizae and, 612, 612f
  formation of, 758–59
  inhibition of flowering, 836, 837f
  lateral, 741
  modified, 743, 743f
  silt, 775, 775f
  structure of, 739–42, 739–42f
Root canal, 891f
Root cap, 730, 730–31f, 739, **739**, 739f, 760f, 810
  columella, 739, 739f
  lateral, 739, 739f
Root hair, 730–31f, **734**, 735, **741**, 770, 770f, 787f
Root-knot nematode, 796, 797f
Root nodule, 787, 787f, 1172
Root pressure, **770**–72
Root-shoot axis, establishment of, 756–57, 756–57f
Root system, 732
Rose, 751, 835, 849
Rosin, 590
Rossmann fold, 45
Rosy periwinkle (*Catharanthus roseus*), 1232, 1232f
Rotifera (phylum), 621f, 623t, 648, 648f
Rough endoplasmic reticulum, 86–87f, **91**, 91f, 93f, 95
Roundup, **336**, 336f
Round window, 980, 981f
Roundworm, 621f, 622t, 635, 646, 646–47f, 882
rRNA. *See* Ribosomal RNA
r-selected population, **1154**, 1154t, 1179
RTF. *See* Resistance transfer factor
Rubber, 37t, 53
Rubella, 1098
Ruber, 849
Rubisco, 201–3, 352, 785
Rubus, 1143f
Ruffini corpuscle, 971t, 973, 973f
Rule of eight. *See* Octet rule
Rumen, 557, 613, 898, 899f
Ruminant, 889, 893, 898, 898–99f
Rumination, 898

Runner, plant, 747, 747f, 849
Runoff, 1184f, 1189, 1189f, 1217, 1241
Rusts (plant disease), 605t, 608

S

Saber-toothed cat, 717t, 1228, 1228f
Saber-toothed-ness, 516, 516f
Saccharomyces, 210t, 610f
Saccharomyces cerevisiae, 603, 610
  genome of, 348f, 352, 493t, 610
  homeodomain proteins of, 16f
Saccule, 978–80, 978f
Sage, 798
Sahelanthropus tchadensis, 723f
St. John's wort, 1164
Salamander, 698, 698t, 701, 888f, 927
Salicylic acid, 802, 804
Saliva, 890, 1025
Salivary gland, 859f, 889f, 890, 900t, 965t
Salivary gland cancer, 417t
Salmonella, 549f
  Ames test, 414, 414f
  type III system in, 553
Saltatory conduction, 946–**47**, 947f
Salt-curing, 47
Salt gland, 1050, 1050f
Salt hunger, 1057
Salt lick, 1057
Salty taste, 975
Samara, 761f
Sand dollar, 622t, 676, 680
Sand flea, 669
Sand fly, 567
Sandpiper, 711t
Sanger, Frederick, 41, 298
SA node. *See* Sinoatrial node
Santa Catalina Mountains, tree species along moisture gradient, 1162–63, 1163f
Sarcoma, **414**, 414t, 417t
Sarcomere, **874**–76, 877f
Sarcoplasmic reticulum, 868f, 878–79, 878f
Sarcopterygii (class), 689, 692–93f, 696–97, 697f
SARS. *See* Severe acute respiratory syndrome
Sartorius muscle, 859f
Satellite image, of world's environments, 1203f
Satiety, 958
Satiety factor, 903
Saturated fatty acid, 53–54, 54f, 815
Sauria (suborder), 703t, 708
Saurischia (order), 703t
Sauropsida, 705f
Savage, Jay, 1234
Savanna, 180f, 1162f, 1191f, 1208, 1208f, **1210**
Scaffold protein, 211, 211f, 295f
Scales, of reptiles, 702
Scaleworm, 660
Scallop, 652, 656–57, 882
Scanning electron microscopy, **82**–83, 83t, 110
Scanning-tunneling microscopy, of DNA, 48, 48f
Scaptomyza, Hawaiian, 482
Scapula, 871f
SCARECROW gene, in *Arabidopsis*, 740, 740f, 810
Scarlet fever, 554
Scarlet kingsnake (*Lampropeltis triangulum elapsoides*), 472f
Scavenger, 180f
Schistosomiasis, 644

Schizogony, **565**
Schizophrenia, 950
Schizosaccharomyces pombe, 348f, 352, 493t
Schleiden, Matthias, 15, 81
Schwann, Theodor, 15, 81
Schwann cells, **941**, 941f, 1091
Schwannoma, 417t
Sciatic nerve, 954f
Sciuridae (family), 511f
Sclera, 983, 984f
Sclereid, 736
Sclerenchyma, 736
Sclerenchyma cells, **736**, 736f
SCN. *See* Suprachiasmatic nucleus
Scolex, 643f, **645**
Scolopendra, 671f
Scorpion, 670, 1170
Scouring rush. *See* Horsetail
Scrapie, 542
Screech owl, 711t
Scrotal cancer, 415
Scrotum, 1068, 1068f
Sculpin, 1242
Scurvy, 904, 904t
Scutellum, **763**, 764f
Scyphozoa (class), 638, 640, 640f
Sea anemone, 622t, 624, 624f, 638, 641, 641f, 1161f, 1173, 1215
Seabird, 711t, 1188
Sea cucumber, 622t, 676, 676f, 679
Sea fan, 641
Seafloor, 1215
Seahorse, 1125
Seal, 719t
Sea level, effect of global warming on, 1224
Sea lily, 679
Sea lion, 719t, 1242, 1242f
Sea mat, 623t
Sea mice, 660
Sea moss, 623t
Sea otter, 1114f, 1168, 1242–43, 1242f
Sea pansy, 641
Sea peach (*Halocynthia auranthium*), 686f
Sea pen, 641
Search image, 444, 444f
Sea slug, 655–56
Sea snake, 709, 1050
Seasonality, 1199, 1205
  variations in solar energy, 1204, 1204f
Sea star, 622t, 676, 676–77f, 679f, 680, 680f, 882, 1031, 1031f
  predation on mussels, 1176, 1176f
Sea turtle, 703t, 882, 1050
Sea urchin, 622t, 676, 676f, 680, 1168, 1242, 1242f
  development in, 501, 501f
  fertilization in, 1083f
Sea walnut, 623t, 641
Sea whip, 641
Sebaceous gland, 860, 1014
Secondary cell wall, **101**, 101f
Secondary chemical compounds, 1169
Secondary consumer, **1190**, 1190f, 1192f
Secondary growth, in plants, **731**, 732–33f, 733, **742**
Secondary immune response, **1027**, 1027f
Secondary induction, **1093**

Secondary lysosome, 94
Secondary metabolite, 798, 799t, 800
Secondary mycelium, 608, 608f
Secondary oocyte, 1073–74, 1073–74f, 1082
Secondary phloem, 731, 733f, 745f, 746
Secondary productivity, **1191**
Secondary radial symmetry, **678**
Secondary sexual characteristics, 1072, **1126**
Secondary spermatocyte, 1068, 1069f
Secondary structure, of proteins, 43, 44f, 45
Secondary succession, **1178**
Secondary tissues, of plant, **731**, **746**
Secondary xylem, 731, 733f, 737, 745f, 746
Second filial generation, 246t, **247**–48, 248f, 250–51, 250–51f
Second Law of Thermodynamics, **146**
Second messenger, **132**, 134, 998
  calcium, 132, 133f, 135
  cAMP, 132, 132–34f, 998, 998f
  cGMP, 129
  IP_3/calcium, 998–99, 999f
Second trimester, 1100
Secretin, 900, 900t, 901f
Secretion, 860, 1046
Secretory phase, of endometrium, **1075**
Secretory protein, 91, 91f, 93f
Secretory vesicle, 92–93f, 119, 119f
Sedge, 845, 849
Seed, 388, 389f, 584, 589, 591f, 593, 595–96, 596f, 732, 760, 848
  adaptive importance of, 760
  dispersal of, 438, 584, 760–62, 845, 849, 1143, 1143f, 1240, 1243
  dormancy in, 760, 814, 814f, 827, 827f
  food reserves in, 763
  formation of, 388, 389f, 760, 760f
  germination of, 388, 389f, 596f, 756, 760, 763, 764f, 808, 814, 824, 827, 840
  imbibition of water, 763
  nutrient storage in, 759, 759f
Seed cache, 1111, 1111f
Seed coat, 589, 591, 595–96, 596f, 760, 760f, 763, 764f, 814, 838
Seed-hoarding bird, 1111, 1111f
Seedless vascular plant, 580, 580f, 585t, 586–88, 586–88f
Seedling, 730
  development of, 731f
  etiolated, 808
  growth of, 763, 764f
Seed plant, 584, 585t, 589–96
Seed traits, in garden pea, 244–45, 246t, 247f, 253–54, 254f
Segmental duplication, 349–50, 350t, 496–99
Segmentation (animals), **627**
  in annelids, 527, 527f, 627, 627f, 658, 659f
  in arthropods, 527, 527f, 627, 627f, 665, 665f
  in chordates, 527, 527f, 627, 627f, 684
  in *Drosophila* development, 395f, 396–97
  evolution of, 527, 527f, 627, 627f
  in insects, 665f
Segment-polarity genes, 395f, **396**
Segregation of traits, **243**
  law of, **251**
Sei whale, 1239, 1239f
Selectable marker, 321, 322f, 327, 327f

I-32   Index

Selection, 437, 437t, 438f, **440–41**, 440f. *See also* Artificial selection; Natural selection
  to avoid predators, 440, 440f
  on color in guppies, 448–49, 448–49f
  directional, 446–47f, **447**
  disruptive, **446**, 446–47f, 481
  forms of, 446–47, 446–47f
  frequency-dependent, **444**, 444f
  group, **1130**
  industrial melanism, 456–57, 456–57f
  interactions among evolutionary forces, 443, 443f
  kin, **1130**–31, 1131f, 1132, 1134
  limits to, 450, 450f
  to match climatic conditions, 441, 441f
  oscillating, **444**, 486, 512
  for pesticide resistance in insects, 441, 441f
  reinforcement of isolating mechanisms, 478–79, 478f
  sexual, **1126**–27, 1126f
  stabilizing, 446–47f, **447**, 477, 486
Selective permeability, **114**
Selenium, bioremediation by fungi, 603
Self, 136, 136f
Self antigen, 1036
Self-fertilization, **244**
Self-incompatibility, in plants, **847**, 847f
  gametophytic, 847, 847f
  sporophytic, 847, 847f
Self-pollination, 595, 844, 846–47, 847f
Self-versus-nonself recognition, **1020**
Semelparity, **1149**
Semen, 1070
Semibalanus balanoides, 1164, 1164f
Semicircular canal, 978–79f, 979, 981f
Semiconservative replication, **288**, 289f
Semidesert, 1208, 1208f
Semilunar valve, 918
Seminal vesicle, 1068f, 1070
Seminiferous tubule, 1069f
Senescence, in plants, 851–52
Sense strand, 306
Sensitive phase, **1112**
Sensitive plant (*Mimosa pudica*), 812–13, 812f
Sensitivity, as characteristic of life, 2, 62–63, 62f
Sensor, **1040**, 1040–41f
Sensory exploitation, **1127**
Sensory hair, **674**, 975, 975f
Sensory information, path of, 970, 970f
Sensory neuron, 869t, **940**, 940f, 961–62f, 963–66
Sensory organs
  development of, 1091
  of insects, 665f
Sensory perception, 970
Sensory receptor, 674, 940, 970–72, 971t
Sensory systems, 969–88
  in plants, 807–27
Sensory transduction, 970, 972, 972f
Sepal, **594**, 595f, 839f, **840**, 840f
Separase, 216–17
Separation layer, 852, 852f
Septation (cell division), 208, 208–9f
Septum
  of annelids, **658**, 659f
  of fungal hyphae, 601–2
  in heart, 916, 916f
Sequence-tagged site (STS), **345**, 345f, **347**, 351
Sequential hermaphroditism, 1062–63
Serine, 41, 42f
Serosa, of gastrointestinal tract, 889, 889f

Serotonin, **950**, 959
Serotonin receptor, 378
Serpentes (suborder), 703t, 708
Serpentine soil, 1163, 1163f
Sertoli cells, 1068, 1069f, 1071, 1071f
Serum, **910**
Sessile animal, 617f, **636**, 657, 669, 669f, 686
Setae, **658**, 659f, 661
Set point, 1040, 1040f
Severe acute respiratory syndrome (SARS), 357, 533t, 540, 540f, **541**
Severe combined immunodeficiency (SCID), 333t
Sewage, 1217
Sex, 1061–78
Sex chromosome, **270**
  of birds, 270, 270t
  of fruit fly, 270, 270t
  of humans, 270, 270t
  of insects, 270, 270t
  nondisjunction involving, 273, 273f
Sex determination, 270, 270t, 1063
  in mammals, 1063, 1063f
  temperature-dependent, **1223**
Sex linkage, **266**, 266f, 268f
Sex ratio, **1145**
  human family, 253, 253t
Sex steroid, **993**, 1009
Sexual development, in vertebrates, 1009
Sexual dimorphism, **1126**, 1126f
Sexual imprinting, **1112**
Sexuality, in eukaryotes, 522
Sexual life cycle, 228–29, 229f
Sexually transmitted disease (STD), 555t, 1076
Sexual receptivity, 1076
Sexual reproduction, **73**, 227–38, **228**, 522, 1062–63. *See also* Meiosis
  in animals, 619t
  evolution of, 73, 237–38
Sexual selection, **1126**–27, 1126f
Shade leaf, 751
Shark, 693t, 695, 695f, 925, 956f, 988, 1032, 1042, 1044, 1049
  teeth of, 695
Sharp-beaked finch (*Geospiza difficilis*), 454–55, 454f
Sheep, cloning of, 394, 394f, 424–25, 424–25f
Shelf fungi, 608
Shell
  of bivalves, 656–57, 657f
  of diatoms, 571, 572f
  of mollusks, 652, 654, 654f, 656, 656f, 1168
  of snail, 653f
  of turtle, 708
Shell pattern, in snails, 435f, 440
Shell shape, in oysters, 461, 461f
Shell-skinned fish, 692, 692f
Shigella, type III system in, 553
Shipworm, 652
Shivering, 1042
Shock, anaphylactic, **1036**
Shoot, 731–32f, 732. *See also* Root-shoot axis
  development of, 764f
  formation of, 758–59
shootmeristemless mutant, in *Arabidopsis*, 758, 758f
Shorebird, 711t, 713
Short-day plant, **834**–35, 834f
  facultative, 835
Short-horned grasshopper, 674
Short interspersed element (SINE), 350

short root mutant, in *Arabidopsis*, 810
Short-term memory, 960
Shotgun sequencing, 347, 347f, 351
Shrew, 719t, 982, 1190f
Shrimp, 668
Shrimp farm, 1232
Shrub, 585t, 593
Shrubland, 1191f
Siamese cat, 256
Siberian tiger, 1227f
"Sick" building, 614
Sickle cell anemia, 45, 249t, 255, **262**, 262f, 264t, 298, 298f
  malaria and, 262, 262f, 445, 445f
Side-blotched lizard (*Uta stansburiana*), 1149, 1149f
Sideroblastic anemia, 269f
Sidneyia, 630f
Sieve area, 738
Sieve cells, 738
Sieve plate, 738, 738f
Sieve tube, 738, 776–77, 777f
Sieve-tube member, 733, 738, 738f
Sight. *See* Vision
Sigma factor, 306, 306f
Sigmoidal growth curve, 1150f, **1151**
Sigmoria, 671f
Signal beacon, 392
Signal sequence, **91**, 91f, 553
Signal transduction pathway, 816
  in plants, 808–9
  in seed germination, 764f
  wound response in plants, 802, 802f
Signature sequence, 519
Sign stimulus, **1106**–7, 1107f
Sildenafil, 126, 951, 1010, 1070
Silica, 586, 796, 1169
Silk, 37t
Silk (spider), 670
Silkworm moth (*Bombyx mori*), 210t, 1009f, 1118–19
Silt root, 775, 775f
Silverfish, 674
Silver fox, domestication of, 459, 459f
Simple epithelium, **860**
  columnar, 860, 861t
  cuboidal, 860, 861t
  squamous, 860, 861t
Simple eye, 504, 505f, **666**
Simple leaf, **749**, 749f
Simple metamorphosis, 674
Simple sequence repeats, 350, 350t
Simpson, O. J., 332f
SINE. *See* Short interspersed element
Singer, S., 108
Single bond, **26**
Single-copy gene, 349
Single nucleotide polymorphism (SNP), 269, **351**
  in human genome, 351, 355, 357
  microarray technology, 355
Single-strand binding protein, 292, 292f, 293f
Sink (plant carbohydrate), **776**–77, 777f
Sinoatrial (SA) node, 917, 919, 919f
Sinosauropteryx, 517f, 712f
Sinus venosus, **915**, 915–16f, 917
Siphonaptera (order), 675f
siRNA. *See* Small interfering RNA
Sister chromatid(s), 212, 212f, 215–17, 231f, 232, 233–36f, 236
Sister chromatid cohesion, 232, 236
Skate, 693t, 695, 988
Skeletal muscle, 856f, 867t, 868, 868f
  actions of, 873, 873f
  lymph movement and, 914
  venous pump, 913, 913f
Skeletal system, 857t, 858f, 871

Skeleton, 685, **856**
  hydrostatic, **871**, 871f
  types of, 871, 871f
  of vertebrates, 689f
Skin, 859f, **1014**
  as barrier to infection, 860, 1014
  of reptiles, 702
  as respiratory organ. *See* Cutaneous respiration
Skin cancer, 191, 415t, 1221
Skink, 709, 709f
Skinner, B. F., 1111
Skinner box, 1111
Skotoptropism, 811
Skull, 856, 858f, 871f, 1091
Slash pine (*Pinus palustris*), 590
Sleep, 959
Sleeping sickness, 1035
Sleep movement, in plants, 813, 813f
Sliding filament model, of muscle contraction, 874, 874–77f, **875**–77
Slightly movable joint, 872, 872f
Slime mold, 520f, 575–76, 576f, 605
  cellular, 575–76, 576f
  plasmodial, 575, 576f
Slipped mispairing, 409–10
S locus, 847, 847f
Sloth, 719t
Slow-twitch muscle fiber, **880**, 881f
Slug (mollusk), 652, 656, 883
Slug (slime mold), 576, 576f
Small interfering RNA (siRNA), **374**, 375f
Small intestine, 859f, 889, 889f, 893, 893–94f, 897f, 900t
  absorption in, 896
  accessory organs to, 895, 895f
  digestion in, 894, 894f
Small nuclear ribonucleoprotein (snRNP), 313–14, 314f, 376
Smallpox, 357t, 533t, 540, 558, 1018, 1018f, 1027f
Small RNA, 374, 374–75f
Smell, 955, 971t, 976, 976f, 1091
Smoking, 931
  cancer and, 223, 415, 415t, 420–21, 420–21f
  cardiovascular disease and, 921
  nicotine addiction, 952–53
Smoking cessation, 953
Smooth endoplasmic reticulum, 86–87f, 91f, **92**
Smooth muscle, 856f, 867–68, 867t
Smuts, 608
Snail, 622t, 635, 652, 653f, 655f, 883
  shell pattern in, 435f, 440
Snake, 703t, 705–6f, 707–9, 709f, 883
  venomous, 708–9, 1170
Snake venom, 39t
Snapdragon, 593, 835, 838, 841
  CYCLOIDIA gene in, 841, 841f
  phantastica mutant in, 748, 748f
Snodgrass, Robert, 527
Snowshoe hare (*Lepus americanus*), population cycles of, 1153–54, 1153f
SNP. *See* Single nucleotide polymorphism
snRNP. *See* Small nuclear ribonucleoprotein
Snuff, 415
Snurp. *See* Small nuclear ribonucleoprotein
Snyder, Evan, 427
Soap, 37t
  antibacterial, 552
Social insects, 1120, 1120f, 1132–33, 1132–33f

Index I-33

Social system
 communication in social group, 1120–21, 1120–21f
 evolution of, 1133–34
Society, **1133**
Socorro isopod, 1140
Sodium, 25, 25f
 in action potential, 944–47, 945–47f
 blood, 1057–58, 1058f
 in cytoplasm and extracellular fluid, 943t, 1044
 extracellular, 1057
 in photoreception, 986
 reabsorption in kidney, 1007, 1048f, 1054–58, 1055f, 1058f
 in resting membrane potential, 942–43, 942–43f
Sodium channel, 109t, 114, 135f
Sodium chloride, 25, 25f, 30f
Sodium hydroxide, 31
Sodium-potassium pump, 39t, 109t, **120**, 121f, 122t, 942, 942–43f, 946
Soft coral, 641
Soft palate, 891, 891f
Soft rot, 553
Soft wood, 590
Soil, 1138
 fertility of, 784
 formation of, 1178
 minerals in, 784
 in photosynthesis, 188
 phytoremediation, 790–92, 790–92f
 plant nutrients in, 781f, 784
 serpentine, 1163, 1163f
Soil pore, 784
Solanum dulcamara, 1143f
Solar energy. *See also* Sunlight
 climate and, 1204, 1204f
 distribution over earth's surface, 1204, 1204f
 seasonal variation in, 1204, 1204f
Soldier fly (*Pteticus trivittatus*), 672f
Solenoid, 211, 211f
Soleus muscle, 881f
Solute, **116**
Solute potential, **769**–70, 769f
Solvent, 28t, 30, 30f, **116**
Somatic cell(s), **228**–29, 229f, 410
Somatic cell embryo, 850f
Somatic DNA rearrangement, **1026**, 1026f
Somatic motor neuron, 879, **940**
Somatic mutation, 410, **1026**
Somatic nervous system, 940f, 963, 963t
Somatosensory cortex, primary, 957, 958f
Somatostatin. *See* Growth hormone-inhibiting hormone
Somatotropin. *See* Growth hormone
Somite, 385, **1090**, 1091f, 1098
Somitomere, **1090**
Sonar, 982
Song, bird's, 1106, 1113, 1113f, 1119, 1124
Songbird, 711t, 713
 declining populations of, 1238, 1238f
 migratory, 1238, 1238f
 woodland, 1238
Song sparrow (*Melospiza melodia*), 1152f
Soot, 415, 415t
s orbital, 22, 22f
Sorghum, 203
Sorocarp, 576f
Sorus, 587, 588f
Sounds
 direction of sound source, 982
 made by insects, 674

Source (plant carbohydrate), **776**–77, 777f
Source-sink metapopulation, **1144**
Sourdough bread, 610
Sour taste, 975
South equatorial current, 1212f
Southern, E. M., 330f
Southern blot, **330**, 330–31f
Sowbug, 669
Soybean (*Glycine max*), 731f, 784, 787, 799f, 800, 835, 851
 transgenic, 337
Soy products, 800
Soy sauce, 603
Spadefoot toad (*Scaphiophus*), 1139
Spallanzani, Lazzaro, 716
Sparrow, 711t, 888
Spastic paraplegia, 269f
Spatial heterogeneity, species richness and, 1198–99, 1198f
Spatial recognition, 959–60
Special connective tissue, **862**, 864–66
Special creation, 64
Speciation, **471**–88, 486f
 adaptation and, 479
 allopatric, **480**, 481f
 geography of, 480–81, 480–81f
 long-term trends in, 487, 487f
 natural selection in, 479
 polyploidy and, 480–81
 rate of, 488
 sympatric, 480–82
Species, **2**, 3f, **510**–11
 clusters of, 482–85
 defining, 477
 effect of global warming on, 1223, 1223f
 endemic, **1230**
 geographic variation within, 472–73, 472f
 hybridization between. *See* Hybridization (between species)
 introduced, 1234t, 1240–41, 1246
 keystone, **1177**, 1177f, 1242f, 1243
 maintenance of genetic distinctiveness, 474–77
 nature of, 472–73
 origin of, 471–88
 sympatric, **472**
Species-area relationship, **1200**, 1236, 1236f
Species concept
 biological, **473**, 477
 ecological, 477
 genealogical, 477
Species diversity, biogeographic patterns of, 1199, 1199f
Species diversity cline, **1199**
Species name, 510–11
Species richness, 1180. *See also* Biodiversity
 causes of, 1198, 1198f
 climate and, 1198, 1198f
 conservation biology, 1227–48
 effects of, 1197, 1197f
 evolutionary age and, 1199
 on island, 1200, 1200f
 predation and, 1199
 productivity and, 1197–99, 1198f
 spatial heterogeneity and, 1198–99, 1198f
 in tropics, 1199, 1199f
Species-specific signal, 1118
Species turnover, **1200**
Specific heat, **29**
 of water, 28t, 29
Specific immune defense, 1018–20

Specific transcription factor, **370**–71, 371f
Speckled wood butterfly (*Pararge aegeria*), 1223f
Spectrin, 109t, 112, 113f
Speech, 957
Spemann, Hans, 1092
Spemann organizer, 397
Sperm, 228f, **1062**, 1069f
 fertilization, 1082–83, 1082–83f
 penetration of egg by, 1082–83, 1082–83f
 of plant, 842f, 848
 production of, 1068, 1069f
Spermatid, 1068, 1069f
Spermatocyte
 primary, 1068, 1069f
 secondary, 1068, 1069f
Spermatogonium, 1068
Spermatophore, 1125
Spermatozoan. *See* Sperm
Sperm competition, **1126**
Sperm count, 1070
Spermicide, 1076, 1076f, 1077t
Sperm nucleus, 840
Sperm whale, 1239, 1239f
S phase, **213**–14, 213f
Sphenisciformes (order), 711t
Sphincter, 892
Sphygmomanometer, 918, 918f
Spicule, **636**, 637f
Spider, 622t, 635, 664f, 670, 670f, 883, 1123, 1170
 poisonous, 670, 670f
Spider's web, 40f
Spiderwort (*Tradescantia*), 227f
Spinach, 835
Spinal cord, 856, 856f, 858f, 869t, 940f, 954f, 955t, 956, 961–62, 961–62f
 injury to, 426, 428, 962
Spinal nerve, 963
Spinal reflex, 961–62, 961–62f
 cutaneous, 962f
Spindle apparatus, 209f, 215, 215–16f, 217, 230, 233, 235f, 601
Spindle checkpoint, 219, 219f, 221, 221f
Spindle microtubules, 209f, 217, 233, 233–34f, 236
Spindle plaque, 601
Spindle pole body, 209f
Spine (plant), 748, 751, 1169
Spiny anteater, 717–18
Spiny fish, 692f, 693
Spiracle, 665–67f, **667**, 674, 927
Spiral cleavage, **626**, 626f
Spirillum, 546
Spirochaete, 549f
Spleen, 858f, 914, 1015f, 1032
Spliceosome, 314, 314f, 376
Sponge, 574, 617, 617f, 621f, 622t, 634–37f, 636–37, 1031
Spongin, **636**, 637f
Spongy bone, 865, 865f
Spongy mesophyll, **750**, 750f, 771f
Spontaneous reaction, 147
Sporangiophore, 607
Sporangium, **576**, 576f, **581**–82, 581f, 583f, 584, 587, 588f, 600f, 606–7f, **607**
Spore
 of bacteria, **546**
 dispersal of, 588f
 of fungi, 599f, 602, 602f, 607f
 of moss, 582, 583f
 of plant, **581**, 581f, 589
Spore mother cell, **581**–82, 581f, 587
Sporocyst, 644, 644f
Sporocyte. *See* Spore mother cell

Sporophyte, **581**, 581f, 582–83, 583f, 586–87, 588f, 589, 591–92, 596, 596f, 606f, 732, 842
Sporophytic self-incompatibility, 847, 847f
Spotted cuscus, 467f
Spotted sandpiper, 1128
Spring overturn, 1217
Springtail, 673
Spruce, 585t, 590, 1178, 1178–79f
Spruce-fir forest, 1141f
Squamata (order), 703t, 705f, 708–9, 709f
Squamous epithelium
 simple, 860, 861t
 stratified, 860, 861t
Squash, 738f
Squid, 652, 653f, 657, 882
Squirrel, 624f, 986
src gene, 422
Src protein kinase, 224f, 422
SRY gene, 270, **1063**, 1063f
Stabilizing fin, 695
Stabilizing selection, 446–47f, **447**, 477, 486
Stain, visualization of cell structure, 83
Stamen, 589, 593f, **594**, 838, 839–40f, **840**, 845–47, 847f
Staminate flower, 846, 847f
Stanley, Wendell, 524
Stapes, 980, 981f
Staphylococcus, 549f
Staphylococcus aureus, penicillin resistance in, 552
Starch, 38t, 54, 56, **57**, 202, 890
Starch grain, 38t
Starfish. *See* Sea star
Starling, 711t, 1117, 1117f, 1143, 1238
Starter culture, 610
Startle reflex, 1100
Stasis, **486**
Statocyst, 971t, 978
Statolith, 978
STD. *See* Sexually transmitted disease
Steam, 27f
Stegosaur, 703t
Stegosaurus, 515f
Stele, **741**
Steller's sea lion, 1242
Stem, 732, 809, 809f, 818
 elongation of, 824
 modified, 747, 747f
 structure of, 744–46, 744–46f
Stem cells, 911
 embryonic, 405f, **426**, 426–27f, 430, 1086
 ethics of stem cell research, 430, 430t
 hemopoietic, 1026
 pluripotent, 429, 429f
 therapeutic applications of, 426–28, 426–27f
 tissue-specific, **427**
Stephens Island wren, 1168, 1240
Stereocilia, 977–79, 977f
Stereoisomer, 56, 56f
Stereotyped behavior, 1106
Sterigma, 608f
Sterilization (birth control), 1078, 1078f
Stern, Curt, 267, 267f
Sternocleidomastoid muscle, 931f
Sternum, 858f
Steroid, 37t, 53, 53f
Steroid hormone, 128t, 129, 902, 993, 995t, 996
 mechanism of action of, 996–97, 996f
 in plants, 825, 825f
 structure of, 996f

I-34 Index

Steroid sulfatase deficiency, placental, 269f
Steward, F. C., 281
Stickleback fish, courtship signaling in, 1107, 1118, 1118f
Stigma, 847, 847–48f
　of euglenoids, 566, 566f
　of flower, 589, 595, 595–96f
　of plants, **840**, 840f
Stimulus, 62, 62f, 970, 1040, 1040–41f, 1106
　conditioned, 1111
　environmental, 970, 970t
　sign, **1106**–7, 1107f
　supernormal, **1107**
　unconditioned, 1110
Stimulus-gated ion channel, 972, 972f
Stimulus-response chain, 1118, 1118f
Stinging nettle, 639
Stipule, 744, 744f, 748, 1174
Stipule scar, 744
Stolon, 747, 747f
Stomach, 859f, 889, 889f, 892–93f, 899f, 900t
　digestion in, 892–93
　innervation of, 965t
　secretion by, 892
　structure and function of, 892
Stomach cancer, 414t
Stomata, 203–4, 203f, **580**, 583–84, **734**, 734f, 737, 744, 750, 750f, 768, 768f, 771f, 772, 782, 796
　mutants in *Arabidopsis*, 734, 734f
　opening and closing of, 772–74, 773f, 827, 827f
Stone canal, 678f
Stone cells, 736f
Stonecrop, 204
Stonecup, 840
Storage polysaccharide, 57
Storage protein, 39t, 40
Stork, 711t
Stramenopile, 564f, **571**–72, 571f
Stratification (seed), 763
Stratified epithelium, 856f, **860**, 861t
　pseudostratified columnar, 861t
　squamous, 860, 861t
Stratum basale, 1014
Stratum corneum, 1014
Stratum spinosum, 1014
Strawberry (*Fragaria ananassa*), 747, 761f, 771f, 849
Strawberry finch, 1113
Stream, 1185, 1191f, 1194, 1194f, 1196f, 1216f
Streptococcus, 549f, 555t
　disease-causing, 554
Streptococcus mutans, 554
Streptococcus pneumoniae, transformation in, 282, 282f
Streptococcus sanguis, 554
Streptomyces, 549f
Streptomycin, 520t
Streptophyta, 518, 526, 526f, 564f, 573
Stress fiber, 98f
Stretch receptor, 869t, 961f, 971t, 974, 974f
Striated muscle, **867**, 874, 874f
Strigiformes (order), 711t
Stroke, 54, **921**
Stroke volume, 920
Stroma, 96, 186–87f, 187
Structural carbohydrates, 58, 58f
Structural DNA, 350, 350t
Structural formula, **26**
Structural isomer, 56
Struthioniformes (order), 711t
STS. *See* Sequence-tagged site
Sturtevant, A. H., 268, 268f

Style, **595**, 595–96f, **840**, 840f, 847f, 848, 848f
Stylet, **646**, 776, 776f
Subcutaneous tissue, 1014
Suberin, 741, 746, 852, 852f
Submucosa, of gastrointestinal tract, 889, 889f, 894f
Subspecies, **472**, 472f
Substance P, **950**
Substrate, 149–50, 150f, 156
Substrate-level phosphorylation, **162**, 162f, 166, 170, 177
Subunit vaccine, **334**, 334f
Succession, **1178**–79
　in animal communities, 1179, 1179f
　interruption by disturbances, 1180, 1180f
　in plant communities, 1178–79, 1178–79f
　primary, **1178**
　secondary, **1178**
Succinate, 171f
Succinate dehydrogenase, 171f
Succinyl-CoA, 170, 171f
Succinyl-CoA synthetase, 171f
Succulent, 774
Sucker
　of fish, 980
　of flatworms, 643f
　of flukes, 644
　of plant, 849
Sucking reflex, 1081f, 1100
Suckling, 1000, 1102
Sucrase, 150f, 900t
Sucrose, 30, 56f, 57, 57f, 150f, 202, 894
　transport in plants, 776–77, 776–77f
Sugar, **55**–56
　isomers of, 56, 56f
　reabsorption in kidney, 1048f
　transport forms of, 57
　transport in plants, 776–77, 776–77f
Sugar beet, 57
Sugarcane, 57, 57f, 203, 210t, 353f
Sulfate respiration, 163
Sulfhydryl group, 36f
Sulfolobus, 519
Sulfur
　in plants, 782, 782t
　requirement for iron uptake, 337
Sulfur bacteria, 163, 196, 196f
Sulfur dioxide, 1219
Sulfuric acid, 1219
Sulphur butterfly (*Colias eurytheme*), 440
Summation, **880**, 880f, **944**, 944f
Summer solstice, 1204f
Summer tanager (*Piranga rubra*), 713f
Sunbird, 845, 1124, 1124f
Sundew (*Drosera*), 751, 788
Sunflower (*Helianthus annuus*), 185, 593, 745f, 845, 851
Sunlight, 1138. *See also* Solar energy
　in photosynthesis, 144, 185–86, 188–90
Supercooling, **815**
Superfund site, 790–91
Superior vena cava, 917f, **918**
Super nitric oxide, 936
Supernormal stimulus, **1107**
Superphosphate, **1188**
Supporting cells, **941**
Suppressor T cells, 1019, 1019t, 1021f
Suprachiasmatic nucleus (SCN), 1010
Surface area-to-volume ratio, **81**, 81f, 101, 101f
Surface marker. *See* Cell surface marker
Surface tension, 29, 29f
Surface water, 1184–85
Surinam frog, 1065f

Survival of the fittest, 11
Survival value, **1122**
Survivorship, **1147**
Survivorship curve, 1147, 1147f
Sushi, 572
Suspensor, **388**, 389f, 756–57, 756f, 758f, 759
suspensor mutant, of *Arabidopsis*, 757, 757f
Suspensory ligament, 983–84, 984f
Sustainable agriculture, 1220
Sutherland, Earl, 998
Sutton, Walter, 265
Suture (joint), 872, 872f
Swallowing, 891–92, 891f
Swamp, 1216
Swan, 711t, 1066
Swarming, 1133
Sweat gland, 860, 1014
Sweating, 1042
Sweet potato, 743
Sweet taste, 975
Sweet woodruff, 749f
Swift, 711t, 713
Swim bladder, 696, 697f, 980
Swimmeret, **668**, 669f
Swimming
　cellular, 100
　by fish, 695, 882, 882f
　by terrestrial vertebrates, 882
Swordfish, 1042, 1239
Symbion pandora, 648f
Symbiosis, 562, **604**, 1161f, **1172**
　coevolution and, 1172
　fungi in, 604, 611–14
　prokaryotes in, 557
Sympathetic chain, of ganglia, 964–65, 964f
Sympathetic division, **940**, 940f, 964, 964f
Sympatric speciation, 480–82
Sympatric species, **472**, 473, **1166**, 1166f
Symplesiomorphy, **513**
Symport, 121
Synapomorphy, **513**–14
Synapse, 879, **948**–53
　chemical, **127**, 948
　electrical, 948
　structure of, 948, 948f
Synapsid, **704**, 705f
Synapsis, 230, 230–31f, 232, 234f
Synaptic cleft, 944f, **948**, 948f, 992
Synaptic integration, **951**, 951f
Synaptic signaling, 127, 127f
Synaptic vesicle, **948**, 948f
Synaptonemal complex, 230, 232, 232f, 237, 409
Syncytial blastoderm, **386**, 387f
Synergid, 595–96, 842f, 843, 843f, 848f
Synergist (muscle), **873**
Syngamy, **228**, **522**
Synovial capsule, 872, 872f
Synovial fluid, 872f
Synovial joint, 872, 872f
Synovial membrane, 872f
Synteny, 351–53, 525
　conservation of, **495**
Syphilis, 556, 556f
Systematic phylogeny, 525
Systematics, **512**
　classification and, 514
　molecular, 525, **628**, 630
Systemic acquired resistance, in plants, **804**, 804f
Systemic circulation, 698, **916**–17, 916–17f
Systemic lupus erythematosus, 1036
Systemin, 802, 802f

Systole, **918**
Systolic pressure, **918**, 918f

T

2,4,5-T, 821
Table salt. *See* Sodium chloride
Table sugar. *See* Sucrose
Tadpole, 491, 491f, 701, 1065, 1065f
TAF. *See* Transcription-associated factor
Tagmata, **665**
Tagmatization, 666
Taiga, 1207f, 1208, 1208f, **1211**
Tail
　postanal, **684**, 684f, 689f
　prehensile, 720
　of sperm, 1068, 1069f
Tanager, 1238
Tandem cluster, 349
Tangles, in Alzheimer disease, 960
Tannin, 798
Tapeworm, 622t, 643f, 645, 1062
Taproot system, 743
Taq polymerase, 329, 329f
Tarantula, 670
Target cells, **992**, 992f
Tarsier, 720, 721f
Tasmanian tiger cat, 467f
Tasmanian wolf, 467f
Taste, 971t, 975, 975f
Taste bud, 971t, 975, 975f
Taste pore, 975
TATA box, **308**, 308f, 370, 370f, 372f
Tatum, Edward, 263, 296–97, 406
Tau protein, 960
Taxis, **1116**
Taxol, 799t, 800
Taxon, **510**, 511
Taxonomic hierarchy, 511, 511f, 514, 525
Taxonomy, **510**
　impact of molecular data on, 525–28
Tay-Sachs disease, 259, 264t, 274
T-box, 502
Tbx5 gene, 503
TCE. *See* Trichloroethylene
T cell(s), 910f, 1010, 1019–22, 1021–22f, 1026
　antigen recognition by, 1020
　cytotoxic, 1019–20, 1019t, 1021–22f, 1022, 1028f
　evolution of, 1032f
　helper, 1019–23, 1019t, 1021f, 1023f, 1028f, 1033
　HIV infection of, 536–37, 537f, 1033–34, 1033–34f
　inducer, 1019, 1019t, 1033
　killer, 1035
　suppressor, 1019, 1019t, 1021f
　in surveillance against cancer, 1022, 1022f
　in transplant rejection, 1022, 1022f
T cell receptor, 136f, 1020, 1020t, 1021f, 1023f
Tectorial membrane, 980, 981f
Teeth, 888
　deciduous, 890
　dental caries, 458, **554**, 555t
　development of, 1091
　diet and, 715, 715f
　of horses, 462–63
　of humans, 890
　of mammals, 715
　saber-toothed-ness, 516, 516f
　of sharks, 695
　specialized, 715, 715f
　of vertebrates, 890, 890f

Index **I-35**

Telencephalon, 955t, 956
Telomerase, 401–2, 423
Telomerase inhibitor, 423
Telomere, 350, **401**, 423
Telomere depletion hypothesis, of aging, 401–2, 401f
Telophase
  meiosis I, 231f, 234f, 236
  meiosis II, 231f, 235f, 236
  mitotic, 213, 213f, 216–**17**, 217f, 231f
Telson, **668**, 669f
Temperate deciduous forest, 1189, 1189f, 1191f, 1208, 1208f, 1211, 1211f
Temperate evergreen forest, 1191f, 1208, 1208f, 1211, 1211f
Temperate forest, 1207f
Temperate grassland, 1191f, 1208, 1208f, 1210
Temperate virus, **534**
Temperature, 29
  adaptation to specific range, 1138
  annual mean, 1209, 1209f
  detection of, 971t, 973
  effect on chemical reactions, 26
  effect on enzyme activity, 47, 152, 152f
  effect on flower production, 836, 838f
  effect on oxyhemoglobin dissociation curve, 935, 935f
  effect on photorespiration, 203, 203f
  effect on photosynthesis, 188–89, 189f
  effect on plant respiration, 786
  effect on transpiration, 773–74
Temperature-dependent sex determination, **1223**
Temperature-sensitive allele, 256, 256f
Template strand, 290–91f, 306, 306–7f
Temporal isolation, 473t, 475
Temporal lobe, 957, 957f, 960
Tendon, 862, 863t, 868, 873, 874f
Tendril, 747, 747f
Tentacle, 654, 654f
Teratorn vulture, 1228f
Terebratolina septentrionalis, 663f
Terminal bud, 732f, **744**, 744f
Terminal chiasmata, 233
Terminalia calamansanai, 1143f
Terminus of replication, 208, 208f, 290
Termite, 672f, 675t, 898, 1132–33
Tern, 711t
Terpene, 37t, 53, 53f
Terpenoid, 799t
Terrestrial ecosystem, 1208–11
  animal locomotion on land, 883, 883f
  compared to aquatic ecosystems, 1217
Territorial behavior, 1119, 1124, 1124f
Territoriality, **1124**
Territory, 1106, 1142
Tertiary consumer, 1190f, 1192f
Tertiary structure, of proteins, 43, 44f, 45, 47f
Test, of forams, 575, 575f
Testcross, **252**, 252f
Testis, 858–59f, 993f, 995t, 1068, 1068–71f
Testosterone, 53, 825f, 995t, 996f, 1009, 1068, 1071, 1071t, 1106
Testudines, 691f
Tetanus (disease), 554, 880
Tetanus (sustained muscle contraction), **880**, 880f
Tetany, 880
Tetra (*Astyanax mexicanus*), 505f
Tetracycline, 551
Tetrahedron, 27
Tetrahymena pyriformis, 569
Tetrahymena thermophila, 374
Tetraploid, 480–81, 498f

Tetrapod, **689**
  locomotion in water, 882
  locomotion on land, 883, 883f
T-even phage, 534
Texas fever, 670
Thalamus, 955f, 955t, **956**, 957f, 958, 987
Thalidomide, 1098
Thallus, 757, 757f
Thecodont, 704, 704–6f, 707
Theory, **6**
Therapeutic cloning, **428**–30, 428–29f
Therapsid, 700, 704, 704f, 706f, 717
Therian, 717
Thermal receptor, 988
Thermal stratification, 1216–17, 1217f
Thermal vent, 553, 1215
Thermocline, 1216–17, 1217f
Thermococcus, 520f
Thermodynamics, **144**
  First Law of, **146**
  Second Law of, **146**
Thermophile, 519, 521f, 549f
  extreme, **71**, 73f
Thermoplasma, 520f
Thermoproteus, 520f, 549f
Thermoreceptor, **973**
Thermoregulation, 909, 909f, 958, 1039f, 1040, 1042–43, 1139
  in birds, 713
  in fish, 1042
  in insects, 1042, 1042f
  in lizards, 1017, 1042, 1138–39, 1139f
  in mammals, 714–15
  negative feedback loop, 1040, 1041f, 1042
  in reptiles, 704, 707, 1042, 1207
Thermotoga, 519, 520f
Thermotolerance, in plants, 815, 815f
Thermotropism, 811
Theropod, 712
Theta wave, 959
Thick myofilament, 574, 868, 868f, 874–77, 874–77f
Thigmonasty, **811**
Thigmotropism, **811**
Thin myofilament, 874–77, 874–77f
"Third eye," 708, 1010
Third filial generation, 248, 248f
Third trimester, 1100
Thirst, 920, 958, 1057, 1057f
Thomson, James, 426
Thoracic breathing, 702
Thoracic cavity, **856**, 856f
Thoracic nerves, 954f
Thorn, 751, 796, 1169, 1174
Thorn-shaped treehopper (*Embonia crassicornis*), 672f
Thoroughbred racehorse, 238, 450, 450f
Three-chambered heart, 653f
Three-point cross, **268**
Threonine, 41, 42f
Threshold, action potential, 944
Thrip, 1132
Thrombin, 911f
Thrombocytes. See Platelet(s)
Thrush (bird), 1238
Thrush (disease), 614
Thx5 gene, 502, 502f
Thylakoid, **96**, 97f, 186–87f, 187, 192f, 198–99, 198f
Thymidine kinase gene, 370f
Thymine, 49–50, 49–50f, 284, 284f, 287, 287f
Thymine dimer, 412

Thymus, 858f, 914, 993f, 1010, 1015f, 1032
Thyroid cancer, 417t
Thyroid gland, 858f, 993, 994t, 1004–5
Thyroid hormone, 128t, 129
Thyroid-stimulating hormone (TSH), 994t, 1001–2, 1001f, 1003–5f, 1004
Thyrotropin. See Thyroid-stimulating hormone
Thyrotropin-releasing hormone (TRH), 1003–4, 1003–5f
Thyroxine (T$_4$), 994t, 996–97, 1004–5
  in amphibian metamorphosis, 1005, 1005f
  mechanism of action of, 997, 997f
  regulation of secretion of, 1004, 1004f
Tibia, 699f, 871f
Tibial nerve, 954f
Tick, 670
Tidal volume, **931**
Tiger, 474, 474f, 488f
Tight junction, 128t, 137f, 138, 138f, 385
Tiglon, 474f
Tilman, David, 1197
Timberline, 1207
Times Beach, Missouri, 1218
Tinbergen, Niko, 1106–7, 1107f, 1118, 1122, 1122f
tinman gene, 397
Ti plasmid, **335**, 335f, 336
Tissue, **2**, 3f, **136**, 619, 638, **856**, 856–57f
  evolution of, 624
  primary, 856
Tissue culture, 405f
  plant, 850, 850f
Tissue plasminogen activator, 333
  genetically engineered, 333
Tissue-specific identity marker, 136
Tissue-specific stem cells, **427**
TKCR syndrome, 269f
Tmespiteris, 586
TMV. See Tobacco mosaic virus
TNT. See Trinitrotoluene
Toad (*Bufo*), 698, 698t, 701, 1065
  declining populations of, 1234–35, 1234–35f
  feeding on bees, 1110f
  hybridization between species of, 474
Toadstool, 599, 605t, 608
Tobacco, 825, 836, 836f
Tobacco hornworm (*Manduca sexta*), 798, 798f
Tobacco mosaic virus (TMV), 524, 524f, 532f
Tocopherol. See Vitamin E
Toe, grasping, 720
Tomato (*Lycopersicon esculentum*), 761f, 826, 835, 840
  Flavr Savr, 335
  mineral deficiencies in, 783f
  transgenic, 335–36, 826, 826f
  wound response in, 802, 802f
Tomato hornworm, 336
Tongue, 890–91, 891f, 975, 975f
Tool use, 722, 724, 726
too many mouths mutation, in *Arabidopsis*, 734, 734f
Tooth. See Teeth
Top carnivore, 1193
Top-down effect, 1194, 1196, 1242
Topoisomerase, 292
Topsoil, 781f, **784**, 784f
Topsoil erosion, 1220f
Torsion, **656**
Tortoise, 619t, 703t, 708, 1066f
  Galápagos, 467

Tortoiseshell cat, 271
Totipotency, in plants, 281
Totipotent cells, **393**
Toucan, 711t
Touch, 971t, 973
  plant response to, 811–13
Touch dome ending. See Merkel cells
Toxin, 39t
  plant, 798, 798f, 799t
Trace (xylem and phloem), 745
Trace element, 904
Trachea, 858f, 864f, 891f, **928**, 928–29f
Tracheae, 665f, **667**, 667f, 670–71, 674, 922, 923f, **926**–27
Tracheata, 527
Tracheid, 733, 737, 737f, 772
Tracheole, **667**, 667f
Trade winds, 1205–6, 1205f, 1212
Trail pheromone, 1120, 1120f
Trait, **243**. See also Segregation of traits
  direct transmission of, 242–43
Transcription, **303**, **306**
  errors in, 307
  in eukaryotes, 308–9, 308–9f, 315–16f, 316
  initiation of, 306–7, 306f, 308f, 309, 362, 366–69, 377f
  posttranscriptional modifications, 309, 309f
  in prokaryotes, 306–7, 306–7f, 316, 316f
  termination of, 307
Transcriptional control, **362**, 366–72, 366–72f
  in eukaryotes, 362, 370–72, 370–72f
  in prokaryotes, 362
  regulation of promoter access, 362
Transcription-associated factor (TAF), 370, 370f
Transcription bubble, **307**, 307f
Transcription complex, 372, 372f
Transcription factor, 224, 308, 308f, 373, 399, 501–2, 504, 758, 838
  basal, 370, 370f, 372f
  in cancer, 417, 417t
  dorsal, 396
  E2F, 222f, 417, 418f, 422
  in eukaryotes, 370–72, 370–72f
  specific, **370**–71, 371f
  TFIID, 370, 372f
Transcriptome, 356
Transducin, 135, 135f
Transfection, **416**
Transferrin, 39
Transfer RNA (tRNA), 35f, **302**
  structure of, 303f
  synthesis of, 308
  in translation, 310–12, 310–12f
Transformation
  in bacteria, **282**–83, 282f
  cell, **535**
Transforming principle, 283
Transgenic animals, 338, 338f
Transgenic plants, **336**, 355, 355f, 826
  herbicide resistance in, 336, 339, 355f
  insect resistance in, 336
  nitrogen fixation in, 335–36
  potential risks of, 339
Translation, **303**, 310–12, 310–12f
  elongation stage of, 311
  in eukaryotes, 311, 315–16f, 316
  initiation of, 310–11, 311f, 316
  in prokaryotes, 310, 316, 316f, 548
  "start" and "stop" signals, 303, 305, 310
  termination of, 312, 312f
  translocation step in, 312, 312f

I-36 Index

Translational control, 377f, 378
Translation factor, 378
Translation repressor protein, **378**
Translocation (chromosome), 272, 411, 416
Translocation (phloem transport), **776**
Translocation (translation), **312**
Translocation Down syndrome, 272
Transmembrane protein, 108, 108f, 109t, **112**, 113f, 130, 130f
Transmissible spongiform encephalopathy (TSE), **542**
Transmission electron microscope, **82**, 83t, 110
Transpiration, **737**, **768**, 769–70, 771f, 772
 regulation of rate of, 772–74
 in water cycle, 1184, 1184f
Transplant rejection, 1022, 1022f
Transport disaccharide, 57
Transport protein, 39, 39t, 80, 108f, 111, 111f
Transport vesicle, 92–93f, 99–100, 99f
Transposable element, 237, 350, 350f, 1035
Transposase, **407**, 408f
Transposition, **407**–8, 408f
Transposon, 350, **406–7**, 408f, 409, 411, 500
 dead, 351
 in *Drosophila*, 411, 500
Transverse tubule (T tubule), **878**–79, 878f
Trap-door spider, 670
Traumotropism, 811
Tree, 585t, 593
Tree fern, 587, 587f
Tree finch (*Camarhynchus*), 483, 483f
 large insectivorous tree finch, 454–55, 454f
 small insectivorous tree finch, 454–55, 454f
Tree frog (*Eleutheradactylus coqui*), 491f
Tree kangaroo, 1140f
Tree of life, 520, 520f
Trematoda (class), 644, 644f
Treponema pallidum, 549f, 556
TRH. *See* Thyrotropin-releasing hormone
Triacylglycerol, **53**, 54f
Triazine resistance, 441
Trichinella, 646f, 647
Trichinosis, 646f, 647
Trichloroethylene (TCE), 415
 phytoremediation for, 790–91, 790–91f
Trichogyne, 609f
Trichome, **734**, 735f, 744, 774, 788, 788f, 796
Trichromat, 987
Tricuspid valve, 917f, **918**
Triglyceride, **53**, 54f, 862, 896
Triiodothyronine, 997, 997f, 1005
Trilobite, 453f
Trinitrotoluene (TNT), phytoremediation for, 791
Triple bond, **26**
Triplet binding assay, 304
Triplet code, **304**
Triplet expansion (mutation), 259, 410, 410t
Triple X syndrome, 273, 273f
Trisomy, 210, **272**, 272f
Trisomy 21. *See* Down syndrome
Trivers, Robert, 1130
tRNA. *See* Transfer RNA
Trochophore, **635**, **655**, 655f, 660
Trophic cascade, **1194–96**, 1194–96f
 human effects on, 1195

Trophic level, 180, **1190**–91, 1190f
Trophoblast, **1086**, 1086f, 1089f, 1096, 1098
Tropical ecosystem, 1207
 species richness in, 1199, 1199f
Tropical forest
 destruction of, 1220, 1220f
 productivity of, 1191, 1191f
Tropical monsoon forest, 1208, 1208f
Tropical rain forest, 1205, 1207–8f, 1208, 1210, 1210f, 1231–32, 1233f
 loss of, 1220, 1220f
 productivity of, 1191f
Tropical seasonal forest, 1191f
Tropic hormone, 1001
Tropin, 1001
Tropomyosin, 876f, **878**–79, 878f
Troponin, 876f, **878**–79, 878f
Trout, 882f, 888, 1116
trp operon, 366–68, 367f
trp promoter, 367–68, 367f
trp repressor, 364f, 367–68, 367f
Truffle, 605t, 609
Tryon, Robert, 1108
Trypanosoma cruzi, 567
Trypanosome, 567, 567f
 antigen shifting in, 1035
Trypanosomiasis, 567
Trypsin, 895, 900t
Tryptophan, 42f, 367, 367f, 820f, 993
TSE. *See* Transmissible spongiform encephalopathy
Tsetse fly, 567, 567f
TSH. *See* Thyroid-stimulating hormone
T tubule. *See* Transverse tubule
Tuatara, 703t, 706f, 707–8
Tubal ligation, 1078, 1078f
Tube cell, 848f
Tube cell nucleus, 842f, 848f
Tube feet, 677–79f, **678**–79, 882
Tube nucleus, 596f
Tuber, 747, 747f
Tuberculosis, 554, 554f, 555t
Tube worm, 622t
Tubulin, 98, 98f, 209, 209f, 214, 217
Tularemia, 357t
Tulip, 747
Tulip tree (*Lirioidendron tulipifera*), 749f
Tumor, **413**
Tumor antigen, 1022
Tumor-suppressor gene, 223–24, 224f, 402, 416–**17**, 418–19, 419f
Tuna, 1042
Tundra, 579, 579f, 1191f, 1207f, 1208, 1208f, **1211**
Túngara frog (*Physalaemus*), 1127, 1127f
Tunic, 686f
Tunicate, 685–86f, 686–87
Turbellaria (class), 644
Turgor, 811, **812**, 812f
Turgor movement, 812–13, 812–13f
Turgor pressure, **117**, 769, 769f, 773, 777, 811, 818
Turner syndrome, 273, 273f
Turnip, 743
Turpentine, 590
Turtle, 515f, 691f, 703t, 705–6f, 707–8, 708f, 882
Tutt, J. W., 456–57
Twig snake, 709
Twin-fan worm, 660
Twin studies, 1108
Twitch, **879**–80
Tympanal organ, 666f
Tympanic canal, 980, 981f
Tympanic membrane. *See* Eardrum

Tympanum, **674**
Type III secretion system, 553
Typhoid fever, 554, 555t
Typhus, 554, 555t
Tyrannosaur, 703t
Tyrannosaurus, 515f, 517f
Tyrosinase, 256
Tyrosine, 42f, 993
Tyrosine kinase receptor, 574

U

Ubiquinone, 174, 174f
Ulcer
 duodenal, 893
 gastric, 893
Ulna, 699f, 871f
Ultimate causation, 1106
Ultracentrifuge, 288, 288–89f
Ultrasound, of fetus, 274
Ultraviolet radiation, 190–91, 190f, 1204
 ozone layer and, 1221
Umbilical artery, 1097f
Umbilical cord, 715f, 1064f, 1096–97f, 1101, 1101f
Umbilical vein, 1097f
Unconditioned response, 1110
Unconditioned stimulus, 1110
Unequal crossing over, 409, 409f
Unicellularity, of prokaryotes, 546, 547t
Uniform pool, 927
Uniramous appendages, 527–28, 528f, 668t
Universal Declaration on the Human Genome and Human Rights, 358
University of Wisconsin-Madison Arboretum, 1246f
Unsaturated fatty acid, **53**–54, 54f, 904
Uracil, 49–50, 49–50f, 284, 284f
 in DNA, 412
Urea, 178f, 902, 1049, **1051**, 1051f, 1054, 1055f, 1056
Ureter, 859f, **1052**, 1052f
Urethra, 859f, 1052f, 1068f, 1072f
Urey, Harold C., 66, 68
Uric acid, **1051**, 1051f, 1054
Uricase, 1051
Urinary bladder, 859f, 965t, **1052**, 1052f
Urinary system, 857f, 859f
Urine, 909, 1046, 1048
 concentration of, 1050
 hypertonic, 1050
 hypotonic, 1048
 pH of, 1054
 volume of, 1000f, 1054
Urochordate (subphylum), 686–87, 686f
Urodela (order), 698t, 700–701, 701f
Urogenital tract, as barrier to infection, 1014
Uropod, 668, 669f
Uterine contractions, 1000, 1004, 1043, 1043f, 1101–2, 1101f
Uterine horn, 1073f
Uterine tube. *See* Fallopian tube
Uterus, 859f, 1072, 1073f, 1101f
Utricle, 978–80, 978f
uvr genes, **412**
UVR photorepair system, **412**

V

Vaccination, **1018**, 1018f, 1027
Vaccine, 540. *See also specific pathogens and diseases*
 AIDS, 1034

 DNA, 334, 1035
 malaria, 569
 subunit, **334**, 334f
 trypanosome, 567
Vaccinia virus, 334, 524t
Vacuole, of eukaryotic cells, 101, 101f, 102t
Vagina, 859f, 1072, 1072–73f, 1101f
Vaginal secretions, 1014
Vagus nerve, 965
Valence electron, **24**
Valine, 42f
Vampire bat, 1130
van Beneden, Pierre-Joseph, 228
Vancomycin, 84
van der Waals attractions, 43f, 45
van Helmont, Jan Baptista, 188
Vanilla orchid, 743
van Niel, C. B., 188
var genes, 1035
Variable, **6**
Varicella zoster virus, 533t
Varicose veins, 913
Variety, **472**
Variola major virus, 357t
Variola virus, 533t
Varmus, Harold, 263
Vasa recta, 1053f, 1056
Vascular bundle, 742, 745f, 750f
Vascular cambium, **731**, 731f, 737, 741, **745**, 745f
Vascular plant, **580**, 729f, 732, 732f
 extant phyla of, 584, 585t
 features of, 584, 585t
 seedless, 580, 580f, 585t, 586–88, 586–88f
Vascular tissue, of plants, **388**, 389f, **584**, **732**, 732f, 737–38, 737–38f, **758**
Vas deferens, 859f, 1068–70f, 1070, 1078, 1078f
Vasectomy, 1078, 1078f
Vase sponge, 622t
Vasoconstriction, 909, 909f, **912**, 920, 936, 994, 1000f, 1042
Vasodilation, 909, 909f, **912**, 920–21, 936, 994, 1010, 1017, 1025, 1042
Vasopressin. *See* Antidiuretic hormone
Vector, cloning. *See* Cloning vector
Vegetal pole, **383**, 392, 392f, 1084, 1084f, 1088, 1088f
Vegetarian finch (*Platyspiza*), 10f, 454–55, 454f, 483, 483f
Vegetative development, in plants, 755–63
Vegetative propagation, 747
Vegetative reproduction, in plants, 849
Vein (blood vessel), 858f, **912**–13, 912–13f
 varicose, 913
Vein (leaf), 750, 750f
Veliger, **655**, 655f
Velociraptor, 515f, 517f, 712, 712f
Velvet (antler), 716
Velvet worm, 623t
Venous pump, **913**, 913f
Venous return, 913
Venous valve, **913**, 913f
Venter, Craig, 348
Ventral portion, 624, 624f
Ventral root, **963**
Ventricle (brain), 957f
Ventricle (heart), **915**, 915–16f
 left, 917–18, 917f
 right, 917–18, 917f
Venule, **912**–13

Index I-37

Venus flytrap (*Dionaea muscipula*), 751, 788, 788f, 811, 821
Vernal equinox, 1204f
Vernalization, 834, **836**
Vertebra, 688, 856, 856f, 872f
   embryonic development of, 688f
   neck, 464
Vertebral column, 685f, 688, 689f, 721, 871f, 961
   of fish, 690
Vertebrata (subphylum), 511f, 688
Vertebrate, **619**, 683–726, 685f
   aquatic, gastrulation in, 1088, 1088f
   biological clock in, 1009–10
   brain of, 954–56, 955f
   characteristics of, 688–89, 688–89f
   circulatory system of, 908–9, 909f, 915–17, 915–17f
   cladogram, 513f
   development in, 382–85, 382–85f, 1081–1102, 1085f, 1095f
     axis formation in, 397
   digestive system of, 888–89, 888–89f
     variations in, 898, 898–99f
   evolution of, 461, 688–718, 691f, 694f
   excretory system of, 689f
   eyes of, 464–65, 465f, 504–5, 504–5f, 969f, 983–84, 983–84f
   forelimb of, 13f, 464, 464f
   heterozygosity in, 435
   Hox genes in, 399
   immune regulation in, 1009–10
   immune system of, 1032
   invasion of land by, 699, 699f
   kidneys of, 1047–50
   locomotion in, 882–84
   organization of body of, 856–57, 856–59f
   photoreceptors of, 985–86, 985–86f
   respiratory system of, 915–17, 915–17f
   sexual development in, 1009–10
   skeleton of, 689f
   social systems of, 1133–34, 1134f
   teeth of, 890, 890f
Vertical gene transfer, **499**
Vervet monkey, 1114, 1121, 1121f
Vesicle, **86**, 94
Vessel member, 733, 737, 737f, 772
Vestibular apparatus, 979
Vestibular canal, 980, 981f
Vestibular nerve, 979f
Vestibule (ear), 979f
Vestigial structure, **465**, 465f
VHL gene, 417t
Viagra. See Sildenafil
Vibration sense, 971t
Vibrio cholerae, 85f, 549f, 555t
   phage conversion in, 535
Viceroy butterfly (*Limenitis archippus*), 1171, 1171f
Victoria (Queen of England), 261, 261f
Vicuña, 1239
Villi, **894**, 894f
Vimentin, 99
Vinblastine, 1232f
Vincristine, 1232f
Vinyl chloride, 415
Violet, 841
Viper, 709
Viral disease, plants, 803f
Virdiplantae (kingdom), 518, 522, 525–26, 526f, 580
Vireo, 1238
Virginia creeper (*Parthenocissus quinquefolia*), 749, 749f
Viroid, 542
Virulent virus, **534**

Virus, 82f, **524**, 524f, 531–42
   bacteriophage. See Bacteriophage
   cancer and, 540–41
   classification of, 524
   cloning vector, 325–26
   discovery of, 524
   disease-causing, 531, 533t, 540–51
   DNA, 532, 532f, 533t
   emerging, **541**
   genome of, 532–33
   host range of, **532**
   recombination in, 540–41
   replication of, 532
   RNA, 524, 532, 532f, 533t
   shape of, 524, 524f, 532, 532f
   size of, 524, 524f
   structure of, 532, 532f
   temperate, **534**
   virulent, **534**
Viscera, 867
Visceral mass, **654**
Visceral pleural membrane, **930**
Vision, 971t, 983–87
   binocular, 720, **987**
   black-and-white, 985
   color, 720, 969f, 985–86, 985f
   nearsightedness and farsightedness, 984f
Vision amplification cascade, 135, 135f
Visual cortex, 957
Visual field, 987
Vital capacity, **931**
Vitamin, 153, 904
Vitamin A, 193, 904t, 985, 996
   deficiency of, 337, 337f
Vitamin B$_1$, 904t
Vitamin B$_2$, 904t
Vitamin B$_3$, 904t
Vitamin B$_5$, 904t
Vitamin B$_6$, 904t
Vitamin B$_{12}$, 557, 892
Vitamin C, 904t
Vitamin D, 128t, 129, 904t, 1006, 1006f, 1010
Vitamin E, 904t
Vitamin K, 557, 898, 904t
Vitelline membrane, 1083f
Viviparity, 1064–67, 1064f
Vivipary, **827**, 827f
Vocalization, to attract mate, 474, 475f
Voice box. See Larynx
Volcanic eruption, 1179, 1179f
Volcanic island, 1178
Voltage-gated ion channel, 944–45, 945f, 972f
Volvox, 561f, 573
Vomitoxin, 614
von Frisch, Karl, 1106, 1107f, 1120
von Humbolt, Alexander, 1185
Vorticella, 565f
Vulture, 180f, 711t, 1190

W

Wachtershauser, Gunter, 65
Wadlow, Robert, 1002f
Waggle dance, 1120, 1120f
Waking state, 959
Walking fern (*Asplenium rhizophyllum*), 751
Walking legs, 669f, 670
Walking pattern, 883
Wallace, Alfred Russel, 12
Wall cress. See *Arabidopsis*
Walnut, 749
Walrus, 719t, 1070
Warbler, 711t, 1238
   resource partitioning in, 1166

Warbler finch (*Certhidea*), 454–55, 454f, 483, 483f
Warfarin resistance, in rats, 441
Warm, moist evergreen forest, 1208
Warm receptor, 973
Warning coloration, 1170, 1171f
Wasp, 664f, 675t, 884, 1133, 1170–71, 1175
   parasitoid, 801, 801f
Waste products, transport in blood, 909–10
Water, **27**
   absorption by plants, 770–71, 770–71f
   adhesive properties of, 29, 768, 772
   chemistry of, 27
   cohesive nature of, 28t, 29, 768, 772
   forms of, 27, 27f
   heat of vaporization of, 28t, 29
   hydrogen bonds in, 28, 28f, 30, 30f
   ionization of, 31–32, 31–32f
   locomotion in, 882, 882f
   molecular structure of, 27, 27f
   osmosis, 116–17, 116f
   in photosynthesis, 187–89, 187f, 198–99, 198–99f
   polarity of, 27f
   prebiotic chemistry, 67f
   production in electron transport chain, 172f, 173–74, 174f
   properties of, 28t
   reabsorption in kidney, 1048f, 1054–57, 1055f, 1057–58f
   requirement of living things, 27, 1138, 1185
   shape of water molecule, 27
   soil, 784
   as solvent, 28t, 30, 30f
   specific heat of, 28t, 29
   surface tension of, 29, 29f
   transpiration from leaves, 772
   transport in plants, 768–76, 768–76f
Water availability, 1138–39
Water balance, 1045–47f, 1046–47
   hormonal control of, 1000, 1000f
Water boatman, color form of, 444, 444f
Watercress, 1169
Water current, over respiratory surface, 923
Water cycle, 1184–85, 1184–85f, 1189
   in forest ecosystem, 1185, 1185f
Water-dispersed fruit, 762, 762f
Water flea, 668
Waterfowl, 711t, 713
Water hyacinth (*Eichornia crassipes*), 1241, 1246
Water lily, 734, 774, 775f
Water moccasin, 709
Water mold, 605
Water potential, 768–**69**, 768–69f, 770, 777
Watersheds, of New York City, 1232–33, 1233f
Water storage root, 743, 743f
Water strider, 29f
   body size and egg-laying in, 442, 442f
Water table, 1185
Water-vascular system, **676**, 677–79f, 678–79
Waterwheel (*Aldrovansa vesicular*), 788–89, 789f
Watson, James, 286–87, 286–87f
Wax, 53, 796, 898
W chromosome, 270, 270t
Wear-and-tear hypothesis, of aging, 402

Weasel, 719t
Weaver bird, 1134
Weberian ossicle, 980
Weed, foreign, 1240
Weevil, 672f
Weight, 20
Weinberg, W., 436
Weissmann, Charles, 542
Welwitschia, 585t, 592, 592f
Wenner, Adrian, 1120
Went, Frits, 819, 819f
WEREWOLF gene, in *Arabidopsis*, 740, 740f
Werner, Otto, 402
Werner's syndrome, 402
Wernicke's area, 959–60
Westerlies, 1205–6, 1205f
West Nile fever, 1240
Wetland, 788, **1217**
   productivity of, 1191, 1191t, 1217
Whale, 528, 528f, 714, 719t, 882, 982
   evolution of, 461, 461f
   overexploitation of, 1239, 1239f, 1242–43, 1242f
Whaling industry, 1239, 1239f, 1242–43, 1242f
Wheat (*Triticum*), 744f, 836, 851
   chromosome number in, 210t, 498f
   evolution of, 498f
   genome of, 353f, 355f, 497
   transgenic, 336
Wheel animal. See Rotifera
Whiskers, 714
Whisk fern, 584, 585t, 586, 586f
White blood cells. See Leukocytes
White Cliffs of Dover, 575, 575f
White-crowned sparrow, 1113, 1113f
White fiber, 880
White-fronted bee-eater, 1132
White matter, **941**, 961, 961–62f
"White meat," 880
White rhinoceros, 1140f
White-tailed deer, 1119, 1168
Whooping cough, 554
Whorl (flower parts), 594
Whorl (leaf pattern), 744, 749, 749f
Wildebeest, 180f
Wild geranium (*Geranium maculatum*), 841f
Wild lettuce (*Lactuca*), hybridization between species, 475
Wild type, **268**
Wilkins, Maurice, 286
Willadsen, Steen, 394
Willow, 613, 762, 1211, 1232
Wilmut, Ian, 394, 424, 424–25f
Wilson, Edward O., 1200
Wilting, 117, 773
Wilts (plant disease), 553
Wind
   dispersal of fungal spores by, 602
   fruit dispersal by, 762, 762f, 1143f
   pollination by, 443, 443f, 594–95, 844–45, 846f
Winder, Ernst, 415
Window leaf, 751
Windpipe. See Trachea
Wine, 167f, 181
Wine-making, 603, 610
Wings, 884
   of bats, 884, 884f
   of birds, 884, 884f
   development of, 502f
   evolution of, 503
   of insects, 503, 664, 665f, 673, 673f, 884
   of pterosaurs, 884, 884f
Wing traits, in fruit fly, 268f
Winter habitat, for birds, 1237

Winter solstice, 1204f
Wishbone, 517f, 710
Wiskott-Aldrich syndrome, 269f
Wiwaxia, 630f
wnt pathway, 397
Woese, Carl, 518
Wolf, 467f, 1138, 1139f, 1195, 1211
   captive breeding of, 1247
Wolf spider, 670
Wolverine, 1211
Woman River iron formation, 163
Wood, 590, 594, 731, 737, 737f
Woodland, 1141f, 1191f
Woodpecker, 711t, 713
Woodpecker finch (*Cactospiza pallida*), 10f, 454–55, 454–55f
Woodward, Robert, 800
Woody plant, 594, 731, 742, 744, 851
Word salad, 960
Worker bee, 1062, 1120, 1132–33, 1132f
Workplace, carcinogens in, 415t
Wound healing, 222
Wound response, in plants, 802, 802f
Wrinkling, 1014
Writing, 960

X

X chromosome, **266,** 270, 270t
   of fruit fly, 268f
   human, 269f
   inactivation of, 271, 271f
   nondisjunction involving, 273, 273f
X-gal, 327, 327f
Xianguangia, 630f
X ray(s), 190, 190f
X-ray diffraction pattern
   of DNA, 286, 286f
   of proteins, 43
Xylem, 584, **732, 737,** 745f
   primary, 730f, 733f, 737, 742, 742f, 745, 745f
   secondary, 731, 733f, 737, 745f, 746
   water and mineral transport through, 768–76, 768–76f
XYY genotype, 273

Y

YAC. *See* Yeast artificial chromosome
Yamagiwa, Katsusaburo, 415
Y chromosome, **266,** 270, 270t, 1063, 1063f
   nondisjunction involving, 273, 273f
   segmental duplication on, 499f
   tracing human evolution, 725
Yeast, 63f, 599, **603,** 609–10, 610f
   cell cycle control in, 220–21
   cell division in, 209f
   chromosome number in, 210t
   commercial uses of, 603
   DNA of, 285t
   ethanol fermentation in, 167f, 181, 181f
   experimental systems using, 610
   fermentation pathways in, 610
   genome of, 348f, 493t, 610
   mating factor of, 131
Yeast artificial chromosome (YAC), 323, **347**
Yellow-eyed junco, 1123
Yellow fever, 533t, 540, 1224
Yellow star thistle, 1241
Yellowstone Park, return of wolves to, 1247
Yersinia, type III system in, 553
Yersinia pestis, 357t, 555t
Yew, 585t, 590
Y gene, 366f
Yohoia, 630f
Yolk, 702, 1064–65, 1084f
Yolk plug, 1088, 1088f
Yolk sac, **702,** 702f, 715f, 1066, **1096,** 1096f
Young, Lorraine, 425
Yucca, 594

Z

Z chromosome, 270, 270t
Z diagram, 197, 197f
Zebra, 180f, 719t
Zebra dove, 1179
Zebra finch, 1123
Zebra mussel, 652, 1240
Zebrina pendula, 810f
Z gene, 366f
Zinc, 149, 153, 792, 904
   in plants, 782, 782t
   deficiency of, 783f
Zinc finger motif, **265,** 365f
Zinkernagel, Rolf, 334
Z line, 874–77, 874–77f
Zoecium, **662,** 663f
Zona pellucida, 1082, 1082f
Zone of cell division, **739**–40, 739–40f
Zone of elongation, **739**–40, 739f
Zone of maturation, **739,** 739f, 741–42, 741–42f
Zooplankton, 1193, 1193f, 1196, 1242
Zoospore, 572, 602, 606f
Zooxanthellae, 653f
Zygnematales, 526f
Zygomycetes, 600, 600f, 607, 607f
Zygomycota (phylum), 605, 605f, 605t, 607, 607f
Zygosporangium, **607,** 607f
Zygospore, 573f
Zygote, **228,** 229f, 588, **1062,** 1081
Zymogen, 895

Index I-39